Fermentation Biotechnology for Functional Foods

Fermentation-based processing advanced with global food systems as humanity started to organize the domestication of agriculture from plant and animal sources. These advances emerged across regions of the world from river valley civilizations in Asia (Mesopotamia, India and China) and Africa (Egypt) to Meso-American civilizations. As food crops and animal foods emerged from domestication to advancements in post-harvest processing stages, growth of invisible microbial systems led to modified foods that aligned with local palates of the communities. This in several cases became part of food systems, as it added a mode of preservation and improved the food quality that was preferred by each community. This is how the early genesis of fermented foods from grains, fruits and vegetables to animal meats and milk products became an integral part of early civilizations over 4000 years ago and has now carried over and further developed across many countries in every region of the world. The foundations of these food fermentations are now an integral part of modern food advances as preferred food substrates and the microbial interactions that drive the metabolic processes in defined food matrixes are being scientifically advanced.

This book focuses on bringing together diverse emergences and advancements of fermented foods across different regions of the world and how the metabolic processes associated with fermentation in several cases add health-relevant functional qualities across different food matrixes. This book contains 26 chapters from the Americas, Africa, Europe and Asia, bringing together the fermentation processes of diverse food substrates and their microbial processing and in several cases providing health-targeted functional benefits.

Key Features:

- Discusses the rationale and basis of fermented food systems from across 26 diverse regions of the world with examples from the Americas, Asia, Europe and Africa
- Presents historical plant and animal food substrates and the metabolic processes of their fermentation across traditions of a range of countries
- Covers traditional food fermentations across diverse food substrates that have potentially health-relevant functional benefits
- Reflects and presents the role of historical, cultural and traditional food concepts of a diverse range of countries with many novel examples and how they have influenced health-relevant functional benefits
- Discusses the relevance of metabolic concepts of fermented and functional foods in advancing healthy food concepts

This book is important for food scientists, nutritionists and the health care sector, but beyond this, it is also relevant for a wider global audience interested in a holistic health approach from food systems where examples of fermentation experience can inform new natural processing strategies to improve food quality and health.

Food Biotechnology

Series Editors:
Kalidas Shetty and Dipayan Sarkar

Functional Foods and Biotechnology
Kalidas Shetty, Gopinadhan Paliyath, Anthony Pometto, Robert E. Levin

Functional Foods and Biotechnology: Sources of Functional Foods and Ingredients
Kalidas Shetty and Dipayan Sarkar

Functional Foods and Biotechnology: Biotransformation and Analysis of Functional Foods and Ingredients
Kalidas Shetty and Dipayan Sarkar

Fermentation Biotechnology for Functional Foods

Edited by

Kalidas Shetty and Lena Gálvez Ranilla

CRC Press
Taylor & Francis Group
Boca Raton London New York

CRC Press is an imprint of the
Taylor & Francis Group, an **informa** business

First edition published 2026
by CRC Press
2385 NW Executive Center Drive, Suite 320, Boca Raton FL 33431

and by CRC Press
4 Park Square, Milton Park, Abingdon, Oxon, OX14 4RN

CRC Press is an imprint of Taylor & Francis Group, LLC

ISBN: 978-1-032-99653-0 (hbk)
ISBN: 978-1-032-99652-3 (pbk)
ISBN: 978-1-003-60530-0 (ebk)

DOI: 10.1201/9781003605300

Typeset in Times
by Apex CoVantage, LLC

To all our ancestors across the world whose curiosity and adaptation to local ecologies advanced fermented foods from diverse food substrates

Contents

Foreword ix
Preface xi
About the Editors xiii
List of Contributors xv
Acknowledgments xxi

1 Introduction **1**
Kalidas Shetty and Lena Gálvez Ranilla

2 Diversity of Traditional Fermented Foods in Algeria and Their Microbial Ecosystems **3**
Zahia Benmouna and Houari Benamar

3 Fermented Sweet Potato Products: Production, Health Benefits, and Utilization **17**
Joy Ehiwuogu-Onyibe and Kolawole Banwo

4 Moroccan Fermented Foods and Their Potential Health Benefits **45**
Tarik Lakhlifi, Samia El Oirdi, and Abdelhaq Belhaj

5 Tempeh: Indonesia's Indigenous Health-Promoting, Sustainable, and Affordable Fermented Food **55**
Amadeus Driando Ahnan-Winarno, Lorraine Cordeiro, Wida Winarno, Florentinus Gregorius Winarno, and Hang Xiao

6 Biotechnology for Functional Foods with a Focus on Traditional Fermented Foods and Beverages of Iran **69**
N. Davati and S. Hesami

7 Terasi: Biotechnology of the Indigenous Fish Fermented Products of Indonesia **95**
Widya Agustinah, Saraswati, and Nuri Andarwulan

8 Fermentation Biotechnology and Bioactive Compounds of Douchi, a Chinese Traditional Salt-Fermented Soybean Product **110**
Siyu Yao, Anqianyi Tu, Minghuang Chen, Chenxin Zhou, Siyong You, Zhenhui Pan, Yue Zhang, Caili Fu, and Jie Xu

9 Soybean-Based Fermented Foods of Northeast India: Microbiology, Nutritional Value and Health Benefits **122**
Kirti Rai and Kriti Ghatani

10 Fermentation Biotechnology of Food and Beverage Products in Malaysia **142**
Yu-Hsuan How, Nor Khaizura Mahmud Ab Rashid, Seng-Joe Lim, Lee-Sin Chang, and Liew-Phing Pui

11 Traditional Fermented Foods of Northeast India: Microbial Diversity and Functionality **162**
Shankar Prasad Sha

12 Lactic Acid Bacteria and Biotechnology of Cereal and Legume-Based Fermented Foods from India **175**
Pooja Singh, Ram Kulkarni, Kalidas Shetty, and Vinay Rale

13 **Microbial Diversity for Novel Fermented Foods and Perspectives from Pakistan** 192
Uroosa Ejaz and Muhammad Sohail

14 **Recent Advances and Future Perspectives in the Development of Healthier Yu-Lu, a Chinese Salt-Fermented Fish Sauce** 203
Chenxin Zhou, Wenshuang Xue, Minghuang Chen, Jing Zhou, Ying Li, Huijie Liu, and Ruichang Gao

15 **Tea Plantation and Processing of Traditional Products with Potential Health Benefits in Laos PDR** 212
Somsay Phovisay, Kridsada Unban, Kalidas Shetty, Saisamorn Lumyong, and Chartchai Khanongnuch

16 **Korea Indigenous Fermented Soy Products** 225
Dae Young Kwon, Kim Soon-Hee, Kyung Rhan Chung, Dai-Ja Jang, Su-Jin Jung, Kalidas Shetty

17 **Worldwide Fermented Non-Meat Sausages and Their Importance for Health** 239
Inês Azevedo, Joana Bastos Barbosa, Helena Albano, and Paula Teixeira

18 **Traditional Fermented Meat Products from Portugal** 271
Luís Soares, Joana Gomes, Teresa Bento de Carvalho, Mónica Oliveira, Lourenço Pinto de Rezende, Irene Martín, Joana Bastos Barbosa, and Paula Teixeira

19 **Innovative Potential of Traditional Greek Lactic Acid Fermented Vegetables for Functional Foods** 293
Spiros Paramithiotis

20 **Traditional Turkish Fermented Beverages and Their Health Benefits** 303
Hasan Tanguler and Hande Baltacioglu

21 **Traditional Turkish Fermented Foods and Their Health Benefits** 328
Hande Baltacioglu and Hasan Tanguler

22 **Fermented Foods from Chile and Health Benefits** 342
M. Carolina Henríquez

23 **Biotechnological Advances in Diverse Brazilian Fruit Fermentation** 351
Carolina Saori Ishii Mauro, Maria Thereza Carlos Fernandes, Paulo Terumitsu Saito, Fernanda Silva Farinazzo, Deepak Kumar Verma, and Sandra Garcia

24 **Food Fermentation in Brazil: Historical Aspects, Production Processes, and Health-Relevant Importance** 380
Francieli Begnini, Leidiane Andreia Acordi Menezes, and Cristina Maria Zanette

25 **Traditional Andean Potato-Based Fermented Foods and Their Potential for Health** 396
Lena Gálvez Ranilla

26 **Cacao Fermentation Process in Mexico and Targeting Functional Food Products** 414
María J. Paris, Leonor Lastra-Vargas, Enrique Palou, Emma Mani-López, and Aurelio López-Malo

27 **Traditional Fermented Foods of Mexico and Potential Functional Benefits** 450
Emma Mani-López and Aurelio López-Malo

Index 467

Foreword

As we appreciate our ancestors' wisdom to capture the essence of fermented food across a diverse range of food substrates, we are indebted to scientific advances from hardworking scientists and the fellow young minds in their laboratories. Their efforts advanced the concepts and discoveries for deeper understanding of fermented foods.

Preface

Fermentation-based processing became an essential part of global food systems as soon as humanity started to organize the domestication of agriculture from plant and animal sources. Fermentation as a natural processing and preservation strategy emerged intuitively through daily experiences with food systems across regions of the world from river valley civilizations in Asia (Mesopotamia, India and China) and Africa (Egypt) to Meso-American civilizations. As food crops and animal foods emerged from domestication to advancements in post-harvest processing stages, the growth of invisible microbial systems led to modified foods which could be preserved better and were adapted to local cuisine and taste profiles. These fermentation discoveries became an integral part of food systems, as they added a mode of preservation and improved the food quality that was preferred by each community. This shaped the early genesis of a range of fermented foods globally from grains, fruits and vegetables to animal meats and milk products to evolve and become an integral part of early civilizations over 4000 years and has now carried over and further advanced across many countries in every region of the world.

The foundations of the methods and utility of these food fermentations are now integral parts of modern food advances as more discoveries of food substrates and their microbial interactions have emerged to drive metabolic process in very defined food matrixes. This has led to scientific advances on the microbial and metabolic processes coupled to microbial interactions in each food matrix. Further, microbial processing through fermentation is largely driven by low oxygen-based metabolism through a succession of bacteria and fungi, yet it is mainly influenced by a food matrix produced through cycles of carbon fixation and oxidative metabolism in eukaryotic systems (plant and animal food matrixes). This food matrix and microbial interactions under a low-oxygen environment have shaped many fermented foods that have become an integral part of communities around the world with the evolving diversity of the food matrix over centuries and the associated natural microbial processing. These critical interactions, beyond improvements in food preservation and food safety, also have relevant and increasingly critical health benefits emerging from the microbially processed food matrix. Understanding these fermented food systems and their matrixes and how fermentation processing could contribute to human health through enhanced functional foods is an important area of research that must be advanced with rigorous scientific methods and advances.

Therefore, this book focuses on bringing together the diverse emergence of fermented food from across different regions of the world and how the metabolic processes associated with fermentation in several examples add health-relevant functional qualities across different food matrixes. The understanding of the metabolic processes of fermentation of diverse food substrates from across the world and how they contribute to health-relevant functional benefits leads to the development and improvement of these processes to further advance the quality of global food systems towards health outcomes. Such a metabolic understanding of food matrixes and microbial interactions can integrate a wider diversity of food systems towards sustainable solutions and quality to advance food security and health. This book contains 26 chapters from the Americas, Africa, Europe and Asia, bringing together fermentation processes of diverse food matrixes and their microbial processing with several examples of health-targeted functional benefits.

Kalidas Shetty,
North Dakota State University, USA
Lena Gálvez Ranilla,
Universidad Catolica de Santa Maria, Peru

About the Editors

Dr. Kalidas Shetty is Professor of Plant Sciences at North Dakota State University, Fargo, USA. Dr. Shetty's research interests focus on the critical role of the cellular and metabolic basis of oxygen biology for advancing new innovations in life sciences and especially agricultural and food innovations to advance global food security and health in a sustainable environment. This has relevance to advance scientific, educational and policy strategies to develop climate-resilient health-targeted food security solutions including malnutrition and hunger challenges. He has published over 260 manuscripts in peer-reviewed journals and over 50 as invited reviews and in conference proceedings, with an H-Index of 82 on Google Scholar. In 2004, he was selected by the US State Department as the inaugural Jefferson Science Fellow to advise on scientific issues as they relate to international diplomacy and international development. Dr. Shetty has widely traveled, presenting lectures and seminars in the areas of food biology, climate-resilient healthy food systems for food security and health and food safety in over 50 countries in Asia, Europe, Africa and the Americas. His current passion is to advance research, education capacity building and policy on a sustainable and ecological basis for climate-resilient healthy food systems and food diversity to drive global food security. This vision is based on crop and food diversity, indigenous wisdom, traditional fermentations and new technological innovations in ethnic and indigenous food systems, integrating an understanding of the comparative cellular biochemistry of plant and animal systems and their interactions with microbial systems in diverse ecologies.

Dr. Lena Gálvez Ranilla is a professor of Universidad Catolica de Santa Maria, Arequipa, Peru. Dr Gálvez has a PhD in food science from the University of Sao Paulo (Brazil) with sandwich doctoral research at the University of Massachusetts (Department of Food Science, USA). Currently, she leads the Laboratory of Research in Food Science at Universidad Catolica de Santa María (Peru). She received the prize of the L'Oréal–UNESCO–National Council for Science and Technology (Peru)–National Academy of Sciences (Peru) for Women in Science. Dr. Gálvez's research has been focused on the study of Latin American food biodiversity as a potential source of bioactive compounds with health-relevant functional properties. She has led research projects funded by Peruvian and international institutions aimed at valorizing crop diversity, with a focus on native grains. Currently, Dr. Gálvez's group is aimed at applying omics-based technologies for the characterization and value addition of Peruvian crop diversity from Andean food systems, with relevance for health and food security.

Contributors

Leidiane Andreia Acordi Menezes
Department of Food Science and Technology
Federal University of Santa Catarina
Florianópolis, SC, Brazil

Widya Agustinah
Food Technology Study Program
School of Bioscience, Technology,
and Innovation
BSD Campus
Atma Jaya Catholic University of Indonesia
Tangerang, Indonesia

Amadeus Driando Ahnan-Winarno
Indonesian Tempe Movement
and
Better Nature Ltd.
London, UK

Helena Albano
Escola Superior de Enfermagem de Coimbra
and
Escola Superior Agrária do Instituto Politécnico de
Viana do Castelo
Portugal

Nuri Andarwulan
Department of Food Science and Technology
Southeast Asian Food and Agricultural Science and
Technology (SEAFAST) Center
IPB University
Bogor, Indonesia

Inês Azevedo
Escola Superior de Biotecnologia
CBQF—Centro de Biotecnologia e Química Fina
Laboratório Associado
Universidade Católica Portuguesa
Porto, Portugal

Hande Baltacioglu
Department of Food Engineering
Faculty of Engineering
Niğde Ömer Halisdemir University
Niğde, Türkiye

Kolawole Banwo
Food Microbiology and Biotechnology Unit
Department of Microbiology
University of Ibadan
Oyo State, Nigeria

Joana Bastos Barbosa
Escola Superior de Biotecnologia
CBQF—Centro de Biotecnologia e Química Fina
Laboratório Associado
Universidade Católica Portuguesa
Porto, Portugal

Francieli Begnini
Department of Chemical Engineering
Post Graduate Program in Food Engineering
Federal University of Paraná
Curitiba, PR, Brazil

Abdelhaq Belhaj
Microbial Ecology, Cellular Interactions and Environment
Team
Department of Biology
Faculty of Sciences
Moulay Ismail University
Meknes, Morocco.

Houari Benamar
Department of Biology
Faculty of Natural and Life Sciences
University of Oran1 Ahmed Ben Bella
Oran, Algeria
and
Laboratory of Research in Arid Areas
Faculty of Biological Sciences
University of Science and Technology Houari Boumediene,
USTHB
Algiers, Algeria

Zahia Benmouna
Department of Living and Environment
Faculty of Natural and Life Sciences
University of Sciences and Technology of Oran Mohamed
Boudiaf, USTOMB
Oran, Algeria
and
Laboratory of Research in Arid Areas
Faculty of Biological Sciences
University of Science and Technology Houari Boumediene,
USTHB
Algiers, Algeria

Maria Thereza Carlos Fernandes
Departamento de Ciência e Tecnologia de Alimentos—CCA
Programa de Pós-Graduação em Ciência e Tecnologia de
Alimentos
Universidade Estadual de Londrina
Londrina, PR, Brazil

Teresa Bento de Carvalho
Escola Superior de Biotecnologia
CBQF—Centro de Biotecnologia e Química Fina
Laboratório Associado
Universidade Católica Portuguesa
Porto, Portugal

Lee-Sin Chang
Department of Food Science and Nutrition
Faculty of Applied Sciences
UCSI University
Kuala Lumpur, Malaysia
and
Department of Food Sciences
Faculty of Science and Technology
Universiti Kebangsaan Malaysia
Selangor, Malaysia

Minghuang Chen
College of Biological Sciences and Engineering
Fuzhou University
Fuzhou, Fujian, China

Lorraine Cordeiro
Department of Nutrition
University of Massachusetts
Amherst, Massachusetts, USA

N. Davati
Department of Food Science and Technology
Faculty of Food Industry
Bu-Ali Sina University
Hamedan, Iran

Joy Ehiwuogu-Onyibe
Biotechnology Department
Federal Institute of Industrial Research
Oshodi, Lagos, Nigeria

Uroosa Ejaz
Department of Biosciences
Faculty of Life Sciences
Shaheed Zulfiqar Ali Bhutto Institute of Science and Technology
SZABIST University
Karachi, Pakistan

Samia El Oirdi
Microbial Ecology, Cellular Interactions and Environment Team
Department of Biology
Faculty of Sciences
Moulay Ismail University
Meknes, Morocco

Caili Fu
College of Biological Sciences and Engineering
Fuzhou University
Fuzhou, Fujian, China
and
College of Chemical Engineering
Nanjing Forestry University
Nanjing, China
And
Fujian Putian Sea-100 Food Co., Ltd
Putian, Fujian, PR China

Ruichang Gao
School of Food and Biological Engineering
Jiangsu University
Zhenjiang, Jiangsu, China

Sandra Garcia
Departamento de Ciência e Tecnologia de Alimentos—CCA
Programa de Pós-Graduação em Ciência e Tecnologia de Alimentos
Universidade Estadual de Londrina
Londrina, PR, Brazil

Kriti Ghatani
Department of Food Technology
University of North Bengal
Siliguri, West Bengal, India

Joana Gomes
Escola Superior de Biotecnologia
CBQF—Centro de Biotecnologia e Química Fina
Laboratório Associado Universidade Católica Portuguesa
Porto, Portugal

M. Carolina Henríquez
Escuela de Nutrición y Dietética
Facultad de Farmacia
Universidad de Valparaíso
Valparaíso, Chile
and
Centro de Micro-Bioinnovación
Universidad de Valparaíso
Valparaíso, Chile
and
Centro de Investigación del Comportamiento Alimentario
Escuela de Nutrición y Dietética
Facultad de Farmacia
Universidad de Valparaíso
Valparaíso, Chile

S. Hesami
Department of Biology
Faculty of Science
Bu-Ali Sina University
Hamedan, Iran

Yu-Hsuan How
Department of Food Science and Nutrition
Faculty of Applied Sciences
UCSI University
Kuala Lumpur, Malaysia

Dai-Ja Jang
Korea Food Research Institute
Jeonju, Jeonbuk, Korea

Su-Jin Jung
Clinical Trial Center for Functional Foods
Jeonbuk National University Hospital
Jeonju, Jeonbuk, Korea

Nor Khaizura Mahmud Ab Rashid
Department of Food Science
Faculty of Food Science and Technology
Universiti Putra Malaysia
Selangor, Malaysia

Chartchai Khanongnuch
Research Center for Multidisciplinary Approaches to Miang
Multidisciplinary Research Institute
Chiang Mai University
Mueang, Chiang Mai, Thailand

Ram Kulkarni
Symbiosis School of Biological Sciences
Symbiosis International (Deemed University)
Lavale, Pune, India

Deepak Kumar Verma
Agricultural and Food Engineering Department
Indian Institute of Technology Kharagpur
Kharagpur, West Bengal, India

Tarik Lakhlifi
Microbial Ecology, Cellular Interactions
and Environment Team
Department of Biology
Faculty of Sciences
Moulay Ismail University
Meknes, Morocco

Leonor Lastra-Vargas
Chemical and Food Engineering Department
University of Americas-Puebla
Puebla, Mexico

Ying Li
School of Food and Biological Engineering
Jiangsu University
Zhenjiang, Jiangsu, China

Seng-Joe Lim
Department of Food Sciences
Faculty of Science and Technology
University Kebangsaan Malaysia
Selangor, Malaysia

Huijie Liu
School of Food and Biological Engineering
Jiangsu University
Zhenjiang, Jiangsu, China

Aurelio López-Malo
Chemical and Food Engineering Department
University of Americas-Puebla
Puebla, Mexico

Saisamorn Lumyong
Department of Biology
Faculty of Science
Chiang Mai University
Mueang, Chiang Mai, Thailand

Emma Mani-López
Chemical and Food Engineering Department
University of Americas-Puebla
Puebla, Mexico

Irene Martín
Escola Superior de Biotecnologia
CBQF—Centro de Biotecnologia e Química
Fina—Laboratório Associado
Universidade Católica Portuguesa
Porto, Portugal

Mónica Oliveira
Escola Superior de Biotecnologia
CBQF—Centro de Biotecnologia e Química
Fina—Laboratório Associado
Universidade Católica Portuguesa
Porto, Portugal

Enrique Palou
Chemical and Food Engineering Department
University of Americas-Puebla
Puebla, Mexico

Zhenhui Pan
College of Biological Sciences and Engineering
Fuzhou University
Fuzhou, Fujian, China

Spiros Paramithiotis
Laboratory of Microbiology
Department of Biological Applications and Technology
University of Ioannina
Ioannina, Greece

María J. Paris
Chemical and Food Engineering Department
University of Americas-Puebla
Puebla, Mexico

Somsay Phovisay
Interdisciplinary Program in Biotechnology
The Graduate School
Chiang Mai University
Mueang, Chiang Mai, Thailand
and
Department of Food Science and Technology
Faculty of Agriculture and Forest Resource
Souphanouvong University
Luangprabang, Lao PDR

Lourenço Pinto de Rezende
Escola Superior de Biotecnologia
CBQF—Centro de Biotecnologia e Química Fina—
Laboratório Associado
Universidade Católica Portuguesa
Porto, Portugal

Shankar Prasad Sha
Food Microbiology Laboratory
Department of Botany
Kurseong College
Darjeeling, West Bengal, India

Liew-Phing Pui
Department of Food Science and Nutrition
Faculty of Applied Sciences
UCSI University
Kuala Lumpur, Malaysia

Kirti Rai
Department of Food Technology
University of North Bengal
Siliguri, West Bengal, India

Vinay Rale
Symbiosis Centre for Research and Innovation
Symbiosis International (Deemed University)
Lavale, Pune, India
and
Viva Enviro Agrisciences LLP
Pune, India

Lena Gálvez Ranilla
Laboratory of Research in Food Science
Universidad Catolica de Santa Maria
Arequipa, Peru
and
Escuela Profesional de Ingeniería de Industria Alimentaria,
Departamento de Ciencias e Ingenierías Biológicas y
Químicas
Facultad de Ciencias e Ingenierías Biológicas y Químicas
Universidad Catolica de Santa Maria
Arequipa, Peru

Kyung Rhan Chung
Institute of Food Culture and Science
Yongin, Kyongki, Korea

Carolina Saori Ishii Mauro
Departamento de Ciência e Tecnologia de Alimentos—CCA
Programa de Pós-Graduação em Ciência e Tecnologia de
Alimentos
Universidade Estadual de Londrina
Londrina, PR, Brazil

Saraswati
Department of Food Science and Technology
IPB University
Bogor, Indonesia

Kalidas Shetty
Global Institute of Food Security and International
Agriculture (GIFSIA)
Department of Plant Sciences
North Dakota State University
Fargo, ND, USA

Fernanda Silva Farinazzo
Departamento de Ciência e Tecnologia de Alimentos—CCA
Programa de Pós-Graduação em Ciência e Tecnologia de
Alimentos
Universidade Estadual de Londrina
Londrina, PR, Brazil

Pooja Singh
Symbiosis Centre for Waste Resource Management
Symbiosis International (Deemed University)
Lavale, Pune, India

Luís Soares
Escola Superior de Biotecnologia
CBQF—Centro de Biotecnologia e Química Fina—
Laboratório Associado
Universidade Católica Portuguesa
Porto, Portugal

Muhammad Sohail
Department of Microbiology
University of Karachi
Karachi, Pakistan

Kim Soon-Hee
Institute of Food Culture and Science
Yongin, Kyongki, Korea

Hasan Tanguler
Department of Food Engineering
Faculty of Engineering
Niğde Ömer Halisdemir University
Niğde, Türkiye

Paula Teixeira
Escola Superior de Biotecnologia
CBQF—Centro de Biotecnologia e Química Fina—
Laboratório Associado
Universidade Católica Portuguesa
Porto, Portugal

Paulo Terumitsu Saito
Departamento de Ciência e Tecnologia de Alimentos—CCA
Programa de Pós-Graduação em Ciência e Tecnologia de
Alimentos
Universidade Estadual de Londrina
Londrina, PR, Brazil

Anqianyi Tu
Faculty of Science
University of Waterloo
Waterloo, Canada

Kridsada Unban
Division of Food Science and Technology
Faculty of Agro-Industry
Chiang Mai University
Mueang, Chiang Mai, Thailand

Florentinus Gregorius Winarno
Engineering Committee
Indonesian Academy of Sciences
Jakarta Pusat, Indonesia

Wida Winarno
MBRIO Food Safety
Bogor, Jawa Barat, Indonesia

Hang Xiao
Department of Food Science
University of Massachusetts
Amherst, Massachusetts, USA

Jie Xu
College of Biological Sciences and Engineering
Fuzhou University
Fuzhou, Fujian, China

Wenshuang Xue
School of Food and Biological Engineering
Jiangsu University
Zhenjiang, Jiangsu, China

Siyu Yao
College of Biological Sciences and Engineering
Fuzhou University
Fuzhou, Fujian, China

Siyong You
College of Biological Sciences and Engineering
Fuzhou University
Fuzhou, Fujian, China

Dae Young Kwon
Korea Academy of Science and Technology
Songnam, Kyongki, Korea
and
Institute of Food Culture and Science
Yongin, Kyongki, Korea

Cristina Maria Zanette
Department of Food Engineering
Midwestern State University (UNICENTRO)
Guarapuava, PR, Brazil

Yue Zhang
College of Biological Sciences and Engineering
Fuzhou University
Fuzhou, Fujian, China

Chenxin Zhou
College of Food Science and Engineering
South China University of Technology
Guangzhou, Guangdong, China
and
College of Biological Sciences and Engineering
Fuzhou University
Fuzhou, Fujian, China

Jing Zhou
School of Food and Biological Engineering
Jiangsu University

Acknowledgments

A special thank you to all the contributors who patiently worked with us through the burdens of COVID-19 and personal challenges. We are grateful to all of you and your contributions to advance fermented foods.

Introduction

1

Kalidas Shetty and Lena Gálvez Ranilla

This book brings together diverse culturally relevant and locally developed fermented foods and some beverages from different regions of the world. The focus is on metabolic processes associated with fermentation, with several examples that add health-relevant functional qualities across different food matrixes. The knowledge of the metabolic processes of fermentation of diverse food substrates and their health-relevant functional benefits provides a range of strategies towards the development and improvement of these processes to further advance the quality of global food systems towards health benefits. The metabolic understanding of food matrixes and the associated microbial interactions allows the integration of a wider diversity of food systems towards sustainable solutions, advancing food security and health, especially in terms of modern challenges of non-communicable chronic disease. This book contains 26 chapters from the Americas, Africa, Europe and Asia, bringing together fermentation processes of diverse food matrixes and their microbial processing with several examples of health-targeted functional benefits.

It is increasingly clear that fermentation-based processing is an important part of global food systems that emerged since the domestication of agriculture from plant and animal sources. Fermentation as a natural processing and preservation strategy emerged with cultural and experimental experiences with food systems across regions of the world from river valley civilizations in the Afro-Asiatic regions of Mesopotamia, India, China and Egypt to Meso-American civilizations in North and South America. As food-based sources and their domestication advanced, methods for post-harvest processing and preservation emerged with the growth of invisible microbial systems. These experiences led to modified foods which preserved better and were adapted to the local community culture and choices. These fermentation discoveries were coupled with overall food systems as they added methods of preservation and improved food quality with local adaptation to taste and therefore preference for such foods. This led to the evolution and early genesis of diverse fermented foods globally from a wide range of food matrixes such as grains, fruits, vegetables and animal meats and milk products and has become an integral part of humanity since the dawn of early civilizations over 4000 years ago. This is now being advanced with newer scientific advances across many regions of the world and is discussed across the 26 chapters in this book.

The advances in understanding food fermentations better are now well integrated with improved methods in modern food advances with continuous discoveries of food substrates. The key discoveries include microbial interactions which drive metabolic processes in well-defined food matrixes. This is leading to scientific advances to better define the microbial and metabolic processes in each food matrix and the key microbial interactions associated with each food matrix. Furthermore, microbial processing through fermentation and understanding of succession from the low oxygen-based metabolism of bacteria and yeast to aerobic growth filamentous fungi influencing the food matrix mobilizing through cycles of carbon fixation and oxidative metabolism in plant and animal food matrix is being clarified. The 26 chapters in the book help advance the understanding of these fermented food systems and their food matrix, and several chapters cover how these processes can advance benefits for human health through the development of functional health benefits. The chapters have been organized from our human emergence from Africa to Asia, Europe and the Americas.

Africa: There are three interesting chapters from Africa: "Diversity of Traditional Fermented Foods in Algeria and Their Microbial Ecosystems"; "Moroccan Fermented Foods and their Potential Health Benefits"; and "Fermented Sweet Potato Products: Production, Health Benefits and Utilization". These chapters provide insights into traditional artisanal fermented foods, which remain the main source of nutrition for many rural communities. The explanation for these health benefits relates to the biologically active compounds produced by the microorganisms responsible for fermentation or their probiotic effects. Furthermore, information on fermented sweet potato in Nigeria provides insights on how traditional fermentation can be aligned to modern contemporary needs.

Asia: Due to the large population and diversity of regions, Asia has the widest coverage of the range of fermented foods, across 12 chapters. These include

DOI: 10.1201/9781003605300-1

processes from Indonesia, "Tempeh: Indonesia's Indigenous Health-Promoting, Sustainable, and Affordable Fermented Food", and "TERASI: Biotechnology of The Indigenous Fishery Fermented Products of Indonesia". From another important country in Southeast (SE) Asia, Malaysia, we have "Fermentation Biotechnology of Food and Beverage Products in Malaysia". From the northern region of SE Asia, we have a novel chapter, "Tea Plantation and Processing to Traditional Products with Potential Health Benefits in Laos PDR". From West Asia, we have an excellent chapter with roots in ancient Persia, "Biotechnology for Functional Foods with a Focus on Traditional Fermented Foods and Beverages of Iran". From South Asia, we have a chapter from Pakistan, "Microbial Diversity for Novel Fermented Foods and Perspectives from Pakistan". From India, a sub-continent, we have additional coverage of the traditional fermented foods and more recent soybean-based fermented foods of northeast (NE) India, "Traditional Fermented Foods of Northeast India; Microbial Diversity and Functionality"; "Lactic Acid Bacteria and Biotechnology of Cereal and Legume-Based Fermented Foods from India"; and "Soybean-Based Fermented Foods of Northeast India: Microbiology, Nutritional Value and Health Benefits". From NE Asia, we have two chapters from China, "Fermentation Biotechnology and Bioactive Compounds of Douchi, a Chinese Traditional Salt-Fermented Soybean Product", and "Recent Advances and Future Perspectives in the Development of Healthier Yu-lu, a Chinese Salt Fermented Fish Sauce", and a chapter from the Republic of Korea, "Korean Indigenous Fermented Soy Products". The 12 representative chapters from Asia cover a wide range of regional food matrixes such as plant sources, milk, tea and fish. Interesting insights can be gained from comparative chapters on soybean fermentation from China, India, Indonesia and the Republic of Korea.

Europe: Fermented foods from Europe provide insights largely from the Mediterranean region over five chapters, starting from the Euro-Asian interface region of Turkey with two chapters, "Traditional Turkish Fermented Foods and Their Health Benefits" and "Traditional Turkish Fermented Beverages and Their Health Benefits"; from Greece, "Innovative Potential of Traditional Greek Lactic Acid Fermented Vegetables for Functional Foods"; from the eastern Mediterranean region of Portugal, two chapters, "Worldwide Fermented Non-Meat Sausages and Their Importance for Health" and "Traditional Fermented Meat Products from Portugal". These cover a wide range of plant- and animal-based fermented food and beverages, with good examples of health-relevant functional benefits.

The Americas: The Americas have very interesting representative fermentation processes based on several interesting, localized substrates across six chapters: From Mexico, we have two chapters, "Cacao Fermentation Process in Mexico and Targeting Functional Food Products" and "Traditional Fermented Foods of Mexico from Pre-Hispanic Times and Potential Functional Benefits". These are very interesting processes with novel traditional and indigenous substrates with contemporary scientific advances from fermentation processing to health benefits. From Brazil, there are two chapters, first "Food Fermentation in Brazil: Historical Aspects, Production Processes, and Health-Relevant Importance". This chapter has wide food substrates and fermentation examples, and a second chapter focuses on fruits: "Biotechnological Advances in Diverse Brazilian Fruit Fermentation". From Peru, we have very interesting and rare coverage of fermented potato with roots in the Andean region, "Traditional Andean Potato Based Fermented Foods and Their Potential for Health". From Chile, the focus is more on from grape fermentation to coverage of other rare, localized fruits: "Fermented Foods from Chile and Health Benefits". Overall, the Americas show a wide range of plant-based fermentation from Mexico to Chile that covers contemporary food substrates and fermentation that integrates indigenous knowledge both on food matrixes and fermentation processing, with several examples of health benefits.

In summary, the 26 chapters covers interesting and novel examples from Africa, Asia, Europe and the Americas, extending to a range of plant and animal substrates with both traditional indigenous fermentation processes with a long history of local cultural integration and more contemporary advances. Beyond understanding the processes of fermentation and novel substrates, several chapters highlight the health-relevant functional benefits. This provides a strong foundation to further build extensive examples of fermentation processes and their novel localized food substrates based on localized ecology and climatic preferences from around the world. Into these we can integrate how contemporary scientific processes and methods can be advanced for a deeper understanding of their benefits for improving food processing, preservation and food quality in terms of nutritional quality and health benefits.

Diversity of Traditional Fermented Foods in Algeria and Their Microbial Ecosystems

2

Zahia Benmouna and Houari Benamar

2.1 INTRODUCTION

Algeria has a rich tradition in food preparation and is known to produce many traditional fermented foods from animal or plant origins (Benkerroum 2013). First, nomadic and sedentary populations living in dry desert regions near oases have practiced these kinds of fermentation processes to preserve and stock their food. They have learned from the challenges of daily life to build resilience in their food system and fermentation to offer an economically efficient strategy for food preservation in harsh conditions. They know how to make the most of what nature gives them so sparingly (Ben Danou 1929). Then, these foods become widely consumed and highly appreciated by people living in the big cities due to increasing interest in healthy eating without extensive processing and chemical preservatives. Nowadays, traditional fermented foods play an important role in the economy and food security of these countries. Yet they are still mainly prepared at the household level, sometimes under poor sanitary conditions, and marketed through informal routes (Odunfa and Oyewole 1998; Benkerroum 2013).

The fermentation process is a natural technique, which mainly involves lactic acid bacteria (LAB) and/or yeasts. They are present naturally in the raw material or added as starter cultures in controlled fermentation and allow the disappearance of fungal flora over time and consequently make the product safe for consumption. It is a simple and inexpensive approach for the improvement of safety, nutritional and organoleptic values and the shelf life of food substrates. The wide variety of fermented foods is a result of differences in the most-represented diversity of bacterial species naturally found in the raw ingredients of each geographical region, local environmental conditions and processing procedures (Deegan et al. 2006; Merabti et al. 2015; Abriouel et al. 2017).

Traditional fermented products are a good source of autochthonous strains, which are usually well adapted to a particular food environment and are very useful to develop functional starters and especially suited for small-scale fermentations (Holzapfel 2002). They thus remain beyond any official control of their compliance to national regulatory standards. Sometimes their consumption is predicted to be of public health risk, although such a risk has never been estimated on a scientific basis due to the lack of consumption patterns, epidemiological data and appropriate surveillance programs (Benkerroum 2013). Improvements in the hygienic quality of food are very important to ensure the safety of traditional fermented food products. Some of the strategies that should be introduced to meet this goal are the provision of technical assistance to the producers and the introduction of compositional and hygienic quality payment schemes (Talon et al. 2007). This chapter provides a description of the microbial diversity of famous traditional fermented foods in Algeria and their probiotic potentials, as well as the application of biotechnology to traditional fermented foods and the safety-associated risk of consumption of these foods.

2.2 TYPES OF TRADITIONAL FERMENTED FOODS

During fermentation, microorganisms convert the chemical composition of raw materials, which enriches the nutritional value in some fermented foods and imparts health benefits to consumers (Steinkraus 2002; Farhad et al. 2010; Tamang 2015). This section explores traditional fermented foods from milk, meat, cereals and vegetables known in Algeria that have been subjected to many studies (Table 2.1), depending on the raw materials.

DOI: 10.1201/9781003605300-2

2.2.1 From Milk

Traditional dairy products with strong cultural, medicinal and economic values are historical products with social and economic relevance for rural communities and especially women (Claps and Morone 2011). They are the first traditional fermented foods, with a range of diversity and heterogeneity. Cheese, for example, occupies a very important socio-economic place established in rural and emerging urban environments.

2.2.1.1 Raib

Raib is spontaneously curdled raw milk developed within 24 to 36 h at the ambient temperature. It is prepared from cow milk without draining or churning (Hamama and Bayi 1991). It is a result of the growth of the microbial flora that was originally present in the milk (natural fermentation). From *raib*, Elotmani et al. (2002) identified *Enterococcus faecalis, E. faecium* and *Lactococcus lactis* with anti-*Listeria* activity. Rehaiem et al. (2010) identified *E. faecium*, also with strong anti-*Listeria* activity.

2.2.1.2 Lben

Lben is a traditional Algerian drink usually obtained from cow milk. Other types of milk can be used in the production of *lben*, such as goat, sheep and camel milks, or a mixture of two of these types, fermented for 24–48 h at about 18–24°C depending on the season. The curd after coagulation is brewed in a goatskin bag, the *chekoua*, which is stirred vigorously back and forth (Tantaoui-Elaraki and El Marrakchi 1987). *E. faecium, E. faecalis, E. avium, Leuconostoc mesenteroides, L. lactis* subsp. *cremoris* and *L. lactis* subsp. *lactis* were isolated from lben (Ghalouni et al. 2018). In other studies, Samet-Bali et al. (2012) identified *L. lactis* subsp. *cremoris* and *L. lactis* subsp. *lactis* from this food.

2.2.1.3 Lfrik

Lfrik is a fermented camel milk produced traditionally from raw camel milk. *Lfrik* is made by spontaneous fermentation of raw camel milk at room temperature in about 12 hours. Its acidity is significantly lower than that of *lben*, obtained from traditional fermented cow milk (Ismaili et al. 2017). The authors observed the dominance of LAB in *lfrik*, marked by the species of *L. lactis* subsp. *lactis biovar diacetylactis, Levilactobacillus brevis* and *Streptococcus salivarius* subsp. *thermophilus*.

2.2.1.4 Cheese

2.2.1.4.1 Jben

Jben is a fresh cheese produced by the spontaneous coagulation of milk casein and sometimes by the addition of vegetable products such as artichoke, cardoon flowers, wild thorny plant and pumpkin seeds able to produce curds. The curd is poured into a cloth or a mold to drain for two or more days until the desired texture is obtained. It may be salted or not. This traditional dairy product is very popular, mainly because of its refreshing quality (Hamama and Bayi 1991; Ouadghiri et al. 2005; Mennane et al. 2007).

2.2.1.4.2 Bouhezza

Bouhezza is a fermented soft-ripened cheese produced in several Chaouia regions (northeastern Algeria). Originally, *bouhezza* was the product of the processing of goat and sheep milks, and currently cow milk is used (Boudalia et al. 2020). The cheese is obtained after the transformation of *lben* in a *chekoua*. Draining, salting and refining *bouhezza* are performed simultaneously in the *chekoua* (Zitoun et al. 2011). The main sensory descriptors are lactic acid flavor and odor. The products of proteolysis give *bouhezza* a richness in volatiles and fragrance characteristics, which reflect the uniqueness of this cheese (Medjoudj et al. 2017).

2.2.1.4.3 Klila

Klila is a fermented cheese produced in several regions of Algeria. It is the most popular traditional cheese, and its artisanal manufacturing process is still in use today (Figure 2.1). The processing consists of moderate heating of *lben* until it becomes curdled, and then it is drained in muslin. The cheese obtained is consumed as it stands, fresh or after drying (Leksir et al. 2019). Amira et al. (2020) identified *Lactiplantibacillus plantarum* and *Lacticaseibacillus casei* in this cheese. In another study, Boubekri and Ohta (1996a, 1996b) isolated *Enterococcus, Pediococcus, Streptococcus, Lactobacillus* and *Leuconostoc* with antimutagenicity activity.

In a study on Algerian *klila*, Benamara et al. (2022) noted that the microbial charge of *klila* cheeses can vary a lot quantitatively and qualitatively. In fact, it was clearly influenced by

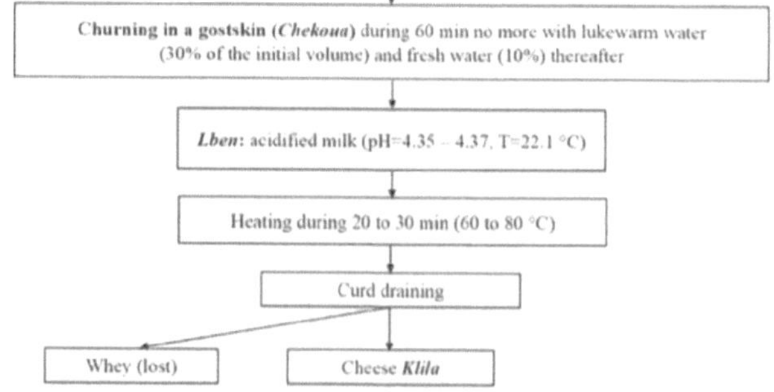

FIGURE 2.1 Raib, lben and klila cheese making (Benamara et al. 2016).

the origin of the milk: ewe, goat or cow. For instance, it is observed that farmers are more attentive to the hygienic conditions of their cows during milking than farmers breeding ewes or goats.

2.2.1.4.4 Rigouta

Rigouta is a fermented cheese that is often prepared by boiling acidified cheese whey from cow milk (Baccouri et al. 2020). Ghrairi et al. (2004) isolated 600 LAB from *rigouta* cheese, of which eight isolates were found to produce anti-*Listeria* activity.

2.2.1.4.5 Other Cheeses

Ighounane is a cheese made in the heights of Djurdjura in Kabylia (Algeria) from colostrum (the first milk of a cow giving birth). *Aghoughlou* is a cheese also made in Kabylia. It is obtained from fresh cow or goat milks coagulated by fig tree latex (*Ficus carica* L.). *Kemariya* or *takemmarite*, which is a traditional cheese made only from goat milk, is made by women according to traditional methods in the regions of M'zab (northern central and northwest of Algeria). *Oudiouan oulli* is a Tuareg cheese similar to cottage cheese served in small portions and eaten fresh or dried. *Ioulsan* is a traditional Algerian cheese prepared by the Tuaregs of Hoggar or Ihaggarren (people from the Algerian Sahara). It is a typical dry cheese (87 to 92% dry matter), obtained by the moderate heating of skimmed *lben*. *Mechouna* is a cheese traditionally prepared with goat or cow milks. The process begins with heat treatment of the milk to boiling. Then, *lben* is added with salt (the amount of *lben* is half that of milk). All the contents are then heated a second time until coagulation and separation of curd and whey (McSweeney et al. 2017; Leksir et al. 2019).

2.2.1.4.6 Dhan *or* Zebda *and* Smen

These are ethnic dairy products, a traditional fermented butter made from whole raw milk (Figure 2.2). These products constitute a significant part of the Algerian diet and represent a

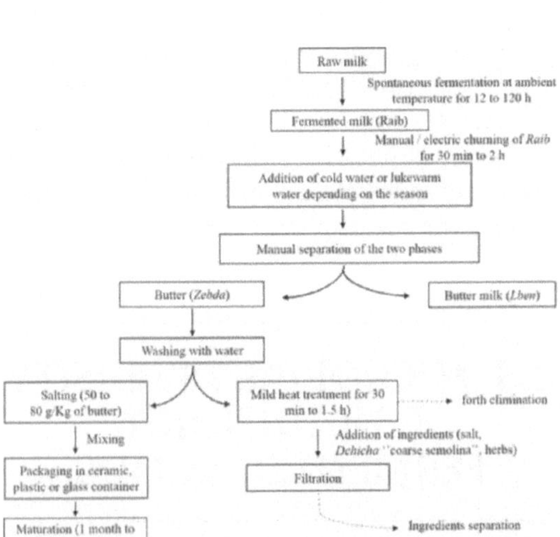

FIGURE 2.2 Diagram showing the different steps of traditional *zebda* and *smen* manufacturing (Boussekine et al. 2020).

gastronomical heritage that needs to be preserved and protected (Jamaly et al. 2011; Boussekine et al. 2020). *Dhan* is a special traditional fermented butter in Algeria (Boussekine et al. 2020). It is also called *zebda*. After the spontaneous coagulation of cow, goat and sheep milks, which are shaken, *dhan el bagra*, *dhan el maaza* and *dhan el naadja* are respectively recovered (Idoui et al. 2009). *Smen* is obtained from *zebda* after washing with salt water (Bensaleh et al. 2011). Several studies have been carried out to isolate LAB from traditional butter (Kacem and Karam 2006; Idoui and Karam 2008; Idoui et al. 2010). The predominance of the genus *Lactococcus* and *Leuconostoc* was observed in *dhan* (Idoui et al. 2009). Bensaleh et al. (2011) identified *S. thermophilus* and *Enterococcus* sp. from *smen* obtained from cow and ewe milks.

2.2.2 From Meat

In Algeria, various traditional meat products have long been known and prepared for the family. They were made to preserve meat when it was available in quantities exceeding immediate needs, while appropriate storage means such as refrigerators and freezers were lacking. Fermented meat is a more stable product that can be kept at room temperature for as long as possible without being spoiled or becoming hazardous to consumers' health (Benkerroum 2013).

2.2.2.1 Gueddid *or* Kadid

This is made without adding starter cultures and relies on the endogenous flora, traditionally selected from the surrounding environment to keep traditional organoleptic qualities and to provide an effective form of natural preservation. It is the most popular traditional dry fermented meat produced in Algeria. It is typically made by salting and drying the meat product (Belgacem et al. 2008). Forty-eight LAB strains have been isolated, of which four exhibited anti-*Listeria* activity. Essid et al. (2009) isolated 17 strains of *L. plantarum* with technological and safety properties.

2.2.2.2 Dried Ossban

This is a traditional dry fermented meat typically prepared from sheep intestine and meat mixed with salt and spices and dried through air exposure to sunlight for several days. Two strains, *P. pentosaceus* and *E. faecium*, were identified from dried *ossban* with resistance to gastrointestinal conditions (acidity and bile salts), susceptibility to clinically relevant antibiotics, production of bacteriocin and probiotic potential (Zommiti et al. 2018a, 2018b).

2.2.3 From Cereals and Vegetables

Cereals are a major component of human food in Africa. Nutritional experts have paid attention to cereal-based foods, and they are frequently complementary foods for young children and serve as dietary staples for adults. Fermented cereal foods play an important socio-economic role in developing countries as well as making a major contribution to the protein requirements of natural populations (Odunfa and Oyewole 1998; Tou et al. 2007; Franz et al. 2014). In Algeria, wheat is the most cultivated, consumed and transformed cereal in the food industry, and it is used in the best-known fermented cereal (Bekhouche et al. 2013; Gourchala et al. 2014; Kalbaza et al. 2018; Merabti et al. 2019).

2.2.3.1 Sourdough

Sourdough is a leavening agent traditionally obtained without the addition of starter microorganisms. Cereal fermentation using sourdough as a natural leavening agent improves bread texture and flavor, retards the staling process of bread and protects from bacterial and mold spoilage; it also improves the nutritional properties of the fermented products (Robert et al. 2009). Mamhoud et al. (2016) identified species belonging to the genera *Lactobacillus* sp. and *Pediococcus* sp. from wheat bread obtained by the sourdough fermentation process.

2.2.3.2 El-hammoum *or* Hamoum

El-hammoum or *hamoum* (Figure 2.3a) is obtained after the fermentation of wheat grains, *Triticum durum* Desf., stored in contact with a wall, *matmora* (Figure 2.3b). This fermentation can last 4–9 years to generate a new stable fermented product called *hamoum* which comes from the word *hmoum*, meaning black, referring to the black and brown color with a special aroma of fermented wheat. After the step of drying under the sun, *el-hammoum* is stored in sealed jars (Bekhouche et al. 2013). It is very much appreciated for its organoleptic and nutritional properties (richness in protein) and its benefits for health, especially by diabetics, because of the reduction of sugars linked to its fermentation (Gourchala et al. 2014). Studies on this food have focused on the identification of LAB and investigation of its technological characteristics and nutritional qualities (Gourchala et al. 2014; Kalbaza et al. 2018).

2.2.3.3 Lemzeiet

This is a traditional couscous manufactured from the fermented wheat *el-hammoum* (Figure 2.4) (Bekhouche et al. 2013). A microbial study on *lemzeiet* showed that the fermentation process allows the disappearance of fungal flora over time through the establishment of LAB, resulting in the characterization of the microbiota of this fermented wheat. It revealed a high level of LAB species and strain diversity during the different stages of maturation with the development of aroma compounds (Merabti et al. 2015). According to a study, *lemzeiet* would be a very interesting raw material for new functional fermented wheat-based products such as couscous and pasta. Therefore, LAB that has potentially imparted health benefits could be included in food processing procedures to increase product appeal and safety. Moreover, identifying LAB strains from fermented wheat *lemzeiet*—in terms of their diversity, functional properties, growth dynamics, and survival—will contribute to improved control of the fermentation process (Becila et al. 2022).

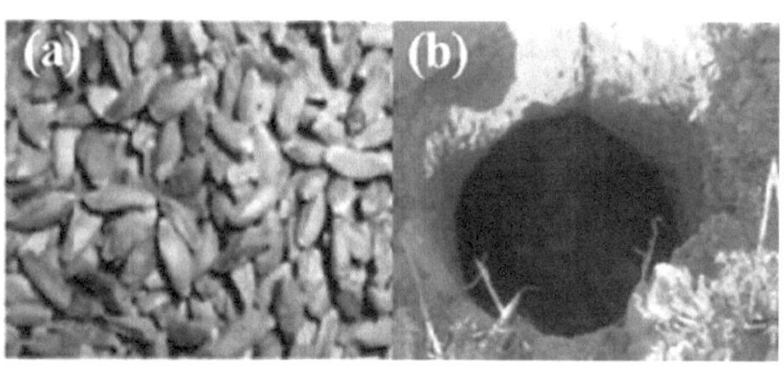

FIGURE 2.3 (a) Fermented wheat, *el-hammoum* (Gourchala et al. 2014); (b) *matmora* (Mokhtari et al. 2016).

FIGURE 2.4 Traditional couscous manufactured from *el-hammoum* (Bekhouche et al. 2013).

2.2.3.4 Traditional Fermented Green Olives

Table olives are an important fermented product in Algeria. The olives are non-alkali treated and are initially brined at higher sodium chloride concentrations (10–15%, w/v), and then they are allowed to undergo spontaneous fermentation (Abouloifa et al. 2019). This fermentation process allows the development of various microorganisms, mainly LAB, enterobacteria, yeasts and molds (Rokni et al. 2015). Abouloifa et al. (2019, 2021) identified *L. brevis, L. pentosus* and *L. plantarum* from traditional fermented green olives with high antifungal activity against *Penicillium* sp. and *Candida pelliculosa* and antibacterial effects against some Gram-positive and Gram-negative bacteria.

2.3 MICROBIAL DIVERSITY OF TRADITIONAL FERMENTED FOODS

Identification of the microorganisms present in traditional fermented foods using molecular biology techniques is highly suggested to understand and preserve their microbial diversity and to select the appropriate strains which could be used as starter cultures for food fermentation. Such strains may be selected on the basis of their performance during production, analysis of health-promoting functional qualities, safety, nutritional benefits and probiotic potential, which contribute to the manufacturing leading to added-value products (Benkerroum and Tamime 2004; Talon et al. 2007; Tamang et al. 2016). Furthermore, exploitation of microbial ecologies allows us to better understand the ecosystem and know which microorganisms are responsible for fermentation (Cocolin et al. 2009). Environmental conditions such as temperature, origin and composition of food substrates, processing and sanitary conditions might have a significant influence on the microbial and physicochemical composition of traditionally made products (Idoui et al. 2010). In the following section, the main microorganisms isolated from traditional fermented food are described (Table 2.1).

2.3.1 Bacteria

Bacteria are the main microorganisms present in traditional fermented food. This is strongly due to their capacity for rapid growth in different environments. This section describes the major groups of bacteria isolated from traditional fermented foods. The LAB group is more studied, probably due to their intervention in the fermentation process and dominance in contributing to functional properties.

2.3.1.1 Lactic Acid Bacteria

Studies usually show the presence of mesophilic LAB, and this group of bacteria has generally recognized as safe (GRAS) status. LAB have rarely been associated with food intoxication or health-relevant infections. Their presence in traditional fermented food is due to their strong association with the fermentation process. They convert most available carbohydrates to lactic acid, with small amounts of acetic acid, resulting in a lowering of the pH (Campbell-Platt 1987).

The main genus isolated from different traditional fermented foods is *Enterococcus*, with different species (Boubekri and Ohta 1996a, 1996b; Elotmani et al. 2002; Bekhouche et al. 2013; Ghalouni et al. 2018; Merabti et al. 2019). The majority of studies focused on the isolation of *Enterococcus* strains from traditional fermented foods (Ghrairi et al. 2004; Belgacem et al. 2008; Rehaiem et al. 2010; Hanchi et al. 2016). In fact, enterococci are the main fermenting microbiota of artisan dairy products from Mediterranean countries, most probably due to the relatively high temperatures found in this area (Franz et al. 2003). Enterococci are most frequently present in many traditional food products. Moreover, several enterococci were found to produce multiple and broad-spectrum bacteriocins (Hanchi et al. 2016; Benmouna et al. 2018, 2020).

The genera *Lactobacillus, Lactococcus, Leuconostoc, Pediococcus, Weissella* and *Streptococcus* were also isolated from traditional fermented foods (Essid et al. 2009; Jamaly et al. 2011; Kalbaza et al. 2018; Mechai et al. 2020). Among the species of *Lactobacillus* isolated from traditional fermented foods (Table 2.1), *L. brevis*, *L. curvatus*, *L. plantarum* and *L. casei* were identified from *klila*, *hammoum* and *lemzeiet* (Kalbaza et al. 2018; Merabti et al. 2019; Amira et al. 2020). From cow, goat and sheep butters and k*lila*, Adjoudj et al. (2020) isolated *L. plantarum, L. casei, L. fermentum* and *L. paracasei.* Strains of *L. brevis, L. pentosus* and *L. plantarum* were identified from traditional fermenting green olives (Abouloifa et al. 2019).

Several authors have mentioned the presence of strains of *Lactococcus* in some traditional fermented food. *L. lactis* was isolated from *lben* (Ouadghiri et al. 2009). *L. lactis* subsp. *cremoris* and *L. lactis* subsp. *lactis* were also isolated from *lben* (Samet-Bali et al. 2012). Other authors identified strains of *Pediococcus* from traditional fermented food (Table 2.1); for example, *P. pentosaceus* and *P. acidilactici* were isolated from cow, goat and sheep butters, *jben*, *klila* and *lemzeiet*, respectively (Bekhouche et al. 2013; Adjoudj et al. 2020; Mechai et al. 2020).

2.3.1.2 Other Bacteria

Traditional fermented foods contain other bacteria and some spoilage microbiota (*Staphylococcus, Pseudomonas* and enterobacteria). Their presence depends on raw materials or on the environment of the manufacturing process (Talon et al. 2007). Idoui et al. (2010) noted that environmental conditions such as temperature, origin and composition of milk and processing and sanitary conditions might have a significant influence on the microbial and physicochemical composition of traditionally made dairy products. Other authors have mentioned that coliforms contaminate traditional fermented foods (Samet-Bali et al. 2012, 2016). Species of *Yersinia* were also identified from traditional fermented milk (Hamama et al. 1992).

2.3.2 Yeast and Mold

The species of yeasts reported for fermented foods were *Candida lusitaniae, C. tropicalis* and *C. albicans.* They were isolated from *raib*, *lben* and *jben* (Hamama and Bayi 1991; Samet-Bali et al. 2012). Samet-Bali et al. (2016) mentioned that yeast and mold are the main microorganisms present in *raib* from cow milk. Some authors also noted the presence of yeast and molds in some fermented foods (Hamama and Bayi 1991; Idoui et al. 2010; Samet-Bali et al. 2016).

2.4 PROBIOTIC POTENTIAL OF TRADITIONAL FERMENTED FOODS

Generally, traditional fermented foods seem to confer a multitude of advantages, not only the preservative characteristics of this food but also various value-added benefits. These are

TABLE 2.1 Main Traditional Fermented Foods and Brief Descriptions of Studies

VERNACULAR NAME	*DEFINITION*	*MICROORGANISM*	*STUDY*	*REFERENCE*
Dairy (*raib, lben, jben* and *rigouta*) and fermented green olives	Fermented food products	*E. casseliflavus* *E. durans* *E. faecalis* *E. faecium* *E. mundtii*	• No resistance acquired against clinically relevant drugs used for enterococcal infections treatment in human medicines. • No hemolytic activity.	Rehaiem et al. (2016)
Artisanal fermented milk	–	*E. durans*	• Active against a wide range of Gram-positive pathogens such as *Listeria* sp. and *A. versicolor* and *Penicillium* commune spores.	Hanchi et al. (2016)
Raib	Traditional fermented milk	*E. faecalis* *E. faecium* *L. lactis*	• Action against *Listeria*. • Bacteriocin-producing LAB.	Elotmani et al. (2002)
	Traditional dairy beverage	*E. faecium*	• Strong anti-*Listeria* activity. • Production of bacteriocin. • Promising potential in biopreservation to enhance the hygienic quality of this dairy product.	Rehaiem et al. (2010)
	Traditional fermented cow milk	Coliforms *Staphylococcus* sp. *S. aureus* Yeast and mold	Hygienic quality.	Samet-Bali et al. (2016)
Raib from goat and cow milks, *bouhezza* and *klila*	Traditional dairy products	*L. plantarum*	• Production of EPS: role as stabilizers, thickeners and/or fat-replacers, and were considered natural biopolymers.	Bachtarzi et al. (2019)
Raib, lben and *jben*	Traditional dairy products	–	• Large case control study: individuals who have a high intake of traditional dairy products were at lower risk for CRC.	El Kinany et al. (2020)
Raib and *lben*	Traditional fermented milks	*Yersinia enterocolitica* *Y. frederiksenii* *Y. intermedia* *Y. kristensenii* *Y. pseudomberculosis*	• pH ranging from 3.9 to 4.9 for *lben* and from 4.1 to 5.1 for *raib*.	Hamama et al. (1992)
Raib, jben and *smen*	Traditional dairy products	*L. brevis* *L. paracasei* *L. plantarum*	• Probiotic potential: tolerated acid gastric conditions (pH 2.0 and pH 3.0) and resistance to bile salts. • Antagonistic activity against food-borne pathogen bacteria. • Sensitive to kanamycin and tetracycline. • Good hydrophobicity and adherence properties.	Jamaly et al. (2011)

(Continued)

TABLE 2.1 (Continued)

VERNACULAR NAME	*DEFINITION*	*MICROORGANISM*	*STUDY*	*REFERENCE*
Raib and *jben*	Traditional fermented milk and fresh cheese	Aerobic flora Fecal coliforms *S. aureus* Enterococci Lactobacilli Leuconostocs Streptococci Yeast and mold	• Microbiological analysis of *raib* and *jben*.	Hamama and Bayi (1991)
Lben	Traditional fermented skimmed milk	*E. durans* *E. faecium* *E. gilvus* *E. hirae* *L. brevis* *L. paracasei* *L. plantarum* *L. rhamnosus* *L. garvieae* *L. lactis* *Ln. citreum* *Ln. kimchii* *Ln. mesenteroides* *Ln. pseudomesenteroides* *P. pentosaceus* *W. cibaria* *W. confusa* *W. paramesenteroides* *W. viridescens*	Identification of LAB.	Ouadghiri et al. (2009)
	Fermented milk	*L. lactis* subsp. *cremoris* *L. lactis* subsp. *lactis Leuconostoc* sp. Coliforms and enterococci Yeasts (*C. albicans*, *C. lusitaniae* and *C. tropicalis*)	• Nutritional value of *Iben*: high mineral content and less fatty and acidic content compared with the industrial *Iben*. • Traditional *Iben* microstructure revealed an open and branched structure, while a compact structure with small pores was observed in industrial *Iben*.	Samet-Bali et al. (2012)
	Traditional fermented cow milk	*E. avium* *E. faecalis* *E. faecium* *L. lactis* subsp. *cremoris* *L. lactis* subsp. *lactis* *Ln. mesenteroides*	• Acidification ability. • Proteolytic activity.	Ghalouni et al. (2018)

(Continued)

TABLE 2.1 (Continued)

VERNACULAR NAME	*DEFINITION*	*MICROORGANISM*	*STUDY*	*REFERENCE*
Lfrik	Fermented camel milk	*Lactobacillus* sp. *L. brevis* *Lactococcus* sp. *L. lactis* subsp. *lactis biovar diacetylactis* *Leuconostoc* sp. *Streptococcus* sp. *Streptococcus salivarius* subsp. *thermophilus*	• Absence of pathogenic flora (*Salmonella* and *S. aureus*) in analyzed *Ifrik* samples due to its inhibition by LAB, which have the ability to produce active substances and bacteriocins, thus acting as bactericidal agents in fermented foods.	Ismaili et al. (2017)
Jben and *klila*	Traditional fermented cheese	*L. casei* subsp. *casei* *L. curvatus* *L. pentosus* *L. plantarum* *L. lactis* subsp. *lactis* *P. acidilactici* *P. pentosaceus*	• Antimicrobial activity against Gram-positive bacteria: *B. cereus*, *B. subtilis*, *E. faecalis*, *S. aureus* and Gram negative bacteria: extended-spectrum beta-lactamases strains, multi-drug resistant strains, *E. coli*, *P. aeruginosa*, *Salmonella* sp., *Shigella* sp. and *Y. enterocolitica*.	Mechai et al. (2020)
Bouhezza	Traditional cheese from raw cow, goat and sheep milks	-	• Classified according to the *Codex Alimentarius* as ripened soft and mid-fat cheese. • The sensory evaluation showed that it was salty, spicy and intense in odor and aroma. • *Bouhezza* was considered a product with satisfactory microbiological quality and was rich in LABs that contribute to its ripening. • Proteolysis activity.	Medjoudj et al. (2017)
			• *Bouhezza* from cow milk was classified as the most satisfactory cheese for the majority of criteria (taste, color and texture).	Boudalia et al. (2020)
Klila	Traditional fermented cheese	*Enterococcus* sp. *Lactobacillus* sp. *Leuconostoc* sp. *Pediococcus* sp. *Streptococcus* sp.	• Antimutagenicity activity of LAB from *klila*: binding to DNA-damaging heterocyclic aromatic amines. LAB were ingested daily by humans.	Boubekri and Ohta (1996a, 1996b)
		L. casei *L. plantarum*	• Encapsulation of probiotic bacteria *L. casei* and *L. plantarum* in sodium alginate, enhanced viability of both isolates.	Amira et al. (2020)
Klila and cow, goat and sheep butters	Traditional fermented food from milks	*L. casei* *L. fermentum* *L. paracasei* *L. plantarum* *P. acidilactici* *P. lolii* *P. pentosaceus*	• Lipolytic and proteolytic activities.	Adjoudj et al. (2020)

(Continued)

TABLE 2.1 (Continued)

VERNACULAR NAME	*DEFINITION*	*MICROORGANISM*	*STUDY*	*REFERENCE*
Rigouta	Traditional cheese	*E. faecalis*	• Antibacterial activity against • *Brochotrix* sp., *L. ivanovii* BUG496, *L. monocytogenes* EGDe and *Serratia* sp.	Ghrairi et al. (2004)
Dhan	Sheep butter	*Lactococcus* sp. *Leuconostoc* sp.	• Capacity of acidification, proteolysis activity and production of EPS.	Idoui et al. (2009)
	Traditional fermented butter	LAB Staphylococci Caseolytic bacteria Lipolytic bacteria Psychotropic bacteria Yeasts	• The influence of environmental conditions such as temperature, origin and composition of milk, processing and sanitary conditions on the microbial and physicochemical composition of traditionally made dairy products.	Idoui et al. (2010)
Smen	Traditional fermented butter	*Enterococcus* sp. *S. thermophilus*		Bensaleh et al. (2011)
Kadid or *gueddid*	Traditional fermented meat	*E. faecium*	• Anti-listerial bacteriocin-producing LAB against *L. innocua* F, *L. ivanovii* BUG 496 and *L. monocytogenes* CECT 4032.	Belgacem et al. (2008)
		L. plantarum	• Technological properties: a good acidifying activity, hydrolysis of casein and antimicrobial activity against *E. coli, P. aeruginosa*, *S. arizonae* and *S. aureus*. • Antibiotic susceptibility to the amikacin, cefuroxime, gentamicin, norfloxacin and streptomycin. • Enzymatic activities. • Selection for use as starter cultures for the production of fermented sausages.	Essid et al. (2009)
Dried *ossban*	Artisanal meat	*E. faecium* *P. pentosaceus*	• Proposed as a probiotic and bioprotective agent for fermented foods and for application in the food and/or feed industries.	Zommiti et al. (2018a, 2018b)
Sourdough fermentation	Wheat bread from sourdough fermentation	*Lactobacillus* sp. *Pediococcus* sp.	• Exploiting the potential of selected autochthonous LAB and extending the use of a sourdough.	Mamhoud et al. (2016)

(Continued)

TABLE 2.1 (Continued)

VERNACULAR NAME	DEFINITION	MICROORGANISM	STUDY	REFERENCE
Hamoum	Traditional fermented wheat	–	• An improvement in the nutritional values of wheat through an increase in proteins, biological values and polyphenols and decrease in starch, total sugars and gliadin fractions. • Interesting perspective on the diet for diabetics and those predisposed to celiac disease.	Gourchala et al. (2014)
		L. brevis *L. plantarum* *L. lactis* subsp. *lactis*	• Acidifying, proteolysis and lipolytic activities.	Kalbaza et al. (2018)
Lemzeiet	Couscous from fermented wheat	*E. faecium* *Lactobacillus* sp. *P. acidilactici* *P. pentosaceus* *Streptococcus* sp. *Weissella* sp.	• Antifungal activity (identification of antifungal compounds). • Identification of aroma compounds.	Bekhouche et al. (2013)
		E. faecium *E. hirae* *L. brevis* *L. buchneri* *L. coryniformis* *L. curvatus* *L. namurensis* *L. rapi* *Leuconostoc* sp. *Ln. fallax* *Ln. mesenteroides* *P. acidilactici* *P. pentosaceus* *W. paramesenteroides*	• Acidifying, enzymatic and antifungal activities and safety. • Inhibited undesirable flora and enhanced the activity of some endogenous enzymes in cereals. • Able to hydrolyze casein. • Amylolytic activity.	Merabti et al. (2019)
Traditional fermented green olives	Traditional fermented food	*L. brevis* *L. pentosus* *L. plantarum*	• High antifungal activity against *C. pelliculosa* and *Penicillium* sp. • Antibacterial effect against some Gram-positive and Gram-negative bacteria. • Autoaggregation, hydrophobicity and antioxidant activities and surface tension. • No acquired or horizontally transferable antibiotic resistance; mucin degradation; or DNase, hemolytic and gelatinase activities.	Abouloifa et al. (2019)

mainly due to their nutritional qualities, a long period of storage and microbial diversity, thus allowing them to be considered functional foods. Several authors have undertaken research into the probiotic potentials of many traditional fermented foods (Jamaly et al. 2011; Gourchala et al. 2014; El Kinany et al. 2020; Kotzekidou 2020). In this context, several investigations have revealed that fermentation processes produce active substances, flavor-enhancing compounds, important enzymes and amino acids (Grajek et al. 2005; Kalui et al. 2010; Franz et al. 2014).

LAB have a GRAS character; others are probiotics or have probiotic potential. They are known to produce anti-microbial substances which lead to a safe and long shelf life of food products. These useful microorganisms which have health benefits are associated with fermentation. During this process, LAB produce lactic acid, known for its inhibitor effect (Anukam and Reid 2009; Rehaiem et al. 2010; Hanchi et al. 2016; Ismaili et al. 2017). Some LAB isolated from traditional fermented food have antifungal and antibacterial effects against Gram-positive and Gram-negative bacteria (Bekhouche et al. 2013; Hanchi et al. 2016; Abouloifa et al. 2019; Mechai et al. 2020). Others were able to produce exopolysaccharides (EPSs), which act as stabilizers, thickeners and/or fat-replacers or viscosity enhancing agents and are considered natural biopolymers (Table 2.1) (Bachtarzi et al. 2019, 2020).

El Kinany et al. (2020), in their large case control study, found that individuals who have a high intake of either modern or traditional dairy products are at lower risk for colorectal cancer (CRC). An important finding of this work is the decreased risk of CRC with consumption of traditional dairy products. In another study, Gourchala et al. (2014) mentioned that *el-hammoum* would be a very interesting possibility for the diet of diabetics and those predisposed to celiac disease.

2.5 RISKS OF TRADITIONAL FERMENTED FOOD CONSUMPTION

The technological and hygienic problems of traditional fermented foods need to be solved in order to reduce losses due to wasteful and insufficient fermentation pathways, poor quality and unstable shelf life of products (Odunfa and Oyewole 1998). The raw material generally hosts an abundant and complex microbial flora whose microbial groups would potentially grow and compete for nutrients (Benkerroum 2013). Some properties and enzymatic activities of the microbiota in traditional fermented foods can generate undesirable effects or even toxic compounds such as biogenic amines, the presence of which should, of course, be avoided (Ladero et al. 2015). Moreover, traditional foods have been reported to be associated with infectious diseases and intoxication (Cosivi et al. 1998; Bendahou et al. 2008). Other risks linked to the consumption of traditional fermented food are the presence of antibiotic-resistant (AR) profiles of bacteria, which may be due in part to the antibiotic commonly used as growth promoters and in therapy (Abriouel et al. 2017).

2.6 APPLICATION OF BIOTECHNOLOGY TO TRADITIONAL FERMENTED FOODS

Data obtained from the studies about the application of biotechnology to traditional fermented food might be helpful to replace widely used chemicals, which were recognized as causes of nutritional problems; waste of products; and losses of particular features of food. Currently, the application of biotechnological research to traditional fermented food seems limited to the identification of microorganisms for starter culture development and the use of bacteria with potential probiotic properties to manufacture functional food. The functional aspects of fermented foods are mostly related to the concept of probiotic bacteria or to the molecules generated during food fermentation, such as bioactive peptides (Leroy and De Vuyst 2014). The use of enterococci in the food industry is currently a subject of debate, as the line between harmless contaminants and pathogenic strains remains unclear (Bertolami and Farnworth 2008). However, they seem to play an important role in fermented dairy products and display a number of useful biotechnological traits in addition to bacteriocin production (Hanchi et al. 2016). Nevertheless, it is clear that the generally accepted safety and technical applicability of *Enterococcus* strains might be evaluated on case-by-case basis. The PCR technique may give information about the bacteriocins produced by bacteria and the safety of bacteria for its application as starter. Hence, the use of LAB with multiple bacteriocin production, such as *E. durans* 61A, is an interesting feature in preventing the growth of undesired bacteria rather than the use of a single bacteriocin producer. *E. durans* 61A is free from virulent genes and could be considered a good candidate for application in food preservation (Hanchi et al. 2016).

In the field of using starter cultures for sourdough fermentation, a novel biotechnological approach has been adopted. This includes the development of unique fermentation processes and the use of LAB starter cultures, aimed at shortening and simplifying the fermentation process at both artisanal and industrial levels. In fact, the use of sourdough leads to an intense and appreciated crust color (Mamhoud et al. 2016). Moreover, the establishment of a starter culture collection bank for the maintenance and characterization of suitable strains as potential starter cultures for small factories is vital, and artisan-type products may be an important source (Benkerroum and Tamime 2004).

Around the world, LAB isolated from traditional fermented foods were used to reduce the trimethylamine off-odor in fish and fishery products. This was applied by using the cell free supernatant of LAB isolated from Korean traditional fermented foods *meju* (fermented soybean), two types of *makgeolli* (rice wine), one type of *kimchi* (Korean fermented fresh sauerkraut) and fermented kelp (Park et al. 2020). According to Yang et al. (2020), metaproteomics is one of the omics methods which is gradually gaining attention in the field of traditional fermented food research. It can reveal the nature of microorganism function at the protein level to enhance an understanding of the traditional fermented food ecosystem. The application of metaproteomic approaches is helpful to allow further control of the fermentation process; regulate flavor metabolism; and improve the safety of traditional fermented food through a deep understanding of the microbial interactions, changes in community function, driving forces and survival strategies of microorganisms in the fermentation process.

2.7 CONCLUSIONS AND FUTURE DIRECTIONS

Fermentation, which was previously a hazardous process, is currently a science to be studied for all the advantages it confers. In the past, the traditional fermented foods were consumed only by nomadic or rural people. Nowadays, their consumption has increased globally, not only for their nutritional qualities but also for their good effects on health. Algeria is rich in traditional fermented foods, such as those originating from milk rather than other sources. Some of them have more than one vernacular name, such as *dhan/zebda/smen*, which mean traditional fermented butter. However, the organoleptic qualities of each depend on the region, the raw material and its microbial diversity. The use of safe bacteria isolated from traditional fermented food as starter in the production of fermented food allows generation of a product with desirable technological properties. Therefore, some of the strategies should be introduced to meet this goal at the farm level for producers.

In the future, thanks to biotechnology methods, more fermented foods with health-promoting properties will become available on the market, with many directed towards consumers with very specific health and metabolism needs.

REFERENCES

Abouloifa, H., Gaamouche, S., Rokni, Y., et al. 2021. Antifungal activity of probiotic *Lactobacillus* strains isolated from natural fermented green olives and their application as food bio-preservative. *Biol. Control* 152:104450.

Abouloifa, H., Rokni, Y., Bellaouchi, R., et al. 2019. Characterization of probiotic properties of antifungal *Lactobacillus* strains isolated from traditional fermenting green olives. *Probiotics Antimicrob. Proteins* 12:683–96.

Abriouel, H., Knapp, C. W., Gálvez, A., et al. 2017. Antibiotic resistance profile of microbes from traditional fermented foods. In *Fermented foods in health and disease prevention*, ed. J. Frias, C. Martinez-Villaluenga, and E. Peñas, 675–704. London: Academic Press.

Adjoudj, F., Guessas, B., Spano, G., et al. 2020. Phenotypic and genotypic characterization of Lactobacilli and Pediococci isolated from traditional Algerian fermentation products. *South Asian J. Exp. Biol.* 10:95–103.

Amira, S., Sifour, M., Ouled-Haddar, H., et al. 2020. Effect of different food stress conditions on the viability of encapsulated *Lactobacillus plantarum* and *Lactobacillus casei* isolated from klila (an Algerian traditional fermented cheese). *J. Microbiol. Biotechnol. Food Sci.* 9:38–43.

Anukam, K. C., and Reid, G. 2009. African traditional fermented foods and probiotics. *J. Med. Food.* 12:1177–84.

Baccouri, O., Boukerb, A. M., Nilly, F., et al. 2020. Draft genome sequence of *Enterococcus faecalis* strain OB15, a probiotic strain recently isolated from Tunisian rigouta cheese. *Microbiol. Resour. Announc.* 9(15).

Bachtarzi, N., Kharroub, K., and Ruas-Madiedo, P. 2019. Exopolysaccharide-producing lactic acid bacteria isolated from traditional Algerian dairy products and their application for skim-milk fermentations. *LWT-Food Sci. Technol.* 107:117–24.

Bachtarzi, N., Speciale, I., Kharroub, K., et al. 2020. Selection of exopolysaccharide-producing *Lactobacillus plantarum* (*Lactiplantibacillus plantarum*) isolated from Algerian fermented foods for the manufacture of skim-milk fermented products. *Microorganisms* 8:1101.

Becila, F. Z., Bouasla, A., Turchiuli, C., et al. 2022. Biochemical and microbiological changes associated with fermentation of durum wheat for Lemzeïet processing, a traditional Algerian fermented food. *Processes* 10:2347.

Bekhouche, F., Merabti, R., and Bailly, J. D. 2013. "Lemzeiet": traditional couscous manufacture from fermented wheat (Algeria): process and technological and nutritional quality. *Afr. J. Food Sci.* 4:167–75.

Belgacem, Z. B., Ferchichi, M., Prévost, H., et al. 2008. Screening for anti-listerial bacteriocin-producing lactic acid bacteria from "gueddid" a traditionally Tunisian fermented meat. *Meat Sci.* 78:513–21.

Benamara, R. N., Benahmed, M., Ibri, K., et al. 2022. Algerian extra hard cheese of klila: a review on the production method, and microbial, organoleptic, and nutritional properties. *J. Ethn. Foods* 9:1–12.

Benamara, R. N., Gemelas, L., Ibri, K., et al. 2016. Sensory, microbiological and physico-chemical characterization of klila, a traditional cheese made in the south-west of Algeria. *Afr. J. Microbiol. Res.* 10:1728–38.

Bendahou, A., Lebbadi, M., Ennanei, L., et al. 2008. Characterization of *Staphylococcus* species isolated from raw milk and milk products (lben and jben) in north Morocco. *J. Infect. Dev. Countries* 2:218–25.

Ben Danou, C. 1929. Quelques notes de laiterie sur l'Algérie. *Le Lait* 9:161–3.

Benkerroum, N. 2013. Traditional fermented foods of north African countries: technology and food safety challenges with regard to microbiological risks. *Compr. Rev. Food Sci. Food Saf.* 12:54–89.

Benkerroum, N., and Tamime, A. Y. 2004. Technology transfer of some Moroccan traditional dairy products (lben, jben and smen) to small industrial scale. *Food Microbiol.* 21:399–413.

Benmouna, Z., Dalache, F., Zadi-Karam, H., et al. 2018. Optimization and some characteristics of bacteriocin produced by *Enterococcus* sp. CM9 collected from Mauritanian camel milk. *Emir. J. Food Agric.* 30:275–82.

Benmouna, Z., Dalache, F., Zadi-Karam, H., et al. 2020. Ability of three lactic acid bacteria to grow in sessile mode and to inhibit biofilm formation of pathogenic bacteria. In *Advances in microbiology, infectious diseases and public health. Advances in experimental medicine and biology*, ed. G. Donelli, 105–14. Berlin: Springer.

Bensaleh, F., Labtar, A., Delorme, C., et al. 2011. Occurrence, isolation and DNA identification of *Streptococcus thermophilus* involved in Algerian traditional butter 'Smen'. *Afr. J. Biotechnol.* 10:17251–7.

Bertolami, M. C., and Farnworth, E. R. 2008. The properties of *Enterococcus faecium* and the fermented milk product Gaio®. In *Handbook of fermented functional foods. Functional foods and nutraceuticals*, ed. E. R. Farnworth, 71–88. Boca Raton, FL: CRC Press.

Boubekri, K., and Ohta, Y. 1996a. Identification of lactic acid bacteria from Algerian traditional cheese, el-klila. *J. Sci. Food Agric.* 70:501–5.

Boubekri, K., and Ohta, Y. 1996b. Antimutagenicity of lactic acid bacteria from el-klila cheese. *J. Sci. Food Agric.* 72:397–402.

Boudalia, S., Boudebbouz, A., Gueroui, Y., et al. 2020. Characterization of traditional Algerian cheese "bouhezza" prepared with raw cow, goat and sheep milks. *Food Sci. Technol.* 40:528–37.

Boussekine, R., Merabti, R., Barkat, M., et al. 2020. Traditional fermented butter smen/dhan: current knowledge, production and consumption in Algeria. *J. Food Res.* 9.

Campbell-Platt, G. 1987. *Fermented foods of the world, a dictionary and guide*. London: Butterworths.

Claps, S., and Morone, G. 2011. *Produits laitiers et fromagers traditionnels de l'Algérie*. CorFilac: Ragusa.

Cocolin, L., Dolci, P., Rantsiou, K., et al. 2009. Lactic acid bacteria ecology of three traditional fermented sausages produced in the north of Italy as determined by molecular methods. *Meat Sci.* 82:125–32.

Cosivi, O., Grange, J. M., Daborn, C. J., et al. 1998. Zoonotic tuberculosis due to *Mycobacterium bovis* in developing countries. *Emerg. Infect. Dis.* 4:59–70.

Deegan, L. H., Cotter, P. D., Hill, C., et al. 2006. Bacteriocins: biological tools for bio-preservation and shelf-life extension. *Int. Dairy J.* 16:1058–71.

El Kinany, K., Deoula, M. M. S., Hatime, Z., et al. 2020. Consumption of modern and traditional Moroccan dairy products and colorectal cancer risk: a large case control study. *Eur. J. Nutr.* 59:953–63.

Elotmani, F., Revol-Junelles, A. M., Assobhei, O., et al. 2002. Characterization of anti-*Listeria* monocytogenes bacteriocins from *Enterococcus faecalis*, *Enterococcus faecium*, and *Lactococcus lactis* strains isolated from Raib, a Moroccan traditional fermented milk. *Curr. Microbial.* 44:10–7.

Essid, I., Medini, M., and Hassouna, M. 2009. Technological and safety properties of *Lactobacillus plantarum* strains isolated from a Tunisian traditional salted meat. *Meat Sci.* 81:203–8.

Farhad, M., Kailasapathy, K., and Tamang, J. P. 2010. Health aspects of fermented foods. In *Fermented foods and beverages of the world*, ed. J. P. Tamang, and K. Kailasapathy, 391–414. New York: CRC Press.

Franz, C. M., Huch, M., Mathara, J. M., et al. 2014. African fermented foods and probiotics. *Int. J. Food Microbiol.* 190:84–96.

Franz, C. M., Stiles, M. E., Schleifer, K. H., et al. 2003. Enterococci in foods-a conundrum for food safety. *Int. J. Food Microbiol.* 88:105–22.

Ghalouni, E., Hassaine, O., and Karam, N. E. 2018. Phenotypic identification and technological characterization of lactic acid bacteria isolated from l'ben, an Algerian traditional fermented cow milk. *J. Pure Appl. Microbiol.* 12:521–32.

Ghrairi, T., Manai, M., Berjeaud, J. M., et al. 2004. Antilisterial activity of lactic acid bacteria isolated from rigouta, a traditional Tunisian cheese. *J. Appl. Microbiol.* 97:621–8.

Gourchala, F., Hobamahoro, A. F., Mihoub, F., et al. 2014. Effect of natural fermentation on the nutritional quality of "el hammoum" durum wheat (*Triticum durum*) fermented product of the Algerian country. *Int. J. Biotechnol. Res.* 4:9–18.

Grajek, W., Olejnik, A., and Sip, A. 2005. Probiotics, prebiotics and antioxidants as functional foods. *Acta Bioch. Pol.* 52(3):665–71.

Hamama, A., and Bayi, M. 1991. Composition and microbiological profile of two Moroccan traditional dairy products: raib and jben. *Int. J. Dairy Technol.* 44:118–20.

Hamama, A., El Marrakchi, A., and El Othmani, F. 1992. Occurrence of *Yersinia enterocolitica* in milk and dairy products in Morocco. *Int. J. Food Microbiol.* 16:69–77.

Hanchi, H., Hammami, R., Fernandez, B., et al. 2016. Simultaneous production of formylated and nonformylated enterocins L50A and L50B as well as 61A, a new glycosylated durancin, by *Enterococcus durans* 61A, a strain isolated from artisanal fermented milk in Tunisia. *J. Agri. Food Chem.* 64:3584–90.

Holzapfel, W. H. 2002. Appropriate starter culture technologies for small-scale fermentation in developing countries. *Int. J. Food Microbiol.* 75:197–212.

Idoui, T., Benhamada, N., and Leghouchi, E. 2010. Microbial quality, physicochemical characteristics and fatty acid composition of a traditional butter produced from cows' milk in east Algeria. *Grasas y Aceites* 61:232–6.

Idoui, T., Boudjerda, J., Leghouchi, E., et al. 2009. Lactic acid bacteria from "sheep's dhan", a traditional butter: isolation, identification and major technological traits. *Grasas y Aceites* 60:177–83.

Idoui, T., and Karam, N. E. 2008. Lactic acid bacteria from jijel's traditional butter: isolation, identification and major technological traits. *Grasas y Aceites* 59:361–7.

Ismaili, M. A., Hamama, A., Saidi, B., et al. 2017. Chemical composition, microbial profile and identification of lactic acid bacteria of Moroccan fermented camel milk "lfrik". *Curr. Res. Nutr. Food Sci. J.* 5:383–90.

Jamaly, N., Benjouad, A., and Bouksaim, M. 2011. Probiotic potential of *Lactobacillus* strains isolated from known popular traditional Moroccan dairy products. *Br. Microbiol. Res. J.* 1:79–94.

Kacem, M., and Karam, N. E. 2006. Physicochemical and microbiological study of "shmen", a traditional butter made from camel milk in the Sahara (Algeria): isolation and identification of lactic acid bacteria and yeasts. *Grasas y Aceites* 57:198–204.

Kalbaza, K., Halima, Z. K., and Karam, N. E. 2018. Identification and major technological characteristics of *Lactococcus* and *Lactobacillus* strains isolated from hamoum, an Algerian fermented wheat. *Afr. J. Biotechnol.* 17:108–17.

Kalui, C. M., Mathara, J. M., and Kutima, P. M. 2010. Probiotic potential of spontaneously fermented cereal based foods-A review. *Afr. J. Biotechnol.* 9:2490–8.

Kotzekidou, P. 2020. Improved traditional fermented foods of the Mediterranean region-health benefits as functional foods. In *Functional foods and biotechnology: biotransformation and analysis of functional foods and ingredients*, ed. K. Shetty and D. Sarkar, 135–50. New York: CRC Press.

Ladero, V., Martín, M. C., Redruello, B., et al. 2015. Genetic and functional analysis of biogenic amine production capacity among starter and non-starter lactic acid bacteria isolated from artisanal cheeses. *Eur. Food Res. Technol.* 241:377–83.

Leksir, C., Boudalia, S., Moujahed, N., et al. 2019. Traditional dairy products in Algeria: case of klila cheese. *J. Ethn. Foods* 6:7.

Leroy, F., and De Vuyst, L. 2014. Fermented food in the context of a healthy diet: how to produce novel functional foods? *Curr. Opin. Clin. Nutr. Metab. Care* 17:574–81.

Mamhoud, A., Nionelli, L., Bouzaine, T., et al. 2016. Selection of lactic acid bacteria isolated from Tunisian cereals and exploitation of the use as starters for sourdough fermentation. *Int. J. Food Microbiol.* 225:9–19.

McSweeney, P. L. H., Ottogalli, G., and Fox, P. F. 2017. Diversity and classification of cheese varieties: an overview. In *Cheese*, ed. P. L. H. McSweeney, P. F. Fox, P. D. Cotter, and D. W. Everett, 781–808. San Diego: Academic Press.

Mechai, A., Debabza, M., and Zouari, S. 2020. Antagonistic activity of lactic acid bacteria isolated from Algerian traditional fermented milks against multi-drug resistant and β-lactamases-producing pathogenic bacteria. *Res. J. Biotechnol.* 15:4.

Medjoudj, H., Aouar, L., Zidoune, M. N., et al. 2017. Proteolysis, microbiology, volatiles and sensory evaluation of Algerian traditional cheese bouhezza made using goat's raw milk. *Int. J. Food Prop.* 20:S3246–65.

Mennane, Z., Khedid, K., Zinedine, A., et al. 2007. Microbial characteristics of klila and jben traditional Moroccan cheese from raw cow's milk. *World J. Dairy Food Sci.* 2:23–27.

Merabti, R., Bekhouche, F., Chuat, V., et al. 2015. A large diversity of lactic acid bacteria species is involved in the fermentation of wheat used for the manufacture of lemzeiet. *Eur. Food Res. Technol.* 241:137–49.

Merabti, R., Madec, M. N., Chuat, V., et al. 2019. First insight into the technological features of lactic acid bacteria isolated from Algerian fermented wheat lemzeiet. *Curr. Microbial.* 76:1095–104.

Mokhtari, S., Kheroua, O., and Saidi, D. 2016. Isolation and identification of lactic acid bacteria from Algerian durum wheat (*Triticum durum*) natural fermented in underground silos matmora "el-hammoum" and their antimicrobial activity again pathogenic germs. *J. Nutr. Health Sci.* 3.

Odunfa, S. A., and Oyewole, O. B. 1998. African fermented foods. In *Microbiology of fermented foods*, ed. B. J. B. Wood, 155–91. London: Blackie Academic and Professional.

Ouadghiri, M., Amar, M., Vancanneyt, M., et al. 2005. Biodiversity of lactic acid bacteria in Moroccan soft white cheese (jben). *FEMS Microbiol. Lett.* 251:267–71.

Ouadghiri, M., Vancanneyt, M., Vandamme, P., et al. 2009. Identification of lactic acid bacteria in Moroccan raw milk and traditionally fermented skimmed milk 'lben'. *J. Appl. Microbiol.* 106:486–95.

Park, S. K., Jo, D. M., Yu, D., et al. 2020. Reduction of trimethylamine off-odor by lactic acid bacteria isolated from Korean traditional fermented food and their *in situ* application. *J. Microbiol. Biotechnol.* 30:1510–5.

Rehaiem, A., Fhoula, I., Slim, A. F., et al. 2016. Prevalence, acquired antibiotic resistance and bacteriocin production of *Enterococcus* spp. isolated from Tunisian fermented food products. *Food Control* 63:259–66.

Rehaiem, A., Martínez, B., Manai, M., et al. 2010. Production of enterocin A by *Enterococcus faecium* MMRA isolated from 'Rayeb', a traditional Tunisian dairy beverage. *J. Appl. Microbiol.* 108:1685–93.

Robert, H., Gabriel, V., and Fontagne-Faucher, C. 2009. Biodiversity of lactic acid bacteria in French wheat sourdough as determined by molecular characterization using species-specific PCR. *Int. J. Food Microbiol.* 135:53–9.

Rokni, Y., Ghabbour, N., Chihib, N. E., et al. 2015. Physico-chemical and microbiological characterization of the natural fermentation of Moroccan picholine green olives variety. *J. Mater. Environ. Sci.* 6:1740–51.

Samet-Bali, O., Ennouri, M., Dhouib, A., et al. 2012. Characterisation of typical Tunisian fermented milk: leben. *Afr. J. Microbiol. Res.* 6:2169–75.

Samet-Bali, O., Felfoul, I., Lajnaf, R., et al. 2016. Hygienic quality of "rayeb", a traditional Tunisian fermented cow's milk. *Int. Food Res. J.* 23:366–9.

Steinkraus, K. H. 2002. Fermentations in world food processing. *Compr. Rev. Food Sci. Food Saf.* 1:23–32.

Talon, R., Leroy, S., and Lebert, I. 2007. Microbial ecosystems of traditional fermented meat products: the importance of indigenous starters. *Meat Sci.* 77:55–62.

Tamang, J. P. 2015. *Health benefits of fermented foods and beverages*. New York: CRC Press.

Tamang, J. P., Watanabe, K., and Holzapfel, W. H. 2016. Review: diversity of microorganisms in global fermented foods and beverages. *Front. Microbiol.* 7:377.

Tantaoui-Elaraki, A., and El Marrakchi, A. 1987. Study of Moroccan dairy products: iben and smen. *Mircen. J.* 3:211–20.

Tou, E. H., Mouquet-River, C., Rochette, I., et al. 2007. Effect of different process combinations on the fermentation kinetics, microflora and energy density of *ben-saalga*, a fermented gruel from Burkina Faso. *Food Chem.* 100:935–43.

Yang, L., Fan, W., and Xu, Y. 2020. Metaproteomics insights into traditional fermented foods and beverages. *Compr. Rev. Food Sci. Food Saf.* 19:2506–29.

Zitoun, O. A., Benatallah, L., Ghennam, E. A., et al. 2011. Manufacture and characteristics of the traditional Algerian ripened bouhezza cheese. *J. Food Agric. Environ.* 9:96–100.

Zommiti, M., Bouffartigues, E., Maillot, O., et al. 2018a. In vitro assessment of the probiotic properties and bacteriocinogenic potential of *Pediococcus pentosaceus* MZF16 isolated from artisanal Tunisian meat "dried ossban". *Front. Microbiol.* 9:2607.

Zommiti, M., Cambronel, M., Maillot, O., et al. 2018b. Evaluation of probiotic properties and safety of *Enterococcus faecium* isolated from artisanal Tunisian meat "dried ossban". *Front. Microbiol.* 9:1685.

Fermented Sweet Potato Products

Production, Health Benefits, and Utilization

3

Joy Ehiwuogu-Onyibe and Kolawole Banwo

3.1 INTRODUCTION

Nature at its best provides solutions to every need of the inhabitants of the earth. The sweet potato is one such excellent provision by nature. The sweet potato (SP) is an herbaceous dicotyledonous flowering plant with edible leaves (heart shaped), stem (vines), and roots, including the root skin. It is a trailing/creeping vine vegetable classified as belonging to the Plantae kingdom, Solanales order, Convolvulaceae family, and of the genus *Ipomoea*. Sweet potato (*Ipomoea batatas)* roots can be round, elongated, long, and tapered. Varieties are visually differentiated based on the root skin colors and flesh colors, which can be white, yellow, beige, orange, brown, and red to purple and their combinations. These colors are possibly defined by types and levels of accumulation of compounds such as phenolic derivatives, for example, anthocyanins and carotenoids. Though described as perennial, SP is now cultivated all year round in some locations where water can be provided during low rainfall periods or the dry season (e.g., by irrigation). The International Potato Center (CIP 2018) listed some advantages of the SP as low risk, high yield, low labor requirements, and low cost. Bergh et al. (2012) described the sweet potato as adaptable to marginal environments; flexible in mixed farming systems, with a short maturation period that allows for more than one crop cycle in one year; and less demanding in amount of rainfall, labor inputs, and soil fertility requirements. This explains why some see the sweet potato as supporting women in farming. The SP is a world crop that is now cultivated in many countries because of its many identified nutritional and health benefits. From the first Food and Agriculture Organization of the United Nations Statistics (FAOSTAT) report of 109 SP-producing countries in 1961, the number of SP-producing countries in the world rose to 127 in 2018, although only 117 countries reported values in tons (FAOSTAT 2020). This number may be higher since some countries may not have been captured in FAOSTAT data. In the class of important vegetables, the sweet potato occupies the second position after the potato.

The history of the sweet potato may date back to the prehistoric era. The SP is reported to have been brought to Europe by Christopher Columbus in 1492 during one of his expeditions and was subsequently introduced to Africa, India, Asia, and the West Indies by Portuguese explorers in the 16th century (Zhang et al. 2009). It was domesticated from the wild in central America and subsequently introduced into Africa and India by the Europeans in the 15th century. Records claim the existence of the SP in China in 1582 (Garcia 2018). The global production of SPs increased from 1961 (98,192,635 tons) up to the early 19th century (2004; 133,755,516 tons) and thereafter declined until recent years (2018; 91,945,358 tons). Out of the 127 countries captured by FAO as sweet potato–producing countries in 2018, 13 countries were in the million-tons (t) production level: China (mainland) 53,009,345, Malawi 5,668,543, Nigeria 4,029,909, United Republic of Tanzania 3,834,779, Ethiopia 1,834,619, Indonesia 1,806,389, Uganda 1,529,608, India 1,400,281, Vietnam 1,374,664, Angola 1,274,871, United States of America 1,241,846, Madagascar 1,088,922, and Rwanda 1,082,364 (Table 3.1). The China Mainland alone accounted for 57.65%, while only four top-producing countries (China, Malawi, Nigeria, United Republic of Tanzania) accounted for 72.37% of world SP production in 2018 (FAOSTAT 2020). While the SP is at the 1st position in the country's crop production for the 2nd 2018 global producer (Malawi), it occupied the 9th position in Nigeria (3rd 2018 SP

DOI: 10.1201/9781003605300-3

TABLE 3.1 2018 Global Sweet Potato Production (tons)

S/N	*COUNTRY*	*VALUE*	*S/N*	*COUNTRY*	*VALUE*
1	China, mainland	53009345	60	Liberia	24179
2	Malawi	5668543	61	Portugal	23470
3	Nigeria	4029909	62	Morocco	18120
4	United Republic of Tanzania	3834779	63	Venezuela (Bolivarian Republic of)	17956
5	Ethiopia	1834619	64	Chile	14758
6	Indonesia	1806389	65	Pakistan	12636
7	Uganda	1529608	66	New Zealand	11030
8	India	1400281	67	Guyana	10082
9	Viet Nam	1374664	68	Congo	9676
10	Angola	1274871	69	Somalia	8564
11	United States of America	1241846	70	Fiji	8542
12	Madagascar	1088922	71	Togo	8446
13	Rwanda	1082364	72	Timor-Leste	8179
14	Kenya	871010	73	Comoros	7862
15	Japan	796500	74	Honduras	7845
16	Brazil	741203	75	Italy	7480
17	Papua New Guinea	728163	76	Tonga	6434
18	Haiti	684581	77	Cabo Verde	5434
19	Mozambique	625868	78	Bolivia (Plurinational State)	4483
20	Democratic Republic of Korea	625000	79	Gabon	4215
21	Burundi	583019	80	Ecuador	3976
22	Cuba	549512	81	Micronesia (Federated States of)	3106
23	Philippines	525634	82	Mauritania	2992
24	Mali	504557	83	Greece	2619
25	Cameroon	410097	84	Eswatini	2466
26	Egypt	387481	85	Dominica	2302
27	Democratic Republic of Congo	384350	86	Palestine	2300
28	Argentina	339359	87	Saint Vincent and the Grenadines	2172
29	Republic of Korea	314248	88	Zimbabwe	2112
30	Peru	275258	89	Bahamas	1885
31	Guinea	265268	90	Barbados	1654
32	Chad	255447	91	Guadeloupe	1521
33	Bangladesh	246816	92	Reunion	1082
34	China, Taiwan Province of	236266	93	Suriname	795
35	Sudan	234392	94	El Salvador	738
36	Zambia	183280	95	Puerto Rico	644
37	Sierra Leone	157033	96	Saint Lucia	607
38	Ghana	151420	97	Trinidad and Tobago	598
39	Niger	129959	98	Mauritius	583
40	Lao People's Democratic Republic	124460	99	New Caledonia	450
41	Equatorial Guinea	107915	100	Yemen	355
42	Solomon Islands	106180	101	Cook Islands	343
43	South Africa	86186	102	Grenada	315

(Continued)

TABLE 3.1 (Continued)

S/N	COUNTRY	VALUE	S/N	COUNTRY	VALUE
44	Uruguay	80794	103	Brunei Darussalam	276
45	Australia	70204	104	Antigua and Barbuda	273
46	Mexico	67486	105	Costa Rica	264
47	Benin	64659	106	Niue	251
48	Burkina Faso	64537	107	Guatemala	194
49	Myanmar	63445	108	Martinique	168
50	Spain	59863	109	Bermuda	160
51	Cote d'Ivoire	54794	110	Belize	150
52	Dominican Republic	52760	111	Saint Kitts and Nevis	149
53	Paraguay	50599	112	Guam	127
54	Jamaica	45410	113	Bhutan	53
55	Sri Lanka	43323	114	China, Hong Kong SAR	46
56	Cambodia	42953	115	Montserrat	30
57	Israel	42650	116	Maldives	4
58	Senegal	39459	117	Cayman Islands	3
59	Malaysia	38664			

Source: FAOSTAT (2020).

producer) crop production, and surprisingly, it was in 11th place in crop production for the highest producer (China Mainland) in the same year. According to FAO statistics, in 1961 and 1962, SP was in the first position for commodity volume in China, with 77,288,439 tons and 83,108,156 tons, respectively, ahead of rice and maize. The output volume increased to above 100 million tons with SP alternating between the number 1 and 2 positions with rice until 1995. It gradually went to position 3 in 1996, 4 in 2001, 6 in 2007 (with 81,274,288 tons, below the 100 million production mark), and continued the decrease in 2015 to 11th place until 2018, while rice maintained its number 1 position. In the case of Nigeria, SP production increased from 1961 (149,000 tons) to 1969 (190,000 tons), decreased in 1970 (113,000 tons), continued the downward trend until 1987 (93,000 tons), increased from 1988 (110,000 tons), reached the million-ton mark in 1995 (1,168,000 tons), and continued with the upward trend until 2018 (4,029,909 tons). Despite this positive record, Nigeria is far below its potential capacity in SP production and utilization. This may be the situation reported with many countries.

Global production increased from 1964 (93,377,412 tons) to 2004 (133,755,516 tons) and decreased from 2005 (130,954,089 tons) to 91,945,358 tons in 2018 (Figure 3.1). These narratives indicate that the world is not at its best capacity in SP production. For an important nutrient crop like the sweet potato, there is a need to reverse this downward production trend by addressing the SP food chain/system challenges, low cultivation and yield, pest/disease prevention and management, raw SP storage and short shelf life, marketability, underutilization and low/no commercial application, disruption of farming due to conflicts/disasters, and general acceptability. An example at hand is Nigeria, where SP is not a key crop for the majority of farmers; rather, it is a secondary crop in mixed farming and majority being targeted at local consumption. It is usually intercropped with primary crops such as yam and maize. Commercial utilization of the SP through diversified usage in consumable products will encourage more attention on SP farming and boost production capacity. SP yield and quality are at times affected by viruses, bacteria, fungi, insects, and nematodes such as sweet potato feathery mottle virus, *Streptomyces ipomoeae*, *Rhizopus stolonifer*, *Cylas formicarius*, and *Meliodogyne incognita*, respectively (Okada et al. 2017). Due to potential high losses impacted by a short life

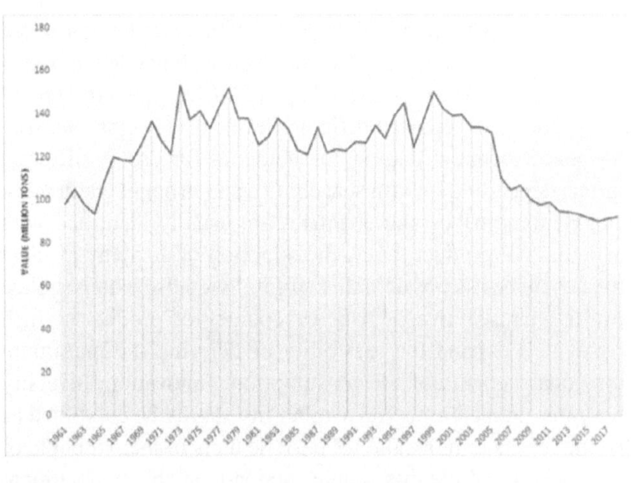

FIGURE 3.1 Global sweet potato production: 1961 to 2017.

Source: Adapted from FAOSTAT (2020).

span and storage challenges, harvesting is sometimes delayed, planned at intervals, and/or spread over a period (e.g., >1 year) to make the roots available even during off seasons. This remains the cheapest storage option for rural and subsistence farmers. However, for this storage option, the availability of cultivation/farming land is an important opportunity cost to be evaluated where accessibility of farming land is a major constraint. Studies indicate a global loss of over 30% of food in most regions of the world (HLPE 2014). The SP contributes significantly to food loss and wastage because of its perishable nature (short fresh produce shelf life), pests and diseases, and inadequate utilization channels.

The value of the SP in human nutrition and survival is reflected in it being named *jinshu*, meaning golden root, during the China famine of 1594, which resulted from large-scale destruction of crops by hurricane. History recorded that the Fujirian governor made farmers widely cultivate sweet potatoes. The cultivated crop saved many from starvation (Zhang et al. 2009; Garcia 2018). The FAO, World Health Organization (WHO), International Fund for Agricultural Development (IFAD), United Nations International Children's Emergency Fund (UNICEF), and World Food Program (WFP) identified the four dimensions of food security as availability, accessibility, utilization, and stability (FAO 2019). Globally, and for every nation, these are indices to track and ensure success in securing food for the populace. The degree of dietary diversity is an important indicator for household or national food security. The SP has great untapped potential to contribute very significantly to achieving individual, household, national, and global food security. To advance the potential of increased utilization of the SP, six varieties of orange fleshed sweet potato (OFSP) were introduced into Africa in 1995 because of its capacity to solve the problem of vitamin A deficiency (CIP 2018). Huge success in this application in households in Africa was reported. The number of OFSP varieties in Africa increased to 42 through breeding by 2016 (Low et al. 2017).

The nutritional (macronutrients and micronutrients such as vitamins and minerals.) composition of raw and processed SP and its production improvement possibilities attest to the strategic position that the SP should occupy in nutrition and food security policies. All SP plant parts, which can be consumed raw, cooked, or processed, are endowed with needed nutrients for humans and animals. Maquia et al. (2013) characterized 44 SP varieties and reported different levels of drought tolerance and hence different degrees of potential for use in different regions depending on rain seasonality and drought challenges. Though some tribes and countries are usually identified with particular SP product types, the adoption of as many product types or dishes as possible that suit individuals or households/families, irrespective of tribe or country of product or dish origin, will expand the feasibility of deriving the maximum nutritional benefits of SP consumption. Innovation, diversity of options, and commercial availability should be combined as driving forces to move the SP into the diets of many people or every person. While this chapter may not be able to document all SP products in the globe, unfermented SP foods produced and consumed in different countries include fresh leaves and roots, cooked leaves and roots (boiled, roasted, baked, grilled), chips (dried, fried, roasted, baked), flour, starch, cake, beverages, donuts, porridge, bread, pie, cookies, biscuits, baby food, puff puff, buns, candy, soup, juice, puree, canned roots, noodles, and salad. There was no data on processed SP products by FAOSAT. This shows that the SP is seriously underutilized, and a lot is lost from the food chain as waste. Efforts to correct this status quo will improve nutrition and food security records. Different varieties are preferentially suited for different products. Oluwole et al. (2015) observed differences in baked crisps produced with orange- and yellow-fleshed sweet potatoes. Apart from variations in nutrients, which may have resulted from differences in varieties, the two produced crisps were rated high but differently in color, taste, crispiness, texture, flavor, mouthfeel, and overall acceptability. The method for crisp production was raw potato root sorting, peeling, washing, cutting, cooking, mashing, mixing with other ingredients (spices, vegetable fat, etc.), dough sheeting and cutting, baking at 120°C, cooling, and packaging (Truong and Ramesh 2010) documented sweet potato puree production methods. One such production technique consisted of SP root washing, peeling, cutting into small sizes, cooking, pulping into puree, canning, or packing in plastic containers and freeze storing. SP roots can be made into flour using FAO (2010) or other methods. FAO steps include sorting, washing without peeling, draining, chipping/slicing into about 3- to 6-mm thickness, drying, milling to flour, and packaging in strong thick black polyethylene bags. Drying can be by sun, solar, or oven drying at a low temperature. SP puree or flour is applied in making bread and other baked or related products (FAO 1994; Gule 2016; Wanjuu et al. 2018; Banwo et al. 2020a). SP bread can be made using SP flour or puree combined with wheat flour, as shown in Figure 3.2. Different formulations are available for good and delicious SP bread production: Wanjuu et al. (2018) used the following composition in bread making– 31.22% SP puree, 48.17% wheat flour, 1.78% sugar, 0.8% salt, 14.27% water, 0.71% yeast, 2.32% fat, 0.36% improver, and 0.36% gluten; FAO (1994) employed the following composition: 100 g SP flour, 230 g wheat flour, 15 g sugar, 1 g salt, 437.5 ml water, 40 g yeast, and 15 g margarine. Gule (2016) adopted the following composition: SP puree 40%, wheat flour 60%, sugar 1%, salt 1.5%, yeast 2%, improvers 0.5%, and shortening fat 4%, while Banwo et al. (2020a) employed the following composition: cassava flour (150 g), sweet potato flour (100 g) and soybean flour (50 g), baking fat (10 g), sugar (30 g), salt (1 g), and egg (1 whole egg), mixed with about 300 mL of water. As shown in Figure 3.2, SP biscuits can be made by thoroughly mixing sieved 300 g flour (sweet potato, wheat, or cassava flour), 4 g salt, 15.5 g baking powder into a mix of 320 g mashed cooked SP roots, 120 g sugar, and 60 g margarine. The dough formed is rolled out, cut to shapes and sizes, placed in oiled tray/pans, and baked. Akinsola et al. (2017) made SP pasta with 200 g SP flour, 30 g carbon methyl cellulose, a pinch of salt, 1 egg, 50 mL water, and 10 mL oil. Some SP root varieties are sweeter than others and hence enjoy differences in preference by individual. SP sweetness can be reduced or eliminated by partial or full fermentation based on the duration of fermentation (see

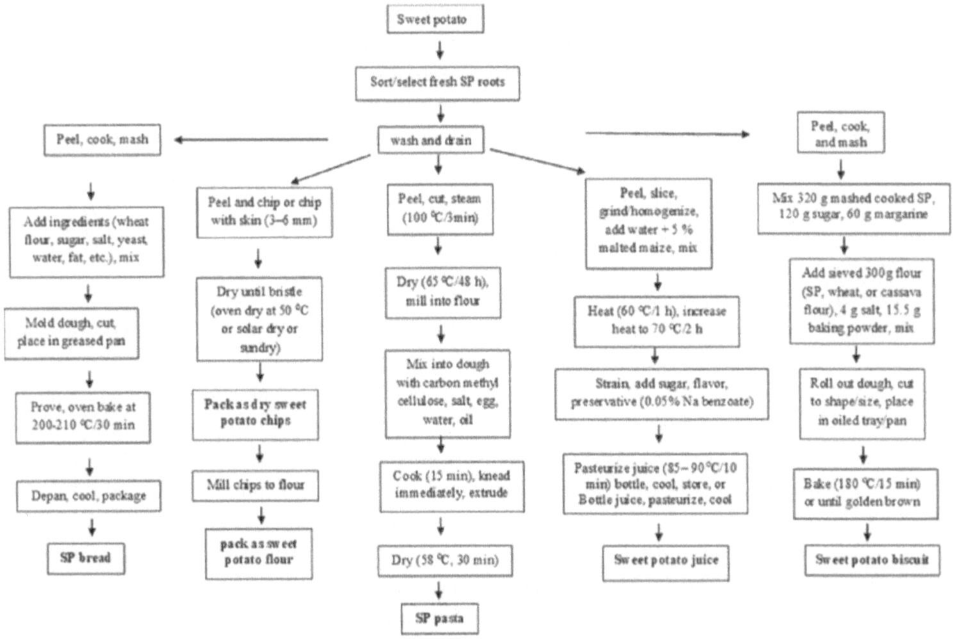

FIGURE 3.2 Unfermented sweet potato product production processes.

Source: Adapted from FAO (1994, 2010), Wireko-Manu et al. (2010), Akinsola et al. (2017), and Wanjuu et al. (2018).

the fermented SP products section). Soaking sliced SP in clean water for 8 to 12 h at room temperature (e.g., 25–32°C) before frying or roasting slightly reduces the sweetness of fried SP chips, making them more enjoyable or acceptable for some consumers.

In all production activities, using applicable good hygiene practice (GHP), good manufacturing practice (GMP), and hazard analysis critical control points (HACCPs) are advised. To ensure quality and an acceptable product, during processing of SP involving peeling or cutting, enzymatic browning can be controlled by cutting/slicing immediately into potable water, blanching (treatment with about 95°C hot water), immersing in food-grade 0.5% sodium bisulphite or 0.5% ascorbic acid, or a combination of any of these treatments, depending on the type of SP (Ajayi et al. 2016; Karim et al. 2016; Oluwole et al. 2016). This inhibits the activity of polyphenol oxidase, which causes enzymatic browning.

3.2 MICROBIAL ECOLOGY OF RAW SWEET POTATO, FERMENTING BROTH, AND FERMENTED PRODUCTS

Microorganisms are living organisms of microscopic size; such organisms are seen with the aid of a microscope. They are organisms ranging between <0.01 μm and >1000 μm in size that are seen using the microscope, such as prions, viruses, bacteria, archaea, fungi (yeast and mold), protozoa, and algae. Anton Leeuwenhoek was the first scientist to discover microscopic organisms' existence in 1675 (Pelczar and Chan 1981). While some microorganisms can be implicated in negative effects on humans, plants, and animals, many play desirable positive roles for the growth, survival, and health benefits of human, animals, and plants. Microorganisms associated with a product or sample are traceable to the plant cultivation location and method, handling and storage, distribution and marketing, raw material, processing, packaging, and the environment of these activities. Researchers have in different studies identified microorganisms in the soil environment of cultivated SP plants, SP roots, stored SP roots, SP root spoilage, fermenting SP roots/vines, and processed SP products. By applying denaturing gradient gel electrophoresis (DGGE) fingerprints, Marques et al. (2014) reported bacterial genus associated with the rhizosphere of three sweet potato types and concluded that the type and number of bacterial populations was affected by the SP age and genotype. Using molecular approaches, these bacterial genus were identified in the SP root rhizosphere: *Pseudomonas, Sphingobium, Acinetobacter, Chryseobacterium, Chitinoghaga, Sphingobacterium, Bacillus, Caulobacter, Phenylobacterium, Hyphomicrobium, Rhizobium, Novosphingobium, Burkholderia, Comamonas, Streptomyces, Stenotrophomonas, Methylophilus, Terrabacter, Nocardioides, Conexibater, Mucilaginibacter, Panteoa, Pseudonocardia, Flavobacterium, Pedobacter, Tumebacillus, Pullulanibacillus, Clostridium, Paenibacillus, Cohnella, Oxalophagus*, and *Streptophyta*. Khan and Doty (2009) applied 16S rRNA gene sequencing to characterize and identify SP endophytic isolates such as *Enterobacter aerogenes, Rahnella aquatilis, Pseudomonas putida, P. fluorescens,*

Rhodanobacter terrae, Stenotrophomonas maltophilia, Xanthomonas retroflexus, and *Phyllobacterium myrsinacearum.* Studying the growth of planted SP, rhizospheric and endophytic bacteria species of SP identified with the aid of the 16S rRNA gene sequence by Tangapo et al. (2018) included rhizospheric bacteria *Methylobacterium platani, Sphingonaonas paucimobilis, S. endophytica, Klebsiella pneumoniae, Enterobacter cloacae, Pseudomonas nitroreducens, Serratia marcencens, Chryseobacterium indologenes, Sphingobacterium multivorum, Exiguobacterium acetyicum, Bacillus mycoides, B. luciferensis, B. acquimaris, B. mojavensis, B. pumilus, B. lincheniformis, B. amyloliquefaciens, B. flexus, B. badius, B. subtilis, B. simplex, B. megaterium, B. aryabhattai, B. cereus, B. acidiceler, B. methylotrophicus, Staphylococcus saprophyticus, Streptomyces abikoensis, S. albus, S. griseus, Arthrobacter ureafaciens, Kocuria rhizophila, Microbacterium laevaniformans, Micrococcus luteus, Paracoccus* spp., and *Bacillus* spp. and endophytic bacteria *Klebsiella pneumoniae, Enterobacter cloacae, Sphingobacterium multivorum, Bacillus mycoides, B. pumilus, B. mojavensis, B. lincheniformis, B. amyloliquefaciens, B. badius, B. subtilis, B. megaterium, Staphylococcus haemolyticus, Streptomyces abikoensis, Arthrobacter ureafaciens, Microbacterium laevaniformans, Bacillus safensis, Pseudomonas putida, Paracoccus cummunis,* and *Pantoea* spp. Tangapo et al. (2018) investigated and recorded that all the endophytic bacteria (in the SP root) could effectively be traced to the plant rhizosphere except *Staphylococcus haemolyticus, Pseudomonas putida, Bacillus safensis, Paracoccus communis,* and *Pantoea* spp.

Different species of microorganisms have been identified in stored SP, which may be related to the subsequent deterioration or spoilage of the SP. Oladoye et al. (2016) used sequenced amplified genes to identify fungi (*Penicillium chrysogenum, Penicillium brevicompactum, Penicillium expansum, Penicillium crustosum, Mucor circinelloides,* and *Rhizopus oryzae*) on stored harvested SP roots. Oladoye et al. (2013) at the molecular level identified *Arthrobacter woluwensis, A. nicotianae, A. methylotrophs, Arthrobacter* sp., *Staphylococcus kloosii, S. sciuri, S. gallinarium, Staphylococcus* sp., *Klebsiella pneumoniae, Klebsiella* sp., *Bacillus* sp., *Rhanella* sp., *Pseudomonas syringae, Lysinibacillus fusiformis,* and *Listeria monocytogenes* on SP four weeks post-harvest storage in plastic containers at 29°C. Oduola et al. (2018) identified *Staphylococcus aureus, Bacillus subtilis, B. licheniformis, Pseudomonas fluorescens, Micrococcus luteus, Aspergillus niger, A. fumigatus, A. terreus, Mucor mucedo, Fusarium* sp., *Penicillium* sp., and *Geotrichum* sp. as bacterial and fungal spoilage organisms on decaying SP purchased from the open market. In another study, fungi *Aspergillus niger, Fusarium oxysporum, Rhizopus stolonifer, Botryodiplodia theobroma,* and *Penicillium* sp. were associated with decaying sweet potato roots (Olaitan 2012). Harter and Weimer (1929) reported SP storage microbes, including *Rhizopus nigricans, Rhizopus tritici, Rhizopus artocarpi, Rhizopus delemar, Rhizopus oryzae, Rhizopus reflexus, Coratostomella fimbriata, Diplodia tubericola, Diporthe batatatis, Plenodomus destruens, Sclerotrum bataticola, Monilochaetes infuscans, Fusarium oxysporum, Mucor racemosus, Altenaria* sp., *Penicillium* sp., *Botrytis cinerea, Epicoccum, Fusarium culmorun, Fusarium acuminatum, Gibberella saubinetti, Trichoderma koningi, Sclerotinia, Fusarium batatatis, Fusarium hyperoxysporum, Fusarium radicicola, Fusarium solani, Fusarium orthoceras, Nectria ipomoeae, Zygorrynchus* sp., *Melanospora* sp., *Aspergillus niger, Sclerotium rolfsii, Cephalothecium* sp., *Fusarium vasinfectum, Pestalozzia* sp., *Verticillium cinnabarinum, Acromoniella* sp., *Macrosporium* sp., *Actinomyces* sp., *Trichosporium* sp., *Sporotrichium* sp., *Pestalozzia* sp., *Neocosmospora vasinfecta,* and *Ceratostoma* spp. Harter et al. (1918) documented that some fungi associated with stored or spoiled SP roots caused spoilage when inoculated into SP, thus establishing the fact that some of these organisms could cause spoilage in SP stored below 2°C. The fungi were *Rhizopus nigricans, Fusarium oxysporum, Fusarium acuminatum, Fusarium culmorum, Mucor racemosus, Penicillium expansum, Altenaria* sp., *Sphaeronema fimbriatum, Sphaeronema infuscans, Spaeronema monilochaetes, Diplodia tubericola, Diplodia gossypina, Diplodia natalensis, Diplodia maclurae, Lasiodiplodia nigra, Diaporthe batatatis, Plenodomus destruens, Slerotium bataticola, Botrytis cinerae, Epicoccum* sp., *Gibberella saubinetii,* and *Trichoderma koningi.* Viruses are also associated with plants, but they are more difficult to culture or grow except in special laboratories. Hegde et al. (2012) recorded viruses causing SP diseases: sweet potato mild mottle virus, sweet potato chlorotic stunt virus, sweet potato feathery mottle virus, sweet potato leaf curl virus, sweet potato mosaic virus, sweet potato latent virus, and sweet potato ring spot virus. Some human gut viruses derived from plant (pepper mild mottle virus, oat blue dwarf virus, grapevine asteroid mosaic-associated virus, maize chlorotic mottle virus, oat chlorotic stunt virus, panicum mosaic virus, and tobacco mosaic virus) were documented by authors who suggested a possible link of some of these viruses with human disease (Scarpellini et al. 2015). More scrutiny of SP and other plant-based products for viral load may be warranted.

Yusuf (2016) assessed microbial population during two Nigerian traditional SP fermentation methods: soaking in cold water and hot water fermentation (parboiling) and observed an initial presence of *Enterobacter cloacae* in cold water fermentation and *Bacillus cereus* in hot water fermentation, followed by a predominance of *Leuconostoc mesenteroides* in both processes. *Lactiplantibacillus plantarum, Staphylococcus aureus, Bacillus cereus, Streptococcus* sp., *Saccharomyces cerevisiae, Aspergillus niger, Rhizopus stolonifer,* and *Penicillium* spp. were isolated during a 24 h steeping of SP in water (Ezeama and Amajor 2015). In a like manner, Amajor et al. (2014) observed the presence of *Staphylococcus aureus, Levilactobacillus brevis, Bacillus cereus, Aspergillus niger,* and *Rhizopus stolonifer* in fermented sun-dried SP flour, while Olawale and Ojokoh (2019) isolated *Bacillus subtilis, Lactiplantibacillus plantarum, Lactococcus lactis, Staphylococcus aureus, Micrococcus* spp., *Mucor mucedo, Aspergillus niger, Aspegillus flavus, Penicillium chrysogenum, Aspergillus fumigatus, Rhizopus oryzae, Altenaria altemate,* and *Geotrichum* spp. during a 72-h solid state fermentation of a combined blend of orange-fleshed

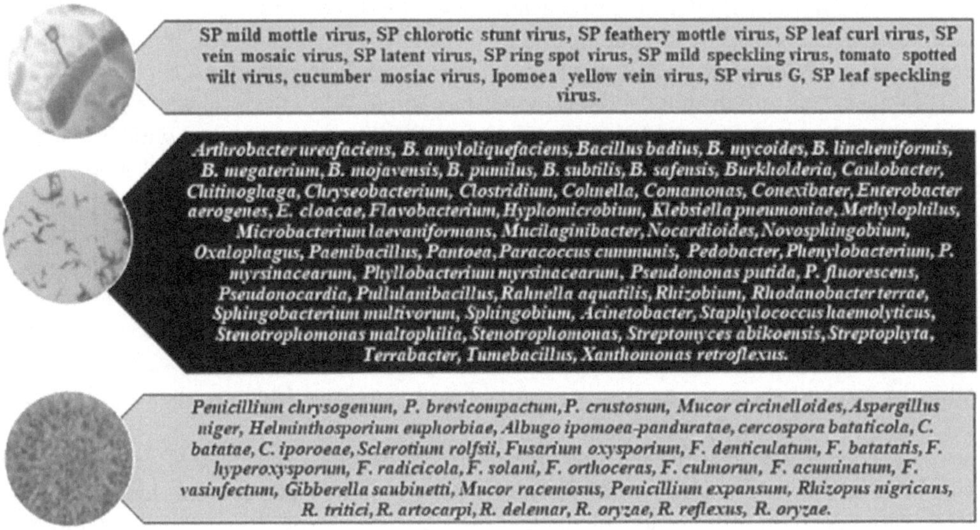

FIGURE 3.3 Some microorganisms associated with the sweet potato.

Source: Harter and Weimer (1929), Khan and Doty (2009), Hegde et al. (2012), Marques et al. (2014), Oladoye et al. (2016), and Tangapo et al. (2018).

SP and beniseed. Some viruses, bacteria, and fungi associated with SP are listed in Figure 3.3.

These listed bacteria and fungi represent only the culturable species or strains which depend on variables such as type of medium, incubation temperature, and aerobic or anaerobic culturing conditions. Some of these SP-associated microorganisms form the identified safe bacteria and fungi for different food processes. Many lactic acid bacteria with origins in plants (root, tuber, fruit, leaf, and flower) have been successfully adapted for starter culture applications. In the same way, non-pathogenic microorganisms from SP plants with the capacity and required attributes can be explored for potential starter culture application.

3.3 FERMENTED SWEET POTATO PRODUCTS

Fermentation is a method of food processing aimed at reducing postharvest losses, aiding in food preservation, improving nutrients/nutrition, adding value to the processed product, enhancing food safety and quality, and increasing product diversification and accessibility (FAO 1998a; Campbell et al. 2017; Banwo et al. 2020a, 2020b; Banwo et al. 2021). Fermented foods are produced and consumed all over the world. In advanced countries, industrialized fermentation processes and technologies abound, while in developing countries like Nigeria, low-cost fermentation procedures and technologies at the household or cottage levels are the case scenarios. Benefits of food fermentation include product preservation by the presence or activities of organic acids and low pH, improved flavor/taste, conversion of inedible food products to edible forms, creating a variety of products, increasing nutrient availability through improved digestibility, improved quality, and safety as a result of toxin reduction (FAO 1998a; Ajayi et al. 2019; Saranraj et al. 2019). In addition, when fermented foods are consumed with the fermenting organisms or their by-products of fermentation, such foods may be considered probiotic and functional in nature (Kádár and Fonseca 2019; Banwo et al. 2020a, 2020b). Saranraj and Ray (2019) defined fermentation as a biotechnological process in which desirable microorganisms are used in the production of value-added products of commercial importance.

Food fermentation is controlled through the manipulation of conditions to produce acceptable safe food. Microbial fermentations are biocatalyzed by extracellular enzymes (released outside the cell into the medium/substrate) or intracellular enzymes (engaged within the cell) of the microorganism. Microorganisms employ different enzymes – amylases, lipases, proteases, and others – during fermentation depending on the available nutrient and prevailing environmental and growth conditions. Fermentation is regarded as a low-cost processing technology. Fermentation can occur in a liquid state or solid state. Liquid state fermentation, also known as submerged fermentation, is a process based on the growth and conversion of substrate in liquid medium by the microorganism(s). In solid state fermentation, microbial growth, substrate conversion, and target product production take place on solid material known as substrate. Lactic acid bacteria (LAB) such as *Lactobacillus acidophilus, Streptococcus lactis,* and *Leuconostoc mesenteroides* generate lactic acid, acetic acid, ethyl alcohol, acetic acid, formic acid, and carbon dioxide from the fermentation of organic compounds, while butyl alcohol bacteria, like *Clostridium acetobulyticum* and *C. beijerinckii*, produce butyric acid, butanol, isopropyl alcohol, ethanol, acetic acid, acetone, formic acid, hydrogen, and carbon dioxide from the fermentation of carbohydrates (Pelczar and Chan 1981; Kádár and Fonseca 2019). Products of carbohydrate fermentation by acetic acid bacteria (e.g., Acetobacter) include acetic acid, gluconic acid, and kojic acid. Species of *Propionibacterium* produce propionic acid, acetic acid, succinic acid, and carbon dioxide from fermentation of carbohydrates (Gonzalez-Garcia et al. 2017; Piwowarek et al. 2018). Fungi like *Saccharomyces cerevisiae* and *Gibberella fujikuroi* (*Fusarium moniliforme*) convert carbohydrates to alcohol and gibberellic acid, respectively, during their utilization of such organic compounds. Alkaline fermentation by *Bacillus* species (e.g., *Bacillus subtilis, Bacillus licheniformis*) leads to the hydrolysis of protein-rich substances into polypeptides, amino acids, and ammonia. Lipolytic conversion of lipids/fat to glycerol and fatty acids also occurs. The degree and duration of fermentation and the type of microbes involved in the process will determine the end product of any fermentation. The application of fermentation in food processing eliminates or reduces the possibility of contamination by pathogenic and non-pathogenic microorganisms because of lowered pH and the production of inhibitory substances – organic acids, phenolic, bacteriocin, antibiotics, other anti-microbial compounds – that are harmful to pathogens and some contaminating nonpathogenic microbes. Rhee et al. (2011) reported the inhibition of inoculated pathogens such as *Clostridium perfringes, Staphylococcus aureus, Salmonella* Typhimurium, *Listeria monocytogenes, Vibrio parahaemolyticus*, and *Escherichia coli* during kimchi fermentation. When the fermentation is from naturally occurring fermenting microorganisms without the addition of any culture, it is referred to as natural fermentation or spontaneous fermentation, while that with the use of an introduced microbial culture or starter culture is called controlled fermentation. In the former fermentation, which proceeds naturally by chance, the product usually varies because of the activities of different microorganisms, while the latter improves product consistency due to the controlled activities of selected known microorganisms (Oguntoyinbo et al. 2011; Banwo et al. 2020a, 2020b).

An array of fermented SP products is produced in different parts of the globe. As room temperature varies from season to season and from location (country) to location, the rate of fermentation varies depending on the prevailing temperature and environmental conditions. During the hot season, the rate of fermentation will be high and the duration of fermentation possibly shorter than during the cold season. The necessity of using temperature devices to maintain a specific temperature during fermentation may be required to ensure process uniformity and product consistency, especially for large-scale productions. Some fermented SP products and a process flow chart are shown in Figure 3.4. Ajayi et al. (2016) documented

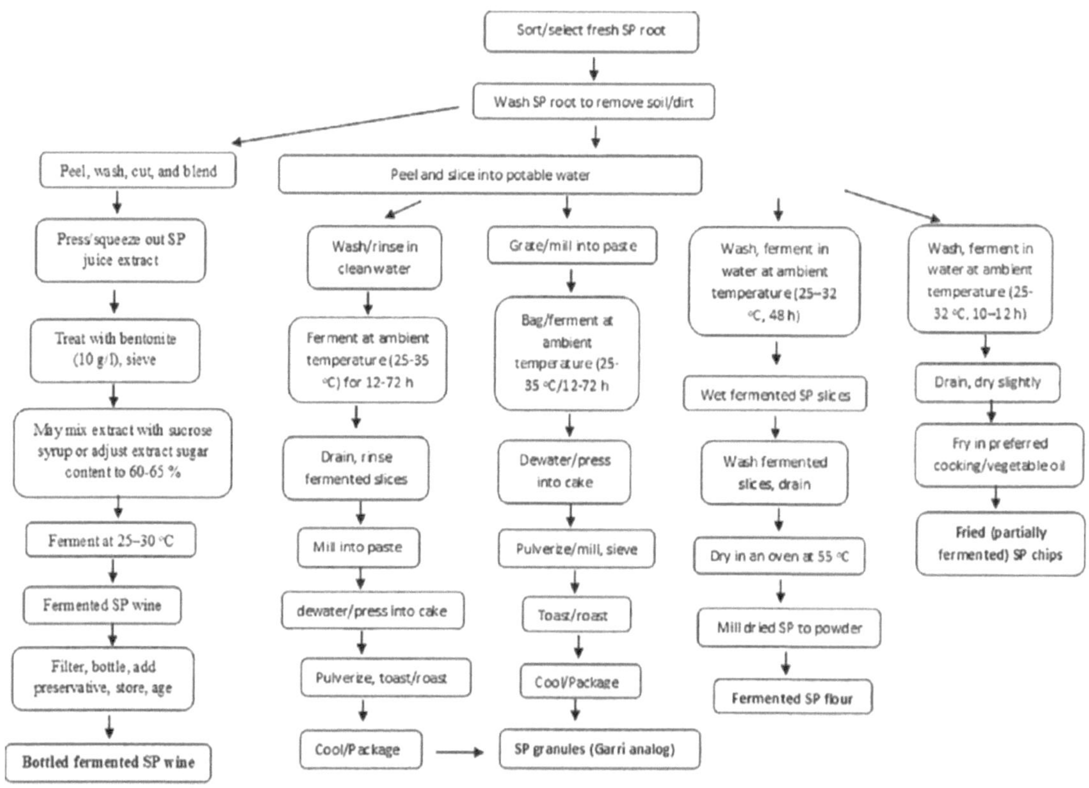

FIGURE 3.4 Fermented sweet potato product production process flow chart.

Source: Adapted from FAO (1998a) and Ajayi et al. (2016); traditional method; homemade method.

a process to produce quality safe SP flour. Garri is a toasted/roasted granule mostly produced from cassava tuber, consumed directly (by soaking in water or chewing) or made into gelatinous dough meal (bolus) form using boiling water. This dough meal is usually served with soup/stew. Ajayi et al. (2016) also made this gel-like dough meal using their produced fermented SP flour, indicating that fermented SP flour can be used for this purpose. An adaptation of the cassava garri production method can be used to produce roasted/toasted SP granules (garri analog). Banwo et al. (2020a) employed the use of composite flour of cassava, sweet potato, and soybean flour in sourdough bread making. Other fermented products of SP include wines, juices, chips, roasted granules, enzymes, organic acids and ethanol, and pickles.

The increasing number of beneficial microorganisms used in various food/microbiological process applications has been documented, especially from fermenting a composite flour of cassava, sweet potato, and soybean flours. Banwo et al. (2020a) in their study isolated lactic acid bacteria and yeasts of functional importance from the composite flour, and these were screened for characteristics vital for the production of sourdough bread. *Lactiplantibacillus plantarum* and *Pichia kudriavzevii* were selected for the controlled fermentation of the composite meal used in the production of the bread. The use of local farm produce such as sweet potato can increase viable revenue for indigenous farmers and develop the economy. The use of quality sweet potato flour as a blend with wheat can be used in bakery and confectionary products that require reduced gluten content. The strains and raw materials used in the sourdough bread production enhance the nutritional contents of the bread (Banwo et al. 2020a, 2020b).

3.4 SWEET POTATO NUTRITION AND HEALTH BENEFITS

The age-old saying "health is wealth" and the current nutrition viewpoint that your health is as good as the food you consume paint a picture of the importance of food and nutrition to life. The problems of improper diets are reflected in the high level of malnutrition (undernutrition, wasting, starvation, stunting, obesity, overweight, and underweight) and non-communicable diseases (such as diabetes, stroke, hypertension, cancer, and heart disease) as noted by FAO, IFAD, UNICEF, WFP, and WHO (FAO 2019). The Sustainable Development Goal (SDG) 2.1 target of ending hunger and ensuring access to food by all and SDG 2.2 target of eliminating all forms of malnutrition is seriously being undermined by current situations of global food insecurity, hunger, undernourishment, economic recession, economic slowdown/downturn, and unemployment. The increasing rate of deaths associated with non-communicable diseases (NCDs) arising from unhealthy food consumption with poor-quality diets reflects the global food and nutritional insecurity. To tackle and reduce the impact of these nutrition-related challenges, there is advocacy for an increased intake of fruits, vegetables, and other forms of dietary fiber which have been discovered to be lacking in many diets around the globe (WHO 2018). Other recommendations include increased intake of potassium to counter the harmful impact of sodium intake, less than 30% fat of total energy intake, less than 5% salt intake, and less than 10% free sugar daily intake (WHO 2018). A balanced daily energy requirement and diets should depend on the age, lifestyle, and activities of an individual. All the dietary requirements for growth, body functions, and maintenance are expected, as advised by nutrition professionals, to be derived from daily nutrition. It therefore means that proper meal planning and diversity, backed up by knowledge of food nutrition content and benefits, is very important to achieve nutritional security.

If properly developed and processed, it is expected that most foods will have the same or more of the health benefits of raw produce. Sweet potato presents a sought-after solution of increasing the availability and affordability of nutritious foods that constitute a healthy diet. The nutritional composition of the sweet potato (Tables 3.2 and 3.3) and its potential to contribute significantly to the dietary requirements of children and adults have been established and documented. In addition to other nutrients, the reported nutritional content of SP products in carbohydrate, sugar, fat, fiber, β-carotene, vitamin A, vitamin C, vitamin E, vitamin K, calcium, magnesium, phosphorus, potassium, and sodium are especially noted (USDA 2019). These macronutrients and micronutrients are essential in human and animal health. Humans and animals derive energy from the metabolism of the consumed macronutrients made of carbohydrate, protein, and fat. Byproducts and compounds of their metabolic processes are also supplied for body activities and functions. Vitamin A (retinol) is needed for vision, proper immune system function, and healthy skin; vitamin C (ascorbic acid) is a very potent antioxidant, has immune regulatory activities, maintains healthy skin/bone/blood vessels, protects and keeps cells healthy, helps wound healing, and has possible antimicrobial and antiviral activities; vitamin E maintains healthy skin and eyes and strengthens the immune system; vitamin K is helpful for normal blood clotting and wound healing; calcium helps build and maintain strong bones and teeth, regulates muscle contractions and heartbeat, helps in blood clotting, and is involved in hormone secretion; magnesium helps in metabolism for energy release and bone health; phosphorus is involved in metabolism, energy release, and building strong bones and teeth; and potassium controls fluids and electrolyte balance and keeps heart muscles functioning properly. These are micronutrients needed in small amounts for the body to function properly and stay in good health. The abundant β-carotene (a bioactive compound and antioxidant) in SP can be converted to retinol (vitamin A) for the body's use (Food and Nutrition Board 2011; USDA 2019). Dietary fiber is known to play an important role in aiding bowel movements, detoxification, cholesterol metabolism/reduction, and weight management. Sweet potato varieties differ in nutrient contents, and the location and method of cultivation also

TABLE 3.2 Nutritional Content per 100 g of Raw Sweet Potato Root, Sweet Potato Leaves, Yam Tuber, and Potato Tuber

RAW SWEET POTATO ROOT		*RAW SWEET POTATO LEAVES*		*RAW POTATO TUBER*		*RAW YAM TUBER*	
Water	77.28 g	Water	86.81 g	Water	79.25 g	Water	69.6 g
Energy	86 kcal	Energy	42 kcal	Energy	77 kcal	Energy	118 kcal
Energy	359 kJ	Energy	175 kJ	Energy	322 kJ	Energy	494 kJ
Protein	1.57 g	Protein	2.49 g	Protein	2.05 g	Protein	1.53 g
Total fat	0.05 g	Total fat	0.51 g	Total fat	0.09 g	Total fat	0.17 g
Ash	0.99 g	Ash	1.36 g	Ash	1.11 g	Ash	0.82 g
Carbohydrate	20.12 g	Carbohydrate	8.82 g	Carbohydrate	17.49 g	Carbohydrate	27.88 g
Fiber, total dietary	3 g	Fiber, total dietary	5.3 g	Fiber, total dietary	2.1 g	Fiber, total dietary	4.1 g
Sugars, total	4.18 g	Calcium	78 mg	Sugars, total	0.82 g	Sugars, total	0.5 g
Sucrose	2.52 g	Iron	0.97 mg	Sucrose	0.17 g	Calcium	17 mg
Glucose	0.96 g	Magnesium	70 mg	Glucose	0.31 g	Iron	0.54 mg
Fructose	0.7 g	Phosphorus	81 mg	Fructose	0.26 g	Magnesium	21 mg
Starch	12.65 g	Potassium	508 mg	Starch	15.29 g	Phosphorus	55 mg
Calcium	30 mg	Sodium	6 mg	Calcium	12 mg	Potassium	816 mg
Iron	0.61 mg	Selenium	0.9 µg	Iron	0.81 mg	Sodium	9 mg
Magnesium	25 mg	Vitamin C	11 mg	Magnesium	23 mg	Zinc	0.24 mg
Phosphorus	47 mg	Thiamin	0.156 mg	Phosphorus	57 mg	Copper	0.178 mg
Potassium	337 mg	Riboflavin	0.345 mg	Potassium	425 mg	Manganese	0.397 mg
Sodium	55 mg	Niacin	1.13 mg	Sodium	6 mg	Selenium	0.7 µg
Zinc	0.3 mg	Pantothenic acid	0.225 mg	Zinc	0.3 mg	Vitamin C	17.1 mg
Copper	0.151 mg	Vitamin B-6	0.19 mg	Copper	0.11 mg	Thiamin	0.112 mg
Manganese	0.258 mg	Folate, food	1 µg	Manganese	0.153 mg	Riboflavin	0.032 mg
Selenium	0.6 µg	Folate, DFE	1 µg	Selenium	0.4 µg	Niacin	0.552 mg
Vitamin C	2.4 mg	Vitamin A, RAE	189 µg	Vitamin C, total	19.7 mg	Pantothenic acid	0.314 mg
Thiamin	0.078 mg	Carotene, beta	2217 µg	Thiamin	0.081 mg	Vitamin B-6	0.293 mg
Riboflavin	0.061 mg	Carotene, alpha	42 µg	Riboflavin	0.032 mg	Folate, total	23 µg
Niacin	0.557 mg	Cryptoxanthin, beta	58 µg	Niacin	1.061 mg	Folate, food	23 µg
Pantothenic acid	0.8 mg	Vitamin A	3778 IU	Pantothenic acid	0.295 mg	Folate, DFE	23 µg
Vitamin B-6	0.209 mg	Lutein + zeaxanthin	14720 µg	Vitamin B-6	0.298 mg	Choline, total	16.5 mg
Folate, total	11 µg	Vitamin K (phylloquinone)	302.2 µg	Folate, total	15 µg	Vitamin A, RAE	7 µg
Folate, food	11 µg	Fatty acids, total saturated	0.111 g	Folate, food	15 µg	Carotene, beta	83 µg
Folate, DFE	11 µg	Fatty acids, total monounsaturated	0.02 g	Folate, DFE	15 µg	Vitamin A	138 IU
Choline, total	12.3 mg	Fatty acids, total polyunsaturated	0.228 g	Choline, total	12.1 mg	Vitamin E (alpha-tocopherol)	0.35 mg
Vitamin A, RAE	709 µg	Tryptophan	0.035 g	Betaine	0.2 mg	Vitamin K (phylloquinone)	2.3 µg
Carotene, beta	8509 µg	Lysine	0.228 g	Carotene, beta	1 µg	Fatty acids, total saturated	0.037 g

(Continued)

TABLE 3.2 (Continued)

RAW SWEET POTATO ROOT		*RAW SWEET POTATO LEAVES*		*RAW POTATO TUBER*		*RAW YAM TUBER*	
Carotene, alpha	7 µg	Methionine	0.086 g	Vitamin A	2 IU	Fatty acids, total monounsaturated	0.006 g
Vitamin A	14187 IU	Cystine	0.047 g	Lutein + zeaxanthin	9 µg	Fatty acids, total polyunsaturated	0.076 g
Vitamin E (alpha-tocopherol)	0.26 mg			Vitamin E (alpha-tocopherol)	0.01 mg	Tryptophan	0.012 g
Tocopherol, beta	0.01 mg			Vitamin K (phylloquinone)	2 µg	Threonine	0.054 g
Tocotrienol, alpha	0.01 mg			Fatty acids, total saturated	0.025 g	Isoleucine	0.052 g
Vitamin K (phylloquinone)	1.8 µg			Fatty acids, total monounsaturated	0.002 g	Leucine	0.096 g
Fatty acids, total saturated	0.018 g			Fatty acids, total polyunsaturated	0.042 g	Lysine	0.059 g
Fatty acids, total monounsaturated	0.001 g			Tryptophan	0.021 g	Methionine	0.021 g
Fatty acids, total polyunsaturated	0.014 g			Threonine	0.067 g	Cystine	0.019 g
Tryptophan	0.031 g			Isoleucine	0.066 g	Phenylalanine	0.071 g
Threonine	0.083 g			Leucine	0.098 g	Tyrosine	0.04 g
Isoleucine	0.055 g			Lysine	0.107 g	Valine	0.062 g
Leucine	0.092 g			Methionine	0.032 g	Arginine	0.127 g
Lysine	0.066 g			Cystine	0.024 g	Histidine	0.034 g
Methionine	0.029 g			Phenylalanine	0.081 g	Alanine	0.063 g
Cystine	0.022 g			Tyrosine	0.048 g	Aspartic acid	0.155 g
Phenylalanine	0.089 g			Valine	0.103 g	Glutamic acid	0.181 g
Tyrosine	0.034 g			Arginine	0.101 g	Glycine	0.053 g
Valine	0.086 g			Histidine	0.035 g	Proline	0.054 g
Arginine	0.055 g			Alanine	0.063 g	Serine	0.081 g
Histidine	0.031 g			Aspartic acid	0.48 g		
Alanine	0.077 g			Glutamic acid	0.351 g		
Aspartic acid	0.382 g			Glycine	0.057 g		
Glutamic acid	0.155 g			Proline	0.063 g		
Glycine	0.063 g			Serine	0.074 g		
Proline	0.052 g						
Serine	0.088 g						

Source: USDA (2019).

contribute to differences in the SP nutrient profile (Bergh et al. 2012; Maquia et al. 2013; Krochmal-Marczak et al. 2014; Abong et al. 2020). Krochmal-Marczak et al. (2014) reported that the Triumph cultivar with white skin and flesh has higher starch, sugars, proteins, vitamin C, ascorbic acid, phosphorus, calcium, and magnesium content than that of the Goldstar and Carmen Rubin cultivars with colored skin and flesh. The orange-fleshed sweet potato is promoted by many, including FAO, UNICEF, and USDA, as a better source of the carotenoids compared to some SP varieties. Bread (125 g) made from 45% OFSP puree containing 15 µg/g beta carotene was estimated to be able to provide 40% of the daily vitamin A requirement for

TABLE 3.3 Nutritional Composition of Boiled Sweet Potato Root, Steamed Sweet Potato Leaves, Fried Sweet Potato Fries, Sweet Potato Paste (per 100 g)

ROOT, BOILED WITHOUT SKIN, SALTED		*LEAVES, STEAMED, SALTED*		*ROOT FRIES, FROZEN, FRIED*		*SWEET POTATO PASTE*	
Water	80.13 g	Water	89.2 g	Water	44.47 g	Water	24.05 g
Energy	76 kcal	Energy	35 kcal	Energy	259 kcal	Energy	294 kcal
Energy	320 kJ	Energy	144 kJ	Protein	2.07 g	Protein	0.41 g
Protein	1.37 g	Protein	2.18 g	Total fat	18.1 g	Total fat	0.04 g
Total fat	0.14 g	Total fat	0.34 g	Carbohydrate	34.05 g	Carbohydrate	75.3 g
Ash	0.63 g	Ash	0.89 g	Fiber, total dietary	5.5 g	Fiber, total dietary	0.8 g
Carbohydrate	17.72 g	Carbohydrate	7.38 g	Sugars, total	12.34 g	Sugars, total	71.58 g
Fiber, total dietary	2.5 g	Fiber, total dietary	1.9 g	Calcium	50 mg	Calcium	9 mg
Sugars, total	5.74 g	Sugars, total	5.48 g	Iron	0.75 mg	Iron	0.25 mg
Sucrose	1.43 g	Calcium	33 mg	Magnesium	25 mg	Magnesium	5 mg
Glucose	0.54 g	Iron	0.63 mg	Phosphorus	56 mg	Phosphorus	10 mg
Fructose	0.43 g	Magnesium	48 mg	Potassium	391 mg	Potassium	70 mg
Maltose	3.34 g	Phosphorus	40 mg	Sodium	140 mg	Sodium	9 mg
Starch	5.22 g	Potassium	312 mg	Zinc	0.36 mg	Zinc	0.07 mg
Calcium	27 mg	Sodium	249 mg	Copper	0.169 mg	Copper	0.033 mg
Iron	0.72 mg	Zinc	0.26 mg	Selenium	0.4 µg	Selenium	0.5 mg
Magnesium	18 mg	Copper	0.033 mg	Vitamin C	6.8 mg	Vitamin C	3.8 mg
Phosphorus	32 mg	Manganese	0.229 mg	Thiamin	0.084 mg	Thiamin	0.017 mg
Potassium	230 mg	Selenium	0.9 µg	Riboflavin	0.088 mg	Riboflavin	0.027 mg
Sodium	263 mg	Vitamin C	1.5 mg	Niacin	0.667 mg	Niacin	0.161 mg
Zinc	0.2 mg	Thiamin	0.112 mg	Vitamin B-6	0.169 mg	Vitamin B-6	0.05 mg
Copper	0.094 mg	Riboflavin	0.267 mg	Folate, total	29 µg	Folate, total	2 µg
Manganese	0.266 mg	Niacin	1.003 mg	Folate, food	29 µg	Folate, food	2 µg
Selenium	0.2 µg	Pantothenic acid	0.2 mg	Folate, DFE	29 µg	Folate, DFE	2 µg
Vitamin C	12.8 mg	Vitamin B-6	0.16 mg	Choline, total	24.9 mg	Choline, total	3.2 mg
Thiamin	0.056 mg	Folate, total	49 µg	Vitamin A, RAE	415 µg	Vitamin A, RAE	236 µg
Riboflavin	0.047 mg	Folate, food	49 µg	Carotene, beta	4980 µg	Carotene, beta	2833 µg
Niacin	0.538 mg	Folate, DFE	49 µg	Lutein + zeaxanthin	8 µg		
Pantothenic acid	0.581 mg	Choline, total	21 mg	Vitamin E	2.55 mg		
Vitamin B-6	0.165 mg	Vitamin A	147 µg	Vitamin K	15.7 µg		
Folate, total	6 µg	Carotene, beta	1725 µg	Fatty acids, saturated	2.416 g		
Folate, food	6 µg	Carotene, alpha	33 µg	Fatty acids, monounsaturated	7.551 g		

(Continued)

TABLE 3.3 (Continued)

ROOT, BOILED WITHOUT SKIN, SALTED		*LEAVES, STEAMED, SALTED*		*ROOT FRIES, FROZEN, FRIED*		*SWEET POTATO PASTE*	
Folate, DFE	6 μg	Cryptoxanthin, beta	45 μg	Fatty acids, polyunsaturated	6.681 g		
Choline, total	10.8 mg	Vitamin A	2939 IU				
Vitamin A	787 μg	Lutein + zeaxanthin	11449 μg				
Carotene, beta	9444 μg	Vitamin E	0.96 mg				
Vitamin A	15740 IU	Vitamin K	108.6 μg				
Vitamin E	0.94 mg	Fatty acids, saturated	0.065 g				
Tocopherol, beta	0.01 mg	Tryptophan	0.02 g				
Vitamin K	2.1 μg	Lysine	0.132 g				
Fatty acids, saturated	0.04 g	Methionine	0.05 g				
Tryptophan	0.02 g	Cystine	0.027 g				
Threonine	0.082 g	Fatty acids, monounsaturated	0.012 g				
Isoleucine	0.082 g	Fatty acids, polyunsaturated	0.134 g				
Leucine	0.121 g						
Lysine	0.081 g						
Methionine	0.041 g						
Cystine	0.013 g						
Phenylalanine	0.099 g						
Tyrosine	0.068 g						
Valine	0.108 g						
Arginine	0.077 g						
Histidine	0.031 g						
Alanine	0.09 g						
Aspartic acid	0.282 g						
Glutamic acid	0.161 g						
Glycine	0.074 g						
Proline	0.072 g						
Serine	0.085 g						

Source: **USDA (2019).**

children (Muzhingi 2016). From personal experience, an average 215-g fresh SP root is currently estimated to have a market price of less than a hundred naira (N 100) in Nigeria and therefore is very affordable even for a low-income earner and can be consumed cooked, roasted, fried, or prepared in any other form for both children and adults. Banwo et al. (2020a) reported in their study the presence of vitamins and minerals essential for normal functioning of the human body from a composite blend which contained sweet potato flour. Ray et al. (2012) reported that purple sweet potato red wine produced by fermentation possessed in 100 mL the following: starch 0.15 g, total sugar 1.35 g, phenol 0.36 g, anthocyanin 55.09 mg, tannin 0.64 mg, lactic acid 1.14 mg; ethanol 9.33% (v/v), and 2,2-diphenyl-1-picrylhydrazyl (DPPH) scavenging activity of 58.95% at a dose of 250 µg/mL.

The specified daily intake for nutrients is shown in Table 3.4. The recommended dietary allowance represents a average daily dietary intake high enough to meet the nutrient requirements of healthy individuals in a group (Food and Nutrition Board 2011). In addition to the nutrients listed here, SP products also contain phenolic acids, steroids, anthocyanins, alkaloids, carotenes, sporamin, flavonoids and tannins, hence conferring the potential on SP products to have hypoglycemic, anti-hypertension, immunoregulatory, antioxidative, anticancer, anti-inflammatory, antimicrobial, anti-obesity, antidiabetes, anti-asthma, anti-ulcer, and detoxification functional qualities (Chandrasekara and Kumar 2016; Nwosisi et al. 2017; Mohanraj 2018; Saranraj et al. 2019). Functional foods are foods or food ingredients that provide additional health benefit(s) above what is normally provided by basic food nutrients (Food Directorate of Health 1998; IFT 1998). Functional foods positively influence the health of the consumer (Ogunremi et al. 2017; Banwo et al. 2020b). SP products can therefore be classified as functional foods if some of the vitamins, minerals, and antioxidants are retained in the final products at concentration that can impact the body functions and/or health of the consumer.

Most processing techniques reduce the moisture content of the raw SP and concentrate nutrients in the final product. During the processing of SP, some nutrients increase, while

TABLE 3.4 Recommended Dietary Allowance/Adequate Intake for Nutrients

	SP 1	*SP 2*	*RECOMMENDED DIETARY ALLOWANCE/ADEQUATE INTAKE*						
Energy (kcal)	192 kcal	259 kcal	**Carbohydrate**: 4–18 y, 45–65% of energy; adults, 45–65% of energy **Fat**: 4–18 y, 25–35% of energy; adults, 20–35% of energy **Protein**: 4–18 y, 10–30% of energy; adults, 10–35% of energy						
			Children	**Male (Age in Years)**			**Female (Age in Years)**		
			4–8 y	9–13	19–30	51–70	9–13	19–30	51–70
Carbohydrate (g)	37.45	34.05	130	130	130	130	130	130	130
Fat (g)	9.39	18.1	ND	ND	ND	ND	ND	ND	ND
Protein (g)	2.27	2.07	19	34	56	56	34	46	46
Sugar (g)	13.58	12.34	ND	ND	ND	ND	ND	ND	ND
Fiber (g)	6	5.5	25	31	38	30	26	25	21
Vitamin C (mg)	7.5	6.8	25	45	90	90	45	75	75
Vitamin A (µg)	457	415	400	600	900	900	600	700	700
Vitamin E (mg)	1.57	2.55	7	11	15	15	11	15	15
Vitamin K (µg)	4.8	15.7	55	60	120	120	60	90	90
Vitamin B6 (mg)	0.186	0.169	0.6	1	1.3	1.7	1	1.3	1.5
β-Carotene (µg)	5478	4980	ND	ND	ND	ND	ND	ND	ND
Calcium (mg)	55	50	1,000	1,300	1,000	1,000	1,300	1,000	1,200
Iron (mg)	0.81	0.75	10	8	8	8	8	18	8
Magnesium (mg)	27	25	130	240	400	420	240	310	320
Manganese (mg)	–	–	1.5	1.9	2.3	2.3	1.6	1.8	1.8
Phosphorus (mg)	62	56	500	1,250	700	700	1,250	700	700
Potassium (mg)	431	391	2,300	2,500	3,400	3,400	2,300	2,600	2,600
Sodium (mg)	154	140	1,000	1,200	1,500	1,500	1,200	1,500	1,500
Zinc (mg)	0.4	0.36	5	8	11	11	8	8	8

Source: SP 1 and SP 2 nutrient values: USDA (2019); dietary allowance/adequate intake: Food and Nutrition Board (2011), National Academy of Sciences (2019). SP 1 = baked orange-fleshed sweet potato, SP 2 = fried orange-fleshed sweet potato, ND = not determined, Kcal = kilocalories, g = grams, mg = milligrams, µg = micrograms.

TABLE 3.5 Effect of Boiling on Nutritional Components of Orange Sweet Potato Leaves/Vines

	GREEN LEAVES		*PURPLE LEAVES*		*PURPLE VINES*	
COMPONENT	*RAW*	*BOILED*	*RAW*	*BOILED*	*RAW*	*BOILED*
Vitamin C (mg/100 g)	482.91	159.11	446.25	188.46	552.99	180.10
β-carotene (μg/100 g)	3208.59	2098.88	3194.58	2199.34	5141.65	2742.42
Protein (%)	20.34	19.89	22.86	19.82	22.60	20.58
Crude fiber (%)	12.48	11.00	12.76	12.23	13.11	11.35

Source: Abong et al. (2016).

some are reduced, depending on the technique employed. For instance, grilling, blanching, and steaming of sliced SP tubers prior to milling into flour increased the flour's carbohydrate content and decreased the fat and protein content of the produced SP flour (Olubunmi et al. 2017). From Tables 3.3 and 3.4, frying appears to significantly decrease β-carotene, potassium, and vitamin A, while other components such as fat, vitamin E, and vitamin K were increased significantly when compared to baking. The method of processing also affects the potential energy yield of SP. A baked SP gave 192 kilocalories, while a fried SP gave 259 kilocalories. The higher energy value for the fried SP is attributable to the oil. When metabolized, the macronutrients fat, carbohydrate, and protein are estimated to give 9, 3.75, and 4 kcal/g of energy, respectively. Abong et al. (2016) reported a significant reduction in protein and carotene and a decrease in vitamin C content of SP leaves when processed by boiling (Table 3.5); however, Tumuhimbise et al. (2009) documented that heating such as boiling, steaming, frying, and baking increases the bioaccessibility of β-carotene in SP. In some documented evidence, fermentation reduced the antinutritive factors of SP and increased some of the nutritional components in the final SP products. Oluwole et al. (2012) reported higher protein and carbohydrate in fermented yellow-fleshed SP flour than the unfermented SP, and Ajayi et al. (2016) observed an increase in carbohydrate, calcium, potassium, sodium, and zinc in 48-h spontaneously fermented SP flour compared to the raw root. The processing method may also affect the glycemic index of SP (Bahado-Singh et al. 2011; Allen et al. 2012). Allen et al. (2012) considered the American Beauregard sweet potato root and documented a low glycemic index of <36 and medium glycemic index of <72 for the raw and cooked (steamed, baked, microwaved) SP, respectively, noting that cooking increased SP glycemic index. Consumption of sweet potatoes can supply needed nutrients and hence be beneficial to people living with diabetes or other related conditions (Allen et al. 2012; CIP 2018).

There is a dearth of reference nutrient profile data for many types and categories of processed SP products in the public domain. Valid comprehensive data from different research studies, product analyses, and evaluation efforts are needed to generate standard data on the nutrient profile of processed sweet potato products (both unfermented and fermented). Some research-generated nutrient data for fermented SP products are represented in Table 3.6.

3.5 PREBIOTIC, PROBIOTIC, AND SYNBIOTIC SWEET POTATO PRODUCTS

Some metabolic dysfunction and illnesses have been attributed to changes in the gut microbiome structure. The gut microflora can be modified with diets/foods containing prebiotics and/or probiotics to improve consumer health. The management and control of the microbiome population and activities are now targeted for the achievement of improved health outcomes for humans and animals and, in a related manner, plants (FAO and WHO 2006; Tona 2017). Dietary compounds such as prebiotics are non-digestible food ingredients or foods that selectively stimulate the growth and/or activities of one or more resident health-beneficial microorganisms (probiotics) in a host/consumer, thus improving the host/consumer's health. Examples of some prebiotics are fructooligosaccharide, resistant starch, inulin, galactooligosaccharide, pectin, and xylooligosaccharide. Probiotics are live microorganisms, which, when consumed in adequate amounts, confer a health benefit(s) on the host (FAO and WHO 2006). Some foods that are rich sources of prebiotics include chicory root, whole grains, onions, garlic, seeds, banana, Jerusalem artichoke, root tubers, leek, and asparagus. Creating synbiotic products containing both prebiotic and probiotic agents will enhance the functional food potentials of the products through synergistic activities (Rowland et al. 1998; Banwo et al. 2020b). Some indicators of an effective prebiotic and/or probiotic are positive modulation of gut microorganisms (e.g., bifidobacteria and lactobacilli), positive modification of metabolic activities in the host system, inhibition of pathogen metabolism/growth/adaptation in the host system, anti-microbial activity, and production of health beneficial compounds (Baroja et al. 2007; Nagpal et al. 2012; Ehiwuogu-Onyibe et al. 2019).

Reports on the prebiotic potential of SP dietary components were previously documented (Lestari et al. 2013; Muchiri and McCartney 2017; Sun et al. 2018; Rodrigues de Albuquerque et al. 2020). These and other reports demonstrated increased acidity and short-chain fatty acid levels; higher bifidobacteria and lactobacilli population; and/or pathogen inhibition by SP fiber, anthocyanin, puree, and flour. Tsai et al. (2013) demonstrated the ability of fermented Chingshey purple

TABLE 3.6 Nutrient Composition of Fermented Sweet Potato Products

PARAMETER	RED WINE	SP FLOUR	SP FLOUR	SP GRANULES
Carbohydrate%		81.85	81.25	81.30
Crude protein%			5.16	1.81
Fat%		0.64	0.68	1.42
Crude fiber%.		2.64	0.76	1.93
Ash%		0.77	2.07	1.19
Starch (g/100 ml)	0.15			
Total sugar (g/100 ml)	1.35			
Phenol (g/100 ml)	0.36			
pH	3.61			4.90
Anthocyanin (mg/100 ml)	55.09			
Tannin (mg/100 ml)	0.64			
Lactic acid (mg/100 ml)	1.14			
Ethanol (% v/v)	9.33			
Total soluble sugar (°Brix)	2.25			
Calcium (mg/100 g)		243.27		
Potassium (mg/100 g)		239.00		
Sodium (mg/100 g)		104.01		
Manganese (mg/100 g)		3.29		
Magnesium (mg/100 g)		2.91		
Iron (mg/100 g)		1.59		
Zinc (mg/100 g)		1.10		
Copper (mg/100 g)		0.59		

Source: SP red wine (Ray et al. 2012); fermented SP flour (Ajayi et al. 2016); fermented SP flour (Ezeama and Amajor 2015); fermented SP granules (garri analog) (Karim et al. 2016).

sweet potato milk rich in gamma amino butyric acid (GABA) to reduce hypertension and improve intestinal microflora balance by increasing fecal populations of *Bifidobacterium* spp. and *Lactobacillus* spp., thereby decreasing the population of *Clostridium perfringens*. These beneficial microorganisms are strategically adapted to the host (human, animal) colon, where resistant starch and other complex substances form the bulk of available nutrients. Their activities and metabolic by-products structure the community of microorganisms that can survive in such an environment, hence their ability to prevent or treat some health conditions (Nagpal et al. 2012; Shimizu 2012). For example, one of the short-chain fatty acids (butyric acid) produced by bifidobacteria during fermentation of prebiotic inulin is confirmed to have therapeutic effects on inflammatory bowel disease in patients (Baroja et al. 2007). Probiotics have been applied for weight management/reduction, immune enhancement, cholesterol reduction, pathogen inhibition, detoxification, peptic ulcers, cancer management, malabsorption, and diarrhea control and treatment (FAO and WHO 2006; Isolauri 2001; Cha et al. 2012; Nagpal et al. 2012; Bordoni et al. 2013). A powerful synergistic effect of a synbiotic (prebiotic and probiotic) product was advanced by Rowland et al. (1998) in a cancer treatment investigation using *Bifidobacterium longum* and inulin.

Probiotic candidates are usually evaluated for functional and technological properties prior to selection and application. A probiotic microorganism should fulfil some necessary criteria: viability, survival through the GIT passage, resistance to gastric juice and bile salt, cell adhesion, genetic stability, survival at low pH, no unacceptable sensory attributes imparted to the product, technological applicability, bacteriocin production, pathogen inhibition, specific observable health benefits, absence of harmful plasmid or transmissible genes, and *in vivo* safety (Conway et al. 1987; Reniero et al. 1992; Del Re et al. 2000; Gopal et al. 2001; FAO and WHO 2006; Banwo et al. 2013a, 2013b; Ehiwuogu-Onyibe et al. 2019; Banwo et al. 2020c). From these properties, it is obvious that not all fermenting microorganisms can be designated as probiotic. Most probiotic microorganisms for humans are selected from *Lactobacillus* and *Bifidobacterium* species. Examples are *Lactobacillus acidophilus*, *Lacticaseibacillus rhamnosus*, *Lacticaseibacillus casei*, *Lactobacillus helveticus*, *Bifidobacterium infantis*, *B. adolescentis*, *B. animalis* subsp *animalis*, *B. animalis* subsp *lactis*, *B. bifidum*, *B. longum*, *B. breve*, and *B. pseudocatenulatum* (Isolauri 2001; Bordoni et al. 2013; Fijan 2014). Some strains of yeasts and other lactic acid bacteria have been applied as probiotics (FAO and WHO

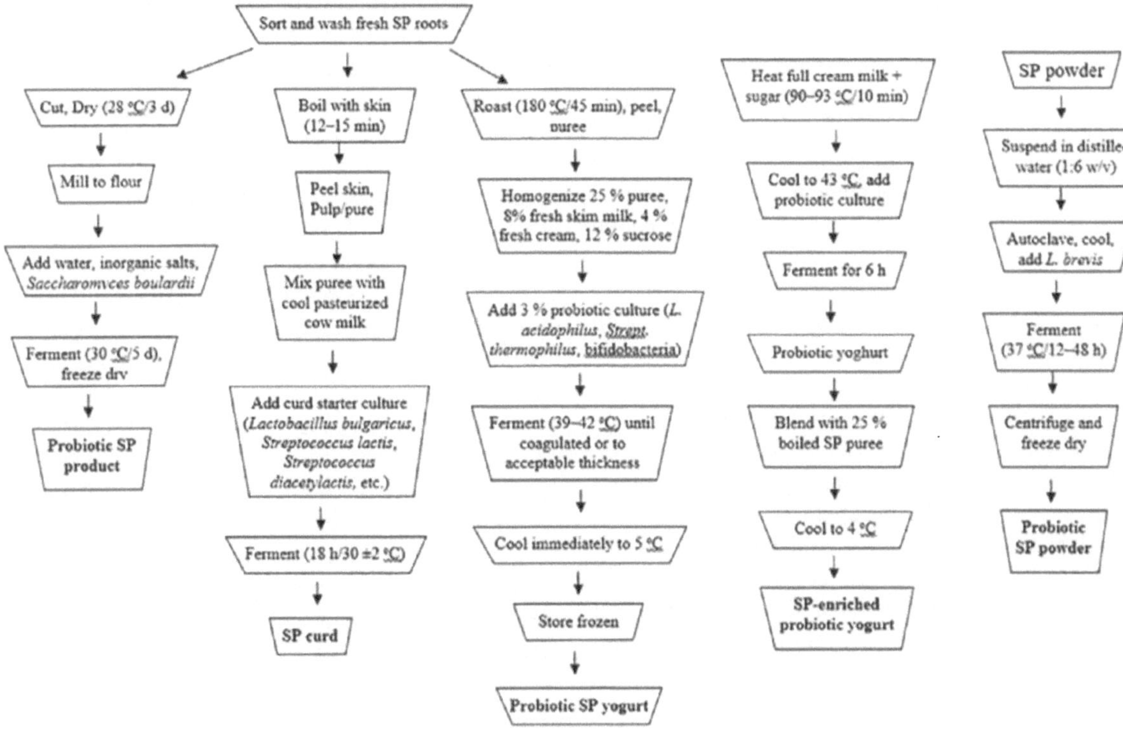

FIGURE 3.5 Probiotic sweet potato product flow process.

Source: Adapted from Hussein and Aumara (2006), Mohapatra et al. (2007), Campbell et al. (2017), Lee et al. (2018), and Muchiri et al. (2020).

2006). Some propionic acid bacteria (PAB) are also used as probiotics. They are known to produce vitamin B_{12} and folic acid, which promote growth and health of the host, regulate the host immunity, protect against pathogens by production of propionic acid and bacteriocin, and scavenge mycotoxins (Piwowarek et al. 2018). Probiotic cultures are used as pure culture(s) on their own, added to products, or used as fermenting or part of the fermenting microorganisms during the production process (Conway et al. 1987; Baroja et al. 2007; Onyibe et al. 2013). Campbell et al. (2017) documented the production of a probiotic-enriched SP product using probiotic *Saccharomyces boulardii* for the fermentation of SP flour (Figure 3.5). In addition to containing a probiotic yeast, the protein, amino acid, and fatty acid profiles of the fermented product were enhanced. Mohapatra et al. (2007) prepared a potential probiotic curd by 18 h fermentation of boiled SP puree (12–16%) and pasteurized (boiled) cow milk with curd starter culture (such as *Lactobacillus bulgaricus*, *Streptococcus lactis*, and *Streptococcus diacetylactis*). Consumption of the produced curd containing live potential probiotic strains (e. g. *Lactobacillus bulgaricus*) may confer a probiotic SP product status. Muchiri et al. (2020) produced orange-fleshed SP–enriched probiotic yoghurt with live cultures of *L. acidophilus*, *S. thermophilus*, and *Bifidobacterium animalis* subsp. *lactis*. A processing technique to produce probiotic *Levilactobacillus brevis* purple sweet potato powder, with antibacterial and antioxidant activities, was given by Lee et al. (2018). In a like manner, Hussein and Aumara (2006) prepared a functional SP yoghurt using probiotic cultures (*L. acidophilus*, *S. thermophilus*, bifidobacteria) through fermentation of roasted SP puree and skim milk (Figure 3.5).

Both unfermented and fermented SP products can be vehicles for probiotic administration in humans and animals. Song et al. (2012) reviewed the application of probiotics in food and agriculture, including non-dairy based food probiotics such as fruit- and vegetable-based, cereal-based, soy-based, and cassava-based food matrixes. Probiotic benefits can also be targeted in plants using applicable beneficial microbes. Commercial probiotic SP products seem not to be very common. Fermented SP products could be next-generation vehicles of probiotics to humans and animals.

3.6 STARTER CULTURE APPLICATION FOR SWEET POTATO FERMENTATION: WHERE ARE WE NOW?

Spontaneous fermentation, also described as natural or uncontrolled fermentation, is one of the oldest food production and preservation methods. It is still the most commonly applied food fermentation technology in many household, cottage, and small-scale fermentation processes, especially in low-income and developing countries. In this uncontrolled method of microbial food substrate conversion, the spontaneity of the fermentation process can expose the final product to possible safety risks arising from contamination and sometimes quality issues because of variation in process parameters. The alternative use of starter culture(s) for fermentation processes offers the advantages of (i) controlled fermentation with reduced or eliminated risk of microbial contamination by the starter culture outcompeting unwanted microorganisms, (ii) reduced fermentation or processing time, (iii) possibility of the application of selected microorganism(s) with added traits for specific roles in the fermentation process or fermented food, (iv) enhanced product quality and nutritional and sensory attribute(s), (v) potential for strain manipulation or improvement, and (vi) prospects for inclusion of probiotic organisms for additional targeted health benefits (Holzapfel 2002; Ogunremi et al. 2017; Banwo et al. 2020b, 2020c).

Starter cultures are viable microorganisms of stable metabolic activities, singly or in defined combinations, used in fermentation processes for specific changes or transformation of the substrate to produce products of desirable quality. The use of starter culture in food fermentation offers the benefit of specific and optimum flavor development. For the targeted advantages of natural adaptability and competitiveness, starter cultures are mostly selected from microorganisms associated with the substrate or from the indigenous microbiota of the product. In some instances, starter cultures from other food fermentations or environments may be effectively and efficiently used. The processes of starter culture isolation to improvement have been discussed by several authors (Hunter-Cevera and Belt 1999; Huang et al. 1999; Monaghan et al. 1999; Vinci and Byng 1999; Ajayi et al. 2016; Ogunremi et al. 2017; Banwo et al. 2020a, 2020b). Some of these activities are presented in Figure 3.6.

Selection of a starter culture is dependent on the target substrate or raw material, fermentation process, and sought-after product. Some characteristics of a good starter culture include rapid growth in fermentation conditions (such as medium/environment, temperature, pH, and anaerobic conditions), ability to outcompete other microorganisms, colonization and dominance in the fermentation environment, capacity to effectively transform the substrate or raw material to the desired product, inhibition of growth of unwanted microbes (pathogens, spoilage organisms, off-flavor–impacting organisms), improvement of color and flavor of the product, enhancement of product safety and shelf life, and no production of undesirable or harmful metabolites. In addition, starter cultures must be safe and pose no risk to consumers. Possessing antimicrobial-resistant genes or transposable elements with genes encoding detrimental functions, such as antibiotic-resistant genes, is an undesirable property for any starter culture. Such undesirable genes may be transferred to other microbes during fermentation or to gut microorganisms of the consumer (human or animal) (Banwo et al. 2013a; Rozman et al. 2020). Heller et al. (1995) demonstrated the possibility of gene transfer between microbes in food microbiota. The generally recognized as safe (GRAS) or presumptively safe species for human and animal consumption status of a starter culture is an obligatory regulatory requirement (EFSA 2017).

FIGURE 3.6 Starter culture development process activities.

One way of making the rich nutrients of the SP available to consumers is through diverse fermented products. Some important industrial products for economic development are biofuels (ethanol), industrial chemicals (organic acids), microbial enzymes, and food additives (flavors, thickeners). The actualization of these production processes requires the use of starter cultures derived from SP, other fermentable substrates, or existing commercial starter cultures. Examples of versatile starter cultures that can be applied for diverse substrates, as well as SP, to produce related end products are *Saccharomyces cerevisiae* for wine production and bread making; *Lactiplantibacillus plantarum* for flour, pickles, and juice; *Lactobacillus bulgaricus* and *Streptococcus thermophilus* for yoghurt; *Saccharomyces cerevisiae* for wine and bioethanol production; *Aspergillus niger* and *Saccharomyces cerevisiae* for distilled liquor; *Aspergillus niger* for glucoamylase; and *Streptomyces* for tetracycline (Azhara et al. 2017; Saranraj and Ray 2019; Saranraj et al. 2019). There is a large existing gap in starter culture commercialization and usage. Hence, the call for goal-driven research, development, investment, and commercial production of starter cultures in Africa is overdue. This is undoubtedly contributing to the slow pace in industrial development of low-income nations. This in turn has created a loophole in maximizing the potential of innovative fermented product design and availability. Developing diverse fermented SP products will ultimately create a market for deploying starter culture applications or vice versa, with subsequent expansion of industrial SP usage. In addition to other factors, fermenting microorganism species and strain variation may determine the final product nutrient composition type and concentration. Starter culture design and application, therefore, present a great opportunity and possibility of producing products with specialized nutrient profiles and health solution potentials (Ogunremi et al. 2017; Lee et al. 2018; Banwo et al. 2020a, 2020b, 2020c; Banwo et al. 2021).

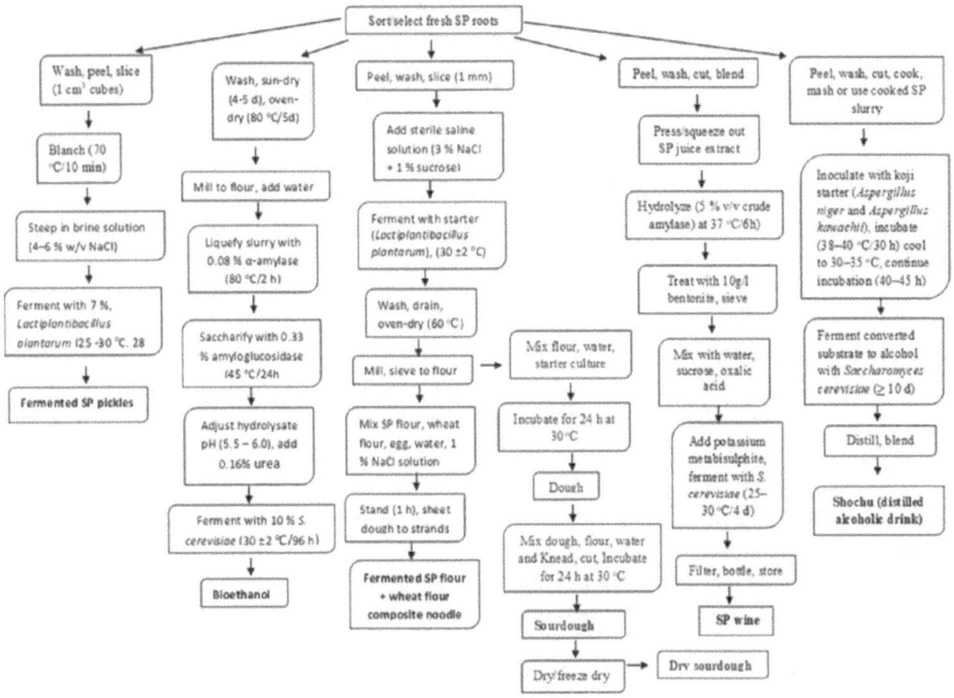

FIGURE 3.7 Starter culture–produced SP product flow process.

Source: Adapted from FAO (1998a), Sato and Sudo (1999), Paramithiotis et al. (2005), Ray and Naskar (2008), Oloo et al. (2014), Sanjib et al. (2014), Yuliana et al. (2018), and Mantzourani et al. (2019).

Application of starter cultures (*Lactiplantibacillus plantarum*, *Leuconostoc mesenteroides*, and *Saccharomyces cerevisiae*) in sweet potato fermentation to positively modify sweet potato flour for an improved composite noodle was demonstrated (Yuliana et al. 2018). Oloo et al. (2014) reported increased beta-carotene retention in SP pickles by *Lactiplantibacillus plantarum* fermentation. Swain et al. (2013) successfully produced bioethanol by solid-state fermentation of sweet potato flour with the application of co-culturing of *Trichoderma* sp. and *Saccharomyces cerevisiae* in a ratio of 1:4. Fermentation of sweet potato leaf with a culture of *Chaetomium globosum* increased the protein content while significantly decreasing the anti-nutrients phytic acid by 57.51%, trypsin inhibitor by 15.31%, and hydrogen cyanide by 61.70% (Jayant et al. 2020). Ray et al. (2012) produced sweet potato red wine with acceptable sensory qualities using the wine yeast *Saccharomyces cerevisiae* as starter culture. Starter culture (*Lactobacillus acidophilus*, *Lactobacillus delbrueckii*, and *Lactobacillus gasseri*)–fermented Chingshey purple sweet potato (CPSP) milk reduced blood pressure in hypertensive rats, thus demonstrating that oral administration of GABA-rich CPSP-fermented milk could be an effective antihypertensive yoghurt drink (Tsai et al. 2013). Ojewumi et al. (2018) produced bioethanol through sweet potato peel fermentation using *Saccharomyces cerevisiae* as the starter culture. Bioethanol production from SP flour was achieved by fermentation using *S. cerevisiae* as a starter culture (Ray and Naskar 2008). Sourdough-based technology is an old but rediscovered fermentation process. The development of novel starter cultures, selected from lactic acid bacteria and yeasts for SP bakery products, through sourdough technology will add value such as better nutrition; prolonged shelf life; and improved flavor, aroma, taste, color, and texture (Chavan and Chavan 2011; Nionelli and Rizzello 2016; Ogunsakin et al. 2017; Banwo et al. 2020a; Olojede et al. 2020). Some microorganisms applied in sourdough production are *Lactiplantibacillus plantarum*, *Levilactobacillus brevis*, *Lacticaseibacillus paracasei*, *Fructilactobacillus sanfranciscensis*, *Lactobacillus bulgaricus*, *Pediococcus pentosaceus*, *Saccharomyces cerevisiae*, and *Pichia kudriavzevii* (Paramithiotis et al. 2005; Ogunsakin et al. 2017; Mantzourani et al. 2019; Banwo et al. 2020a; Olojede et al. 2020; Banwo et al. 2021). This has the potential of speeding up industrialization of SP usage. Figure 3.7 shows some starter culture–driven production process flow charts. Industrial application of these and many other starter culture–based processes will expand the commercial production and easy accessibility of improved SP-process–based types of starter cultures.

3.7 PRODUCT SAFETY

Ready-to-eat crops and foods present a huge concern for contamination and negative health repercussions. SP is classified as a vegetable, and it is a product that some people consume raw because of its low level of toxicant and antinutrient contents (Antia et al. 2006; OECD 2010). A safe product is free of hazard and risk to the consumer (human or animal). A hazard is a biological, chemical, or physical agent in, or in the condition of, food with the potential to cause an adverse health effect. Examples are viruses, bacteria, fungi, parasites, natural chemicals (e.g., aflatoxin), human-made chemicals (heavy metals from fertilizers/irrigation water, pesticides, fungicides, insecticides, harmful preservatives, and other chemical additives), and physical substances (bone, stone, glass, wood). Biological hazard agents are viruses (Hepatitis A virus), bacteria (enterohaemorrhagic *Escherichia coli*, *Listeria*, *Vibrio cholerae*, *Salmonella*, *Campylobacter*), fungi (*Aspergillus* sp.), and parasites (*Taenia solium*, *Echinococcus*, *Ascaris*, *Cryptosporidium*, *Giardia*, *Entamoeba histolytica*), among others, which can cause harm to human or animal health, usually due to infection or by toxins produced in food products (Przyrembel 2008). An adverse health effect signifies a change in morphology, physiology, growth, development, reproduction, or life span of an organism, system, or (sub)population that results in an impairment of functional capacity, an impairment of the capacity to compensate for additional stress, or an increase in susceptibility to other influences (FAO and WHO 2006; Przyrembel 2008). While risk can be defined as the possibility, chance, or probability that there will be harm or the experience of an adverse health effect if exposed to a hazard, safety is the state of being safe: the condition of being protected from harm or other non-desirable outcomes or recognized hazards. Food safety is, therefore, the assurance that food will not cause harm to the consumer when it is prepared and/or eaten according to instructions or specifications and used for the intended purpose (FAO and WHO 2006; Ogunremi et al. 2017; Banwo et al. 2020b).

What are the possible sources of hazards that require attention in SP production/products? This list includes raw sweet potato produce and processed SP product (unfermented and fermented) quality and safety, packaging material quality, fermenting microorganism safety, equipment and processing environment quality, operating personnel, and input chemical/reagent quality. To achieve quality consistently, a producer or organization must be proactive and internalize the concept of quality at every stage of operation. Some practical principles and quality systems will be applicable depending on the level of operation under consideration. Quality is defined by the customer, and a product is of good quality if it meets the customer's needs and expectations. Continuous research, dissemination of data and information, use of risk analysis in risk management for food safety decisions, innovative product design, and responsive manufacturing and distribution that put consumers' lives first are the way forward. Concerted efforts to adopt the best global practice in SP food design, production, approval, monitoring, and post-marketing surveillance are critical to consistently ensuring SP product safety (OECD 2010; Hegde et al. 2012).

Food quality and safety control begin with good hygiene practice at all levels of food production. GHP may be defined as the process of ensuring that manufactured products are safe for consumption, free from harmful microorganisms, and of good keeping quality during their shelf life. The best food safety practice will be to immediately process freshly harvested SP or store harvested SP in a temperature-controlled environment (rooms, stores, and warehouses). Good agricultural practice (GAP) provides guidelines on growing, harvesting, transportation, and storage of crops. Questions about soil and environment quality, irrigation water quality, fertilizer type and quality, pesticide and fungicide safety and usage, and chemical application type and levels for various purposes must always be conscientiously answered, since what is used in any process is invariably carried over through the product to the consumer. Proper practices such as GAP, GHP, and current good manufacturing practice (CGMP) indicate proper sorting and discarding of damaged, infected, discolored, rotten, or spoiled products during harvest and before use or processing to reduce or eliminate the risk posed by some of these contaminants. As specified by CODEX (2019), the maximum allowable level of heavy metals in root and tuber vegetables, such as SP, includes arsenic 0.1 mg/kg, lead 0.1 mg/kg, and cadmium 0.1 mg/kg. Also, the maximum levels allowed for canned vegetables are lead 0.1 mg/kg and for canned foods 250 mg/kg and acrylonitrile 0.02 mg/kg. Ideally, animal feed will need to comply with the same food safety rules that require food to be fit and safe for human consumption, since animals that consume such food will be consumed by humans. Food virology is another area of food quality control, gaining more attention because of the implication of foodborne viruses in human and animal infections. Control measures for viruses at every step of the food chain are therefore of importance to food safety. Virus detection and characterization are performed with biological and molecular-based (e.g., PCR, next-generation sequencing) techniques (OECD 2010; Hegde et al. 2012; Boscha et al. 2018).

Raw SP, like other plant foods, contains antinutrients that are known to reduce the bioavailability and absorbance of some nutrients during digestion. Oxalates, phytates, trypsin inhibitors, and tannins are some antinutrients that have been detected in SP tubers and leaves (OECD 2010). Oxalate reduces calcium availability, and trypsin inhibitors inhibit protein degradation by trypsin. Various processing methods such as soaking, blanching, steaming, boiling, roasting, frying, and fermentation are known to reduce antinutrients to varying degrees, thus improving the bioavailability and absorption of targeted nutrients (Wanasundera and Ravindran 1992; Hotz and Gibson 2007). In an investigation, Wanasundera and Ravindran (1992) observed a 40–50%, 20–25% and 12–15% reduction of oxalates by the boiling, steaming, and baking of yam, respectively. Abubakar et al. (2010) reported low phytate and tannin compared to oxalate antinutrient contents (mg/100 g) in cooked porridge, boiled SP, and fried SP: phytate-porridge 0.57, boiled SP 0.88, fried SP 0.72; tannin-porridge 0.63, boiled SP 0.68, fried SP 0.33; oxalate-porridge 178.27, boiled SP 167.15 and fried SP 148.28. Oboh et al. (1989) evaluated 49 varieties of raw SP tubers and detected phytate, tannins, and oxalate in mean values of 0.01 g/100 g, 0.71 g/100 g, and 0.71 g/100 g, respectively.

With respect to fermented SP products, fermenting microorganisms, starter cultures, and probiotics should not harbor acquired or transferable antibiotic-resistant gene(s), plasmids, viruses, or transposons that can negatively impact consumer resident microorganisms and health structure. Product quality and safety are not chance occurrences; they happen by deliberate design, purposeful implementation, and sustained activities to guarantee results (FAO 1998b). HACCP is a common quality and safety management system recommended for food handling, food processing, food packaging, and food distribution establishments. GHP, GAP, and CGMP are some prerequisite programs for successful HACCP applications and implementation. The HACCP concept is defined as a scientific, systematic preventive approach to ensuring food safety through the identification, evaluation, and control of food safety hazards. The HACCP plan must be documented: identification of food safety hazard(s) that need control at every critical control point (CCP), determining control measures, establishing critical limits for each control measure, putting monitoring procedures in place, instituting the necessary corrections and corrective actions, defining responsibilities and authorities, and planning for records of monitoring (FAO 1998b). The system works on the following seven principles.

- Conduct hazard analysis of the process; identify potential hazards and measures for their control.
- Determine the critical control points of the process that will eliminate/minimize hazards or risks.
- Establish critical limits which must be met to ensure that each CCP is under control.
- Establish a monitoring system to demonstrate that the CCP(s) is under control.
- Establish the corrective action to be taken when monitoring indicates that a CCP is moving out of control.
- Establish verification procedures to confirm that the HACCP system is working effectively.
- Establish documentation for procedures and records.

Using the production flow process for ready-to-eat roasted SP granules (garri analog) in Figure 3.4 as an example, the potential hazards and poor-quality issues include (not exhaustive):

1. spoilt, discolored, infected roots
2. contaminated process water
3. mud, sand, filth on roots
4. pathogenic and thermophilic microorganisms from water, the environment, machines and tools (such as graters and roasters)
5. under/overfermentation of mash
6. contamination from handlers (poor hygiene, pathogenic organism–carrying personnel)
7. dirt, dust, and particles from an unclean environment
8. contamination by insects and rodents
9. over-roasted or burnt granules
10. contamination from rusted or dirty equipment (grater, roaster)
11. pathogenic microorganisms from materials (e.g., troughs, sacks, and trays)

TABLE 3.7 Ready-to-Eat Roasted Sweet Potato Granules (Garri Analog) Production Process Potential Critical Control Points

CCP 1 (Raw material)	SP roots should be fresh, of the right variety, of the right maturity/age at harvest, devoid of defects, devoid of spoilage, devoid of discoloration. Acceptable raw material should not be stored beyond established post-harvest storage period before processing.
CCP 2 (Peeling and washing)	Peeling should be performed with a clean, effective implement/machine to reduce wastage and ensure proper peeling of all roots. Only potable water of determined quality (pH, conductivity, microbial level, free of pathogens, etc.) should be used for washing. Potable water should be colorless, odorless, and tasteless (COT). The washing process should thoroughly remove mud, dirt, soil/sand, particles, and coloration, resulting in clean SP roots for grating.
CCP 3 (Grating/mashing)	The grating machine should always be clean, appropriately disinfected or sterilized, and free of pathogens and rusts. Effective grating/mashing to acceptable appearance and particle size is a critical process requirement.
CCP 4 (Fermentation, dewatering, pressing)	Dewatering and fermentation should be carried out in a clean environment with clean disinfected/sterilized materials (troughs, sacks). Established fermentation temperature and duration are critical control points that should be monitored and adhered to. At the end of fermentation and dewatering, cake pH, moisture content, acidity level, and pathogen content are some monitoring parameters.
CCP 5 (Roasting)	Roasting equipment should be made of acceptable materials, clean, rust free, and free of pathogenic/thermophilic microorganisms. Roasting should be performed in a very clean environment at a predetermined temperature/heat degree and to a specified moisture level (e.g., about 10%). Avoid overheating and charring/burning of granules.
CCP 6 (Cooling and packaging)	Establish and monitor cooling method to avoid heaping roasted granules that may occasion trapping of moisture or overheating, which may result in overcooking granules. Granule size specification should be adhered to and sieving/milling instituted if necessary.
	Packaging material type and packaging method/equipment necessary to protect the packaged product (roasted SP granules) is a critical control to maintain quality and safety throughout shelf life.
CCP 7 (Storage)	A clean, insect- and rodent-free environment with the right temperature and humidity is an important requirement during storage.

12. moisture absorption by the product (improper sealing or moisture-permeable packaging material)
13. weevil infestation resulting from incorrect processing and/or faulty packaging

The CCPs (Table 3.7) for this production process are some suggested steps to which controls must be enforced to eliminate, reduce, or prevent the occurrence of each identified potential hazard or quality problem. The producer is expected to set limits or allowable specifications for the CCPs and ensure compliance.

The impact of packing on the quality and safety of SP products cannot be undermined, especially having in mind the sensitivity of target SP nutrients to atmospheric oxygen/air, heat, light, moisture, and water (Nasir et al. 2016). The function of good packaging in preserving products from deterioration and microorganism attack cannot be overemphasized. In a laboratory-scale investigation of fried SP crisps stored at ambient temperature (30 ± 2°C), Nasir et al. (2016) confirmed laminated aluminum foil offered a better barrier and better carotene retention than high-density polyethylene against rancidity and microbial degradation. Packaging materials should be selected and designed to safeguard product integrity (such as quality, hygiene, nutrients, and organoleptic properties) until the expiration of the packaged product. In consonance with food safety concerns, food and feed fraud is a current phenomenon that impacts safety and public health. Food or feed fraud refers to intentional and economically motivated adulteration of foods. To mitigate these challenges, producers, regulators, consumers, and the public must be aware and responsive to these occurrences (such as adulteration, substitution, addition, tampering, misrepresentation of food and food ingredients, unapproved enhancement, false claims, and misleading label statements) along the food chain and in their environment (Spink et al. 2019; European Union 2020).

3.8 IMPROVING ANIMAL FARMING WITH SWEET POTATO

Livestock or farm animals can be categorized into ruminant (such as cattle, goat, buffalo, and sheep) and non-ruminant (such as chicken, pig, and horse) animals based on their pattern of feeding and digestion. Fish, snail, and rabbit farming are also popular businesses in some localities. While ruminants are herbivores, non-ruminants are monogastric (Tona 2017). In many countries like Nigeria, livestock feeds are imported; consequently, livestock farmers constantly complain of the prohibitive cost of inputs, which ultimately leads to a high cost of farmed animal products. Local production of feeds using locally available raw materials is preferred as a solution to this persistent challenge. Most farm animal feeds are based on cereals (e.g., maize). Research and expanded use of alternative local feed raw materials will reduce costs and may improve efficiency. Sweet potato roots, vines, and residues have been utilized in animal feed production. Seath et al. (1947) showed verified evidence of very good performance of various parts of SP as feed for cattle rearing. They pointed out that SPs are good carbohydrate feed for dairy animals, SP vines as a supplement pasture for cows increases milk production, fresh SP was valuable as silage in dairy rations, dehydrated sweet potatoes tubers were palatable, and the high carotene content increased the vitamin A value of the butterfat. Olorunnisomo (2007) concluded that mixing sun-dried SP forage and roots in equal proportions improved nutrient utilization, reduced the cost of live weight gain, and maximized economic returns for sheep. The effectiveness of SP feed was corroborated by González et al. (2002), who suggested that SP root meal can replace 75% of the cereal components in pig feeding. Jayant et al. (2020) experimented on feeding fingerlings with fermented SP leaf meal and confirmed that fermented sweet potato leaf meal can replace 100% of de-fatted rice bran in the diet of *Labeo rohita* fingerlings. In a similar manner, Yacout et al. (2016) studied SP performance as a possible sheep feed. They reported anti-nutritional compounds and pesticide residue reduction by sun-drying sweet potato and SP silage and observed that feeding with LAB-inoculated SP vine silage was an economical, good roughage for lactating sheep. Ensiling is a method of preservation for plant materials, which is based on microbial fermentation of carbohydrates, adopted for animal nutrition and farming. Sweet potato silage for animal farming has been in practice. The mixture of sweet potato root (SPR) and sweet potato vine (SPV) was well preserved (with good smell) for several months as feed for pigs (Giang et al. 2004). SP silage methods for quality feeds have been documented (FAO 2001; Giang et al. 2004; CGIAR 2017). In a review report, Paneque Ramirez (1992) concluded that cooked and mashed sweet potato may totally replace maize for fattening pigs when supplemented with protein, and dehydrated sweet potato can replace up to 50% of maize in diets for poultry and pigs.

In raw or processed forms, identifiable SP products, residues, or "waste" should be made readily available for use as feedstock. Local or national product development and industrial production of SP or SP-supplemented feeds should be given better attention to maximize the accruable benefits. As research in human microbiome communities for improved nutrition and health is expanding, research into the animal microbiome as a means of harnessing information to aid in providing innovative products targeted at improved livestock nutrition and health must be advocated. SP probiotic products fit into this vision of future innovative feeds.

3.9 FUTURE PERSPECTIVES

More than 2 billion people of the world do not have access to safe, nutritious, and sufficient food (FAO 2019). The FAO recognized four dimensions of food security: availability, accessibility, utilization, and stability. Amid food and nutritional insecurity, a lot of wastage along the food chain is reported all over the globe. The worst affected is the African continent, where most of the foods produced are not refrigerated and huge annual food loses are common. A solution to the high energy cost required to run and maintain cold chains or alternative cheap energy sources for cold systems will greatly move this trajectory towards a wastage reduction trend. Another option to eliminate food wastage is prompt conversion of foods into nutritious food products. At present, sweet potato products, production processes, and technological development are at various stages of completion and commercialization. Many are still in the early stage of research and development, others are at advanced stages, and few are commercialized at the industrial scale in some developed and developing countries. Serious attention and concerted efforts in diversification of SP utilization will improve the current state of local, national, and global food security. Farmers' productivity and economic status will also be enhanced; this will motivate them to cultivate and produce more. Research and refining of processes to benefit and rescue humankind from the challenges of wastage, hunger, malnutrition, poverty, and economic limitations should be continuous development. The slow pace in translating results of laboratory investigation to successful and profitable industrial applications has been the reason for delayed commercialization of many scientific breakthroughs. If all the laboratory-developed SP food products were successfully translated to industrial products and commercialized, the world would have substantial availability of nutritious products, and the global population would have more options to counter nutritional insecurity.

The Food and Agriculture Organization of the United Nations Statistics displayed in Figure 3.8 are an indication that much can still be achieved in many countries and continents to boost SP production and availability. In real terms, continental contributions to the 91,945,356-ton total were: Asia 66% (60,671,837 t), Africa 29.1% (26,728,240 t), North America 2.9% (2,655,422 t), South America 1.7% (1,547,108 t), Oceania 0.23% (206,667 t), and Europe 0.15% (136,082 t). If in 2018 the Chinese Mainland contributed 87.4% of the 66% for the continent with the highest production of SP (Asia), it can be interpreted to mean that the remaining Asian countries can

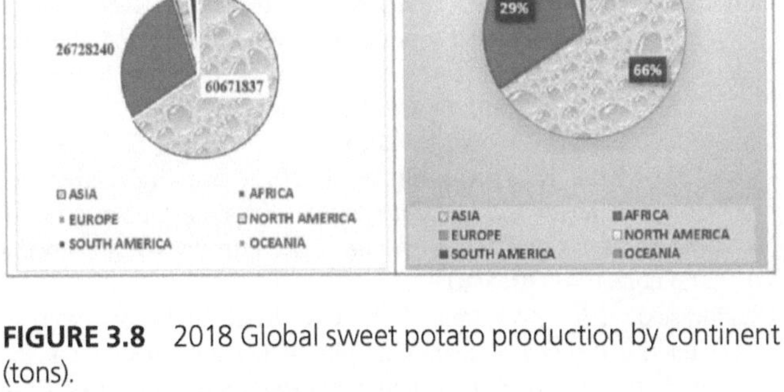

FIGURE 3.8 2018 Global sweet potato production by continent (tons).

Source: Adapted from FAOSTAT (2020).

raise their contributions. The other continents and the countries of those continents could achieve higher outputs. More worrisome is the absence of data on processed SP products in FAOSTAT, a possible indication of its limited contribution to national trade volume and its absence in global trade activities. The industry of sweet potato products (unfermented and fermented) is still at the level of discovery and evaluation. The OECD (2010) identified very limited international trading in unprocessed SP roots, with the United States, China, Israel, and France being the main exporters. The potential positive roles of SP in advancing food security, nutrition and health, income generation, climate change mitigation, and sustainable food systems have been highlighted (Low et al. 2017; CIP 2018). The findings in this chapter agree with the CIP (2018) on underdeveloped markets and potential products as major SP challenges. Fermentation is a viable technique in the quest for SP produce utilization, diversification, development of new innovative food products, preservation, and storage. Efforts at generation and collation of data for the unfermented SP product nutrition profile seem to be progressively developing, but that of fermented SP products is still awaiting research, product development, data generation, and documentation. In other words, nothing significant in available reference data is visible in the public domain pertaining to the diversity of possibilities in fermented SP products for industrial application. Just as the landscape for cassava has changed as a result of the increasing use of cassava derivatives for innovative food products, the current position of SP production and application can greatly expand in the near future.

Areas being addressed (and still needing attention) to drive enhanced SP production and utilization output include improved root and dry matter yield, increased nutritional value (e.g., beta carotene, vitamin A, and anthocyanins), changes in maturity period and more acceptable specific sensory attributes, disease and pest resistance, and tolerance to stress (salinity, drought, nutrient depleted soil, flood, and acid soils). Most SP cultivar or variety improvements are being done through breeding with less need for safety protocols. Discussions on strain and species improvements, as applicable to microorganisms or plants, point to the fact that the potential for modifications to suit specific desired traits or properties is a continuous and progressive need. The recent emergence of effective scientific tools has opened new avenues for this approach. Current trends in gene sequencing, genomics, proteomics, metagenomics, bioinformatics, and related metabolic tools attest to the wide array of innovative possibilities. Countries can leverage resources to increase both scientific and technological research, innovation, and industrial actualization of research results to enhance sustainable production and processing of nutritious SP food to meet growing global demand. Biotechnology research is capital intensive, and an increased government and grant/donor agency commitment to science and technology is needed to create avenues for starter culture discovery, improvement, optimization, integration, standardization, and industrial application.

FAO (2017) reported one-third of global food production leads to wastage and loss. To counter the devastating impact of national and global food wastage, it is recognized by all, including FAO, that food processing and distribution must be given prominent attention in the transformation of food systems. It is also obvious that SP products (unfermented and fermented) can contribute immensely to the nutritional security and improved health of growing populations. The future is still waiting for national and international investments and actions to fast-track the much-expected contribution of SP, especially fermented SP products, to the national and global economy and nutrition.

REFERENCES

Abong, G. O., Chepkoech, M. and Oguya, G. 2016. Nutritional value and acceptability of lactic acid fermented sweetpotato leaves utilized as vegetables. In The Markets, Processing, and Utilization Community of Practice Meeting Dar Es Salam, 14–16 March, Tanzania.

Abong, G. O., Muzhingi, T., Michael Wandayi Okoth, M. W., Ng'ang'a, F., Ochieng', P. E., Mbogo, D. M., Malavi, D., Akhwale, M. and Ghimire, S. 2020. Phytochemicals in Leaves and roots of selected Kenyan orange fleshed sweet potato (OFSP) varieties. *International Journal of Food Sciences* 2020:1–11. https://doi.org/10.1155/2020/3567972.

Abubakar, H. N., Olayiwola, I. O., Sanni, S. A. and Idowu, M. A. 2010. Chemical composition of sweet potato (*Ipomea batatas* Lam) dishes as consumed in Kwara state, Nigeria. *International Food Research Journal* 17:411–416.

Ajayi, O. I., Ehiwuogu-Onyibe, J., Oluwole, O. B., Jegede, A. A., Salami, T. A., Asieba, G. O., Chiedu, I. E., Suberu, Y. L., Aba, E. M., Dike, E. N., Ajuebor F. N. and Elemo, G. N. 2016. Production of fermented sweet potato flour using indigenous starter cultures. *African Journal of Microbiology Research* 10(41):1746–1758.

Ajayi, O. I., Okedina, T. A., Samuel, A. E., Asieba, G. O., Jegede, A. A., Onyemali, C. P., Ehiwuogu-Onyibe, J., Lawal, A. K. and Elemo, G. N. 2019. Evaluation of starter culture fermented sweet potato flour using FTIR spectra and GC-MS chromatogram. *African Journal of Microbiology Research* 13:1–13.

Akinsola, A. O., Idowu, O. A., Laniran, A. M., Ojubanire, B. A. and Oke, E. K. 2017. Development, evaluation and sensory quality of orange fleshed sweet potato (*Ipomoea batatas* Lam) extruded pasta products. *Croatian Journal of Food Technology, Biotechnology and Nutrition* 12:83–89.

Allen, J. A., Corbitt, A. D., Maloney, K. P., Butt, M. S. and Truong, V.-D. 2012. Glycemic index of sweet potato as affected by cooking methods. *The Open Nutrition Journal* 6:1–11.

Amajor, J. U., Oti, E., Ekeledo, N., Omodamiro, R., Amajor, E. E. and Aniedu, C. 2014. Studies on the characteristic properties of fermented sundried orange-fleshed sweet potato flour. *Nigerian Food Journal* 32 (1):45–53.

Antia, B. S., Akpan, E. J., Okon, P. A. and Umoren, I. U. 2006. Nutritive and anti-nutritive evaluation of sweet potatoes (*Ipomoea batatas*) leaves. *Pakistan Journal of Nutrition* 5:166–168.

Azhara, S. H. M., Abdullaa, R., Jamboa, S. A., Marbawia, H., Gansaua, J. A., Faika, A. A. M. and Rodrigues, K. F. 2017. Yeasts in sustainable bioethanol production: A review. *Biochemistry and Biophysics Reports* 10:52–61.

Bahado-Singh, P. S., Riley, C. K., Wheatley, A. O. and Lowe, H. I. 2011. Relationship between processing method and the glycemic indices of ten sweet potato (*Ipomoea batatas*) cultivars commonly consumed in Jamaica. *Journal of Nutrition and Metabolism* 584832. https://doi.org/10.1155/2011/584832.

Banwo, K., Adesulu-Dahunsi, A. T. and Dahunsi, O. S. 2020c. Molecular methods in fermented foods. In Sankaranarayanan, A., Amaresan, N. and Dhanasekaran, D. (Eds.). *Fermented Food Products*. CRC Taylor and Francis Group. Pp. 365–380. ISBN: 13: 978-0-367-22422-6.

Banwo, K., Fasuyi, T. and Olojede, A. O. 2021. Potentials of *Lactobacillus plantarum* and *Pichia kudriavzevii* in co-fermentation of sourdough from millet. *International Journal of Food Science and Technology* 56:857–864. https://doi.org/10.1111/ijfs.14729.

Banwo, K., Osagbemi, O., Ajao, O. and Sanni, A. 2020a. Sourdough bread from the blend of cassava, sweet potato, and soybean flours using *Lactobacillus plantarum* and *Pichia kudriavzevii*. *Acta Alimentaria* 49:441–450.

Banwo, K., Ogunremi, O. R. and Sanni, A. I. 2020b. Fermentation biotechnology of African traditional foods. In Shetty, K. and Sarkar, D. (Eds.). *Functional Foods and Biotechnology: Biotransformation and Analysis of Functional Foods and Ingredients*. CRC Taylor and Francis Group. Pp. 102–124. ISBN: 978-0-367-43522-6.

Banwo, K., Sanni, A. I. and Tan, H. R. 2013a. Technological properties and probiotic potential of *Enterococcus faecium* strains isolated from cow milk. *Journal of Applied Microbiology* 114:229–241.

Banwo, K., Sanni, A. I. and Tan, H. R. 2013b. Functional properties of *Pediococcus* species isolated from traditional fermented cereal gruel and milk in Nigeria. *Food Biotechnology* 27 (1):14–38.

Baroja, L. M., Kirjavainen, P. V., Hekmat, H. and Reid, G. 2007. Anti-inflammatory effects of probiotic yoghurt in inflammatory bowel disease patients. *Clinical and Experimental Immunology* 149:470–479.

Bergh, K., Orozco, P., Gugerty, M. K. and Anderson, C. L. 2012. *Sweet Potato Value Chain: Nigeria EPAR Brief No. 220*. EVANS School Policy Analysis and Research.

Bordoni, A., Amaretti, A., Leonardi, A., Boschetti, E., Danesi, F., Matteuzzi, D., Roncaglia, L., Raimondi, S. and Rossi, M. 2013. Cholesterol-lowering probiotics: *In vitro* selection and *in vivo* testing of bifidobacteria. *Applied Microbiology and Biotechnology* 9:8273–8281.

Boscha, A., Gkogkab, E., Le Guyaderc, F. S., Loisy-Hamond, F., Leee, A., van Lieshoutf, L., Marthig, B., Myrmeli, M., Sansomj, A., Schultzk, A. C., Winklerl, A., Zuberm, S. and Phistern, T. 2018. Foodborne viruses: Detection, risk assessment, and control options in food processing. *International Journal of Food Microbiology* 285:110–128.

Campbell, C., Nanjundaswamy, A. K., Njiti, V., Xia, Q. and Chukwuma, F. 2017. Value-added probiotic development by high-solid fermentation of sweet potato with *Saccharomyces boulardii*. *Food Science and Nutrition* 5:633–638.

CGIAR. 2017. Sweetpotato silage manual for smallholder farmers. In CGIAR Research Program on Roots, Tubers and Bananas (RTB) Funded by the European Union and Technical Support of IFAD. Pp. 1–24. http://www.rtb.cgiar.org/endure.

Cha, M. K., Lee, D. O., An, H. M., Lee, S. W., Shin, S. H., Kwon, J. H., Kim, K. J. and Ha, N. J. 2012. Antiviral activity of *Bifidobacterium adolescentis* SPM1005-A on human papillomavirus type 16. *BMC Medicine* 10:72. http://www.biomedcentral.com/1741-7015/10/72.

Chandrasekara, A. and Kumar, T. J. 2016. Roots and tuber crops as functional foods: A review on phytochemical constituents and their potential health benefits. *International Journal of Food Science* 2016:1–15. http://dx.doi.org/10.1155/2016/3631647.

Chavan, R. S. and Chavan, S. R. 2011. Sourdough technology—A traditional way for wholesome foods: A review. *Comprehensive Reviews in Food Science and Food Safety* 10:170–182.

CIP. 2018. The origin and importance of sweetpotato. In *Everything You Ever Wanted to Know About Sweetpotato: Reaching Agents of Change ToT Manual*. Lima: International Potato Center.

CODEX. 2019. General standard for contaminants and toxins in food and feed. *Codex Alimentarius International Food Standards*, CXS 193-1995.

Conway, P. L., Gorbach, S. L. and Goldin, B. R. 1987. Survival of lactic acid bacteria in the human stomach and adhesion to intestinal cells. *Journal of Dairy Science* 70:1–12.

Del Re, B., Sgorbati, B., Miglioli, M. and Palenzona, D. 2000. Adhesion, autoaggregation and hydrophobicity of 13 strains of *Bifidobacterium longum*. *Letters in Applied Microbiology* 31:438–442.

EFSA. 2017. The European Food Safety Authority (EFSA) Panel on Biological Hazards 2017. Scientific Opinion on the update of the list of QPS-recommended biological agents intentionally added to food or feed as notified to EFSA. *EFSA Journal* 15 (3):4664, 177 pp. https://doi.org/10.2903/j.efsa.2017.4664.

Ehiwuogu-Onyibe, J., Ayodele, O., Mordi, J. and Elemo, G. 2019. *In vivo* safety and hypolipidemic effect of *Bifidobacterium adolenscentis* CH2 in female albino rats. *African Journal of Microbiology Research* 13:195–205.

European Union. 2020. Agri-food fraud. https://ec.europa.eu/food/safety/food-fraud_en.

Ezeama, C. F. and Amajor, J. U. 2015. Microbiological profile, shelf stability and some physicochemical properties of fermented flour from different sweet potato cultivars. *Journal of Food Technology Research* 2 (2):33–45.

FAO. 1994. *Utilising Root Crops*. Bangkok: Food and Agriculture Organization of the United Nations. fao.org.

FAO. 1998a. *Fermented Fruits and Vegetables. A Global Perspective*. FAO Agricultural Services Bulletin No. 134. Rome: Food and Agriculture Organization of the United Nations.

FAO. 1998b. *Food Quality and Safety Systems—A Training Manual on Food Hygiene and the Hazard Analysis and Critical Control Point (HACCP) System*. Rome: Food and Agriculture Organization of the United Nations.

FAO. 2001. Sweet potato root silage for efficient and labor-saving pig raising in Vietnam. http://www.fao.org/tempref/upload/Agrippa/554_en.doc.

FAO. 2010. *Making Sweet Potato Chips and Flour*. Food and Agriculture Organization of the United Nations. www.fao.org/3/ca4510en

FAO. 2017. The future of food and agriculture-Trends and challenges. http://www.fao.org/3/a-i6881e.pdf.

FAO. 2019. *FAO, IFAD, UNICEF, WFP and WHO. The State of Food Security and Nutrition in the World 2019. Safeguarding Against Economic Slowdowns and Downturns*. Rome: FAO.

FAO and WHO. 2006. *Probiotics in Food: Health and Nutritional Properties and Guidelines for Evaluation*. Rome: World Health Organization and Food and Agricultural Organization of the United Nations. Pp. 1–56.

FAOSTAT. 2020. *FAO Statistical Database*. Food and Agricultural Organization of the United States. https://www.fao.org

Fijan, S. 2014. Microorganisms with claimed probiotic properties: An overview of recent literature. *International Journal of Environmental Research and Public Health* 11:4745–4767.

Food and Nutrition Board. 2011. *Recommended dietary intakes*. Food and Nutrition Board, Institute of Medicine, National Academy of Science, USA.

Food Directorate of Health. 1998. *Nutraceuticals/Functional Foods and Health Claims on Foods Final Policy*. Therapeutic Products Programme and the Food Directorate from the Health Protection Branch. Food Directorate of Health, Canada.

Garcia, M. P. 2018. Challenging national narratives: On the origins of sweet potato in China as global commodity during the early modern period. In Perez Garcia, M. and De Sousa, L. (Eds.). *Global History and New Polycentric Approaches. Palgrave Studies in Comparative Global History*. Singapore: Palgrave Macmillan. Pp. 53–80. https://doi.org/10.1007/978-981-10.4053-5_4.

Giang, H. H., Le, V. L. and Ogle, B. 2004. Evaluation of ensiling methods to preserve sweet potato roots and vines as pig feed. *Livestock Research for Rural Development* 16:Art. #45. Retrieved November 18, 2023, from http://www.lrrd.org/lrrd16/7/gian16045.htm.

González, C., Díaz, I., León, M., Vecchionacce, H., Blanco, A. and Ly J. 2002: Growth performance and carcass traits in pigs fed sweet potato (*Ipomoea batatas* [Lam.] L) root meal. *Livestock Research for Rural Development* 14:Article #54. Retrieved November 18, 2023, from http://www.lrrd.org/lrrd14/6/gonz146.htm.

Gonzalez-Garcia, R. A., McCubbin, T., Navone, L., Stowers, C., Nielsen, L. K. and Marcellin, E. 2017. Microbial propionic acid production. *Fermentation* 3 (2):21. https://doi.org/10.3390/fermentation3020021.

Gopal, P. K., Prasad, J., Smart, J. and Gill, H. S. 2001. In vitro adherence properties of *Lactobacillus rhamnosus* DR20 and *Bifidobacterium lactis* DR10 strains and their antagonistic activity against an enterotoxigenic *Escherichia coli*. *International Journal of Food Microbiology* 67:207–216.

Gule, S. 2016. Use of orange fleshed sweet potato to produce commercially viable bakery items. In Rebranding OFSP for Health and Wealth. Marketing, Processing and Utilization Community of Practice, International Potato Centre Proceedings of Community of Practice Meeting, 14–16 March, Dar es Salaam.

Harter, L. L. and Weimer, J. L. 1929. *A Monographic Study of Sweet-Potato Diseases and Their Control*. Technical Bulletin No. 99. Washington, DC: United States Department of Agriculture.

Harter, L. L., Weimer, J. L. and Adams, J. M. R. 1918. Sweet-potato storage-rots. *Journal of Agricultural Research* XV (6):337–368.

Hegde, V., Misra, R. S. and Jeeva, M. L. 2012. Sweet potato diseases: Diagnosis and management. *Fruit, Vegetable, and Cereal Science and Biotechnology* 6:65–78.

Heller, K. J., Gels, A. and Never, H. 1995. Behaviour of genetically modified microorganisms in yoghurt. *System and Applied Microbiology* 18:504–509.

HLPE. 2014. Food losses and waste in the context of sustainable food systems. In A Report by High Level Panel of Experts on Food Security and Nutrition of the Committee on World Food Security, Rome.

Holzapfel, W. H. 2002. Appropriate starter culture technologies for small-scale fermentation in developing countries. *International Journal of Food Microbiology* 75:197–212.

Hotz, C. and Gibson, R. S. 2007. Traditional food-processing and preparation practices to enhance the bioavailability of micronutrients in plant-based diets. *The Journal of Nutrition* 137 (4):1097–1100. https://doi.org/10.1093/jn/137.4.1097.

Huang, L., Stevens-Miles, S. and Lingham, R. B. 1999. Screening for activities. In Demain A. L. and Davies J. E. (Eds.). *Manual of Industrial Microbiology and Biotechnology*, 2nd edition. Washington, DC: American Society for Microbiology. Pp. 21–28.

Hunter-Cevera, J. C. and Belt, A. 1999. Isolation of cultures. In Demain A. L. and Davies J. E. (Eds.). *Manual of Industrial*

Microbiology and Biotechnology, 2nd edition. Washington, DC: American Society for Microbiology. Pp. 3–20.

Hussein, G. A. M. and Aumara, I. E. 2006. Preparation and properties of probiotic frozen yoghurt made with sweet potato and pumpkin. *Arab Universities Journal of Agricultural Sciences* 14 (2):679–695.

IFT. 1998. Functional foods: Their role in disease prevention and health promotion. The Institute of Food Technologists' Expert Panel on Food Safety and Nutrition in Food Technology. *Food Technology* 52 (11):63–70.

Isolauri, E. 2001. Probiotics in human disease. *The American Journal of Clinical Nutrition* 73:1142S–1146S.

Jayant, M., Sahu, N. P., Deo, A. D., Gupta, S., Garg, C. K. and Valappil, R. K. 2020. Nutritional evaluation of fermented sweet potato leaf meal as a replacer of deoiled rice bran in the diet of *Labeo rohita* fingerlings. *Journal of Experimental Zoology India* 23 (1):61–70.

Kádár, Z. and Fonseca, C. 2019. Bio-products from sugar-based fermentation processes. In Bastidas-Oyanedel, J.-R. and Schmidt, J. E. (Eds.). *Biorefinery*. Springer Nature. https://doi.org/10.1007/978-3-030-10961-5_12.

Karim, O. R., Adebanke, B. M., Akintayo, O. A. and Awoyale, W. 2016. Physical, chemical and sensory properties of cassava (Manihot esculenta)—sweet potato (*Ipomoea batatas*) gari. *Ukrainian Journal of Food Science* 4 (2):276–289.

Khan, Z. and Doty, S. 2009. Characterization of bacterial endophytes of sweet potato plants. *Plant Soil* 322:197–207. https://doi.org/10.1007/s11104-009-9908-1.

Krochmal-Marczak, B., Sawicka, B., Supski, J., Cebulak, T. and Paradowska, K. 2014. Nutrition value of the sweet potato (*Ipomoea batatas* (L.) Lam) cultivated in south—eastern Polish conditions. *International Journal of Agronomy and Agricultural Research* 14:169–178.

Lee, J. H., Velmurugan, P., Park, J. H., Chang, W. S., Park, Y. J. and Oh, B. T. 2018. Photo-fermentation of purple sweet potato (*Ipomoea batatas* L.) using probiotic bacteria and LED lights to yield functionalized bioactive compounds. *3 Biotech* 8 (7):300. https://doi.org/10.1007/s13205-018-1327-7.

Lestari, L. A., Soesatyo, M. H. N. E., Iravati, S. and Harmayani, E. 2013. Characterization of Bestak sweet potato (*Ipomoea batatas*) variety from Indonesian origin as prebiotic. *International Food Research Journal* 20 (5):2241–2245.

Low, J. W., Nwanga, R. O. M., Adrade, M., Carey, E. and Ball, A.-M. 2017. Tackling vitamin A deficiency with biofortified sweetpotato in sub-Saharan Africa. *Global Food Security* 14:23–30.

Mantzourani, I., Terpou, A., Alexopoulos, A., Bezirtzoglou, E. and Plessas, S. 2019. Assessment of ready-to-use freeze-dried immobilized biocatalysts as innovative starter cultures in sourdough bread making. *Foods* 8 (1):40. https://doi.org/10.3390/foods8010040.

Maquia, I., Muocha, I., Naico, A., Martins, N., Gouveia, M., Andrade, I., Goulao, L. F. and Ribeiro, A. I. 2013. Molecular, morphological, and agronomic characterization of the sweet potato (*Ipomoea batatas* L.) germplasm collection from Mozambique: Genotype selection for drought prone regions. *South African Journal of Botany* 88:142–151.

Marques, J. M., da Silva, T. F., Vollu, R. E., Blank, A. F., Ding, G.-C., Seldin, L. and Smalla, K. 2014. Plant age and genotype affect the bacterial community composition in the tuber rhizosphere of field-grown sweet potato plants. *FEMS Microbiology Ecology* 88:424–435.

Mohanraj, R. 2018. Sweet potato: Bioactive compounds and health benefits. In Mérillon, J.-M. and Ramawat, K. G. (Eds.). *Bioactive Molecules in Food, Reference Series in Phytochemistry*. Springer International Publishing. https://doi.org/10.1007/978-3-319-54528-8_62-1.

Mohapatra, S., Panda, S. H., Sahoo, S. K., Sivakumar, P. S. and Ray, R. C. 2007. β-Carotene-rich sweet potato curd: Production, nutritional and proximate composition. *International Journal of Food Science and Technology* 42:1305–1314.

Monaghan, R. L., Gagliardi, M. M. and Streicher, S. L. 1999. Culture preservation and inoculum development. In Demain A. L. and Davies J. E. (Eds.). *Manual of Industrial Microbiology and Biotechnology*, 2nd edition. Washington, DC: American Society for Microbiology. Pp. 29–48.

Muchiri, M. and McCartney, A. L. 2017. *In vitro* investigation of orange fleshed sweet potato prebiotic potential and its implication on human gut health. *Functional Foods in Health and Disease* 7 (10):833–848.

Muchiri, M. N., McCartney, A. L. and Methven, L. 2020. Sensory profile and consumer preference of novel probiotic yoghurt enriched with orange sweet potato (*Ipomoea batatas*). *African Journal of Food Agriculture Nutrition Development* 20:16471–16489.

Muzhingi, T. 2016. Effect of baking on the β-carotene content of orange fleshed sweetpotato (*Ipomoea batatas*) puree bread and OFSP flour bread. In Rebranding OFSP for Health and Wealth. Marketing, Processing and Utilization Community of Practice. International Potato Centre Proceedings of Community of Practice Meeting, 14–16 March, Dar es Salaam.

Nagpal, R., Kumar, A., Kumar, M., Behare, P. V., Jain, S. and Yadav, H. 2012. Probiotics, their health benefits and applications for developing healthier foods: A review. *FEMS Microbiology Letters* 334:1–15.

Nasir, S., Olatunde, G., Adebowale, A. R. and Ayelaagbe, I. 2016. Optimization of deep-fat frying process for sweetpotato crisps from different varieties. In Rebranding OFSP for Health and Wealth. Marketing, Processing and Utilization Community of Practice. International Potato Centre Proceedings of Community of Practice Meeting, 14–16 March, Dar es Salaam.

National Academy of Sciences 2019. *Dietary Reference Intakes for Sodium and Potassium*. Washington, DC: National Academies of Sciences, Engineering, and Medicine. The National Academies Press. https://doi.org/10.17226/25353.

Nionelli, L. and Rizzello, C. G. 2016. Sourdough-based biotechnologies for the production of gluten-free foods. *Foods (Basel, Switzerland)* 5 (3):65.

Nwosisi, S., Nandwani, D. and Ravi, R. 2017. Bioactive compounds in organic sweet potato. *Journal of Advances in Molecular Biology* 1 (2):81–90. https://dx.doi.org/10.22606/jamb.2017.12001.

Oboh, S., Ologhobo, A. and Tewe, O. 1989. Some aspects of the biochemistry and nutritional value of the sweet potato (*Ipomoea batatas*). *Food Chemistry* 31 (1):9–18.

Oduola, A. A., Awojobi, K. O. and Adeyemo, S. M. 2018. Microorganisms associated with post harvest spoilage of sweet potatoes in Ile-Ife, Nigeria. *Journal of Microbiology Research* 8 (1):1–8.

OECD 2010. *OECD Environment, Health and Safety Publications Series on the Safety of Novel Foods and Feeds No. 20. Consensus Document on Compositional Considerations for New Varieties of SWEET POTATO [Ipomoea batatas (L.) Lam.]: Key Food and Feed Nutrients, Anti-nutrients, Toxicants and Allergens*. Paris: Organisation for Economic Co-operation and Development.

Ogunremi, O. R., Banwo, K. and Sanni, A. I. 2017. Starter-culture to improve the quality of cereal-based fermented foods: Trends in selection and application. *Current Opinion in Food Science* 13:38–43.

Ogunsakin, A. O., Vanajakshi, V., Anu-Appaiah, K. A., Vijayendra, S. V. N., Walde, S. C., Banwo, K., Sanni, A. I. and Prabhasankar,

P. 2017. Evaluation of functionally important lactic acid bacteria and yeasts from Nigerian sorghum as starter cultures for gluten-free sourdough preparation. *LWT—Food Science and Technology* 82:326–334.

Oguntoyinbo, F. A., Tourlomousis, P., Gasson, M. J. and Narbad, A. 2011. Analysis of bacterial communities of traditional fermented west African cereal foods using culture independent methods. *International Journal of Food Microbiology* 145:205–210.

Ojewumi, M. E., Job A. I., Taiwo, O. S., Obanla, O. M., Ayoola, A. A., Ojewumi, E. O. and Oyeniyi, E. A. 2018. Bio-conversion of sweet potato peel waste to bioethanol using *Saccharomyces cerevisiae*. *International Journal of Pharmaceutical and Phytopharmacological Research* 8 (3):46–54.

Okada, Y., Kobayashi, A., Tabuchi, H. and Kuranouchi, T. 2017. Review of major sweetpotato pests in Japan, with information on resistance breeding programs. *Breeding Science* 67:73–82. https://doi.org/10.1270/jsbbs.16145.

Oladoye, C. O., Connerton, I. F., Kayode, R. M. O., Omojasola, P. F. and Kayoed, I. B. 2016. Biomolecular characterization, identification, enzyme activities of molds and physiological changes in sweet potatoes (*Ipomea batatas*) stored under controlled atmospheric conditions. *Biomedicine and Biotechnology* 17 (4):317–332.

Oladoye, C. O., Olaoye, O. A. and Connerton, I. F. 2013. Isolation and identification of bacteria associated with spoilage of sweet potatoes during post harvest storage. *International Journal of Agriculture and Food Science* 3 (1):10–15.

Olaitan, O. O. 2012. Bio-deterioration of sweet potato (*Ipomoea batatas* Lam) in storage, inoculation-induced quality changes, and control by modified atmosphere. *Journal of Applied Science and Environmental Management* 16 (2):189–193.

Olawale, K. M. and Ojokoh, A. O. 2019. Effects of fermentation and extrusion on the proximate compositions and organoleptic properties of sweet potato (*Ipomoea batatas*) and Beniseed (*Sesamum indicum*) blends. *South Asian Journal of Research in Microbiology* 5 (4):1–12.

Olojede, O. A., Sanni, A. I. and Banwo, K. 2020. Rheological, textural and nutritional properties of gluten-free sourdough made with functionally important lactic acid bacteria and yeast from Nigerian sorghum. *LWT—Food Science and Technology* 120:108875.

Oloo, B. O., Shitandi, A., Mahungu, S., Malinga, J. B., Ogata and B. Rose 2014. Effects of lactic acid fermentation on the retention of beta-carotene content in orange fleshed sweet potatoes. *International Journal of Food Studies* 3:13–33.

Olorunnisomo, O. A. 2007. A cost-benefit analysis of sweet potato production for sheep feeding in the southwest of Nigeria. *Livestock Research for Rural Development* 19:Article #80. http://www.lrrd.org/lrrd19/6/olor19080.htm.

Olubunmi, A. A., Abraham, I. O., Mojirade, L. A., Afolake, O. B. and Kehinde, O. E. 2017. Development, evaluation and sensory quality of orange fleshed sweet potato (*Ipomoea batatas* Lam) extruded pasta products. *Croatian Journal of Food Technology, Biotechnology and Nutrition* 12 (1–2):83–89.

Oluwole, O. B., Kosoko, S. B., Owolabi, S. O., Onyibe, J. E. and Jegede, A. A. 2015. Comparative evaluation of the quality parameters of baked potato crisps: Yellow-fleshed and orange-fleshed. *ECronicon Nutrition* 1.5:245–253.

Oluwole, O. B., Kosoko, S. B., Owolabi, S. O., Salami, M. J., Elemo, G. N. and Olatope, S. O. A. 2012. Development and production of fermented flour from sweet potato (*Ipomea batatas* L.) as a potential food security product. *Journal of Food Science and Engineering* 2:257–262.

Oluwole, O., Odediran, O., Alagbe, G., Eboagwu, I., Jegede, A., Ogundeji, K., Olokoshe, A., Asieba, G. and Onyibe, J. 2016. Polyphenol oxidase activity and inhibition in white yam (*Dioscorea rotundata* var Laasirin) chips as African fries for human consumption. *Journal of Food Science and Engineering* 6:43–50.

Onyibe, J. E., Oluwole, O. B., Ogunbanwo, S. T. and Sanni, A. I. 2013. Antibiotic susceptibility profile and survival *of Bifidobacterium adolescentis* and *Bifidobacterium catenulatum* of human and avian origin yoghurt. *Nigerian Food Journal* 31:73–83.

Paneque Ramirez, G. 1992. Cultivation harvesting and storage of sweet potato products. www.fao.org; https://www.google.com/search?q=FAO+SWEET+POTATO&rlz=1C1CHZL_enNG873NG873&oq=FAO+SWEET+POTATO&aqs=chrome.

Paramithiotis, S., Chouliaras, Y., Tsakalidou, E. and Kalantzopoulos, G. 2005. Application of selected starter cultures for the production of wheat sourdough bread using a traditional three-stage procedure. *Process Biochemistry* 40:2813–2819.

Pelczar, M. J. and Chan, E. C. S. 1981. Microbiology of foods. In Vastyan, J. E., Vinnicombe, A. T. and Amar, J. S. (Eds.). *Elements of Microbiology*. McGraw-Hill Inc. Pp. 620–641.

Piwowarek, K., Lipińska, E., Hać-Szymańczuk, E., Kieliszek, M. and Ścibisz, I. 2018. *Propionibacterium* spp.—source of propionic acid, vitamin B12, and other metabolites important for the industry. *Applied Microbiology and Biotechnology* 102:515–538.

Przyrembel, H. 2008. What is a benefit in relation to food consumption? *Toxicology Letters* 180 (2):72–74.

Ray, R. C. and Naskar, S. K. 2008. Bioethanol production from sweet potato (*Ipomoea batatas* L.) by enzymatic liquefaction and simultaneous fermentation (SSF) process. *Dynamic Biochemistry, Process Biotechnology and Molecular Biology* 2 (1):47–49.

Ray, R. C., Panda, S. K., Swain, M. R. and Sivakumar, P. S. 2012. Proximate composition and sensory evaluation of anthocyanin-rich purple sweet potato (*Ipomoea batatas* L.) wine. *International Journal of Food Science and Technology* 47:452–458.

Reniero, R., Cocconcelli, P., Bottazzi, V. and Morelli, L. 1992. High frequency of conjugation in *Lactobacillus* mediated by an aggregation-promoting factor. *Journal of General Microbiology* 138:763–768.

Rhee, S. J., Lee, J. E. and Lee, C. H. 2011. Importance of lactic acid bacteria in Asian fermented foods. *Microbial Cell Factories* 10 (Suppl 1):S5. https://doi.org/10.1186/1475-2859-10-S1-S5

Rodrigues de Albuquerque, T. M., Borges, C. W. P., Cavalcanti, M. T., Lima, M. S., Magnani, M. and Souza, E. L. 2020. Potential prebiotic properties of flours from different varieties of sweet potato (*Ipomoea batatas* L.) roots cultivated in Northeastern Brazil. *Food Bioscience* 36:10061. https://doi.org/10.1016/j.fbio.2020.100614.

Rowland, I. R., Rumney, C. J., Coutts, J. T. and Lievense, L. C. 1998. Effects of *Bifidobacterium longum* and inulin on gut bacterial metabolism and carcinogen-induced aberrant crypt foci in rats. *Carcinogenesis* 19:281–285.

Rozman, V., Lorbeg, P. M., Accetto, T. and Matijašić, B. B. 2020. Characterization of antimicrobial resistance in lactobacilli and bifidobacteria used as probiotics or starter cultures based on integration of phenotypic and *in silico* data. *International Journal of Food Microbiology* 314. https://doi.org/10.1016/j.ijfoodmicro.2019.108388.

Sanjib, K. P., Himjyoti, D. C., Lata, M. P. and Kumar, G. V. 2014. Process standardization, characterization and storage study of a sweet potato (*Ipomoea batatas* L.) wine. *International Food Research Journal* 21 (3):1149–1156.

Saranraj, P., Behera, S. S. and Ray, R. C. 2019. Traditional foods from tropical root and tuber crops: Innovations and challenges. In Galanakis, C. M. (Ed.). *Innovations in Traditional Foods*. Woodhead Publishing. Pp. 159–191.

Saranraj, P. and Ray, R. C. 2019. Tropical fruit wines: Health aspects. *International Journal of Microbiological Research* 10 (3):94–110.

Sato, K. and Sudo, S. 1999. Small scale solid-state fermentations. In Demain A. L. and Davies J. E. (Eds.). *Manual of Industrial Microbiology and Biotechnology*, 2nd edition. Washington, DC: American Society for Microbiology. Pp. 61–79.

Scarpellini, E., Ianiro, G., Attili, F., Bassanelli, C., De Santis, A. and Gasbarrini, A. 2015. The human gut microbiota and virome: Potential therapeutic implications. *Digestive and Liver Disease* 47:1007–1012.

Seath, D. M., Rusoff, L. L., Miller, G. D. and Branton, C. 1947. *Utilizing Sweet Potatoes as Feed for Dairy Cattle*. Louisiana Bulletin No. 423. LSU Agricultural Experiment Station Reports. http://digitalcommons.lsu.edu/agexp/710.

Shimizu, M. 2012. Functional food in Japan: Current status and future of gut-modulating food. *Journal of Food and Drug Analysis* 20 (Suppl. 1):213–216.

Song, D., Ibrahim, S. and Hayek, S. 2012. Recent application of probiotics in food and agricultural science. *INTECH Open Access*. http://dx.doi.org/10.5772/50121.

Spink, J., Embarek, P. B., Savelli, C. J. and Bradshaw, A. 2019. Global perspectives on food fraud: Results from a WHO survey of members of the International Food Safety Authorities Network (INFOSAN). *NPJ Science of Food* 3:12. https://doi.org/10.1038/s41538-019-0044-x.

Sun, H., Zhang, P., Zhu, Y., Lou, Q. and He, S. 2018. Antioxidant and prebiotic activity of five peonidin-based anthocyanins extracted from purple sweet potato (*Ipomoea batatas* (L.) Lam.). *Scientific Reports* 8 (1):5018. https://doi.org/10.1038/s41598-018-23397-0.

Swain, M. R., Mishra, J. and Thatoi, H. 2013. Bioethanol production from sweet potato (*Ipomoea batatas* L.) flour using co-culture of *Trichoderma* sp. and *Saccharomyces cerevisiae* in solid-state fermentation. *Brazilian Archives of Biology and Technology* 56 (2):171–179.

Tangapo, A. M., Astuti, D. I. and Aditiawati, P. 2018. Dynamics and diversity of cultivable rhizospheric and endophytic bacteria during the growth stages of cilemba sweet potato (*Ipomoea batatas* L. var. *cilemba*). *Agriculture and Natural Resources* 52:301–316.

Tona, G. O. 2017. Chapter 7: Current and future improvements in livestock nutrition and feed resources. In Yucel, B. and Taskin, T. (Eds.). *Animal Husbandry and Nutrition*. Pp. 147–169. http://dx.doi.org/10.5772/intechopen.73088.

Truong, V.-D. and Ramesh, Y. A. 2010. Sweet potato purees and powders for functional food ingredients. In Ray, R. C. and Tomlins, K. I. (Eds.). *Post Harvest Aspects in Food*. New York: Nova Science Publishers. Pp. 117–161.

Tsai, C.-C., Chiu, T.-H., Ho, C.-Y., Lin, P.-P. and Wu, T.-Y. 2013. Effects of anti-hypertension and intestinal microflora of spontaneously hypertensive rats fed gamma aminobutyric acid-enriched Chingshey purple sweet potato fermented milk by lactic acid bacteria. *African Journal of Microbiology Research* 7 (11):932–940.

Tumuhimbise, G. A., Namutebi, A. and Muyonga, J. H. 2009. Microstructure and in vitro beta carotene bioaccessibility of heat processed orange fleshed sweet potato. *Plant Foods and Human Nutrition* 64 (4):312–318. https://doi.org/10.1007/s11130-009-0142-z.

USDA 2019. United States Department of Agriculture FoodData Central. https://fdc.nal.usda.gov/.

Vinci, V. A. and Byng, G. 1999. Strain improvement by nonrecombinant methods. In Demain A. L. and Davies J. E. (Eds.). *Manual of Industrial Microbiology and Biotechnology*, 2nd edition. Washington, DC: American Society for Microbiology. Pp. 103–113.

Wanasundera, J. P. D. and Ravindran, G. 1992. Effects of cooking on the nutrient and antinutrient contents of yam tubers (*Dioscorea alata* and *Dioscorea esculenta*). *Food Chemistry* 45 (4):247–250.

Wanjuu, C., Abong, G., Mbogo, D., Heck, S., Low, J. and Muzhingi, T. 2018. The physiochemical properties and shelf-life of orange-fleshed sweet potato puree composite bread. *Food Science and Nutrition* 6:1555–1563.

WHO 2018. *Healthy Diet*. WHO fact sheet No. 394. World Health Organization. www/whoint/mediacentre/factsheets/fs394/en.

Wireko-Manu, F. D., Ellis, W. O. and Oduro, I. 2010. Production of a non-alcoholic beverage from sweet potato (*Ipomoea batatas* L.). *African Journal of Food Science* 4 (4):180–183.

Yacout, M. H., Khayyal A. A., Shwerab, A. M. and Khalel, M. S. 2016. Introduce sweet potato vines as good roughage for small ruminants. *ECRONICON Veterinary Science* 2 (4):184–204.

Yuliana, N., Nurdjanah, S., Setyani, S., Sartika, D., Martiansari, Y. and Nabila, P. 2018. Effect of fermentation on some properties of sweet potato flour and its broken composite noodle strand. *American Journal of Food Technology* 13:48–56.

Yusuf, A. B. 2016. Microbiological assessment of traditional fermentation methods used in the production of flour from sweet potato in Nigeria. In 4th World Congress and Expo on Applied Microbiology, 19–21 September, Las Vegas. Applied Microbiology Open Access 2016 2:3(Suppl).

Zhang, L., Wang, Q., Liu, Q. and Wang, Q. 2009. Sweet potato in China. In Loebenstein G. and Thottapilly G. (Eds.). *The Sweetpotato*. Springer. Pp. 325–358. https://doi.org/10.1007/978-1-4020-9475-0_15.

Moroccan Fermented Foods and Their Potential Health Benefits

4

Tarik Lakhlifi, Samia El Oirdi, and Abdelhaq Belhaj

4.1 INTRODUCTION

Fermentation is one of the oldest methods of food preparation in the world. It was primarily used by ancient humans to preserve perishable foods. Today, fermentation refers to the use of microorganisms (bacteria, yeast, and molds) to achieve desirable food features in fermented food (Nout 2014). Food fermentation can be classified into two categories: aerobic fermentation, including alkaline and fungal fermentation on the one hand, and anaerobic fermentation, including alcoholic and acidic fermentation on the other hand (Malo and Urquhart 2016; Nout 2014; Şanlier et al. 2019). Moreover, considering the way fermentation microorganisms are utilized, two types of fermentation can be considered. The first type is natural fermentation, in which the fermentation process is achieved by the autochthonous (indigenous/native) microflora of the raw food materials or the processing environment (Tamang et al. 2015; Voidarou et al. 2021). This type of fermentation is also called "wild fermentation" or "spontaneous fermentation" (Dimidi et al. 2019). The second type is controlled fermentation, in which the fermentation process is achieved by the addition of defined strains, known as "starter cultures" (Rezac et al. 2018; Voidarou et al. 2021).

In traditional fermented foods, the fermentation process is spontaneous and uncontrolled. These foods have been prepared and consumed for thousands of years and are strongly associated with culture and tradition (Chaves-López et al. 2014; Petrova et al. 2021). The tradition of the fermentation process has been handed down from parent to child for centuries. Fermented foods vary from region to region depending on geography, climatic conditions, types of raw materials, preparation methods, and surrounding environment (Anagnostopoulos and Tsaltas 2019; Tamang et al. 2020).

Traditional fermented foods are valuable and popular for several reasons, such as diet enrichment through the development of a diversity of flavors and textures; shelf life extension of food products; biological enrichment of food with protein, essential amino acids, and vitamins; and detoxification of foods from undesired or toxic components, such as phytic acid, cyanide, and oxalic acid (Malo and Urquhart 2016). Beyond these features, it is increasingly understood that fermented foods also provide various potential health benefits. These health benefits include anti-hypertensive, anti-diabetic, anti-carcinogenic, anti-pathogenic, anti-inflammatory, anti-atherosclerotic, and anti-obesity activity (Anagnostopoulos and Tsaltas 2019; Marco et al. 2021; Şanlier et al. 2019; Voidarou et al. 2021).

In Morocco, traditional fermented foods represent an important part of the human diet and an integral part of culture and traditions. Indeed, these foods play a significant role in the economy and food security in the country (Benkerroum and Tamime 2004; Benkerroum 2013). Traditional Moroccan fermented foods are produced at the household level for daily domestic consumption or at the level of local associations/private individuals to be marketed.

Morocco has a wide range of traditional fermented foods that are unique and of potential health and functional benefits. This chapter describes the main characteristics of the most common traditional fermented foods of Morocco, prepared using indigenous microorganisms to ferment different raw materials of animal and plant origin. Their potential health and functional benefits are also highlighted.

4.2 FERMENTED DAIRY PRODUCTS

Morocco has a variety of traditional fermented dairy foods that are highly appreciated by consumers. The most popular of them are raib, lben, smen, jben, zabda baldya, and klila.

DOI: 10.1201/9781003605300-4

4.2.1 Raib

Raib is a common dairy product in Morocco obtained by spontaneous fermentation of raw milk from cow, goat, or a blend of both. To prepare raib, raw milk is collected in earthenware jars or plastic containers and left to ferment spontaneously at room temperature until coagulation takes place. This process can take up to 24–72 h depending on the temperature during the four seasons of the year. Raib may be a finished product, consumed as such, or an intermediate for the production of other traditional products (Figure 4.1). Raw milk used for raib production contains diverse and complex microbial flora. The main microorganisms involved in fermentation belong to lactic acid bacteria (Benkerroum 2013) (Table 4.1), which make the product safe for human consumption.

4.2.2 Lben

Lben is a popular dairy product obtained after churning fermented milk (raib). Figure 4.1 shows the traditional stages of the manufacture of lben. Raw milk (cow, goat, or a blend of

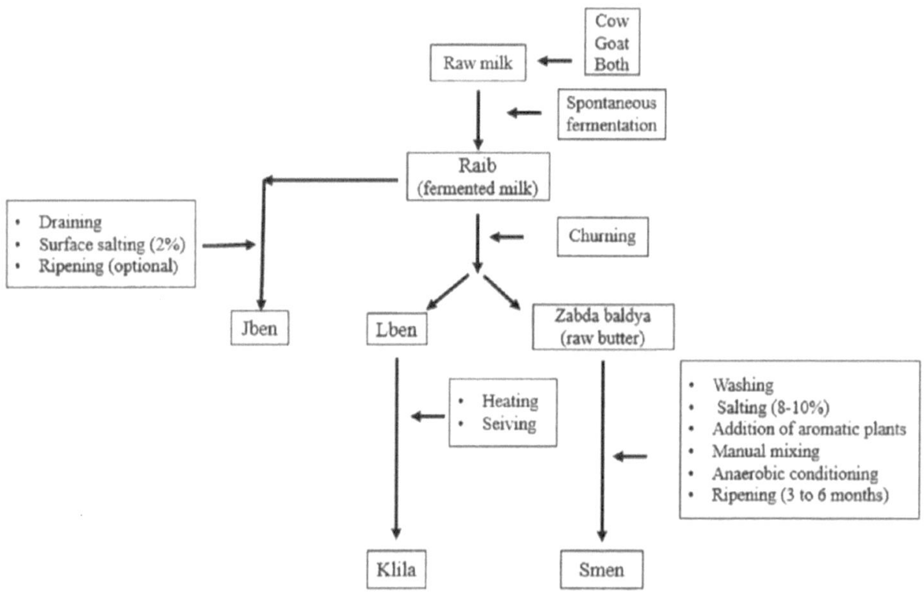

FIGURE 4.1 Generalized flowchart for manufacture of traditional fermented dairy products in Morocco. Finished products are shown in yellow.

TABLE 4.1 Main Microorganisms Involved in Traditional Fermented Dairy Products of Morocco (Benkerroum and Tamime 2004; Benkerroum 2013)

DAIRY PRODUCT	*PREDOMINANT MICROORGANISMS*	*GROWTH TEMPERATURE*	*TYPE OF FERMENTATION*
Raib	*Lc. lactis*; *Lc. lactis* ssp. *lactis* biovar *diacetylactis*; *Leuconostoc mesenteroides; Leuconostoc mesenteroides* subsp. *mesenteroides*.	Mesophilic	Spontaneous
Lben	*Leuconostoc mesenteroides* subsp. *lactis*; *Leuconostoc mesenteroides* subsp. *cremoris; Lactococcus lactis* subsp. *lactis*: *Lactococcus lactis* subsp. *lactis* biovar *diacetylactis*; *Saccharomyces cerevisiae*; *Kluyveromyces marxianus* var. *marxianus*.	Mesophilic	Spontaneous
Zabda baldya	Same as lben from which it derives but dominated by *Lactococcus lactis* subsp. *cremoris* and subsp. *lactis*.	Mesophilic	Spontaneous
Smen	Lactobacilli; yeasts.	Mesophilic	Spontaneous
Klila	*Pediococcus acidilactis*; *Lactobacillus confusus*; *Enterococcus faecalis*; *Enterococcus faecium*	Mesophilic	Spontaneous
Jben	*Leuconostoc mesenteroides* subsp. *lactis*; *Lactococcus lactis* subsp. *lactis* and subsp. *lactis* biovar *diacetylactis*; and *Lacticaseibacillus casei* subsp. *casei*	Mesophilic	Spontaneous

both) is left to sour spontaneously at room temperature until coagulation. Then, the curdled milk (raib) is churned for 45–60 min to be separated into lben and raw butter, called zabda baldya. Traditionally, churning is performed in a skin bag called the chekwa, made from a goat or lamb in one piece. The chekwa, filled with approximately 10 L of curdled milk, is hung on a roof or a wooden tripod and vigorously shaken until hearing the butter lumps.

Mesophilic lactic acid bacteria (LAB) are the main group of microorganisms responsible for the fermentation and aroma development in lben. The most predominant species are *Leuconostoc mesenteroides* subsp. *lactis* and subsp. *Cremoris* and *Lactococcus lactis* subsp. *Lactis* and *subsp. lactis* biovar *diacetylactis* (Benkerroum and Tamime 2004). Yeasts, including *Saccharomyces cerevisiae* and *Kluyveromyces marxianus* var. *marxianus*, are also found in lben (Benkerroum and Tamime 2004).

The strong flavor that characterizes lben can be associated with several compounds, such as diacetyl, acetaldehyde, ethanol, and acetoin (Benkerroum and Tamime 2004). However, other aroma compounds that may play a role in lben's flavor need to be identified and quantified.

4.2.3 Zabda Baldya

Zabda baldya literally means "butter of the country". It is a raw butter that is separated from lben after churning fermented milk (raib). Traditionally, a limited amount (approximately 10 %) of warm water (45–50°C) is added at the end of the churning period to accentuate the coalescence of fat globules, which increases the yield of zabda baldya. This raw butter has a strong diacetyl flavor and a softer body compared to that of industrially produced butter. The main microorganisms involved in fermentation are the same as lben, from which it derives, but dominated by *Lactococcus lactis* subsp. *cremoris* and subsp. *lactis* (Benkerroum 2013). Zabda baldya can be consumed as such or used for the production of another traditional product called smen.

4.2.4 Smen

Smen is a very popular dairy product in Morocco obtained from raw butter (zabda baldya). To prepare smen, zabda baldya is washed with warm water, decanted, and then washed with fresh salted water. This process is repeated several times until the rinsing water is clear, indicating that zabda baldya is free from any remaining lben that may cause spoilage later. The butter is then salted (approximately 8 to 10 g/100 g), well mixed, and conditioned in earthenware pots. The pots are entirely filled and compacted to promote anaerobiosis conditions; then they are stored in the dark in a cool place (13–15°C) for 3 to 6 months to ripen. In order to prevent mold growth and enhance the flavor of smen, ground leaves of aromatic plants, especially thyme or rosemary, can be added either to the raw butter (zabda baldya) during the manufacturing process or to the surface of smen just before its packaging. Ripened smen is characterized by a strong rancid and salty taste.

During the ripening period, lipolysis is the main mechanism that determines the typical flavor of smen (Benkerroum and Tamime 2004). The addition of salt and release of free fatty acids in smen create an unfavorable environment for microbial growth. Therefore, resistant microorganisms with lipolytic activity such as mesophilic lactobacilli are mainly responsible for smen ripening. Yeasts may also play an active role in the maturation of smen because of their known high lipolytic activity. However, the identification of these microorganisms at the species level, using molecular methods, is required.

4.2.5 Klila

Klila is a traditional dairy product obtained from lben. During storage, lben sours, and its pH reaches low levels after 3 to 4 days. The soured lben is heated until the separation of curd and whey. The aqueous phase is decanted, and the curd (klila) is collected. Klila is consumed as white fresh cheese. The heat treatment improves the microbiological safety of klila. The predominant microorganisms in klila are *Pediococcus acidilactis*, *Lactobacillus confusus*, *Enterococcus faecalis*, and *Enterococcus faecium* (Abd El-Salam and Benkerroum 2006).

4.2.6 Jben

Jben is a common dairy product obtained from raw milk (cow, goat, or a blend of both). Figure 4.1 shows the traditional procedure for jben manufacturing. Raw milk, collected in an earthenware container, is fermented spontaneously at room temperature until coagulation. The curdled milk (raib) is then placed in a muslin cloth bag that is hung to allow curd draining. The firmness of jben is achieved within approximately 10 days. The cheese (jben) is then emptied from the muslin bag, cut into pieces of approximately 200 g, and salted on the surface. The predominant microflora of jben is lactic acid bacteria, represented by *Leuconostoc mesenteroides* subsp. *lactis*, *Lactococcus lactis* subsp. *lactis* and subsp. *lactis* biovar *diacetylactis*, and *Lacticaseibacillus casei* subsp. *casei* (Benkerroum and Tamime 2004; Benkerroum 2013).

4.3 FERMENTED MEAT PRODUCTS

The consumption and preparation of fermented meat products have been a part of the deep-seated culture in Morocco for thousands of years. Indeed, the average consumption of red meat has grown significantly and reached 17.2 kg/capita/year in 2019 (MAPM 2021). The majority of these fermented meat products have been prepared for family or religious feasts or to preserve meat when it is available in large quantities, as is the case with the religious feast Al Adha, which traditionally involves sacrificing a sheep or goat for each family.

4.3.1 Keddid

Keddid is a Moroccan meat product, produced mostly by each family or in small-scale enterprises using a traditional fermentation method (Figure 4.2). Meat is cut into strips, about the size of a finger, and then spices (salt, garlic, cumin, coriander seed, and ground ginger) are added to the meat. After mixing well, meat strips are put in the refrigerator to macerate for one night. The next day, they are hung for fermentation and drying at the ambient temperature in open air for 8–12 days. To be sure that the keddid is ready for use; it is cut into two pieces; the inside of the meat must be very dry. The final aspect of keddid is shown in Figure 4.3(c). Before its use, keddid should be desalted by immersion in water to make it tender and soft. For consumption, it can be used as an ingredient in some dishes like couscous or can be further processed into khlii.

It should be noted that the fermentation of keddid is accomplished by microorganisms of the natural meat microflora. Bennani et al. (1995) demonstrated that the high counts of lipolytic microorganisms may give evidence that they are involved in the biochemical processing of keddid. According to Bennani et al. (1995) the pH and the A_w of this product are as follows: pH 5.2 to 5.3 and A_w 0.50 to 0.65.

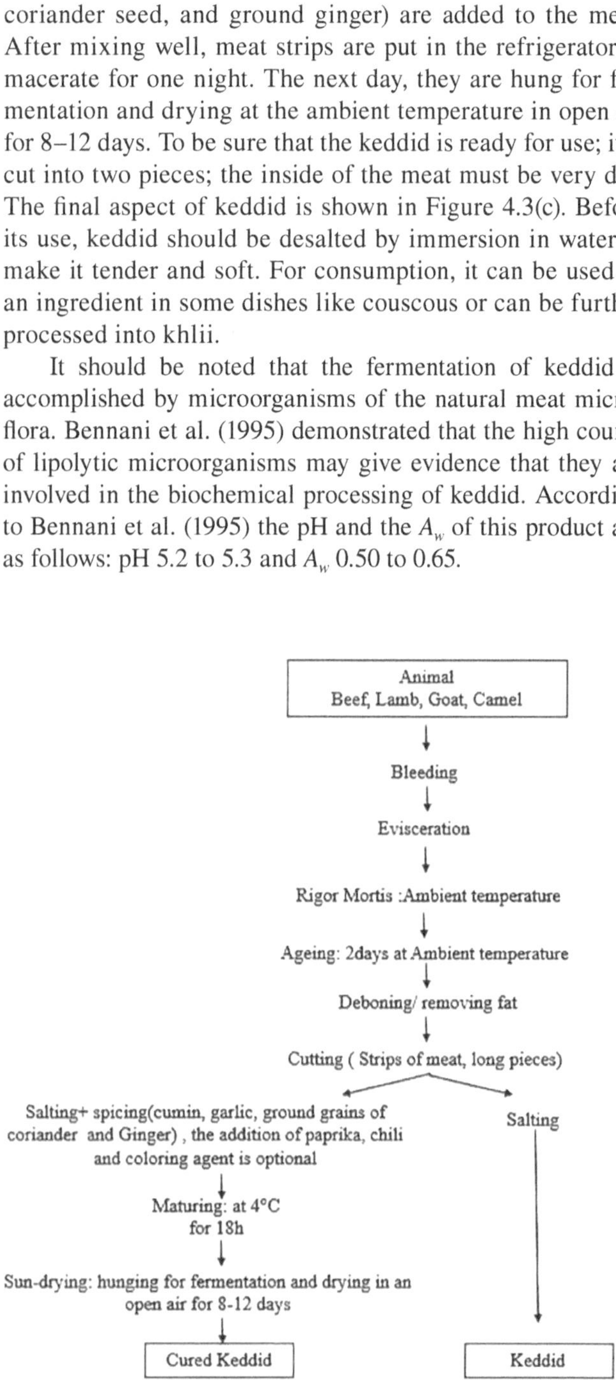

FIGURE 4.2 Traditional process to produce two keddid variants.

FIGURE 4.3 Illustration of some traditional Moroccan fermented foods: (a) fermented capers; (b) fermented lemon; (c) keddid; (d) kourdass.

4.3.2 Khlii

Khlii or khlea is a culinary preparation specific to Maghrebian gastronomy, especially used in Moroccan cuisines, based on meat with aromatics candied in fat. A variant consists of using olive oil instead of fat (diet khlii). The tradition of preparing khlii was started in Fes (a city in northeastern Morocco) before spreading across Morocco and North Africa. This is the reason Fes is still called "The Capital of Khlii" in Morocco. It is rare to find a house in this city without this traditional product or at least its byproduct, called agrish, which refers to the mixture that settles to the bottom of the pot after cooking khlii. Khlii can be used as an ingredient to spice up the taste of certain dishes (legumes and Moroccan pancakes), and it is particularly appreciated in Morocco with fried eggs. Every house used to prepare khlii, but at the moment, most go for buying it ready-made from the market. The preparation process of khlii has been described in previous published studies (Benkerroum 2013; Gagaoua and Boudechicha 2018). The preparation diagram for khlii is given in Figure 4.4. There are three types of khlii.

4.3.2.1 *Classic Khlii*

This khlii is obtained by the traditional method that involves cooking/boiling keddid for many hours in a mixture of water and animal fat until the complete evaporation of water. It is a lengthy process but worth the effort to obtain authentic khlii.

4.3.2.2 *Light/Diet Khlii*

Due to the consumer demand for healthy products, it is wise to prepare "diet khlii". The method is based on steaming dried meat, then preserving it in olive oil instead of animal fat.

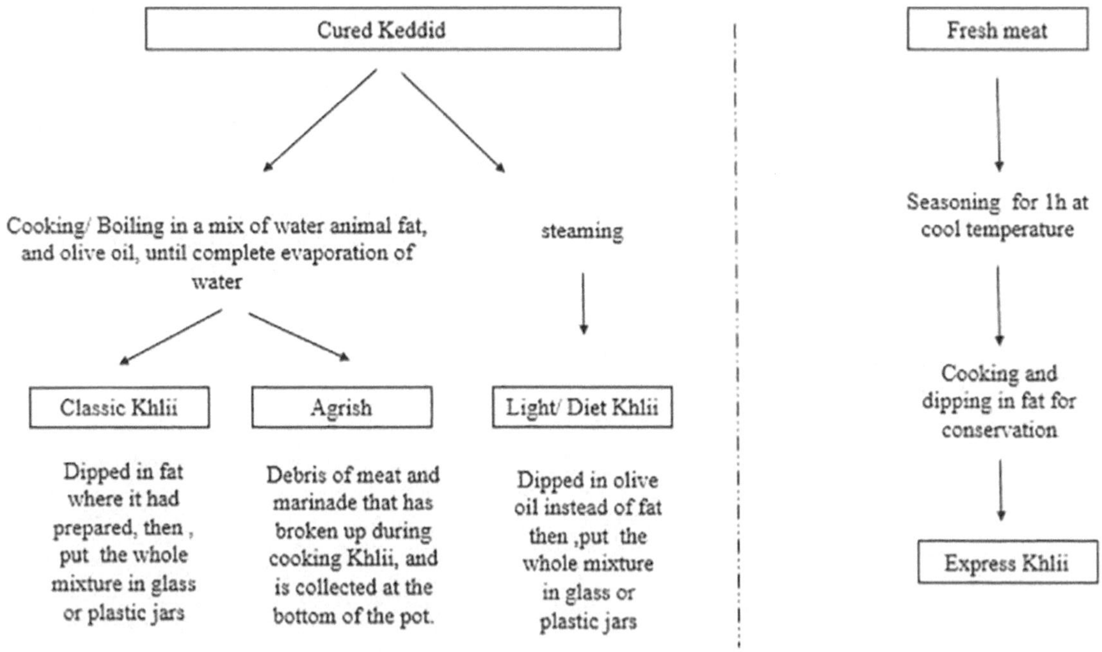

FIGURE 4.4 Traditional process to produce khlii and its variants.

4.3.2.3 Express Khlii

This is prepared by using fresh meat strips, rather than dried, which are marinated in a marinade (sharmula in Moroccan dialect) for 1 h, cooked, and dipped in animal fat for preservation. In the past, khlii was kept in earthenware jars called khabyate; currently it is placed in glass or plastic Tupperware.

4.3.3 Boubanita–Boubanite

Boubanita or boubanite is a culinary specialty typical of the city of Rabat in Morocco, which is prepared mainly at Aïd al-adha. It is made from a lamb's belly stuffed with meat. To prepare boubnita, lamb meat is cut into small cubes, seasoned with spices (coriander, cinnamon, cumin, ginger, chili, paprika, olive oil, and salt), and stuffed into cleaned lamb rumen. Then the two openings are tied, and it is left to dry and ferment in the shade. The final product is consumed with vegetables.

4.3.4 Kourdass

This is a sausage-like product of southeastern Morocco made with lamb stomach, prepared from offal using the intestines, liver, lung, and spleen. These constituents are sliced and marinated with salt, black pepper, cumin, hot pepper, and coriander. A piece of stomach stuffed with the mixture of previously prepared and marinated offal must be rolled up with the intestine (the rough side of the stomach toward the exterior). Then every roll is hung for fermentation and sun-drying in the open air. Kourdass is served with couscous or other Moroccan dishes (Figure 4.3(d)).

As in Maghreb countries, stuffing meats inside casing material (usually animal intestines), followed by fermentation, is one of the most popular forms of the consumption of meat products in the West (Franciosa et al. 2018). It should be noted that the natural microflora of meat are involved in Moroccan fermented meat products. Studies in this context must be done to have an idea of the dominant microorganisms in this kind of product.

4.4 MAHIA

Mahia, literally meaning "water of life", is a traditional alcoholic beverage. It was historically produced by the Moroccan Jewish population before their emigration in the second half of the 20th century. Nowadays, Mahia is still produced and consumed in Morocco, especially in rural areas. Mahia is obtained from the distillation of fermented figs. The traditional procedure for making mahia is shown in Figure 4.5. Dried figs are placed in a plastic barrel containing water and left to ferment spontaneously at the ambient temperature for 1 to 3 months. The fermented juice is then transferred to a pressure cooker and put on low heat. The distilled mahia comes out of a tube connected to the pressure cooker. Mahia is collected and stored in a new barrel until it is sold. The microorganisms involved in mahia are the natural microflora, particularly yeasts, able to metabolize fermentable sugars from figs and produce ethanol and carbon dioxide. No data are available on the characterization of these microorganisms and the chemical composition of mahia.

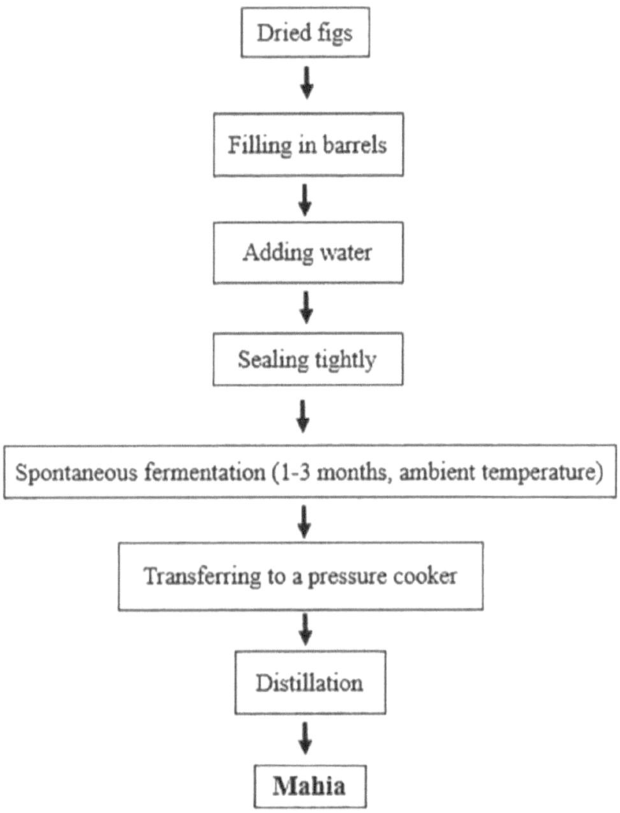

FIGURE 4.5 Flowchart for the manufacture of mahia.

4.5 FERMENTED FRUITS AND VEGETABLES

Fermented vegetables and fruits are broadly consumed as side dishes or as ingredients for several dishes in various regions of the world, particularly in the Mediterranean area. Each country or region produces and consumes its own unique fermented vegetables. Indeed, the fermentation of products of plant origin is done to preserve seasonal products for a long period and to benefit from their nutritional and organoleptic quality. In Morocco, a multitude of fermented vegetables and fruits such as lemons, olives, onions, peppers, carrots, gherkins, capers, and so on are available in various communities. The protocols of some vegetables and fruits traditionally fermented in Morocco are described in the following sections.

4.5.1 Pickled Peppers, Capers, Onions, Carrots, and Gherkins

Pickled peppers, capers (Figure 4.3(a)), onions carrots, and gherkins are very popular in Morocco. For their preparation, these foods are sorted, washed, occasionally cut, and submerged in a brine solution in a glass container, which is then tightly sealed and kept in a cool dry place for 4–6 weeks to allow fermentation. Another method consists of adding vinegar to the brine to ensure successful fermentation.

4.5.2 Table Olives

Morocco has advanced to the international level with an offer of quality olives and is now the third-largest international exporter (MAPM 2021). Despite developments in the process of the production of table olive, fermented products are generally still produced at the family or village scale with traditional methods. The olive harvest begins in mid-September when the fruit has reached its maximum size. Generally, this step is done by hand, which allows obtaining a good quality product. Olives are placed in perforated plastic baskets to have good aeration and avoid the risk of increase in temperature at the beginning of fermentation. This fruit contains a phenolic compound, oleuropeine, that gives it a bitter taste. The fermentation of olives is a process that consists of creating optimal conditions in the olive vats for the development of lactic acid bacteria. These bacteria consume the sugars that diffuse into the brine and produce lactic acid. At the end of the fermentation process, the olives lose their bitterness, acquire the desired organoleptic characteristics, and change color.

The first operation consists in removing leaves and damaged fruits. After washing (removal of dust), the olives are sorted, manually opened with knives, or slightly crushed with a stone. Then the injured olives are covered with saline solution (5 g/100 mL water), which is replaced daily with freshly prepared brine until the elimination of the bitter taste. In the last step, olives are covered with strong brine (12 g/100 mL water) and a thin layer of olive oil. Finally, the container is tightly sealed and placed at 15°C–20°C to promote the growth of mesophilic lactic acid bacteria (Benkerroum 2013). Ghabbour et al. (2011) indicated that *Lactiplantibacillus plantarum, Lactiplantibacillus pentosus*, and *Pediococcus pentosaceus* are the predominant species that naturally occur in the traditional fermentation of green olives.

4.5.3 Fermented Lemons

The most common technique to make Moroccan fermented lemons, msayer or msir, is given here. Thin-skinned lemons (Meyer lemons) are generally used since they require less time for fermentation and for releasing bitterness.

After washing and drying the lemons, each unit is cut into quarters from the top (nipple), while the rest is kept attached to the base, filled with coarse salt, and placed into a glass jar previously washed with hot water. The lemons are pressed as tightly as possible in the jar to allow the leakage of the lemon juice and the dissolving of salt. After tightly sealing the container, it is left in a cool and dry place for 2 months (Figure 4.3(b)). Salt exerts an osmotic pressure that leads to an exuding of lemon juice and creates an acidic salty brine that should totally cover the lemons. These conditions ensure the process of fermentation by acid- and salt-tolerant microorganisms. Lemon when pickled may be stored for more than a year at room temperature and can be used in various ways, such as whole or chopped in small pieces, in, for example, tajines, soups, stews, and fish-based dishes.

Bousmaha et al. (2006) showed that the dominant species are lactic acid bacteria and yeasts. Aayah et al. (2010) showed that yeasts, represented by *Candida parapsislosis* and unclassified Saccharomycetales, are the predominant microbial group.

4.6 MOROCCAN SOURDOUGH

Sourdough is one of the oldest forms of cereal fermentation, dating back to ancient Egypt. It is used primarily for baking purposes, and its main function was to leaven the dough to produce more aerated bread until it was replaced by baker's yeast. The sourdough matrix consists of a sharp and acidic taste mixture of both flour and water, which is fermented by a complex microbial ecosystem containing LAB and yeasts, which allow obtaining bread with improved crumb structure, volume, texture, and sensory properties as well as nutritional value. Broadly speaking, LAB and yeasts have an important role in sourdough fermentation. Dominant LAB were shown to belong to the genera *Lactobacillus, Leuconostoc, Lactococcus, Pediococcus,* and *Weissella*, and dominant yeast genera are *Saccharomyces* and *Candida*. The most competitive species reported in the literature are *L. plantarum, L. fermentum, S. cerevisiae*, and *I. orientalis* (Vogelmann et al. 2009).

During the bread-making process, the yeast releases gas (CO_2) that leavens the dough, and the LAB generates lactic acid that gives taste and delays the deterioration of the bread (Behera et al. 2016). In Morocco, each family makes their own bread using the traditional sourdough ferment. After the kneading operation, a portion of the leaven is kept for further use in the next few days; that is, the mother dough represents the natural microbial inoculum for the following doughs. The preparation of traditional sourdough is based on mixing sieved white flour and sieved strong whole meal flour with an equal amount of water and in some cases with the addition of salt and sugar (Figure 4.6).

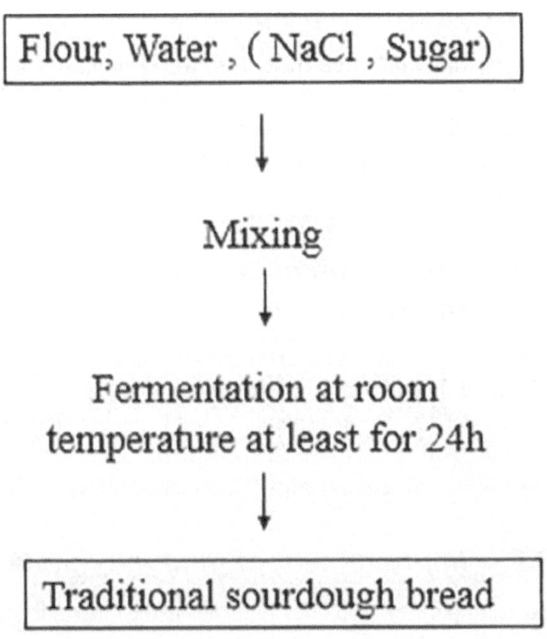

FIGURE 4.6 Traditional process to prepare Moroccan sourdough.

4.7 TRADITIONAL CIDER VINEGAR

Apple vinegar is a natural product widely used in food and traditional medicine, as it contains many bioactive compounds, and it may be defined as a sour liquid containing mainly acetic acid and water. It is generally used as a condiment and preservative and is obtained by acetic fermentation. Microorganisms involved in the bioconversion of vinegar are mainly yeasts and acetic acid bacteria (AAB) (Mounir et al. 2018).

To prepare Moroccan cider vinegar, it is necessary to put pieces of apple in a glass jar, close it with a clean cloth that allows the passage of oxygen to support the acetic fermentation by the acetic bacteria, and leave the jar in a dark place at room temperature for 40 days. After this period, it is necessary to proceed to the filtration stage to obtain pure cider vinegar. A gelatinous film develops on top of the vinegar, referred to as "mother vinegar", which can be used as a base to start making future jars of vinegar.

4.8 POTENTIAL HEALTH BENEFITS OF MOROCCAN FERMENTED FOODS

Beyond their preservative and organoleptic attributes, Moroccan fermented foods are increasingly understood for their health and functional benefits. The potential health promotion by these fermented products is discussed in the following.

4.8.1 Health Benefits due to Modifying the Food Constituents

During fermentation, microbial activity results in the enrichment and/or removal of certain compounds, which affects the nutritional composition of the final fermented product (Marco et al. 2021). Microorganisms are able to break down high-calorie monosaccharides and disaccharides present in food raw materials via catabolic pathways. Breaking down certain sugars can lead to reducing the glycemic index (Capurso and

Capurso 2020; Wolever 2017) and improving the tolerability of foods containing sugars, such as the lactose in dairy products and fructans in wheat (Nyyssölä et al. 2020). Microbial catabolism during fermentation can result in the hydrolysis of macromolecules, including polysaccharides, fats, and proteins, thereby increasing their digestion (Joye 2019; Poutanen et al. 2009; Wagenknecht et al. 1961). Furthermore, fermentation can lead to the removal of toxic or undesirable food components such as phytic acids in cereals (Ojha et al. 2018). In the fermentation of polyphenol-containing foods, LAB growth increases the bioavailability of flavonoids, tannins, and other active compounds (Septembre-Malaterre et al. 2018; Sun et al. 2011).

4.8.2 Health Benefits due to Bioactive Compounds

The potential health benefits of fermented foods can be also attributed to the bioactive compounds produced during fermentation (Marco et al. 2017; Melini et al. 2019). Lactic acid (or lactate) is one of these compounds, synthesized by LAB. This organic acid seems to have anti-inflammatory activity by reducing pro-inflammatory cytokine secretion of toll-like receptor (TLR)-activated macrophages and dendritic cells (Iraporda et al. 2015). Lactate also shows antioxidant properties by reducing the reactive oxygen species in intestinal epithelial cells (Kahlert et al. 2016).

Other important health-promoting compounds are bioactive peptides with several functions. Some peptides enhance the absorption of different minerals, including phosphorus, calcium, magnesium, and iron (Dimitrov et al. 2015). Bacteriocins, known for their anti-pathogenic activity, can also be produced in fermented products (Simova et al. 2008). Other peptides have angiotensin-converting enzyme (ACE-I)-inhibitory activity, which results in an overall antihypertensive effect (Donkor et al. 2007). Cyclic dipeptides, such as cyclo(leucyloprolyl) and cyclo(phenylalanyl-prolyl), possess antitumor, antimicrobial, and antiviral activities (Ivanov et al. 2021).

One of the important functional compounds is conjugated linoleic acid, which has athero-protective activity and may be involved in colon cancer prevention and weight loss (Lehnen et al. 2015). Furthermore, the biosynthesis of vitamins such as folate (B9), riboflavin (B2), vitamin B12, and vitamin K can also occur during fermentation (Şanlier et al. 2019; Tarvainen et al. 2019). Other interesting compounds that can be produced by fermentation-associated microorganisms are exopolysaccharides (EPSs). These compounds are known for their anticancer and immunomodulatory activity (Liu et al. 2011; Wang et al. 2014).

4.8.3 Health Benefits due to Probiotic Effects

In general, fermented foods are labeled probiotic since supposedly probiotic microorganisms have been isolated or detected in the fermentation process or the final product. It should be noted that these potential probiotic properties have been partially confirmed and stem from *in vitro* activity or analysis in animal models. According to the International Scientific Association for Probiotic and Prebiotic (ISAPP), probiotics are defined as "living microorganisms that create health benefits in the host when taken in sufficient amounts" (Marco et al. 2021).

The process of fermentation destroys many of the pathogenic microorganisms and chemicals in food and produces beneficial bacteria which in turn produce enzymes that aid digestion. In general, most fermented foods contain viable cells (probiotics) between 10^6 and 10^9 cells/g or cells/mL (Marco et al. 2017). These microorganisms not only provide nutritional health benefits but also maintain a delicate balance between the gastrointestinal (GI) tract and immune system. LAB are among the most important and widely found microorganisms in food fermentation, serving essential functions in fermented dairy, meat, cereal, and vegetable products. Besides LAB, species of AAB are solely responsible for some fermented foods, such as vinegar.

4.8.3.1 Protection against Infection

It is suggested that viable lactic acid bacteria interfere with the colonization and subsequent proliferation of foodborne pathogens, thus preventing the manifestation of infection. The beneficial effects of lactic acid bacteria have also been attributed to their ability to inhibit the growth of pathogens, either directly or through the production of antibacterial substances. Antibiotics, which are used to kill pathogenic bacteria, also disrupt the intestinal flora, leading to diarrhea and other intestinal disorders. Fermented foods have been reported to be effective in the prevention of various gastrointestinal infections (Hasan et al. 2014). Probiotic bacteria may counteract the inflammatory process by stabilizing the gut microbial environment and the intestine's permeability barrier (Isolauri et al. 2001).

4.8.3.2 Intestinal pH Balance

A healthy large intestine (or colon) has a slightly acidic pH. Naturally, fermented foods contain active Lactobacilli bacteria, and when consumed, these lactic acid bacteria produce acids, which help to maintain the pH of the large intestine at a healthy level.

4.8.3.3 Prevention of Toxic Effects of Mycotoxins

Inhibition of mycotoxin absorption in the gastrointestinal tract can be modulated by mycotoxin binding agents such as yeast and LAB derived from fermented food products. The mechanism of detoxification by yeast is due to the adhesion of mycotoxins to cell wall components (Hasan et al. 2014).

4.8.3.4 Improvement of Immunity System

The consumption of probiotic organisms in the form of fermented foods is claimed to offer several functional properties, including stimulation of the immune system. It suggests that probiotics can enhance both specific and nonspecific immune responses. It is well established that the intestinal microflora affects the functioning of the immune system, so it is reasonable to conclude that probiotics could provide a means of improving immune status. However, the mechanism of modulation of the immune system is not entirely known. According to Ashraf and Shah (2014), probiotics that can be delivered via fermented products could improve the gut mucosal immune system by increasing the number of IgA $^{+}$ and cytokine-producing cells in the effector site of the intestine.

4.8.3.5 Control of Blood Cholesterol and Hyperlipidemia

Previous studies indicated that fermented food products have an hypocholesterolemic effect. There have been some human studies that suggest that blood cholesterol levels can be reduced by the consumption of probiotic-containing fermented products. According to Hasan et al. (2014), *L. acidophilus* has the ability to lower serum cholesterol levels. This supports the health-promoting properties of traditional fermented products based on *L. acidophilus* or other lactic acid bacteria, as high cholesterol is one of the major risk factors for cardiovascular diseases.

LAB-producing bile salt hydrolase enzyme (BSH) prevents the reabsorption of bile salts, which forces the liver to use cholesterol to form bile salts, thus lowering blood cholesterol levels (Miremadi et al. 2014).

4.9 CONCLUSIONS AND FUTURE DIRECTIONS

Morocco has a rich diversity of traditional fermented foods that are highly appreciated by consumers. The potential health benefits of these foods may be attributed either to the bioactive compounds produced during fermentation or to the direct action of microorganisms, mainly LAB and yeasts, in the consumer's digestive tract (probiotic effect). However, there is a need for future studies on various aspects of these traditional fermented products, such as the determination of biomarkers for these products affecting health benefits, characterization of fermentation-associated microorganisms and the relationships between them, bioaccessibility of microbial metabolites, physicochemical characterization, and safety concerns related to these products. Also needed are clinical trials to determine the quantitative and consistent effects of Moroccan fermented foods on human health. One of the important challenges facing scientists and the food industry in the future will be to allow the large-scale production of these traditional products without losing their unique flavors, textures, and beneficial traits. Moroccan fermented foods can play an interesting role in the development of new fermented products. Indeed, the production of innovative and functional foods with novel properties will lead to final products with an added value even higher than what is already currently available.

REFERENCES

Aayah, H., Ahansal, L., Guyot, J. P., Ibnsouda, S., Chevallier, I., and A. Boussaid. 2010. Characterisation of the dry salted process for the production of the msayer, a traditional lemon aromatising condiment. *LWT-Food Science and Technology* 43(3): 568–572.

Abd El-Salam, M. H., and N. Benkerroum. 2006. North African Cheeses. In *Brined Cheeses*, ed. A. Y. Tamime, 139–187. Oxford: Blackwell Publishing.

Anagnostopoulos, D. A., and D. Tsaltas. 2019. Fermented Foods and Beverages. In *Innovations in Traditional Foods*, ed. M. G. Charis, 257–291. Elsevier.

Ashraf, R., and N. P. Shah. 2014. Immune system stimulation by probiotic microorganisms. *Critical Reviews in Food Science and Nutrition* 54(7): 938–956.

Behera, S. S., Ray, R. C., and R. C. Ray. 2016. Sourdough Bread. In *Bread and Its Fortification*, ed. C. M. Rosell, J. Bajerska, and A. F. El Sheikha, 53–67. Boca Raton: CRC Press.

Benkerroum, N. 2013. Traditional fermented foods of North African Countries: Technology and food safety challenges with regard to microbiological risks. *Comprehensive Reviews in Food Science and Food Safety* 12(1): 54–89.

Benkerroum, N., and A. Y. Tamime. 2004. Technology transfer of some Moroccan traditional dairy products (Lben, Jben and Smen) to small industrial scale. *Food Microbiology* 21(4): 399–413.

Bennani, L., Zenati, Y., Faid, M., and M. Ettayebi. 1995. Physico-chemical and microbiological characteristics of a dried salted meat product (Kaddid) in Morocco. *Zeitschrift für Lebensmittel-Untersuchung und Forschung* 201(6): 528–532.

Bousmaha, L., Ouhssine, M., and M. El Yachiouia. 2006. Fermentation du citron par inoculation microbienne. *Afrique Science: Revue Internationale des Sciences et Technologie* 2(1).

Capurso, A., and C. Capurso. 2020. The Mediterranean Way: Why elderly people should eat wholewheat Sourdough bread—A little known component of the Mediterranean diet and healthy food for elderly adults. *Aging Clinical and Experimental Research* 32: 1–5.

Chaves-López, C., Serio, A., Grande-Tovar, C. D., Cuervo-Mulet, R., Delgado-Ospina, J., and A. Paparella. 2014. Traditional fermented foods and beverages from a microbiological and nutritional perspective: The Colombian heritage. *Comprehensive Reviews in Food Science and Food Safety* 13(5): 1031–1048.

Dimidi, E., Cox, S. R., Rossi, M., and K. Whelan. 2019. Fermented foods: Definitions and characteristics, impact on the gut microbiota and effects on gastrointestinal health and disease. *Nutrients* 11(8): 1806.

Dimitrov, Z., Chorbadjiyska, E., Gotova, I., Pashova, K., and S. Ilieva. 2015. Selected adjunct cultures remarkably increase the content of bioactive peptides in Bulgarian white brined cheese. *Biotechnology & Biotechnological Equipment* 29(1): 78–83.

Donkor, O. N., Henriksson, A., Singh, T. K., Vasiljevic, T., and N. P. Shah. 2007. ACE-inhibitory activity of probiotic yoghurt. *International Dairy Journal* 17(11): 1321–1331.

Franciosa, I., Alessandria, V., Dolci, P., Rantsiou, K., and L. Cocolin. 2018. Sausage fermentation and starter cultures in the era of molecular biology methods. *International Journal of Food Microbiology* 279: 26–32.

Gagaoua, M., and H. R. Boudechicha. 2018. Ethnic meat products of the North African and Mediterranean countries: An overview. *Journal of Ethnic Foods* 5(2): 83–98.

Ghabbour, N., Lamzira, Z., Thonart, P., Cidalia, P., Markaoui, M., and A. Asehraou. 2011. Selection of oleuropein-degrading lactic acid bacteria strains isolated from fermenting Moroccan green olives. *Grasas y Aceites* 62(1): 84–89.

Hasan, M. N., Sultan, M. Z., and M. Mar-E-Um. 2014. Significance of fermented food in nutrition and food science. *Journal of Scientific Research* 6(2): 373–386.

Iraporda, C., Errea, A., Romanin, D. E., Cayet, D., Pereyra, E., Pignataro, O., Sirard, J. C., Graciela L. G., Analia G. A., and M. Rumbo. 2015. Lactate and short chain fatty acids produced by microbial fermentation downregulate proinflammatory responses in intestinal epithelial cells and myeloid cells. *Immunobiology* 220(10): 1161–1169.

Isolauri, E., Sütas, Y., Kankaanpää, P., Arvilommi, H., and S. Salminen. 2001. Probiotics: Effects on immunity. *The American journal of clinical nutrition* 73(2): 444–450.

Ivanov, I., Petrov, K., Lozanov, V., Hristov, I., Wu, Z., Liu, Z., and P. Petrova. 2021. Bioactive compounds produced by the accompanying microflora in Bulgarian yoghurt. *Processes* 9(1): 114.

Joye, I. 2019. Protein digestibility of cereal products. *Foods* 8: 199.

Kahlert, S., Junnikkala, S., Renner, L., Hynönen, U., Hartig, R., Nossol, C., Barta-Böszörményi, A., Dänicke, S., Souffrant, W. B., Palva, A., Rothkötter, H. J., and J. Kluess. 2016. Physiological concentration of exogenous lactate reduces antimycin a triggered oxidative stress in intestinal epithelial cell line IPEC-1 and IPEC-J2 in vitro. *PLoS One* 11(4): e0153135.

Lehnen, T. E., da Silva, M. R., Camacho, A., Marcadenti, A., and A. M. Lehnen. 2015. A review on effects of conjugated linoleic fatty acid (CLA) upon body composition and energetic metabolism. *Journal of the International Society of Sports Nutrition* 12(1): 1–11.

Liu, C.-F., Tseng, K.-C., Chiang, S.-S., Lee, B.-H., Hsu, W.-H., and T.-M. Pan. 2011. Immunomodulatory and antioxidant potential of *Lactobacillus* exopolysaccharides. *Journal of the Science of Food and Agriculture* 91(12): 2284–2291.

Malo, P. M., and E. A. Urquhart. 2016. Fermented Foods: Use of Starter Cultures. In *Encyclopedia of Food and Health,* ed. B. Caballero, P. M. Finglas, and F. Toldrá, 681–685. Academic Press.

MAPM. 2021. https://www.agriculture.gov.ma/fr/programme/exportation (accessed December 4, 2023).

Marco, M. L., Heeney, D., Binda, S., Cifelli, C. J., Cotter, P. D., Foligné, B., Gänzle, M., et al. 2017. Health benefits of fermented foods: Microbiota and beyond. *Current Opinion in Biotechnology* 44: 94–102.

Marco, M. L., Sanders, M. E., Gänzle, M., Arrieta, M. C., Cotter, P. D., De Vuyst, L., Hill, C., et al. 2021. The International Scientific Association for Probiotics and Prebiotics (ISAPP) consensus statement on fermented foods. *Nature Reviews Gastroenterology and Hepatology* 18(3): 196–208.

Melini, F., Melini, V., Luziatelli, F., Ficca, A. G., and M. Ruzzi. 2019. Health-promoting components in fermented foods: An up-to-date systematic review. *Nutrients* 11(5): 1189.

Miremadi, F., Ayyash, M., Sherkat, F., and L. Stojanovska. 2014. Cholesterol reduction mechanisms and fatty acid composition of cellular membranes of probiotic Lactobacilli and Bifidobacteria. *Journal of Functional Foods* 9: 295–305.

Mounir, M., Fauconnier, M. L., Afechtal, M., Thonart, P., Alaoui, M. I., and F. Delvigne. 2018. Aroma profile of pilot plant-scale produced fruit vinegar using a thermotolerant Acetobacter pasteurianus strain isolated from Moroccan cactus. *Acetic Acid Bacteria* 7(1).

Nout, M. J. R. 2014. Food technologies: Fermentation. In *Encyclopedia of Food Safety: Foods, Materials, Technologies and Risks*, ed. S. Holt, and R. Phadnis, 168–177. Academic Press.

Nyyssölä, A., Ellilä, S., Nordlund, E., and K. Poutanen. 2020. Reduction of FODMAP content by bioprocessing. *Trends in Food Science & Technology* 99: 257–272.

Ojha, P., Adhikari, R., Karki, R., Mishra, A., Subedi, U., and T. B. Karki. 2018. Malting and fermentation effects on antinutritional components and functional characteristics of sorghum flour. *Food Science & Nutrition* 6(1): 47–53.

Petrova, P., Ivanov, I., Tsigoriyna, L., Valcheva, N., Vasileva, E., Parvanova-Mancheva, T., Arsov, A., and K. Petrov. 2021. Traditional Bulgarian dairy products: Ethnic foods with health benefits. *Microorganisms* 9(3): 1–21.

Poutanen, K., Flander, L., and K. Katina. 2009. Sourdough and cereal fermentation in a nutritional perspective. *Food Microbiology* 26(7): 693–699.

Rezac, S., Kok, C. R., Heermann, M., and R. Hutkins. 2018. Fermented foods as a dietary source of live organisms. *Frontiers in Microbiology* 9: 1785.

Şanlier, N., Gökcen, B. B., and A. C. Sezgin. 2019. Health benefits of fermented foods. *Critical Reviews in Food Science and Nutrition* 59(3): 506–527.

Septembre-Malaterre, A., Remize, F., and P. Poucheret. 2018. Fruits and vegetables, as a source of nutritional compounds and phytochemicals: Changes in bioactive compounds during lactic fermentation. *Food Research International* 104: 86–99.

Simova, E. D., Beshkova, D. M., Dimitrov, Z. P., and Z. I. Simov. 2008. In vitro and in situ bacteriocin activity of lactic acid bacteria from Bulgarian dairy products and methods for making of *Lactobacillus* protective fermented milks with bacteriocin inhibitory substances. *Bulgarian Journal of Agricultural Science* 14: 28–42.

Sun, B., Neves, A. C., Fernandes, T. A., Fernandes, A. L., Mateus, N., De Freitas, V., Leandro, C., and M. I. Spranger. 2011. Evolution of phenolic composition of red wine during vinification and storage and its contribution to wine sensory properties and antioxidant activity. *Journal of Agricultural and Food Chemistry* 59(12): 6550–6557.

Tamang, J. P., Cotter, P. D., Endo, A., Han, N. S., Kort, R., Liu, S. Q., Mayo, B., Westerik, N., and R. Hutkins. 2020. Fermented foods in a global age: East meets West. *Comprehensive Reviews in Food Science and Food Safety* 19(1): 184–217.

Tamang, J. P., Thapa, N., Tamang, B., Rai, A., and R. Chettri. 2015. Microorganisms in Fermented Foods and Beverages. In *Health Benefits of Fermented Foods and Beverages*, ed. J. P. Tamang, 1–110. New York: CRC Press/Taylor & Francis.

Tarvainen, M., Fabritius, M., and B. Yang. 2019. Determination of vitamin K composition of fermented food. *Food Chemistry* 275: 515–522.

Vogelmann, S. A., Seitter, M., Singer, U., Brandt, M. J., and C. Hertel. 2009. Adaptability of lactic acid bacteria and yeasts to sourdoughs prepared from cereals, pseudocereals and cassava and use of competitive strains as starters. *International Journal of Food Microbiology* 130(3): 205–212.

Voidarou, C., Antoniadou, M., Rozos, G., Tzora, A., Skoufos, I., Varzakas, T., Lagiou, A., and E. Bezirtzoglou. 2021. Fermentative foods: Microbiology, biochemistry, potential human health benefits and public health issues. *Foods* 10(1):1–27.

Wagenknecht, A. C., Mattick, L. R., Lewin, L. M., Hand, D. B., and K. H. Steinkraus. 1961. Changes in soybean lipids during tempeh fermentation. *Journal of Food Science* 26(4): 373–376.

Wang, K., Li, W., Rui, X., Chen, X., Jiang, M., and M. Dong. 2014. Characterization of a novel exopolysaccharide with antitumor activity from *Lactobacillus plantarum* 70810. *International Journal of Biological Macromolecules* 63: 133–139.

Wolever, T. M. S. 2017. Yogurt is a low—glycemic index food. *The Journal of Nutrition* 147(7): 1462–1467.

Tempeh

Indonesia's Indigenous Health-Promoting, Sustainable, and Affordable Fermented Food

5

Amadeus Driando Ahnan-Winarno, Lorraine Cordeiro, Wida Winarno, Florentinus Gregorius Winarno, and Hang Xiao

5.1 INTRODUCTION

Tempeh, a staple source of protein, was first documented in Indonesia in the 1600s. Tempeh is composed of soybeans fermented by *Rhizopus* spp. mold, which forms a cake that can be cooked like meat. Over the past decades, tempeh has gained significant scientific attention. Since 2015, the number of research papers on tempeh has increased up to six times compared to the previous years, mostly related to nutrition, health benefits, sustainability, fermentation process, and application (Ahnan-Winarno et al. 2021). Tempeh is a rich source of protein, fiber, minerals, vitamins, and bioactive compounds, leading to exponential growth in its use as a natural and nutritious plant-based meat alternative in the meat analogue market. The global meat analogue market is projected to increase to USD 3.5 billion by 2026 (Boukid 2021).

However, the adoption of tempeh as a mainstream food ingredient globally has been hindered by the general unfamiliarity with its benefits, applications, and taste—even among Indonesians, where tempeh is locally consumed and is often seen as a less preferred food or a 'food of the poor'. With the rise of new tempeh-based food products and more positive brand images, Indonesians are beginning to appreciate tempeh's potential in the local and global food market. This chapter will present the benefits and challenges of tempeh; its potential as a fermented food and plant-sourced protein in the food market; and the vision of tempeh as food processing biotechnology that has a larger application to the production of nutritious, sustainable, and affordable foods.

5.2 DEFINITION

There have been two regulations found standardizing the definition of tempeh. According to the CODEX Alimentarius Commission, tempeh (*tempe* in Bahasa Indonesia) is described as a compact, white, cake-formed product, prepared from dehulled boiled soybeans through solid-state fermentation with *Rhizopus oligosporus, Rhizopus oryzae,* and *Rhizopus stolonifer* (Food and Agriculture Organization-World Health Organization 2017). According to the National Standardization Agency of Indonesia (Badan Standardisasi Nasional), tempeh is made from soybeans or other ingredients fermented by a 'tempeh starter culture'. While both standards provide slightly different amount of limits in proximate content, only the Indonesian standard provides limits for maximum heavy metal content. Besides the standard definitions of tempeh, other non-soybean tempehs fermented by other organisms have been consumed in Indonesia, such as velvet bean (*Mucuna pruriens*), koro (*Phaseolus vulgaris*), and koro kratok (*Phaseolus lunatus*) (Ariani et al. 2016; Higasa et al. 1996; Pugalenthi et al. 2005; Paredes-López and Harry 1989; Rochín-Medina et al. 2015; Pertiwi et al. 2020). Tempehs made using *Rhizopus delemar* and *Rhizopus arrhizus* have also been found (Khasanah et al. 2018; Vebliza et al. 2018).

DOI: 10.1201/9781003605300-5

FIGURE 5.1 Sliced raw tempeh.

FIGURE 5.2 Tempeh fermented in banana leaf.

5.3 HISTORY

The earliest documentation of tempeh in an ancient Javanese manuscript, *Serat Centhini*, written in the 1600s and published in 1815 with the help of the Surakarta Kingdom from Central Java, was documented by Shurtleff and Aoyagi (1984, 2020). According to Winarno (2021), Indonesian food scientist, historian, and author of several publications on tempeh,

> *Jamu* is one of the oldest traditional sources of wisdom in Indonesia that indicates how to cure or prevent health issues and has something to do with the use of functional foods, including tempeh. It would be useful to understand traditional uses of tempeh in future research.

Tempeh was first documented as 'sambal lethok' or 'sambal tumpang', a curry-like dish typically made of tempeh; overripe tempeh; coconut milk; shallots; chili peppers; tofu; and boiled vegetables like bean sprouts, watercress, and long beans, consumed with rice and rice crackers. Winarno (2021) noted that while

> we may not be able to find historical records of tempeh in scientific journals, it is well-documented in old cookbooks such as one released in 1947 with various recipes of tempeh dishes. It is important to preserve the information that tempeh was first documented in Klaten, Central Java, Indonesia, and came in the form of sambal tumpang that has been popular in Yogyakarta. The overripe tempeh found in the dish has been considered as a traditional flavoring component.

The word *tempe* in Bahasa Indonesia was hypothesized to come from the word *tumpi*, a Javanese sago flour-based food that resembles tempeh (Purwadaria et al. 2016). The first use of the term 'tempeh' and documentation of tempeh in other countries was recorded by Prinsen and Geerligs in a Dutch article published in 1896, and consistent use of the term 'tempeh' was found in a 1950 English article by Van Veen and Schaefer (Shurtleff and Aoyagi 2020; van Veen and Sohaefer 1950). Winarno (2021) commented that "for generations upon generations, Indonesians have eaten tempeh, but they are not familiar with the functional nature of tempeh as a food. Greater awareness in Indonesia and worldwide is needed on the health benefits of tempeh."

5.4 SPREAD AND EVOLUTION

Tempeh has experienced major negative branding; for example, the first president of Indonesia, Soekarno, said in his national speech, "Don't be a tempeh nation!" This likely referred to the context that traditional tempeh production involved dehulling with feet. The negative expression of 'mental tempe' also has been widely spread as a symbol of not having a strong mentality, likely referring to tempeh's texture.

As a staple protein source, especially on Java and Bali islands, tempeh has been consumed fried, boiled, steamed, or grilled. Fried tempeh has been especially popular in Indonesia for its taste, although one might categorize tempeh into fries, which have been considered an unhealthy category of food in Indonesia. Tempe bongkrek (made of coconut pressed cake) has been historically well known for the food poisoning outbreak it caused due to the toxin resulting from the contamination of *Burkholderia gladioli* pathovar *cocovenenans* (*B. cocovenans*). The bacterium can produce bongkrekic acid, a mitochondrial toxin that can cause malaise; dizziness; jaundice; and, in extreme cases, shock, coma, and death (Anwar et al. 2017). Besides this, there was only one other outbreak related to tempeh in North Carolina, which was poisoning caused by the growth of *Salmonella enterica* Paratyphi B variant L(+) tartrate(+) coming from contaminated *Rhizopus* spp. culture from Indonesia on unpasteurized tempeh (Griese et al. 2013).

Recently, there have been initiatives in leveraging tempeh as nutritious food, such as the emergence of the Indonesian Tempe Movement since 2014, with efforts to prioritize tempeh to be registered as a UNESCO Intangible Cultural Heritage and the introduction of modernized methods of producing tempeh by the directives of entities like Rumah Tempe Indonesia (Ahnan-Winarno et al. 2019; Astawan and Maskar 2019).

The perception of tempeh in Indonesia has also become more positive, according to a survey conducted by the Indonesian Tempe Movement and Big Change Agency in 2021 on an online survey platform involving 109 participants from several provinces in Indonesia. The compositions of respondents were 58% 18–24 years old, 40% 24–34 years old, 1% 35–44 years old, and 1% 55–64 years old; 38.03% professionals, 42.45% students, 2.82% housewives, and 16.9% others (e.g., entrepreneurs, freelancers). Thirty-five percent were in Jakarta; 21% in West Java; 15% in Banten; 10% in Central Java; 7% in Yogyakarta; and 3% in East Java, Kalimantan, Bali, and Sumatera, and they were 57.75% female and 42.25% male.

FIGURE 5.3 Fried tempeh and white rice drizzled with sweet soy sauce.

What first came to their minds when they heard the word 'tempeh' were soybean (22.6%), delicious (16.98%), and protein (14.15%). Ninety-five percent (95%) of the respondents liked tempeh, 9.2% of them did not like tempeh, while the rest did not like nor dislike tempeh. The reasons they liked tempeh were because tempeh was perceived as delicious (59.14%), healthy (12.9%), and rich in protein (9.68%). The reasons they did not like tempeh were due to tempeh having plain flavor, simply not liking the taste, or preferring tofu instead. Ninety-three-point five percent of the respondents said that they knew that tempeh was highly nutritious.

In a 2020 study by Fibri and Frost, Indonesian millennials were proud of tempeh and preferred traditional over modern tempeh, which was amplified by the knowledge of the use of local ingredients. According to Winarno (2021),

> tempeh production, which was largely a home-based process, has been changed by the Indonesian Tempeh Movement which introduced new forms of tempeh preparation that addressed hygiene habits, sanitation, and generally food safety concerns. Large scale producers started producing tempeh in hygienic conditions soon after the Indonesian Tempeh Movement's promotion of the need for food safety in tempeh production in 2015.

Since the first foreign documentation of tempeh was found in 1896 in a Dutch article by Prinsen and Geerligs, tempeh has been found to be documented in a 1950 English article by Van Veen and Schaefer and was identified to have been introduced in Japan by 1912, India by 1936, Suriname in 1936, the United States by 1958, and Zambia by 1971 (Shurtleff and Aoyagi 2020).

In Indonesia, new tempeh products have been increasing in the past years, sold especially through online platforms. Some of the most easily found products are tempeh chips, dry seasoned tempeh crunches, fresh tempeh, and other non-conventional tempeh products such as protein balls and cookies. In the rapidly growing global meat analogue market, several tempeh-related plant-based companies can be found worldwide in 2021, embracing the tempeh fermentation as a biotechnology process. Better Nature is innovating on tempeh fermentation and selling ready-to-cook plain and seasoned meat alternatives in the UK, Germany, and France markets. Lightlife and Tofurky are selling tempeh as a food ingredient in US supermarkets, starting from Massachusetts and Oregon, respectively. Kema Foods is starting a shelf-stable tempeh snack business from California. Planit Protein is preparing to launch a home tempeh fermentation system that allows the production of mixed-grain–based tempehs in various shapes in the US. Schouten from the Netherlands has been selling tempeh as a food ingredient in Europe and now is expanding to India. There have been many other conventional and/or smaller tempeh businesses around the world, for example, Cultura Umami in Chile; Holy Tempeh in Israel; and Rusto's Tempeh in Japan, South Korea, China, Ecuador, and Mexico. Tempeh also has been sold as ready-to-eat dishes in food trucks or restaurants such as Java Cove in Washington DC, US, as well as Warung Tempeh and Pino's Warung in London.

5.5 NUTRITION

Indonesians have used tempeh as a primary source of protein, thus positioning tempeh in place of meat in the Indonesian diet. Research on the nutritional content of tempeh has found that besides the high protein content, other nutrients such as fiber, calcium, iron, vitamin B12, and bioactive compounds make tempeh a nutrient-dense food. As more research on tempeh is conducted, these analyses could shape the understanding of tempeh as a nutritious plant-based source of protein that has specific health benefits (Ahnan-Winarno et al. 2021). Winarno (2021) proposed that "tempeh as a weaning food holds nutritional potential due to the fermentation process that contributes to the gut microbiota." She noted that at the International Conference on Tempeh, leading food scientists and food technologists led by the Indonesian Academy of Sciences (AIPI) made a formal declaration that tempeh is good as complementary food (Winarno 2021).

Compared to beef, a popular source of protein worldwide, tempeh contains similar amounts of energy, protein, and iron; significantly higher amounts of fiber and calcium; and significantly lower amounts of saturated fat and sodium (Ahnan-Winarno et al. 2021). Additionally, tempeh could contain vitamin B12, which is considered to be sourced only from animal products, due to the co-fermentation of vitamin B12–producing bacteria, for example, *Propionibacterium freudenreichii, Klebsiella pneumoniae,* and *Citrobacter freundii* (Keuth and Bisping 1994; Liem et al. 1977; Okada 1989). These microorganisms typically originate from soil and natural water. Unlike tempehs in other countries with hygienic production systems like in Canada, tempehs made traditionally in Indonesia are usually prepared near rivers, using natural water sources, and/or in a floorless structure and have been found to contain vitamin B12 (0.34–52–2.44 μg/84 g) (Liem et al. 1977). Further analysis in the types of vitamin B12 (e.g. methylcobalamine, cyanocobalamin, etc.) in tempeh shall be conducted and followed by *in vivo* and clinical analyses to determine the bioavailablity of tempeh vitamin B12.

According to Winarno (2021), "the presence of other microorganisms is not only related to the cleanliness of making tempeh, but the temperature, humidity, and other environmental conditions which allow for higher quality tempeh." The biodiversity of tempeh starters in Indonesia should be given special consideration, as

> Indonesians use many *Rhizopus* species for making tempeh. Maybe initially only one type of species was sent to the U.S. for production of tempeh. However, the diversity of tempeh starters in Indonesia is important for making tempeh in various altitudes with different environmental conditions. Indonesia has one of the highest levels of biodiversity in the world which is well reflected in the tempeh starters and perhaps tempehs derived from starters from Indonesia have a special niche market for high quality tempeh.
>
> (Winarno 2021)

Besides macro- and micronutrients, tempeh contains bioactive compounds that come from soybeans as the main ingredient. The most-researched bioactive compounds in tempeh have been soy isoflavones, which include glycosides (genistin, daidzin, glycitin) and aglycones (genistein, daidzein, glycitein). These compounds have been linked to potential health benefits around cancer inhibition, cognitive function, lung health, cardiovascular health, liver health, type 2 diabetes mellitus, skeletal muscle recovery, and malnutrition (Ahnan-Winarno et al. 2021).

As a fermented product, tempeh could be among the sources of isoflavones with the highest bioavailability. Whole soybean foods contained higher concentrations of isoflavones compared to 'second-generation soy foods' such as soy noodles, burgers, and hotdogs (Baiano 2010). Isoflavones in tempeh and miso, as fermented soy foods, were found to be more bioavailable compared to those of unfermented soy foods like soy protein and soy milk derivatives (Baiano 2010). Among other fermented soy foods, like tofu and bean curd sheets, tempeh contained significantly higher isoflavone content in either raw or cooked forms (Haron et al. 2016).

The higher amount and bioavailability of bioactive compounds in tempeh could be due to the fermentation process that can free and/or transform soy isoflavones. Food matrix, fiber, or sugar-bound isoflavones can be freed, for example, by the enzymatic activity occurring during the tempeh fermentation process (Wang and Murphy 1996; Berghofer et al. 1998; Borges et al. 2016; Esaki et al. 1994; Kuligowski et al. 2017; Rochín-Medina et al. 2015; Sánchez-Magana et al. 2014). This process resulted in increases in antioxidative activity, reducing power, and lipid peroxidation inhibitory activity, which have been linked to favorable health effects through reduction of inflammation and cell oxidation (Ahmad et al. 2015; Chang et al. 2009; Xiao et al. 2016). Winarno (2021) encourages more "research on the use of black soybeans in tempeh as this may result in a tempeh product with higher phytonutrients and antioxidant properties than other types of soybeans."

5.6 POTENTIAL HEALTH BENEFITS

Compared to other popular health foods like berries, tempeh is under-researched; there are approximately 68 papers related to the health benefits of tempeh, with a low number of clinical and population studies. Nonetheless, these papers provide a foundation for establishing a literature base on the potential health benefits of tempeh and directions for future research.

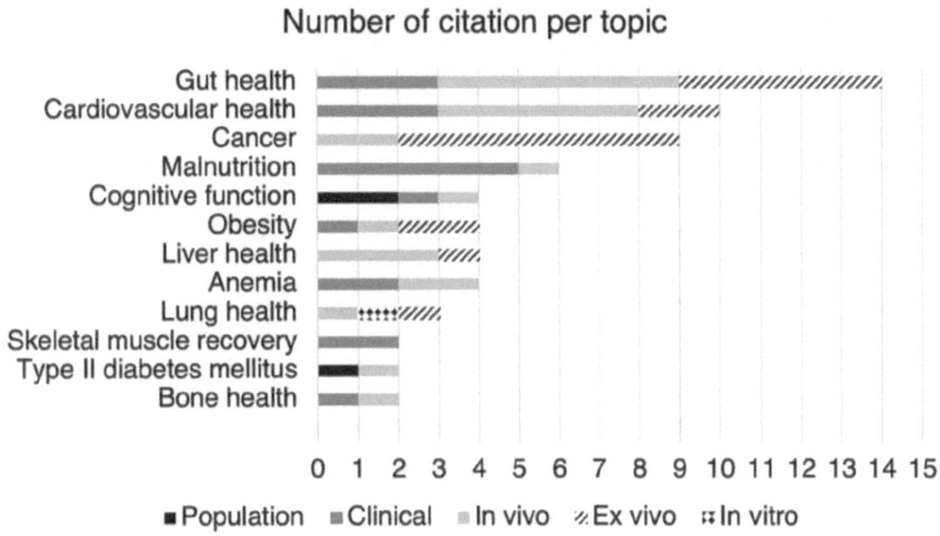

FIGURE 5.4 Number of studies on the health benefits of tempeh (Ahnan-Winarno et al. 2021).

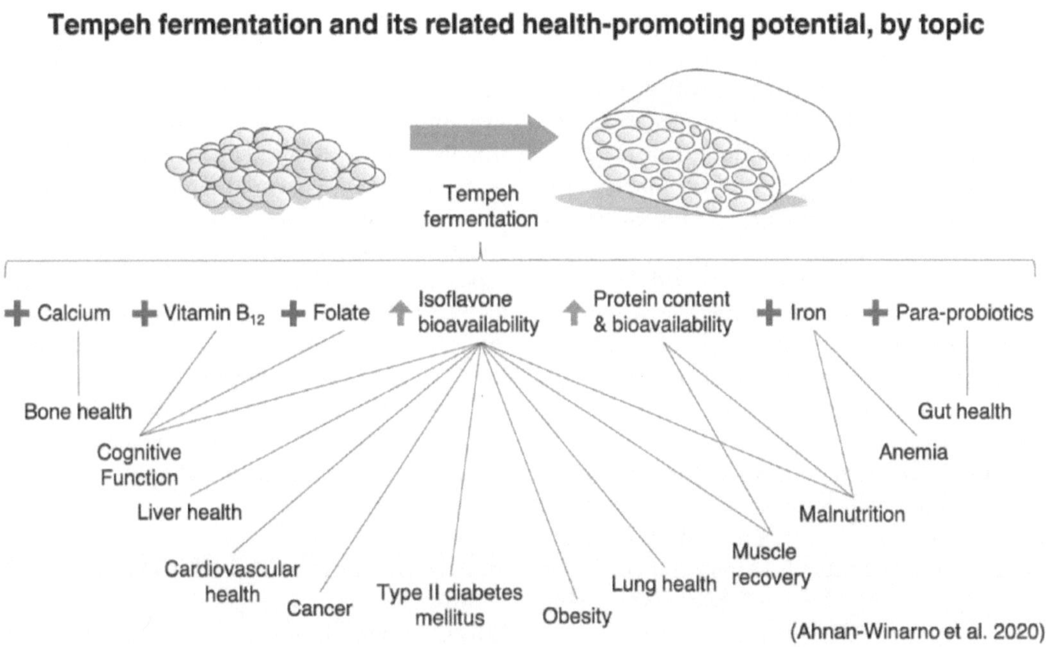

FIGURE 5.5 Potential health benefits of tempeh (Ahnan-Winarno et al. 2021).

The health benefit topics that have been linked to tempeh, sorted from the most to the least number of researched benefits, include gut health, cardiovascular health, cancer, malnutrition, cognitive function, obesity, liver health, anemia, lung health, skeletal muscle recovery, type II diabetes mellitus, and bone health (Figure 5.4) (Ahnan-Winarno et al. 2021). Gut health was a leading topic, with a total of 14 research articles with 6 *in vivo* studies (the highest), 5 *in vitro* experiments, and 3 clinical studies (Ahnan-Winarno et al. 2021). Cognitive function was researched in two population-based experiments and type II diabetes mellitus in one study (Ahnan-Winarno et al. 2021). Research on tempeh and malnutrition had the most clinical reports (five), and cancer-related articles had the highest number of *in vitro* experiments (seven) (Ahnan-Winarno et al. 2021).

5.6.1 Gut Health

In terms of gut health, tempeh supplementation resulted in beneficial effects by altering gut microbiota composition, antibody production, and short chain fatty acids (Ahnan-Winarno et al. 2021). Tempeh supplementation in human studies (100 g/person per day for 2 weeks in eight healthy males and females

20–23 years of age) increased the fecal number of *Akkermansia muciniphila* (Stephanie et al. 2017). In rats, tempeh supplementation was correlated with increases in cecal numbers of *A. muciniphila, Bifidobacterium,* and *Lactobacillus*; cecal propionate and acetate levels; and decreases in cecal numbers of *Enterobacteriaceae* (Yang et al. 2018), lithocholic acid (a risk factor of colon cancer) and IgA (index of intestinal immune function) (Utama et al. 2013), and gut transit time (Matsuo 1995b). Tempeh supplementation was also reported to alleviate diarrhea severity among 6- to 24-month-old Indonesian children ($N = 304$) (Partawihardja 1990), shorten diarrhea duration among malnourished children in Kenya ($N = 56$) (Kalavi et al. 1996), and show anti-adhesion activity against *E. coli* in *in vitro* experiments (Kiers et al. 2002; Mo et al. 2012; Roubos-van den Hil et al. 2010).

5.6.2 Cardiovascular Health

In humans, supplementation of tempeh has shown hypolipidemic effects among hyperlipidemia subjects. Tempeh supplementation decreased the levels of low-density lipoprotein (LDL) and total cholesterol and increased high-density lipoprotein (HDL) and triglycerides in women with hyperlipidemia ($N = 41$) (Afifah et al. 2020). Similar results, except for the decreased triglyceride level, were observed in male and female subjects ($N = 51$). Supplementation of germinated tempeh drink (105 g per day, three times a day) decreased systolic blood pressure (Ansarullah et al. 2017).

In rats, supplementation of tempeh showed beneficial effects in hyperglycemic, normal, and hypertensive rats. In hyperglycemic rats, supplementation of tempeh reduced serum total cholesterol, triglycerides, free fatty acid, and LDL levels and increased HDL levels (Huang et al. 2018). In normal rats, supplementation of okara tempeh decreased bile acid as well as liver and plasma cholesterol levels (Masako Matsuo and Hitomi 1993). In hypertensive rats, tempeh supplementation decreased blood urea nitrogen level, which was a predictor of mortality in decompensated heart failure patients (Aronson et al. 2004).

5.6.3 Cancer

Research on tempeh and its association with cancer has been limited to one animal study and *in vitro* studies using cancerous cell lines. Tempeh extract has shown inhibitory activities on T47D breast cancer cells (IC50 value of 196.066 ± 15.956 µg/mL) (Yuliani et al. 2016), Caco-2 human colon adenocarcinoma cells (Hsu et al. 2009), Hep3B human carcinoma cells (IC50 = 150.2 mg/mL) (Cheng et al. 2011), HuH-7 human hepatocellular carcinoma cells (apoptosis at 600 to 700 µM supplementation concentrations) (Matsuo et al. 1997), and MCF-7 breast cancer cells (IC50 values ranging from 8.70 to 5.2 µg/mL) (Athaillah et al. 2019; Muzdalifah et al. 2018). In mouse myeloma cells, tempeh extract showed higher inhibitory activity than that of soy (Kiriakidis et al. 1997). In chicken chorioallantois membrane assays, supplementation of tempeh isoflavone extracts showed inhibition in angiogenesis by 24.42%–75.09% (Kiriakidis et al. 2005). In rats treated with 1,2-dimethylhydrazine dihydrochloride, tempeh supplementation (600 mg/kg body weight for 12 weeks) significantly reduced the number of aberrant crypt foci in the colon.

5.6.4 Malnutrition

Tempeh supplementation showed potential in improving physiological parameters against malnutrition in numerous clinical studies. In undernourished Indonesian patients aged 1–10 years ($N = 30$), tempeh flour-based food intake (containing 13.6 g of tempeh per day for 14 days) increased serum iron and hemoglobin levels similarly to the gold standard, WHO's F100 milk-based formula (Iva et al. 2012). In Indonesian toddlers ($N = 70$), tempeh-based biscuit (providing 4.81 g of protein per day per subject) supplementation increased weight, height, and blood hemoglobin levels for 12 weeks (Fatmah 2018). In Indonesian underweight children ($N = 19$), supplementation with a tempeh–jicama mixture (100 g/day) increased body weight and blood albumin levels compared to the biscuit control (Symond et al. 2016). In iron-deficient pregnant women ($N = 252$), a tempeh-dominant diet during pregnancy seemed to improve any decrease in blood hemoglobin, ferritin, and iron that are experienced due to higher demands compared to no intervention (Wijaya-Erhardt et al. 2011). In underweight children under 5 years ($N = 46$), supplementation of tempeh nuggets for 30 days increased energy intake compared to no supplementation (Permatasari et al. 2018). In rats, 30 days of tempeh-based formula increased weight gain and serum albumin levels compared to the control of commercial products used to address malnutrition (Khasanah et al. 2018).

Further research is needed to elucidate the health-beneficial effects of tempeh supplementation in humans regarding prevention and treatment of malnutrition.

5.6.5 Cognitive Function

Effects of tempeh on cognitive function have been reported in animal, human, and population studies, especially in terms of memory improvement. Tempeh flour supplementation in Indonesian elders ($N = 15$) improved the scores of the Mini-Mental State Examination (MMSE) and Hopkins Verbal Learning Test (HVLT) (Kridawati et al. 2019). In population studies in Indonesia, tempeh consumption also resulted in better memory test scores ($N = 719$) and better immediate memory recall ($N = 142$) compared to consumption of tofu (Hogervorst et al. 2008; Hogervorst et al. 2011). Similar memory improvement was also observed in rats supplemented with tempeh isoflavone extract at 40 mg/kg, in addition to observed better cognitive biomarkers like reduced acetylcholinesterase levels compared to that of unfermented soybeans (Ahmad et al. 2014). The cognitive effects of tempeh have been linked to its isoflavone and vitamin B12 content (Mo et al. 2013; Ahmad et al. 2014).

5.6.6 Obesity

Tempeh's potential in addressing obesity has been researched within the scope of serum biomarkers only, focusing on cholesterol and malondialdehyde (MDA) as an oxidative stress marker (Ahnan-Winarno et al. 2021). In a human study, daily consumption of a tempeh-based drink for three months significantly decreased total cholesterol, LDL, and MDA levels (Astawan et al. 2018). In rats, supplementation of tempeh for one month improved serum MDA levels (Harun et al. 2017). Better profiles of serum triacyl glyceride, HDL, and LDL levels were observed after high-GABA tempeh supplementation for six weeks in rats (Astawan et al. 2018).

5.6.7 Liver Health

In vitro and *in vivo* studies showed hepatoprotective effects of tempeh supplementation, correlated with tempeh's antioxidant content and changes in enzyme activities. In human hepatocellular carcinoma HuH-7 cells, exposure to antioxidant extracts of tempeh (at 800 μM of feed concentration) resulted in cytotoxic, apoptotic, and cell growth inhibitory activities (Matsuo et al. 1997). In mice, supplementation of lyophilized tempeh (1000 mg/kg feed ad libitum) reversed alcohol-induced liver damage accompanied with reduced aminotransferase, alanine aminotransferase, cholesterol, triglyceride, malondialdehyde, and nitric oxide levels (Yusof et al. 2013). In rats, tempeh supplementation (21% (w/w) of concentration in diet, ad libitum) lowered the level of thiobarbituric acid, a lipid peroxidation biomarker (Murata et al. 1971), while okara tempeh supplementation (50% (w/w) ad libitum for 36 days) decreased glutathione peroxidase activity (Matsuo 1995a). In hypertensive rats, tempeh–natto supplementation (204 to 208 g/kg) increased liver DPPH scavenging activity and ACE inhibitory activity (Chang et al. 2009).

5.6.8 Lung Health

The potential beneficial effects of tempeh on lung health have been limited to *in vitro* and *in vivo* studies. Black bean tempeh milk extract exhibited inhibitory activity toward human carcinoma cells Hep 3B (IC50 = 150.2 mg/mL) but not against human carcinoma cells HeLa, Hep G2, or CL-1, nor normal human lung fibroblast cells MRC-5 (Cheng et al. 2011). In hypertensive rats, tempeh–natto supplementation was correlated with a decrease in angiotensin-converting enzyme (ACE) activity (Chang et al. 2009). These effects were hypothesized to be due to tempeh's phenolic content, which was mainly isoflavones (Cheng et al. 2011).

5.6.9 Skeletal Muscle Recovery

Clinical studies have shown the potential benefits of tempeh intake on muscle recovery. In Indonesian tuberculosis patients ($N = 128$), daily intake of tempeh for two months coupled with baseline therapy improved body weight, handgrip strength, and walking test scores (Setiawan 2016). The improvements were not associated with the nutritional content of tempeh, suggesting that the effects might come from tempeh's antioxidant content. In Indonesian student athletes ($N = 18$), tempeh drink intake after exercise lowered the serum creatine kinase level, a muscle damage indicator, as well as maximal strength at 24 hours post exercise compared to whey and a placebo (Jauhari et al. 2013). These potential effects could be due to branched chain amino acids in the tempeh drink (Setiawan 2016; Howatson et al. 2012; Jackman et al. 2010).

5.6.10 Bone Health

Tempeh contains 81.6 to 94.35 mg of calcium, which can fulfill 6.80% to 7.86% of the RDA per 85-g portion (Food and Drug Administration 2019; National Institute of Health 2020; United States Department of Agriculture 2019). Tempeh supplementation showed better calcium absorption compared to that of whole milk containing an equal calcium amount in postmenopausal Malay women ($N = 20$) (Haron et al. 2010), as well as compared to that of unfermented soybeans in rats (Watanabe et al. 2008). Calcium content and bioavailability are critical for maintaining healthy bone mass and functionality (Cashman 2002).

5.7 SUSTAINABILITY

Compared to animal-sourced protein, that is, beef, mutton, lamb, pork, chicken, fish, egg, milk, and cheese, tempeh demonstrated better energy and greenhouse gas (GHG) efficiency for the same amount of protein (Ahnan-Winarno et al. 2021). These findings were obtained by comparing the results of life cycle analyses of each source of protein. Tempeh was more than four times more energy efficient and more than 20 times more GHG efficient compared to beef (Ahnan-Winarno et al. 2021) (Figures 5.6 and 5.7).

More life cycle assessments need to be done to directly compare tempeh's sustainability to other plant-based alternative proteins, based on the CO_2 equivalent of GHG emission. Tempeh has the potential to be as sustainable or more so compared to extruded vegetable meat alternatives, whose protein delivery efficiency energy GHG ranged from 89 to 162 g protein/kg CO_2 equivalent (Detzel et al. 2021). Compared to the increasingly popular ingredient of soy protein isolate (SPI), tempeh theoretically has better sustainability in terms of lower contributions to global warming, water depletion, and fossil depletion, parameters for which SPI was less sustainable than or only as sustainable as unprocessed chicken, pork, and beef (Berardy et al. 2015). SPI was superior in terms of land use given that SPI was obtained as a utilized by-product of soymeal from the soy oil industry (Berardy et al. 2015).

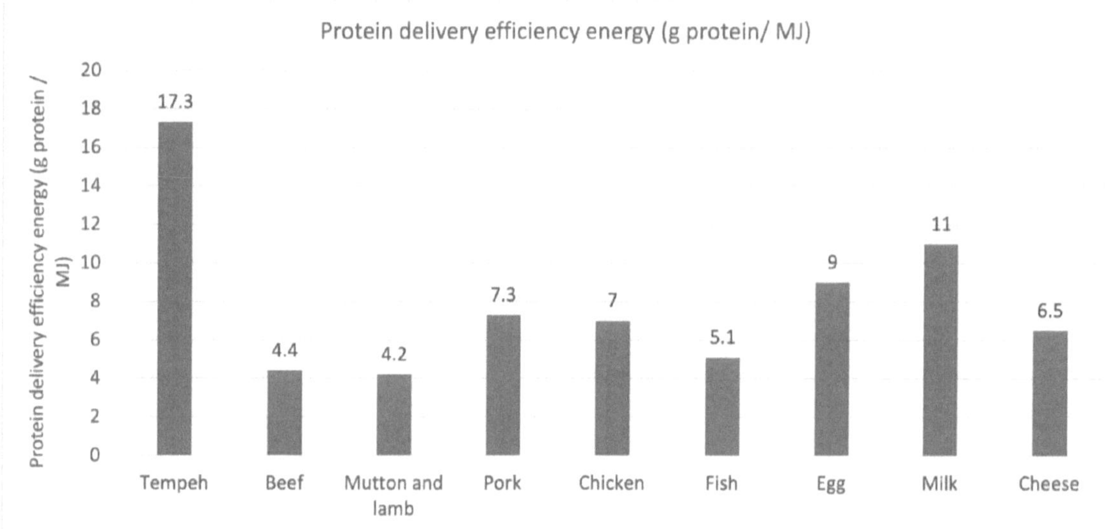

FIGURE 5.6 Protein delivery efficiency energy of tempeh compared to other common sources of protein (Ahnan-Winarno et al. 2021; González et al. 2011).

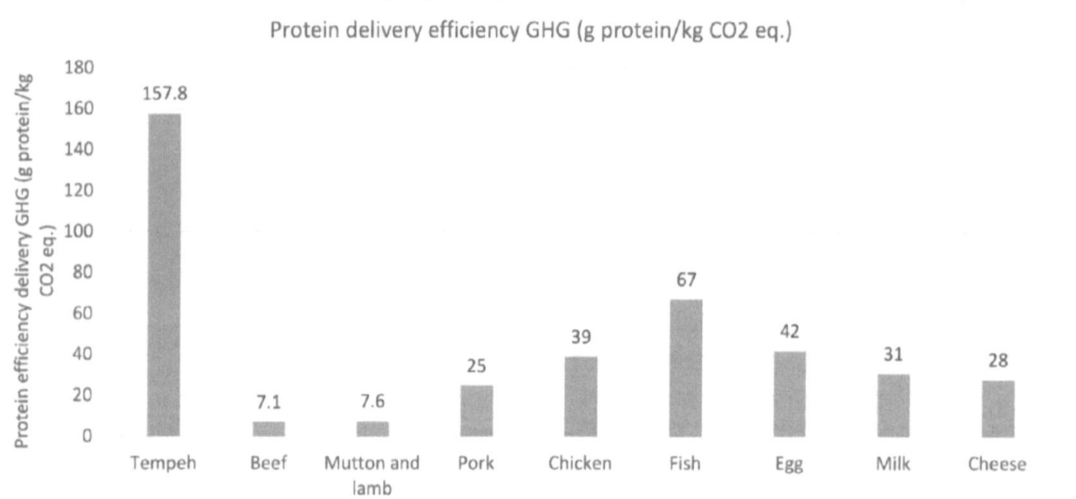

FIGURE 5.7 Protein delivery efficiency energy GHG of tempeh compared to other common sources of protein (Ahnan-Winarno et al. 2021; González et al. 2011).

5.8 AFFORDABILITY

While traditional tempeh has been very affordable in Indonesia (about six times cheaper than beef and almost ten times cheaper than milk for the same amount of protein), tempeh could become more expensive depending on its production method. Hygienic tempeh produced in a modern facility in Indonesia could be five times more expensive than traditionally prepared tempeh and more expensive than chicken and eggs for the same amount of protein. In the United States, where tempeh is commonly produced in modern facilities, the price of tempeh could range from being similar to that of pork to higher than beef for the same amount of protein, all higher than the price of chicken and eggs. To position tempeh as an affordable source of protein, careful thought needs to be put into production optimization to ensure a lower cost.

5.9 OPPORTUNITIES AND CHALLENGES

Tempeh as a plant-sourced protein offers significant health and environmental benefits that are critical given persistent malnutrition, the rise in non-communicable diseases, and climate

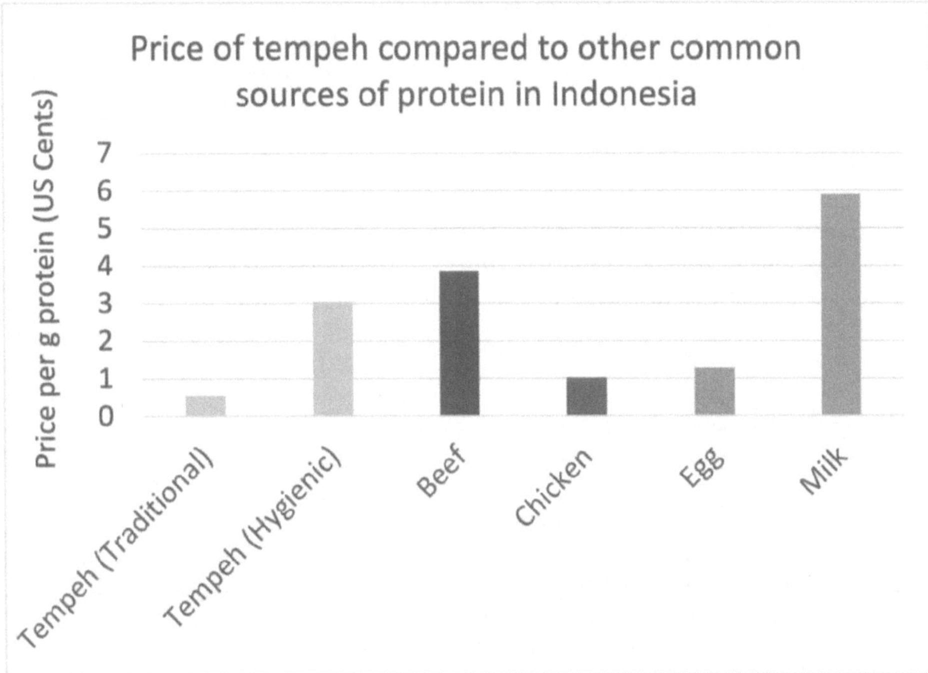

FIGURE 5.8 Price of tempeh per g of protein compared to other popular sources of protein in Indonesia (Ahnan-Winarno et al. 2021).

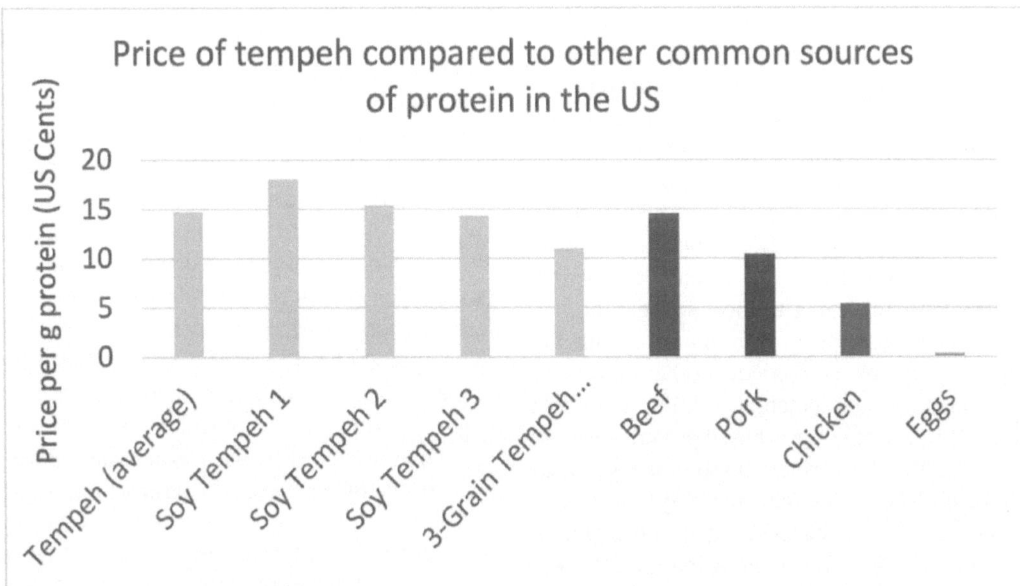

FIGURE 5.9 Price of tempeh per gram of protein compared to those of other popular protein sources in the US (Ahnan-Winarno et al. 2021).

change. Tempeh and tempeh fermentation is a gift from Indonesia to the world, with its potential to produce nutritious, sustainable, and affordable foods. The adoption and expansion of tempeh use globally could enable new ways of utilizing local crops as meat alternatives to meet the demand for new approaches in food consumption to tackle the climate crisis.

Beyond soy tempeh as a food, the results of tempeh fermentation still hold the potential to produce enhanced sensorial properties, health benefits, sustainability, and affordability. Tempeh fermentation shall be seen as a food processing biotechnology in which every change in fermentation parameters (e.g., substrate, conditions, durations) could result in enhancement of one or many aspects of food. A systematic approach could be adopted to promote cost- and time-effective enhancement for such optimization. For example, orderly or cross-sectional multi-omics (genomics, transcriptomics, proteomics, metabolomics) could be done to screen and select the best tempeh fermentation fungal strain, co-inoculation partner organisms, fermentation temperature, time, and humidity, post-harvest treatment such as pasteurization settings, and new applications such as meat alternatives for optimum sensorial properties and nutritional profiles. Evidence-based research on the health benefits of tempeh can be applied to the process of optimizing tempeh fermentation to produce food products most suitable for improving or ameliorating gut health, heart disease, cancer, malnutrition, and other health conditions mentioned previously.

Improving access to tempeh as a nutritious, sustainable, and affordable source of protein requires addressing challenges in terms of domestic and international reputation, supply chain, global production system, plastic use, and further elucidating health benefit potentials.

The 'food of the poor' image that has been instilled among Indonesians might have contributed to the slow spread of the global adoption of tempeh compared to tofu, the latter of which was presented to global cuisine initially by the Chinese with a higher sense of pride in the food. As Indonesians develop national pride regarding tempeh and expertise in marketing local foods, new initiatives have emerged to produce higher-quality tempeh products, such as supermarket-friendly tempeh chips in Indonesia and frozen tempeh for export.

The emergence of new tempeh-related businesses around the world has fueled the momentum to engage tempeh in the mainstream food market while also providing international validation of this food source for Indonesians. In terms of the domestic supply chain, the soybean supply in Indonesia highlights the domination of soybean imports from the United States, which has impacted farmers and tempeh producers through price and supply fluctuations. This concern became more apparent when tempeh producers complained when the soybean price was too high, and the farmers complained when the price was low. According to local soybean farmers in Grobogan, Central Java, soybean farming is tedious, less productive, and less profitable compared to other crop farming, such as corn and rice. One potential solution is to open new market channels in which tempeh-based products are sold as value-added products such as protein powder or vegan meat alternatives that focus on health and sustainability values. Such new market channels would enable better and higher pricing of the end products and allow larger profit margins for both soybean farmers and tempeh producers to gradually improve production quality.

Improving international sales channels would greatly benefit local growers and producers with the potential for larger profit margins; however, hurdles like quality standardization and regulation need to be tackled as upfront investments in time and cost. Therefore, farmer- and tempeh producer–friendly investment schemes should be developed simultaneously with expansion of international markets for tempeh. Tempeh is a traditional food that is sold in local markets and has been exempted from supermarket-level food safety regulations. Traditional tempeh in Indonesia can be made in a regular kitchen and sold wrapped in banana leaf on push carts on the same day of harvest, not requiring shelf-life longer than 2–3 days. Therefore, improvements in food safety would benefit local consumers as well as conditioning tempeh for international markets, resulting in improved quality of tempeh in the market. One successful example is the Rumah Tempe Indonesia, which has an international food safety-standardized facility and has exported tempeh to South Korea with optimal production and financial outcomes.

The traditional tempeh production system in Indonesia greatly benefits from the climate, having room temperature and humidity consistently suitable for tempeh fermentation. Outside of Indonesia, especially in non-tropical areas, producing tempeh could be challenging. Larger investments are needed to establish the tropical climate conditions for tempeh fermentation and for food safety regulations. As a result, the typical tempeh sold in overseas markets has been pasteurized in plastic bags and produced in modern facilities or commercial kitchens. While the benefits of such processing include the convenience of a longer shelf life, which is typically up to three months in refrigeration, the flavor and culinary experience of fresh unpasteurized tempeh deteriorates. Fresh tempeh could be promoted in these countries if the turnover rate of tempeh sales was as high as that of commercially produced tempeh. Planit Protein is a good example of a company that has been addressing this issue by selling a simplified tempeh-making apparatus.

Plastic use in tempeh production raises a concern for sustainability and environmentally sound production. Traditionally made tempeh is fermented and sold in a cover made from leaves, while commercial tempeh is typically sold wrapped in plastic, which allows for a better seal and facilitates pasteurization for shelf-life improvement. However, plastic use creates an environmental concern due to the single-use nature of tempeh's plastic wrap/packaging. To address this, numerous tempeh producers have developed new methods of tray fermentation, in which tempeh fermentation is conducted in trays instead of poked plastic bags or folded leaves. Adoption of plastic-free tempeh fermentation at a large production scale is needed to prevent employment of single-use plastics in the tempeh industry, particularly as it scales up for supply to markets outside of Indonesia. Winarno (2021) argues that

> besides the starter cultures, *laru* leaves have traditionally been used to make tempeh. When we compare traditional Indonesian tempeh to international tempeh, the use of plastic in the fermentation process of international tempehs is a limiting factor for sustainability. In fact, the use of *laru* leaves in preparation of tempeh is sustainable and better for the planet. Tempeh fermented in *laru* leaves is also perceived as more delicious in Indonesia, [sic] therefore could contribute to taste/sensorial improvements.

Simultaneously, addressing food safety concerns regarding leaf-wrapped tempeh to be sold in smaller, local markets should be a priority of national food safety agencies. This includes the potential growth of some pathogens that should be considered, such as *Salmonella* spp., *Cronobacter sakazakii*,

Listeria monocytogenes, and *Bacillus cereus*, together with the growth of *Rhizopus* spp., without decreasing the quality of tempe (Nuraida and Krusong 2014).

In addition to nutritious, sustainable, and affordable aspects, the potential health benefits provide a critical opportunity for the growth of the tempeh industry. More research is needed to establish causation in health effects of tempeh, with clinical trials being a priority area for future studies. The scope, quality, and rigor of research on the health benefits of tempeh are also essential to elucidating its potential for health benefits and inclusion in dietary recommendations. Expanding the breadth of studies (i.e., *in vitro, in vivo*, clinical, intervention, and population studies) would provide a solid foundation for positioning tempeh as a functional food, defined as a food that promotes certain health benefits beyond its ability to fulfill nutritional needs. In terms of identity protection and promotion, tempeh's culturally and historically rich background should be translated into international recognitions, including as a UNESCO Intangible Cultural Heritage and Geographical Indication(s) by World Intellectual Property Organization (WIPO) covering types of traditional tempeh and tempeh dishes.

In summary, tempeh has the potential to be one of the premier foods for health, with a sustainable approach in the production and supply chain. Future steps in tempeh production can be made to reduce its carbon footprint, promote a predominantly plant-based diet, and promote other environmentally friendly approaches in food production for population health.

REFERENCES

Afifah, D. N., N. Nabilah, G. T. Supraba, S. N. Pratiwi, Nuryanto, M. Sulchan. 2020. The effects of tempeh gembus, an Indonesian fermented food, on lipid profiles in women with hyperlipidemia. *Current Nutrition & Food Science* 16(1):56–64. https://doi.org/10.2174/1573401314666180807112549

Ahmad, A., K. Ramasamy, S. M. Jaafar, A. B. A. Majeed, V. Mani. 2014. Total isoflavones from soybean and tempeh reversed scopolamine-induced amnesia, improved cholinergic activities and reduced neuroinflammation in brain. *Food and Chemical Toxicology* 65:120–128. https://doi.org/10.1016/j.fct.2013.12.025

Ahmad, A., K. Ramasamy, A. B. A. Majeed, V. Mani. 2015. Enhancement of β-secretase inhibition and antioxidant activities of *tempeh*, a fermented soybean cake through enrichment of bioactive aglycones. *Pharmaceutical Biology* 53(5):758–766. https://doi.org/10.3109/13880209.2014.942791

Ahnan-Winarno, A., W. Winarno, M. Nanere. 2019. Can tempeh be greenly marketed by the Indonesian Tempeh Movement (ITM)? *Proceedings of the 1st Sampoerna University-AFBE International Conference*, SU-AFBE 2018, 6–7 December 2018, Jakarta. https://doi.org/10.4108/eai.6-12-2018.2286267

Ahnan-Winarno, A. D., L. Cordeiro, F. G. Winarno, J. Gibbons, H. Xiao. 2021. Tempeh: A semicentennial review on its health benefits, fermentation, safety, processing, sustainability, and affordability. *Comprehensive Reviews in Food Science and Food Safety* 20(2):1717–1767. https://doi.org/10.1111/1541-4337.12710

Ansarullah, A., Hardinsyah, S. A. Marliyati, M. Astawan. 2017. The effect of Tempe drink intervention on blood pressure in hypertension and hypercholesterolemia subjects. *Jurnal Gizi Dan Pangan* 12(2):101–108. https://doi.org/10.25182/jgp.2017.12.2.101-108

Anwar, M., A. Kasper, A. R. Steck, J. G. Schier. 2017. Bongkrekic acid—A review of a lesser known mitochondrial toxin. *Journal of Medical Toxicology* 13(2):173–179. https://doi.org/10.1007/s13181-016-0577-1

Ariani, S. R. D., S. Matsjeh, Mustofa, B. Purwono. 2016. Quantitative analysis of L-dopa (l-3,4-dihydroxyphenylalanine) content in extracts of various products from velvet bean (Mucuna pruriens (L.) DC.) Tempe processing. *Oriental Journal of Chemistry* 32(6):2921–2927. https://doi.org/10.13005/ojc/320611

Aronson, D., M. A. Mittleman, A. J. Burger. 2004. Elevated blood urea nitrogen level as a predictor of mortality in patients admitted for decompensated heart failure. *The American Journal of Medicine* 116(7):466–473. https://doi.org/10.1016/j.amjmed.2003.11.014

Astawan, M., Y. S. Mardhiyyah, C. H. Wijaya. 2018. Potential of bioactive components in Tempe for the treatment of obesity. *Jurnal Gizi Dan Pangan* 13(2):79–86. https://doi.org/10.25182/jgp.2018.13.2.79-86

Astawan, M., D. H. Maskar. 2019. Rebranding Tempe from traditional food to be more acceptable for the 21st century consumers. *Annals of Nutrition and Metabolism* 75:7–7.

Athaillah, Z. A., D. Muzdalifah, A. Lestari, A. F. Devi, L. Z. Udin, N. Artanti, H. N. Lioe. 2019. Phenolic compound profile and functionality of aqueous overripe Tempe extracts. *Current Research in Nutrition and Food Science* 7(2):382–392. https://doi.org/10.12944/CRNFSJ.7.2.08

Baiano, A. 2010. Effects of processing on isoflavone content and profile in foodstuffs: A review. In M. J. Thompson (Ed.), *Isoflavones: Biosynthesis, Occurrence and Health Effects* (pp. 111–135). Nova Science Publishers, Inc.

Berardy, A., C. Costello, T. Seager. 2015. Life cycle assessment of soy protein isolate. *Proceedings of the International Symposium on Sustainable Systems and Technologies*, Dearborn, MI, 3.

Berghofer, E., B. Grzeskowiak, N. Mundigler, W. B. Sentall, J. Walcak. 1998. Antioxidative properties of faba bean-, soybean- and oat tempeh. *International Journal of Food Sciences and Nutrition* 49(1):45–54. https://doi.org/10.3109/09637489809086403

Borges, C. W. C., M. C. Carrão-Panizzi, J. M. G. Mandarino, J. B. da Silva, S. Benedetti, E. I. Ida. 2016. Contents and bioconversion of β-glycoside isoflavones to aglycones in the processing conditions of soybean tempeh. *Pesquisa Agropecuária Brasileira* 51(3):271–279. https://doi.org/10.1590/S0100-204X2016000300009

Boukid, F. 2021. Plant-based meat analogues: From niche to mainstream. *European Food Research and Technology* 247(2):297–308. https://doi.org/10.1007/s00217-020-03630-9

Cashman, K. D. 2002. Calcium intake, calcium bioavailability and bone health. *British Journal of Nutrition* 87(S2):S169–S177. https://doi.org/10.1079/BJN/2002534

Chang, C.-T., C.-K. Hsu, S.-T. Chou, Y.-C. Chen, F.-S. Huang, Y.-C. Chung. 2009. Effect of fermentation time on the antioxidant activities of tempeh prepared from fermented soybean using Rhizopus oligosporus. *International Journal of Food Science and Technology* 44(4):799–806. https://doi.org/10.1111/j.1365-2621.2009.01907.x

Cheng, K.-C., J.-T. Lin, W.-H. Liu. 2011. Extracts from fermented black soybean milk exhibit antioxidant and cytotoxic activities. *Food Technology and Biotechnology* 49(1):111.

Chung, Y. C., C. K. Hsu, Y. T. Huang, Y. F. Lin. 2009. Mechanisms study on the antihypertensive effect of water extract from natto-tempeh. *New Biotechnology* 25:S206. https://doi.org/10.1016/j.nbt.2009.06.148

Detzel, A., M. Krüger, M. Busch, I. Blanco-Gutiérrez, C. Varela, R. Manners, J. Bez, E. Zannini. 2021. Life cycle assessment of animal-based foods and plant-based protein-rich alternatives: An environmental perspective. *Journal of the Science of Food and Agriculture* 102(12):5098–5110. https://doi.org/10.1002/jsfa.11417

Esaki, H., H. Onozaki, T. Osawa. 1994. Antioxidative activity of fermented soybean products. In M. T. Huang, T. Osawa, C. T. Ho, R. T. Rosen (Eds.), *Food Phytochemicals for Cancer Prevention I: Fruits and Vegetables* (Vol. 546, pp. 353–360). Amer Chemical Soc.

Fatmah. 2018. Mocaf tempeh dates biscuit for the improvement of nutritional status of underweight children. *International Journal of Geomate* 14(49):155–161.

Fibri, D. L. N., M. B. Frost. 2020. Indonesian millennial consumers' perception of Tempe—And how it is affected by product information and consumer psychographic traits. *Food Quality and Preference* 80:103798. https://doi.org/10.1016/j.foodqual.2019.103798

Food and Agriculture Organization-World Health Organization. 2017. *Regional Standard for Tempe*. FAO-WHO Codex Alimentarius.

Food and Drug Administration. 2019, April 1. *CFR—Code of federal regulations title 21*. https://www.accessdata.fda.gov/scripts/cdrh/cfdocs/cfcfr/cfrsearch.cfm?fr=101.12

González, A. D., B. Frostell, A. Carlsson-Kanyama. 2011. Protein efficiency per unit energy and per unit greenhouse gas emissions: Potential contribution of diet choices to climate change mitigation. *Food Policy* 36(5):562–570. https://doi.org/10.1016/j.foodpol.2011.07.003

Griese, S. E., A. T. Fleischauer, J. K. MacFarquhar, Z. Moore, C. Harrelson, A. Valiani, S. E. Morrison, D. Sweat, J.-M. Maillard, D. Griffin, D. Springer, M. Mikoleit, A. E. Newton, B. Jackson, T.-A. Nguyen, S. Bosch, M. Davies. 2013. Gastroenteritis outbreak associated with unpasteurized tempeh, North Carolina, USA. *Emerging Infectious Diseases* 19(9):1514–1517. https://doi.org/10.3201/eid1909.130334

Haron, H., T. Shaari, C. B. Keng. 2016. Effects of different cooking methods on isoflavone content in Malaysian soy-based dishes. *Sains Malaysiana* 45:1329–1335.

Haron, H., S. Shahar, K. O. O'Brien, A. Ismail, N. Kamaruddin, S. A. Rahman. 2010. Absorption of calcium from milk and tempeh consumed by postmenopausal Malay women using the dual stable isotope technique. *International Journal of Food Sciences & Nutrition* 61(2):125–137. https://doi.org/10.3109/09637480903348080

Harun, I., H. Susanto, A. Rosidi. 2017. The Tempe giving decrease malondialdehyde [MDA] level and increase the activities of superoxide dismutase enzyme [SOD] on rats with high physical activities. *Jurnal Gizi Dan Pangan* 12(3):211–216. https://doi.org/10.25182/jgp.2017.12.3.211-216

Higasa, S., Y. Negishi, Y. Aoyagi, T. Sugahara. 1996. Changes in free amino acids of Tempe during preparation with velvet beans (Mucuna pruriens). *Journal of the Japanese Society for Food Science and Technology-Nippon Shokuhin Kagaku Kogaku Kaishi* 43(2):188–193.

Hogervorst, E., F. Mursjid, D. Priandini, H. Setyawan, R. I. Ismael, S. Bandelow, T. B. Rahardjo. 2011. Borobudur revisited: Soy consumption may be associated with better recall in younger, but not in older, rural Indonesian elderly. *Brain Research* 1379:206–212. https://doi.org/10.1016/j.brainres.2010.10.083

Hogervorst, E., T. Sadjimim, A. Yesufu, P. Kreager, T. B. Rahardjo. 2008. High tofu intake is associated with worse memory in elderly Indonesian men and women. *Dementia and Geriatric Cognitive Disorders; Basel* 26(1):50–57.

Howatson, G., M. Hoad, S. Goodall, J. Tallent, P. G. Bell, D. N. French. 2012. Exercise-induced muscle damage is reduced in resistance-trained males by branched chain amino acids: A randomized, double-blind, placebo controlled study. *Journal of the International Society of Sports Nutrition* 9(1):20. https://doi.org/10.1186/1550-2783-9-20

Hsu, C. K., Y. P. Yu, Y. C. Chung. 2009. Effect of tempeh on the intestinal microbiota and colon cancer in rats. *New Biotechnology* 25:S205. https://doi.org/10.1016/j.nbt.2009.06.145

Huang, Y.-C., B.-H. Wu, Y.-L. Chu, W.-C. Chang, M.-C. Wu. 2018. Effects of tempeh fermentation with Lactobacillus plantarum and Rhizopus oligosporus on streptozotocin-Induced type II diabetes mellitus in rats. *Nutrients* 10(9):1143. https://doi.org/10.3390/nu10091143

Indonesian Tempe Movement. 2021. *Tempethon 2021*. https://www.youtube.com/playlist?list=PL0oRfdG-EHNq09tBOoTQaB-V1PXRN0-bbZ

Iva, T., E. P. Prawirohartono, L. A. Lestari. 2012. Efek F100 dan formula tepung tempe terhadap kadar serum Fe dan hemoglobin pada anak gizi kurang. *Jurnal Gizi Klinik Indonesia* 9(1):25–33. https://doi.org/10.22146/ijcn.15373

Jackman, S. R., O. C. Witard, A. E. Jeukendrup, K. D. Tipton. 2010. Branched-chain amino acid ingestion can ameliorate soreness from eccentric exercise. *Medicine & Science in Sports & Exercise* 42(5):962–970. https://doi.org/10.1249/MSS.0b013e3181c1b798

Jauhari, M., A. Sulaeman, H. Riyadi, I. Ekayanti. 2013. Effect of administering tempeh drink on muscle damage recoveries after resistance exercise in student athletes. *Pakistan Journal of Nutrition* 12(10):924–928.

Kalavi, F. N. M., N. M. Muroki, A. M. Omwega, R. K. N. Mwadime. 1996. Effect of Tempe-yellow maize porridge and milk-yellow maize porridge on growth rate, diarrhoea and duration of rehabilitation of malnourished children. *East African Medical Journal* 73(7):427–431.

Keuth, S., B. Bisping. 1994. Vitamin B12 production by Citrobacter freundii or Klebsiella pneumoniae during tempeh fermentation and proof of enterotoxin absence by PCR. *Applied and Environmental Microbiology* 60(5):1495–1499.

Khasanah, M., W. Sjamsuridzal, A. Oetari, I. Santoso, G. Roosheroe. 2018. Phylogenetic analyses based on ITS regions of rDNA identified five Rhizopus strains from tempeh as R. delemar and R. oryzae. In T. Mart, D. Triyono, I. T. Anggraningrum, K. A. Sugeng, R. Yuniati (Eds.), *Proceedings of the 3rd International Symposium on Current Progress in Mathematics and Sciences 2017 (ISCPMS2017)* (Vol. 2023, pp. 020141–1). Amer Inst Physics. https://doi.org/10.1063/1.5064138

Kiers, J. L., M. J. R. Nout, F. M. Rombouts, M. J. A. Nabuurs, J. van der Meulen. 2002. Inhibition of adhesion of enterotoxigenic Escherichia coli K88 by soya bean Tempe. *Letters in Applied Microbiology* 35(4):311–315. https://doi.org/10.1046/j.1472-765X.2002.01182.x

Kiriakidis, S., O. Hogemeier, S. Starcke, F. Dombrowski, J. C. Hahne, M. Pepper, H. C. Jha, N. Wernert. 2005. Novel tempeh (fermented soyabean) isoflavones inhibit in vivo angiogenesis in the chicken chorioallantoic membrane assay. *British Journal of Nutrition* 93(3):317–323. https://doi.org/10.1079/BJN20041330

Kiriakidis, S., S. Stathi, H. C. Jha, R. Hartmann, H. Egge. 1997. Fatty acid esters of sitosterol 3 beta-glucoside from soybeans and

Tempe (fermented soybeans) as antiproliferative substances. *Journal of Clinical Biochemistry and Nutrition* 22(3):139–147.

Kridawati, A., T. B. W. Rahardjo, E. Hogervorst. 2019. Tempe, isoflavones and cognitive function. *Annals of Nutrition and Metabolism* 75:376–376.

Kuligowski, M., K. Pawłowska, I. Jasińska-Kuligowska, J. Nowak. 2017. Isoflavone composition, polyphenols content and antioxidative activity of soybean seeds during tempeh fermentation. *CyTA—Journal of Food* 15(1):27–33. https://doi.org/10.1080/19476337.2016.1197316

Liem, I. T., K. H. Steinkraus, T. C. Cronk. 1977. Production of vitamin B-12 in tempeh, a fermented soybean food. *Applied and Environmental Microbiology* 34(6):773–776.

Matsuo, M. 1995a. In vivo antioxidant action of a new foodstuff, okara Tempe, in rats fed oxidized oil. *Nippon Nogeikagaku Kaishi-Journal of the Japan Society for Bioscience Biotechnology and Agrochemistry* 69(2):169–171.

Matsuo, M. 1995b. Utilization of dietary fiber in a new foodstuff, okara-Tempe, by microflora in rats. *Nippon Nogeikagaku Kaishi-Journal of the Japan Society for Bioscience Biotechnology and Agrochemistry* 69(2):163–167.

Matsuo, M., E. Hitomi. 1993. Suppression of plasma cholesterol elevation by okara Tempe in rats. *Bioscience, Biotechnology, and Biochemistry* 57(7):1188–1190. https://doi.org/10.1271/bbb.57.1188

Matsuo, M., N. Nakamura, Y. Shidoji, Y. Muto, H. Esaki, T. Osawa. 1997. Antioxidative mechanism and apoptosis induction by 3-hydroxyanthranilic acid, an antioxidant in Indonesian food tempeh, in the human hepatoma-derived cell line, HuH-7. *Journal of Nutritional Science and Vitaminology* 43(2):249–259.

Mo, H., S. Kariluoto, V. Piironen, Y. Zhu, M. G. Sanders, J.-P. Vincken, J. Wolkers-Rooijackers, M. J. R. Nout. 2013. Effect of soybean processing on content and bioaccessibility of folate, vitamin B12 and isoflavones in tofu and Tempe. *Food Chemistry* 141(3):2418–2425. https://doi.org/10.1016/j.foodchem.2013.05.017

Mo, H., Y. Zhu, M. J. R. Nout. 2012. In vitro digestion enhances anti-adhesion effect of Tempe and tofu against Escherichia coli. *Letters in Applied Microbiology* 54(2):166–168. https://doi.org/10.1111/j.1472-765X.2011.03189.x

Murata, K., H. Ikehata, Y. Edani, K. Koyanagi. 1971. Studies on the nutritional value of tempeh: Part II. Rat feeding test with tempeh, unfermented soybeans, and tempeh supplemented with amino acids. *Agricultural and Biological Chemistry* 35(2):233–241. https://doi.org/10.1080/00021369.1971.10859904

Muzdalifah, D., Z. A. Athaillah, A. F. Devi, L. Z. Udin. 2018. Cytotoxicity of ethanol overripe tempe extract against MCF-7 breast cancer cell and its antioxidant activity. In S. N. A. Jenie, A. A. Dwiatmoko, M. A. Fitriady (Eds.), *Proceedings of the 4th International Symposium on Applied Chemistry 2018* (Vol. 2024, p. 020037). Amer Inst Physics. https://doi.org/10.1063/1.5064323

National Institute of Health. 2020, March 26. *Office of dietary supplements—Calcium.* https://ods.od.nih.gov/factsheets/Calcium-HealthProfessional/

Nuraida, L., Krusong, W. 2014. Starter cultures. *Indigenous Fermented Foods of Southeast Asia*, pp. 112–138.

Okada, N. 1989. Role of microorganisms in tempeh manufacture—Isolation of vitamin-b12 producing bacteria. *JARQ-Japan Agricultural Research Quarterly* 22(4):310–316.

Paredes-López, O., G. I. Harry. 1989. Changes in selected chemical and antinutritional components during tempeh preparation using fresh and hardened common beans. *Journal of Food Science* 54(4):968–970. https://doi.org/10.1111/j.1365-2621.1989.tb07923.x

Partawihardja, I. S. 1990. *Effect of tempeh supplementation on growth rate of children aged 6 to 24 months with diarrhea* [PhD Dissertation]. Dissertation, University of Indonesia.

Permatasari, O., R. Murwani, M. Z. Rahfiludin. 2018. Tempe nuggets provision improves energy adequacy and protein intake in underweight underfive children. *Current Research in Nutrition and Food Science* 6(1):89–96. https://doi.org/10.12944/CRNFSJ.6.1.09

Pertiwi, M. G. P., Y. Marsono, R. Indrati. 2020. In vitro gastrointestinal simulation of tempe prepared from koro kratok (Phaseolus lunatus L.) as an angiotensin-converting enzyme inhibitor. *Journal of Food Science and Technology-Mysore* 57(5):1847–1855. https://doi.org/10.1007/s13197-019-04219-1

Pugalenthi, M., V. Vadivel, P. Siddhuraju. 2005. Alternative food/feed perspectives of an underutilized legume Mucuna pruriens var. Utilis—A review. *Plant Foods for Human Nutrition* 60(4):201. https://doi.org/10.1007/s11130-005-8620-4

Purwadaria, H. K., D. Fardiaz, L. B. S. Kardono, A. McElhatton. 2016. Tempe from traditional to modern practices. In *Modernization of Traditional Food Processes and Products* (pp. 145–160). Springer.

Rochín-Medina, J. J., R. Gutiérrez-Dorado, L. M. Sánchez-Magaña, J. Milán-Carrillo, E. O. Cuevas-Rodríguez, S. Mora-Rochín, A. Valdez-Ortiz, C. Reyes-Moreno. 2015. Enhancement of nutritional properties, and antioxidant and antihypertensive potential of black common bean seeds by optimizing the solid state bioconversion process. *International Journal of Food Sciences and Nutrition* 66(5):498–504. https://doi.org/10.3109/09637486.2015.1052377

Roubos-van den Hil, P. J., H. A. Schols, M. J. R. Nout, M. H. Zwietering, H. Gruppen. 2010. First characterization of bioactive components in soybean Tempe that protect human and animal intestinal cells against enterotoxigenic Escherichia coli (ETEC) infection. *Journal of Agricultural and Food Chemistry* 58(13):7649–7656. https://doi.org/10.1021/jf101379y

Sánchez-Magana, L. M., E. O. Cuevas-Rodríguez, R. Gutiérrez-Dorado, A. E. Ayala-Rodríguez, A. Valdez-Ortiz, J. Milán-Carrillo, C. Reyes-Moreno. 2014. Solid-state bioconversion of chickpea (Cicer arietinum L.) by Rhizopus oligosporus to improve total phenolic content, antioxidant activity and hypoglycemic functionality. *International Journal of Food Sciences and Nutrition* 65(5):558–564. https://doi.org/10.3109/09637486.2014.893284

Setiawan, B. 2016. *The effect of fermented soybean (tempeh) supplementation among active pulmonary tuberculosis patients with standard therapy in Indonesia* [PhD Dissertation]. Justus Liebig University Giessen.

Shurtleff, W., A. Aoyagi. 1984. *History of tempeh.* Soyinfo Center.

Shurtleff, W., A. Aoyagi. 2020. *History of tempeh and tempeh products (1815–2020): Bibliography and sourcebook.* Soyinfo Center.

Stephanie, S., N. K. Ratih, S. Soka, A. Suwanto. 2017. Effect of tempeh supplementation on the profiles of human intestinal immune system and gut microbiota. *Microbiology Indonesia* 11(1):2.

Symond, D., F. Oenzil, E. Darwin, N. I. Lipoeto. 2016. Efikasi suplemenasi formula tempe bengkuang terhadap kadar albumin dan z-skor berat badan menurut umur (bb/u) pada anak gizi kurang. *Jurnal Gizi Dan Pangan* 11(1):Article 1. https://doi.org/10.25182/jgp.2016.11.1.%p

United States Department of Agriculture. 2019. *USDA FoodData Central.* https://fdc.nal.usda.gov

Utama, Z., Y. Okazaki, H. Tomotake, N. Kato. 2013. Tempe consumption modulates fecal secondary bile acids, mucins, Immunoglobulin A, enzyme activities, and cecal microflora and organic acids in rats. *Plant Foods for Human Nutrition* 68(2):177–183. https://doi.org/10.1007/s11130-013-0357-x

van Veen, A. G., G. Sohaefer. 1950. The influence of the tempeh fungus on the soya bean. *Documenta Neerlandica et Indonesica de Morbis Tropicis* 2:270–281.

Vebliza, Y., W. Sjamsuridzal, A. Oetari, I. Santoso, I. G. Roosheroe. 2018. Re-identification of five strains of Rhizopus arrhizus from tempeh based on ITS regions of rDNA sequence data. In T. Mart, D. Triyono, I. T. Anggraningrum, K. A. Sugeng, R. Yuniati (Eds.), *Proceedings of the 3rd International Symposium on Current Progress in Mathematics and Sciences 2017 (ISCPMS2017)* (Vol. 2023, pp. 020167–1). Amer Inst Physics. https://doi.org/10.1063/1.5064164

Wang, H.-J., P. A. Murphy. 1996. Mass balance study of isoflavones during soybean processing. *Journal of Agricultural and Food Chemistry* 44(8):2377–2383. https://doi.org/10.1021/jf950535p

Watanabe, N., H. Aoki, K. Fujimoto. 2008. Fermentation of soybean by Rhizopus promotes the calcium absorption ratio in rats. *Journal of the Science of Food and Agriculture* 88(15): 2749–2752. https://doi.org/10.1002/jsfa.3403

Wijaya-Erhardt, M., S. Muslimatun, J. G. Erhardt. 2011. Fermented soyabean and vitamin C-rich fruit: A possibility to circumvent the further decrease of iron status among iron-deficient pregnant women in Indonesia. *Public Health Nutrition* 14(12):2185–2196. https://doi.org/10.1017/S1368980011000954

Winarno, F., G. 2021. *Pengetahuan, Kearifan Lokal, Pangan dan Kesehatan*. Jakarta: Gramedia Pustaka Utama.

Xiao, Y., J. Fan, Y. Chen, X. Rui, Q. Zhang, M. Dong. 2016. Enhanced total phenolic and isoflavone aglycone content, antioxidant activity and DNA damage protection of soybeans processed by solid state fermentation with Rhizopus oligosporus RT-3. *RSC Advances* 6(35):29741–29756.

Yang, Y., T. Kameda, H. Aoki, D. E. Nirmagustina, A. Iwamoto, N. Kato, N. Yanaka, Y. Okazaki, T. Kumrungsee. 2018. The effects of Tempe fermented with Rhizopus microsporus, Rhizopus oryzae, or Rhizopus stolonifer on the colonic luminal environment in rats. *Journal of Functional Foods* 49:162–167. https://doi.org/10.1016/j.jff.2018.08.017

Yuliani, S. H., E. P. Istyastono, F. D. O. Riswanto. 2016. The cytotoxic activity on T47D breast cancer cell of genistein-standardized ethanolic extract of tempeh-a fermented product of soybean (Glycine max). *Oriental Journal of Chemistry* 32(3):1619–1624.

Yusof, H. M., N. M. Ali, S. K. Yeap, W. Y. Ho, B. K. Beh, S. P. Koh, K. Long, S. Abdul Aziz, N. B. Alitheen. 2013. Hepatoprotective effect of fermented soybean (nutrient enriched soybean tempeh) against alcohol-induced liver damage in mice. *Evidence-Based Complementary and Alternative Medicine* 2013:1–8. https://doi.org/10.1155/2013/274274

Biotechnology for Functional Foods with a Focus on Traditional Fermented Foods and Beverages of Iran

6

N. Davati and S. Hesami

6.1 INTRODUCTION

Iran is the largest country in the Middle East with climate diversity. Depending on the diverse climate of Iran, various agricultural products are produced in different regions of this country. In the northwest and west of Iran, winters are very cold; however, the temperature is mild during spring and fall and hot during the summer. In the south of Iran, the temperature is very high in the summer. The main agricultural products of Iran include wheat, rice, and barley. Due to the diverse climate in Iran, various dairy products are produced from different livestock such as cattle, ewes, and goats (Pak and Farajzadeh 2007; Meybodi et al. 2016). Iran is a country with many traditional dishes and a long history in preparing ancient foods that continues to contemporary times. Among these, fermented foods have a special place in the Iranian people's diet. Iranian people consume fermented foods, which are present in different tastes and forms depending on the effects of microorganisms which are naturally present or intentionally added and the types and quantities of ingredients of foods. The people of Iran have used methods of producing fermented foods since ancient times. Fermented products as functional foods have beneficial properties such as high nutritional value, antioxidant and antimicrobial activities, and prevention of diseases (allergic reactions, cardiovascular, cancer, and gastrointestinal disorders) (Tamang et al. 2016; Vasiee et al. 2018b). Traditional fermented dairy products are the main source of potential probiotic microorganisms, in particular lactic acid bacteria (LAB). LAB is a group of fastidious, non-spore forming, gram-positive, catalase negative, facultative anaerobic microorganisms in the form of cocci or rods, and they are able to reduce the pH of fermented products (Moraes et al. 2013). The main Iranian fermented foods are presented in Table 6.1.

6.2 FERMENTED BEVERAGES AND FOODS OF IRAN

In this chapter, Iranian local fermented foods which are produced through the fermentation process are included:

1. Vegetable-based products
2. Dairy-based products
3. Cereal-based products
4. Cereal and dairy–based products
5. Fish-based products

6.2.1 Vegetable-Based Products

In Iran, the preparation of fermented vegetables in brine (pickled vegetables) in households or factories is popular. Pickled vegetables are usually served as appetizers along with every meal. The ingredients of these products are mostly cucumbers, cauliflower, carrots, garlic, eggplant, pepper, and cabbage.

6.2.1.1 *Winter Salad*

Winter salad is one of the most famous types of pickled vegetables in tomato juice and is produced mainly in the Khorasan Razavi province located in the northeastern of Iran.

DOI: 10.1201/9781003605300-6

TABLE 6.1 Some Iranian Traditional Fermented Foods

SOURCE	*PRODUCT NAME*	*INGREDIENTS*	*MICROORGANISM*	*REFERENCES*
Vegetables and fruits	*Winter salad*	Carrots, tomato, cabbage, cauliflower, garlic, angelica, chili pepper, salt, and black pepper	*Levilactobacillus brevis, Lactobacillus japonicus, Lactobacillus pentosus, Lactobacillus senmaizukei, Lactiplantibacillus plantarum, Leuconostoc mesenteroides*	Hesami and Davati (2018) Davati and Hesami (2021b)
	Verjuice	Unripe green grapes, salt	*Saccharomyces cerevisiae*	Kaya et al. (2020)
	Pickled garlic	Red grape vinegar, garlic	*S. cerevisiae, Pediococcus ethanolidurans, L. plantarum, L. brevis*	Jung et al. (2011) Pure and Pure (2016)
Dairy	*Lighvan*	Raw milk of ewe and goat, rennet, salt	*Lactococcus garvieae, L. plantarum, S. parauberis, Enterococcus faecium, Enterococcus faecalis, Lactococcus lactis, Escherichia coli, Lacticaseibacillus paracasei, Lactococcus raffinolactis, Lactococcus plantarum, Streptococcus thermophilus, Staphylococcus haemolyticus, Ligilactobacillus salivarius, Macrococcus caseolyticus*	Edalatian et al. (2012a)
	Koozeh	Raw milk of ewe and goat, rennet, salt	*L. garvieae, Leu. mesenteroides, Debaryomyces hansenii, Streptococcus parauberis, L. lactis, Warcupia* spp. *Vibrio tapetis, L. raffinolactis, S. uberis, S. thermophilus, Carnobacterium maltomaricum, Lactobacillus curvatus, Celerinatantimonas diazotrophica, Lacticaseibacillus casei, L. plantarum, Limosilactobacillus fermentum*	Edalatian et al. (2012b) Hassanzadazar et al. (2012)Tavakoli et al. (2017)
	Poosti	Raw milk of ewe and goat, salt, lamb rennet	*L. plantarum, L. paracasei, Lactobacillus acidophilus, L. brevis, Lactobacillus buchneri*	Hemmatian et al. (2015) Abdali et al. (2019)
	Khiki	Raw milk of ewe and goat, salt, lamb rennet	*L. plantarum, L. paracasei, L. casei, E. faecium, Enterococcus durans*	Parsaeimehr et al. (2019)
	Siahmazgi	Raw milk of ewe and goat, salt, lamb rennet	*L. plantarum, L. brevis, Lactobacillus parabuchneri, E. faecium, Pediococcus parvulus, E. durans*	Partovi et al. (2015) Partovi et al. (2019)
	Doogh	Yogurt, water, salt	*S. thermophilus, Lactobacillus bulgaricus*	Dana et al. (2011)
	Yogurt	Milk, starter culture	*S. thermophilus, L. bulgaricus*	Han et al. (2014)
	Kashk	Yogurt, salt, water	*S. thermophilus, L. bulgaricus*	Soltani and Güzeler (2013)
Cereals	*Sangak*	Whole wheat flour, water, salt, sourdough	*L. plantarum, L. delbrueckii, Leu. mesenteroides, S. cerevisiae*	Khoshakhlagh et al. (2014) Rahnama et al. (2015) Meybodi et al. (2016)
	Barbari	Flour, water, salt, sourdough, oil		
	Taftoon	Soft white wheat flours, water, salt, sourdough		
	Lavash	Flour, water, salt, sourdough, soda		

(Continued)

TABLE 6.1 (Continued)

SOURCE	PRODUCT NAME	INGREDIENTS	MICROORGANISM	REFERENCES
Dairy and cereals	*Horreh*	Kardeh (*Biarum carduchorum*), wheat flour, *doogh*, rice, onion, ghee	*L. fermentum, L. plantarum, L. brevis, Weissella cibaria, E. faecium, E. faecalis, Leuconostoc citreum, Leu. mesenteroides subsp. mesenteroides; Pediococcus pentosaceus*	Vasiee et al. (2018a)
	Kashk-e zard	Wheat flour, salt, yogurt, *doogh*, spices, garlic	*lactobacillus, Enterococcus, Pediococcus*	Mashak and Moradi (2019)
	Tarkhineh	Wheat grout, turnip, *doogh*, salt, sourdough	*Lactobacillus nagelii, Lactobacillus bifermentans, Leuconostoc cermoris, Lactobacillus fructosus, L. fermentum, Lactobacillus intestinalis, Lactobacillus agilis, Lactobacillus acidipiscis, L. plantarum, L. pentosus, L. brevis, Lactobacillus diolivorans, E. faecium, E. faecalis, Leu. citreum, Pediococcus acidilactici, P. pentosaceus*	Tabatabaei Yazdi et al. (2012) Vasiee et al. (2018b)
Fish	*Mahyaveh*	Sardines (*Sardinella* sp.) or anchovies (*Stolephorus* sp.), salt, mustard, cumin, coriander, fennel seeds, black pepper, thyme, water	*L. plantarum, Lactobacillus alimentaria, Micrococcus luteus, Staphylococcus lentus, Candida* sp., *Saccharomycopsis* sp., *Pichia* sp.,	Zarei et al. (2012) Taheri et al. (2014)

It is usually consumed as an appetizer along with food and is produced based on traditional recipes through spontaneous fermentation. For winter salad production, white cabbage is one of main components, along with raw garlic, cauliflower, marjoram, salt, carrot, and other spices which are fermented in the tomato juice under anaerobic conditions (Davati and Hesami 2021b). The ripening of winter salad is distinguished by its specific texture, taste, flavor, and color. White cabbage is a rich source of carbohydrates (4.18–5.51%), dietary fiber (1.9–2.9%), proteins (1.27–1.37%), vitamin C (0.03–0.04%), fat (0.06–0.20%), and minerals (0.3–0.7%) (Souci et al. 2000). The presence of salt in tomato salad inhibits the growth of food-borne pathogenic bacteria. The air evacuation stimulates the growth of LAB at the beginning of the fermentation process. The production of CO_2 by heterofermentative LAB and oxygen removal through cellular metabolism in vegetables tissues lead to more anaerobic conditions, resulting in more growth of LAB. The production of antimicrobial metabolites by LAB leads to inhibition or death of food-borne microorganisms (Harris 1998; Hutkins 2006).

6.2.1.1.1 The Preparation Method for Winter Salad

A diagram of the manufacturing process of winter salad is shown in Figure 6.1. In this protocol, the chopped vegetables are completely immersed in tomato juice containing salt to create anaerobic conditions and prevent undesirable color and flavor changes (Davati and Hesami 2021b).

6.2.1.1.2 Microbiology of Winter Salad

According to Hesami and Davati (2018), the biodiversity of fermented pickled vegetables in tomato juice using the next-generation sequencing (NGS) technique identified *Lactobacillus* (77.12%), *Pediococcus* (7.51%), *Leuconostoc* (5.61%), *Acinetobacter* (1.20%), *Enterobacter* (1.00%), *Erwinia* (0.35%), *Dickeya* (0.35%), and unclassified bacteria (2.79%). At the species level, *Lactobacillus acidifarinae* (2.32%), *Lactobacillus brevis* (currently *Levilactobacillus brevis*) (19.32%), *Lactobacillus japonicus* (15.71%), *Lactobacillus pentosus* (13.25%), *Lactobacillus senmaizukei* (10.26%), *Lactobacillus plantarum* (currently *Lactiplantibacillus plantarum subsp. plantarum*) (4.6%), *Leuconostoc mesenteroides* (2.65%), and unclassified bacteria (17.26%) were identified. Based on previous studies, *L. plantarum, L. pentosus, Leu. mesenteroides*, and *L. brevis* were identified as probiotic bacteria.

6.2.1.1.3 Microbial Community Changes of Winter Salad during Ripening Period

Davati and Hesami (2021b) isolated 120 strains from winter salad during ripening and studied the microbial community changes of this product during ripening period (Figure 6.2). They indicated that 44% *Lactobacillus* (*L. japonicus, L. alimentarius, L. senmaizukei, L. pentosus, L. plantarum, L. parabrevis, L. brevis*, and *L. acidifarinae*), 7% *Pediococcus* (*P. cellicola, P. pentosaceus, P. parvulus, P. stilesii* and *P. argentinicus*), 13% *Leuconostoc* (*Leu. palmae* and *Leu. mesenteroides*),

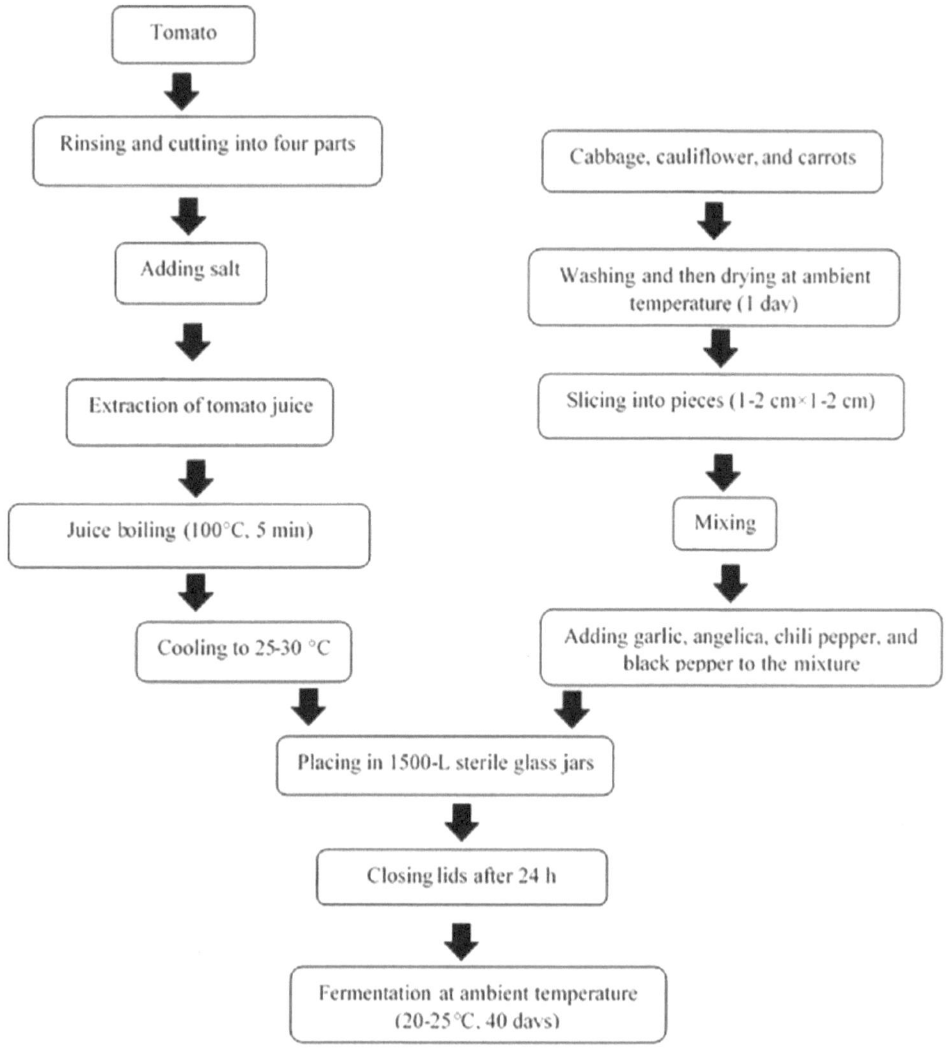

FIGURE 6.1 Winter salad production.

Source: Reprinted with permission from Davati and Hesami (2021b).

10% *Enterobacter soli*, 4% *Acinetobacter johnsonii*, and 22% unknown isolates were dominant at the beginning of the fermentation process. In the last stage of fermentation, 84% *Lactobacillus* (*L. japonicus, L. brevis, L. senmaizukei, L. pentosus, L. acidifarinae, L. alimentarius, L. parabrevis,* and *L. plantarum*), 4% *Leuconostoc* (*Leu. palmae* and *Leu. mesenteroides*), 4% *Pediococcus* (*P. argentinicus, P. cellicola, P. stilesii, P. parvulus,* and *P. pentosaceus*), 0.5% *E. soli*, 0.2% *A. johnsonii*, and 7.3% unknown isolates were dominant. The changes in the microbial community along with acid production during the ripening of winter salad are shown in Figure 6.2. Their study showed that the major bacterial species in the ripening of winter salad are *L. brevis, L. japonicus, L. pentosus, L. senmaizukei, L. plantarum,* and *Leu. mesenteroides.* During fermentation, the numbers of *L. japonicus* (23%) and *L. plantarum* (8.7%) increased until the 10th day and gradually declined until the 40th day. The number of *L. brevis* (23%), *L. pentosus* (15.5%), and *L. senmaizukei* (12%) increased until the 40th day. The numbers of *Leu. mesenteroides* (10%) increased until the 5th day, then declined sharply by the 10th day and subsequently saw a slight decline by the 40th day. A sharp reduction of *E. soli* was detected by the 5th day and then continued with a slight decline (0.5%) until the 40th day.

The phylogenetic relationships indicated high biodiversity in the microbial community of winter salad (Figure 6.3).

6.2.1.1.4 Physicochemical Properties of Winter Salad

In the early stage of fermentation, the pH decreased from 4.34 to 3.17 in the final stage of fermentation. During fermentation, the salt content (from 6% to 11.12(% and titratable acidity (from 0.69% to 1.99%) gradually increased. Other changes in the physicochemical properties of winter salad during fermentation are shown in Table 6.2 (Davati and Hesami 2021b).

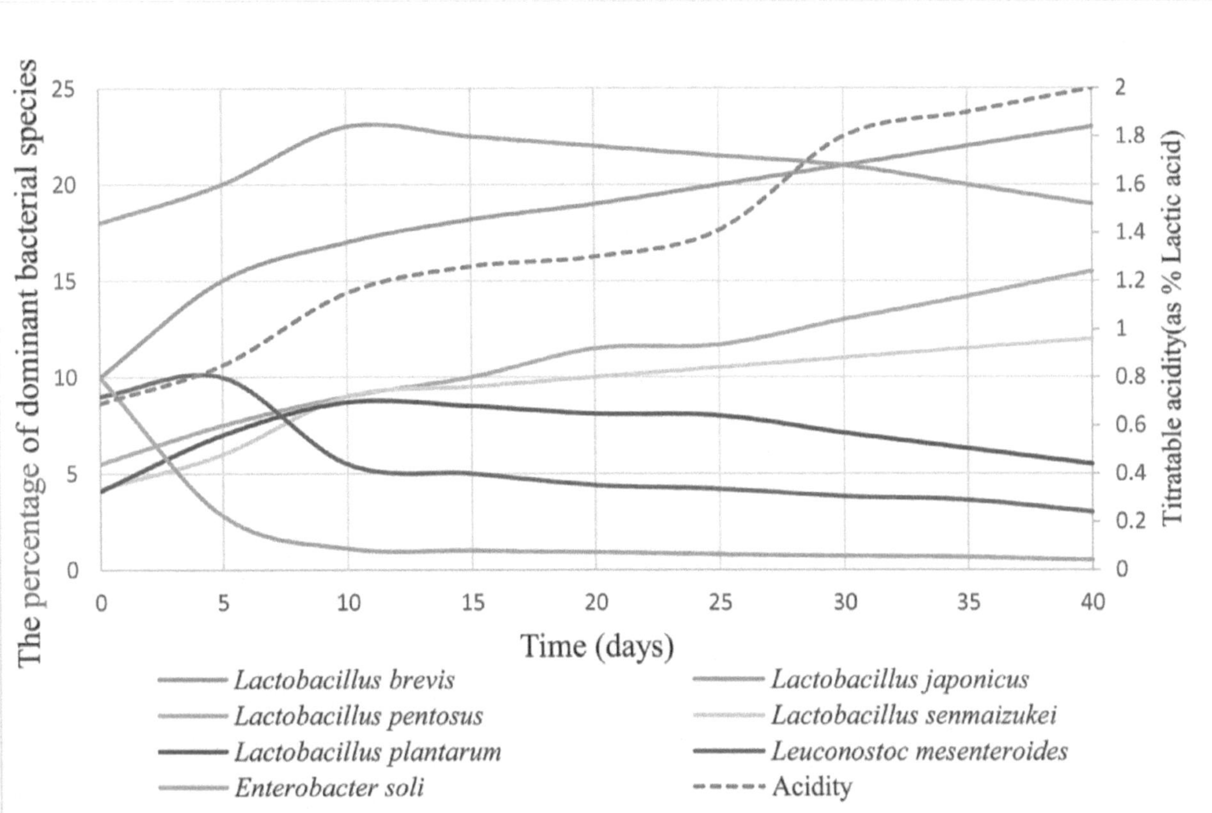

FIGURE 6.2 Changes in microbial dynamics and acidity during the ripening period of winter salad. *Lactobacillus brevis* (currently *Levilactobacillus brevis*); *Lactobacillus plantarum* (currently *Lactiplantibacillus plantarum*).

Source: Reprinted with permission from Davati and Hesami (2021b).

6.2.1.2 Verjuice

Verjuice, with a unique aroma and taste, is used as a food flavoring (Aminian et al. 2003). It is usually served as a salty and sour side dish along with vegetable salads, lunch, and snacks in the Middle East and the Mediterranean and is mainly produced based on homemade recipes. *Verjuice* appears to have therapeutic and nutritional properties such as serum lipid-lowering agents and antimicrobial and antioxidant activity properties (Karapinar and Sengun 2007; Setorki et al. 2010; Mousa-Al-Reza et al. 2011).

6.2.1.2.1 Preparation Method of Verjuice

In different countries, *verjuice* can be prepared from different kinds of grapes with or without fermentation. In Iran, *verjuice* is usually produced using spontaneous fermentation of juice from unripe *Asgari* grapes (green grapes). A preparation diagram of *verjuice* is shown in Figure 6.4.

6.2.1.2.2 Microbiology of Verjuice

According to Kaya et al. (2020), *S. cerevisiae* was identified as the most dominant microorganism in *verjuice* made of fermented unwashed unripe grapes.

6.2.1.2.3 Chemical Properties of Verjuice

The nutritional value of *verjuice* is almost the same as that of green grapes. According to the USDA Food Data Central, raw green grapes contain approximately 81% water, 18% carbohydrate, 0.16% fat, 1% protein, and B complex vitamins (thiamine, pantothenic acid, riboflavin, folate, and niacin), vitamin C, vitamin E, and vitamin K. Also, *verjuice* contains a variety of mineral elements, including iron, magnesium, phosphorous, sodium, manganese, potassium, and zinc.

6.2.1.3 Pickled Garlic

Garlic is the main flavor ingredient in some fermented vegetables, such as winter salad, pickled cucumber, and kimchi (Farnworth 2008). Moreover, garlic alone is used for fermentation, and *seven-year-old pickled garlic* is one of the most common fermented garlic products in Iran. This product is a nutritious seasoning known for its unique odor and sour taste. In addition to removing the strong taste and odor of raw garlic, its color changes from white to brown during seven-year fermentation (Pure et al. 2017).

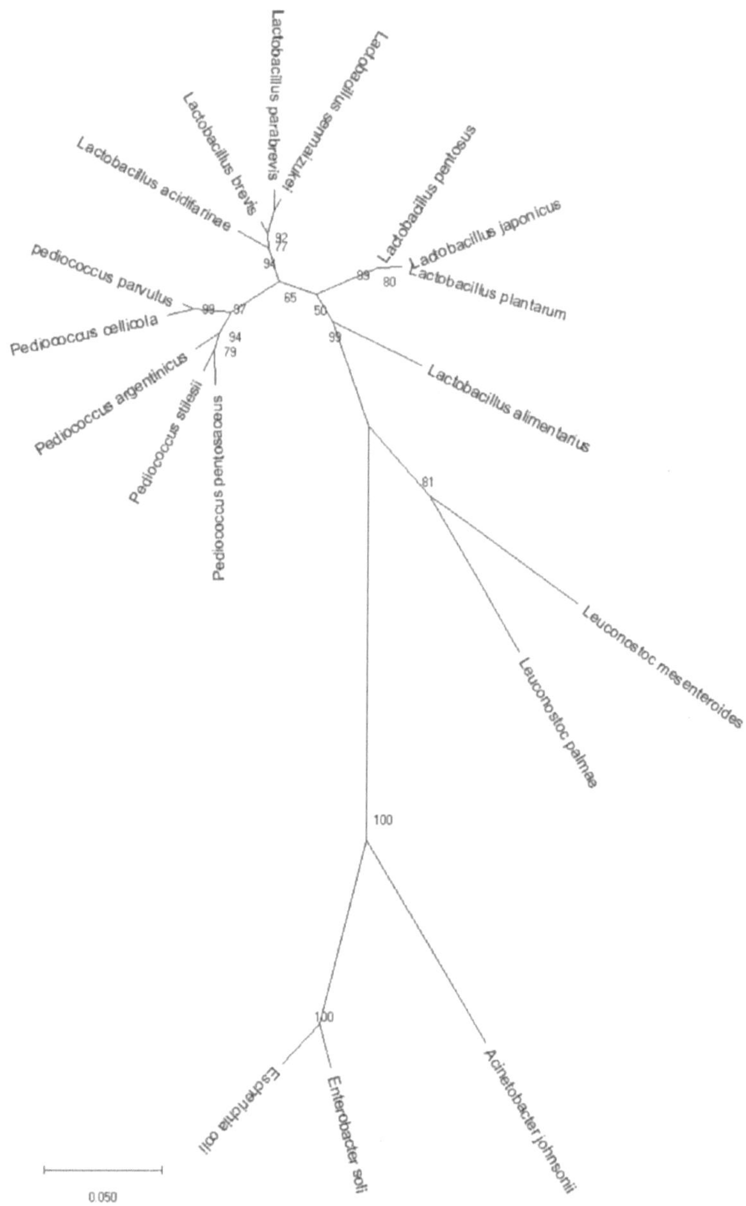

FIGURE 6.3 Phylogenetic tree based on 16S rRNA gene sequence analysis of microbial community from winter salad. *Lactobacillus brevis* (currently *Levilactobacillus brevis*); *Lactobacillus plantarum* (currently *Lactiplantibacillus plantarum*).

Source: Reprinted with permission from Davati and Hesami (2021b).

TABLE 6.2 Changes in the Physicochemical Properties of Winter Salad during Ripening Time

	RIPENING TIME (DAYS)								
CHARACTERISTICS	*1*	*5*	*10*	*15*	*20*	*25*	*30*	*35*	*40*
pH	4.34	4.07	3.83	3.62	3.52	3.32	3.22	3.20	3.17
% acidity	0.69	0.74	1.07	1.24	1.30	1.41	1.76	1.94	1.99
Salt (g/100g)	6	6.11	6.22	6.34	8.28	9.53	10.05	10.29	11.12
Protein (g/100g)	6.11	6.12	6.12	6.12	6.12	6.12	6.12	6.12	6.12
Ash (g/100g)	2.11	2.11	2.11	2.10	2.10	2.10	2.10	2.10	2.10
Fat (g/100g)	0.24	0.23	0.23	0.23	0.22	0.22	0.22	0.22	0.22

Source: Data from Davati and Hesami (2021b).

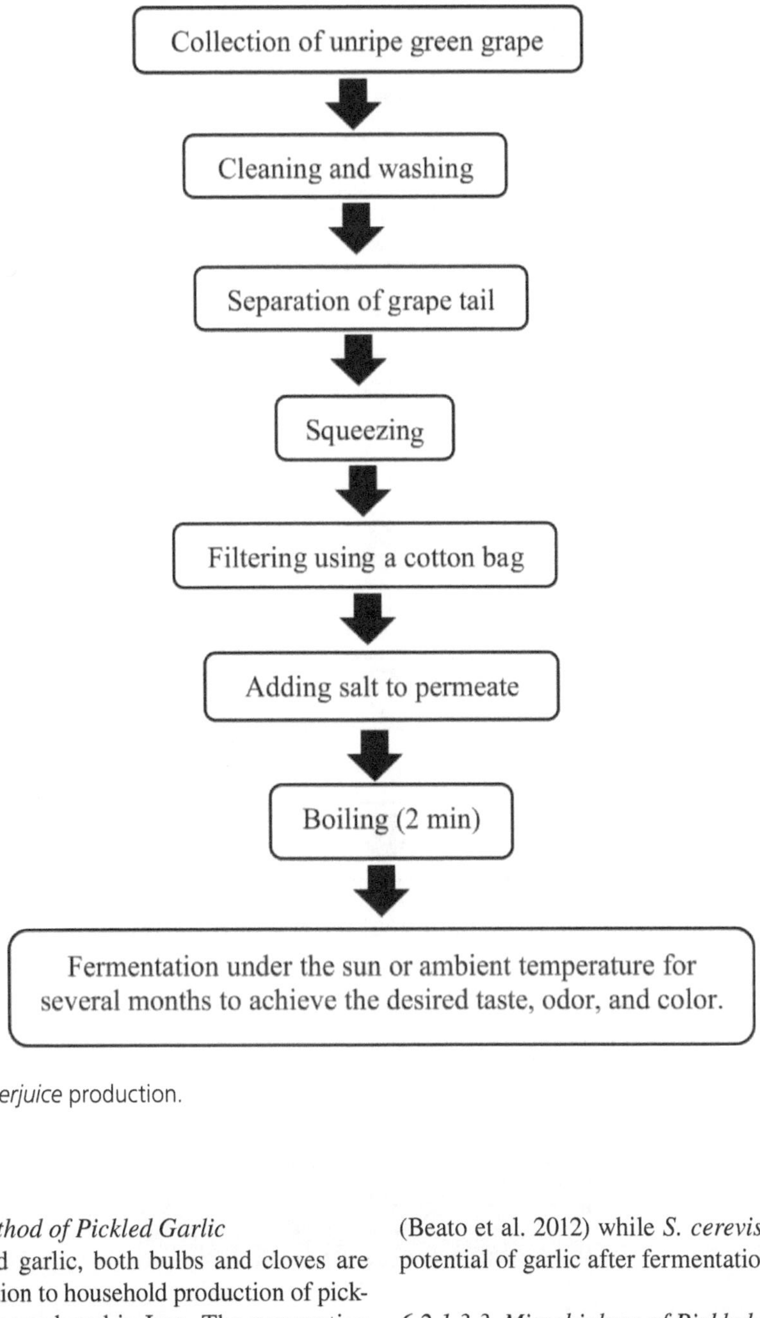

FIGURE 6.4 Diagram of *verjuice* production.

6.2.1.3.1 Preparation Method of Pickled Garlic

For seven-year-old pickled garlic, both bulbs and cloves are traditionally used. In addition to household production of pickled garlic, it is industrially produced in Iran. The preparation diagram for pickled garlic is shown in Figure 6.5.

6.2.1.3.2 Chemical Properties of Pickled Garlic

According to Casado et al. (2004), the chemical properties of pickled garlic were reported as follows: titratable acidity: 0.70–2.66%, water: 86.89%, protein: 3.35%, salt: 2.39–7.40%, and dietary fiber contents: 2.1%. In terms of wet weight, other components included ash: 2.65–8.40%, sugars: 2.21–4.22%, and fat: 0.21–0.35%. Moreover, there were high nutritional value compounds such as lysine, sulfur-containing amino acids, α-tocopherol: 0.36–2.53 mg/kg, thiamine: 0–0.055 mg/kg, ascorbic acid: 0–47.9 mg/100 g, and riboflavin: 0.013–0.032 mg/kg. According to previous studies, the amount of organosulfur (OS) compounds in garlic changes during fermentation. For example, *L. pentosus* reduced the OS content of garlic (Beato et al. 2012) while *S. cerevisiae* increased the reducing potential of garlic after fermentation (Jung et al. 2011).

6.2.1.3.3 Microbiology of Pickled Garlic

According to Jung et al. (2011), *S. cerevisiae* was identified as the dominant microorganism in fermented black garlic, which was preheated before fermentation. In addition to yeasts, LAB including *L. plantarum* (currently *Lactiplantibacillus plantarum subsp. plantarum*), *P. ethanolidurans,* and *L. brevis* (currently *Levilactobacillus brevis*) have been identified in pickled garlic according to Nikfarjam et al. (2021).

6.2.2 Dairy Products

There is a wide range of dairy products, including various cheeses, *kashk* (curd), yogurt, and *doogh* in Iran. Along with increasing the shelf life, the fermentation process enhances the taste, odor, and digestibility of products.

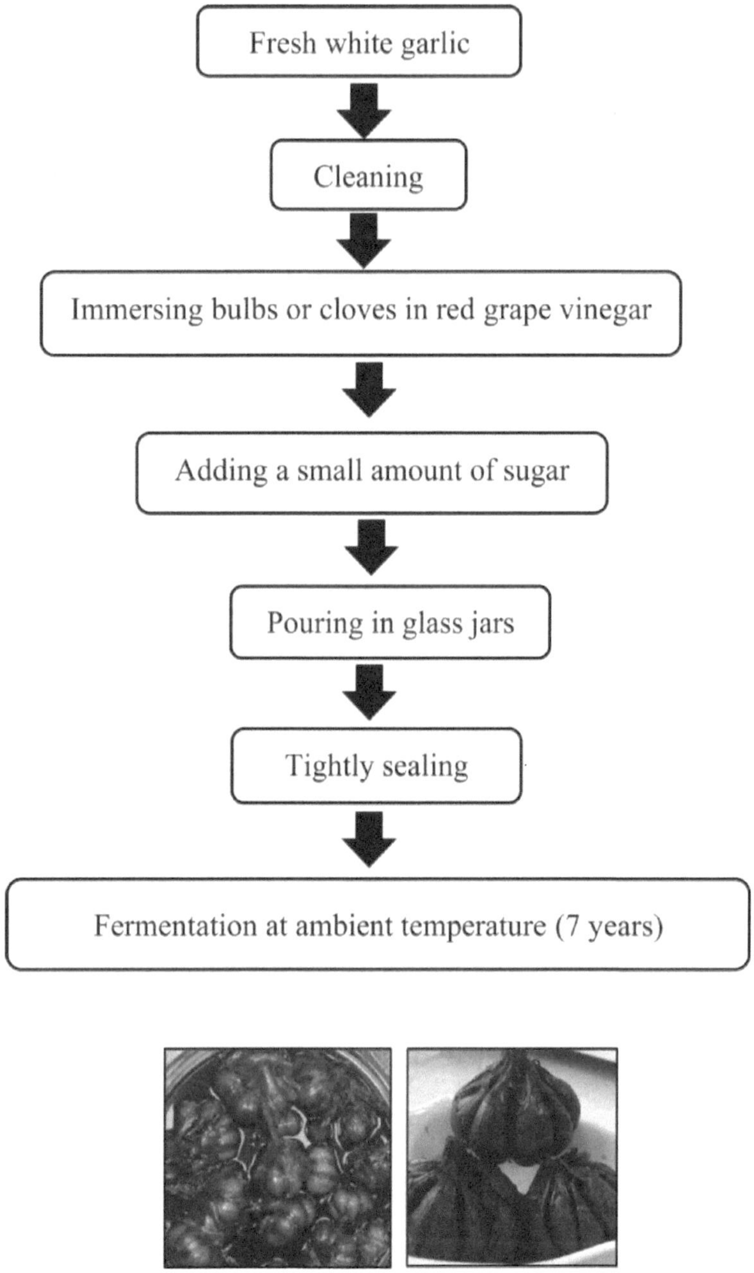

FIGURE 6.5 Diagram of pickled garlic production.

Source: Reprinted with permission from Pure and Pure (2016).

6.2.2.1 Kashk

Kashk derives from the word *khushk* (meaning drying), which implies that this product is produced through the drying process (Noori et al. 2013). *Kashk* is a rich, creamy, slightly salty, nutritious, and flavorful product and as a concentrated whitish-yellow product is prepared in dried or semi-solid liquid form in the different regions of Iran. *Kashk* is manufactured in three forms, dried *kashk*, traditional liquid *kashk*, and industrial liquid *kashk*. Dried *kashk*, a rich source of calcium and protein with high nutritional value, is produced by removing water from local yogurts under sun-drying (Ebrahimi et al. 2011; Dehkordi et al. 2014). The dried form of *kashk* must be soaked and softened to be rehydrated before being used. This product has a long shelf life without spoilage or reduction in nutritional value (Soltani and Güzeler 2013; Dehkordi et al. 2014).

6.2.2.1.1 Preparation Method of Traditional and Industrial Dried Kashks

For the preparation of dried *kashk*, there are two methods available: traditional and industrial processes. In the traditional method (Figure 6.6A), water is added to high-fat yogurt and transferred into a goat or sheepskin as container that is

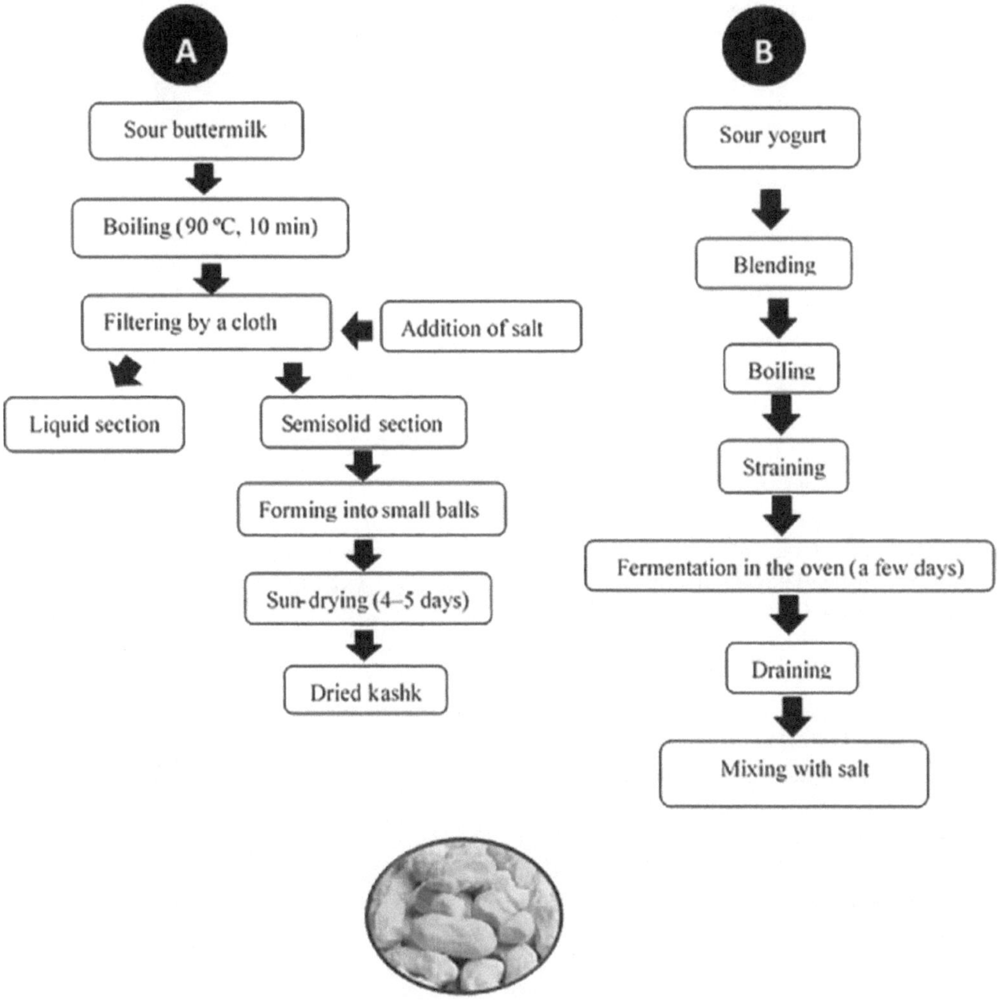

FIGURE 6.6 Diagram of traditional (A) and industrial (B) *dried* kashk manufacturing.

Source: Adapted from Ezzatpanah (2020).

hanging from a stable place. Then this container is pushed back and forth until the buttermilk separates from the butter. The buttermilk is boiled afterward and dehydrated to obtain *kashk*, which is gradually dried under sunlight (Walker 2000). In the industrial method (Figure 6.6B), sour yogurt is blended, then boiled and filtered. The product is fermented in the oven for a few days; then the moisture is drained and mixed with salt to make *kashk*. The remaining drained liquid can be cooked further, thickened with wheat starch, and used for the preparation of *ghara ghoroot* (*black kashk* or *black concentrated whey*) (Walker 2000).

6.2.2.1.2 Preparation Method of Traditional and Industrial Liquid Kashks

Traditional liquid *kashk* is prepared from dried *kashk*. First, dried *kashk* is washed with water several times. Then *kashk* is minced and mixed with water, 0.5–1% of salt, and spices (dill or mint). The volume of water used is up to five times the *kashk* amount. The product is then filtered and transferred into the separator. After standardization of fat in the product, the separated solid section is homogenized and heated (85°C, 15 s). After packaging in glass jars, product is stored at 4 ± 1°C (Meybodi et al. 2016). For industrial liquid *kashk* preparation, the standardized milk with fat ratio of about 0.4–0.6% is pasteurized (90–95°C, 5 min) and then cooled down to 43 ± 1°C. At the next step, 1–2% of starter culture containing *S. thermophilus* and *L. bulgaricus* is added to milk and incubated at 37°C for 6–7 h. The manufactured yogurt is homogeneous at 50–55°C, and the dry matter percentage is increased to 18–19% using a separator. The concentrated yogurt is blended with 0.8–1% salt and 1–2% whey powder, subsequently heated for 10 min at 88°C, and cooled down. Then, the final product is placed in glass jars and stored at 4 ± 1°C (Soltani and Güzeler 2013).

6.2.2.1.3 Chemical Properties of Dried Kashk

Dried *kashk* has a long shelf life of 4 years due to the processes of concentration, salting, heating, and fermentation during its production. The chemical composition of dried *kashk* with pH = 4.27 is presented in Table 6.3.

TABLE 6.3 Chemical Properties of Traditional Dried *Kashk*

CHARACTERISTICS	AMOUNT (%W/W)
Moisture	14.21
Ash	12.25
Fat	9.17
NaCl	9.77
Protein	51.74
Acidity (lactic acid)	1.4

Source: Data from Soltani and Güzeler (2013).

6.2.2.2 *Yogurt*

Yogurt as a fermented product containing LAB is popular in Iran and is flavored by adding dried vegetable powders such as garlic, pennyroyal, tarragon, mint, cucumber, grilled eggplant, and shallot (Meybodi et al. 2016). The manufacturing diagram for set yogurt is shown in Figure 6.7.

6.2.2.3 Doogh

According to the codex standard, *doogh* is defined as a fermented drinking yogurt that is traditionally produced in Iran (Esfandiari et al. 2013). It is a delicious beverage that is made of yogurt, water, and salt and seasoned with dried vegetables based on different tastes that are optionally added. In addition to its remarkable sensory properties, *doogh* is thought to have health benefits for humans (Soltani and Güzeler 2013).

6.2.2.3.1 Preparation Method of Doogh

Traditionally, *doogh* is prepared by dilution of yogurt and blending in a bag made of goat or sheep skin that is called a *mashk* (Soltani and Güzeler 2013). Based on a local recipe, it is made of a mixture containing milk (1 liter); water (2 liters); yogurt (5 tablespoons); salt (1 tablespoon); sugar (1/4 tablespoon); and dried aromatic vegetables like mint, dill, oregano, and thyme followed by incubation at 39–45°C for 7 days. Carbon dioxide gas is naturally produced in *doogh* due to microbial fermentation. However, recently to increase the bubble tissue of *doogh*, it is carbonated (Kiani et al. 2008). Regarding the Iranian National Standard Organization (ISIRI) statement, *dooghs* are categorized into four groups in dairy product factories (Soltani and Güzeler 2013):

1. Noncarbonated and nonheated
2. Noncarbonated and heated
3. Carbonated and nonheated
4. Carbonated and heated

The heating process is done to prevent the growth of microorganisms and to increase the product shelf life. The carbonating and heating processes are usually performed after fermentation. The industrial production method of *doogh* is shown in Figure 6.8.

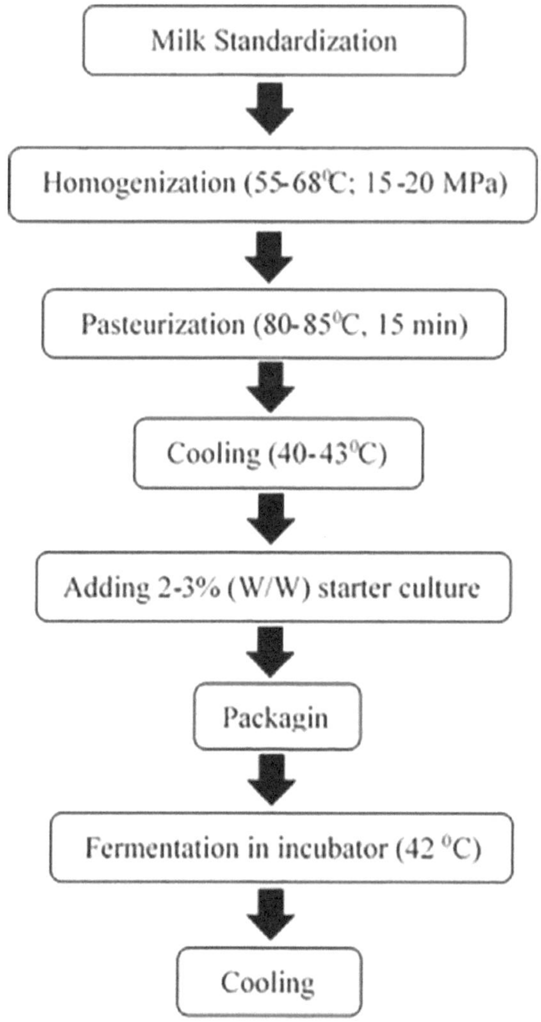

FIGURE 6.7 Diagram of set yogurt manufacturing.
Source: Adapted from Meybodi et al. (2016).

In the process of yogurt dilution to make acidic beverages like *doogh*, the casein gel particles are separated; as a result, it causes excessive loss of stability and more sediment in the particles (Lucey et al. 1999). Phase separation in *doogh* occurs mainly due to the aggregation of casein (Foroughinaia et al. 2007). For *doogh* production, the addition of salt plays an important role in increasing phase separation (Köksoy and Kılıç 2003). Therefore, the consumer must shake the product before use. To avoid phase separation and increase the stabilization of texture, researchers have used several gums such as tragacanth, pectin, gellan, and guar in *dooghs* (Foroughinaia et al. 2007; Azarikia and Abbasi 2010; Kiani et al. 2010).

6.2.2.3.2 Microbiology of Local Yogurt and Doogh

Industrial yogurt and *doogh* are mainly produced by two bacterial species, *S. thermophilus* and *L. bulgaricus*. However, research has shown that the microbial flora of local yogurt and *doogh* in Iran are more diverse due to non-compliance with hygienic production (poor hygiene practice). Even *L.*

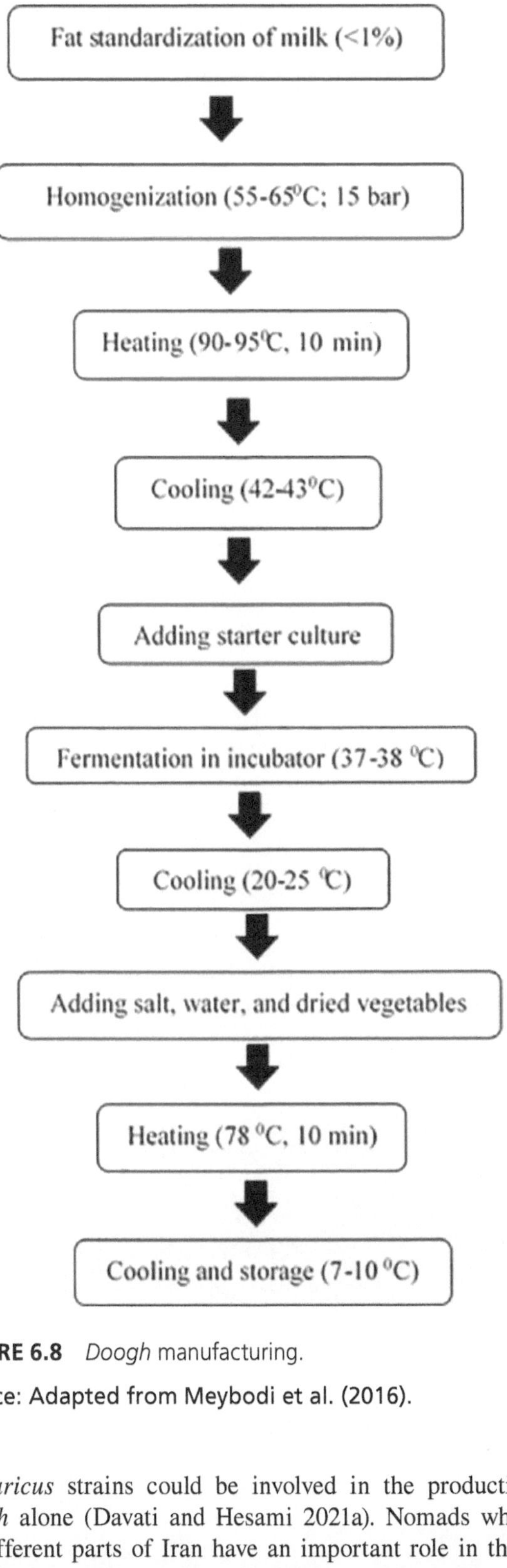

FIGURE 6.8 *Doogh* manufacturing.

Source: Adapted from Meybodi et al. (2016).

bulgaricus strains could be involved in the production of *doogh* alone (Davati and Hesami 2021a). Nomads who live in different parts of Iran have an important role in the production of traditional fermented dairy products. Davati and Hesami (2021a) studied the microbial changes during fermentation of local *doogh* and yogurt prepared from ewe milk by nomads using 16S rRNA-based NGS analysis. They showed that the most dominant microorganisms are from the *Enterobacteriaceae* family (49.16%) in ewe milk (Figure 6.9). The phylogenetic tree representing the bacterial community of ewe milk showed that there is a great biodiversity of bacteria in ewe milk supplied by nomads (Figure 6.10). However, non-LAB were overtaken later by *Lactobacillus delbrueckii* (99.9%) in local yogurt and *doogh*. In their study, this result is surprising because *Streptococcus* was in very low abundance for *doogh* and yogurt. A reason could be the rapid autolysis of *Streptococcus* spp. in yogurt and *doogh*. The cell-free supernatants (CFSs) of these products (*doogh* and yogurt) showed antimicrobial properties against food-borne pathogens.

In their study, the heat map of the 25 most different operational taxonomic units (OTUs) showed that the microbial composition of yogurt was extremely similr to that of *doogh* (Figure 6.11). According to Sayevand et al. (2018), the dominant bacteria identified in the traditional yogurts and *dooghs* from Esfahan, Hamadan, Chaharmahal, and Bakhtiari provinces in Iran were *L. acidophilus, Lactobacillus helveticus, Lactobacillus kefiranofaciens*, and *Lactobacillus amylovorus*. Therefore, the microbial composition structure of each local fermented food is dependent on geographical location and product preparation method.

6.2.2.4 Cheeses

No one knows how the first cheese was made. According to an ancient story, an Arab merchant put his milk in a bag made of a sheep's stomach while traveling on a day trip in the desert. The rennet in the sheep's stomach, along with the heat of the sun, turned the milk into cheese and whey (Classen et al. 1994). The different sensory properties and variety of traditional cheeses made from raw milk depend on their microbial flora. Commonly, commercial LAB strains are not used to produce traditional dairy products, and the organoleptic properties of these products are mainly dependent on the natural LAB strains available in local milk. Due to pH reduction by LAB, whey extraction from the curd is increased, causing the microbial spoilage to decrease (Partovi et al. 2015). The flavor of cheese, which is very important to the consumer, depends on the balance between the various non-volatile and volatile components released during the aging period (Fox and Wallace 1997). Flavor formation is because of several metabolic pathways such as casein proteolysis, lipolysis, and glycolysis during cheese aging. The free amino acids released by casein degradation contribute to the flavor formation of the cheese. In addition to amino acids, esters, ketones, free fatty acids, and lactones, which are produced by lipid oxidation reactions and lipolysis, participate in the formation of flavors (Siek et al. 1971). There are types of native Iranian cheeses based on the region of production, type of milk, processing method, and additives, which include *lighvan, siahmazgi, koozeh, khiki*, and *poosti* cheeses. These are well-known Iranian cheeses that have high fat content, and their specific organoleptic properties make them distinctive.

6.2.2.4.1 Poosti *and* Khiki *Cheeses*

Poosti cheese has traditionally been produced from the raw milk of ewes and goats alone or mixed by the Qashqai and Bakhtiari nomads in southwestern Iran and in the mountains

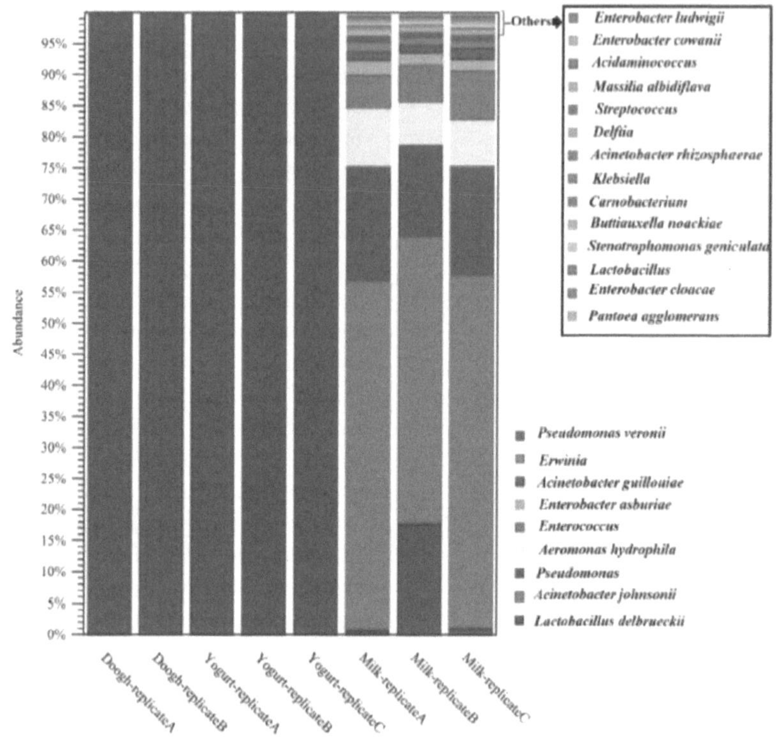

FIGURE 6.9 The dominant microbial community for milk, yogurt, and *doogh*.

Source: Reprinted with permission from Davati and Hesami (2021a).

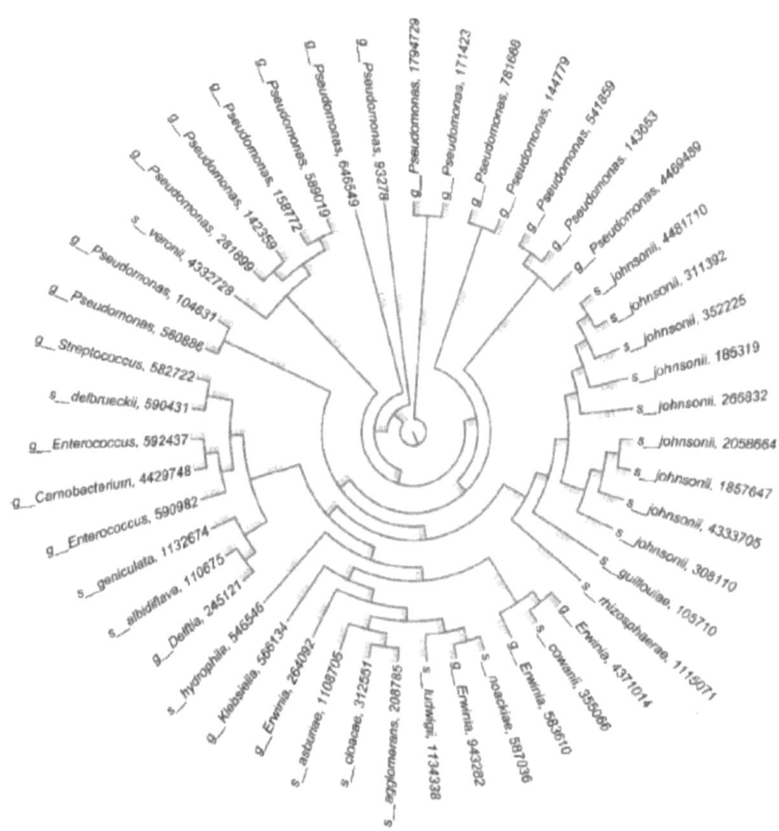

FIGURE 6.10 The phylogenetic tree of 16S rRNA gene sequences from the microbial community of ewe milk.

Source: Reprinted with permission from Davati and Hesami (2021a).

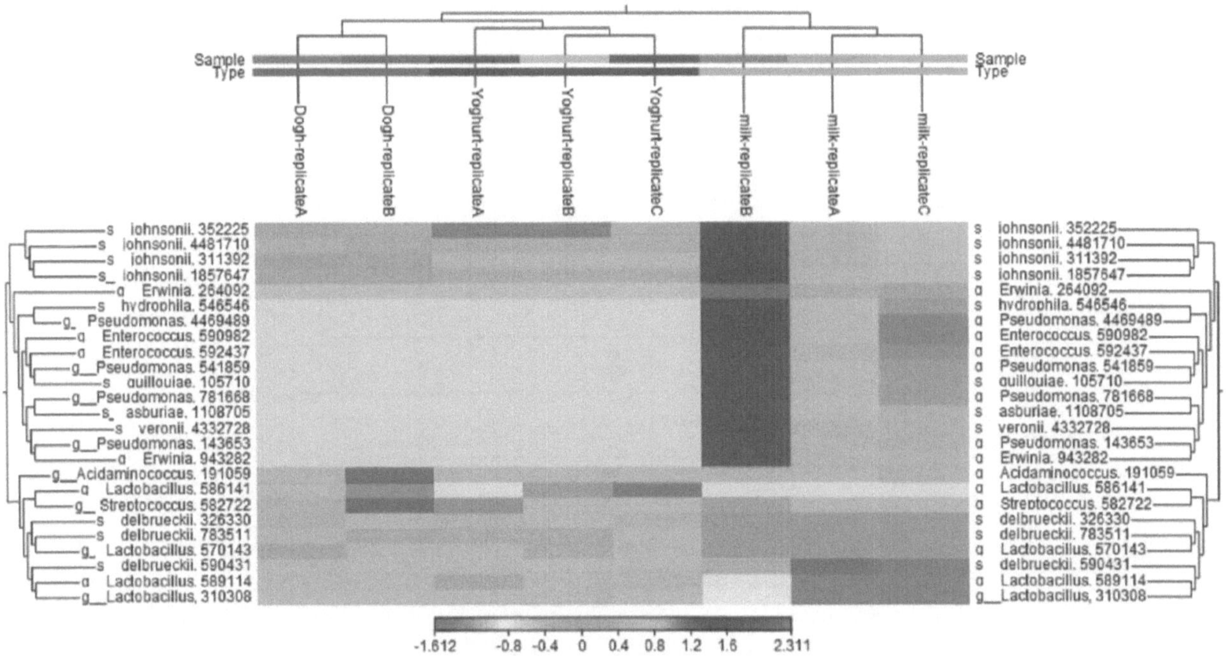

FIGURE 6.11 Heat map displaying different OTU distribution patterns in yogurt, milk, and *doogh*.

Source: Reprinted with permission from Davati and Hesami (2021a).

of the Sangesar area in the Semnan province for a long time (Hemmatian et al. 2015). *Khiki* cheese is another traditional cheese produced by the residents of the Semnan and Kerman provinces (Parsaeimehr et al. 2019).

6.2.2.4.1.1 Preparation Method of Poosti *and* Khiki *Cheeses* *Mashk* is made of goat or sheep skin, which is used as a container for the preparation and preservation of both of *khiki* and *poosti* cheeses. The production methods of *khiki* and *poosti* cheeses are similar. *Khiki* cheese is produced using ruminant milk like that of goats and sheep without a heating process and starter cultures, and the ripening stages of *khiki* cheese in *mashk* last 6 months (Parsaeimehr et al. 2019). For skin container preparation, the intact and healthy skin of a goat or sheep without rupture is selected, and its endoderm is cleaned by scraping with a sharp knife. Then, the cleaned skin is washed and dried to prevent the growth of microorganisms under the sunlight. To whiten, yogurt and salt are poured over the dried skin (Hemmatian et al. 2015). The method of *poosti* cheese production is shown in Figure 6.12.

6.2.2.4.1.2 Microbiology of Poosti *Cheese* Hemmatian et al. (2015) reported that during the ripening of *poosti* cheese (60 days), the total count of bacteria, fungi, and coliform decreased; however, the LAB population increased (Table 6.4). According to Abdali et al. (2019), *Lactobacillus paracasei* (*Lacticaseibacillus paracasei*), *L. plantarum* (currently *Lactiplantibacillus plantarum subsp. plantarum*), *L. acidophilus*, *L. buchneri*, and *L. brevis* (currently *Levilactobacillus brevis*) are responsible for the ripening of *poosti* cheese. Moreover, they revealed that *poosti* cheese contains probiotic bacteria with antioxidative properties.

6.2.2.4.1.3 Microbiology of Khiki *Cheese* According to Parsaeimehr et al. (2019), the dominant *Lactobacillus* species, including *L. paracasei* (*Lacticaseibacillus paracasei*) (32.7%), *L. plantarum* (currently *Lactiplantibacillus plantarum subsp. plantarum*) (53.6%), and *Lactobacillus casei* (*Lacticaseibacillus casei*) (13.7%), are responsible for the ripening of *khiki* cheese. In addition, the most dominant *Enterococcus* species were *E. faecium* and *E. durans*.

6.2.2.4.1.4 Physicochemical and Micro-Structural Properties of Poosti *Cheese* According to Hemmatian et al. (2015), the moisture content and fat percentage of *poosti* cheese decreased, and acidity, salt, and ratio of total nitrogen/dry matter (TN/DM) increased during 60 days of ripening (Table 6.5). The microstructure of this cheese was changed due to molecular motion (salt and water) in texture and microbial community activity during the aging period. SEM images showed that the casein network density, number, and depth of pores, as well as roughness, increased during the ripening period.

6.2.2.4.2 Siahmazgi *Cheese*

Siahmazgi cheese is traditionally prepared in *siahmazgi*, a village in the Gilan province in the north of Iran. This cheese is manufactured using raw fresh ewe's milk or a mixture of goat's milk (33%) and ewe's milk (67%) obtained from local native herds (Farahani et al. 2014). The good quality, particular taste, and strong texture of this cheese have made it very popular among consumers in the region.

6.2.2.4.2.1 Preparation Method of Siahmazgi *Cheese* To produce *siahmazgi* cheese, milk clarifying is usually done through a cotton bag fastened around a copper bowl, and the

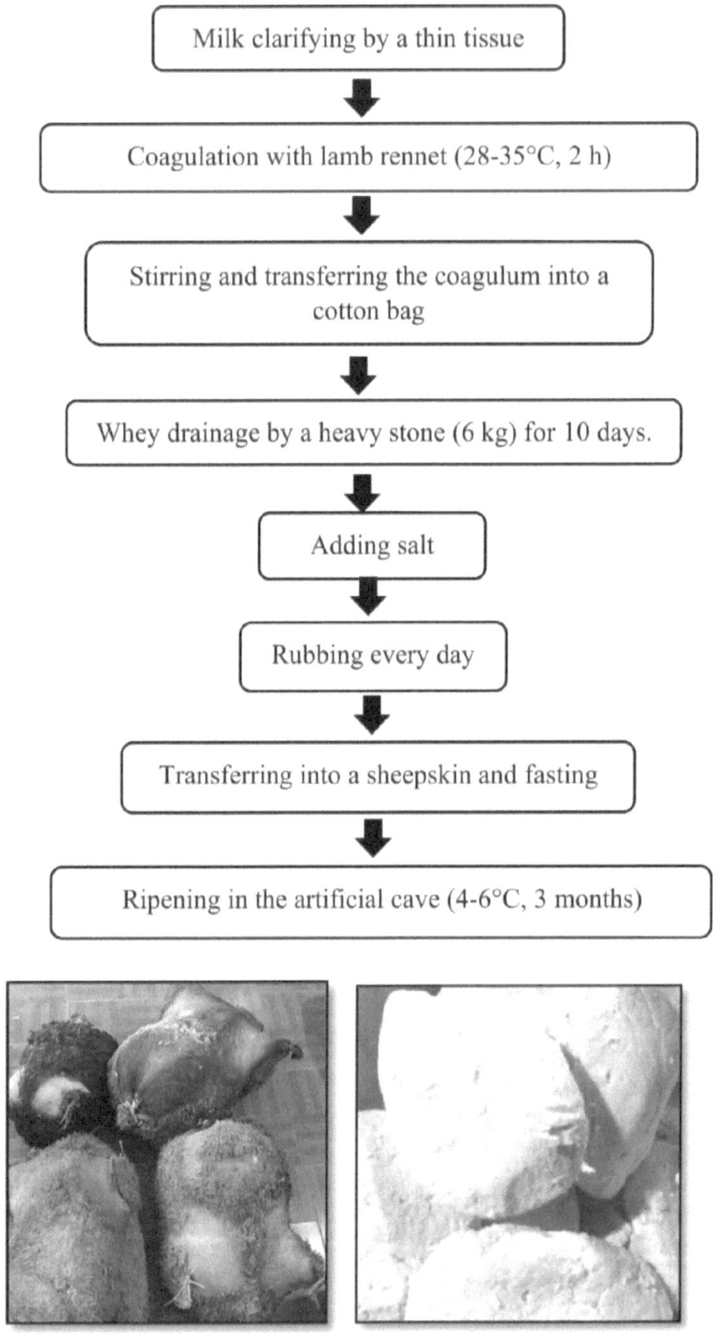

FIGURE 6.12 *Poosti* cheese making process.

TABLE 6.4 Microbial Changes of *Poosti* Cheese during the Aging Period

	TIME (DAYS)			
	CURD	*1*	*30*	*90*
	+	++++	+++	++
Fungi	+++	++	–	–
E. coli	+	+	–	–
Coliform	+	+	–	–
LAB	+	++	+	+++

Source: Data from Hemmatian et al. (2015).

TABLE 6.5 Chemical Properties of *Poosti* Cheese for the Ripening Period

	RIPENING TIME (DAYS)		
CHARACTERISTICS	*1*	*30*	*60*
Acidity (%W/W)	0.87	1,48	1.54
Salt (%W/W)	3.52	6.54	7.76
Fat (%W/W)	24.76	27.66	24.23
WSN/TN [1] (%W/W)	35.89	40.66	39
TN/DM [2] (%W/W)	10.27	27.19	29
Moisture (%W/W)	39.17	27.53	20.41

Source: Data from Hemmatian et al. (2015).

Note: 1: Water-soluble nitrogen to total nitrogen; 2: total nitrogen to dry matter

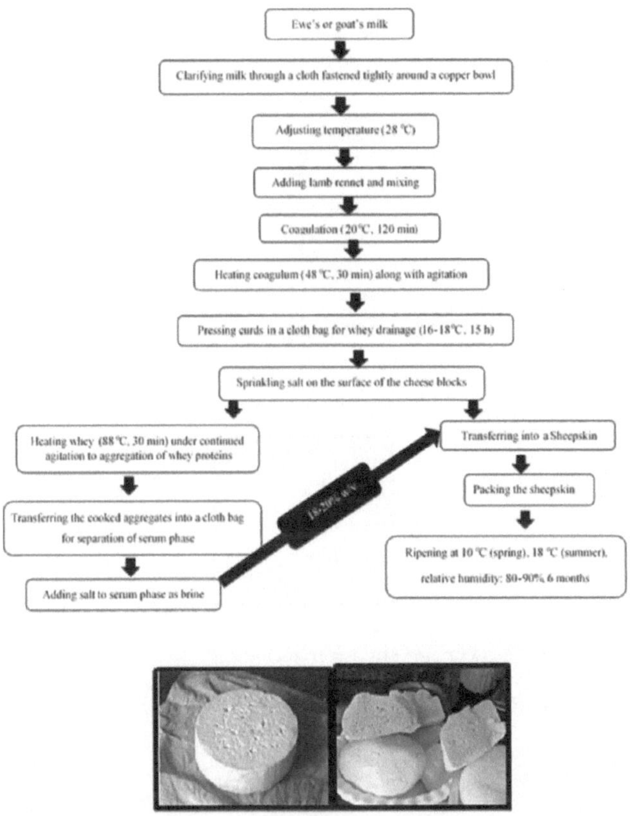

FIGURE 6.13 *Siahmazgi* cheese making process

Source: Adapted from Farahani et al. (2014).

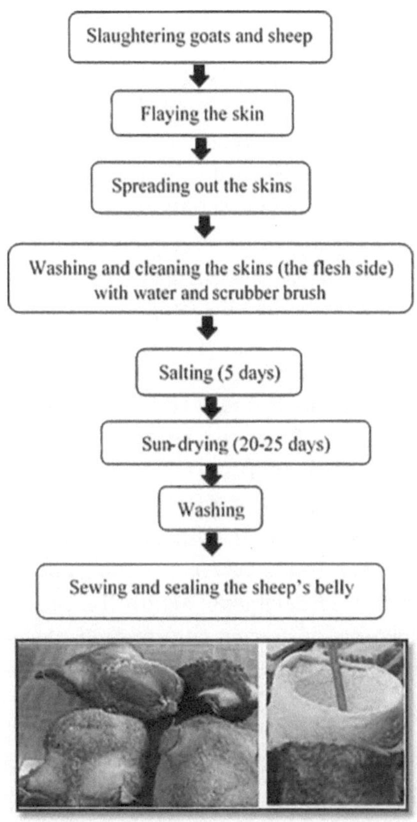

FIGURE 6.14 Sheepskin container preparation.

Source: Adapted from Farahani et al. (2014).

stomach juice of ruminant mammals is used as a coagulant agent. The method of *siahmazgi* cheese production is shown in Figure 6.13. For stomach juice extraction, a suckling lamb stomach is dried for 3–4 days. In the next step, it is placed in a pot containing hot water (90–100°C) and gently pressed by hand until its juice is extracted (Farahani et al. 2014). To prepare the sheepskins for *siahmazgi* cheese, like *mashk*, the intact and clean sheepskins are washed, then salted and dried under direct sunlight for 20–25 days. The sheepskins are rewashed and scrubbed. The cheese is mixed with salt, placed into the sheepskin container, and left to mature (Farahani et al. 2014). A flowchart of the preparation method of the sheepskin is presented in Figure 6.14.

6.2.2.4.2.2 Microbiology of Siahmazgi *Cheese* Partovi et al. (2015) studied the microbial community of *siahmazgi* cheese obtained from raw milk of ewe and goat. Using biochemical tests and 16S rRNA region sequencing, the isolated LAB responsible for cheese ripening were identified as *L. brevis* (currently *Levilactobacillus brevis*), *L. parabuchneri*, *L.*

plantarum (currently *Lactiplantibacillus plantarum subsp. plantarum*), *E. faecium, P. parvulus,* and *E. duran.* Other researchers demonstrated that *E. coli, Staphylococcus,* coliforms, molds, and yeast were present in *siahmazgi* cheese made from raw milk; however, these contaminants were not observed in samples made from pasteurized milk (Moradi et al. 2019). Partovi et al. (2019) reported that all *L. plantarum* strains had significant antibacterial activity against *Listeria monocytogenes* and *Salmonella typhimurium.*

6.2.2.4.2.3 Physicochemical and Sensory Properties of Siahmazgi *Cheese* The changes in physicochemical properties during the ripening of siahmazgi cheese are shown in Table 6.6 (Farahani et al. 2014). They classified this cheese in the group of semi-hard and high-fat cheeses because the elasticity is higher than the stickiness. In another study, Moradi et al. (2019) examined the sensory and physicochemical properties of *siahmazgi* cheese made of pasteurized and unpasteurized milk on days 0, 30, and 60. Based on their results, dry matter, fat content, and pH of both cheeses decreased for 60 days, while the protein content and salt percentage of both cheeses significantly increased. The texture of *siahmazgi* cheese was significantly changed on the 60th day; the other sensory properties, including color, aroma, and taste, remained unchanged.

6.2.2.4.3 Koozeh *Cheese*

Koozeh is one of the most famous traditional Iranian cheeses. This cheese is made of raw sheep's milk or a mixture of ewe's and goat's milk in the northwestern and western provinces of Iran. For the preparation of *koozeh* cheese, old cheese-making protocols are used without adding a starter. The specific taste and aroma of *koozeh* cheese have increased consumer interest (Edalatian et al. 2012b). The acid production and ripening of starter culture–free cheeses derived from raw milk rely on the activity of their indigenous LAB (Wouters et al. 2002).

6.2.2.4.3.1 Preparation Method of Koozeh *Cheese* *Koozeh* cheese is produced in different regions of Iran using similar methods. Mashak and Rooshani (2019) have shown that the *koozeh* cheese of Kurdistan province has high nutritional values and is rich in calcium, phosphorus, and protein. The main steps of *koozeh* cheese preparation are presented in Figure 6.15.

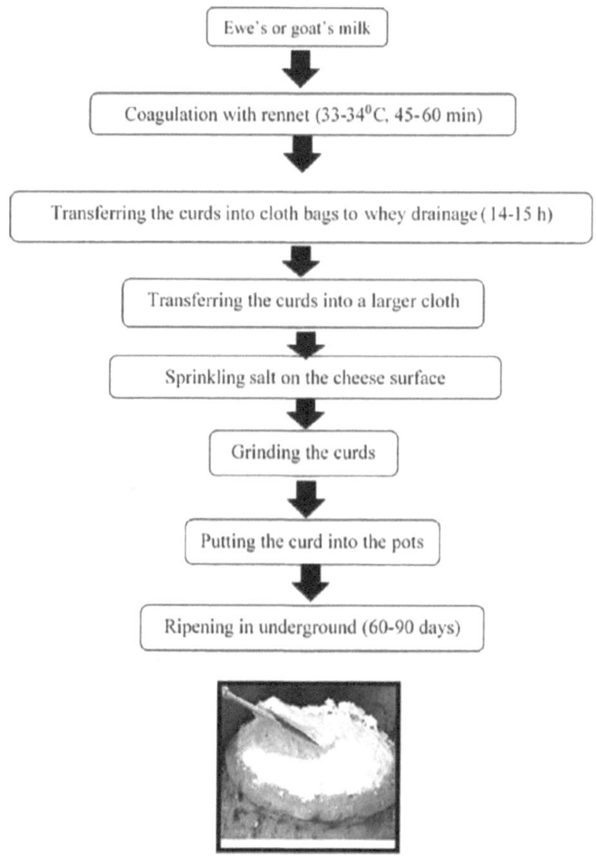

FIGURE 6.15 *Koozeh* cheese manufacturing.
Source: Adapted from Edalatian et al. (2012b).

TABLE 6.6 Physicochemical Properties of *Siahmazgi* Cheese during the Ripening Period

	RIPENING TIME (DAYS)						
CHARACTERISTICS	3	30	60	90	120	150	180
Dry matter (%w/w)	41.53	50.10	59.81	61.08	62.76	62.65	59.95
Fat (%w/w)	18.12	22.62	29.12	31.75	30.12	31.87	31
Total nitrogen	3.11	3.28	3.60	3.33	3.82	3.71	3.40
Protein (%w/w)	19.83	20.94	22.96	21.30	24.37	23.73	21.71
Ash (%w/w)	2.40	5.74	7.50	8.03	8.26	7.04	7.24
Salt (%w/w)	0.51	4.01	5.44	5.71	6.49	5.46	5.65
WSN	0.11	0.14	0.14	0.14	0.17	0.18	0.17
pH	7.25	5.11	5.07	4.78	4.57	4.46	4.46
Titrable acidity (Dornic degree)	77.50	80.25	92.00	122.50	125.00	134.50	135.25

Source: Data from Farahani et al. (2014).

6.2.2.4.3.2 Microbiology of Koozeh *Cheese* Edalatian et al. (2012b) studied the microbial diversity of *koozeh* cheese using DGGE[1] analysis. The results showed that the dominant bacteria in *koozeh* cheese belonged to *L. curvatus, L. garvieae, Leu. mesenteroides, S. parauberis, L. lactis, L. raffinolactis, D. hansenii, S. uberis, C. diazotrophica, C. maltomaricum, V. tapetis, S. thermophilus,* and *Warcupia* spp. Several studies have shown that *koozeh* cheese, as a functional nutritious food, has probiotic properties. According to Mehdizadeh et al. (2019), the *koozeh* cheese matrix can be applied as a suitable carrier of probiotic bacteria such as *Bifidobacterium bifidum.* Hassanzadazar et al. (2012) isolated 28 *Lactobacillus* strains from *koozeh* cheese and investigated their probiotic properties. In their study, *L. casei* (*Lacticaseibacillus casei*), as a native probiotic species, could tolerate low pH and bile salts and showed antagonistic activity against *Lis. monocytogenes.* In another study, Tavakoli et al. (2017) confirmed the probiotic abilities of *L. casei, L. plantarum* (currently *Lactiplantibacillus plantarum subsp. plantarum*), and *Lactobacillus fermentum* (currently *Limosilactobacillus fermentum*) isolated from *koozeh* cheese.

6.2.2.4.4 Lighvan *Cheese*

Liqhvan cheese is one of the most famous Iranian cheeses and is traditionally produced mainly in Lighvan village located in East Azerbaijan, Iran. It is one of the unique brine cheeses in terms of its sour taste and its semi-hard texture. The ripening process is important to develop the flavor and texture characteristics of *lighvan* cheese (Kafili et al. 2013; Mirzaei et al. 2008). Despite the rapid growth of industrial cheese production in Iran, *lighvan* cheese is still prevalent in different regions of Iran and is preferred (Abdi et al. 2006).

6.2.2.4.4.1 Preparation Method of Lighvan *Cheese* *Lighvan* cheese is prepared by using raw ewe's milk or a combination of ewe's and goat's milk. It is ripened for 3 to 4 months at 10 ± 2°C in 10 to 12% salt brine. *Lighvan* is a starter culture–free cheese, and as a result, the composition of indigenous microflora plays a key role during ripening of this cheese (Mirzaei 2011). The tools used to make *lighvan* cheese are handmade and used to make traditional cheese. While pasteurized milk is recommended for making cheese, many traditional cheesemakers believe that cheese made from raw milk has a more desirable taste due to the presence of indigenous microflora and enzymes in raw milk. The production method of *lighvan* cheese is summarized in Figure 6.16.

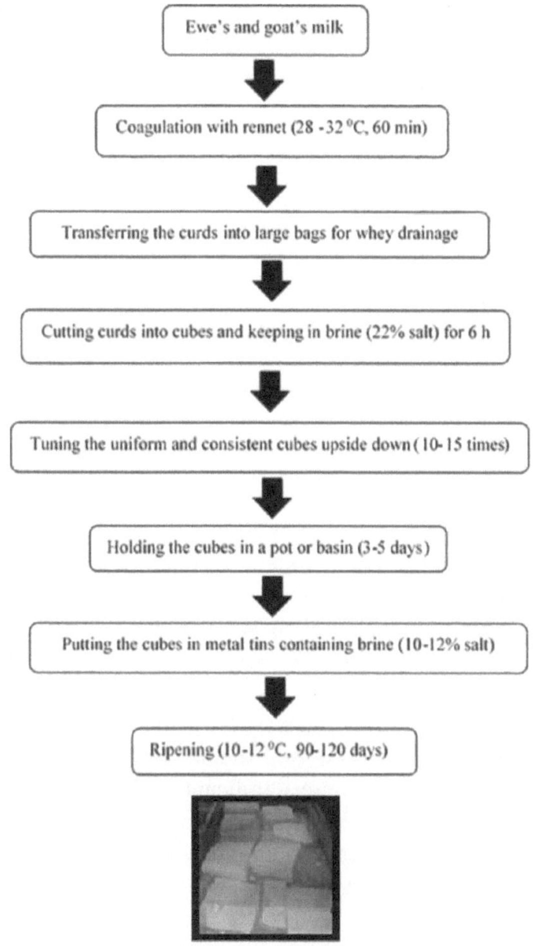

FIGURE 6.16 *Lighvan* cheese manufacturing

Source: Adapted from Edalatian et al. (2012a).

6.2.2.4.4.2 Microbiology of Lighvan *Cheese* Edalatian et al. (2012a) investigated the biodiversity of LAB of *lighvan* cheese during the ripening stage and isolated *L. garvieae, L. plantarum* (currently *Lactiplantibacillus plantarum subsp. plantarum*), *S. parauberis, E. faecalis, L. lactis, E. faecium, E. coli, L. plantarum, L. paracasei* (*Lacticaseibacillus paracasei*), *L. raffinolactis, S. thermophilus, S. haemolyticus, Lactobacillus salivarius* (*Ligilactobacillus salivarius*), and *M. caseolyticus.* The LAB evaluation showed that *Lactobacillus* and *Lactococcus* dominated in the first stage; however, they were replaced by *Enterococcus* at the end of ripening. The most predominant species were *E. faecium, L. lactis* ssp. *lactis, E. faecalis,* and *L. plantarum.* The authors stated that there is high biodiversity among 38 *E. faecium* isolates from *lighvan* cheese.

6.2.2.4.4.3 Physicochemical Changes of Lighvan *Cheese during Ripening* Shahab Lavasani et al. (2012a, 2012b) studied changes in the organoleptic and physicochemical properties of *lighvan* cheese during ripening. They demonstrated that dry matter, total nitrogen, and fat values of *lighvan* cheese decreased during the aging period. However, salt content, lipolysis, and trichloroacetic acid-soluble nitrogen values (TCA-SN) increased during the ripening period (90 days). The ripening period was the important factor affecting the cheese's organoleptic properties. Furthermore, they showed that significant amounts of microbial metabolites are transferred into the brine; therefore, their effect on the organoleptic characteristics of the final product is limited. Shahab Lavasani (2021) employed *Bifidobacteria* to produce probiotic *lighvan* cheese and compared the proteolysis profile and aroma development of this probiotic cheese with traditional *lighvan* cheese. They detected that the proteolysis profile of traditional

TABLE 6.7 Volatile Compounds (AU[1] × 10^5) from *Lighvan* Cheese during the Ripening Period

	RIPENING PERIOD (DAYS)			
CHARACTERISTICS	*1*	*30*	*60*	*90*
Esters	3.4	20.4	36.8	75.3
Aromatic	4.5	35.7	47.7	54.8
Ketones	2.0	33.8	45.7	53.3
Acids	62.9	122.0	130.2	223.6
Aldehydes	11.1	13.0	17.2	28.1
Alcohols	29.6	32.2	42.5	28.4

Source: Data from Aminifar et al. (2010).

Note: [1]AU: Area unit.

lighvan cheese was affected by *Bifidobacterium lactis* subsp. *animalis*; however, *Bifidobacteria* did not affect the sensory characteristics of probiotic *lighvan* cheese. In another study, Aminifar et al. (2010) studied the microstructural, compositional, and textural properties of *lighvan* cheese through the ripening stage. These researchers reported that the salt-in-moisture ratio increased, but the moisture content of the cheese decreased during the storage period. Changes in the proteolysis level were significant, and this affected the hardness of cheese during the ripening period. This study demonstrated that the contribution of water-soluble nitrogen/total nitrogen (WSN/TN), the number of pores, and their size significantly increased, and a casein gel network with lots of pores and homogenous texture was formed in *lighvan* cheese as a result of ripening. According to Aminifar et al. (2014), by applying the SPME–GC–MS[2] technique, the volatile products responsible for the flavor formation of *lighvan* cheese were recognized. They reported that esters, aldehydes, ketones, cyclic aromatic compounds, and carboxylic acids increased during the ripening period (Table 6.7), and carboxylic acids with 48.22% of total compounds were the dominant volatile compounds in this cheese. Moreover, the proteolysis of *lighvan* cheese during this period was monitored by these researchers. Using the SDS-PAGE[3] technique, they indicated that degradation of β- and α-S casein during the first month of cheese ripening was more significant than in the other stages. They realized that the reduction of the proteolysis rate after 30 days could be correlated with the inhibitory effect of salt on bacterial growth and proteolytic enzyme activities.

6.2.2.4.4.4 Safety of Lighvan *Cheese* The specific flavor of *lighvan* cheese is attributed to its indigenous microbial community derived from raw milk (Abdi et al. 2006). Since *lighvan* cheese is prepared from raw milk without a heating process (Mirzaei 2011), pathogenic microorganisms such as *Enterococci* with virulent genes and fungi may grow. Joghataei et al. (2017) evaluated the safety and antimicrobial activity of *Enterococci* isolated from *lighvan* cheese. They isolated 23 *Enterococci* sensitive to ampicillin and vancomycin, and further virulent genes (gelE, ace, esp, asa1, vanA, cpd, and vanB) in some of these isolates were detected. Attar et al. (2018) reported that the frequency of bacteriocin structural genes like Lactococcin A, Lactococcin B, Lactococcin M, Lactococcin GQ, and Nisin were 1, 6, 7, 6, and 1 out of 12 *L. lactis* isolated from *lighvan* cheese, respectively. They stated that the presence of bacteriocin producers such as *L. lactis* in *lighvan* cheese can inhibit the growth of pathogenic bacteria. Furthermore, according to Nayyeri et al. (2017), the antifungal property of *L. plantarum* strains isolated from *lighvan* cheese against *Penicillium expansum* and *Rhodotorula mucilaginosa* was confirmed.

6.2.3 Cereal-Based Products

6.2.3.1 Bread

The quality of bread has been noted based on various factors, such as:

1. Sensory features, such as its volume, form, color, texture, and taste.
2. Nutritional values, such as iron, calcium, and vitamin levels.
3. Health consideration, including the presence of prohibited additives and microbial contamination in products.

Traditional Iranian breads are mostly popular because of their taste, diversity, and quality. Nonetheless, urban bread quality has diminished in some cases. Furthermore, due to the application of sodium bicarbonate, rural breads are of high quality, because they are mostly associated with sourdough, in which baker's yeast and lactobacilli are abundant. The indigenous LAB in sourdough as native starter cultures play an important role in increasing the shelf life, flavor, and quality of bread (Mollakhalili Meybodi et al. 2015). There are four forms of bread in Iran, *barbari, sangak, taftoon*, and *lavash* (Faridi et al. 1983), described in the following.

6.2.3.1.1 Barbari *Bread*

Barbari, a flat bread, is extensively consumed in the northwestern and northern regions of Iran (Ghanbari and Farmani 2013) and is 70 to 80 cm in length and 25 to 30 cm in width with a thickness of nearly 3.5 mm. It is produced through blending all the ingredients (flour, water, and oil) together to obtain an appropriate consistency, followed by fermentation for about 2 h. The balls are prepared from the dough, then flattened, and it is allowed to rest for 20 min. Next, the flattened balls are placed on a surface to create a brown color and polished following baking. Final baking and proof time are about 15 and 8–12 min, respectively (Meybodi et al. 2016).

6.2.3.1.2 Sangak *Bread*

Sangak is a triangular-shaped flat bread with good flavor as well as high nutritional value. This bread with high digestibility (containing high fiber) is produced through baking on a hot bed containing small rubble in the oven. This bread is baked using whole wheat flour with a thickness of 0.5–1 cm. The temperature of ovens for the traditional baking of *sangak* bread is 250–350°C, and the bread is produced in a few minutes. The dough of *sangak* bread is characterized by lower consistency and higher moisture content than others (Najafi et al. 2012; Meybodi et al. 2016). According to Movahhed (2021), increasing the amount of sourdough in bread samples reduced pH, staling, the growth of fungi, and total bacterial count, as well as increasing acidity, moisture, ash, volume, fiber content, and organoleptic properties compared to bread without sourdough. Analysis of the microbial community from 18 samples of fresh dough and sourdough obtained from 14 bakeries indicated that *Leuconostoc* and *Lactobacillus* spp. made up 77% of the total bacteria of the dough. These bacteria are necessary to achieve the favorable aroma and taste for the bread; however, the leavening of the dough is caused by *S. cerevisiae* (Azar et al. 1977; Meybodi et al. 2016).

6.2.3.1.3 Taftoon *Bread*

This Iranian flat bread is made from soft wheat flour. *Taftoon* is commonly round and is about 400 to 500 mm in length, 400 to 500 mm in width, and has a thickness of 2 mm. The ingredients of *taftoon* are water = 78–90 L, flour = 100 kg, salt = 90–130 g, and yeast = 450 to 650 g, and it is prepared like other types of bread (Salehifar and Shahedi 2010; Meybodi et al. 2016).

6.2.3.1.4 Lavash *Bread*

Lavash as a traditional Iranian bread is produced by white wheat; it is flat and is used for preparing sandwiches. It is lightly baked and consumed as a healthy snack or an appetizer (Fazeli et al. 2004; Mortazavi and Sadeghi 2011). *Lavash* has an oval shape and is 30–40 cm in width and 60–70 cm in length, with a thickness of 2–3 cm. Yeast, water soda, flour, and salt are its ingredients (Movahed et al. 2011), which are mixed completely, followed by fermentation for 1–3 h. A roller is used to flatten the balls of the dough on a wooden surface. The bread is then heated at 200 to 250°C for nearly 5 min (Meybodi et al. 2016).

6.2.4 Cereal and Dairy–Based Products

Many people in the Middle East consume fermented milk-cereal mixtures (Ibanoglu et al. 1996). Cereal and dairy–based fermented foods often contain probiotic bacteria. As the Food and Agriculture Organization/World Health Organization (FAO/WHO) stated, the production of antimicrobial substances *in vitro* is an important factor by which new probiotic strains should be characterized (Vasiee et al. 2018b).

6.2.4.1 Kashk-e Zard

Kashk-e zard is commonly used in the southeastern region of Iran (Sistan and Balouchestan provinces) and still mainly prepared by households and is also sold in the market by informal routes. Its nutritional value is high, and it is a good source of bioavailable vitamins, minerals, and protein. Other similar products in different countries are *kushuk* (Iraq), *kishk* (Lebanon and Egypt), *tarhana* (Turkey and Greece), *atole* (Scotland) *talkuna* (Finland), *madeer-oggt* (Saudi Arabia), *tahanya* (Hungary), and *kichk* (India), (Mashak et al. 2014).

6.2.4.1.1 Preparation Method of Kashk-e Zard

Kashk-e zard is produced through the combination of yoghurt, cereal flour (mainly wheat flour), salt, vegetables, and spices and is fermented for some days. It is commonly a mixture of 35% wheat flour and 65% yoghurt (Mashak et al. 2014). The production method of *kashk-e zard* is presented in Figure 6.17.

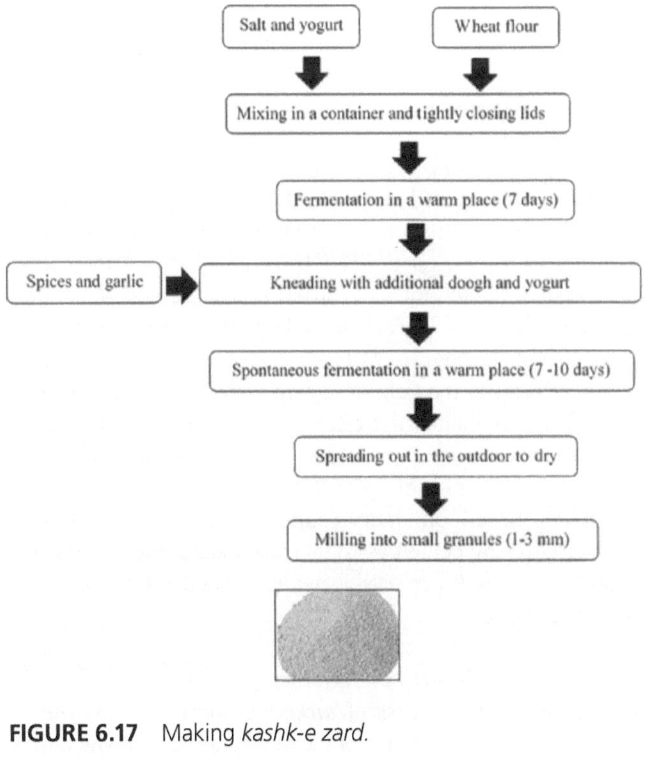

FIGURE 6.17 Making *kashk-e zard.*

Source: Adapted from Mashak et al. (2014).

6.2.4.1.2 Microbiology of Kashk-e Zard

Lactobacillus as a dominant genus in the ripening of *kashk-e zard* was identified. Furthermore, in addition to *Lactobacillus*, there are probiotic strains of *Enterococcus* and *Pediococcus* in *kashk-e-zard* (Mashak and Moradi 2019). Mashak (2016) showed the antimicrobial effect of the *Lactobacillus* strains collected from *kashk-e zard* and tarkhineh against *Staphylococcus aureus*, *E. coli*, *S. typhimurium*, and *Lis. monocytogenes* based on the agar well diffusion assay. She reported that *tarkhineh* and *kashk-e zard* have probiotic effects, deserving more investigation.

6.2.4.2 Tarkhineh

Tarkhineh, a fermented product based on dairy and cereal products, is consumed in countries such as Egypt, Iran, Turkey, Greece, Iraq, Finland, and Hungary with various names like *tarhana*, *kushuk*, *talkhuna*, and *kishk* (İbanoğluet al. 1999; Tamang et al. 2005). In winter, *tarkhineh* in the western region of Iran (Kurdistan, Kermanshah, and Hamedan) is consumed as soup because of its natural taste and therapeutic properties (Ebrahimi et al. 2014; Mashak et al. 2014), and it is otherwise generally consumed (Vasiee et al. 2018b). Fermented milk products, such as *doogh*, are highly nutritious foods including remarkable levels of fat, protein, and amino acids that are limited in cereal (methionine and lysine). On the other hand, wheat flour has some minerals (Mn, Cu, and Fe) that are deficient in milk (Tamime et al. 1999). Therefore, *tarkhineh* is rich in organic acids, minerals, and free amino acids, making it healthy for old people, patients, and children. Furthermore, it has a high level of vitamins like riboflavin, thiamine, vitamin B12, niacin, folic acid, and pantothenic and ascorbic acid (Ibanoglu et al. 1999; Ekıncı 2005; Ekinci and Kadakal 2005). Also, LAB in *doogh* help to absorb nutrients that otherwise would be poorly digestible or indigestible (Lahtinen et al. 2011).

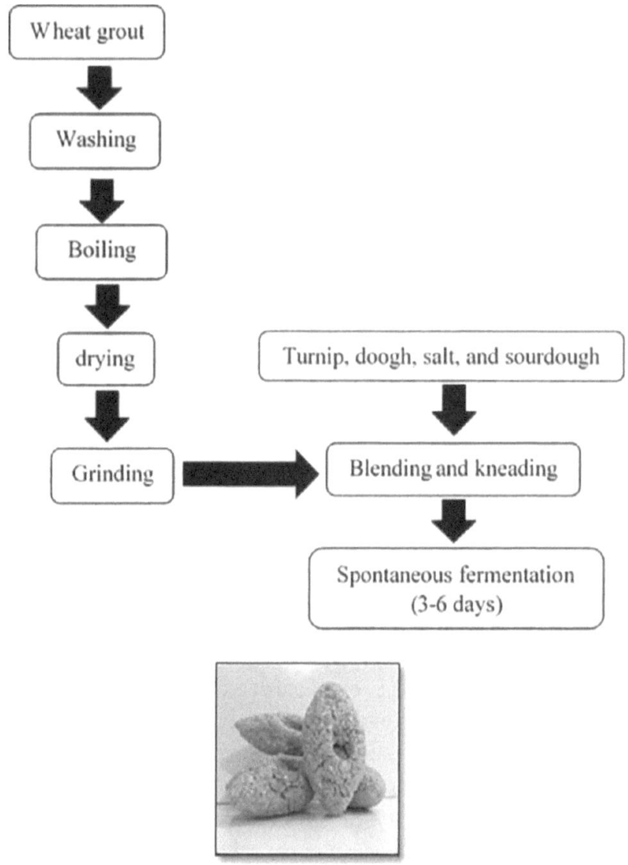

FIGURE 6.18 Flow diagram for traditional *tarkhineh* production.
Source: Adapted from Vasiee et al. (2018b).

6.2.4.2.1 Preparation Method of Tarkhineh

Tarkhineh is prepared with wheat grout, *doogh*, turnip, salt, vegetables, sourdough, and spices, followed by fermentation with yeasts and LAB derived from *doogh* for 3 to 6 days. Following fermentation, it is dried under the sun. *Tarkhineh* can be kept for 1 to 2 years in cold and dry conditions (Vasiee et al. 2018b). A diagram of *tarkhineh* production is presented in Figure 6.18. In the field of drying *tarkhineh*, Ghaitaranpour et al. (2013) evaluated the effect of wheat boiling time, grout particle size (fine, medium, coarse), drying time (30–900 min), and temperature (70, 80, 90°C) on the physical properties of *tarkhineh*. Based on their results, these variables affected *tarkhineh* drying behavior. To increase the production rate and improve *tarkhineh*'s color, it is recommended to boil for 1 hour and dry at 80°C.

6.2.4.2.2 Microbiology of Tarkhineh

The fermentation process of *tarkhineh* mostly occurs due to the natural microorganisms of the raw materials, in which their activity and growth are affected during the fermentation and impacted by environmental conditions, particularly salt content and temperature. According to Tabatabaei Yazdi et al. (2012), *L. bifermentans* (21.3%), *Leu. cermoris* (6%), *L. nagelii* (67%), *L. intestinalis* (0.9%), *L. fructosus* (1.45%), *L. agilis* (0.9%), *L. fermentum (currently Limosilactobacillus fermentum)* (1%), and *L. acidipiscis* (0.9%), isolated from *tarkhineh* on the third day of fermentation, are responsible for fermentation and ripening of *tarkhineh* and *doogh* as a main source for these LAB. In another study, Vasiee et al. (2018b) identified 75 isolates from *tarkhineh*. Most strains belonged to the *Lactobacillus* genus and were identified as *L. pentosus* (12%), *L. fermentum (currently Limosilactobacillus fermentum)* (22.66%), *L. diolivorans* (1.33%), *L. brevis* (currently *Levilactobacillus brevis*) (10.66%), and *L. plantarum* (currently *Lactiplantibacillus plantarum subsp. plantarum*) (25.33%). The strains from other genera were *Leuconostoc citreum* (5.33%), *Enterococcus* (*faecium* and *faecalis*) (18.66%), and *Pediococcus* (*acidilactici* and *pentosaceus*) (3.99%). All strains of *L. brevis* and *L. plantarum* underwent rep-PCR typing, and *L. plantarum* and *L. brevis* isolates were found with high intra-species diversity.

6.2.4.2.3 Chemical Characteristics of Tarkhineh

Vasiee et al. (2018b) indicated that fat, protein, salt, and ash contents in *tarkhineh* did not change significantly (P value < 0.01) during the fermentation period (3 days). However, pH (from 4.5 to 3.8) and lactic acid content (from 2.2% to 5.6%) changed significantly (P value < 0.01) during this period. Mashak et al. (2014) reported that the low pH value (4.91 ± 0.06) as well as low moisture content (3.03 ± 0.16%) of the dried *tarkhineh* provided bacteriostatic activity against food spoilage and pathogenic microorganisms and increased the shelf life of the product.

6.2.4.2.4 Functionality and Probiotic Properties of Tarkhineh

Tarkhineh as a functional food can be a source of probiotic bacteria such as *L. fermentum, L. plantarum* (Vasiee et al. 2014), *P. acidilactici* (with high probiotic potential) (Vasiee et al. 2020), and *Enterococcus* (Mashak and Moradi 2019). Kiani et al. (2021a) showed that *tarkhineh* can be used in potato chips as a protective product for probiotic cells. They assessed the morphological features, survival rate, and sensory assessment of the evaluated probiotic cells using *tarkhineh* in potato chips throughout the storage period. Two formulations of *tarkhineh*, plain and turmeric, were used using a simple spraying approach to cover the probiotic cells in potato chips. They demonstrated elliptical to spherical (480 to 770 mm) probiotic drops. According to storage stability data, probiotic cells added to plain and turmeric *tarkhineh* during 4 months of storage at 4°C showed good protection ability, with nearly 3.70 and 2.85 log reductions in CFU/g, respectively. Moreover, probiotic potato chips in comparison with commercial and non-probiotic potato chips showed probiotic product characteristics, like good quality and superior sensory features throughout the storage period (Kiani et al. 2021a). According to morphological assessment using scanning electron microscopy, drops with different formulations were found with almost spherical to elliptical forms, and the *tarkhineh* texture protected the probiotic cells. Furthermore, microscopy images showed integration of *tarkhineh* particles into network-like structures in drops entrapping the probiotic cells. In addition, SEM images indicated several tiny pores on the surface of all the *tarkhineh* formulations. Kiani et al. (2021b) showed that *tarkhineh* as a novel microencapsulation matrix improved the sensory and quality features of potato chips containing probiotic *L. lactis* KUMS-T18. Potato chips generated with plain and turmeric *tarkhineh* during incubation at 4°C showed well-protected probiotic cells with nearly 4.52 and 3.46 log reductions in CFU/g, respectively. However, probiotic potato chips in comparison with commercial and non-probiotic potato chips exhibited the characteristics of probiotic products, like superior sensory and quality properties. Therefore, probiotic *L. lactis* that are covered with *tarkhineh* ingredients are applicable to produce probiotic potato chips (Kiani et al. 2021a). In recent years, cancer is increasing worldwide, and *tarkhineh* consumption as a functional food has potential for cancer prevention. Mehrabian et al. (2012) studied the anticancer and antimutagenic effects of LAB collected from *tarkhineh* using the Ames test. In this study, only four strains of *Lactobacilli* showed antimutagenic activity. The inhibitory rate in these four strains at its highest and lowest level was 60.38 and 39.37%, respectively. In addition, the inhibitory rate of mutation in the strains with the existence of microsomes were markedly different (P < 0.05) compared to the absence of microsomes. The addition of the microsome of rat liver (S9) could help to prove the anticancer properties of four *Lactobacilli* strains. Aflatoxins are the most important mycotoxin produced by fungi, and some strains of LAB can control mutagenic factors like aflatoxin. It seems that this is because of physical attachment between LAB and mutagens. According to Tajabadi et al. (2011), *Lactobacillus* strains isolated from *tarkhineh* could sequester aflatoxin by binding the toxin. These isolates are recommended to be used in the food industry as a native probiotic after more safety assessment.

6.2.4.3 Horreh

Horreh is a dairy-cereal based food that is traditionally produced in regions in the southwest of Iran, such as Khoozestan and Fars, and commonly consumed as a nutritious seasoning (Vasiee et al. 2018a).

6.2.4.3.1 Preparation Method of Horreh

For *horreh* preparation, wheat flour is mixed with *kardeh* (*Biarum carduchorum*) and kneaded, followed by adding *doogh* and keeping for 2 days in a warm environment. Then, onion (fried in ghee) and steamed cooked rice are blended with the prepared mixture. *Horreh* is produced based on spontaneous fermentation started by indigenous microorganisms (Figure 6.19). The biodiversity of the microbial community in *horreh* is dependent mostly on the applied ingredients (Vasiee et al. 2018a).

6.2.4.3.2 Microbiology of Horreh

According to Vasiee et al. (2018a), LAB including *W. cibaria, Enterococcus (faecalis* and *faecium), L. fermentum* (currently *Limosilactobacillus fermentum*), *L. brevis* (currently

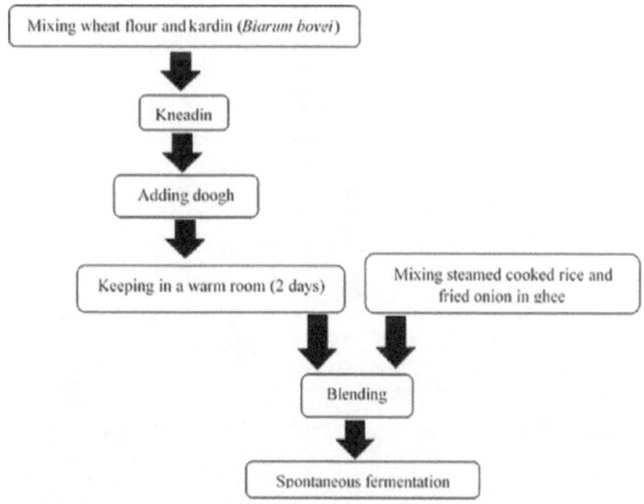

FIGURE 6.19 *Horreh* production.

Source: Adapted from Vasiee et al. (2018a).

Levilactobacillus brevis) and *Lactobacillus plantarum* (currently *Lactiplantibacillus plantarum subsp. plantarum*) *Leuconostoc* (*mesenteroides* and *citreum* subsp. *mesenteroides*), and *P. pentosaceus* are responsible for *horreh* ripening. Some *L. plantarum* and *L. fermentum* isolates indicated high antibacterial activity and hydrophobicity. They are appropriate candidates to be used as new probiotic isolates in animal or human foods.

6.2.5 Fish-Based Products

6.2.5.1 Mahyaveh

Mahyaveh, a fermented fish sauce, is used in the south of Iran (Zarei et al. 2012). It is mainly prepared based on the family tradition, type of available raw materials, and climate conditions of the area. Thus, wide variations are observed in the ratio of raw materials and preparation methods.

6.2.5.1.1 Preparation Method of Mahyaveh

Mahyaveh includes anchovies (*Stolephorus* sp.) or sardines (*Sardinella* sp.), mustard (*Brassica juncea* L.), salt, and water. *Mahyaveh* is produced from dried or fresh fish (Zarei et al. 2012). A diagram of *mahyaveh* preparation is shown in Figure 6.20.

6.2.5.1.2 Microbiology of Mahyaveh

According to Taheri et al. (2014), the number of total viable cells in the first week of preparation was 8.7 ± 102 CFU/g, which reached 7.2 ± 102 CFU/g in the following 2 months. *L. alimentaria* and *L. plantarum* (currently *Lactiplantibacillus plantarum subsp. plantarum*) were 59.9% of the total LAB. *Staphylococcus* sp. and *Micrococcus* sp. made up 70% of the overall bacterial community. *M. luteus* exhibited a remarkable difference compared to other *Micrococcus* species ($P < 0.05$). The number of *S. lentus* was higher than that of other related species such as *Staphylococcus* spp. *Pichia* sp., *Saccharomycopsis* sp., *Candida* sp., and *Mucor* were identified as the dominant fungal genera. Elevated levels of pathogenic bacteria, spore-producing bacteria and coliforms were not detected in *mahyaveh*, confirming the *mahyaveh* safety however, more studies are essential.

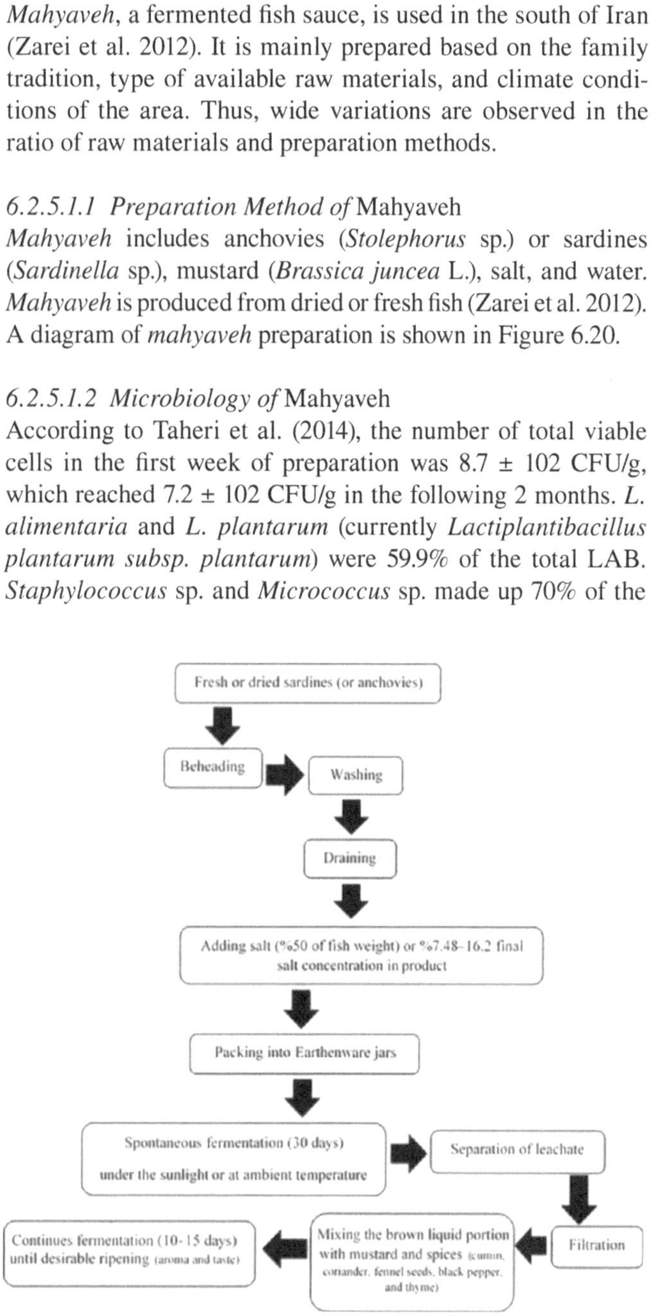

FIGURE 6.20 *Mahyaveh* production.

Source: Adapted from Taheri et al. (2014).

Zarei et al. (2012) studied the microbial quality of Iranian fish sauce samples, and the microbial analysis results of *mahyaveh* samples from five different regions were reported as follows (log CFU/ml): aerobic mesophilic bacteria = 4.71 ± 3.65, *Enterobacteriaceae* = 3.41 ± 2.03, LAB = 4.13 ± 3.05, and halophilic bacteria = 3.66 ± 2.24 (Zarei et al. 2012).

LAB are the dominant microorganisms in several fermented products prepared from fish (Olympia et al. 1992; Ostergaard et al. 1998). LAB usually start the fermentation process and cause a reduction in pH. The presence of salt, organic acids (lactic acid), and low pH can be the most important factors for the preservation of fermented fish products (Ijong and Ohta 1995; Sanni et al. 2002).

6.2.5.1.3 Chemical Properties of Mahyaveh

The results of physicochemical property analysis of *mahyaveh* samples received from five different regions are: pH = 6.28 ± 0.11, %NaCl = 11.7 ± 0.45, total volatile basic nitrogen = 3098 ± 197 (mg/kg), histamine = 2662 ± 413 (mg/kg), putrescine = 5.44 ± 0.89 (mg/kg), spermidine = 1096 ± 451 (mg/kg), and tyramine = 11.5 ± 3.06 (mg/kg) (Zarei et al. 2012). Zarei et al. (2012) reported the range of 5.4–7.55 for the pH of *mahyaveh* from various locations and the range of 7.48–16.2% for the levels of NaCl. The high biogenic amine levels are the main problem with this product. The high histamine level of Iranian fish sauce may be due to the high bacterial counts, particularly LAB and *Enterobacteriaceae*, in this product. Such bacteria can produce hazardous levels of histamine in a very short time when the fish is stored at high temperatures (Omura et al. 1978). As far as we know, there were no reports of poisoning because of a high intake of *mahyaveh* in the studied locations. The concentration of histamine is high in most of the samples because these products are commonly used in small amounts and are not eaten regularly, so they are associated with a relatively low prevalence of histamine poisoning. Nonetheless, high amounts of this product should not be consumed on a regular basis (Zarei et al. 2012).

6.3 CONCLUSIONS AND FUTURE DIRECTIONS

There are a wide range of various fermented products in the world that have their own specific aroma, taste, and texture. Even though in Iran, as in other parts of the world, fermented foods have been largely marginalized with the advent of new technologies, recently there has been a resurgence of interest in traditional fermented foods, basically due to the proposed health benefits of fermented foods. Temperature diversity in different regions of Iran and diversity of climatic regions have led to the production of various agricultural products as well as various livestock products, especially dairy products. Fermented products provide a significant part of the diet in Iran, and people prefer fermented foods that are produced with various flavors, shapes, and sensory properties in different parts of Iran. The sensory properties of fermented foods and beverages are dependent on the chemical composition of raw materials, intrinsic microbial flora found in raw materials, climatic conditions, production method, and geographical location of these products. Various studies have shown that native fermented products in each region have their own unique intrinsic microbial strains, which can be distinctive among similar products from different regions of the world. The indigenous microbial community (mainly LAB) of local fermented foods, produced by spontaneous fermentation, is responsible for increasing the shelf life and specific sensory characteristics of these products. To take advantage of the beneficial properties of these products in the future, their intrinsic strains should be isolated, identified, and maintained; then the technological and probiotic properties of the microbial isolates need to be evaluated. Selected strains of LAB with high technological abilities can be recommended for fermentation industries, with a focus on the dairy industry. To select suitable microbial strains, the microbial communities of fermented products should be accurately identified by applying more advanced techniques to identify even unculturable strains. Nowadays, NGS techniques as one of the independent-culture methods can be used for more comprehensive and accurate understanding of microbial communities in the ecosystems of fermented products. However, studies on dynamic changes in the microbial communities of Iranian fermented products and their health benefits have received very little attention from researchers, and few publications have been available for many products in recent years. In this chapter, we have referred to some fermented food products in Iran. Further studies will be required to investigate the unique nutritional composition and dynamic changes in the microbial communities of these products.

NOTES

1. Denaturiwwng gradient gel electrophoresis.
2. Solid-phase microextraction–gas chromatography–mass spectrometry.
3. Sodium dodecyl sulfate polyacrylamide gel electrophoresis.

REFERENCES

Abdali, H., Saei-Dehkordi, S.S., Mobini-Dehkordii, M. and Abtahi-Froushani, S.M., 2019. First isolation and molecular detection of autochthonous potential probiotic lactobacilli isolates from Iranian traditional Poosti cheese and their antioxidative activity. *Biological Journal of Microorganism*, *8*(32), pp.25–39.

Abdi, R., Sheikh-Zeinoddin, M. and Soleimanian-Zad, S., 2006. Identification of lactic acid bacteria isolated from traditional Iranian Lighvan cheese. *Pakistan Journal of Biological Sciences*, *9*(1), pp.99–103.

Aminian, B., Masoumpour, S., Sadegh, A.A. and Omrani, G.H., 2003. Unripe grape juice (Verjuice) as a lipidlowering agent: Fact or fiction. *Archives of Iranian Medicine*, *6*(1), pp.32–34.

Aminifar, M., Hamedi, M., Emam-Djomeh, Z. and Mehdinia, A., 2010. Microstructural, compositional and textural properties during ripening of Lighvan cheese, a traditional raw sheep cheese. *Journal of Texture Studies*, *41*(4), pp.579–593.

Aminifar, M., Hamedi, M., Emam-Djomeh, Z. and Mehdinia, A., 2014. Investigation on proteolysis and formation of volatile compounds of Lighvan cheese during ripening. *Journal of Food Science and Technology*, *51*, pp.2454–2462.

Attar, M.A., Yavarmanesh, M., Mortazavi, A., Dovom, M.R.E. and Najafi, M.B.H., 2018. Antibacterial effects of Lactococcus lactis isolated from Lighvan cheese regarding the recognition of Nisin, Lacticin and Lactococcin structural genes. *LWT*, *89*, pp.186–191.

Azar, M., Ter-Sarkissian, N. and Ghavifek, H., 1977. Microbiological aspects of Sangak bread. *Journal of Food Science and Technology*, *14*(6), pp.251–254.

Azarikia, F. and Abbasi, S., 2010. On the stabilization mechanism of Doogh (Iranian yoghurt drink) by gum tragacanth. *Food Hydrocolloids*, *24*(4), pp.358–363.

Beato, V.M., Sánchez, A.H., de Castro, A. and Montaño, A., 2012. Effect of processing and storage time on the contents of organosulfur compounds in pickled blanched garlic. *Journal of Agricultural and Food Chemistry*, *60*(13), pp.3485–3491.

Casado, F.J., López, A., Rejano, L., Sánchez, A.H. and Montaño, A., 2004. Nutritional composition of commercial pickled garlic. *European Food Research and Technology*, *219*(4), pp.355–359.

Classen, C., Howes, D. and Synnott, A., 1994. *Aroma: The cultural history of smell.* Taylor & Francis, pp.135–180.

Dana, M.G., Yakhchali, B., Salmanian, A.H. and Jazii, F.R., 2011. High-level acetaldehyde production by an indigenous Lactobacillus strain obtained from traditional dairy products of Iran. *African Journal of Microbiology Research*, *5*(25), pp.4398–4405.

Davati, N. and Hesami, S., 2021a. 16S rRNA metagenomic analysis reveals significant changes of microbial compositions during fermentation from ewe milk to doogh with antimicrobial activity. *Food Biotechnology*, *35*(3), pp.179–198.

Davati, N. and Hesami, S., 2021b. Bacterial community analysis of winter salad during fermentation, and its antimicrobial properties. *International Food Research Journal*, *28*(6), pp.1285–1297.

Dehkordi, F.S., Yazdani, F., Mozafari, J. and Valizadeh, Y., 2014. Virulence factors, serogroups and antimicrobial resistance properties of Escherichia coli strains in fermented dairy products. *BMC Research Notes*, *7*, pp.1–8.

Ebrahimi, M.T., Ouwehand, A.C., Hejazi, M.A. and Jafari, P., 2011. Traditional Iranian dairy products: A source of potential probiotic lactobacilli. *African Journal of Microbiology Research*, *5*(1), pp.20–27.

Ebrahimi, M.T., Shariatpanahi, M., Jafari, P. and Sadeghi, S., 2014. Inhibitory effect of Lactobacilli (isolated from Tarkhineh Dough) on E. coli and Lis. monocytogenes colonization. *International Journal of Biosciences (IJB)*, *5*(10), pp.29–36.

Edalatian, M.R., Habibi Najafi, M.B., Mortazavi, A. and Mayo, B., 2012a. The biodiversity and evolution of lactic flora during ripening of the Iranian semisoft Lighvan cheese. *International Journal of Dairy Technology*, *65*(1), pp.81–89.

Edalatian, M.R., Najafi, M.B.H., Mortazavi, S.A., Alegría, Á., Nassiri, M.R., Bassami, M.R. and Mayo, B., 2012b. Microbial diversity of the traditional Iranian cheeses Lighvan and Koozeh, as revealed by polyphasic culturing and culture-independent approaches. *Dairy Science & Technology*, *92*, pp.75–90.

Ekinci, R. and Kadakal, C., 2005. Determination of seven water-soluble vitamins in tarhana, a traditional Turkish cereal food, by high-performance liquid chromatography. *ACTA Chromatographica*, *15*, p.289.

Ekıncı, R., 2005. The effect of fermentation and drying on the water-soluble vitamin content of tarhana, a traditional Turkish cereal food. *Food Chemistry*, *90*(1–2), pp.127–132.

Esfandiari, Z., Badiey, M., Mahmoodian, P., Sarhangpour, R., Yazdani, E. and Mirlohi, M., 2013. Simultaneous determination of sodium benzoate, potassium sorbate and natamycin content in Iranian yoghurt drink (Doogh) and the associated risk of their intake through Doogh consumption. *Iranian Journal of Public Health*, *42*(8), p.915.

Ezzatpanah, H., 2020. Traditional food and practices for health: Iranian dairy foods. In *Nutritional and health aspects of food in south Asian countries* (pp.275–287). Academic Press.

Farahani, G., Ezzatpanah, H. and Abbasi, S., 2014. Characterization of Siahmazgi cheese, an Iranian ewe's milk variety: Assessment of physico-chemical, textural and rheological specifications during ripening. *LWT-Food Science and Technology*, *58*(2), pp.335–342.

Faridi, H.A., Finney, P.L. and Rubenthaler, G.L., 1983. Iranian flat breads: Relative bioavailability of zinc. *Journal of Food Science*, *48*(1), pp.107–110.

Farnworth, E.R.T. ed., 2008. *Handbook of fermented functional foods*. CRC Press.

Fazeli, M.R., Shahverdi, A.R., Sedaghat, B., Jamalifar, H. and Samadi, N., 2004. Sourdough-isolated Lactobacillus fermentum as a potent anti-mould preservative of a traditional Iranian bread. *European Food Research and Technology*, *218*, pp.554–556.

Foroughinaia, S., Abbasi, S. and Hamidi Esfahani, Z., 2007. Effect of individual and combined addition of salep, tragacantin and guar gums on the stabilisation of Iranian Doogh. *Iranian Journal of Nutrition Sciences & Food Technology*, *2*(2), pp.15–25.

Fox, P.F. and Wallace, J.M., 1997. Formation of flavor compounds in cheese. *Advances in Applied Microbiology*, *45*, pp.17–86.

Ghaitaranpour, A., Yazdi, F.T., Behbahani, B.A., Mortazavi, A. and Mohebbi, M., 2013. The effect of wheat boiling time, bulgur particle size, drying time and temperature on some physical properties of hot air dried Tarkhineh. *Scientific Journal of Pure and Applied Sciences*, *2*(4), pp.175–182.

Ghanbari, M. and Farmani, J., 2013. Influence of hydrocolloids on dough properties and quality of Barbari: An Iranian leavened flat bread. *Journal of Agricultural Science and Technology*, *15*(3), pp.545–555.

Han, X., Zhang, L., Yu, P., Yi, H. and Zhang, Y., 2014. Potential of LAB starter culture isolated from Chinese traditional fermented foods for yoghurt production. *International Dairy Journal*, *34*(2), pp.247–251.

Harris, L.J., 1998. The microbiology of vegetable fermentations. *Microbiology of Fermented Foods*, pp.45–72.

Hassanzadazar, H., Ehsani, A., Mardani, K. and Hesari, J., 2012. Investigation of antibacterial, acid and bile tolerance properties of lactobacilli isolated from Koozeh cheese. *Veterinary Research Forum*, *3*(3), p.181.

Hemmatian, M., Aminifar, M. and Attar, F., 2015. Characterization of Poosti cheese, a traditional raw sheep cheese during ripening: Physicochemical, microbial and micro-structural aspects. *Nutrition and Food in Health and Disease*, *2*(2), pp.39–48.

Hesami and Davati. (2018). Identification of microbial diversity of fermented pickled vegetable with tomato juice using Next-Generation Sequencing technique. *Journal of Food Science and Technology*, *15*(83), pp.15–27.

Hutkins, R.W., 2006. Fermented vegetables. *Microbiology and Technology of Fermented Foods*, pp.223–259.

Ibanoglu, S., Ainsworth, P. and Hayes, G.D., 1996. Extrusion of tarhana: Effect of operating variables on starch gelatinization. *Food Chemistry*, *57*(4), pp.541–544.

İbanoğlu, Ş., İbanoğlu, E. and Ainsworth, P., 1999. Effect of different ingredients on the fermentation activity in tarhana. *Food Chemistry*, *64*(1), pp.103–106.

Ijong, F.G. and Ohta, Y., 1995. Microflora and chemical assessment of an Indonesian traditional fermented fish sauce "Bakasang". *Journal of the Faculty of Applied Biological Science, Hiroshima University*, *34*(2), pp.95–100.

Joghataei, M., Yavarmanesh, M. and Dovom, M.R.E., 2017. Safety evaluation and antibacterial activity of enterococci isolated from Lighvan cheese. *Journal of Food Safety*, *37*(1), p.e12289.

Jung, Y.M., Lee, S.H., Lee, D.S., You, M.J., Chung, I.K., Cheon, W.H., Kwon, Y.S., Lee, Y.J. and Ku, S.K., 2011. Fermented garlic protects diabetic, obese mice when fed a high-fat diet by antioxidant effects. *Nutrition Research*, *31*(5), pp.387–396.

Kafili, T., Emam Djomeh, Z. and Mayo, B., 2013. Physiological biodiversity of Lactobacillus strains isolated during traditional Iranian lighvan cheese manufacturing. *International Journal of Food Properties*, *16*(1), pp.9–17.

Karapinar, M. and Sengun, I.Y., 2007. Antimicrobial effect of koruk (unripe grape—Vitis vinifera) juice against Salmonella typhimurium on salad vegetables. *Food Control*, *18*(6), pp.702–706.

Kaya, Z., Unluturk, S., Martin-Belloso, O. and Soliva-Fortuny, R., 2020. Effectiveness of pulsed light treatments assisted by mild heat on Saccharomyces cerevisiae inactivation in verjuice and evaluation of its quality during storage. *Innovative Food Science & Emerging Technologies*, *66*, p.102517.

Khoshakhlagh, K., Hamdami, N., Shahedi, M. and Le-Bail, A., 2014. Quality and microbial characteristics of part-baked' Sangak bread packaged in modified atmosphere during storage. *Journal of Cereal Science*, *60*(1), pp.42–47.

Kiani, A., Nami, Y., Hedayati, S., Elieh Ali Komi, D., Goudarzi, F. and Haghshenas, B., 2021a. Application of Tarkhineh fermented product to produce potato chips with strong probiotic properties, high shelf-life, and desirable sensory characteristics. *Frontiers in Microbiology*, *12*, p.657579.

Kiani, A., Nami, Y., Hedayati, S., Jaymand, M., Samadian, H. and Haghshenas, B., 2021b. Tarkhineh as a new microencapsulation matrix improves the quality and sensory characteristics of probiotic Lactococcus lactis KUMS-T18 enriched potato chips. *Scientific Reports*, *11*(1), p.12599.

Kiani, H., Hossein Kiani, Mousavi, S.M.A. and Emam-Djomeh, Z., 2008. Rheological properties of Iranian yoghurt drink, Doogh. Department of Food Science, Technology and Engineering,

Faculty of Biosystem Engineering, University of Tehran, Iran. *International Journal of Dairy Science*, *3*(2), pp.71–78.

Kiani, H., Mousavi, M.E., Razavi, H. and Morris, E.R., 2010. Effect of gellan, alone and in combination with high-methoxy pectin, on the structure and stability of doogh, a yogurt-based Iranian drink. *Food Hydrocolloids*, *24*(8), pp.744–754.

Köksoy, A. and Kılıç, M., 2003. Effects of water and salt level on rheological properties of Ayran, a Turkish yoghurt drink. *International Dairy Journal*, *13*(10), pp.835–839.

Lahtinen, S., Ouwehand, A.C., Salminen, S. and von Wright, A. eds., 2011. *Lactic acid bacteria: Microbiological and functional aspects*. CRC Press.

Lucey, J.A., Tamehana, M., Singh, H. and Munro, P.A., 1999. Stability of model acid milk beverage: Effect of pectin concentration, storage temperature and milk heat treatment. *Journal of Texture Studies*, *30*(3), pp.305–318.

Mashak, Z., 2016. Antimicrobial activity of lactobacillus isolated from Kashk-e-Zard and Tarkhineh, two Iranian traditional fermented foods. *International Journal of Enteric Pathogens*, *4*(2), pp.7–34692.

Mashak, Z. and Moradi, B., 2019. Identification of probiotic strains of enterococcus and pediococcus isolated from two Iranian traditional fermented dairy–cereal food products (Kashk-e-Zard and Tarkhineh). *Journal of Food Microbiology*, *6*(1), pp.79–89.

Mashak, Z. and Rooshani, J., 2019. The survey of chemical and microbial characteristics of traditional Koozeh-Cheese (Koopeh) in Kurdistan province. *New Findings in Veterinary Microbiology*, *2*(1), pp.67–80.

Mashak, Z., Sodagari, H., Mashak, B. and Niknafs, S., 2014. Chemical and microbial properties of two Iranian traditional fermented cereal-dairy based foods: Kashk-e-Zard and Tarkhineh. *International Journal of Biosciences*, *4*(12), pp.124–133.

Mehdizadeh, T., Sheikhkanloui Milan, H. and Mojaddar Langroodi, A., 2019. Viability of Bifidobacterium bifidum and its effect on the microbial, chemical and sensorial characteristics of traditional Koozeh cheese. *Iranian Journal of Nutrition Sciences & Food Technology*, *13*(4), pp.51–60.

Mehrabian, S., Tajabadi-Ebrahimi, M., Abbas-Ahmadi, M. and Bahrami, H., 2012. Study of antimutagenic and anticancer effect of lactic acid bacteria isolated from Tarkhineh by Ames Test. *Journal of Arak University of Medical Sciences*, *15*(7), pp.72–79.

Meybodi, N.M., Ebrahimi, M.T. and Mortazavian, A.M., 2016. Ethnic fermented foods and beverage of Iran. *Ethnic Fermented Foods and Alcoholic Beverages of Asia*, pp.309–322.

Mirzaei, H., 2011. Microbiological changes in Lighvan cheese throughout its manufacture and ripening. *African Journal of Microbiology Research*, *5*(13), pp.1609–1614.

Mirzaei, H., Khosroshahi, A.G. and Karim, G., 2008. The microbiological and chemical quality of traditional Lighvan cheese (white cheese in brine) produced in Tabriz, Iran. *Journal of Animal and Veterinary Advances*, *7*(12), pp.1594–1599.

Mollakhalili Meybodi, N., Mohammadifar, M.A. and Feizollahi, E., 2015. Gluten-free bread quality: A review of the improving factors. *Journal of Food Quality and Hazards Control*, *2*(3), pp.81–85.

Moradi, Z.K., Nasiri, N., Tabari, E., and Kioumarsi, H., 2019. Investigating and comparing the physicochemical, microbial and sensory characteristics of traditional and pasteurized Siahmazgi cheese. *Saudi Journal of Nursing and Health Care*, *2*(11), pp.357–361.

Moraes, P.M., Perin, L.M., Silva Júnior, A. and Nero, L.A., 2013. Comparison of phenotypic and molecular tests to identify lactic acid bacteria. *Brazilian Journal of Microbiology*, *44*, pp.109–112.

Mortazavi, S.A. and Sadeghi, A., 2011. Investigating the sourdough potential for enhance microbiological shelf life and roasty aroma of traditional Lavash bread. *African Journal of Biotechnology*, *10*(47), pp.9673–9679.

Mousa-Al-Reza, H., Ziba, R., Zakieh, K., Rad Tania, S. and Mohammad-Mahdi, S., 2011. The effect of verjuice on serum lipid levels in mice rendered atherosclerosis. *Journal of Biologically Active Products from Nature*, *1*(4), pp.229–235.

Movahhed, S., 2021. Effect of Sourdough on microbial, chemical and organoleptic characteristics of Sangak bread. *Journal of Food Science and Technology (Iran)*, *18*(110), pp.37–47.

Movahed, S., Rooshenas, G.H. and Ahmadichenarbon, H., 2011. Evaluation of the effect of liquid sourdough method on dough yield, bread yield and organoleptic properties Iranian Lavash bread. *World Applied Sciences Journal*, *15*(7), pp.1054–1058.

Najafi, M.A., Rezaei, K., Safari, M. and Razavi, S.H., 2012. Use of sourdough to reduce phytic acid and improve zinc bioavailability of a traditional flat bread (Sangak) from Iran. *Food Science and Biotechnology*, *21*, pp.51–57.

Nayyeri, N., Edalatian Dovom, M.R., Habibi Najafi, M.B. and Bahreini, M., 2017. A Preliminary study on antifungal activity of lactic acid bacteria isolated from different production stages of Lighvan cheese on Penicillium expansum and Rhodotorula mucilaginosa. *Journal of Food Measurement and Characterization*, *11*, pp.1734–1744.

Nikfarjam, A.K., Mahdian, E., Ghavidel, R.A. and Karazhyan, R., 2021. Probiotic potential of Lactobacillus strains isolated from Iranian traditional pickled garlic and using them in fermented olive. *Journal of BioScience and Biotechnology*, *10*(1), pp.41–51.

Noori, A., Keshavarzian, F., Mahmoudi, S., Yousefi, M. and Nateghi, L., 2013. Comparison of traditional Doogh (yogurt drinking) and Kashk characteristics (Two traditional Iranian dairy products). *European Journal of Experimental Biology*, *3*(6), pp.252–255.

Olympia, M., Ono, H., Shinmyo, A. and Takano, M., 1992. Lactic acid bacteria in fermented fishery product,"burong bangus". *Journal of Fermentation and Bioengineering*, *73*(3), pp.193–197.

Omura, Y., Price, R.J. and Olcott, H.S., 1978. Histamine—forming bacteria isolated from spoiled skipjack tuna and jack mackerel. *Journal of Food Science*, *43*(6), pp.1779–1781.

Ostergaard, A., Embarek, P.K.B., Yamprayoon, J., Wedell-Neergaard, C., Huss, H.H. and Gram, L., 1998. Fermentation and spoilage of Som Fak, a Thai low-salt fish product. *Tropical Science*, *38*, pp.105–112.

Pak, A. and Farajzadeh, M., 2007. Iran's integrated coastal management plan: Persian Gulf, Oman Sea, and southern Caspian Sea coastlines. *Ocean & Coastal Management*, *50*(9), pp.754–773.

Parsaeimehr, M., Khazaei, M., Jebellijavan, A. and Staji, H., 2019. The isolation and identification of dominant lactic acid bacteria by the sequencing of the 16S rRNA in traditional cheese (Khiki) in Semnan, Iran. *Journal of Human, Environment, and Health Promotion*, *5*(1), p.15.

Partovi, R., Gandomi, H. and Basti, A.A., 2019. Safety aspects of Lactobacillus plantarum strains isolated from Siahmazgi cheese. *Ankara Üniversitesi Veteriner Fakültesi Dergisi*, *66*(4), pp.337–342.

Partovi, R., Gandomi, H., Akhondzadeh Basti, A., Noori, N., Nikbakht Borujeni, G. and Kargozari, M., 2015. Microbiological and chemical properties of Siahmazgi cheese, an Iranian Artisanal Cheese: Isolation and identification of Dominant Lactic Acid bacteria. *Journal of Food Processing and Preservation*, *39*(6), pp.871–880.

Pure, A.E., Mofidi, S.M.G., Keyghobadi, F. and Pure, M.E., 2017. Chemical composition of garlic fermented in red grape vinegar and kombucha. *Journal of Functional Foods*, *34*, pp.347–355.

Pure, A.E. and Pure, M.E., 2016. Antioxidant and antibacterial activity of kombucha beverages prepared using banana peel, common nettles and black tea infusions. *Applied Food Biotechnology*, *3*(2), pp.125–130.

Rahnama, F., Mohammadzadeh Milani, J. and Gohari Ardabili, A., 2015. Improved quality characteristics of Sangak bread by response surface optimisation of farinograph and extensograph traits of doughs formulated with fenugreek gum. *Quality Assurance and Safety of Crops & Foods*, *7*(4), pp.413–421.

Salehifar, M. and Shahedi, M., 2010. Effects of oat flour on dough rheology, texture and organoleptic properties of Taftoon bread. *Journal of Agricultural Science and Technology, 9*, pp.227–234.

Sanni, A.I., Asiedu, M. and Ayernor, G.S., 2002. Microflora and chemical composition of momoni, a Ghanaian fermented fish condiment. *Journal of Food Composition and Analysis*, *15*(5), pp.577–583.

Sayevand, H.R., Bakhtiary, F., Pointner, A., Remely, M., Hippe, B., Hosseini, H. and Haslberger, A., 2018. Bacterial diversity in traditional doogh in comparison to industrial doogh. *Current Microbiology*, *75*, pp.386–393.

Setorki, M., Asgary, S. and Eidi, A., 2010. Effects of acute verjuice consumption with a high-cholesterol diet on some biochemical risk factors of atherosclerosis in rabbits. *Medical Science Monitor: International Medical Journal of Experimental and Clinical Research*, *16*(4), pp.BR124–130.

Shahab Lavasani, A., 2021. Proteolysis characteristics and aroma development during ripening of functional Iranian Lighvan cheese. *Journal of Food Processing and Preservation*, p.e15615.

Shahab Lavasani, A., Ehsani, M.R., Mirdamadi, S. and Ebrahim Zadeh Mousavi, M.A., 2012a. Changes in physicochemical and organoleptic properties of traditional Iranian cheese Lighvan during ripening. *International Journal of Dairy Technology*, *65*(1), pp.64–70.

Shahab Lavasani, A., Ehsani, M.R., Mirdamadi, S. and Mousavi, S.M. 2012b. Study of proteolysis and lipolysis of probiotic Lighvan cheese. *International Journal of AgriScience*, *2*(4), pp.341–352.

Siek, T.J., Albin, I.A., Sather, L.A. and Lindsay, R.C., 1971. Comparison of flavor thresholds of aliphatic lactones with those of fatty acids, esters, aldehydes, alcohols, and ketones. *Journal of Dairy Science*, *54*(1), pp.1–4.

Soltani, M. and Güzeler, N., 2013. The production and quality properties of liquid kashks. *Gida*, *38*(1), pp.1–7.

Souci, S.W., Fachmann, W. and Kraut, H., 2000. *Food composition and nutrition tables* (6th Ed.). Medpharm GmbH Scientific Publishers.

Tabatabaei Yazdi, F., Alizadeh Behbahani, B., Mohebbi, M., Mortazavi, S.A. and Ghaitaranpour, A., 2012. Identification of lactic acid bacteria isolated from Tarkhineh, a traditional Iranian fermented food. *Science Journal of Microbiology*, *1*(7), pp.152–159.

Taheri, A., Jalalinezhad, S., Hosseini, S.V., Ahmadi, A. and Nasery, F., 2014. Analysis of bacterial community in Mahyaveh, an Iranian traditional fish sauce. *Pajoohandeh Journal*, *19*(5), pp.273–280.

Tajabadi, E.M., Jafari, P., Bahrami, H. and Hashemi, M., 2011. Evaluation of Aflatoxin B1 reduction in present of lactobacilli isolated from Tarkhineh and fermented milk of Tarkhineh by ELISA. *Journal of Microbial Biology*, *3*, pp.43–48.

Tamang, J.P., Shin, D.H., Jung, S.J. and Chae, S.W., 2016. Functional properties of microorganisms in fermented foods. *Frontiers in Microbiology*, *7*, p.578.

Tamang, J.P., Tamang, B., Schillinger, U., Franz, C.M., Gores, M. and Holzapfel, W.H., 2005. Identification of predominant lactic acid bacteria isolated from traditionally fermented vegetable products of the Eastern Himalayas. *International Journal of Food Microbiology*, *105*(3), pp.347–356.

Tamime, A.Y., Barclay, M.N., McNulty, D. and O'Connor, T.P., 1999. Kishk-a dried fermented milk/cereal mixture. 3. Nutritional composition. *Le Lait*, *79*(4), pp.435–448.

Tavakoli, M., Hamidi-Esfahani, Z., Hejazi, M.A., Azizi, M.H. and Abbasi, S., 2017. Characterization of probiotic abilities of Lactobacilli isolated from Iranian Koozeh traditional cheese. *Polish Journal of Food and Nutrition Sciences*, *67*(1).

Vasiee, A., Alizadeh Behbahani, B., Tabatabaei Yazdi, F., Mortazavi, S.A. and Noorbakhsh, H., 2018a. Diversity and probiotic potential of lactic acid bacteria isolated from horreh, a traditional Iranian fermented food. *Probiotics and Antimicrobial Proteins*, *10*, pp.258–268.

Vasiee, A., Falah, F., Behbahani, B.A. and Tabatabaee-Yazdi, F., 2020. Probiotic characterization of Pediococcus strains isolated from Iranian cereal-dairy fermented product: Interaction with pathogenic bacteria and the enteric cell line Caco–2. *Journal of Bioscience and Bioengineering*, *130*(5), pp.471–479.

Vasiee, A.R., Mortazavi, A., Tabatabaei-yazdi, F. and Dovom, M.R., 2018b. Detection, identification and phylogenetic analysis of lactic acid bacteria isolated from Tarkhineh, Iranian fermented cereal product, by amplifying the 16s rRNA gene with universal primers and differentiation using rep-PCR. *International Food Research Journal*, *25*(1).

Vasiee, A.R., Yazdi, T.F., Mortazavi, A. and Edalatian, M.R., 2014. Isolation, identification and characterization of probiotic Lactobacilli spp. from Tarkhineh. *International Food Research Journal*, *21*(6), p.2487.

Walker, H. ed., 2000. Milk: Beyond the dairy. In *Proceedings of the Oxford Symposium on Food and Cookery 1999. Oxford Symposium* (p. 70). Prospect Books.

Wouters, J.T., Ayad, E.H., Hugenholtz, J. and Smit, G., 2002. Microbes from raw milk for fermented dairy products. *International Dairy Journal*, *12*(2–3), pp.91–109.

Zarei, M., Najafzadeh, H., Eskandari, M.H., Pashmforoush, M., Enayati, A., Gharibi, D. and Fazlara, A., 2012. Chemical and microbial properties of Mahyaveh, a traditional Iranian fish sauce. *Food Control*, *23*(2), pp.511–514.

Terasi

Biotechnology of the Indigenous Fish Fermented Products of Indonesia

7

Widya Agustinah, Saraswati, and Nuri Andarwulan

7.1 INTRODUCTION TO *TERASI*

Fermented fish products have played an important role in the diets of the people of East and Southeast Asia. Fermented fish products are described as the products of freshwater and marine finfish, shellfish, and crustaceans that are processed with salt to cause fermentation and thus prevent spoilage (Ruddle and Ishige 2010). In the East Asia region, eating large amounts of rice is an inexpensive source of vegetable protein, amino acids, and energy. Therefore, salty side dishes or condiments that facilitate the consumption of rice are preferred as vital foodstuffs. Fermented fish products are easy to prepare and cook, have a long shelf life, and give an umami and salty taste to vegetable dishes (Mizutani et al. 1992).

The Southeast Asia region is proposed as the center of origin of fermented fish products since the largest variety of fermented fish products and their principal dietary importance occur in the region. The development of fermented fish products in that region is related to the ecological zonal alignment of rice cultivation, the seasonal behavior of fish stocks, and the presence of salt production centers. Throughout Southeast Asia, fermented fish products are crucial for the economically poorer population, which consumes them daily in relatively large amounts (Ruddle and Ishige 2010).

Various kinds of fermented fish products can be found in various regions in Indonesia. Raw materials for the fermentation process can be whole fish or ground fish and viscera from marine and freshwater fish species. There are several fermented fish products in Indonesia, such as *peda, jambal roti, kecap ikan, terasi, ikan tukai, bekasang, bekasam, naniura, picungan,* and *cincalok* (Irianto and Giyatmi 1998). Of the various types of fermented fish products in Indonesia, *terasi* is one of the most popular. *Terasi* has been used for centuries as a flavor enhancer in various types of Indonesian cuisine. *Terasi* existed in Indonesia even before the country was established and was used before the rise and fall of the Cirebon and Sunda kingdoms located on Java Island. *Terasi* is a fermented product of shrimp, fish, or the mixture of both materials and is mainly available in the form of paste. *Terasi* can be obtained from the fermentation process with or without salt added (Prihanto and Muyasyaroh 2021). Related products are found throughout Asia, such as *belacan* in Malaysia, *kapi* in Thailand and Cambodia, *bagoong-alamang* in the Philippines, *mam-ruoc* or *mam-tom* in Vietnam, *saewo-jeot* in Korea, and *ngapi-seinsa* or *hmyinnga-pi* in Myanmar (Yoshida 1998). Some fermented shrimp pastes in Southeast Asia are produced without the addition of salt, and the step of sun drying is important for extending the shelf life. However, salting this kind of paste enhances both the taste and shelf life (Ruddle and Ishige 2010).

The color of *terasi* varies from reddish brown to black depending on the raw materials used and the natural microbes present during fermentation. Shrimp-based *terasi* (*terasi udang*) is generally red, while fish-based *terasi* (*terasi ikan*) is usually a dull grey color (Prihanto and Muyasyaroh 2021). *Terasi udang* is much more favored than *terasi ikan* for most Indonesians (Hajeb and Jinap 2013). The raw materials for *terasi udang* are small shrimp-like/planktonic crustaceans (*Acetes, Mesopodopsis, Lucifer, Mysis*, or *Atya*), usually called *udang rebon*, whilst the raw materials for *terasi ikan* are small anchovies (*Stolephorus* sp. or *Engraulis* sp.) or splendid ponyfish (*Leiognathus* sp.) (Karim et al. 2014; Huda 2012; Hajeb and Jinap 2013). A review by Hajeb and Jinap (2013) stated that *Acetes* species are the most widely used small shrimp to produce fermented shrimp products in Southeast Asian countries,

DOI: 10.1201/9781003605300-7

in which *Acetes indicus*, *A. erythraeus*, *A. vulgaris*, and *A. japonicus* are the most commonly used species. In Indonesia, *A. japonicus* and *A. sibogaesibogae* are commonly utilized as the raw materials for *terasi udang* production.

The production of *terasi* is spread along Sumatra and Java Island (Huda 2012; Ruddle and Ishige 2010). The centres of shrimp paste production in Indonesia are primarily located in South Bangka Regency, East Lampung Regency, Cirebon, Purworejo Regency, Tuban Regency, Jember Regency (Puger Subdistrict), Madura Island, and Lombok Island. In addition, *Terasi* is produced in almost all areas of the northern coast of Java Island (Prihanto and Muyasyaroh 2021)

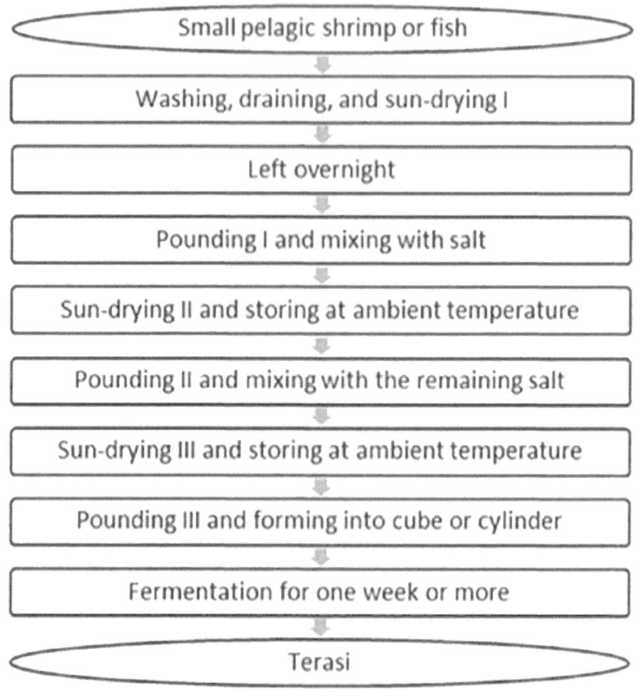

FIGURE 7.1 Flow diagram of *terasi* production.

7.2 *TERASI* PRODUCTION, PACKAGING, AND SHELF LIFE

The process of making *terasi* depends on the area of production and is closely related to the recipes developed by the ancestors of the locals. In general, the production process usually consists of drying, milling/pounding, forming, and fermentation stages. The drying and milling processes are repeatedly performed. In the fermentation process with salt addition, about 2–15% salt can be added to the raw material.

Traditional *terasi* production with salt is started by washing, draining, and drying of fish/shrimp until the raw material is half-dried. During drying, impurities, such as small fish, mussel, shells, and corals, are removed. Then the half-dried raw material is left overnight at ambient temperature and pounded the next day. In the first round of pounding, salt is added (around half of the total salt required during processing). The mixture of salt and pounded shrimp/fish is sun-dried and subsequently kept in a container at ambient temperature for 2–3 days. The second pounding step is applied, and the remaining salt is added to reach the desired concentration. After that, the pounded mixture is sun-dried again for 5–6 hours and kept at ambient temperature for 2–3 days. A further pounding step may be applied to get fine particles. The dough is formed in cubes or cylinders of 1 kg weight and subsequently fermented for a week or more at ambient temperature (Huda 2012). Heating *terasi* using an oven at 100°C for a certain period could be implemented to produce ready-to-use *terasi* (Handayani et al. 2020; Sari and Rinto 2011). The oven heating procedure was reported to extend the shelf life of ready-to-use *terasi* up to a minimum of 60 days at room temperature. A flow diagram of *terasi* production can be seen in Figure 7.1.

In principle, the processing method of *terasi* with salt and other ingredients is like the processing method of *terasi* with salt only, but in the second pounding step, the salt solution is mixed with coconut/palm sugar. Tamarind can also be added in this step. The addition of other ingredients is expected to improve the taste and flavor as well as to accelerate the fermentation process (Irianto and Giyatmi 1998). The coconut sugar addition is usually applied by producers in the Cirebon area. Junianto (2011) reported that the concentration of coconut sugar added in the production of *terasi* varied between producers in Cirebon, ranging from 1–12%. Burying small pelagic shrimp in beach sand for 2 days is performed during *terasi* production in the Belitung area. This method is carried out to hasten the anaerobic fermentation process (Prihanto and Muyasyaroh 2021).

So far, most of the *terasi* producers from various regions in Indonesia are categorized as small and medium-scale industries. The primary packaging materials usually used to wrap *terasi* can be pandan leaves, dried banana leaves, and oil-resistant paper. Polyethylene plastic is also commonly used as both primary and secondary packaging. Fungal growth during distribution and storage becomes difficult to avoid in traditional packaged *terasi* (Hidayat and Mahanani 2018; Cahyo et al. 2016; Prihanto and Muyasyaroh 2021). The use of proper packaging material can maintain *terasi* quality and improve shelf life. Cahyo et al. (2016) evaluated the effect of different packaging materials (aluminum foil and high-density polyethylene (HDPE) plastic) on *terasi* powder shelf life. The shelf life of *terasi* was assessed by accelerated shelf-life testing (ASLT) methods. Results showed that the use of aluminum foil can maintain the quality of *terasi* five to six times longer than HDPE. The quality of *terasi* powder packaged by HDPE and stored at 30°C can last up to 3 months. The observed parameters in their study included moisture content, aroma, color, and ash content.

Several studies summarized by Amir and Mahdi (2017) reported the use of rhodamine B as a red colorant in *terasi* production. That adulteration practice was found in several areas of terasi production centres. The prohibited colorant is often used in the production of *terasi* made from under-utilized fish. The main reason for rhodamine utilization is due to the lack of knowledge by producers about regulations related to food additives and the dangers of rhodamine B. Currently, several research groups have developed the use of angkak and dragon fruit peel extract as natural colorants in *terasi* production (Indriati and Andayani 2012; Permatasari et al. 2018; Hestiani et al. 2019).

7.3 CHEMICAL COMPOSITION OF *TERASI*

7.3.1 Proximate Composition, Salt, and Water Activity

As previously explained, *terasi udang* (fermented shrimp-based paste) is more popular for consumers in Indonesia because it generally has preferable sensory qualities. Along with that fact, most studies related to the analysis of the chemical composition of *terasi* from Indonesia used *terasi udang* as the sample. The chemical composition of *terasi* is strongly affected by ingredients used (types of raw material, salt percentage, and other ingredients), method of production, seasonal variation, and so on. As a consequence, those factors will yield varied unique taste profiles, aromas, and overall sensory characteristics (Damanik Ambarita et al. 2019). Ali et al. (2019) reported the proximate composition of *terasi udang* from six different producers in Madura Island, Indonesia. The moisture, protein, fat, and ash content of the observed samples ranged from 20.35–48.39%, 24.10–48.43%, 3.43–5.93%, and 21.27–29.69%, respectively. The pH and water activity values of the observed samples were about 7–8 and 0.73–0.80. In general, the types of raw materials used by various *terasi* producers in Madura are the same, that is, small pelagic shrimp (*Acetes* sp.). However, the methods applied vary. The concentration of salt used ranges from 5–10%. They also apply different pre-treatment processes and fermentation periods. But in principle, all applied methods were carried out through spontaneous fermentation techniques consisting of several stages, namely the addition of salt, drying, grinding, and fermentation at ambient temperature. Furthermore, Ali et al. (2019) added that the moisture, protein, fat, and ash content of fresh small pelagic shrimp (*Acetes* sp.) were respectively 65.45%, 20.89%, 5.83%, and 4.66%.

Ambarita et al. (2019) observed the chemical composition of *terasi* from six different regions (West Java, Jakarta, North Sumatera, Central Java, West Kalimantan, and East Java) in Indonesia. As *terasi* samples were collected from different regions which might have different ingredients and production processes, it was not surprising that the proximate composition, water activity, and salt content were significantly different ($p < 0.05$) among the observed terasi samples. The main material used in the observed *terasi* was *rebon*/small pelagic shrimp (*Acetes* sp.), except for *terasi* from West Java, which used small fish, but the type of fish used was not specified. The ranges for moisture, protein, fat, ash, carbohydrate, and salt content were respectively 27.01–37.41, 22.63–42.83, 2.19–4.46, 11.93–43.42, 3.15–7.37, and 4.60–32.83 in g/100 g fresh weight. The lowest protein content was found in *terasi* from West Java, while the highest was in *terasi* from Central Java. In addition, *terasi* from Central Java also showed the highest fat content. *Terasi* from West Kalimantan had the lowest salt and ash content, while the highest salt and ash content was found in *terasi* from West Java. The range for water activity of all samples was 0.662–0.796.

Terasi udang from Cirebon-West Java is usually processed with the addition of coconut/palm sugar. Junianto (2011) reported that the concentration of coconut sugar added in the production of *terasi* varied between producers in Cirebon, ranging from 1–12%. The moisture, protein, ash, and total sugar of *terasi udang* from three different producers in Cirebon (Pesisir Selatan, Cangkol, and Samadikun Village) ranged from 33–34.5%, 12.31–30.19%, 14.5–33.5%, and 1.59–6.49%, respectively. *Terasi* from Samadikun Village in Cirebon had the lowest total sugar because it used the lowest amount of coconut sugar (1%). The addition of sugar to *terasi* production is expected to increase *terasi* yield or improve its flavor. Pongsetkul et al. (2014) stated that the production of *kapi* (a *terasi*-like product in Thailand) also often adds root crops, flour, or sugar to improve the taste and flavor. A study by Sumardianto et al. (2019) reported the effect of different sugar concentrations on the physicochemical, microbial, and sensory characteristics of *terasi udang*. The materials used in their study were small pelagic shrimp, salt (5%), and coconut sugar (0, 7.5, and 10%). The best treatment was found in the processing of terasi with 10% coconut sugar. The best treatment resulted in *terasi udang* with the following characteristics: protein content at 29.08%, water content at 34.11%, sugar content at 15.37%, pH at 6.97, L^* value at 46.52, salt content at 13.63%, glutamic acid at 23115.83 mg/kg, BAL <10 log CFU/ml, and overall sensory score of 8.

Differences in the origin of shrimp as raw materials have been reported to have an influence on the quality of *terasi*. Romadhon et al. (2018) found that grade 1 *terasi* made from 100% pond shrimp in Pati Regency, Central Java, had the highest protein content (42.78%), fat (5.33%), and glutamic acid (40.50%). Grade 3 *terasi* that used fishery shrimp showed lower protein and glutamic acid content, as well as higher ash content. As previously discussed, the oven heating of *terasi* is often applied to produce ready-to-use/instant *terasi*. Sari and Rinto (2011) observed that the oven heating treatment on *terasi udang* from Banyuasin Regency, south Sumatra Province (100°C, 15–60 minutes), could increase the protein content from 18.62–24.28% to 43.2–45.66%. Furthermore, oven-heated *terasi* had an ash content of 33.43–35.78% and a fat content of 1.08–2.46%. Oven heating at 100°C for 60 minutes was chosen by Sari and Rinto (2011) as the best treatment because it produces *terasi* that meets the SNI 2716:2016 criteria for *terasi udang* (National Standarization Agency of Indonesia 2016).

TABLE 7.1 Quality and Food Safety Requirements of *Terasi Udang* According to SNI 2716:2016

TESTED PARAMETER	*UNIT*	*REQUIREMENT*
a. Sensory	–	Min. 7*
b. Chemical parameter		
• Moisture content	%	Max. 45**
• Acid insoluble ash	%	Max. 35***
• Salt content	%	Max. 10****
• Protein content	%	Max. 1.5
	%	12–20
	%	Min. 20
c. Microbial contamination		
• *Escherichia coli*	APM/g	<3
• *Salmonella*	per 25 g	negative

Notes: * for all sensory parameters (appearance, aroma, taste, texture)
** *terasi in paste form*
*** *dried terasi in block form*
**** *dried terasi in powder or granule form*

Until recently, the Indonesian National Standardization Agency had only set quality requirements for *terasi udang*. Quality and food safety requirements for *terasi udang* can be seen in Table 7.1.

Kim et al. (2014) reported that *terasi udang*-like product from Brunei Darussalam (*belacan*) and Korea (*saewo-jeot*) contained several important nutritional components. The observed fermented shrimp pastes had moisture content at 27–48%, protein content at 22–30%, fat content at 0.6–4.9%, and ash content at 19–46%. Moreover, the observed sample had mineral levels in the following order: Na (5069–13668 mg/100 g) > Ca (931–1457 mg/100 g) > P (513–757 mg/100 g) > K (345–548 mg/100 g) > Mg (193–438 mg/100 g) > Fe (2.09–2.26 mg/100 g) > Zn (1.34–2.22 mg/100 g). Pongsetkul et al. (2014) also observed the characteristics of *kapi* (*terasi udang*-like product) from several producers in Thailand. They found that the moisture content of *kapi* varied from 33.79 to 52.54%. Based on dry matter, protein constituted the major component (29.44–53.17%). Salt content was in the range of 22.77–35.47%, contributing to the high ash content of all samples (33.80–50.50%). Fat contents of samples were about 1.41 to 3.67%. Moreover, a wide range of carbohydrate content (4.90–32.48%) was recognized because some producers added carbohydrate sources (sugar or flour) in the production of *kapi*. Like the characteristics of *terasi* in Indonesia, *kapi* had an a_w value that ranged from 0.669–0.774.

In addition to small pelagic shrimp, fish can also be used as raw material for *terasi* production. Karim et al. (2014) conducted a study observing the effects of using different raw materials to the content of glutamic acid and the quality of *terasi*. The material used were fresh anchovies (*Stolephorus*), fresh splendid ponyfish (*Leiognathus* sp.), and small pelagic shrimp (*Acetes* sp.). Salt was added at concentration of 2%. The mixture of raw material and salt was fermented for 20 days at ambient temperature. *Terasi* made from small pelagic shrimp (*terasi udang*) showed the highest organoleptic scores for appearance, aroma, and taste (7.7–7.8 out of 9), indicating better sensory quality compared to anchovy and splendid ponyfish fish. Texture scores were similar across all samples (≈7.4–7.5), with no significant differences. Additionally, all samples received the maximum score of 9.0 in the mold category, indicating similarly high microbial quality across treatments. *Terasi udang* was reported to contain glutamic acid of 12.56%, protein content of 35.10%, moisture of 40.05%, total plate count of 1.74 × 10^3 colony/g, a_w of 0.84, salt content of 0.64%, and pH of 6.28. In contrast, protein content in anchovy- and ponyfish-based terasi was lower, i.e. 25.91 ± 0.39% and 28.20 ± 0.21%, respectively, while their moisture and glutamic acid levels ranged between 43.71–45.69% and 10.75–11.95%. *Terasi udang* has a bright and attractive red color, which is associated with the astaxanthin in shrimp. The findings of Karim et al. (2014) were in accordance with the study of Hestiani et al. (2019), showing that *terasi* derived from ponyfish (*Leiognathus equulus*) after 15 days of fermentation had a protein content of 28.56%. They used 5% salt in *terasi* production from ponyfish.

7.3.2 Amino Acid Composition

An observation concerning the amino acid profile of *terasi udang* from Madura Island was conducted by Ali et al. (2020). They reported that *terasi udang* from 2, 4, and 6 weeks fermentation contained complete essential amino acids (EAAs), indicating the nutritional value and the quality of the protein contained in *terasi*. They found a relatively high amount of EAAs (Met, Thr, Val, His, Iso, Try, Phe, Arg, Lys, Leu) than non-essential amino acids (NEAAs) (Ser, Gly, Val, Leu, Asp, Glu, Ala, Pro, Cys, Tyr, Asp, Cys, Glu) both in the raw material *Acetes* sp. and in *terasi* products.

The EAA profile was dominated by lysine, valine, and leucine, while the predominant NEAAs were glutamic acid, aspartic acid, and alanine. The concentration of amino acids in *terasi* was relatively low compared to the raw material. This can be due to the degradation of amino acids into simpler compounds during fermentation. These findings were similar to the study of Anggo et al. (2014), which reported the same tendency of the EAA and NEAA pattern in *terasi udang*. Anggo et al. (2014) also found that a longer fermentation time (up to 32 days) caused a significant decrease in amino acid concentration, primarily glutamic acid.

Data on amino acid composition of *terasi* from six different regions in Indonesia were provided in a study by Ambarita et al. (2019). Generally, total amino acids found in *terasi udang* from Indonesia ranged from 10.7–36.0 g/100 g dry matter, while total amino acids in *terasi ikan* originating from West Java was about 6.7 g/100 g dry matter, which is associated with its lower protein content (22.63%) when compared to other observed samples. The observed samples contained about 1–6 g/100 g dry matter, with glutamic acid as the most dominant amino acid in *terasi*. The other predominant amino acids found in the observed *terasi* samples were glycine, alanine, valine, leucine, and lysine. Citrulline was also detected in a considerable amount (0.51–3.0 g/100 g dry matter). Mizutani et al. (1992) added that taurine, ornithine, and citrulline are typical amino acids which also could be found in both *terasi udang* and *terasi ikan*. Ornithine and citrulline are decomposition products of arginine. Moreover, L-pyroglutamic acid and gamma-aminobutyric acid, the secondary decomposition products of glutamic acid, could be detected in *terasi udang* and *terasi ikan*. But actually, the conversion of glutamic acid was not as high as that of other amino acids, such as arginine.

In accordance with the study of *terasi* from Indonesia, *terasi*-like products from Brunei Darussalam and Korea reported by Kim et al. (2014) also showed similar tendency. The most abundant amino acid was glutamic acid in all pastes, followed by aspartic acid and leucine. Total amino acids in *belacan* from Brunei Darussalam and *saewo-jeot* from South Korea were respectively 27.1 and 18.4 g/100 g. The proportion of amino acids related to taste (sweet, bitter, meaty, and tasteless) was similar for each paste in their study. Glutamic acid, cysteine, and methionine contributed to the meaty taste. Aspartic acid, serine, arginine, threonine, proline, valine, isoleucine, and lysine are considered tasteless amino acids, while alanine, glycine, histidine, threonine, and tryptophan are known to contribute to sweet and bitter tastes. Pongsetkul et al. (2015) found the range of total amino acids in 11 *kapi* products from Thailand was about 23.77–68.95 mg/g sample, with glutamic and aspartic acid as the major amino acids found in all samples. Glycine, leucine, and lysine were also found to a high extent in all samples.

7.3.3 Fatty Acid and Organic Acid Composition

Study related to the fatty acid profile of *terasi* from various regions in Indonesia is still limited. However, Ali et al. (2020) documented the fatty acid profile of fresh *Acetes* sp. as well as *terasi* produced with different fermentation periods (2, 4, and 6 weeks). The saturated fatty acids (SFAs) of the observed samples were about 11.10–13.14 mg/g, while monounsaturated fatty acids (MUFAs) and polyunsaturated fatty acids (PUFAs) were about 8.26–18.33 and 8.11–13.19 mg/g. Palmitic, palmitoleic, eicosapentaenoic (EPA), and docosahexaenoic acid (DHA) were observed as major fatty acids found in both fresh small pelagic shrimp and *terasi* products. Saturated fatty acid levels were relatively stable during 6 weeks of fermentation, while MUFAs were detected in the highest concentration in the early stage of fermentation and decreased along the fermentation period. PUFAs, especially the omega-3 and 6 groups, had a significant increase after 6 weeks, particularly ALA, EPA, and DHA.

Kim et al. (2014) reported that *terasi*-like products from different countries (Brunei Darussalam and South Korea) had SFA, MUFA, and PUFA content ranging from 23.74–31.16%, 18.99–21.63%, and 47.21–57.27% of total fatty acids. Dominant fatty acids found in fermented shrimp pastes were palmitic, palmitoleic, oleic, EPA, and DHA. EPA content in the observed samples reached 21.73–26.92%, while DHA content was 21.84–23.74%. According to those results, *terasi* can be determined as a potential source of essential omega-3 fatty acids. Furthermore, Kim et al. (2014) also documented the cholesterol content of fermented shrimp pastes, which were about 25.29–46.99%. Montano et al. (2001) reported that Philippines shrimp paste had a SFA of 47.3%, MUFA of 18.0%, and PUFA of 34.7%. Coinciding with the findings of Ali et al. (2020) and Kim et al. (2014), Peralta et al. (2005) reported that the contents of palmitic acid, EPA, and DHA were high in salt-fermented shrimp paste.

In addition to the commonly known fatty acids content in shrimp paste, Wang et al. (2019) found the presence of branched chain fatty acids (BCFAs) in Asian fermented shrimp paste (1.63 ± 0.72%). Habitual *nātto* and/or fermented seafood consumption could support BCFA intakes in Asian countries. Milk is well known as a big contributor to the BCFA intake in the world, especially in milk-producing countries. BCFAs are mostly saturated fatty acids with a terminal propan-2-yl (isopropyl) group (iso) or butan-2-yl (sec-butyl) group (anteiso). Food BCFAs originate primarily from bacterial action. Wang et al. (2019) summarized that BCFAs may support gut health by reducing the incidence of necrotizing enterocolitis as well as reducing pro-inflammatory markers in a human intestinal cell.

Yoshida (1998) summarized that typical organic acids which could be found in fermented shrimp paste in Southeast Asia were pyroglutamic, acetic, malic, and succinic acid. Acetic acid was observed as dominant organic acid in fermented shrimp paste; its concentration in Bangladesh *nappi* reached 2941.12 mg/100 g. A study by Mizutani et al. (1987) showed that acetic acid content in Indonesian *terasi* was 440.36 mg/100 g, while lactic and succinic acid content reached 152.6 and 95.92 mg/100 g. Similar to these findings, Mizutani et al. (1992) found that acetic, lactic, and succinic acid were detected in considerable amounts in Asian fermented shrimp pastes. The observed total organic acids reached 1.47 ± 0.9%. When referring to a recent study conducted by Ambarita et al. (2019), the acetic acid content in *terasi* samples from various regions in Indonesia was around 0.63–1.5 g/100 g dry matter. Other kinds of short chain fatty acids which could be detected in *terasi* samples from Indonesia were propionic acid (0.055–0.11 g/100 g), isobutyric acid (0.009–0.24 g/100 g), butyric acid (0.013–0.13 g/100 g), and isovaleric acid (0.051–0.64 g/100 g).

7.4 MICROBIOLOGICAL ASPECTS OF *TERASI*

7.4.1 Microbial Diversity of *Terasi*

The manufacturing process in each region determines the flavor of *terasi* as a result of naturally occurring microbial growth and enzyme activities. The length of sun exposure in the drying process and the fermentation time are the contributing factors in determining the similar yet unique flavors and aromas of the final product that was produced in each region of Indonesia (J. Pongsetkul et al. 2015; Murwani et al. 2016). Other factors, such as the geographical condition and raw material quality, could also influence the final product characteristics in each region (Arfarita and Chukeatirote 2015). Fish or shrimp that are caught in different ecological settings in each location have different characteristics (Murwani et al. 2016). The fresher and more uniform the ingredients are, the higher the quality of the *terasi* produced.

The color of *terasi* is reddish-brown to grey to black depending on the raw materials and the presence of indigenous microbes in the fermentation process (Prihanto and Muyasyaroh 2021). *Terasi* paste made from fermented shrimp exhibits a red color. Shrimp contain a carotenoid pigment compound, astaxanthin, which can undergo non-enzymatic browning and change to a brown-reddish color upon heating, such as sun-drying (Murwani et al. 2016). The protease enzyme also helps in releasing the free form of astaxanthin into the product and darkening the color of *terasi* (Prihanto et al. 2021a). The non-enzymatic browning reaction produces certain compounds which would improve the taste and aroma of *terasi*. When fish is used as a substrate of *terasi* fermentation, the color of *terasi* is typically dull grey. A coloring agent is commonly used to improve the color of the fermented fish *terasi*/*terasi ikan* and, therefore, attract more consumers.

The high salt environment of 10–17% salt concentration during *terasi* fermentation could select the growth of halophilic bacteria. The presence of indigenous halophilic bacteria in *terasi*, such as *Bacillus* sp. and *Micrococcus* sp., could originate from the fish, salt in the fermentation process, and contamination from the environment (Surono and Hosono 1994b). The large microbial diversity in *terasi*, which is due to the traditional manufacturing process, could be potentially used to differentiate the originality of *terasi* from each region and can be further standardized as a geographical indication of each region. Table 7.2 shows several unique traditional ways of producing *terasi* which could affect the microbial diversity in the final product (Prihanto and Muyasyaroh 2021).

The quality of *terasi* can also be determined from the flavor and nutritional profile. *Terasi* is commonly used in the local cuisines because its distinct taste and strong aroma can induce appetite, increase food intake, and improve consumer satisfaction (Murwani et al. 2016). The unique flavor and aroma of *terasi* are the results of the fermentation process, which was carried out predominantly by the halophilic lactic

TABLE 7.2 Various Methods of *Terasi* Manufacturing Process in Indonesia as Potential Sources of Microbial Diversity (Summarized from Murwani et al. 2016 and Prihanto and Muyasyaroh 2021)

		SOURCE OF MICROBIAL DIVERSITY	
REGION OF PRODUCER	*LOCAL NAME*	*SEAFOOD TYPE*	*UNIQUE PROCESSING METHOD*
Belitung Regency: Sijuk and Tanjungpandan districts	*Belacan*	Small shrimp or *calok* (*Atya* sp.)	Burying shrimp in beach sands for 2 days to initiate spontaneous anaerobic fermentation in the sands Packaging of terasi in woven pandanus leaves
South Bangka: Toboali area	Toboali terasi	Small white shrimp (1–2 cm in length)	
Lombok island	Lombok terasi	Shrimp, fish, and mixed	
Puger Kulon Village	Puger terasi	Planktonic shrimp (*Acetes* sp.)	
East Java: Tuban area	Tuban terasi	Shrimp	Pre-treatment before fermentation: Sun-drying of the shrimp for 1 day, pulverizing with a wood mortar, sun-drying of the dough, re-grinding with a wood mortar, and packing in a cylindrical or cubic shape. Fermentation in the packaging occurs in approximately 2 weeks
Central Java: Pekalongan area	Pekalongan terasi	Shrimp	Adding salt and food color
Central Java: Batang area	Batang terasi	Shrimp	Without adding salt
West Java: Cirebon area	Cirebon terasi	Shrimp	Drying for 1 day before fermentation

acid bacteria (LAB) and indigenous enzyme-producing bacteria. These microbes can degrade the proteins and fats in the raw materials and produce amino acids; fatty acids; other carbonyl compounds; ammonia; amines; and simple sulfur compounds, such as sulfides, mercaptans, and disulfides (Prihanto and Muyasyaroh 2021). The presence of LAB is important to inhibit the growth of spoilage bacteria due to the production of lactic acid that reduces the pH value temporarily to 4.5 during fermentation (Kobayashi et al. 2003). The growth of proteolytic bacteria is also responsible for the transformation of proteins to amino acids, which can improve the nutritional value of the fermented fish or shrimp paste. Essential amino acids, such as lysine and leucine, and non-essential amino acids, such as glutamic acid and alanine, are reported as the predominant free amino acids in Thai traditional fermented shrimp paste (Kleekayai et al. 2016).

7.4.2 Enzymatic Activities of *Terasi* Microbial Flora

Terasi is fermented anaerobically with the primary presence of halophilic or halotolerant bacteria. The high salt concentration in Indonesian *terasi* creates high osmotic pressure that causes a selection of halophilic LAB growth in the final product, such as the *Tetragenoccocus* species (previously classified in the genus *Pediococcus*). Two predominating types of halophilic LAB isolates have been found from Cirebon *terasi*, *Tetragenococcus halophilus* and *T. muriaticus*, which are both homofermentative, catalase-negative, Gram-positive cocci and able to grow at 18% NaCl and pH 8.5 but not at pH 4.2 (Kobayashi et al. 2003). *T. halophilus* and *T. muriaticus* were found to be monosaccharide and oligosaccharide fermenters, respectively. Unlike *T. muriaticus*, which could commonly produce histamine and was associated with histamine accumulation in fermented seafood, the isolated strain of *T. muriaticus* from Cirebon *terasi* did not produce histamine (Kobayashi et al. 2003).

Terasi fermentation involves the activity of several enzymes, such as protease, lipase, chitinase, chitosanase, and fibrinase (Prihanto et al. 2021a). Since protease serves as the key enzyme in *terasi* fermentation and determines the quality of the final product, almost 99% of the bacterial isolates from *terasi* can produce protease enzyme (Arfarita and Chukeatirote 2015). The activity of intracellular aminopeptidase enzyme in the proteolytic system of LAB is important in the release of amino acids as the major precursors for volatile compound formation. *T. halophilus* isolates from Thai fish sauce (*nam pla*) were reported to have aminopeptidase activity and were associated with the volatile compound formation during fermentation (Udomsil et al. 2010).

Gram-positive bacteria isolated from Cirebon and Tuban *terasi* were *Bacillus thuringiensis* strain S2–3 and *B. subtilis* strain S4–5. Both strains had high proteolytic activity and were resistant to ampicillin and susceptible to several antibiotics, such as chloramphenicol, erythromycin, kanamycin, tetracycline, vancomycin, and bacitracin (Chukeatirote et al. 2015). The *B. subtilis* UBT7 isolate from *terasi* served as a potential source of a thermophilic and neutral extracellular protease enzyme which showed an optimum activity at 50°C and pH 7 (A. Prihanto et al. 2016). The ability to synthesize and excrete protease was essential for the bacteria's survival in extreme conditions, such as the low water activity of the dried solid paste form and high salt concentration of *terasi*. The release of protease was important to hydrolyze the protein to amino acids and provide energy for microbial growth. However, an exception was found in Batang *terasi* manufactured without salt addition (Table 7.2). The presence of a higher number of volatile compounds in Batang *terasi*, as compared to Pekalogan and Kendal *terasi*, which involved salt addition in the processing, was presumably correlated with the growth of mesophilic proteolytic bacteria during fermentation (Murwani et al. 2016).

The microflora of *terasi*, particularly from the Bagansiapi-api region in North Sumatera, were dominated by *Bacillus* sp and cocci bacteria. They were composed of *B. brevis*, *B. pumilus*, *B. megaterium*, *B. coagulans*, *B. subtilis*, and *Micrococcus kristinae*, which were all aerobic and salt tolerant to 10% NaCl (Surono and Hosono 1994a). *Bacillus* sp. isolated from *terasi* could grow in the temperature range of 10–50°C and show esterase (C4) and esterase lipase (C8) activity. However, all isolates were weakly proteolytic. Lipolytic activity on the long-chain fatty acids was only present in *B. pumilus*, *B. coagulans*, and *M. kristinae*. The presence of these lipolytic bacteria contributed to the development of cheesy and ammoniacal odors in *terasi* due to the production of low molecular weight fatty acids, amines, and ammonia, respectively. In addition, the presence of endogenous proteolytic enzymes in the fish muscle protein was also important to initiate muscle protein denaturation prior to the bacterial activity and release short-chain saturated fatty acids that were firmly bound to the fish proteins (Surono and Hosono 1994b).

The presence of halophilic *Bacillus* sp and *Micrococcus* sp. could be found in the *terasi* starter and final product in the amount of up to 5 log CFU/g. Other microorganisms, such as *Pseudomonas* sp., *Kurthia* sp., and *Sporolactobacillus* sp., could also be found in the final product, which originated from contamination during *terasi* manufacturing process (Surono and Hosono 1994b). *Bacillus, Lactobacillus, Corynebacterium,* and *Staphylococcus* were commonly found as indigenous microflora in Malaysian fermented shrimp paste (*belacan* and *cincalok*), while *T. halophilus* was the indigenous microflora in Filipino fermented shrimp paste (*bagoong alamang*) (Nunal et al. 2016).

Other enzymatic activities with potential industrial and medical therapy applications were also reported from the bacterial isolates of fermented shrimp and fish paste *terasi*. Chitinolytic activity was shown from the supernatant of bacterial isolate T5a1 and JB4, which was isolated from Sidoarjo (East Java) shrimp *terasi*. The activity of T5a1 chitinase crude enzyme fraction was optimum at pH 7.0 and temperature 40°C and required calcium (Ca^{2+}) and ferric (Fe^{3+}) ions to increase its activity (Noviendri et al. 2008), while the JB4 chitinase enzyme fraction was optimum at pH 8.0 and temperature 40°C and needed monovalent cations (NH_4^+ and Na^+) as the activator (Noviendri et al. 2006). The isolate of *B. cereus* HMRSC30 from *terasi* had the ability to produce a low-cost chitosanase and could potentially be used simultaneously with chitin extraction under optimized conditions (Cahyaningtyas et al. 2021).

Novel fibrinolytic enzymes were extracted from *Bacillus* sp. nov. SK006 isolate from fermented shrimp paste (Hua et al. 2008). This fibrinolytic enzyme had a high specific activity (11.2 U/mg) which was optimum at 30°C and pH 7.2, was stable below 40°C between pH 5.0 and 11.0, and could degrade fibrin clots. The bacterial isolates from fermented anchovy (*Stolephorus* sp.) paste *terasi* were reported to show higher amylase activity (7.7 U/mg) than protease activity (1.6 U/mg) (Oktavia et al. 2018). The optimum production time to achieve the highest protease and amylase activity was 36 h at 37°C and 28°C, respectively.

Prihanto et al. (2021b) reported the growth of L-asparaginase-producing bacteria in shrimp *terasi*. The highest activity of extracellular L-asparaginase (14 U/mg) was produced by *B. cereus* strain T4 isolate. *B. cereus* was abundant in the soil and plants, and it was generally known as an opportunistic pathogen by its ability to produce enterotoxin and emetic toxin. *B. cereus* was also detected in Indian fermented fish paste, *tungtap*, at more than 10^2 CFU/g, along with *Staphylococcus aureus* and Enterobacteriaceae as the pathogenic contaminants (Thapa et al. 2004).

The growth of *B. cereus* in *terasi* might be associated with contamination from the soil and air during sun drying and fermentation. However, further investigation is required to confirm the strain specificity and pathogenicity of the *B. cereus* isolate from *terasi*. Fermented food could provide a different environment that could inhibit the growth of pathogenic bacteria and select the growing microorganisms in the final product. For example, *Klebsiella pneumoniae* is known as an opportunistic pathogen which can produce enterotoxins and cause pneumoniae and liver abscesses in humans. The growth of *K. pneumoniae* in tempeh, the Indonesian mold-fermented soybean, is associated with vitamin B12 production. It was confirmed that *K. pneumoniae* isolates from tempeh were genetically different from pathogenic isolates by the ERIC-PCR method (Ayu et al. 2014) and did not produce enterotoxins (Keuth and Bisping 1994).

7.4.3 The Importance of *Terasi* Starters

Starter cultures are usually added to initiate the fermentation process and speed up the process to obtain certain quality of the final product. The utilization of approved starter cultures along with the appropriate monitoring of fermentation procedures, correct storage conditions, and knowledge of the microorganisms are important to reduce the risk of contamination during food fermentation, therefore ensuring the safety and quality of the fermented product (Allwood et al. 2021). *Terasi* is produced by a spontaneous fermentation which utilizes endogenous bacteria from the raw materials. This process results in the varying quality of *terasi* due to many aspects, such as variation of the raw materials, processing steps, and storage condition, as previously described. The addition of starter culture in *terasi* making could control the fermentation process and improve the safety, nutritional quality, sensory characteristics, and functional properties of *terasi*.

Suitable candidates for the starter culture development of *terasi* production are LAB and hydrolytic enzyme-producing bacteria since they are the major types of bacteria found in *terasi* and other fermented shrimp or fish paste in Asian countries. *L. plantarum* SB7 and *B. amyloliquefaciens* BC9 isolates from mare's milk have been applied as a starter in shrimp *terasi* fermentation (Prihanto et al. 2021a). *Terasi* made using either or both bacteria showed a shorter fermentation period and good quality, such as darker color, higher protein content, high glutamic acid level, more variety of volatile alcohols, and improved sensory properties. The use of *Pediococcus halophilus* (which has been renamed to *T. halophilus*) as a starter in anchovy *terasi* at 10^6 CFU/mL showed an ability to shorten the fermentation period by producing a stable amount of amino acids since day 14 of fermentation (Lestari et al. 2017).

Starter culture composition could also be developed from the isolates of natural microflora from fermented shrimp paste. Three strains of natural microflora, *Staphylococcus saprophyticus*, *B. subtilis*, and *L. murinus*, were isolated from Cirebon *terasi* and further optimized as the starter in the making of semi-wet *terasi* (Sato et al. 2020). Semi-wet *terasi* had higher water activity and thus a shorter shelf life than dry *terasi*. The use of these three strains in the optimized mixture ratio of 2:1:2, respectively, was effective in suppressing the growth of indigenous microbes in *terasi* by exhibiting antimicrobial activities against Gram-positive and Gram-negative bacteria and showing a different metabolite profile than the product of natural fermentation without a starter addition. In another study, the mixture of three isolated strains from Filipino fermented shrimp paste, *bagoong alamang*, that is, *Corynebacterium* sp., HT13, *B. subtilis* LB15, and *Lb. plantarum* LB4, were developed into a starter culture for the making of the same fermented shrimp paste with antibacterial activities against several foodborne pathogens, increased soluble protein content, and improved sensory properties as compared to fermented shrimp paste made without a starter culture (Nunal et al. 2016). Fermentation occurred at a faster rate due to the elimination of unwanted microorganisms and indigenous microbes. All studies demonstrated the improvement of safety, nutrition, and sensory qualities of fermented shrimp pastes which were made by the addition of a mixed starter culture.

The major concerns for the safe consumption of fermented food are the contamination of pathogenic and toxin-producing microorganisms and production of biogenic amines in a high-protein substrate. Some contaminating and possible toxin-forming bacteria were reported in *terasi*. *B. cereus* was isolated from *terasi* produced in East Java and suspected to grow in *terasi* due to contamination of the soil and air during the traditional *terasi* manufacturing process (Prihanto et al. 2021b). The ability of *B. cereus* in producing enterotoxins was shown when the bacteria reached levels higher than 10^4 CFU/g. The contamination of *B. cereus* in other fermented food, such as Korean fermented soybean paste (*doenjang*) and Japanese *koji* or *miso,* was found at low levels, and no toxins were detected (Allwood et al. 2021). Other studies did not find the presence of contaminating pathogens, such as *B. cereus, S. aureus*, and Enterobacteriaceae, in Cirebon, Malang, and Tuban *terasi* (Arfarita and Chukeatirote 2015). There were also no reports of food safety outbreaks by the consumption of *terasi*; therefore, *terasi* was considered safe for consumption. The findings, however, indicated the need to develop appropriate starter cultures and control the hygiene of the manufacturing process to ensure the quality and safety of *terasi*.

Biogenic amines, which are heat stable and can be found in various foods of animal and plant origins, including fermented foods, are associated with spoilage and hygiene conditions of food processing (Shalaby 1996). A high concentration of biogenic amines in food is usually found in decomposed, spoiled, or organoleptically unacceptable products due to bacterial contamination. Biogenic amines are also considered a natural antinutritive factor which could have harmful effects on human health, such as histamine poisoning, tyramine toxicity, and carcinogenic effects, upon consumption of high amounts. The synthesis of biogenic amines from microorganisms occurs when free amino acids are available in the substrate and subjected to decarboxylation by the production of amino acid decarboxylases to remove the α-carboxyl group and form the corresponding amines. These biogenic amines include tyramine, cadaverine, putrescine, histamine, tryptamine, and beta-phenylethylamine.

The formation of biogenic amines in fermented shrimp paste during storage is dependent on several factors, such as shrimp species, microbial flora, packaging type, storage temperature, and the use of antimicrobial agents (Cai et al. 2015). Several free amino acids in food, such as histidine, tyrosine, ornithine, lysine, phenylalanine, and tryptophan, are the precursors of biogenic amine formation. Some microorganisms are known as producers of amino acid decarboxylase, such as certain *Enterobacteriaceae, Clostridium, Lactobacillus, Streptococcus, Micrococcus*, and *Pseudomonas* species (Shalaby 1996). Some of these microorganisms have been isolated from *terasi,* such as *M. kristinae* and *Pseudomonas* sp. (Surono and Hosono 1994b). The presence of proteolytic bacteria in *terasi* and the resulting high concentration of free amino acids in *terasi* (Prihanto et al. 2021a) might also potentially contribute to the synthesis of biogenic amines by microorganisms. However, the microorganisms' capability to produce biogenic amines is strain dependent (Lestari et al. 2017).

The halophilic LAB strain of *T. muriaticus* which was isolated from Cirebon *terasi* was confirmed to lack histamine-producing properties (Kobayashi et al. 2003). Bacterial isolates from Indian fermented fish paste, *tungtap*, did not show the ability to decarboxylate amino acids, such as tyrosine, lysine, histidine, and ornithine (Thapa et al. 2004). *Terasi* made from fermented anchovy (*Stolephorus* sp.) using *P. halophilus* showed a lower level of lysine as the precursor of cadaverine formation (Lestari et al. 2017) and no decarboxylase activity. Several strains of pedioccocci were not able to produce amines (Bover-Cid and Holzapfel 1999). The cell-free supernatant of other LAB at certain concentrations, such as *P. acidophilus, Leuconostoc mesentroides* subsp. *cremoris, Lactococcus lactis* subsp. *lactis*, and *Streptococcus thermophilus*, could reduce the growth of some pathogens and the production of histamine, ammonia, and other biogenic amines (Özogul et al. 2017). These results indicated that the selection of starter was important to provide protection and specific interaction with the indigenous microbes in *terasi* to prevent biogenic amine formation.

Some reports showed that only low levels of biogenic amines were detected in fermented shrimp or fish paste. Damanik Ambarita et al. (2019) documented the biogenic amine profile of terasi samples obtained from six different areas in Indonesia, as explained previously (total biogenic amine range 0.99–6.10 g/100g). Putrescine (0.31–2.10 g/100 g dry matter) and cadaverine (0.49–3.00 g/100 g) were the major biogenic amines detected in all samples. Putrescine is formed by the decarboxylation of ornithine and arginine, whilst cadaverine resulted from the decarboxylation of lysine. Histamine, which was detected in trace amounts (0.05–0.12 g/100 g dry matter), is well known as a derivative of histidine. Tryptamine is an indolamine metabolite of tryptophan. Beta-phenylethylamine is a derivative of phenylalanine, while tyramine is a by-product of tyrosine breakdown. Histamine was detected in Thai fermented shrimp paste (*kapi*) in low amounts in the range of 20–54 mg/kg (Daroonpunt et al. 2016). Cai et al. (2015) reported that the fermented shrimp paste produced by the addition of 0.1% starter culture and 18% salt showed low levels of biogenic amines (<50 mg/kg) upon storage for 80 days at room temperature. The maximum limit of histamine that could exhibit toxicity is 200 mg/kg, according to Codex Alimentarius (Daroonpunt et al. 2016). The low histamine content also indicated the correct handling and storage of raw materials and the proper processing of the fermented shrimp paste. Optimally, fresh shrimp should be immediately mixed with salt after being caught and before reaching the shore (Daroonpunt et al. 2016). Sorting small shrimp at the shore during *terasi* making is important to select a fresh and uniform size of shrimp and discard small fish, insects, clams, and other non-shrimp materials, as they would lower the quality of *terasi* produced. Further, the addition of natural antioxidants, such as the combination of tea polyphenols and 6-gingerol, could significantly inhibit the production of biogenic amines, maintain more amino acids in the fermented shrimp paste, and extend shelf life to 160 days under ambient temperature storage (Cai et al. 2015).

Several bacterial isolates from *terasi* and other fermented shrimp or fish pastes may be classified as generally recognized as safe (GRAS) microorganisms and serve as potential candidates for the development of *terasi* starter inocula to produce certain metabolites with health benefits. Enzymes and amino acids are examples of primary metabolites from bacterial isolates which play an important role in improving the sensory properties and nutritional value of fermented shrimp or fish pastes, including *terasi*. Enzymes with well-known industrial applications, such as protease, lipase, and amylase, and other enzymes with health-related application, such as chitinase, chitosanase, L-asparaginase, and fibrinase, have been extracted and characterized from various bacterial isolates of *terasi*, as described in the previous section.

Chitinase has been intensively studied in the production of bioactive compounds. Chitin is the major polysaccharide component in the crustacean shell, which is built from N-acetylglucosamine monomers. Chitosan is a product of partial de-acetylation of chitin and has a higher solubility at a neutral pH than chitin. Bacterial chitinase and chitosanase were used to hydrolyze chitin and chitosan into low molecular weight oligomers which were reported to show biological activities, such as antimicrobial, antioxidant, antitumor, antihypertensive, and immunostimulant activity (Puspita et al. 2017). L-asparaginase (L-asparagine amidohydrolase, E.C.3.5.1.1), which hydrolyzes the essential amino acid asparagine needed for leukemic cell growth, has been used as a chemotherapeutic enzyme with antineoplastic activity (Batool et al. 2016). Crude L-asparaginase has been produced by *B. cereus* isolates from *terasi* (Prihanto et al. 2021b). Fibrinolytic enzyme extracted from a bacterial isolate from fermented shrimp paste had the ability to degrade fibrin clots by forming active plasmin from plasminogen or direct fibrinolysis (Hua et al. 2008). The impairment of fibrinolytic mechanisms was associated with the pathogenesis of atherosclerosis and increased risk of cardiovascular diseases.

The growing bacteria in fermented shrimp pastes, such as *terasi*, Thai *kapi*, and Myanmar *ngapi*, are also capable of producing antibacterial compounds naturally. Bacteriocins from *kapi* and *ngapi* are produced by *B. amyloliquefaciens* and *Bacillus* sp. isolates, respectively. *Virgibacillus salexigens* has been isolated and genetically identified from shrimp *terasi* as a Gram-positive, spore-forming, and strictly aerobic bacterium. It was reported that a *V. salexigens* strain P2 from *terasi* produced a Class IId bacteriocin compound, which was characterized as having a one-peptide and non-pediocin-like linear structure, 5.3 kDa of monoisotropic mass weight, stability at high temperatures and various pH treatments, and a broad spectrum of inhibiting other Gram-positive bacteria, including *Listeria monocytogenes* (Kobayashi et al. 2016). These bacteriocins could be potentially applied in food industries as naturally derived food additives to replace synthetic preservatives, following nisin A, the Class I bacteriocin, which has received an FDA approval and been added to the European food additive list.

Bacterial isolates, such as *Bacillus* sp. and some LAB species, from various fermented foods are known for their ability to produce gamma-aminobutyric acid (GABA), a non-protein amino acid with bioactive functions. *B. thuringiensis* and *Lc. lactis*, from Korean fermented shrimp paste (*jeotgal*) have been isolated and characterized for their ability to produce high concentrations of GABA (Vo and Park 2019). A high concentration of glutamate in fermented shrimp and/or fish paste is an essential precursor for GABA accumulation through the activity of glutamic acid decarboxylase; therefore, an isolated strain of GABA-producing bacteria could be developed into a promising starter culture in *terasi* making with increased GABA content.

7.5 FLAVOR COMPOUNDS AND MICROBIAL ACTIVITIES IN *TERASI*

Peptides, amino acids, other nitrogenous compounds, and volatile compounds are the result of microbial and enzymatic activities during the fermentation process which influence the taste and aroma of a fermented seafood product (J. Pongsetkul et al. 2015). The production of these compounds is generally derived from protein and lipid degradation through autolytic reactions, bacterial enzyme activities, and chemical reactions during fermentation and storage (Kleekayai et al. 2016; Prihanto et al. 2021a). Both endogenous and exogenous enzymes are involved in the production of bioactive peptides and free amino acids, including essential amino acids, in fermented fish or shrimp products. The content of free amino acids can vary within fermentation due to protein decomposition and transformation of amino acids to amines, volatile acids, and other nitrogenous compounds through bacterial and enzymatic activities, including transamination and oxidative deamination (Kleekayai et al. 2016).

The concentration of amino acids in the final product determines sensory acceptance. The amino acid composition varies in different fermented salted shrimp pastes and is dominated by glutamate/glutamine and aspartate/asparagine amino acids (J. Pongsetkul et al. 2015). *Terasi* is confirmed as a rich source of amino acids, such as glutamic acid, aspartic acid, leucine, and arginine, with amounts above 20,000 mg/kg (Prihanto et al. 2021a). Glutamic acid influences the perception of meaty and umami flavors, while asparagine is associated with sweetness (J.-S. Kim et al. 2005). The presence of histidine, leucine, isoleucine, methionine, phenylalanine, and tryptophan is related to bitter taste, while that of glycine, alanine, serine, threonine, and valine is related to sweet taste (Daroonpunt et al. 2016).

The content of *kokumi* peptide, γ-glutamyl-valyl-glycine (γ-glu-val-gly), in fermented shrimp paste was reported by Miyamura et al. (2014). This peptide was presumably perceived through the calcium-sensing receptor (CaSR) in humans and showed sensory attributes of continuity, mouthfulness, and thick flavor. *Kokumi* is considered the sixth taste because its sensation cannot be explained by the five basic tastes alone. The presence of *kokumi* substances is important as a flavor modifier and enhancer in food, although the substances themselves do not have a taste (Kuroda and Miyamura 2015). It is expected that a potent *kokumi* peptide could be detected in fermented shrimp paste which is already used as a condiment in many Asian countries. The content of γ-glu-val-gly in

Indonesian *terasi* was five times higher than that in Chinese *xiajiang* and Filipino *bagoong* fermented shrimp paste (Miyamura et al. 2014). The variation of *kokumi* peptide concentration could be affected by the shrimp species, microbial flora, and fermentation condition.

The release of volatile compounds in *terasi* occurs upon heating and cooking. The volatile compounds of *terasi* are grouped into nitrogen-containing compounds, sulfur-containing compounds, alcohols, carbonyl compounds, hydrocarbons, fatty acids, and ester compounds. These compounds are also found in fermented fish and shrimp paste from other countries, including China and Thailand (Li et al. 2013; J. Pongsetkul et al. 2015). Different types of volatile compounds are found in different regions of shrimp *terasi* manufacturing areas along the northern coast of Central Java, such as Demak, Jepara, Pati, Pekalongan, Batang, and Kendal (Murwani et al. 2016). Batang *terasi* is considered unique in the manufacturing process since salt is not added in the process. It contains unique volatile compounds not found in other salted *terasi*, such as 1-docen-3-yne and cyclododecane (hydrocarbons), 2-hydroxy-4,6-cyclooctadien-1-one (carbonyl), ethane, 1,2-bis-methylthio (sulfur), 3-heptanol, 3,5-dimethyl, 1,3-benzendiol, 5-methyl, 1-tetradecanol, 11-hexadecen-1-ol (alcohol), 2,5-dimethylpyrazine, 2–(2-hydroxyethyl) piperidine, 1-phenyl-2-methylenaminopropane, 2-phenyl-N-benzilidine ethylamine (nitrogen), and ethyl laurate (ester). Some compounds, such as trimethylamine, indole, S-containing compounds, acetic acid, esters, and phenols, are continually generated in stinky mandarin fish, while the aldehyde level decreases during fermentation (Li et al. 2013). Trimethylamine, which gives the fishy odor characteristic, is the product of microbial reduction of trimethylamine oxide and decreases upon prolonged fermentation (Kleekayai et al. 2016).

A study by Ali et al. (2020) reported the volatile compound profiles of *terasi udang* from six different producers on Madura Island. Volatile compounds in *terasi udang* are formed by autolytic and bacterial degradation of proteins and lipids as well as due to chemical reactions through fermentation. In addition to identifying the volatile compounds, Ali et al. (2020) also quantified the relative amount of each detected volatile compound by referring to the peak area. A total of 59 compounds were detected in fresh shrimp and *terasi* samples, including alcohol compounds (8.25–10.85%), carbonyls (26.65–31.51%), acids (15.42–17.49%), esters (3.12–9.87%), hydrocarbons (4.48–9.27%), N-containing compounds (20.87–24.78%), S-containing compounds (6.16–7.58%), and phenolic compounds (0–1.5%). There were only 45 volatile compounds found in fresh shrimp. The longer the fermentation period applied, the more types of volatile compounds were detected. Fermentation of *terasi* for 2, 4, and 6 weeks resulted in as many as 46, 55, and 57 volatile compounds, respectively. Volatile compounds varied depending on shrimp/fish species and manufacturing conditions as well as the strains of microorganisms involved.

Nitrogen-containing compounds, especially pyrazine derivatives, are the major volatile compounds and the important contributors of flavor development in fermented shrimp paste (*Ka-pi*) in Thailand (Kleekayai et al. 2016; J. Pongsetkul et al. 2015). In accordance with a study by Pongsetkul et al. (2015), nitrogen-containing compounds, especially pyrazine derivatives such as 2-methylpyrazine, 2,5-dimethylpyrazine, 2,6-dimethylpyrazine, 2-ethypyrazine, 2,3-dimethylpyrazine, 2,3,5-trimethylpyrazine, tetramethylpyrazine, and 2,6-dimethyl-3-propylpyrazine, seemed to be the major volatile components in *terasi udang*. Pyrazines were recognized to contribute to nutty, roasted, and toasted aromas in many foods. Pyrazines are formed by the Maillard reaction through Strecker degradations from various nitrogen sources such as amino acids. Pongsetkul et al. (2015) also summarized that alcohol derivatives derive from lactic acid bacterial fermentation and the degradation of oxidized lipid. Aldehydes, such as benzaldehydes, are more likely generated from lipid oxidation during fermentation. Branched short-chain aldehydes or aromatic aldehydes might result from deamination of amino acids. Additionally, alcohols, aldehydes, and ketones are typically derived from lipid oxidation and the degradation of amino acids and carbohydrate derivatives such as monosaccharides or high chain length oligosaccharides, whereas aromatic ring–derived compounds such as phenolics can be generated from lipids oxidation, sugar, and protein degradation. Sulfuric compounds in fermented shrimp paste originate from enzymatic or heat protein degradation and might be associated with S-containing amino acids (Varlet and Fernandez 2010). Various hydrocarbon compounds, such as alkanes, alkenes, and alkadienes, are produced through fat oxidation during fermentation (Karahadian and Lindsay 1989).

A different profile of N-containing compounds was detected in *terasi*. 2,5-dimethylpyrazine was only found in Batang *terasi* (Murwani et al. 2016) and not in salted *terasi*. Pyrazines are the products of the Maillard reaction through Strecker degradation, which used amino acids as the precursor of nitrogen sources and reducing sugars as carbon sources and where a slightly high-pH environment was favorable (pH 7.2–8.4). The formation of pyrazines also resulted from microbial activities during fermentation. The presence of pyrazines contributed to the roasted nutty odor in Thai fermented shrimp paste (Kleekayai et al. 2016) and burnt and sweet odors in Vietnamese fish sauce (*nuoc-mam*) (J. Pongsetkul et al. 2015).

Indole, a product of tryptophan catabolism, was found in all types of *terasi* produced along the northern coast of Central Java (Murwani et al. 2016). Indole was also found in Chinese traditional fermented shrimp paste (Fan et al. 2017) and some Thai fermented shrimp paste (Kleekayai et al. 2016). The production of indole has been associated with shrimp spoilage; therefore, it can be used as a quality (freshness) indicator of raw materials in fermented shrimp paste manufacturing. The rate of tryptophan decomposition might vary, as it is affected by which microorganisms were present during fermentation (J. Pongsetkul et al. 2015).

Dimethyl trisulfide was the major S-containing compounds in Central Java province *terasi* in Indonesia (Murwani et al. 2016). Dimethyl disulfide was found in Thai fermented shrimp paste. Both dimethyl disulfide and dimethyl trisulfide concentrations, which are responsible for an undesirable fecal note odor, increased with the fermentation time and generated the distinctive aroma of the final product (Kleekayai et al. 2016). The formation of S-containing compounds was presumably associated with the degradation of S-containing amino acids, such as cysteine and methionine, by microbial activities and high temperature (Varlet and Fernandez 2010). The strong aroma of *terasi* and other fermented shrimp paste was due to the low threshold values of S-containing compounds, that is, between 0.16–12 and 0.005–0.01 ppb, for dimethyl disulfide and dimethyl trisulfide, respectively (Kleekayai et al. 2016). It was reported that the addition of *T. halophilus* isolate in fish sauce fermentation could reduce the dimethyl disulfide and biogenic amine content of a fish sauce product (Udomsil et al. 2011).

The formation of alcohol, aldehydes, and ketones was generally derived from the lipid oxidation reaction and degradation of amino acids and sugars (Kleekayai et al. 2016). Protein was hydrolyzed into peptide and amino acids, which were further degraded into other components, such as hydrocarbons, esters, and alcohols. Alcohols were detected in large amounts in fish paste but were lower in shrimp paste. 2-butanol and 3-methyl-butanol were the major alcohol compounds in Thai salted shrimp paste, while phenolic compounds and longer carbon chains of alcohol groups were dominantly found in *terasi* from Central Java. Butanol derivatives might be produced through LAB fermentation from amino acids, such as phenylalanine, valine, and leucine (Udomsil et al. 2010). A longer carbon chain of an alcohol group, such as 1-hexanol, 1-penten-3-ol, 1-octen-3-ol, and benzene ethanol, might be derived from the degradation products of lipid oxidation (J. Pongsetkul et al. 2015).

Phenol-type compounds are aromatic compounds which create an undesirable aroma in seafood. Phenolic compounds were detected in high amounts in the initial fermentation, and their content decreased to a minimum level with the fermentation time of shrimp paste (Kleekayai et al. 2016). Toluene and benzene derivatives are the typical phenolic compounds in fish and shellfish paste and can be detected in low amounts in Indonesian and Thailand fermented shrimp paste, *terasi* and *kapi* (Kleekayai et al. 2016; Murwani et al. 2016).

Aldehyde compounds are the products of amino acid biotransformation by microbial activities through the branched-chain amino acids biosynthesis pathways of leucine, valine, and isoleucine. Branched short-chain aldehydes or aromatic aldehydes are commonly found in fermented products. 3-methylbutanal and 2-methylbutanal re associated with a sweet, caramel-like odor in soy sauce and a meaty odor note in fish sauce (Udomsil et al. 2010). The content of 3-methylbutanal decreases along with the fermentation time of the Thai shrimp paste *kapi*, yet it is not detected in Indonesian *terasi*. Benzaldehyde is associated with the favorable almond-like odor in fish sauce. It is detected in some Central Java *terasi* and Thai *kapi*. It is also the major aldehyde produced by *T. halophilus* isolate in soy sauce (Udomsil et al. 2010). The formation of benzaldehyde in *terasi* could be closely associated with the aminopeptidase activity of *T. halophilus* isolate from *terasi* in producing desirable flavor in the fermented product (Kobayashi et al. 2003). Benzaldehyde could also be generated either from the Strecker degradation of phenylalanine or the biotransformation pathway of *Lactiplantibacillus plantarum* through the conversion of phenylalanine to phenylpyruvic acid and subsequent transformation to benzaldehyde during fermentation (Kleekayai et al. 2016). Some aldehydes are also produced from the oxidation of oleic and linoleic acids (Prihanto et al. 2021a).

Ketone components, such as dectatriene-2-one, cyclooctadiene-1-one, and 2-tridecanone, were detected in *terasi*, while only 2-heptanone was found in *kapi*. Ketones could be generated either from the Maillard reaction or microbial enzymatic activity on lipids and/or amino acids (Kleekayai et al. 2016). Hexadecane, 1,11-dodecadiene, and 2-tetradecene were some other hydrocarbons detected in Central Java terasi (Murwani et al. 2016). The formation of hydrocarbon during *terasi* fermentation is the result of microbial activity on lipids.

The aerobic halophiles in *terasi* showed lipolytic activities, which could hydrolyze the fats into low molecular weight fatty acids (Surono and Hosono 1994b). These volatile fatty acids are also responsible for the acidic aroma of *terasi* (Murwani et al. 2016). This is further supported by findings from the Filipino fermented fish sauce product *patis* (Sanceda et al. 1992). The aerobic fermentation process in the production of fish sauce produces significantly higher concentrations of volatile acids than the anaerobic one, and this is characterized by a sharper and stronger acidic, less sweet, more rancid, pungent, and cheesy odor in aerobically fermented fish sauce.

Short-chain fatty acids, such as 2-methylbutanoic acid and 2-methylpropanoic acid, can be found in fermented shrimp paste as the products of the microbial metabolism of isoleucine and valine, respectively. 2-methylbutanoic acid, which gives sweet and cheesy odor notes, was detected in *terasi* and *kapi* in high amounts, while 2-methylpropanoic acid, which gives a pungent butter-fat–like odor, was only found in *kapi* in lower amounts (Kleekayai et al. 2016). These acids can also be converted into fruity esters, such as butanoate esters. However, the corresponding ester component in *terasi* was not detected. As longer chains of fatty acids, such as hexadecanoic acid and tetradecanoic acid, were found in Central Java *terasi*, the corresponding ester components could also be detected, such as hexadecanoic acid methyl ester and other ester components, such as ethyl linoleic and ethyl laurate (Murwani et al. 2016). Ester compounds are generated from the esterification of alcohols and carboxylic acid, which were associated with esterase and esterase lipase activities in *terasi* starter (Surono and Hosono 1994b).

7.6 CONCLUSION

Terasi is one of the most popular fermented seafood products in Indonesia. It has been used for centuries as a flavor enhancer in various types of Indonesian cuisine. *Terasi* can be obtained from a fermentation process with or without salt addition. The chemical composition of *terasi* is strongly affected by the ingredients used, methods of production, seasonal variation, and so on. Therefore, these factors will yield a varied unique taste profile, aromas, and overall sensory characteristics. Besides being known as a source of umami amino acids, shrimp paste is reported to be rich in essential amino acids and omega-3 long-chain saturated fatty acids. Studies also show that fermented shrimp paste (including *terasi*) can be a source of branched-chain fatty acids, which can benefit human health. The unique flavor and aroma of *terasi* are the results of a fermentation process carried out predominantly by halophilic LAB and indigenous enzyme-producing bacteria. In addition to the role of developing *terasi* taste and flavor, several enzymes derived from *terasi* isolates have the potential to be developed for industrial and therapeutic applications.

Studies concerning the relationship of flavor characteristics, geographical indications, microbiome profile, and bioactive components of *terasi* are limited but promising for further exploration. *Terasi* has roots in the ancestral heritage of Indonesia and is commonly used in Asian cuisines. It has the potential to achieve global market expansion due to its unique and rich umami flavor profile. However, the traditional methods applied by farmers and small-scale producers limit their ability to produce products with consistent quality. The implementation of good manufacturing practice and the addition of selected starter cultures can be key factors to improve the sensory and nutritional qualities of *terasi*.

REFERENCES

Ali, Mahrus, Joni Kusnadi, Aulanni'am, and Yunianta. 2020. "Amino Acids, Fatty Acids and Volatile Compounds of Terasi Udang, an Indonesian Shrimp Paste, During Fermentation." *AACL Bioflux* 13 (2): 938–50.

Ali, Mahrus, Yunianta, Aulanni'Am, and Joni Kusnadi. 2019. "Characteristic of Acan: A Traditional Shrimp Paste of Maduranese, Indonesia." *IOP Conference Series: Earth and Environmental Science* 230 (1): 8–14. https://doi.org/10.1088/1755-1315/230/1/012048.

Allwood, Joanne G., Lara T. Wakeling, and David C. Bean. 2021. "Fermentation and the Microbial Community of Japanese Koji and Miso: A Review." *Journal of Food Science* 86 (6): 2194–207. https://doi.org/10.1111/1750-3841.15773.

Amir, Nursinah, and Chanif Mahdi. 2017. "Evaluation of Rhodamine B Use in Shrimp Paste Product Marketed in Makassar City." *Jurnal IPTEKS PSP* 4 (8): 128–33.

Anggo, Apri Dwi, Fronthea Swastawati, and Widodo Farid Ma'ruf. 2014. "The Quality of Organoleptic and Chemically in Rebon Shrimp Paste to Different of Salt Concentration and Duration Fermentation." *Jurnal Pengolahan Hasil Perikanan Indonesia* 17 (1): 53–59. https://doi.org/10.17844/jphpi.v17i1.8137.

Arfarita, Novi, and Ekachai Chukeatirote. 2015. "Characterization of Protease-Producing Bacteria Isolated from Terasi." *Journal of Biological Researches* 21 (1): 18–23. https://doi.org/10.23869/89.

Ayu, Eveline, Antonius Suwanto, and Tati Barus. 2014. "Klebsiella Pneumoniae from Indonesian Tempeh Were Genetically Different from That of Pathogenic Isolates." *Microbiology Indonesia* 8 (1): 9–15. https://doi.org/10.5454/mi.8.1.2.

Batool, Tahira, Essam A. Makky, Muna Jalal, and Mashitah M. Yusoff. 2016. "A Comprehensive Review on L-Asparaginase and Its Applications." *Applied Biochemistry and Biotechnology* 178 (5): 900–23. https://doi.org/10.1007/s12010-015-1917-3.

Bover-Cid, Sara, and Wilhelm Heinrich Holzapfel. 1999. "Improved Screening Procedure for Biogenic Amine Production by Lactic Acid Bacteria." *International Journal of Food Microbiology* 53 (1): 33–41. https://doi.org/10.1016/S0168-1605(99)00152-X.

Cahyaningtyas, Hilmi Amanah Aditya, Wasana Suyotha, Benjamas Cheirsilp, and Shigekazu Yano. 2021. "Statistical Optimization of Halophilic Chitosanase and Protease Production by Bacillus Cereus HMRSC30 Isolated from Terasi Simultaneous with Chitin Extraction from Shrimp Shell Waste." *Biocatalysis and Agricultural Biotechnology* 31 (January): 101918. https://doi.org/10.1016/j.bcab.2021.101918.

Cahyo, Mohammad Firdaus N., Sri Hastuti, and Iffan Maflahah. 2016. "Penentuan Umur Simpan Terasi Instan Dalam Kemasan." *Agrointek* 10 (1): 55. https://doi.org/10.21107/agrointek.v10i1.2026.

Cai, Luyun, Shucheng Liu, Lijun Sun, Yaling Wang, Hongwu Ji, and Jianrong Li. 2015. "Application of Tea Polyphenols in Combination with 6-Gingerol on Shrimp Paste of during Storage: Biogenic Amines Formation and Quality Determination." *Frontiers in Microbiology* 6 (September): 981. https://doi.org/10.3389/fmicb.2015.00981.

Chukeatirote, Ekachai, Novi Arfarita, Piyanuch Niamsup, and Anittaya Kanghae. 2015. "Phenotypic and Genetic Characterization of Bacillus Species Exhibiting Strong Proteolytic Activity Isolated from Terasi, an Indonesian Fermented Seafood Product." *Journal of Northeast Agricultural University (English Edition)* 22 (4): 15–22. https://doi.org/10.1016/S1006-8104(16)30014-9.

Damanik, Ambarita, Mery Tambaria, Katleen Raes, and Bruno De Meulenaer. 2019. "The Sensory Acceptance and Profile of Indonesian Sambal-Terasi: Impact of Terasi Type and Concentration." *International Journal of Gastronomy and Food Science* 16 (July): 100149. https://doi.org/10.1016/j.ijgfs.2019.100149.

Daroonpunt, Rungsima, Masataka Uchino, Yoshimasa Tsujii, Machiko Kazami, Daiki Oka, and Somboon Tanasupawat. 2016. "Chemical and Physical Properties of Thai Traditional Shrimp Paste (Ka-Pi)." *Journal of Applied Pharmaceutical Science* 6 (5): 58–62. https://doi.org/10.7324/JAPS.2016.60509.

Fan, Yan, Li'ang Yin, Yong Xue, Zhaojie Li, Hu Hou, and Changhu Xue. 2017. "Analyzing the Flavor Compounds in Chinese Traditional Fermented Shrimp Pastes by HS-SPME-GC/MS and Electronic Nose." *Journal of Ocean University of China* 16 (2): 311–18. https://doi.org/10.1007/s11802-017-3194-y.

Hajeb, P., and S. Jinap. 2013. "Fermented Shrimp Products as Source of Umami in Southeast Asia." *Journal of Nutrition & Food Sciences* 1 (S10). https://doi.org/10.4172/2155-9600.s10-006.

Handayani, Baiq Rien, Bambang Dipokusumo, Wiharyani Werdiningsih, and Sitta Fitri Rahmadhina. 2020. "Partial Properties of Ready-to-Use Shrimp Paste Affected by Heating Time." *Current Research on Biosciences and Biotechnology* 1 (2): 57–61. https://doi.org/10.5614/crbb.2019.1.2/jthm4041.

Hestiani, Asnani, and K.T. Isamu. 2019. "The Effect of Fermentation Time on Sensory Value, Proximate Composition, and Total Bacteria Shrimp Paste of the Bete-Bete Fish (Leiognathus Equulus)." *Journal of Fisheries Processing Technology* 2 (2): 267–73.

Hidayat, Taufik, and Retno Sari Mahanani. 2018. "Inovasi Teknologi Pengemasan Terasi Udang Puger 'Hj. Ariyatun.'" In *Seminar Nasional Hasil Penelitian Dan Pengabdian Masyarakat Politeknik Negeri Jember*, 16–19 Jember.

Hua, Ying, Bo Jiang, Yoshinori Mine, and Wanmeng Mu. 2008. "Purification and Characterization of a Novel Fibrinolytic Enzyme from Bacillus Sp. Nov. SK006 Isolated from an Asian Traditional Fermented Shrimp Paste." *Journal of Agricultural and Food Chemistry* 56 (4): 1451–57. https://doi.org/10.1021/jf0713410.

Huda, Nurul. 2012. "Indonesian Fermented Fish Products." In *Handbook of Animal-Based Fermented Food and Beverage Technology*, edited by Y.H. Hui and E. Özgül Evranuz, 758–65. CRC Press. https://doi.org/10.1201/b12084-47.

Indriati, Ninoek, and Fairdiana Andayani. 2012. "Pemanfaatan Angkak Sebagai Pewarna Alami Pada Terasi Udang." *Jurnal Pascapanen Dan Bioteknologi Kelautan Dan Perikanan* 7 (1): 11. https://doi.org/10.15578/jpbkp.v7i1.65.

Irianto, G, and Giyatmi. 1998. "Traditional Fermented Fish Products in Indonesia." In *Fish Utilization in Asia and the Pacific*, edited by D.G. James, 67–75. Bangkok: FAO of the UN Regional Office for Asia and the Pacific.

Junianto. 2011. "Studi Karakteristik Pengolahan Terasi Cirebon Dalam Upaya Mendapatkan Perlindungan Indikasi Geografi." *Jurnal Akuatika* 2 (1): 1–14.

Karahadian, C., and R.C. Lindsay. 1989. "Role of Oxidation in the Formation and Stability of Fish Flavors." In *Flavor Chemistry: Trends and Developments*, edited by R. Teranishi, R.G. Buttery, and F. Shahidi, 60–75. Washington, DC: American Chemical Society.

Karim, F., F. Swastawati, and A. Anggo. 2014. "The Effect of Different Raw Material to Glutamic Acid Content in Fish Paste." *Jurnal Pengolahan Dan Bioteknologi Hasil Perikanan* 3 (4): 51–58.

Keuth, S., and B. Bisping. 1994. "Vitamin B12 Production by Citrobacter Freundii or Klebsiella Pneumoniae during Tempeh Fermentation and Proof of Enterotoxin Absence by PCR." *Applied and Environmental Microbiology* 60 (5): 1495–99. https://doi.org/10.1128/aem.60.5.1495-1499.1994.

Kim, Jin-Soo, Fereidoon Shahidi, and Min-Soo Heu. 2005. "Tenderization of Meat by Salt-Fermented Sauce from Shrimp Processing by-Products." *Food Chemistry* 93 (2): 243–49. https://doi.org/10.1016/j.foodchem.2004.09.022.

Kim, Young-Boong, Yun-Sang Choi, Su-Kyung Ku, Dai-Ja Jang, Hajah Hasnah Binti Ibrahim, and Ki Bong Moon. 2014. "Comparison of Quality Characteristics between Belacan from Brunei Darussalam and Korean Shrimp Paste." *Journal of Ethnic Foods* 1 (1): 19–23. https://doi.org/10.1016/j.jef.2014.11.006.

Kleekayai, Thanyaporn, Surapong Pinitklang, Natta Laohakunjit, and Worapot Suntornsuk. 2016. "Volatile Components and Sensory Characteristics of Thai Traditional Fermented Shrimp Pastes During Fermentation Periods." *Journal of Food Science and Technology* 53 (3): 1399–410. https://doi.org/10.1007/s13197-015-2142-3.

Kobayashi, Takeshi, Tri Winarni Agustini, Ratna Ibrahim, Kaeko Kamei, Akihiro Kondo, Michika Kajiwara, Yoshihiro Ooka, Hidetoshi Nakamura, Takeshi Terahara, and Chiaki Imada. 2016. "Production of Bacteriocin by Virgibacillus Salexigens Isolated from 'Terasi': A Traditionally Fermented Shrimp Paste in Indonesia." *World Journal of Microbiology and Biotechnology* 32 (3): 47. https://doi.org/10.1007/s11274-015-1991-2.

Kobayashi, Takeshi, Michika Kajiwara, Mita Wahyuni, Toshihide Kitakado, Naoko Hamada-Sato, Chiaki Imada, and Etsuo Watanabe. 2003. "Isolation and Characterization of Halophilic Lactic Acid Bacteria Isolated from 'Terasi' Shrimp Paste: A Traditional Fermented Seafood Product in Indonesia." *The Journal of General and Applied Microbiology* 49 (5): 279–86. https://doi.org/10.2323/jgam.49.279.

Kuroda, Motonaka, and Naohiro Miyamura. 2015. "Mechanism of the Perception of 'Kokumi' Substances and the Sensory Characteristics of the 'Kokumi' Peptide, γ-Glu-Val-Gly." *Flavour* 4 (1): 11. https://doi.org/10.1186/2044-7248-4-11.

Lestari, S.D., Herpandi, and G.R.R. Simamora. 2017. "Effects of Different Pediococcus Halophilus Level and Fermentation Time on Chemical Properties of Fermented Anchovy Paste 'Terasi Ikan.'" *IOP Conference Series: Materials Science and Engineering* 193 (April): 012004. https://doi.org/10.1088/1757-899X/193/1/012004.

Li, Chunping, Jiajia Wu, Yan Li, and Zhiyuan Dai. 2013. "Identification of the Aroma Compounds in Stinky Mandarin Fish (Siniperca Chuatsi) and Comparison of Volatiles During Fermentation and Storage." *International Journal of Food Science & Technology* 48 (11): 2429–37. https://doi.org/10.1111/ijfs.12254.

Miyamura, Naohiro, Motonaka Kuroda, Yumiko Kato, Junko Yamazaki, Toshimi Mizukoshi, Hiroshi Miyano, and Yuzuru Eto. 2014. "Determination and Quantification of a Kokumi Peptide, γ-Glutamyl-Valyl-Glycine, in Fermented Shrimp Paste Condiments." *Food Science and Technology Research* 20 (3): 699–703. https://doi.org/10.3136/fstr.20.699.

Mizutani, T., A. Kimizuka, K. Ruddle, and N. Ishige. 1987. "A Chemical Analysis of Fermentedfish Products and Discussion of Fermented Flavors in Asian Cuisines." *Bulletin of the National Museum of Ethnology* 12 (3): 801–64.

Mizutani, Tadashi, Akimitsu Kimizuka, Kenneth Ruddle, and Naomichi Ishige. 1992. "Chemical Components of Fermented Fish Products." *Journal of Food Composition and Analysis* 5 (2): 152–59. https://doi.org/10.1016/0889-1575(92)90031-E.

Montano, N., G. Gavino, and V.C. Gavino. 2001. "Polyunsaturated Fatty Acid Contents of Some Traditional Fish and Shrimp Paste Condiments of the Philippines." *Food Chemistry* 75: 155–58.

Murwani, Retno, Hera Surya Adhi Putra, Henggar Widiyanto, Agus Trianto, and Ambariyanto Ambariyanto. 2016. "Shrimp Paste 'Terasi' Volatile Compounds from Northern Coast of Central Java." *Jurnal Teknologi* 78 (4–2): 187–92. https://doi.org/10.11113/jt.v78.8203.

National Standarization Agency of Indonesia. 2016. "SNI No 2716-2016 Terasi Udang." Jakarta.

Noviendri, Dedi, Ekowati Chasanah, and Yusro Nuri Fawzya. 2006. "Karakterisasi Enzim Kitinase Yang Diproduksi Oleh Isolat Bakteri JB4 Dari Terasi." *Jurnal Pascapanen Dan Bioteknologi Kelautan Dan Perikanan* 1 (2): 85–91. https://doi.org/10.15578/jpbkp.v1i2.390.

Noviendri, Dedi, Yusro Nuri Fawzya, and Ekowati Chasanah. 2008. "Karakteristik Dan Sifat Kinetika Enzim Kitinase Dari Isolat Bakteri T5a1 Asal Terasi." *Jurnal Pascapanen Dan Bioteknologi Kelautan Dan Perikanan* 3 (2): 123–29. https://doi.org/10.15578/jpbkp.v3i2.15.

Nunal, Sharon N., Christine Marie C. Florece, and Michael Andre Treyes. 2016. "Antagonistic Activities against Foodborne Pathogens and Use as Functional Starter Culture of Bacterial Strains Isolated from Fermented Shrimp Paste." *Philippine Jounal of Natural Sciences* 21 (1): 26–36.

Oktavia, Yulia, Shanti Dwita Lestari, Susi Lestari, Herpandi, and Miftahul Jannah. 2018. "Optimasi Waktu Inkubasi Produksi Protease Dan Amilase Isolat Bakteri Asal Terasi Ikan Teri Stolephorus Sp." *Jurnal Ilmu Dan Teknologi Kelautan Tropis* 10 (3): 719–25. https://doi.org/10.29244/jitkt.v10i3.18840.

Özogul, Fatih, Nurten Toy, Yesim Özogul, and Imen Hamed. 2017. "Function of Cell-Free Supernatants of Leuconostoc, Lactococcus, Streptococcus, Pediococcus Strains on Histamine Formation by Foodborne Pathogens in Histidine Decarboxylase Broth." *Journal of Food Processing and Preservation* 41 (5): e13208. https://doi.org/10.1111/jfpp.13208.

Peralta, E.M., H. Hatate, D. Watanabe, D. Kawabe, H. Murata, Y. Hama, and R. Tanaka. 2005. "Antioxidative Activity of Philippine Salt-Fermented Shrimp and Variation of Its Constituents During Fermentation." *Journal of Oleo Science* 54: 553–58.

Permatasari, Amelia Ayu, Sumardianto Sumardianto, and Laras Rianingsih. 2018. "Perbedaan Konsentrasi Pewarna Alami Kulit Buah Naga (Hylocereus Polyrhizus) Terhadap Warna Terasi Udang Rebon (Acetes Sp.)." *Jurnal Teknologi Hasil Pertanian* 11 (1): 39. https://doi.org/10.20961/jthp.v11i1.29094.

Pongsetkul, Jaksuma, Soottawat Benjakul, Punnanee Sampavapol, Kazufumi Osako, and Nandhsha Faithong. 2014. "Chemical Composition and Physical Properties of Salted Shrimp Paste (Kapi) Produced in Thailand." *International Aquatic Research* 6 (3): 155–66. https://doi.org/10.1007/s40071-014-0076-4.

Pongsetkul, Jaksuma, Soottawat Benjakul, Punnanee Sampavapol, Kazufumi Osako, and Nandhsha Faithong. 2015. "Chemical Compositions, Sensory and Antioxidative Properties of Salted Shrimp Paste (Ka-Pi) in Thailand." *International Food Research Journal* 22 (4): 1454–65.

Prihanto, Asep A., Abdul Jaziri, and Ima Perwira. 2016. "Purification and Characterization of Neutral Protease from Bacillus Substilis UBT7 Isolated from Terasi, Indonesian Fermented Fish." *Biosciences, Biotechnology Research Asia* 13 (3): 1409–13. https://doi.org/10.13005/bbra/2283.

Prihanto, Asep A., and Hidayatun Muyasyaroh. 2021. "The Indonesian Fermented Food Product Terasi: History and Potential Bioactivities." *Systematic Reviews in Pharmacy* 12 (2): 378–84. https://doi.org/10.31838/srp.2021.2.52.

Prihanto, Asep A., Rahmi Nurdiani, Yoga D. Jatmiko, Muhamad Firdaus, and Titis S. Kusuma. 2021a. "Physicochemical and Sensory Properties of Terasi (an Indonesian Fermented Shrimp Paste) Produced Using Lactobacillus Plantarum and Bacillus Amyloliquefaciens." *Microbiological Research* 242 (January): 126619. https://doi.org/10.1016/j.micres.2020.126619.

Prihanto, Asep A., Rahmi Nurdiani, Happy Nursyam, Hidayatun Muyasharoh, and L. Hayati Royani. 2021b. "Terasi, an Indonesian Fermented Shrimp Paste, a New Reported Source of L-Asparaginase-Producing Bacteria." *Research Journal of Biotechnology* 16 (1): 178–81.

Puspita, Indun Dewi, Ari Wardani, R.R. Oki Arum Puspitasari, Prihati Sih Nugraheni, M.G.S. Muhammad Prima Putra, Sri Pudjiraharti, and Ustadi Ustadi. 2017. "Occurrence of Chitinolytic Bacteria in Shrimp Rusip and Measurement of Their Chitin-Degrading Enzyme Activities." *Biodiversitas Journal of Biological Diversity* 18 (3): 1275–81. https://doi.org/10.13057/biodiv/d180354.

Romadhon, Romadhon, Laras Rianingsih, and Apri Dwi Anggo. 2018. "Antibactery Activities from Some Quality Rate of Rebon Shrimp Paste." *Jurnal Pengolahan Hasil Perikanan Indonesia* 21 (1): 68. https://doi.org/10.17844/jphpi.v21i1.21263.

Ruddle, Kenneth, and Naomichi Ishige. 2010. "On the Origins, Diffusion and Cultural Context of Fermented Fish Products in Southeast Asia." In *Globalization, Food and Social Identities in the Asia Pacific Region*, edited by James Farrer. Tokyo: The Sophia University Institute of Comparative Culture.

Sanceda, Norlita G., Tadao Kurata, Yoshikatsu Suzuki, and Nobuhiko Arakawa. 1992. "Oxygen Effect on Volatile Acids Formation during Fermentation in Manufacture of Fish Sauce." *Journal of Food Science* 57 (5): 1120–35. https://doi.org/10.1111/j.1365-2621.1992.tb11277.x.

Sari, D.I., and S.A. Rinto. 2011. "Karakteristik Terasi Jembret Instan Dengan Perbedaan Lama Waktu Pengeringan." *Jurnal Ilmu-Ilmu Perikanan Dan Budidaya Perairan* 6 (1): 15–26.

Sato, Arisa, Dea Indriani Astuti, Sastia Prama Putri, and Eiichiro Fukusaki. 2020. "Quality Improvement of Semi-Wet Terasi by Optimizing the Starter Culture Ratio of Controlled Fermentation." *HAYATI Journal of Biosciences* 27 (4): 320. https://doi.org/10.4308/hjb.27.4.320.

Shalaby, Ali R. 1996. "Significance of Biogenic Amines to Food Safety and Human Health." *Food Research International* 29 (7): 675–90. https://doi.org/10.1016/S0963-9969(96)00066-X.

Sumardianto, Ima Wijayanti, and Fronthea Swastawati. 2019. "Karakteristik Fisikokimia Dan Mikrobiologi Terasi Udang Rebon Dengan Variasi Konsentrasi Gula Merah." *Jurnal Pengolahan Hasil Perikanan Indonesia* 22 (2): 287–98.

Surono, Ingrid S., and Akiyoshi Hosono. 1994a. "Chemical and Aerobic Bacterial Composition of 'Terasi', a Traditional Fermented Product from Indonesia." *Food Hygiene and Safety Science (Shokuhin Eiseigaku Zasshi)* 35 (3): 299–304. https://doi.org/10.3358/shokueishi.35.299.

Surono, Ingrid S., and Akiyoshi Hosono. 1994b. "Microflora and Their Enzyme Profile in 'Terasi' Starter." *Bioscience, Biotechnology, and Biochemistry* 58 (6): 1167–69. https://doi.org/10.1271/bbb.58.1167.

Thapa, Namrata, Joydeb Pal, and Jyoti Prakash Tamang. 2004. "Microbial Diversity in Ngari, Hentak and Tungtap, Fermented Fish Products of North-East India." *World Journal of Microbiology and Biotechnology* 20 (6): 599–607. https://doi.org/10.1023/B:WIBI.0000043171.91027.7e.

Udomsil, Natteewan, Sureelak Rodtong, Yeung Joon Choi, Yanglin Hua, and Jirawat Yongsawatdigul. 2011. "Use of Tetragenococcus Halophilus as a Starter Culture for Flavor Improvement in Fish Sauce Fermentation." *Journal of Agricultural and Food Chemistry* 59 (15): 8401–8. https://doi.org/10.1021/jf201953v.

Udomsil, Natteewan, Sureelak Rodtong, Somboon Tanasupawat, and Jirawat Yongsawatdigul. 2010. "Proteinase-Producing Halophilic Lactic Acid Bacteria Isolated from Fish Sauce Fermentation and Their Ability to Produce Volatile Compounds." *International Journal of Food Microbiology* 141 (3): 186–94. https://doi.org/10.1016/j.ijfoodmicro.2010.05.016.

Varlet, V., and X. Fernandez. 2010. "Sulfur-Containing Volatile Compounds in Seafood: Occurrence, Odorant Properties and Mechanisms of Formation." *Food Science and Technology International* 16 (6): 463–503.

Vo, Thi Thu-Thao, and Jong-Hyun Park. 2019. "Characteristics of Potential Gamma-Aminobutyric Acid-Producing Bacteria Isolated from Korean and Vietnamese Fermented Fish Products." *Journal of Microbiology and Biotechnology* 29 (2): 209–21. https://doi.org/10.4014/jmb.1811.09072.

Wang, Dong Hao, Yupeng Yang, Zhen Wang, Peter Lawrence, Randy W. Worobo, and J. Thomas Brenna. 2019. "High Levels of Branched Chain Fatty Acids in Nātto and Other Asian Fermented Foods." *Food Chemistry* 286 (January): 428–33. https://doi.org/10.1016/j.foodchem.2019.02.018.

Yoshida, Yoshiko. 1998. "Umami Taste and Traditional Seasonings." *Food Reviews International* 14 (2–3): 213–46. https://doi.org/10.1080/87559129809541158.

Fermentation Biotechnology and Bioactive Compounds of Douchi, a Chinese Traditional Salt-Fermented Soybean Product

8

Siyu Yao, Anqianyi Tu, Minghuang Chen, Chenxin Zhou, Siyong You, Zhenhui Pan, Yue Zhang, Caili Fu, and Jie Xu

8.1 INTRODUCTION

As one of the most important crops, the soybean is a main source of high-quality proteins for humans, while douchi is a famous fermented black bean product that originated in China with a long consumption history. Douchi contains a high quantity of proteins, various amino acids, lactic acid, magnesium, calcium, phosphorus, and a variety of vitamins.

The raw materials of douchi are mainly soybeans and specifically black beans, supplemented by wine, salt, and various spices. The special flavor and bioactive compounds of douchi are produced by the fermentation of soy protein by *Aspergillus*, *Mucor*, related proteases, or other microorganisms (Wu et al. 2011; Hu et al. 2014). Douchi has been becoming popular throughout the world due to its special qualities, such as unique flavors; the ability to improve health benefits; and value as medicinal food (Caplice and Fitzgerald 1999; Granito et al. 2002). Therefore, products similar to Chinese douchi are now popular in several countries, for instance, Japanese natto (Murooka and Yamshita 2008; Suzuki et al. 2023), tempeh in Indonesia (Utami et al. 2016; Su et al. 2023), kinema and libi ippa in Bhutan (Sugawara et al. 1998; Kumari et al. 2023), tungrymbai and bekang in India (Chettri and Tamang 2015), chongkukjang or cheonggukjang in Korea (Chang 2018), pepok in northern Myanmar, sieng in Cambodia and Laos (Chettri and Tamang 2015), and thuo nao in Thailand and Laos (Chukeatirote et al. 2018).

Douchi for human consumption has many benefits. Recently, the ability of douchi to promote appetite (Reseland et al. 2001), lower the number of inflammatory factors in the body (Jung et al. 2016), enhance the structure of intestinal flora (Dong et al. 2014), and prevent cervical thrombosis (Yuan et al. 2012) has been confirmed. Additionally, douchi may have medicinal attributes with antioxidant activities, antidiabetic properties, and hypotensive activity (Chen et al. 2012; Hou et al. 2023).

Therefore, researchers have extensively studied and continuously innovated fermentation technologies for douchi. This chapter summarizes the existing fermentation biotechnologies for douchi, the effects of fermentation on functional bioactive compounds, and the recent research progress. It provides a theoretical basis for the development and industrial production of douchi.

8.2 DIVERSITY AND PRODUCTION

Due to various procedures of making douchi in different regions of China, several kinds of douchi are produced. This is based on the types of microorganisms used in fermentation, which is the most important classification method. Based on this, douchi products are divided into *Mucor*-type (e.g., Yongchuan douchi), *Aspergillus*-type (e.g., Liuyang douchi and Yangjiang douchi), *Rhizopus*-type (e.g., Indonesian tempeh), and bacterial-type (e.g., Qianxikou and Babao douchi) douchi (Chen et al. 2011). Alternatively, douchi products are also divided into semen sojae praeparatum (SSP) and salted douchi according to whether salt is used during fermentation. As mentioned, Liuyang douchi,

DOI: 10.1201/9781003605300-8

Japanese natto, and tempeh in Indonesia all belong to the SSP category, while Yongchuan douchi, Babao douchi, and others fall into the salted douchi category. Salted douchi has high moisture content and is suitable for cooking dishes. Depending on the moisture content, there are dry douchi, wet douchi, and water douchi (Wang et al. 2022).

Although there are many kinds of douchi, the production processes have three stages in general. First, the pretreatment stage, including sorting, soaking, cooking, and other processes, makes the soybean more suitable for the growth and reproduction of subsequent microorganisms. This is followed by koji-making by natural or artificial inoculation, also known as the pre-fermentation stage. In this stage, the microorganisms utilize the nutrients in soybean to rapidly grow and reproduce while secreting various enzymes such as protease and glucoamylase that directly affect the fermentation results and quality. Subsequently, the macromolecules are decomposed into micro-molecule peptides and amino acids mediated by the enzymes, which produce a variety of flavor substances and functional bioactive compounds. This stage is known as the post-fermentation stage. Last, spices are added to form the characteristic color, flavor, and nutrients in douchi.

This chapter specifically introduces the production processes of the four types of douchi made by different microorganisms. Figure 8.1 shows a flowchart of the fermentation process to produce the four types of douchi (Liu et al. 2020).

Aspergillus oryzae and *Aspergillus egyptiacus* are the most widely used microorganisms in *Aspergillus*-type douchi fermentation. Soybeans with a full seed matrix are selected and soaked with water until the beans expand and have few wrinkles. The soaked soybeans are cooked under a certain temperature and pressure, then cooled down to 40°C. After that, *Aspergillus* spp. is inoculated in precooked soybeans. Koji is then formed after ventilation; then the beans with salt are mixed and fermented for a few days (Figure 8.1a) (Zhang et al. 2007; He et al. 2016). The representative products of *Mucor*-type douchi are Chongqing Yongchuan and Yongchuan douchi. Yongchuan douchi has the highest thrombolytic enzyme activity among all the products, and its health benefits to the heart and cerebral vessels can be as good as Japanese natto (Feng et al. 2010). Like *Aspergillus*-type douchi, *Mucor*-type douchi is made by screening, soaking, cooking, cooling, and koji making, then thoroughly mixing, followed by the post-fermentation process (Figure 8.1b). Thus, the product has the characteristics of rich fragrance, bright color, and pure flavor (Wang and Li 2012). Bacteria-type douchi is fermented by bacteria, where natto (Japan) and water douchi (China) serve as common examples fermented by the main microorganisms of *Bacillus subtilis* and *Tetragenococcus* (Liu

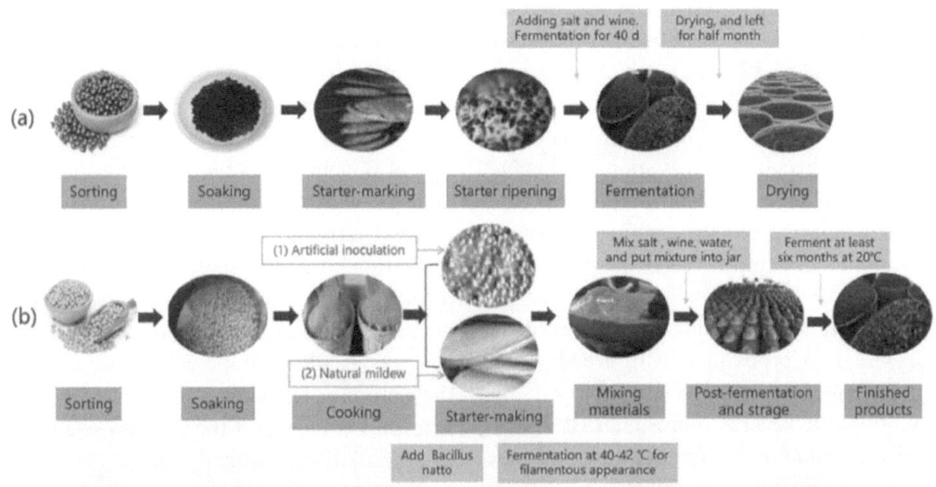

FIGURE 8.1 Flowchart of fermentation process to produce (a) *Aspergillus*-type, (b) *Mucor*-type, (c) bacteria-type, and (d) *Rhizopus*-type douchi.

Source: Adapted with permission from ref (Liu et al. 2020). Copyright 2024 Informa UK Limited.

TABLE 8.1 Fermentation Process of Partial Douchi

TYPE OF DOUCHI	*REPRESENTATIVE PRODUCTS*	*TIME OF KOJI-MAKING*	*TEMPERATURE OF KOJI-MAKING*	*TIME OF FERMENTATION*	*FERMENTATION TEMPERATURE AND METHOD*	*REFERENCES*
Mucor-type	Yongchuan douchi	10–12 d	15–18°C	10–12 months	20°C, natural	(Li, Luo, et al. 2020)
	Tongchuan douchi	15–21 d	15–18°C	10–12 months	20°C, natural	(Zhao 2020)
Aspergillus -type	Yangjiang douchi	4–5 d	26–30°C	About 40 days	30–45°C, heat preservation	(Zhao 2020)
	Liuyang douchi	2–4 d	28–32°C	About 7 days	35–55°C, heat preservation	(Zhao 2020)
Bacterial-type	Natto	About 11 h	38–42°C	1–2 d	5°C, pure fermentation	(Kobayashi et al. 2017)
Rhizopus-type	Tempeh	/	/	1–3 d	30–37°C, pure fermentation	(Setiawan et al. 2016)

et al. 2018; Wang et al. 2022). In the past, bacterial-type douchi was made by natural fermentation. Soybeans were wrapped and steamed in straw, then fermented by natural *Bacillus subtilis*. Nowadays, the natural *Bacillus subtilis* has been replaced by standardized pure strains. The production process of bacterial-type douchi is shown in Figure 8.1c, after the pre-treatment, pre-fermentation, and post-fermentation stages, respectively, where ingredients are added to increase the flavors in the post-fermentation stage (Kobayashi et al. 2017). Tempeh is the representative soybean product fermented by *Rhizopus*, which has a soft, slimy texture and is shaped like a white cake. First, the soybeans are soaked in water, followed by peeling and cooking. Then the acidified matrix is added with bacteria such as spores or hyphae cultured by *Rhizopus* in liquid or solid form. Finally, soybeans are put into the container and fermented at a certain temperature for several days to obtain mature tempeh (Setiawan et al. 2016) (Figure 8.1d).

Based on these descriptions, the major differences of the fermentation processes of four types of douchi are the koji-making stage and post-fermentation stage. In Table 8.1, the specific process conditions of several representative douchi products are summarized, including the temperature and time of koji preparation in different types of douchi, as well as aspects such as fermentation temperature and time.

8.3 RESEARCH STATUS OF FERMENTATION TECHNOLOGY

The fermentation of Chinese traditional douchi mostly adopted natural fermentation methods. However, this is greatly affected by the environment and seasonality, for instance, complex microorganism population, long production periods, high salt concentration, and unstable quality. The production cost of enterprises increases due to natural fermentation, seriously affecting the expansion and supply of the production scale of douchi. Consequently, an increasing number of studies have focused on pure fermentation, double strain koji-making, post-fermentation technology exploration, and low-salt douchi.

Pure fermentation combines modern microecology with traditional natural fermentation methods. The fermentation temperature, strain, humidity, and time are controlled throughout the process, which is being advanced to improve product quality, reduce food safety risks, and shorten the production cycle. Yang et al. (2019) successfully used high-efficiency fermentation strains to prepare pure fermented SSP and obtained an optimal process for pure fermented SSP.

Nowadays, the optimization of the fermentation technology of douchi is mostly as follows: utilizing response surface analysis with indicators of functional components to achieve optimal fermentation process conditions. To shorten the fermentation time and reduce the salinity, Gao and Liu (2010) determined the optimal conditions of *Mucor* G-1 koji-making using protease activity as the index. Yeast, *Saccharomyces rouxii*, was added in the early stage of post-fermentation; then the optimum fermentation process of fermented tempeh was determined based on the sensory evaluation index. Li, Chen, et al. (2020) determined the optimal fermentation conditions with amino acid nitrogen content and fibrinolytic enzyme activity as indexes. Similarly, Shen et al. (2020) used soy isoflavone content as an index to obtain the optimal fermentation conditions by using efficient fermentation strains from shuidouchi.

Additionally, 12–18% of salt was added during the post-fermentation period to prevent microbial over-fermentation and product decay, but it has been proved that long-term ingestion of high-salt food is harmful to physical health. Therefore, low-salt production of douchi became an important health relevant trend. As a result, some scholars have researched the process of low-salt fermentation to produce douchi as a substitute. Liang et al. (2020) found increased levels of total acid (TA), amino state nitrogen (AN), and free amino acids (FAAs) in douchi samples by fermenting soybeans in lower salt and higher temperature, which was an effective way to solve the low-salt fermentation of douchi.

The dynamic changes of microbial species, quantity, and flora during the traditional fermentation of douchi are the research focus. Polymerase chain reaction denaturing gradient gel electrophoresis (PCR-DGGE) is one of the main molecular biological methods to examine the genetic diversity of microbial communities, which are being widely used in the microbial diversity study of douchi. Guo et al. (2020) analyzed community dynamic changes of food-borne pathogenic bacteria during bacterial-type douchi fermentation using the PCR-DGGE technique. In recent years, second-generation sequencing technology has developed rapidly for use in such studies (Melanie et al. 2015; Deng et al. 2021). Compared with traditional molecular biological methods, high-throughput sequencing technology quickly and accurately identifies microorganism species in fermented food, helping to provide detailed information on microbial diversity (Roh et al. 2010; Zhao 2020). Meanwhile, the rapid development of electronic tongue technology has eliminated the influences of subjective factors and provided rapid digital evaluation of taste indices, which has broad application prospects in the filed of traditional fermented food (Wang et al. 2020). In the future, single-cell expansion technologies will be introduced to gain insight into the metabolic properties and potentials of microorganisms during fermentation of douchi.

8.4 EFFECT OF FERMENTATION ON THE QUALITY OF DOUCHI

8.4.1 Study on the Nutritional Characteristics of Douchi

The contents of protein, fat, and other nutrients in douchi are significantly lower, while the contents of riboflavin (vitamin B_2), retinol (vitamin A), and others are obviously higher than in unfermented soybeans (Zhao 2001; Jia et al. 2019). During the fermentation process, the proteases and lipases produced by microbes degrade proteins and fats into small molecule water-soluble proteins, free amino acids, and fatty acids in soybeans, which explains the significantly reduced levels of proteins and fats in douchi. Moreover, low molecular weight proteins could be potentially absorbed directly into the intestinal mucosa without being digested, which is beneficial for patients with impaired digestion and digestive dysfunction. On the other hand, microbes also use soybean nutrients to produce other active substances, especially vitamin B; hence, the concentrations of retinol, riboflavin, and vitamin B_{12} are significantly increased in douchi (Jia et al. 2019). Retinol is essential to promote the formation of photoreceptor pigments within visual cells and maintaining normal visual responses. Riboflavin is the main component of intracellular dehydrogenase that participates in various processes of oxidative respiration. Lack of riboflavin influences biological oxidation within the body and can cause metabolic disorders. Vitamin B_{12} (cobalamin) promotes erythrocyte development and maturation and maintains the body's hematopoiesis and nervous system function.

It is observed that soybeans are rich in mineral contents. However, most of them are sequestered in the form of phytate, which is the compound salts of potassium, calcium, and magnesium of inositol phosphates. A range of 70~80% of the phosphorus in soybean is not capable of being used by the human body, and about 60% is discharged from the body. Insoluble calcium is formed when calcium is combined with phytic acid, and approximately 70–80% of insoluble calcium is not absorbed by the body and remains in the stool. Insoluble iron is found when iron interacts with phytates, which results in a 7% iron absorption rate in soybeans. Phytates also combine with zinc to form insoluble salts, which reduces the utilization rate. During fermentation, phytates are reduced by 15–20% due to the active phytase secreted by microorganisms, which hydrolyzes phytates into inositol and phosphate. Thus, the solubility and utilization of minerals increases by two- to three-fold and 30–50%, respectively (Song 2003).

8.4.2 Effects of Fungal Diversity on the Flavor of Douchi

The growth and propagation of fungi affect the formation of taste, flavor, and texture of douchi. Fungi produce various amino acids by breaking down proteins, and these amino acids play a major role in the production of flavor compounds (Peng et al. 2013). Recently, more and more studies have identified the structure and composition of microbial community in douchi (Chen et al. 2011; Tamang et al. 2016).

Wang et al. (2020) identified the fungal communities in different types of douchi from two regions by high-throughput sequencing technology and evaluated their taste quality digitally with electronic tongue technology. The dried and wet douchi samples collected from Lichuan county in China were named LCD and LCW, respectively, while the dried and wet douchi samples collected from Badong county in China were named BDD and BDW, respectively. Figure 8.2 analyzed the fungal microbial composition of the four kinds of douchi samples. *Ascomycota* was the dominant phylum that existed at the phylum level in all samples, followed by *Basidiomycota* (Figure 8.2a). The average relative abundance of *Basidiomycetes* was the highest in the LCW group and the lowest in the BDD group, while *Ascomycetes* was the opposite. At the genera level (Figure 8.2b), there were a total of 14 fungal genera in fermented soybeans, with an average relative content greater than 1.00%. In order from the highest to the lowest content, there were *Debaryomyces*, *Aspergillus*, *Fusarium*, *Pichia*, *Wallemia*, *Candida*, *Trichosporon*, *Saccharomyces*, *Penicillium*, *Cladosporium*, *Xeromyces*, *Endothia*, *Colletotrichum*, and *Scopulariopsis*. From Figure 8.2a–b, the relative abundance of dominant fungi was even different in various samples from the same group. Different bacterial genera were identified with redundancy analysis (RDA) (Figure 8.2c). In addition, from Figure 8.2d, it can be seen that five dominant fungal genera (*Saccharomyces, Debaryomyces*, *Pichia*, *Aspergillus*, and *Fusarium*) led the main differences in the overall fungal structure of samples from different groups. Further evidence suggested that differences in dominant genera were the main factor causing overall differences in the faunal structure of different groups of douchi samples.

Also, Wang et al. (2020) discussed the effect of dominant fungi on the flavor quality of douchi (Figure 8.3). There were four fungal genera (*Trichosporon, Debaryomyces*, *Candida,* and *Cladosporium*) that had great effects on five basic flavors (umami, sourness, saltiness, astringency, and bitterness). According to the authors' analysis, *Debaryomyces* (positively correlated with sourness) and *Trichosporon* (positively correlated with umami) were beneficial for the taste quality of douchi, while *Cladosporium* (positively correlated with umami) and *Candida* (negatively correlated with umami) had a negative effect on the taste quality of douchi (Wang et al. 2020).

8.5 HEALTH-RELATED BIOACTIVE COMPONENTS OF DOUCHI

During the fermentation process, douchi produced bioactive compounds, such as polypeptide, free soybean isoflavones, melanoidins, soyasaponins streptokinase, and soybean oligosaccharides. These active substances significantly improved the nutritional and health functions of the product. Existing research results showed that douchi had many beneficial effects, for instance, anti-cancer, anti-oxidation, anti-thrombotic, anti-hypertension, anti-hyperlipidemia, anti-hyperglycemic, preventing senile dementia, and improving the intestinal environment (Figure 8.4) (Liu et al. 2003; He et al. 2006; Jia et al. 2019).

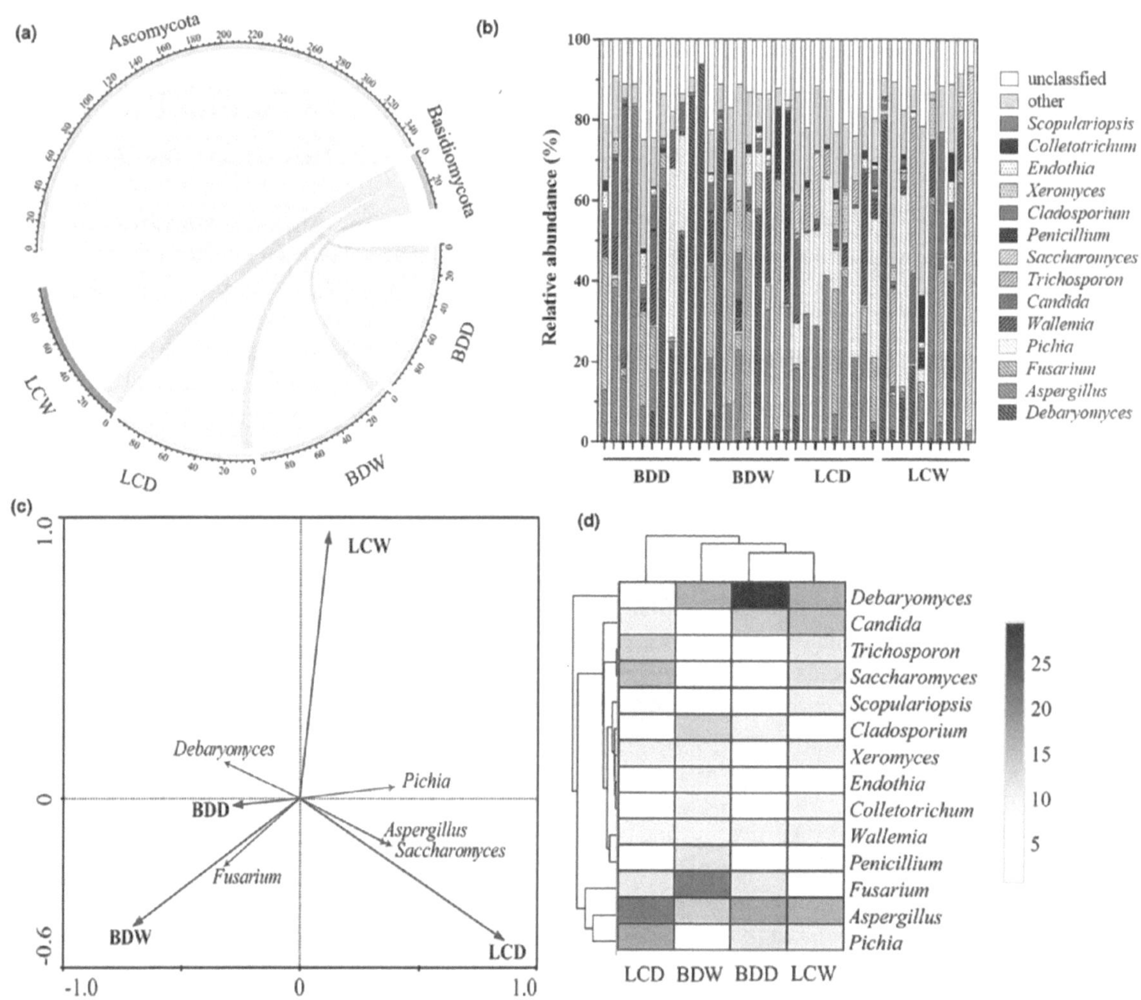

FIGURE 8.2 Analysis of fungal and microbial composition in different douchi samples. Relative abundance of fungi at the (a) phylum and (b) genus level, (c) biordered graph of redundancy analysis (RDA), (d) heat maps of dominant fungi.

Source: Adapted with permission from Wang et al. (2020).

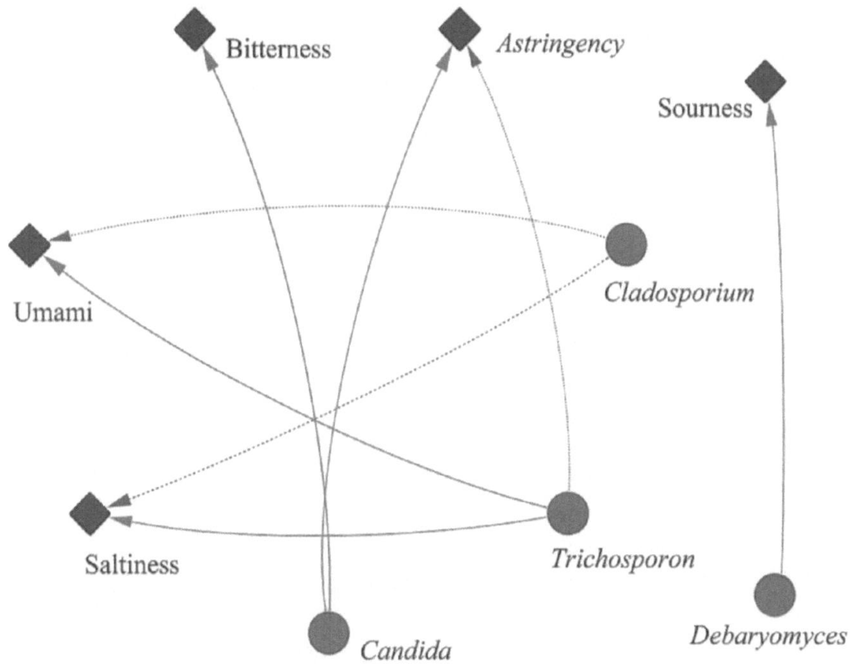

FIGURE 8.3 Network diagram of the relationship between dominant microorganisms and taste indexes. Red and green lines indicate positive and negative correlation, respectively.

Source: Adapted with permission from Wang et al. (2020).

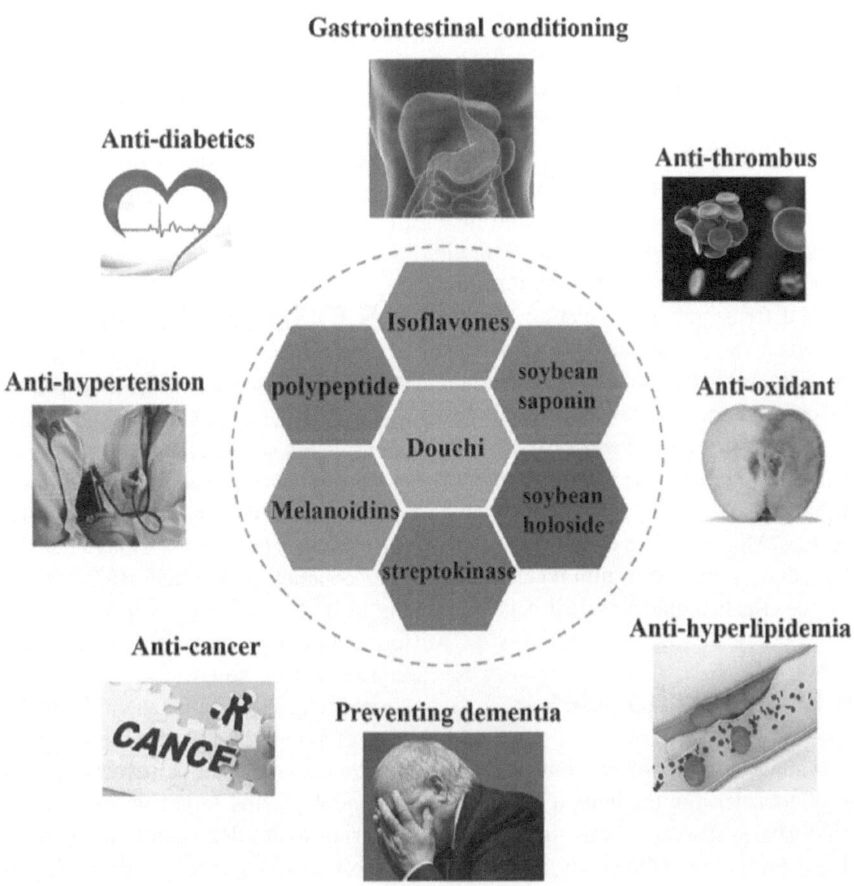

FIGURE 8.4 Bioactive compounds and health functions of douchi.

8.5.1 Soybean Isoflavone

Soybean isoflavone is one of the most attractive functional components in douchi and is also known as phytoestrogen because of its estrogen-like effects. Studies showed that soybean isoflavone has anti-cancer activity (Li and Wang 2020), antioxidant activity (Mishra et al. 2019), hypoglycemic activity (Li and Huang 2008), activity improving osteoporosis in a rat model (Yu et al. 2019), hypolipidemic activity (Liu et al. 2014), anti-inflammatory activity (Keila et al. 2018), and other physiological functions. Cui et al. (2007) determined the contents of total isoflavones in soybean and douchi by ultraviolet spectrophotometry. The results showed that the relative content of total isoflavones in douchi increased after fermentation. However, Song et al. (2002) suggested that fermentation treatment had no contribution to the total content of isoflavones, while the isoflavones of glycoside-type soybeans were partially converted to free-type soybean isoflavones under the effect of β-glucosidase, thus improving the antioxidant properties of douchi.

The main extraction methods of soybean isoflavones included ultrasonic-assisted extraction, microwave-assisted extraction, and ultra-high-pressure extraction. Isoflavones have low water solubility, which limits their high bioactivity in the body. Therefore, improving the bioavailability of soybean isoflavones will be a future research direction (Zhang et al. 2020).

8.5.2 Polypeptides

Soybean polypeptides and a derived oligopeptide mixture are obtained from the decomposition of soybean proteins by protease or microorganism. Studies indicated composition of amino acids was the same as the soybean proteins, but soybean polypeptide had more functional characteristics than soybean proteins. Its physiological functions included lowering cholesterol, lowering blood pressure, antioxidation, regulating the immune system, promoting microbial fermentation, regulating insulin, and promoting fat metabolism (Sun and Jiang 2004). Cheng et al. (2005) pointed out that the contents of low and medium molecular peptides in douchi were increased by 3–8 times after microbial action. Zhang (2003) found that *Aspergillus*-type douchi contained polypeptides that inhibited the activity of angiotensin-converting enzyme (ACE), which is positively correlated with increased blood pressure. Moreover, Wu et al. (2017) demonstrated that the ACE inhibitor promoted insulin regeneration in diabetic rats to achieve the effects of diabetes treatment.

8.5.3 Soybean Oligosaccharides

Oligosaccharides, also known as bifidobacterium factors, increase the number of bifidobacteria in the human intestines, which allows the human digestive system to assist in reducing blood pressure, enhancing the immunity of the body, and reducing the production of intestinal toxic substances. Meanwhile, it helps to lower blood pressure and enhance immunity (Zhang et al. 2015). The oligosaccharides in douchi consist of two types, soybean oligosaccharides in raw materials, such as raffinose, sucrose, stachyose, and another synthesized by transglycosylation of glycoside transferase or obtained from the hydrolysis of polysaccharides by the hemicellulosic enzymes, such as galactanase and mannase, during fermentation (Ji et al. 2016). The concentration of oligosaccharides decreased after fermentation; this might be caused by the usage of oligosaccharides for microbial growth (Wang 2008). Currently, there are many extraction methods for oligosaccharides, including water extraction, alkali extraction, alcohol extraction, membrane separation, enzymatic hydrolysis, the microwave method, and the ultrasonic-assisted method. Microwave- and ultrasonic-assisted methods are the ideal extraction methods since they have high extraction efficiency and do not destroy the structure and properties of oligosaccharides (Zhang et al. 2015).

8.5.4 Soyasaponins

Soyasaponins are a class of important secondary metabolites that are formed during the growth of soybeans. In recent years, multiple studies showed that soyasaponins had lipid-lowering, weight loss, anti-thrombotic, anti-diabetic, anti-inflammatory, antiviral, and immunomodulatory effects and anti-cancer benefits (Wen et al. 2020; Chen et al. 2020). Past studies indicated that it was difficult to separate and extract soyasaponins since the chemical composition of soyasaponins was complex and its concentration in soybean was only 0.6%. At present, there are three feasible methods for soybean saponin extraction, n-butanol extraction, macroporous resin adsorption, and lead salt precipitation. With the deepening research on the physiological activity and function of soyasaponins, the application prospects have become more and more extensive in the fields of food, medicine, and cosmetics.

8.5.5 Streptokinase

Streptokinase is a serine protease produced by *Bacillus subtilis* during the fermentation of douchi. This is a novel thrombolytic agent with a strong effect of dissolving thrombo. Studies indicated that it hydrolyzed thrombus cross-linked fibrin directly and activated plasmin of the body to improve the efficiency of thrombolysis (Han 2015; Dong et al. 2019). Many studies have been conducted on douchi streptokinase, including the screening of high-producing strains, optimization of the traditional fermentation process, isolation, and purification of streptokinase (Zhou, Yang, et al. 2017). Future studies for douchi streptokinase should focus on the molecular levels, enzymatic characteristics, and thrombolytic mechanisms. Douchi thrombolytic enzymes can be separated or purified using high-tech methodology and target the use fibrinolytic enzymes as raw materials to develop healthy food with multi-functions with no side effects. This is targeted to achieving the clinical application of douchi streptokinase in the prevention and treatment of thrombotic diseases.

8.5.6 Melanoidins

Melanoidins in douchi are a kind of polymer mixture with complex structure and different degrees of polymerization produced by the Maillard reaction during fermentation. They are weakly acidic and easy to hydrolyze under acidic or alkaline conditions but not degraded by digestive enzymes (Han et al. 2019). Melanoidins not only have a strong ability to scavenge free radicals and show antioxidant activity but also improve the effect of glucose tolerance and inhibit the formation of nitrosamine (Zhang et al. 2015). Nowadays, research on melanoidins has focused on antioxidant activity and optimization of extraction methods. Kong et al. (2019) optimized ultrasonic extraction conditions for melanoidins by a single-factor experiment and orthogonal experiment with water as an extractant.

8.5.7 Biogenic Amines

A biogenic amine is a kind of low molecular weight organic compound within the amino group. It has positive effects when it exists in the human body in a small amount, such as improving immunity, promoting intelligence, and improving vascular activity (Kalac 2009). But excessive concentrations have many negative effects on the human body, for instance, hypotension, vomiting, allergy, headache, other physiological manifestations, and even toxic consequences (Pegg and Anthony 2013). Figure 8.5 shows the schematic diagram of generation and a degradation method that applies the amine-degrading-enzymes (ADEs) of biogenic amines (Li and Lu 2020). The biogenic amines in douchi were mainly exogenous biogenic amines (produced by the transamination of ketones or aldehydes) that have three necessary conditions for production: free amino acids, amine-positive microorganisms, and a suitable environment (Tabanelli et al. 2019). Biogenic amines found in douchi included cadaverine, putrescine, tryptamine, spermidine, 2-phenylethylamine, histamine, tyramine, and spermine, which were degraded to hydrogen peroxide, acetaldehyde, and ammonia in combination with ADEs and O_2. There were also other methods to inhibit the production of biogenic amines in the fermentation of douchi, for example, controlling the number and species of microorganisms involved in fermentation. Tan et al. (2019) found that biogenic amines had strain specificity if produced by different molds, and *Actinomucor elegans* was a suitable strain to produce rapid and safe douchi fermentation with low biogenic amine content. Finally, reducing free amino acids in the raw soybean also inhibited the generation of biogenic amines; however, this method will affect the quality and flavor of douchi.

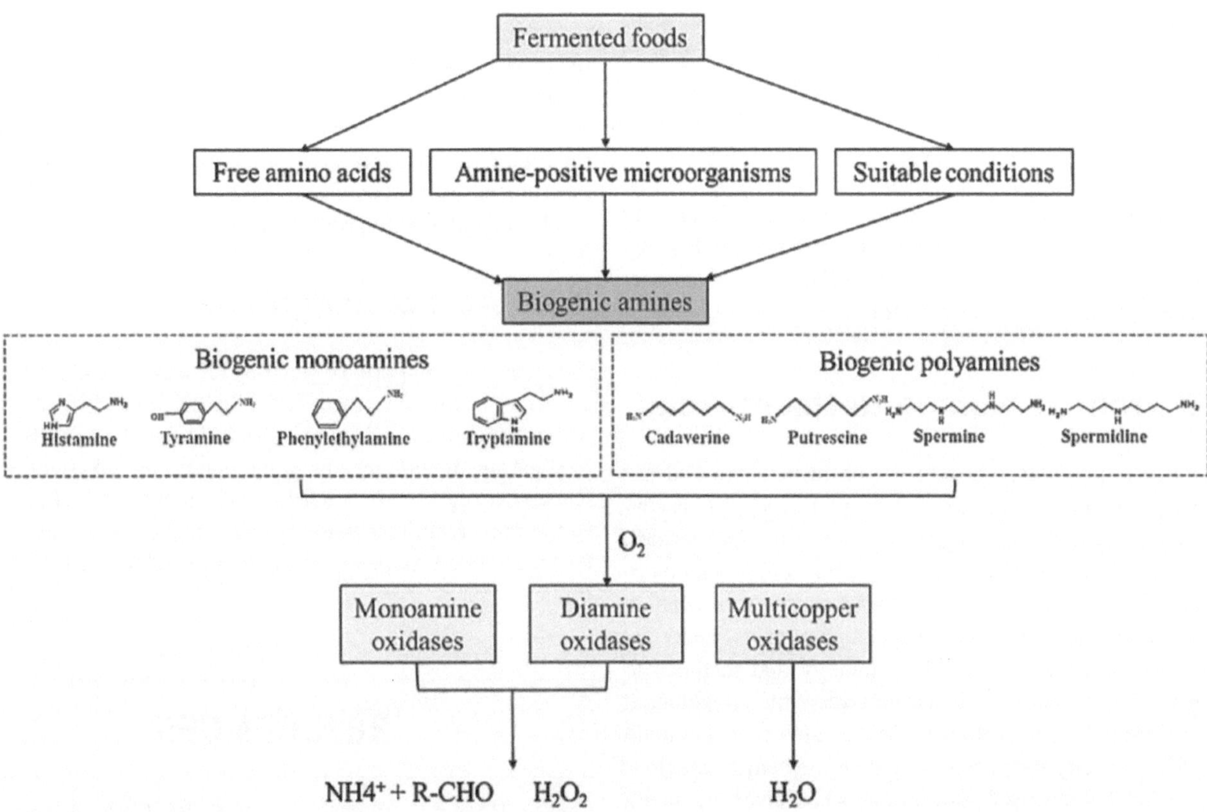

FIGURE 8.5 Schematic diagram of generation and degradation of biogenic amines.

Source: Adapted with permission from Li and Lu (2020). Copyright 2024 Elsevier.

8.6 DEVELOPMENT STRATEGIES FOR DOUCHI

8.6.1 From High Salt to Low Salt

The high salt content essential to douchi's traditional production process has reinforced its role as a condiment in Chinese cuisine, while simultaneously limiting its development as a standalone food product. Thus, it is necessary to solve the technical difficulties of low-salt fermentation. Recently, in methods of low-salt fermentation, some substances (chloride, amino acid, spices) have been used to replace salt. Otherwise, the fermentation temperature is raised to slow down the fermentation process, which plays an important role in anticorrosion (Hoang et al. 2016; Zhou, Liao, et al. 2017). Low-salt heat-preservation fermentation promoted the hydrolysis of proteins and increased the umami flavor of the fermented douchi (Liang et al. 2020), which advanced the transition of douchi in the staple food direction.

8.6.2 From Seasoning to Fast-Moving Consumer Goods

Currently, douchi products on the market are monotonous in taste and form and have great potential for market development. In the future, the development of douchi should conform to the concept of nutrition, health, and convenience. The development of douchi should be carried out by using innovative food processing technology, and it must be processed into a variety of forms, such as snack food with a unique flavor and diverse taste, special foods for the elderly and children, and breakfast for vegetarians, to improve the product quality and sale volume of douchi, aiming to fulfill the need for product upgrading, transformation, and diversity in processing.

8.6.3 From Seasoning to Health Food

As mentioned previously, douchi is rich in nutrition and has bioactive compounds with health care functions, such as soybean isoflavones, thrombolytic enzymes, oligosaccharides, and melanoidins. Therefore, treating douchi only as a condiment greatly reduces its value. Further research can focus on the separation, extraction, and pharmacology of its functional components by using new food technology and methods to develop a series of health products and explore the full potential of the nutritional value of douchi. At the same time, it should strengthen publicity for people to realize the health functions of douchi so that douchi health products can enter the health industry. As an emerging service industry, the health industry has great development potential, which is expected to improve the sales volume and quality of douchi.

8.7 CONCLUSIONS

Douchi is a form of fermented soybean food with unique flavor, high nutritional value, and rich functional bioactive compounds. Douchi has four categories depending on the type of microorganism involved in fermentation or two types according to the amount of salt or water added. Although the production processes of douchi, such as soybean pretreatment, koji making, and post-fermentation stage, are roughly the same, there are some differences in the latter two stages that lead to different kinds of douchi. Compared with the raw soybean, proteins and fats are reduced in douchi products, while vitamins are significantly increased. However, the proteins in douchi are easier to digest and absorb after fermentation while improving the utilization of minerals in the human body. The generation of taste and flavor of douchi is closely related to the growth and propagation of fungi during fermentation. The difference in dominant genera is the primary factor that causes the difference of the whole faunal structure of different groups of douchi samples. Thus, different types of douchi have distinct flavors (sour, bitter, astringent, salty, and umami) due to differences in dominant genera. There are many bioactive compounds formed during fermentation, such as polypeptides, free soybean isoflavones, melanoidins, soyasaponins, streptokinase, and soybean oligosaccharides, which are all positively related to health improvement and have a variety of physiological health care functions. Meanwhile, fermentation produces unhealthy substances like biogenic amines, but it is a promising way to degrade biogenic amines into hydrogen peroxide, acetaldehyde, and ammonia by biogenic amine oxidase. Douchi is mainly a seasoning in China due to high salt concentration, which hinders its development into a staple food. In the future, the research must focus on developing douchi in the direction of low salinity, fast-moving consumer goods) (FMCG), and healthy food to improve the quality and sales of douchi.

8.7.1 Acknowledgments

This work was supported by the Key Research and Development Project of Hainan Province (ZDYF2022XDNY335), the project of Amway China Co., Limited (Grant No. Am20220229RD), the Science and Technology Tackling Plan Project (Agriculture) of Rugao City (grant no. SRG(21)1040), and the 5th High-Level Entrepreneurship and Innovation Team Project of Putian City (Added-Value Processing of Predominant Aquatic Products in Putian City).

REFERENCES

Caplice, E., and G. F. Fitzgerald. 1999. Food fermentations: role of microorganisms in food production and preservation. *International Journal of Food Microbiology* 50 (1–2):131–149. doi: 10.1016/S0168-1605(99)00082-3.

Chang, H. C. 2018. Healthy and safe Korean traditional fermented foods, kimchi and chongkukjang. *Journal of Ethnic Foods* 5 (3):161–166. doi: 10.1016/j.jef.2018.08.003.

Chen, J. B., H. Ullah, Z. Zheng, X. F. Gu, C. H. Su, L. Y. Xiao, X. L. Wu, F. Xiong, Q. Li, and L. Y. Zha. 2020. Soyasaponins reduce inflammation by downregulating MyD88 expression and suppressing the recruitments of TLR4 and MyD88 into lipid rafts. *BMC Complementary Medicine and Therapies* 20 (1):167. doi: 10.1186/s12906-020-2864-2.

Chen, T., S. Xiong, S. Jiang, M. Wang, Q. Wu, and W. Hua. 2011. Molecular identification of microbial community in Chinese Douchi during post-fermentation process. *Food Science and Biotechnology* 20 (6):1633–1638. doi: 10.1007/s10068-011-0225-0.

Chen, T. T., S. Y. Jiang, S. Q. Xiong, M. J. Wang, D. C. Zhu, and H. Wei. 2012. Application of denaturing gradient gel electrophoresis to microbial diversity analysis in Chinese Douchi. *Journal of the Science of Food and Agriculture* 92 (10):2171–2176. doi: 10.1002/jsfa.5604.

Cheng, L. J., S. X. Zhao, J. G. Cao, and H. Z. Liang. 2005. Analysis on the nitrogen components in Chinese traditional fermented soybean products. *China Brewing* (7):47–49. doi: 10.3969/j.issn.0254-5071.2005.07.016. (In Chinese).

Chettri, R., and J. P. Tamang. 2015. Bacillus species isolated from Tungrymbai and Bekang, naturally fermented soybean foods of India. *International Journal of Food Microbiology* 197:72–76. doi: 10.1016/j.ijfoodmicro.2014.12.021.

Chukeatirote, E., K. Niraphai, U. Sardsud, and S. Popluechai. 2018. Identification of antagonistic bacteria isolated from Thai fermented soybean (Thua Nao) for biocontrol of lasiodiplodia theobromae. *Acta Phytopathologica et Entomologica Hungarica* 53 (1):19–28. doi: 10.1556/038.53.2018.004.

Cui, L., Y. Huang, W. Zhan, Y. Dou, and C. Wei. 2007. Effect of fermentation process on total content of isoflavones in soybean. *Soybean Science* (4):588–590+596. doi: 10.3969/j.issn.1000-9841.2007.04.026. (In Chinese).

Deng, G., Y. Liu, P. Li, L. Y. Liao, H. L. Yin, L. W. Jiang, Y. Y. Qin, and H. Xiao. 2021. *Food Science* 42 (24):61–66. doi: 10.7506/spkx1002-6630-20200902-017. (In Chinese).

Dong, K., Y. H. Chen, D. Y. Shen, and X. F. Pei. 2019. Research on screening, identification and growth performance of high-yield Douchi fibrinolytic enzyme strain from fermented soybeans. *China Condiment* 44 (9):85–89. doi: 10.3969/j.issn.1000-9973.2019.09.017. (In Chinese).

Dong, S. Q., X. L. Wu, B. G. Wang, S. J. Li, Y. L. Wang, and X. U. Feng. 2014. The effect of the main bacteria isolated from Douchi on biochemical indexes and the intestinal flora in mice. *Science and Technology of Food Industry* 35 (7):343–347. doi: 10.13386/j.issn1002-0306.2014.07.067. (In Chinese).

Feng, J. L., J. Y. Li, L. Xin Lu, and T. L. Li. 2010. Changes in angiotensin I-converting enzyme inhibitory activities during the ripening of Douchi (a Chinese traditional soybean product) fermented by various starter cultures. *International Journal of Food Properties* 13 (3):512–524. doi: 10.1080/10942910802688176.

Gao, Y., and X. Liu. 2010. Subsection-fermented process of subsalt-fermented Douchi using multiple strains. *Transactions of the Chinese Society of Agricultural Engineering* 26 (s1):369–373. doi: 10.3969/j.issn.1002-6819.2010.z1.065.

Granito, M., J. Frias, R. Doblado, M. Guerra, M. Champ, and C. Vidal-Valverde. 2002. Nutritional improvement of beans (Phaseolus vulgaris) by natural fermentation. *European Food Research and Technology* 214 (3):226–231. doi: 10.1007/s00217-001-0450-5.

Guo, Y., X. R. Huang, Z. J. Li, J. X. Zeng, Y. Tao, and Y. J. Wu. 2020. Dynamics of bacterial flora of bacterial water-Douchi during pure-culture intensive fermentation. *China Brewing* 39 (7):25–29. doi: 10.11882/j.issn.0254-5071.2020.07.006. (In Chinese).

Han, L. 2015. Research on the thrombolytic effect of Douchi fibrinolytic enzyme. *Modern Food* (23):18–19. doi: 10.3969/j.issn.1007-3582.2015.23.007. (In Chinese).

Han, Y., Y. Zhao, M. S. Xu, Y. Yao, N. Wu, H. Y. Du, and Y. G. Tu. 2019. Research progress on the melanoidins of maillard reaction product. *Science and Technology of Food Industry* 40 (9):339–345. doi: 10.13386/j.issn1002-0306.2019.09.058. (In Chinese).

He, G. Q., J. Huang, R. Liang, C. D. Wu, and R. Q. Zhou. 2016. Comparing the differences of characteristic flavour between natural maturation and starter culture for Mucor-type Douchi. *International Journal of Food Science and Technology* 51 (5). doi: 10.1111/ijfs.13077.

He, J., Z. W. Huang, J. J. Wu, and Z. E. Zhou. 2006. A study on the antioxidant activity of Aspergillus-type Douchi melanoidin. *Acta Agriculturae Universitatis Jiangxiensis* (5):776–779. doi: 10.3969/j.issn.1000-2286.2006.05.029. (In Chinese).

Hoang, N. X., S. Ferng, C. H. Ting, W. H. Huang, Y. Y. Robin, and C. K. Hsu. 2016. Optimizing the initial moromi fermentation conditions to improve the quality of soy sauce. *LWT-Food Science and Technology* 74:242–250. doi: 10.1016/j.lwt.2016.07.049.

Hou, H., W. Zhou, L. Guo, S. Jia, X. Zhang, and L. Wang. 2023. Effects of characteristics of Douchi during rapid fermentation and antioxidant activity using different starter cultures. *Journal of the Science of Food and Agriculture* 103:2459–2472. doi: 10.1002/jsfa.12419.

Hu, H. P., D. C. Liu, L. J. Yin, Y. Q. Cheng, and L. X. Ding. 2014. Screening and production performance of poredominant strains during Douchi post-fermentation. *Food Science* 35 (17):146–152. doi: 10.7506/spkx1002-6630-201417029. (In Chinese).

Ji, Y. P., X. Chen, and R. M. Shen. 2016. Optimization of ultrasonic-assisted extraction of Douchi oligosaccharides by response surface methodology. *China Brewing* 35 (5):129–133. doi: 10.11882/j.issn.0254-5071.2016.05.027. (In Chinese).

Jia, F., X. Gua, C. He, Q. L. Li, and X. L. Wang. 2019. Research progress in nutritional and functional ingredients of the traditional fermented soybean products. *China Brewing* 38 (4):1–6. doi: 10.11882/j.issn.0254-5071.2019.04.001. (In Chinese).

Jung, A. R., S. H. Ahn, I. S. Park, S. Y. Park, and K. Kim. 2016. Douchi (fermented Glycine max Merr.) alleviates atopic dermatitis-like skin lesions in NC/Nga mice by regulation of PKC and IL-4. *BMC Complementary and Alternative Medicine* 16 (1):416. doi: 10.1186/s12906-016-1394-4.

Kalac, P. 2009. Recent advances in the research on biological roles of dietary polyamines in man. *Journal of Applied Biomedicine* 7 (2):65. doi: 10.32725/jab.2009.007.

Keila, K. M. Ada, A. G. Janet, S. Arlette, O. S. Sergio, and M. Cristina. 2018. Peptides and isoflavones in gastrointestinal digests contribute to the anti-inflammatory potential of cooked or germinated desi and kabuli chickpea (Cicer arietinum L.). *Food Chemistry* 268:66–76. doi: 10.1016/j.foodchem.2018.06.068.

Kobayashi, K., Y. Horii, S. Watanabe, Kubo Y, K. Koguchi, Y. Hoshi, K. I. Matsumoto, and K. Soda. 2017. Comparison of soybean cultivars for enhancement of the polyamine contents in the fermented soybean natto using Bacillus subtilis (natto). *Bioscience, Biotechnology, and Biochemistry* 81 (3):587–594. doi: 10.1080/09168451.2016.1270738.

Kong, F. L., R. B. Kuang, J. C. Zheng, D. Q. Lin, and Z. M. Hao. 2019. Optimization of ultrasonic extraction technology for aspergillus-type Douchi melanoidin. *Modern Food* (24):77–78+83. doi: 10.16736/j.cnki.cn41-1434/ts.2019.24.028. (In Chinese).

Kumari, R., N. Sharma, S. Sharma, S. Samurailatpam, S. Padhi, S. P. Singh, and A. Kumar Rai. 2023. Production and characterization of bioactive peptides in fermented soybean meal produced using proteolytic Bacillus species isolated from kinema. *Food Chemistry* (421):136130. doi: 10.1016/j.foodchem.2023.136130.

Li, B. B., and S. L. Lu. 2020. The Importance of amine-degrading enzymes on the biogenic amine degradation in fermented foods: a review-ScienceDirect. *Process Biochemistry* 99:331–339. doi: 10.1016/J.PROCBIO.2020.09.012.

Li, N., and Q. B. Huang. 2008. Isoflavones in fermented soybean and their pharmacological effects and clinical application. *Modern Chinese Medicine* (7):18–19. doi: 10.13313/j.issn.1673-4890.2008.07.009. (In Chinese).

Li, S., and J. Wang. 2020. Research progress on the clinical application of soybean isoflavones. *Soybean Science* 39 (4):633–640. doi: 10.11861/j.issn.1000-9841.2020.04.0633. (In Chinese).

Li, W., S. B. Luo, Z. R. Qiu, D. L. Lai, H. W. Wang, and H. Y. Suo. 2020. Microbial community structure and dynamic succession in traditional fermentation of mucor-type Yongchuan Douchi. *Food and Fermentation Industries* 46 (23):60–67. doi: 10.13995/j.cnki.11-1802/ts.024566. (In Chinese).

Li, Y., Y. H. Chen, X. W. Liu, and X. F. Pei. 2020. Optimization of technological conditions for fermented soya beans by bacillus velezensis using response surface methodology. *China Condiment* 45 (2):49–53. doi: 10.3969/j.issn.1000-9973.2020.02.012. (In Chinese).

Liang, R., X. P. Zhu, H. L. Li, X. Huang, R. R. Cui, and D. M. Su. 2020. Effect of salt and temperature on the quality of mucor-type Douchi. *Journal of Ankang University* 32 (4):116–123. doi: 10.16858/j.issn.1674-0092.2020.04.025. (In Chinese).

Liu, L. B., X. Q. Chen, L. L. Hao, G. F. Zhang, Z. Jin, C. Li, Y. Z. Yang, J. J. Rao, and B. C. Chen. 2020. Traditional fermented soybean products: processing, flavor formation, nutritional and biological activities. *Critical Reviews in Food Science and Nutrition* (4):1–19. doi: 10.1080/10408398.2020.1848792.

Liu, L. L., Z. T. Wang, C. B. Ding, L. I. Shuang, L. I. Wei, W. C. Liu, and Y. N. Zheng. 2014. Determination of soy isoflavone aglycone content of edamame prepared by different processing methods and its hypoglycemic and hypolipidemic activity for diabetic mice. *Natural Product Research and Development* 26 (10):1659–1663. doi: 10.16333/j.1001-6880.2014.10.025. (In Chinese).

Liu, X., Q. Yu, X. F. Pei, H. C. Liu, and X. Xu. 2003. A study on the modulation functions of Shansong soybean oligosaccharide on the structure of intestinal flora. *Chinese Journal of Microecology* (1):13–15. doi: 10.3969/j.issn.1005-376X.2003.01.004. (In Chinese).

Liu, Y., H. L. Song, and H. Z. Luo. 2018. Correlation between the key aroma compounds and gDNA copies of bacillus during fermentation and maturation of natto. *Food Research International (Ottawa, Ont.)* 112:175–183. doi: 10.1016/j.foodres.2018.06.033.

Melanie, S., U. Z. Ijaz, D. Rosalinda, H. Neil, W. T. Sloan, and Q. Christopher. 2015. Insight into biases and sequencing errors for amplicon sequencing with the Illumina MiSeq platform. *Nucleic Acids Research* 43 (6):e37. doi: 10.1093/nar/gku1341.

Mishra, B. K., S. Hati, and S. Das. 2019. Bio-nutritional aspects of Tungrymbai, an ethnic functional fermented soy food of Khasi Hills, Meghalaya, India. *Clinical Nutrition Experimental* 26:8–12. doi: 10.1016/j.yclnex.2019.05.004.

Murooka, Y., and M. Yamshita. 2008. Traditional healthful fermented products of Japan. *Journal of Industrial Microbiology and Biotechnology* 35 (8):791–798. doi: 10.1007/s10295-008-0362-5.

Pegg, Anthony E. 2013. Toxicity of polyamines and their metabolic products. *Chemical Research in Toxicology* 26 (12):1782–1800. doi: 10.1021/tx400316s.

Peng, H., G. Z. Hu, G. J. Chen, H. Y. Suo, and J. Q. Kan. 2013. Changes of basal components and biogenic amines in the preparation of Yongchuan Douchi koji. *Food and Fermentation Industries* 39 (3):163–166. doi: 10.13995/j.cnki.11-1802/ts.2013.03.042. (In Chinese).

Reseland, J. E., U. Syversen, I. Bakke, G. Qvigstad, L. G. Eide, O. Hjertner, J. O. Gordeladze, and C. A. Drevon. 2001. Leptin is expressed in and secreted from primary cultures of human osteoblasts and promotes bone mineralization. *Journal of Bone and Mineral Research* 16 (8):1426–1433. doi: 10.1359/jbmr.2001.16.8.1426.

Roh, S. W., K. H. Kim, Y. D. Nam, H. W. Chang, E. J. Park, and J. W. Bae. 2010. Investigation of archaeal and bacterial diversity in fermented seafood using barcoded pyrosequencing. *The ISME Journal* 4 (1):1–16. doi: 10.1038/ismej.2009.83.

Setiawan, B., S. K. Thamtam, I. R. A. Jati, R. C. Purwestri, D. Nohr, and H. K. Biesalski. 2016. The influence of traditional stir-frying with oil on acceptability, antioxidant activities, nutrients, and the phytic acid content of fermented soybean (tempeh). *Nutrition and Food Science* 46 (2):259–271. doi: 10.1108/NFS-09-2015-0105.

Shen, D. Y., K. Dong, Z. W. Fan, J. X. Chen, X. He, and X. F. Pei. 2020. Optimization of pure breed fermentation conditions for soybean isoflavones produced by fermented soybeans with response surface methodology. *China Condiment* 45 (6):31–36. doi: 10.3969/j.issn.1000-9973.2020.06.008. (In Chinese).

Song, Y. S. 2003. Study on changes of nutrimental composition and active composition in lobster sauce. *Science and Technology of Food Industry* (7):79–80+64. doi: 10.3969/j.issn.1002-0306.2003.07.034. (In Chinese).

Song, Y. S., B. W. Zhang, Z. H. Hao, and Y. S. Chi. 2002. Study on the affects of fermentation on lobster sauce antioxidize activity. *Food Science* (8):263–267. doi: 10.3321/j.issn:1002-6630.2002.08.081. (In Chinese).

Su, H. K., W. C. Chen, J. H. Lu, H. R. Chao, Y. F. Liang, S. Haruka, W. L. Hsu, M. L. Wu, and M. H. Tsai. 2023. The effects of using Tempeh as a supplement for type 2 diabetes. *Food Science and Nutrition* :1–9. doi: 10.1002/fsn3.3319.

Sugawara, E., T. Suzuki, and Y. Yoshida. 1998. The comparison of pyrazine compounds in non-salted fermented soybean products. *Japanese Society for Food Science and Technology* 4 (1):85–88. doi: 10.3136/fsti9596t9798.4.85.

Sun, X. H., and W. Q. Jiang. 2004. Soybean peptides functional properties and application. *Machinery for Cereals Oil and Food Processing* (5):7–10. doi: CNKI:SUN:NJSP.0.2004-07-014. (In Chinese).

Suzuki, K., M. Nakamura, N. Sato, K. Futamura, K. Matsunaga, and A. Yagami. 2023. Nattokinase (Bac s 1), a subtilisin family serine protease, is a novel allergen contained in the traditional Japanese fermented food natto. *Allergology International* 72 (2):279–285. doi: 10.1016/j.alit.2022.11.010.

Tabanelli, G., C. Montanari, and F. Gardini. 2019. Biogenic amines in food: a review of factors affecting their formation. *Encyclopedia of Food Chemistry* 1:337–343. doi: 10.1016/B978-0-08-100596-5.21833-7.

Tamang, J. P., D. H. Shin, S. J. Jung, and S. W. Chae. 2016. Functional properties of microorganisms in fermented foods. *Frontiers in Microbiology* 7:578. doi: 10.3389/fmicb.2016.00578.

Tan, Y., R. F. Zhang, G. J. Chen, S. S. Wang, C. F. Li, Y. Xu, and J. Q. Kan. 2019. Effect of different starter cultures on the control of biogenic amines and quality change of douchi by rapid fermentation. *LWT-Food Science and Technology* 109:395–405. doi: 10.1016/j.lwt.2019.04.041.

Utami, R., C. H. Wijaya, and H. N. Lioe. 2016. Taste of water-soluble extracts obtained from over-fermented Tempe.

International Journal of Food Properties 19 (9):2063–2073. doi: 10.1080/10942912.2015.1104509.

Wang, J. M., and L. Y. Li. 2012. Optimization enzyme hydrolysis conditions of new enzymatic fermented bean curd by response surface analysis. *Sichuan Food and Fermentation* 48 (2):44–48. doi: 10.3969/j.issn.1674-506X.2012.02-012.

Wang, N., W. X. Chen, W. P. Zhang, and J. S. Zhang. 2022. Community diversity of different bacterial-type of Guizhou Douchi using high throughput sequencing. *Food and Fermentation Industries* 48 (20):85–90+104. doi: 10.13995/j.cnki.11-1802/ts.030524. (In Chinese).

Wang, Q. S. 2008. Research on separation and purification of soybean oligosaccharides from sweet slurry in soybean sheet production. Doctor, Zhejiang University Cnki. (In Chinese).

Wang, Y. R., F. S. Xiang, Z. D. Zhang, Q. C. Hou, and Z. Guo. 2020. High-throughput sequencing-based analysis of fungal diversity and taste quality evaluation of Douchi, a traditional fermented food. *Food Science and Nutrition* 8 (12):6612–6620. doi: 10.1002/fsn3.1953.

Wen, L. W., Q. L. Zhang, G. L. Qing, S. P. Jia, P. H. Ji, Q. Chen, H. X. Wang, and J. Sun. 2020. Research advances on biological activities and application of saponins. *Animal Husbandry and Feed Science* 41 (4):90–96. doi: 10.12160/j.issn.1672-5190.2020.04.018. (In Chinese).

Wu, W., Y. L. Zhu, and L. Yuan. 2017. Protective effect of ACEI/ARB on T2DM. *Chinese Journal of Clinical Rational Drug Use* 10 (21):163–165. doi: 10.15887/j.cnki.13-1389/r.2017.21.082. (In Chinese).

Wu, Y. J., W. Meng, Y. Z. Li, S. Zhan, and Y. J. Wu. 2011. Production of bacterial fermented Douchi by pure fermentation. *China Brewing* (9):56–60. doi: 10.3969/j.issn.0254-5071.2011.09.016. (In Chinese).

Yang, C. P., B. R. Xu, L. Y. Huang, M. Ren, and K. Cai. 2019. Optimization of preparation technology of SSP by pure breed fermentation. *Journal of Guiyang College of Traditional Chinese Medicine* 41 (3):54–57. doi: 10.16588/j.cnki.issn1002-1108.2019.03.012. (In Chinese).

Yu, H., P. Chang, B. Y. Chen, R. Wei, L. N. Ma, S. J. Niu, L. Gan, Y. H. Mao, and G. Y. Cheng. 2019. Melioration and antioxidation effect of complexes of soy isoflavones and bovine colostrum on osteoprosis of ovariectomized rats. *Science and Technology of Food Industry* 40 (15):284–291. doi: 10.13386/j.issn1002-0306.2019.15.047. (In Chinese).

Yuan, J., J. Yang, Z. H. Zhuang, Y. L. Yang, L. Lin, and S. H. Wang. 2012. Thrombolytic effects of Douchi fibrinolytic enzyme from Bacillus subtilis LD-8547 in vitro and in vivo. *BMC Biotechnology* 12 (1):36. doi: 10.1186/1472-6750-12-36.

Zhang, J. H. 2003. Studies on the fermentation mechanism and healthy function of aspergillus-type Douchi. Doctor, China Agricultural University, Cnki. (In Chinese).

Zhang, J. H., E. Tatsumi, J. F. Fan, and L. T. Li. 2007. Chemical components of aspergillu-type Douchi, a Chinese traditional fermented soybean product, change during the fermentation process. *International Journal of Food Science and Technology* 42 (3):263–268. doi: 10.1111/j.1365-2621.2005.01150.x.

Zhang, X. M., L. Zhang, F. H. Gao, J. Y. Tang, and X. Li. 2015. Nutrition research status and development prospect of Douchi. *South China Agriculture* 9 (4):51–53. doi: 10.19415/j.cnki.1673-890x.2015.04.018. (In Chinese).

Zhang, Y. J., T. Lan, Q. Chu, K. X. Dong, S. L. Wang, Y. Zhang, and Z. J. Zheng. 2020. Progress in preparation technology and functional activity of soybean isoflavones. *Journal of Food Safety and Quality* 11 (17):5964–5970. doi: 10.19812/j.cnki.jfsq11-5956/ts.2020.17.029. (In Chinese).

Zhao, D. A. 2001. Nutrition and health protection functions of soybean fermented food. *China Brewing* (4):9–13. doi: 10.3969/j.issn.0254-5071.2001.04.004. (In Chinese).

Zhao, W. P. 2020. Quality characteristics and microbial community evolution of aspergillus-type Douchi during fermentation. Master, Jiangxi Normal University Cnki. doi: 10.27178/d.cnki.gjxsu.2020.000183. (In Chinese).

Zhou, T., T. Liao, X. H. Gu, S. Y. Jin, and H. Li. 2017. The inhibition of spices on the dominant bacteria and protease activity during low-salt fermentation of Douchi. *Modern Food* 17 (17):111–113. doi: 10.16736/j.cnki.cn41-1434/ts.2017.17.034. (In Chinese).

Zhou, X. L., Y. C. Yang, L. Liu, and Y. Zhang. 2017. Research on screening and fermentation properties of fibrinolysin producing strains from fermented soya beans. *China Condiment* 42 (7):41–45. doi: 10.3969/j.issn.1000-9973.2017.07.009. (In Chinese).

Soybean-Based Fermented Foods of Northeast India

9

Microbiology, Nutritional Value and Health Benefits

Kirti Rai and Kriti Ghatani

9.1 INTRODUCTION

Soybean (*Glycine max*) is a legume like peas, beans and lentils, and it belongs to the family *Leguminosae* (which includes some 500 genera and more than 12,000 species) in the subfamily *Papilionideae*. Soybean was first cultivated and used from the beginning of China's agricultural age (Da Silva & de Majo, 2022).

The most widely accepted theory about the origin of the soybean is that it was first grown in the eastern half of north China in about the 11th century BC, and by the end of the first century AD, it had reached north, central and south China; Korea; and Japan. In the 10th century AD, the soybean was first introduced in India from China and transported through the Himalayan routes and via Burma (now Myanmar) by traders and merchants from Indonesia. Soybeans are mainly grown in Asia, South and North America and have been an important component in the daily diet of Asian people. Due to large-scale trading of soybeans today, it is now grown in different parts of India, including Himachal Pradesh, the Kumaon Hills of Uttar Pradesh (now Uttaranchal), Eastern Bengal, the Khasi Hills of Meghalaya, Manipur, the Naga Hills of Nagaland and many parts of central India covering Madhya Pradesh. Today, the Indian subcontinent is the secondary centre for the cultivation of soybeans after China (Hymowitz, 1990). Soybean is a source of plant-based protein, and apart from proteins, it is also rich in components such as lipids, vitamins, minerals, free sugars, flavonoids, isoflavones, saponins and peptides that have a high nutritional value (Kim et al., 2006). They also contain antioxidants and phytonutrients and are regarded as one of the most important crops in the world due to their major health benefits to humans. Soybean products are considered a good substitute for animal protein. The nutritional value of fermented soybean nearly equals that of animal protein because it contains all the necessary and nutritious components which are essential for human health. All except sulphur amino acids such as methionine and cysteine which are present in animal protein are absent in fermented soybean. Also, in addition to high protein content, soybeans contain high levels of unsaturated fatty acids, dietary fibre, isoflavones and minerals, which possess a number of health benefits (Bolla, 2015). There are worldwide varieties of soybean products such as boiled soybeans, roasted soybean, miso, soy cheese, soy mayonnaise, soy milk, soy sauce, soy yogurt, tamari, tempeh, textured soy protein (TSP) or textured vegetable protein (TVP) and tofu. In the northeastern regions of India, soybeans are cultivated, grown and then processed into many different types of products (Tamang, 2015). Soybeans in this region are of traditional importance, and the art of making soybean products by the process of fermentation has been passed down by people's ancestors. The northeastern regions of India are home to a distinct number of ethnic tribes, and each of these tribes is accustomed to producing different soybean products by employing their own indigenous methods and processes (Tamang, 2015). Soybean is usually planted from the third week of June to the middle of July and harvested in November, and it requires well-drained, fertile loamy soil with a pH range between 6.0 and 7.5 (Tamang, 2015). Soybean is usually consumed in two forms: the unfermented soybean forms like roasted and fried soybeans seeds, soybean oil, soybean powder, soy sauce and the fermented soybean forms such as soybean pickle and different ethnic fermented soybean products. Fermented soybean products are common among the tribes of northeastern India and include foods such as *kinema* (Sikkim),

DOI: 10.1201/9781003605300-9

FIGURE 9.1 Soybean seeds.

bekang (Mizoram), *hawaijar* and *chagempomba* (Manipur), *tungrymbai* (Meghalaya), *akhuni/aakhone/axone* (Nagaland) and *peruyyan* (Arunachal Pradesh). These fermented soybean foods are similar to each other in texture (sticky) and taste (slightly alkaline), and they all have a strong ammonic odour, which is their signature trait even though the preparation methods are different for each (Tamang, 2015). These fermented products form an important staple dietary component and also serve as a major protein source to the people of the corresponding ethnic tribes. Fermentation is the commonly used method to produce variations of soybean products and also to increase their shelf life. Fermentation improves the physicochemical and sensory quality of soybean products, where *Bacillus subtilis* and *Aspergillus* are the microorganisms commonly used for fermentation (Jayachandran & Xu, 2019). Soybean fermentation results in the release of bioactive peptides by the action of proteolytic enzymes produced by the microorganisms involved in the fermentation process (Sanjukta et al., 2015). Bioactive peptides in fermented soybean have been studied for various therapeutic properties such as antioxidant (Sanjukta et al., 2015), antihypertensive (Zhang et al., 2006), anti-tumour (Jung et al., 2006) and anti-diabetic (Kwon et al., 2011) and are also known to prevent atherosclerosis (Tsai et al., 2009). Fermentation is a simple technique that can significantly help in improving the texture, nutrition and flavour of the soybean (Granito et al., 2002). The fermentation process helps to enhance bioactive compounds for major health benefits (Sourabh et al., 2015). Bioactive compounds are nutritive constituents that are found in small quantities in foods and provide health benefits beyond the basic nutritional value of the product. During the fermentation process, the biofunctional properties of the soybean increase due to the increase in free isoflavones and peptides in the soybean (Cho et al., 2011). Fermentation can also result in the reduction of antinutritional components such as proteinase inhibitors, phytic acid, urease and oxalic acids (Reddy & Pierson, 1994; Egounlety & Aworh, 2003).

9.2 FERMENTED FOODS

Fermentation is one of the most feasible, economical and widely used methods in the world for different types of food preparation processes, and it is also defined as a technology in which the growth and metabolic activities of microorganisms are used to preserve foods (Nuraida, 2015; Terefe, 2016; Wilburn & Ryan, 2017). Food fermentation can be divided into two categories: aerobic fermentation (fungal and alkaline) and anaerobic fermentation (alcoholic and lactic acid) (Nout, 2014). During the fermentation process, the microorganisms that are used break down the fermentable carbohydrates into end products such as organic acid, carbon dioxide, alcohol (Kim et al., 2016) and antimicrobial metabolites such as bacteriocins that help ensure food safety by killing or inhibiting foodborne pathogens, if present in the food product (Nout, 2014). Fermentation also increases the shelf life of foods which are highly perishable in nature (Nuraida, 2015; Terefe, 2016). Fermentation also increases the organoleptic properties of food, the digestibility of proteins and carbohydrates and the bioavailability of vitamins and minerals (Altay et al., 2013; Hwang et al., 2017).

9.2.1 Types of Fermented Foods

Fermented foods can be defined as those foods or beverages that are produced by the controlled growth of microorganisms, which releases different enzymes during the process, that in turn helps in the conversion of food components (Marco et al., 2017). Food fermentation has widely used two main methods to carry out the process. First, some foods are fermented naturally, also called "wild ferments" or "spontaneous ferments". By this method, the microorganisms essential for the process to commence are present naturally in the raw material or in the processing environment, such as sauerkraut, kimchi and certain fermented soy foods and products. Second, foods are fermented with the help of the addition of starter cultures, also known as "culture-dependent ferments", like kefir, kombucha and natto (Rezac et al., 2018). Culture-dependent ferments are performed by a process referred as "backslopping", which includes adding a small amount of a previously fermented batch to the fresh raw food, as in the case of preparation of sourdough bread (Marco et al., 2017). The starter culture used for the initiation of the fermentation process can be natural (backslopping) or a selected commercial starter, which can also enhance the overall characteristics of the final fermented product (Yann & Pauline, 2014). Fermented food can be milk

TABLE 9.1 Fermented Foods of the World

FERMENTED FOOD	*DESCRIPTION*	*RAW MATERIALS*	*ORGANISMS USED*	*HEALTH BENEFITS*	*REFERENCES*
Yoghurt, kefir, cheese	Milk based	Skimmed/ fortified milk	*Streptococcus thermophilus, Lactobacillus delbrueckii/ Lactobacillus bulgaricus, Lactococcus* sp., *Lactobacillus* sp. *or Streptococcus* sp.	Rich source of calcium, vitamin B2, vitamin B6 and vitamin B12; boosts immune system; aids in digestion; contains probiotics	(De, 1980; Aneja, 1989),(Johnson, 2017)
Kimchi	Vegetable based	Cabbage, salt, spices	Lactic acid bacteria (LAB)	Abundant vitamins, minerals and antioxidants; boosts cell growth and muscle growth; improved cardiovascular health and digestive system; blood clotting	(Cheigh et al., 1994), (Lee, 1997)
Sauerkraut	Vegetable based	Shredded cabbage, salt	*Klebsiella, Enterobacter, Lactobacillus brevis, Lactiplantibacillus plantarum, Leuconostoc mesenteroides*	Promotes healthy bowel flora; rich source of iron, vitamin C, copper, potassium, folate and manganese	(Orgeron II et al., 2016)
Sausages	Meat based	Animal meat	*Lactobacillus* sp.	Contains proteins and collagen that strengthen skin tissue and make it more youthful; source of vitamin B12; production of red blood cells (RBCs) in body	(Oster, 2011)
Idli	Cereal based	Fermented black lentils and rice	*Lactobacillus* sp.	Increases absorption of vital materials from gastrointestinal tract and enhances vitamin B content	(Ansari & Pourjafar, 2022)
Gundruk	Vegetable based	*Rayosaag*	*Pediococcus, Lactobacillus* sp.	Aids milk production in lactating mothers; cures diarrhoea and stomach indigestion	(Battcock & Azam-Ali, 1998)
Bread	Cereal based	Flour	*Saccharomyces cerevisiae*	Rich source of fibre and lower in fats; good for digestion and overall health; reduces cholesterol and blood pressure	(Mondal & Datta, 2008)

based, vegetable based, soybean based, fruit based, meat based, cereal based and so on. Some popular fermented foods of the world are kefir, kombucha, sauerkraut, natto, tempeh, miso, kimchi, sausages and yoghurt. These fermented food products also contain essential live microorganisms which have many health-benefiting properties for the human body, and they are called probiotics. Fermented foods exhibit beneficial effects on health by reducing blood cholesterol levels; increasing immunity; protecting against pathogens; fighting carcinogenesis, osteoporosis, diabetes, obesity, allergies and atherosclerosis; and alleviating the symptoms of lactose intolerance (Farhad et al., 2010). Several epidemiological studies have proved that consumption of fermented soy foods reduces serum cholesterol and triacylglycerol concentrations, thereby helping the body against type 2 diabetes (Anderson et al., 1998). It also helps to improve bone health (Berenjian et al., 2011). Furthermore, some studies have shown that a diet rich in soy can have neuroprotective, fibrinolytic and anticancer activity (Cao et al., 2009; Dabbagh et al., 2014; Chen et al., 2019). The health benefits associated with fermented foods are often attributed to the bioactive peptides that are synthesized in the microbial degradation of proteins involved in fermentation (María Hebert et al., 2010; Walther & Sieber, 2011; Otağ & Hayta, 2013).

9.2.2 Soybean-Based Fermented Food Products of Northeast India

There are more than 12,000 varieties of food that contain soybean protein; hence it is consumed by a large number of people all over the world (Montgomery, 2003). Soybean-based fermented foods of the world include natto, *koji* and miso from Japan; tempeh from Indonesia; *douchi* and tofu from China; *thuanao* from Thailand; and *kochujang, kanjang, doengjang* and *chungkukjang* from Korea (Jayachandran & Xu, 2019). The northeast region of India comprises the states of Arunachal Pradesh, Assam, Manipur, Mizoram, Tripura, Sikkim, Meghalaya and Nagaland. The climate and altitude variations in northeastern India such as landscape, favourable climatic conditions, communities and geographical and ecological diversity have resulted in a great range of ecological habitats (Bhutani, 2008).

Northeast India not only has unique natural diversity but is also home to a large number of Indigenous and immigrant tribal and ethnic communities. More than 200 different tribes with distinct dialects, customs, beliefs, heritage and socio-religious traditions are settled in the northeastern states of India (Bhutani, 2008). Some of the tribes linked with traditional fermented food making in northeast India are the Meitei community of Manipur; the Khasi and Jaintia tribespeople of Meghalaya; the Sema Naga tribe of Nagaland; the Apatani tribes in Arunachal Pradesh; the Mizo people of Mizoram; and the Nepali community in the regions of Nepal, Sikkim, Darjeeling and Bhutan (Tamang, 2015). The traditional knowledge of fermented food making has been generated, refined and passed down from the ancestors by ethnic and indigenous communities.

9.2.2.1 Kinema

Kinema is one of the most important traditional foods of the Nepali community in the regions of Nepal, Sikkim, Darjeeling and Bhutan in the northeast of India. The word *kinema* is derived from *kinambaa* in the Limboo dialect (Limboo being one of the major ethnic Nepali communities); *ki* means fermented, and *nambaa* means flavour (Tamang, 2024). *Kinema* is a type of fermented food which is made from whole soybeans, which are harvested and grown by the locals living in that region. The texture of *kinema* is somewhat like the natto of Japan as well as tempeh from Indonesia since both are fermented forms of soybean. It is sticky and slimy in nature, with a strong ammoniacal smell and taste (Tamang, 2015). Various types of microorganisms are involved in the fermentation; the predominantly used strains are *Bacillus* sp., *Aspergillus* sp. and *Lactobacillus* sp. During the traditional production of *kinema*, small, dry soybean seeds are used. They are first washed and soaked overnight in water. The soaked soybean seeds are then taken out and put into a cooking container which contains fresh water and then boiled for 2–3 hours until they are soft. Excess water is drained off, and the cooked soybean seeds are cracked lightly to split the cotyledons. This practice of cracking the cooked seeds of soybeans is observed only during *kinema* production and not in other similar fermented soybean foods. This is probably done to increase the surface area for speedy fermentation by the aerobic spore-forming bacteria *Bacillus* sp. (Tamang, 2003). After that, approximately 1% of firewood ash is added directly to the cooked soybeans and mixed thoroughly to maintain the taste, nature and condition of the product. Then the soybean is placed in a basket lined with locally grown fresh fern (*Glaphylopteriolopsise rubescens*). The basket is then covered with a jute bag and left to ferment naturally at 25–40°C for 1–3 days. If the process of fermentation is complete, there is an appearance of a white mass and typical *kinema* flavour with a strong odour. During the summer season, the fermentation time may be 1–2 days, whereas in winter, it may require 2–3 days. *Kinema* is eaten as a side-dish curry with boiled rice and cooked with vegetables such as onions, chilies, tomatoes, garlic and ginger. Several species of *Bacillus* have been isolated from *kinema*, including *Bacillus subtilis*, *Bacillus licheniformis*, *Bacillus cereus*, *Bacillus circulans*, *Bacillus thuringiensis* and *Bacillus sphaericus* at an average load of 108 colony-forming units (CFU/g) (Sarkar et al., 1994). However, *B. subtilis* was the dominant functional bacterium in *kinema* (Sarkar & Tamang, 1994; Tamang & Nikkuni, 1996). Besides *Bacillus*, *Enterococcus faecium* (10^7 CFU/g) and the yeasts *Candida parapsilosis* and *Geotrichum candidum* (10^4 CFU/g) have also been reported in *kinema* (Sarkar & Tamang, 1994; Tamang, 2003).

9.2.2.2 Hawaijar

Hawaijar is a popular traditional fermented soybean food of the Meitei community of Manipur (Tamang, 2015). The process of preparation and the texture of the food product are

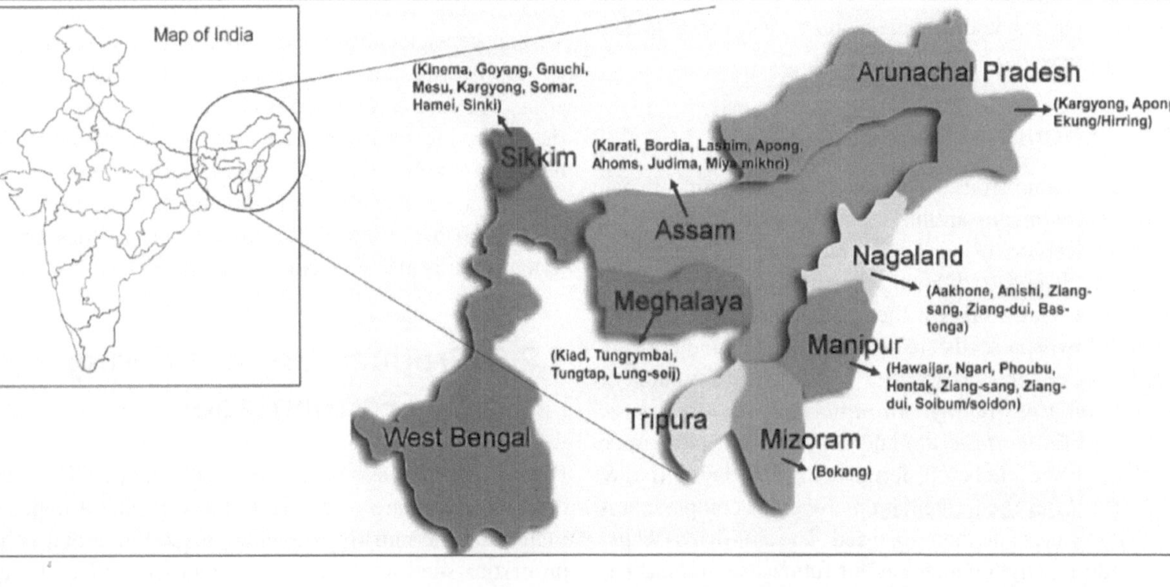

FIGURE 9.2 Map of India showing the northeastern regions along with the traditional fermented foods of each place (Das et al., 2021).

similar to those of *kinema*. *Hawaijar* is also made from small-seeded soybeans that are washed and boiled in an open container for 2–3 hours and drained. They are then cooled down and packed in baskets, which are again lined with the leaves of the fig plant (*Ficus hispida*), locally called *asseeheibong* in the Meitei language, or banana leaves can also be used. Then the basket is closed with an airway and kept in a warm place, usually at a 25–40°C in the kitchen or the sun. The soybeans are allowed to naturally ferment for 3–5 days. After this, *hawaijar* is ready to be consumed. It is usually eaten as a paste which is mixed with salt and chilli. *Chagempomba* is also mainly prepared by the Meitei people of Manipur. They add various kinds of seasonal vegetables, spices and flavourings to *hawaijar* and then cook it. It has a characteristic pungent strong smell, and it is eaten with steamed rice (Jeyaram et al., 2008).

9.2.2.3 Tungrymbai

Tungrymbai is a fermented soybean product which is exclusively made by the local Khasi and Jaintia tribespeople of Meghalaya in northeast India. It is a sticky food that exhibits a unique flavour and texture. The preparation of *tungrymbai* is done by using traditional technology by the local tribal people (Sekar & Mariappan, 2007). Soybean seeds are collected, cleaned, washed and soaked in water for approximately 4–6 hours (Agrahar-Murugkar & Subbulakshmi, 2006). The seed coat of the soybeans is then normally removed before cooking by rubbing the soaked seeds gently. The soaked soybeans are cooked for about 1–2 hours and then cooled. They are then packed with fresh leaves of *Clinogyne dichotoma* (locally called *lamet*), placed inside a basket and covered with a thick cover such as a cloth. The covered basket is left to ferment naturally for 3–5 days. After the fermentation process is complete, the *tungrymbai* is taken out and mashed well. The mashed *tungrymbai* is poured into a container with water and boiled until the water evaporates, with continuous stirring. It is mixed with fried onion, garlic, ginger, chilli, ground black sesame (locally called *til*) and salt. A thick curry is made out of it and served as a side dish with steamed rice. Pickle is also made from *tungrymbai*.

9.2.2.4 Akhuni/Aakhone/Axone

Akhuni gets its name from two words from the Sumi dialect, *axone*, where *axo* means aroma or smell, and the word *ne* or *nhe*, meaning deep or strong. Therefore, it means "the strong-smelling dish". It is a traditional fermented soybean food of the Sema Naga tribe living in the southern part of the state of Nagaland. Soybean seeds are selected, soaked and cooked, and then the beans are wrapped in fresh leaves of banana or *Phrynium pubinerve* (family: *Marantaceae*) or *Macaranga indica* (family: *Euphorbiaceae*) and kept in a hot place such as a fireplace in the kitchen to ferment for 5–7 days (Mao & Odyuo, 2007). After the fermentation process is complete, it is then made into cakes and then sun dried. The sun-dried *akhuni* is then stored in airtight containers for future use. *Akhuni* has a bitter, smoked taste to it and can be mixed with seasonal vegetables to make a stew or smoked pork, dried fish or dried beef. It is also used to make chutney and pickle by adding onions, chillies, garlic and tomatoes. In Nagaland, *akhuni* is considered a common fermented food product and prepared year-round by the tribespeople (Tamang, 2015).

9.2.2.5 Peruyyan

Peruyaan is a traditional fermented soybean food of the Apatani tribes in Arunachal Pradesh in northeast India. The word *peruyaan* is derived from the Apatani dialect, where *perun* means beans and *yannii* means packing in leaves (Tamang et al., 2009). For the traditional preparation of *peruyaan*, soybean seeds are selected, washed and cooked for 2–3 hours until they become soft, and then the excess water is drained. They are then cooled for some time and kept in a basket lined with fresh ginger leaves, locally called *takiyannii*. The basket is loosely covered with a covering such as a thick cloth or the same ginger leaves. It is then kept in a warm place, probably beside a fireplace in the kitchen, for 3–5 days. After the fermentation process is complete, the stickiness of the product is examined, and if the product is considered sticky enough, then it is regarded as ready for consumption. *Peruyaan* is consumed mostly as a side dish with steamed rice, but it is also sometimes mixed with hot water; chillies, locally called *tero*; and salt and directly consumed without frying or cooking.

9.2.2.6 Bekang

Bekang is also a traditional fermented soybean food commonly consumed by the Mizo people of Mizoram in the northeast part of India. The traditional method of preparation of *bekang* is similar to that of *kinema* except for the cracking of the seeds. The soybeans are selected, cleaned and soaked in water for 10–12 hours, and excess water is then drained out. The beans are then boiled for 2–3 hours in an open pot until the beans become soft. Excess water is again drained off, and then it is wrapped in fresh leaves of *Calliparpa aroria* (family: Verbanaceae), locally called *nuhlhan*, or in leaves of *Phrynium* sp. (family: Merantaceae), locally known as *hnahthial* (Tamang et al., 2009). The wrapped beans are kept inside a basket and then placed in a warm place where they are allowed to ferment naturally for 3–4 days. The end product is a sticky soybean with an ammoniacal odour. *Bekang* is consumed as it is or made into curry with the addition of salt, chillies, onions and tomatoes. It is also consumed as a side dish with steamed rice.

9.2.3 Products Formed during Soy Fermentation

During the process of fermentation, many different types of components are also released which play a major role in increasing the nutritive qualities of food item and enhancing properties such as texture, taste and shelf life. The qualitative and quantitative composition of soybean components is drastically changed by physical and enzymatic processes during the fermentation process (Baek et al., 2008). Soybean fermentation results in the release of peptides by the action of proteolytic enzymes produced by the microorganisms involved during fermentation. Bioactive peptides are inactive within the sequence of the parent protein, and they are released on enzymatic hydrolysis during the fermentation process and gastrointestinal digestion. The large protein, lipid and carbohydrate molecules present in the raw soybean are broken down by enzymatic hydrolysis during fermentation to small molecules such as peptides, amino acids, fatty acids and sugars, which are responsible for the unique sensory and functional properties of the final products. The degradation of these lipids and carbohydrates proceeds rapidly during the initial stage of fermentation, because these are used as the major energy sources for the microorganisms.

However, after the initial stage of fermentation, the soy proteins are rapidly degraded, and then only a small amount of crude protein remains as a result of long-term fermentation (Yamabe et al., 2007). Soybean isoflavones are major components responsible for bioactive functions such as lowering the risks of cancers of the breast, prostate, and colon; cardiovascular diseases; and osteoporosis (Ain et al., 2022; Chatterjee et al., 2018; Martino et al., 2011). Fermented soybean products have also shown negligible human IgE immunoreactivity, which may be due to the hydrolysis of immunoreactive soybean proteins during fermentation (Song et al., 2008). Bioactive peptides derived from food proteins have positive health properties for the human body. Peptides are released by the hydrolysis of soybean proteins by microbial proteases, which exhibit therapeutic properties such as anti-hypertensive (Zhang et al., 2006), antioxidant (Sanjukta et al., 2015), hypocholesterolaemic, chemopreventive and antidiabetic properties (Kwon et al., 2011). Bioactive peptides in fermented soybeans are either formed by the hydrolysis of soybean protein or produced by the starter culture. Peptides also improve the functional properties of fermented foods and can act as a natural alternative to various synthetic drugs (Sanjukta et al., 2015; Rai et al., 2015). Fermentation also helps to increase the antioxidant capacity in fermented food, and many studies have shown that fermentation with different species of microorganisms could potentially improve the antioxidant capacity, vitamin B2, vitamin B12, and gamma-aminobutyric acid (GABA). GABA is responsible for normal regulation of the central nervous system. The levels of these vitamins and GABA are significantly increased in fermented soy products (Huang et al., 2017).

9.3 HEALTH BENEFITS OF FERMENTED SOYBEAN FOODS

Fermented soybean products have been prepared since ancient times, and the people of those times only used the fermentation process because they believed that it increased shelf life, kept the food fresh and also improved the overall taste of the food. However, in the present, fermented foods have been a part of the daily diet of different communities of people all over the

TABLE 9.2 Soybean-Based Fermented Foods of Northeast India

FERMENTED SOYBEAN FOOD PRODUCTS	ORIGIN OF FOOD (IN INDIA)	FERMENTATION MICROORGANISM	HEALTH BENEFITS	REFERENCES
Kinema	Sikkim, Darjeeling, Nepal, Assam	*Bacillus subtilis, B. licheniformis, B. cereus, B. circulans, B. thuringiensis, B. sphaericus, Geotrichum candidum*	Increases calcium absorption in body; decreases bone loss	(Tanimoto et al.,2001), (Sarkar & Tamang, 1994), (Tamang, 2003)
Hawaijar	Manipur	*B. subtilis, B. licheniformis, B. amyloliquefaciens, B. cereus, Staph. aureus, Staph. sciuri, Alkaligenes* sp., *Provedencia rettgeri, Proteus mirabilis*	Antioxidant properties; source of vitamin B2 and vitamin B12; organoleptic properties	(Jeyaram et al., 2008), (Singh et al., 2014)
Tungrymbai	Meghalaya	*B. subtilis, B. licheniformis, B. pumilus*	Acidifying capacity; degradation of antinutritive factors; antimicrobial activities	(Chettri & Tamang, 2015), (Omizu et al., 2011), (Agrahar-Murungkar & Subbulakshmi, 2006)
Akhuni	Nagaland	*B. subtilis, Proteus mirabilis*	Nutraceutical properties	(Singh et al., 2014)
Peruyyan	Arunachal Pradesh	*B. subtilis, B. licheniformis, B. amyloliquefaciens, Vagococcus lutrae, Ped. acidilactici*	Nutraceutical properties	(Singh et al., 2014)
Bekang	Mizoram	*B. subtilis, B. brevis, B. circulans, B. coagulans, B. licheniformis, B. pumilus, B. Sphaericus, Lysinibacillus fusiformis*	Antioxidant properties in body	(Chettri & Tamang, 2015), (Mishra et al., 2019)

world. Much research has proved the potential health benefits of soybean-based fermented foods, and the health benefits can be attributed to the different microorganisms that are present and which are also responsible for the fermentation process.

9.3.1 Poly-γ-Polyglutamic Acid

Poly-γ-polyglutamic acid (γ-PGA) is an amino acid polymer which can be synthesized by Gram-positive bacteria (Yao et al., 2022) and some Gram-negative bacteria (Candela et al., 2009) during the fermentation process. Polyglutamic acid (PGA) is a biodegradable, non-immunogenic and unusual anionic homopolyamide which is made of D and L glutamic acid units. PGA can be differentiated into two isoforms, α-PGA and γ-PGA, depending on the attachment of the carboxy group to α and γ, respectively (Shih & Van, 2001). Some of PGA-producing bacteria are *Bacillus subtilis, B. anthracis, B. licheniformis, B. thuringensis, B. cereus, B. pumilus, B. amyloliquefaciens, B. mojavensis, B. atrophaeus, B. megaterium, Staphylococcus epidermidis, Natrialba aegyptiaca, Lysinibacillus sphaericus* and *Fusobacterium nucleatum* (Cachat et al., 2008; Meerak et al., 2008; Candela et al., 2009; Cao et al., 2009).

Research was conducted where many species of *Bacillus* from *kinema*, a locally fermented soybean food from Sikkim, was successfully isolated (Chettri et al., 2016). The isolated strains were identified as *Bacillus subtilis, B. licheniformis, B. pumulis, B. sphaericus* and *B. cereus*. The molecular characterization also revealed that the PGA-producing bacteria in *kinema* were *B. subtilis* and *B. licheniformis*. PGA produced by *Bacillus* spp. in many fermented soybean products gives a sticky texture to the product (Urushibata et al., 2002; Nishito et al., 2010), such as *natto* from Japan (Nagai, 2012; Kada et al., 2013), *chungkokjang* from Korea (Lee et al., 2010), *tungrymbai* and *bekang* from India (Chettri & Tamang, 2014), and *thaunao* from Thailand (Chunhachart et al., 2006).

Chettri (2012) previously isolated some strains of *Bacillus* from *tungrymbai* and *bekang*. These cultures included *Bacillus subtilis, B. pumilus, B. licheniformis, B. cereus, B. coagulans, B. circulans* and *B. brevis*. The screening of the production of poly-glutamic acid by *Bacillus* sp. was then carried out (Chettri & Tamang, 2014). Results showed that *B. subtilis* TS1:B25 in *tungrymbai* and *B. subtilis* BT: B9 in *bekang* showed the highest production of γ-PGA among all the other strains tested, which proved that *B. subtilis* is a more efficient producer of γ-PGA than the other species of *Bacillus*.

Mucilaginous or sticky substances that can be visible on the surface of soybeans during soybean fermentation are actually due to the production of poly-γ-glutamic acid produced by *Bacillus* spp. (Chettri et al., 2016; Li et al., 2022). Gram-positive *Bacillus* spp. are a natural source of PGA production (Chettri & Tamang, 2015; Kharnaior & Tamang, 2021; Tamang et al., 2021). Studies have been conducted on screening PGA-producing *Bacillus* from fermented soybean foods such as *peruyyan*. The isolates were screened based on their high stickiness (>20 cm) and identified up to the species level. *B. safensis* subsp. *safensis* PY1–19, isolated from *peruyyan*, showed 78.9% glutamic acid, the ability to produce phytase and antagonistic properties against pathogenic bacteria with no haemolytic activity. This strain was further used as a starter culture for the preparation of *peruyaan* in the laboratory and then examined for PGA production. The resulting product showed high stickiness and viscosity and production of PGA of about 11 g/L. It also showed phytase production, which helps in the reduction of phytic acid, a major anti-nutritive factor found in soybeans (Pariyar et al., 2022).

Recent studies on PGA-rich fermented soybean foods have also shown the ability to suppress increased postprandial blood glucose levels (Araki et al., 2020), regulating blood pressure owing to their ability to inhibit angiotensin-converting enzyme (ACE) activity (Lee et al., 2014), and they are associated with a lower incidence of certain diseases that lower the risk of mortality (Chan et al., 2021). They also have major applications in medical fields (wound healing, treatment of dental caries, bone regeneration, bone improvement, drug delivery) and food industries (flavouring agents, thickeners, packaging material) (Nair et al., 2021).

Poly-γ-PGA has been found to increase calcium absorption in the body and decrease bone loss in human beings (Tanimoto et al., 2001). It also helped to increase the bioavailability of calcium by increasing solubility and intestinal absorption. This increase in calcium solubility is due to the inhibition of the production of insoluble calcium phosphate. High-PGA foods have been associated with an anti-obesity effect, lowering blood glucose and reduction of cardiovascular diseases (Lee et al., 2014; Kobori et al., 2022).

Women in a post-menopausal phase were seen to have better intestinal absorption when they were given a single dose of γ-PGA. People suffering from lower basal absorptive capacity also particularly benefitted from γ-PGA (Tanimoto et al., 2001).

9.3.2 Saponins

Omizu et al. (2011) found that *kinema* has a high content of Group B saponins (naturally occurring phytochemicals with at least one glycosidic linkage (C-O-sugar bond) at C-3 between aglycone and a sugar chain, which are usually found in legumes, oats and spinach). During the study, saponin composition and content in seeds of raw dry soybean and fermented soybean foods of India, including *kinema*, *bekang* and *tungrymbai*, were investigated by liquid chromatography-tandem mass spectrometry analysis. The results showed that these fermented soybean foods contained large amounts of Group B saponins, which contributed to health-promoting effects.

Leterme (2002) conducted clinical studies which suggested that the presence of saponins in food decreased blood lipids and lowered the blood glucose response. Saponins directly interact with cholesterol in our body and decrease its absorption from the gut by forming an insoluble complex and hence inhibit the cholesterol absorption or bile acid reabsorption in the body, which is again a beneficial mechanism for lowering blood cholesterol in humans. Also, saponins reduce the risk of coronary heart disease by lowering plasma cholesterol levels (Sidhu & Oakenfull, 1986).

Saponins also reduce the risk of cancer because they have the ability to knock out tumour cells from the body. These anticarcinogenic properties involve mechanisms such as acid and neutral sterol metabolism, cytotoxicity of cancerous cells, immune modulatory effects and normalization of carcinogen-induced cell proliferation. These cytotoxic and growth inhibitory effects were seen in colon tumour cells in cultures which resulted due to the interactions with the free sterols or cholesterol present in the cell membrane and changes in its permeability (Rao & Sung, 1995; Du et al., 2014).

Recent reviews have also enlightened us about the potential use of different classes of phytochemicals in the treatment of obesity. Among these classes of phytochemicals, several saponins have been demonstrated to lower body weight and serum lipid levels (Williams et al., 2013). Studies also showed that saponins show haemolytic action towards red blood cells and can be toxic if given intravenously (Milgate & Roberts, 1995).

Tamang & Sarkar (1996) studied the influence of processing on soybean lipids during the production of *kinema*. They identified the lipids, namely campesterol, stigmasterol and 13-sitosterol, and also found that their content increased during fermentation by 50–61% by using capillary column gas chromatography (GC) analysis of petroleum ether extracts of lyophilized samples obtained from unfermented and fermented beans during a 2-day fermentation at 37°C in the experiment.

9.3.3 Anti-Thrombotic Activity

According to the medical dictionary, antithrombotic means "preventing or interfering with the formation of thrombin", where thrombin/thrombus means a stationary blood clot along the wall of a blood vessel which causes frequent vascular obstruction in human veins and arteries. Antithrombotic agents or drugs are used therapeutically to prevent formation and for treatment of these dangerous blood clots which may otherwise cause strokes, myocardial infarction, venous syndromes, deep vein thrombosis and resultant pulmonary emboli (McCarthy & Rinella, 2012). Fibrinolysis is a process which helps to prevent these blood clots and plays a crucial role in preventing blockages in blood vessels over a period of time. Fibrinolysis is an enzymatic breakdown of the fibrin present in blood clots which prevents it from remaining in place and growing, thereby allowing the body to clear fragments of clots safely. It avoids many risks such as strokes and damage to the heart which otherwise can be caused by big floating clots (Marieb & Hoehn, 2013). According to the World Health Organization (WHO) in 2022, CVDs were the leading cause of death, with an estimated 17.9 million lives lost each year.

Singh et al. (2014) studied traditional fermented foods of northeast India and their associated microorganisms for fibrinolytic activity. For the process, samples of traditional fermented foods from northeast India were obtained from different markets in northeast India, such as fermented soybean foods (*hawaijar, bekang, akhuni, tungrymbai*), fermented fish food (*hentak, shedal, ngapi*) and fermented pork meat (*sa'um*). Ten grams of each sample were then transferred to a sterile stomacher bag containing 90 ml of physiological saline (0.1% w/v bacteriological peptone, 0.85% w/v NaCl) and homogenized in a Stomacher 400-Circulator, Seward, UK, at 250 x *g* for 3 min. This homogenate was then divided into two portions, one for screening fibrinolytic activity and the other for microbial isolation. For reliable quantification of fibrinolytic activity, a simple spectrophotometric assay was also established and validated by comparing with a conventional fibrin plate assay. The results showed that traditionally fermented soybean and fish products from northeast India possessed fibrinolytic activity. The fibrinolytic activity in fermented soybean products originated from microorganisms, mostly from *Bacillus* spp., and in fermented fish products, it was of endogenous origin.

The presence of microbial fibrinolytic enzymes in naturally fermented soybean foods imparts anti-thrombotic properties to the food. Fibrinolytic enzymes of microbial origin have cost-effective production, has broad biochemical diversity, allow genetic manipulation and have few or no side effects (Diwan et al., 2021). Nattokinase (NK) is a potent microbial fibrinolytic enzyme which is produced by *Bacillus subtilis* in natto, a fermented soybean food from Japan (Yao et al., 2022). NK is also a potent blood-clot dissolving protein used for the treatment of CVDs (Weng et al., 2017). Apart from *B. subtilis* from natto, many microbial fibrinolytic enzyme-producing bacteria were isolated from naturally fermented soybean foods, such as *B. subtilis* DC27, *B. amyloliquefaciens* DC-4 and *B. mojavensis* LY-06 from douchi from China (Hu et al., 2019; Yang et al., 2020; Li et al., 2022) and *B. amyloliquefaciens*, *B. subtilis* and *B. licheniformis* from *cheonggukjang* from Korea (Ryu et al., 2017: Yao et al., 2017). Furthermore, *B. subtilis*, *B. amyloliquefaciens* and *Vagococcus carniphilus*, isolated from naturally fermented soybean foods from northeast India such as *tungrymbai*, *hawaijar* and *bekang*, were also reported to show fibrinolytic activities (Singh et al., 2014).

Fibrinolytic enzyme-producing bacteria from northeastern fermented soybean foods such as *kinema* and *peruyyan* were screened (Kharnaior et al., 2023). Also, the samples were assessed for their therapeutic properties such as antioxidant, anti-inflammatory, anti-obesity and anti-diabetic properties and flavones and isoflavones. Out of the total isolates, *Bacillus subtilis* (ten strains), *B. velezensis* (three strains), *B. inaquosorum* (two strains) and *B. halotolerans* (one strain) showed fibrinolytic enzyme activity ranging from 700 to 1230.61 U/mL, with a blood clot-degrading property of more than 50%. A remarkable DPPH radical scavenging property (82.93 ± 7.32%) was observed from the samples. Fermented soybean foods showed anti-thrombotic, antioxidant, anti-inflammatory, anti-diabetic and anti-obesity properties, indicating their therapeutic value.

9.3.4 Increased Nutraceutical Potential along with Antioxidant Properties

During the fermentation process, the shelf life of food and the organoleptic properties are increased along with nutraceutical potential due to the action of the microorganisms that produce many enzymes and secondary metabolites. A nutraceutical is a pharmaceutical-grade or alternative standardized nutrient which exhibits many physiological benefits (Sarris et al., 2016). Nutraceutical potential means the ability of products derived from food sources to provide extra health benefits other than just the basic nutritional value available in the food. These secondary metabolites and bioactive compounds present in the food impart antimicrobial and antioxidant properties (Vijayendra & Halami, 2015; Tangkhawanit & Siriamornpun, 2019).

Kinema showed an increase in the level of total content of amino acids, riboflavin and niacin, which in turn possess cholesterol-lowering effects. *Kinema* is also rich in linoleic acid, has antioxidant activities (Sanjukta et al., 2015) and contains all essential amino acids in it (Sarkar & Tamang, 1995). *Kinema* is considered a functional food because it has health-promoting factors which benefit the human body. It is a rich source of essential amino acids, antioxidants, vitamin B complexes and proteins and also has low cholesterol content; hence, *kinema* is also sometimes referred to as a "functional food" (Tamang, 2010). It also has a large quantity of plant protein and hence is also regarded as a cheap source of protein available to vegetarians. Glycosidic isoflavones are phenolic compounds found in soybean and contribute as the main source of biological activities in fermented soybean food. They are hydrolysed with the help of β-glucosidase produced by the microorganisms in soybean fermentation, which in turn increases the aglycone isoflavones and antioxidant activity (Kwak et al., 2007; Samruan et al., 2012).

Moktan et al. (2008) evaluated the antioxidant activities of methanolic extract of *kinema*, fermented with *Bacillus subtilis* DK-W1 (MTCC 1747), and cooked non-fermented (CNF) soybean by methods like DPPH (2,2-diphenyl-1-picrylhydrazyl) radical scavenging activity, $Fe3^{+}$-reducing power, $Fe2^{+}$-chelating activity and activity in linoleic acid emulsion system. *Kinema* and cooked non-fermented soybean were prepared in a laboratory and then dried separately for 24 hours at 60°C in a hot-air oven and then ground to powder. It was then extracted using methanol for 5 hours at room temperature with continuous stirring. This extraction process was repeated thrice, and the extracts were filtered through Whatman no.1 paper. The filtrates were then concentrated at 40°C under vacuum and freeze-dried using an Eyela freeze-dryer, model FDU–506. The lyophilized extracts were then stored in a desiccator at 4°C. Before using it, the lyophilized extract was dissolved in methanol. The stable free radical 2, 2-diphenyl-1-picrylhydrazyl method was used to measure the antioxidant activity of the extracts in terms of hydrogen-donating or radical-scavenging ability (Sánchez-Moreno et al., 1998). According to the method, a 0.1-ml methanolic solution of the lyophilized extract (10–50 mg ml^{-1}) was added to 2.9 ml of 60-μM methanolic solution of DPPH. The mixture was then shaken immediately and allowed to stand at room temperature in the dark. The decrease in absorbance at 517 nm was then measured using a Systronics-type 118 spectrophotometer at 10-min interval up to 60 min until a plateau was reached. The results showed that the total phenolic content and reducing power of *kinema* were 144% and 147%, respectively, which was higher than that of CNF soybeans (3.3 mg/g dry weight), and at all the tested concentrations, *kinema* was found to be a better free radical scavenger, which increased in a time- and dose-dependent manner compared to the CNF soybean (Moktan et al., 2008).

Reports have shown that *B. subtilis* (MTTC5480) that was isolated from *hawaijar* showed higher anti-human immunodeficiency virus (HIV) potential than the other reported microRNAs against human immunodeficiency virus. The further purification of the dsRNA fraction with validation experiments confirmed the suppression of HIV-1 viral production in a human PBMC model, where there was effective inhibition of HIV-1 at the early stage before integration rather than post-integration transcription. Hence it is proposed that the dsRNA fraction from *B. subtilis* MTCC5480 targeted the HIV1 genomic RNA at two stages of the retroviral cycle and gave twofold effects. First, it inhibited the reverse transcription at the early stage of infection (before nuclear import and integration), and second, it suppressed translation at the post-integration stage (Labala et al., 2020).

9.3.5 Increased Nutritional Value

After fermentation, the nutritional value of the soybean is essentially improved. The total amino acids, free amino acids and mineral contents are exponentially increased in value, along with phytosterols, riboflavin, niacin and major antioxidants (Sarkar & Tamang, 1995). The phytosterols are similar to cholesterol and serve as components of membranes of plants and also act as an essential micronutrient for human body (Zaloga, 2015). They have a cholesterol-lowering effect on the body, and the fermentation process breaks complex proteins down into easily digestible amino acids.

Fermentation undoubtedly improves the quality of soybeans in various ways like increasing digestibility, enhancing nutrition and also increasing the isoflavones content (Xu et al., 2015). Also, due to the reactions such as the enzymatic breakdown of larger macromolecules present in soybean such as carbohydrates, proteins and lipids into sugars, peptides, amino acids and fatty acids, the sensory and functional properties are also increased gradually.

Gamma-aminobutyric acid (GABA) is an anxiety neurotransmitter and also responsible for the regulation of the central nervous system (CNS). Soybean fermentation can also improve the antioxidant capacity, vitamin B2, aglycone, beta-carotene and GABA (Huang et al., 2017).

Hawaijar has also been promoted to be used for commercial purposes. During the preparation process, the soaked and cooked soybean is placed in a bamboo basket after draining off the water portion. This greasy portion is then washed with lukewarm water until it becomes non-greasy. The drained water is believed to help in curing tuberculosis and to be good for women to consume (Devi & Suresh Kumar, 2012).

9.3.6 Prevention of Non-Alcoholic Fatty Liver Disease

Non-alcoholic fatty liver disease (NAFLD) is a disease that affects the liver in humans, and it is a major health issue in many developed countries (Chang et al., 2019). The main function of the liver is to modulate lipid metabolism and also to maintain lipid homeostasis, but due to NAFLD, problems like obesity and insulin resistance can be caused (Lonardo et al., 2015). NAFLD also causes a broad range of liver disorders such as simple hepatic steatosis (intrahepatic accumulation of triglycerides) and severe steatohepatitis, which is associated with hepatocyte damage, chronic inflammation and fibrosis, which may ultimately lead to liver failure (Lonardo et al., 2015).

Tungrymbai's microflora consist of mainly *Lactobacillus* sp., and 16S rDNA sequence analyses reported the strains to be *Levilactobacillus brevis* (K38A and RD10) and *Limosilactobacillus fermentum* (K20) (Mishra et al., 2019). Thokchom and Joshi (2012) reported that *tungrymbai* had antibiotic resistance and probiotic properties. It could also prevent intracellular lipid accumulation, impaired lipid oxidation and hepatic steatosis (Dihingia et al., 2018). It was found that soy isoflavones found in fermented soybean food decrease hepatic levels of malondialdehyde (MDA) and increase hepatic antioxidant enzymes (Ibrahim et al., 2008).

An experiment was conducted by Liu et al. (2017) to determine the effects of soy isoflavone on hepatic fat accumulation in non-alcoholic fatty liver disease rats induced by a high-fat diet (HFD). The results showed that soy isoflavone reduced HFD-induced lipid accumulation in the liver, increased serum ALT and improved liver lobule structure. Also, the expression of SREBP-1c and FAS was lower, whereas the protein level of the peroxisome proliferator-activated receptor alpha (PPAR-α) was higher in two soy isoflavone groups than that of the HFD group, thereby alleviating hepatic steatosis and delaying the progression of NAFLD via inhibiting lipogenesis and promoting fatty acid oxidation in the liver.

9.3.7 Anti-Diabetic Effects

Any defect in insulin secretion or insulin action can lead to a sickness called diabetes mellitus (DM). It is a metabolic disease which can be classified as type I and type II, where type I is insulin-dependent diabetes, where the pancreas cannot secrete insulin, and type II is non-insulin dependent, caused by an imbalance in insulin secretion as well as blood sugar absorption (Hameed et al., 2015). It was reported that fermented soybean foods contain many functional components such as isoflavonoids, unsaturated fatty acids and peptides, which provide protection against many chronic diseases like obesity, type II diabetes, hypertension and carcinogenesis (Kwon et al., 2011).

Many soybean and soy products are shown to lower the negative effects of DM and its prevention. Soy products are rich in isoflavones and exhibits anti-diabetic effect which thereby inhibit the glucose uptake into brush border membrane vesicles (Bhathena & Velasquez, 2002).

Also, it was reported that soy proteins may improve obesity and diabetes by reducing insulin resistance and affect lipid synthesis and breakdown in adipose tissue (Palacios-González et al., 2014).

Skeletal muscle cells in the human body are important glucose-responsive tissue that helps in glucose homeostasis and energy balance in addition to the storage and utilization of glucose molecules in the body (Evans et al., 2019). They also account for about 80% of postprandial glucose uptake from the blood, and hence the improper functioning of skeletal muscle is associated with diabetes (Merz & Thurmond, 2011). An *in vitro* study on the L6 cell line (an immortalized rat skeletal muscle cell line) was used in examining the anti-diabetic effects of the fermented soybean food *hawaijar. Hawaijar* showed a decrease in high glucose-mediated oxidative stress and altered glucose metabolism, therefore increasing glucose utilization in high-glucose treated culture L6 myotubes. Also, *in vivo* studies on type II diabetes (T2D)–induced rats were conducted. Its administration in the rats caused a reduction in body weight, fasting blood glucose (FBG) levels, glycohemoglobin (GHb), insulin resistance (IR), an improvement in glucose tolerance and no toxic effects. Both the cell culture and animal studies demonstrated the upregulation of phospo-PI3K/phospo-AKT/phospo-AMPK/GLUT4 signalling pathways and the alleviation of the impaired glucose metabolism associated with diabetes (Das et al., 2022).

9.3.8 Formation of Bioactive Peptides

During the fermentation process, many bioactive peptides are released, such as antioxidant peptides, antimicrobial peptides and antidiabetic peptides. They are either formed by the hydrolysis of soybean proteins or produced by the starter culture used during the fermentation (Sanjukta & Rai, 2016). Oxidation of biomolecules in the human body can lead to the death of cells along with tissue damage. Also, the free radicals present in the body such as superoxides, hydrogen peroxide, hydroxyl radicals, lipid hydroperoxides and peroxyl radicals can cause cancer, diabetes, arteriosclerosis and arthritis (Halliwell & Gutteridge, 2015). Antioxidants works as molecules that capture these free radicals within their structure and prevent oxidation and associated cellular damage. It has been found that soybeans fermented using proteolytic *B. subtilis*

MTCC5480 show high DPPH radical scavenging and power-reducing activity (Sanjukta et al., 2015).

During fermentation, the beta-glucosidic bond in the soybean is cleaved, and the isoflavone glycosides are transformed into the free isoflavone aglycone, which is a bioactive compound (Liu et al., 2022). The most abundant form of isoflavones is glycosides (genistin and daidzin), which are converted to the corresponding aglycones (genistein and daidzein) during the fermentation of *tungrymbai* (Mishra et al., 2019). They are then hydrolysed when ingested in the digestive tract of the human body by the intestinal hydrolytic enzymes and microbial glucosidases. This reaction helps in increasing the water solubility and bioavailability in the body (Hsiao et al., 2020). The isoflavones produced also possess properties such as anti-inflammatory, anti-cancer, anti-obesity and anti-diabetes properties; gut biota regulation; and osteoporosis prevention (Chen et al., 2014).

Baruzzi et al. (2011) also reported that *Bacillus* species, especially *B. subtilis*, that are mainly found in traditional fermented soybeans were able to produce many antimicrobial compounds such as proteins, enzymes, lipopeptides and bacteriocin. These are helpful against bacteria and fungi and also act against pathogens such as *Campylobacter* sp., *Clostridium botulinum, Listeria monocytogenes, Salmonella typhimurium, Staphylococcus aureus* and *Pseudomonas aeruginosa*.

Moktan et al. (2008) found that the increase in polyphenols and isoflavones are responsible for increase in the antioxidant property of the fermented soybeans. Amadou et al. (2010) found that antioxidant peptides isolated from *Lactiplantibacillus plantarum* Lp6 fermented soybean protein meal hydrolysate using a Sephadex G-15 gel filtration column and gel filtration chromatography exhibited strong antioxidant activity.

Fermented soybeans, besides being a good source of proteins, are rich in isoflavonoids and peptides (Senrung et al., 2017). They help in reducing cholesterol, help to slow down the progression of atherosclerosis (a disease of the arteries) and enhance overall intestinal and digestive tract health. Apart from this, they also prevent inflammation and lubricate joints associated with arthritis due to the production of hyaluronic acid during fermentation. Hence the consumption of fermented soybeans increases bone strength and improves the functioning of the heart. Fermented soybeans are easier to digest than unfermented soybeans (Gopikrishna et al., 2021).

9.3.9 Anti-Cancer Effects

According to the World Health Organization, cancer is a generic term for a large group of diseases that can affect any part of the body. There is rapid creation of abnormal cells that grow beyond their usual boundaries and which can then invade adjoining parts of the body and spread to other organs; the latter process is referred to as metastasis. This can be prevented by staying healthy and following proper dietary measures. Studies with fermented soy milk beverages were conducted, and the results showed the suppression of proliferation of the human colon cancer cell lines HT-29 and Caco-2 (Lai et al., 2013), and it inhibited the generation of reactive oxygen species and inhibited the growth of oestrogen-receptor positive MCF-7 human breast cancer in mice (Zhu et al., 2006).

Kim and co-workers conducted research where they administered γ-PGA with a molecular weight (2×10^6 Da) orally to C57BL/6 mice bearing major histocompatibility complex (MHC) class I-deficient tumours. This resulted in inducing natural killer (NK) cell-mediated antitumor immunity in the mice (Kim et al., 2007). MHC class I-deficient tumours can be fatal in humans. The mechanism behind antitumor immunity in mice was due to the activation of NK cells by γ-PGA rather than a direct cytotoxic effect. The results also led to the conclusion that γ-PGA (2000kDa) was better in stimulating antitumor effects against B16 tumours in C57BL/6 mice compared to β-glucan (curdlan), an immune-modulating antitumor agent which is known to activate NK cells. This research highlighted the potential use of γ-PGA in cancer immunotherapy.

The consumption of soybeans and their products may also have a significant effect on the prevention of cancer and regulation of metalloproteinase (MMP) levels due to its high isoflavone content and phenolic composition (Lima et al., 2017). Also, soybean cotyledon contains a protein called lunasin which has been reported as an anticancer and anti-inflammatory agent (Zhu, 2018). Various studies dealing with animal models and epidemiologic analyses have shown that soy protein could inhibit the growth of prostate, colon and mammary cancer cells (Badger et al., 2005). Experiments were done on 7, 12-dimethylbenz(a)anthracene (DMBA) and methylcholanthrene treated (MCA) fibroblast NIH/3T3 cell lines to analyse the anti-cancer properties of soybean active peptides (Singh et al., 2014). Also, studies on anticancer properties of lunasin on mouse models with DMBA-induced mammary cancer showed that soy proteins like lunasin can be used as a regular component in our diets to reduce the risk of cancer (de Lumen, 2005).

Rayaprolu et al. (2013) reported that soybean meal contained certain bioactive peptides which inhibit liver (HepG-2), colon (HCT-116) and lung (NCL-H1299) cancer cell lines at the rate of 70%, 73% and 68%, respectively. Hwang and his co-workers reported that breast cancer in the MCF7 cell line was inhibited by activating the TGFβ pathway using extract of fermented soybeans (Hwang et al., 2011).

9.3.10 Probiotic Properties

Studies have shown that regular consumption of probiotics improves immunological and gastrointestinal functions and reduces the risk of cancer and cardiovascular diseases (Hirayama & Rafter, 2000; Floch et al., 2011, Nagpal et al., 2012).

Probiotics have a wide range of applications in the treatment of several diseases, including diarrhoea, relief of chronic bowel inflammatory diseases, immunomodulation, lowering of serum cholesterol, minimized risk of colon cancer, improvement of lactose digestion and reduction of allergies and adverse effect on intestinal microbiota (Saad et al., 2013).

Thokchom and Joshi (2012) reported that *tungrymbai* had antibiotic resistance and probiotic properties. It was also reported that lactic acid bacteria, *Lactobacillus* sp. and *Lactococcus* sp. isolated from *tungrymbai*, can be considered a potent probiotic due to their acid and bile tolerance as well as their antibacterial activity against both gram-positive and gram-negative bacteria. Also, the high degree of hydrophobicity of LAB strains isolated from fermented soybean products probably indicates the potential of adhesion to gut epithelial cells of the human intestine, advocating for their probiotic character (Chettri & Tamang, 2014).

9.3.11 Antihypertensive Effects

Blood pressure imbalance is a major reason for various types of cardiovascular problems. Lactic acid bacteria have major health benefits like weight maintenance, preventing cardiovascular disease, anti-diabetic properties, cholesterol reduction and maintaining blood pressure levels (Mathur et al., 2020). A study showed that the peptides derived from soy milk fermented by lactic acid *Bacillus* acted as an antihypertensive agent (Tsai et al., 2006). In the study, the soy milk was first fermented with *Bacillus*, and then various factors such as systolic blood pressure (SBP) and total peptide content were analysed. At the end, the results showed that an increase in peptide levels could bring a reduction in the SBP; hence it can maintain blood pressure and keep it under control.

9.3.12 Miscellaneous Health Benefits

Singh and Singh (2014) conducted an experiment on weaned and sexed Swiss albino SPF mice. The mice (both male and female) were placed into two groups where the first group was fed a control diet (feed containing cooked soy flour at 1%) and the second group was fed an *akhuni* diet (feed containing 1% *Akhuni*). After 38 days of recording the weight of mice, the pregnant female mice were separated. After the parturition period, the babies born from both groups were counted, and when the baby mice from the *akhuni* group reached the 15th day, they were again divided into two groups. One group was given the control group diet, and the other was given the *akhuni* diet again for 21 days. The mice of both groups were weighed weekly to record their body weight gain. The results revealed that there was no difference between weight gain patterns of male and female mice fed on the control diet. However, *akhuni* feeding led to a significantly higher weight of mice and larger litter size than the control diet group of mice. The babies born of the *akhuni* group mice also had 5% better survival, attained more weight at the weaning stage and had significantly improved immune response. *Akhuni* and its related product *kinema* have also been shown to exert similar a growth-promoting effect in piglets (Singh et al., 2009)

Essential fat-soluble vitamin K, also called menaquinone-7 (MK-7), is very important for the blood clotting process. Soybean fermentation results in the production of menaquinone-7, which is in high demand, and many industries just extract menaquinone-7 during the fermentation process for making nutritional products (Hu et al., 2023). The tribal people dwelling in the northeastern part of India include soybean-based fermented food products as a daily part of their diet. Many studies have also indicated that the protein content of these fermented food products had increased significantly as compared to its raw form after the fermentation process. The reason for the increase is due to the action of microorganisms (naturally inoculated) during the process of fermentation of products such as *tungrymbai*, which showed an increase in the amounts of proteins, fat, folic acid and minerals (Agrahar-Murugkar & Subbulakshmi, 2006); *kinema*, which showed increased proteins, total amino acids, free amino acids and mineral contents, along with riboflavin and niacin; and *hawaijar*, with increased proteins, carbohydrates and minerals (Sarkar et al., 1994; Tamang et al., 2012; Bhandari & Thapaliya, 2023). All these nutrient analyses were reported on a dry matter basis.

9.4 IMPROVEMENT OF SOYBEAN-BASED FERMENTED FOODS

Traditional fermented food products are generally made by indigenous methods and processes by the local inhabitants of the given region. More than half of the traditional fermented foods are made at the household level for the consumption of family members. Fermented food products made by indigenous methods are healthy and tasty, but the methods are also very laborious, and one cannot guarantee an excellent product without any major defects. The production of traditional fermented soybean products adopts natural fermentation technology. The fermentation process is difficult to control, and the quality of the product is largely dependent on factors such as geographical location, environment, weather and season. The preparation of fermented soybean products requires the exposure of minimally treated soybean and other supplements for a long period of time. Thus, contamination can take place in those raw materials during the period due to inadequate hygiene. Thus, the improvement of product quality and safety without affecting the original characteristics of the product is yet another challenge. Also, there might be problems in storage, or sometimes the product may not turn out as expected. Hence, to overcome such problems, some improvements should be undertaken.

The solid-state fermentation process for *Bacillus amyloliquefaciens*, *Lactobacillus* sp., *Saccharomyces cerevisiae* and *B. amyloliquefaciens* was analysed to evaluate the impact of fermentation on the nutritional quality and bioactivity of soybean meal (SBM). The part remaining from the processing of soybean-based food is called soybean meal, and it is an important and cheap protein source for food and animal feed.

The results showed significant enhancement in the nutritional quality as well as bioactivity by removing the protein and carbohydrate-based anti-nutritional factors (ANFs) and allergens in the fermented food (Chi & Cho, 2016).

Soybean is also one of the top eight allergenic foods that contribute to almost 90% of food allergies in the world (Hefle et al., 1996; Kabourek & Taylor, 2003; Cordle, 2004). The nutritive quality and bioactivity of soybean meal was also investigated by solid-state fermentation using *Neurosporacrassa* (NC) or *Neurosporacrassa* and *Saccharomycescerevisiae* (NCSC) together. *Neurosporacrassa* (CGMCC 3088) is a fungus that produces high active cellulase and carotenoid, which significantly improves the nutritional value of fermented soybean meal (FSM) and also decreases the content of anti-nutritional factors such as phytic acid and trypsin inhibitors (Ren et al., 2016). It was also found that the total protein content was increased by 23.4%, and the antigenic immunogenicity of the antigen decreased by 88.7% after the fermentation process with *Saccharomyces cerevisiae* (SC) (Song et al., 2008). Also, soybean meal fermented by *Neurosporacrassa* was an excellent source of functional proteins, with protein digestibility and low allergenicity. It could be applied to food materials such as biscuits, bread and steamed bread (Li et al., 2020).

9.4.1 Microbial Strain Improvement Techniques

1. First, the microorganisms that are solely responsible for the fermentation process should be isolated, identified and cultured well in an advanced laboratory, and then the strains should be preserved well for future use as a starter culture. These microorganisms can also be improved by genetic engineering techniques. Lactic acid bacteria are important microorganisms used in the food industry. LABs associated with fermentation of soybean foods include species of *Lactobacillus, Lactococcus, Enterococcus, Pediococcus, Leuconostoc* and *Streptococcus* (Tamang et al., 2015).
   Nowadays, there is a constant need for new strains to work along with the starter culture for improved fermentation processes as well as to yield a good-quality product with nutritional value. Usually new strains can be obtained by screening of wild-type strains from natural sources (Johansen, 2018). Classical strain improvement techniques can be combined with latest techniques of gene transfer (conjugation, transduction and natural transformation) that can help produce strains with improved properties in many genetic backgrounds such as enhancement of bacteriophage resistance, elimination of traits undesirable in any or specific food fermentation and improvement of stress tolerance by the strain. An alternative approach is strain improvement using techniques of recombinant DNA technology (Johansen, 1999; Derkx et al., 2014). Technologies like recombineering (recombination-mediated genetic engineering) (Van Pijkeren & Britton, 2014) and use of the CRISPR-Cas9 system (Al Amin et al., 2018) allow us to precisely modify the desired strain at the single nucleotide level in a genome of interest.
   There can also be improvements in the overall quality of the fermented foods by the manipulation of many physical factors such as temperature, pH and aeration, as these factors mainly influence the growth and activity of microorganisms during the fermentation process. They can also impact the quality of the final product in a favourable manner. Sometimes the raw materials used for the fermentation process can also be replaced by other raw materials of good quality when available. All these processes and methods should also be tested in advanced laboratories prior to implementation on a large scale because positive results are necessary to carry on the process at a higher level.
2. In the Himalayan region of India, two varieties of soybean are usually found; they are called yellow and black varieties of soybean. In a study by Sanjukta et al. (2015), some local cultivars of these soybean seeds were purchased from the local market at Gangtok, Sikkim. They were then fermented using the bacterial strain *B. subtilis* MTCC 5480 (MH6B18) which had been isolated from *hawaijar* and *B. subtilis* MTCC 1747 which had been isolated from *kinema* and procured from microbial-type culture collection (MTCC). They were then processed and fermented. Water and methanol extracts of fermented soybean were also prepared by adding soybean powder to the respective solvents (1:10, w/v) and shaking at 28 ± 2°C for 4 hours. The methanol extracts were filtered using a Whatman No.1 filter paper, whereas the water extracts were centrifuged at 8000 g for 20 min, and the supernatant was stored at 4°C. All extracts were further analysed for antioxidant assays (reducing potential, total antioxidant activity, DPPH radical scavenging and superoxide radical anion scavenging assays). The results of the experiment showed the effectiveness of water extract of yellow soybean fermented with proteolytic *B. subtilis* MTCC 5480 to scavenge free and superoxide radicals, high reducing power potential and enhancement of antioxidant effects due to peptides as well as polyphenols. The increase in free phenolic acids in fermented soybean is probably due to β-glucosidase production during fermentation (Bhanja et al., 2009), the radical scavenging activity is due to the deleterious role of free radicals in biological systems during oxidative stress and the increase in antioxidant activity is due to the increase in free polyphenols (Moktan et al.,

2008; Ren et al., 2012; Lien et al., 2015; Xiao et al., 2016). In this study, higher antioxidant activity was found in water extracts in comparison with methanolic extracts of soybean. Overall, the product had potential as an additive for functional foods and nutraceuticals to reduce oxidative stress. Hence, the yellow variety of soybean can be preferred over black during the production process.

3. A large variety of fermented soybean products require a considerable amount of salt additive (up to 20%) to control water activity, increase shelf life and influence the microbial growth and rate of fermentation. In short, it helps to control the safety and sensory qualities of the product. As a consequence, concerns about high salt intake associated with the increased consumption of fermented soybean products are on the rise because high salt intake is one of the causative agents of hypertension. One of the common strategies to tackle this situation is to find the lowest salt concentration where the growth of the fermenting microorganisms and the flavour and texture of the fermented soybean products will not be compromised. Also, in setting up a maximum amount of salt per serving target, is necessary to engage soybean fermentation manufacturers to reduce salt use in fermented soybean products (Liu et al., 2022). Some innovative fermentation processes have been developed to produce low-salt fermented soybean products such as *douchi* (Wang et al., 2007).
4. *Bacillus cereus* is a pathogenic endospore-forming and enterotoxin-excreting bacterium that causes diarrhoea and emesis (Zhou et al., 2014). In addition to pathogenic microorganisms, fermented soybean products may also be contaminated with many mycotoxins such as aflatoxins (AFs), ochratoxin A (OTA), zearalenone (ZEA), fumonisins (FBs) and rhizoxins which can threaten human health (Tian et al., 2022). Clearly, specific guidelines and regulations on pathogenic microorganisms and mycotoxins should be set up for fermented soybean products in order to guarantee food safety. The development of rapid detection methods to monitor pathogenic microorganisms and mycotoxins throughout the whole production process of fermented soybean products would be an effective control to avoid contamination and ensure food safety.
5. Biogenic amines (BAs) are low molecular weight nitrogenous organic bases which have biological activities commonly produced during the soybean fermentation processes. They perform physiological effects, like acting as neurotransmitters, regulating body temperature and stomach pH, secreting gastric acid, aiding immune response and helping in cell growth and differentiation. However, excessive consumption can also cause toxic effects, such as headaches, nausea, cardiac palpitations and abnormal blood pressure (Ladero et al., 2010). Therefore, there should be some reasonable limits imposed in order to ensure safety during consumption. Such limits and guidelines should be kept in mind by present food manufacturing practices and companies. There should be progress and the development of new analytical and microbiological technologies that keep BA concentrations below dangerous limits (Ladero et al., 2010).
6. Last, the improvement or renovation of fermentation equipment for steaming, cooling, mixing and standardized production can reduce labour intensity and ensure the stability and safety of process and product quality.

9.5 CONCLUSION

Fermented food remains one of the best sources of many nutritional components, and it has also been an integral part of the human diet over a long period of time. Soybeans have various proven health benefits and have been a source of essential nutrients, minerals, vitamins and amino acids. The qualitative and quantitative composition of soybean components is changed by physical and enzymatic processes during the fermentation process; hence the organoleptic functional properties of the food product are also significantly improved. The ethnic soybean fermented foods of northeast India are also proven to provide many health benefits to the human body. Many fermented food improvement research and techniques have also been experimented to enhance the productivity and nutritional value of the fermented food. Studies have shown that soybean fermented foods help in the prevention of non-alcoholic fatty liver disease and diabetes mellitus and contain bioactive components which have antimicrobial and antioxidant properties. Fermented soybean foods also have increased nutritional and nutraceutical potential along with probiotic potential. The increase of the protein content in the fermented soybean product is the most significant factor. Fermented foods such as *kinema* have been proven to possess poly-γ-polyglutamic acid, which increases calcium absorption in the body, decreases bone loss in human beings and increases the bioavailability of calcium by increasing the solubility and intestinal absorption. Reports have also shown that *B. subtilis* (MTTC5480) that was isolated from *hawaijar* showed higher anti-HIV potential than the other reported microRNAs against human immunodeficiency virus. Other miscellaneous health benefits include an essential fat-soluble, vitamin K, also called menaquinone-7, which is very important for the blood clotting process and healthy bone formation in our body and is found in fermented foods. Saponins present in fermented soybeans help to decrease blood lipids, lower the blood glucose response and prevent cancer. Fermented soybean products contain isoflavones that possess properties such as anti-inflammatory, anti-cancer, anti-obesity and anti-diabetes properties; gut biota regulation; and osteoporosis prevention. Also, they exhibit anti-thrombotic activity in the human body. Due to all these health benefits, the popularity of soybean fermented foods has been increasing day by day, and their consumption has practically increased. However, the union of modern science and technological knowledge with traditional technology can be helpful in the production of products with better nutritional factors that are beneficial to human health and which in turn can help uplift the rural masses, especially in the northeastern states of India.

REFERENCES

Agrahar-Murugkar, D., & Subbulakshmi, G. (2006). Preparation techniques and nutritive value of fermented foods from the Khasi tribes of Meghalaya. *Ecology of Food and Nutrition*, *45*(1), 27–38.

Ain, H. B. U., Saeed, F., Kashif, M., Tufail, T., & Suleria, H. A. R. (2022). Nutritional Aspects of Soybean (Glycine Max). In *Bioactive Compounds from Multifarious Natural Foods for Human Health* (pp. 3–19). Apple Academic Press.

Al Amin, N., Ahmad, N., Nan, W., Xiuming, F., Nan, W., Xiaoxue, B., & Piwu, W. (2018). An efficient transient assay for CRISPR CAS9 system delivering targeted mutation using synthetic oligo SgRNA in soybean (Glycine max). *Pakistan Journal of Botany*, *50*(6), 2223–2230.

Altay, F., Karbancıoglu-Güler, F., Daskaya-Dikmen, C., & Heperkan, D. (2013). A review on traditional Turkish fermented non-alcoholic beverages: Microbiota, fermentation process and quality characteristics. *International Journal of Food Microbiology*, *167*(1), 44–56.

Amadou, I., Gbadamosi, O. S., Shi, Y., Kamara, M. T., & Jin, S. (2010). Identification of antioxidative peptides from Lactobacillus plantarum Lp6 fermented soybean protein meal. *Research Journal of Microbiology*, *5*(5), 372–380.

Anderson, J. W., Blake, J. E., Turner, J., & Smith, B. M. (1998). Effects of soy protein on renal function and proteinuria in patients with type 2 diabetes, *American Journal of Clinical Nutrition*, *68*, 1347S–1353S.

Aneja, R. P. (1989). World survey of traditional milk products. India and neighboring countries including the Himalayan region. *FAO Manuscript*, *6*, 123–126.

Ansari, F., & Pourjafar, H. (2022). Influence of fermentation and germination on food bioactives. In *Retention of Bioactives in Food Processing* (pp. 277–308). Cham: Springer International Publishing.

Araki, R., Fujie, K., Yuine, N., Watabe, Y., Maruo, K., Suzuki, H., & Hashimoto, K. (2020). The possibility of suppression of increased postprandial blood glucose levels by gamma-polyglutamic acid-rich *natto* in the early phase after eating: A randomized crossover pilot study. *Nutrients*, *12*(4), 915.

Badger, T., Segrin, C., Meek, P., Lopez, A. M., Bonham, E., & Sieger, A. (2005). Telephone interpersonal counseling with women with breast cancer: Symptom management and quality of life. *Oncology Nursing Forum*, *32*(2).

Baek, L. M., Park, L. Y., Park, K. S., & Lee, S. H. (2008). Effect of starter cultures on the fermentative. *Korean Journal of Food Science and Technology*, *40*(4), 400–405.

Baruzzi, F., Quintieri, L., Morea, M., & Caputo, L. (2011). Antimicrobial compounds produced by Bacillus spp. and applications in food. *Science Against Microbial Pathogens: Communicating Current Research and Technological Advances*, *2*(1), 1102–1111.

Battcock, M., & Azam-Ali, S. (1998). *Fermented Fruits and Vegetables: A Global Perspective*. Italy: Food and Agriculture Organization of the United Nations.

Berenjian, A., Mahanama, R., Talbot, A., Biffin, R., Regtop, H., Valtchev, P., & Dehghani, F. (2011). Efficient media for high menaquinone-7 production: Response surface methodology approach. *New Biotechnology*, *28*(6), 665–672.

Bhandari, A., & Thapaliya, S. (2023). Comparison of physicochemical properties of biscuits supplemented with soy, kinema, and buckwheat flour. *Nepal Journal of Biotechnology*, *11*(1), 16–26.

Bhanja, T., Kumari, A., & Banerjee, R. (2009). Enrichment of phenolics and free radical scavenging property of wheat *koji* prepared with two filamentous fungi. *Bioresource Technology*, *100*(11), 2861–2866.

Bhathena, S. J., & Velasquez, M. T. (2002). Beneficial role of dietary phytoestrogens in obesity and diabetes. *American Journal of Clinical Nutrition*, *76*, 1191–1201.

Bhutani, K. K. (Ed.). (2008). *Herbal wealth of North-East India: Database and appraisal*. Department of Natural Products, National Institute of Pharmaceutical Education and Research.

Bolla, K. N. (2015). Soybean consumption and health benefits. *International Journal of Scientific & Technology Research*, *4*(7), 50–53.

Cachat, E., Barker, M., Read, T. D., & Priest, F. G. (2008). A *Bacillus thuringiensis* strain producing a polyglutamate capsule resembling that of *Bacillus anthracis*. *FEMS Microbiology Letters*, *285*, 220–226.

Candela, T., Moya, M., Haustant, M., & Fouet, A. (2009). *Fusobacterium nucleatum*, the first Gram-negative bacterium demonstrated to produce polyglutamate. *Canadian Journal of Microbiology*, *55*(5), 627–632.

Cao, X. H., Liao, Z. Y., Wang, C. L., Yang, W. Y., & Lu, M. F. (2009). Evaluation of a lipopeptide biosurfactant from Bacillus natto TK-1 as a potential source of anti-adhesive, antimicrobial and antitumor activities. *Brazilian Journal of Microbiology*, *40*, 373–379.

Chan, E. W. C., Wong, S. K., Kezuka, M., Oshiro, N., & Chan, H. T. (2021). *Natto* and *Miso*: An overview on their preparation, bioactive components and health-promoting effects. *Food Research*, *5*(3), 446–452.

Chang, Y., Cho, Y. K., Cho, J., Jung, H. S., Yun, K. E., Ahn, J., & Ryu, S. (2019). Alcoholic and nonalcoholic fatty liver disease and liver-related mortality: A cohort study. *Official Journal of the American College of Gastroenterology*, *114*(4), 620–629.

Chatterjee, C., Gleddie, S., & Xiao, C. W. (2018). Soybean bioactive peptides and their functional properties. *Nutrients*, *10*(9), 1211.

Cheigh, H. S., Park, K. Y., & Lee, C. Y. (1994). Biochemical, microbiological, and nutritional aspects of *Kimchi* (Korean fermented vegetable products). *Critical Reviews in Food Science & Nutrition*, *34*(2), 175–203.

Chen, J. M., Zhang, C. H., & Tian, A. I. (2014). Study of the antibacterial peptides produced by *Bacillus amyloliquefaciens* KN-BL-1 and its fermented soybean meal. *China Biotechnology*, *34*(10), 61–66.

Chen, Z., Li, W., Santhanam, R. K., Wang, C., Gao, X., Chen, Y., . . . & Chen, H. (2019). Bioactive peptide with antioxidant and anticancer activities from black soybean [Glycine max (L.) Merr.] byproduct: Isolation, identification and molecular docking study. *European Food Research and Technology*, *245*, 677–689.

Chettri, R. (2012). Microbiological evaluation of *turangbai* and *bekang* ethnic fermented soybean foods of North East India (Doctoral dissertation, University of North Bengal).

Chettri, R., Bhutia, M. O., & Tamang, J. P. (2016). Poly-γ-glutamic acid (PGA)-producing Bacillus species isolated from *Kinema*, Indian fermented soybean food. *Frontiers in Microbiology*, *7*, 971.

Chettri, R., & Tamang, J. P. (2015). Bacillus species isolated from *tungrymbai* and *bekang*, naturally fermented soybean foods of India. *International Journal of Food Microbiology*, *197*, 72–76.

Chi, C. H., & Cho, S. J. (2016). Improvement of bioactivity of soybean meal by solid-state fermentation with *Bacillus amyloliquefaciens* versus *Lactobacillus spp.* and *Saccharomyces cerevisiae*. *LWT-Food Science and Technology*, *68*, 619–625.

Cho, K. M., Lee, J. H., Yun, H. D., Ahn, B. Y., Kim, H., & Seo, W. T. (2011). Changes of phytochemical constituents (isoflavones, flavanols, and phenolic acids) during *cheonggukjang* soybeans fermentation using potential probiotics *Bacillus subtilis* CS90. *Journal of Food Composition and Analysis*, *24*(3), 402–410.

Chunhachart, O., Itoh, T., Sukchotiratana, M., Tanimoto, H., & Tahara, Y. (2006). Characterization of γ-glutamyl hydrolase produced by Bacillus sp. isolated from Thai thua-nao. *Bioscience, Biotechnology, and Biochemistry*, *70*(11), 2779–2782.

Cordle, C. T. (2004). Soy protein allergy: Incidence and relative severity. *Journal of Nutrition, 134*(5), 1213S.

Dabbagh, F., Negahdaripour, M., Berenjian, A., Behfar, A., Mohammadi, F., Zamani, M., . . . & Ghasemi, Y. (2014). Nattokinase: Production and application. *Applied Microbiology and Biotechnology*, *98*, 9199–9206.

Das, A., Gujre, N., Devi, R. J., & Mitra, S. (2021). A review on traditional ecological knowledge and Its role in natural resources management: North East India, a cultural paradise. *Environmental Management*, 1–22.

Das, D., Sarkar, S., Dihingia, A., Afzal, N. U., Wann, S. B., Kalita, J., & Manna, P. (2022). A popular fermented soybean food of Northeast India exerted promising antihyperglycemic potential via stimulating PI3K/AKT/AMPK/GLUT4 signaling pathways and regulating muscle glucose metabolism in type 2 diabetes. *Journal of Food Biochemistry*, *46*(12), e14385.

Da Silva, C. M., & de Majo, C. (2022). *The age of the soybean: An environmental history of soy during the great acceleration*. The White Horse Press.

de Lumen, B. O. (2005). Lunasin: A cancer-preventive soy peptide. *Nutrition Reviews*, *63*(1), 16–21.

De, S. (1980). *Outlines of dairy technology*. Oxford University Press.

Derkx, P. M., Janzen, T., Sørensen, K. I., Christensen, J. E., Stuer-Lauridsen, B., & Johansen, E. (2014). The art of strain improvement of industrial lactic acid bacteria without the use of recombinant DNA technology. *Microbial Cell Factories*, *13*(1), 1–13.

Devi, P., & Suresh Kumar, P. (2012). Traditional, ethnic and fermented foods of different tribes of Manipur. *Indian Journal of Traditional Knowledge*, *11*(1), 70–77.

Dihingia, A., Bordoloi, J., Dutta, P., Kalita, J., & Manna, P. (2018). Hexane-isopropanolic extract of Tungrymbai, a North-East Indian fermented soybean food prevents hepatic steatosis via regulating AMPK-mediated SREBP/FAS/ACC/HMGCR and PPARα/CPT1A/UCP2 pathways. *Scientific Reports*, *8*(1), 10021.

Diwan, D., Usmani, Z., Sharma, M., Nelson, J. W., Thakur, V. K., Christie, G., & Gupta, V. K. (2021). Thrombolytic enzymes of microbial origin: A review. *International Journal of Molecular Sciences*, *22*(19), 10468.

Du, C. Y., Zhou, Y., Song, C., Wang, Y. P., Jie, Z. G., He, Y. L., & Shi, Y. Q. (2014). Is there a role of surgery in patients with recurrent or metastatic gastrointestinal stromal tumours responding to imatinib: A prospective randomised trial in China. *European Journal of Cancer*, *50*(10), 1772–1778.

Egounlety, M., & Aworh, O. C. (2003). Effect of soaking, dehulling, cooking and fermentation with Rhizopus oligosporus on the oligosaccharides, trypsin inhibitor, phytic acid and tannins of soybean (Glycine max Merr.), cowpea (Vigna unguiculata L. Walp) and groundbean (Macrotyloma geocarpa Harms). *Journal of Food Engineering*, *56*(2–3), 249–254.

Evans, P. L., McMillin, S. L., Weyrauch, L. A., &Witczak, C. A. (2019). Regulation of skeletal muscle glucose transport and glucose metabolism by exercise training. *Nutrients*, *11*(10), 2432.

Farhad, M., Kailasapathy, K., & Tamang, J. P. (2010). Health aspects of fermented foods. *Fermented Foods and Beverages of the World*, 391–414.

Floch, M. H., Walker, W. A., Madsen, K., Sanders, M. E., Macfarlane, G. T., Flint, H. J., & Brandt, L. J. (2011). Recommendations for probiotic use—2011 update. *Journal of Clinical Gastroenterology*, *45*, S168–S171.

Gopikrishna, T., Suresh Kumar, H. K., Perumal, K., & Elangovan, E. (2021). Impact of Bacillus in fermented soybean foods on human health. *Annals of Microbiology*, *71*(1), 1–16.

Granito, M., Frias, J., Doblado, R., Guerra, M., Champ, M., & Vidal-Valverde, C. (2002). Nutritional improvement of beans (*Phaseolus vulgaris*) by natural fermentation. *European Food Research and Technology*, *214*, 226–231.

Halliwell, B., & Gutteridge, J. M. (2015). *Free radicals in biology and medicine*. Oxford University Press.

Hameed, I., Masoodi, S. R., Mir, S. A., Nabi, M., Ghazanfar, K., & Ganai, B. A. (2015). Type 2 diabetes mellitus: From a metabolic disorder to an inflammatory condition. *World Journal of Diabetes*, *6*(4), 598.

Hefle, S. L., Nordlee, J. A., & Taylor, S. L. (1996). Allergenic foods. *Critical Reviews in Food Science & Nutrition*, *36*(S1), 69–89.

Hirayama, K., & Rafter, J. (2000). The role of probiotic bacteria in cancer prevention. *Microbes and Infection*, *2*(6), 681–686.

Hsiao, Y. H., Ho, C. T., & Pan, M. H. (2020). Bioavailability and health benefits of major isoflavone aglycones and their metabolites. *Journal of Functional Foods*, *74*, 104164.

Hu, P., Peng, C., Zhang, B., Hu, X., Milon, R. B., & Ren, L. (2023). Enhancing menaquinone-7 biosynthesis through strengthening precursor supply and product secretion. *Bioprocess and Biosystems Engineering*, 1–12.

Hu, Y., Yu, D., Wang, Z., Hou, J., Tyagi, R., Liang, Y., & Hu, Y. (2019). Purification and characterization of a novel, highly potent fibrinolytic enzyme from *Bacillus subtilis* DC27 screened from *Douchi*, a traditional Chinese fermented soybean food. *Scientific Reports*, *9*(1), 1–10.

Huang, G., Cai, W., & Xu, B. (2017). Improvement in beta-carotene, vitamin B2, GABA, free amino acids and isoflavones in yellow and black soybeans upon germination. *LWT—Food Science and Technology*, *75*, 488–496.

Hwang, J., Kim, J. C., Moon, H., Yang, J. Y., & Kim, M. (2017). Determination of sodium contents in traditional fermented foods in Korea. *Journal of Food Composition and Analysis*, *56*, 110–114.

Hwang, J. S., Yoo, H. J., Song, H. J., Kim, K. K., Chun, Y. J., Matsui, T., & Kim, H. B. (2011). Inflammation-related signaling pathways implicating TGFβ are revealed in the expression profiling of MCF7 cell treated with fermented soybean, *chungkookjang*. *Nutrition and Cancer*, *63*(4), 645–652.

Hymowitz, T. (1990). Soybeans: The success story. *Advances in New Crops*, 159–163.

Ibrahim, W. H., Habib, H. M., Chow, C. K., & Bruckner, G. G. (2008). Isoflavone-rich soy isolate reduces lipid peroxidation in mouse liver. *International Journal for Vitamin and Nutrition Research*, *78*(45), 217–222.

Jayachandran, M., & Xu, B. (2019). An insight into the health benefits of fermented soy products. *Food Chemistry*, *271*, 362–371.

Jeyaram, K., Singh, W. M., Premarani, T., Devi, A. R., Chanu, K. S., Talukdar, N. C., & Singh, M. R. (2008). Molecular identification of dominant microflora associated with '*Hawaijar*'—a traditional fermented soybean (*Glycine max* (L.)) food of Manipur, India. *International Journal of Food Microbiology*, *122*(3), 259–268.

Johansen, E. (1999). *Generic engineering| modification of bacteria* (pp. 917–921). Elsevier.

Johansen, E. (2018). Use of natural selection and evolution to develop new starter cultures for fermented foods. *Annual Review of Food Science and Technology*, *9*, 411–428.

Johnson, M. E. (2017). A 100-year review: Cheese production and quality. *Journal of Dairy Science*, *100*(12), 9952–9965.

Jung, K. O., Park, S. Y., & Park, K. Y. (2006). Longer aging time increases the anticancer and antimetastatic properties of *doenjang*. *Nutrition*, *22*(5), 539–545.

Kabourek, J. L., & Taylor, S. L. (2003). Soyfoods and allergies: Separating fact from fiction. *The Soy Connection*, *11*(2), 1–6.

Kada, S., Ishikawa, A., Ohshima, Y., & Yoshida, K. I. (2013). Alkaline serine protease AprE plays an essential role in poly-γ-glutamate production during *natto* fermentation. *Bioscience, Biotechnology, and Biochemistry*, *77*(4), 802–809.

Kharnaior, P., Das, M., & Tamang, J. P. (2023). Therapeutic and anti-thrombotic properties of some naturally fermented soybean foods of the Eastern Himalayas. *Fermentation*, *9*(2), 91.

Kharnaior, P., & Tamang, J. P. (2021). Bacterial and fungal communities and their predictive functional profiles in *kinema*, a naturally fermented soybean food of India, Nepal and Bhutan. *Food Research International*, *140*, 110055.

Kim, B., Hong, V. M., Yang, J., Hyun, H., Im, J. J., Hwang, J., & Kim, J. E. (2016). A review of fermented foods with beneficial effects on brain and cognitive function. *Preventive Nutrition and Food Science*, *21*(4), 297.

Kim, S. L., Berhow, M. A., Kim, J. T., Chi, H. Y., Lee, S. J., & Chung, I. M. (2006). Evaluation of soyasaponin, isoflavone, protein, lipid, and free sugar accumulation in developing soybean seeds. *Journal of Agricultural and Food Chemistry*, *54*(26), 10003–10010.

Kim, T. W., Lee, T. Y., Bae, H. C., Hahm, J. H., Kim, Y. H., Park, C., & Poo, H. (2007). Oral administration of high molecular mass poly-γ-glutamate induces NK cell-mediated antitumor immunity. *The Journal of Immunology*, *179*(2), 775–780.

Kobori, M., Araki, R., & Hashimoto, K. (2022). Health functions of fermented soybean product natto: Poly-gamma-glutamic acid reduces the postprandial blood glucose levels. In *Washoku. Japanese food* (pp. 87–99). American Chemical Society.

Kwak, C. S., Lee, M. S., & Park, S. C. (2007). Higher antioxidant properties of Chungkookjang, a fermented soybean paste, may be due to increased aglycone and malonylglycoside isoflavone during fermentation. *Nutrition Research*, *27*(11), 719–727.

Kwon, D. Y., Hong, S. M., Ahn, I. S., Kim, M. J., Yang, H. J., & Park, S. (2011). Isoflavonoids and peptides from *meju*, long-term fermented soybeans, increase insulin sensitivity and exert insulinotropic effects in vitro. *Nutrition*, *27*(2), 244–252.

Labala, R. K., Velhal, S., Bhagat, S., Patel, V., & Jeyaram, K. (2020). Small double-stranded RNA with anti-HIV activity abundantly produced by Bacillus subtilis MTCC5480 isolated from fermented soybean. *International Journal of Biological Macromolecules*, *161*, 828–835.

Ladero, V., Calles-Enríquez, M., Fernández, M., & A Alvarez, M. (2010). Toxicological effects of dietary biogenic amines. *Current Nutrition & Food Science*, *6*(2), 145–156.

Lai, L. R., Huang, S. C., Chou, H. Y., & Chun, C. (2013). Effect of lactic fermentation on the total phenolic, saponin and phytic acid contents as well as anti-colon cancer cell proliferation activity of soymilk. *Journal of Bioscience and Bioengineering*, *115*(5), 552–556.

Lee, C. H. (1997). Lactic acid fermented foods and their benefits in Asia. *Food Control*, *8*(5–6), 259–269.

Lee, H., Chang, M. J., & Kim, S. H. (2010). Effects of poly-γ-glutamic acid on serum and brain concentrations of glutamate and GABA in diet-induced obese rats. *Nutrition Research and Practice*, *4*(1), 23.

Lee, N. R., Go, T. H., Lee, S. M., Jeong, S. Y., Park, G. T., Hong, C. O., & Son, H. J. (2014). In vitro evaluation of new functional properties of poly-γ-glutamic acid produced by *Bacillus subtilis* D7. *Saudi Journal of Biological Sciences*, *21*(2), 153–158.

Leterme, P. (2002). Recommendations by health organizations for pulse consumption. *British Journal of Nutrition*, *88*(S3), 239–242.

Li, S. H., Zhao, W., Li, P. L., Min, G., Zhang, A. X., Zhang, J. L., & Liu, J. K. (2020). Effects of different cultivars and particle sizes of non-degermed millet flour fractions on the physical and texture properties of Chinese steamed bread. *Cereal Chemistry*, *97*(3), 661–669.

Li, Y., Tang, X., Chen, L., Xu, X., & Li, J. (2022). Characterization of a nattokinase from the newly isolated bile salt-resistant *Bacillus mojavensis* LY-06. *Foods*, *11*(16), 2403.

Lien, D. T. P., Tram, P. T. B., & Toan, H. T. (2015). Effects of extraction process on phenolic content and antioxidant activity of soybean. *Journal of Nutrition & Food Sciences*, *3*(1–2), 33–38.

Lima, A., Oliveira, J., Saude, F., Mota, J., & Ferreira, R. B. (2017). Proteins in soy might have a higher role in cancer prevention than previously expected: Soybean protein fractions are more effective mmp-9 inhibitors than non-protein fractions, even in cooked seeds. *Nutrients, 9*, 201.

Liu, H., Zhong, H., Leng, L., & Jiang, Z. (2017). Effects of soy isoflavone on hepatic steatosis in high fat-induced rats. *Journal of Clinical Biochemistry and Nutrition*, *61*(2), 85–90.

Liu, L., Chen, X., Hao, L., Zhang, G., Jin, Z., Li, C., & Chen, B. (2022). Traditional fermented soybean products: Processing, flavor formation, nutritional and biological activities. *Critical Reviews in Food Science & Nutrition*, *62*(7), 1971–1989.

Lonardo, A., Ballestri, S., Marchesini, G., Angulo, P., & Loria, P. (2015). Nonalcoholic fatty liver disease: A precursor of the metabolic syndrome. *Digestive and Liver Disease*, *47*(3), 181–190.

Mao, A. A., & Odyuo, N. (2007). Traditional fermented foods of the Naga tribes of Northeastern India. *Indian Journal of Traditional Knowledge*, *6*(1), 37–41.

Marco, M. L., Heeney, D., Binda, S., Cifelli, C. J., Cotter, P. D., Foligné, B., & Hutkins, R. (2017). Health benefits of fermented foods: Microbiota and beyond. *Current Opinion in Biotechnology*, *44*, 94–102.

María Hebert, E., Saavedra, L., & Ferranti, P. (2010). Bioactive peptides derived from casein and whey proteins. *Biotechnology of lactic acid bacteria: Novel applications*, 233–249. Blackwell Publishing.

Marieb, E. N., & Hoehn, K. (2013). The cardiovascular system: Blood vessels. *Human Anatomy & Physiology*, 703–720.

Martino, H. S. D., Cardoso, L. D. M., Ribeiro, S. M. R., Dantas, M. D. S., Piovesan, N. D., & De Mejía, E. (2011). Nutritional and bioactive compounds of soybean: Benefits on human health. In *Soybean and health* (pp. 465–488). IntechOpen.

Mathur, H., Beresford, T. P., & Cotter, P. D. (2020). Health benefits of lactic acid bacteria (LAB) fermentates. *Nutrients*, *12*(6), 1679.

McCarthy, E. M., & Rinella, M. E. (2012). The role of diet and nutrient composition in nonalcoholic fatty liver disease. *Journal of the Academy of Nutrition and Dietetics*, *112*(3), 401–409.

Meerak, J., Yukphan, P., Miyashita, M., Sato, H., Nakagawa, Y., & Tahara, Y. (2008). Phylogeny of γ-polyglutamic acid-producing Bacillus strains isolated from a fermented locust bean product manufactured in West Africa. *The Journal of General and Applied Microbiology*, *54*(3), 159–166.

Merz, K. E., & Thurmond, D. C. (2011). Role of skeletal muscle in insulin resistance and glucose uptake. *Comprehensive Physiology*, *10*(3), 785–809.

Milgate, J., & Roberts, D. C. K. (1995). The nutritional & biological significance of saponins. *Nutrition Research*, *15*(8), 1223–1249.

Mishra, B. K., Hati, S., & Das, S. (2019). Bio-nutritional aspects of *Tungrymbai*, an ethnic functional fermented soy food of Khasi Hills, Meghalaya, India. *Clinical Nutrition Experimental*, *26*, 8–22.

Moktan, B., Saha, J., & Sarkar, P. K. (2008). Antioxidant activities of soybean as affected by Bacillus-fermentation to kinema. *Food Research International*, *41*(6), 586–593.

Mondal, A., & Datta, A. K. (2008). Bread baking–A review. *Journal of Food Engineering*, *86*(4), 465–474.

Montgomery, K. S. (2003). Soy protein. *Journal of Perinatal Education*, *12*(3), 42–45.

Nagai, T. (2012). Overview of studies on *Bacillus subtilis* (*natto*) bacteriophages and the prospects. *JARQ*, *46*, 305–310.

Nagpal, R., Kumar, A., Kumar, M., Behare, P. V., Jain, S., & Yadav, H. (2012). Probiotics, their health benefits and applications for developing healthier foods: A review. *FEMS Microbiology Letters*, *334*(1), 1–15.

Nair, P., Navale, G. R., & Dharne, M. S. (2021). Poly-gamma-glutamic acid biopolymer: A sleeping giant with diverse applications and unique opportunities for commercialization. *Biomass Conversion and Biorefinery*, *13*(6), 4555–4573.

Nishito, Y., Osana, Y., Hachiya, T., Popendorf, K., Toyoda, A., Fujiyama, A., & Sakakibara, Y. (2010). Whole genome assembly of a natto production strain Bacillus subtilis natto from very short read data. *BMC Genomics*, *11*, 1–12.

Nout, M. J. R. (2014). Food technologies: Fermentation. In *Encyclopedia of food safety, volume 3: Foods, materials, technologies and risks* (pp. 168–177). Academic Press.

Nuraida, L. (2015). A review: Health promoting lactic acid bacteria in traditional Indonesian fermented foods. *Food Science and Human Wellness*, *4*(2), 47–55.

Omizu, Y., Tsukaoto, C., Chettri, R., & Tamang, J. P. (2011). Determination of saponin contents in raw soybean and fermented soybean foods of India. *Journal of Scientific & Industrial Research*, *70*, 533–538.

Orgeron II, R. P., Corbin, A., & Scott, B. (2016). Sauerkraut: A probiotic superfood. *Functional Foods in Health and Disease*, *6*(8), 536–543.

Oster, K. V. (2011). *The complete guide to preserving meat, fish, and game: Step-by-step instructions to freezing, canning, and smoking*. Atlantic Publishing Company.

Otağ, F. B., & Hayta, M. (2013). Effects of heat treatment and fermentation on the formation and activity of bioactive peptides in foods. *GIDA-Journal of Food*, *38*(5), 307–314.

Palacios-González, B., Zarain-Herzberg, A., Flores-Galicia, I., Noriega, L. G., Alemán-Escondrillas, G., Zariñan, T., & Tovar, A. R. (2014). Genistein stimulates fatty acid oxidation in a leptin receptor-independent manner through the JAK2-mediated phosphorylation and activation of AMPK in skeletal muscle. *Biochimica et Biophysica Acta (BBA)-Molecular and Cell Biology of Lipids*, *1841*(1), 132–140.

Pariyar, P., Yaduvanshi, P. S., Raghu, P., & Tamang, J. P. (2022). Screening of poly-glutamic acid (PGA)-producing Bacillus species from Indian fermented soybean foods and characterization of PGA. *Fermentation*, *8*(10), 495.

Rai, A. K., Sanjukta, S., & Jeyaram, K. (2015). Production of Angiotensin I converting enzyme inhibitory (ACE-I) peptides during milk fermentation and its role in treatment of hypertension. *Critical Reviews in Food Science and Nutrition*, *57*(13), 2789–2800.

Rao, A. V., & Sung, M. K. (1995). Saponins as anticarcinogens. *The Journal of Nutrition*, *125*(suppl_3), 717S–724S.

Rayaprolu, S. J., Hettiarachchy, N. S., Chen, P., Kannan, A., & Mauromostakos, A. (2013). Peptides derived from high oleic acid soybean meals inhibit colon, liver and lung cancer cell growth. *Food Research International*, *50*(1), 282–288.

Reddy, N. R., & Pierson, M. D. (1994). Reduction in antinutritional and toxic components in plant foods by fermentation. *Food Research International*, *27*(3), 281–290.

Ren, S. C., Liu, Z. L., & Wang, P. (2012). Proximate composition and flavonoids content and in vitro antioxidant activity of 10 varieties of legume seeds grown in China. *Journal of Medicinal Plants Research*, *6*(2), 301–308.

Ren, Z. Q., Deng, Z. Y., Song, L. W., Yang, J. Y., Liu, R., & Fan, Y. W. (2016). Study on fermented by *Neurospora crassa* to improve structural style of soybean meal nutrition. *Science and Technology of Food Industry*, *7*(12), 222–225.

Rezac, S., Kok, C. R., Heermann, M., & Hutkins, R. (2018). Fermented foods as a dietary source of live organisms. *Frontiers in Microbiology*, *9*, 1785.

Ryu, M. S., Yang, H. J., Kim, J. W., Jeong, S. J., Jeong, S. Y., Eom, J. S., & Jeong, D. Y. (2017). Potential probiotics activity of Bacillus spp. from traditional soybean pastes and fermentation characteristics of Cheonggukjang. *Korean Journal of Food Preservation*, *24*(8), 1168–1179.

Saad, N., Delattre, C., Urdaci, M., Schmitter, J. M., & Bressollier, P. (2013). An overview of the last advances in probiotic and prebiotic field. *LWT-Food Science and Technology*, *50*(1), 1–16.

Samruan, W., Oonsivilai, A., & Oonsivilai, R. (2012). Soybean and fermented soybean extract antioxidant activities. *International Journal of Nutrition and Food Engineering*, *6*(12), 1134–1137.

Sánchez-Moreno, C., Larrauri, J. A., & Saura-Calixto, F. (1998). A procedure to measure the antiradical efficiency of polyphenols. *Journal of the Science of Food and Agriculture*, *76*(2), 270–276.

Sanjukta, S., & Rai, A. K. (2016). Production of bioactive peptides during soybean fermentation and their potential health benefits. *Trends in Food Science & Technology*, *50*, 1–10.

Sanjukta, S., Rai, A. K., Muhammed, A., Jeyaram, K., & Talukdar, N. C. (2015). Enhancement of antioxidant properties of two soybean varieties of Sikkim Himalayan region by proteolytic Bacillus subtilis fermentation. *Journal of Functional Foods*, *14*, 650–658.

Sarkar, P. K., & Tamang, J. P. (1994). The influence of process variables and inoculum composition on the sensory quality of *kinema*. *Food Microbiology*, *11*(4), 317–325.

Sarkar, P. K., & Tamang, J. P. (1995). Changes in the microbial profile and proximate composition during natural and controlled fermentations of soybeans to produce *kinema*. *Food Microbiology*, *12*, 317–325.

Sarkar, P. K., Tamang, J. P., Cook, P. E., & Owens, J. D. (1994). *Kinema*—a traditional soybean fermented food: Proximate composition and microflora. *Food Microbiology*, *11*(1), 47–55.

Sarris, J., Murphy, J., Mischoulon, D., Papakostas, G. I., Fava, M., Berk, M., & Ng, C. H. (2016). Adjunctive nutraceuticals for depression: A systematic review and meta-analyses. *American Journal of Psychiatry*, *173*(6), 575–587.

Sekar, S., & Mariappan, S. (2007). Usage of traditional fermented products by Indian rural folks and IPR. *Indian Journal of Traditional Knowledge*, *70*, 533–538.

Senrung, A., Anal, S. J. L., Joshi, P., Malik, G., Jha, S., & Avasthy, S. (2017). Indigenous fermented food habits of Manipur and Nagaland. *North East India: The Untapped Tourism Industry*, 53–59.

Shih, L., & Van, Y. T. (2001). The production of poly-(γ-glutamic acid) from microorganisms and its various applications. *Bioresource Technology*, *79*(3), 207–225.

Sidhu, G. S., & Oakenfull, D. G. (1986). A mechanism for the hypocholesterolaemic activity of saponins. *British Journal of Nutrition*, *55*(3), 643–649.

Singh, B. R., Ngullie, E., & Bhatt, B. P. (2009). Effect of *axone* (fermented soybean) feeding on growth performance, intestinal physiology and humoral response of suckling and grower pigs in Nagaland state of North Eastern Hill Region. In *Proceedings of 13th biennial animal nutrition conference December* (pp. 122–123).

Singh, B. R., & Singh, R. K. (2014). Nagamese fermented soybean (Axone) reduces age at first parturition, increases litter size and modulate immune response in mouse model. *International Journal of Probiotics & Prebiotics*, *9*(1/2), 49.

Singh, T. A., Devi, K. R., Ahmed, G., & Jeyaram, K. (2014). Microbial and endogenous origin of fibrinolytic activity in traditional fermented foods of Northeast India. *Food Research International*, *55*, 356–362.

Song, Y. S., Frías, J., Martinez-Villaluenga, C., Vidal-Valdeverde, C., & de Mejia, E. G. (2008). Immunoreactivity reduction of soybean meal by fermentation, effect on amino acid composition and antigenicity of commercial soy products. *Food Chemistry*, *108*(2), 571–581.

Sourabh, A., Amit, K. R., & Chauhan, A. (2015). Health-related issues and indigenous fermented products. *Indigenous Fermented Foods of South Asia*, *7*, 309.

Tamang, J. P. (2003). Native microorganisms in the fermentation of *kinema*. *Indian Journal of Microbiology*, *43*(2), 127–130.

Tamang, J. P. (2010). Diversity of fermented foods. *Fermented Foods and Beverages of the World*, *1*, 41–83.

Tamang, J. P. (2015). Naturally fermented ethnic soybean foods of India. *Journal of Ethnic Foods*, *2*(1), 8–17.

Tamang, J. P. (2024). Unveiling kinema: Blending tradition and science in the Himalayan fermented soya delicacy. *Journal of Ethnic Foods*, *11*(1), 29.

Tamang, J. P., & Nikkuni, S. (1996). Selection of starter cultures for the production of *kinema*, a fermented soybean food of the Himalaya. *World Journal of Microbiology and Biotechnology*, *12*(6), 629–635.

Tamang, J. P., & Sarkar, P. K. (1996). Microbiology of mesu, a traditional fermented bamboo shoot product. *International Journal of Food Microbiology*, *29*(1), 49–58.

Tamang, J. P., Chettri, R., & Sharma, R. M. (2009). Indigenous knowledge of Northeast women on production of ethnic fermented soybean foods. *Indian Journal of Traditional Knowledge*, *8*(1), 122–126.

Tamang, J. P., Kharnaior, P., Pariyar, P., Thapa, N., Lar, N., Win, K. S., & Nyo, N. (2021). Shotgun sequence-based metataxonomic and predictive functional profiles of Pe poke, a naturally fermented soybean food of Myanmar. *PLoS One*, *16*(12), e0260777.

Tamang, J. P., Tamang, N., Thapa, S., Dewan, S., Tamang, B., Yonzan, H., & Kharel, N. (2012). Microorganisms and nutritional value of ethnic fermented foods and alcoholic beverages of North East India. *Indian Journal of Traditional Knowledge*, *11*(1), 7–25.

Tamang, J. P., Thapa, N., Tamang, B., Rai, A., & Chettri, R. (2015). Microorganisms in fermented foods and beverages. *Health Benefits of Fermented Foods and Beverages*, *7*, 1–10.

Tangkhawanit, E., & Siriamornpun, S. (2019). Study of extraction and processing effect on bioactive compounds and nutritional value of soybean for utilization of soybean meal as an natural additive in functional food products (Doctoral dissertation, Mahasarakham University).

Tanimoto, H., Mori, M., Motoki, M., Torii, K., Kadowaki, M., & Noguchi, T. (2001). Natto mucilage containing poly-γ-glutamic acid increases soluble calcium in the rat small intestine. *Bioscience, Biotechnology, and Biochemistry*, *65*(3), 516–521.

Terefe, N. S. (2016). *Food fermentation. Reference module in food science*. Elsevier.

Thokchom, S., & Joshi, S. R. (2012). Antibiotic resistance and probiotic properties of dominant lactic microflora from Tungrymbai, an ethnic fermented soybean food of India. *Journal of Microbiology*, *50*(3), 535–539.

Tian, F., Woo, S. Y., Lee, S. Y., Park, S. B., Im, J. H., & Chun, H. S. (2022). Mycotoxins in soybean-based foods fermented with filamentous fungi: Occurrence and preventive strategies. *Comprehensive Reviews in Food Science and Food Safety*, *21*(6), 5131–5152.

Tsai, J. S., Lin, Y. S., Pan, B. S., & Chen, T. J. (2006). Antihypertensive peptides and γ-aminobutyric acid from prozyme 6 facilitated lactic acid bacteria fermentation of soymilk. *Process Biochemistry*, *41*(6), 1282–1288.

Tsai, T. Y., Chu, L. H., Lee, C. L., & Pan, T. M. (2009). Atherosclerosis-preventing activity of lactic acid bacteria-fermented milk–soymilk supplemented with *Momordica charantia*. *Journal of Agricultural and Food Chemistry*, *57*(5), 2065–2071.

Urushibata, Y., Tokuyama, S., & Tahara, Y. (2002). Characterization of the Bacillus subtilis ywsC gene, involved in γ-polyglutamic acid production. *Journal of Bacteriology*, *184*(2), 337–343.

Van Pijkeren, J. P., & Britton, R. A. (2014). Precision genome engineering in lactic acid bacteria. *Microbial Cell Factories*, *13*(1), 1–10.

Vijayendra, S. V. N., & Halami, P. M. (2015). Health benefits of fermented vegetable products. *Health Benefits of Fermented Foods and Beverages*, 297–324.

Walther, B., & Sieber, R. (2011). Bioactive proteins and peptides in foods. *International Journal for Vitamin and Nutrition Research*, *81*(2), 181.

Wang, L. J., Li, D., Zou, L., Dong Chen, X., Cheng, Y. Q., Yamaki, K., & Li, L. T. (2007). Antioxidative activity of Douchi (a Chinese traditional salt-fermented soybean food) extracts during its processing. *International Journal of Food Properties*, *10*(2), 385–396.

Weng, Y., Yao, J., Sparks, S., & Wang, K. Y. (2017). Nattokinase: An oral antithrombotic agent for the prevention of cardiovascular disease. *International Journal of Molecular Sciences*, *18*(3), 523.

Wilburn, J. R., & Ryan, E. P. (2017). Fermented foods in health promotion and disease prevention: An overview. *Fermented Foods in Health and Disease Prevention*, 3–19.

Williams, D. J., Edwards, D., Hamernig, I., Jian, L., James, A. P., Johnson, S. K., & Tapsell, L. C. (2013). Vegetables containing phytochemicals with potential anti-obesity properties: A review. *Food Research International*, *52*(1), 323–333.

World Health Organisation. (2022). Cardiovascular diseases (CVDs). https://www.who.int/news-room/fact-sheets/detail/cardiovascular-diseases-(cvds)

Xiao, Y., Fan, J., Chen, Y., Rui, X., Zhang, Q., & Dong, M. (2016). Enhanced total phenolic and isoflavone aglycone content, antioxidant activity and DNA damage protection of soybeans processed by solid state fermentation with *Rhizopus oligosporus* RT-3. *RSC Advances*, *6*(35), 29741–29756.

Xu, L., Du, B., & Xu, B. (2015). A systematic, comparative study on the beneficial health components and antioxidant activities of commercially fermented soy products marketed in China. *Food Chemistry, 174*, 202–213.

Yamabe, S., Kobayashi-Hattori, K., Kaneko, K., Endo, H., & Takita, T. (2007). Effect of soybean varieties on the content and composition of isoflavone in rice-koji miso. *Food Chemistry, 100*(1), 369–374.

Yang, H., Yang, L., Li, X., Li, H., Tu, Z., & Wang, X. (2020). Genome sequencing, purification, and biochemical characterization of a strongly fibrinolytic enzyme from *Bacillus amyloliquefaciens* Jxnuwx-1 isolated from Chinese traditional *Douchi*. *The Journal of General and Applied Microbiology*, *66*(3), 153–162.

Yann, D., & Pauline, G. (2014). Usefulness of natural starters in food industry: The example of cheeses and bread. *Food and Nutrition Sciences*, *5*(17), 1679–1691.

Yao, M., Yang, Y., Fan, J., Ma, C., Liu, X., Wang, Y., & Sun, Q. (2022). Production, purification, and functional properties of microbial fibrinolytic enzymes produced by microorganism obtained from soy-based fermented foods: Developments and challenges. *Critical Reviews in Food Science and Nutrition*, 1–26.

Yao, Z., Liu, X., Shim, J. M., Lee, K. W., Kim, H. J., & Kim, J. H. (2017). Properties of a fibrinolytic enzyme secreted by *Bacillus amyloliquefaciens* RSB34, isolated from Doenjang. *Journal of Microbiology and Biotechnology*, *27*(1), 9–18.

Zaloga, G. P. (2015). Phytosterols, lipid administration, and liver disease during parenteral nutrition. *Journal of Parenteral and Enteral Nutrition*, *39*, 39S–60S.

Zhang, J. H., Tatsumi, E., Ding, C. H., & Li, L. T. (2006). Angiotensin I-converting enzyme inhibitory peptides in Douchi, a Chinese traditional fermented soybean product. *Food Chemistry*, *98*(3), 551–557.

Zhou, G., Bester, K., Liao, B., Yang, Z., Jiang, R., & Hendriksen, N. B. (2014). Characterization of three *Bacillus cereus*strains involved in a major outbreak of food poisoning after consumption of fermented black beans (*douchi*) in Yunan, China. *Foodborne Pathogens and Disease*, *11*(10), 769–774.

Zhu, Y. (2018). Anticancer effects of soybean bioactive components and anti-inflammatory activities of the soybean peptide Lunasin (Doctoral dissertation, Universite de Liege, Belgium).

Zhu, Y., Wang, A., Liu, M. C., Zwart, A., Lee, R. Y., Gallagher, A., & Clarke, R. (2006). Estrogen receptor alpha positive breast tumors and breast cancer cell lines share similarities in their transcriptome data structures. *International Journal of Oncology*, *29*(6), 1581–1589.

Fermentation Biotechnology of Food and Beverage Products in Malaysia

10

Yu-Hsuan How, Nor Khaizura Mahmud Ab Rashid, Seng-Joe Lim, Lee-Sin Chang, and Liew-Phing Pui

10.1 INTRODUCTION

Fermented foods are widely popular in Southeast Asia countries. Most of the Southeast Asian countries are tropical countries, where food products such as meat, fish, fruit, and vegetables are perishable. Hence, fermentation techniques are adopted for food preservation (Nuraida 2015; Pakwan et al. 2020). Living in the same sub-region, similar substrates or raw materials are found across Southeast Asian countries as compared to Western countries due to availability and accessibility. Nevertheless, the sensory characteristics may vary with the origin of the raw material, starter cultures, geographical area, and fermentation techniques (Pakwan et al. 2020; Tamang et al. 2020; Endo et al. 2014).

The difference in religion and cultures across Southeast Asia countries also influenced the type of raw materials or end products from fermentation, such as restrictions in consuming alcohol, pork, beef, or any animal products (Tamang et al. 2020). Southeast Asia consumed more fish, as most of the Southeast Asian countries are located along the sea. Besides that, plant materials are also more commonly used in their diet as compared to dairy products in Southeast Asia. Hence, Southeast Asian fermented products are derived from fish, soybean, palm, rice, and fruits (Endo et al. 2014; Swain et al. 2014; Raji et al. 2017; Surono 2016).

Malaysia is a multi-racial nation that consists of Malay, Chinese, Indian, and other ethnic groups. As the time progresses over generations, the food and cuisines from different groups have integrated to form a Malaysian food identity. Although there is cross-cultural integration in some of the traditional foods to make fusion based-cuisines, Malaysians also embraced their heritage and traditions in which many traditional foods and cuisines from different ethnic groups are still preserved, including fermented food or beverage products (Ishak et al. 2019; Perry 2017).

Fermented soybean cake (tempeh), fermented fish (*ikan pekasam*), and fermented durian pulp (*tempoyak*) are often found in Malaysia are traditional Malay or Javanese food (Ezzat et al. 2021; Raji et al. 2017; Ahnan-Winarno et al. 2021). On the other hand, fermented bean curd (*sufu*) and fermented soybean paste (*taucu*) originating from China are widely found in Malaysia due to the migration of Chinese people (Wilson 1995; Liu et al. 2006). Malaysian Indians also adapted traditional fermented products from India such as yoghurt drink (*lassi*), which is commonly served in Indian cuisine restaurants in Malaysia (Sudheendra et al. 2018). Besides that, rice wine (*tapai*) and palm wine (*tuak*) are traditional wines of Indigenous people, mostly from Sabah and Sarawak, East Malaysia (Endo et al. 2014; Law et al. 2011).

10.2 TYPES AND TECHNIQUES OF FERMENTATION

10.2.1 Types of Fermentation

Different types of fermentation are involved in producing different types of fermented foods. However, a single fermented food could also be produced from more than one type of fermentation, as some fermented foods may involve multiple microorganisms (Malo and Urquhart 2015). There are four main types of fermentation that are commonly used in both industrial and homemade processes: acetic acid fermentation, lactic acid fermentation, alcohol fermentation, and alkaline fermentation. These fermentation types are highly dependent

DOI: 10.1201/9781003605300-10

on the types of food matrix and microorganisms used in the fermentation process (Anal 2019). As the end products such as lactic acid, acetic acid, alcohol, and amino acids are produced from different types of fermentation, they often contribute to the taste and flavor of the fermented foods.

The types of fermentation for fermented food products commonly found in Malaysia are listed in Table 10.1. Among the different types of fermentation, lactic acid fermentation is found to be used across different types of food substrates, including fruits, dairy, and seafood, while fermented soybean products found in Malaysia are shown to mainly use the alkaline-fermented process. On the other hand, fruits in Malaysia are mostly fermented into wine or vinegar through alcoholic and alkaline fermentation.

10.2.2 Fermentation Techniques

Traditional fermentation techniques can be dated back to thousands of years ago, where they first originated from South Asia. People were producing fermented food, beverages, and condiments traditionally at small-scale levels since the beginning of organized community living. Fermented food production started to shift towards large-scale production only after the late 1800s, when food industrialization began (Ray and Joshi 2014). However, many traditional fermented foods, beverages, or condiment products are still produced traditionally at a small scale in Southeast Asia (Law et al. 2011; Tamang et al. 2020).

Open fermentation is often considered a traditional method for food fermentation where the fermentation is carried out in an open environment or with an open lid (Li et al. 2014). Several traditional fermented food products produced with open fermentation are widely found in Southeast Asia (Table 10.2). This type of fermentation raises safety concerns, as it is mostly carried out in a non-sterile environment. However, open fermentation allows the substrate to be recycled, obtaining high yield from a substrate; has low energy consumption; and is non-labor intensive. Hence, it is explored by the industry to produce low-cost and sterile fermented food products (Li et al. 2014).

On the other hand, some of them are also produced under full or partial closed fermentation, where the fermentation of the substrate occurs in closed containers due to the requirement of anaerobic conditions (Table 10.2). However, the current modern technology also allows aerobic fermentation to be carried out in closed fermentation with a constant supply of sterile oxygen. It is widely used in the industry due to its simple operation and is a safe and hygienic process (Ingledew and Lin 2011).

Interestingly, fermented products produced with the same substrate, such as original *sufu* and grey *sufu* from soybean, used different fermentation techniques to produce unique products, flavors, and aroma (Ma et al. 2013; Han et al. 2001). Furthermore, different wines were also found to be produced using different fermentation techniques in terms of open or closed fermentation for different substrates.

Fermentation can also be categorized into submerged fermentation and solid-state fermentation (Behera and Ray 2019). As compared to solid-state fermentation, submerged fermentation is mostly used in industry. However, many traditional fermented food products found in Malaysia are produced with solid-state fermentation (Table 10.2). Submerged fermentation is often used in the production of beverages or condiments in Malaysia such as fruit wines, vinegars, and dairy drinks (Lynch et al. 2019; Sudheendra et al. 2018). Nevertheless, there are also fermented beverages or condiments such as fermented rice wine (*tapai*) and fermented soybean paste (*taucu*) that involve both solid-state and submerged fermentation in their production process (Law et al. 2011; Surono 2016).

TABLE 10.1 Types of Fermentation for Fermented Food, Beverages, or Condiments Found in Malaysia

TYPE OF FERMENTATION	*SUBSTRATES*	*FERMENTED PRODUCTS*	*REFERENCES*
Lactic acid fermentation	Durian	Fermented durian pulp (*tempoyak*)	Swain et al. (2014)
	Fish	Fermented fish (*ikan pekasam*)	Ezzat et al. (2021)
	Milk	Fermented milk (*lassi*)	Behare et al. (2010)
Alcoholic fermentation	Coconut	Fermented coconut/palm wine (*tuak*)	Anal (2019)
	Rice	Fermented rice wine (*tapai*)	
Acetic acid fermentation	Papaya	Papaya vinegar	Kong et al. (2018)
	Soursop	Soursop vinegar	Chin et al. (2017)
	Pineapple	Pineapple vinegar	Mohamad et al. (2015)
	Rambutan	Rambutan vinegar	Mokhtar et al. (2016)
	Dokong	*Dokong* vinegar	
Alkaline fermentation	Soybean	Fermented soybean cake (tempeh)	Zainal Abidin et al. (2020)
		Fermented soybean paste (*taucu*)	Gopikrishna et al. (2021)
		Fermented bean curd (*sufu*)	
	Duck egg	Fermented duck egg (pidan)	Wang and Fung (1996)

TABLE 10.2 Fermentation Techniques Used for Fermented Food, Beverages, or Condiments Found in Malaysia

SUBSTRATE	*FERMENTED PRODUCT*	*FERMENTATION TECHNIQUE*		*REFERENCES*
Duck egg	Fermented duck egg (*pidan*)	Open fermentation	Solid-state fermentation	Wang and Fung (1996)
Soybean	Fermented soybean curd (*sufu*)	Open fermentation	Solid-state fermentation	Fan et al. (2020), Han et al. (2001)
	Fermented soybean curd (*grey sufu*)	Closed fermentation	Solid-state fermentation	Ma et al. (2013)
	Fermented soybean cake (tempeh)	Open fermentation	Solid-state fermentation	Ahnan-Winarno et al. (2021), Radita et al. (2017)
	Fermented soybean paste (*taucu*)	Closed fermentation	Solid-state and submerged fermentation	Nandiyanto et al. (2018), Surono (2016)
Coconut	Fermented palm wine (*tuak*)	Open fermentation	Submerged fermentation	Santiago-Urbina and Ruíz-Terán (2014), Lynch et al. (2019)
Durian	Fermented durian pulp (*tempoyak*)	Closed fermentation	Solid-state fermentation	Surono (2016), Swain et al. (2014)
Rice	Fermented rice wine (*tapai*)	Closed fermentation	Solid-state and submerged fermentation	Endo et al. (2014), Law et al. (2011)
Fish	Fermented fish (*ikan pekasam*)	Closed fermentation	Solid-state fermentation	Ezzat et al. (2021)
Milk	Fermented milk (*lassi*)	Closed fermentation	Submerged fermentation	Sudheendra et al. (2018), Rasane et al. (2017)
Fruits	Fermented fruit vinegars	Closed fermentation	Submerged fermentation	Lynch et al. (2019), Luzón-Quintana et al. (2021)

10.3 FERMENTED FOODS, BEVERAGES, AND CONDIMENTS FOUND IN MALAYSIA

10.3.1 Food Products

10.3.1.1 Ikan Pekasam *(Fermented Fish)*

Fermented fish are categorized into two groups: widely used fish sauce and the less popular ones, such as fermented fish with salt and carbohydrates. Fermented fish is locally known as *ikan pekasam* in Malaysia, *plaa-som* in Thailand, *burong dalag* in the Philippines, *narezushi* in Japan, and *sikhae* in Korea. *Ikan pekasam* is produced commonly in the northern states of Malaysia (Perak, Kedah, and Perlis), and the production is generally on a small scale. *Ikan pekasam* is commonly deep-fried or wrapped in banana leaves and roasted on charcoal or a wood fire (Ismail and Zain 1978). Fried *ikan pekasam,* together with fried chilies and onions, is served as a side dish when eaten with hot rice (Ezzat et al. 2015).

Ikan pekasam is a traditional delicacy in Malaysia of fermented fish mixed with roasted rice, tamarind, and salt ingredients. The most common freshwater fish used are tilapia and feather back (*belida*), carp (*terubul, temperas, mengkawan*, and *loma*), Javanese carp (*lampam jawa*), barb (*mata merah*), giant goramy (*kalui*), goramy (*sepat padi*), snakeskin goramy (*sepat siam*), marble goby (*ketutu*), perch (*kepar*), and snakehead (*toman*) (Department of Fisheries 2013; Syahputra et al. 2012; Huda 2012). The fish is combined with a minimum of 10% salt and ground roasted uncooked rice (Syahputra et al. 2012) before being allowed to ferment for a minimum of two weeks to one month.

There are two methods of *ikan pekasam* production, natural and acid-assisted fermentation. The *pekasam* production starts with evisceration and thorough cleaning of the fish, which is then mixed with coarse salt (approximately 30%) and left to ferment for two to three days. After the salting process or first stage fermentation, the fish is washed a few times and tossed to remove excess water. When the fish is fully drained, it is mixed with ground roasted rice and kept in a tightly sealed jar to ferment (second stage) for approximately 14 days. After fermentation, it is ready to be packed and distributed. For acid-assisted fermentation, a mixture of dried slices of the *Garcinia atroviridis* fruit or tamarind (*Tamarindus indica*) pulp is added to the fish before the second stage of fermentation (Syahputra et al. 2012).

Ikan pekasam is considered to undergo natural fermentation, with the addition of a carbohydrate source from ground roasted rice. The available carbohydrate is degraded by the natural presence of lactic acid bacteria (LAB) to produce lactic acid during fermentation (Riebroy et al. 2004). However, some *ikan pekasam* producers add tamarind or dried slices of the *Garcinia atroviridis* fruit as natural acidulants (Syahputra et al. 2012) to decrease the pH rapidly (Riebroy et al. 2004). The natural pH of tamarind is pH 1.3 to 2.1 (Abukakar et al. 2008; Ahmed et al. 2007), and *G. atroviridis* fruit is pH 2.1–2.5 (Siripongvutikorn et al. 2009). To date, both natural and acid-assisted fermentation are carried out depending on the *ikan pekasam* producers.

The chemical composition of *ikan pekasam* may vary from producer to producer, and some may have higher salt concentrations. However, the salt content must be more than 10%, according to the Malaysian Food Act and Regulation (1983). The moisture and salt content of *ikan pekasam* is reported within the range of 57 to 73% and 10 to 16%, respectively (Hassan 1980). The major protein component accounts for 15.0 to 25.0% of the dry matter, with 3.0 to 8.0% fat and 6.0 to 14.0% ash. Mansor et al. (2014) reported 0.1 to 0.4% lactic acid and a pH of 4.8 to 5.2 of *ikan pekasam*. This chemical composition of *ikan pekasam* is a suitable medium for most natural microflora, especially lactic acid bacteria, to initiate fermentation.

Fermented *ikan pekasam* also contains organic acid. The most dominant organic acid found is lactic acid, followed by succinic acid, acetic acid, formic acid, malic acid, oxalic acid, citric acid, L-pyroglutamic acid, and fumaric acid. The presence of acetic and formic acids in *ikan pekasam* indicates that the fermentation is heterofermentative (Ezzat et al. 2021). Besides that, the dominant aromatic compounds in *ikan pekasam* are carboxylic acids, giving a rancid and "goaty" off-flavor to the product (Tan et al. 2017).

Lactic acid bacteria that have been responsible for *ikan pekasam* fermentation are *Lactiplantibacillus plantarum, Levilactobacillus brevis, Lacticaseibacillus casei, Corynebacterium vitaeruminis, Streptococcus anginosus,* and *Staphylococcus carnosus* subsp. *carnosus* (Muryany et al. 2017; Tan et al. 2017; Mansor et al. 2014). Among these lactic acid bacteria, *Lactiplantibacillus plantarum* is indicated as the potential probiotic in which *L. plantarum* fulfilled several criteria of probiotics. For instance, it has antagonism against common pathogenic bacteria, has no hemolytic activity, is tolerant to lower pH (pH 3 and 5), can hydrolyze bile salts, and is susceptible to most antibiotics (Muryany et al. 2017).

Despite having good bacteria, *ikan pekasam* may have some food safety issues related to biogenic amines. Ezzat et al. (2015) reported that *ikan pekasam*'s total biogenic amine content did not exceed the tolerance level of 1000 ppm, which is considered safe to consume. Exceeding the tolerance level could lead to scombroid fish poisoning in humans. Cadaverine and putrescine were found as significant biogenic amines in *ikan pekasam*. Therefore, the combination of these two compounds could be used as an acceptability index for fermented fish products.

10.3.1.2 Tempeh (Fermented Soybean Cake)

Tempeh has been a staple food in the Javanese community for more than 300 years. The earliest mention of tempeh was in a manuscript published in 1815, referred to as Serat Centini, which included the teaching of Javanese cultures and philosophy (Ahnan-Winarno et al. 2021). Although tempeh is currently widely consumed by people in different countries, it was once a food consumed by lower-income people as a source of protein due to its economic value (Radita et al. 2017). Besides that, the unique sensory profile of tempeh is also highly favored by the locals in Indonesia and Malaysia (Ahnan-Winarno et al. 2021; Zainal Abidin et al. 2020). Figure 10.1 displays tempeh served in restaurants in Malaysia.

Tempeh is mainly produced using soybean; however, it could also be produced with other seeds, grains, nuts, or beans such as rice bran, lentils, barley, peanuts, oats, quinoa, chickpeas, cowpea, sunflower seeds, millet, velvet beans, fava beans, winged beans, black beans, white beans, mung beans, jack beans, and corn grits (Ahnan-Winarno et al. 2021; Radita et al. 2017). The main microorganisms used for tempeh fermentation are *Rhizopus microsporus* var. *oligosporus* (Ahnan-Winarno et al. 2021). Aside from *R. oligosporus*, other lactic acid bacteria were also identified in traditional Malaysian tempeh such as *Enterococcus faecium, Leuconostoc mesenteroides* ssp. *mesenteroides*, and *Lactobacillus delbrueckii* ssp. *delbrueckii* by API 20 Strep, API 50 CHL kit, and 16S rRNA gene sequences (Pisol et al. 2015). *Enterococcus faecium* was also identified in Malaysian tempeh through sodium dodecyl sulphate-polyacrylamide gel electrophoresis (SDS-PAGE) (Moreno et al. 2002).

Tempeh is produced through solid-state fermentation with a final product of minimum 15% (w/w) protein content, 7% (w/w) fat content, 2.5% (w/w) fiber content, and maximum

FIGURE 10.1 Tempeh served in Malaysian Malay cuisine restaurant.

65% (w/w) moisture content (Ahnan-Winarno et al. 2021). The fermentation process of tempeh was found to reduce the ash content while increasing the protein and amino acids content in soybeans. However, these vary with the type of soybean (Handoyo and Morita 2006). The protein content in tempeh was also reported to be higher than dairy products such as yogurt (Song et al. 2008). Tempeh is generally white in color due to the mycelium from *Rhizopus* spp., with a nutty taste (Handoyo and Morita 2006). The main volatile compounds found in tempeh are undecane, furan, and dimethyl disulfide, which contribute to the nutty taste and flavor of tempeh (Kustyawati et al. 2017).

There are nine main steps in the production of tempeh: soaking or fermenting, dehulling, washing, boiling, draining/drying, cooling, inoculating, incubating, and packing (Pisol et al. 2015). However, traditional tempeh from different locations of Indonesia can be produced using different sequences of steps. The soaking of beans in water is to allow the hulls to be peeled off easily, while the occurrence of natural acidification by microorganisms is to inhibit unwanted pathogens (Ahnan-Winarno et al. 2021). The duration of soaking can vary between 6 and 24 hours (Pisol et al. 2015).

The soybean is then dehulled for hygienic purposes. The dehulled beans are boiled for 20 to 30 minutes to further remove pathogens (Ahnan-Winarno et al. 2021). The moisture content in tempeh is then reduced through the draining or drying process (Pisol et al. 2015). As tempeh is produced through solid-state fermentation, water activity of 1.0 with 62% humidity is required for fermentation (Ahnan-Winarno et al. 2021). *Rhizopus* spp. is inoculated into the substrate and stored in packaging with restricted airflow. Traditional tempeh is wrapped in banana leaves prior to incubation. After the inoculation, the packed substrate is fermented at 25 to 38°C for 18 to 72 hours to support the growth of the fungi (Ahnan-Winarno et al. 2021; Pisol et al. 2015).

The fermentation process of tempeh improves the bioavailability of nutrients and reducing anti-nutrients that were present in the soybean. Also, the soaking step of producing tempeh decreases phytates during natural fermentation (Ahnan-Winarno et al. 2021). The enzymes released from the microorganism during fermentation are responsible for improving bioactive compounds such as amino acids, fatty acids, isoflavones aglycones, crude fiber, ash, vitamin B12, and phenolic acids (Bavia et al. 2012). They also contribute to the breakdown of anti-nutrients such as phytate, trypsin inhibitor, oxalate, cyanogenic glycoside, and oligosaccharides. These bioactive compounds in tempeh are reported to possess antioxidant, anti-hypertensive, anti-diabetic, anti-obesity, and anti-cancer activity (Ahnan-Winarno et al. 2021; Wan Saidatul Syida et al. 2018).

Furthermore, bacteriocins produced by the microorganisms present in tempeh could also exert antimicrobial properties (Moreno et al. 2002). The isoflavone aglycones in tempeh also have the potential to improve human cognitive, liver, cardiovascular, lung, and muscle health, as well as preventing cancer, type 2 diabetes, obesity, and malnutrition (Ahnan-Winarno et al. 2021). An amount of calcium in tempeh comparable to that of milk was found to be absorbed by postmenopausal women. This allows tempeh to be a non-dairy source of calcium for women with low bone mass (Haron et al. 2010).

Aside from consuming fresh, tempeh is also fried to produce different aroma compounds (Kustyawati et al. 2017). It can also be seasoned with salt or garlic to improve shelf life and organoleptic and antimicrobial properties (Astawan et al. 2016). In recent years, tempeh has also been transformed into different types of food ingredients and products such as soy protein isolate, burger patties, and energy bars. The high sensory acceptable score in these products reflects its potential in commercialization (Zainal Abidin et al. 2020; Vital et al. 2018; Wan Saidatul Syida et al. 2018).

10.3.1.3 Sufu *(Fermented Bean Curd)*

Sufu is a fermented bean curd that originated in China with a long history dating back to the Han Dynasty (260 BCE–220 AD). However, its manufacturing process was not recorded until the Wei Dynasty (220–265 AD) and became more widely popular during the Ming Dynasty (1368–1644), with more records of its processing methods. In the Qing Dynasty, *sufu* was considered a health-promoting food, and current studies show that the high protein content in *sufu* is more easily digested by children and patients (Han et al. 2001). Although *sufu* originated in China, it was later introduced to people in Southeast Asia after the mass migration of Chinese people to Southeast Asia (Lockard 2013).

Sufu is also referred to as "Chinese cheese", as it has a creamy cheese texture with a unique fermented soy flavor (Liu et al. 2006). *Sufu* can be categorized into three types based on the microorganisms used: mold fermented, bacteria fermented, and natural fermented. The common species used for mold-fermented *sufu* are *Geotrichum* spp., *Mortierella* spp., *Rhizopus* spp., *Actinomucor* spp., and *Mucor* spp. (Cheng et al. 2009; Jang et al. 2021). On the other hand, *Bacillus* spp. such as *B. pumilus*, *B. megaterium*, *B. stearothermophilus*, and *B. firmus* are used for the inoculation of bacteria-fermented *sufu* (Gopikrishna et al. 2021; Wan et al. 2020). Natural fermentation only involves the action of naturally present microorganisms in the soybean without additional inoculation (Li et al. 2020).

Sufu can also be classified into four main types: red *sufu*, white *sufu*, grey *sufu*, and other types of *sufu*. White *sufu* is considered the original *sufu* and consists of mixtures such as salt, sugar, flour paste, spices, and alcoholic beverage. On the other hand, red *sufu* is developed with the same mixture as white *sufu* with the addition of a natural red colorant known as *angkak*, a byproduct produced from fermented cooked rice. Instead of adding colorant, grey *sufu* is produced using leftover whey from tofu making, salt, and spices with the inoculation of both bacteria and mold. Aside from these three main types of *sufu*, other ingredients such as vegetables, rice, bacon, and alcohol can also be incorporated in the making of *sufu* (Han et al. 2001). Figure 10.2 shows commercial original white *sufu* in chili and sesame oil that can be purchased in Malaysia's grocery stores.

FIGURE 10.2 Commercial white *sufu* in chili and sesame oil found in Malaysian grocery store.

Generally, the basic proximate composition of *sufu* is 25–44% moisture content, 20–27% fat content, 24–35% protein content, and 30–45% salt content. *Sufu* is also a mildly acidic product with a pH ranging from 5.11–6.46. Free amino acids produced during fermentation contribute to the flavor of *sufu*. The main free amino acids found in *sufu* are glutamic acid; however, phenylalanine, leucine, valine, and aspartic acid are also largely identified in *sufu*. The taste of *sufu* was found to be primarily bitter, followed by salty and umami taste (Xie et al. 2019).

Aside from its similarity in texture, the production steps of *sufu* are also similar to cheese manufacturing (Jang et al. 2021). There are four main steps in manufacturing *sufu*: tofu production, pehtze production, brining, and ripening (Li et al. 2020). In tofu production, soybeans are soaked in water, ground into a slurry, and filtered to produce soymilk (Han et al. 2001). The soymilk is boiled, added with salt, and left to stand at approximately 80°C for 15 minutes until it is coagulated (Li et al. 2020). The coagulated tofu is left at the preferred temperature and humidity with or without the inoculation of desired microorganisms (Ma et al. 2013). The incubation duration to form mycelium on the pehtze could last from 2 days up to 15 days. In the brining process, pehtze is added with salt and allowed to absorb for 5 to 12 days until the salt content reaches 15 to 30% (Li et al. 2020; Han et al. 2001). The salted pehtze is then added with different mixtures depending on the types of *sufu* and left to ripen in a sealed jar at room temperature for 2 to 6 months (Han et al. 2001).

The proteins in *sufu* are hydrolyzed into bioactive compounds such as peptides and amino acids by the microorganisms during fermentation (Han et al. 2001). The quantity and types of essential amino acids are also reported to be comparable to the amino acids found in eggs and milk proteins. As a high-protein plant-based food, *sufu* is a potential healthy meat substitute to replace animal proteins in the diet, as it is cholesterol-free (Xie et al. 2019). Aside from cholesterol-lowering properties, *sufu* has also been reported to be anti-hypertensive, exerting an influence on lowering blood pressure, reducing cardiovascular and cancer risk (Sanjukta and Rai 2016).

Isoflavone glycoside is the main compound found in beans that are converted into bioactive aglycones by the microorganisms in *sufu*. Bioactive aglycones have been shown to reduce cancer risks for breast and prostate cancer and improve bone health as well as cardiovascular health. It is reported that low fermentation temperature (13°C) and low salt content (15%) contribute to the increase of isoflavone aglycone formation in *sufu* (Cheng et al. 2011). A high concentration of butyric acid was also found in *sufu* after fermentation. It is a fatty acid known for its immunomodulatory activity in managing gastrointestinal tract diseases (Ma et al. 2013).

Apart from the beneficial compounds, biogenic amines such as histamine, tyramine, putrescine, cadaverine, tryptamine, and β-phenylethylamine that are harmful to human health have also been detected in *sufu* (Fan et al. 2020). Nevertheless, low salt, the sterilization process, the inclusion of selected microorganisms for fermentation instead of natural fermentation, and the addition of spices such as ginger powder have been shown to reduce biogenic amines. It is hypothesized that the sterilization process and ginger powder were able to reduce unwanted microorganisms in *sufu*, leading to the reduction of biogenic amines (Fan et al. 2020; Li et al. 2020).

10.3.2 Fermented Beverages and Condiments

10.3.2.1 Lassi *(Yoghurt Drink)*

Lassi is a refreshing fermented dairy drink that originated from India. The origin of *lassi* dates back to the Paleolithic and Neolithic eras in India (Sudheendra et al. 2018; Abesinghe et al. 2020). However, it is also often found in Southeast Asian countries such as Thailand and Malaysia (Chee et al. 2020) (Figure 10.3). *Lassi* is a drinkable yoghurt that is also referred to as buttermilk or stirred curd, as it is mainly derived from *dahi* (traditional Indian curd) and diluted with water (Abesinghe et al. 2020; Sharma 2016). It is considered a yoghurt drink due to the similar bacteria used for fermentation, which are *Streptococcus thermophilus* and *Lactobacillus bulgaricus* (Panesar 2011; Upadhyay and Kumar 2017; Sarkar 2008).

Lassi is mainly made from cow or buffalo milk (Panesar 2011). The process of *lassi* production is separated into nine steps: standardization, pasteurization, homogenization and cooling, inoculation, fermentation, curd breaking, sugar and flavor addition, blending, and filling (Upadhyay and Kumar 2017). Cow or buffalo milk is standardized on milk fat and milk solids non-fat (MSNF) in the range of 0.5 to 4.5% milkfat and 8.5 to 10% MSNF (Chandan and O'Rell 2013). The

FIGURE 10.3 Original *lassi* served at Malaysian Indian cuisine restaurant.

standardized raw milk is pasteurized at 85 to 95°C for 5 to 30 mins to remove microorganisms naturally present in the milk (Rasane et al. 2017; Sarkar 2008). Homogenization and cooling are then carried out on pasteurized milk prior to inoculation (Chandan and O'Rell 2013; Rasane et al. 2017).

Different microorganisms are inoculated for the fermentation of *lassi*, such as *S. thermophilus*, *L. bulgaricus*, *Lactobacillus acidophilus*, *Lactococcus lactis* subsp. *lactis* and *cremoris*, and *Leuconostoc lactis* (Upadhyay and Kumar 2017; Panesar 2011; Rasane et al. 2017; Sarkar 2008). The milk is fermented until curd (*dahi*) is formed. However, the inoculation concentration, fermentation time, and fermentation temperature may vary with different starter cultures and methods. After the curd is set, the coagulum is broken through agitation (Chandan and O'Rell 2013). Other ingredients such as sugar (8 to 12% of total concentration) and flavoring are then added and blended with the fermented milk. The *lassi* is then put in a bottle and stored at 4°C (Rasane et al. 2017).

Lassi is a protein-based probiotic beverage (Chee et al. 2020; Ranadheera et al. 2017). The combination of milk and probiotics, as well as the fermentation process, contributes to the health benefits of *lassi* (Sharma 2016; Abesinghe et al. 2020; Padghan et al. 2018). *Lassi* that is produced using exopolysaccharide (EPS)-producing LAB strains may exert health benefits such as lowering pathogen adhesion, interacting with immune reactors, and strengthening the gut epithelial barrier (Behare et al. 2010).

During the fermentation process of *lassi*, the proteins in raw milk are broken down into bioactive peptides, vitamins, and enzymes which possess potential health effects (Rasane et al. 2017; Padghan et al. 2018). These bioactive compounds in *lassi* have also been found to display antioxidant activity (Upadhyay and Kumar 2017; Padghan et al. 2018). Furthermore, the conjugated linoleic acid produced from the hydrolysis of milk fat during the production of *lassi* could also exhibit anti-carcinogenic properties (Sarkar 2008).

Different flavors of *lassi* are served in restaurants. It is usually blended with fruits such as mango, strawberry, pineapple, and banana. Nowadays, other ingredients such as carrot, dried figs, dried dates, ginger extract, and turmeric extract are also fortified in *lassi*. Fortified *lassi* has higher phenolic content and antioxidant activity as compared to non-fortified lassi. Furthermore, *lassi* with dried fruit pieces added has been shown to improve the probiotic cell count and sensory evaluation of the dairy beverage. *Lassi* blended with juices such as carrot also displays high organoleptic properties (Upadhyay and Kumar 2017; Sudheendra et al. 2018; Patel et al. 2015). Moreover, *lassi* has also been carbonated to improve the refreshing quality and shelf life (Ravindra et al. 2014).

10.3.2.2 Tuak *(Palm Wine)*

Wine can be made from other fruits or combinations of fruits, which gives a hint of various pleasant and enhanced flavors. Figure 10.4 indicates the general steps in production of fruit wine and fruit vinegar. Fruit wines are alcoholic beverages produced from alcoholic fermentation of fruits. The alcoholic fermentation usually occurs through yeast fermentation, in which the sugars present in fruit are converted into ethanol under anaerobic conditions. Yeast metabolizes sugar to produce energy in the form of adenosine triphosphate (ATP) and yields ethanol (CH_3CH_2OH) and carbon dioxide (CO_2) along the way (van Maris et al. 2006).

The common yeast used in alcoholic fermentation is the brewer's yeast, *Saccharomyces cerevisiae*. Other fermentative yeasts were also found to be able to produce alcoholic beverages, including genera such as *Hansenula*, *Candida*, *Zygosaccharomyces*, *Schizosaccharomyces*, *Hanseniaspora*, and *Brettanomyces* (Romano et al. 2003). These fermentative yeasts have different tolerances and responses towards ethanol stress and stimulate different mechanisms for fermentation, as well as producing different secondary metabolites and thus yielding different aromas and flavors of fruit wine.

Traditionally, the most popular fruit wine that is available in Malaysia, Indonesia, and the Philippines is palm wine, locally known as *tuak* (Malaysia and Indonesia) or *tubâ* (the Philippines). Palm wine can be made from the sap of various types of palm trees such as date palms (*Phoenix dactylifera*), sugar palms (*Borassus flabellifer Linn.*), and coconut palms (*Cocos nucifera L.*). The common source of this alcoholic beverage is from coconut nectar (*nira*) (Figure 10.5), which is obtained by cutting off the inflorescences of the unopened spadix from the coconut palm (Asghar et al. 2019).

The nectar is a liquid containing sugar (approximately 10–20%), with neutral pH and translucent and oyster white-colored liquid, which in its natural form is non-alcoholic. However, the nectar, if kept for a certain duration, through

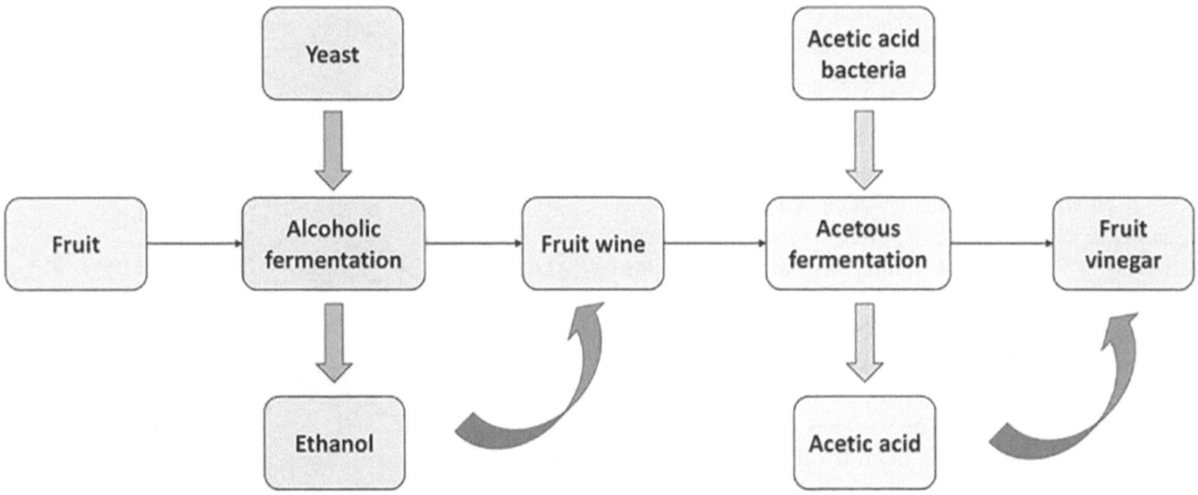

FIGURE 10.4 General steps in the preparation of fruit wine and fruit vinegar through alcoholic and acetous fermentation.

FIGURE 10.5 Coconut palm tree with unopened spadix (left) containing nectar (right).

means of spontaneous fermentation, that is, fermentation by the action of natural microflora (not inoculated), will be naturally converted into an alcoholic beverage, *tuak*, Depending on the duration of the spontaneous fermentation, usually between 24 and 48 hours, the sugar present will be converted into alcohol, where the alcohol content could go up to 5–6% (Law et al. 2011). The sweetness of the beverage decreases over time. Palm wine has mild alcoholic characteristic and malty and floral-fruity notes, with off-colored milky appearance (Chandrasekhar et al. 2012). Upon further fermentation, palm wine will turn sour due to acetate fermentation by acetic acid bacteria, which converts the alcohol into acetic acid that contains 4 to 7% acetic acid, that is, vinegar (Law et al. 2011; Chandrasekhar et al. 2012). Its highly fermentable nature poses preservation problems, leading to unacceptable taste and aroma, which is usually considered "spoiled".

Besides palm wine, colored fruit that contain natural pigments with high antioxidant capacity are gaining popularity among wine drinkers. Tropical fruits that are available in abundance all year round can be used to produce fruit wine either for household consumption or commercial purposes. Dragon fruit, for instance, is widely cultivated in the Philippines and is processed into dragon fruit wine during the harvesting season (Dimero and Tepora 2018). Meanwhile, soursop fruit, widely grown in the southern part of Malaysia, is used to produce soursop wine (Ho et al. 2020). Thailand, on the other hand, produces mangosteen wine, lychee wine, and passionfruit wine that are exported to the United States (Zouhali-Worrall 2008). In Thailand, there are several vineyards across the country, among them Prachuap Khiri Khan province, Nakhon Ratchasima province, and Chiang Rai province, while in Vietnam, most of the vineyards are located in southern Vietnam. These vineyards, as the name suggests, grow grapes for winemaking. The wineries in these countries produce red wines, white wines, and rose wines.

Another common type of wine available in this region, although not from fruit, is rice wine, which is called *sake* in Japan, *cheongju* in Korea, and *shaosingjiu* in China. Rice, which contains starch, can be fermented by adding *ragi* (yeast). The rice is soaked and steamed prior to the addition of *ragi*. Since starch is not fermentable in its native form, it will first go through a saccharification process by the enzymes (such as amylase) produced by yeast to break the starch into simpler fermentable sugars, such as glucose (Ray et al. 2016). The monosaccharides are then fermented by the yeast to produce alcohol. Due to the high carbohydrate content of rice and microbial (*ragi*) inoculation, the fermentation process is more controlled (compared to that of palm wine) and thus is able to produce high concentrations of alcohol, ranging from 15–25%, producing a strong alcoholic characteristic with a sweet taste.

10.3.2.3 Taucu *(Fermented Soybean Paste)*

Taucu is a traditional fermented soybean paste from Indonesia and Malaysia (Wilson 1995; Tamang et al. 2020). However, it is also commercially found in markets across Southeast Asia (Gopikrishna et al. 2021). A few centuries ago, Hokkien natives from China came to present-day Malaysia and Indonesia for trading and business during the Ming dynasty. The Hokkien communities have since existed in many Southeast Asian countries, including Malaysia, Indonesia, Singapore, Brunei, the Philippines, and Thailand (Ding 2016). Although *taucu* is a traditional condiment in Indonesia and Malaysia (Figure 10.6), it is reported to originate from China, as the pronunciation of *taucu* is from the Hokkien Chinese dialect (Surono 2016).

Taucu is produced specifically from yellow soybean with various mold and bacteria species such as *Rhizopus* spp., *Bacillus* spp., *Lactobacillus* spp., *Lactococcus* spp., *Leuconostoc* spp., *Streptococcus* spp., *Weissella* spp., *Hansenula* spp., *Aspergillus oryzae*, *Zygosaccharomyces soyae*, *Enterococcus hermanniensis*, and *Pediococcus acidilactici* (Tamang et al. 2020; Surono 2016). The production of *taucu* is considered an alkaline and high-salt fermentation (Nuraida 2015; Tamang et al. 2020). *Taucu* is produced with at least 60 kg of soybeans, either homemade or by micro-scale industries in order to obtain the maximum profit (Nandiyanto et al. 2018).

FIGURE 10.6 Commercial *taucu* found in Malaysian grocery store.

As compared to the process of making tempeh and *sufu*, *taucu* has more steps involved: soaking, washing and filtering, peeling, boiling, filtering and cooking, yeast fermentation, drying and grinding, brine fermentation, cooking, and packing. The first three steps (soaking, washing, peeling) are to remove the hulls from the soybean (Surono 2016). This part is necessary as the cellulose in the hulls may hinder the growth of the microorganisms. The dehulled beans are then boiled for one to two hours in order to soften the beans. The boiled beans are cooled, and roasted rice and yeast are added for fermentation. The enzymes from the microorganisms hydrolyze the macronutrients in the soybeans during the fermentation process (Nandiyanto et al. 2018).

The fermentation process can be carried out either on a tray (open system) or in a container (closed system) for three to five days. The fermented soybean is dried and mashed prior to brine fermentation (Surono 2016). The mashed soybean has 20% of saltwater with lactic acid bacteria added and is incubated for 20 to 30 days. The role of salt in this process is to preserve the product by inhibiting unwanted microorganisms and enhancing taste, aroma, and flavor (Nandiyanto et al. 2018). Both lactic acid bacteria fermentation and alcohol fermentation occur simultaneously during this process, where lactic acid bacteria contribute to the production of lactic acid, while alcohol is formed by the yeast. Last, the twice-fermented soybeans are cooked, with sugar and water added, and packed in bottles (Nandiyanto et al. 2018; Surono 2016).

The yellow-colored soybean paste has a meaty flavor; hence, it is often used as a flavoring agent for Indonesian and Chinese cuisine in Southeast Asia (Usman Pato 2003). Many flavor compounds in *taucu* are produced through the fermentation process. One of the amino acids broken down from protein in the fermentation is glutamic acid, which is the main contributor to the flavor of *taucu*. Aside from proteins, lipids are also hydrolyzed into fatty acids by microorganisms during fermentation. The fatty acids are combined with the main products of fermentation such as alcohol, acetic acids, and organic acids to form esters, which contribute to the aroma of the *taucu* (Nandiyanto et al. 2018). *Taucu* has been reported to exert antioxidant activity against 2,2-diphenyl-1-picrylhydrazyl (DPPH) radicals from its ethyl acetate and ethanol extracted contents, which contain flavonoids, monoterpenoids, and sesquiterpenoids (Mustarichie et al. 2017).

10.3.2.4 Tempoyak *(Fermented Durian Pulp)*

Tempoyak is a ubiquitous fermented plant-based product in Malaysia made from durian (*Durio zibethinus*) pulp, in which durian is an essential seasonal fruit cultivated in tropical Asian countries. It is crucial for its economic value, exotic aromas, and flavors. Durian has a short shelf life of two to three days after harvest or during the storage and retail sale phase. The pulp undergoes rapid physiochemical changes, which lead to dehiscence and softening. Therefore, a rapid decline in the value of durian requires important value-added product development (Paull and Ketsa 2014).

Low-quality durian pulp obtained from the cracked or over-ripened durian is usually used for *tempoyak*, allowing for the use of durians that would otherwise be discarded (Gandjar 1999). Thus, fermentation of over-ripe and low-quality durian pulp is an efficient and cheap method of preserving highly perishable climacteric fruits, especially when proper refrigeration or post-harvest handling facilities are inadequate in certain areas. *Tempoyak* is widely consumed in Malaysia and Indonesia but is not yet wholly commercialized. *Tempoyak* is considered an acid-fermented condiment, consumed with fresh vegetables or as an ingredient in certain local dishes. One of the most well-known dishes is called *gulai tempoyak* (Abdul Aris et al. 2015), in which the *tempoyak* is added to coconut milk sauce with any protein-based food like fish.

Tempoyak is mainly produced during the durian fruiting season. *Tempoyak* is still domestically produced for self-consumption or as a small business. The pulp of over-ripe, cracked, or unripe durians is pooled, and the durian pulp used may be obtained from different cultivars (Chuah et al. 2016). Durian pulp is traditionally fermented through spontaneous and uncontrolled processes to produce *tempoyak* (acid and salt-fermented durian) (Amin et al. 2004; Wasnin et al. 2014). The main intention of the fermentation process is to retain a high degree of the characteristic sensory qualities of the durian.

Tempoyak has variable quality due to the inconsistent indigenous microflora, typical for naturally fermented fruits (Chen et al. 2013; Nyanga et al. 2007). For this type of fermentation, the process relies on the natural microflora already present in the pulp. The microflora may originate from various sources, such as from the environment (soil, rotting vegetation, retail environment), and can be due to cross-contamination from other cracked durians (Papalexandratou et al. 2011).

The making of *tempoyak* begins with the collection of durians either from one or various cultivars that are suitable to produce *tempoyak*. The durian pulp will be separated from the seeds and placed in a covered container. Then, durian pulp constitutes a soft custard with or without a certain percentage of salt added before fermentation. For *tempoyak* with salt, a range of 1.3 to 3% (w/w) of salt can be added (Amin et al. 2004). Next, it will be fermented in a sealed container, or some even use a pottery container. This process will take at least seven days to a month at the ambient temperature (Merican 1977; Chuah et al. 2016). There are no starter cultures and preservatives added, as it is considered spontaneous fermentation. The *tempoyak* or fermented durian product has a strong odor and is yellow, with a soft texture and sour taste.

The chemical composition of *tempoyak* is as follows: 6.28% carbohydrates, 8.10% protein, 6.65% fat, 1.83% ash, and 74.95% moisture (Pato and Surono 2013). Sugar content is essential as the substrate for fermentation, and the main sugar present in *tempoyak* is glucose, followed by fructose, maltose, and sucrose (Wasnin et al. 2014). Meanwhile, *tempoyak* also contains lactic acid, acetic acid, citric acid, propionic acid, and succinic acid (Chuah et al. 2016; Wasnin et al. 2014). The pH of *tempoyak* depends on the fermentation temperature and time. Therefore, it varies from one production to another since it is an uncontrolled fermentation. *Tempoyak* obtained from a local market in Malaysia was between pH 4.00 and 4.20 (Leisner et al. 2001), while other studies reported that the pH of *tempoyak* ranges from 3.96 to 4.08 (Amin et al. 2004; Wasnin et al. 2014).

Volatile compounds are also produced and usually increase in concentration during fermentation. Neti et al. (2011) found that the spontaneous fermentation of durian pulp still contained volatile sulfur constituents and lacked a fruity aroma after eight days of storage at 30°C. They reported the sulfur compounds (diethyl trisulfide, N dimethylthioisophynil-3-amino, and dipropyl trisulfide) decreased significantly, with no substantial change in ester compounds (ethyl 2-methyl butanoate and ethyl propanoate) at the end of fermentation.

Spontaneous lactic acid fermentation is a technique used in the production of *tempoyak*. This technique is highly dependent on the natural microorganisms present in the raw material and the processing environment itself. Several researchers have also reported that the native microflora found on fruit skins and utensils used during the fermentation process and other environmental resources can grow extensively (7 to 9 $\log_{10}$ CFU/g) and are responsible for fermentation (Nyanga et al. 2007; Fleet 2003).

The dominant lactic acid bacteria in *tempoyak* varies according to the region (Table 10.3). In the central part of Malaysia (Kuala Lumpur and Selangor), the dominant lactic acid bacteria are *Lactiplantibacillus plantarum* and *Lacticaseibacillus casei* (Khalil et al. 2018; Leisner et al. 2001, 2002; Adnan and Tan 2007), whereas *Fructobacillus durionis* and *Lactiplantibacillus plantarum* (Chuah et al. 2016) are predominant in the north and *Lentilactobacillus buchneri* and

TABLE 10.3 Microflora Diversity in *Tempoyak* from Various States in Malaysia

STATE	*LACTIC ACID BACTERIA*	*IDENTIFICATION METHODS*	*REFERENCES*
Selangor (commercial *tempoyak*)	*Lactiplantibacillus plantarum* (dominant), *Levilactobacillus brevis*, *Leuconostoc mesenteroides*, *Liquoriactobacillus mali*, *Limosiactobacillus fermentum* and an unidentified *Lactobacillus* sp.	SDS-PAGE and API 50 CHL	Leisner et al. (2001, 2002)
Kuala Lumpur (commercial *tempoyak*)	*Lacticaseibacillus casei*	API 50 CHL	Adnan and Tan (2007)
Penang	*Fructobacillus durionis*, *Lactiplantibacillus plantarum* (dominant), *Fructilactobacillus fructivorans*, *Leuconostoc dextranicum*, *Lactobacillus collinoides* and *Lacticaseibacillus paracasei*	16S rRNA gene sequencing	Chuah et al. (2016)
Selangor (commercial *tempoyak*)	*Lactiplantibacillus plantarum*, *Limosilactobacillus fermentum*, *Lactobacillus crispatus*, *Limolactobacillus reuteri, and Lactobacillus pentosus*	API 50 CHL and 16S rRNA gene sequences	Khalil et al. (2018)
Terengganu (commercial *tempoyak*)	*Lentilactobacillus buchneri*, *Lactiplantibacillus plantarum*, *Levilactobacillus brevis* and *Lactobacillus acidophilus.*	API 50 CHL	Salleh et al. (2021)
Terengganu (commercial *tempoyak*)	*Lacticaseibacillus paracasei* DSM 2649, *Lentilactobacillus buchneri* DSM 20057T, *Lactobacillus parabuchneri* DSM 57069, *Lacticaseibacillus paracasei* DSM 20020 and *Lactobacillus farcimini* CIP 103136T.	MALDI-TOF/MS	

Lacticaseibacillus paracasei (Salleh et al. 2021) in the east coast region. The microflora diversity present in durian may be due to different geographical locations, fruit cultivars, and developmental stages.

During *tempoyak* fermentation, the dominant bacteria changed from early fermentation to the end of the fermentation. In the initial fermentation stage, *Leuconostoc mesenteroides* and *Fructobacillus durionis* (heterofermentative) are dominant. As fermentation proceeded, *F. durionis* remained dominant. However, towards the end of fermentation, *L. plantarum* (homofermentative) became the dominant species (Chuah et al. 2016). Hence, *tempoyak* fermentation involved both homofermentative and heterofermentative lactic acid bacteria.

Another distinctive characteristic of lactic acid bacteria isolated from *tempoyak* is fructophilic characteristics. *F. durionis* and *L. brevis* are reported for having fructophilic characteristics. This characteristic refers to a specific group of lactic acid bacteria that prefers fructose and sucrose as carbohydrate substrates. It has usually been isolated from fructose-rich environments such as fruits, flowers, and honey (Neveling et al. 2012; Endo et al. 2009).

Some lactic acid bacteria isolated from *tempoyak* also have value as probiotics: *L. plantarum, L. fermentum, L. crispatus, L. reuteri,* and *L. pentosus*. They exhibited high survivability to acid (pH 3.0) and bile salt conditions. The strains can also produce exopolysaccharides, auto-aggregate, and have cell surface hydrophobicity. In addition, they showed good inhibitory activity against the tested pathogens with high antioxidant activity and a good ability to reduce cholesterol (Khalil et al. 2018; Ahmad et al. 2018). The probiotic properties increase the value of *tempoyak* as a fermented condiment with a beneficial effect on health.

10.3.2.5 Fruit Vinegar

Vinegar has been used as a preservative, condiment, and healthy drink with or without the presence of probiotics, the health-relevant beneficial bacteria. It is a common fact that vinegar has tremendous nutritional benefits that can improve human health status, such as anti-visceral obesity properties, improved digestive system function, appetite stimulation, antioxidant activities, reduced cholesterol level, and regulation of blood pressure, as reviewed in many studies (Ho et al. 2017). Production of vinegar from fruits is a traditional practice since ancient times, and today, fruit vinegar is widely available commercially.

Malaysia is a country that has a tropical climate with warm temperature and rainfall throughout the year, which creates luxuriant plant life and allows abundant growth of a wide variety of tropical fruits. Seasonal tropical fruits that are highly cultivated are rambutan, mangosteen, durian, pulasan, langsat, and duku, while non-seasonal tropical fruits are those such as pineapple, banana, papaya and guava (Department of Agriculture Malaysia 2019). According to the Malaysian Agricultural Research and Development Institute (MARDI), the production of tropical fruits in Malaysia is increasing every year due to the increase in land area and productivity, which highly contributes to the agricultural gross domestic product

(GDP) (Abu Dardak 2019). This gives rise to the transformation and production of these agricultural fruits into various value-added products with a longer shelf life, such as wine and vinegar.

As shown in Figure 10.4, fruit that contains sugar undergoes the first fermentation process—alcoholic fermentation, in which the sugar is converted into ethanol by the actions of a starter culture, such as yeast, to yield fruit wine under anaerobic conditions (Ho et al. 2020) (as described in Section 10.3.2.2). Then the solution is further treated with a second fermentation—acetate fermentation, to produce fruit vinegar (Ho et al. 2021), in which the ethanol from the fruit wine produced is oxidized to acetic acid by the action of the acetic acid bacteria under aerobic conditions (Ho et al. 2017). Common starter cultures used in vinegar production include *Acetobacter aceti*, *Gluconacetobacter entanii*, *Gluconobacter oxygens*, and *Komagataeibacter europaeus* (Gomes et al. 2018). Production of vinegar from the four main genera *Acetobacter*, *Gluconacetobacter*, *Gluconobacter*, and *Komagataeibacter* is favorable, as these bacteria have high activity to convert ethanol into acetic acid and high acid tolerance—they survive under relatively high acetic acid concentrations released into the fermentative medium (Gomes et al. 2018). Production of vinegar from fruit uses the sugar naturally present in the fruit, which generally produces vinegar with good organoleptic properties (Gokirmakli et al. 2019).

Conventionally, fruit vinegars are usually imported into Malaysia, mainly from Europe, China, and the United States (Abdul Latiff et al. 2018). However, in recent years, adoption of online platforms, social media sales, and marketing approaches by traders and the general public offer greater exposure and convenience of shopping and new product penetration into the market. Consumers can now produce fruit vinegars easily at home; therefore fruit vinegar made from tropical fruits is no longer a niche market. There are some established brands in the market such as Eu Yan Sang, Signature Market, and VinegPlus, while imported vinegars include Kraft Heinz and Bragg. Common fruit vinegars that are widely consumed are made from apple, grape, strawberry, raspberry, blueberry, blackcurrant, and cranberry.

Meanwhile, in Southeast Asia, the tropical fruits utilized to produce vinegar vary depending on the "signature" fruit harvested from the countries. For instance, banana vinegar comes from Thailand (Boonsupa et al. 2019), sugarcane/cane vinegar from the Philippines (Beltran-Milan 2016), and persimmon vinegar from China (Zou et al. 2018), while plum and cherry vinegars are abundantly available in Japan (Matsuura et al. 2008). In Malaysia, the production of coconut vinegar and palm vinegar has a long history in which the vinegar produced has low acidity, musty flavor, and a unique aftertaste (Mokhtar et al. 2016). Due to the abundance of the highly perishable tropical fruits, attempts have been made to produce vinegars from fruits such as papaya (Kong et al. 2018), soursop (Chin et al. 2017), pineapple (Mohamad et al. 2015), rambutan, and *dokong* (Mokhtar et al. 2016), leading to the increase and expansion of the tropical fruit vinegar market.

10.4 EMERGING FERMENTED FOODS IN SOUTHEAST ASIA

In this ever-changing world, communities are actively embracing the rapid Internet development and digitalization trends, which bring conveniences and a wellspring of information. Consumers are looking beyond "prebiotic" and "probiotic" foods, in which the health benefits of these "superfoods" have been described and highlighted. The ever-increasing consumer demand as well as the need for product variation has brought new fermented products to the market. Over the years, more exotic fermented products are seen on shelves and online portals. Among these are tropical fruit wines and vinegars. One that is prominent is durian-based wines, in which the complex flavors of durian, with more than 200 volatile compounds and abundant esters, give a peculiar taste and aroma profile to the beverage. An interesting study conducted by Lu et al. (2015) revealed that alcoholic fermentation of durian using *Torulaspora delbrueckii* Biodiva and *Saccharomyces bayanus* EC-1118 produced durian wine with high acetyl esters and alcohols, specifically isoamyl alcohol and 2-phenylethyl alcohol—which impart fruity notes, whereas the sulfur-containing volatiles that contribute the onion-like odor were metabolized or declined to trace levels.

Other than durian, there are various fermented beverages being developed from tropical fruits. Among them are soursop wine (Ho et al. 2020) and vinegar (Ho et al. 2017, 2021), fermented papaya beverage (Kong et al. 2018), pineapple vinegar (Mohamad et al. 2015), coconut water vinegar (Othaman et al. 2014), rambutan vinegar and *dokong* vinegar (Mokhtar et al. 2016), banana vinegar (Boonsupa et al. 2019), and many others. One of the reasons that these fermented beverages from tropical fruits are produced is the fact that these tropical fruits are highly perishable, with a short shelf life.

Therefore, these perishable and short shelf-life fruits that are seasonal, resulting in oversupply and underuse during the peak season, can now be converted to fermented products. In order to prevent postharvest losses and reduce wastage, the approach taken is to convert them into wine, vinegars, or fermented beverages to prolong their shelf life (Lim et al. 2021). It is also fascinating to find that upon fermentation, bioactivities such as total phenolic content and antioxidant activities are found to increase compared to those of the raw tropical fruit itself. This shows that the secondary metabolites generated from the microbes are beneficial to human health. Qualitative studies conducted by Abdul Latiff et al. (2018) and Mokhtar et al. (2016) indicated that consumers preferred natural vinegar over synthetic vinegar and accepted vinegar made from tropical fruits as long as the quality, safety, and nutritional content of the products are well specified. A variety of wines and vinegars from different local raw ingredients (substrates) has emerged, and they have been marketed as niche and premium products.

Currently, the research direction is progressing toward sustainable utilization and management of food waste by reduction and utilization of waste material, such as fruit peels, in which attention has been given to address several elements of the Sustainable Development Goals (SDGs). Sustainable practices are used to recover valuable compounds from the waste material, convert it into value-added tropical food products, and provide nutrition to the public. Tropical fruit is a high-value agricultural commodity that contributes to the national economy, and the usage of waste material will generate interest from investors to promote sustainable economic growth. Fruit peels, which are usually discarded during the processing or consumption of the fruit, contain valuable bioactive compounds and natural pigments which can be utilized by fermentative microorganisms to metabolize the organic substances present in fruits. The production of vinegars had explored fruit peels that are underutilized such as banana peel (Fatima and Mishra 2015), pineapple peel (Raji et al. 2012), persimmon peel (Kim et al. 2003) and many others. The outcome is promising as it can be a value added organic products.

Kombucha is a fermented beverage produced using sweetened tea as the substrate and fermented by a symbiotic culture of bacteria and yeasts (SCOBY), producing an acidic yet fragrant taste, with a slight tangy sensation. The primary metabolic pathway involves the hydrolysis of sugar (disaccharides) by yeast invertase, converting into its monosaccharides, and subsequently converted into ethanol and organic acids (Jayabalan et al. 2014). During the fermentation, the SCOBY will produce a layer of pellicle of cellulose that floats on the beverage, which is usually used as starter culture for subsequent fermentation (Amarasekara et al. 2020). Kombucha is typically composed of ethanol and organic acids (gluconic acids, glucuronic acids, lactic acids, and acetic acids as the dominant acid), proteins, polyphenols, minerals (Cu, Fe, Mn, Ni, and Zn) and vitamins (B1, B2, B6, B12, and vitamin C) (Mousavi et al. 2020). The alcohol content of kombucha must be maintained at less than 0.5% alcohol by volume (ABV) to be commercially sold as an alcohol-free drink (Chan et al. 2021). Kombucha has been on the market for many years now. However, it was mostly produced by homebrewers in the past, where it is only consumed within the household or being sold to the local community. Quite recently, local kombucha manufacturers and bottlers have emerged in Malaysia and Southeast Asian countries, producing interesting kombucha varieties for the market. Various local ingredients ae used to spice up the taste and aroma of kombucha, such as passionfruit, blue pea flower, pomelo, *kedondong*, and many others.

Aside from research on using novel substrates for fermentation, some studies were performed using novel cultures for the fermentation. One of the interesting studies performed in Malaysia was using mushroom mycelia (*Pleurotus pulmonarius* and *Volvariella volvacea*) in combination with yeast (*S. cerevisiae*) for the alcoholic fermentation process (Mat Isham et al. 2019; Ho et al. 2020). The mushrooms were found to contain alcohol dehydrogenase (ADH) enzyme, which could convert sugars into ethanol. Aside from that, alcoholic fermentation using yeast is always limited by the inhibition effect of the ethanol towards the yeast itself, usually at concentrations of approximately 15% ethanol. Mushroom, on the other hand, being a more complex organism, has higher tolerance towards ethanol and thus was able to produce higher ethanol content (Ho et al. 2020). Mushrooms were also reported to contain amylase enzyme, which will be able to break down complex starches into fermentable sugars, thus giving a higher fermentation yield. However, a study by Mat Isham et al. (2019) found that yeast has a much shorter lag phase and could adapt to the fermentation environment in a shorter period (within 24 hours), compared to that of mushroom mycelia, which needed up to 120 hours to produce a stable amount of ethanol. Therefore, the combination of mushroom mycelia and yeast was recommended to produce the maximum amount of ethanol in a shorter time (Ho et al. 2020).

With the advancement of technology and knowledge, more fermented products are emerging and being introduced to the market. Traditionally fermented products are now produced at the industrial scale, with better control of parameters and thus higher safety and quality consistency. These products are available in retail malls, which makes it more convenient for consumers to obtain such products. The closer links between academia and industry also encourage more research outputs and innovative product development from universities and research institutions being translated into products on the market through industry collaboration.

10.5 REGULATION AND CONCERNS

Fermented foods are like other food products that need to be regulated under specific food laws and regulations. Food regulation is crucial in food production to ensure quality and safety and protect against adulterated or misbranded food. In Malaysia, Food Regulation 1985 is used in all food production, including fermented foods. The fermented foods listed in the regulation are cheese and cheese products, cultured milk and fermented milk, fish paste, *belacan*, fish sauce, *cincalok*, *pekasam*, fermented soybean product, vinegar, and fruit wine (Malaysian Food Regulations 1985).

From the fermented products listed, further elaboration on the regulation emphasizes local fermented products, as compiled in Table 10.4. This regulation is a requirement for the fermented food producer to comply regarding the quality and safety aspects of the product. This compliance is mainly related to the product composition and minimizing any chemical or microbiological contamination issues.

TABLE 10.4 Standard of Malaysian Fermented Products Based on Food Regulation 1985

REGULATION NO.	*FERMENTED PRODUCT*	*PRODUCT DESCRIPTION*
162	Fish paste	(1) Fish paste shall be the fish product in the form of paste obtained by salt fermentation of fish but shall not include belacan. (2) Fish paste - *(a)* shall contain not less than - (i) 15 per cent of salt; and (ii) 30 per cent of protein; *(b)* shall not contain more than - (i) 40 per cent of water; and (ii) 25 per cent of ash; and *(c)* shall be clean and wholesome and shall not contain other extraneous matter. (3) Fish paste may contain permitted preservative, permitted coloring substance and permitted flavor enhancer.
163	*Belacan*	(1) *Belacan* shall be the fish product in the form of paste obtained by salt fermentation of fresh shrimp or prawn or both. (2) *Belacan* - *(a)* shall contain not less than - (i) 15 per cent of salt; and (ii) 30 per cent of protein; *(b)* shall not contain more than- (i) 40 per cent of water; and (ii) 30 per cent of ash; and *(c)* shall be clean and wholesome and shall not contain any extraneous matter. (3) *Belacan* may contain permitted preservative, permitted coloring substance and permitted flavor enhancer.
164	Fish sauce	(1) Fish sauce shall be the fish product in the form of liquid prepared from fresh fish, other than shell-fish, with salt fermentation and includes *budu*. (a) For the purpose of these Regulations, *budu* shall be the fish product obtained by salt fermentation of anchovies of *Stolephorus* species or mixture of anchovies with other small fish. (2) Fish sauce - *(a)* shall contain not less than - (i) 15 per cent of salt; and (ii) 5 per cent of protein; *(b)* may contain other food; and (c) shall be clean and wholesome and shall not contain other extraneous matter. (3) Fish sauce may contain permitted preservative, caramel as a coloring substance and permitted flavor enhancer.
166	*Cincalok*	(1) Cincalok shall be the fish product obtained by salt fermentation of fresh shrimp of *Acetes* species with the addition of rice or other fermentable carbohydrates. (2) Cincalok - *(a)* shall contain not less than- (i) 10 per cent of protein; and (ii) 10 per cent of salt; *(b)* shall contain not more than 15 per cent of ash; (c) may contain other food; and (d) shall be clean and wholesome and shall not contain other extraneous matter. (3) Cincalok may contain permitted preservative, permitted coloring substance and permitted flavor enhancer.
170	*Pekasam*	(1) *Pekasam* shall be the fish product obtained by the fermentation of fish with the addition of carbohydrates. It may contain condiments and ingredients necessary to achieve the desired flavor and shall not less than 10 per cent of salt. (2) *Pekasam* may contain permitted preservative. permitted coloring substance and permitted flavor enhancer.

(Continued)

TABLE 10.4 (Continued)

REGULATION NO.	FERMENTED PRODUCT	PRODUCT DESCRIPTION
221	Fermented soya bean product	(1) Fermented soya bean product shall be the soya bean, whole or ground, that have been fermented with harmless microorganism. It may contain other wholesome food. The product shall not be a sauce and includes *tauju* or *sufu*, and *tauceong* or *taucu*. (2) For the purposes of these Regulations- *(a)* *tauju* or *sufu* shall be the product obtained by fermenting soya bean curd with cultures of *Actinomucor elegans* or other harmless organism; and *(b)* *tauceong* or *taucu* shall be the product, whole or ground, obtained by fermenting soya beans with *Aspergillus* species followed by brine fermentation. (3) Fermented soya bean product may contain permitted preservative, permitted coloring substance and permitted flavor enhancer.
334	Vinegar	(1) Vinegar shall be a liquid product prepared from the alcoholic fermentation and subsequent acetous fermentation of any suitable food. (2) Vinegar- *(a)* shall contain not less than 4 per cent w/v of acetic acid; and *(b)* shall not contain any mineral acid. (3) Vinegar may contain permitted preservative, caramel as a coloring substance and spices as permitted flavoring substance.
367	Fruit wine	(1) Fruit wine shall be the product of the partial or complete alcoholic fermentation of the juice, or of the juice and other portions of any fruit other than grapes and includes the food for which a standard is prescribed in regulation 368 to 371. It shall not contain more than 15 per cent v/v of alcohol. (2) Fruit wine-- *(a)* may contain-- (i) urea and yeast; (ii) fructose, glucose, glucose syrup, sugar; (iii) carbon dioxide, nitrogen, oxygen; and (iv) volatile acidity calculated as acetic acid, not including preservative acids, in a proportion not exceeding 1.2 g per liter; and *(b)* shall not contain potable water. (3) Fruit wine may contain permitted preservative and permitted food conditioner, including polyvinylpyrrolidone in a proportion not exceeding 60 mg per liter. (4) No package of fruit wine shall be labelled with the word "wine" unless the name of the fruit from which the wine is made is conjoined in uniform lettering of not less than 10 point with the word "wine".

Malaysia Food Regulation 1985 emphasizes not only food composition but also microbial cultures for fermented products. Even though most of the Malaysian local fermented foods are produced by natural or spontaneous fermentation, it is crucial to have proper regulations or guidelines to assess the safety of a food product added with cultures (Malaysian Food Regulations 1985).

Regulation 26A in Malaysia Food Regulation 1985 on microbial cultures for food fermentation explicitly applies to probiotic cultures. This regulation states a probiotic culture shall comply with the following criteria: has been evaluated for its safety for humans and beneficial to health as demonstrated in human studies, resistant to gastric acidity, resistant to bile acid, adheres to mucus and/or human epithelial cells and cell lines, able to hydrolyze bile salts, has a long history of consumption, has a clear taxonomy identification (genus, species, strain), has non-pathogenic strains, has non-toxigenic strains, and is free of antibiotic-resistant genes (Malaysian Food Regulations 1985).

The European Food Safety Agency (EFSA) has a guideline stated as Food Cultures in the European Union countries in which the named cultures must have specific criteria such as no presence of virulence factors, toxic metabolites, transmissible antibiotic resistance, or resistance to antimicrobials (Laulund et al. 2017). Similarly, in the United States, generally recognized as safe (GRAS) criteria are applied for cultures used in food products. All these regulations and guidelines certify that the cultures or probiotics used in fermented foods have proven beneficial health effects and no safety issues for the consumer.

10.6 CONCLUSION AND FUTURE DIRECTIONS

There are many unique fermented foods and beverages in Southeast Asia due to the available distinctive food substrates that are found in this region, ranging from seafood to tropical fruits. The fermentation methods and characterizations of the traditional Southeast Asian fermented food products discussed in this chapter have been well researched; nevertheless, more *in vitro* and *in vivo* research could be carried out on these fermented products in terms of their health benefits for consumers. Furthermore, nutritive ingredients or mixed cultures could be incorporated into traditional fermented products to increase their value. Innovation can also be carried out on traditional fermented products to improve their application and accessibility. As many traditional fermented products in Southeast Asia are produced at home and sold locally, the safety aspect of homemade fermented products remains a concern. Hence, it is also crucial to regulate and provide guidelines for home-based fermenters for better food safety.

REFERENCES

Abdul Aris, M. H., H. Y. Lee, N. Hussain, H. Ghazali, W. N. Nordin, and N. A. Mahyudin. 2015. Effect of Vietnamese coriander (*Persicaria odorata*), turmeric (*Curcuma longa*) and asam gelugor (*Garcinia atroviridis*) leaf on the microbiological quality of gulai tempoyak paste. *International Food Research Journal* 22, no. 4: 1657–1661.

Abdul Latiff, Z. A., S. I. Mokhtar, S. W. Soon, and M. A. Ayob. 2018. Study on the customer perception of the tropical fruit vinegar in Kelantan and Kuala Lumpur. *Journal of Food Products Marketing* 24, no. 3: 263–279.

Abesinghe, A. M. N. L., H. Priyashantha, P. H. P. Prasanna, M. S. Kurukulasuriya, C. S. Ranadheera, and J. K. Vidanarachchi. 2020. Inclusion of probiotics into fermented buffalo (Bubalus bubalis) milk: An overview of challenges and opportunities. *Fermentation* 6, no. 4: 121.

Abu Dardak, R. 2019. *Trends in Production, Trade and Consumption of Tropical Fruit in Malaysia.* Taipei: FFTC Agricultural Policy Platform.

Abukakar, M. G., A. N. Ukwuani, and R. A. Shehu. 2008. Phytochemical screening and antibacterial activity of *Tamarindus indica* pulp extract. *Asian Journal of Biochemistry* 3, no. 2: 134–138.

Adnan, A. F. M., and I. K. Tan. 2007. Isolation of lactic acid bacteria from Malaysian foods and assessment of the isolates for industrial potential. *Bioresource Technology* 98, no. 7: 1380–1385.

Ahmad, A., W. B. Yap, N. T. Kofli, and A. R. Ghazali. 2018. Probiotic potentials of *Lactobacillus plantarum* isolated from fermented durian (tempoyak), a Malaysian traditional condiment. *Food Science and Nutrition* 6, no. 6: 1370–1377.

Ahmed, J., H. S. Ramaswamy, and K. C. Sashidhar. 2007. Rheological characteristics of tamarind (*Tamarindus indica L.*) juice concentrates. *LWT—Food Science and Technology* 40, no. 2: 225–231.

Ahnan-Winarno, A. D., L. Cordeiro, F. G. Winarno, J. Gibbons, and H. Xiao. 2021. Tempeh: A semicentennial review on its health benefits, fermentation, safety, processing, sustainability, and affordability. *Comprehensive Reviews in Food Science and Food Safety* 20, no. 2: 1717–1767.

Amarasekara, A. S., D. Wang, and T. L. Grady. 2020. A comparison of kombucha SCOBY bacterial cellulose purification methods. *SN Applied Sciences* 2, no. 2: 1–7.

Amin, A., Z. Jaafar, and L. K. Ng. 2004. Effect of salt on tempoyak fermentation and sensory evaluation. *Journal of Biological Sciences* 4, no. 5: 650–653.

Anal, A. K. 2019. Quality ingredients and safety concerns for traditional fermented foods and beverages from Asia: A review. *Fermentation* 5, no. 1: 2021.

Asghar, M. T., Y. A. Yusof, M. N. Mokhtar, M. E. Ya'acob, H. Mohd. Ghazali, L. S. Chang, and Y. N. Manaf. 2019. Coconut (*Cocos nucifera L.*) sap as a potential source of sugar: Antioxidant and nutritional properties. *Food Science and Nutrition* 35: 1–11.

Astawan, M., J. Hermanianto, Suliantari, and G. S. P. Sugiyanto. 2016. Application of vacuum packaging to extend the shelf life of fresh-seasoned tempe. *International Food Research Journal* 23, no. 6: 2571–2580.

Bavia, A. C. F., C. E. da Silva, M. P. Ferreira, R. S. Leite, J. M. G. Mandarino, and M. C. Carrão-Panizzi. 2012. Chemical composition of tempeh from soybean cultivars specially developed for human consumption. *Food Science and Technology* 32, no. 3: 613–620.

Behare, P. V., R. Singh, S. K. Tomar, R. Nagpal, M. Kumar, and D. Mohania. 2010. Effect of exopolysaccharide-producing strains of *Streptococcus thermophilus* on technological attributes of fat-free lassi. *Journal of Dairy Science* 93, no. 7: 2874–2879.

Behera, S. S., and R. C. Ray. 2019. Forest bioresources for bioethanol and biodiesel production with emphasis on mohua (*Madhuca latifolia*) flowers and seeds. In *Bioethanol Production from Food Crops*, ed. R. C. Ray and S. Ramachandran, 233–247. London: Academic Press.

Beltran-Milan, L. 2016. Effects of the different types of container on the characteristics of sugarcane vinegar. *International Journal of Education and Research* 4, no. 1: 235–248.

Boonsupa, W., W. Pimda, K. Sreeninta, C. Yodon, S. Nawarat, B. Bou-on, and P. Hemwiphat. 2019. Development of fermented banana vinegar: Chemical characterization and antioxidant activity. *Journal of Food Health and Bioenvironmental Science* 12, no. 1: 21–27.

Chan, M., H. Sy, J. Finley, J. Robertson, and P. N. Brown. 2021. Determination of ethanol content in kombucha using headspace gas chromatography with mass spectrometry detection: Single-laboratory validation. *Journal of AOAC International* 104, no. 1: 122–128.

Chandan, R. C., and K. O'Rell. 2013. Principles of yogurt processing. In *Manufacturing Yogurt and Fermented Milks*, ed. R. C. Chandan and A. Kilara, 239–260. Chichester: John Wiley & Sons.

Chandrasekhar, K., S. Sreevani, P. Seshapani, and J. Pramodhakumari. 2012. A review on palm wine. *International Journal of Research in Biological Sciences* 2, no. 1: 33–38.

Chee, Y. F., S. Kaur, and R. W. K. Yap. 2020. Association between dietary patterns and cardiometabolic risks in Malaysian punjabi adults. *Malaysian Journal of Medicine and Health Sciences* 16: 106–115.

Chen, Y. S., H. C. Wu, S. F. Pan, B. G. Lin, Y. H. Lin, W. C. Tung, Y. L. Li, C. M. Chiang, and F. Yanagida. 2013. Isolation and characterization of lactic acid bacteria from yan-taozih (pickled peaches) in Taiwan. *Annals of Microbiology* 63, no. 2: 607–614.

Cheng, Y. Q., Q. Hu, L. Te Li, M. Saito, and L. J. Yin. 2009. Production of Sufu, a traditional Chinese fermented soybean food, by fermentation with *Mucor flavus* at low temperature. *Food Science and Technology Research* 15, no. 4: 347–352.

Cheng, Y. Q., Y. P. Zhu, Q. Hu, L. Te Li, M. Saito, S. X. Zhang, and L. J. Yin. 2011. Transformation of isoflavones during sufu (a traditional Chinese fermented soybean curd) production by fermentation with *Mucor flavus* at low temperature. *International Journal of Food Properties* 14, no. 3: 629–639.

Chin, W. H., A. Mat Lazim, S. Fazry, U. K. Hj Hussain Zaki, and S. J. Lim. 2017. Effects of fermentation time and pH on soursop

(*Annona muricata*) vinegar production towards its chemical compositions. *Sains Malaysiana* 46, no. 9: 1505–1512.

Chuah, L. O., A. K. Shamila-Syuhada, M. T. Liong, A. Rosma, K. L. Thong, and G. Rusul. 2016. Physio-chemical, microbiological properties of tempoyak and molecular characterisation of lactic acid bacteria isolated from tempoyak. *Food Microbiology* 58: 95–104.

Department of Agriculture Malaysia. 2019. *Fruit Crop Statistic Malaysia 2019*. Putrajaya.

Department of Fisheries. 2013. *Annual Fisheries Statistics*. Putrajaya.

Dimero, F. N., and T. F. Tepora. 2018. Processing and development of dragon fruit wine. *International Journal of Environment, Agriculture and Biotechnology* 3, no. 5: 1943–1947.

Ding, P. S. 2016. Introduction. In *Southern Min (Hokkien) as a Migrating Language*, ed. P. S. Ding, 1–18. Singapore: Springer.

Endo, A., Y. Futagawa-Endo, and L. M. Dicks. 2009. Isolation and characterization of fructophilic lactic acid bacteria from fructose-rich niches. *Systematic and Applied Microbiology* 32, no. 8: 593–600.

Endo, A., T. Irisawa, L. Dicks, and S. Tanasupawat. 2014. Fermented foods: Fermentations of East and Southeast Asia. In *Encyclopedia of Food Microbiology: Second Edition*, ed. R. C. Ray and S. Ramachandran, 1, 846–851. London: Academic Press.

Ezzat, M. A., M. H. Ghazali, K. Roselina, and D. Zare. 2021. Organic acid composition and consumer acceptability of fermented fish. *Food Research* 5: 262–271.

Ezzat, M. A., D. Zare, R. Karim, and H. M. Ghazali. 2015. Trans- and cis-urocanic acid, biogenic amine and amino acid contents in ikan pekasam (fermented fish) produced from Javanese carp (*Puntius gonionotus*) and black tilapia (*Oreochromis mossambicus*). *Food Chemistry* 172: 893–899.

Fan, X., X. Lv, L. Meng, M. Ai, C. Li, F. Teng, and Z. Feng. 2020. Effect of microwave sterilization on maturation time and quality of low-salt sufu. *Food Science and Nutrition* 8, no. 1: 584–593.

Fatima, B., and A. A. Mishra. 2015. Optimization of process parameter for the production of vinegar from banana peel and coconut water. *International Journal of Science, Engineering and Technology* 3, no. 3: 817–823.

Fleet, G. H. 2003. Yeasts in fruit and fruit products. In *Yeasts in Food*, ed. T. Boekhout and V. Robert, 267–287. Cambridge: Woodhead Publishing.

Gandjar, J. 1999. Fermentations of the far east. In *Encyclopedia of Food Microbiology*, ed. R. K. Robinson, C. A. Batt, and P. D. Patel, 767–773. New York: Academic Press.

Gokirmakli, Ç., N. H. Budak, Z. B. Guzel-Seydim, A. C. Seydim, B. E. Filiz, and I. D. Karakulak. 2019. Antioxidant properties of strawberry vinegar. *International Journal of Food Engineering* 5, no. 3: 171–174.

Gomes, R. J., M. de F. Borges, M. de F. Rosa, R. J. H. Castro-Gómez, and W. A. Spinosa. 2018. Acetic acid bacteria in the food industry: Systematics, characteristics and applications. *Food Technology and Biotechnology* 56, no. 2: 139–151.

Gopikrishna, T., H. K. Suresh Kumar, K. Perumal, and E. Elangovan. 2021. Impact of *Bacillus* in fermented soybean foods on human health. *Annals of Microbiology* 71, no. 1: 1–16.

Han, B. Z., F. M. Rombouts, and M. J. R. Nout. 2001. A Chinese fermented soybean food. *International Journal of Food Microbiology* 65, no. 1–2: 1–10.

Handoyo, T., and N. Morita. 2006. Structural and functional properties of fermented soybean (tempeh) by using *Rhizopus oligosporus*. *International Journal of Food Properties* 9, no. 2: 347–355.

Haron, H., S. Shahar, K. O. O'Brien, A. Ismail, N. Kamaruddin, and S. A. Rahman. 2010. Absorption of calcium from milk and tempeh consumed by postmenopausal Malay women using the dual stable isotope technique. *International Journal of Food Sciences and Nutrition* 61, no. 2: 125–137.

Hassan, Z. 1980. Pekasam—A fermented fish product (in Peninsular Malaysia). In *Seminar on Modernization of Malaysian Cottage Food Industries*. Serdang: Universiti Pertanian Malaysia.

Ho, C. W., L. S. Chang, S. K. Syed Muzni, S. Fazry, A. Lazim, U. K. Hj Hussain Zaki, and S. J. Lim. 2021. Functional beverage production using acetous fermentation of soursop: Physicochemical, toxicity and organoleptic properties. *Food Bioscience* 39: 100812.

Ho, C. W., A. M. Lazim, S. Fazry, U. K. H. Hussain Zaki, and S. J. Lim. 2017. Varieties, production, composition and health benefits of vinegars: A review. *Food Chemistry* 221: 1621–1630.

Ho, C. W., A. M. Lazim, S. Fazry, U. K. H. Hussain Zaki, S. Massa, and S. J. Lim. 2020. Alcoholic fermentation of soursop (*Annona muricata*) juice via alternative fermentation technique. *Journal of the Science of Food and Agriculture* 100, no. 3: 1012–1021.

Huda, N. 2012. Malaysian fermented fish products. In *Handbook of Animal-Based Fermented Food and Beverage Technology*, ed. Y. H. Hui and E. Ö. Evranuz, 734–757. Boca Raton: CRC Press.

Ingledew, W. M. M., and Y. H. Lin. 2011. Ethanol from starch-based feedstocks. In *Comprehensive Biotechnology*, ed. M. Moo-Young, 37–49. Oxford: Pergamon.

Ishak, N., M. S. M. Zahari, S. A. Talib, and H. M. Hanafiah. 2019. The influence of biculturalism/integration attributes on ethnic food identity formation. *Journal of Ethnic Foods* 6, no. 1: 1–13.

Ismail, M. S., and A. Zain. 1978. Utilization of tilapia, a trash freshwater fish. In *Proceedings of the Indo-Pacific Fishery Commission*, 303–309. Bangkok: FAO-UN.

Jang, C. H., J. Oh, J. S. Lim, H. J. Kim, and J. S. Kim. 2021. Fermented soy products: Beneficial potential in neurodegenerative diseases. *Foods* 10, no. 3: 1–17.

Jayabalan, R., R. V. Malbaša, E. S. Lončar, J. S. Vitas, and M. Sathishkumar. 2014. A review on kombucha tea-microbiology, composition, fermentation, beneficial effects, toxicity, and tea fungus. *Comprehensive Reviews in Food Science and Food Safety* 13, no. 4: 538–550.

Khalil, E. S., A. Manap, M. Yazid, S. Mustafa, A. M. Alhelli, and P. Shokryazdan. 2018. Probiotic properties of exopolysaccharide-producing *Lactobacillus* strains isolated from tempoyak. *Molecules* 23, no. 2: 398.

Kim, S.-K., G.-D. Lee, and S.-K. Chung. 2003. Monitoring on fermentation of persimmon vinegar from persimmon peel. *Korean Journal of Food Science and Technology* 35, no. 4: 642–647.

Kong, C. T., C. W. Ho, J. W. A. Ling, A. Lazim, S. Fazry, and S. J. Lim. 2018. Chemical changes and optimisation of acetous fermentation time and mother of vinegar concentration in the production of vinegar-like fermented papaya beverage. *Sains Malaysiana* 47, no. 9: 2017–2026.

Kustyawati, M. E., O. Nawansih, and S. Nurdjanah. 2017. Profile of aroma compounds and acceptability of modified tempeh. *International Food Research Journal* 24, no. 2: 734–740.

Laulund, S., A. Wind, P. M. F. Derkx, and V. Zuliani. 2017. Regulatory and safety requirements for food cultures. *Microorganisms* 5, no. 2: 1–14.

Law, S. V., F. Abu Bakar, D. Mat Hashim, and A. Abdul Hamid. 2011. Popular fermented foods and beverages in Southeast Asia. *International Food Research Journal* 18, no. 2: 475–484.

Leisner, J. J., M. Vancanneyt, R. Gulam, B. Pot, K. Lefebvre, A. Fresi, and L. K. Tee. 2001. Identification of lactic acid bacteria constituting the predominating microflora in an acid-fermented condiment (tempoyak) popular in Malaysia. *International Journal of Food Microbiology* 63: 149–157.

Leisner, J. J., M. Vancanneyt, K. Lefebvre, K. Vandemeulebroecke, B. Hoste, N. Euras Vivalta, G. Rusul, and J. Swings. 2002. *Lactobacillus durianis* sp. *nov.*, isolated from an acid-fermented condiment (tempoyak) in Malaysia. *International Journal of Systemic and Evolutionary Microbiology* 52: 927–931.

Li, D., Y. Ma, J. Liang, R. Shi, J. Wang, S. Guo, and X. Li. 2020. Effects of different production technologies (fermented strains and spices) on biogenic amines in sufu fermentation. *Journal of Food Processing and Preservation* 44, no. 8: 1–10.

Li, T., X. Bin Chen, J. C. Chen, Q. Wu, and G. Q. Chen. 2014. Open and continuous fermentation: Products, conditions and bioprocess economy. *Biotechnology Journal* 9, no. 12: 1503–1511.

Lim, S. J., L. S. Chang, S. Fazry, W. A. Wan Mustapha, and A. S. Babji. 2021. Functional food & ingredients from seaweed, edible bird's nest and tropical fruits: A translational research. *LWT—Food Science and Technology* 151: 112164.

Liu, D. M., L. Li, X. Q. Yang, S. Z. Liang, and J. S. Wang. 2006. Survivability of *Lactobacillus rhamnosus* during the preparation of soy cheese. *Food Technology and Biotechnology* 44, no. 3: 417–422.

Lockard, C. A. 2013. Chinese migration and settlement in southeast asia before 1850: Making fields from the sea. *History Compass* 11, no. 9: 765–781.

Lu, Y., D. Huang, P. R. Lee, and S. Q. Liu. 2015. Effects of cofermentation and sequential inoculation of *Saccharomyces bayanus* and *Torulaspora delbruckii* on durian wine composition. *International Journal of Food Science and Technology* 50, no. 12: 2653–2663.

Luzón-Quintana, L. M., R. Castro, and E. Durán-Guerrero. 2021. Biotechnological processes in fruit vinegar production. *Foods* 10, no. 5: 945.

Lynch, K. M., E. Zannini, S. Wilkinson, L. Daenen, and E. K. Arendt. 2019. Physiology of acetic acid bacteria and their role in vinegar and fermented beverages. *Comprehensive Reviews in Food Science and Food Safety* 18, no. 3: 587–625.

Ma, Y. L., J. H. Wang, Y. Q. Cheng, L. J. Yin, and L. Te Li. 2013. Some biochemical and physical changes during manufacturing of grey sufu, a traditional Chinese fermented soybean curd. *International Journal of Food Engineering* 9, no. 1: 45–54.

Malaysian Food Regulations. 1985. Food Regulations 1985. *Food Act 1983*. Putrajaya: Ministry of Health.

Malo, P. M., and E. A. Urquhart. 2015. Fermented foods: Use of starter cultures. In *Encyclopedia of Food and Health*, ed. B. Caballero, P. M. Finglas, and F. Toldra, 681–685. Oxford: Academic Press.

Mansor, A., A. A. Kahar, A. Husin, S. A. Rahman, D. L. A. Razak, S. A. Sharifudin, M. A. Othaman, N. Y. A. Rashid, T. G. Hun, and K. Long. 2014. Distribution and characterization of indigenous microbes from Malaysian fermented fish products. In *International Conference of Beneficial Microbes (ICOBM 2014)*, 228–231. Penang: Universiti Sains Malaysia.

Mat Isham, N. K., N. Mokhtar, S. Fazry, and S. J. Lim. 2019. The development of an alternative fermentation model system for vinegar production. *LWT—Food Science and Technology* 100: 322–327.

Matsuura, R., H. Moriyama, N. Takeda, K. Yamamoto, Y. Morita, T. Shimamura, and H. Ukeda. 2008. Determination of antioxidant activity and characterization of antioxidant phenolics in the plum vinegar extract of cherry blossom (*Prunus iannesiana*). *Journal of Agricultural and Food Chemistry* 56, no. 2: 544–549.

Merican, Z. 1977. Malaysian tempoyak. In *Handbook of Indigenous Fermented Foods*, ed. K. H. Steinkraus, 148. New York: Marcel Dekker.

Mohamad, N. E., S. K. Yeap, K. L. Lim, H. M. Yusof, B. K. Beh, S. W. Tan, W. Y. Ho, S. A. Sharifuddin, A. Jamaluddin, K. Long, N. M. A. Nik Abd Rahman, and N. B. Alitheen. 2015. Antioxidant effects of pineapple vinegar in reversing of paracetamol-induced liver damage in mice. *Chinese Medicine (United Kingdom)* 10, no. 1: 1–10.

Mokhtar, S. I., F. Zakaria, S. W. Soon, Z. Ariff, A. Latiff, and J. Campus. 2016. Study on the nutritional values and customer acceptance of *Lansium domesticum* & *Nephelium lappaceum* newly fermented natural fruit vinegars in Malaysia. *Asia Pacific Conference on Advanced Research* 2, no. 2: 63–74.

Moreno, M. R. F., J. J. Leisner, L. K. Tee, C. Ley, S. Radu, G. Rusul, M. Vancanneyt, and L. De Vuyst. 2002. Microbial analysis of Malaysian tempeh, and characterization of two bacteriocins produced by isolates of *Enterococcus faecium*. *Journal of Applied Microbiology* 92, no. 1: 147–157.

Mousavi, S. M., S. A. Hashemi, M. Zarei, A. Gholami, C. W. Lai, W. H. Chiang, N. Omidifar, S. Bahrani, and S. Mazraedoost. 2020. Recent progress in chemical composition, production, and pharmaceutical effects of kombucha beverage: A complementary and alternative medicine. *Evidence-Based Complementary and Alternative Medicine* no. 4397543: 1–14.

Muryany, I. M., I. M. Salwany, A. R. Ghazali, H. L. Hing, and N. R. Fadilah. 2017. Identification and characterization of the lactic acid bacteria isolated from Malaysian fermented fish (pekasam). *International Food Research Journal* 24, no. 2: 868.

Mustarichie, R., S. Megantara, W. Indriyati, and A. Zuchrotun. 2017. Antioxidant activity of tauco ethanol extract and its fractions. *Asian Journal of Pharmaceutical and Clinical Research* 10: 57–61.

Nandiyanto, A. B. D., R. Ismiati, J. Indrianti, and A. G. Abdullah. 2018. Economic perspective in the production of preserved soybean (tauco) with various raw material quantities. *IOP Conference Series: Materials Science and Engineering* 288: 1–6.

Neti, Y., I. D. Erlinda, and V. G. Virgilio. 2011. The effect of spontaneous fermentation on the volatile flavor constituents of durian. *International Food Research Journal* 18: 625–631.

Neveling, D. P., A. Endo, and L. M. Dicks. 2012. Fructophilic *Lactobacillus kunkeei* and *Lactobacillus brevis* isolated from fresh flowers, bees and bee-hives. *Current Microbiology* 65: 507–515.

Nuraida, L. 2015. A review: Health promoting lactic acid bacteria in traditional Indonesian fermented foods. *Food Science and Human Wellness* 4, no. 2: 47–55.

Nyanga, L. K., M. J. R. Nout, T. H. Gadaga, B. Theelen, T. Boekhout, and M. H. Zwietering. 2007. Yeasts and lactic acid bacteria microbiota from masau (*Ziziphus mauritiana*) fruits and their fermented fruit pulp in Zimbabwe. *International Journal of Food Microbiology* 120, no. 1–2: 159–166.

Othaman, M. A., S. A. Sharifudin, A. Mansor, A. A. Kahar, and K. Long. 2014. Coconut water vinegar: New alternative with improved processing technique. *Journal of Engineering Science and Technology* 9, no. 3: 293–302.

Padghan, P., M. Bimlesh, and S. Rajan. 2018. In-vivo studies of antioxidant activity of fermented milk (Lassi) by *Lactobacillus acidophilus* and standard dahi culture. *Journal of Pharmacognosy and Phytochemistry* 7, no. 2: 25–30.

Pakwan, C., T. Chitov, P. Chantawannakul, M. Manasam, S. Bovonsombut, and T. Disayathanoowat. 2020. Bacterial compositions of indigenous Lanna (Northern Thai) fermented foods and their potential functional properties. *PLoS One* 15, no. 11: 1–12.

Panesar, P. S. 2011. Fermented dairy products: Starter cultures and potential nutritional benefits. *Food and Nutrition Sciences* 2: 47–51.

Papalexandratou, Z., G. Vrancken, K. De Bruyne, P. Vandamme, and L. De Vuyst. 2011. Spontaneous organic cocoa bean box fermentations in Brazil are characterized by a restricted species diversity of lactic acid bacteria and acetic acid bacteria. *Food Microbiology* 28, no. 7: 1326–1338.

Patel, P., V. Patel, and R. Subhash. 2015. Probiotic lassi preparation and sensory evaluation using *L. paracasei* immobilized dry fruit pieces. *International Journal of Fermented Foods* 4, no. 1–2: 23–33.

Pato, U. 2003. Bile and acid tolerance of lactic acid bacteria isolated from dadih and their antimutagenicity against mutagenic heated tauco. *Asian-Australasian Journal of Animal Sciences* 16, no. 11: 1680–1685.

Pato, U., and I. S. Surono. 2013. Bile and acid tolerance of lactic acid bacteria isolated from tempoyak and their probiotic potential. *International Journal of Agricultural Technology* 9, no. 7: 1849–1862.

Paull, R. E., and S. Ketsa. 2014. Durian: Postharvest quality-maintenance guidelines. *Fruit, Nut, and Beverage Crops* 1–4.

Perry, M. S. 2017. Feasting on culture and identity: Food functions in a multicultural and transcultural Malaysia. *3L: Language, Linguistics, Literature* 23, no. 4: 184–199.

Pisol, B., N. Abdullah, K. A. Khalil, and L. Nuraida. 2015. Isolation and identification of lactic acid bacteria from different stages of traditional Malaysian tempeh production. *Malaysian Journal of Microbiology* 11, no. 4: 358–364.

Radita, R., A. Suwanto, N. Kurosawa, A. T. Wahyudi, and I. Rusmana. 2017. Metagenome analysis of tempeh production: Where did the bacterial community in tempeh come from? *Malaysian Journal of Microbiology* 13, no. 4: 280–288.

Raji, M. N. A., S. Ab Karim, F. A. C. Ishak, and M. M. Arshad. 2017. Past and present practices of the Malay food heritage and culture in Malaysia. *Journal of Ethnic Foods* 4, no. 4: 221–231.

Raji, Y. O., M. Jibril, I. M. Misau, and B. Y. Danjuma. 2012. Production of vinegar from pineapple peel. *International Journal of Scientific Research in Science and Technology* 3, no. 2: 656–666.

Ranadheera, C. S., J. K. Vidanarachchi, R. S. Rocha, A. G. Cruz, and S. Ajlouni. 2017. Probiotic delivery through fermentation: Dairy vs. non-dairy beverages. *Fermentation* 3, no. 4: 1–17.

Rasane, P., R. Kailey, and S. K. Singh. 2017. Fermented indigenous Indian dairy products: Standards, nutrition, technological significance and opportunities for its processing. *Journal of Pure and Applied Microbiology* 11, no. 2: 1199–1213.

Ravindra, M. R., K. J. Rao, B. S. Nath, and C. Ram. 2014. Carbonated fermented dairy drink—effect on quality and shelf life. *Journal of Food Science and Technology* 51, no. 11: 3397–3403.

Ray, M., K. Ghosh, S. Singh, and K. Mondal. 2016. Folk to functional: An explorative overview of rice-based fermented foods and beverages in India. *Journal of Ethnic Foods* 3, no. 1: 5–18.

Ray, R. C., and V. K. Joshi. 2014. Fermented foods: Past, present and future. In *Microorganisms and Fermentation of Traditional Foods*, ed. R. C. Ray and V. K. Joshi, 1–36. Boca Raton: CRC Press.

Riebroy, S., S. Benjakul, W. Visessanguan, K. Kijrongrojana, and M. Tanaka. 2004. Some characteristics of commercial Som-fug produced in Thailand. *Food Chemistry* 88, no. 4: 527–535.

Romano, P., C. Fiore, M. Paraggio, M. Caruso, and A. Capece. 2003. Function of yeast species and strains in wine flavour. *International Journal of Food Microbiology* 86, no. 1–2: 169–180.

Salleh, F., M. N. Lani, N. A. Kamaruding, T. Z. T. Chilek, and N. Ismail. 2021. Lactic acid bacteria producing sorbic acid and benzoic acid compounds from fermented durian flesh (tempoyak) and their antibacterial activities against foodborne pathogenic bacteria. *Applied Food Biotechnology* 8, no. 2: 121–132.

Sanjukta, S., and A. K. Rai. 2016. Production of bioactive peptides during soybean fermentation and their potential health benefits. *Trends in Food Science and Technology* 50: 1–10.

Santiago-Urbina, J. A., and F. Ruíz-Terán. 2014. Microbiology and biochemistry of traditional palm wine produced around the world. *International Food Research Journal* 21, no. 4: 1261–1269.

Sarkar, S. 2008. Innovations in Indian fermented milk products—A review. *Food Biotechnology* 22, no. 1: 78–97.

Sharma, N. 2016. Identification and evaluation of in vitro probiotic attributes of novel and potential strains of lactic acid bacteria isolated from traditional dairy products of north-west Himalayas. *Journal of Clinical Microbiology and Biochemical Technology* 2: 018–025.

Siripongvutikorn, S., C. Thongraung, W. Usawakesmanee, T. Buatoom, and P. Thammarutwasik. 2009. Development of instant garcinia (*Garcinia atroviridis*) Tom-Yum mix as a high acid seasoning. *Journal of Food Processing and Preservation* 33, no. 1: 74–86.

Song, Y. S., J. Frias, C. Martinez-Villaluenga, C. Vidal-Valdeverde, and E. G. de Mejia. 2008. Immunoreactivity reduction of soybean meal by fermentation, effect on amino acid composition and antigenicity of commercial soy products. *Food Chemistry* 108, no. 2: 571–581.

Sudheendra, C. V., T. Madhavi, G. Gopikrishna, and K. S. Rajiv. 2018. Study of microbial changes in probiotic and synbiotic lassi during storage. *The Pharma Innovation Journal* 7, no. 2: 94–97.

Surono, I. S. 2016. Ethnic fermented foods and beverages of Indonesia. In *Ethnic Fermented Foods and Alcoholic Beverages of Asia*, ed. J. P. Tamang, 341–382. New Delhi: Springer.

Swain, M. R., M. Anandharaj, R. C. Ray, and R. Parveen Rani. 2014. Fermented fruits and vegetables of asia: A potential source of probiotics. *Biotechnology Research International* 2014: 1–19.

Syahputra, B. S. A., J. H. Ali, and M. K. H. Omar. 2012. Ikan pekasam. *Jurnal Ilmiah Abadi Ilmu* 825–832.

Tamang, J. P., P. D. Cotter, A. Endo, N. S. Han, R. Kort, S. Q. Liu, B. Mayo, N. Westerik, and R. Hutkins. 2020. Fermented foods in a global age: East meets West. *Comprehensive Reviews in Food Science and Food Safety* 19, no. 1: 184–217.

Tan, W. C., S. J. Lim, and W. A. W. Mustapha. 2017. Pencirian bakteria asid laktik dan sebatian aroma ikan pekasam. *Sains Malaysiana* 46, no. 3: 439–448.

The Malaysian Food Act and Regulation. 1983. *Food Act 1983*. Putrajaya: Ministry of Health.

Upadhyay, S., and P. Kumar. 2017. Preparation of carrot lassi. *The Pharma Innovation Journal* 6, no. 8: 8242.

van Maris, A. J. A., D. A. Abbott, E. Bellissimi, J. van den Brink, M. Kuyper, M. A. H. Luttik, H. W. Wisselink, W. A. Scheffers, J. P. van Dijken, and J. T. Pronk. 2006. Alcoholic fermentation of carbon sources in biomass hydrolysates by *Saccharomyces cerevisiae*: Current status. *Antonie van Leeuwenhoek, International Journal of General and Molecular Microbiology* 90, no. 4: 391–418.

Vital, R. J., P. Z. Bassinello, Q. A. Cruz, R. N. Carvalho, J. C. M. De Paiva, and A. O. Colombo. 2018. Production, quality, and acceptance of tempeh and white bean tempeh burgers. *Foods* 7, no. 136: 1–9.

Wan, H., T. Liu, C. Su, X. Ji, L. Wang, Y. Zhao, and Z. Wang. 2020. Evaluation of bacterial and fungal communities during the fermentation of Baixi sufu, a traditional spicy fermented bean curd. *Journal of the Science of Food and Agriculture* 100, no. 4: 1448–1457.

Wang, J., and D. Y. C. Fung. 1996. Alkaline-fermented foods: A review with emphasis on pidan fermentation. *Critical Reviews in Microbiology* 22, no. 2: 101–138.

Wan Saidatul Syida, W. K., A. Noriham, I. Normah, and M. Mohd Yusuf. 2018. Changes in chemical composition and amino acid content of soy protein isolate (SPI) from tempeh. *International Food Research Journal* 25, no. 4: 1528–1533.

Wasnin, R. M., M. S. A. Karim, and H. M. Ghazali. 2014. Effect of temperature-controlled fermentation on physico-chemical properties and lactic acid bacterial count of durian (*Durio zibethinus Murr.*) pulp. *Journal of Food Science and Technology* 51, no. 11: 2977–2989.

Wilson, L. A. 1995. Soy foods. In *Practical Handbook of Soybean Processing and Utilization*, ed. D. R. Erickson, 428–459. Massachusetts: Academic Press.

Xie, C., H. Zeng, J. Li, and L. Qin. 2019. Comprehensive explorations of nutritional, functional and potential tasty components of various types of sufu, a Chinese fermented soybean appetizer. *Food Science and Technology* 39: 105–114.

Zainal Abidin, N. A., Z. Mohd Zin, M. A. A. Abdullah, N. D. Rusli, and M. K. Zainol. 2020. Physicochemical properties and sensory acceptance of *Canavalia ensiformis* tempeh energy bar. *Food Research* 4, no. 5: 1637–1645.

Zou, B., G. Xiao, Y. Xu, J. Wu, Y. Yu, and M. Fu. 2018. Persimmon vinegar polyphenols protect against hydrogen peroxide-induced cellular oxidative stress via Nrf2 signalling pathway. *Food Chemistry* 255: 23–30.

Zouhali-Worrall, M. 2008. A vintner's vision: Wine from mangosteens. *Fortune Small Business*, December 2. https://money.cnn.com/2008/12/02/smallbusiness/thai_wine.fsb/index.htm

Traditional Fermented Foods of Northeast India

11

Microbial Diversity and Functionality

Shankar Prasad Sha

11.1 INTRODUCTION

Food fermentation is one of the oldest traditional technologies for the production of edible products in the development of human civilization (Tamang et al. 2016; Sha et al. 2018). Fermented foods are defined as food products prepared by people using their indigenous knowledge of food fermentation from locally available raw materials from plant or animal sources either naturally or by adding starter culture(s) containing functional microorganisms which modify the substrates biochemically and organoleptically into edible products that are socially and culturally acceptable to consumers (Tamang et al. 2016). Fermented beverages and alcoholic drinks are socially and culturally acceptable products for consumption, drinking, entertainment, and religious purposes as well as customary practices (Sha et al. 2016). Drinking of alcoholic beverages and distilled alcohol are widespread and interest enhancing the pleasure of eating due to some nutritional role (Tamang et al. 2016). The consumption of alcoholic drinks in India is mentioned in the *Ramayana*, dating from 300–75 BC (Tamang et al. 2016). Ethnic alcoholic drinks have strong ritualistic importance among the ethnic people in Asia and Africa, where social activities require the provision and consumption of appreciable quantities of alcohol, and wine has a deep-rooted cultural history for European as well as Mediterranean ethnic people (Sha et al. 2018; Tamang and Samuel 2010). Although a diverse range of alcoholic products is available, a general scheme for their production can be presented as: (i) selection of the raw material; (ii) processing of the raw material to give a fermentable extract; (iii) alcoholic fermentation by yeast, principally by strains of *Saccharomyces cerevisiae*; (iv) distillation of the fermented material to give the distillate product; and (v) post-distillation processing (Sha et al. 2018). About ten different categories of global alcoholic beverages have been reported by Tamang et al. (2016).

To understand the ethno-microbiology, both cultivable and uncultivable microorganisms from fermented food and beverages may be identified using culture-dependent and culture-independent methods. This will help to document a complete profile of native functional and non-functional microorganisms and also to study both inter- and intra-species diversity within a particular genus or among genera (Yan et al. 2013). Molecular identification is emerging as an accurate and reliable identification tool for identification of both culture-dependent and culture-independent microorganisms from fermented foods (Dolci et al. 2015). Due to some limitations, the culture-dependent methods may not detect the whole microbial community in foods. However, culture-independent methods extracting whole genomic DNA directly from small (<1 g) samples of fermented food may detect the whole microbial communities in food samples (Puerari et al. 2015). Culture-independent methods, including pyrosequencing, PCR-denaturing gradient gel electrophoresis (DGGE) analysis, and recently next-generation sequencing such as high-throughput metagenomic amplicon sequencing, may serve to give more insights into the microbial ecology of natural food fermentation with increased accuracy in a relatively short period of time (Ercolini 2004; Chen et al. 2014; Puerari et al. 2015; Tamang et al. 2016).

Analysis of the internal transcribed spacer (ITS) region has been widely applied in explorations of the diversity of yeasts associated with various traditional fermented foods (Caggia et al. 2001; Las Heras-Vazquez et al. 2003), which

DOI: 10.1201/9781003605300-11

may provide a fast and easy means of accurate identification at the species level (Esteve-Zarzoso et al. 1999) due to higher sequence variation (Iwen et al. 2002; Korabecna 2007). PCR-DGGE analysis is the most commonly used among culture-independent fingerprinting techniques and is based on the separation of amplicons (PCR-products) of the same size but with different sequences of 16SrRNA and 26SrRNA amplicons (~~Cocolin~~ et al. 2000; Ercolini 2004). The PCR-DGGE approach is used to investigate yeast diversity during commercial wine fermentation (Cocolin et al. 2001). Recently, bioinformatics approaches have been used to characterize microbial genomes directly from metagenomes, allowing strain-level identification during the process and genomic comparison (Eren et al. 2015; Scholz et al. 2016). Rapid evolution in high-throughput sequencing techniques has enabled researchers to have increased accuracy, high-throughput sequencing tools with reasonably low cost and in a relatively short period of time (Cocolin et al. 2013; Mayo et al. 2014).

Some researchers have reported microbial communities in traditionally prepared starters cultures and traditional alcoholic drinks of some countries by using PCR-DGGE analysis, such as Chinese amylolytic starter *yaoqu/hongqu* (Lv et al. 2013) and *daqu* (Chen et al. 2014); the sorghum-based alcoholic beverage of Benin, *tchoukoutou* (Greppi et al. 2013); and *chicha*, the ethnic alcoholic beverage of Brazil (Puerari et al. 2015). Metagenomic studies using high-throughput sequencing techniques on various fermented milk products have shown a realistic view of the microbial community structure involved in natural fermentation (Dobson et al. 2012). There are eight states located in the region of India commonly known as the northeast (www.northeasttourism.gov.in). All eight states in the northeast have diverse types of traditionally prepared amylolytic starters prepared by different linguistic ethnic groups of people, including *marcha* from Sikkim, *humao* from Assam, *hamei* from Manipur, *chowan* from Tripura, *thiat* from Meghalaya, *khekhrii* from Nagaland, *dawdim* from Mizoram, and *phut* from Arunachal Pradesh. These starter cultures, except *khekhrii*, are traditionally prepared from soaked rice with wild herbs and previously prepared starter powder (1–2%) as an inoculum (back-sloping). Then mixtures are ground in a wooden mortar with the addition of water to make a thick dough which is kneaded into round to flattened balls/cakes of different sizes and shapes, covered with fern fronds/paddy straws/jute sags, and fermented at room temperature for 1–3 days; then fresh balls/cakes are sun-dried for a few days (Anupma et al. 2018; Sha et al. 2018). *Khekhrii/khrie* from Nagaland is prepared by naturally fermenting germinated sprouted rice grains and then sun-dried to use as a dry starter culture to prepare *zutho*, a local alcoholic beverage.

This chapter focuses on various traditional fermented foods and alcoholic beverages of northeast India as well their advanced microbial diversity and functionality.

Traditional practice of sub-culturing by back-sloping method serves to preserve the essential microbes consisting consortia of yeasts, molds, and bacteria, in the form of dry, fattened, or round balls called amylolytic starters that will be used for the production of different types of alcoholic beverages (Tamang et al. 2016). Traditionally prepared alcoholic beverages have consortia of mixed microbiota representing filamentous molds, yeast, and bacteria; hence many researchers have studied the fungal, yeast, and bacterial populations in northeastern fermented foods, commonly based on culture-dependent techniques, including biochemical phenotypic and culture-dependent 16S rRNA sequencing, and identified yeasts such as *Candida glabrata, C. tropicalis, Clavispora lusitaniae, Issatchenkia* sp., *Pichia anomala, P. ranongensis, P. burtonii, Saccharomycopsis fbuligera, Sm. capsularis, Saccharomyces cerevisiae, Sacch. Bayanus;* molds such as *Absidia corymbifera, Amylomyces rouxii, Botryobasidium subcoronatum, Mucor circinelloides forma circinelloides, Mucor hiemalis, Rhizopus oryzae, Rhi. microsporus, Rhi. chinensis, Rhi. stolonifer,* and *Xeromyces bisporus,* and bacteria *Lpb. plantarum, B. circulans, A. pasteurianus, Bacillus amyloliquefaciens, B. sporothermodurans, B. subtilis, Pediococcus pentosaceus, Lactobacillus bifermentans, Lv. brevis, Weissella confusa, W. paramesenteroides,* and *Acetobacter orientalis* (Sha et al. 2017; Tamang et al. 2016).

Introduction of culture-independent methods and its application in food microbiology studies (Sha et al. 2018; Dolci et al. 2015; Tamang et al. 2016) has been a great tool for researchers to profile the microbial diversity of fermented foods using PCR-DGGE, and High-throughput sequencing (HTS) (Sha et al. 2018; Tamang et al. 2016). Rapid evolution in next-generation sequencing (NGS) technologies has enabled researchers to have increased accuracy and high throughput with reasonably low cost in a relatively short period of time and have shown the wide microbial community structure responsible for fermentation and product development (Sha et al. 2018; Jung et al. 2012; Tamang et al. 2016). These cutting-edge technologies are used to investigate the wide microbial community structure of ethnic fermented foods and alcoholic beverages from northeast India.

11.2 NORTHEASTERN FERMENTED FOODS AND ALCOHOLIC BEVERAGES

Traditional practice of sub-culturing by back-sloping and preservation of important indigenous consortia of microorganisms for alcoholic beverages production in northeast Asia including the Himalayan regions of India, Nepal, Bhutan, and China is the worth important wisdom of the ethnic people for centuries (Tamang et al. 2016). Ethnic fermented foods have been prepared for many centuries, and preparation processes vary from country to country. In northeast India, fermented foods and alcoholic beverages are been a fundamental part of the ethnic heritage. One of the oldest traditional and most cost-effective methods for producing and preserving food is fermentation (Sha et al. 2013; Tamang et al. 2016; Sha et al. 2018).

These ethnic foods have been consumed from time immemorial, as the natural process of fermentation increases the shelf life, decreases volume, shortens the cooking time, and enhances the nutritive value as compared to fresh foods. The process of fermentation is a slow decomposition reaction of organic substances like complex carbohydrates rapidly broken down by microbial enzymes that convert complex substances into simpler forms such as simple sugars first and then to products such as alcohols or organic acids, vitamins, and antibiotics. Fermentation boosts flavor, aroma, and digestibility and exerts health-promoting effects. Fermented food preparation requires starter cultures that are added to the substrate or material to start the process of fermentation. Lactic acid (LA) fermentation of foods like milk, cereals, vegetables, and fruits is a major tradition to preserve and improve the nutritional and sensory qualities of food. Many lactic acid bacteria (LAB) have been reported from various traditional naturally fermented foods and observed to be the predominant microflora of fermented products (Tamang et al. 2016). Fermented foods are classified on the basis of the substrate used as fermented meat and fish products, fermented milk products, fermented vegetable and fruit products, fermented cereal-legume/legume products, and fermented cereal products.

11.2.1 Fermented Legume-Based Products

11.2.1.1 Kinema

Kinema is one of the important types of fermented food among the ethnic communities in the eastern Himalayan regions of Nepal, Darjeeling hills, and Sikkim in India. *Kinema* is a whole-soybean fermented food with a sticky texture, gray tan in color flavorful and highly proteinaceous in nature. It is like *natto,* a soybean-based fermented food from Japan. The heat-resistant spore-forming bacterium *Bacillus subtilis,* lactic acid bacteria such as *Enterococcus faecium,* and a few types of yeast like *Candida parapsilosis* and *Geotrichum candidum* have been recovered from *kinema.* However, *Bacillus subtilis* is the dominant microflora in *kinema* fermentation, followed by *Enterococcus faecium.* It is observed that rich microbial diversity in various sources, particularly soybeans and leaves as wrapping materials, harnesses indigenous microbiota for spontaneous fermentation of *kinema* (Tamang et al. 2016).

11.2.1.2 Tungrymbai/Tung Rymbái

Tungrymbai/tung rymbái is a soybean fermented food traditionally prepared by the Khasi and Jaiñtia people of Meghalaya, northeast India. (Sohliya et al. 2009; Tamang et al. 2016). The soybeans are washed and cleaned, then boiled until they are cooked. After cooking, the excess water is drained off, and the beans are allowed to cool on a bamboo tray. The soybeans are then transferred to a bamboo basket lined with fresh slamet (*Phrynium pubinerve*) leaves. The leaves are then wrapped over the cooked beans, and hot charcoal is placed on top before more leaves are added once again. The entire basket is then placed inside a jute bag to ferment for three to four days by a fireplace. When the fermentation is complete, the beans are taken out and crushed with a mortar and pestle (Sohliya et al. 2009; Tamang et al. 2016). *Bacillus subtilis* and *Bacillus licheniformis* are the major microbial communities of *tungrymbai* (Sohliya et al. 2009; Tamang et al. 2016).

11.2.2 Fermented Milk-Based Products

Milk and milk products are the major food components in the dietary culture of ethnic people living in different geographical areas of India. Cow milk is commonly consumed throughout the Himalayan regions, and buffalo milk is more popular in the foothills of the western Himalayas. Yak (*chauri* in Nepali and *kno*, the female yak, in the Tibetan language) is commonly used in the subalpine and alpine Himalayas as a source for milk, meat, skin, and hair in the highlands above elevations of 2100 m. Consumption of goat milk is uncommon in the Sikkim Himalayas.

The Sikkim Himalayan people prepare and consume a variety of ethnic fermented milk products made from cow milk and yak milk, such as *dahi*, *mohi*, *gheu*, *soft chhurpi*, *chhu*, *somar*, *maa*, *philu*, and *shyow*. Some of the milk products, notably *dahi* (curd), *mohi* (buttermilk), and *gheu* (butter), are familiar in all regions of the Himalayas; *chhurpi*, *chhu*, *philu*, and others are confined mostly to the Tibetans. *Somar* is exclusively prepared and consumed by the Sherpa of Nepal and Sikkim living at high altitudes (Tamang et al. 2016).

11.2.2.1 Dahi

Dahi (curd) is a popular fermented milk product of the Himalayas and is also used for the preparation of a number of other ethnic milk products like *gheu*, *mohi*, and soft *chhurpi chhu*. *Dahi* is the Nepali and Hindi word; the Tibetans call it *shyow* (Tamang et al. 2016). Curd has probiotic lactic acid bacteria. Probiotics are live beneficial lactic acid bacteria and yeasts that exert positive health effects when consumed by humans. Fresh milk from a cow or yak is boiled in a vessel, cooled to room temperature, and transferred to a hollow wooden vessel or container locally known as a *theki*. A small quantity of previously prepared *dahi* (called *mau* in Nepali) serves as a source of inoculum and is added to the boiled and cooled milk. This is left for fermentation for one to two days in summer and two to four days in winter at room temperature. The duration of fermentation depends on the season as well as on the geographical location. Lactic acid bacteria include *Limosilactobacillus fermentum, Paracasei* subsp. *pseudoplantarum, Lactococcus lactis,* and *Lactis* subsp. *cremoris*, and yeasts include *Saccharomycopsis* spp. and *Candida* spp.

11.2.2.2 Ghee

Gheu or *ghee* is prepared butter from cow or buffalo milk with a typical flavor and aroma and is produced by churning *dahi*. It is a common milk product of India and is consumed by the entire Indian population. It is important commercially as well as for various other functions. Lactic acid bacteria *Lactococcus lactis, Lactobacillus acidophilus*, and *Lc. lactis* subsp. *cremoris are* associated with *ghee* (Tamang et al. 2016).

11.2.2.3 Hard Chhurpi

Hard-variety *chhurpi* is mostly prepared from yak milk at high altitudes (2100–4500 m) and has characteristic gumminess and chewiness, and it is commonly eaten as a masticator (Tamang et al. 2016). Two types of hard *chhurpi* are prepared from yak milk: common *chhurpi* and *dudh chhurpi* that is highly popular in the Darjeeling, Sikkim, and Bhutan areas of Himalayas. By making *chhurpi* yak milk can be preserve for future use in the form of square or round milk cake. The lactic acid bacteria *Lactobacillus farciminis*, *Lb. paracasei* subsp. *paracasei*, *Weisella confuses*, and *Lb. bifermentans* are involved in *chhurpi*.

11.2.2.4 Soft Chhurpi

Soft-variety *chhurpi* or *kachcha chhurpi* (Tamang et al. 2016) is a soft, cheese-like Himalayan ethnic fermented milk product. It has a rubbery texture with a slightly sour taste and aroma. It is consumed mostly in Sikkim, the Darjeeling hills, and Bhutan. The lactic acid bacteria involved are *Lactobacillus farciminis*, *Lv. brevis*, *Lb. alimentarius*, *Lb. salivarius*, *Lactococcus lactis* subsp. *cremoris*, *Saccharomycopsis* sp., and *Candida* sp. (Tamang et al. 2016)

11.2.2.5 Somar

Somar is an ethnic Himalayan fermented milk product consumed by the Sherpa tribes of Nepal and Sikkim living in the high mountains. It is prepared from cow or yak milk. *Somar* is a soft paste with a bitter taste and a strong flavor. The lactic acid bacteria are *Lactobacillus paracasei* subsp. *pseudoplantarum* and *Lactococcus lactis* subsp. *cremoris* (Tamang et al. 2016).

11.2.3 Fermented Vegetables

11.2.3.1 Gundruk

Gundruk is an ethnic fermented leafy vegetable, a dry and acidic product indigenous to the Nepali living in the Himalayan regions of India, Nepal, and Bhutan. *Gundruk* is mainly produced during November to March, when large quantities of fresh leaves of mustard, *rayo-sag* (local variety of mustard), radish, and other vegetables pile up, much more than people could consume fresh. The word *gundruk* is derived from the Newari (Newar/Pradhan), one of the major ethnic communities of Nepali (origin), word *gunnu*, meaning "dried taro (*Colocasia*) stalk." *Gundruk* is similar to the fermented acidic vegetable products of other countries, such as *kimchi* from Korea, sauerkraut from Germany, *sunki* from Japan, *pak-gard-dong* from Thailand, and *suan-cai* from China. The pH and acidity (as lactic acid) in *gundruk* are 4.0–4.3 and 0.8%–1.0%, respectively. In *gundruk*, 90% of the organic acids consist of lactic and acetic acids, and other organic acids are citric, malic, and acetic acids. In mustard *gundruk*, free amino acids, mostly glutamic acid, alanine, leucine, lysine, and threonine, remarkably increase with a decrease in asparagine, glutamine, histidine, and arginine during fermentation. *Gundruk* prepared from leaves of mustard comprises cyanides (15.7%) and isothiocyanates (8.5%) as the main flavor components, followed by alcohols (12.3%). The lactic acid bacteria involved are *Limosilactobacillus fermentum, plantarum, Lb. casei, Lb. casei* subsp. *pseudoplantarum*, and *Pediococcus pentosaceus* (Tamang et al. 2016).

11.2.3.2 Sinki

Sinki, a fermented radish taproot product, is prepared by pit fermentation in the Himalayas. It is usually prepared when radish taproots are in plentiful supply. When the leaves of the radish are fermented, it is called *gundruk*, and when the taproot is fermented, it is called *sinki*. The lactic acid bacteria involved are *Limosilactobacillus fermentum, Lb. casei*, and *Leuconostoc fallax* (Tamang et al. 2016).

11.3 AMYLOLYTIC STARTERS AND ALCOHOLIC BEVERAGES

11.3.1 Amylolytic Starters

11.3.1.1 Amou/perokkushi

Amou/perokkushi is an amylolytic starter for the preparation of rice-based alcoholic beverages in Assam by the Deori and Bodo communities, respectively (Das et al. 2017). They identified the amylolytic fungi, based on the sequencing of their internal transcribed spacer (ITS) regions, as *Amylomyces rouxii* and *Rhizopus oryzae*, and both strains showed the ability to break down and saccharify starch (Polysaccharides) in initial stages. The DGGE-NGS investigation showed the dominance of *Saccharomyces cerevisiae* yeast and *Lactococcus lactis* LAB in overall fermentation using simple sugars from the initial saccharification of starch (Das et al. 2017).

11.3.1.2 Bakhar

Bakhar is a starter culture used to make *pachwai*, rice wine, in the eastern part of India and contains *Rhizopus* sp., *Mucor* sp., and at least one species of yeast (Das et al. 2017). Ginger and other plant materials are dried, ground, and added to rice flour. Water is added to make a thick paste, and a small round cake of 1.0–1.5 cm in diameter is formed and inoculated with powdered cakes from previous batches. The cakes are then wrapped in leaves, allowed to ferment for three days, and then sun-dried (Das et al. 2017; Sha et al. 2019).

11.3.1.3 Hamei

Hamei is an ethnic amylolytic mixed dry, round to flattened starter of Manipur in India. It is very similar to *marcha* (Tamang 2010). *Hamei*, an ethnic amylolytic starter from Manipur, is used for the preparation of alcoholic beverages from glutinous rice and is very interesting because of its unique flavor and aroma. Yeast communities of *hamei* were identified by phenotypic (biochemical characterization) and molecular tools such as restriction digestion patterns generated from the PCR amplified internal transcribed spacer region along with the 5.8S rRNA gene (ITS1–5.8S-ITS2) which included the yeasts *Saccharomyces cerevisiae*, *Pichia anomala*, *Trichosporon* sp., *Candida tropicalis*, *Pichia guilliermondi*, *Candida parapsilosis*, *Torulaspora delbrueckii*, *Pichia fabianii*, and *Candida montana* (Jeyaram et al. 2008). The genetic diversity of the industrially important *S. cerevisiae* group isolated from *hamei* was investigated using pulsed field gel electrophoresis (PFGE) (Tamang et al. 2007; Jeyaram et al. 2008).

11.3.1.4 Huamo

Huamo is a traditionally prepared rice-based ethnic amylolytic starter from Assam and is commonly used for the preparation or fermentation of *judima* and *xaaj* (Tamang et al. 2016). *Humao* is prepared by using the locally available glutinous rice, bark, and roots of wild plants (Chakrabarty et al. 2014). During the preparation of *huamo*, rice is first washed and powdered in a wooden *okhari* along with the bark and roots of wild plants, and a few old *humao* starters are mixed properly with clean water to make a paste. Then the paste is used to make different-sized round to flat cake-like starters on a mat or carpet, fermented for one to two days, sun-dried, and then stored at room temperature for further use. The fermented product is dominated by yeasts and lactic acid bacteria, has a unique aroma, and is sweet in taste due to the high content of sugar.

11.3.1.5 Ipoh

Ipoh is a traditionally fermented amylolytic starter from Arunachal Pradesh used for the traditional fermentation of *apong* and *ennog*, popular mild alcoholic beverages (Tamang 2010). It is prepared through various processes of washing and grinding glutinous rice into fine powder and mixing the powder of leaves, bark, and seeds of locally available plant species *Veronia cinerea* and *Clerodendron viscosum*. Then this mixture is mixed properly in a large container (*dekchi*) and made into paste by using previously stored rice water; spread on clean bamboo; and made into circular, disc-shaped small cakes or biscuits. The cakes are then carefully kept to dry out completely either in the attic above the fireplace of traditional houses or in a cool dry place for four to five days for fermentation and then sun-dried. After drying, they are stored for further use. The major microorganisms involved in *ipoh* are yeast populations (Tiwari and Mahanta 2007).

11.3.1.6 Khekhrii

Khekhrii is a traditionally prepared unique type of ethnic starter from Nagaland used to prepare a local alcoholic beverage, *zutho* (Tamang 2010). During the traditional preparation, unhulled glutinous rice is washed, soaked in water for two to five days, kept and covered with *Khreihenyii* leaves, and allowed to germinate for three to four days in summer and five to six days in winter. After partial germination when the rice sprout is about half an inch in length, the sprouted rice is exposed to the sun for drying, powdered, and again sun-dried and stored for further use (Jamir and Rao 1990; Jamir and Deb 2014).

11.3.1.7 Keem

Keem is a traditionally prepared barley-based ethnic amylolytic starter from Himachal Pradesh and is commonly used for the preparation of *soor*, which is commonly consumed as a mild alcoholic beverage on various occasions (Rana et al. 2004). During traditional preparation, chopped fresh twigs of *Cannabis sativa* (8 kg), 5 kg leaves of *Sapindus mukorossi*, and 10–15 kg total of different plant species are dried in the shade for a few days and then powdered and mixed properly with about 50 kg of barley flour. For the desired quantity of this dry mixture, a sufficient quantity of *jayaras* (a compound prepared by keeping finely chopped leaves and tender parts of *Dicliptera roxburghiana, Zanthoxylum armatum, Leucas lanata*, and *Melia azedarach*) is added to a bigger vessel, kept overnight, and made into dough for round cakes of about 1–2 kg weight. Many oval-shaped cakes are prepared and kept on a plant bed (*sathar*) made up of 15 different tender shoots of *Pinus roxburghii* and *Cannabis sativa* alternating between the cakes and incubated in a closed room. The prepared starters are allowed to remain undisturbed for 24 days. On the 25th day of incubation, the room is opened, and the cake is placed upside down and allowed to remain there for another 12 days for fermentation. Cakes of *keem* are then taken out, sun-dried, and used in preparation of the local alcoholic beverage known as *soor* (Rana et al. 2004).

11.3.1.8 Mod pitha

Mod pitha is a traditionally fermented amylolytic starter from Assam and Arunachal Pradesh used for the traditional preparation of *sujen*, which is consumed as a mild alcoholic beverage (Deori et al. 2007). For the traditional preparation of *mod pitha*, glutinous rice grains (*saol*) and a handful each of cleaned leaves, fronds, bark, and bulbs of various plant parts are put in a round bamboo tray (*saloni*) and exposed for sun drying for a day. The rice grains (*saol*, 4–5 kg) are soaked in water for 2 h, cleaned and mixed with the dried plant parts, and ground in a wooden grinder (*dheki*). The ground rice powder is taken out and sieved in a round bamboo tray (*saloni*), the coarse part is returned to the wooden grinder (*dheki*) for grinding, and this process is continued until a fine powder is obtained. Old *mod pitha* (two- to three-day-old starter) is added to the mass during grinding. Powdered glutinous rice is put into a container (*soriya*); then water is added to make a sticky paste, and small round to flattened cakes (2–3 cm in diameter and 0.1 to.04 cm in thickness) are prepared. *Mod pitha* is used for the fermentation of *xaaj*, a beverage from Assam (Deori et al. 2007).

11.3.1.9 Marcha

Marcha is a ball-like amylolytic starter used to ferment starchy materials into a fermented beverage in Nepal, Bhutan, and the Darjeeling hills and Sikkim in India (Tamang and Sarkar 1995; Tamang 2010). During preparation of marcha, glutinous rice is soaked, excess water is removed, pounded, wild herbs, old *marcha* (~1 %) are added in the mixture, then made into thick paste by adding water, then round or flate cakes are made and covered with wild fern leaves and allow it for fermentation up to 1–2 days, then sun-dried it and stored for further use (Tamang et al. 1996). Kobayashi et al. (1961) reported *Rhizopus oryzae, Mucor praini*, and *Absidia lichtheimi* in *marcha* samples collected from Sikkim. Hesseltine and Ray (1988) isolated *Mucor* and *Rhizopus* sp. in *marcha*. Tamang and Sarkar (1995) identified the microorganisms in *marcha* from the Darjeeling Hills and Sikkim as *Pediococcus pentosaceus, Saccharomycopsis fibuligera, Pichia anomala, Mucor circinelloides*, and *Rhizopus chinensis*. Batra and Millner (1974) reported *Hansenula anomala* var. *schneggii* (*Pichia anomala*) in *marcha*. In Bhutan, *marcha* is called *chang-poo*, in which *Saccharomycopsis*, *Penicillium* sp., and *Aspergillus* sp. were reported (Uchimura et al. 1990). Microbial profiles of amylolytic starters from India, Nepal, and Bhutan are filamentous molds like *Mucor circinelloides, Mucor hiemalis, R. chinensis*, and *R. stolonifer* variety *lyococcus* (Tamang et al. 1988) and yeasts *S. cerevisiae, S. bayanus, Saccharomycopsis fibuligera, Sm. capsularis, Pichia anomala, P burtonii*, and *Candida glabrata* (Tamang and Sarkar 1995; Shrestha et al. 2002; Tsuyoshi et al. 2005; Jeyaram et al. 2011).

11.3.1.10 Ranu Dabai

Ranu dabai is an amylolytic starter culture from Assam (Ghosh and Das 2004). During the preparation of *ranu dabai*, six steps are involved: washing rice and storing wash water. After cleaning, glutinous rice on a *soop* (a flat traditionally prepared tray generally made of sliced bamboo) is placed in a vessel (made of metal/clay) for washing. Clean water is poured into it, mixed, and drained off. The discarded wash water is stored in a container for future use. Mixing and grinding: In this step, the traditional wooden husking machine *dhiki* is used for grinding. The freshly collected plant materials and grains are chopped and ground properly and taken out onto a *soop*. Glutinous rice is taken in the *dhiki* and partially powdered, and 3–4 large old tablets of *ranu dabai* are added for 10 kg of rice. After some time, the pastes of various plants are also added and mixed properly. The powdered mixture is placed in a large *dekchi* and made into paste using the previously collected washed rice water. Clean gunny bags are then spread on the floor under shade. These tablets are completely handmade. The standard size is about 4.5–7 cm in diameter, which is kept in rows on the gunny bags, where these tablets are kept for 40–60 min. The size of *ranu* cakes vary from 1.5–15 cm in diameter.

11.3.1.11 Ranu Goti

Ranu goti is prepared by some ethnic communities of central India for the preparation of alcoholic beverages such as *handia* (Kumar and Rao 2007). During the perpetration of *ranu goti*, first, glutinous rice is washed and soaked, excess water is drained off, and then it is powdered with help of *dhiki*. The rice powder is mixed with powdered roots, leaves, bark, rhizomes, and seeds of about 20–25 plant species in a ratio of 2:1 with clean water, and small pieces of cakes are made. These *goti* are kept for incubation in a bamboo basket under closed conditions. Then, after incubation, the *ranu goti* are taken out of the bamboo basket and exposed to the sun for drying for 7–8 days, then used in the preparation of the local alcoholic beverage *handia* (Kumar and Rao 2007).

11.3.1.12 Thiat

Thiat is an amylolytic starter from Meghalaya used to ferment the alcoholic beverage *kiad* (Tamang et al. 2016). During the preparation of *thiat*, first, the glutinous rice is washed, soaked, and then powdered. The rice powder is mixed with khaw-*iang-/hawiang* plants leaves with clean water, then small pieces of *thiat* cakes are made in size ranging from 4–5 cm in diameter and 0.8–1.0 cm in thickness, and are kept for fermentation in *malieng* and covered by *sla-pashor.* After fermentation, the *thiat* is sun-dried and used as dry starter for alcohol production (Amati and Begum 2007).

11.3.1.13 Vekur Pitha

Vekur pitha is a traditionally prepared ethnic amylolytic starter culture from Assam and is commonly used for the traditional preparation of *ahom*, which is consumed as a mild alcoholic beverage during various ceremonies (Saikia et al. 2007). For the traditional preparation of amylolytic starter culture, *vekur pitha* glutinous rice grains (*saol*) and the leaves of a few wild plants are used. The plant parts and additive ingredients serve as source of the yeast *Saccharomyces cerevisiae*. The leaves of plants are collected from the wild and exposed to natural sunlight for two to three days. Sun-dried leaves are powdered and mixed with the powdered rice grains in a vessel containing a few ml of clean water. Here, old powdered *pitha* commonly called *ghai pitha* is mixed with freshly prepared *pitha* as source of yeast microflora. The semi-solid *pitha* is mixed with the required ingredients and rolled into a plate-disc shape, wrapped with fresh leaves of *Musa paradisiaca*, and kept in anaerobic environment over fire heat. The fire heat is maintained at 90–180 cm height for five to six days to dry until it gets hard. Oval-shaped dried *pitha* containing yeast inoculum, rice powder, and plant material is known as *vekur pitha*, which is preserved in natural conditions for future use for the preparation of various alcoholic beverages. *Saccharomyces cereviceae* is the major yeast that plays a vital role in the fermentation of *vekur pitha* (Saikia et al. 2007).

11.3.1.14 Xaj-Pitha

Bora et al. (2016) reported that *xaj-pitha*, a rice-based ethnic amylolytic starter culture from Assam, is used to prepare local alcoholic beverages. The microbial community of *xaj-pitha,* by an NGS approach, revealed amylase producers, such as *Rhizopus delemar, Mucor circinelloides,* and *Aspergillus* sp. Ethanol producers fungal strains such as *Candida glabrata, Debaryomyces hansenii, Ogataea parapolymorpha, Wickerhamomyces ciferrii, Saccharomyces cerevisiae, Meyerozyma guilliermondii* and *Dekkera bruxellensis* (Bora et al. 2016). Some opportunistic contaminants were also reported in *xaj-pitha*. The bacterial population was dominated by LAB such as *Lactobacillus plantarum, Levilactobacillus brevis, Weissella cibaria, Lactococcus lactis Leuconostoc lactis, Weissella para mesenteroides,* and *Leuconostoc pseudomesenteroides* (Bora et al. 2016).

11.3.2 Ethnic Alcoholic Beverages

11.3.2.1 Atingba

Atingba is a popular traditionally prepared alcoholic beverage from Manipur, prepared from rice (Jeyaram et al. 2009). The Meitei community of Manipur mainly consumes it as a food beverage on several occasions. For the preparation of *atingba*, rice is cooked first, and then its water is removed and it is allowed to cooled to room temperature. Powdered *hamei* (the starter culture for *atingba*) is mixed properly with cooked rice in a ratio of 5 cakes/10 kg of rice. The mixture is then placed in earthen pots, which are covered with *hangla* leaves (*Alocasia* sp.) and then allowed to mature for three to four days of fermentation in summer and six to seven days in winter. This process is then followed by two to three days of submerged fermentation in an earthen pot to produce the final alcoholic product, *atingba*. It is then distilled to give a clear liquor alcoholic beverage called *yu* in Manipur. There are various types of yeasts and filamentous fungi that are responsible for the fermentation of rice (substrate) to *atingba* (product) such as the yeasts *Saccharomyces cerevisiae, Pichia anomala, montana, C. parapsilosis, Pichia guilliermondii, Torulaspora delbrueckii, P. fabianii*, and *Candida tropicalis* and molds like *Mucor* sp. and *Rhizopus* sp., whereas some important LAB are *Pediococcus pentosaceus* and *Levilactobacillus brevis* that play a vital role in flavor and texture development (Jeyaram et al. 2008).

11.3.2.2 Bhaati Jaanr

Bhaati jaanr is an ethnic rice-based mild alcoholic food beverage fermented by *marcha* in the Eastern Himalayan regions of Nepal, India, and Bhutan (Tamang 2010). During preparation, first rice is saccharified for one to two days in an earthen pot at room temperature, and once saccharification is achieved, the vessel is made airtight and is allowed to ferment for two to three days in summer and seven to eight days in winter. The major microflora involved in *bhatti jaanr* saccharification and fermentation are filamentous fungi (*Rhizopus chinensis, M. hiemalis, Mucor circinelloides, R. stolonifer,* and var. *lyococcus*), yeasts (*Candida glabrata, Saccharomyces cerevisiae,* and *S. bayanus*), and lactic acid bacteria (*Pediococcus pentosaceus, Lactobacillus bifermentans,* and *Lb. brevis*). These microflora are responsible for the development of flavor and acidity in the product. The pH, acidity, ethanol content, and moisture content of the *Bhaati jaanr* are 3.5, 0.24%, 5.9%, and 83.4%, respectively. *Bhaati jaanr* is consumed as a staple food directly in Sikkim and Darjeeling (Tamang 2010).

11.3.2.3 Chhang

Chyang or *lugri* is a mild alcoholic, foamy, translucent beverage, which is prepared by traditional fermentation. It is prepared by using the substrate barley (*Hordeum nulum*), locally known as *sherokh* in Ladakh (Bhatia et al. 1977). *Chyang* has a sweet-sour taste and aromatic flavor. During *chhang* preparation, first barley grains are cooked over a slow fire in water just sufficient for absorption, and after cooking, the mixture is spread on a blanket or burlap mat to remove the excess water. The cooked barley grains at the lukewarm stage are mixed with starter culture, *phab*, at a ratio of 1g/kg of barley. These mixtures are placed in drill bags, mostly in 20-kg batches, and then tightly packed. These mixtures are then packed in gunny bags to maintain the temperature around 30–35°C, which is required for fermentation of barley into *chyang* after seven to eight days of fermentation (Bhatia et al. 1977). Microorganisms that play a significant role in the fermentation process of *chhang* are the yeasts *Saccharomyces cerevisiae* and *S. uvarum* (Batra 1986). *Chyang* is one of the popular mild alcoholic beverages traditionally prepared and consumed by the people of Ladakh (Bhatia et al. 1977).

11.3.2.4 Kodo Ko Jaanr

Kodo ko jaanr is one of the most popular ethnic fermented finger millet (*Eleusine coracana*) beverages of the Himalayan regions of India, with a mild alcoholic (4.8%) and sweet taste (Tamang et al. 2016). *Kodo ko jaanr* has several synonyms as used by different ethnic groups of the Himalayan region, such as *chyang* (Tibetan, Ladakhi, Drupka), mandokpenaa *thee* (Limboo), and *mong chee* (Lepcha) (Tamang et al. 2016). During its production, finger millet seeds are cleaned, washed, and cooked for about 30 min; excess water is drained off; and cooked millet is spread on a bamboo mat for cooling. About 1–2% of powdered *marcha* is sprinkled over the cooked seeds, mixed thoroughly, and packed in a bamboo basket lined with fresh fern (*Thelypteris erubescens*), then covered with jute cloth and fermented at room temperature for two to four days. The saccharified mass is transferred into an earthen pot or bamboo basket, made airtight, and fermented for three to four days during summer and five to seven days in winter at room temperature for alcohol production. Freshly fermented *kodo ko jaanr* is placed into a bamboo-made vessel locally called a *toongbaa,* and lukewarm water is added up to its rim and left for 10–15 min. Then, the milky white extract of *jaanr* is sipped through a narrow bamboo straw, called *pipsing*, which has a hole in the side near the bottom to avoid the passage of grits. Water is added twice or thrice after sipping of the extract. Consumption of fermented finger millet beverages in exclusively decorated bamboo or wood vessels called *toongbaa* is unique in the Himalayas (Tamang et al. 1996). *Kodo ko jaanr* liquor is believed to be a good tonic for ailing persons and post-natal women. After consumption, the residue or grits of *kodo ko jaanr* is used as fodder for pigs and cattle. This is a good example of total utilization of a substrate as food and fodder, and the discarded grits contain nutrient used as animal feed.

Marcha is used as an amylolytic starter supplement for all functional microorganisms in *kodo ko jaanr* fermentation (Thapa and Tamang 2004). Mycelial molds have roles only in the initial phase of fermentation, mostly in saccharification of the substrates. Yeasts *Pichia anomala*, *Saccharomyces cerevisiae*, *Candida glabrata*, and *Saccharomycopsis fibuligera* and LAB *Pediococcus pentosaceus* and *Lactobacillus bifermentans* have been recovered in *kodo ko jaanr* samples. The population of filamentous molds that originated in *marcha* declines daily during *in situ* fermentation of *kodo ko jaanr* and finally disappears after the fifth day (Thapa and Tamang 2004).

11.3.2.5 Sujen

It is a mild alcoholic beverage popular among the Deori, an ethnic community in Assam (Deori et al. 2007). It is also considered pure and used as a holy water by Deoro priests during various festivals and ceremonies. During *sujen* preparation, first a natural starter called *mod pitha* is prepared, and then *sujen* is fermented (Deori et al. 2007). Several types of plant species are used for the preparation of the *mod pitha* starter culture. Five kg of glutinous rice is soaked for about 2 hours in water, cleaned, and then mixed properly with the dried plant parts and ground in a *dheki,* a wooden grinder, along with old *mod pithas* starter cultures. The ground starter culture powder is placed in a vessel for fermentation for a duration of 7–15 days. After fermentation, it is diluted for consumption (Das et al. 2017).

11.4 HEALTH BENEFITS OF FERMENTED FOODS AND BEVERAGES

Many fermented foods also contain live microorganisms that may improve gastrointestinal health and provide other health benefits, including lowering the risk of type 2 diabetes and cardiovascular diseases. The number of organisms in fermented foods can vary significantly, depending on how products were manufactured and processed, as well as conditions and duration of storage (Wu et al. 2019). In addition, we review the metabolic changes that take place within the food matrix during probiotic fermentation, as reported by Chung et al. (2018). Probiotics are defined as live microorganisms that, when administered in adequate amounts, confer a health benefit on the host (Tamang et al. 2016). Probiotic organisms used in foods must have the ability to resist gastric juices and exposure to bile and be able to proliferate and colonize the digestive tract (Anandharaj et al. 2014). Bioactive compounds present in plant-based foods and their metabolites derived from gut microbiota and endogenous metabolism represent thousands of chemical structures of potential interest for human nutrition and health. Probiotics in combination with prebiotics give more productive results in the host by improving the survival of beneficial live microbes or activating the metabolism of health-promoting bacteria in the gastrointestinal tract (Anandharaj et al. 2014). Ingestion of probiotic bacteria has the potential to stabilize the immunological barrier in the gut mucosa by reducing the generation of local inflammatory cytokines. Most evidence for a role of aberrant gut microflora in inflammatory bowel disease, however, derives from experimental animal models. Transgenic mice with targeted deletion of the T-cell receptor (TCR α) spontaneously develop colitis in response to the gut microflora (Wu et al. 2019). If organized gut-associated lymphoid tissue is removed from the mice by appendicectomy at neonatal age, but not later, tolerance to gut microflora develops, and the mice do not develop colitis, indicating that the initial colonization pattern may determine subsequent immunological processes. The limited availability of controlled data in human inflammatory disease calls for more research on the effects of specific probiotic strains on the distinct forms of inflammatory bowel disease and their complications (Wu et al. 2019).

11.4.1 Cholesterol-Lowering Activity of Lactic Acid Bacteria

There are different fermented foods and alcoholic beverages which have lactic acid bacterial populations. This combination of live microbial organisms and metabolites contributes to a wide range of health benefits (Chung et al. 2018). Cardiovascular disease (CVD) is one of the leading causes of death in the eastern world, with high levels of serum cholesterol being a major risk factor for the disease. Diet can play a major role in the management of serum cholesterol levels and thus one's risk of developing CVD (Wang et al. 2018). It has been shown that milk and especially fermented milks are able to reduce serum cholesterol levels in animal trials. While cholesterol reduction varied from one grain to another, these differences did not reflect the country of origin of the grain; Yugoslavian grains had both the highest and lowest levels of cholesterol reduction (Marsh et al. 2014). Bile salt hydrolase deconjugates bile acids, and, as deconjugated bile salts is less soluble and less efficiently reabsorbed from the intestinal lumen, this leads to increased bile salt excretion in the feces (Wu et al. 2019). BSH deconjugation contributes to cholesterol-lowering abilities, as cholesterol is utilized in bile acid synthesis by *Lactiplantibacillus plantarum* (Wu et al. 2019).

11.4.2 Suppression of Pathogens

One of the main ways through which probiotics-containing food products can exert beneficial effects is altering the gut microbiota. This can be done either through the introduction of new species or strains into the gastrointestinal tract or by promoting the growth of beneficial microbes which are already present. Some examples are presented here. In multiple studies, consumption of kefir or kefiran in an animal model has been associated with an increase in microbes thought of as beneficial, such as *Lactobacillus* and *Bifidobacterium*, while simultaneously decreasing harmful microbial species such as *Clostridium perfringens* (Wu et al. 2019). Specific strains of *Lactobacillus* isolated from kefir have been shown to adhere to Caco-2 cells and inhibit the adherence of *Salmonella typhimurium* and *Escherichia coli* O157:H7. Many *Clostridium* species are present as commensal members of the intestinal microbiome (Chung et al. 2018). However, imbalances of the microbiota may lead to certain infections caused by these microorganisms, mainly *Clostridium butyricum*, *Clostridium difficile*, and *Clostridium perfringens*. The antimicrobial activity and probiotic potential of *Bifidobacterium* and *Lactobacillus* against various types of *Clostridium* suppress their growth and improve gut microbiota (Chung et al. 2018). The perturbation of infant gut microbiota caused by cesarean delivery is partially restored by exclusive breastfeeding. *Bifidobacterium* is known for its health-promoting properties and ability to stimulate the immune response (Quagliariello et al. 2016).

11.4.3 Antibacterial and Antifungal Properties

The fermented food–associated microbiome has shown a wide range of antibacterial and antifungal activities (Sha et al. 2017, 2018). Many microbial cultures isolated from fermented foods showed a wide range of antibacterial and antifungal activities against pathogenic bacterial and fungal species and were found to have antimicrobial activity equal to ampicillin, azithromycin, ceftriaxone, amoxicillin, and ketoconazole against many of these species (Rodrigues et al. 2018).

11.4.4 Antitumor Effects

The cell free fraction of kefir has shown antitumor activity *in vitro*, when it was observed to have a dose-dependent antiproliferative effect on the gastric cancer cell line (Meng et al. 2018). In addition to the promotion of cell death in cancerous cells, antimutagenic effects have been demonstrated in studies with known carcinogens such as methylmethanosulphate, methy-lazoxymethanol, sodium azide, aflatoxin B1, and 2-aminoanthracene, as indicated by the Ames test (Kirjavainen et al. 2001). Interleukin-17A (IL-17A) is a pro-inflammatory cytokine produced by T_{H17} cells that participates and contributes to host defense and autoimmune disease. Authors recently reported the antitumor properties of the probiotic strain of *Lacticaseibacillus casei* BL23 in mice and T_{H17}cells. The beneficial effects of yoghurt consumption on health and on the improvement of the mucosal immune system are well established, as is the diet-associated risk of colon cancer. Anticancer activity of buttermilk against colon cancer cells was reported by (Meng et al. 2018). There are various differential protective effects of red wine polyphenol extracts on colon carcinogenesis. Yogurt starter cultures *Streptococcus thermophilus* and *Lactobacillus bulgaricus* modulate the immune response in a mouse model of dextran sulfate sodium–induced colitis, which could play an important role in potential cancer treatment (Meng et al. 2018).

11.4.5 Wound-Healing Activity

The antimicrobial properties of kefir may lead to its use for non-traditional applications. Indeed, when rats bearing open wounds inoculated with *S. aureus* were treated with a gel made from kefir grains, it was found that the wounds healed at a much faster rate than was observed in control rats that received no treatment or rats that received a traditional treatment of 5 mg/kg neomycin-clostebol emulsion (Ong et al. 2019). *Lactiplantibacillus plantarum* aids in wound healing and suppresses *Staphylococcus aureus* infection at wound sites, reported by Ong et al. (2019). Furthermore, a study using a rabbit model for contaminated open wounds also found that gel made from kefir grain resulted in quicker healing times and clearing of infection (Ong et al. 2019). Beneficial lactic acid probiotic bacteria *Limosilactobacillus reuteri* stimulate oral wound healing (Ghatani and Tamang 2017; Sha et al. 2019). Current research into original therapies to treat intestinal inflammation is focusing on no-drug therapies. KLD is a mixture of krill oil (KO), probiotic *Limosilactobacillus reuteri*, and vitamin D (VitD3). Krill oil, vitamin D and *Limosilactobacillus reuteri* cooperate to reduce gut inflammation (Ong et al. 2019).

11.4.6 Immunomodulatory Effects

Probiotics in fermented foods have attracted considerable attention lately as treatment options for immune diseases. The immunomodulatory activity of *Lactiplantibacillus plantarum* was reported by Meng et al. (2018). Reports mentioned that when young rats inoculated intra-duodenally with cholera toxin (CT) were fed kefir, the levels of anti-CT IgA in the serum increased, as did the secretion levels of anti-CT IgA in the Peyer's patches, the mesenteric lymph nodes, the spleen, and the intestinal lamina propria compared to CT alone (Wu et al. 2019). A purified neutral exopolysaccharide (EPS) was obtained from *Lactiplantibacillus plantarum* consisting of glucose and galactose in an approximate molar ratio of 2.13:1.06 and had a molecular weight of 1.34×10^5 Da. This exopolysaccharide showed immunomodulatory activity reported by Wu et al. (2019) and mmunomodulatory activity of polysaccharides from soybean residue fermented with *Morchella esculenta,* is a potent component for human hepatocellular and cervical carcinoma treatment as reported by Meng et al. (2018). Dairy *Propionibacterium* (PAB) are used as a ripening starter in combination with lactic acid bacteria for dairy products such as Swiss-type cheese, which showed immunomodulatory potential, as reported by (Foligné et al. 2016).

11.4.7 Anti-Allergenic Effects

Allergic diseases have been on the rise in the developed world for decades, leading to higher incidences of conditions such as asthma and food allergies (Zhu et al. 2011). Many allergies, especially those related to food, are developed early in life, with most food allergies developing within the first two years of life. Although most food allergies developed early in life do not persist, some can become lifelong conditions (Zhu et al. 2011). Recent research work has shown that an increasingly important factor in determining if a child develops allergic disease, be it food allergy or asthma, is the level of complexity and the specific organisms present in the gut microbiota (Zullo and Ciafardini 2019). Higher levels of *Bifidobacterium* and group 1 lactobacilli (obligate heterofermentative lactobacilli such as *L. acidophilus, L. delbrueckii*, and *L. helveticus*) in the gut of infants have been associated with a lower incidence of allergic disease later in life (Zhu et al. 2011), and both kefir and kefiran have been observed to exert these effects on the gut microbiota in animal trials. Supplementation with *Bifidobacterium* has been shown to influence the intestinal microbiota of weaning infants by reducing levels of *Bacteroides* and has been associated with lower incidence of food allergies (Meng et al. 2018; Ghatani and Tamang 2017). Studies with antibiotics in the early life period have also highlighted the importance of appropriate microbial stimulation of the immune system for protection against asthma development (Zhu et al. 2011).

11.4.8 Health Benefits of Yeast

Although the majority of commercialized probiotic microbes are bacteria such as *lactobacilli* and *bifidobacteria,* there are some yeast species and strains that have been recognized to have probiotic properties, such as *Saccharomyces boulardii* (Wen et al. 2008; Sha et al. 2016, 2019). *S. boulardii* has been shown to improve the symptoms of *Clostridium difficile*–associated diarrhea as well as reducing inflammation and altering the immune state and reactions in the gut, leading to its adoption as a treatment for *C. difficile* diarrhea (Wen et al. 2008). Some yeasts from kefir have also shown immunomodulatory activities. *K. marxianus* has been shown to have the ability to adhere to Caco-2 cells (Wen et al. 2008). The health-enhancing yeast *Saccharomyces cerevisiae* is in two types of commercial products for animal nutrition (Wen et al. 2008). Probiotic potential, antioxidant activity, and phytase production were observed in *Saccharomyces cerevisiae* var. *boulardii* and *Kluyveromyces fragilis, Saccharomyces cerevisiae*, and *Pichia membranifaciens*, as reported by (Menezes et al. 2020). *Candida diddensiae, Nakazawaea wickerhamii, Candida adriatica, Wickerhamomyces anomalus,* and *Yamadazyma terventina* were used in this study and compared with the reference yeast *Saccharomyces boulardii* (Sha et al. 2019). The live oil-borne yeast strains synthesize polyunsaturated fatty acids (PUFAs) in quantities greater than those found in olive oil, which provide health benefits (Zullo and Ciafardini 2019). The consortia of yeast, molds, and lactic acid bacteria play an important role in fermentation as well as probiotics (Sha et al. 2017, 2018; Tamang et al. 2016). Next-generation sequencing tools are an important tool to investigate the complete microbiome of any fermented foods and alcoholic beverages (Sha et al. 2019; Tamang et al. 2016) and their relevance to provide health benefits.

11.5 CONCLUSION

Traditional fermented foods and alcoholic beverages are attractive alternatives to fulfil the nutritional and food requirements of a large number of people. In this chapter we have compiled all the information regarding traditional fermented foods and alcoholic beverages of northeast India and their methods of preparation, consumption, sales, ethnic significance, microbial diversity, and functionality. These fermented foods and alcoholic beverages are related to the livelihood of the ethnic tribes of the entire region of northeast India and north Bengal. Due to the indigenous knowledge of ethnic tribes and the new methods introduced by immigrants arriving from other states of India, a wide variety of fermented foods are produced in north Bengal. These fermented foods and alcoholic beverages have the potential to grow into a small to medium-sized industry if proper scientific and technical support is extended to the existing indigenous practices of home-based fermentations. Ancient people throughout the world are real bio-technologists, having independently discovered the value of fermenting food as a sustainable and cost-effective means of preservation, improving nutritional quality as well as enhancing sensory properties. The fermentation of dairy products, pulses, cereals, and other substrates to produce fermented foods and mild alcoholic beverages with health-promoting properties is indigenous to many regions of the world. The ethnic communities of northeast India and north Bengal used to ferment various raw materials naturally by using their home-based traditional knowledge and produce; hence ethnic people are using a mechanism of rural biotechnology for the preservation of their agricultural harvest as well as for the production of various fermented foods and alcoholic beverages for their consumption and livelihood.

Fermentation is the breakdown of polysaccharides like starch and sugar by bacteria and yeast and is an ancient technique of preserving food and preparing fermented foods. Common Indian fermented foods include *dahi, mohi, churpii, gundruk, kinema, sinki, khalpi,* and *haria*. Fermentation not only gives fermented products a unique sensory signature but can also improve the nutritional value and digestibility of foods in several ways. The process of fermentation can produce vitamins, antioxidants, and molecules that lower blood pressure and inflammation. Beyond nutritional value, fermented foods which contain live bacteria may influence the intestinal microbiota. Fermented foods and alcoholic beverages are sources of live beneficial microorganisms known as probiotics, with several health-promoting activities, as shown in Figure 11.1. These microbes have various beneficial roles such as anti-allergic, anti-cancer, cholesterol-lowering, and immunomodulatory properties. The wide range of potential health-promoting effects of fermented foods could lead to further expansion of the popularity of traditional fermented food products that are manufactured using microorganisms. A complex interactive relationship between food and the gut microbiota exists and is receiving intense scrutiny to provide information on how the interaction can produce either positive or negative effects on human health. In the current state of human health sciences, precision nutrition and personal medicine dominate our future direction based on studies of human health. Hence, these fermented foods and mild alcoholic beverages are playing a significant role in promoting human health.

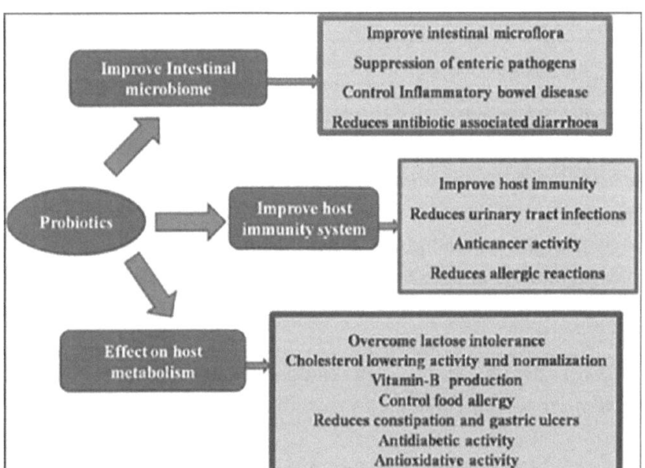

FIGURE 11.1 Health benefits of probiotics associated with fermented foods and alcoholic beverages.

REFERENCES

Amati, H. and Begum, S.S. 2007. *Kiad*-a popular local liquor of Pnar tribe of Jaintia hills district, Meghalaya. *Journal of Traditional Knowledge*, *6*(1), 133–135. https://www.cabidigitallibrary.org/doi/full/10.5555/20073117495

Anandharaj, M., Sivasankari, B. and Rani, R.P. 2014. Effects of probiotics, prebiotics, and synbiotics on hypercholesterolemia: A review. *Chinese Journal of Biology*, *2014*, 572754. https://doi.org/10.1155/2014/572754

Anupma, A., Pradhan, P., Sha, S.P. and Tamang, J.P. 2018. Traditional skill of ethnic people of the Eastern Himalayas and North-East India in preserving microbiota as dry amylolytic starters. *Indian Journal of Traditional Knowledge*, *17*, 184–190.

Batra L.R. 1986. Microbiology of some fermented cereals and grain legumes of India and vicinity. In: Hesseltine CW, Wang HL (eds), *Indigenous fermented food of non-Western origin*. Cramer, Berlin, pp. 85–104.

Batra, L.R. and Millner, P.D. 1974. Some Asian fermented foods and beverages and associated fungi. *Mycologia*, *66*, 942–950. https://doi.org/10.1080/00275514.1974.12019699

Bhatia, A.K., Singh, R.P. and Atal, C.K. 1977. 'Chhang' the fermented beverage of Himalayan folk. *Indian Food Packer*, *4*, 1–8.

Bora, S.S., Keot, J., Das, S., Sarma, K. and Barooah, M. 2016. Metagenomics analysis of microbial communities associated with a traditional rice wine starter culture (*Xaj-pitha*) of Assam, India. *3Biotech*, *6*(2), 153. https://doi.org/10.1007/s13205-016-0471-1

Caggia, C., Restuccia, C., Pulvirenti, A. and Giudici, P. 2001. Identification of *Pichia anomala* isolated from yogurt by RFLP of the ITS region. *International Journal of Food Microbiology*, *71*, 71–73. https://doi.org/10.1016/S0168-1605(01)00556-6.

Chakrabarty, J., Sharma, G.D. and Tamang, J.P. 2014. Traditional technology and product characterization of some lesser-known ethnic fermented foods and beverages of North Cachar Hills District of Assam. *Indian Journal of Traditional Knowledge*, *13*(4), 706–715.

Chen, B., Wu, Q. and Xu, Y. 2014. Filamentous fungal diversity and community structure associated with the solid state fermentation of Chinese Maotai-flavor liquor. *International Journal of Food Microbiology*, *179*, 80–84. https://doi.org/10.1016/j.ijfoodmicro.2014.03.011

Chung, H.J., Sim, J.H., Min, T.S. and Choi, H.K. 2018. Metabolomics and lipidomics approaches in the science of probiotics: A review. *Journal of Medicinal Food*, *21*(11), 1086–1095. https://doi.org/10.1089/jmf.2017.4175

Cocolin, L., Alessandria, V., Dolci, P., Gorra, R. and Rantsiou, K. 2013. Culture independent methods to assess the diversity and dynamics of microbiota during food fermentation. *International Journal of Food Microbiology*, *167*, 29–43. https://doi.org/10.1016/j.ijfoodmicro.2013.05.008

Cocolin, L., Bisson, L.F. and Mills, D.A. 2000. Direct profiling of the yeast dynamics in wine fermentations. *FEMS Microbiology Letter*, *189*(1), 81–87. https://doi.org/10.1111/j.1574-6968.2000.tb09210.x

Cocolin, L., Manzano, M., Aggio, D., Cantoni, C. and Comi, G. 2001. A novel polymerase chain reaction (PCR)-denaturing gradient gel electrophoresis (DGGE) for the identification of *Micrococcaceae* strains involved in meat fermentations: Its application to naturally fermented Italian sausages. *Meat Science*, *58*, 59–64. https://doi.org/10.1016/S0309-1740(00)00131-5

Das, A.J., Miyaji, T. and Deka, S.C. 2017. Amylolytic fungi in starter cakes for rice beer production. *Journal of General and Applied Microbiology*, *63*(4), 236–245. https://doi.org/10.2323/jgam.2016.11.004

Deori, C., Begum, S.S. and Mao, A.A. 2007. Ethnobotany of Sujen – A local rice beer of Deori tribe of Assam. *Journal of Traditional Knowledge*, *6*(1), 121–125. https://www.cabidigitallibrary.org/doi/full/10.5555/20073117510

Dobson, A., Cotter, P.D., Ross, R.P. and Hill, C. 2012. Bacteriocin production: a probiotic trait? *Applied and Environmental Microbiology*, *78*(1), 1–6. https://doi.org/10.1128/AEM.05576-11

Dolci, P., Alessandria, V., Rantsiou, K. and Cocolin, L. 2015. Advanced methods for the identification, enumeration, and characterization of microorganisms in fermented foods. In: W.H. Holzapfel (Eds.) *Advances in Fermented Foods and Beverages*. London: Elsevier, 157–176. https://doi.org/10.1016/b978-1-78242-015-6.00007-4.

Ercolini, D. 2004. PCR-DGGE fingerprinting: Novel strategies for detection of microbes in food. *Journal of Microbiological Methods*, *56*, 297–314. doi:10.1016/J.MIMET.2003.11.006.

Eren, A.M., Esen, O.C., Quince, C., Vineis, J.H., Morrison, H.G., Sogin, M.L. and Delmont, T.O. 2015. Anvi'o: an advanced analysis and visualization platform for 'omics data. *Peer J*, *3*, e1319. https://doi.org/10.7717/peerj.1319

Esteve-Zarzoso, B., Belloch, C., Uruburu, F. and Querol, A. 1999. Identification of yeasts by RFLP analysis of the 5.8S rRNA gene and the two ribosomal internal transcribed spacers. *International Journal of Systematic and Evolutionary Microbiology*, *49*, 329–337. https://doi.org/10.1099/00207713-49-1-329.

Foligné, B., Parayre, S., Cheddani, R., Famelart, M.-H., Madec, M.-N., Plé, C., Breton, J., Dewulf, J., Jan, G. and Deutsch S.-M. 2016. Immunomodulation properties of multi-species fermented milks. *Food Microbiology*, *53*, 60–69. https://doi.org/10.1016/j.fm.2015.04.002

Ghatani, K. and Tamang, B. 2017. Assessment of probiotic characteristics of lactic acid bacteria isolated from fermented yak milk products of Sikkim, India: Chhurpi, shyow, and khachu. *Food Biotechnology*. *31*, 210–232. https://doi.org/10.1080/08905436.2017.1335212

Ghosh, C. and Das, A.P. 2004. Preparation of rice beer by the tribal inhabitants of tea gardens in Terai of West Bengal. *Indian Journal of Traditional Knowledge*, *3*(4), 373–382.

Greppi, A., Rantisou, K., Padonou, W., Hounhouigan, J., Jespersen, L., Jakobsen, M. and Cocolin, L. 2013. Yeast dynamics during spontaneous fermentation of mawè and tchoukoutou, two traditional products from Benin. *International Journal of Food Microbiology*, *165*(2), 200–207. https://doi.org/10.1016/j.ijfoodmicro.2013.05.004

Hesseltine, C.W. and Ray, M.L. 1988. Lactic acid bacteria in murcha and *ragi*. *Journal of Applied Microbiology*, *64*(5), 395–401. https://doi.org/10.1111/j.1365-2672.1988.tb05096.x

Iwen, P.C., Hinrichs, S.H. and Rupp, M.E. 2002. Utilization of the internal transcribed spacer regions as molecular targets to detect and identify human fungal pathogens. *Medical Mycology*, *40*(1), 87–109. https://doi.org/10.1080/mmy.40.1.87.109

Jamir, B. and Deb, C.R. 2014. Studies on some fermented foods and beverages of Nagaland, India. *International Journal of Fermented Foods, 3*(2), 127–138. https://doi.org/10.5958/2321-712X.2014.01314.3

Jamir, N. S. and Rao, R.R. 1990. Fifty new or interesting medicinal plants used by the Zeliang of Nagaland (India). *Ethnobotany*, *2*(1), 11–18.

Jeyaram, J., Anand Singh, Th., Romi, W., Ranjita Devi, A., Mohendro Singh, W., Dayanidhi, H., Rajmuhon Singh, N. and Tamang, J.P. 2009. Traditional fermented foods of Manipur. *Indian Journal of Traditional Knowledge*, *8*(1), 115–121.

Jeyaram, K., Singh, W.M., Capece, A. and Romano, P. 2008. Molecular identification of yeast species associated with '*Hamei*'-a traditional starter used for rice wine production in Manipur, India. *International Journal of Food Microbiology*, *124*(2), 115–125. https://doi.org/0.1016/j.ijfoodmicro.2008.02.029

Jeyaram, K., Tamang. J.P., Capece, A. and Patrizia, R.P. 2011. Geographical markers for *Saccharomyces cerevisiae* strains with similar technological origins domesticated for rice-based ethnic fermented beverages production in North East India. *Antonie van Leeuwenhoek*, *100*, 569–578. https://doi.org/10.1007/s10482-011-9612-z

Jung, M.J., Nam, Y.D., Roh, S.W. and Bae, J.W. 2012. Unexpected convergence of fungal and bacterial communities during fermentation of traditional Korean alcoholic beverages inoculated with various natural starters. *Food Microbiology*, *30*(1), 112–123. https://doi.org/10.1016/j.fm.2011.09.008

Kirjavainen, P.V., Apostolou, E., Arvola, T., Salminen, S.J., Gibson, G.R. and Isolauri, E. 2001. Characterizing the composition of intestinal microflora as a prospective treatment target in infant allergic disease. *FEMS Immunology & Medical Microbiology, 32*(1), 1–7. https://doi.org/10.1111/j.1574–695X.2001.tb00526.x

Kobayashi, Y., Tubaki, K. and Soneda, M. 1961. Several moulds and a yeast used for brewing native beer (*Kodok Jar*) among the Sikkimese of India. *Journal of Japanese Botany*, *36*, 321–331.

Korabecna, M. 2007. The variability in the fungal ribosomal DNA (ITS1, ITS2, and 5.8 SrRNA gene): its biological meaning and application in medical mycology. *Communicating Current Research Education Top Trend Application Microbiology*, *2*, 783–787.

Kumar, V. and Rao, R.R. 2007. Some interesting indigenous beverages among the tribals of Central India. *Indian Journal of Traditional Knowledge*, *6*(1), 141–143.

Las Heras-Vazquez., Javier, F., Mingorance-Cazorla, L., Clemente-Jimenez, J.M. and Rodriguez, V.F. 2003. Identification of yeast species from orange fruit and juice by RFLP and sequence analysis of the 5.8S rRNA gene and the two internal transcribed spacers. *FEMS Yeast Research*, *3*(1), 3–9. https://doi.org/10.1111/j.1567-1364.2003.tb00132.x

Lv, X.C., Huang, X.L., Zhang, W., Rao, P.F. and Ni, L. 2013. Yeast diversity of traditional alcohol fermentation starters for *Hong Qu* glutinous rice wine brewing, revealed by culture-dependent and culture-independent methods. *Food Control*, *34*, 183–190. https://doi.org/10.1016/J.FOODCONT.2013.04.020

Marsh, A.J., Hill, C., Ross, R.P. and Cotter, P.D. 2014. Fermented beverages with health-promoting potential: past and future perspectives. *Trends in Food Science & Technology, 38*(2), 113–124. https://doi.org/10.1016/j.tifs.2014.05.002

Mayo, B., T.C.C. Rachid, C., Alegría, Á., M.O Leite, A., S Peixoto, R. and Delgado, S. 2014. Impact of next generation sequencing techniques in food microbiology. *Current Genomics*, *15*(4), 293–309.

Menezes, A.G.T., Ramos, C.L., Cenzi, G., Melo, D.S., Dias, D.R. and Schwan, R.F. 2020. Probiotic potential, antioxidant activity, and phytase production of indigenous yeasts isolated from indigenous fermented foods. *Probiotics and Antimicrobial Proteins*, *12*, 280–288. https://doi.org/10.1007/s12602-019-9518-z

Meng, Y., Li, B., Da Jin, M.Z., Lu, J. and Huo, G. 2018. Immunomodulatory activity of Lactobacillus plantarum KLDS1. 0318 in cyclophosphamide-treated mice. *Food & Nutrition Research*, *62*, 10.29219. https://doi.org/10.29219/fnr.v62.1296

Ong, J.S., Taylor, T.D., Yong, C.C., Khoo, B.Y., Sasidharan, S., Choi, S.B., Ohno, H. and Liong, M.T. 2019. *Lactobacillus plantarum* USM8613 aids in wound healing and suppresses Staphylococcus aureus infection at wound sites. *Probiotics and Antimicrobial Proteins*, *12*, 125–137. https://doi.org/10.1007/s12602-018-9505-9

Puerari, C., Magalhães-Guedes, T.M. and Schwan, R.F. 2015. Physicochemical and microbiological characterization of chicha, a rice-based fermented beverage produced by Umutina Brazilian Amerindians. *Food Microbiology*, *46*, 210–217. https://doi.org/10.1016/j.fm.2014.08.009

Quagliariello, A., Aloisio, I., Bozzi Cionci, N., Luiselli, D., D'Auria, G., Martinez-Priego, L., Pérez-Villarroya, D., Langerholc, T., Primec, M., Mičetić-Turk, D. and Di Gioia, D. 2016. Effect of bifidobacterium breve on the intestinal microbiota of coeliac children on a gluten free diet: a pilot study. *Nutrients*, *8*(10), 660. https://doi.org/10.3390/nu8100660

Rana, T.S., Datt, B. and Rao, R.R. 2004. Soor: a traditional alcoholic beverage in Tons valley, Garhwal Himalaya. *Indian Journal of Traditional Knowledge*, *3*(1), 59–65.

Rodrigues, P.A., Ferrari, R.G. and Conte-Junior, C.A. 2018. Application of molecular tools to elucidate the microbiota of seafood. *Journal of Applied Microbiology*, *124*(6), 1347–1365. https://doi.org/10.1111/jam.13701

Saikia, B., Tag, H. and Das, A.K. 2007. Ethnobotany of foods and beverages among the rural farmers of Tai Ahom of North Lakhimpur district, Asom, *Indian Journal of Traditional Knowledge*, *6*(1), 126–132.

Scholz, M., Ward, D.V., Pasolli, E., Tolio, T., Zolfo, M. and Asnicar, F. 2016. Strain-level microbial epidemiology and population genomics from shotgun metagenomics. *Nature Methods*, *13*, 435–438. https://doi.org/10.1038/nmeth.3802

Sha, S.P., Ghatani, K. and Tamang, J.P. 2013. Dalbari, a traditional pulses based fermented food of West Bangal. *International Journal of Agriculture and Food Science and Technology*, *4*, 6–10.

Sha, S.P., Anupma, A., Pradhan, P., Prasad, G.S. and Tamang, J.P. 2016. Identification of yeasts by PCR-mediated DGGE in *marcha*, an ethnic amylolytic starter of India. *Journal of Ethnic Foods*, *3*, 292–296. https://doi.org/10.1016/j.jef.2016.11.009

Sha, S.P., Jani, K., Sharma, A., Anupma, A., Pradhan, P., Shouche, Y. and Tamang, J.P. 2017. Analysis of bacterial and fungal communities in Marcha and Thiat, traditionally prepared amylolytic starters of India. *Scientific Reports*, *7*(1), 10967. https://doi.org/10.1038/s41598-017-11609-y

Sha, S.P., Suryavanshi, M.V. and Tamang, J.P. 2019. Mycobiome diversity in traditionally prepared starters for alcoholic beverages in India by high-throughput sequencing method. *Frontiers in Microbiology*, *10*, 348. https://doi.org/10.3389/fmicb.2019.00348

Sha, S.P., Suryavanshi, M.V., Jani, K., Sharma, A., Shouche, Y. and Tamang, J.P. 2018. Diversity of yeasts and molds by culture-dependent and culture-independent methods for mycobiome surveillance of traditionally prepared dried starters for the production of Indian alcoholic beverages. *Frontiers in Microbiology*, *9*, 2237. https://doi.org/10.3389/fmicb.2018.02237

Shrestha, H.N.K., Nand, K. and Rati, E.R. 2002. Microbiological profile of *murcha* starters and physico-chemical characteristics of *poko*, a rice based traditional food products of Nepal. *Food Biotechnology*, *16*, 1–15. https://doi.org/10.1081/FBT-120004198

Sohliya, I., Joshi, S.R., Bhagobaty, R.K. and Kumar, R. 2009. Tungrymbai – A traditional fermented soybean food of the ethnic tribes of Meghalaya. *Indian Journal of Traditional Knowledge*, *8*, 559–561.

Tamang, J.P. 2010. *Himalayan Fermented Foods: Microbiology, Nutrition, and Ethnic Values*. New York, NY: CRC Press, Taylor & Francis Group. ISBN: 9781420093247.295.

Tamang, J.P. and Samuel, D. 2010. Dietary culture and antiquity of fermented foods and beverages. In: J.P. Tamang and K. Kailasapathy (Eds.) *Fermented Foods and Beverages of the World*. New York, NY: CRC Press, Taylor & Francis Group, 1–40. ISBN: 97814 20094954.

Tamang, J.P. and Sarkar, P.K. 1995. Microflora of marcha: an amylolytic fermentation starter. *Microbios*, *81*, 115–122.

Tamang, J.P., Watanabe, K. and Holzapfel, W.H. 2016. Review: Diversity of microorganisms in global fermented foods and beverages. *Frontiers in Microbiology*, *7*, 377. https://doi.org/10.3389/fmicb.2016.00377

Tamang, J.P., Sarkar, P.K. and Hesseltine, C.W. 1988. Traditional fermented foods and beverages of Darjeeling and Sikkim – A review. *Journal Science Food Agricultural 44*(4), 375–385. https://doi.org/10.1002/jsfa.2740440410

Tamang, J.P., Thapa, S., Tamang, N. and Rai, B. 1996. Indigenous fermented food beverages of Darjeeling hills and Sikkim: process and product characterization. *Journal of Hill Research*, *9*(2), 401–411.

Thapa, S. and Tamang, J.P. 2004. Product characterization of kodo ko jaanr: Fermented finger millet beverage of the Himalayas. *Food Microbiology*, *21*(5), 617–622. https://doi.org/10.1016/j.fm.2004.01.004

Tiwari, S.C. and Mahanta, D. 2007. Ethnological observations on fermented food products of certain tribes of Arunachal Pradesh. *Indian Journal of Traditional Knowledge*, *6*(1), 106–110.

Tsuyoshi, N., Fudou, R., Yamanaka, S., Kozaki, M., Tamang, N., Thapa, S. and Tamang, J.P. 2005. Identification of yeast strains isolated from Marcha in Sikkim, a microbial starter for amylolytic fermentation. *International Journal of Food Microbiology*, *99*(2), 135–146. https://doi.org/10.1016/j.ijfoodmicro.2004.08.011

Uchimura, T., Kojima, Y. and Kozaki, M. 1990. Studies on the main saccharifying microorganism in the Chinese starter of Bhutan "Chang poo". *Journal of the Brewing Society of Japan*, *85*(12), 881–887. https://doi.org/10.6013/jbrewsocjapan1988.85.881

Wang, J., Wu, T., Fang, X., Min, W. and Yang, Z. 2018. Characterization and immunomodulatory activity of an exopolysaccharide produced by *Lactobacillus plantarum* JLK0142 isolated from fermented dairy tofu. *International Journal of Biological Macromolecules*, *115*, 985–993. https://doi.org/10.1016/j.ijbiomac.2018.04.099

Wen, L., R.E. Ley, P.Y. Volchkov, et al., 2008. Innate immunity and intestinal microbiota in the development of Type 1 diabetes. *Nature, 455*(7216), 1109. https://doi.org/10.1038/nature07336

Wu, Y., J. Wan, U. Choe, et al., 2019. Interactions between food and gut microbiota: impact on human health. *Annual Review of Food Science and Technology, 10*, 389–408. https://doi.org/10.1146/annurev-food-032818-121303

Yan, Y., Y. Qian, F. Ji, J. Chen, B. Han. 2013. Microbial composition during Chinese soy saucekoji-making based on culture dependent and independent methods. *Food MIbrobiology, 34*, 189–195. https://doi.org/10.1016/j.fm.2012.12.009

Zhu, Y., T.M. Luo, C. Jobin, H.A. Young. 2011. Gut microbiota and probiotics in colon tumorigenesis. *Cancer Letters, 309*(2), 119–127. https://doi.org/10.1016/j.canlet.2011.06.004

Zullo, B.A. and Ciafardini, G. 2019. Evaluation of physiological properties of yeast strains isolated from olive oil and their in vitro probiotic trait. *Food Microbiology, 78*, 179–187. https://doi.org/10.1016/j.fm.2018.10.016

Lactic Acid Bacteria and Biotechnology of Cereal and Legume-Based Fermented Foods from India

12

Pooja Singh, Ram Kulkarni, Kalidas Shetty, and Vinay Rale

12.1 INTRODUCTION

The field of biotechnology pioneered with the practice of food fermentation, a practice that is followed traditionally in many cultures across the globe since many centuries. Subsequently the study of the processes at play, agents responsible, presence and role of specific microflora, and use and developments in the fermentation technology further expanded and strengthened this field, making it more profitable and productive. Most of the fermented food production and preservation practices adhered to the age-old traditions being passed through generations and hence many such practices failed to evolve and sustain in the current food security crisis. According to the International Scientific Association for Probiotics and Prebiotics (ISAPP), fermented foods are defined as "foods made through desired microbial growth and enzymatic conversions of food components" (Marco et al. 2021). Hence the key players in this process are the varied microbial communities, consisting of multiple genera and taxa, that establish in the food being fermented. A mere change in the texture or properties of the food matrix does not necessitate this shift. The specificity and clarity of the microflora in the fermented food is primarily responsible for the food to be categorized in the "probiotic food" category. The process of food fermentation has been well established to increase the probiotic and organoleptic properties of the food items by the varied enzymes, acids, and other metabolites produced by the beneficial fermenting microbes, shifting many fermented products into the "functional food" category as well. Hence a detailed study of the microflora of the fermented food is necessarily pertinent. Additionally, many fermented foods have been studied to contain antimicrobial compounds that aid in increasing the shelf life of the product and hence require a detailed analysis (Achi and Ukwuru 2015). Although many popular fermented food like cheese, beer, and wine are well studied (Sicard and Legras 2011), not much is known in detail about the traditional food preparations made from cereals and legumes, especially those that are prepared from the locally available raw materials. Cereal based food is the primary, non-dairy, fermented food category in India. Since microbial action forms the core of the fermented food industry, understanding the microbial involvement in the process and their limitations associated with different foods assumes importance. Extensive work in this field has been done by the group of Tamang et al., especially on the fermented food from the hilly regions of North-east India (Tamang 2022a). This calls for more intense research on the biotechnological aspects of fermented foods especially the lesser-known processes prevalent in the Indian subcontinent, including the study of millets for fermented food preparations. Although widely grown and available in the Indian subcontinent, not much has been documented on the fermentation food potential of different millets. Fermentation, being essentially a microbiological activity, all the prominent and profound attributes of fermented food is dependent on the type, number, and succession of microbial community in the specific food matrix. Whether food has been prepared by natural fermentation (wild ferments), or by using a primary starter culture (culture dependent ferments), the physiological behavior of microorganisms is the crux of the process. As a thumb rule, members of the lactic acid bacteria (LAB) have been implicated in this process and are presumed to impart numerous organoleptic and health benefits to the relevant food substrate (Dimidi et al. 2019).

This chapter specifically explores the Indian fermented foods sourced from cereals and legumes and the challenges

DOI: 10.1201/9781003605300-12

associated with their commercial production. We explore the various aspects of metabolomics of lactobacilli and enterococci to understand and elucidate their participatory role. An interesting aspect highlighted in this chapter is the scope of bioprocess engineering in increasing the probiotic and palatable potential of these foods with value added nutritional qualities. This will enable the fermented food to be selective (targeted towards specific health needs) and generic (mass appeal with the benefit of addressing nutritional food crisis) at the same time.

12.2 CEREALS AND LEGUMES AS SUBSTRATE FOR FERMENTED FOOD PREPARATION

The history of fermented food preparations dates to the Vedic and post Vedic period in the history of India, wherein preparations like *medaka* (spiced rice beer) and *prasanna* (spiced barley or wheat beer) have been mentioned in the ancient texts (Ray et al. 2016). These and other such preparations were known to be made under rustic local conditions in earthen pots using locally available materials. A thorough knowledge of the existing ecosystem and requirements for a successful fermentation process were passed down through generations in the native population. Although the processes were well managed and the organoleptic and nutritional value of the products well established, the technicalities were not studied scientifically in much detail and the products were to a large extent dependent on the environmental conditions and a natural fermentation progression. Nonetheless, these technologies have been well integrated in the Indian culinary art and tradition from time immemorial and have been successfully passed down through generations in common households. More importantly the knowledge transfer was not merely perfunctory, but pragmatic home-grown approach to achieve an economical means of food preservation in the absence of refrigeration and other cold storage processes.

India, with a history of over 800 years of civilization, has over 1000 different food and beverages preparations and the territorial "ethno-microbiological" knowledge plays a vital part in the passing down of this practice over multiple generations (Tamang 2022b). Apart from milk, cereals and legumes have been used since generations for making many savory food items in the Indian subcontinent, especially in the Southern parts of India. Wheat, rice, millets, and other cereals have all been documented to be used for the preparation of a variety of fermented foods, primarily in combination with some legumes that make the fermentation processes easier, and the products more wholesome. Many fermented foods are made using nutritionally diverse substrates, for example, mix of cereals with legumes, to counter the deficiency of any vital component that may decrease the overall nutritional quotient of the fermented food. Cereals contain high levels of methionine and cysteine but are generally deficient in the amino acid lysine. On the other hand, legumes are deficient in sulfur containing amino acids. Mixing of legumes with cereals compensates for this deficiency and makes the fermented food more nutritional, wholesome and a potential functional food (Achi and Ukwuru 2015). Conversion of a "fermented food" to the category of a "functional food" further increases its nutritional value and commercial importance. Role of microbial consortia on fermented foods and their impact on the final food quality has been well established. A complete understanding of the microbiology of such foods now forms a vital aspect of the food production sector.

Several fermented foods, including many sweet dishes, are prepared from cereals mostly wheat and rice as well as millets, based on their availability in the region. Table 12.1 lists a number of cereal based fermented foods (with or without legumes) prevalent and mostly known from the Indian subcontinent in the Western and the Southern states (Table 12.1 a) and Northern and Eastern States (Table 12.1 b), and also highlights the prominent microflora that has been studied so far from these food items (Blandino et al. 2003; Das et al. 2012; Kanwar et al. 2007; Kumar et al. 2013; Rawat et al. 2018; Bhalla and Bhalla 2007; Roy et al. 2007; Ray et al. 2016; Senapati et al. 2016; Singh et al. 2019; Tamang et al. 2016a; Yonzan and Tamang 2010; Thapa and Tamang 2020). Most of the beverages made using fermented cereals have been found to have rice as their substrate (Table 12.1). There are many other lesser-known fermented foods, especially from the North and North-eastern states, with variations in their substrate, or processing time and methods, that yet remain to be scientifically studied.

As evident from Table 12.1, a variety of local traditional foods are prevalent in the hill states of India, including Himachal Pradesh in the North and other North-Eastern states. Consumption of fermented foods is an integral part of their lifestyle. While wheat is the primary substrate for fermented foods in the North, primarily Himachal Pradesh, Eastern and Southern India mostly use rice based fermented foods. Soybean, however, is the widely used ingredient in the North-Eastern States, where use of cereal in making fermented local items is limited. In fact, Soybean based fermented food is very specific to this region of India and has close similarity with similar soybean based fermented food of other East Asian countries including Nepal, Bhutan and Japan (Tamang et al. 2020). Although the Eastern and North-eastern regions have a rich cultural tradition of fermented foods, very little information is available on the microbiology of numerous fermented foods and beverages from the hill states of Himachal and other North-eastern states of India, especially those sourced from cereals and legumes. Most of the fermented foods prepared from cereals are based on the process of natural fermentation and the use of a specific starter culture in such cases has not been manifested. However, as mentioned in Table 12.1, there are reports on the use of some beneficial starter cultures for enhancing the quality of few foods and beverages prepared using various cereals (for example *Malera, Balam and Phab*).

TABLE 12.1A Fermented Foods and Prominent Microflora from Western and Southern India

NAME	*CEREAL*	*LEGUME*	*REGION/STATE*	*STARTER CULTURE SOURCE (IF ANY)*	*PROMINENT MICROFLORA*
Idli	Rice	Urad Daal	Southern India	Natural	*Leuconostoc mesenteroides, Streptococcus faecalis*
Dosa	Rice	Urad Daal	Southern India	Natural	*Leuconostoc mesenteroides, Enterococcus faecalis*
Dhokla	Rice	Bengal Gram	Western India	Curd	*Lactobacillus fermentum, Leuconostoc mesenteroides, Enterococcus faecalis*
Koozhu	Millet and Broken rice	–	Tamil Nadu	Curd	*Enterococcus faecalis, Pediococcus acidilactici, Weissella paramesenteroides, leuconostoc mesenteroides, Lactobacillus fermentum, Lactiplantibacillus plantarum*
Kallappam	Rice (+ coconut)	–	Southern India	Toddy	*Lactobacillus fermentum, Lactiplantibacillus plantarum*
Ambali	Millet (Ragi) and Rice	–	Karnataka	Buttermilk	*Lactobacillus fermentum, Leuconostoc mesenteroides, Enterococcus faecalis*
Pazhaiya soru	Rice	–	Southern India	Curd	*Enterococcus faecalis, Pediococcus acidilactici*
Mor kuzhambhu	–	Gram flour	Tamil Nadu	Buttermilk	*Weissella paramesentroides*
Rabdi	Millet	–	Rajasthan	Buttermilk	*Bacillus* sp., *Micrococcus* sp.
Jilebi	Wheat	Gram flour	Western India	Natural/Curd	*Lentilactobacillus buchneri, Enterococcus faecalis* *Lactobacillus fermentum, Streptococcus lactis*
Anarshe	Rice	–	Western and Northern India	*Marcha*	LAB and yeasts like *Hansenula anomala* and *Mucor rouxianus*
Ambeli	Rice & Ragi	–	Central India	Natural	*Leuconostoc mesenteroides, Lactobacillus fermentum, Enterococcus faecalis*

TABLE 12.1B Fermented Foods and Prominent Microflora from Northern and Eastern India

NAME	*CEREAL*	*LEGUMES*	*REGION/STATE*	*STARTER CULTURE SOURCE (IF ANY)*	*PROMINENT MICROFLORA*
Kulu	Wheat flour	–	Himachal Pradesh	Buttermilk	–
Kadi	–	Gram flour	Himachal Pradesh	Buttermilk/ Curd	–
Siddu/Khobli	Wheat flour supplemented with opium seeds/walnut	Black gram	Himachal Pradesh	Malera	*Saccharomyces cerevisiae, Candida valida*
Chilra or Lwar	Wheat/barley mixed with buckwheat flour	–	Himachal Pradesh	Treh	*Saccharomyces cerevisiae, Debaromyces hansenii, Lactobacillus acidophilus, Leuconostoc mesenteroides, Lactococcus lactis*
Marchu	Wheat flour	–	Himachal Pradesh	Malera	–
Pinni	Wheat flour	–	Himachal Pradesh	Buttermilk	–
Sez	Rice	–	Himachal Pradesh	Balam	*Bacillus* sp., *Saccharomycopsis fibuligera, Kluveromyces maxianus*

(Continued)

TABLE 12.1B (Continued)

NAME	*CEREAL*	*LEGUMES*	*REGION/STATE*	*STARTER CULTURE SOURCE (IF ANY)*	*PROMINENT MICROFLORA*
Jhan chang	Barley	–	Himachal Pradesh	Phab	–
Babru	Wheat	Black gram	Himachal Pradesh	Natural	*Lactiplantibacillus plantarum, Lactococcus lactis, Saccharomyces cerevisiae, Debaromyces hansenii*
Gulgule	Wheat flour	–	Himachal Pradesh	Malera	*Saccharomyces cerevisiae, Saccharomyces fermentati, Leuconostoc mesenteroides, Lactococcus lactis*
Seera	Wheat flour	–	Himachal Pradesh	Natural	*Saccharomyces cerevisiae, Cryptococcus laurentii* and *Torulospora delbrueckii, Bacillus* sp., *Leuconostoc* sp. *Lactobacillus amylovorus*
Bhaturu	Wheat	–	Himachal Pradesh	Malera	*Lactiplantibacillus plantarum, Leuconostoc.* sp., *Lactobacillus mesenteroides, Saccharomyces cerevisiae*
Selroti	Rice & Wheat	–	Himachal Pradesh & Sikkim	Natural/milk	*Saccharomyces cerevisiae, Saccharomyces kluyveri, Pichia burtonii, Leuconostoc mesenteroides, Enterococcus faecium, Pediococcus pentosaceus, Lactobacillus curvatus, Debaryomyces hansenii, Zygosaccharomyces rouxii*
Manna	Wheat	–	Himachal Pradesh	–	–
Thuktal	Barley	–	Himachal Pradesh	–	*Lactobacillus acidophilus, Leuconostoc mesenteroides, Lactobacillus fermentum, Lactobacillus buchneri, Lactiplantibacillus plantarum, Lactococcus lactis, Streptococcus lactis, Enterococcus faecalis.*
Aet, aktori, baari, babroo, bhatooru, chhura, mande/ manna, malpude, tcung, shunali	Wheat		Himachal Pradesh	–	*Lactobacillus plantarum, Lactobacillus acidophilus, Leuconostoc mesenteroides, Lactobacillus lactis*
Khatta	–	Black gram	Himachal Pradesh	–	–
Mangjangkori	Buckwheat bran	–	Himachal Pradesh	Natural	–
Tchhoso roti	Kodra flour	–	Himachal Pradesh	–	–
Tiskori	Wheat bran	–	Himachal Pradesh	Natural	–
Sang	Wheat grain	–	Himachal Pradesh	–	–
Borhe	–	Black gram	Himachal Pradesh	Natural	–
Sepubari	–	Black gram	Himachal Pradesh	Natural	–
Madrah	–	Kidney beans	Himachal Pradesh	Yoghurt	–
Bedvin roti	–	Black gram	Himachal Pradesh	Natural	–

(Continued)

TABLE 12.1B (Continued)

NAME	*CEREAL*	*LEGUMES*	*REGION/STATE*	*STARTER CULTURE SOURCE (IF ANY)*	*PROMINENT MICROFLORA*
Hawaijar	–	Soybean	Hills of North-Eastern India	Natural	*Bacillus subtilis, Bacillus licheniformis*
Torani	Rice	–	Orissa	Yoghurt	–
Taotjo	Wheat/Rice	Soybean	East India	Natural	*Aspergillus oryzae*
Wadi	–	Black gram	Punjab, and West Bengal	Natural	*Lactobacillus fermentum, Leuconostoc mesenteroides*
Wari	–	Black bean & Soybean	Uttar Pradesh	Yoghurt	*Streptococcus thermophilus, Lactobacillus bulgaricus*
Chakuli	Rice	Black gram	Orissa	Natural	–
Enduri pitha, Chhuchipatra pitha, Podo pitha	Rice	Black gram	Orissa	Natural/ Yoghurt	–
Chitou	Rice	Black gram/ Urad Daal	Orissa	Natural	*Lactiplantibacillus plantarum, Bacillus* sp. *Tetragenococcus halophilus, Pediococcus pentosaceus*
Amriti	–	Black gram	West Bengal	Natural	–
Panta Bhat, Poita bhat, Pokhalo	Rice	–	West Bengal, Assam, Orissa	Natural/ Yoghurt	*Streptococcus faecalis, Streptococcus thermophilus, Microbacterium flavum, Lactobacillus bul-garicus, Lactobacillus casei, Pediococcus acidilactici, Saccharomyces* sp.
Kinema	–	Soybean	Sikkim, West Bengal	Natural	*Enterococcus faecium, Bacillus subtilis*
Masyaura	–	Black gram/ Green gram	Sikkim, West Bengal	Natural (aided by colocasia)	*Pediococcus pentosaceous, Pediococcus acidilactici Lactobacillus* sp.
Bari	–	Soybean	Sikkim	Natural	–
Tungrymbai		Soybean	Meghalaya	Natural	*Enterococcus faecium, Bacillus licheniformis*
Axone	–	Soybean	N-E India	Natural	*Bacillus subtilis*
Bekang	–	Soybean	Mizoram	Natural	*Bacillus subtilis, Brevibacillus brevis, Bacillus circulans, Lysinbacillus luciformis*
Aagya, Peron namsing, Yanni perung, Chukchoro, soybean (libi) chhurpi	–	Soybean	Arunachal Pradesh	Natural	*Bacillus* sp.
Peruyyan	–	Soybean	Arunachal Pradesh	Natural	*Bacillus subtilis*
Akhuni	–	Soybean	Manipur	–	*Bacillus subtilis*, LAB
Bekanthu	–	Soybean	Nagaland	–	*Bacillus subtilis*, LAB
Yu	Rice	–	Manipur	*Hamei*	*Saccharomyces cerevisiae, Aspergillus niger, Bacillus* sp.

(Continued)

TABLE 12.1B (Continued)

NAME	*CEREAL*	*LEGUMES*	*REGION/STATE*	*STARTER CULTURE SOURCE (IF ANY)*	*PROMINENT MICROFLORA*
Atingba	Rice	–	Manipur	*Hamei*	*Saccharomyces cerevisiae, Aspergillus niger, Bacillus* sp.
Apong	Rice	–	Assam & Arunachal Pradesh	*Ipoh* or *epop*	*Saccharomyces cerevisiae, Hanseniaspora* sp., *Kloeckera* sp., *Pischia* sp., *Candida* sp.
Zutho	Rice	–	Nagaland	*Khekhrii* or *Grist*	*Saccharomyces. cerevisiae, Rhizopus* sp.
Jou	Rice	–	North-East India	*Amao* or *Angkur*	*Saccharomyces cerevisiae*
Judima	Rice	–	North-East India	*Umhu* or *Humao*	*Pediococcus pentosaceus, Bacillus circulans, Bacillus catarosporous, Bacillus pumilus, Bacillus firmus, Debaryomyces hansenii, Saccharomyces cerevisiae*
Bhaati jaanr	Rice	–	North-East India	*Marcha*	*Hansenula anomala, Rhizopus* sp., *Pichia* spp., *LAB, Pediococcus, Lactobacillus bifermentous*
Chhang/Sra	Rice	–	Himachal Pradesh	Phab	–
Sur	Millet (Ragi)	–	Himachal Pradesh	Natural	–
Rice *jann*	Rice	–	Uttar Pradesh, Himachal Pradesh	Balam/Balma	–
Haria	Rice	–	West Bengal, Orissa, Jharkhand	Bakhar	*Saccharomyces cerevisiae, Lactobacillus fermentum, Bifidobacterium* sp.

Marcha is another community specific starter culture, used to make alcoholic beverage from cereals and millets in some regions in the eastern Himalayas. In fermented foods involving mixed grains and raw materials, for example mix of cereal and legumes, the microbiological processes are markedly different. It has been observed that in legume and cereal composite fermented products, appropriate lactic acid fermentation enhances the safety and shelf life of the product and has a profound effect on palatability as well (Budhwar et al. 2020).

12.3 MICROBIOLOGY OF CEREAL AND LEGUME BASED FERMENTED FOODS

The microorganisms and microbial processes associated with fermenting and fermented foods play a critical role in the nutrient composition, taste, texture and the type and number of beneficial metabolites in the products. Hence a clear understanding of the associated microflora and their specific metabolic attributes forms the foundation of any fermentation-based industry. Although culture dependent studies form the basis of the analysis of microflora in fermented foods, culture independent methods and omics technologies give a clear picture of the dynamic shift and microbial successions at play. Nevertheless, a "core microbiota" is often found to be the dominant and central to the functioning of the whole community (Marco et al. 2021). As seen in Table 12.1, microflora of fermented foods is both diverse and complex, with different substrates harboring and supporting different dominant microorganisms. The prominent organisms found in fermented foods with cereal as substrate include members of *Leuconostoc mesenteroides, Limosilactobacillus fermentum, Lactiplantibacillus plantarum, Lactobacillus acidophilus, Enterococcus faecalis and Weissella paramesentroides. Enterococcus faecium, L. mesenteroides, L. fermentum, Lactobacillus bulgaricus, Streptococcus thermophilus, Pediococcus pentosaceous* and *Pediococcus acidilactici* are found to form the prominent microflora of fermented food with legume as substrate. On the other hand, *Bacillus subtilis* is functionally dominant in foods prepared using soybean (Tamang 2015; Tamang et al. 2020). However, the diversity of microflora in a fermented food is not limited to these organisms and there is a dynamic variability that exists in different food items based on a number of factors, the type of substrate being a prominent one. Fermentation processes with rice as substrate mostly have been found to be dominated by *Lactobacillus bulgaricus* and *Streptococcus thermophilus*, while *L. planetarum* and *Pediococcus* sp. are mostly found in fermentation with maize as substrate (Achi and Ukwuru 2015).

Cereal based foods contain carbohydrates suitable for the growth of the starter culture and the establishment and proliferation of indigenous microflora. Cereals contain non digestible carbohydrates and water-soluble fibers that aid in the growth of bacteria primarily lactobacilli and bifidobacteria. Other organisms prevalent in such foods include species of *Enterococcus*, *Streptococcus* and *Pediococcus* among others. Multiple microbial strains offer the added advantage of greater beneficial effects than a single prevalent strain and hence the study of all specific organisms at play, during and post fermentation process, assumes importance. This specific microbial profiling will also aid in "designing" of the starter culture, used for the fermentation process, as per the target health condition or target population. Microbial profiling of the starter cultures and final fermented food has been undertaken for items from different regions of India. *Malera* and *Phab* are the two starter cultures used in the preparation of many cereals based traditional recipes, including for example *bhaturoo*, *gulgule* and *Jhang chang* from the Himalayan states in India and neighboring countries. Bacterial Denaturing Gradient Gel Electrophoresis (DGGE) profile targeting the V3 region of 16S rRNA of these two traditional starter cultures revealed the dominance of *Bifidobacterium* sp. in both the starter cultures along with the presence of Enterobacteriaceae members such as *Klebsiella* sp. and a pathogenic strain of *Dickeya chrysanthemi* (Sharma et al. 2019). Conventional techniques are insufficient to map the diversity of metabolic changes taking place during the process of fermentation and hence tools like metabolomics play a vital role to decipher the intricacies of this aspect and understand the specific metabolites imparting beneficial properties to the fermented foods. The myriad of compounds, like for example phenolics, flavones, sterols, and oryzonol, get enriched during rice fermentation. Compounds like tocopherols and tocotrienols, formed during rice fermentation, have been shown to exhibit anti-inflammatory cell signalling properties. Nutritious and antimicrobial properties of *Haria* were attributed to the presence of different malto-oligosaccharides (G5-G2), such as maltose (G2), maltotriose (G3, 28.1 mg/g), and maltotetrose (G4, 20.1 mg/g). Further the immune-stimulatory, antioxidant, and antimutagenic activities were attributed to several pyranose derivatives like 2,3,4,5-tetra-O-acetyl-1-deoxy-b-D-glucopyranose, b-D-mannopyranose pentaacetate, b-D-galactopyranose pentaacetate, and 1,2,3,6-tetra-O-acetyl-4-O-formyl-D-glucopyranose. However, studies like these are limited and also not available for the diverse food preparations prevalent across the Indian subcontinent (Ghosh et al. 2015). *Kinema*, a naturally fermented soybean-based food of Northeast India has numerous health benefits and is widely being studied for its microbiological properties. Study of bacteriocin from *Lactiplantibacillus plantarum* isolated from *Kinema* revealed antimicrobial properties against *Kocuria rhizophila* (Goel et al. 2020). *Bacillus*, prominently present in *kinema*, was found to be responsible for the reduction in anti-nutritive compounds during the fermentation process and a possible increase in free amino acids (Tamang et al. 2016b). Presence of both antimicrobial and probiotic properties makes *Kinema* a promising nutrient and protein rich probiotic food. Like other soybean based food, including *akhuni, bekang, hawaijar, peruyaan*, and *tungrymbai*, *kinema* microflora is also dominated by *Bacillus* sp., including *B. subtilis, B. cereus, B. licheniformis, B. circulans, B. thuringiensis*, and *B. sphaericus*. *Kinema* is primarily consumed in the hilly areas of Sikkim and Darjeeling in India, and other neighbouring countries like Nepal and Bhutan. Although difference in the microflora between *kinema* from different regions has been observed, dominance of multiple *Bacillus* sp., including *B. subtilis* followed by *B. licheniformis*, was found to be constant (Kharnaior and Tamang 2021). Nevertheless, the presence of these organisms plays a vital role in the texture of the end product and is responsible for the myriad benefits associated with these soyabean rich foods.

12.3.1 LAB and Fermentation

Lactic acid bacteria are the most commonly and successfully used functional starter culture in the food fermentation industry. One of the major metabolic changes brought about by the lactic acid bacteria during fermentation of cereals is reduction in the pH. This is mainly attributed to the production of lactic acid, as well as to a lesser extent acetic acid. Apart from increasing the sourness of the final fermented food, production of acids by LAB during cereal fermentation also stimulates the inherent proteolytic enzymes in the cereal matrix. Additionally, LAB also have their own proteolytic systems owing to their property of being auxotrophic for many amino acids. Although these proteolytic enzymes have been mostly characterized for the hydrolysis of casein in milk, some studies have also been carried out on their involvement in the protein degradation during sourdough fermentation (Di Cagno et al. 2002). Amino acids and peptides released during proteolysis are metabolized by the fermenting lactobacilli, further contributing to the production of several aroma chemicals including alcohols, carbonyls, and esters (Liu et al. 2008). Since LAB do not have a complete TCA cycle, apart from lactate, pyruvate formed because of sugar consumption is also converted into other metabolites including diacetyl, acetoin, and acetic acid. In case of heterofermentative LAB, ethanol and CO_2 additionally arise because of the phosphoketolase pathway. Collectively, all these small organic molecules affect the sensory attributes of the cereals and other foods fermented by LAB. Starch, an important polysaccharide in cereals is hydrolyzed by the action of α-amylases during the fermentation, releasing maltose and maltodextrins. In addition to the amylolytic activity of the cereal enzymes, some lactobacilli also contribute their own α-amylases (Oguntoyinbo and Narbad 2012). Lactobacilli further have the ability to utilize the α-glucans originating from starch hydrolysis. The genes encoding the required proteins such as the ABC transporter, α-glucosidases, maltose phosphorylase, and β-phosphoglucomutase are organized in maltodextrin operon in some lactobacilli. Amongst the most abundant fructooligosaccharides and sugars in the cereals, sucrose can be utilized by many lactobacilli mostly through the phosphotransferase or the ABC transport systems (Ganzle and Follador 2012).

Many LAB have the ability to produce exopolysaccharides (EPS) that have significant impact on the texture of the fermented cereal products. Additionally, EPS also have several biological impacts on the consumer, including promotion of probiotic microorganisms, by serving as prebiotics, producing short-chain fatty acids, affecting energy metabolism, immunomodulation and assistance in intestinal colonization (Ripari 2019). The EPS broadly fall into two categories, *viz*., homopolysaccharides (HoPS) and heteropolysaccharides (HePS). Biosynthesis of two subtypes of HoPS—glucans and fructans—is governed by glucansucrases and fructansucrases, respectively, with sucrose as the substrate. Many of these enzymes and the encoding genes have been characterized from bacterial genera including *Streptococcus*, *Leuconostoc*, and *Lactobacillus* with many of the constituent species being cereal-associated (Tieking and Gänzle 2005; van Hijum et al. 2006). The biosynthesis of HePS, on the other hand, requires numerous genes organized in operons in LAB (Deo et al. 2019). Due to the complex metabolic and regulatory networks that generate the building blocks for HePS biosynthesis, the HePS yields are generally lower than those of HoPS (Zeidan et al. 2017).

Certain LAB have an ability to produce B-group vitamins and thus have the potential of improving the nutritional value of the foods upon fermentation. *Lactiplantibacillus plantarum* strains are especially known to produce riboflavin and biotin. Since the strains resistant to roseoflavin are overproducers of riboflavin, *Lactobacillus plantarum* strains were isolated from durum wheat flour. When these strains were used for bread-making through sourdough fermentation or pasta-making through a pre-fermentation step, it resulted in 2–3-fold increases in the riboflavin content in pasta and bread (Capozzi et al. 2011). Some strains of *Lactiplantibacillus plantarum*, *Lactococcus lactis*, and *Streptococcus thermophilus* are also able to produce folic acid (Capozzi et al. 2012). One such *Lactiplantibacillus plantarum* strain isolated from injera fermented dough was able to improve folate status in the folate-deficient rats (Tamene et al. 2019). Although *Lactobacillus* and *Bifidobacterium* are the most commonly associated microorganisms with fermented foods (Ray et al. 2016), the diversity of microflora was found to vary with the substrate and depends on several factors including temperature attained during fermentation, duration of fermentation, pH, inoculum size, and the nature of raw material. Depending on the conditions and type of organisms in the microflora, a succession of biochemical changes in the substrate takes place, accumulating many intra and extra cellular enzymes and metabolites, like butyrate, that collectively strengthen gut microbiota, immune health and impart other health beneficial effects.

12.4 FUNCTIONALITY AND NATURE OF CEREAL AND LEGUME-BASED FERMENTED FOODS

Most of the fermented foods are prepared by natural fermentation process carried out by select group of organisms that get developed in the food being fermented. This microbial process of selection and the subsequent microbial action is defined and steered by the environmental conditions that determine the odor, flavor, texture, nutritional properties, and palatability of fermented food. Not only the presence of specific microorganisms, but the succession of their enrichment in the substrate is also vital for the development of specific flavor and texture in the product. Hence, duplicating the precisely similar microbial culture environment and processes using external inoculum is difficult. The naturally developed culture in the fermenting food is in turn responsible for offering additional antimicrobial properties to the product. Protein rich starter raw materials are especially prone to being contaminated by pathogenic microbial flora and hence maintaining, monitoring and use of a starter inoculum for these is especially critical. More clear understanding and adaption of factors like these, enhances the consumer compatibility and acceptance, especially if dealing with specific groups like infants, malnourished people, or patients with specific dietary limitations including those with cardiac or celiac diseases. Fermentation and fermenting microorganisms bring about several changes in the fermented food including bio-enrichment of vitamins, increased antioxidant activity, peptides, and enzymes production, increase in saponin and isoflavone levels and polyglutamic production (Sharma et al. 2020). At the same time, undesirable metabolites and antinutritive compounds like tannins, phytates, polyphenols, and mycotoxins, as well as pathogens if any, are concomitantly removed during the fermentation process. Apart from the usual nutritional benefits derived from cereals, the cereal based fermented food items also contain many antimicrobials, therapeutics, and enhanced nutritive potential. Presence of water-soluble fibers such as beta-glucan and galacto- and fructo-oligosaccharides support the growth of multiple microorganisms, primarily LAB and yeast, that play the key role in producing many metabolites, including short chain fatty acids, that lower the pH and prevent growth of harmful microflora. Further, presence and activity of LAB acts as a natural way to concentrate and enhance nutrients and enrich the fermented product with various therapeutic components like probiotics, prebiotics, antioxidants, phenolics, maltooligomers, antimicrobials, and other bioactive substances (Tamang et al. 2021).

Compared to milk and milk products, cereals have lower consumer acceptance in terms of their nutritional qualities and sensory properties. Deficiency of essential amino acids like lysine, low availability of starch and presence of antinutrients, like polyphenols, tannins, and phytic acids, were prominent reasons for their low acceptance (Sharma et al. 2020). Cereal based food specifically contain many antinutritive factors like lectins, phytic acids, polyphenols, and trypsin inhibitors that get reduced during fermentation making the food more digestible and nutritious (Samtiya et al. 2020). Cereals contain large amount of phytic acid as the *in-planta* reserves of phosphorus which, on the other hand, act as antinutrients for humans since it interferes with the absorption of minerals such as calcium, zinc, and iron. Reduction of fructans in wheat, and raffinose, stachyose and verbascose in soybeans and legumes, brings about a lowering the glycaemic index of the fermented food, thereby increasing its acceptability (Nyyssola et al. 2020). Hence, fermentation was found to be the most low-cost beneficial process through which not only the nutritional value and organoleptic properties, but also the acceptability of cereals and legumes can be greatly enhanced. Enzymatic degradation of phytates by the fermenting microorganisms offer a natural process of reduction in the level of this antinutritional compound, concomitantly elevating the levels of elements such as calcium, soluble iron, and zinc (Nkhata et al. 2018). A few studies have shown the presence of phytase activity in lactic acid bacteria from fermented cereals (Reale et al. 2007). Additionally, these cereal phytases are also active at the lower pH achieved through lactic acid bacterial fermentation, thus contributing significantly towards phytate reduction in the final fermented product (Ganzle 2020). Low pH, primarily due to the many organic acids, as well as production of antimicrobial agents like hydrogen peroxide with antifungal fatty acids, diacetyl, flavin nucleotides, phenyllactic acid, bacteriocins, and bacteriocin-like products by the fermenting microbial flora especially LAB, result in an antimicrobial environment for the many pathogenic microorganisms that are otherwise present in cereals and legumes (Papadimitriou et al. 2016). Non starch polysaccharides, like beta-glucan, in cereal-based foods increase the nutritional component and probiotic properties of the fermented product and have been reported to contain vitamin E, selenium, linolenic acid and folate that offer many cardiovascular health benefits (Beck et al. 2010). Phenolics are an important class of secondary metabolites in the cereals and are nutritionally beneficial because of their large number of bioactivities. During fermentation of cereals, the enzymatic activities of LAB chemically transform phenolics. The main effects include release of free phenolic acids from the complex polyphenols, their conversion into simpler compounds, and concomitant increase in the antioxidant and free-radical scavenging activities (Skrajda-Brdak et al. 2019; Rodríguez et al. 2009). Besides being a rich source of vitamins especially B1, B5, and C, maize based fermented foods contain highest amounts of phenolic compounds that have promising cancer preventive properties due to antioxidative features. Additionally, maize contains cryptoxanthine, a compound reported to prevent lung cancer, and homocysteine, known for beneficial cardiac properties (Yuan 2003). With the current scenario of environmental deterioration, the above-mentioned beneficial properties call for the incorporation of such health-targeted foods in daily diet regime in order to combat the body's exposure to carcinogenic compounds.

Fermented food products prepared from cereals and legumes were studied to have numerous health benefits as compared to their raw substrates. These have been established to have higher digestibility, increased level of amino acids such as lysine, are low in fats/cholesterol, and provide added health benefits along with being a rich source of proteins, minerals. They often find themselves categorized as functional foods due to their innumerable health benefits (Charalampopoulos et al. 2002). About 100 g of *Panta bhat* for example, commonly eaten in West Bengal, was reported to contain 850 mg calcium, 839 mg potassium, 303 mg sodium, and 73.91 mg iron. Another fermented food, *chitou* from Orissa, India, was found to contain 31.7 mg sodium, 23.2 g total carbohydrate, 13.5 mg potassium, 2.1 g protein, and 1.1 g dietary fiber, along with vitamin A, B-complex, calcium, folate, iron, niacin, riboflavin, and thiamine (Ray and Swain 2013). However, different substrates offer varied nutritional advantages in the fermented products and hence many fermented foods find their mention in special diet prescribed for specific ailments or health conditions and nutritionally challenged groups like infants or feeding and weaning mothers. *Koozhu*, a broken wheat and millet based fermented food from the state of Tamil Nadu in India, given to children at a weaning age, has been found to contain increased levels of thiamine, riboflavin, and niacin. *Wadi,* made from black gram has been found to have soluble nitrogen, free amino acids, enhanced proteolytic activity, and B-vitamins including thiamine, riboflavin, and cyanocobalamine. Beverages like *bhaati jaanr* (Table 12.1 b) are recommended for postnatal recovering women and ailing people for gaining back their lost strength (Ray et al. 2016). Fermented soybean-based food like *tungrymbai*, *bekang*i and *kinema* (Table 12.1 b) have been reported to have an increased amount of free amino acids, bioactive peptides, minerals, Vit B complex as well as group B saponins, and exhibit free radical scavenging, antioxidant, and antiviral properties. These can hence be used as low cost natural source of bioactive peptides and antioxidants especially for the undernourished and economically challenged population (Chettri and Tamang 2014; Omizu et al. 2011; Sharma et al. 2020). *Bacillus licheniformis* from *kinema* (Table 12.1 b) was found to be an active source of these bioactive peptides, many of which have been found to have antioxidant and antithrombotic activities (Kumari et al. 2022). Saponins, implicated in the prevention of hypercholesterolemia, and polyglutamic acid, both have been found to be present in the *Bacillus* rich *kinema* (Chettri et al. 2016; Tamang et al. 2016b). In one study, *kinema* extract was found to exhibit a better free radical scavenging and higher antioxidant activity with 44% inhibition of linoleic acid peroxidation as compared to 36% inhibition by the raw soybean extract (Moktan et al. 2008). Similar antioxidant property was also found in *bekang* and *tumgrymbai*, other soybean based fermented food reported to contain *Bacillus* as the prominent microorganism at work. Bioactive peptides have also been known to have antithrombic and antihypertension properties and play a vital role as an immunomodulatory agent (Sanjukta et al. 2016). *Selroti* (Table 12.1 b) is generally recommended as a part of diet for minimizing dyslipidemia and cardiometabolic risks (Yonzan and Tamang 2010). *Panta bhat* (Table 12.1 a), commonly consumed in West Bengal, is being recognized world-wide for its nutritional superiority owing to the numerous gastrointestinal benefits including benefits in ulcers, colitis, Crohn's disease, irritable bowel syndrome and celiac disease (Choi et al. 2014). *Haria* (Table 12.1 b) has numerous digestive benefits and has antioxidant activity. *Apong* (Table 12.1 b) has been found to be beneficial for reducing and in prevention of kidney stones. *Jou* (Table 12.1 b) consumed by the Boro tribe has been known to prevent jaundice and urinary disorders. *Zutho* (Table 12.1 b) has been found to be effective in diabetes treatment, wound healing, and in the lowering of cholesterol (Das et al. 2012; Ray et al. 2016). *Judima* (Table 12.1 b) has been found to have multiple beneficial properties, including hypolipidemic, hypotensive, anti-inflammatory, antiaging, antiallergic, antioxidant, antibacterial, antifungal, antispasmodic, hepatoprotective, neuroprotective, and antidiabetic (Chakrabarty et al. 2014). A *Saccharomyces* strain isolated from *Khambir*, a wheat based leavened bread made in the Leh region, exhibited anti-inflammatory activity towards induced ulcers, and hence can be used in anti-ulcer therapies (Banik et al. 2019). The numerous proteinases, present in the fermenting microorganisms, including LABs, are responsible for the great diversity of biologically active peptides being produced. Compilation of these have been done to curate a data base, FermFooDB, that maintains a list of the numerous types of bioactive peptides present in various fermented food (Chaudhary et al. 2021). However, milk based fermented food are more widely studied and reports on bioactive peptides from cereal based food are limited. Table 12.2 lists some of the known physiological and health benefits of fermented foods primarily those inhabited by LAB (Cerbo et al. 2016; Tamang et al. 2016b).

As described in Table 12.2, production of organic acids, decrease in pH, and production of many volatile and other compounds by LAB (like bacteriocins, carbon dioxide, ethanol, hydrogen peroxide, and diacetyl), all play a prominent role in establishing the antimicrobial nature of fermented foods, inhibiting growth and proliferation of pathogenic organisms, and increasing their nutritional value, health benefits, and shelf life. Multiple factors like composition of the substrate (type of cereal or specific combinations with legumes), processing conditions, type of inoculum, growth characteristics of starter cultures, stability of the strains during storage and nutritional and sensory characteristics of the product, are all important factors that determine the user acceptance of the final fermented product (Achi and Ukwuru 2015). Additionally, starter inoculum, water content of substrate during fermentation, temperature, duration of fermentation, and environmental factors during fermentation, all have a definite bearing on the success of the process. Pre-fermentation conditions, especially germination, is a crucial step for fermentation process as it results in significant lowering of the antinutritive factors such as tannin

TABLE 12.2 Health and nutritional benefits of LAB in fermented foods

Nutritional and Health benefits of LAB in fermented food
Source of beneficial metabolites—Increase in vitamins, minerals, enzymes, free amino acids, bioactive peptides, polyglutamic acids
Antimicrobial activity—Competition with pathogens for binding site, Secretion of antimicrobial compounds by LAB, Inhibition of many pathogenic bacteria including *Helicobacter pylori, E. coli* and *Pseudomonas aeruginosa*, activity against periodontitis and gingivitis causing microbes
Antioxidant and Detoxifying activity—Prevention in lipid peroxidation, reduction in oxidation reduction potential, Production of antioxidants, Inactivation of mycotoxins and many bacterial toxins, Removal of toxicity causing metals, Presence of radical scavenging compounds such as 1,1-diphenyl-2-picryl hydrazyl (DPPH) and 2–2-azino-bis (3-ethylbenzo-thiazoline-6-sulfonic acid (ABTS), Reduction of phytic acids
Promotion of gut health—Enhanced nutrient absorption and maintenance of healthy gut microflora, Protection against gastrointestinal disorders
Enhancement in immune health—Potential adjuvants, Adhesion to the gut mucosa, Enhancement of natural killer cell cytotoxicity, Induction of interferon gamma production and cytokine expression
Prevention of cancer/Antileukemic and Antitumor effects—Inactivation of cancerous compounds, Regulation of cell apoptosis and cell differentiation, Antimutagenic and anti-cancerous compounds
Enhance Heart health—Lowering of serum cholesterol levels, Antilipidemic compounds, Antihypertensive peptides, Lower cholesterol assimilation, Cholesterol reduction, Enhanced production of bile salt hydrolase, Saponins like DDMP (2,3-dihydro-2,5-dihydroxy-6-metyl-4*H*-pyran-4-one) prevent hypercholesterolemia, Lower hypertriacylglycerolaemia
Kidney Disease therapy—High urease activity and ability to degrade biogenic amines and oxalate
Antidiabetic activity—Inhibition of a- and b-glucosidase activity
Antiarthritic effect—Downregulation of Th1 type inflammatory response, decrease in proinflammatory cytokines, Upregulation of anti-inflammatory cytokines, presence of Vitamin like K2 and minerals like Mg, Ca and P enhance bone health
Prevention of vascular diseases—Enhanced Fibrinolytic activity by the production of enzymes such as natto kinase, Activation of LDL receptors by isoflavones,
Antithrombotic—Presence of bioactive peptides

and phytic acids (Budhwar et al. 2020). The buildup of ecological niche and the succession of multitude of microorganisms' aid in the release of various enzymes and compounds bringing about a change in the grain matrix and altering the nature of substrate. Since the probiotic and other health beneficial properties of fermented food items are largely dependent on the type of fermenting microbial population, manipulation of this microflora with human beneficial microbes is being researched.

12.5 ENTEROCOCCI IN FERMENTATION INDUSTRY

Enterococci, although indicators of fecal contamination, play a positive and prominent role in many fermented food processes and preparations (M'hir et al. 2012). Enterococci have the ability to produce bacteriocin, including those from Type 1 (cytolysin), Type 2 (enterocin AS-48), Type 3 (bacteriocin) 31 and Type 4 (enterocin 1071A and enterocin 1071B) class (Moreno et al. 2006). The ability of *Enterococcus* to produce bacteriocins, and their properties to enhance the organoleptic properties of fermented foods, are two important characteristic that justify and promote their role in the food technology sector. However, their use for fermentation is mostly limited primarily to fermentations of dairy, meat, and vegetable substrate and only limited reports exist on their presence, use and beneficial activities in cereal and legume based fermented foods (Moreno et al. 2006). One such product where enterococci are being studied is in the preparation of sourdough. Sourdough is a slow fermented baked product obtained from many cereal flours. Containing a stable culture of LAB and yeast in flour and water, sourdough can also be made using enterococci in starter culture. It was found that the acids produced by enterococci acidified the dough, thus reducing the colonization of molds and rope bacteria, thereby protecting food from spoilage (Corsetti et al. 2007). Compared to other dough, highest amount of free amino acids were recorded in sourdough fermented with *Enterococcus faecium* in the starter culture (Collar et al. 1992). Free amino acids in turn affect the growth of yeast and subsequent production of flavor compounds. *Enterococcus faecium* has been demonstrated *in vitro* to exhibit proteolytic activity against gluten. This causes a decrease of the gliadin (Prolammins) concentration from 45 to 18 g/kg. The gluten content can be further reduced using multiple strains to bring the cereal based fermented food product to near gluten free (GF) status (M'hir et al. 2008).

Some strains of enterococci have been found to be present in some cereal and legume based fermented foods and beverages used in many countries. This includes *pozol* found in Mexico (Ben Omar and Ampe 2000), *bushera* from Uganda (Muyunja et al. 2003), *miso* from Japan (Onda et al. 2002), *doenjang* from Korea (Kim et al. 2009), *hussuwa* in Sudan (Yousif et al. 2005) and *turhana and boza* from Turkey (Sengen et al. 2009; Todorov 2010) among others. *Idli,* widely prevalent in South India and Southeast Asian countries has been found to contain *Enterococcus faecalis* along with *Leuconostoc mesenteroides.* These LAB were reported to be involved in acid production, flavour formation, and leavening of the batter (Nout 2009), as well as anti-listerial properties due to production of bacteriocin (Vijayendra et al. 2010). *Enterococcus faecium* was studied for its adhesive properties, acid and bile tolerance and resistance to gastrointestinal juice. It was also found to have high hydrophobicity (high adhesive potential), and high tolerance to acidic pH and bile salts, attributes that establishes its probiotic potential (Rehaiem et al. 2014). Further, it was found to be positive for genes encoding for enterocins, a feature that would enhance its role in the production of probiotic fermented foods. Some other soybean based fermented food, like *kinema, peruyaan, bekang* and *tungrymbai/turangbai* have also been found to harbour strains of *Enterococci*, including *E. faecalis, E. faecium, E. durans* and *E. raffinosus,* although these strains in soybean based fermented food are not dominant (Tamang et al. 2015). Biochemical and biotechnological properties of different *Enterococcus* strains need to be studied in more detail to fully unravel the probiotic potential of this class of microorganisms. Although cereal based fermented foods have multiple health benefits, there is increasing research on the adverse health effects of beta glucan and the benefits of low beta-glucan food products. As a result, there is growing research trend to elucidate nutritional benefits of millet-based fermented foods that are low in beta-glucan and provide the benefits of a fermented food at the same time.

12.6 MILLET BASED FERMENTED FOODS

Millets, one of the major grains grown in India, form a staple food and integral part of the local Indian food culture. Every region in the Indian subcontinent, is known for a variety of local millets, prepared traditionally in different ways. Major millets like white pearl millets (Bajra), sorghum (Jowar), finger millet (Ragi) and minor millets including Foxtail millet (Kakum), Kodo Millet (Kodon), and barnyard millet (Sanwa), have been cultivated and consumed in India since long periods. Millets were, and are, a part of the staple diet of primarily rural Indian population. However, they have recently gained much attention due their highly diverse nutritive attributes and GF nature that have been reported to give them a dietary and nutritional advantage over other cereals and legumes. Packed with beneficial dietary fibers, substantially high nutrients, and a low gluten content, they are a promising and extremely beneficial alternative to the use of wheat flour, especially for people with low or no tolerance to gluten, and those suffering from celiac disease (Kumar et al. 2018). They are being touted as a superfood and their inclusion in national health policies of India is being highly advocated (Kumar et al. 2018). Millets have been reported to act as an excellent source of calcium, protein (7–12%), micronutrients and phytochemicals and are being promoted for their superior nutritionally rich cereal properties (Saini et al. 2021). Budhwar et al. (2020) describe and discuss the various flavonoids and phenolic acids present in cereals, especially millets, and the bioactive properties imparted by them. In the hills of Darjeeling, Sikkim and Arunachal Pradesh, finger millet, locally called kodo, is fermented to a high calorie and mineral rich alcoholic beverage (Tamang 2022a). Fermentation of GF flour enhances the nutritional and digestibility quotient of the GF food and incorporates many nutritional components such as essential amino acids like lysine, methionine, and tryptophan (Zannini et al. 2012). One of the most widely studied and promising fermentation process of GF fermented food is the preparation of sourdough bread. Sourdough technology is one of the oldest processes in cereal food biotechnology, known for over 5000 years to improve the texture and flavor of baked cereals (Zannini et al. 2012). Although not a traditional bread of Indian origin, this technology is gaining wide acceptance in the Indian market and is a promising field for the study of manipulations in the fermentation process for enhanced product quality. Sourdough bread has been prepared primarily from a heterofermentative mix of strains of LAB using wheat flour and water. However, sourdough prepared from rice, oats, buckwheat, teff, amaranth, maize, millet, and quinoa are now being studied for GF fermented products and their microbiological health benefits (Zannini et al. 2012). Efficient and stable starter culture forms the basis of a successful sourdough process. Manipulation of starter culture hence is an interesting aspect that can be explored to create desirable characteristics in the product. Inoculum developed on wheat/rye has been used to carry the process in rice flour and it was found that the final established microflora in rice flour was diverse from that of the starter culture thus establishing the fact that microflora competition and establishment is substrate specific, and the type of GF flour determines the final microbiota (Moroni et al. 2010). Study of GF sourdough reveals novel LAB not normally reported in traditional sourdough made from wheat and rye. These include LAB species like *Pediococcus pentosaceus, Lactobacillus graminis, Lactobacillus gallinarum* and *Lactobacillus sakei* not normally found in traditional sourdough (Moroni et al. 2009). The use of specific microbial species as starter culture has been explored to bring out the desired changes in the GF fermented product. Use of *Lactiplantibacillus plantarum* as starter culture to give a softer bread (Moore et al. 2008), use of exopolysaccaride producing LAB strain to replace hydrocolloids (Galle et al. 2011), use of *Weissella cibaria* to produce prebiotic molecule gluco-oligosaccharides in the GF bread being made using sorghum and quinoa flour (Schwab et al. 2008) are some of the interventions that gave a superior fermented GF product (Zannini et al. 2012). Use of efficient LAB also brings about gluten detoxification in many flours and hence this technology can be investigated in depth to bring about desired changes in other fermented food. Creating high quality GF fermented products with high palatability, by use of an efficient and stable starter culture and fermentation conditions manipulations will help strengthen the millet based fermented food market. Earlier GF products have been produced by means of either use of different GF flour, by addition of ingredients or additives or by the use of technologies like enzymatic reduction or high hydrostatic pressure processing for reducing gluten content in food. (Zannini et al. 2012). These approaches either tend to increase the price of the food or add additional allergic entities to the product and at the same time diminish the natural food status of the product. Hence the use of natural processes to make the food GF and at the same time increasing the palatability and nutritional component using desired fermentation, is a challenge and the need of the hour.

In the scenario of rising global unemployment, severe food shortage due to pandemic and the subsequent heightened food security issues, it becomes even more imperative that fermented millet-based foods are extensively promoted to endure food security and nutritional needs of the COVID-19 pandemic affected Indian population. Additionally, the role of millets in boosting and strengthening immune system and supporting healthy gut microflora has currently been proposed and is being researched (Muthamilarasan and Prasad 2021). It is hence advisable to bring in clarity and invite policies and incentives to attract research focus around fermented millet-based foods. Additionally, nutritive value of individual cereal and millet-based food and products need to be studied thoroughly to establish their use across all age groups and in different health conditions.

12.7 TECHNOLOGICAL ADVANCEMENTS AND FUTURE CONSIDERATIONS

Although a lot of research has revolved around different fermented foods, there are innumerable fermented foods prepared from cereals and legumes that are yet to be explored in detail. Further, it would be beneficial to make fermented food items a relevant and essential part of food systems with national importance and copyright the details of the process to preserve this rich heritage of India. Various interventions can be done to enhance the efficiency, nutritional potential, health benefits, acceptability, awareness, and availability of this class of food items.

12.7.1 Study of Microbiological Processes and Stabilization of Starter Inoculum

Microorganisms are the key to a successful fermentation process. A deep understanding of the process controlling microbes, their presence *in situ*, the influence of environmental factors on the process in enriching and favoring metabolically performing microbes, all are the most pertinent aspects. How to maintain a robust, active, and efficient starter culture and how to reduce the preparation/pretreatment/fermentation time are few of the microbiologically relevant issues that need attention in the development of a successful bioprocess. Naturally, the effect of starter culture on the quality of fermented foods is specific and determinant. Fermentation of pearl millet by *Lactobacillus acidophilus* resulted in insignificant rise in thiamine content as compared to when the same substrate was fermented using *Saccharomyces cerevisiae* and *Saccharomyces diastaticus* (Arora et al. 2010). In another report, use of *Lactiplantibacillus plantarum* gave increased gamma amino butyric acid (GABA) content in the prepared sourdough when used for the fermentation of buckwheat and quinoa while *Lactococcus lactis* gave enhanced GABA yield during amaranth flour and chickpea fermentation (Coda et al. 2010).

Another aspect that needs to be further researched is the increase in the antimicrobial potential of LABs and the enhancement in the antimicrobial nature of fermented foods. This is especially relevant in the current scenario of increasing cases of multidrug resistant strains of pathogenic bacteria. One of the biggest challenges in the making of a safe GF fermented food is contamination with the notorious mycotoxins that not only cause economical losses by reducing the shelf life of the product but are also harmful health wise especially for patients of celiac disease. Although LABs have been demonstrated to reduce mycotoxins in the fermented food, the exact mechanism by which LABs exhibit anti-mycotoxin nature needs thorough investigation to make the process technologically more reproducible and economical at the industrial level. Two theories that have been proposed are toxin binding effect (Turbic et al. 2002) and enzymatic detoxification (Zinedine et al. 2005). A study of 39 food types from Northeast India revealed the presence of pathogens such as *Clostridium perfringens, Clostridium botulinum, Clostridium difficile, Staphylococcus aureus, Proteus mirabilis,* and *Listeria monocytogenes* especially in soybean-based fermented food. Analysis of these pathogens revealed the presence of toxins like diarrheal toxins, and pathogenic factors like hemolysin, urease and multidrug resistance, in many of these pathogens (Keisam et al. 2019). Such wide prevalence necessitates more detailed studies on the toxicity of microorganisms associated with various fermented food. Replacing natural fermentation with a controlled, formulated and well-defined starter culture can to some extent reduce the chances of contamination by pathogenic microflora and lead to a consistent and safe fermented end product. Additionally, maintaining an inoculum such that it creates unfavorable conditions for pathogenic microorganisms in the substrate during the fermentation process is a beneficial approach that needs more detailed research. Analysis of pathogenic metabolic markers in the starter cultures and fermented food can be a rapid assessment for the presence of unwanted microorganisms. Figure 12.1 details the beneficial and profitable features of starter culture that need to be further investigated and adopted to enhance the properties, profitability, nutritional quotient, and consumer acceptance of fermented foods.

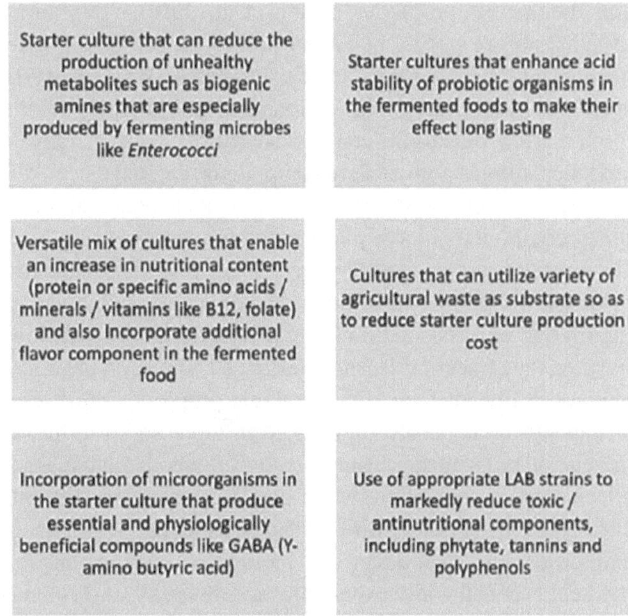

FIGURE 12.1 Recommendations for research on starter culture.

Similar such studies on traditional Indian foods needs to be carried out using millets as GF substrate, primarily to explore the use of specific microbial strains in starter culture for the incorporation of physiologically beneficial metabolites, as well as to reduce the glycaemic/or insulinaemic response in GF fermented food (Zannini et al. 2012).

12.7.2 Study of Substrate and Pre-Fermentation Treatments

Substrate plays a pivotal role in determining the type of microorganisms that develop and populate the fermented foods. Each cereal substrate harbours and supports different microflora and hence the organoleptic and nutritional benefits of each cereal based fermented product varies. Composite substrates are being promoted due to the combined enhanced benefits observed from them. These mainly involve the use of millets with or without legumes and other cereals. Sindhu and Khetarpaul (2001), in their experiment, constructed a substrate comprised of barley, coprecipitate, green gram and tomato (2:1:1:1) and fermented the same. Two different fermentation starter culture and processes were followed: (i) Single culture fermentation using starter culture of *Lacticaseibacillus casei* and *L. plantarum* and (ii) sequential culture fermentation using *Saccharomyces boulardii* (25^{0}C; 24 hours), *Lacticaseibacillus casei* (37^{0}C; 24 hours), *Saccharomyces boulardii* (25^{0}C; 24 hours), and *L. plantarum* (37^{0}C; 24 hours). The sequential culture process was found to give a superior highly digestible fermented product with higher reduction in phytic acid and polyphenols. In another similar study, millet flour, whey powder, and tomato pulp were mixed (2:1:1) and the fermented product was found to have unique and enhanced nutritional and therapeutic properties (Arora et al. 2010). Pretreatment of substrate, or pre-fermentation processes, are also a determining factor for microbiological functions. Many processes such as soaking, germination, roasting, and malting prior to fermentation, relaxes the compact structure of starch, activates enzymes, dilutes antinutritive compounds, and have been documented to increase the nutritive value and reduce antinutritional compounds in many fermented foods (Ray et al. 2016). Germination especially plays a critical role in establishing favorable conditions for microbial colonization and in activating myriad enzymes facilitating multiple metabolic pathways. Further, the effect of this, and other such mixing, on the microbiological attributes and the metabolic properties of different fermented foods need to be studied in more detail to make a variety of tailor-made fermented food items available across all classes of population. More detailed studies on the benefits and drawbacks of different pre-fermentation treatments and processing techniques are required to further enhance the health benefits and expand the acceptability of fermented foods. Specific processing conditions play an important part in gluten detoxification as well, a feature that is especially beneficial and critical for patients with celiac disease. This includes temperature of fermentation, water content in dough mixture and the duration of fermentation as was observed by when using a pool of specific LABs that acted on wheat gluten proteins thereby combining the goodness of wheat flour with the attributes of GF food thereby minimizing risk of toxicity of the fermented food (De Angelis et al. 2006). GF fermented food are generally low in folate (B11) and hence strategies to increase folate content, a major nutraceutical, is a prime area to be addressed. Use of yeast and incorporation of folate rich flours like quinoa and amaranth are promising aspects (Schoenlechner et al. 2010). Hence working out a combination of millet and legume mix, for making a defined wholesome functional fermented product with consistent characteristics and rich in pre and probiotic properties, is a promising aspect of fermentation technology to be explored.

12.7.3 Increasing the Economic Value of Fermented Food

Fermented foods are now being developed to be functional foods with a strong probiotic/or biogenic activity. Maintaining the viable beneficial bacterial count to the recommended levels (10^6 cfu/g at the time of consumption) along with adequate sensory characteristics is the key to maintaining the market and economic value of the product. Further, additional prebiotic characteristics (like oligosaccharides) in a probiotic fermented food would greatly enhance its market value and hence more diverse availability, study, and awareness about synbiotics would be beneficial. The need for increased shelf life of fermented products was heightened in the current pandemic scenario when it was not feasible for the consumers to step out at regular frequent intervals for their daily needs. Further, even though challenging, making one wholesome fermented product with multiple health benefits and increased nutritive value, targeting different medically challenged individuals and catering to different age groups and economic strata, would be an ideal successful market strategy. All the technological advancements that can be done on fermented food has to be without any compromise on the traditional desirable attributes of the food item and without any commercial price inflation.

12.7.4 Compilation and Dissemination of Knowledge Base

Ignorance and low market exposure hinders the success of any product. Making the knowledge of the fermented food available to non-local masses and at the same time preserving the knowledge as right of indigenous population, will help in promoting and preserving indigenous products and processes. However, developing local capabilities and rural industrialization, instead of shifting production to urban areas will ensure the preservation of core capabilities and intricacies of the process. Hence, efforts to target the rural sector for preserving and promoting the knowledge and practice of fermented food assumes importance.

12.8 CONCLUSIONS

Preparation of fermented foods is both an art and science for the purpose of food preservation. South and Southeast Asia has been the pioneer in this field with numerous fermented delicacies included in their traditional cuisine. Cereals and legumes have been used since generations for preparation of fermented foods and have gained widespread attention. With increasing evidence of the expanding health benefits of millet-based foods, it is imperative to study about the various millet based fermented foods, the microbiological processes in the same, the dominant and lesser-known microorganisms from the various food products and the scientific and technological factors involved in their production and preservation. The aim of any biotechnological approach towards fermented foods should be to preserve and promote the intricacies of the traditional methods as much as possible. The exact process and methodology used for the preparation is vital in giving the fermented foods their characteristic flavor, texture, sensory and dietary characteristics and hence these need to be documented and disseminated scientifically. Study and manipulation of starter cultures and metabolomics in the mixed substrates are few of the most prominent and upcoming research areas related to fermented foods. Use of low-cost substrates, use of controlled and fixed starter cultures, reuse of by products, increasing the safety and shelf life of fermented food using low-cost methods, are all strategies that will aid in circular economy and enable an overall sustainability to this sector.

12.8.1 Acknowledgments

The authors thank the Provost, Faculty of Medical and Health Sciences and Registrar, Symbiosis International (Deemed University), for their support.

REFERENCES

Achi, O.K., Ukwuru, M. 2015. Cereal-based fermented foods of Africa as functional foods. *International Journal of Microbiology and Application* 2(4):71–83.

Arora, S., Jood, S., Khetarpaul, N. 2010. Effect of germination and probiotic fermentation on nutrient composition of barley-based food mixtures. *Food Chemistry* 119(2):779–784.

Banik, A., Mondal, J., Rakshit, S., Ghosh, K., Sha, S.P., Halder, S.K., Ghosh, C., Mondal, K.C. 2019. Amelioration of cold-induced gastric injury by a yeast probiotic isolated from traditional fermented foods. *Journal of Functional Foods* 59:164–173.

Beck, E.J., Tapsell, L.C., Batterham, M.J., Tosh, S.M., Huang, X.F. 2010. Oat betaglucan supplementation does not enhance the effectiveness of energy-restricted diet in overweight women. *British Journal of Nutrition* 103(8):1212–1222.

Ben Omar, N., Ampe, F. 2000. Microbial community dynamics during production of the Mexican fermented maize dough Pozol. *Applied and Environmental Microbiology* 66:3664–3673.

Bhalla, S., Bhalla, T.C. 2007. Traditional foods and beverages of Himachal Pradesh. *Indian Journal of Traditional Knowledge* 6(1):17–24.

Blandino, A., Al-Asceri, M.E., Pandiella, S.S., Cantero, D., Webb, C. 2003. Cereal based fermented foods and beverages. *Food Research International* 36(6):527–543.

Budhwar, S., Sethi, K., Chakraborty, M. 2020. Efficacy of germination and probiotic fermentation on underutilized cereal and millet grains. *Food Production, Processing and Nutrition* 2:1–17.

Capozzi, V., Menga, V., Digesu, A.M., De Vita, P., van Sinderen, D., Cattivelli, L., Fares, C., Spano, G. 2011. Biotechnological production of vitamin B2-enriched bread and pasta. *Journal of Agricultural and Food Chemistry* 59(14):8013–8020.

Capozzi, V., Russo, P., Dueñas, M.T., López, P., Spano, G. 2012. Lactic acid bacteria producing B-group vitamins: a great potential for functional cereals products. *Applied Microbiology and Biotechnology* 96(6):1383–1394.

Cerbo, A., Palmieri, B., Aponte, M., Morales-Medina, J.C., Iannitti, T. 2016. Mechanisms and therapeutic effectiveness of lactobacilli. *Journal of Clinical Pathology* 69(3):187–203.

Chakrabarty, J., Sharma, G.D., Tamang, J.P. 2014. Traditional technology and product characterization of some lesser-known ethnic fermented foods and beverages of north Cachar hills district of Assam. *Indian Journal of Traditional Knowledge* 13:706–715.

Charalampopoulos, D., Wang, R., Pandiella, S.S., Webb, C. 2002. Application of cereals and cereal components in functional foods: a review. *International Journal of Food Microbiology* 79(1–2):131–141.

Chaudhary, A., Bhalla, S., Patiyal, S., Raghava, G.P., Sahni, G. 2021. FermFooDb: a database of bioactive peptides derived from fermented foods. *Heliyon* 7(4):e06668.

Chettri, R., Bhutia, M.O., Tamang, J.P. 2016. Poly-γ-glutamic acid (PGA)-producing *Bacillus* species isolated from Kinema, Indian fermented soybean food. *Frontiers in Microbiology* 7:971.

Chettri, R., Tamang, J.P. 2014. Functional properties of *Tungrymbai* and *Bekang*, naturally fermented soybean foods of North East India. *International Journal of Fermented Foods* 3(1):87–103.

Choi, J.S., Kim, J.W., Cho, H.R., Kim, K.Y., Lee, J.K., Ku, S.K., Sohn, J.H. 2014. Laxative effects of fermented rice extract (FRe) in normal rats. *Toxicology and Environmental Health Science* 6:155e63.

Coda, R., Rizzello, C.G., Gobbetti, M. 2010. Use of sourdough fermentation and pseudo-cereals and leguminous flours for the making of a functional bread enriched of gamma-aminobutyric acid (GABA). *International Journal of Food Microbiology* 137(2–3):236–245.

Collar, C., Mascaros, A.F., Benedito, De Barber, C. 1992. Amino acid metabolism by yeasts and lactic acid bacteria during bread fermentation. *Journal of Food Science* 57:1423–1427.

Corsetti, A., Settani, L., Lopez, C.C., Felis, G.E., Mastrangelo, M., Suzzi, G. 2007. A taxonomic survey of lactic acid bacteria isolated from wheat (*Triticium durum*) kernels and non-conventional flours. *Systemic and Applied Microbiology* 30:561–571.

Das, A., Raychaudhuri, U., Chakraborty, R. 2012. Cereal based functional food of Indian subcontinent: a review. *Journal of Food Science and Technology* 49:665e72.

De Angelis, M., Rizzello, C.G., Fasano, A., Clemente, M.G., Simone, C.D., Silano M, Vincenzi MD, Losito, I., Gobbetti, M. 2006. VSL#3 probiotic preparation has the capacity to hydrolyze gliadin polypeptides responsible for Celiac Sprue probiotics and gluten intolerance. *Biochimica et Biophysica Acta (BBA)—Molecular Basis of Disease* 1762(1):80–93.

Deo, D., Davray, D., Kulkarni, R. 2019. A diverse repertoire of exopolysaccharide biosynthesis gene clusters in *Lactobacillus* revealed by comparative analysis in 106 sequenced genomes. *Microorganisms* 7(10):444.

Di Cagno, R., De Angelis, M., Lavermicocca, P., De Vincenzi, M., Giovannini, C., Faccia, M., & Gobbetti, M. 2002. Proteolysis by sourdough lactic acid bacteria: effects on wheat flour protein fractions and gliadin peptides involved in human cereal intolerance. *Applied and Environmental Microbiology* 68(2):623–633.

Dimidi, E., Cox, S.R., Rossi, M., Whelan, K. 2019. Fermented foods: definitions and characteristics, impact on the gut microbiota and effects on gastrointestinal health and disease. *Nutrients* 11(8):1806.

Galle, S., Schwab, C., Arendt, E.K., Ganzle, M.G. 2011. Structural and rheological characterisation of heteropolysaccharides produced by lactic acid bacteria in wheat and sorghum sourdough. *Food Microbiology* 28(3):547–553.

Ganzle, M., Follador, R. 2012. Metabolism of oligosaccharides and starch in lactobacilli: a review. *Frontiers in Microbiology* 3:33219.

Ganzle, M.G. 2020. Food fermentations for improved digestibility of plant foods–an essential ex situ digestion step in agricultural societies? *Current Opinion in Food Science* 32:124–132.

Ghosh, K., Ray, M., Adak, A., Dey, P., Halder, S.K., Das, A., Jana, A., Parua Mondal, S., Das Mohapatra, P.K., Pati, B.R., Mondal, K.C. 2015. Microbial, saccharifying and antioxidant properties of an Indian rice based fermented beverage. *Food Chemistry* 168:196e202.

Goel, A., Halami, P.M., Tamang, J.P. 2020. Genome analysis of Lactobacillus plantarum isolated from some Indian fermented foods for bacteriocin production and probiotic marker genes. *Frontiers in Microbiology* 11(40):1–12.

Kanwar, S.S., Gupta, M.K., Katoch, C., Kumar, R., Kanwar, P. 2007. Traditional fermented foods of Lahaul and Spiti area of Himachal Pradesh. *Indian Journal of Traditional Knowledge* 6(1):42–45.

Keisam, S., Tuikhar, N., Ahmed, G., Jeyaram, K. 2019. Toxigenic and pathogenic potential of enteric bacterial pathogens prevalent in the traditional fermented foods marketed in the Northeast region of India. *International Journal of Food Microbiology* 296:21–30.

Kharnaior, P., Tamang, J.P. 2021. Bacterial and fungal communities and their predictive functional profiles in *kinema*, a naturally fermented soybean food of India, Nepal and Bhutan. *Food Research International* 140:110055.

Kim, T.W., Lee, J.H., Kim, S.E., Park, M.H., Chang, H.C., Kim, H.Y. 2009. Analysis of microbial communities in doenjang, a Korean fermented soybean paste, using nested PCR-denaturing gradient gel electrophoresis. *International Journal of Food Microbiology* 131:265–271.

Kumar, A., Tomer, V., Kaur, A., Kumar, V., Gupta, K. 2018. Millets: a solution to agrarian and nutritional challenges. *Agriculture and Food Security* 7(1):1–15.

Kumari, R., Sanjukta, S., Sahoo, D., Rai, A.K. 2022. Functional peptides in Asian protein rich fermented foods: production and health benefits. *Systems Microbiology and Biomanufacturing* 2:1–13.

Liu, M., Nauta, A., Francke, C., Siezen, R.J. 2008. Comparative genomics of enzymes in flavor-forming pathways from amino acids in lactic acid bacteria. *Applied and Environmental Microbiology* 74(15):4590–4600.

M'hir, S., Aldric, J.M., El Mejdoub, T., Destain, J., Mejri, M., Hamdi, M., Thonart, P. 2008. Proteolytic breakdown of gliadin by *Enterococcus faecalis* isolated from Tunisian fermented dough. *World Journal of Microbiology and Biotechnology* 24:2775–2781.

M'hir, S., Minervini, F., Di Cagno, R., Chammem, N., Hamdi, M. 2012. Technological, functional and safety aspects of enterococci in fermented vegetable products: a mini-review. *Annals of Microbiology* 62(2):469–481.

Marco, M.L., Sanders, M.E., Gänzle, M, Arrieta, M.C., Cotter, P.D., Vuyst, L.D., Hill, C., Holzapfel, W., Lebeer, S., Merenstein, D., Reid, G., Wolfe, B.E., Hutkins, R. 2021. The International Scientific Association for Probiotics and Prebiotics (ISAPP) consensus statement on fermented foods. *Nature Reviews: Gastroenterology and Hepatology* 18:196–208. https://doi.org/10.1038/s41575-020-00390-5

Moktan, B., Saha, J., Sarkar, P.K. (2008). Antioxidant activities of soybean as affected by *Bacillus*-fermentation to kinema. *Food Research International* 41(6), 586–593.

Moore, M.M., Dal Bello, F., Arendt, E.K. 2008. Sourdough fermented by Lactobacillus plantarum FST 1.7 improves the quality and shelf life of gluten-free bread. *European Food Research and Technology* 226(6):1309–1316.

Moreno, M.F., Sarantinopoulos, P., Tsakalidou, E., De Vuyst, L. 2006. The role and application of enterococci in food and health. *International Journal of Food Microbiology* 106(1):1–24.

Moroni, A.V., Arendt, E.K., Morrissey, J.P., Dal Bello, F. 2010. Development of buckwheat and teff sourdoughs with the use of commercial starters. *International Journal of Food Microbiology* 142(1–2):142–148.

Moroni, A.V., Dal Bello, F., Arendt, E.K. 2009. Sourdough in gluten-free bread-making: an ancient technology to solve a novel issue? *Food Microbiology* 26(7):676–684.

Muthamilarasan, M., Prasad, M. 2021. Small millets for enduring food security amidst pandemics. *Trends in Plant Science* 26(1):33–40.

Muyunja, C.M.B.K., Naruhus, J.A., Treimo, J., Langsrudt, T. 2003. Isolation, characterization and identification of lactic acid bacteria from Bushura: a Ugandan traditional fermented beverage. *International Journal of Food Microbiology* 80:201–210.

Nkhata, S.G., Ayua, E., Kamau, E.H., Shingiro, J.B. 2018. Fermentation and germination improve nutritional value of cereals and legumes through activation of endogenous enzymes. *Food Science & Nutrition* 6(8):2446–2458.

Nout, M.J.R. 2009. Rich nutrition from the poorest cereal fermentations in Africa and Asia. *Food Microbiology* 26:685–692.

Nyyssola, A., Ellilä, S., Nordlund, E., Poutanen, K. 2020. Reduction of FODMAP content by bioprocessing. *Trends in Food Science and Technology* 99:257–272.

Oguntoyinbo, F.A., Narbad, A. 2012. Molecular characterization of lactic acid bacteria and in situ amylase expression during traditional fermentation of cereal foods. *Food Microbiology* 31:254–262.

Omizu, Y., Tsukaoto, C., Chettri, R., Tamang, J.P. 2011. Determination of saponin contents in raw soybean and fermented soybean foods of India. *Journal of Scientific and Industrial Research* 70:533–538.

Onda, T., Yanagida, F., Uchimura, T., Tsuji, M., Ogino, S., Shinohara, T., Yokotsuka, K. 2002. Widespread distribution of the bactreiocin-producing lactic acid cocci in miso-paste products. *Journal of Applied Microbiology* 92695–705.

Papadimitriou, K., Alegria, A., Bron, P.A., De Angelis, M., Gobbetti, M., Kleerebezem, M., Lemos, J.A., Linares, D.M., Ross, P., Stanton, C., Turroni, F., Sinderen, D., Varmanen, P., Ventura, M., Zuniga, M., Tsakalidou, E., Kok, J. 2016. Stress physiology of lactic acid bacteria. *Microbiology and Molecular Biology Reviews* 80:837–890.

Rawat, K., Kumari, A., Kumar, S., Kumar, R., Gehlot, R. 2018. Traditional fermented products of India. *International Journal of Current Microbiology and Applied Science* 7(4):1873–1883.

Ray, M., Ghosh, K., Singh, S., Mondal, K.C. 2016. Folk to functional: an explorative overview of rice-based fermented foods and beverages in India. *Journal of Ethnic Foods* 3(1):5–18.

Ray, R.C., Swain, M.R. 2013. Indigenous fermented foods and beverages of Odisha, India: an overview. In *Indigenous fermented foods of South Asia*, ed. V.K. Joshi. CRC Press. pp. 1–30.

Reale, A., Konietzny, U., Coppola, R., Sorrentino, E., & Greiner, R. 2007. The importance of lactic acid bacteria for phytate degradation during cereal dough fermentation. *Journal of Agricultural and Food Chemistry* 55(8):2993–2997.

Rehaiem, A., Belgacem, Z.B., Edalatian, M.R., Martínez, B., Rodríguez, A., Manai, M., Guerra, N.P. 2014. Assessment of potential probiotic properties and multiple bacteriocin encoding-genes of the technological performing strain *Enterococcus faecium* MMRA. *Food Control* 37:343–350.

Ripari, V. 2019. Techno-functional role of exopolysaccharides in cereal-based, yogurt-like beverages. *Beverages* 5(1):16.

Rodríguez, H., Curiel, J.A., Landete, J.M., de las Rivas, B., de Felipe, F.L., Gómez-Cordovés, C., Mancheno, J.M., Muñoz, R. 2009. Food phenolics and lactic acid bacteria. *International Journal of Food Microbiology* 132(2–3):79–90.

Roy, A., Moktan, B., Sarkar, P.K. 2007. Traditional technology in preparing legume-based fermented foods of Orissa. *Indian Journal of Traditional Knowledge* 6(1):12–16.

Saini, S., Saxena, S., Samtiya, M., Puniya, M., Dhewa, T. 2021. Potential of underutilized millets as Nutri-cereal: an overview. *Journal of Food Science and Technology* 58(12):4465–4477.

Samtiya, M., Aluko, R.E., Dhewa, T. 2020. Plant food anti-nutritional factors and their reduction strategies: an overview. *Food Production Processing and Nutrition* 2(1):1–14.

Sanjukta, S., Rai, A.K. 2016. Production of bioactive peptides during soybean fermentation and their potential health benefits. *Trends Food Science and Technology* 50:1–10.

Satish Kumar, R., Kanmani, P., Yuvaraj, N., Paari, K.A., Pattukumar, V., Arul, V. 2013. Traditional Indian fermented foods: a rich source of lactic acid bacteria. *International Journal of Food Sciences and Nutrition* 64(4):415–428.

Schoenlechner, R., Wendner, M., Siebenhandl-Ehn, S., Berghofer, E. 2010. Pseudocereals as alternative sources for high folate content in staple foods. *Journal of Cereal Science* 52(3):475–479.

Schwab, C., Mastrangelo, M., Corsetti, A., Ganzle, M. 2008. Formation of oligosaccharides and polysaccharides by *Lactobacillus reuteri* LTH5448 and *Weissella cibaria* 10 M in sorghum sourdoughs. *Cereal Chemistry* 85(5):679–684.

Senapati, A.K., Ann, A., Raj, A., Gupta, A., Sharma, A., Neopany, B., Panmei, C., Diwedi, D.H., Dev, R., Angchok, D., Bakar, F.A., Chye, F.Y., Rapsang, G.F., Vyas, G., Devi, G.A.S., Prajapati, J.P., Sim, K.Y., Targais, K., Reddy, L.V.A., Swain, M.R., Reza, S., Zaman., M.Z., Garg, N., Singh, N.S., Sharma, N., Ray, R.C., Thorat, S.S., Pinto, S.V., Gautam, S., Thokchom, S., Joshi, S.R., Khomdram, S., Stobdan, T. 2016. Diversity of indigenous fermented foods and beverages of South Asia. In *Indigenous Fermented Foods of South Asia*, ed. V. Joshi, 69–106. CRC Press.

Sengen, I.Y., Nielsen, D.S., Karapinar, M., Jakobsen, M. 2009. Identification of Lactic acid bacteria isolated from Tarhana, a traditional Turkish fermented food. *International Journal of Food Microbiology* 135(2):105–111.

Sharma, R., Garg, P., Kumar, P., Bhatia, S.K., Kulshrestha, S. 2020. Microbial fermentation and its role in quality improvement of fermented foods. *Fermentation* 6(4):106.

Sharma, V., Attri, S., Mahajan, R., Goel, G. 2019. Microbial profiling of Malera and Phab: starters used for preparing traditional fermented foods and beverages in Himachal Pradesh, India. *Current Nutrition & Food Science* 15(7):707–711.

Sicard, D., Legras, J.L. 2011. Bread, beer and wine: yeast domestication in the *Saccharomyces* sensu stricto complex. *Comptes rendus Biologies* 334(3):229–236.

Sindhu, S.C., Khetarpaul, N. 2001. Probiotic fermentation of indigenous food mixture: effect on antinutrients and digestibility of starch and protein. *Journal of Food Composition and Analysis* 14(6):601–609.

Singh, C., Kumar, V., Sharma, M., Singh, A.K., Kumar, A. 2019. Study of Physicochemical Properties of Fermented Wheat Flour (Sheera). *Journal of Agricultural Engineering and Food Technology* 6(4):271–274.

Skrajda-Brdak, M., Konopka, I., Tańska, M., Czaplicki, S. 2019. Changes in the content of free phenolic acids and antioxidative capacity of wholemeal bread in relation to cereal species and fermentation type. *European Food Research and Technology* 245(10):2247–2256.

Tamang, J.P. 2015. Naturally fermented ethnic soybean foods of India. *Journal of Ethnic Foods* 2:8–17.

Tamang, J.P. 2022a. Dietary culture and antiquity of the Himalayan fermented foods and alcoholic fermented beverages. *Journal of Ethnic Foods* 9:30.

Tamang, J.P. 2022b. "Ethno-microbiology" of ethnic Indian fermented foods and alcoholic beverages. *Journal of Applied Microbiology* 133(1):145–161.

Tamang, J.P., Cotter, P.D., Endo, A., Han, N.S., Kort, R., Liu, S.Q., Mayo, B., Hutkins, R. 2020. Fermented foods in a global age: East meets West. *Comprehensive Reviews in Food Science and Food Safety* 19(1):184–217.

Tamang, J.P., Jeyaram, K., Rai, A.K., Mukherjee, P.K. 2021. Diversity of beneficial microorganisms and their functionalities in community-specific ethnic fermented foods of the Eastern Himalayas. *Food Research International* 148:110633.

Tamang, J.P., Shin, D.H., Jung, S.J., and Chae, S.W. 2016b. Functional properties of microorganisms in fermented foods. *Frontiers in Microbiology* 7:578.

Tamang, J.P., Thapa, N., Tamang, B., Rai, A., Chettri, R. 2015. Microorganisms in fermented foods and beverages. In *Health benefits of fermented foods and beverages*, ed. J. Tamang, 1–110. CRC Press.

Tamang, J.P., Watanabe, K., Holzapfel, W.H. 2016a. Review: diversity of microorganisms in global fermented foods and beverages. *Frontiers in Microbiology* 7:377.

Tamene, A., Baye, K., Kariluoto, S., Edelmann, M., Bationo, F., Leconte, N., Humblot, C. 2019. *Lactobacillus plantarum* P2R3FA isolated from traditional cereal-based fermented food increase folate status in deficient rats. *Nutrients* 11(11):2819.

Thapa N., Tamang J.P. 2020. Ethnic Fermented Foods and Beverages of Sikkim and Darjeeling Hills (Gorkhaland Territorial Administration). In *Ethnic fermented foods and beverages of India: science history and culture*, ed. J. Tamang, 479–537. Springer.

Tieking, M., Gänzle, M.G. 2005. Exopolysaccharides from cereal-associated lactobacilli. *Trends in Food Science & Technology* 16(1–3):79–84.

Todorov, S.D. 2010. Diversity of bacteriocinogenic lactic acid bacteria isolated from boza, a cereal-based fermented beverage from Bulgaria. *Food Control* 21:1011–1021.

Turbic, A., Ahokas, J.T., Haskard, C.A. 2002. Selective in vitro binding of dietary mutagens, individually or in combination, by lactic acid bacteria. *Food Additives and Contaminants* 19(2):144–152.

van Hijum, S.A., Kralj, S., Ozimek, L.K., Dijkhuizen, L., and van Geel-Schutten, I.G. 2006. Structure-function relationships of glucansucrase and fructansucrase enzymes from lactic acid bacteria. *Microbiology and Molecular Biology Reviews* 70(1): 157–176.

Vijayendra S.V.N., Rajashree K., Halami P.M. 2010. Characterization of a heat stable anti-listerial bacteriociin produced by vancomycin sensitive Enterococcus faecium isolated from idli batter. *Indian Journal of Microbiology* 50:243–246.

Yonzan, H., Tamang, J.P. 2010. Microbiology and nutritional value of selroti, an ethnic fermented cereal food of the Himalayas. *Food Biotechnology* 24(3):227–247.

Yousif, N.M.K., Dawyndt, P., Abriouel, H., Wijaya, A., Shillinger, U., Vancanneyt, M., Swings, J., Dirar, H.A., Holzapfel, W.H., Franz, C.M.A.P. 2005. Molecular characterization, technological properties and safety aspects of enterococci from Husuwa, an Africain fermented sorghum product. *Journal of Applied Microbiology* 98:216–228.

Yuan, J.M., Stram, D.O., Arakawa, K., Lee, H.P., Yu, M.C. 2003. Dietary cryptoxanthin and reduced risk of lung cancer: the Singapore Chinese Health Study. *Cancer Epidemiology Biomarkers & Prevention* 12(9):890–898.

Zannini, E., Pontonio, E., Waters, D.M., Arendt, E.K. 2012. Applications of microbial fermentations for production of gluten-free products and perspectives. *Applied Microbiology and Biotechnology* 93(2):473–485.

Zeidan, A.A., Poulsen, V.K., Janzen, T., Buldo, P., Derkx, P.M., Øregaard, G., Neves, A.R. 2017. Polysaccharide production by lactic acid bacteria: from genes to industrial applications. *FEMS Microbiology Reviews* 41(Supp_1):S168–S200.

Zinedine, A., Faid, M., Benlemlih, M. 2005. In vitro reduction of aflatoxin b1 by strains of lactic acid bacteria isolated from Moroccan sourdough bread. *International Journal of Agriculture and Biology* 7(1):67–70.

Microbial Diversity for Novel Fermented Foods and Perspectives from Pakistan

13

Uroosa Ejaz and Muhammad Sohail

13.1 INTRODUCTION

The term "fermentation" comes from the Latin word "fermentum", which means the natural decomposition of complex organic materials into a simpler form facilitated by microbial catalysis (Steinkraus 1983). It is one of the traditional methods to preserve food (alongside salting and drying) and for culinary enjoyment and food quality modification (Xiang et al. 2019). Fermented beverages and foods are enzymatically altered by microorganisms, which results in a better taste and smell of food. Beyond preservation, fermentation provides new and unique variations in foods and improves food palatability (Rodgers 2008). The types of fermented foods are driven by raw materials, environmental conditions, taste preferences, availability of food sources and, ultimately, new technological development (Rezac et al. 2018).

13.2 BRIEF HISTORICAL PERSPECTIVE

The history of fermented foods dates back thousands of years. The evidence suggests that as early as 6000 BC, the civilizations of the Fertile Crescent in the Middle East used fermented vegetables, meat and milk (Ali et al. 2009). However, studies emerged in the 19th century to understand the mechanism of the process and to develop better equipment. The art of fermentation has now become a well-established science where carefully selected raw materials are fermented by using microorganisms with desirable attributes by adopting various techniques (Azam et al. 2017). At present, more than 3500 fermented vegetable, meats, fish, tea, pulses, cereals and dairy products are available (Xiang et al. 2019).

13.3 MICROORGANISMS INVOLVED IN FERMENTATION OF FOOD PRODUCTS

Fermented foods cannot be prepared without microorganisms either naturally present in the food or added externally in pure or mixed form. Their presence in fermented beverages and food varies according to the preparation methods, types of raw materials, surrounding environment, climactic factors and geography (Tamang et al. 2020). From a list of microorganisms (Table 13.1) which are well known for the production of fermented beverages and food, examples can be such as *Bifidobacterium*, *Lactococcus*, *Enterococcus*, *Streptococcus*, *Lactobacillus*, *Penicillium chrysogenum*, *P. roqueforti*, *Aspergillus sojae*, *Aspergillus oryzae*, *Candida humilis*, *Candida krusei* and *Saccharomyces cerevisiae* (Xiang et al. 2019). In the western world and Asia, most fermented foods are reliant on lactic acid bacteria (LAB) to mediate the fermentation procedure (Ahmed et al. 2013), and several genera, including *Leuconostoc*, *Streptococcus*, *Pediococcus*, and *Lactobacillus* are predominant in manufacturing fermented vegetables, meat and dairy products (Rezac et al. 2018; Kok and Hutkins 2018). Furthermore, acetic acid bacteria are also involved in the fermentation of kombucha, vinegar, beans and cocoa (De Roos and De Vuyst 2018; Meersman et al. 2013). *Aspergillus* sp., *Penicillium* sp. and other molds are also used in food manufacturing, such as cheese, while yeasts are employed for the production of bread, wine and beer (Kok and Hutkins 2018). Indeed, for preparing many foods, yeast and bacteria are used in combination to produce the desired product (Sieuwerts et al. 2018; Liu et al. 2017).

 DOI: 10.1201/9781003605300-13

TABLE 13.1 Microorganisms Involved in Fermentation of Food

MICROORGANISM		*FERMENTED PRODUCT*	*REFERENCE*
Bacteria	Lactic acid bacteria	Blue cheese	(Xiang et al. 2019)
	Staphylococcus sp. SK1-1-5, *Virgibacillus* sp. SK37, and *Virgibacillus* sp. SK33	Fermented anchovy sauce	(Yongsawatdigul et al. 2007)
	Pediococcus and *Lactobacillus*	Fermented narezushi (modern Japanese sushi)	(Koyanagi et al. 2011)
	Pediococcus pentosaceus ZY40 GY23, *Lactobacillus plantarum*, *Staphylococcus*, *Macrococcuss*, *Enterococcus*, *Vagococcus*, *Lactococcus*, *Latilactobacillus sakei*	Grass carp sausages, *chouguiyu* (a traditional Chinese fermented-fish product)	(Ngasotter et al. 2020)
	Lactic acid bacteria including *Lactobacillus pentosus* and *Lactiplantibacillus plantarum*	Fermented Thai fish product som-fak	(Paludan-Müller et al. 2002)
	Staphylococcus carnosus and *Staphylococcus xylosus*	Fermented sausages of Vallo di Diano (southern Italy)	(Casaburi et al. 2007)
	Coagulase-negative cocci (CNC, *Kocuria* spp. and *Staphylococcus* spp.) and lactic acid bacteria (LAB, *Lactobacillus* spp.)	Fermentation and ripening of sausages	(Corbière Morot-Bizot et al. 2006)
Fungi	*Aspergillus* species (eg, *Aspergillus versicolor*), *Penicillium* species (*P. verrucosum*, *P. commune*, *P. brevicompactum*), *P. nalgiovense*, *P. camemberti*, *P. roqueforti*	Mold-ripened cheeses, brie and camembert cheese	(Sanjukta and Rai 2016)
Yeast	*Saccharomyces*, *Trichosporon*, *Cryptococcus*, *Debaryomyces*, *Rhodotorula*, *Yarrowia*, *Pichia*, *Kluyveromyces*, and *Candida*	Fermented meat products	(Dillon and Board 1989)

13.4 BENEFITS OF FERMENTED FOODS

The effective use of fermentation techniques can maximize the utilization of bio-resources and hence can aid in addressing the global food crisis (Yu and Bogue 2013). The awareness of consumers regarding diet–disease relationships has renewed calls for health-promoting benefits in fermented food products. Fermentation transforms raw food material into new products which enhance gut health properties (due to the presence of probiotics), as well as increasing nutritional value (due to the enrichment or generation of certain bioavailable nutrients during fermentation) (Gobbetti et al. 2010; Granato et al. 2010). Microorganisms utilize the raw materials and produce bioactive metabolites and nutrients with desirable bioavailability and molecular size (e.g. amino acids and peptides) (Ghosh et al. 2013). During fermentation, large proteins are converted into active peptides by enzymatic hydrolysis, such as the production of milk-derived peptides with immunomodulatory, antimicrobial, mineral binding, anti-hypertensive and antioxidant effects (Udenigwe and Mohan 2014).

Fermented foods of an innovative or traditional nature possess quality and nutritional advantages as they offer health functionalities to specific groups of consumers (such as vegetarians or those on cholesterol-restricted or lactose-intolerant diets) and also serve as a source of better nutrition to the general population. The microorganisms used for fermentation purposes actually function as "microbial factories" which not only produce desired bioactives and nutrients but also consume undesired substances; for example, bacteria break down lactose during fermentation by producing β-galactosidase, which results in the production of lactose-reduced or lactose-free milk products (Martins et al. 2013).

The consumption of fermented foods offers many advantages as compared to non-fermented foods, such as reduction of gastrointestinal disorders; inhibition of pathogenic bacteria; degradation of bacterial and plant toxins (e.g. cyanogenic glycosides); increase of glucose tolerance; improvements in digestibility and decreased risks of various diseases and illness, including skin problems, liver disorders (Higashikawa et al. 2010), bone deformities (Narva et al. 2004), respiratory illnesses (Makino et al. 2010), periodontitis (Adegboye et al. 2012), type 2 diabetes (O'Connor et al. 2014), arthritis and cardiovascular diseases (Sonestedt et al. 2011).

13.4.1 Bioactive Compounds

Fermented foods carry various bioactive compounds which offer many beneficial health effects. These are small molecules which are generated during fermentation (Martins et al. 2011). Phenolic compounds are the well-known bioactive components with immunomodulating activities.

The variety or cultivar of the plant also affects the production of bioactive compounds. Many non-nutritive phytochemicals exhibit disease-fighting properties (Dueñas et al. 2005). Ryan et al. (2011) reported that fermentation of different rice brans by *Saccharomyces boulardii* released different metabolites. Keller et al. (2013) observed an increased amount of epicatechin gallate, stearamide, linoleamide and dodecanamide in tea leaves fermented by *Eurotium cristatum* as compared to non-fermented green tea. Bioactive compounds including non-saponins and saponins (ginsenosides) in red ginseng roots help in the regulation of insulin and blood glucose levels (Oh et al. 2014).

Simple phenols (e.g. phenolic acids) and complex phenolics (e.g. ellagitannins, lignans and chalcones) are produced during many fermentation processes (Callemien and Collin 2010). Based on the structure, phenolic acids are divided into two types: hydroxycinnamic and hydroxybenzoic acids. The most commonly found hydroxycinnamic acids include syringic, vanillic, protocatechuic, hydroxybenzoic and gallic acids, whereas caffeic acids, ferulic acid, p-coumaric acid and sinapic acid can be included in the category of hydroxybenzoic acids (Martins et al. 2011).

Physical and chemical methods are reported to produce bioactive compounds from food, but the use of fermentation techniques for the extraction/production of bioactive compounds provides certain advantages, including the production of high-quality and bioactive extracts. These bioactive compounds are produced as secondary metabolites during fermentation (Sanjukta and Rai 2016). Barrios-Gonzalez et al. (2005) reported that exhaustion of key nutrients like nitrogen, phosphate or carbon sources results in the production of bioactive compounds.

13.5 FERMENTED FOOD PRODUCTS

Fermented beverages and foods may vary by the intentional application of microbes utilized, fermentation time and food type (Table 13.2). There are number of fermented foods available in today's global food markets, as shown in Table 13.3.

TABLE 13.2 Types of Foods Based on Fermentation Time

FOOD	*FERMENTATION*	*REFERENCES*
Sourdough bread	Fermentation in the preparation phase	(Corsetti and Settanni 2007; (Hansen and Schieberle 2005)
Fermented tea leaves		(Lv et al. 2013)
Chocolate		(Crafack et al. 2014)
Coffee		(Lee et al. 2015)
Wine	Fermentation of the final food product	(Lee et al. 2009)
Beer		(Grassi et al. 2014)
Kombucha		(Chu and Chen 2006)
Kimchi		(Cho et al. 2006)
Yogurt		(Soukoulis et al. 2007)

TABLE 13.3 Fermented food products (Xiang et al. 2019)

FOOD	*FERMENTED PRODUCT*
Soy based	Wadi, papad, idli, dosa, dholka, amrita, yellow soybean paste, tauchu, tamari, gochujang, doubanjiang, doenjang, chunjang, cheonggukjang, natto, miso, tofu, tempeh, soybean sauce and soy sauce
Tea based	Miang, leppet-so, bai-ming, kombucha, pu-erh tea
Barley, maize, rice based	Pozol, sake, amazake, ogi, kvass, vinegar, sorghum beer, kurosu, vinegar, som moo, nem chua, pepperoni, salami, chorizo, jamón ibérico
Meat based	Bosnian sudzuk, pepperoni, beef sticks, salami, sausages, chin som mok
Fish based	Shiokara, soy sauce, shidal, surströmming, shrimp paste, rakfisk, jeotgal, hákarl, garum, fish sauce, faseekh, bagoong
Dairy based	Shuba, dahi, kumis, kefir, cheese, cultured milk products like: whey vinegar, yogurt, skyr, smetana, crème fraîche, yogurt, skyr, smetana, filmjölk, quark
Honey based	Venigar, metheglin, mead
Fruit based	Nata de pina, nata de coco, pickling (mango pickle, lime pickle, lemon pickle), asinan, mangga, burong, atchara, brandy, perry, cider, vinegar, wine
Vegetable based	Potato beer, mushrooms, oiso baegi, oiji, oigee, mootsanji, szechwan cabbage, vanilla, tempoyak, tang-chai, sunki, si-sek-chai, santol, sambal tempo-jak, sajur asin, phonlami-dong, phak-dong, paw-tsay, pak-siam-dong, naw-mai-dong, kong-chai, kiam-cheyi, jeruk, dalok, burong mangga, asinan, gundruk, sauerkraut, Indian pickle, mixed pickle, kimchi
Grain based	Vinegar, kishk, bouza, seketeh, merissa, zambian opaquemaize beer, malawa beer, busaa (maize beer), pitokaffir beer, burukutu, vodka, dosa, idli, grainwhisky, malt whisky hisky, rice wine, sowans, sourdough, sikhye, sake, ogi, murri, makgeolli, kvass, injera, gamju, bread choujiu, beer, amazake

13.5.1 Fermented Vegetable Products

Archeological records suggest that from the period 960 to 1279 CE, fermented vegetables were commonly produced and consumed in China (Tamang et al. 2020). These vegetables are rich in vitamins, antioxidants, minerals and dietary fibers and therefore remain in demand for natural nutraceutical markets. Fermented vegetables such as peppers, green peas, olives, tomatoes, cauliflowers, carrots, sauerkraut and kimchi are traditional foods. Lactic acid fermentation of salted white cabbage resulted in the popular fermented vegetable sauerkraut (Lu et al. 2003). Korean kimchi products include nabak kimchi (cut radish and Chinese cabbage), dongchimi (whole radish kimchi with water), baik kimchi (beachu kimchi with water), kakdugi, yeolmoo kimchi, tongbaechu kimchi and beachu kimchi, as well as mulkimchi (with water) and ordinary kimchi (without additional water) (Xiang et al. 2019). All these kimchi products are formed by lactic acid bacterial fermentation and thus contain significant amounts of minerals, vitamins, dietary fibers and lactic acid bacteria (LAB) (10^7–10^9 CFU g^{-1}). Furthermore, kimchi products have anticancer, anti-obesity, anti-bacterial, anti-inflammatory, anti-aging, antioxidant and probiotic effects (Park et al. 2014).

13.5.2 Cereals and Breads

The production of bread has been described in ancient scriptures dated back to 10,000 BCE in Egypt (Tamang et al. 2020). Fermentation of a rye flour matrix and wheat with LAB was found to decrease the glycemic index of the obtained breads (Maioli et al. 2008; Juntunen et al. 2003). On the other hand, cereals are rich source of bioactive phytochemicals, that is, phytosterols (e.g. stanols and sterols), phenolics (e.g. phenolic acids and lignans) and vitamins (e.g. folate, vitamin E and thiamine). Cereals such as wheat, sorghum, rye, rice, oats, millet, maize and barley have been used as substrates for fermentation (Xiang et al. 2019). In contrast to unfermented cereals, fermented cereals have higher bioavailability of minerals and lower anti-nutritional effects, and they tend to be more palatable (Katina et al. 2005).

13.5.3 Fermented Meats

Fermentation has long been used to preserve meat worldwide. It has been part of deep-rooted culture in Europe for thousands of years (Tamang et al. 2020). Fermentation of meat involves complex microbiological, chemical and physical events, resulting in products with unique nutritional value, texture, flavor and color. Molds, yeasts and bacteria (such as Staphylococci, Micrococci and lactic acid bacteria) are involved in the fermentation of meat (Singh et al. 2012).

13.5.4 Fermented Soybeans

The use of fermented soybean paste was first recorded in China around 2,200 years ago (Tamang et al. 2020). Fermented soy products have higher nutritional value, unique flavor and significant amounts of fatty acids and amino acids. Mold, bacteria and yeasts are used in pure form or in combinations to produce fermented soy products such as natto, tempeh, miso, tamari sauce or soy sauce (Shin and Jeong 2015).

13.5.5 Fermented Seafood

Fermented seafoods are composed of fermented whole fish or derived by-products. The use of fermented fish started in Southeast Asia around 200 BCE to 200 CE (Tamang et al. 2020). Streptococci, Pediococci and Lactobacilli are typically used for the fermentation of fish (Zang et al. 2020).

13.5.6 Yogurt

Yogurt has been produced for over 10,000 years. It is a well-known fermented milk vehicle because it possesses advantages in nutritional value over non-fermented milk owing to the presence of health-promoting microorganisms and health benefits (Issazadeh et al. 2013).

13.5.6.1 Nutrient Profiles

Yogurt is a rich source of various micro- and macronutrients such as vitamin B12, phosphorus, magnesium, calcium, digestible carbohydrates and high-quality proteins. Plain yogurt is relatively low in fat and glycemic index and high in protein, making it an excellent meal or snack accompaniment. An average 170-g portion of popular low-fat fruit-flavored yogurt provides 30% of vitamin B12, 21–25% of riboflavin, 26% of phosphorus, 20–24% of calcium and 6% of potassium and magnesium (Kok and Hutkins 2018). Yogurt consists of high-quality proteins called whey proteins representing around 20% and casein proteins constituting 80% of the dairy proteins (McGregor and Poppitt 2013). The milk fats present in yogurt provide a pleasant texture and impart richness to yogurts (Chandan 2007). Yogurt is also a rich source of fat-soluble vitamins, K, E, D and A (Tunick and Van Hekken 2015) and conjugated linoleic acid, which has anticancer properties (Serafeimidou et al. 2012)

13.5.6.2 Proto-Cooperation

Yogurt is produced by a process called proto-cooperation (Azam et al. 2017), which is a process that produces yogurt by mutual stimulation of growth of *Streptococcus thermophilus* and *Lactobacillus delbrueckii* subsp. *bulgaricus* at levels ranging from $<10^4$ to 10^9 cfu g^{-1} or mL^{-1} (Rezac et al. 2018). *S. thermophilus* produces nutrients (carbon dioxide and formate), and *L. delbrueckii bulgaricus* utilizes these nutrients and produces cell wall–bound proteases (Sieuwerts et al. 2008). Horizontal gene transfer, particularly transfer of a gene cluster for the production of sulfur-containing amino acids from *L. bulgaricus* to *S. thermophilus*, also takes place during yogurt formation

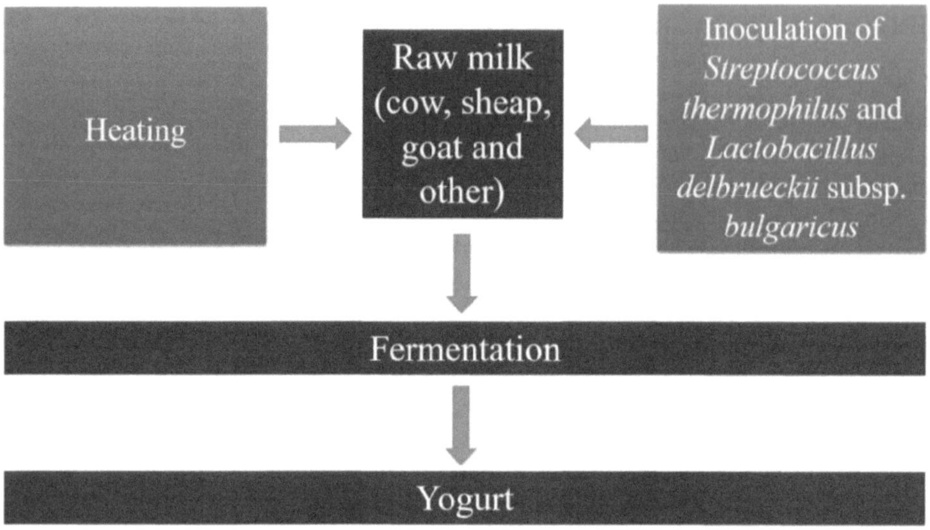

FIGURE 13.1 Flowchart showing yogurt formation from milk.

and supports combined growth (Bolotin et al. 2004). In several studies, other microorganisms, including *Lactobacillus* sp. and *Bifidobacterium* sp., were also enumerated (Rezac et al. 2018). In essence, fermentation of milk by lactic acid bacteria in a closed incubated environment results in the production of yogurt (Figure 13.1).

13.5.6.3 Advantages of Yogurt Consumption

Yogurt, as a nutrient-dense fermented food, contains various bioactive metabolites, macro-and micro-nutrients and bacterial cultures which provide many health benefits (Sodini et al. 2004). According to some epidemiological studies, consumption of yogurt has been shown to improve health and to reduce the risk of disease. In one large cohort study, less long-term weight gain was associated with yogurt consumption (Mozaffarian et al. 2011). In another large cohort study, it was concluded that consumption of a yogurt-rich diet was inversely associated with metabolic syndrome (Babio et al. 2015) and resulted in a lower risk of developing bladder cancer (Larsson et al. 2008). The European Prospective Investigation into Cancer and Nutrition (EPIC) cohort study of European adults reported that combined consumption of fermented milk, yogurt and cheese reduced the risk of diabetes (Sluijs et al. 2012). Similarly, the EPIC-Italy cohort study suggested reduced risk of colorectal cancer with the consumption of yogurt (Pala et al. 2011). Results of the Swedish Malmo Diet and Cancer cohort study showed that consumption of fermented milk reduces the risk of cardiovascular diseases (Sonestedt et al. 2011). Additionally, the Quebec family study revealed that consumption of yogurt improved body composition and provided health benefits independently of diet quality (Panahi et al. 2018). Individuals who are lactose intolerant are able to consume yogurt without developing symptoms (Kok and Hutkins 2018). It is suggested that the microbes in yogurt, including *L. delbrueckii bulgaricus* and *S. thermophilus*, have a protective effect against lactose (Oak and Jha 2019; Shaukat et al. 2010). Furthermore, the release of branched-chain amino acids during fermentation by *S. thermophilus* helped in the maintenance of muscle function, protein synthesis and lean body mass (McGregor and Poppitt 2013).

13.5.7 Probiotics

Beneficial bacteria present in fermented food are called probiotics and, when administered in adequate amounts, confer a health benefit to the host. Probiotic products have acquired a rapidly growing market worldwide. In recent years, probiotics have been used to treat various diseases *in vitro* and *in vivo* (Zhao et al. 2019). Probiotic bacteria maintain the microbial gastrointestinal balance (Zhao et al. 2019). Particularly, yogurt is used as a vehicle for probiotic delivery. Yogurt and lactic acid bacterial species have shown promising health effects on the digestive and immune systems; for example, they provide protection against colon cancer, *Helicobacter pylori* infection, inflammatory bowel disease, diarrheal diseases and constipation (Parvez et al. 2006; Adolfsson, Meydani, and Russell 2004). Lactic acid bacteria are able to produce bacteriocins in response to environmental stimuli, which modulates gut microbiota populations (Mokoena 2017). Incorporation of *L. paracasei* ssp. *tolerans*, a probiotic species, into yogurt showed beneficial effects against inflammatory response and pathogens (Korbekandi et al. 2015; Maragkoudakis et al. 2006). In addition, *Bifidobacterium bifidum* strains can inhibit pathogenic flora such as *Clostridium difficile* (Savadogo and Traore 2012). Probiotic milk culture of *Lacticaseibacillus* casei and *Lactobacillus acidophilus* was capable of delaying the growth of *Listeria innocua, Enterococcus faecalis, E. faecium* and *Staphylococcus aureus* (Millette et al. 2007). Sarvari et al. (2014) demonstrated that the survival of yogurt bacteria and probiotics is dependent on the strain, storage temperature and time.

13.6 USE OF FERMENTED FOODS IN ASIAN COUNTRIES

Fermented vegetables and fruits are associated with several cultural and social aspects of different races in Asian countries. Some foods of Southeast Asia are famous in other parts of the world (Swain et al. 2014). Fermented food products are generally produced by traditional methods at the village or family scale in Southeast Asia. However, there is a need to give more attention to improve the quality and safety of food (Waché et al. 2018). Nuoc-mam is a famous Asian fermented fish. However, safety-related concerns are associated with this food product, as high amounts of biogenic amines are detected in nuoc-mam (Waché et al. 2018).

Fermented fish is generally the most widely consumed food in many parts of Asia. It is available in two forms. Fish can be mixed with 15% to 25% salt to create fish pastes and fish sauces. Fish fermentation technology is a traditional method, and many of the products are prepared at home in Asia. They are regularly consumed as flavor-enhancing condiments and side dishes. Many of these products are region specific and include mehiawah from Middle East Asia; yucha from China; ngari, hentak, tungtak and shidal from India; sukuti and sidra from India and Nepal; nga pi from Myanmar; budu from Malaysia; pindang from Indonesia; jeot kal from Korea; shottsuru, bonito flakes, izushi (fermented sushi) and shiokara from Japan; nam pla and pla ra from Thailand; and patis from Philippines (Tamang et al. 2020). Furthermore, fish sauces are used as a condiment throughout East Asia. They are most commonly associated with Cambodia, Myanmar, Malaysia, Laos, Vietnam, Thailand and other South Asian regions (Thongthai and Gidberg 2005). In contrast, sauces prepared from fermented soybean paste are more commonly used in China, Korea and Japan (Tamang et al. 2020).

Each region or country in East Asia produces and consumes its own unique fermented vegetables. Lactic acid bacteria are mostly used for the fermentation of vegetables; thus, they are often considered possible sources of probiotic-like microbes. The list of fermented vegetables includes carrot, celery, cauliflower, turnips, beets, cucumbers, radishes and napa cabbage (Tamang et al. 2020). The representative fermented vegetables are burong mustala in the Philippines; sayur asin in Indonesia; pak-gard-dong and noa-mai-dong in Thailand; gundruk, sinki, mesu, soibum and soidon in India and Nepal; nukaduke and sunki in Japan, pao-tsai and suan-tsai in China and Taiwan; and kimchi in Korea. All of these products are made by LAB, with *Weissella* spp., *Pediococcus* spp., *Leuconostoc* spp. and *Lactobacillus* spp. as the major species involved in the fermentation (Tamang et al. 2008).

Wine has not been part of the culture or tradition in Asia. Usually, fruits, including grapes, are consumed directly without fermentation. Instead, alcoholic beverages are usually made from potatoes and cereals that contain few fermentable carbohydrates. Traditionally, dry amylolytic starters are used to convert starches to sugars (Shah et al. 2017; Tamang et al. 2016). Fermented tea such as miang is frequently consumed in Thailand, whereas kombucha, fuzhuan brick and puer tea are more commonly consumed in China (Tanasupawat et al. 2007). Pidan (or century egg) is a fermented duck egg mainly consumed in China (Ganasen and Benjakul 2010).

13.7 FERMENTED FOODS IN PAKISTAN

Pakistan, situated in South Asia, has coordinates between 70°E and 30°N and covers a total land and sea area of approximately 796,095 km. The country borders Afghanistan in the northwest, China in the north, Iran in the west, India in the east and the Arabian Sea in the south. The application of fermentation techniques in Pakistan and other developing countries of the region is highly desirable, as the low-cost technology has introduced a nutritionally reliable and safer class of fermented foods in the south (Soomro and Masud 2012). The conventional fermentation being adopted in Pakistan as a food-processing operation enhances the shelf life, aroma, texture and nutritional status of the fermented products. Fermented foods prepared at the household or domestic level in Pakistan have a simple and generic procedure. This is among the reasons that specialty fermented food products of the region have greater popularity among people owing to their unique nutritional and organoleptic attributes.

Paneer is a native coagulated dairy product. It is a soft, non-rennet, non-fermentative and un-ripened type of cheese in the South Asian region. Paneer is prepared from buffalo milk in Khyber Pakhtunkhwa, Pakistan, and extensively utilized in cooked meat and vegetable dishes. Fermented fruits and vegetable are also used in Pakistan. Pickles are one of the delightful cuisines of Pakistan and are the product of lactic acid fermentation. They have very good spicy flavor and are consumed routinely with food (Akhtar et al. 2016).

Cereal grains, as the staple diet of people of Pakistan, are considered an energy-dense dietary source of fiber, protein, carbohydrates and a range of essential micronutrients. Cereals in Pakistan, like other countries of South Asia, have a range of applications in indigenous fermented foods. Andrassay, molten lava cake (doli ki roti), pancakes preserved in syrup (jalebi) and bread and leavened bread (naan, also known as kulcha in some regions) are among the widely consumed cereal-based fermented food preparations (Akhtar et al. 2016).

Flat bread or naan is a cheap source of carbohydrates and protein for the people of South Asia (India, Iran, Afghanistan and Pakistan). Naan is prepared with sugar, salt and refined wheat flour, while product fermentation is carried out either with heterofermentative cultures of *Lactobacilli* spp. or yeasts to improve qualitative and nutritional features, increase product digestibility and ease fermentation. Another fermentative product, andrassay, is native to the Kasur district of Punjab,

TABLE 13.4 Fermented Food Products of Pakistan (Akhtar et al. 2016; Soomro and Masud 2012; Bhatia and Khetarpaul 2012)

REGIONS OF CONSUMPTION IN PAKISTAN	FOOD	MICROORGANISM	SUBSTRATE
Punjab, Sindh	Mango pickle	Indigenous micro flora	Green mango
Southern areas	Doli ki roti	Indigenous microflora	Fine flour
In all areas of Pakistan	Naan	Yeast	Wheat flour
Khyber Pakhtunkhuwa, northern areas	Paneer	Lactic acid bacteria	Fresh buffalo/cow milk
Southern, central areas of Punjab	Dahi	Lactic acid bacteria	Cow/buffalo/goat milk

Pakistan. Kasuri andrassay can be manufactured by three combinations: 1) butter oil, sugar and khoya; 2) butter oil, jaggery and wheat flour and 3) butter oil, sugar and wheat flour.

Another product that is unique in taste, jalebi, is an Asian desert prepared from refined white flour (maida). It is also called funnel cake and is fried and immersed in sugar syrup. Doli ki roti, like fried puri, is found in western Pakistan. It is prepared by blending cereals with legumes. It consists of 52.7–62.6 mg/100 g of Ca, 8.7–10.6 mg/100 g of iron, 3.8–4.7 % of ash and 14.5–17.1 % protein (Bhatia and Khetarpaul 2012).

Dahi is one of the dairy products found in the Indo-Pak sub-continent, having comparable features to yogurt. It accounts for 70% of total fermented dairy products. In Pakistan, lassi (butter milk) is made from dahi, which has a refreshing flavor and taste. A mixed culture of lactobacilli, *L. d. bulgaricus* and *S. thermophilus* is used in the preparation of dahi (Soomro and Masud 2012; Soomro et al. 2001).

However, there is a need to approach innovative strategies such as optimization of fermentation process, use of more sophisticated pure cultures and product safety, ensuring consistent and ideal utilization of locally adoptable and acceptable nourishing food commodities in the country.

13.7.1 Conclusion

In addition to traditional and historical knowledge, there is a growing body of scientific literature which shows that many substances are present in fermented foods, including microorganisms, bioactive compounds and health-promoting nutrients. Yogurt in various forms is one of the dominating fermented products which contains probiotics. The generally perceived healthy image, acceptability and popularity of yogurt make it an ideal snack or meal accompaniment. Probiotics are active microorganisms that are good for human health, especially for the human digestive system, and can be delivered through a wide range of fermented foods and beverages, including fermented foods commonly consumed in Pakistan. There are several questions to be answered in future studies of fermented beverages and food, such as the safety of fermentation under different conditions, portion sizes of fermented beverages and food necessary to elicit health benefits and determining biomarkers of fermented food health benefits. The diversity of fermented foods of Pakistan and Asia must be optimized for quality and health benefits based on these research needs.

REFERENCES

Adegboye, Amanda R.A., Lisa B. Christensen, Poul Holm-Pedersen, Kirsten Avlund, Barbara J. Boucher, and Berit L. Heitmann. 2012. "Intake of Dairy Products in Relation to Periodontitis in Older Danish Adults." *Nutrients* 4 (9): 1219–29. https://doi.org/10.3390/nu4091219.

Adolfsson, Oskar, Simin Nikbin Meydani, and Robert M. Russell. 2004. "Yogurt and Gut Function." *American Journal of Clinical Nutrition* 80 (2): 245–56. https://doi.org/10.1093/ajcn/80.2.245.

Ahmed, Sarifuddin, Krushna Chandra Dora, Sreekanta Sarkar, Supratim Chowdhury, and Subha Ganguly. 2013. "Quality Analysis of Shidal—A Traditional Fermented Fish Product of Assam, North-East India." *Indian Journal of Fisheries* 60 (1): 117–23.

Ali, Farrah Wahida Othman, Nazamid Bin Saari, Fatimah Abu Bakar, A.S. Abdulamir, Abdulkarim Sabo Mohammed, Yazid Abdul Manap, and Anwarul Hidayah Zulkifli. 2009. "Novel, Practical and Cheap Source for Isolating Beneficial γ-Aminobutyric Acid-Producing Leuconostoc NC5 Bacteria." *Research Journal of Medical Sciences* 3 (4): 146–53.

Azam, Mariya, Mashkoor Mohsin, Hira Ijaz, Ume Ruqia Tulain, Muhammad Adnan Ashraf, Ahad Fayyaz, Zain Ul Abadeen, and Qindeel Kamran. 2017. "Lactic Acid Bacteria in Traditional Fermented Asian Foods." *Pakistan Journal of Pharmaceutical Sciences* 30 (5): 1803–14.

Babio, Nancy, Nerea Becerra-Tomás, Miguel Ángel Martínez-González, Dolores Corella, Ramon Estruch, Emilio Ros, Carmen Sayón-Orea, et al. 2015. "Consumption of Yogurt, Low-FatMilk, and Other Low-Fat Dairy Products Is Associated with Lower Risk of Metabolic Syndrome Incidence in an Elderly Mediterranean Population1–3." *Journal of Nutrition* 145 (10): 2308–16. https://doi.org/10.3945/jn.115.214593.

Barrios-Gonzalez, J., F.J. Fermandez, A. Tomasini, and A. Mejia. 2005. "Secondary Metabolites Production by Solid-State Fermentation." *Malaysian Journal of Microbiology* 1 (November 2014): 1–6. https://doi.org/10.21161/mjm.110501.

Bhatia, Anshu, and Neelam Khetarpaul. 2012. "'Doli Ki Roti'—an Indigenously Fermented Indian Bread: Cumulative Effect of Germination and Fermentation on Bioavailability of Minerals." *Indian Journal of Traditional Knowledge* 11 (1): 109–13.

Bolotin, Alexander, Benoît Quinquis, Pierre Renault, Alexei Sorokin, S. Dusko Ehrlich, Saulius Kulakauskas, Alla Lapidus, et al. 2004. "Complete Sequence and Comparative Genome Analysis of the Dairy Bacterium Streptococcus Thermophilus." *Nature Biotechnology* 22 (12): 1554–58. https://doi.org/10.1038/nbt1034.

Callemien, D., and S. Collin. 2010. "Structure, Organoleptic Properties, Quantification Methods, and Stability of Phenolic Compounds in Beer-A Review." *Food Reviews International* 26 (1): 1–84. https://doi.org/10.1080/87559120903157954.

Casaburi, Annalisa, M. Conception Aristoy, Silvana Cavella, Rossella Di Monaco, Danilo Ercolini, Fidel Toldrá, and Francesco Villani. 2007. "Biochemical and Sensory Characteristics of Traditional Fermented Sausages of Vallo Di Diano (Southern Italy) as Affected by the Use of Starter Cultures." *Meat Science* 76 (2): 295–307. https://doi.org/10.1016/j.meatsci.2006.11.011.

Chandan, Ramesh C. 2007. "Manufacturing Yogurt and Fermented Milks." *Manufacturing Yogurt and Fermented Milks* 60 (3): 1–364. https://doi.org/10.1002/9780470277812.

Cho, Jinhee, Dongyun Lee, Changnam Yang, Jongin Jeon, Jeongho Kim, and Hongui Han. 2006. "Microbial Population Dynamics of Kimchi, a Fermented Cabbage Product." *FEMS Microbiology Letters* 257 (2): 262–67. https://doi.org/10.1111/j.1574-6968.2006.00186.x.

Chu, Sheng Che, and Chinshuh Chen. 2006. "Effects of Origins and Fermentation Time on the Antioxidant Activities of Kombucha." *Food Chemistry* 98 (3): 502–7. https://doi.org/10.1016/j.foodchem.2005.05.080.

Corbière Morot-Bizot, Stéphanie, Sabine Leroy, and Régine Talon. 2006. "Staphylococcal Community of a Small Unit Manufacturing Traditional Dry Fermented Sausages." *International Journal of Food Microbiology* 108 (2): 210–17. https://doi.org/10.1016/j.ijfoodmicro.2005.12.006.

Corsetti, Aldo, and Luca Settanni. 2007. "Lactobacilli in Sourdough Fermentation." *Food Research International* 40 (5): 539–58. https://doi.org/10.1016/j.foodres.2006.11.001.

Crafack, Michael, Hanna Keul, Carl Emil Eskildsen, Mikael A. Petersen, Sofie Saerens, Andreas Blennow, Mathias Skovmand-Larsen, et al. 2014. "Impact of Starter Cultures and Fermentation Techniques on the Volatile Aroma and Sensory Profile of Chocolate." *Food Research International* 63: 306–16. https://doi.org/10.1016/j.foodres.2014.04.032.

Dillon, Vivian M., and R.G. Board. 1989. "A Medium for Detecting Sulphite-binding Yeasts in Meat Products." *Letters in Applied Microbiology* 8 (5): 165–67. https://doi.org/10.1111/j.1472-765X.1989.tb00239.x.

Dueñas, Montserrat, Dolores Fernández, Teresa Hernández, Isabel Estrella, and Rosario Muñoz. 2005. "Bioactive Phenolic Compounds of Cowpeas (Vigna Sinensis L). Modifications by Fermentation with Natural Microflora and with Lactobacillus Plantarum ATCC 14917." *Journal of the Science of Food and Agriculture* 85 (2): 297–304. https://doi.org/10.1002/jsfa.1924.

Ganasen, Palanivel, and Soottawat Benjakul. 2010. "Physical Properties and Microstructure of Pidan Yolk as Affected by Different Divalent and Monovalent Cations." *LWT—Food Science and Technology* 43 (1): 77–85. https://doi.org/10.1016/j.lwt.2009.06.007.

Ghosh, Debasree, Dipti K. Chattoraj, and Parimal Chattopadhyay. 2013. "Studies on Changes in Microstructure and Proteolysis in Cow and Soy Milk Curd during Fermentation Using Lactic Cultures for Improving Protein Bioavailability." *Journal of Food Science and Technology* 50 (5): 979–85. https://doi.org/10.1007/s13197-011-0421-1.

Gobbetti, Marco, R. Di Cagno, and M. de Angelis. 2010. "Functional Microorganisms for Functional Food Quality." *Critical Reviews in Food Science and Nutrition* 50 (8): 716–27. https://doi.org/10.1080/10408398.2010.499770.

Granato, Daniel, Gabriel F. Branco, Adriano Gomes Cruz, José de Assis Fonseca Faria, and Nagendra P. Shah. 2010. "Probiotic Dairy Products as Functional Foods." *Comprehensive Reviews in Food Science and Food Safety* 9 (5): 455–70. https://doi.org/10.1111/j.1541-4337.2010.00120.x.

Grassi, Silvia, José Manuel Amigo, Christian B. Lyndgaard, Roberto Foschino, and Ernestina Casiraghi. 2014. "Beer Fermentation: Monitoring of Process Parameters by FT-NIR and Multivariate Data Analysis." *Food Chemistry* 155: 279–86. https://doi.org/10.1016/j.foodchem.2014.01.060.

Hansen, A., and P. Schieberle. 2005. "Generation of Aroma Compounds During Sourdough Fermentation: Applied and Fundamental Aspects." *Trends in Food Science and Technology* 16 (1–3): 85–94. https://doi.org/10.1016/j.tifs.2004.03.007.

Higashikawa, Fumiko, Masafumi Noda, Tomokazu Awaya, Kazuhiro Nomura, Hirotaka Oku, and Masanori Sugiyama. 2010. "Improvement of Constipation and Liver Function by Plant-Derived Lactic Acid Bacteria: A Double-Blind, Randomized Trial." *Nutrition* 26 (4): 367–74. https://doi.org/10.1016/j.nut.2009.05.008.

Issazadeh, Khosro, Morteza Azizollahi Ali Abadi, Reza Kazemi Darsanaki, Fatemeh Alikhani, Hassan Dadras, and Ahmad Tajehmiri. 2013. "Isolation, Identification and Analysis of Probiotic Properties of Lactobacillus Spp from Traditional Yoghurts in North of Iran." *Journal of Pure and Applied Microbiology* 7 (4): 2965–71.

Juntunen, Katri S., David E. Laaksonen, Karin Autio, Leo K. Niskanen, Jens J. Holst, Kari E. Savolainen, Kirsi Helena Liukkonen, Kaisa S. Poutanen, and Hannu M. Mykkänen. 2003. "Structural Differences between Rye and Wheat Breads but Not Total Fiber Content May Explain the Lower Postprandial Insulin Response to Rye Bread." *American Journal of Clinical Nutrition* 78 (5): 957–64. https://doi.org/10.1093/ajcn/78.5.957.

Katina, K., E. Arendt, K.H. Liukkonen, K. Autio, L. Flander, and K. Poutanen. 2005. "Potential of Sourdough for Healthier Cereal Products." *Trends in Food Science and Technology* 16 (1–3): 104–12. https://doi.org/10.1016/j.tifs.2004.03.008.

Keller, Amy C., Tiffany L. Weir, Corey D. Broeckling, and Elizabeth P. Ryan. 2013. "Antibacterial Activity and Phytochemical Profile of Fermented Camellia Sinensis (Fuzhuan Tea)." *Food Research International* 53 (2): 945–49. https://doi.org/10.1016/j.foodres.2013.04.023.

Kok, Car Reen, and Robert Hutkins. 2018. "Yogurt and Other Fermented Foods as Sources of Health-Promoting Bacteria." *Nutrition Reviews* 76: 4–15. https://doi.org/10.1093/nutrit/nuy056.

Korbekandi, H., Abedi, D., Maracy, M., Jalali, M., Azarman, N., and Iravani, S. 2015. "Evaluation of Probiotic Yoghurt Produced by Lactobacillus Paracasei Ssp. Tolerans." *Journal of Food Biosciences and Technology* 5 (1): 37–44. http://jfbt.srbiau.ac.ir/article_6080_ed8292fc998edb4bc5056a4cda91c5a9.pdf.

Koyanagi, T., M. Kiyohara, H. Matsui, K. Yamamoto, T. Kondo, T. Katayama, and H. Kumagai. 2011. "Pyrosequencing Survey of the Microbial Diversity of 'Narezushi', an Archetype of Modern Japanese Sushi." *Letters in Applied Microbiology* 53 (6): 635–40. https://doi.org/10.1111/j.1472–765X.2011.03155.x.

Larsson, Susanna C., Swen Olof Andersson, Jan Erik Johansson, and Alicja Wolk. 2008. "Cultured Milk, Yogurt, and Dairy Intake in Relation to Bladder Cancer Risk in a Prospective Study of Swedish Women and Men." *American Journal of Clinical Nutrition* 88 (4): 1083–87. https://doi.org/10.1093/ajcn/88.4.1083.

Lee, Jang Eun, Geum Sook Hwang, Cherl Ho Lee, and Young Shick Hong. 2009. "Metabolomics Reveals Alterations in Both Primary and Secondary Metabolites by Wine Bacteria." *Journal of Agricultural and Food Chemistry* 57 (22): 10772–83. https://doi.org/10.1021/jf9028442.

Lee, Liang Wei, Mun Wai Cheong, Philip Curran, Bin Yu, and Shao Quan Liu. 2015. "Coffee Fermentation and Flavor—An Intricate and Delicate Relationship." *Food Chemistry* 185: 182–91. https://doi.org/10.1016/j.foodchem.2015.03.124.

Liu, Youzhong, Sandrine Rousseaux, Raphaëlle Tourdot-Maréchal, Mohand Sadoudi, Régis Gougeon, Philippe Schmitt-Kopplin, and Hervé Alexandre. 2017. "Wine Microbiome: A Dynamic World of Microbial Interactions." *Critical Reviews in Food Science and Nutrition* 57 (4): 856–73. https://doi.org/10.1080/10408398.2014.983591.

Lu, Z., F. Breidt, V. Plengvidhya, and H.P. Fleming. 2003. "Bacteriophage Ecology in Commercial Sauerkraut Fermentations." *Applied and Environmental Microbiology* 69 (6): 3192–3202. https://doi.org/10.1128/AEM.69.6.3192-3202.2003.

Lv, Hai Peng, Ying Jun Zhang, Zhi Lin, and Yue Rong Liang. 2013. "Processing and Chemical Constituents of Pu-Erh Tea: A Review." *Food Research International* 53 (2): 608–18. https://doi.org/10.1016/j.foodres.2013.02.043.

Maioli, Mario, Giovanni Mario Pes, Manuela Sanna, Sara Cherchi, Mariella Dettori, Elena Manca, and Giovanni Antonio Farris. 2008. "Sourdough-Leavened Bread Improves Postprandial Glucose and Insulin Plasma Levels in Subjects with Impaired Glucose Tolerance." *Acta Diabetologica* 45 (2): 91–96. https://doi.org/10.1007/s00592-008-0029-8.

Makino, Seiya, Shuji Ikegami, Akinori Kume, Hiroshi Horiuchi, Hajime Sasaki, and Naoki Orii. 2010. "Reducing the Risk of Infection in the Elderly by Dietary Intake of Yoghurt Fermented with Lactobacillus Delbrueckii Ssp. Bulgaricus OLL1073R-1." *British Journal of Nutrition* 104 (7): 998–1006. https://doi.org/10.1017/S000711451000173X.

Maragkoudakis, Petros A., Christos Miaris, Pavel Rojez, Nikos Manalis, Flora Magkanari, George Kalantzopoulos, and Effie Tsakalidou. 2006. "Production of Traditional Greek Yoghurt Using Lactobacillus Strains with Probiotic Potential as Starter Adjuncts." *International Dairy Journal* 16 (1): 52–60. https://doi.org/10.1016/j.idairyj.2004.12.013.

Martins, Eliane M. Furtado, Afonso Mota Ramos, Ellen Silva Lago Vanzela, Paulo César Stringheta, Cláudia Lúcia de Oliveira Pinto, and José Manoel Martins. 2013. "Products of Vegetable Origin: A New Alternative for the Consumption of Probiotic Bacteria." *Food Research International* 51 (2): 764–70. https://doi.org/10.1016/j.foodres.2013.01.047.

Martins, Silvia, Solange I. Mussatto, Guillermo Martínez-Avila, Julio Montañez-Saenz, Cristóbal N. Aguilar, and Jose A. Teixeira. 2011. "Bioactive Phenolic Compounds: Production and Extraction by Solid-State Fermentation. A Review." *Biotechnology Advances* 29 (3): 365–73. https://doi.org/10.1016/j.biotechadv.2011.01.008.

McGregor, Robin A., and Sally D. Poppitt. 2013. "Milk Protein for Improved Metabolic Health: A Review of the Evidence." *Nutrition and Metabolism* 10 (1): 1–13. https://doi.org/10.1186/1743-7075-10-46.

Meersman, Esther, Jan Steensels, Melissa Mathawan, Pieter Jan Wittocx, Veerle Saels, Nore Struyf, Herwig Bernaert, Gino Vrancken, and Kevin J. Verstrepen. 2013. "Detailed Analysis of the Microbial Population in Malaysian Spontaneous Cocoa Pulp Fermentations Reveals a Core and Variable Microbiota." *PLoS One* 8 (12): e81559. https://doi.org/10.1371/journal.pone.0081559.

Millette, M., C. Dupont, D. Archambault, and M. Lacroix. 2007. "Partial Characterization of Bacteriocins Produced by Human Lactococcus Lactis and Pediococccus Acidilactici Isolates." *Journal of Applied Microbiology* 102 (1): 274–82. https://doi.org/10.1111/j.1365–2672.2006.03044.x.

Mokoena, Mduduzi Paul. 2017. "Lactic Acid Bacteria and Their Bacteriocins: Classification, Biosynthesis and Applications against Uropathogens: A Mini-Review." *Molecules* 22 (8): 1255. https://doi.org/10.3390/molecules22081255.

Mozaffarian, Dariush, Tao Hao, Eric B. Rimm, Walter C. Willett, and Frank B. Hu. 2011. "Changes in Diet and Lifestyle and Long-Term Weight Gain in Women and Men." *New England Journal of Medicine* 364 (25): 2392–2404. https://doi.org/10.1056/nejmoa1014296.

Narva, Mirkka, Riikka Nevala, Tuija Poussa, and Riitta Korpela. 2004. "The Effect of Lactobacillus Helveticus Fermented Milk on Acute Changes in Calcium Metabolism in Postmenopausal Women." *European Journal of Nutrition* 43 (2): 61–68. https://doi.org/10.1007/s00394-004-0441-y.

Ngasotter, Soibam, David Waikhom, Susmita Mukherjee, Manoharmayum Shaya Devi, and Asem Sanjit Singh. 2020. "Diversity of Lactic Acid Bacteria (LAB) in Fermented Fish Products: A Review." *International Journal of Current Microbiology and Applied Sciences* 9 (5): 2238–49. https://doi.org/10.20546/ijcmas.2020.905.255.

Oak, Sophia J., and Rajesh Jha. 2019. "The Effects of Probiotics in Lactose Intolerance: A Systematic Review." *Critical Reviews in Food Science and Nutrition* 59 (11): 1675–83. https://doi.org/10.1080/10408398.2018.1425977.

O'Connor, Laura M., Marleen A.H. Lentjes, Robert N. Luben, Kay Tee Khaw, Nicholas J. Wareham, and Nita G. Forouhi. 2014. "Dietary Dairy Product Intake and Incident Type 2 Diabetes: A Prospective Study Using Dietary Data from a 7-Day Food Diary." *Diabetologia* 57 (5): 909–17. https://doi.org/10.1007/s00125-014-3176-1.

Oh, Mi Ra, Soo Hyun Park, Sun Young Kim, Hyang Im Back, Min Gul Kim, Ji Young Jeon, Ki Chan Ha, et al. 2014. "Postprandial Glucose-Lowering Effects of Fermented Red Ginseng in Subjects with Impaired Fasting Glucose or Type 2 Diabetes: A Randomized, Double-Blind, Placebo-Controlled Clinical Trial." *BMC Complementary and Alternative Medicine* 14: 1–7. https://doi.org/10.1186/1472-6882-14-237.

Pala, Valeria, Sabina Sieri, Franco Berrino, Paolo Vineis, Carlotta Sacerdote, Domenico Palli, Giovanna Masala, et al. 2011. "Yogurt Consumption and Risk of Colorectal Cancer in the Italian European Prospective Investigation into Cancer and Nutrition Cohort." *International Journal of Cancer* 129 (11): 2712–19. https://doi.org/10.1002/ijc.26193.

Paludan-Müller, C., R. Valyasevi, H.H. Huss, and L. Gram. 2002. "Genotypic and Phenotypic Characterization of Garlic-Fermenting Lactic Acid Bacteria Isolated from Som-Fak, a Thai Low-Salt Fermented Fish Product." *Journal of Applied Microbiology* 92 (2): 307–14. https://doi.org/10.1046/j.1365-2672.2002.01544.x.

Panahi, Shirin, Caroline Y. Doyon, Jean Pierre Després, Louis Pérusse, Marie Claude Vohl, Vicky Drapeau, and Angelo Tremblay. 2018. "Yogurt Consumption, Body Composition, and Metabolic Health in the Québec Family Study." *European Journal of Nutrition* 57 (4): 1591–1603. https://doi.org/10.1007/s00394-017-1444-9.

Park, Kun Young, Ji Kang Jeong, Young Eun Lee, and James W. Daily. 2014. "Health Benefits of Kimchi (Korean Fermented Vegetables) as a Probiotic Food." *Journal of Medicinal Food* 17 (1): 6–20. https://doi.org/10.1089/jmf.2013.3083.

Parvez, S., K.A. Malik, S. Ah Kang, and H.Y. Kim. 2006. "Probiotics and Their Fermented Food Products Are Beneficial for Health." *Journal of Applied Microbiology* 100 (6): 1171–85. https://doi.org/10.1111/j.1365-2672.2006.02963.x.

Rezac, Shannon, Car Reen Kok, Melanie Heermann, and Robert Hutkins. 2018. "Fermented Foods as a Dietary Source of Live Organisms." *Frontiers in Microbiology* 9: 396129. https://doi.org/10.3389/fmicb.2018.01785.

Rodgers, Svetlana. 2008. "Novel Applications of Live Bacteria in Food Services: Probiotics and Protective Cultures." *Trends in Food Science and Technology* 19 (4): 188–97. https://doi.org/10.1016/j.tifs.2007.11.007.

Roos, Jonas De, and Luc De Vuyst. 2018. "Acetic Acid Bacteria in Fermented Foods and Beverages." *Current Opinion in Biotechnology* 49. Elsevier Ltd: 115–19. https://doi.org/10.1016/j.copbio.2017.08.007.

Ryan, Elizabeth P., Adam L. Heuberger, Tiffany L. Weir, Brittany Barnett, Corey D. Broeckling, and Jessica E. Prenni. 2011. "Rice Bran Fermented with Saccharomyces Boulardii Generates Novel Metabolite Profiles with Bioactivity." *Journal of Agricultural and Food Chemistry* 59 (5): 1862–70. https://doi.org/10.1021/jf1038103.

Saeed Akhtar, Majid Hussain, Tariq Ismail, and Muhammad Riaz. 2016. "Ethnic Fermented Foods of Pakistan." In *Ethnic Fermented Foods and Alcoholic Beverages of Asia*, 119–37. http://dx.doi.org/10.1007/978-81-322-2800-4_5.

Sanjukta, Samurailatpam, and Amit Kumar Rai. 2016. "Production of Bioactive Peptides during Soybean Fermentation and Their Potential Health Benefits." *Trends in Food Science and Technology* 50: 1–10. https://doi.org/10.1016/j.tifs.2016.01.010.

Sarvari, F., A.M. Mortazavian, and M.R. Fazeli. 2014. "Biochemical Characteristics and Viability of Probiotic and Yogurt Bacteria in Yogurt during the Fermentation and Refrigerated Storage." *Applied Food Biotechnology* 1 (1): 55–61. https://doi.org/10.22037/afb.v1i1.7125.

Savadogo, A., and A.S. Traore. 2012. "La Flore Microbienne et Les Propriétés Fonctionnelles Des Yaourts et Laits Fermentés." *International Journal of Biological and Chemical Sciences* 5 (5): 2057. https://doi.org/10.4314/ijbcs.v5i5.28.

Serafeimidou, Amalia, Spiros Zlatanos, Kostas Laskaridis, and Angelos Sagredos. 2012. "Chemical Characteristics, Fatty Acid Composition and Conjugated Linoleic Acid (CLA) Content of Traditional Greek Yogurts." *Food Chemistry* 134 (4): 1839–46. https://doi.org/10.1016/j.foodchem.2012.03.102.

Shah, N.T., B. Umrethia, and T.P. Shah. 2017. "Preliminary Pharmacognostic and Phyto-Chemical Evaluation of Darbha (Desmostachya Bipinnata Stapf.)." *Journal of Ayurveda and Integrated Medical Sciences* 2 (3): 75–83.

Shaukat, Aasma, Michael D. Levitt, Brent C. Taylor, Roderick MacDonald, and Tatyanna A. Shamliyan. 2010. "Annals of Internal Medicine NIH Conference Systematic Review: Smoking Cessation Intervention Strategies For." *Annals of Internal Medicine* 152 (12): 797–803.

Shin, Donghwa, and Doyoun Jeong. 2015. "Korean Traditional Fermented Soybean Products: Jang." *Journal of Ethnic Foods* 2 (1): 2–7. https://doi.org/10.1016/j.jef.2015.02.002.

Sieuwerts, Sander, Frank A.M. De Bok, Jeroen Hugenholtz, and Johan E.T. Van Hylckama Vlieg. 2008. "Unraveling Microbial Interactions in Food Fermentations: From Classical to Genomics Approaches." *Applied and Environmental Microbiology* 74 (16): 4997–5007. https://doi.org/10.1128/AEM.00113-08.

Sieuwerts, Sander, Peter A. Bron, and Eddy J. Smid. 2018. "Mutually Stimulating Interactions between Lactic Acid Bacteria and Saccharomyces Cerevisiae in Sourdough Fermentation." *LWT—Food Science and Technology* 90: 201–6. https://doi.org/10.1016/j.lwt.2017.12.022.

Singh, V.P., V. Pathak, and Akhilesh K. Verma. 2012. "Fermented Meat Products: Organoleptic Qualities and Biogenic Amines-A Review." *American Journal of Food Technology* 7 (5): 278–88. https://doi.org/10.3923/ajft.2012.278.288.

Sluijs, Ivonne, Nita G. Forouhi, Joline W.J. Beulens, Yvonne T. Van Der Schouw, Claudia Agnoli, Larraitz Arriola, Beverley Balkau, et al. 2012. "The Amount and Type of Dairy Product Intake and Incident Type 2 Diabetes: Results from the EPIC-InterAct Study." *American Journal of Clinical Nutrition* 96 (2): 382–90. https://doi.org/10.3945/ajcn.111.021907.

Sodini, Isabelle, Florent Remeuf, Cyme Haddad, and Georges Corrieu. 2004. "The Relative Effect of Milk Base, Starter, and Process on Yogurt Texture: A Review." *Critical Reviews in Food Science and Nutrition* 44 (2): 113–37. https://doi.org/10.1080/10408690490424793.

Sonestedt, Emily, Elisabet Wirfält, Peter Wallström, Bo Gullberg, Marju Orho-Melander, and Bo Hedblad. 2011. "Dairy Products and Its Association with Incidence of Cardiovascular Disease: The Malmö Diet and Cancer Cohort." *European Journal of Epidemiology* 26 (8): 609–18. https://doi.org/10.1007/s10654-011-9589-y.

Soomro, Aijaz Hussain, and Tariq Masud. 2012. "Probiotic Characteristics of Lactobacillus Spp. Isolated from Fermented Milk Product Dahi." *Food Science and Technology Research* 18 (1): 91–98. https://doi.org/10.3136/fstr.18.91.

Soomro, Aijaz Hussain, Tariq Masud, and Kiran Anwaar. 2001. "Role of Lactic Acid Bacteria (LAB) in Food Preservation and Human Health—A Review." *Pakistan Journal of Nutrition* 1 (1): 20–24. https://doi.org/10.3923/pjn.2002.20.24.

Soukoulis, C., P. Panagiotidis, R. Koureli, and C. Tzia. 2007. "Industrial Yogurt Manufacture: Monitoring of Fermentation Process and Improvement of Final Product Quality." *Journal of Dairy Science* 90 (6): 2641–54. https://doi.org/10.3168/jds.2006–802.

Steinkraus, Keith H. 1983. "Lactic Acid Fermentation in the Production of Foods from Vegetables, Cereals and Legumes." *Antonie van Leeuwenhoek* 49: 337–48.

Swain, Manas Ranjan, Marimuthu Anandharaj, Ramesh Chandra Ray, and Rizwana Parveen Rani. 2014. "Fermented Fruits and Vegetables of Asia: A Potential Source of Probiotics." *Biotechnology Research International* 2014: 1–19. https://doi.org/10.1155/2014/250424.

Tamang, Buddhiman, Jyoti P. Tamang, Ulrich Schillinger, Charles M.A.P. Franz, Michael Gores, and Wilhelm H. Holzapfel. 2008. "Phenotypic and Genotypic Identification of Lactic Acid Bacteria Isolated from Ethnic Fermented Bamboo Tender Shoots of North East India." *International Journal of Food Microbiology* 121 (1): 35–40. https://doi.org/10.1016/j.ijfoodmicro.2007.10.009.

Tamang, Jyoti Prakash, Paul D. Cotter, Akihito Endo, Nam Soo Han, Remco Kort, Shao Quan Liu, Baltasar Mayo, Nieke Westerik, and Robert Hutkins. 2020. "Fermented Foods in a Global Age: East Meets West." *Comprehensive Reviews in Food Science and Food Safety* 19 (1): 184–217. https://doi.org/10.1111/1541-4337.12520.

Tamang, Jyoti Prakash, Koichi Watanabe, and Wilhelm H. Holzapfel. 2016. "Review: Diversity of Microorganisms in Global Fermented Foods and Beverages." *Frontiers in Microbiology* 7: 181961. https://doi.org/10.3389/fmicb.2016.00377.

Tanasupawat, Somboon, Amnat Pakdeeto, Chitti Thawai, Pattaraporn Yukphan, and Sanae Okada. 2007. "Identification of Lactic Acid Bacteria from Fermented Tea Leaves (Miang) in Thailand and Proposals of Lactobacillus Thailandensis Sp. Nov., Lactobacillus Camelliae Sp. Nov., and Pediococcus Siamensis Sp. Nov." *Journal of General and Applied Microbiology* 53 (1): 7–15. https://doi.org/10.2323/jgam.53.7.

Thongthai, C., and A Gidberg. 2005. "Asian Fish Sauce as a Source of Nutrition." In *Asian Functional Food*, 215–65. https://doi.org/10.1201/9781420028119.

Tunick, Michael H., and Diane L. Van Hekken. 2015. "Dairy Products and Health: Recent Insights." *Journal of Agricultural and Food Chemistry* 63 (43): 9381–88. https://doi.org/10.1021/jf5042454.

Udenigwe, Chibuike C., and Aishwarya Mohan. 2014. "Mechanisms of Food Protein-Derived Antihypertensive Peptides Other than ACE Inhibition." *Journal of Functional Foods* 8 (1): 45–52. https://doi.org/10.1016/j.jff.2014.03.002.

Waché, Yves, Thuy Le Do, Thi Bao Hoa Do, Thi Yen Do, Maxime Haure, Phu Ha Ho, Anil Kumar Anal, et al. 2018. "Prospects for Food Fermentation in South-East Asia, Topics from the Tropical Fermentation and Biotechnology Network at the End of the Asifood Erasmus+project." *Frontiers in Microbiology* 9: 2278. https://doi.org/10.3389/fmicb.2018.02278.

Xiang, Huan, Dongxiao Sun-Waterhouse, Geoffrey I.N. Waterhouse, Chun Cui, and Zheng Ruan. 2019. "Fermentation-Enabled Wellness Foods: A Fresh Perspective." *Food Science and Human Wellness* 8 (3): 203–43. https://doi.org/10.1016/j.fshw.2019.08.003.

Yongsawatdigul, J., S. Rodtong, and N. Raksakulthai. 2007. "Acceleration of Thai Fish Sauce Fermentation Using Proteinases and Bacterial Starter Cultures." *Journal of Food Science* 72 (9): M382–89. https://doi.org/10.1111/j.1750-3841.2007.00532.x.

Yu, Huan, and Joe Bogue. 2013. "Concept Optimisation of Fermented Functional Cereal Beverages." *British Food Journal* 115 (4): 541–63. https://doi.org/10.1108/00070701311317838.

Zang, Jinhong, Yanshun Xu, Wenshui Xia, and Joe M. Regenstein. 2020. "Quality, Functionality, and Microbiology of Fermented Fish: A Review." *Critical Reviews in Food Science and Nutrition* 60 (7): 1228–42. https://doi.org/10.1080/10408398.2019.1565491.

Zhao, Wenbin, Yuheng Liu, Maria Latta, Wantong Ma, Zhengrong Wu, and Peng Chen. 2019. "Probiotics Database: A Potential Source of Fermented Foods." *International Journal of Food Properties* 22 (1): 197–216. https://doi.org/10.1080/10942912.2019.1579737.

14 Recent Advances and Future Perspectives in the Development of Healthier Yu-Lu, a Chinese Salt-Fermented Fish Sauce

Chenxin Zhou, Wenshuang Xue, Minghuang Chen, Jing Zhou, Ying Li, Huijie Liu, and Ruichang Gao

14.1 INTRODUCTION

Fish sauce is a kind of delicious and nutritious traditional condiment, which is popular in Japan, coastal China, and Southeast Asia. Fish sauce is made by natural fermentation at an appropriate temperature by mixing a certain proportion of salt with low-value fish, shrimp, or aquatic product processing waste materials. This process mostly relies on the joint action of various enzymes and microorganisms inside and outside the fish (Liu 1989). Fish sauce is highly nutritionally valued and rich in a variety of amino acids, including eight essential amino acids, unsaturated fatty acids, calcium, magnesium, zinc (Jiang et al. 2019a), and other minerals and trace elements, and it has high nutritional value. With the development of the modern food industry and the improvement of people's living standards, the consumption of fish sauce is rising. To meet the demand for fish sauce, many countries are increasing the production level of fish sauce. At present, fish sauce is produced in many countries and regions, such as Thailand, Japan, and France. Among them, Thailand is a major country of fish sauce consumption and production, and its fish sauce processing technology is the most advanced in the world.

Fermentation is an ancient method of food processing and preservation that can enhance the shelf life, flavor, and nutritional value of food products (Marco et al. 2017). However, traditional fish sauce production has a lengthy fermentation process and inconsistent sensory quality, affecting the overall quality of the product. To address these issues, a new bacterial starter culture can be introduced to reduce the fermentation time and ensure a consistent quality of fish sauce. In the past, higher amounts of salt were used in traditional fish sauce production to prevent the growth of harmful bacteria during fermentation. However, modern research has shown that long-term excessive salt consumption can lead to health issues such as hypertension, heart disease, and other ailments (Aliño et al. 2010). Therefore, a low-salt fermentation process should be used to reduce the salt concentration, which meets the dietary needs of consumers with low sodium. For example, Huang (2017) desalted abalone visceral fish sauce by electrodialysis, which solved the high salt content in naturally fermented fish sauce. Zhou et al. (2017) used the by-products of freshwater fish as raw materials to produce low-salt fish sauce by splitting salt stages and replacing salt with soy sauce yeast in the later stages. However, the resulting flavor of fish sauce is not as good as that of the traditional fermentation method, and therefore there is still a need to find more effective methods to improve the flavor of fish sauce. Microorganisms play an important role in the fermentation process of fish sauce. This chapter provides a comprehensive review of the microbial composition present in fish sauce and its influence on the flavor profile of fish sauce. Also, the production of biogenic amines in the fermentation process of fish sauce and the degradation of biogenic amines by microorganisms are summarized, since biogenic amines are one of the factors that influence the safety of fish sauce.

DOI: 10.1201/9781003605300-14

14.2 MICROORGANISMS IN FISH SAUCE FERMENTATION

The salinity content in fish sauce is relatively high, typically ranging between 25% and 30%. Hence, to accommodate the hyper-saline environment, the microorganisms present in fish sauce are predominantly halophilic or salt-tolerant species, which play a crucial role in the decomposition of raw materials and the formation of flavor. To date, numerous microorganisms have been identified within fish sauce, including but not limited to *Bacillus*, *Staphylococcus*, *Pediococcus*, *Halophilic*, extremely red halophilic Archaea, and halophilic *Lactobacillus*. The halophilic lactic acid bacteria in fish sauce can not only produce lactic acid rapidly, reduce pH, and inhibit the growth of miscellaneous bacteria and spoilage bacteria but also facilitate the formation of flavor substances. *Tetragenococcus muriaticus* is a moderately halophilic lactic acid bacterium, which was isolated from Thai fish sauce (Chokchai et al. 2017). Wu et al. (2018) screened five types of salt-tolerant bacteria from fermented fish sauce. After 16S rRNA high-throughput sequencing, the researchers found that strains JL-B, JL-C, JL-D, and JL-E were all *Tetracoccus halophilus*, and strain JL-F was *Staphylococcus epidermidis*. *Tetracoccus halophilus* is a common halophilic lactic acid bacterium found in fish sauce and soy sauce. It can grow under the condition of 25% salt concentration and has a significant impact on the formation of fish sauce flavor substances. Furthermore, researchers have also isolated lactic acid bacteria with antibacterial activity from fermented fish, capable of producing antibiotics and other substances that inhibit the growth of pathogenic and spoilage bacteria, ensuring the safety of the fermented products. Zeng et al. (2014) discovered *Lactiplantibacillus plantarum* with antibacterial properties against *Listeria monocytogenes*, *Staphylococcus aureus*, and *Escherichia coli* from a traditional Chinese low-salt fermented fish product. In addition to the aforementioned findings, Rungsima et al. (2016) identified a novel species of Gram-positive, spore-forming, rod-shaped *Bacillus* from Thai fish sauce, which they named *Bacillus piscicola sp. nov*. Furthermore, Udomsil et al. (2015) isolated *Staphylococcus* sp. CMC5-3-1 from Thai fish sauce capable of imparting a dark chocolate flavor to the product. A strain of *Kodamaea ohmeri* M8 with biogenic amine-degrading capabilities was also isolated from naturally fermented fish sauce (Yang et al. 2012).

In recent years, researchers have conducted studies on the variations in microbial composition throughout the different fermentation processes of traditional fish sauce. Wang and Duan (2019) analyzed the bacterial community in the heat preserved fermented fish sauce and found that the dominant bacteria in the initial stage of fish sauce fermentation were *Psychrobacter*, *Streptococcus*, and *Lactobacillus*, while the dominant bacteria were *Streptococcus*, *Lactobacillus*, *Staphylococcus*, *Acinetobacter*, and *Pseudomonas*, which indicated that the dominant bacteria would change with different fermentation stages of fish sauce. Huang et al. (2010) also found that the dominant bacteria of traditional fermented fish sauce were lactic acid bacteria and yeast. Considering the changes of various physicochemical indexes and free amino acids, researchers found that the changing trend of lactic acid bacteria and yeast was closely related to the changes of multiple physical, chemical indexes, and free amino acids.

The microbial species of fish sauce in different fermentation periods are not consistent, and the microbial composition of fish sauce products in different regions is also different due to various raw materials and production methods. For example, *Bacillus*, *Corynebacterium*, a small number of *Pseudomonas*, and extremely red halophilic Archaea are the main strains involved in Thai fish sauce (Ventosa et al. 1998), while *Lactiplantibacillus plantarun* and *Enterococcus faecium* are the dominant strains involved in the fermentation of mahyaveh (Karparvar et al. 2019).

14.3 MICROBIAL STARTER OF LOW-SALT FISH SAUCE

In recent years, many halophilic or salt tolerant microorganisms, such as *Virgibacillus sp.* SK37 (Sinsuwan et al. 2012), *Halobacterium sp.* SP1 (1) (Akolkar et al. 2010), and *Virgibacillus sp.* SK33 (Sinsuwan et al. 2008), have been reported as fish sauce starter cultures. The protease secreted by these microorganisms can accelerate the fermentation process of fish sauce and improve the flavor and quality of fish sauce by promoting the hydrolysis of fish protein. Traditional fish sauce has the problem of high salt content; hence, researchers have improved the traditional fish sauce fermentation technology, which used a low-salt fermentation process to reduce the salt content, to adapt to the health need of consumers. However, the low-salt fermentation process resulted in a less distinct flavor. Therefore, enhancing the quality and flavor of fish sauce while ensuring reduced salt levels has become a key issue in fish sauce research.

With the isolation of more and more microorganisms, researchers screened out some bacteria adapted to the low-salt environment to improve the rapid brewing process of fish sauce and improve the flavor of low-salt fish sauce. Kim et al. (2016) reported that the addition of *Aspergillus oryzae* significantly improved the flavor of low-salt fish sauce and increased the contents of total nitrogen, amino nitrogen, and free amino acids. Gao et al. (2020) isolated a cold-adapted protease-producing strain, *Planococcus maritimus* XJ2, from traditional fermented shrimp sauce and used it for low-salt fish sauce fermentation. The results showed that the fish sauce inoculated with *Planococcus maritimus* XJ2 significantly shortened the fermentation time and had better flavor compared to the control group. Zhou et al. (2017) used a by-product of freshwater fish as raw material to produce low-salt fish sauce by adding salt in stages and salt-tolerant soy sauce yeast in the later stage, which not only greatly shortened the fermentation time but also reached the national first-class fish sauce standard. Chen et al. (2020b) isolated a *Lodderomyces elongisporus* strain from fish protein hydrolysate, and the growth was inhibited by increasing the salt concentration. When the content of NaCl was higher than 10%, the growth of yeast was greatly inhibited. Researchers studied fermentation broth with yeast added. Compared with the control group, the yeast could produce alcohols and pyrazines, which could give a rose fragrance, roasted nut fragrance, and barbecue flavor to the fermentation broth. This provides a theoretical basis for the strain as a starter culture to improve the flavor of rapid-fermented low-salt fish sauce.

Exogenous lysine and mixed starter culture also significantly influence the fermentation of low-salt fish sauce. Lysine, a sweet-tasting amino acid, can, to some extent, enhance the flavor formation of fish sauce (Zheng et al. 2019). Zheng et al. (2019) used *Halobacillus campisalis*, *Halobacillus faecis*, *Bacillus aquimaris*, and *Bacillus hwajinpoensis* as the mixed starter cultures and exogenous lysine for the fermentation of 15% low-salt fish sauce. The results indicated an increase in the amino acid nitrogen content of the fish sauce, enriching its nutritional value. However, reports on the impact of exogenous lysine on the quality of low-salt fermented fish sauce are currently scarce.

14.4 PROTEASE IN FISH SAUCE FERMENTATION

14.4.1 Protease from Fish

Fish viscera or tissues are rich in protease, such as tryptase, chymotrypsin, cathepsin, and so on (Tao et al. 2008). Ketnawa et al. (2015) isolated peptidase from the viscera of cultured giant catfish, which exhibited optimal protease activity at pH 8.0 and 70°C, maintaining over 50% of its activity at the highest salt concentration (30% w/v). This suggests that peptidase from the viscera of giant catfish has potential applications in food processing at high temperatures and high salt concentrations. Villalba et al. (2013) conducted purification and characterization studies on trypsin from the viscera of vermiculated sailfin catfish. It was found that this enzyme displayed maximal activity at pH 9.0 and 50°C, but its activity progressively decreased with increasing NaCl concentrations. Klomklao et al. (2004) found that in addition to some digestive enzymes, there was also high protease activity in the tuna spleen, which was identified as tryptase like serine protease.

During the fermentation process of fish sauce, the addition of an appropriate amount of fish viscera and enzymes can accelerate the decomposition of proteins, thereby achieving the goal of reducing the fermentation time. In recent years, many scholars globally fermented fish sauce by adding fish viscera rich in protease. Gildberg (2001) added Atlantic cod intestines rich in proteases to Arctic small sea fish, and the results showed that by adding 5~10% enzyme-rich cod sausage into surimi for 6 months, the protein recovery rate of fish raw material reached 60%, and the fermentation time of fish sauce was significantly shortened. Zhong et al. (2015) used lizard fish *Saurida wanieso* viscera rich in protease as raw material to ferment fish sauce at pH 11.0 and 40°C. The results showed that the concentrations of umami amino acids and sweet amino acids reached 11.95 and 20.33 mg/mL, respectively. Klomklao and coworkers (2005) added tuna spleen to the sardine fish sauce fermentation process; although the fermentation time was shortened, the color and flavor were not good. Compared with traditional fermentation, the total nitrogen and amino acid nitrogen reached the commercial index in a short time and accelerated the fermentation process by adding proteases to fish during fermentation. However, some flavor substances did not form due to the shorter fermentation time, resulting in a poor flavor of fish sauce.

14.4.2 Protease from Microorganisms

The fermentation process of fish sauce depends not only on the endogenous enzymes in natural materials but also on the exogenous enzymes secreted by microorganisms. The high-salt environment of fish sauce can inhibit the growth of some spoilage bacteria. However, it also inhibits the protease activity of fish itself and reduces the fermentation ability of fish sauce, which affects the flavor of fish sauce. Therefore, it has become a hot research direction to find microorganisms that can secrete salt-tolerant protease with high hydrolysis activity in fish sauce production.

The protease activity of halophilic bacteria usually increases as the salinity increases in a certain range, but it is very low or even inactivated under lower salinity conditions. This is related to the amino acid composition and structure of functional proteins expressed by halophilic bacteria that adapt to high-salt environment (Zhao and Tang 2020). Sinsuwan et al. (2007) isolated a strain of salt-tolerant serine alkaline protease from Thai fish sauce, *Virgibacillus* sp. SK33. Compared with the control group, the protease still had the activity of hydrolyzing casein in the presence of 25% NaCl, and the activity of hydrolyzing casein to the synthetic substrate increased four times, which indicated that the protease produced by *Virgibacillus* sp. SK33 had a strong ability of proteolysis at high salt concentration. Xiao et al. (2015) isolated *P. citrinum* YL-1 from traditional Chinese fish sauce, which was also capable of secreting salt-tolerant serine protease. Upon purification of the protease, it was identified as an alkaline protease that retained 18% of its enzymatic activity at a high salt concentration of 30% NaCl, enabling it to hydrolyze fish protein in a high-salt environment. A protease-producing strain, *Virgibacillus halodenitrificans* SK1-3-7, was isolated from fish sauce. This protease exhibited sodium chloride activity and detergent stability, showing high stability in 0~2 moL/L NaCl and various organic solvents (Montriwong et al. 2015). A high fibrinolytic strain *Virgibacillus sp.*SK1-3-7 was isolated from fermented fish sauce by Montriwong et al. (2012). The enzyme secreted by this strain is serine protease, which can be activated by 0.02 moL/L $CaCl_2$ and 0.15 moL/L NaCl.

Some halophilic Archaea also have the ability to secrete proteases. Gao et al. (2017) conducted purification and characteristic studies on the extracellular protease of the halophilic Archaeon *Halogranum rubrum*. The research indicated that this enzyme was an alkaline, salt-tolerant protease, exhibiting optimal catalytic activity at pH 8.0 and 50°C. Also, the enzymes with high hydrolytic activity for silver carp myosin and calcium ions can enhance the hydrolytic activity of the enzyme, which makes them potential starter enzymes for low-salt fish sauce.

14.5 FLAVOR FORMED DURING FERMENTATION OF FISH SAUCE

The flavor of fish sauce is formed through the combined action of fish endogenous protease or other enzymes and microorganisms, which gradually decompose and ferment the proteins and fats in fish meat, forming taste substances such as free amino acids, oligopeptides, and free fatty acids. These substances then undergo a series of complex reactions to further form various volatile and non-volatile flavor compounds (Ning et al. 2018). Flavor is an important quality characteristic of fish sauce. Traditional fish sauce fermentation often struggles to ensure consistency in product quality and flavor, whereas the modern rapid fermentation processes for fish sauce may result in poor flavor. Therefore, it is necessary to combine microbial fermentation with fish sauce fermentation techniques in future research.

Jiang et al. (2017) pointed out that the protease secreted by halophilic bacteria not only promotes the fermentation process of fish sauce but also contributes to the formation of volatile odor. Halophilic lactic acid bacteria promote the formation of volatile fatty acids and accelerate the nonenzymatic browning process of fish sauce. Udomsil et al. (2010) screened out *Tetragenococcus halophilus* MRC5–5–2 and MS33 from fish sauce with high intracellular aminopeptidase activity, capable of producing benzaldehyde and 2-methylpropanol, which could contribute to the formation of meaty and almond flavors in the product. The same researcher (Udomsil et al. 2017) also added *Tetragenococcus halophilus* MS33 and protease-producing strain *Virgibacillus sp.*SK37 to anchovy in order to improve the concentrations of glutamate and volatile compounds in fish sauce and the overall acceptability of fish sauce. Li et al. (2016) inoculated *Pediococcus pentosaceus* with high protease and peptidase activities during the fermentation of fish sauce and found that this significantly increased the content of flavorful amino acids, markedly enhancing the umami taste and aroma of the fish sauce. In addition, yeast also plays an essential role in fermentation products (Yang et al. 2017). It can reproduce in large numbers in the acidic environment created by lactic acid bacteria and has strong aroma production ability. Wang et al. (2015) used pond fish as raw materials to prepare fish sauce by adding solid bran koji fermentation and adding *Saccharomyces rouxii* and *Saccharomyces globosum* for post-ripening and aroma enhancement. The results showed that *Saccharomyces rouxii* and *Saccharomyces globosum* in the post-ripening stage of fish sauce not only contributed to the production of alcohol, polyols, and other wine aroma substances but also promoted the production of a variety of esters with the participation of ethanol, giving the products a certain ester content. It also participated in the formation of organic acids and amino acids, which make fish sauce have an obvious mellow flavor.

The biotransformation and metabolism of halophilic microorganisms can also effectively reduce the production of bad flavor in the process of fish sauce fermentation. Fukami et al. (2004) inoculated *Staphylococcus xylose* from chub mackerel sauce, which not only weakened the fishy, sour, and other bad smells of fish sauce but also enhanced the characteristic flavor and improved the flavor of fish sauce effectively. Wang et al. 2014) significantly improved product quality by adding *Kodamaea ohmeri* M8 during fermentation, which enhanced the flavor of umami, cheese, and meat and weakened the fishy, ammonia, and putrid smell.

14.6 BIOGENIC AMINES IN FISH SAUCE FERMENTATION AND THEIR DEGRADATION

A biogenic amine is a kind of nitrogen-containing substance with a small molecular weight, which widely exists in fermented products like sausage, fermented aquatic products, beer, and wine. According to the chemical structures, biogenic amines can be divided into the aliphatic group, such as putrescine, cadaverine, spermidine; aromatic group, such as tyramine, phenylethylamine, octopamine; and heterocyclic group, such as histamine, tryptamine (Chen and Zhou 2016). Common biogenic amines include spermine, putrescine, tryptamine, phenylethylamine, cadaverine, histamine, tyramine, and spermidine (Xie et al. 2020). In marine foods, fish like tuna, sardine, and mackerel are rich in histidine, which has high histamine levels and is prone to histamine poisoning (Liu et al. 2017). There are also high biogenic amines in dried salted fish, canned fish, salted fish, or packaged fish. Wu et al. (2016) analyzed 43 samples of commercially available dried salted fish. The results showed that putrescine and cadaverine had the highest content in all biogenic amines, with 14% of the samples exceeding 100 mg/kg of histamine, and the highest histamine content was found in dried pomfret, reaching 347.79 mg/kg. Additionally, 6.98% of the samples had tyramine content exceeding 100 mg/kg, and 23.26% of the samples had high total biogenic amine content (greater than 600 mg/kg).

During the fermentation process of fish sauce, amine-producing bacteria will hydrolyze a large amount of protein into free amino acids, which form biogenic amines through decarboxylation, amination, and transamination of aldehydes and ketones (Fausto et al. 2016). Biogenic amines are common in fish sauce. A total of 35 commercial fish sauce products were analyzed, and 32 fish sauce products contained 7 kinds of biogenic amines, 2 fish sauce products contained 6 kinds of biogenic amines (excluding tryptamine), and 1 fish sauce product contained 5 kinds of biogenic amines (excluding histamine and spermidine) (Jiang 2014). The contents of histamine, putrescine, cadaverine, and tyramine in fish sauce were higher than other biogenic amines (Reiter et al. 2008), while the contents of phenylethylamine, spermine, and spermidine were lower (Sanceda et al. 1996; Tsai et al. 2006). Jiang et al. (2014) analyzed the biogenic amines in 35 commercial fish sauce samples and found that putrescine, cadaverine, histamine, and tyramine were the main biogenic amines (more than 100 mg/kg) in the fish sauce, while tryptamine, spermine, and spermidine were the secondary biogenic amines (less than 25 mg/kg).

The normal physiological activities of the human body need a small amount of biogenic amines to eliminate free radicals, enhance metabolic activity, and control blood pressure. However, excessive intake of biogenic amines causes harm to the human body. Among the common biogenic amines, histamine is the most harmful to the human body, followed by tyramine. The symptoms of biogenic amine poisoning include nausea, headache, dyspnea, palpitation, hypertension, rash, hypotension, and abdominal spasms (Becker et al. 2001; Anlı and Bayram 2008) and even death in severe cases. Biogenic amines can be decomposed into aldehydes, ammonia, and hydrogen through oxidative deamination by amino acid oxidases. Studies have suggested that some halophilic bacteria have the ability to degrade biogenic amines. At present, one of the most effective ways to prevent the formation and accumulation of biogenic amine in fish sauce is to inoculate biogenic amine oxidase-producing microorganisms. Kim et al. (2019) isolated a non-biogenic amine producing *Tetragonococcus halophilus* strain (MJ4). Compared with the negative control group, strain MJ4 significantly inhibited the formation of ptomaine during sau-jeot (shrimp sauce) fermentation, which indicated that *Tetragonococcus halophilus* MJ4 could be used as a starter to inhibit the formation of biogenic amines. Wang et al. (2014) isolated and screened a strain of *Kodamaea ohmeri* with anti-histamine activity from traditional fermented fish sauce, which can degrade more than 75% histamine and more than 65% tyramine in fish sauce. The strain SWA25, selected from fermented fish sauce by Jiang (2014), was used as a new starter culture in the fermentation of yellow crucian carp sauce. Compared with the control group, the contents of histamine, tyramine, cadaverine, tryptamine, phenethylamine, putrescine, and total biogenic amines in fish sauce inoculated with *H. shantousis* SWA25 were decreased by 64.5%, 59.2%, 71.0%, 63.4%, 68.2%, 22.0%, and 55.3%, respectively, which had little effect on spermidine and spermine. Takura et al. (2019) isolated *Tetragenococcus halophilus* from fermented fish food and confirmed that it carried the aspartate decarboxylase gene. Using this strain as a fish sauce starter can not only prevent the accumulation of biogenic amines in fish sauce products but also transform aspartic acid into alanine to obtain a softer-tasting product.

14.7 BIOACTIVE COMPONENTS OF FISH SAUCE

As a traditional condiment, fish sauce not only has good flavor but also is rich in nutrients, including all essential amino acids, bioactive peptides, taurine, and octopamine. Many researchers have also done relevant research on the content and activity of functional components in fish sauce.

14.7.1 Bioactive Peptides

A bioactive peptide is a kind of substance with special physiological activity whose molecular weight is less than 6 kDa. It is composed of two or more amino acids in different composition and arrangement (Qian and Zhang 2016). Most marine bioactive peptides are obtained by enzymatic hydrolysis of marine organisms. Marine bioactive peptides have different biological activities due to their different structure, amino acid composition, and sequence, including anti-cancer, antioxidant, antihypertensive, immunomodulatory, antibacterial, hypoglycemic activities, and so on. Najafian and Babji (2018) isolated and identified two new antioxidant peptides, Lue-Asp-Asp-Pro-Val-Phe-Ile-His (LDDPVFIH) and Val-Ala-Ala-Gly-Arg-Thr-Asp-Ala-Gly-Val-His (VAAGRTDAGVH), from fermented anchovy extract. The results indicated that LDDPVFIH had high antioxidant activity, and the hydrophobic amino acids (lle and Leu), acidic (Asp), and alkaline (His) amino acids in the peptide sequence were the reasons for the high antioxidant activity of fermented anchovy extract. Sasaki et al. (2013) extracted and purified a new angiotensin-I converting enzyme inhibitor from squid sauce and identified it as Leu-Ala-Arg. In addition, the researchers also found that Leu-Ala-Arg has an antihypertensive effect and can reduce the systolic blood pressure of spontaneously hypertensive rats.

Peptides in fish sauce have an important influence on the taste of the fish sauce. Miyamura et al. (2016) analyzed four kinds of commercial fish sauce fermented with koji and found that there were Kokumi peptide, γ-Glu-Val-Gly, in four kinds of fish sauce, and the content of γ-Glu-Val-Gly in three kinds of fish sauce ranged from 1.7 to 5.7 mg/L. The results showed that γ-Glu-Val-Gly existed widely in all kinds of fermented fish sauce. Park et al. (2002) analyzed the Vietnamese fish sauce nuoc mam and found that the complex taste of the fish sauce was related to many peptides produced during fermentation. A total of 20% of total nitrogen was in the form of polypeptides in the fish sauce, and 17 oligopeptides, including dipeptides, tripeptides, and tetrapeptides, were isolated from it as well.

14.7.2 Octopamine and Taurine

Octopamine is a homologous vertebrate hormone norepinephrine, with the chemical structure of p-hydroxybenzene-β-hydroxyethylamine (Pan and Chen 1999). Researchers found octopamine in various tissues of most invertebrates since it was first discovered in the salivary gland of octopus in 1951 (Erspamer and Boretti 1951). As a marine bioactive substance, octopamine participates in many physiological activities in the human body, such as regulating human metabolism, suppressing appetite, improving attention, and so on. Studies have shown that octopamine is a natural β-3-adrenergic receptor agonist (Carpéné et al. 1999), which has potential application value in the treatment of obesity and type II diabetes (Arch 2002). The content of octopamine in fish sauce is relatively high, about 0.06%~0.1% (Rong et al. 2008), and it is the main source of natural octopamine for human. Rong et al. (2008) measured the octopamine content in Chinese fish sauce and anchovy sauce at 566 and 1055 μg/mL, respectively. Qu et al. (2011) fermented fish sauce with jumbo flying squid as raw material. The results suggested that the octopamine content could reach 6~7 mg/mL after 15~20 days of brewing, which was much higher than that of commercial anchovy sauce.

Taurine is a small molecular sulfur-containing amino acid with non-protein structure. It mainly exists in various tissues and organs of animals in the form of small peptides or free amino acids, which play an important role in maintaining the stability of tissues and cells *in vivo*. Taurine is abundant in aquatic products. Studies have shown that taurine has many biological activities, such as regulating cell osmotic pressure, as an antioxidant, reducing cholesterol, nerve conduction, promoting brain development, and so on. At the same time, taurine is also widely used as a nutritional enhancer in all kinds of food and infant food. Taurine is also an important active ingredient in fish sauce. Jiang et al. (2019b) analyzed the nutrition of mackerel sauce produced in Chaoshan, China, and found that its taurine content was very high, up to 1.13 mg/mL. Chen et al. (2020a) determined the content of taurine in grass carp sauce by HPLC, which was 14.29 ± 0.38 mg/mL, and the content of taurine in eel sauce was 1.21 ± 0.24 mg/mL.

The biological activities of octopamine and taurine give them important application value in health food or medicine applications. Therefore, changing the new fermentation technology of fish sauce and finding efficient and convenient extraction methods to extract octopamine and taurine from fish sauce are research directions to improve the economic benefits of the fish sauce–processing industry in the future.

14.8 CONCLUSION

As a traditional Chinese condiment, fish sauce, or yu-lu, is deeply enjoyed by consumers because of its delicious flavor and rich nutrition. With in-depth study of the fermented fish sauce by researchers globally, a certain amount of microbial information has been obtained. Microorganisms, as an indispensable part of fish sauce, have important effects on the fermentation process. Protease from fish viscera or microorganisms not only shortens the fermentation time but also promotes the formation of fish sauce flavor substances and improves the nutritional value of fish sauce. At present, more and more researchers are advancing fish sauce fermentation to improve the quality of the fish sauce for the industry and consumers. At the same time, more and more low-salt fish sauce starters will be developed to meet the requirements of low salt and rapid fermentation. Biogenic amine is an important factor affecting the safety of fish sauce. Hence, researchers will continue to pay attention to microorganisms with biogenic amine degradation ability and evaluate their safety in fish sauce production.

REFERENCES

Akolkar, A. V., D. Durai, and A. J. Desai. 2010. *Halobacterium sp.* SP1(1) as a starter culture for accelerating fish sauce fermentation. *Journal of Applied Microbiology* 109 (1):44–53. https://doi.org/10.1111/j.1365-2672.2009.04626.x.

Aliño, M., R. Grau, A. Fuentes, and J. M. Barat. 2010. Characterisation of pile salting with sodium replaced mixtures of salts in dry-cured loin manufacture. *Journal of Food Engineering* 97 (3):434–439. https://doi.org/10.1016/j.jfoodeng.2009.10.039.

Anlı, R. E., and M. Bayram. 2008. Biogenic amines in wines. *Food Reviews International* 25 (1):86–102. https://doi.org/10.1080/87559120802458552.

Arch, J. R. S. 2002. β3-Adrenoceptor agonists: potential, pitfalls and progress. *European Journal of Pharmacology* 440 (2–3):99–107. https://doi.org/10.1016/S0014-2999(02)01421-8.

Becker, K., K. Southwick, J. Reardon, R. Berg, and J. Newton MacCormack. 2001. Histamine poisoning associated with eating tuna burgers. *JAMA: The Journal of the American Medical Association* 285 (10):1327–1330. https://doi.org/10.1001/jama.285.10.1327.

Carpéné, C., J. Galitzky, E. Fontana, C. Atgié, M. Lafontan, and M. Berlan. 1999. Selective activation of β_3-adrenoceptors by octopamine: comparative studies in mammalian fat cells. *Naunyn-Schmiedeberg's Archives of Pharmacology* 359 (4):310–321. https://doi.org/10.1007/PL00005357.

Chen, H., and Q. S. Zhou. 2016. Biogenic amines in fish sauce: a review. *Meat Research* 30 (1):40–45. https://doi.org/10.15922/j.cnki.rlyj.2016.01.009. (In Chinese).

Chen, L. L., C. Q. Bai, M. L. Yuan, Y. Jiang, and L. Zhao. 2020a. Determination of octopamine and taurine in fish sauce by HPLC. *Food and Fermentation Industries* 46 (2):261–265. https://doi.org/10.13995/j.cnki.11-1802/ts.022179. (In Chinese).

Chen, R. L., M. J. Li, S. Y. Feng, H. Yu, X. B. Fang, and X. E. Chen. 2020b. Screening of aroma-producing yeast strains from fish protein hydrolysate and research on their aroma-production and fermentation characteristics. *Food Science and Technology* 45 (1):13–18. https://doi.org/10.13684/j.cnki.spkj.2020.01.003. (In Chinese).

Chokchai, C. N., S. Rodtong, J. Yongsawatdigul, and J. L. Steele. 2017. Draft genome sequences of *Tetragenococcus muriaticus* strains 3MR10-3 and PMC-11-5 isolated from Thai fish sauce during natural fermentation. *Genome Announcements* 5 (15): e00198–17. https://doi.org/10.1128/genomeA.00198–17.

Erspamer, V., and G. Boretti. 1951. Identification and characterization, by paper chromatography, of enteramine, octopamine, tyramine, histamine and allied substances in extracts of posterior salivary glands of octopoda and in other tissue extracts of vertebrates and invertebrates. *Fundamental and Clinical Pharmacology* 88 (3):296–332. https://doi.org/10.1016/S1876–0813(08)60032–9.

Fausto, G., Ö. Yesim, S. Giovanna, T. Giulia, and Ö. Fatih. 2016. Technological factors affecting biogenic amine content in foods: a review. *Frontiers in Microbiology* 7 (1371):1218. https://doi.org/10.3389/fmicb.2016.01218.

Fukami, K., Y. Funatsu, K. Kawasaki, and S. Watabe. 2004. Improvement of fish-sauce odor by treatment with bacteria isolated from the fish-sauce mush (moromi) made from frigate mackerel. *Journal of Food Science* 69 (2):45–49. https://doi.org/10.1111/j.1365-2621.2004.tb15514.x.

Gao, R. C., T. Shi, X. D. Liu, M. Q. Zhao, H. L. Cui, and L. Yuan. 2017. Purification and characterisation of a salt-stable protease from the halophilic archaeon halogranum rubrum. *Journal of the Science of Food and Agriculture* 97 (5):1412–1419. https://doi.org/10.1002/jsfa.7879.

Gao, R. C., J. Zhou, W. J. Leng, T. Shi, W. G. Jin, and L. Yuan. 2020. Screening of a Planococcus bacterium producing a cold-adapted protease and its application in low-salt fish sauce fermentation. *Journal of Food Processing and Preservation* 44 (8). https://doi.org/10.1111/jfpp.14625.

Gildberg, A. 2001. Utilisation of male Arctic capelin and Atlantic cod intestines for fish sauce production—evaluation of fermentation conditions. *Bioresource Technology* 76 (2):119–123. https://doi.org/10.1016/S0960-8524(00)00095-X.

Huang, S. F. 2017. Electrodialysis desalination process in the fish sauce made from abalone viscera. *Academic Periodical of Farm Products Processing* (5):19–21. https://doi.org/10.16693/j.cnki.1671-9646(X).2017.03.006. (In Chinese).

Huang, Z. Y., Z. W. Zhu, Q. X. Zeng, Y. S. Zhang, and Y. S. Jian. 2010. Study on the trends of microorganisms in traditional fermentation fish sauce. *Food and Fermentation Industries* 36 (07):18–22. https://doi.org/10.13995/j.cnki.11-1802/ts.2010.07.037. (In Chinese).

Jiang, H. J., M. Wei, and Y. Wang. 2019a. Determination of 13 elements in fish sauce by ICP-OES with microwave digestion. *China Condiment* 44 (10):165–167. https://doi.org/10.3969/j.issn.1000-9973.2019.10.037. (In Chinese).

Jiang, J. J., Q. Jia, W. L. Lin, L. L. Liang, and Z. J. Lai. 2019b. Nutritional analysis and physiological activity evaluation of Pneumatophorus japonicus Sauce. *Agricultural Science and Technology* 20 (1):49–54. https://doi.org/10.16175/j.cnki.1009-4229.2019.01.008.

Jiang, J. J., L. L. Liang, W. L. Lin, and J. Y. Lin. 2017. Research progress of characteristic volatile compounds of fish sauce. *Journal of Fisheries of China* 41 (6):984–992. https://doi.org/10.11964/jfc.20170310765.

Jiang, W. 2014. Screening, identification and application of a new salt tolerant and efficient biogenic amine degrading bacterium. Doctor, Ocean University of China, Cnki. (In Chinese).

Jiang, W., Y. Xu, C. S. Li, X. Y. Dong, and D. F. Wang. 2014. Biogenic amines in commercially produced Yulu, a Chinese fermented fish sauce. *Food Additives and Contaminants: Part B* 7 (1):25–29. https://doi.org/10.1080/19393210.2013.831488.

Karparvar, N., H. G. Safaei, A. Derakhshandeh, F. Hemmati, and S. M. Mazloomi. 2019. Isolation and identification of lactic acid bacteria from a traditional fermented fish sauce (Mahyaveh) in Fars Province, Iran. *International Journal of Nutrition Sciences* 4 (1):49–53. https://doi.org/10.30476/ijns.2019.81437.1005.

Ketnawa, S., O. M. Alvarez, S. Benjakul, and S. Rawdkuen. 2015. Extraction and biochemical characterization of peptidases from giant catfish viscera by Aqueous Two-Phase system. *Journal of Food Biochemistry* 39 (4):429–438. https://doi.org/10.1111/jfbc.12145.

Kim, B. M., J. H. Park, D. S. Kim, Y. M. Kim, J. Y. Jun, I. H. Jeong, S. Y. Nam, and Y. M. Chi. 2016. Effects of rice koji inoculated with Aspergillus luchuensis on the biochemical and sensory properties of a sailfin sandfish (Arctoscopus japonicus) fish sauce. *International Journal of Food Science and Technology* 51 (8):1888–1899. https://doi.org/10.1111/ijfs.13162.

Kim, K. H., S. H. Lee, B. H. Chun, S. E. Jeong, and C. O. Jeon. 2019. *Tetragenococcus halophilus* MJ4 as a starter culture for repressing biogenic amine (cadaverine) formation during saeu-jeot (salted shrimp) fermentation. *Food Microbiology* 82:465–473. https://doi.org/10.1016/j.fm.2019.02.017.

Klomklao, S., S. Benjakul, and W. Visessanguan. 2004. Comparative studies on proteolytic activity of splenic extract from three tuna species commonly used in Thailand. *Journal of Food Biochemistry* 28 (5):355–372. https://doi.org/10.1111/j.1745-4514.2004.05203.x.

Klomklao, S., S. Benjakul, W. Visessanguan, H. Kishimura, and B. K. Simpson. 2005. Effects of the addition of spleen of skipjack tuna (Katsuwonus pelamis) on the liquefaction and characteristics of fish sauce made from sardine (Sardinella gibbosa). *Food Chemistry* 98 (3):440–452. https://doi.org/10.1016/j.foodchem.2005.06.013.

Li, M. R., X. J. Wang, S. Duan, and F. Y. Wu. 2016. Metabolic behavior and transcriptomic analysis of a *Pediococcus pentosaceus* strain in simulated fish sauce. *Modern Food Science and Technology* 32 (8):54–63. https://doi.org/10.13982/j.mfst.1673-9078.2016.8.009. (In Chinese).

Liu, G. M., H. Liu, J. Min, C. Y. Li, Z. J. Zeng, X. Y. Han, and M. J. Cao. 2017. Evaluation methods and control measures of biogenic amines in seafood. *Journal of Chinese Institute of Food Science and Technology* 17 (8):1–11. https://doi.org/10.16429/j.1009-7848.2017.08.001. (In Chinese).

Liu, P. Z. 1989. Some technical measures to improve the quality of fish sauce. *Chinese Condiment* (3):17-20+32. https://doi.org/CNKI:SUN:ZGTW.0.1989-03-005. (In Chinese).

Marco, M. L., D. Heeney, S. Binda, C. J. Cifelli, P. D. Cotter, B. Foligné, M. Gänzle, R. Kort, G. Pasin, and A. Pihlanto. 2017. Health benefits of fermented foods: microbiota and beyond. *Current Opinion Biotechnology* 44:94–102. https://doi.org/10.1016/j.copbio.2016.11.010.

Miyamura, N., M. Kuroda, Y. Kato, J. Yamazaki, T. Mizukoshi, and H. Miyano. 2016. Quantitative analysis of γ-glutamyl-valyl-glycine in fish sauces fermented with Koji by LC/MS/MS. *Chromatography* 37 (1):39–42. https://doi.org/10.15583/jpchrom.2015.034.

Montriwong, A., S. Kaewphuak, S. Rodtong, S. Roytrakul, and J. Yongsawatdigul. 2012. Novel fibrinolytic enzymes from *Virgibacillus halodenitrificans* SK1-3-7 isolated from fish sauce fermentation. *Process Biochemistry* 47 (12):2379–2387. https://doi.org/10.1016/j.procbio.2012.09.020.

Montriwong, A., S. Rodtong, and J. Yongsawatdigul. 2015. Detergent-stable salt-activated proteinases from *Virgibacillus halodenitrificans* SK1-3-7 isolated from fish sauce fermentation. *Applied Biochemistry and Biotechnology* 176 (2):505–517. https://doi.org/10.1007/s12010-015-1591-5.

Najafian, L., and A. S. Babji. 2018. Purification and identification of antioxidant peptides from fermented fish sauce (Budu). *Journal of Aquatic Food Product Technology* 28 (1):1–11. https://doi.org/10.1080/10498850.2018.1559903.

Ning, Y. C., Z. F. Wu, and P. F. Weng. 2018. Research progress in the utilization of microorganisms to promote fermentation process and quality of traditional fish sauce. *Journal of Fisheries of China* 42 (9):1497–1509. https://doi.org/10.11964/jfc.20170310761. (In Chinese).

Pan, C. P., and F. H. Chen. 1999. Research progress on mechanism of octopamine and its receptor. *Chinese Journal of Pesticide Science* (3):1–7. https://doi.org/CNKI:SUN:NYXB.0.1999-03-000. (In Chinese).

Park, J. N., K. Ishida, T. Watanabe, K. I. Endoh, K. Watanabe, M. Murakami, and H. Abe. 2002. Taste effects of oligopeptides in a Vietnamese fish sauce. *Fisheries Science* 68 (4):921–928. https://doi.org/10.1046/j.1444-2906.2002.00511.x.

Qian, L., and Z. J. Zhang. 2016. Research advance on the preparation of bioactive peptides from edible and medicinal fungi. *Northern Horticulture* (18):193–195. https://doi.org/10.11937/bfyy.201618048. (In Chinese).

Qu, Y. H., M. Wang, Z. D. Liu, X. J. Chen, and S. S. Chen. 2011. Study on fish sauce brewed from the viscera of Dosidicus gigass. *Chinese Journal of Marine Drugs* 30 (1):50–52. https://doi.org/10.13400/j.cnki.cjmd.2011.01.001. (In Chinese).

Reiter, R. J., D. X. Tan, M. J. Jou, A. Korkmaz, L. C. Manchester, and S. D. Paredes. 2008. Biogenic amines in the reduction of oxidative stress: melatonin and its metabolites. *Neuro Endocrinology Letters* 29 (4):391–8. https://doi.org/10.1007/s11064-008-9745-2.

Rong, Y. P., T. T. Zhu, S. S. Chen, J. Z. Wang, and Y. H. Qu. 2008. Analysis of adrenoceptor agonist-octopamine in fish sauce by HPLC. *Chinese Journal of Marine Drugs* 27 (1):43–45. https://doi.org/10.3969/j.issn.1002-3461.2008.01.010. (In Chinese).

Rungsima, D., I. Takashi, K. Takuji, O. Moriya, and T. Somboon. 2016. Bacillus piscicola sp nov., isolated from Thai fish sauce (Nam-pla). *International Journal of Systematic and Evolutionary Microbiology* 66 (3):1151–1155. https://doi.org/10.1099/ijsem.0.000851.

Sanceda, N. G., T. Kurata, and N. Arakawa. 1996. Accelerated fermentation process for the manufacture of fish sauce using histidine. *Journal of Food Science* 61 (1):220–222. https://doi.org/10.1111/j.1365-2621.1996.tb14764.x.

Sasaki, T., M. Koudou, T. Michihata, S. Nakamura, and T. Enomoto. 2013. Purification and antihypertensive activity of a novel angiotensin-I converting enzyme inhibitory peptide from fish sauce, Ishiru. *Japanese Journal of Complementary and Alternative Medicine* 10 (1):45–49. https://doi.org/10.1625/jcam.10.45.

Sinsuwan, S., S. Nawong, S. Rodtong, N. Raksakulthai, and J. Yongsawatdigul. 2008. Characterization of *Virgibacillus sp.* SK33 cell-bound proteinases and its application as a starter culture for fish sauce fermentation. *Journal of Biotechnology* 136:S745–S745. https://doi.org/10.1016/j.jbiotec.2008.07.1774.

Sinsuwan, S., S. Rodtong, and J. Yongsawatdigul. 2007. Production and characterization of NaCl-activated proteinases from *Virgibacillus sp.* SK33 isolated from fish sauce fermentation. *Process Biochemistry* 43 (2):185–192. https://doi.org/10.1016/j.procbio.2007.11.015.

Sinsuwan, S., S. Rodtong, and J. Yongsawatdigul. 2012. Hydrolytic activity of *Virgibacillus sp.* SK37, a starter culture of fish sauce fermentation, and its cell-bound proteinases. *World Journal of Microbiology and Biotechnology* 28 (8):2651–2659. https://doi.org/10.1007/s11274-012-1075-5.

Takura, W., I. Satomi, T. Yuya, W. Jun, M. Yoshinobu, T. Yuichiro, and S. Yuki. 2019. Isolation of halophilic lactic acid bacteria possessing aspartate decarboxylase and application to fish sauce fermentation starter. *International Journal of Food Microbiology* 292:137–143. https://doi.org/10.1016/J.IJFOODMICRO.2018.12.013.

Tao, H. L., Z. W. Zhu, Q. X. Zeng, and J. J. Jiang. 2008. Current researches of rapid process fermentation in fish sauce. *Food Research and Development* 29 (3):161–165. https://doi.org/10.3969/j.issn.1005–6521.2008.03.052. (In Chinese).

Tsai, Y. H., C. Y. Lin, L. T. Chien, T. M. Lee, C. I. Wei, and D. F. Hwang. 2006. Histamine contents of fermented fish products in Taiwan and isolation of histamine-forming bacteria. *Food Chemistry* 98 (1):64–70. https://doi.org/10.1016/j.foodchem.2005.04.036.

Udomsil, N., S. Chen, S. Rodtong, and J. Yongsawatdigul. 2017. Improvement of fish sauce quality by combined inoculation of *Tetragenococcus halophilus* MS33 and *Virgibacillus sp.* SK37. *Food Control* 73:930–938. https://doi.org/10.1016/j.foodcont.2016.10.007.

Udomsil, N., S. Rodtong, S. Tanasupawat, and J. Yongsawatdigul. 2010. Proteinase-producing halophilic lactic acid bacteria isolated from fish sauce fermentation and their ability to produce volatile compounds. *International Journal of Food Microbiology* 141 (3):186–194. https://doi.org/10.1016/j.ijfoodmicro.2010.05.016.

Udomsil, N., S. Rodtong, S. Tanasupawat, and J. Yongsawatdigul. 2015. Improvement of fish sauce quality by strain CMC5-3-1: a novel species of *Staphylococcus sp. Journal of Food Science* 80 (9):2015–2022. https://doi.org/10.1111/1750–3841.12986.

Ventosa, A., J. J. Nieto, and A. Oren. 1998. Biology of moderately halophilic aerobic bacteria. *Microbiology and Molecular Biology Reviews* 62 (2):504–544. https://doi.org/10.0000/PMID9618450.

Villalba, A. G. V., J. C. R. Suárez, E. M. V. Soto, G. G. Sánchez, G. C. Ruiz, and R. P. Aguilar. 2013. Trypsin from viscera of vermiculated sailfin catfish, *Pterygoplichthys disjunctivus*, Weber, 1991: its purification and characterization. *Food Chemistry* 141 (2):940–945. https://doi.org/10.1016/j.foodchem.2013.03.078.

Wang, H., X. J. Fu, W. Wu, Y. Wu, J. L. Ren, Q. L. Lin, and Z. H. Li. 2014. Effect of Omer Kodak yeast on the degrading of biogenic amine in fish sauce. *Journal of Chinese Institute of Food Science and Technology* 14 (8):137–141. https://doi.org/10.16429/j.1009-7848.2014.08.018. (In Chinese).

Wang, L., L. Yang, and Y. F. Chen. 2015. Study on post-mature process of fish sauce of pond smelt. *China Condiment* 40 (9):41–44+50. https://doi.org/10.3969/j.issn.1000-9973.2015.09.010. (In Chinese).

Wang, X. J., and S. Duan. 2019. 16S rDNA-based analysis of bacterial community in fish sauce during temperature-controlled fermentation. *Science and Technology of Food Industry* 40 (12):112–119+127. https://doi.org/10.13386/j.issn1002-0306.2019.12.019. (In Chinese).

Wu, R. L., X. Ouyang, S. Duan, X. X. Duan, M. M. Zhai, L. J. Zhang, X. Y. Ling, and X. Y. Gao. 2018. Isolation, identification and fermentation characteristics of salt-tolerant bacteria in fish sauce. *China Brewing* 37 (5):108–113. https://doi.org/10.11882/j.issn.0254-5071.2018.05.021. (In Chinese).

Wu, Y. Y., Y. F. Chen, L. H. Li, X. Q. Yang, S. L. Yang, W. L. Lin, Y. Q. Zhao, and J. C. Deng. 2016. Study on biogenic amines in various dry salted fish consumed in China. *Journal of Ocean University of China* 15 (4):681–689. https://doi.org/10.1007/s11802-016-2958-0.

Xiao, Y. Z., D. K. Wu, S. Y. Zhao, W. M. Lin, and X. Y. Gao. 2015. Statistical optimization of alkaline protease production from *Penicillium citrinum* YL-1 under solid-state fermentation.

Preparative Biochemistry and Biotechnology 45 (5):447–462. https://doi.org/10.1080/10826068.2014.923450.

Xie, S. Y., B. Sun, L. Liu, T. Y. Li, and X. Cong. 2020. Effect of salt reduction on biogenic amine formation in Northeast farm sauce. *Food Science*:1-8. https://doi.org/10.7506/spkx1002-6630-20200105-048. (In Chinese).

Yang, L. K., X. J. Fu, Y. B. Hu, Q. Z. Zhou, and B. Li. 2012. Screening and characterization of biogenic amine degrading bacteria from fish sauce. *Food Science* 33 (11):158-162. https://doi.org/CNKI:SUN:SPKX.0.2012-11-033. (In Chinese).

Yang, Y. P., M. Y. Yu, H. L. Zhou, Q. Y. Zhuo, and L. Gao. 2017. Research status of fermented fish and its modern technology. *Science and Technology Vision* (27):3–4. https://doi.org/10.3969/j.issn.2095–2457.2017.27.002. (In Chinese).

Zeng, X. F., W. S. Xia, J. S. Wang, Q. X. Jiang, Y. S. Xu, Y. Qiu, and H. Y. Wang. 2014. Technological properties of *Lactobacillus plantarum* strains isolated from Chinese traditional low salt fermented whole fish. *Food Control* 40:351–358. https://doi.org/10.1016/j.foodcont.2013.11.048.

Zhao, Z. F., and S. C. Tang. 2020. Progress in the research of halophiles and their extracellular proteases. *Food Science and Technology* 45 (3):23–27+35. https://doi.org/10.13684/j.cnki.spkj.2020.03.009. (In Chinese).

Zheng, Z. Y., J. Zhou, L. Yuan, and R. C. Gao. 2019. Effects of mixed starter cultures and exogenous L-Lys on fermentation quality of fish paste. *Food Science* 40 (12):108–114. https://doi.org/10.7506/spkx1002-6630-20180709-119. (In Chinese).

Zhong, C., M. Nakanishi, J. T. Geng, E. Okazaki, M. J. Cao, W. Y. Weng, and K. Osako. 2015. Comparison of non-volatile taste-active components in fish sauce produced from lizardfish Saurida wanieso viscera under different conditions. *Fisheries Science* 81 (3):581–590. https://doi.org/10.1007/s12562-015-0865-8.

Zhou, M., M. L. Yuan, L. L. Chen, L. Zhao, Z. Z. Wu, and D. L. Jing. 2017. Study on the biochemical characteristics of low-salt fish sauce with by-products of freshwater fish. *China Condiment* 42 (1):18–22. https://doi.org/10.3969/j.issn.1000–9973.2017.01.005. (In Chinese).

Tea Plantation and Processing of Traditional Products with Potential Health Benefits in Laos PDR

15

Somsay Phovisay, Kridsada Unban, Kalidas Shetty, Saisamorn Lumyong, and Chartchai Khanongnuch

15.1 INTRODUCTION

The fermented tea leaves of *Camellia sinensis* var. *assamica* are important part of traditional food systems of the Lao People's Democratic Republic (Lao PDR, also known as Laos). Laos is a small landlocked country in Southeast Asia and shares borders with China to the north, Vietnam to the east, Cambodia to the south, and Thailand and Myanmar to the west. Laos covers a surface area of 236,800 km^2, including plains along the Mekong riverbanks in the middle and south of the country. The northern terrain is mostly mountainous, characterized by dense jungles and rugged mountains. With a population of approximately 7.5 million people, Laos has the smallest population in Southeast Asia. Its nominal GDP was reported at 18,827.15 million USD in 2021 (The World Bank 2021). Agriculture forms the core of Laos' economy, accounting for 61.44% of its GDP, followed by 12.94% for industry and services at 25.62% in 2019 (Statista 2023). Within the agricultural sector, Laos' tea industry holds significant potential for growth and contribution to the local economies. The country's ancient and wild tea resources represent valuable assets. There are opportunities for expanding tea production, investing in tea processing, and developing high-value tea products. However, the tea sector must fully leverage these comparative advantages and also strategically plan to protect these advantages from ecological and sustainability challenges. Major challenges include the protection of ancient and wild tea resources; ensuring sustainable tea production; and fostering equitable and mutually beneficial relationships among stakeholders such as tea farmers, processors, and traders (Earth Systems 2016).

The northern region of Laos is considered one of the tea homelands, evidenced by ancient tea plantations and a long history of tea culture and consumption. This region encompasses mountains roughly spanning latitude 20°N to 30°N (Van Driem 2019). In ancient times, native people produced infusion tea, commonly known as beverage tea, on a small scale in their households. These products have been widely found in northern provinces and have evolved into commercial products today. Additionally, certain ethnic groups in Xayaboury province fermented tea leaves to create a product known as miang (Wilson 2021). The leaves of *Camellia sinensis* var. *assamica*, commonly referred to as the Assamica variety, have been fermented by local people for a long time. Similar fermentation processes have been observed in southern China, Myanmar, and northern Thailand, and the fermented product is used for chewing or as a food ingredient. In Laos, local people have employed their traditional wisdom to ferment the Assamica variety of tea leaves into miang using various methods such as bamboo tubes, bamboo baskets, and burying them in pots (Ban et al. 2017; Van Driem 2019). The miang fermentation process transforms steamed-fresh leaves into a local delicacy through biotransformation, resulting in a distinct sour taste and flavor (Tamang 2010; Xiao et al. 2015). Miang leaves are also rich in bioactive compounds with health benefits, including antimicrobial and antioxidant activities (Phromrukachat et al. 2010; Sun et al. 2018). Moreover, lactic acid bacteria isolated from miang have shown probiotic ability and produced extracellular bacteriocin, suggesting potential health benefits (Cao et al. 2019; Unban et al. 2019). Like other fermented foods, the properties of fermented tea are influenced by the microbial community, geographical conditions, regional varieties, and fermentation processes, all of which impact the final product qualities (Xiao

DOI: 10.1201/9781003605300-15

et al. 2015). In this chapter, all obtainable literature on tea production in Laos was critically reviewed, focusing on scientific investigation regarding tea production and traditional tea products in Laos and how they contribute to existing advances by highlighting the tea's unique characteristics, exploring growth opportunities, addressing challenges, and examining the cultural and health aspects of tea.

15.2 HISTORY OF TEA PRODUCTION IN LAO PDR

The primordial origins of tea are in lush mountains roughly between 20°N and 30°N (Figure 15.1). This zone covers Nepal, Sikkim, Bhutan, Assam, northeastern India, southeastern Tibet, northern Myanmar, Yunnan, Sichuan, northern Thailand, Laos, and Vietnam (Van Driem 2019). Commercially, *Camellia sinensis* is categorized into two main species: *Camellia sinensis* var. *sinensis* and *Camellia sinensis* var. *assamica* (Figure 15.2). The *sinensis* variety provides small leaves and slower-growing bushes that can survive in colder climates and is used in the production of white tea and green tea. On the other hand, the *assamica* variety has large leaves and taller trees that are quick growing and well suited to warm tropical climates. It is used in the production of fermented teas such as black tea, pickled tea, laphet, and miang (Ban et al. 2017). The *assamica* variety is distributed in the evergreen lands covering valleys and forests in the northern part of Laos.

The origin of ancient tea production and trading in Laos has a legendary history that dates back to the Lanxang kingdom (1353–1707), where tea and other non-timber forest products were traded among feudal states in mainland Southeast Asia. During the French colonial period (1893–1953), introduced tea varieties were brought from Vietnam to grow in the fertile volcanic soils of the Bolaven Plateau in southern Laos in 1920 (Pedersen et al. 2016). In 1928, M. du Pasquier, an agronomist and tea expert, studied the endemic wild tea

FIGURE 15.1 The tea homeland (Van Driem 2019).

FIGURE 15.2 Tea varieties A. *Camellia sinensis* var. *assamica* and B. *Camellia sinensis* var. *sinensis* (Ban et al. 2017).

in the northern part of Laos to identify and assess its quality. The study confirmed the exceptional quality of the wild tea, which was found to be identical to the tea variety in Shan state of Myanmar. Subsequently, a French tea research station was established in Xiengkhouang province to provide tea nurseries and conduct experimental processing of high-quality green and black teas for commercial purposes (Marseille 1990).

Tea trading between northern Laos and China is believed to have existed since at least the 7th century, involving feudal states in South China and Southeast Asia. This exchange has continued and expanded quantitatively up to the present day (Wu et al. 2011). However, commercial tea production was disrupted during the French colonial period, and it was only after independence that the tea sector began to thrive again in the 1990s, with further expansion in the 2000s, coinciding with the flourishing of pu-erh tea in China. There is a demand for *Camellia sinensis* var. *assamica* tea, which naturally grows in Laos. In 2008, the demand for pu-erh tea in China declined, resulting in a temporary decline in exports of Lao tea to China. However, the tea market has since recovered, and China remains the largest destination for tea produced in Laos.

Tea is considered a crucial plant in the northern mountains of Laos, where it has been traditionally processed on a small scale using tea leaves from forests or domestically cultivated *Camellia sinensis* (Figure 15.3). Local people have been using *Camellia sinensis* to make the infusion that is commonly known as beverage tea. However, in some parts of western and northern Laos, tea leaves are processed into a fermented product called miang, which is consumed by chewing or eating. The tea product has attracted the attention of Europeans since

FIGURE 15.3 Wild tea processing in northern Laos.

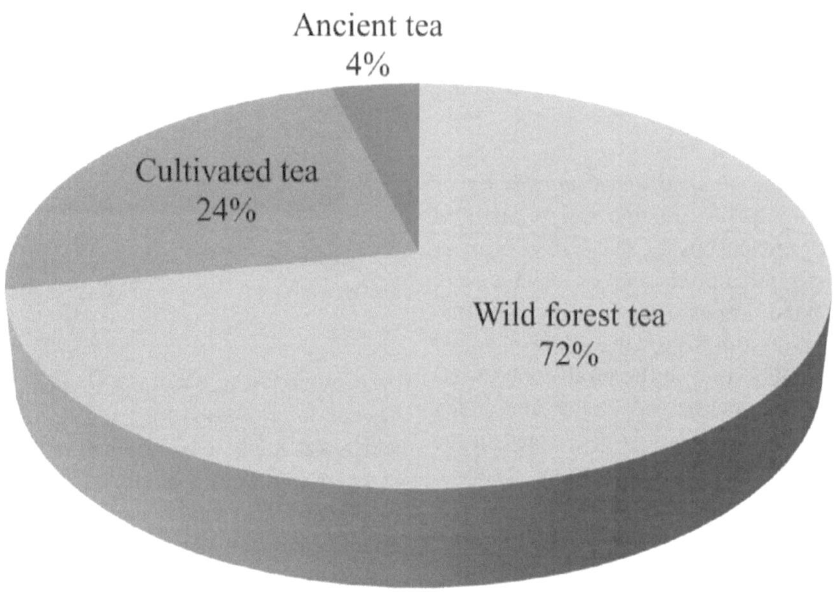

FIGURE 15.4 Percentage of tea resources (Earth Systems 2016).

the French colonial era in the 19th century (Royal Botanic Gardens 1887). Despite this, tea trading has remained a relatively unknown local specialty.

15.3 RECENT TEA PRODUCTION AND STRUCTURE OF THE TEA SECTOR IN THE LAO PDR

According to the National Agriculture and Forestry Research Institute, tea resources in Laos can be classified into three main categories based on the production system (Table 15.1): wild forest tea, cultivated forest tea (known locally as "ancient tea"), and commercially cultivated tea (introduced strains). A study conducted across the nine northern provinces in 2010–2011 identified around 8,500 households engaged in tea cultivation. Another study, as seen in Figure 15.4, estimated the current tea resources to be approximately 12,000 hectares, with 72% consisting of wild forest tea, 24% commercially cultivated tea, and 4% ancient tea (Earth Systems 2016).

In Laos, the majority of tea products are processed into beverages, following market trends. This includes green tea and mao-cha (rough tea) used for producing pu-erh (a fermented tea), as well as other finished teas such as white tea, black tea, yellow tea, red tea, green tea, and oolong tea (Figure 15.5A). However, there is a traditional fermented tea called miang, the same word used in northern Thailand, that holds significance among ethnic groups in the northwest regions of Laos (Figure 15.5B). These regions were a part of the ancient Lanna Kingdom, and the tradition of miang has been deeply rooted in their culture for centuries. Miang has become an integral part of their cultural heritage and continues to be cherished to this day (Mougne et al. 1982).

According to the official national report (NAFRI 2019), the recorded tea plantation area in Laos was 5,430 hectares (Figure 15.6) and showed the dynamic nature of tea plantation areas in different provinces of Laos. Some provinces,

TABLE 15.1 Tea Varieties and Production

DESCRIPTION	*VARIETY*	*MANAGEMENT*	*MARKET CONDITIONS*
Wild tea (wild tea trees and wild tea tree gardens)	*Camellia sinensis* var. *assamica*	• Native tea varieties growing naturally in the forest/natural gardens • Local people communally manage and harvest the wild tea leaves • Organic/natural agriculture • Tea plantation was developed from native wild tea/ancient trees	• High market demand on international market and China • Price premiums
Ancient tea (cultivated forest tea)	*Camellia sinensis* var. *assamica*	• Tea plantation was planted from wild tea seedlings in tea gardens over 100 years ago • Tea trees are old and usually located in remote evergreen forests • Villagers communally manage tea trees as NTFP (natural product)	• High market demand, especially in China • High price
Cultivated tea (introduced strains)	*Camellia sinensis* var. *sinensis*	• The system of investment was land concession/contract farming • Tea plantation areas were managed by private households/domestic or foreign companies • Engagement local workers to harvest the tea trees • Use of fertilizers • Seedlings from the wild tea and neighboring countries (China and Vietnam)	• Lower price • Local and international markets

A B

FIGURE 15.5 The majority of tea products in Laos PDR. A: Beverage teas; B: fermented tea or miang. **(Source: European Chamber of Commerce and Industry in Lao PDR 2017).**

such as Phongsaly and Huaphanh, have experienced consistent growth, while others have witnessed fluctuations or emerged as new tea cultivation areas. Overall, the tea industry in Laos has shown positive development and plays a significant role in the agricultural sector of the country. However, it is worth noting that knowing the estimated tea plantation area is challenging due to the diverse tea production systems in Laos, as well as the presence of wild tea in forested areas, which poses difficulties in data collection and accuracy of statistics. Despite these challenges, approximately 8,055 tons of fresh tea leaves were harvested from an area of 4,195 hectares. This harvest was estimated to yield around 1,650 tons of dried tea product.

The prices of Lao tea vary based on the region and the harvest seasons. In Phongsaly province, for example, all three types of tea (wild forest tea, ancient tea, and cultivated tea) have specific seasons that greatly influence demand, price, and the timing of production. The main tea picking seasons in northern Laos are divided into three periods according to national studies (NAFRI 2018). The first period is the beginning of the dry season, which typically occurs around March to May. Tea harvested during this period is popular in the Chinese market and commands the highest price. The second period is the rainy season, which takes place from June to September. Tea harvested during this period is generally less popular in the

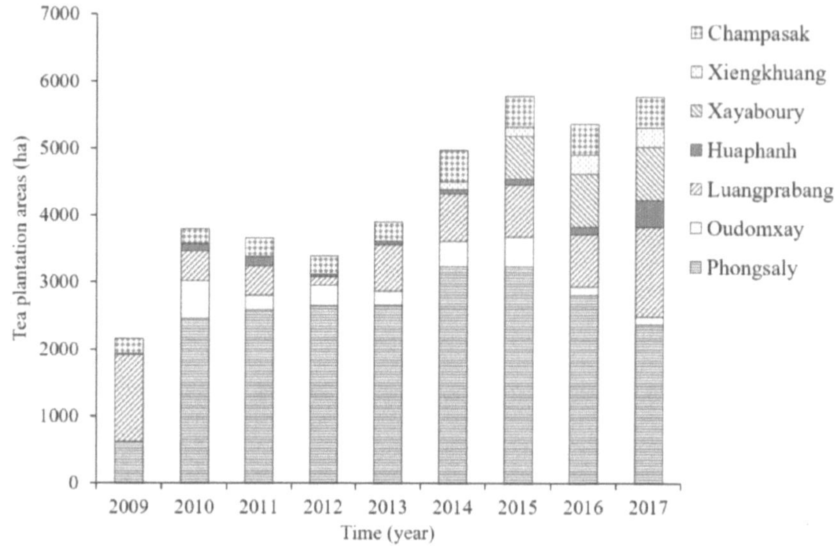

FIGURE 15.6 Tea plantation area by province.

(Source: Department of Agriculture and PAFOs 2019).

Chinese market. The third period is the autumn season, which occurs after the rain, typically from October to November. Tea produced during this season is of higher quality than tea harvested during the rainy season, although still inferior to dry season tea. The prices of tea during the autumn season may be like those of the rainy season, but there is a higher demand, indicating that farmers are able to gain more benefits during this period (Wilson 2021).

15.4 ROLE AND SOCIOCULTURAL RELATIONS OF THE TEA SECTOR

Miang culture, or eating fermented tea, has been preserved among minority tribes who settled in northwestern Laos. These regions were once part of the ancient Lanna Kingdom, and they have maintained a long-standing relationship with this traditional fermented tea as food for hundreds of years. Miang culture involves the consumption of tea for pleasure, often in the form of chewing and direct consumption in solid form. It holds significant importance in the ethnic social customs of welcoming, along with the offering of cigarettes and drinking water in everyday life. Furthermore, the miang fermented tea-consuming culture plays a vital role in the ceremonial practices and special occasions of Southeast Asia. For example, during marriage parties or other traditional ceremonies, miang is packaged in small parcels and distributed to invite people to participate. This cultural practice highlights the significance of miang in fostering social connections and community engagement (Sasaki 2008). For instance, miang was put into small packages then distributed to other people to invite people to a marriage party or other traditional ceremony (Van Driem 2019). The miang product is commonly found in local markets in Xayaboury province. It is used like a chewing condiment and is often enjoyed during various traditional occasions, including welcoming ceremonies, funeral ceremonies, and by laborers during the harvest season. In the past, miang served as an important commodity for exchange between people living in lowland and hill communities (DAFOs 2020). Any study on the roots of fermented tea must consider the practices and traditions of the ancestors of the linguistic Tai families, who were Austroasiatic people. They lived in various countries across Asia, sharing the same roots as the ancestors of the Tai Lue people. The Tai Lue community has settled in the tea heartland of Sipsongpanna, which includes northern Thailand, Myanmar's Shan state, and northern Laos (Brassett et al. 2006). Therefore, the hill tribes of Austroasiatic origin were former communities of the Lanna Kingdom. These areas included northern Thailand (Chiang Mai, Chiang Rai, Phayao, Lampang, Lamphun, Mae Hong Son, Phrae, and Nan provinces), as well as parts of present-day Laos. Consequently, the miang culture was transferred to the next generation (Preechapanya 1996). Xayaboury province in Laos is a well-known region for miang fermented tea culture, covering the northwestern part of the country. It shares a border with the northeast Lanna region of Thailand, which includes Chiang Rai, Phayao, Nan, and Uttaradit provinces.

15.5 TEA PROCESSING FOR TRADITIONAL PRODUCTS

The tea industry in Laos PDR emerged as a notable contributor to the overall economy, contributing between US$6 and $10 million to the Lao economy in 2015, involving over 8,000 households comprising about 44,000 individuals (Pedersen et al. 2016). The cultivation and acquisition of tea resources are predominantly undertaken by smallholder farmers and local communities, particularly within upland regions inhabited by minority ethnic groups. Women play crucial roles in various aspects, including cultivation, collection, processing, and trade. This engagement is characterized by diverse arrangements, with certain smallholders forming cooperatives to facilitate processing, while others establish direct collaborations with processors and traders, often constrained by limited market accessibility (European Chamber of Commerce and Industry in Lao PDR 2017). The emergence of processing facilities to encourage tea production is readily evident, particularly evident in Phongsaly province. Over the years, the province has witnessed a substantial increase in the number of factories, growing from merely 3 in 2010 to a minimum of 11 by 2016. Among these, enterprises from Lao, Malaysian, and Chinese origins are prominent. Notably, similar processing infrastructures are dispersed across various other tea-growing regions throughout the country. For instance, the Tea Company's processing plants in Xayaboury, Xiengkhuang, and Borkeo province represent this trend. While a subset of processors maintains modest plantations to ensure a consistent supply of fresh tea for processing, the majority rely on collaborative partnerships with tea producers to access resources.

Within the framework of tea processing and consumption in Laos, a clear delineation emerges, categorizing these practices into two primary domains: beverage tea and eating tea, also known as miang, which is produced through fermentation. Beverage tea constitutes the more conventional form, meticulously processed with the main intention of being brewed and consumed as a liquid beverage. This type of tea holds paramount significance within the commercial tea sector, catering to widespread drinking preferences. On the other hand, eating tea, or miang fermented tea, represents a distinct departure and is focused on the customary act of chewing tea for traditional edible consumption.

15.5.1 Beverage Teas

The classification of diversified beverage tea types predominantly depends on the specific processing techniques employed. Based on the conditions of tea processing in Lao PDR, a fundamental categorization distinguishes beverage teas into two primary groups, mao-cha (rough tea), which is normally used for crafting pu-erh (fermented tea), and other finished teas, encompassing a range of varieties, including white, black, yellow, red, green, and oolong teas. Particularly, the end product called mao-cha is predominantly traded to Chinese intermediaries, serving as the foundational material for pu-erh tea production in China. This classification illuminates the intricate landscape of tea varieties, showcasing a combination of traditional and innovative methods employed in tea production within the area (Earth Systems 2016).

Mao-cha (an intermediate product), representing tea in its initial processed state, often precedes further processing or aging to develop into specific tea types such as pu-erh and black tea. The process involves plucking fresh tea leaves, followed by withering to reduce moisture, fixation through heat, gentle rolling to release essential oils, and ultimately drying to reduce moisture content. The resulting rough tea leaves exhibit diverse sizes and qualities, offering potential for subsequent processing or aging to manifest distinct flavors and characteristics in the final tea product.

Among finished tea products (Figure 15.7), oolong tea occupies a unique position as a semi-oxidized variety, offering

FIGURE 15.7 The finished products of some Lao tea brands.

a rich blend of flavors and aromas that bridge the gap between green and black teas in terms of oxidation levels. After plucking, leaves undergo withering to reduce moisture, followed by rolling to break cell walls and induce controlled oxidation. This oxidation process ranges from light to moderate and culminates in firing to arrest further oxidation, resulting in a dynamic tea profile encompassing floral, fruity, and toasty notes. Distinct in its rarity and delicacy, yellow tea possesses a subtly oxidized and gently processed character, yielding an exceptional flavor profile. Echoing green tea's processing, leaves undergo withering and are either pan-fired or steamed to halt oxidation. An additional step termed "smothering" or "sealing yellow" is introduced, involving leaf wrapping to facilitate gradual oxidation. This method imparts the tea's distinctive mellow and nuanced taste. Meanwhile, green tea, celebrated for its unoxidized nature and fresh, grassy flavors, is promptly heated through pan-firing or steaming upon plucking to prevent oxidation. Subsequent rolling shapes the leaves before they are dried, a minimal process that preserves the tea's innate green hue and delicate flavor profile. In contrast, black tea, characterized by full oxidation, boasts robust flavors and a dark appearance. After plucking, leaves undergo withering to reduce moisture, followed by rolling to promote oxidation. The leaves are allowed to fully oxidize, culminating in deep flavors and dark hues. A firing stage halts the oxidation process, followed by sorting into different grades.

While tea processing has incorporated modern technology to meet market demands and increase production, remnants of local tea processing traditions endure within the Phounoy ethnic group. These traditional practices persist among native villagers, who cultivate ancient tea leaves dating back over 400 years (Figure 15.8). In Phongsaly province, favorable geographical and climatic conditions like China's Yunnan province have facilitated tea cultivation. Notably, Longjing village, inhabited by the Phounoy people, boasts a centuries-old tea garden with ancient tea trees. The tea revolution, characterized by advancements in cultivation and cross-border trade, has extended to Phongsaly County in northern Laos from China's Yunnan province. The presence of ancient tea trees in Longjing village has spurred Chinese efforts to establish a cross-border tea base. This collaboration has elevated agricultural technologies among Phongsaly farmers, leading to increased income and bolstering sustainable livelihoods. Overall, Phongsaly's tea industry, leveraging its unique geographic advantages and cross-border trade, has spurred positive changes within the agricultural sector, benefiting local farmers and fostering regional economic growth (Guo and Qin 2020).

15.5.2 Eating Tea, or Miang Fermented Tea

During the time of the Lanna Kingdom in the 13th century, the people of the kingdom migrated from Xishuangbanna or Sipsongpanna in Yunnan Province, China, to the southern regions. In this migration, the culture and commerce of the Lanna people were influenced by the presence of Maing fermented tea, a significant aspect of their heritage. These areas encompassed northern Thailand, northeastern Myanmar, and the northwestern part of Laos (Preechapanya 1996). One notable traditional food of northwestern Laos is this fermented tea, known as miang. Unlike black tea, which is typically consumed as a beverage, miang is consumed by chewing and eating. Similar fermented tea products can be found in various countries, but they are known by different names, such as suancha or yancha in China, laphet in Myanmar, and goishicha and awaban-cha in Japan (Kato et al. 1994; Naba et al. 1999; Reichart et al. 1988). The process of fermentation transforms the chemical properties, organic acid qualities, flavor, and taste of the tea leaves through microbial activities. This fermentation process also leads to the generation of metabolites that offer potential health benefits (Huang et al. 2016; Mo et al. 2008). Overall, miang has deep cultural roots in northwestern Laos and even into contemporary times is cherished as an important ethnic food. The historical connections between the Lanna Kingdom and the migration from Sipsongpanna (Yunnan Province, China) have influenced the development and consumption of miang in the region, while similar fermented tea traditions can be found in neighboring countries with their own unique names and variations.

FIGURE 15.8 Small-scale production of traditional beverage tea in north Laos.
(Source: Earth Systems 2016).

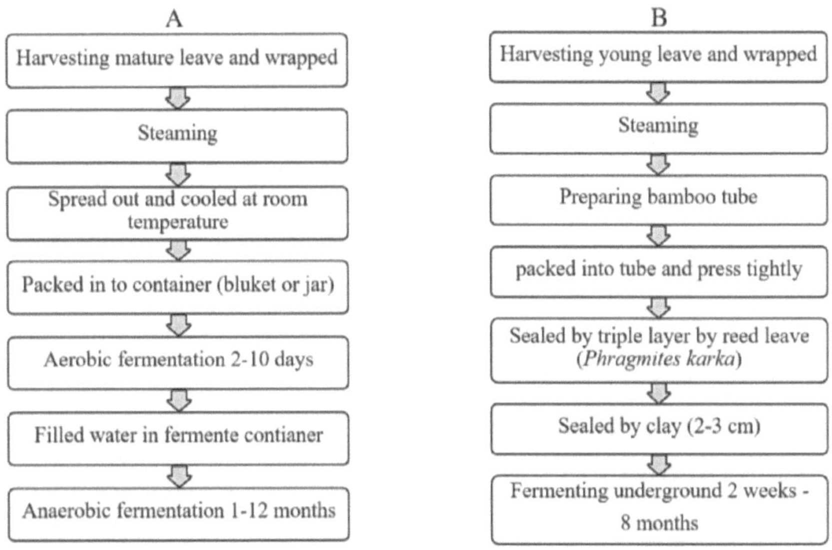

FIGURE 15.9 Traditional miang processing based on ethnic group. A: Laolum (Yuan/Lue); B: Khmue (Luar).

15.5.2.1 Miang Fermented Tea Processing among the Ethnic Groups of Laos

Eating tea, or miang, refers to a traditional fermented tea consumed through chewing and eating, a practice established by the native tribes of Laos. The foundational culture of miang has primarily been transmitted through the ancestral lineage of linguistic Tai communities (Van Driem 2019). Three group names have been used to classify ethnic minorities in Laos based on ethno-linguistic family and customary habitats: Laolum (Tai-Kadai), Laothoeng or Khmue (Mon-Khmer), and Laosung (Sino-Tibetan and Hmong-Myan) (Ethnic Minorities and Indigenous People 2019). Nowadays, the miang process has been found in the sociocultural communities of Lue-Yuan (Laolum) and Luar (Khmue) in the northwest of Laos. However, the miang fermentation process is diverse across each tribe, using their local wisdom to apply the fermentation process according to their environment (Figure 15.9). Although miang products have been found in many places in Laos, but the main areas of miang production are the villages in Saysathan and Xienghorn districts.

15.5.2.2 Miang Fermentation Processing of the Lue-Yuan Ethnic Group (Laolum)

The Lue and Yuan ethnic groups share a common cultural and linguistic history with the Lao-Tai people. They migrated from the southeast of China to the northern part of Laos during the 13th century. Both ethnic groups preferred to settle in lowland areas close to rivers for rice cultivation and animal husbandry (Lao Front for National Construction 2015). Currently, the miang fermentation process followed by these ethnic groups is like the filamentous fungi growth-based process (FFP) described in eastern Lanna (Khanongnuch et al. 2017). This similarity can be attributed to the fact that the people in this region share the same ancestral roots as the people of Lanna. The miang fermentation process is commonly observed in the Xienghorn and Muangkhop districts (Figure 15.9A). During this process, the villagers gather fully grown tea leaves and use bamboo rope to bundle them into hand-sized bundles. These bundles of tea leaves are then steamed for a period of 1 to 2 hours and left to cool naturally at room temperature. Following the cooling process, the steamed tea leaves are placed in baskets and allowed to undergo aerobic fermentation for approximately 1 week, promoting the growth of filamentous fungi. After this initial fermentation stage, the bunches of steamed tea leaves are tightly packed into containers to undergo anaerobic fermentation. The duration of fermentation can vary, ranging from 1 month to as long as a year, depending on the preferences of individual producers (DAFOs 2020).

15.5.2.3 Miang Fermentation Processing of the Luar Ethnic Group (Khmue)

According to the information from community leaders, the Luar ethnic group originally lived in Chiang Mai, Lamphun, and Lampang provinces in Thailand, but civil unrest and conflict disrupted their livelihoods. Thus, they migrated and inhabited the new region of the modern-day Saysathan district. Thus, people in this region have had a lifestyle relationship to miang fermented chewing tea culture for a long time. Miang is traditionally fermented in underground conditions for several weeks to months and commonly found in every household in Saysathan. The achieved products were normally used for personal consumption, commercial purposes, and sometimes as a bartering item instead of cash in rural areas (Lao Front for National Construction 2015; Pattaratuma 2008). The miang fermenting process is without filamentous fungi (Figure 15.9B). First, young tea leaves are tied by bamboo rope and steamed completely for 1–2 hours to develop a soft texture. After that, steamed tea leaves undergo cooling for 6–12 hours or overnight. Then steamed tea leaves are transferred and pressed tightly inside a bamboo tube. Subsequently, they are sealed in triple layers using reed leaves and sealed again tightly using clay (2–3 cm thick). Finally, the bamboo tubes are put underground for fermentation with moist soil for a week to several months with watering once a week in the dry season (DAFOs 2020).

15.6 TRADITIONAL TEA PRODUCT CHARACTERISTICS AND HEALTH BENEFITS

Beverage tea is a rich source of various compounds with associated health benefits, including potent antioxidants like catechins that reduce oxidative stress, support heart health, and aid in weight management, along with epigallocatechin gallate (EGCG) found in green tea, which exhibits antioxidant properties and has the potential to prevent cancer growth and inflammation (Sinija and Mishra 2008; Thielecke and Boschmann 2009; Chacko et al. 2010). Tea contains lower amounts of caffeine than coffee but can enhance mental alertness and physical performance (Glade 2010; Samoggia and Rezzaghi 2021). L-theanine promotes relaxation without causing sleepiness, reduces stress, and potentially enhances cognitive function (Kim et al. 2019). A range of polyphenols, including catechins, offers broad-spectrum antioxidant benefits that protect cells from free radical damage and reduce inflammation related to chronic diseases (Mandel et al. 2011; Musial et al. 2020). Flavonoids are associated with a decreased risk of stroke and improved blood vessel function through regular tea consumption. Additionally, theaflavins and thearubigins, primarily found in black tea, play a significant role in enhancing its antioxidant effects and have the potential to promote cardiovascular health by improving cholesterol levels and reducing blood pressure (Riemersma et al. 2001; Butt et al. 2014; Dias et al. 2014).

In the examination of different processes, significant findings have revealed that caffeine content has a sequence that was observed as follows: black tea > oolong tea > green tea > fresh tea leaves. This suggests that caffeine levels are increased by the fermentation process, as demonstrated by Lin et al. (1998). However, concerning epigallocatechin gallate and total catechins, a different ranking emerged: green tea > oolong tea > fresh tea leaf > black tea. This indicates that higher levels of these beneficial compounds are retained in green tea in comparison to teas that undergo more extensive fermentation (Lin et al. 1998; Zuo et al. 2002). Furthermore, an investigation into the biological functions of tea extracts from various tea types was carried out. Remarkably, protective effects against DNA damage were exhibited by pu-erh tea extracts, and they displayed superior capabilities in scavenging hydroxyl radicals when compared to other teas. The order of effectiveness in suppressing nitric oxide was as follows: pu-erh tea > black tea > green tea > oolong tea. Moreover, the potential of tea extracts to induce apoptosis (cell death) revealed that the highest potential for inducing apoptosis was associated with green tea, followed by oolong tea, black tea, and pu-erh tea (Wong et al. 2022). This result highlights the variations in caffeine, catechins, and biological functions among different tea types, shedding light on their diverse health benefits and potential applications in preventing chronic diseases (Zuo et al. 2002).

In the case of traditional fermented tea like miang, microbial properties played a crucial role in the fermentation process, influencing texture, taste, and health benefits. Miang fermentation process was carried out with native microflora, mainly by lactic acid bacteria (LAB). This process preserved food, enriched nutrient content, and involved key LAB species like *L. plantarum* and *L. pentosus*, which were fundamental microbes typically engaged in the miang process. (Chaikaew et al. 2017). LAB diversity was influenced by factors like tannin concentration, processing techniques, and fermentation duration. During fermentation, various LAB species appear, including those with lower acid tolerance that were observed in the initial stage (Plengvidhya et al. 2007). *Lactobacillus* species from traditional fermented tea showed probiotic potential, namely *L. plantarum*, which was a common LAB in naturally fermented tea (Cao et al. 2019). Some strains of *L. fermentum* from miang exhibited antibacterial effects against pathogenic bacteria such as *L. monocytogenes, Staphylococcus aureus, Salmonella typhi,* and *Shigella sonnei.* The antibacterial mechanism of *L. fermentum* involves breaking down the bacterial cell membranes, leading to the shrinking or cracking of the pathogenic cells (Klayraung and Okonogi 2009). *Candida ethanolica* was the primary yeast in the north Thailand miang process, followed by *Pichia manshurica* and *Pichia occidentalis*. These yeasts had a significant impact on miang's characteristics and quality. They exhibited tannase activity, tannin tolerance, and crucial behaviors for successful fermentation. (Kanpiengjai et al. 2016). The taste and flavor of tea were formed by the fatty acids and amino acids in the leaves, influenced by their freshness. During the tea leaf fermentation, microbes played a crucial role in generating aroma compounds that contribute to its distinct flavor. The flavor profile could be classified into two categories: one arises from enzymatic activity and the second through microbial processes during fermentation. This process leads to the distinct ester aroma found in miang. The complex and diverse taste of miang is a result of both natural leaf decomposition and microbial interactions (Kawakami et al. 1987; Huang et al. 2016).

The color characteristics of miang underwent changes during its processing. After being steamed, the leaves turned yellow-green and became soft. Enzymatic oxidation during fermentation contributed to these color changes. Pigments like theaflavins and thearubigins degraded, leading to the observed color transformation (Kodchasee et al. 2021). During FFP miang fermentation, a significant reduction in hardness was observed, and it remained relatively stable during prolonged fermentation. The decrease in hardness during fermentation was attributed to the activity of β-mannanase (Unban et al. 2020). Moreover, the texture of FFP miang was influenced by *Aspergillus* species, which generated a variety of enzymes, including cellulase, pectinases, xylanase, proteases, lipase, and α-amylases. These enzymes enhanced the biotransformation process, resulting in improvements in miang's texture (Kodchasee et al. 2021).

Miang's health benefits stem from its bioactive elements, which remain robust even after microbial fermentation, retaining high concentrations of beneficial phenolic compounds, such as catechins, epicatechins, and caffeine (Phromrukachat et al. 2010). Extracted compounds from miang were found to inhibit the growth of pathogenic bacteria like *Streptococcus mutans*, *Vibrio cholerae*, and *Salmonella enterica* serovar Typhimurium (Unban et al. 2019). The health benefits of miang are enhanced by its antimicrobial and antioxidant properties. The antioxidant activity is influenced by various flavonoids and catechins like EGCG, EGC, ECG, EC, and C, obtained from the rich content of these phenolic compounds known for their antioxidant abilities. Notably, no significant difference in antioxidant activity was found between sour miang and astringent miang, but prolonged fermentation in the sour miang process showed an increasing trend in antioxidant activity (Unban et al. 2019).

15.7 CHALLENGE AND OPPORTUNITIES IN SUSTAINABLE DEVELOPMENT

In the tea production sector of Laos, there are two key significant challenges. First, safeguarding the genetic diversity of ancient areas and forest tea is paramount. Second, it's crucial to make sure that small-scale farmers and local communities who grow tea have secure rights to their land, as emphasized in reports (Phouyyavong 2018; Earth Systems 2016). This emphasis highlights the importance of preserving and effectively managing these invaluable genetic resources to support sustainable livelihoods and empower producers. Furthermore, it is advantageous for tea production to align with environmentally friendly practices. In the certain regions of Laos, especially in Xienghorn district of Xayaboury province, agroforestry system has been successfully implemented. This approach not only conserves soil quality and biodiversity and efficiently manages water resources but also addresses climate change concerns effectively. Laos has also founded a unique tea agroforestry technique (Figure 15.10A), similar to miang gardens in Thailand, where this approach offers various benefits, including an enriched environment conducive to various plant species such as traditional herbs (Sukkho et al. 2022). The roots of tea plants fortify the soil, and fallen leaves enhance soil quality through decomposition. Additionally, the garden's trees provide shade, shielding tea plants from excessive heat and aiding in the absorption of carbon dioxide from the atmosphere, thereby mitigating the effects of climate change. This cultivation approach intelligently manages water resources, preventing wastage and promoting efficient water retention in the soil (Sasaki et al. 2021; Piriyatachagul et al. 2020). The practice not only proves beneficial for nature but also creates economic opportunities by facilitating the sale of items derived from both ranges of plants and trees.

On the other hand, the ancient tea system has undergone a transformation in some regions. Driven by the demand from Chinese purchasers for tea shoots used in beverage production, villagers in these areas have shifted from agroforestry to conventional tea gardens. This transition has introduced certain drawbacks, most notably the reduction of larger tea plants and the conversion of integrated forest areas into single-crop tea fields. This shift underscores the importance of conserving the agroforestry tea system for the future, aligning with the findings in the Xaysathan district of Xayaboury province (Figure 15.10B). Additionally, miang gardens encounter challenges, notably a reduction in plantations, primarily due to replacement with economically viable crops such as rubber trees (Figure 15.10C). Efforts to commercialize tea processing have been actively pursued, especially in the Xayaboury province. The establishment of a green tea investment project in Saysathan, as reported by the Provincial Agriculture and Forestry Offices (PAFOs 2020), facilitated the purchase of fresh tea leaves from villagers by Chinese merchants.

15.8 FUTURE PROSPECTS

The tea sector, particularly forest tea and ancient tea-producing land, holds significant importance in Laos for multiple reasons. First, forest tea and associated ancient tea carry cultural

FIGURE 15.10 Miang plantation area in Xienghorn district, Laos.

Notes: A: Integration of miang garden.
B: The integrated miang garden was transformed into a single crop.
C: Reduction of miang plantations regarding the replacement by rubber trees.

and historical significance, contributing to the community with their traditional flavors and connecting them to their heritage. These teas are produced using time-honored fermentation techniques passed down through generations, preserving the art of tea-making and contributing to the cultural identity of local communities. Additionally, the cultivation and production of forest tea and associated ancient tea play a vital role in conserving the tea environment in Laos. Tea plantations are typically established within forested areas, promoting the preservation of natural ecosystems and biodiversity. Forest tea cultivation often integrates shade-grown techniques, allowing tea plants to thrive beneath tall trees. This approach helps sustain the integrity of forest ecosystems, safeguard soil health, and preserve wildlife habitats.

Specifically, scientific studies have advanced to understand traditional fermented tea such as miang fermentation in bamboo tubes with underground incubation and the scientifically recognized health benefits of consuming miang, and these involve several key areas of research. The research purposes are to provide clarity on the fermentation mechanism and explore the potential advantages of miang consumption. The following major trends in this field can guide future research: exploring the history and cultural background of miang culture with ethnic groups in Laos. This research will shed light on the traditional practices and cultural significance of miang, providing a broader context for its fermentation process. Investigating the changes in microbial population and bioactive compounds during the underground fermentation of miang in bamboo tubes in Laos: Understanding the dynamics of microbial communities and their impact on the fermentation process will contribute to the optimization and control of miang production. Evaluating the chemical composition and conducting sensory tests of miang at different stages of fermentation: Analyzing the chemical constituents and sensory attributes of miang throughout the fermentation process will provide insights into the flavor development, nutritional content, and overall quality of the final product. By focusing on these research areas, we can enhance our understanding of the miang fermentation process, its microbial dynamics, chemical composition, sensory characteristics, and potential health advantages. This knowledge will contribute to the scientific advances and further development of miang as a valuable and sustainable fermented food product with health benefits.

REFERENCES

Ban, Q., Wang, X., Pan, C., Wang, Y., Kong, L., Jiang, H., and Y. Li. 2017. Comparative analysis of the response and gene regulation in cold resistant and susceptible tea plants. *PLoS One, 12*(12), e0188514.

Brassett, C., Brassett, P., and M. Lu. 2006. *The Tujia Language* (Vol. 455). Lincom Europa.

Butt, M. S., Imran, A., Sharif, M. K., Ahmad, R. S., Xiao, H., Imran, M. H. A. R., and H. A. Rsool. 2014. Black tea polyphenols: a mechanistic treatise. *Critical Reviews in food Science and Nutrition, 54*(8), 1002–1011.

Cao, Z., Pan, H., Li, S., Shi, C., Wang, S., Wang, F., Ye, P., Jia, J., Ge, C., Lin, Q., and Z. Zhao. 2019. In vitro evaluation of probiotic potential of lactic acid bacteria isolated from Yunnan De'ang pickled tea. *Probiotics Antimicrob Proteins, 11*(1), 103–112.

Chacko, S. M., Thambi, P. T., Kuttan, R., and I. Nishigaki. 2010. Beneficial effects of green tea: a literature review. *Chinese Medicine, 5*, 13.

Chaikaew, S., Baipong, S., Sone, T., Kanpiengjai, A., Chui-Chai, N., Asano, K., and C. Khanongnuch. 2017. Diversity of lactic acid bacteria from Miang, a traditional fermented tea leaf in northern Thailand and their tannin-tolerant ability in tea extract. *Journal of Microbiology, 55*(9), 720–729.

DAFOs. 2020. Traditional tea products in Xayaboury province [Unpublished report]. District Agriculture and Forestry Offices. Ministry of Agriculture and Forestry, Xayaboury, Lao PDR.

Department of Agriculture and PAFOs. 2019. Tea production statistics in Lao PDR [Unpublished report]. Ministry of Agriculture and Forestry, Xayaboury, Lao PDR.

Dias, P. M., Changarath, J., Damodaran, A., and M. K. Joshi. 2014. Compositional variation among black tea across geographies and their potential influence on endothelial nitric oxide and antioxidant activity. *Journal of Agricultural and Food Chemistry, 62*(28), 6655–6668.

Earth Systems. 2016. Industry Position Paper Sustainable Development of the Lao Tea Sector. Retrieved March 8, 2023, from https://www.mrlg.org/wp-content/uploads/2019/09/Industry-Position-Paper-Sustainable-Development-of-the-Lao-Tea-Sector.pdf

Ethnic minorities and indigenous people. 2019. Open Development Laos. Retrieved May 5, 2023, from https://laos.opendevelopmentmekong.net/topics/ethnic-minorities-and-indigenous-people/#ref-2457-1

European Chamber of Commerce and Industry in Lao PDR. 2017. Lao Tea Sector Industry Report, Vientiane, Capital, October. Retrieved May 5, 2023, from https://eccil.org/?ddownload=7032

Glade, M. J. 2010. Caffeine-not just a stimulant. *Nutrition (Burbank, Los Angeles County, Calif.), 26*(10), 932–938.

Guo, J. W., and Q. Qin. 2020. Laos Phongsaly Tea Cross-Border Agriculture Culture Dissemination and Interaction Study. *International Workshop on Advances in Social Sciences (IWASS 2020)*, Francis Academic Press, UK.

Huang, Y., Xiao, X., Cong, L., Wu, M., and Y. Yao. 2016. A fermented tea with high levels of gallic acid processed by anaerobic solid-state fermentation. *LWT—Food Science and Technology, 71*, 260–267.

Kanpiengjai, A., Chui-Chai, N., Chaikaew, S., and C. Khanongnuch. 2016. Distribution of tannin-'tolerant yeasts isolated from Miang, a traditional fermented tea leaf (*Camellia sinensis* var. *assamica*) in northern Thailand. *Internation Journal of Food Microbiology, 238*, 121–131.

Kato, M., Tamura, A., Omori, M., Nanba, A., Miyagawa, K., Nishmura, O., and W. Kameda. 1994. Changes of flavor during manufacturing process of Japanese fermented tea (Goishi-cha) and its characteristic. *Journal of Home Economics of Japan, 45*(6), 527–532.

Kawakami, M., Chairote, G., and A. Kobayashi. 1987. Flavor constituents of pickled tea, Miang, in Thailand. *Agricultural and Biological Chemistry, 51*(6), 1683–1687.

Khanongnuch, C., Unban, K., Kanpiengjai, A., and C. Saenjum. 2017. Recent research advances and ethno-botanical history of Miang, a traditional fermented tea (*Camellia sinensis* var. *assamica*) of northern Thailand. *Journal of Ethnic Foods, 4*(3), 135–144.

Kim, S., Jo, K., Hong, K. B., Han, S. H., and H. J. Suh. 2019. GABA and L-theanine mixture decreases sleep latency and improves NREM sleep. *Pharmaceutical Biology*, *57*(1), 64–72.

Klayraung, S., and S. Okonogi. 2009. Antibacterial and antioxidant activities of acid and bile resistant strains of Lactobacillus fermentum isolated from Miang. *Brazilian Journal of Microbiology*, *40*, 757–766.

Kodchasee, P., Nain, K., Abdullahi, A. D., Unban, K., Saenjum, C., Shetty, K., and C. Khanongnuch. 2021. Microbial dynamics-linked properties and functional metabolites during Miang fermentation using the filamentous fungi growth-based process. *Food Bioscience*, *41*, 100998.

Lao Front for National Construction. 2015. *The Identity of the Prai Ethnic*. Department of Ethnic Affairs, Vientiane, pp. 1–2. (Lao version).

Lin, J. K., Lin, C. L., Liang, Y. C., Lin-Shiau, S. Y., and I. M. Juan. 1998. Survey of catechins, gallic acid, and methylxanthines in green, Oolong, Pu-erh, and black teas. *Journal of Agricultural and Food Chemistry*, *46*(9), 3635–3642.

Mandel, S. A., Amit, T., Weinreb, O., and M. B. Youdim. 2011. Understanding the broad-spectrum neuroprotective action profile of green tea polyphenols in aging and neurodegenerative diseases. *Journal of Alzheimer's Disease*, *25*(2), 187–208.

Marseille, A. J. E. 1990. The origins of Shan tea with white tips from Bao Loc (Haut Donnaï) South Vietnam. *Peninsula*, *21*(20), 99–117.

Mo, H., Zhu, Y., and Z. Chen. 2008. Microbial fermented tea–a potential source of natural food preservatives. *Trends in Food Science and Technology*, *19*(3), 124–130.

Mougne, C., MacLennan, R., and S. Atsana. 1982. Smoking, chewing and drinking in Ban Pong, northern Thailand. *Social Science and Medicine*, *16*(1), 99–106.

Musial, C., Kuban-Jankowska, A., and M. Gorska-Ponikowska. 2020. Beneficial properties of green tea catechins. *International Journal of Molecular Sciences*, *21*(5), 1744.

Naba, A., Nyein N. M., Win, S. Y., and K. Miyagawa. 1999. Post-heated and fermented edible teas and their dried forms. *Journal of Home Economics of Japan*, *50*(6), 639–646.

NAFRI. 2018. Tea production and market study in Northern Laos. National Agriculture and Forestry Research Institute, Ministry of Agriculture and Forestry, Vientiane, Lao PDR.

NAFRI. 2019. Tea plantation statistics in Lao PDR. National Agriculture and Forestry Research Institute. Ministry of Agriculture and Forestry, Vientiane, Lao PDR.

PAFOs. 2020. Green tea investment project in Saysathan district [Unpublished report]. Provincial Agriculture and Forestry Office, Department of Agriculture and Forestry, Xayaboury, Lao PDR.

Pattaratuma, A. 2008. Lua. *Journal of Forest Management*, *2*(3), 87–92. Thai version. http://frc.forest.ku.ac.th/frcdatabase/bulletin/Document/Forest_management_journal003_006.pdf

Pedersen O. S., Carroll M., Chen L., and S. Yang. 2016. *Phou San Wild Tea Xieng Khouang Province, LAO PDR From Early Days to Current Production and Market Development* (pp. 5–6). Agro-Biodiversity Project, Ministry of Agriculture and Forestry, Vientiane, Laos PDR.

Phouyyavong, K. 2018. *Unlucking the full potential of Lao PDR's tea sector*. National Agriculture and Forestry Research Institute, Agriculture and Forestry Policy Research Centre, Vientiane, Laos. Retrieved March 8, 2023, from https://www.malica.org/content/download/4633/34246/version/1/file/6.+Unlocking+the+full+potential+of+Lao+PDR%E2%80%99s+tea+sector.pdf

Phromrukachat, S., Tiengburanatum, N., and J. Meechui. 2010. Assessment of active ingredients in pickled tea. *Asian Journal of Food and Agro-Industry*, *3*(3), 312–318.

Piriyatachagul, G., Phuwanatwichit, T., Mangkhang, C., and A. Sarobol. 2020. "Miang" culture: The community resources management through" design-based learning" for self-reliance of highland communities in the upper Northern Thailand. *Higher Education Studies*, *10*(3), 63–71.

Plengvidhya, V., Breidt, F., Jr., Lu, Z., and H. P. Fleming. 2007. DNA fingerprinting of lactic acid bacteria in sauerkraut fermentations. *Applied and Environmental Microbiology*, *73*(23), 7697–7702.

Preechapanya, P. 1996. *Indigenous Ecological Knowledge about the Sustainability of Tea Gardens in the Hill Evergreen Forest of Northern Thailand*. Bangor University.

Reichart, P., Philipsen, H., Mohr, U., Geerlings, H., and S. Srisuwan. 1988. Miang chewing in northern Thai villagers. *Tropical and Geographical Medicine*, *40*(1), 39–44.

Riemersma, R. A., Rice-Evans, C. A., Tyrrell, R. M., Clifford, M. N., and M. E. J. Lean. 2001. Tea flavonoids and cardiovascular health. *QJM*, *94*(5), 277–282.

Royal Botanic Gardens, K. 1887. *Bulletin of Miscellaneous Information* (Vol. 1887). H.M. Stationery Office, London.

Samoggia, A., and T. Rezzaghi. 2021. The consumption of caffeine-containing products to enhance sports performance: an application of an extended model of the theory of planned behavior. *Nutrients*, *13*(2), 344.

Sasaki, A. 2008. Changes in the management system of the resources in the 'Miang tea gardens': a case study of PMO village, northern Thailand. *Tropics*, *17*(3), 271–280.

Sasaki, A., Kanzaki, M., Mochizuki, K., Choocharoen, C., and P. Preechapanya. 2021. Aboveground biomass and carbon sequestration potential of tea and shade trees in Miang tea gardens, an agroforestry system in Northern Thailand. *Tropics*, *29*(4), 105–119.

Sinija, V. R., and H. N. Mishra. 2008. Green tea: health benefits. *Journal of Nutritional & Environmental Medicine*, *17*(4), 232–242.

Statista. 2023. Distribution of employment by economic sector from 2009 to 2019. Retrieved March 8, 2023, from https://www.statista.com/statistics/804983/employment-by-economic-sector-in-laos/

Sukkho, T., Khanongnuch, C., Lumyong, S., Ruangsuriya, J., Pattananandecha, T., Apichai, S., Ogata, F., Kawasaki, N. and C. Saenjum. 2022. Local wisdom and diversity of medicinal plants in cha Miang forest in Mae kampong village, Chiang Mai, Thailand, and their potential for use as osteoprotective products. *Plants*, *11*(11), 1492.

Sun, H., Chen, Y., Cheng, M., Zhang, X., Zheng, X., and Z. Zhang. 2018. The modulatory effect of polyphenols from green tea, oolong tea and black tea on human intestinal microbiota in vitro. *Journal of Food Science and Technology*, *55*(1), 399–407.

Tamang, J. P. 2010. Diversity of fermented foods. In *Fermented Foods and Beverages of the World*, 41–84, CRC Press, Taylor and Francis Group.

Thielecke, F., and M. Boschmann. 2009. The potential role of green tea catechins in the prevention of the metabolic syndrome–a review. *Phytochemistry*, *70*(1), 11–24.

Unban, K., Khatthongngam, N., Shetty, K., and C. Khanongnuch. 2019. Nutritional biotransformation in traditional fermented tea (Miang) from north Thailand and its impact on antioxidant and antimicrobial activities. *Journal of Food Science and Technology*, *56*(5), 2687–2699.

Unban, K., Kochasee, P., Shetty, K., and C. Khanongnuch. 2020. Tannin-tolerant and extracellular Tannase producing *Bacillus* isolated from traditional fermented tea leaves and their probiotic functional properties. *Foods*, *9*(4).

Van Driem, G. L. 2019. *The Tale of Tea: A Comprehensive History of Tea from Prehistoric Times to the Present Day*. Brill.

Wilson, A. 2021. *Tea Sector Development in Lao PDR*. Lao PDR: LURUS.

Wong, M., Sirisena, S., and K. Ng. 2022. Phytochemical profile of differently processed tea: a review. *Journal of Food Science, 87*(5), 1925–1942.

The World Bank. 2021. GDP (current US$)—Lao PDR. Retrieved March 8, 2023, from https://data.worldbank.org/indicator/NY.GDP.MKTP.CD?locations=LA

Wu, Y., Chanhda, H., Zhang, X., Yoshida, A., and C. Wu. 2011. Tea industry development and land utilization along the China-Laos border: a case study of Komen Village in Laos. *African Journal of Business Management, 5*(11), 4328.

Xiao, P., Huang, Y., Yang, W., Zhang, B., and X. Quan. 2015. Screening lactic acid bacteria with high yielding-acid capacity from pickled tea for their potential uses of inoculating to ferment tea products. *Journal of Food Science and Technology, 52*(10), 6727–6734.

Zuo, Y., Chen, H., and Y. Deng. 2002. Simultaneous determination of catechins, caffeine and gallic acids in green, Oolong, black and Pu-erh teas using HPLC with a photodiode array detector. *Talanta, 57*(2), 307–316.

Korea Indigenous Fermented Soy Products

16

Dae Young Kwon, Kim Soon-Hee, Kyung Rhan Chung, Dai-Ja Jang, Su-Jin Jung, Kalidas Shetty

16.1 INTRODUCTION

Food is born in the process of solving the following three challenges (Kwon et al. 2023; Mandelblatt 2012). First, the challenge of satisfying hunger must be addressed, and the food must be edible and safe to consume to prevent anyone from getting sick or dying from it. Second, how can we make food more delicious? The initial task for humans was to satisfy their hunger, but once this was accomplished, they wanted to enjoy their food. Third, how can we safely eat leftovers later? Food is not always available, with a short shelf life, such as wild meat, fruits, and vegetables, which spoil quickly. Every region or nation in the world develops food by addressing these concerns, and the process of solving these three issues is called food development. Food is made according to the conditions and environment of the region, and each region has its unique characteristics, which is what creates ethnic food. It is the same in Korea, with its predominantly mountainous terrain, cultivated various grasses and relied on wild rice and barley in the fields, where rice farming or the cultivation of beans in the fields flourished (Kwon et al. 2025). These geographical and agricultural characteristics gave rise to Korea's unique cuisine.

16.1.1 Korean Cuisine

A distinctive feature of Korean cuisine is the unique presentation of bap (bap is made by boiling rice or barley with soy and sorghum and others) and banchan (side dishes) (see figure in Kwon et al. 2023). That is, while most ethnic cuisines focus on what to eat, Korean cuisine is a culture of "With what to eat bap?" Korean ancestors always pondered how to eat bap deliciously as the main staple grain and enhance its taste (mat, 맛) by adding condiments from plants collected from the fields and mountains (Kwon et al. 2023). The staple food in Korean cuisine is bap, but it is always eaten with banchans rather than alone. Therefore, banchans are important to enjoy a delicious meal. Jang (fermented soy products) was always eaten with kimchi on the bapsang (meal table), and they used to be served to enjoy bap and kuk (like soup or broth) more deliciously. This was called banchan culture, and it is a food culture that does not exist in any other country in the world. Usually, people scoop up a spoon of bap with a spoon, put it in their mouth, and pick up one of the banchans with chopsticks to eat. To make it so, they adjusted the 'kan' with salt or 'jang' to make it feel good and comfortable. This is called delicious state of 'banchan' or 'kuk', 'kan is right' (Jang et al. 2016; Kim et al. 2023; Kwon et al. 2023). An important concept in Korean cuisine is kan, and when the salinity of food is appropriate, it is expressed as 'kan is right' or 'kan is suitable'; salinity of food becomes appropriate (Kwon et al. 2023). This proper kan is the point why Korean food has different ethnic root from China, unless using frying oil or sugar as elements to make food delicious based on agriculture system.

16.1.2 Jang: Fermented Soybean Products

Chongkukjang is a fermented food made by placing boiled soybeans on rice straw and leaving them in a warm place at 40–50 °C for 2–3 days (Figure 16.1) (Park 2018; Shin and Jeong 2015). Jang (醬) is an essential element of Korean cuisine, used to season other foods and enhance their favor. While the concept may be similar to sauces or dressings in Western cuisine, those are typically made by simply mixing liquid or powdery seasonings. In contrast, Korean jang is a fermented food, in which the original ingredients are transformed into new metabolites through the fermentation process. The resulting food has a taste and aroma completely different from the original. Jang can be classified according to the ingredients and fermentation method (Jang et al. 2016; Shin and Jeong 2015). Chongkukjang (fermented soy bean), doenjang (fermented soy paste), and

DOI: 10.1201/9781003605300-16

FIGURE 16.1 Chongkukjang (a) and meju (b) was prepared with the same procedure from boiled soybeans. However, meju was made by drying after making brick-shaping, while chongkukjang was non-dried moistured product. Chongkukjang is a fermented food made by placing boiled soybeans on rice straw and leaving them in a warm place at 40–50 °C for 2–3 days. Meju, dried fermented soybean brick, was a food intermediate to make kanjang, doenjang, and kochujang by adding saline solution for further fermentation.

kanjang (fermented soy sauce) are made using only soybeans as the main ingredient, while kochujang is made by adding red chili powder to soybeans and fermenting them. Additionally, jang can be divided according to whether or not salt is added (Shin 2022; Shin and Jeong 2015). Chongkukjang is fermented without salt, resulting in a short fermentation time and a short shelf-life, typically lasting only a few days. In contrast, jangs fermented with salt have a slower fermentation process due to the presence of salt and can be fermented for 3 months to 1 year. Salted jang has a longer shelf life and can be stored for more than a year. Here are some additional characteristics of each type of jang. Chongkukjang is believed to be the first type of jang (prototype of jang, just like *geotjori* in kimchi (Kwon et al. 2025) to have been made because it is the simplest form of fermentation (Kim et al. 2023). During the fall harvest, soybeans were boiled, and the leftovers were kept in the room to prevent mice from eating them. After a few days, the appearance and smell of the food changed, and it resembled spoiled beans. However, it was considered a waste to throw it away. Not only does chongkukjang not cause stomach discomfort, it also seems to taste even better when cooked in a soup or dish with broth, known as 'kuk' in Korean cuisine. Despite its initial discovery being accidental, chongkukjang gradually became a deliberately made food due to its unique taste and texture. Unlike other jangs, chongkukjang has seasonal restrictions as spoilage occurs faster after fermentation in warmer seasons. In fact, chongkukjang was not initially categorized as a jang until the early Chosun Dynasty and was simply called 'Chongkuk', as it was mainly eaten by itself rather than used as a seasoning for other dishes. Over time, it was classified as a jang and renamed chongkukjang (Kwon et al. 2019).

FIGURE 16.2 To enjoy the bap (rice), chongkukjang was cooked around 100 °C with vegetables and some meat as a chigae like a stew. It was called as chongkukjang chigae, which was very common like as kimchi chigae [5].

The main way of consuming chongkukjang is by boiling it with vegetables to make kuk or chigae (Figure 16.2). Microorganisms present in the rice straw laid under the boiled soybeans are naturally transferred to the soybeans and inoculated for natural fermentation. *Bacillus subtilis* is the main bacterium responsible for the fermentation of chongkukjang (Shin 2022).

16.1.3 Role of Jang in Korean Cuisine

In many Asian countries, particularly Korea, the traditional diet consists mostly of grains and vegetables, making protein deficiency a common problem. This is why Koreans ate a lot of soybean products to supplement their diet. Soybeans is suggested to have come from Manchuria (Chavan 2005; Tamang 2015) and have been eaten in Korea for a long time. Koreans added protein to their diet by eating roasted soybeans directly or using them to make tofu. In addition, soybeans are also used to make food be delicious by making fermented soybean products to adjust kan (Jang et al. 2016; Kim et al. 2023). In short, soybeans are a very important food in Korean cuisine.

Korean food is often described as having the "taste of a mother's hand", while Chinese food is often characterized as having a "fire taste" (Kwon et al. 2023). The former refers to the taste born of a mother's sacrifice and sincerity to feed her family despite difficult economic conditions, while the latter emphasizes the taste of food cooked in oil at high temperatures. Jang was developed as a way to create delicious food even in situations where high heat and sugar, which are elements that enhance flavor, cannot be used in ancient times (Kim et al. 2023). As mentioned earlier, Koreans season almost all their food with jangs, meaning that jangs play a vital role in determining the taste of Korean cuisine. Jang plays a role in enhancing the flavor of banchans. Given the importance of jang in Korean cuisine, it is no surprise that Koreans place great emphasis on making it delicious after boiling. Jang was first prepared accidentally by serendipity finding when boiled soybean was stored for later consumption, but since then it has become an indispensable ingredient in Korean cuisine (Kim et al. 2023).

16.1.4 Chongkukjang and Similars

Chongkukjang, one type of jang, is one of Korea's most well-known fermented jang. Chongkukjang is proto-type fermented product produced from soybeans. It is made by fermenting soybeans after boiling. Unlike doenjang, kanjang and kochujang which require several months of fermentation (Park 2018; Shin and Jeong 2015), chongkukjang only takes 2–4 days to ferment. After harvesting soybeans in Autumn, they are washed, boiled, and left to cool. They are then placed in the warm part of a room on a basket or clay pot covered with straw, then covered with a blanket to control the temperature (40–45 °C), and left to ferment for 2–3 days (Kim et al. 2023; Park 2018; Shin and Jeong 2015). The fermenting organism is *Bacillus*, as most of the strains involved in the fermentation of Korean chongkukjang come from straw. Chongkukjang is a classic example of a fermented soybean food that does not contain salt. Since chongkukjang has been eaten in Korea for a long time, it makes sense that the food has several thousands of years of history. A bacteria, *Bacillus subtilis* from the straw, is naturally inoculated to prepare chongkukjang (Park 2018; Shin 2022; Shin and Jeong 2015).

Japanese natto, a type of chongkukjang that originated in Korea, has also been around for several thousands of years. Natto is an improved version of chongkukjang after inoculating the strains that has become a global food. Natto was allowed to be eaten on its own, while chongkukjang is usually eaten as a chongkukjang kuk or chigae (Park 2018; Shin 2022). There are also foods similar to Korean chongkukjang found in South and Southeast Asia. Some examples are douchi from Sichuan (China), kinema from Nepal and India, and fermented soybeans from countries on the IndoChinese Peninsula, including Thai thuanao (Tamang 2015). These fermented foods are mainly produced through *Bacillus* fermentation. In countries such as Indonesia and Malaysia, which were formerly part of the Santara dynasty (Chavan 2005), there is also a food that closely resembles chongkukjang called 'tempeh'. However, the microorganism involving in tempeh is the mold such as *Rhizopus* or *Aspergillus*, and yeast such as *Sacharomyces*. Of course, bacteria such as *Lactobacillus* also play a role in producing tempeh, but the majority of microbial growth and fermentation takes place through mold or yeast (Shurtleff and Aoyagi 2010). The same principle of fermentation applies to chongkukjang in northern countries and tempeh in southern countries, it is sufficient to understand that mold and yeast were mainly responsible for the fermentation of chongkukjang in humid countries located further to the south. The variation in chongkukjang based dishes between countries can be explained by the differences in each culture's ethnic foods.

16.1.5 Wrong Stories of Chongkukjang

Although chongkukjang is an indigenous fermented soybean food that has existed in Korea for thousands of years (Chung 2018; Kwon et al. 2019), some scholars have been distorting the truth and spreading false information about chongkukjang. Representative misconceptions are: "chongkukjang was first created during war, as it could be prepared quickly" (Lee 1991a), "chongkukjang was introduced to Korea via the Qing dynasty" (National Institute of Korean Language 1999), and "The first documented record of chongkukjang in Korea is in Jungbosallimkyongje(增補山林經濟), written by Yu JL(柳重臨) in 1760 (the 36th year of King Yeongjo's reign)" (Donga Encyclopedia 1982; Park 2015).

These misconceptions are why Professor Lee Seong-woo (Lee 1991a), a scholar based on Chinese characters (Park 2015) was involved in publishing various dictionaries as mentioned above. He also claims that chili peppers came into Korea from Japan during the Japanese invasion of Korea, and distorted the history of all Korean foods to justify this claim (Lee 1991b). Thus, he is the one who distorted the history of kimchi, kochujang and chonggukjang, and talked about our food history based on literature based only on Chinese characters, as if all Korean food's history might be derived from China.

In this chapter, the history of chongkukjang is reviewed based on the science in terms of literature and agricultural science, food science and biological science and also reviewed health benefits of chongkukjang. The authors also aim to correct unscientific misconceptions about the history, origins, and value of chongkukjang.

16.2 SOYBEAN AND PULSE

Pulses (beans, peas and lentils) belong to Leguminosae family are an important Nutritiontional sources for people around the world from thousands of years. There are various kinds of pulses on Earth, such as soybeans (*Glycine* spp) in Korea and Manchuria (now it is north-east China, but was in Korea in ancient times), lentils (*Lens culinaris* Medik) in India, chickpeas (*Cicer arietinum* L.) in Egypt and Mesopotamia. Sometimes legumes and pulses are used interchangeably, in fact, the terms "legumes" and "pulses" are references to the same, but all pulses are considered legumes but not all legumes are considered pulses (Singh 2017).

Pulses have existed for ten millions of years before humans appeared on earth (Fukuda 1933; Hymowitz 1970; Kang et al. 2014; Kim et al. 2010). However, these pulses were not spread to other continents in the world and were only growing in localized areas. Each seed of any pulse is a food that animals or birds like, and once they eat them, the seed including germ get easily digested in their bellies. The birds can fly far across the continents, but they cannot typically spread the seeds. That is why certain areas can only be the original land of wild varieties of pulses including beans when humans first discovered them (Kwon et al. 2023). Based on similar rationale, each soybean species grows only in a specific area because soybeans cannot be transplanted without human intervention.

Through analyzing the genome of soybean, it is estimated that soybeans have been growing on earth, particularly in the Korean Peninsula including Manchuria, for some million years before humans started cultivating them about ten thousand years ago (Kang et al. 2014; Kim et al. 2010). Nevertheless, since pulses, especially soybeans are not grains and eating them raw may cause food poisoning due to the presence of protease inhibitors such as a trypsin inhibitor, they were not widely spread by humans before use of fire for cooking. Through cooking and heat, protease inhibitor is easily deactivated by heat, and can be eaten widely after boiling. In this way, the beans were boiled or germinated to eat, like as chongkukjang or tofu and soybean sprout (*kongnamul*). And soybean sometimes the remaining cooked beans were stored in a room for several days or stored under the roof for drying after making blocks (Figure 16.1). Most of the leftover boiled beans decompose and cannot be eaten, however, some in winter smell strange even after a few days, but it is fine to eat without getting stomach discomfort. This is the primitive chongkukjang and meju, In other words, chongkukjang or meju started where beans were grown and where beans were boiled and used as food. It cannot be said that chongkukjang can be born in a soybean-free land like China.

16.2.1 Native Country of Soybean

Tracing the origin of such rarely propagated pulses or crops has been accomplished through following four analyses: first, classical biological analysis based on documentary analyses since long ago (Fukuda 1933; Hymowitz 1970); second, it can sometimes be identified through literature records such as Chinese books (Jeminyosul (齊民要術, *Qiminyaoshu*) by Ka S. H (賈思勰) 532–549) and many reports (Lee 2022); third, carbonized soybean as a relic of soybeans on Korean Peninsula (Lee and Park 2006); and finally recent developments in genomic analysis of soybeans (Kang et al. 2014; Kim et al. 2010).

As a biological approach, the distribution of wild species is used as an essential indicator for deducing the birthplace of crops in botany. Russian botanist Vavilov stated that the region with abundant crop variation and primitive dominance is the place of plant origin (Vavilov 1926). Carbonized soybean, a relic of soybeans, have been frequently excavated from the Neolithic Age to the Bronze Age sites on the Korean Peninsula (Lee and Park 2006). Even dried soybean relics, which are estimated to have been around from the 3rd-century AD, have been found, and it is presumed to be dried boiled soybean which would be 'meju' (Park 2018; Shin and Jeong 2015) [and see following text] or chongkukjang (Chung 2018). Recent developments in genome analysis technology have scientifically confirmed such claims (Kim et al. 2010). Korean soybeans were found to be located in the early part of the evolutionary tree, like as African in human evolutionary tree (Kwon et al. 2023).

The results of the analyses through these four methods, academia defines the place where wild, intermediate, and cultivated soybeans grow in one place as the plant's place of origin, and the Korean Peninsula and Manchuria as the territory of Kochosun and Kokuryo correspond to this (Fukuda 1933; Hymowitz 1970; Kwon et al. 2017). Kochosun and Kokuryo has its origin on Liao River-based culture. In these respects, Korea, Manchuria and the Korean Peninsula are considered to be the origins of soybeans (Kwon et al. 2017). These insights showed that soybeans were native to Korea.

Even though the origin of the beans is not China but Manchuria (Kang et al. 2014; Kim et al. 2010), some people claim that chongkukjang originated in China because it appears in some Chinese documents (Jeong and Min 2019). It is, however, an incorrect claim since it is made without understanding the gastronomical, agricultural and scientific backgrounds behind them. If we read the Chinese character book carefully, we can see that Chongkukjang has rather entered China. It was not until 2,000 years ago that bean products, such as primitive chongkukjang, were introduced into China. In Bakmulji (博物誌) (Jeong and Min 2019), a compendium written by Zhang Hua (張華, 232–300 AD) during the Chin (晉) dynasty, a method for making chongkukjang (*si*, 豉) in foreign countries is described, claiming that the recipe was passed down from barbarians by Kang-Baik (康伯).

Thus, soybeans were cultivated and became common food for Koreans gastro-geographically (Kim and Ryu 2018), and Koreans have been able to develop soybean-based foods such as chongkukjang, doenjang, kanjang, and tofu since ancient times.

16.3 BASICS FOR LITERATURE REVIEW OF CHONGKUKJANG

16.3.1 Key Words Represent Chongkukjang

Gastro-geographically, chongkukjang is simply found where soybean was grown and where soybean was cooked as a food. The actual food material called chongkukjang has been around for a long time. In a country with letters, of course, letters representing chongkukjang will occur later. Nevertheless, Kochoson and Kokuryo had no choice but to analyze ancient literature data by examining the records of Chinese characters because they did not have their own letters. In Chinese characters, word '*si* (豉)' appear as Chinese characters representing chongkukjang many times (Chung 2018). Here, in ancient times the '*si*' expresses today's chongkukjang and meju at the same time. Originally chongkukjang and meju were prepared with the same procedure after boiling, however, meju was made by drying after crushing boiled soybean and making brick-shaped blocks, while chongkukjang was non-dried moisture-rich product (Figure 16.1). Chongkukjang was eaten by boiling as a chongkukjang chigae (Figure 16.2), however, meju was not eaten right way. Meju was used for further fermenting to kanjang and doenjang by adding saline solution (Shin and Jeong 2015).

In addition to this, the fact that chongkukjang is recorded in Chinese character dictionaries for Koreans like Hunmongjahoe (訓蒙字會, Naegakmunkobon, by Choi SJ (崔世珍), 1527) demonstrates that Koreans have been eating the dish for a very long time, because the food itself must have existed prior to the point when written records were first kept. There were many names for chongkukjang in pure Korean such as 'jyeonkuk (젼국)', 'cheongkuk (청국)', 'chyeonkuk (쳔국)', and 'chyeongkuk (쳥국)' in Hunmongjahoe. A closer examination reveals that there is no logic to these coined words, and they were simply created by different people based on personal preference.

Early records such as Hunmongjahoe show that chongkukjang was originally not classified as jang (Kwon et al. 2019), a type of sauce to enjoy food, but just a type of food with boiled bean. Therefore, since chongkukjang did not have the suffix word "jang (醬)" in Korean before 18th century, the word simple '*si* (豉)' and Korean words such as "jeonkuk or jeonkyuk (전국, 젼국)", "chongkuk or chyongkyuk (청국, 쳥국)" indicate chongkukjang. In documents after 1443, when Korean script (Hankul, ᄒᆞᆫ글) was invented, '*si*' is written in Korean script, which shows that chongkukjang was simply pronounced 'jyeonkuk (젼국)', 'chyeongkuk (쳥국)', or 'chyeonkuk (쳔국)' (Hunmunjahoe), (Sauikyeongheombang, 四醫經驗方, 1608), and (Donguibogam, 東醫寶鑑, by Heo J (許俊), 1610) (Chung 2018; Kwon et al. 2019). It is quite scientific that our ancestors did not include chongkukjang in a jang such as kanjang and doenjang. Even today, chongkukjang, especially natto is used to eat as food like nutritional source rather than to season or taste food.

Although there were pure Korean words for chongkukjang, it does not appear in written records as Hankul even after the advent of Hankul. Instead, 'cheongkuk' was recognized as a type of 'jang' and gradually became known as chongkukjang as Chinese characters by adding the character 'jang (醬)' as a suffix from that point onwards, and when it became possible to be called and to write records in Hankul, Chinese characters were coined to refer to each Hankul sound. For this reason, many other different Chinese characters which refer to chongkukjang as jeonsijang (煎豉醬), jeonkukjang (戰國醬), chongkukjang (靑局醬), or chyeongkukdyang/chongkukjang (靑麴醬), and chongtanjang (淸湯醬) (Ojuyeonmun-jangjeon-Sanko, 五洲衍文長箋散稿, by Lee KG (李圭景), 1788–1849); Bakhaetongko (博海通攷, beginning of the eighteenth century); Saryubakhae (事類博解, by Lee G (李公), 1855) (Chung 2018), do appear in Chinese characters' books because there were no official Chinese characters for chongkukjang (Chung 2018). Nevertheless, it appears that different individuals developed their own characters for it based on their interpretation of the word's meaning, which led to the variations and mistakes.

Thus when we study literature reviews of chongkukjang from ancient records, we must study "*si*" character firstly regardless whether it represents chongkukjang or meju, and then it is necessary to study the Chinese characters, which are the expressions of the Hankul chongkukjang in detail after the 18th century.

16.3.2 Literature Basics Related to Chongkukjang

There are many records over thousands of years which refer to chongkukjang using the word '*si*' in Chinese character records (Chung 2018). However, looking at such records does not present an accurate picture of the history of chongkukjang. This can only be achieved through an examination of the records about "*si*". Korea is a Liao River-based culture that has its origins in Kochosun and Kokuryo. Chung reviewed chongkukjang's history very comprehensively and scientifically (Chung 2018). The painting in Deokheungri Cave (408 AD) carrying the 'yeomsi (鹽豉)' and '*si*' referred to salted meju or chongkukjang were found (Chung 2018; Woo 2018). Three Kingdoms History (三國史記, Kim BS (金富軾), 1145), recording the period from the first century to 668 AD during Kokuryo, Baekje (百濟), and Silla (新羅), described an incident the food, yeomsi, was not available, because the road to Woongjin was blocked off, and Haedong History (海東繹史, by Han CY (韓致奫), 1800s), describing the customs during the Balhae (渤海) period (698–926 AD), recorded '*si*' included in a list of traditional wedding ceremony foods (Chung 2018). The fact that chongkukjang was around during the Kokuryo, Baekje, and Silla periods, alongside evidence that Chaekseong-*si* (柵城豉), was highly regarded in the customs of Balhae, which was founded by migrants from Kokuryo, is strong evidence that the soybean were eaten in Korea as chongkukjang before the Three Kingdom period in the first century. It proves that the claim in New Tang's History (新唐書, by Ku YS (歐陽修), Song G(宋祁) et al., China, 1044–1060) that '*si*' came from *Dongi* (東吏) is factual, and if Balhae's Chaekseong-*si* was already well known at this time, it is likely that the food already existed in Korea thousands of years before the publication of Keupchipyun (急就篇, by 史遊, 40 BC), the first Chinese document that contains a record of '*si*'.

Chung's paper (Chung 2018) also uses records contained in ancient documents to make the case how chongkukjang was used in Korean bapsang (Kwon et al. 2023) during the Koryo kingdom. Based on records contained in Haedong History, Koryo Dynasty History (高麗史, by Kim JS (金宗瑞) and Chung IJ (鄭麟趾), 1613) and Dongkukyisangkukjip (東國李相國集, Lee GB (李奎報), 1241), chongkukjang was put into kuk during the Koryo Kingdom (916–1392), and also notes that chongkukjang was an important part of providing famine relief to the starving people in Koryo Dynasty period. Considering that chongkukjang is still put in making kuk or chigae (Figure 16.2), it is the same as eating chongkukjang in the past and now. There are also many references to chongkukjang and '*si*' in documents from the Chosun dynasty (Chung 2018; Kwon et al. 2019) which demonstrate that it was a very common food consumed among people.

As in chongkukjang rather than *si*, references written in Korean script begin to appear in documents about 100 years after the advent of Hankul, with the first record found in Hunmongjahoe. In addition to Hunmongjahoe, 'jyeonkukjyang' in Eumsikdimibang (閨壼是議方, by Jang Kye-hyang (張桂香), 1670), 'chyeongkukdyang' in Bakhaetongko, and 'chyeongkukjyang' in Jongniuikwe (整理儀軌, 1796) were used. 'Cheongtangjang (淸湯醬)' in Siuijeonseo (是議全書, 1890) is referred to as because it was mainly eaten as a chigae (湯, see Figure 16.2).

It seems that even in the 18th century, there should be many people who incorrectly talked about chongkukjang, just as today's Chinese scholars talking about its wrong history, and spreading misinformation about chongkukjang due to their lack of knowledge; the widespread misconception is that chongkukjang was developed during wartime because of easy preparation steps same as it is today. Regarding this phenomenon, Lee (李圭景) in Oju-Yeonmunjangjeon-Sanko, wrote that "Somebody claimed that '*si*' is currently called jeonkukjang, and was given this name because it can be made overnight and was easy to prepare in the palace during times of war. However, there is no evidence to support this claim (豉者。今俗所謂戰國醬也。此物成於一夜之間。而戰國時。軍中以易制而食。名之云。蓋無稽之說也)" (Chung 2018; Kwon et al. 2019). The book also states that chongkukjang only appeared after the Han (漢) dynasty (則蓋自漢以來。始有之也). It is a very easy and groundless argument that the some Chinese character in the chongkukjang contains the word 'war (戰)' and that the chongkukjang was born in relation to the war. Because the chongkukjang is called 'jeonkukjang in Korean, they simply used the Chinese character 'jeonkukjang' by borrowing the sound 'jeon (戰, war)' without any meaning. It appears that this sentence was used as the basis for the belief that chongkukjang was created during wartime, without giving it a broad interpretation. However, the theory that chongkukjang was called jeonkukjang (戰國醬) because it was easy to prepare during times of war is unsupported by any evidence.

16.4 HISTORY OF CHONGKUKJANG

16.4.1 Birth of Chongkukjang

As with the history of all foods, ethnic food develops through the migration and progress of mankind (*Homo sapiens*) to some geographical different places, and the struggle of finding taste based on agricultural and ecological background. As mentioned earlier, the direction of food development is how to eat all living things that are available in the natural environment without dying and without an upset stomach?, how can they eat it deliciously?, and how can they eat it later? During these struggles, scientific wisdom is accumulated and developed (Kwon et al. 2023; Mandelblatt 2012). Manchuria, including the Korean Peninsula, is the origin of soybeans, so Koreans advanced substantial efforts to use them efficiently and deliciously. For this reason, chongkukjang, like other fermented foods such as kimchi (Kwon et al. 2025), is a food that the ancestors of the Korean people found by accidental discovery after endless efforts. Geo-historically this was a serendipitous finding to the ancient people who cultivated soybean. In other words, the first food found in a country with beans is chongkukjang, which can never occur in a place with no beans. In a way, chongkukjang is inevitably born in a region with beans gastro-geographically.

Soybeans cannot be eaten raw because they contain a trypsin inhibitor. It can be eaten, once boiled, when the trypsin inhibitor is inactivated, and does not cause indigestion. Therefore, soybeans are harvested in Autumn and eaten once boiled, so they are often left in the open while boiling beans. Usually, the leftover of boiled beans cannot be eaten because of decaying, however, in the cold winter it can be eaten after a few days. This is called as chongkukjang because it is the simplest form of fermentation. The appearance and smell of the cooked beans changed, and it resembled spoiled beans, but it was considered as a waste to throw it away. Not only does chongkukjang not cause stomach upset, it also seems to taste even better. Despite its initial discovery being accidental, chongkukjang gradually became a deliberately made food due to its unique taste and texture. It is not so high technology adapted from outside country, but the result of serendipity from wisdom of ancient peoples naturally.

16.4.2 Historical Backgrounds

Many papers showed that Korea has different track from China in terms of biological difference and human migration, civilization and history, language, culture and food (Kwon et al.

2025; Woo 2018). Korea and people of Korea have their own agriculture and culture with more than 5000 years of history. Although Korea is close to China in geographic and ethnologic terms, the country possesses its own unique migration history and culture. Old Koreans migrated different ways from China and constructed different culture. Korea was inhabited by an Altaic people (of Mongolian descendants) and developed its own distinct language (Woo 2018). Chinese culture can be represented by "Huanghe Civilization" (黃河文明, Yellow River culture), while Korean culture is a "Liaohe Civilization" (遼河文明, Yoha River culture) from the plains of Manchuria (Woo 2018). These two cultures differ in biological, archaeological, linguistic, and cultural terms. Also difference between two civilization is evident even when you look at relics. The progression of Korean history passes through Kochosun and the Three Kingdom period of Kokuryo, Baekje, and Silla, which was followed by Koryo (高麗) and the Chosun (朝鮮) dynasty. During this period, the country remained separate from China, which led to its distinct history and culture. The Korean language is also part of the Altaic family which includes Manchuria (formerly Kokuryo), Mongolia, Japan, Hungary, and Finland (Robbeets et al. 2021). This linguistic system is quite different from that of China and other Indo-European languages. Korea also has its own unique food culture which greatly differs from China (Kwon et al. 2025; Robbeets et al. 2021). Furthermore genetic analysis of human genome, languages, foods and cultures are very conservative and not affected much even from invasion by neighboring countries. That is the reason why food culture is not changed even when Korea was invaded from China or Mongolia, or even Japan. It is totally not correct that source of Korean food culture is mainly from China or Korea food history is affected by China (Kwon et al. 2025).

16.4.3 Agricultural Backgrounds

Although China was developed on the basis of its vast plains centering on the Yellow River and the Yangtze River, ancient Korea had another agricultural environment under different geographical conditions than China. By starting to cultivate rice, barley and sorghum, they were obtain basic nutrition, however, they always wondered how to eat bap (rice) deliciously (Kwon et al. 2025). In China, oil was easily obtained from animals such as pigs and fish, and plants, so the food culture developed by frying or roasting food in oil and allowing it to be eaten for later. Gastro-historically, Korea did not have much oil unlike China, so it had to find other ways to taste food rather than taste it with oil. In fact, there was an easy way to get oil in Korea as well. However, there was no need to squeeze too much oil from plant in Korea. When chilies were harvested in China, oil was first extracted, but when peppers were harvested in Korea, they were dried and ground it to make powder. If oil was needed in Korea, soybean oil could be obtained as much as possible from soybeans, but they never have got the oil from soy.

Soybean, which is native to Korea, is an important resource for replenishing proteins that are likely to be lacking in Korean diets. They had to eat beans somehow to supplement protein as tofu, chongkukjang, and later they started cultivating beans because they had to make doenjang, kanjang, with soybean.

16.4.4 Literature Backgrounds

As described previously, the Chinese character '*si* (豉)' was originally used to refer to chongkukjang. Accordingly, analysis of the history of chongkukjang should focus on this character (*si*), as it was the original word used for chongkukjang prior to the advent of Korean script. There are hundreds records on '*si*' in Chinese book and Korean book written in Chinese characters (Chung 2018).

The oldest Chinese record available is Kupchwipyun (急就篇, China, BC 40) from the Han (漢) dynasty. In this document, there is a reference to a dish called 'yeomsi', which is chongkukjang as painted in Deokheungri Cave, while in Solmunhaeja (說文解字, by Hur S (许慎), China, 100 AD), a dish called baeyeomyusuk (配鹽幽菽) is described which is made by seasoning soybeans with salt and sealing them away for a long time. '*Si*' is also referred to as *baeyeomyusuk* in Chinese book, Bonchokangmok (本草綱目, by Li SJ (李時珍), China, 1518–1593). When put together, this evidence suggests that '*si*' existed in China as early as 40 BC and that it was introduced to China by a person called Kang-Baik from a barbarian tribe like as Korea. The barbarian referred to here cannot speak of any other country except *Dongi* (東吏), Northeast barbarian in Manchuria, who appears in Kochosun (古朝鮮) or Kokuryo (高句麗) which represent ancient Korea. And also this barbarian can be from Southeastern country such as Indonesia or Malaysia (*Namman*, 南蠻). However, Yonkamlyuham (淵監類函 by Chang Y (張英)) states that the food is originally from Shandong Province (山東, Northeast China), it is likely that the origin is closer to Korea. In biological, Chinese chongkukjang ferments through the same *Bacillus* bacteria as Korea, while tempeh similar to chongkukjang ferments through mold or yeast. This suggests that this old document is rooted in truth. Accordingly, it can be extrapolated that chongkukjang originally came from a foreign land and was introduced to China more than 2000 years ago. This can also be inferred from the fact that a kind of chongkukjang called '*si*' has been around for a long time in China as well (Kwon et al. 2019).

16.5 DISTORTIONS ABOUT CHONGKUKJANG

16.5.1 Other Misconceptions About Chongkukjang

Chongkukjang was first eaten thousands of years ago in Korea, where soybeans began to be grown and cultivated, and entered China in about 2000 years ago for consumption, and subsequently soybeans entered to be cultivated in Southeast Asia or Southern Asia and developed into soybean fermented foods like as tempeh, thuanao or kinema (Tamang 2015). However, all others are misconceptions of chongkukjang such as "chongkukjang is fermented with mold and the suggestion that it refers to mold in the word chongkukjang (淸麴醬, the letter 麴 means mold), chongkukjang was originally referred as 'jang (醬)' and '*si*' is only meju". All these misconceptions of chongkukjang came in the process of rationalizing the main misconception: "it was made to consume the beans quickly during the war with the Qing Dynasty" (Donga Encyclopedia 1982; National Institute of Korean Language 1999). Lee Seong-woo (Lee 1991a, 1991b; Park 2015) and his followers mainly created these distortions based on the belief that chongkukjang must have been written in Chinese characters as jeonsijang (煎豉醬), jeonkukjang (戰國醬), chongkukjang (靑局醬), or chongkukjang (靑國醬), and refuse to recognize records written in '*si* (豉)' or Korean script (젼국, 쳥국, 청국, 전국). In fact, linguistically, none of this is true when there are so many Chinese characters with the same object.

Moreover, they began to make the name of chongkukjang into another wrong Chinese characters, chongkukjang (淸麴醬), with the intention of adding science in addition to the original wrong Chinese characters of chongkukjang (靑國醬) by modifying some Chinese characters to the endeavor with introducing the character for blue mold (kuk, 麴) which was involved in fermentation. In science, this is very clumsy because chongkukjang is not fermented by mold but bacteria (菌). These kinds of unscientific misinformation have been widely disseminated across a number of websites, including the Encyclopedia without scientific verification of the truth (Park 2015; Donga Encyclopedia 1982; National Institute of Korean Language 1999), and have been compounded by the fact that people are easily persuaded by the arguments of scholars in Chinese characters with scientific evidence.

Another serious misconception is that chongkukjang first emerged from the Jungbosallimkyongje after the 18th century. Lee most likely wrote this entry after seeing records of jeonsijang (煎豉醬) and jeonkukjang (戰國醬) in Jungbosallimkyongje (Lee 1991a). However, this reference is very likely to say this, assuming that chongkukjang was spread from other countries without considering the scientifically created environment of its original invention. This author makes no mention of '*si*', the original word that was used to refer to chongkukjang in written records (Chung 2018).

Even if it is correct that the words jeonsijang and jeonkukjang refer to chongkukjang, the statement that the first reference to chongkukjang in Korea was in Jungbosallimkyongje is clearly and totally incorrect. Jeonkukjang (戰國醬) is mentioned in a number of documents produced before Jungbosallimkyongje, for example, Hujaesonsaengbyuljip (厚齋先生別集, by Kimkan (金幹), 1646–1732), and Chisaengyoram (治生要覽, by Kang-wa (强窩), 1691), and both jeongukjang and chyeonggukjang also feature in Bakhaetongko. Despite the fact that the Chinese character cheong (淸) in chongkukjang, which represents the Qing dynasty, was never used in any documents from the Chosun dynasty, Lee even claims that chongkukjang was given its name because it was introduced to Korea via the Qing dynasty (Lee 1991a). Without even conducting a proper examination of these ancient documents, author has perhaps advanced unscientific information that has distorted the truth about Korean food culture. The author also makes no mention of the fact that jyeonkukjang was written in Korean script in Eumsikdimibang (1670), which was released long before the publication of Jungbosallimkyongje. There is no evidence in ancient records to support these entries in the Encyclopedia of Korean Culture (Lee 1991a, 1991b). Claims must be advanced with scholarly insights. Since publications sponsored by government organization (The Academy of Korea Studies) contain this inaccurate information, the same wrong claims are more harmful when repeated in privately printed materials. Most of the misunderstandings about chongkukjang have been made and sustained by scholars familiar with Chinese characters. Furthermore, since many of these scholars are university professors or authors of books, even though they lack scientific relevance of food, people tend to regard them as right. Recently, a paper was published that translated ancient documents on the history of cheonggukjang and presented this as evidence to refute high-profile false theories about food (Chung 2018; Kwon et al. 2019). It was a very well-written paper based on written evidence (Chung 2018).

16.5.2 The Reasons Why There Is So Much Misinformation About Chongkukjang

The purpose of telling about the reasons for having so many misinformation about chongkukjang is to avoid making the same errors in other foods and food systems to advance the scientific value of indigenous food systems.

First, the main reason is that people who talk about the history of chongkukjang discuss the history by relying only on Chinese characters written in China or Korea without scientifically knowing the biological characteristics such as the type and origin of soybeans, the agricultural characteristics of soybean use, and the food characteristics of chongkukjang fermentation.

This is why caution is required when speaking about Korean history as seen through Chinese character records. Focusing purely on analyzing Chinese character records while ignoring what had actually happened and speaking in Korean language and context will naturally lead to a lot of built in bias and errors using old Chinese character literature. This means the circumstances surrounding such records need to be taken into account to obtain an accurate picture of Korean food culture scientifically. However, since Korea was liberated from Japan, a number of scholars including Lee Seong-woo have adopted this approach without a complete understanding about the nature of foods; Korea's food culture and history have been severely distorted (Ju 1999). In scientific, anthropologic, and agricultural terms, birth of chongkukjang occurs naturally and inevitably in original countries with soybeans.

Second, gastro-geographically there is its own philosophy and science in all foods in the world, and the philosophy or science of Korean foods must be understood which have never traveled around the world, and they have advanced locally without knowing the changes in language and how it is written. Not all ethnic foods in the world have been spread from other countries by technology transfer, but been rather spontaneously developed during survival strategies of local communities by eating living things in that local regional ecology. Therefore it is false rationale that Korean food would have come from China, forgetting that it developed naturally and independently based on local ecology and conditions.

As stated previously, Korean culture is Liao River-based (Woo 2018), while China is rooted in Yellow River culture. Therefore, Korean food or culture should not be similar to or never be derived from Chinese food, but should be looked for the similarities in Altaic attributes such as from Yojin (Jurchen), Keoran (Khitan), Malkal (Gokturks), Balhae, and Mongols. Also, the etymology of Korean words can be found in Altaic tribes. Although languages that belong to this family still exist in Hungary, Finland, and Mongolia, many of these tribes have unfortunately disappeared (Robbeets et al. 2021). The roots of the Korean language should be understood through Korean experience which was based on these tribes. In other words, only attempting to trace the food history or etymology of a word like chongkukjang in Chinese book has limited scientific value and relevance.

Third, the authors do not know the principle of generation or use of Chinese characters, and are convinced that already written Chinese characters will have expressed the exact objective, so they think Chinese characters are flawless. Chinese characters are created and added even though they know that they cannot express accurate objects or facts when they are made.

In particular, when translating Korean native words that are not in China into Chinese characters, it is necessary to understand and interpret that they can be translated in various ways. It is likely that it was simply translated by copying the sounds of chongkukjang in the Korean languages (젼시 or 뎐시, 전국 or 젼국, 청국 or 쳥국) and recorded them as jeonsijang (煎豉醬), jeonkukjang (戰國醬), chongkukjang (靑局醬), or chongkukjang (靑國醬). Since the sound is borrowed and written in Chinese characters, any meaning should not be given or added to the translated Chinese characters. If the old literature is written in many Chinese characters, mistakes occur a lot because it is almost impossible to accurately convey the meaning and sound.

Fourth, because Chinese characters are meaningful characters in the world, inevitably most misunderstandings could have taken place during the interpretation. Therefore, there is a lot of room for interpretation in various ways. However, because of difficulties to learn, people who use Chinese characters are envious of them as great people like having superior attitude to knowing Chinese characters. Sometimes, even if people who know Chinese characters can misinterpret due to the limitations of their scientific knowledge and they easily believe that those who speak Chinese characters will tell the truth.

Once the word 'chongkukjang' is written in Chinese characters, many scholars try to interpret the meaning of Chinese characters first. In fact, it is one of the easiest way to talk about the history of chongkukjang by interpreting the meaning of Chinese characters rather than gastro-historical thinking of history or culture scientifically. Even-though it is the easiest but wrong way, when speaking in the meaning of Chinese characters, they are often envied as a learned scholar. By exploiting this phenomenon, they create many wrongly interpreted theories in this way. Thus wrong history that chongkukjang was first made during war, as it could be prepared quickly, and chongkukjang was introduced to Korea via the Qing dynasty was creation from spread of misinformation of this fermented food. This is the reason why many misconceptions were made when they trace the history of chongkukjang by examining records of chongkukjang (淸國醬) or similar Chinese characters, instead of focusing on records about '*si*'.

Fifth, for thousands of years, Eastern philosophy has had no climate to refute his teacher's theory or philosophy. There has already been a climate in the field of science and technology that recognized that discovering theories or advancing findings different, was considered outperforming our ancestors, and as result in the field of humanities, there was a cultural climate that followed the teacher's theory with a convincing bias. Rather, even if there is data that contradicts the teacher's theory, there is tendency to waste a lot of time refuting it and solidifying the teacher's theory.

Despite the existence of clear records that refer to chongkukjang (豉) or kimchi (菹) (Kwon et al. 2025) thousands of years ago, the misconception that only the Chinese character-based words chongkukjang (淸國醬) and chimchae (沈菜) refer to these foods in documents led some to believe that chongkukjang have only been around in Korea for 500 years. Despite a series of literature that chongkukjang had roots in Korea, they ignored these data and literature and did not accept their arguments, but rather created various unscientific hypotheses to rationalize their false beliefs with confirmatory bias. At one time, Korea treated these scholars as scholars who knew our food culture and history well. This distortion is the same as in kimchi, kochujang and chilli (Kwon et al. 2025).

Finally, although King Sejong created Hangul in 1443 and wanted it to be widely used by the people, the *yangban* (aristocrats) and bureaucrats treated Hangul as a commoner or women language, and insisted on writing Chinese characters, preventing Hangul from systematically developing. If the chongkukjang analysis in Hangul had been steadily used and developed, these Chinese characters could not have distorted Hangul language and writing.

The chongkukjang called as 'jyeonkuk', 'chyeonkuk', or 'chyeongkuk' were referred to as '*si*' in written records prior to the advent of Korean script, while a number of different characters including 煎豉醬, 戰國醬, 靑局醬, and 靑麴醬 were written after the introduction of Korean script because a new Chinese character expression was required to match the phonetic sound of chongkukjang. This is why Lee's claim

(Lee 1991a, 1991b) that chongkukjang did not exist in Korea until these character-based words appeared in documents is false, and this also explains why the history of foods such as red chili, kimchi, and chongkukjang have been misinterpreted. Advancing history of these food needs careful scientific analysis.

16.5.3 The Reason Why 'Si (豉)' Refer to Meju Nowadays

In Korean dictionaries that clearly indicate the meaning of Chinese characters in Korean, such as Hunmongjahoe and Naehun (內訓, by Sohyewanghu (昭惠王后), 1475), '*si*' is written as chongkukjang, but most of today's dictionaries only indicate it as meju (dried fermented soybean block) (Moon 1938, 1957). That was one of the reasons why they distorted the chongkukjang's history. So, does '*si*' really refer to only meju?

Meju, called as 'myejo', 'myojo', 'myoju' or 'myeoju' in ancient times in Hunmongjahoe and Naehun (Chung 2018; Kwon et al. 2019), is process for dried fermented soybean block for fermenting soybean, is a food intermediate for the purpose of making kanjang, doenjang, and kochujang (Shin 2022; Shin and Jeong 2015). Meju and chongkukjang are fundamentally similar to boiling beans and placing them in a room. Fermentation took place in meju during drying from the surface into the surface of cracks which was formed during drying, while chongkukjang is fermented in a wet state by bacteria. So, meju cannot be eaten in that state, but fermented for a long time by mold or yeast, and then used to make other jangs, whereas chongkukjang is a food that can be eaten immediately. So the character '*si*' was used to refer to both chongkukjang and meju in documents produced prior to the advent of Korean script in 1443, while the phonetic pronunciation of chongkukjang was jyeonkuk, chyeongkuk, or chyeonkuk in early developed book as Hunmongjahoe (Chung 2018).

While the original name for chongkukjang was cheongkuk or jeonkuk, at some point of publishing Jungbosalimkyongje, the suffix 'jang' was added to create the word that is used today. Yu (柳重臨), the author of Jungbosalimkyongje began to view chongkukjang as another type of jang alongside doenjang, kanjang, and kochujang, and the names chyeongkukdyang and chyeongkukjyang emerged, and because there was a need to put this name into Chinese characters, 'jang' was added as a suffix and the word was recorded as jeonsijang (煎豉醬), jeonkukjang (戰國醬), chongkukjang (靑局醬), or chongkukjang (靑麴醬) from the 1600s, especially in Jungbosallimkyongje. As chongkukjang was written in different Chinese characters, the character 'si (豉)' originally representing both chongkukjang and meju at the same time gradually began to remain as characters representing only meju exclusively (Choi 2017).

There are a number of references that '*si*' referred to meju (myejo and myojo): Yeosaseoseonhae (女四書諺解, by Lee DS (李德壽), 1737), Jungbosallim-kyongje, Uihwi (宜彙, 1871), Yeoyudangjeonseo (與猶堂全書, by Chung YY (丁若鏞), 1762–1836), and Mulmyeongko (物名攷, by 丁若鏞, 1820), and referred to as myeoju/myoju: Jaryujuseok (字類註釋, Chung YY (鄭允容), 1856) and Jajeonseokyo (字典釋要, Ji SY (池錫永), 1909) (Chung 2018; Kwon et al. 2019). Despite these, as Yu began to use chongkukjang in Chinese characters (煎豉醬, 戰國醬) in the Jeungbosallimkyongje, the '*si*' began to be misunderstood that it was no longer cheonggukjang, but only meju. Particular distinction was drawn between chongkukjang (煎豉醬) and meju (豉) in Jungbosallimkyongje. Even in kanjang and doenjang, because there is no Chinese expression in pure Korean, it is often incorrectly marked with '*si*' when printing doenjang after the 1600s. Uirimchwalyo (醫林撮要, Yang YS (楊禮壽), 1500–1600s) and Donguibogam (東醫寶鑑) state that '*si* (豉)' refers to meju (末醬), while some later authors may have mixed up the expressions for chongkukjang and meju by writing that '*si*' refers to doenjang in Chosunmussangsinsikyorijebeop (朝鮮無雙新式料理製法, by Lee YK (李用基), 1924) and Hanilseonsinokpyeon (韓日鮮新玉篇, 1935), or both meju and doenjang in Hanhandaesajeon (漢韓大事典, 1966) (Chung 2018).

Those who do not know these scientific facts conclude that '*si*' only represents meju and ignore the '*si*' representing chongkukjang in ancient texts. However, those who claim that all records of the character '*si*' after the seventeenth century refer only to meju is unsound from a literary perspective and entails the risk of mistranslation. It is likely that some writers continued to prefer to use the original expression '*si*' for chongkukjang. Food science analysis is required to determine whether each use of the character '*si*' refers to chongkukjang or meju.

16.6 THE HEALTHY BENEFITS OF CHONGKUKJANG

As fermented foods are now believed to be healthy in most countries, there is greater interest in the functions of these foods. In Korea, a lot of research is being conducted into the functions of chongkukjang (Jung et al. 2022; Park 2018). In fact, the nutritional value of chongkukjang is mostly derived from the nutritional value of soybeans, which has been well studied. There has been a concern about a protein shortage in a vegetable-oriented diet like Korea, but this problem has been solved by eating soybeans. Soybeans are often problematic in terms of nutrition, such a lack of essential amino acids as sulfur-containing amino acids, and a lack of some vitamins (Yang et al. 2011). However, during the fermentation of chongkukjang, lots of metabolites that are not found in soybeans are generated, and the metabolites produced in this way are beneficial substances that have a healthy effect on our body (Yang et al. 2011). Generally, vitamin K_2 and vitamin B_{12} are rarely present in soybeans (Park 2018). However, they are produced by microorganisms during fermentation (Kurosu 2011). In addition, isoflavones, which are functional components of soybeans, exist as β-glucoside conjugates and have a low absorption rate. However, when soybeans are fermented into chongkukjang, they are converted into aglycones such as daidzein, glycitein, and genistein by the β-glucosidase of microorganisms, increasing the absorption rate, further enhancing physiological activity (Yang et al. 2011).

There are many records in ancient literature that chongkukjang is good for various symptoms, probably these records were written after epidemiological observations. Ancient records in Sikuisimkam (食醫心鑑, by Jam E (昝殷), 800) show that chongkukjang was prescribed for indigestion and stroke patients. There are also record that claim chongkukjang was used to treat illnesses in Hyangyakjipseongbang (鄉藥集成方, by Yu HT (兪孝通) et al., 1433). Yonkamlyuham (淵鑒類函, by Chang Y (張英), 1637–1708) repeats the content about '*si*' found in Bakmulji claiming that '*si*' originated from Shandong Province is good for the body. In medical books, Hyangyakgukeupbang (鄉藥救急方, 1236) and Hyangyakjipseongbang (鄉藥集成方, 1433) also state that putting chongkukjang to *kuk* can help to relieve fever and also that '*si*' was used to treat other illnesses (Kwon et al. 2019). Since life science was not developed at the time, there is a lack of scientific descriptions on what diseases and how the chongkukjang was good for, and only description for what symptoms it was good for.

Chongkukjang is rich in dietary fiber and has a low glycemic index value due to indigestible saccharides, making it an excellent food for blood glucose management and preventing arteriosclerosis by reducing triglycerides and total cholesterol absorption in the intestine and promoting lipid excretion through feces (Lee 2005). Chongkukjang effectively prevented diabetic complications, regulated enzymes associated with glucose metabolism in the liver, and increased insulin sensitivity in diabetic animal models (Kwon et al. 2007; Park et al. 2013). Moreover, chongkukjang improves insulin resistance by suppressing α-glucosidase activity in the small intestine (Choi et al. 2016). This is consistent with clinical observation, that the consumption of chongkukjang for eight weeks significantly reduced fasting blood glucose, total cholesterol, and low density cholesterol levels (Shin et al. 2011). The presence of trypsin inhibitor in chongkukjang enhances insulin secretion and sensitivity, which helps to manage blood glucose levels. Protease, a proteolytic enzyme in chongkukjang, suppressed cholesterol synthesis and showed an antithrombotic effect by reducing the inhibitory activity of the 3-hydroxy-3-methylglutaryl coenzyme-A reductase in the body. Clinical test show chongkukjang consumption exhibited an anti-arteriosclerosis effect on overweight and obese adults (Back et al. 2011). Moreover, chongkukjang consumption reduced the ratio of apolipoprotein B (Apo A) and apolipoprotein A1 (apo A1), an index protein associated with atherosclerosis.

Chongkukjang contains large amounts of isoflavone, glycans, peptides, dietary fiber, and microorganisms, which can control inflammation and oxidative stress. When chongkukjang is fermented by *Bacillus subtilis*, it produces thread-like mucoid, poly-γ-glutamic acid (γ-PGA), which exhibits immunomodulatory properties and promotes immunological function (Park 2018). Chongkukjang has considerable levels of γ-PGA, which enhances the brain's insulin sensitivity and improves memory (Jeong et al. 2021) and γ-PGA stimulates interferon-gamma and toll-like receptor 4 secretion, which improves natural killer-cell activity, leading to collective improvements in the immune system (Chang et al. 2005).

As we mentioned, many metabolites are produced during chongkukjang fermentation, which act as a healthy compound in animal model systems and potentially humans. Chongkukjang appeared superior to boiled soybeans in regulating blood pressure as valine and tyrosine produced during fermentation inhibit angiotensin-converting enzyme activity. In hypertensive rats, the administration of chongkukjang effectively reduced blood pressure (Yang et al. 2003). Clinically, the consumption of chongkukjang pills (about 35 g/day) improved obesity and lipid metabolism (apolipoprotein B) (Byun et al. 2016). In particular, Apo A1, an index of lipid metabolism, increased in men and apolipoprotein B significantly decreased in women, confirming the improvement effect of arteriosclerosis indexes (Byun et al. 2016). Furthermore, the administration of chongkukjang in obesity-induced animal models showed an anti-obesity effect and improved lipid metabolism (Choi et al. 2016; Kim et al. 2013; Lee et al. 2011; Soh et al. 2008). Chongkukjang helps to improve allergies (atopic dermatitis) caused by immune sensitivity (Baek et al. 2015; Cho et al. 2019). In an atopic dermatitis animal model, the administration of chongkukjang suppressed the expressions of serum immunoglobulin E, Th (T cell helper) 2 cytokine interleukin-4, and itching-related interleukin-31 cytokines (Kim et al. 2013) that appeared during an excessive immune response. Moreover, chongkukjang pills (27 g/d of fresh chongkukjang) taken for 12 weeks significantly alleviated allergic reactions to histamine Baek et al. 2015).

With the development of life science, it is well known that microbiomes play various roles in the intestine and maintain the health of our body as a signaling messenger. Microorganisms in fermented foods, especially lactic acid bacteria, play a number of roles as probiotics. Microorganisms in chongkukjang are very diverse and large amount: 100 billion germs per 100 g of chongkukjang, which results in beneficial effects on the body and eases intestinal movement (An et al. 2023). *Bacillus* sp. in chongkukjang enhances intestinal function, and insoluble dietary fiber shortens the colon transit time, improving overall intestinal health. In addition, chongkukjang promotes intestinal peristaltic motion and increases Bifidobacteria, which helps suppress the proliferation of harmful bacteria in the intestines (An et al. 2023; Jeong et al. 2020). Furthermore, chongkukjang is known to relieve memory impairment induced by Alzheimer's disease and cerebral ischemia by modulating the gut microbiome brain axis (Jeong et al. 2021).

When consuming chongkukjang, there was a lot of controversy over whether it is better to eat it raw like natto or eat it boiled like chongkukjang chigae (stew). Many people said that it is better to eat it raw by establishing the logic that boiling chongkukjang will kill the probiotics (microbiomes) and it will not have functional benefits for health. As the purpose of eating chongkukjang is to enjoy the bap (rice), it would be better for Koreans to eat it with chongkukjang chigae or stew (Figure 16.2). This is the same as in kimchi chigae. And recent research results indicate that even though microorganisms in fermented foods were killed by boiling in chigae, but bioactive substances including nucleic acids still have health effects as postbiotics like eating raw, this seem to end this controversy. It has been announced that fermented foods work healthily as postbiotics even if they are boiled without being eaten raw (Salminen et al. 2011).

Currently, there is lot of research conducted on what genes are expressed and how they affect our bodies when we eat traditional fermented foods like as kimchi and chongkukjang. In addition, many studies are being conducted on traditional fermented foods at the nutri-epigenomics level. As for the cause of these effects, differences in gene expression in the liver have been observed after eating chongkukjang (Soh et al. 2008). In other words, chongkukjang has an impact on human health by causing changes in genetic expression and also influencing metabolism (Park et al. 2010).

Is chongkukjang safe to eat in deed? Although chongkukjang have been eaten for thousands of years, a study shed light on the mechanism that makes them safe for consumption (Chang 2018). When chongkukjang is fermented, the beneficial bacterium *Bacillus* grows first, and as this bacteria produces certain substances called as bacteriocin, this protein kills off other harmful microorganisms, and allowing the fermentation to proceed. This is the same in kimchi fermentation.

16.7 CONCLUSION

Analyzing ancient documents to understand ethno-geographic and ethno-historical perspective is conducive to providing a more accurate understanding of Korean food culture. However, Korean food culture can only be understood by starting from the assumption that Chinese character records are incomplete or contain a number of errors. Chongkukjang is an indigenous fermented soybean food that has existed in Korea for thousands of years. However, representative false claims are that chongkukjang was first created during war, as it could be prepared quickly and chongkukjang was introduced to Korea via the Qing dynasty. In other words, even though chongkukjang has existed as a fermented soybean food in Korea for almost 2000 years, a number of false claims about chongkukjang have been made, mostly by scholars who are familiar with Chinese characters but know nothing about conceptual rationale of natural science or food science towards understanding of ethno-geographic and ethno-historical perspectives.

Before Hankul was created, chongkukjang was referred to as jyeonkuk (전국), cheongkuk (청국), chyeonkuk (천국), or chyeongkuk (청국) and was written as *'si* (豉)' in official Chinese character records. After the advent of Korean script, chongkukjang was recognized as a kind of jang alongside doenjang, kanjang, and kochujang, which is why the character 'jang (醬)' was added to the end. Accordingly, people created words like 戰國醬, 青局醬, and 青麴醬 by selecting Chinese characters that represent the pronunciation of jyeonkuk, cheongkuk, chyeonkuk, and chyeongkuk and adding the suffix 'jang' to the end of each. Instead of looking at the history of chongkukjang through the character *'si'*, they claim that the only references to chongkukjang in ancient texts are words written in Chinese characters. This theory is rooted in the belief that the history of chongkukjang lies in the numerous variations of Chinese characters made to refer to the dish (煎豉醬, 戰國醬, 青局醬, 青麴醬, 清國醬 and 清麴醬), while ignoring records of *'si'*, the character that was originally used for chongkukjang. But those who do not know these scientific facts conclude that chongkukjang is called jeonkukjang (戰國醬) because its short fermentation period allows it to be prepared quickly during times of war, but also that another name for the dish is chongkukjang (清國醬) because it was introduced from the Qing dynasty. However, there is no evidence in ancient records to support this claims and this claim is very non-scientific.

Even though there are clear differences between the origins and foods of Korea and China, some scholars seek to study the history of Korean culture and foods through documents written in Chinese characters, as there are very few such records written in Korean script. They believe that records written in Chinese characters can account for Korean food culture. Korean food culture is very different from that of China. Instead of placing blind faith in records written in Chinese script to shed light on Korean food culture, it makes more sense to consider which parts of Korean culture were recorded in this way. Many errors like this have been committed in Korean food culture, and they cause great harm. Correcting such misinformation is essential for rectifying the history of Korean food culture. The foundations of these insights can then help us better advance food science and health benefits of chongkukjang based on better defined ethno-geographical and ethno-historical perspectives and deeper insights.

16.7.1 Acknowledgements

This works was done with the support of Sunchang Institute of Jang in part.

16.7.2 Author Contributions

DYK conceptualized and designed manuscript structure and collect the data or wisdom in terms of history and science with the help of SHK, DJJ, and SJJ contributed in surveying many documents, KRC surveyed old literature books and documents, and KS contributed to conceptualization and writing. SHK contributed to writing & editing.

REFERENCES

An, F., J. Wu, Y. Feng, G. Pan, et al. 2023. A systematic review on the flavor of soy-based fermented foods: Core fermentation microbiome, multisensory flavor substances, key enzymes, and metabolic pathways. *Comprehensive Reviews in Food Science and Food Safety* 22:2773–801.

Back, H. I., S. R. Kim, J. A. Yang, M. G. Kim, et al. 2011. Effects of Chungkookjang supplementation on obesity and atherosclerotic indices in overweight/obese subjects: A 12 week, randomized, double blind placebo controlled clinical trial. *Journal of Medicinal Food* 14:532–7.

Baek, H. I., S. Y. Jung, K. C. Ha, H. M. Kim, et al. 2015. Effect of Chongkuk-jang on histamine-induced skin wheal response: A randomized, double-blind, placebo-controlled trial. *Journal of Ethnic Foods* 2:52–7.

Byun, M., O. Yu, Y. S. Cha, and T. Park. 2016. Korean traditional Chungkookjang improves body composition, lipid profiles and atherogenic indices in overweight/obese subjects: A double-blind, randomized, crossover, placebo-controlled clinical trial. *European Journal of Clinical Nutrition* 70:1116–22.

Chang, H. C. 2018. Healthy and safe Korean traditional fermented foods, kimchi and Chongkukjang. *Journal of Ethnic Food* 5:161–6.

Chang, J. H., Y. Y. Shim, S. H. Kim, K. M. Chee, and S. K. Cha. 2005. Fibrinolytic and immunostimulating activities of Bacillus spp. strains isolated from Chungkuk-jang. *Korean Journal of Food Science Technology* 37:255–60.

Chavan, S. P. 2005. *Santara Dynasty: Jainism in Southern Karnataka up to AD 1565*. 1–323. DK Print World.

Cho, B. O., J. Y. Shin, J. S. Kim, D. N. Che, et al. 2019. Soybean fermented with *Bacillus amyloliquefaciens* (Cheonggukjang) ameliorates atopic dermatitis-like skin lesion in mice by suppressing infiltration of mast cells and production of IL-31 cytokine. *Journal of Microbiol BioTechnology* 29:828–37.

Choi, J. H., P. Pichiah, M. J. Kim, and Y. S. Cha. 2016. Cheonggukjang, a soybean paste fermented with *B. licheniformis*-67 prevents weight gain and improves glycemic control in high fat diet induced obese mice. *Journal of Clinical Biochemistry and Nutrition* 59: 31–8.

Choi, Y. J. 2017. A study on the characteristics of the Korean sauce 'Shi (豉)' in ancient Korean literature. *Journal of East Asian Society for Diet Life* 27:684–93.

Chung, K. R. 2018. The history of Chongkukjang (in Korean). *Journal of Korea Contents Society* 18:647–55.

Donga Encyclopedia. 1982. (doopedia: http://www.doopedia.co.kr/doopedia/master/master.do?_method=view&MAS_IDX=101013000860952), Seoul, Korea.

Fukuda, Y. 1933. Cytogenetical studies on the wild and cultivated Manchurin soybean (Glycine L.). *Japanese Botany* 6:489–506.

Hymowitz, T. 1970. On the domestication of the soybean. *Economic Botany* 24:408–21.

Jang, D. J., A. J. Lee, S. A. Kang, S. M. Lee, and D. Y. Kwon. 2016. Does siwonhan-mat represent delicious in Korean foods? *Journal of Ethnic Food* 3:159–62. https://doi.org/10.1016/j.jef.2016.06.002.

Jeong, D. Y., J. W. Daily, G. H. Lee, M. S. Ryu, et al. 2020. Short-term fermented soybeans with *Bacillus amyloliquefaciens* potentiated insulin secretion capacity and improved gut microbiome diversity and intestinal integrity to alleviate Asian Type 2 diabetic symptoms. *Journal of Agriculture and Food Chemistry* 68:13168–78. https://doi.org/10.1021/acs.jafc.9b07962.

Jeong, D. Y., M. S. Ryu, H. J. Yang, and S. Park. 2021. γ-PGA-rich chungkookjang, short-term fermented soybeans: Prevents memory impairment by modulating brain insulin sensitivity, neuro-inflammation, and the Gut–Microbiome–Brain Axis. *Foods* 10:221. https://doi.org/10.3390/foods10020221.

Jeong, Y., and K. D. Min. 2019. The study on the newly found Joseon's printed book of Bakmulji (博物志). 59: 95–117. https://doi.org/10.17004/jrcn.2019.59.004.

Ju, Y. H. 1999. Kimchi: Descendant of Changachi and Chanji. *Pulmuwon Newsletter* 69:18.

Jung, S. J., S. W. Chae, and D. H. Shin. 2022. Fermented foods of Korea and their functionalities. *Fermentation* 8:645. https://doi.org/10.3390/fermentation8110645.

Kang, J. Y., S. K. Kim, M. Y. Kim, P. Lestari, et al. 2014. Genome sequence of mungbean and insights into evolution within *Vigna* species. *Nature Communication* 5:5443. https://doi.org/10.1038/ncomms6443.

Kim, J., J. N. Choi, J. H. Choi, Y. S. Cha, et al. 2013. Effect of fermented soybean product (Cheonggukjang) intake on metabolic parameters in mice fed a high-fat diet. *Molecular Nutrition Food Research* 57:1886–91.

Kim, M., and A. Ryu. 2018. Fermented soybean and foodways of the three kingdoms period. *HankukSangkosaHakbo* 5:165–87.

Kim, M. Y., S. Lee, K. Van, T. H. Kim, et al. 2010. Whole-genome sequencing and intensive analysis of the undomesticated soybean (*Glycine soja* Sieb. and Zucc.) genome. *PNAS* 107:22032–7.

Kim, S. H., J. Ko, D. Y. Kwon, and D. J. Jang. 2023. Korean fermented soybean product, the result of endeavors of ancients for the best taste of Korean diet. *Journal of Ethnic Food* 10:33. https://doi.org/10.1186/s42779-023-00183-61.

Kurosu, M. 2011. Biologically active molecules from soybeans. In *Soybean and Health*, ed. by H. El-Shemy, 208–16. IntechOpen, Rijeka, Croatia.

Kwon, D. Y., K. R. Chung, and D. J. Jang. 2019. The history and science of Chongkukjang, a Korean fermented soybean product. *Journal of Ethnic Food* 6: 5. https://doi.org/10.1186/s42779-019-0004-8.

Kwon, D. Y., J. S. Jang, S. M. Hong, J. E. Lee, et al. 2007. Long-term consumption of fermented soybean-derived Chungkookjang enhances insulinotropic action unlike soybeans in 90% pancreatectomized diabetic rats. *European Journal of Nutrition* 46:44–52.

Kwon, D. Y., S. H. Kim, K. R. Chung, J. W. Daily, and S. Park. 2023. Science and philosophy of Korea traditional foods (K-food). *Journal of Ethnic Food* 10:26. https://doi.org/10.1186/s42779-023-00194-3.

Kwon, D. Y., C. H. Lee, K. R. Chung, J. W. Daily, and S. Park. 2025. Scientific knowledge and wisdom of kimchi: A blessing Korean. *Journal of Ethnic Food* 12:1–11. https://doi.org/10.1186/s42779-025-00269-3.

Kwon, T. W., C. H. Lee, D. H. Shin, D. Y. Kwon, et al. 2017. *Soybean Storytelling*. Committee for the Establishment of a Korean Soybean Museum. Sikanyon, Seoul, Korea.

Lee, C. H. 2022. The onset of agriculture and Northeast Asian neolithic farm. In *Korean Food and Foodways: The Root of Health Functional Foods*, ed. C. H. Lee, 51–70. Springer, Singapore.

Lee, J. J., A. R. Kim, H. Lee, C. H. Kim, et al. 2011. Effects of soybean. Cheonggukjang and doenjang on serum cholesterol level and weight reduction in rats fed a high-fat/high-cholesterol diet. *Korean Journal of Food Preservation* 18: 226–35.

Lee, P. 2005. The Effect of Fermented soybean powder on improvement of constipating patients receiving maintenance Hemo dialysis; Kyung-Pook National University, Daegu.

Lee, S. W. 1991a. Chongkukjang. In *Encyclopedia of Korean Culture*, Academy of Korea Studies, vol. 22. 93, Seoul.

Lee, S. W. 1991b. Kochu (red pepper chilli). In *Encyclopedia of Korean Culture*, Academy of Korea Studies, vol. 2. 643–5, Seoul.

Lee, Y. H. and T. S. Park. 2006. Origin of legumes cultivation in Korean peninsula by viewpoint of excavated grain remains and genetic diversity of legumes. *Korean Journal of Agriculture History* 5:1–31.

Mandelblatt, B. 2012. Geography of food. In *Oxford Handbooks Online*, ed. J. M. Pilcher, 1–23. https://doi.org/10.1093/oxfordhb/9780199729937.013.0009.

Moon, S. Y. (文世榮). 1938. *Chosuneo Sajeon* (조선어사전). Korean Language Dictionary Publisher, Seoul.

Moon, S. Y. 1957. *Keun Sajeon (큰사전, Great Korean Dictionary)*. Korean Language Society, Seoul.

National Institute of Korean Language. 1999. *Standard Korean Language Dictionary*. Doosan Dong-A, Seoul.

Park, J. 2015. Chongkukjang is food in the war because of easy fermentation (in Korean). *Premium Chosun*, March 4. http://premium.chosun.com/site/data/html_dir/2015/03/04/2015030400161.html

Park, M. K., I. H. Cho, S. Lee, H. K. Choi, et al. 2010. Metabolite profiling of Cheonggukjang, a fermented soybean paste, during fermentation by gas chromatography-mass spectrometry and principal component analysis. *Food Chemistry* 22:1313–19.

Park, S. 2018. Cheongkukjang. In *Korean Functional Foods: Composition, Processing and Health*, ed. K. Y. Park, D. Y. Kwon, K. W. Lee, and S. Park, 145–64. Springer, New York.

Park, S., S. Kang, H. J. Kim, D. Y. Kwon and J. W. Daily. 2013. Standardized chungkookjang, short-term fermented soybeans with *Bacillus lichemiformis*, improves cognitive function and insulin resistance in β-amyloid-infused diabetic rats. *FASEB Journal* 27:1079.

Robbeets, M., R. Bouckaert, M. Conte, A. Savelyev, et al. 2021. Triangulation supports agricultural spread of the Transeurasian languages. *Nature* 599:616–21. https://doi.org/10.1038/s41586-021-04108-8.

Salminen, S., M. C. Collado, A. Endo, H. Colin, et al. 2011. The international scientific association of probiotics and prebiotics (ISAPP) consensus statement on the definition and scope of postbiotics. *Nature Reviews* 18:649–67.

Shin, D. H. 2022. History of Jang and its origin. In *Overview of Korean Jang (Fermented Soybean Products) Manufacturing*, ed. D. H. Shin and D. Y. Kwon, 12–32. Korean Jang Association, Seoul.

Shin, D. H. and D. Y. Jeong. 2015. Korean traditional fermented soybean products: Jang, *Journal of Ethnic Food* 2:2–7. https://doi.org/10.1016/j.jef.2015.02.002

Shin, S. K., J. H. Kwon, Y. J. Jeong, S. M. Jeon, et al. 2011. Supplementation of cheonggukjang and red ginseng cheonggukjang can improve plasma lipid profile and fasting blood glucose concentration in subjects with impaired fasting glucose. *Journal of Medicinal Food* 14:108–13.

Shurtleff, W. and A. Aoyagi. 2010. *History of Soybeans and Soyfoods in Southeast Asia (13th Century to 2010) Extensively Annotated Bibliopgrphy and Sourcebook*. Soyinfo Center, Lafayette, CA.

Singh, N. 2017. Pulses: An overview. *Journal of Food Science Technology* 54:853–7. https://doi.org/10.1007/s13197-017-2537-4.

Soh, J. R., D. H. Shin, D. Y. Kwon and Y. S. Cha. 2008. Effect of cheonggukjang supplementation upon hepatic acyl-CoA synthase, carnitine palmitoyltransferase I, acyl-CoA oxidase and uncoupling protein 2 mRNA levels in C57BL/6J mice fed with high fat diet. *Genes Nutrition* 2:365–9.

Tamang, J. P. 2015. Naturally fermented ethnic soybean foods of India. *Journal of Ethnic Food* 2:8–17.

Vavilov, N. I. 1926. *The Origin of Cultivated Crops, Origin and Geography of Cultivated Plants*. Translated by Doris Löve. Cambridge University Press, Cambridge.

Woo, S. H. 2018. *Origin of Kochosun civilization and Lyoha culture*. Series of Kochosun Civilization, vol. 3. Knowledge Industry, Seoul.

Yang, H. J., S. Park, V. Pak, K. R. Chung and D. Y. Kwon. 2011. Fermented soybean products and their bioactive components. In *Soybean and Health*, ed. H. El-Shemy, 21–58. IntechOpen, Rijeka.

Yang, J., S. Lee, and Y. Song. 2003. Improving effect of powders of cooked soybean and chongkukjang on blood pressure and lipid metabolism in spontaneously hypertensive rats. *Journal of Korea Society of Food Science Nutrition* 32: 899–905.

Worldwide Fermented Non-Meat Sausages and Their Importance for Health

17

Inês Azevedo, Joana Bastos Barbosa, Helena Albano, and Paula Teixeira

17.1 INTRODUCTION

Fermented sausages are part of the daily diet of many individuals around the world. Variations in ingredients and cooking recipes and various processing conditions as well as habits and customs of different countries lead to the production of diverse types of fermented sausages (Bis-Souza *et al.* 2019; Flores and Piornos 2021; Franciosa *et al.* 2018; Gardini *et al.* 2001). Traditional sausages, mainly made with meat, fat, spices, water, and salt, can have a huge variety depending on their origin. Although these meat products contribute to the intake of different nutrients, potential adverse effects on health have been attributed to their consumption (Bou *et al.* 2017; Font-i-Furnols and Guerrero 2022b; Godfray *et al.* 2018; Holck *et al.* 2017; Leroy *et al.* 2022). Apart from their variety, consumers' preferences are constantly changing. Therefore, the food industry must be active and perceptive in developing new products allowing alternative nutritional options to be offered (Adise *et al.* 2015; Flores and Piornos 2021; Pernu *et al.* 2020; Sha and Xiong 2020).

Vegetarian, vegan, and flexitarian diets are on the rise, and sausages made with different ingredients, such as fish, seafood, and meat analogues, are emerging worldwide. Using modified technologies, companies are presenting innovative meat-like products. A meat analogue is generally made from vegetarian ingredients, and most of them are soy-based or gluten-based, but they are also made from pulse proteins (Font-i-Furnols and Guerrero 2022a; Kumar *et al.* 2017). Meat analogues' texture, flavour, colour, and other properties depend entirely on the ingredients used. Century-old recipes have been modified using materials such as wheat gluten, soy protein, mushrooms, pulses, and flavouring additives to produce a final product tasting like meat or seafood. It has been a considerable challenge for food producers to develop products that resemble meat while being, at the same time, appealing, healthy, and flavourful alternatives (Hoogstraaten *et al.* 2022; Tziva *et al.* 2020).

Overall, meat substitutes may be classified as plant-based (such as soy, gluten, and pulses), cell-based (*in vitro* or cultured meat), and fermentation-based (mycoprotein). Recent developments incorporate other protein sources, such as microalgae proteins extracted from Spirulina, and insect proteins (Altmann *et al.* 2018; Onwezen *et al.* 2021; Trotta *et al.* 2022).

17.2 NON-MEAT SAUSAGE TRENDS

Non-meat sausages are claimed to be healthier since they generally contain less animal fat or cholesterol, salt, and oil, and calorie quantities can be easily adjusted using the same rules used for making traditional sausages. The biggest problem associated with this kind of product lies in the fact that it is quite challenging to mimic meat flavour (Smart and Pontes 2022; Chen *et al.* 2022; McClements and Grossmann 2021). However, it must be accepted that if the meat is not used, the product will have a different taste and will never compete with traditionally made meat products. This does not imply that the product lacks a distinctive character. It is not designed to imitate or compete with meat sausages, but instead to fulfill a specific role within targeted niche markets (Boukid 2021; Chen *et al.* 2022; He *et al.* 2020).

By definition, a traditional sausage is composed of meat stuffed into a casing, together with other ingredients, such as bread, water, olive oil, and spices (Silva *et al.* 2019). As a popular alternative to meat products, vegetarian sausages are available

DOI: 10.1201/9781003605300-17

on the market with ingredients including typically a plant or fungal protein source. Soy, wheat, chickpea, pea protein, mycoprotein, and other plants, such as corn, potato, pepper, onion, garlic, and so on mixed with herbs, spices, salt, vegetable oils, and additives (clotting agents, stabilisers, pH regulators, and antioxidants) are some ingredients that can be found in these substitute products (Pernu *et al.* 2020; Tremlova *et al.* 2022). Most commercialised vegetarian sausages, such as veggie hot dogs, are emulsified since it is easy to place all ingredients into a grinder or a mixer machine and then stuff the resulting paste into a casing. Controlling the texture of non-emulsified sausages in a food processor is challenging. Still, when the texture is under control, it is fairly easy to add an infinite number of materials, and hundreds of recipes can be created (Bedin *et al.* 2018; Birke Rune *et al.* 2022; Yesuraj *et al.* 2022).

Many controversies have revolved around these meat alternatives since public opinion declares that the word "meat" should not be used in these products' labelling. Different names are being studied in order to meet consumers' needs, including "rolls", "patties", and "portions" for products like sausages, burgers, and steaks, respectively. In the USA, the National Cattlemen's Beef Association has been in conflict since February 2018 about whether plant-based products should be called "meat" since they do not originate from animals, which may confuse the consumer. For example, in France, the use of meat-like products, such as vegetarian sausage, was excluded from vegetarian food labelling once it was known that it might delude consumers. All products made from soy, tofu, and other vegetables which are considered meat substitutes may not include words such as "burger", "sausage", or "ham" on their labels (Sha and Xiong 2020). In December 2019, the Plant-Based Foods Association (PBFA) published Voluntary Standards for the Labeling of Meat Alternatives in the United States. However, the Food and Drug Administration (FDA) is still trying to establish label guidelines for plant-based foods, which were expected to be published as drafts or finals by the end of December 2022.

17.3 NON-MEAT SAUSAGES WORLDWIDE

In addition to consumer trends, factors such as cost, availability, environmental sustainability, suitability of new products, and their functional proprieties are the main drivers for the demand for new protein sources (Haque *et al.* 2016; Lima *et al.* 2022; Kumar *et al.* 2022). Nowadays, more and more proteins of different origins and their functionalities, such as water and oil holding abilities, solubility, emulsification, foaming, and gelation are being studied since they are vital for meat analogue structure development. At present, the main sources of vegetable proteins incorporate oilseeds (soybeans, canola, cottonseed, peanut, sunflower seed, sesame, safflower, and flaxseed), cereal (wheat, corn, rice, barley, oats, sorghum, and amaranth), pulses (beans, chickpeas, cluster beans, lentils, lupines, and peas), and leaves (alfalfa, lucerne, tobacco, mulberry bush, grass, sugar cane, sugar beet, and clovers) (Fu *et al.* 2022; Kyriakopoulou *et al.* 2018; Zahari *et al.* 2022).

Taking a cursory look around the World Wide Web, hundreds of non-meat sausages can be found, and it can be seen that there have never been as many plant-based sausages on the market as there are today. Table 17.1 presents examples of plant-based sausages produced worldwide. This table was compiled based on a web search using the keywords "vegan/vegetarian sausages/products/brands in 'country name'", and only those with fermentative ingredients, such as yeasts and starter cultures, were included.

17.4 MEAT ANALOGUES USED IN NON-MEAT SAUSAGES

17.4.1 Soy Protein

Due to their unique properties and low prices, at the moment, most meat analogues are based on soy protein. Other oilseed crops and proteins generated through fermentation, based on several substrates and microorganisms, are also being integrated into meat analogue manufacturing (Dekkers *et al.* 2018; Hashempour-Baltork *et al.* 2020; Sun *et al.* 2021). For over 50 years, soy protein has been used as a nutritional and functional food ingredient (Jooyandeh 2011; Qin *et al.* 2022; Ravani and Sharma 2022). Soy proteins are considered a good source of essential amino acids, offering innumerable functional benefits and a healthy diet. Additionally, soy proteins are believed to be appropriate for most diets since they contain no cholesterol and are lactose free (Bond 2010; Guo 2009; He *et al.* 2020; Jooyandeh 2011; Kapoor *et al.* 2014; Singh and Sit 2022). Soy-based analogues are mainly produced to mimic traditional foods in appearance, colour, flavour, and texture (Bohrer 2019; Hoogstraaten *et al.* 2022; McClements and Grossmann 2021).

17.4.2 Wheat Protein

Wheat protein gluten, a byproduct obtained from starch isolation from wheat flour, is frequently used due to its binding, dough-forming, and leavening qualities (Chen *et al.* 2022; Day *et al.* 2006; Godschalk-Broers *et al.* 2022). Gluten, the dough-forming protein of wheat flour, is the key to the exceptional ability of wheat to generate leavened products. Although known for almost 300 years, gluten has only been used as a unique vegetable protein of commercial significance in the past five decades. Since the need for plant-sourced proteins with different functional properties is quickly growing, the high protein content, exclusive viscoelastic characteristics, thermosetting, and water absorption properties of wheat gluten offer food scientists and technologists opportunities for innovative product design (Chen *et al.* 2022; Day 2011). Being rich in glutamine and cheaper than soy, gluten is finding its way to inclusion in meat analogues (Kumar *et al.* 2017). The main problem associated with wheat proteins is gluten-related disorders due to people with genetically predisposed conditions. For example, celiac disease is a chronic inflammatory intestinal disease triggered by gluten and is estimated to affect 1% of people worldwide (Anzani *et al.* 2020; Elli *et al.* 2015; Gazikalović *et al.* 2021; Pi *et al.* 2022). New gluten-free products are being developed in order to work around this health challenge (Jnawali *et al.* 2016; Starowicz *et al.* 2022; Szpicer *et al.* 2022). Wheat can be found in a large number of sausages (Table 17.1), whether in flour, starch, protein, or gluten form, usually combined with other sources of vegetable protein such as peas or soy.

17.4.3 Pulse Proteins

About 20 leguminous species are commonly used as dry grains for human nutrition. Peas, lentils, lupines, chickpeas, and other beans play a vital role in the human diet due to their high protein content and beneficial nutritional value and also because they are low cost and popular (Chandler and McSweeney 2022; Klupšaitė and Juodeikienė 2015). Along with wheat protein gluten, the most promising for meat analogue application is still pea protein due to its high moisture extrusion (Berrios *et al.* 2022; Kim *et al.* 2021; Vatansever *et al.* 2020). Several studies indicate that these proteins present weaker gelling capacities than soy, with the exception of chickpeas (Berghout *et al.* 2015; Grasso *et al.* 2022; Osen *et al.* 2014). Providing energy, dietary fibre, protein, minerals, and vitamins required for human health, they are the best-known pulse proteins for emulsification, foam stabilisation, and gel formation. Playing an important role in food formulation and processing, they are highly recommended as a potential supplement in a vast number of food applications. Pulse protein concentrates or production of isolates is booming since they enhance the nutritional value of food products (Fernando 2022; Nadeeshani *et al.* 2022). Table 17.1 shows that some brands use exclusively, for example, pea protein, flour, or starch, either combined with soy or wheat. Overall, the products described in Table 17.1 indicate that most meat alternatives are prepared with soy, wheat, and/or pea protein combinations to get the desired structure.

17.5 NON-MEAT FERMENTED SAUSAGE MICROBIOLOGICAL CHARACTERISTICS AND THEIR ROLE IN FOOD SAFETY

Whether they are traditional or innovative, the characteristics of fermented sausages are known to be highly dependent on the ingredient and raw material quality. Also, due to particular processing conditions, ripening, and existing microbial population, all of these products have specific microbiota (Belleggia *et al.* 2022a; Belleggia *et al.* 2022b; Belleggia *et al.* 2022c; Ferrocino *et al.* 2022; Yang *et al.* 2022).

In this chapter, the major aim is to know the emerging worldwide meat alternatives to traditional sausages and understand more about their microbial ecology. In addition, it is also intended to understand their potential role in food safety since they are becoming market relevant, and little is known about their microbiological and chemical characteristics (Azevedo *et al.* 2020; Bedin *et al.* 2018; Silva *et al.* 2019).

In Italy, Bedin *et al.* 2018 developed suitable recipes to prepare food products that mimic the shape and texture of traditional würstel and mortadella, focusing on the Italian market. For this, they faced several challenges, such as finding ways to maintain similar characteristics of traditional foods while fulfilling the consumer's requests and enlarging the market share of the food industries. To avoid confusion and misunderstanding, and since the primary goal of this research was to mimic these products with vegan-allowed ingredients and proteins of vegetal origin, the authors renamed them "mimic-mortadella" and "mimic-würstel". Microbiological analyses were carried out to determine the level of safety of the novel products, and food safety of the mimic-würstel and mimic-mortadella was respected in alignment and agreement with the food regulations when compared to traditionally processed meat-based products (Borch *et al.* 1996). In the sector of traditional salami production, bacterial contamination usually occurs after slaughtering since the internal muscles of the healthy animals are uncontaminated until that moment. Animal muscles' chemical and physical composition nurture the development and colonisation of an elevated type of microorganism, mostly bacteria; even though some are necessary for fermentation and aroma production, but some are human pathogens. These non-meat products respected food safety regulations when compared to their meat analogues. Still, the greatest challenge was trying to fulfil consumers' requests as well as eliminating issues related to shelf life and expected texture as a result of animal protein substitution. Related to the challenges faced in this study, the authors first concentrated on using different proteins as the basis to create a clear structure, identifying wheat gluten proteins as the best solution since this plant-based substitute is accepted in vegan formulations and helps to preserve the elasticity of the product. Both non-meat products obtained were marketable regarding texture and organoleptic properties. In conclusion, the biggest challenge is to handle consumer requests, but on the other hand, animal protein replacement can raise questions related to shelf life and expected texture.

In Portugal, the authors of a preliminary study (Silva *et al.* 2019) microbiologically characterised nine different formulations of *alheira* from five different producers. The isolates obtained were characterised using various phenotypic and biochemical tests, and their susceptibility to different antimicrobials and the presence of virulence factors were also investigated. Although lactic acid bacteria were the predominant microbiota, pathogenic bacteria such as coagulase-positive staphylococci, *Listeria monocytogenes* and *Salmonella* spp., and other indicator organisms were also found. Several virulence factors were detected among different groups, presenting a high incidence of β-hemolytic isolates and also resistance to several antimicrobials, leading to a large number of multiresistant isolates. In short, microbial hazards of traditional *alheira* were also found in these new formulations, which is not surprising since the possibility of cross-contamination always exists. It is crucial to warn consumers to safely cook these products since some common cooking practices of *alheira* may not guarantee a sufficient inside temperature that can eradicate potentially present pathogens (Campos *et al.* 2013; Felício *et al.* 2011). Given a sufficient environment for bacterial contact, horizontal transmission of antimicrobial-resistant genes or virulence determinants is a matter of concern. The results obtained in the study by Silva *et al.* (2019) refer to the need to analyse higher numbers of products and inform consumers of the need for safe cooking time/temperatures for correct preparation of these products.

Following the previously mentioned study, Azevedo *et al.* (2020) characterised more innovative products concerning their microbiological and chemical safety. Therefore, 14 innovative products and 8 corresponding meat analogues acquired in the Portuguese market and made from codfish, mushrooms, tofu, soy, and vegetables were analysed. *Enterobacteriaceae*, *Enterococcus*, lactic acid bacteria, yeasts, and moulds were the main bacteria found in innovative *alheiras,* but no sulphite-reducing *Clostridium* spores, *Escherichia coli*, *L. monocytogenes*, *Salmonella* spp., or *Staphylococcus aureus* were detected. In this study, the authors concluded that no biological threat was detected in these innovative products even though they were produced by the same companies. Still, some of the traditional analogues have shown concerning microbiological status in terms of food safety due to the presence of *E. coli* and *L. monocytogenes*. Due to this, an important aspect to consider in future confirmation studies is understanding why traditional products are more contaminated than innovative *alheiras*. Although produced in the same facilities and eventually under the same conditions, it is undisputed that some ingredients used in their manufacture are different, and meats used in traditional *alheiras* might be more contaminated. Also, producing conventional *alheiras* in higher amounts can be one of the answers to diverse contamination types found and, consequently, to different associated hazards (Azevedo *et al.* 2020).

In Belgium, Geeraerts *et al.* (2020) explored bacterial diversity in a selection of packaged meat product alternatives. Bologna sausage, chorizo, and salami produced without meat, in a total of 15 samples (8 vegetarian, s6ix vegan imitations of meat products, and 1 insect sample), were acquired at different supermarket groups and one bio-shop in Brussels. In this study, bacterial counts ranging from <2.0 to 8.7 log colony-forming units per gram (cfu/g) were detected in these vegetarian and vegan imitations of fermented meat sausages, with some differences between the products. The bacterial load of the edible insect product was generally higher than those in vegan and vegetarian products. Several isolates were collected using different selective agar media, classified, and identified using (GTG)-5-PCR fingerprinting, followed by gene sequencing. *Latilactobacillus sakei* was the main isolate collected from vegetarian products, while *Lb. sakei*, *Enterococcus faecium*, and *Carnobacterium divergens* were mainly found in vegan products. The insect product mostly contained *Ent. faecium*, *Macrococcus caseolyticus*, and *Cronobacter sakazakii*. These LAB species, specifically *Lb. sakei* and *Ent. faecium*, were the most prevalent microorganisms found, both of which have been associated with food spoilage (Geeraerts *et al.* 2017, 2018; Dušková *et al.* 2015, 2016). Despite these food quality and safety concerns, *Lb. sakei* has potential use as a starter culture, and *Ent. faecium* can be a bioprotective culture as well as flavour producing in meat products (Barbosa *et al.* 2014; Favaro and Todorov 2017; Kaškonienė *et al.* 2017). As for future studies, the authors emphasised the need to evaluate how this divergent variability in bacterial loads can be explained. They also intend to find, when prolonged storage is needed, whether the bacteria encountered may lead to spoilage or food safety concerns. Also, related to food safety, the occasional presence of relatively high enterococci levels deserves extra attention. Even though they are lactic acid bacteria, they are also important nosocomial pathogens and might cause bacteremia, endocarditis, and other infections. Some strains are resistant to many antibiotics, and studies conducted on the incidence of virulence factors among enterococcal strains isolated from foods revealed that some harbour virulent traits. The use of metagenomic, meta-transcriptomic, and meta-metabolomic analyses is suggested as a way to offer additional information and to gain better insight into the variable microbial communities that are correlated with this category of food products (Franz *et al.* 2011).

In Finland, a significant food safety concern is emerging since *Clostridium botulinum* is being found in numerous outbreaks in the raw materials used for preserved vegetables. Vegetarian sausages are becoming popular as alternative meat products, and their preparation involves limited additives and salt for products with several months of shelf life. In a study by Pernu *et al.* 2020, 74 (8 frozen and 66 chilled) packaged vegetarian sausage samples from seven producers were acquired in Finland and Germany. These vacuum-packaged products' shelf life varied from less than 2 weeks to 6 months, and some of these products contained detailed instructions for cooking. In some cases, these instructions included heating temperature and time, but some products had just a suggestion of heating method without time indications, and other products were advised to be served either heated or cold. The main ingredients used were soy (soy protein or tofu), wheat protein, vegetable oil, sugar, spices, salt, corn, wheat, or potato starch, and additives such as pH regulators, emulsifiers, stabilisers, and antioxidants were also commonly included but not identified in more detail. *C. botulinum* is a major food safety concern due to its ability to produce a highly potent neurotoxin and resistant endospores. This pathogen was found in 32% of the analysed vegetarian sausages, suggesting that these products can be a source of botulism (Pernu *et al.* 2020). Vegetarian sausages have become a popular source of plant protein and alternatives to meat products. While vegetarian sausages have not been linked to botulism, numerous outbreaks due to preserved vegetables suggest a frequent occurrence of *C. botulinum* spores in the raw material (Date *et al.* 2011; Jalava *et al.* 2011; King *et al.* 2009; Lindström *et al.* 2006; Read and Grundy 2012; Sobel *et al.* 2004). The main food safety concern is that it might grow and produce toxins at refrigeration temperatures. For this reason, the safety of these sausages relies mainly on heat treatment and chilling storage. The product formulation of vegetarian sausages involves limited NaCl and preservatives, and the shelf life may be several months. The highest cell counts (1200 spores/kg) were observed for *C. botulinum* Group II in products with a remaining shelf life of 6 months at the time of purchase, leading to the conclusion that vacuum-packaged vegetarian sausage products frequently contain *C. botulinum* spores and may possess a high risk of *C. botulinum* growth and toxin production (Pernu *et al.* 2020). Even though chilled storage below 3°C and thorough reheating before consumption are warranted, moderate heat treatments will not eliminate *C. botulinum* Group I and Group II spores, and long shelf life may support spore germination, outgrowth, and toxinogenesis (Graham *et al.* 1997). Therefore, vegetarian sausage product safety relies on multiple hurdles controlling growth and toxin production and certainly not simply on toxin inactivation during cooking. The authors suggest subsequent challenge studies where product packs might be inoculated, and shelf life tests will be mandatory to determine the growth and toxic potential of *C. botulinum* in vegetarian sausages and the length of safe shelf life (Pernu *et al.* 2020).

17.6 NON-MEAT FERMENTED SAUSAGE PHYSICOCHEMICAL CHARACTERISTICS

Similar to what happens in the microbiological evaluation of these innovative products, studies concerning chemical characterisation are also scarce. A study conducted by Bayram and Bozkurt (2006) revealed that bulgur is proper as a meat replacement for the production of a vegetarian sucuk, a dry, spicy, fermented sausage that is eaten from the Balkans to the Middle East and Central Asia. Bulgur sucuk's main quality parameters were similar to those of meat sucuk, such as moisture content and pH decrease along the ripening period. The initial moisture content decreased to almost half the value during the ripening period. This might be due to water evaporation as a result of the low relative humidity of the ripening cabinet and low pH. In order to control fermentation, taste, texture, and colour, the bulgur sucuk pH was crucial. The considerable pH decrease in the first 3 days of ripening was mainly due to the bacteria's organic acid production. The high pH values found in the sucuk initial batter allow the growth of most bacteria, yeasts, and moulds, so pH must be reduced. Acid production will be highly conditioned by the type and amount of added carbohydrates, lactic acid bacteria present, drying level, and temperature. This product should find good acceptance among consumers due to its lack of nitrites/nitrates, cholesterol, dangerous microorganisms, chemical compounds, and animal fat and low price. Also, the probability of nitrosamines, biogenic amines, and cancerous compound formation or presence seems to be very low in this product (Bayram and Bozkurt 2006).

The use of starfruit dietary fibre concentrate as a different ingredient in Vienna sausages was investigated by Vivar-Vera *et al.* (2018). Different amounts of starfruit fibre were used in a restricted mixture layout to assess trim force, shrinkage, colour, residual nitrite, moisture, and polyphenol contents on sausages. This study demonstrated that this ingredient in a mixture with pork or turkey meat has a reducing effect on nitrites, moisture, trim force, and shrinkage, although an increase in polyphenol concentration was observed as the starfruit fibre amount rose. Optimisation was achieved with a mixture of pork/turkey and between 7.4 to 8.4% of starfruit fibre. The developed product possessed low residual nitrite content, shrinkage, red colour, high total dietary fibre content, and antioxidant polyphenols. In conclusion, it seems clear that Vienna-type sausages enhanced with starfruit fibre are an excellent approach to achieving prospective functional meat products with high antioxidant dietary fibre content and reduced nitrites.

Mousavi *et al.* (2019) investigated the effects of different levels of tofu on the nutritional value and physicochemical properties of sausages, aiming to explore its potential as a non-meat ingredient. Sausages prepared with 25, 50, and 75% of tofu were studied, and higher moisture and residue content was observed in 75% added tofu. Still, protein content results revealed that as tofu levels increased, the protein content decreased. The best sausage creation included 25% of tofu since it exhibited an exceptional nutritional composition with the highest protein content and lowest fat content compared to the other formulations.

The objective of the study performed by Kamani *et al.* (2019) was to partially and completely replace chicken meat with plant proteins in sausage preparations and then compare their characteristics with the whole meat sample. The outcomes demonstrated that mixed plant proteins, such as soy protein isolate and gluten, improved emulsion stability and minimised cooking loss and shrinkage, but poor elasticity and gel quality were also observed.

Azevedo *et al.* (2020) also performed chemical characterisation of innovative and traditional *alheiras*. Concerning pH and water activity values, no differences were observed between innovative products and their traditional analogues, and nitrites, nitrates, and biogenic amines were found to be within acceptable limits for these kinds of products. Lactic acid was found in all analysed samples, but malic and succinic acid seemed to be predominant in innovative products.

Sha and Xiong (2020) argued about the many supposed advantages of plant-based meat alternatives compared to animal meat and meat products. The authors claimed that meat alternatives often hold more salt than the products they intend to replace, and thus, it has been a challenge to reduce sodium content, promoting health. Another hurdle described is related to the lack of a clean label, where a large number of ingredients are found in these innovative products, including preservatives, stabilisers, and colourants, that are not usually added in meat products (Berman 2019). This raises the question of whether these products are indeed more nutritious and healthier than analogous meat products.

Recently, Keerthana Priya *et al.* (2022) developed a vegan sausage using novel ingredients such as jackfruit and banana florets. This meatless sausage showed good sensory and physico-chemical properties, was rich in fibre and protein, and proved a good alternative in various value-added products, helping in the environmental reduction effects of our food system.

17.7 NON-MEAT FERMENTED SAUSAGE SENSORIAL TRAITS

In the study by Bayram and Bozkurt (2006), bulgur sucuk consumer satisfaction scores related to its sensory properties, like flavour, colour, and cutting ease were evaluated by qualified panellists. For the formulation of this product, olive oil was used in order to obtain a healthy vegetarian sucuk. In meat products, fat is a major contributor to the flavour, texture, mouthfeel, and juiciness of the final product. Fat allows ease of mixture, enabling continuous moisture, which is essential for proper fermentation and aromatisation of fermented products. Fat reduction may influence product suitability since it promotes sensory properties. In several studies, olive oil and other vegetable oils have been included as being suitable in various meat products, replacing animal fat in order to reduce saturated fatty acid and cholesterol content in these products (Bloukas *et al.* 1997; Kayaardi and Gök 2004; Kılıç and Özer 2017; Muguerza *et al.* 2002; Öztürk-Kerimoğlu *et al.* 2021). This ingredient, bulgur, can be found in more than 250 different types of food due to its physical, textural, nutritional, and functional properties. Its chewiness, adhesiveness, water-holding capacity, resistance to mould contamination and insect attack, appealing taste, easy preparation, and long shelf life make it a very good option for a meat substitute.

Tahmasebi *et al.* (2016) investigated sesame and walnut paste, corn flour, and pigeon pea flour effects using response surface methodology (RSM) to establish the ideal formulation of sausages made using the previous ingredients in order to maximise emulsion stability, minimise cooking loss, and improve textural properties. When increasing pigeon pea and corn flour amounts, high emulsion stability and textural properties, such as hardness, gumminess, and deformability, were observed. Nevertheless, when these flours in increased proportions were combined with walnut paste, a slight reduction was observed in cohesiveness, and the results showed high levels of hardness. These vegetable ingredients improved gel-network formation while revealing enhanced emulsion stability parameters and high textural property promotion. Microstructural images of sausages with different formulations revealed interactions between the protein matrix and various vegetable ingredients, as well as the presence of fat globules, which may also influence product parameters. Scanning electron microscopy images showed compact structures with low sponginess due to combining vegetable ingredients and their interaction with meat protein during cooking, which generates a constant and stable protein matrix. Through these results, they concluded that RSM is a promising method to improve sausage formulation for the addition of vegetable ingredients.

Neville *et al.* (2017) developed hybrid meat analogues where a fraction of meat was replaced by other sustainable protein sources. Consumers tested unique formulations of hybrid beef burgers and pork sausages in order to obtain satisfaction score analogues with both meat and meat-free commercial products. Samples of commercial meat, meat-free, and hybrid products were presented to consumers, and the sensory attributes perceived were identified. The results showed that hybrid products were appreciated by consumers, although hybrid sausages presented better general acceptability when compared to hybrid burgers. Meat-free products showed substantially lower acceptability when compared with meat and hybrid products, where no meaningful discrepancies were observed. They also found that imitating a "meaty flavour" and "meaty colour" in hybrid products is vital to increase tolerability among consumers mainly used to meaty foods. It seems clear that this hybrid model helps to link the tolerability gap among mainly meat consumers when it comes to meat and meat-free products. This hypothesis, through enhancing the experience of consumers with meat substitution, may support the conversion of meat eaters to a meat-reduced diet. The authors intend to follow up on their studies in order to understand if, although these hybrid products were found to be acceptable, the consumers intend to buy them.

As reported by Vivar-Vera *et al.* (2018), adding starfruit fibre to Vienna-type sausages made from a pork and turkey mixture resulted in potentially functional meat products with good sensory acceptance by consumers, while also helping to meet daily recommended fibre intake. Good taste, colour, and texture qualities were described when pork/turkey was mixed with starfruit dietary fibre concentrate in Vienna sausages.

The main challenge Bedin *et al.* (2018) faced in their study was to find a proper recipe for würstel and mortadella imitations. Trying to create a good structure, they used different proteins and recognised that wheat gluten was the best solution found not only because it is accepted in vegan formulations but also because it helps to keep elasticity. The developed products are viable in terms of texture but also in organoleptic and sensory texture evaluation. Overall, although the requests of consumers can be fulfilled, animal protein replacement can cause concerns related to shelf life and expected texture.

In Mousavi *et al.*'s (2019) study, as previously reported, among all treatments, 25% tofu was suggested to be incorporated in sausages. According to sensory evaluation, 25% added tofu showed global approval qualities and high scores related to colour, texture, juiciness, and flavour. High fat content found in the 50 and 75% formulations was due to the low emulsification qualities of soy lecithin to mix fat and water. During the cooking process, this formulation became oily, enhancing the sausage's instability.

Analysing the sensory evaluation results in Kamani *et al.*'s (2019) study, meat-free sausages were highly appreciated in terms of texture, odour, colour, and global approval. The global acceptance of chicken meat replacement with plant proteins is promising, and an 80 to 100% replacement might be considered. Nevertheless, further studies must be done in order to improve gel-forming characteristics, which is the greatest obstacle in meat-free sausage production.

17.8 NON-MEAT FERMENTED SAUSAGE FUNCTIONAL CHARACTERISTICS

Like the microbiological and chemical traits of non-meat fermented sausages, little is known about these innovative products related to their functional characteristics. From what is known, traditional fermented sausages have a long tradition in consumers' habits, and these products have specific microbiota typical of the region/area from where they are produced. Even though many typically fermented meat products are still manufactured with traditional technologies, not using starter cultures, in some industries, the use of starter cultures is becoming more frequent since it provides an additional tool in foodborne pathogens prevention as well as enhancing the competitiveness of the starter organisms (Franciosa *et al.* 2018). The introduction of starter cultures has become essential in such a way that allows shortened ripening period, ensures colour development, enhances flavour, and improves food safety (Ren *et al.* 2022; Li *et al.* 2023). Actually, a starter culture should be qualified enough to allow fermentation, colonise the product, and dominate other microorganisms from the beginning until the end of the process (Cocolin *et al.* 2006; Gradinarska *et al.* 2022; Li *et al.* 2022). However, using commercially available starter cultures may result in a loss of organoleptic characteristics, with consequent loss of flavour and aroma. For this reason, in some countries, there are still artisanal sausages that are made relying on unknown microbiota since they are preferred by most consumers. These spontaneously fermented sausages possess unique characteristics and are frequently superior when compared to regulated fermentations inoculated with industrial starters (Franciosa *et al.* 2018). During the fermentation process, two broad groups are mainly prevalent, lactic acid bacteria (LAB) and coagulase-negative cocci (CNC). Lactic acid bacteria are usually more numerous than CNC during fermentation and ripening, staying more stable in cured products. Facultatively heterofermentative lactobacilli, such as *Lb. sakei* and *Lb. curvatus*, generally prevail, and within CNC, *Staphylococcus xylosus* dominates (Aquilanti *et al.* 2016). In the study by Geeraerts *et al.* (2020), the authors described some of these microorganisms as being dominant in some alternative meat products. In vegetarian products, *Lb. sakei* was the main isolate that remained, while *Lb. sakei*, *Ent. faecium*, and *Carnobacterium divergens* were mainly found in vegan products. *Enterococcus faecium*, *M. caseolyticus*, and *C. sakazakii* were the main microbiota in insect products. These microbial cultures can increase fermented products' safety by means of rapid matrix acidification or due to the production of antimicrobial substances such as bacteriocins. These compounds comprise a group of peptides with bactericidal or bacteriostatic activity against some food spoilage and food poisoning bacteria such as *Bacillus* spp., *Clostridium* spp., *Staphylococcus* spp., and *Listeria* spp. Nisin, pediocin, sakacin, curvacin, plantaricin, and bacteriolysins are some examples of bacteriocins that are effective in several pathogenic species control, such as *L. monocytogenes*, *S. aureus*, *Campylobacter* spp., *E. coli*, *C. perfringens*, and *Bacillus cereus* (Arbulu *et al.* 2022; Ferrocino *et al.* 2022; Laranjo *et al.* 2019). This means that, similar to traditional fermented sausages, bacteriocin research in innovative products is also feasible, and we might identify novel bacteriocins and bacteriocin-producing strains for specific applications. The main compromise is finding the best-suited bacteriocin/bacteriocin producer to control spoilage/pathogenic microorganisms often present in this kind of product. Bacteriocin-producing lactic acid bacteria active against *L. monocytogenes*, *Ent. faecalis*, and *Clostridium sporogenes* were found in innovative *alheiras* (unpublished data from Azevedo *et al.*)

17.9 PROBIOTICS IN NON-MEAT FERMENTED SAUSAGES

According to Pimentel *et al.* 2021, probiotic products represent an appealing economic point for vegan consumers who demand products with health benefits. Probiotics are microorganisms that, when consumed in adequate amounts, will benefit the host (FAO/WHO 2002). In the study by Pimentel *et al.* (2021), the effect of probiotics on the technological and sensory properties of various vegan products was explored, and a summary of several other studies conducted in order to evaluate probiotics' health benefits can be found. These probiotic products may improve the immune system (Agraib *et al.* 2022; Kristensen *et al.* 2016; Yadav *et al.* 2022), manage diabetes (Ahmadian *et al.* 2022; Madempudi *et al.* 2022;

Razmpoosh *et al.* 2019), show anticarcinogenic properties (Dasari *et al.* 2017; Pop *et al.* 2022), and improve overall well-being. Based on these benefits, the food industry is being compelled to produce new products containing probiotics and food researchers to study their effects on human health and their specific characteristics. Besides their effectiveness and proven healthiness, probiotic cultures can easily adapt to different food matrices, so the development of products that are nutritionally stable and/or add value should be encouraged (Açik *et al.* 2020; Behera and Panda 2020; Grom *et al.* 2019; Tangyu *et al.* 2019).

For years, probiotics were mainly centred on dairy products, but the rise of lactose-intolerant, vegan, and hypercholesterolemic individuals demanded drastic changes (Cosme *et al.* 2022; Nguyen *et al.* 2019). Based on vegetables such as almonds, coconut, oats, rice, chickpeas, and soybeans, the food industry was able to suggest vegetable matrices as potential probiotic agents (Rincon *et al.* 2020). According to Min *et al.* (2019), fermented foods are selected as non-diary food matrices once studies have shown that fermentation allows superior viability of probiotic culture survival. In developing innovative probiotic non-dairy products, it is essential to consider factors such as viability and stability of the probiotic bacteria, resistance to pH, temperature and other stresses, adequate water content, carbohydrate consumption, and metabolite production from fermented or non-fermented foods. Even though non-dairy food matrices have shown the ability to sustain the growth of probiotics, being good alternatives to offer benefits without milk proteins, lactose, cholesterol, or saturated fat, at present, most probiotic food markets are dairy based (Cosme *et al.* 2022).

Probiotics may modify product composition, colour, and acidity, but the main problem relies on the survival of the probiotic culture, which is dependent on processing steps, food matrix, and storage conditions. Also, even when products with proper probiotic survival, physicochemical characteristics, technological properties, and sensory acceptance are obtained, the strain source may compromise vegan status since most of those presently available are not isolated from vegetable matrices (Pimentel *et al.* 2021).

Probiotics used in food products require evidence of their effectiveness and safety. This might be achieved through scientific studies that validate a history of safe use, lack of adverse incidents, non-existence of virulence and pathogenic issues, absence of substances and/or metabolites that might cause human health risks, absence of antibiotic resistance, and susceptibility to at least two antibiotics (FAO/WHO 2006). A relatively large collection of lactic acid bacteria species are registered as generally recognised as safe (GRAS) and include the most-used probiotic species in supplements or food products (WGO 2017). Along with *Bifidobacterium* genus, species of *Lactobacillus* are the leading used probiotic cultures. *Saccharomyces* yeast and some *Streptococcus* and *Bacillus* species are also used, and their incorporation into non-dairy foods has been studied (Abid *et al.* 2022; Alves *et al.* 2016).

17.10 PREBIOTICS IN NON-MEAT FERMENTED SAUSAGES

A lot is heard about probiotics' value and how they are crucial to our gut microbiome and good digestion promotion, but prebiotics are as important as probiotics to gut health. Therefore, not being informed about them can lead us to unconsciously not consume enough in our diet, which will weaken any probiotic action. Prebiotics are food elements that allow beneficial microorganisms, such as bacteria and fungi, towards beneficial growth or functional action. These nutritional prebiotics are usually nondigestible fibre compounds not digested in the gastrointestinal tract's upper part and promote beneficial bacteria growth by acting as a substrate. They must withstand host digestion and adsorption before fermentation by more than one species of the local microbiota (Preidis *et al.* 2011; Preidis and Versalovic 2014). Prebiotics, predominantly polysaccharides and oligosaccharides, are abundant in high-carbohydrate foods. These compounds may be found in many fruits, vegetables, and whole grains, such as apples, bananas, berries, tomatoes, asparagus, green vegetables, leeks, soybeans, wheat, peas, and beans. Through label reading, they might be found if words such as galactooligosaccharides, fructooligosaccharides, oligofructose, chicory fibre, and inulin are searched. Proven benefits include, besides feeding our good gut bacteria, support of calcium, faster food fermentation during digestion, and favouring healthy gut cells (Câmara *et al.* 2020).

Dietary fibres' value in health is well known, and some benefits are directly associated with prebiotic use. Seeds, nuts, and mushrooms are good sources of prebiotic compounds, which have enhanced industry interest in these ingredients due to their possible use as non-meat protein sources. Several studies have evaluated meat replacement by non-meat protein sources in meat products. Leonard *et al.* (2019) used up to 36% of the total ingredient composition of lupin flour for meat replacement within beef sausage and concluded that carbohydrate levels were increased, but fat and protein content decreased. A lupin flour increase also improved emulsion stability and cooking yield, while the strength of the textural structure became weaker. After sensory analysis, they observed that they could incorporate lupin flour for up to 12% of the total content with no significant effect on consumer acceptability. At higher amounts (up to 36%), a negative impact was observed in consumer acceptability in flavour and texture terms.

Another study by Kamani *et al.* (2019) investigated the suitability of different plant proteins to wholly and partially replace chicken meat in sausage. They determined that 100% replacement with plant proteins is promising since sensory evaluation demonstrated that the overall acceptability of meat-free sausages was like the full meat product. Despite this, they suggested more studies and future research that might improve gel-forming characteristics, which have been shown to be the main obstacle in meat-free sausage manufacturing.

In a study by Wang *et al.* (2019) using *Lentinula edodes*, also known as shiitake mushroom, replacing lean pork meat in sausages has proven to be a feasible option. Moisture contents, total dietary fibre, and total phenolic contents were superior when using this altered sausage. Also, methionine, glutamic acid, and cysteine were enhanced compared to the control. Nevertheless, this alternative reduced the protein content and energy levels of the modified sausages. The use of mushrooms induced a slight darkening and softening of the sausage texture. From this sensory perspective, the best formulation was replacing 25% of lean pork meat with *Lentinula edodes*. Through these results, investigators concluded that this substitute is a potential alternative for reducing lean pork meat in sausages.

In a study by Gad EL Rab *et al.* (2019), where the main aim was to produce a new low-cost product with noteworthy nutritional value, two formulas replaced beef meat in sausages with 10% milled flaxseeds and 20% chickpea flour. According to sensory evaluations, these replacements were found to be the best palatable proportions. Both preparations were evaluated for physicochemical qualities, consumer acceptance, mineral content, and chemical composition, but also amino and fatty acid profiles. Results have proven that flaxseed and chickpea incorporation considerably decreased cooking damage percentage while boosting cooking profit percentage and water holding capability. The sensory assessment revealed that, in general acceptability, no significant differences were reported between the control sample and the two prepared formulas. Adding flaxseeds enhanced raw fibre %, carbohydrates, and unprocessed fat as well as dietary minerals calcium (Ca), iron (Fe), zinc (Zn), and phosphorus (P) while increasing the fatty acid profile and methionine levels. In comparison, the chickpea alternative reduced fat content while improving raw fibre and carbohydrate content as well as linolenic acid and total essential amino acid percentage. Additionally, using flaxseeds and chickpeas in beef sausage preparation reduced the final costs by 8.3 and 16%, respectively.

17.11 CONCLUSION

Due to the increase in demand for vegetarian, vegan, and flexitarian diets over the past years, consumers' search for products with high functional and nutritional values has increased substantially. Demand for products not derived from animals (i.e., vegetarian and/or vegan) has increased globally. Additionally, the current abundance of products based on beans, seeds, and nuts is mainly due to consumers' demands for plant-based alternatives, not only for dietary reasons, lifestyle, or health concerns but also due to the concerns associated with environmental sustainability. This dynamic process and change in consumer preferences have made the overall population more health conscious, and they have demanded beneficial health-targeted values for food and its sustainability in the food chain. As a result, the food industry is compelled to advance functional foods, providing innovative foods that possess higher quality while adding value based on their functional properties. Despite the known advantages, the biggest hurdle associated with these innovative products lies in the fact that meat alternatives are currently much more expensive when compared to their meat counterparts, representing an economic challenge to vegetarian and vegan consumers.

Despite that, meat analogues present a more sustainable method of production when compared to traditional meat production systems requiring smaller natural sources volumes. Several studies have been done incorporating a broad range of ingredients to improve the physicochemical, nutritive, textural, and sensory properties of these meatless products. In the near future, more studies should be conducted in order to increase recognition and support of these products through consumer acceptability evaluation and improve perception of the beneficial effects of consuming meat analogues. New formulations and innovations related to these products should be supported since meat analogues are increasingly considered a potential alternative to real meat.

Several studies related to the microbiological characteristics of these innovative products show that their microbiota are highly dependent on ingredients, raw material quality, processing conditions, ripening, and existing microbial population. Although there are few studies related to the chemical characterisation of these products, which indicated that the parameters were within acceptable limits for these kinds of foodstuffs, there are many challenges to overcome. Currently, these non-meat alternative products are still poorly studied. For this reason, several studies should be taken into consideration, not only to verify their functional and nutritional value but also to confirm that they are safe from a microbiological and chemical point of view.

17.11.1 Acknowledgments

The authors would like to thank the scientific collaboration under the *Fundação para a Ciência e a Tecnologia* (FCT) project UID/Multi/50016/2019 and the project "Cleanlabel+: Innovative natural, nutritious and consumer-oriented "cleanlabel" foods" (POCI-01–0247-FEDER-046080) co-financed by *Fundo Europeu de Desenvolvimento Regional* (FEDER) by *Programa Operacional Competitividade e Internacionalização* (POCI). Financial support for authors Inês Azevedo and Joana Bastos Barbosa was provided by a doctoral fellowship SFRH/BPD/125275/2016 (FCT) and a post-doctoral fellowship SFRH/BPD/113303/2015 (FCT), respectively.

TABLE 17.1 Origin and Ingredients of Available Fermented Non-Meat Sausages Worldwide

ORIGIN	*BRAND*	*PRODUCT*	*INGREDIENTS*	*REFERENCE*
Australia	Made with Plants	Vegan Bangers And Mash	Vegan mash (45%) (potato, soy milk powder (soybean), salt, white pepper), plant-based sausages (35%) (plant-based mince (33%) (water, vegetable protein (soy), vegetable oil, cocoa butter, thickeners (461, 412), vegetable protein powder, vegetable protein extract (soy), smoke flavour, yeast extract, beetroot powder, salt, natural colour (150a)), garlic, parsley, black pepper), gravy (15%) (water, gravy powder (thickener (1442), maltodextrin, salt, natural flavours, sugar, vegetable powder, tomato powder, vegetable gums (415, 412), natural colours (150a, 160c), herbs)), peas (5%)	https://madewithplants.com.au/
	Quorn	Quorn Sausage Rolls	Wheat flour (calcium carbonate, iron, niacin, thiamin), mycoprotein (23%), water, vegetable oil, seasoning {textured wheat protein [wheat gluten, firming agent (516), preservative (223)], salt, free range egg white powder, whey powder (from milk), wheat flour (calcium carbonate, iron, niacin, thiamin), yeast extract, dextrose, thickener (pectins), dried onion, herbs, sunflower oil, barley malt extract, yeast, acidity regulator (503), white pepper extract, herb extract, colour (172)}, onion, canola oil	https://www.quorn.com.au/
	V2 Food	V2 Plant Based Sausages	Water, soy protein, vegetable oils, thickeners (methylcellulose, modified cornstarch, carrageenan), flavours (contains glutamic acid), colours (caramelised sugar, beetroot powder), onion, dehydrated garlic, salt, yeast extract, herbs (parsley, thyme), minerals (zinc, iron), vitamins (niacin (B3), pyridoxine (B6), cobalamin (B12)), antioxidants (tocopherol, ascorbic acid)	https://www.v2food.com/
Belgium	The Vegan Butcher's Choice	Vegan Garlic Salami	Vital wheat gluten, water, sunflower oil, organic nutritional yeast, miso paste, brown sugar, apple cider vinegar, spices, salt	https://shop.veganbutcher.be/
		Vegan Italian Sausage	Vital wheat gluten, water, extra virgin olive oil, onion, garlic, silken tofu, tomato paste, white miso paste, nutritional yeast, spices, liquid smoke, sugar, salt	
Brazil	SuperBom	Pea-based Smoked Sausage	Water, sodium caseinate, powdered egg white, isolated pea protein, pea starch, sunflower oil, pea fiber, casein, salt, hyposodium salt, yeast extract, sugar, calcium chloride firming agent, stabilizer -zantes (carrageenan, xanthan gum and locust bean gum), thickeners (sodium alginate and methylcellulose), maltodextrin, hydrolyzed corn protein, spices, aromatic herbs, aroma similar to natural sausage, radish, carrot and bell pepper, natural smoke aroma, iron and zinc minerals, vitamins A, B9, and B12	https://superbom.com.br
		Pea Sausage	Water, sodium caseinate, powdered egg white, isolated pea protein, pea starch, sunflower oil, pea fiber, salt, hyposodium salt, yeast extract, sugar, carrageenan and xanthan gum stabilizers, maltodextrin, hydrolyzed protein corn, spices, aromatic herbs, aroma identical to natural sausage, radish, carrot and pepper concentrates, iron and zinc minerals, vitamins A, B9, and B12	
		Traditional Pea Sausage	Water, sodium caseinate, powdered egg white, isolated pea protein, pea starch, sunflower oil, pea fiber, casein, salt, hyposodium salt, yeast extract, sugar, calcium chloride firming agent, stabilizer -zantes (carrageenan, xanthan gum and locust bean gum), thickeners (sodium alginate and methylcellulose), maltodextrin, hydrolyzed corn protein, spices, aromatic herbs, aroma similar to natural sausage, radish, carrot and bell pepper, natural smoke aroma, iron and zinc minerals, vitamins A, B9, and B12	

(Continued)

TABLE 17.1 (Continued)

ORIGIN	BRAND	PRODUCT	INGREDIENTS	REFERENCE
Canada	Sol Cuisine	Veggie Breakfast Sausages	Filtered water, soy protein, expeller pressed sunflower oil, yeast extract, cellulose gum, evaporated cane syrup, signature superfoods blend (fava bean powder, maca powder*, lucuma powder*, mesquite powder*), spices, sea salt, onion, vitamin blend (potassium chloride, maltodextrin, magnesium oxide, ferric orthophosphate, niacinamide, zinc oxide, cyanocobalamin, D-calcium pantothenate, copper gluconate, pyridoxine hydrochloride, thiamine mononitrate, riboflavin) (*organic)	https://solcuisine.com/
	Yves Veggie Cuisine®	Jumbo Veggie Dogs	Water, soy protein isolate, vital wheat gluten, tofu (water, soybeans, magnesium chloride). Contains 2% or less of: spices, natural flavor, vitamin & mineral blend (dipotassium phosphate, dimagnesium phosphate, ferric orthophosphate, zinc oxide, niacinamide, vitamin B12, calcium pantothenate, thiamine hydrochloride, pyridoxine hydrochloride, riboflavin, iron oxide), extractives of spices, wheat starch, yeast extract, red beet powder (color), carrageenan	http://yvesveggie.com/
		The Good Dog	Water, soy protein isolate, vital wheat gluten, tofu (water, soybeans, magnesium chloride). Contains 2% or less of: cane sugar, wheat starch, yeast extract, salt, ground mustard seed, carrageenan, natural flavor, tomato lycopene extract (color), rice flour, ferric orthophosphate (iron), vitamin B12, extractives of spices	
		Veggie Dogs	Water, soy protein isolate, vital wheat gluten. Contains 2% or less of: spices, natural flavor, dehydrated onion, vitamin and mineral blend (dipotassium phosphate, dimagnesium phosphate, ferric orthophosphate, zinc oxide, niacinamide, vitamin B12, calcium pantothenate, thiamine hydrochloride, pyridoxine hydrochloride, riboflavin, iron oxide), extractives of spices, beet powder and paprika (color), wheat starch, maltodextrin, yeast extract, carrageenan, sugar, salt	
		Veggie Tofu Dogs	Water, soy protein isolate, vital wheat gluten, tofu (water, soybeans, magnesium chloride). Contains 2% or less of: wheat starch, yeast extract, salt, sugar, canola oil, ground mustard seed, carrageenan, seasoning (maltodextrin, salt, spice, dehydrated onion), natural flavor, vitamin and mineral blend (dimagnesium phosphate, ferric orthophosphate, zinc oxide, niacinamide, vitamin B12, calcium pantothenate, thiamine hydrochloride, pyridoxine hydrochloride, riboflavin, iron oxide), paprika, extractives of spices, beet powder (color), salt blend (sea salt, potassium chloride, magnesium chlorides, magnesium sulfates)	
	Zoglos	Incredible HotDogs	Water, wheat gluten, isolated soy protein, soybean or, salt, methylcellulose, carrageenan, spices, onion, vitamin mix (potassium phosphate, magnesium oxide, niacinamide, ferric phosphate, zinc oxide, calcium pantothenate, thiamine mononitrate, pyridoxine hydrochloride, riboflavin, copper sulphate, folic acid, cyanocobalamin), garlic, paprika, flavours, locust bean gum, yeast extract, maltodextrin, xanthan gum, potassium chloride, smoke flavour, iron oxide, flavours, calcium phosphate, citric acid, disodium guanylate, and inosinate	https://zoglos.com/
		Meatless Franks	Water, wheat gluten, isolated soy protein, soybean or, salt, methylcellulose, carrageenan, spices, onion, vitamin mix (potassium phosphate, magnesium oxide, niacinamide, ferric phosphate, zinc oxide, calcium pantothenate, thiamine mononitrate, pyridoxine hydrochloride, riboflavin, copper sulphate, folic acid, cyanocobalamin), garlic, paprika, flavours, locust bean gum, yeast extract, maltodextrin, xanthan gum, potassium chloride, smoke flavour, iron oxide, flavours, calcium phosphate, citric acid, disodium guanylate, and inosinate	

(Continued)

TABLE 17.1 (Continued)

ORIGIN	*BRAND*	*PRODUCT*	*INGREDIENTS*	*REFERENCE*
Germany	Gut Bio	Tofu Sausages	Potable water, wheat protein* (27%), tofu (soy bean*, potable water, solidifying agents: magnesium chloride (E-511) and calcium sulfate (E-516) (2%), vegetable no. hydrogenated*, onion*, sea salt, condiment*, raw sugar*, yeast*, thickening agent: carob seed flour* (E-410), spices*, garlic*, celery* (* Ingredients from ecological agriculture)	https://www.aldi.pt
	Taifun	Grill Cracker	Tofu* 80% (soybeans* 55%, water, coagulant: magnesium chloride, calcium sulfate), cold-pressed sunflower oil*, oats*, vegetable broth* (sea salt, yeast extract*, sunflower oil*, leek*, carrots*, celery*, mace*, nutmeg*, parsley*), white pepper*, coriander*, black pepper*, garlic*, ginger*, nutmeg*, oregano*, parsley*, mace* (* from controlled organic cultivation)	https://www.taifun-tofu.de/
		Grilled Roast	Tofu* 62% (soybeans* 55%, water, coagulant: magnesium chloride, calcium sulfate), brown rice*, wheat protein*, cold-pressed sunflower oil*, vegetable broth* (sea salt, yeast extract*, sunflower oil*, leek*, carrots*, celery*, mace*, nutmeg*, parsley*), onions*, marjoram*, caraway*, black pepper*, mace*, nutmeg*, ginger*, turmeric* (* from controlled organic cultivation)	
	Vantastic Foods	Sim Sala Mi Vegan smoked salami in one piece	Seitan (76%) (water, wheat protein), yeast extract, flavor, onions, soy sauce (water, soybeans, sea salt), spices (contains mustard), salt, thickener: agar agar, locust bean gum, wheat starch, coloring: iron oxides	https://www.vantastic-foods.com/
		Vegan Bockwurst sausages	Seitan (water, wheat protein) 80%, rapeseed oil, yeast extract, salt, onions, flavoring, thickener: locust bean gum, agar-agar, xanthan; spices, wheat starch, coloring: iron oxide; beechwood smoke	
		Vegan Frying Sausages	Seitan* (water, wheat protein*) 80%, sunflower oil*, onion*, yeast extract*, salt, wheat starch*, spices*, thickener: locust bean gum* (* = certified organic)	
	Veggyness	Garliç Sujuk	Seitan* (water, wheat protein*) (70%), textured wheat protein* (wheat protein*, wheat flour*), high oleic sunflower oil*, yeast extract*, paprika*, garlic* (2%), spices*, onion*, rock salt, thickening agent: guar gum and locust bean gum, acidifying agent: vegan lactic acid, paprika extract**organic ingredient	
		Spacebar Chorizo vegan	Seitan* (67%): (water, wheat protein*), red pepper*, coconut fat*, onion*, yeast extract*, rock salt, spices* (contains celery* and mustard*) thickening agent: locust bean gum*, beech wood smoke (* certified organic)	https://www.veggyness.com/
		Spacebar Hem	Seitan* (86%): (water, wheat protein*), coconut oil*, hemp seed* (2.00%), rock salt, yeast extract*, spices*, onion*, thickening agent: locust bean gum* and guar gum*, beech wood smoke (* certified organic)	
		Vegan Merguez	Seitan* (72.00%): (water, wheat protein*), onions*, high oleic (high oleic) sunflower oil*, spices*, red peppers*, yeast extract*, rock salt, thickening agent: agar-agar*, guar gum* and locust bean gum*, smoke** (*certified organic,**natural beech wood smoke)	
		Vegane Bockwurst	Seitan (water, wheat protein) (80%), rapeseed oil, yeast extract, salt, onions, flavor, thickeners: locust bean gum, agar-agar and xanthan; spices, wheat starch, coloring agent: iron oxide; beechwood smoke	

(Continued)

TABLE 17.1 (Continued)

ORIGIN	BRAND	PRODUCT	INGREDIENTS	REFERENCE
		Vegane Bratwurst	Seitan (water, wheat protein) (72%), rapeseed oil, onions, yeast extract, rock salt, spices, flavorings, thickeners: agar-agar, locust bean gum	https://www.veggyness.com/
		Vegane Bratwurst Chorizo	Seitan* (64.00%): (water, wheat protein*), red peppers*, coconut fat*, onion*, yeast extract, spices* (contains celery and mustard), rock salt, thickening agent: locust bean gum*, smoke** (*certified organic,**natural beech wood smoke)	
		Vegane KaLeWu	Seitan (water, wheat protein) (80%), rapeseed oil, yeast extract, salt, onions, flavor, thickeners: locust bean gum, agar-agar and xanthan; spices, wheat starch, coloring agent: iron oxide; beechwood smoke	
		Vegane Sausage Rostbratwurst	Seitan* (81.00%): (water, wheat protein*), high oleic (high oleic) sunflower oil*, yeast extract*, rock salt, wheat starch*, onion*, spices*, thickening agent: carob seed flour (* certified organic)	
		Vegane Winzi-Weenies	Seitan* (82%): (water, wheat protein*), high oleic (high oleic) sunflower oil*, yeast extract*, stone salt, spices* (contains mustard*), onions* thickening agent: locust bean gum*, paprika extract*, smoke** (* certified organic,** natural beech wood smoke)	
	Viana	Original Currywurst—Tofu Queen	Sauce (60%): (tomato paste*, drinking water, raw cane sugar*, sunflower oil*, apple cider vinegar*, sea salt, curry*, thickener: locust bean gum*, yeast*), vegetarian sausage (40%): tofu (soy beans*, drinking water, Coagulant magnesium chloride), drinking water, wheat protein*, sunflower oil*, sea salt, spices*, raw cane sugar*, thickener: locust bean gum*, celery* (* controlled organic cultivation)	http://www.viana.de/
	Vitaquell	Tofu-Dings Feingewürzt	Tofu 41% (water, soybeans, coagulant nigari (magnesium chloride), calcium sulfate), carrot juice, egg white, rapeseed oil, onions, herbs, potato starch, sea salt, yeast extract, natural flavor, brandy vinegar, spices (contain mustard seeds)	https://www.vitaquell.de/
		Tofu-Wurst Gartenkräute	Tofu 42% (water, soybeans, coagulant nigari (magnesium chloride), calcium sulfate), egg white, rapeseed oil, onions, oil seed mixture (linseed, sesame, pumpkin seeds, sunflower seeds, almonds, mustard seeds), herbs, potato starch, sea salt, cane sugar, yeast extract	
		Vegetarische Brat-Dings	Tofu (water, soybeans, coagulant nigari (magnesium chloride), calcium sulfate), rapeseed oil, water, egg white, herbs, spices, sea salt, yeast extract, mustard seed, brandy vinegar, raw cane sugar, natural flavor, smoke flavor	
		Vegetarische Curry-Brat-Ding	Water, egg white, rapeseed oil, sweet lupine flour, corn starch, onions, potato starch, herbs, spices, barley malt, mustard seeds, curry (0.4%), brandy vinegar, sunflower oil, carrots, parsnips, rice flour, yeast extract, sea salt, natural flavor, freshly developed smoke made of natural woods	
		Vegetarische Vier Weiße	Tofu 32% (water, soybeans, nigari coagulant (magnesium chloride, calcium sulfate), rapeseed oil, water, palm oil, egg white, onions, stabilizer gum arabic, sea salt, yeast extract, potato starch, spices, herbs, smoke flavor, cane sugar, lemon oil	
	Wheaty	Bavarian Style	Seitan* (water, wheat protein*) 82%, high oleic sunflower oil*, rock salt, yeast extract*, onion*, spices*, thickening agent guar gum*, parsley*, lemon* (* certified organic)	https://www.wheaty.com/
		Chorizo	Seitan* (water, wheat protein*) 78%, coconut fat*, onion*, yeast extract*, soy sauce* (water, soybeans*, sea salt), red bell pepper*, garlic*, rock salt, spices*, paprika extract*, acidifying agent: vegan lactic acid (* certified organic)	

(Continued)

TABLE 17.1 (Continued)

ORIGIN	BRAND	PRODUCT	INGREDIENTS	REFERENCE
		Country Style	Seitan* (water, wheat protein*) 86%, high oleic sunflower oil*, coconut fat*, rock salt, spices* (contains celery* and mustard*), yeast extract*, thickening agent locust bean gum* and guar gum*, onion*, red bell pepper*, paprika extract*, natural beech wood smoke (* certified organic)	https://www.wheaty.com/
		Vegan Merguez	Seitan* (water, wheat protein*) 72%, onion*, high oleic sunflower oil*, spices*, red bell pepper*, yeast extract*, rock salt, thickening agent: agar-agar*, guar gum* and locust bean gum*, natural beech wood smoke (* certified organic)	
		Hot Dog/Tiny Hot Dog	Seitan* (water, wheat protein*) 82%, high oleic sunflower oil*, yeast extract*, rock salt, spices* (contains mustard*), onion*, thickening agent locust bean gum*, paprika extract*, natural beech wood smoke (* certified organic)	
		Herbalicious	Seitan* (water, wheat protein*) 81%, high oleic sunflower oil*, yeast extract*, rock salt, wheat starch*, onion*, spices*, thickening agent locust bean gum* (* certified organic)	
		Mini-Herby	Seitan* (water, wheat protein*) 81%, high oleic sunflower oil*, yeast extract*, rock salt, wheat starch*, onion*, spices*, thickening agent locust bean gum* (* certified organic)	
		Red Hot Chili Peppers	Seitan* (water, wheat protein*) 82%, coconut fat*, spices* (contains celery* and mustard*), yeast extract*, rock salt, onion*, thickening agent locust bean gum*, chili* (0,3%), paprika extract*, natural beech wood smoke (* certified organic)	
		Red Sausages	Seitan* (water, wheat protein*) 73%, onion*, high oleic sunflower oil*, yeast extract*, rock salt, paprika*, spices*, thickening agent: agar-agar*, guar gum* and locust bean gum*, natural beech wood smoke (* certified organic)	
		Vegan BBQ Mix	Seitan* (water, wheat protein*) 74%, coconut fat*, red bell pepper*, high oleic sunflower oil*, rock salt, yeast extract*, spices* (contains celery* and mustard*), onion*, oat fiber*, wheat starch*, potato starch*, thickening agent xanthan* and locust bean gum, natural beech wood smoke (*certified organic)	
		Vegan farmer crackers	Seitan* (water, wheat protein*) 86%, high oleic sunflower oil*, coconut fat*, rock salt, spices* (contains celery* and mustard*), yeast extract*, thickener: locust bean gum* and guar gum*, onions*, Red pepper*, pepper extract*, beech wood smoke (* from controlled organic cultivation)	
		Vegan Sucuk	Seitan* (water, wheat protein*) (70%), textured wheat protein* (wheat protein*, wheat flour*), high oleic sunflower oil*, yeast extract*, paprika*, garlic* (2%), spices*, onion*, rock salt, thickening agent:(guar gum* and locust bean gum*), acidifying agent: vegan lactic acid, paprika extract* (* certified organic)	
Ireland	Denny	Meat Free Sausages	Water, rehydrated textured soya and wheat protein (16%), soya protein, wheat protein, salt, soya bean oil, natural flavoring), rusk (wheat), coconut oil, soya protein concentrate (4%), wheat starch, chicory root fiber, natural flavorings, stabilizer: methyl cellulose; salt, yeast extract, sodium alginate, coloring foods: beetroot, safflower; spice extracts: black pepper, ginger, nutmeg; herbs, spices, flavoring	https://www.denny.ie/

(Continued)

TABLE 17.1 (Continued)

ORIGIN	BRAND	PRODUCT	INGREDIENTS	REFERENCE
	Eden	Eden Vegan Sausages	Water, soya, sunflower oil, potato starch, maize semolina, rice semolina, cellulose, tomato puree, spices, salt, rusk [wheat flour (wheat flour, calcium carbonate, iron, niacin, thiamin), salt], maltodextrin, tapioca starch, stabilisers: sodium triphosphate, sodium diphosphate, yeast extract, barley malt extract, spice extracts, rape oil, preservative: sodium sulphite, beetroot juice powder, natural flavourings, antioxidant: sodium ascorbate, thickener: guar gum, smoke flavouring	https://store.edenfoods.com/
	Glas	Perfectly Plant-based Sausages	Water, soya flour, potato starch, textured soya, sunflower oil, maize semolina, rice semolina, soya protein, spices, wheat gluten, salt, cellulose, tomatoes, tapioca starch, yeast extract, maltodextrin, wheat flour, soybean oil, rapeseed oil, barley malt extract, spice extract, natural flavors, beetroot juice powder, thickener: guar gum, smoke flavoring, antioxidant: sodium ascorbate, calcium carbonate, iron, niacin, thiamin, filled into sodium alginate casing	https://glasfoods.ie/
Israel	Tivall Food Industries Ltd.	Cocktail Sausage	Rehydrated wheat and soya proteins (46%), water, vegetable oil (sunflower, rapeseed in varying proportions), egg white powder, potato starch, dextrose, hydrolyzed vegetable protein (soya, maize), garlic powder, salt, yeast extract, stabilisers (guar gum, xanthan gum), flavouring, spices, maltodextrin, colour (iron oxide), vitamins & minerals (tricalcium phosphate, vitamin C, vitamin B3, zinc oxide, ferric diphosphate, vitamin E, vitamin B5, vitamin B6, vitamin B2, vitamin B1, folic acid, vitamin B12)	http://www.trialiafoods.com.au/
		Smoked Frankfurters	Rehydrated wheat and soya proteins (46%), water, vegetable oil (sunflower, rapeseed in varying proportions), egg white powder, potato starch, dextrose, hydrolysed vegetable protein (soya, maize), garlic powder, salt, yeast extract, stabilizers (guar gum, xanthan gum), flavouring, spices, maltodextrin, colour (iron oxide), vitamins & minerals (tricalcium phosphate, vitamin C, vitamin B3, zinc oxide, ferric diphosphate, vitamin E, vitamin B5, vitamin B6, vitamin B2, vitamin B1, folic acid, vitamin B12)	
Italy	Veghiamo—Deli Slices	Black Pepper	Water, gluten from wheat* 32%, durum wheat flour* 8%, chickpea flour*, extra virgin olive oil*, red beetroot powder*, natural flavoring, tamari sauce* (water, soybean seeds*, salt, shochu* (water, rice*, salt), toasted soybean seeds*), black pepper*1.5%, soft wheat flour*, sourdough, coconut oil*, salt, mixture of spices and aromatic herbs* (*organic ingredients)	https://www.felsineoveg.com/
		Carpaccio	Water, gluten from wheat*34%, durum wheat flour* 8%, sunflower oil*, natural flavorings, coloring: betanin, citric acid, salt, soft wheat flour*, pea protein, sourdough, white pepper powder*, garlic powder* (*organic ingredients)	
		Classic	Water, gluten from wheat* 31%, durum wheat flour* 8%, chickpea flour*, extra virgin olive oil*, natural flavorings, red beetroot powder*, soft wheat flour*, salt, sourdough, sunflower oil*, pepper powder*, garlic powder* (*organic ingredients)	
		Cured Ham Flavored	Water, gluten from wheat 31%, natural flavorings, durum wheat flour 7%, sunflower oil, coloring: beetroot red, pea protein, salt, soft wheat flour, sourdough, white pepper powder, garlic powder	
		Lupins	Water, lupin flour* 27%, gluten from wheat* 23%, sunflower oil*, wheat starch*, natural flavoring, salt, apple vinegar*, tamari sauce* (water, soybean seeds*, salt, shochu* (water, rice*, salt), toasted soybean seeds*), coconut oil*, chickpea flour*, durum wheat flour*, dehydrated onion powder*, sourdough, mixture of spices and aromatic herbs*, acidity regulator: sodium bicarbonate (*organic ingredients)	

(Continued)

TABLE 17.1 (Continued)

ORIGIN	*BRAND*	*PRODUCT*	*INGREDIENTS*	*REFERENCE*
		Pepperoni Flavored	Water, gluten from wheat 35%, natural flavorings, durum wheat flour 7%, chickpea flour, extra virgin olive oil, coloring: beetroot red, salt, soft wheat flour, chili pepper powder, sourdough, sunflower oil, garlic powder	https://www.felsineoveg.com/
		Salmon Flavored	Water, gluten from wheat 31%, coloring concentrate from paprika, durum wheat flour 6%, natural flavorings, chickpea flour, extra virgin olive oil, soft wheat flour, salt, sourdough, sunflower oil, white pepper powder, garlic powder	
		Smoked Flavored	Water, gluten from wheat* 34%, durum wheat flour* 8%, chickpea flour*, natural flavorings, extra virgin olive oil*, tamari sauce* (water, soybean seeds*, salt, shochu* (water, rice*, salt), toasted soybean seeds*), coconut oil*, soft wheat flour*, sourdough, salt, mixture of spices and aromatic herbs* (*organic ingredients)	
		Smoked Ham Flavored	Water, gluten from wheat 30%, durum wheat flour 7%, chickpea flour, natural flavorings, extra virgin olive oil, coloring: beetroot red, flavor, smoke flavorings, soft wheat flour, salt, sourdough, sunflower oil, pepper powder, garlic powder	
		Spicy	Water, gluten from wheat* 32%, durum wheat flour* 8%, chickpea flour*, extra virgin olive oil*, natural flavoring, tamari sauce* (water, soybean seeds*, salt, shochu* (water, rice*, salt), toasted soybean seeds*), mixture of spices and aromatic herbs*, coconut oil*, soft wheat flour*, sourdough, salt (*organic ingredients)	
		Truffle	Water, gluten from wheat* 32%, durum wheat flour* 8%, chickpea flour*, extra virgin olive oil*, natural flavorings, truffle dices (*Tuber aestivum*)* 2%, tamari sauce* (water, soybean seeds*, salt, shochu* (water, rice*, salt), toasted soybean seeds*), coconut oil*, soft wheat flour*, sourdough, salt, mixture of spices and aromatic herbs* (*organic ingredients)	
		Turmeric & Ginger	Water, gluten from wheat* 32%, durum wheat flour* 9%, chickpea flour*, sunflower oil*, natural flavoring, salt, ginger powder*0,4%, turmeric powder* 0,3%, chili pepper powder*, sourdough, pepper powder*, garlic powder* (*organic ingredients)	
Japan	Terra Foods	Vegan Marude Sausage	Soy protein, okara, devil's root powder, salt, beet sugar, yeast extract, spices, vegetable powder (onion, celery), maltodextrin/calcium hydroxide, trehalose	https://www.terrafoods.shop/
New Zeland	Tonzu	Organic Vegan Sausages—Curry Spice	Certified organic activated whole soybeans, organic sunflower oil, organic buckwheat, organic tapioca flour, curry spices, organic onion flakes, organic quinoa, organic rice flour, psyllium husks, sea salt, vegan yeast extract, tomato powder, cocoa, ground nutmeg, pure nigari	https://tonzu.co.nz/
		Organic Vegan Sausages—Garlic & Chilli	Certified organic activated whole soybeans, filtered water, organic sunflower oil, organic tapioca flour, organic garlic granules (2.8%), organic rice flour, New Zealand sea salt, tomato powder, yeast extract, mixed herbs, organic cocoa, chilli powder (0.2%), ground nutmeg, pure nigari	
		Organic Vegan Sausages—Italian Herb	Certified organic activated whole soybeans, filtered water, organic sunflower oil, organic tapioca flour, organic onion flakes, organic rice flour, New Zealand sea salt, yeast extract, pure nigari, mixed herbs (0.4%), organic cocoa, organic garlic granules, ground nutmeg	

(Continued)

TABLE 17.1 (Continued)

ORIGIN	BRAND	PRODUCT	INGREDIENTS	REFERENCE
		Organic Vegan Sausages—Sage & Onion	Certified organic activated whole soybeans, filtered water, organic sunflower oil, organic tapioca flour, organic onion flakes (2.7%), organic rice flour, sage (0.9%), New Zealand sea salt, yeast extract, tomato powder, organic cocoa, ground nutmeg, pure nigari	
Poland	Meatless Meat/ Bezmiesnymiesny	Meatless Classic Kabanos	Water, vegetable proteins (wheat, pea), rapeseed oil, coconut oil, inactive yeast flakes, soybeans, soy sauce, Kłodawa salt, black pepper, natural spices, spice extracts, rice flour, dextrose, natural flavors, fiber (bamboo, pea, plantain), maltodextrin	https://www.bezmiesnymiesny.pl/
		Meatless Chilli Kabanos	Water, vegetable proteins (wheat, pea), rapeseed oil, coconut oil, inactive yeast flakes, soybeans, soy sauce, salt, natural spices (including chili flakes), spice extracts, rice flour, dextrose, natural flavors, fiber (bamboo, pea, plantain), maltodextrin	
		Meatless Country Sausages	Water, vegetable proteins (wheat, pea), rapeseed oil, tofu, inactive yeast flakes, soy flour, soybeans, Kłodawa salt, soy sauce, natural spices (including dried onion, dried garlic), smoke flavor, fibers (bamboo, pea, plantain)	
		Meatless Italian Sausages	Water, vegetable proteins (pea, wheat), rapeseed oil, inactive yeast flakes, soy flour, soybeans, Kłodawa salt, soy sauce, natural spices (including dried tomatoes, smoked paprika, oregano), smoke flavor, fibers (bamboo, pea, plantain), spice extracts	
		Meatless Onion Kabanos	Water, vegetable proteins (wheat, pea), rapeseed oil, coconut oil, inactive yeast flakes, soya, soy sauce, Kłodawa salt, natural spices (including dried garlic, dried onion), spice extracts, rice flour, dextrose, natural flavors, fibers (bamboo, pea, plantain), maltodextrin	
		Meatless Pepperoni	Water, wheat protein, tomato concentrate, rapeseed oil, inactive yeast flakes, soybean, potato starch, fiber (bamboo, pea, plantain), natural spices (including sweet pepper, hot pepper, smoked pepper), mustard, Kłodawa salt	
		Meatless sausages	Water, wheat protein, rapeseed oil, inactive yeast flakes, soybean, Kłodawa salt, natural spices, fibers (bamboo, pea, plantain), glucose, flavors (including smoke flavor), spice extracts, carboxymethylcellulose, starch (rice, potato)	
		Meatless Sausages Curry	Water, vegetable proteins (pea, wheat), rapeseed oil, apple juice, inactive yeast flakes, soy flour, soy, Kłodawa salt, soy sauce, glucose, sugar, natural spices (including turmeric, sweet pepper, coriander), fiber (bamboo, pea, plantain), spice extracts	
		Meatless Spanish Sausages	Water, vegetable proteins (pea, wheat), rapeseed oil, inactive yeast flakes, soy flour, soybeans, Kłodawa salt, soy sauce, natural spices (including cumin, chili pepper), smoke flavor, fibers (bamboo, pea), plantain), spice extracts	
		Meatless White Sausages	water, vegetable proteins (wheat, pea), rapeseed oil, inactive yeast flakes, soybean, soy sauce, Kłodawa salt, black pepper, natural spices (including marjoram), modified corn starch, soy flour, glucose, aromas, sugar, fibers (bamboo, pea, plantain)	
	Polsoja	Breakfast Sausages	Water, soy protein, vegetal oil, wheat protein (gluten), methyl-cellulose (binder), corn starch (e1422), spice aromas and extracts, salt, yeast extract, vegetal proteins	http://www.polsoja.com.pl/

(Continued)

TABLE 17.1 (Continued)

ORIGIN	BRAND	PRODUCT	INGREDIENTS	REFERENCE
Portugal	Biodharma	Salsichão	Tofu (soy beans, nigari), cornmeal, arrowroot flour, beer yeast flakes, seaweed agar, olives, onion, garlic, sweet pepper, curry, mustard, extra virgin olive oil, tamari, water	https://biodharma.pt/
	Casa da Prisca	Soy Alheira	Soybean (70%), wheat bread (wheat flour type 65, water, salt, yeast), olive oil, salt, spices, garlic, starter crops, thickener (E407), stabilizer (E508; E516), water, water regulator acidity (E260; E270), dextrose, antioxidant (E330, E334)	https://www.casadaprisca.pt/
	Irmãos Oliveira	Vegetables and Mushrooms Alheira	Mushrooms 36% (Shiitake*), water, wheat bread* (wheat flour, water, natural yeast, salt) vegetables 16% (onion*, pepper*, tomato*, garlic*, parsley*), olive oil* salt and spices. Non-edible cellulose casing (* biological ingredients)	https://www.irmaosoliveiras.pt/
	Nobre	Vegalia Soy Sausages	Water, sunflower oil, soy protein (7%), egg albumin, starch, salt, flavorings, smoke aroma, yeast extract, gelling agents (guar gum, xanthan gum), spices	https://www.nobre.pt/
		Vegalia Tofu Sausages	Water, sunflower oil, tofu (7%) (water, soya, coagulants: nigari (water, magnesium chloride), calcium chloride), egg albumin, soya protein, potato starch, salt, flavorings, smoke flavor, yeast extract, gelling agents (guar gum, xanthan gum), spices, fine herbs (0.2%)	
	Próvida	Tofu Alheira	Tofu* (38%) (soy*, water, natural magnesium nigari-chloride], whole wheat bread* (whole and white wheat flour*, water, sea salt, natural yeast (wheat flour* and water)), seitan* (wheat flour*, wheat gluten*, water), pepper*, garlic powder*, soybean sauce* (water, soybean*, whole wheat, sea salt and koji-aspergillus oryzae), parsley*, coriander*, cumin*, fennel* and ginger* Non-edible casing (* organic production)	https://www.provida.pt/
	Topitéu	Wild Alheira	Mushrooms (45%), wheat bread (wheat flour, water, salt, yeast) (31%), cooking syrup (11%), green asparagus (6%), extra virgin olive oil (5%), spices (1%), salt (1%). Edible casing	https://topiteu.pt/
	Veg In	Vegan Alheira	Water, bread baked in a wood oven (wheat flour, water, yeast, salt), protein and vegetable fibers (wheat gluten, soy flour, isolated soy protein, wheat flour, rice bran), sunflower oil, olive oil, red pepper mass (red pepper, salt), spices, vinegar, onion, salt, garlic paste (garlic, salt) and natural smoke	http://www.vegin.pt/
Spain	Bio Veggie Sausages (Ahimsa)	Cheese	Water, sunflower oil*, wheat gluten*, oat flakes*, Gouda cheese*(5.8%), textured soya*, onion*, wheat flour*, egg albumin*, sea salt, spices*, brewer's yeast* (contains wheat*), vegetable stock* (contains celery*) (*Organically grown).	https://www.ecologicosahimsa.com/
		Chorizo Tofu and Seitan	Water, wheat gluten* (seitan) (22.3%), sunflower oil*, tofu* (11.07%) (soybeans*, water, food grade coagulant) calcium sulphate)), spices* (sweet paprika*, sea salt, cane sugar*, onion*, thyme*, garlic*, corn starch*, rosemary*), textured soya*, tomato*, brewer's yeast*(contains wheat*), spices*, cornstarch*, thickeners (xanthan gum and carrageen), hummus (common salt, natural hummus aromas), emulsifier (soy lecithin) (*Organically grown)	
		Fine herbs	Water, sunflower oil*, wheat gluten*, oat flakes*, textured soya*, onion*, tomato*, wheat flour*, egg albumin*, sea salt, brewer's yeast* (contains wheat*), spices*, parsley*(0.5%), sweet paprika* (*Organically grown)	
		German style	Water, wheat gluten*, Tofu* (water, soybeans*, food grade coagulant [calcium sulphate]), sunflower oil*, brewer's yeast* (contains wheat*), sea salt, corn starch*, spices*, soy sauce*(contains wheat*), sweet cayenne pepper*, thickener (xantana gum), garlic*, thickener (carrageen), apple vinegar*, hummus (common salt, natural hummus aromas), emulsifier (soy lecithin) (*Organically grown)	

(Continued)

TABLE 17.1 (Continued)

ORIGIN	BRAND	PRODUCT	INGREDIENTS	REFERENCE
		Smoked	Water, sunflower oil*, wheat gluten*, soya protein*, tomato*, yeast* (contains wheat*), soy sauce* (contains wheat*), hummus aroma (salt, hummus aroma) (0.5%), sea salt, thickener (carrageen), yeast extract* (contains wheat*), emulsifier (soy lecithin) (*Organically grown)	
		Tofu and Seitan	Wheat gluten*(seitan) (25.6%), sunflower oil*, tofu*(12.8%) (soybeans*, water, nigari), brewer's yeast*, spices*(contains celery*), sea salt, corn starch*, emulsifier/soya lecithin), thickener (xanthan gum), hummus (common salt, natural hummus aromas), thickener (carrageen) (*Organically grown)	
	Biosurya	Salsicha vegetal de tofu y quinoa	Water, tofu* (23.6%) (water, soybeans*, stabilizer: calcium sulfate), wheat gluten*, cooked quinoa* (11.3%) (water, quinoa*, sea salt), sunflower oil*, carrot*, toasted onion*, vegetable broth* (sea salt*, rice flour*, yeast extract*, onion*, garlic*, sunflower oil*), onion*, sea salt, starch corn*, garlic*, thickeners: xanthan gum and carrageenan, sweet paprika*, emulsifier: soy lecithin** From organic farming	https://www.biosurya.com/
	Hijas del Sol	Alemana	Water, wheat gluten*, sunflower oil*, tofu* (water, soy beans*, coagulant [calcium sulfate]), beer yeast* (wheat*), carrots*, sea salt, soy sauce* (soy*, whole wheat (gluten)*, *Aspergillus oryzae*, water), onion*, corn starch*, spices* (bay leaf*, apple blossom*, garlic*, onion*, corn starch*, sweet pepper* pepper black*), thickener (xanthan gum), smoke (common salt, natural smoke flavors), garlic*, thickener (carrageenan), apple cider vinegar*, emulsifier (soy lecithin).* From organic agriculture	https://www.elcorteingles.es/
		Chipolata	Water, wheat gluten*, sunflower oil*, tofu* (water, soy beans*, coagulant [calcium sulfate]), beer yeast* (contains wheat*), sea salt, soy sauce* (soy beans)*, whole wheat*, *Aspergillus oryzae*, water), corn starch*, thickener (xanthan gum), parsley*, thickener (carrageenan), white pepper*, black pepper*, emulsifier (soy lecithin). Ecological cultivation	
		Seitan	Water, wheat gluten* (seitan*) (26%), sunflower oil*, tofu (water, soy*, stabilizer [calcium sulfate]), yeast* (contains wheat* [gluten]), sea salt, starch corn*, garlic*, thickener (xanthan gum), white pepper*, gelling agent (carrageenan), emulsifier (soy lecithin*).* from organic farming	
Sweden	Hänsans Kök	Plant-Based Chorizo	Water, wheat protein (10.1%), vegetable oil (rapeseed, sunflower), onion, stabilizer (methylcellulose, carrageenan, guar gum, xanthan gum), soy protein (3.5%), starch (wheat, potato), potato protein, yeast extract, dried red pepper, salt, spices (chili, dried garlic), garlic, vinegar, flavorings (lycopene), pea fiber, spice extract (jalapeno, paprika/pepper)	https://www.halsanskok.se/
		Plant-Based Sausage	Water, wheat protein (10.3%), soy protein (5.3%), spices (dried onion, dried garlic, black pepper, mustard, pepper, nutmeg/muscat), onion, stabilizers (methylcellulose, xanthan gum, guar gum, carrageenan), starch (wheat, potato), potato protein, vegetable oil (rapeseed, sunflower), salt, yeast extract, beetroot powder, rice flour, hydrolyzed soy protein, aromas, vinegar, smoked maltodextrin, lycopene, pea protein	
	Peas of Heaven	Chorizo	Water, rapeseed oil, pea protein, Stabilizer (carrageenan, methylcellulose), corn starch, spices, salt, paprika, chili, garlic, vegetable broth, citrus fibers, dextrose, aroma, yeast extract, acidity regulator (malic acid), sugar, vitamin B12, iron, riboflavin	https://peasofheaven.com/

(Continued)

TABLE 17.1 (Continued)

ORIGIN	*BRAND*	*PRODUCT*	*INGREDIENTS*	*REFERENCE*
		Hot Dog	Water, rapeseed oil, pea protein, Stabilizer (carrageenan, methylcellulose), corn starch, salt, spices, onion, paprika, vegetable broth, citrus fibers, dextrose, aroma, yeast extract, acidity regulator (malic acid), sugar, vitamin B12, iron, riboflavin	
		Perfect Bratwurst	Water, rapeseed oil, pea protein, pea flour, starch (pea, potato), salt, thickening agents (methylcellulose, processed eucheuma seaweed), spices, garlic, citrus fibers, caramel powder, acidity regulator (tartaric acid), yeast extract, grape sugar, aroma, stabilizer (E450)	
		Perfect Chorizo	Water, rapeseed oil, pea protein, pea flour, starches (pea, potato), salt, thickener (methylcellulose, processed *Eucheuma* seaweed), spices, chili, garlic, paprika, citrus fibers, acidity regulator (tartaric acid), yeast extract, dextrose, aroma, stabilizer (E450), Sugar	
		Prinskorv	Water, rapeseed oil, pea protein, Stabilizer (carrageenan, methylcellulose), corn starch, salt, spices, onion, paprika, vegetable broth, citrus fibers, dextrose, aroma, yeast extract, acidity regulator (malic acid), sugar, vitamin B12, iron, riboflavin	
Switzerland	Migros	V-Love Plant-Based Bratwurst	Water, rapeseed oil, potato protein (contains sulfites), pea protein, thickeners: carrageenan, methyl cellulose, processed *Eucheuma* algae, locust bean gum, konjac, spice extract, iodized table salt, seasoning, spices, starch, flaxseed flour, enzymes	https://www.migros.ch/
		V-Love Plant-Based Grilled Sausage	Water, rapeseed oil, pea protein, thickeners: carrageenan, methyl cellulose, processed *Eucheuma* algae, locust bean gum and konjac, potato protein (contains sulfites), table salt, spices, glucose, coloring: paprika extract, starch, yeast extract, seasoning, flavor, flaxseed flour, enzymes	
	Vegusto	No Moo, Vegi-sausage Wiener	Water, wheat protein (gluten), coconut oil, sunflower oil, spices (contains celery), rock salt, rice starch, yeast, potato starch, rice flour, binder (guar gum, xanthan), almond butter, Colourants (radish and carrot concentrate), binder (carrageen), flavouring (vegetable), natural smoke flavour, lactic acid (vegetable)	https://www.vegusto.ch/
		Vegi-sausage, Bianca	Water, wheat protein (gluten), coconut oil, sunflower oil, spices (contains mustard), breadcrumbs (wheat, table salt (not iodized), yeast), rock salt, rice starch, natural smoke flavour, lemon juice, binder (guar gum, xanthan), vegetable fibres	
		Vegi-sausage, Country-style	Water, wheat protein (gluten), sunflower oil, spices, yeast, rock salt, wheat flour, pear syrup, rice starch, natural smoke flavour, lactic acid (vegetable), paprika oil, binder (guar gum, xanthan)	
		Vegi-sausage, Curry	Water, wheat protein (gluten), sunflower oil, spices (contains celery), rock salt, breadcrumbs (wheat, table salt (not iodized), yeast), rice starch, curry 0.7%, carrot and apple concentrate	
		Vegi-sausage, Farmhouse-style	Water, wheat protein (gluten), sunflower oil, coconut oil, spices (contains celery), rock salt, breadcrumbs (wheat, table salt (not iodized), yeast), rice starch, Colourants (radish and carrot concentrate), antioxidant (ascorbic acid), binder (guar gum, xanthan), natural smoke flavour	
	Vegusto (cont.)	Vegi-Sausage with No-Muh	Water, rice flour, pea protein, corn flour, coconut oil, sunflower oil, yeast, plant fibres, spices, potato starch, rock salt, flavouring (vegetable), rice starch, almond butter, fruit juice, binder (carrageen), natural smoke flavour, Colourants (radish and carrot concentrate, beta-carotene), antioxidant (ascorbic acid)	https://www.vegusto.ch/
		Vegi-sausage, Onion	Water, wheat protein (gluten), sunflower oil, spices (contains celery), rock salt, breadcrumbs (wheat, table salt (not iodized), yeast), rice starch, onions 1.6%	

(Continued)

TABLE 17.1 (Continued)

ORIGIN	BRAND	PRODUCT	INGREDIENTS	REFERENCE
		Vegi-sausage, Rosso	Water, wheat protein (gluten), coconut oil, sunflower oil, spices, breadcrumbs (wheat, table salt (not iodized), yeast), rice starch, dried tomatoes, rock salt, natural smoke flavour, Colourants (radish and carrot concentrate), stabilizer (ascorbic acid)	
The Netherlands	Schoutenfood	Frankfurter	Water, 14% vegetable proteins (soy, wheat gluten), rapeseed oil, onion, free range egg white, vinegar, natural flavoring, rice bran, yeast extract, dextrose, salt, wheat flour, smoked water, extracts (onion, garlic, spice), sugar, mineral (ferrous fumarate), vitamin B12, color (E172)	https://www.schoutenfood.com/
		Smoked Sausage	Water, 14% vegetable proteins (soy, wheat gluten), rapeseed oil, onion, free range egg white, vinegar, natural flavoring, rice bran, yeast extract, dextrose, salt, wheat flour, smoked water, extracts (onion, garlic, spice), sugar, mineral (ferrous fumarate), vitamin B12, color (E172)	
	Unox	Vegetarian Smoked Sausage	Water, vegetable oils 15.9% (sunflower, coconut), thickeners (guar gum, sodium alginate, xanthan gum, cellulose, carrageenan, methylcellulose), protein powder 4.1%, pea fiber, pea protein 1.5%, aroma, spices, salt, hydrolyzed corn protein, food acid: citric acid, yeast extract, mushroom extract. aroma, dextrose, glucose syrup, smoke flavor, sugar, caramel powder, antioxidants (ascorbic acid, alpha-tocopherol), colors (iron oxide, anthocyanin), stabilizer: E450	https://www.unox.nl/
United Kingdom	BonSan	Bonsan Organic Kofu Grill Sausages	Tofu*80% (water, soya beans*, coagulant: kombucha*[black tea* with sugar* fermented into vinegar] and nigari [magnesium chloride]), sunflower oil*, sea salt, tamari* (soya sauce)*, curry spices* 1.3% (contains celery*, mustard*, garlic*), sultanas*, yeast flakes*, thickening agent: locust bean gum*, corn starch*.*certified organic ingredients	https://www.bonsan.co.uk/
	Cauldron Foods	Vegetarian Cumberland Sausages	Rehydrated textured vegetable protein (45%) (water, soya protein, potato starch, wheat gluten, stabilizer: dicalcium phosphate), onion, rapeseed oil, cumberland seasoning (5%) (yeast extract, salt, black pepper, sugar, rosemary, barley malt extract, bay, carrot, leek, rapeseed oil), dried free range egg white, soya protein, stabilizer: methylcellulose	https://www.cauldronfoods.co.uk/
		Vegetarian Lincolnshire Sausages	Rehydrated textured vegetable protein (45%) (water, soya protein, potato starch, wheat gluten, stabilizer: dicalcium phosphate), water, onion, rapeseed oil, lincolnshire seasoning (5%) (yeast extract, salt, potassium chloride, herbs, fructose, white pepper, rusk (wheat flour, salt, raising agent: ammonium bicarbonate) barley malt extract, carrot powder, leek powder, sage extract, nutmeg extract), dried free range egg white, soya protein, stabilizer: methylcellulose	
	GoodLife	Good Life Mighty Non-Meaty Sausages	Water, mushrooms (15%), red onions (13%), lentils (12%), soy protein, spinach (4.5%), fried onions (onions, sunflower oil), fortified wheat flour (wheat flour, calcium carbonate, iron, niacin, thiamin), rusk (fortified wheat flour (wheat flour, calcium carbonate, iron, niacin, thiamin), water, salt, raising agent (ammonium carbonates)), potato flakes, gram flour, pea protein, yeast extract, rapeseed oil, garlic, stabiliser (methyl cellulose), salt, yeast, sage, coriander, nutmeg, thyme, black pepper, acidity regulator (citric acid), sausages coated in sodium alginate casing	https://www.goodlife.co.uk/
	Heck Foods	Heck Meat Free Chipolatas	Water, pea flour, meat-free base (thickener (methyl cellulose), pea fibre, modified starch, flavour, beetroot powder (maltodextrin, beetroot juice concentrate, acidity regulator (citric acid)), tea extract, spice extracts, smoke flavour), textured vegetable protein (pea protein isolate, pea flour), pea protein (2%), seasoning (salt, spice, herb, yeast extract, preservative (sodium sulphite), antioxidant (ascorbic acid), spice extract), citrus fibre, sunflower oil. Filled into calcium alginate casing	https://www.heckfood.co.uk/

(Continued)

TABLE 17.1 (Continued)

ORIGIN	BRAND	PRODUCT	INGREDIENTS	REFERENCE
		Vegan Italia Chipolatas	Water, pea flour, meat-free base (thickener (methyl cellulose), pea fibre, modified starch, flavour, beetroot powder (maltodextrin, beetroot juice concentrate, acidity regulator (citric acid)), tea extract, spice extracts), pea protein (4%), oven dried tomato (3%), basil (3%), seasoning (rice flour, salt, sugar, leek powder, preservative (sodium sulphite), antioxidant (ascorbic acid), yeast extract, spice, acidity regulator (citric acid), spice extract, natural flavouring, herb extract), citrus fibre, vegan mozzarella-style cheese (1%) (water, coconut oil (25%), modified potato starch, gluten free oat fibre, maize starch, salt, natural flavourings, modified maize starch, tricalcium citrate, thickeners (carrageenan, guar gum), acidity regulators (lactic acid, sodium lactate), colour (mixed carotenes)), garlic (1%), sunflower oil. Filled into calcium alginate casing	
	Just WholeFoods	Organic & Vegan Sausage Mix	Extured soya protein*, wholemeal breadcrumbs* (wheat*, yeast, salt), brown rice flour*, couscous*, onion*, vegetable bouillon* (sea salt, rice flour*, vegetables* [onion*, carrots*, parsnip*], yeast extract, sunflower oil*, parsley*, turmeric*), tomato powder*, celery powder*, leek*, sea salt, parsley*, black pepper*, sage*.*certified organic ingredient	https://www.justwholefoods.co.uk/
	Linda McCartney Foods	Vegetarian Red Onion & Rosemary Sausages	Rehydrated textured soya protein (62%), red onion (19%), water, seasoning (red onion powder, kibbled red onion, salt, sugar, yeast extract, rosemary, sunflower oil, white pepper, ginger), soya protein concentrate, rapeseed oil, stabilizer: methyl cellulose	https://lindamccartneyfoods.co.uk/
	No Meat	No Porkies Cumberland Sausages	Water, rehydrated textured soya protein (31%) (textured soya protein, water), sunflower oil, thickener: methyl cellulose; pea fibre, flavouring, modified potato starch, modified maize starch, salt, black pepper, stabiliser: triphosphates; dried sage, ground coriander, yeast extract, beetroot powder (maltodextrin, beetroot juice from concentrate, acidity regulator: citric acid), tea extract, dried parsley, white pepper, garlic extract, preservative: sodium sulphite; ground nutmeg, antioxidant: ascorbic acid; black pepper extract, nutmeg extract, coriander extract, smoke flavouring	https://www.no-meat.co.uk/
	Suma	Suma Sausage Mix—Vegan	Whole wheat flour, palm oil, textured soya protein (21%), thickener: methylcellulose; maltodextrin; seasoning (sugar, potato starch, yeast extract, rapeseed oil, herb & spice extracts), hydrolysed maize protein), salt, onion powder, raising agent: ammonium carbonate; sage, thyme, pepper & beetroot juice powder	www.suma.coop
	Symington's Ltd	Granose Lincolnshire Sausage Mix	Dried textured soya protein (45%), breadcrumbs (32%) [fortified wheatflour (wheatflour, calcium carbonate, iron, niacin, thiamin), yeast, salt], flavourings (contain barley), wheat gluten, palm oil, dried parsley, onion powder, stabiliser (methylcellulose), dried sage, ground black pepper	https://www.sainsburys.co.uk/
	The Tofoo	Sage & onion sizzlers	Tofu* (60%) (water, soya beans*, nigari), maize flour*, water, rapeseed oil*, nutritional yeast*, onion powder*(1%), sea salt, sage*(0.5%), ground linseed*, ground white pepper*, ground nutmeg* (* organic ingredients)	https://tofoo.co.uk/
	VBites	Garlic Sausg Slices	Water, wheat gluten, soya protein, coconut oil, rapeseed oil, garlic powder, potato starch, salt, yeast extract, natural flavoring, dried yeast, sugar, thickener: carrageenan, onion powder, preservative: potassium sorbate, black pepper, color: iron oxide	https://www.vbites.com/
		Lincolmmmshire Sausgs	Water, wheat gluten, rapeseed oil, onion, soya protein, whole meal wheat rusk, sugar, salt, dried yeast, dried herbs, yeast extract, thickener: methyl cellulose; natural flavoring, preservative: potassium sorbate; garlic powder, barley malt extract	

(Continued)

TABLE 17.1 (Continued)

ORIGIN	*BRAND*	*PRODUCT*	*INGREDIENTS*	*REFERENCE*
		Meat-Free Oregano & Basil Sausages	Water, wheat gluten, vegetable oil (rapeseed), soya protein, whole meal wheat rusk, herbs, salt, dried yeast, yeast extract, sugar, natural flavoring, thickener: methyl cellulose, garlic powder, onion powder, herb & spice extracts, malt extract, black pepper	
		Meat-Free Mini Pork Sausages	Water, wheat gluten, vegetable oil (rapeseed), onion, whole meal wheat rusk, soya protein, dried yeast, yeast extract, salt, herbs & spices, natural flavoring, thickener: methyl cellulose; sugar, preservative: potassium sorbate, garlic powder, malt extract (contains barley)	
		Meat-Free Pork and Apple Sausage Rolls	Pork style sausage (water, wheat gluten, onion, vegetable oil (rapeseed), whole meal wheat rusk, soya protein, dried yeast, yeast extract, salt, dried spices, natural flavoring, sugar, preservative: potassium sorbate, thickener: methyl cellulose, garlic powder, malt extract, color: iron oxide). Puff pastry (wheat flour, non-hydrogenated vegetable oils & fats, water, salt, emulsifier: mono- and diglycerides of fatty acids)	
		Meat-Free Pork Sausage Rolls	Pork style sausage (water, wheat gluten, onion, vegetable oil (rapeseed), whole meal wheat rusk, soya protein, dried yeast, yeast extract, salt, dried spices, natural flavoring, sugar, preservative: potassium sorbate, thickener: methyl cellulose, garlic powder, malt extract, color: iron oxide). Puff pastry (wheat flour, non-hydrogenated vegetable oils & fats, water, salt, emulsifier: mono- and diglycerides of fatty acids)	
		Sage & Marj Sausgs	Water, wheat gluten, rapeseed oil, soya protein, whole meal wheat rusk, herbs, salt, dried yeast, yeast extract, sugar, natural flavoring, thickener: methyl cellulose, preservative: potassium sorbate, garlic powder, onion powder, herb & spice extracts, barley malt extract, black pepper	
USA	*365 Everyday Value*	Meatless Breakfast Patties	Water, textured soy protein, textured vegetable protein (soy flour, caramel color), seasoning (spices, organic evaporated cane sugar, paprika, sea salt, autolyzed yeast extract, citric acid), expeller pressed canola oil, wheat gluten, methylcellulose	https://eu.wholefoodsmarket.com/
	Amy's	Meatless Veggie Sausages	Organic onions, organic mushrooms, organic bulgur wheat, wheat gluten, organic celery, organic carrots, organic tofu (filtered water, organic soybeans, magnesium chloride), organic long-grain red rice, organic quinoa, organic oats, expeller pressed high oleic safflower and/or sunflower oil, organic green lentils, organic unbleached wheat flour, organic bell peppers, filtered water, sea salt, organic potatoes, organic garlic, spices, organic flax seed meal, organic extra virgin olive oil, hickory smoke flavor, organic agave syrup, black pepper, organic cane sugar, yeast	https://www.amys.com
	Beyond Meat	Beyond Breakfast Sausage®—Classic	Water, pea protein, expeller-pressed canola oil, refined coconut oil, calcium alginate casing, natural flavors, yeast, rice protein, chicory root fiber, methylcellulose, salt, tapioca syrup solids, tapioca dextrose, apple extract, psyllium husk fiber, yeast extract, potassium chloride, pomegranate extract, vinegar, lemon juice concentrate, pea fiber, sunflower lecithin, beet powder (for color), spices, carrot	https://www.beyondmeat.com/
		Beyond Breakfast Sausage®—Classic Patties	Water, pea protein, expeller-pressed canola oil, refined coconut oil, natural flavors, inactivated yeast, rice protein, methylcellulose, yeast extract (niacin (vitamin B3), pyridoxine hydrochloride (vitamin B6), thiamin hydrochloride (vitamin B1), riboflavin (vitamin B2), folic acid (vitamin B9), cyanocobalamin (vitamin B12)), apple extract, salt, pomegranate extract, vinegar, lemon juice concentrate, sunflower lecithin, beet juice extract (for color), carrot	

(Continued)

TABLE 17.1 (Continued)

ORIGIN	BRAND	PRODUCT	INGREDIENTS	REFERENCE
		Beyond Breakfast Sausage®—Spicy Patties	Water, pea protein, expeller-pressed canola oil, refined coconut oil, natural flavors, inactivated yeast, rice protein, methylcellulose, yeast extract (niacin (vitamin B3), pyridoxine hydrochloride (vitamin B6), thiamin hydrochloride (vitamin B1), riboflavin (vitamin B2), folic acid (vitamin B9), cyanocobalamin (vitamin B12)), apple extract, salt, spices, pomegranate extract, vinegar, lemon juice concentrate, smoked paprika, sunflower lecithin, beet juice extract (for color), carrot.	
	El Burrito	Non-GMO Soy Longaniza™	Water, textured soy protein (soy flour), distilled vinegar, seasoning blend (paprika, salt, spices, lime juice powder (corn syrup solids, lime juice solids, lime oil), dextrose, garlic powder, spices extractives), soybean oil, xanthan gum, yeast extract, extractives of paprika, caramel color	https://elburrito.com
	Field Roast	Apple & Maple	Filtered water, vital wheat gluten, expeller pressed safflower oil, unsulfured dried apples, yeast extract (yeast, salt, sugar), wheat protein isolate (wheat gluten, trisodium phosphate, malic acid, L-cysteine), pure maple syrup, onion powder, barley malt extract, garlic, spices, sea salt, black pepper, ground ginger, nutmeg, natural smoke flavor	https://fieldroast.com/
		Caramelized Onions & Beer	Vital wheat gluten, filtered water, caramelized onions (yellow onions, sugar, apple juice concentrate), expeller pressed safflower oil, organic expeller pressed palm fruit oil, beer (malted barley, hops, yeast, water), yeast extract (yeast, salt, sugar), garlic, pea protein, barley malt extract, onion powder, granulated garlic, natural flavor (potato maltodextrin, gum arabic, dried brewer's yeast, natural flavor, organic sunflower oil), spices, cultured cane sugar, vinegar, sea salt, caraway seeds, red pepper, black pepper	
		Classic Recipe	Water, pea protein, canola oil, spices (black pepper, white pepper, sage, rosemary, marjoram) fava bean protein, brown rice protein, methylcellulose, vinegar, salt, cane sugar, dried garlic, yeast extract, rice concentrate, flavor, beet powder (color)	https://fieldroast.com/
		Classic Smoked	Filtered water, vital wheat gluten, expeller pressed safflower oil, yeast extract (yeast, salt, sugar, natural flavor), organic expeller pressed palm fruit oil, barley malt extract, tomato paste, apple cider vinegar, paprika color, spices, sea salt, onions, wheat flour, garlic, natural smoke flavor, celery seed, paprika oleoresin (color)	
		Italian Garlic & Fennel	Filtered water, vital wheat gluten, expeller pressed safflower oil, eggplant, onions, wheat protein isolate (wheat gluten, trisodium phosphate, malic acid, l-cysteine), yeast extract (yeast, salt, sugar), garlic, barley malt extract, onion powder, dried red bell peppers, red cooking wine (red wine, water, salt), fennel seed, dehydrated garlic, sea salt, spices	
		Smoked Apple & Sage	Filtered water, vital wheat gluten, expeller pressed safflower oil, unsulfured dried apples, yukon gold potatoes, yeast extract (yeast, salt), onion powder, barley malt extract, garlic, spices, sea salt, yeast, rubbed sage, natural smoke flavor	
		Spicy Mexican Chipotle	Vital wheat gluten, filtered water, expeller pressed safflower oil, onions, garlic, wheat protein isolate (wheat gluten, trisodium phosphate, malic acid, L-cysteine), apple cider vinegar, yeast extract (yeast, salt, sugar), chipotle peppers, onion powder, granulated garlic, brown sugar, sea salt, barley malt extract, spices, black pepper, paprika oleoresin (color), chili de arbol peppers, cumin, oregano	
	Gardein	Original breakfast saus'age patties	Water, soy protein concentrate, canola oil, vital wheat gluten, soy protein isolate, contains 2% or less of: yeast extract, methylcellulose, organic cane sugar, potato starch, sea salt, spices, salt, paprika, garlic powder, natural flavors, malt extract (malted barley, water), onion powder, annatto (color), oleoresin paprika (color), spice extract. Contains wheat and soy	https://www.gardein.com/

(Continued)

TABLE 17.1 (Continued)

ORIGIN	BRAND	PRODUCT	INGREDIENTS	REFERENCE
		Sliced italian saus'age	Water, soy protein concentrate, canola oil, fractionated palm oil shortening, vital wheat gluten, soy protein isolate, contains 2% or less of: yeast extract, methylcellulose, organic cane sugar, potato starch, sea salt, spices, fennel, paprika, garlic powder, salt, natural flavors, dried garlic, malt extract (malted barley, water), lactic acid	
		Spicy breakfast saus'age patties	Water, soy protein concentrate, canola oil, vital wheat gluten, soy protein isolate, contains 2% or less of: yeast extract, methylcellulose, organic cane sugar, potato starch, sea salt, spices, salt, paprika, garlic powder, natural flavors, malt extract (malted barley, water), onion powder, red bell pepper flakes, annatto (color), oleoresin paprika (color), spice extract. Contains wheat and soy	
	Lightlife® Foods, Inc	Gimme Lean® Sausage	Water, soy protein concentrate, soy flour, tapioca starch, soy sauce (water, soybeans, salt, wheat), less than 2% of: soy protein isolate, natural flavors (from vegetable sources), wheat gluten, cellulose gum, evaporated cane sugar, sea salt, spices, soy milk powder, barley malt extract, torula yeast, beet powder, salt, yeast extract	https://lightlife.com/
		Plant-Based Bratwurst Sausages	Water, pea protein, canola oil, modified cellulose (from plant fiber), less than 2% of natural flavors, cane sugar, salt, tapioca starch, citrus fiber, fava bean protein, brown rice protein, beet powder (color), yeast extract, dried torula yeast, dehydrated lemon peel, smoked sugar. In a calcium alginate casing	https://lightlife.com/
		Plant-Based Breakfast Links	Water, pea protein, canola oil, brown rice protein, less than 2% of natural flavors, spice, modified cellulose (from plant fiber), cane sugar, vinegar, beet powder (color), salt, yeast extract. In a calcium alginate casing	
		Plant-Based Italian Sausage	Water, pea protein, canola oil, modified cellulose (from plant fiber), less than 2% of natural flavors, cane sugar, salt, tapioca starch, citrus fiber, fava bean protein, brown rice protein, dried red bell peppers, beet powder (color), yeast extract, dried torula yeast, paprika extract (color). In a calcium alginate casing	
		Smart Dogs®	Water, soy protein isolate, soybean oil, evaporated cane sugar, pea protein isolate, tapioca starch, salt, potassium chloride, yeast extract, carrageenan, dehydrated garlic, natural flavor, natural smoke flavor, xanthan gum, fermented rice flour, guar gum, oleoresin paprika (color), vital wheat gluten	
		Smart Dogs® Jumbo	Water, soy protein isolate, soybean oil, evaporated cane sugar, pea protein isolate, tapioca starch, salt, potassium chloride, yeast extract, carrageenan, dehydrated garlic, natural flavor, natural smoke flavor, xanthan gum, fermented rice flour, guar gum, oleoresin paprika (color), vital wheat gluten	
		Smart Sausage® Chorizo	Water, textured soy protein concentrate, soybean oil, egg whites, soy protein isolate, less than 2% of: potato starch, yeast extract, sea salt, wheat gluten, paprika, cellulose gum, dried garlic, salt, spices, fermented rice flour, oleoresin paprika (color), natural flavor (from plant sources), barley malt extract, dried onions	
		Smart Sausage® Italian	Water, textured soy protein concentrate, soybean oil, egg whites, dried vegetables (onions, garlic, tomato, red and green bell peppers), potato starch, soy protein isolate, wheat gluten, cellulose gum, sea salt, yeast extract, salt, tapioca starch, barley malt extract, spices, natural flavor (from plant sources), evaporated cane syrup, calcium phosphate	

(Continued)

TABLE 17.1 (Continued)

ORIGIN	BRAND	PRODUCT	INGREDIENTS	REFERENCE
	MorningStar Farm	Incogmeato™ Original Bratwurst	Water, soy protein concentrate, canola oil, wheat gluten, palm oil, methylcellulose. Contains 2% or less of potato starch, salt, onion powder, spices, sugar, yeast extract, fruit juice (color), natural flavors, sodium alginate, cultured dextrose for freshness, sunflower lecithin, guar gum, citric acid	https://www.morningstarfarms.com/
		Incogmeato™ Italian Sausage	Water, soy protein concentrate, canola oil, wheat gluten, palm oil, methylcellulose. Contains 2% or less of yeast extract, potato starch, spices, salt, fruit juice (color), sodium alginate, natural flavors, cultured dextrose for freshness, dextrose, paprika (color), sugar, sunflower lecithin, garlic powder, guar gum, corn oil, citric acid, onion powder, vegetable juice concentrate (color), soy sauce (water, soybeans, salt)	
		MorningStar Farms® Corn Dogs	Water, wheat flour, sugar, wheat gluten, yellow corn meal, yellow corn flour, corn oil. Contains 2% or less of salt, dextrose, leavening (sodium acid pyrophosphate, sodium bicarbonate), methylcellulose, brown sugar (sugar, molasses), spices, yeast extract, onion powder, pea protein isolate, natural flavors, carrageenan, garlic powder, xanthan gum, red beet juice concentrate (color), paprika extract (color), paprika (color), dried yeast	https://www.morningstarfarms.com/
		MorningStar Farms® Hot & Spicy Sausage Patties	Water, wheat gluten, soy protein concentrate, corn oil, soy flour, egg whites, sodium caseinate, modified tapioca starch. Contains 2% or less of soy protein isolate, soybean oil, lactose, spices, methylcellulose, autolyzed yeast extract, natural and artificial flavors, sodium tripolyphosphate, salt, modified corn starch, hydrolyzed wheat gluten, caramel color, disodium inosinate, crushed red pepper, whey, hydrolyzed corn gluten, red bell peppers, hydrolyzed soy protein, potassium chloride, dextrose, onion powder, disodium guanylate, tetrasodium pyrophosphate, tricalcium phosphate, hydrolyzed wheat protein, sodium hexametaphosphate, succinic acid, niacinamide, monosodium phosphate, lactic acid, brewer's yeast, torula yeast, calcium phosphate, soy lecithin, iron (ferrous sulfate), magnesium carbonate, vitamin B1 (thiamin mononitrate), vitamin B6 (pyridoxine hydrochloride), vitamin B2 (riboflavin), vitamin B12	
		MorningStar Farms® Maple Flavored Sausage Patties	Water, wheat gluten, soy protein concentrate, sugar, corn oil, soy flour, egg whites, sodium caseinate, modified tapioca starch. Contains 2% or less of soy protein isolate, lactose, soybean oil, natural and artificial flavors, methylcellulose, autolyzed yeast extract, spices, sodium tripolyphosphate, salt, modified corn starch, hydrolyzed wheat gluten, disodium inosinate, caramel color, whey, hydrolyzed corn gluten, hydrolyzed soy protein, potassium chloride, dextrose, onion powder, disodium guanylate, tetrasodium pyrophosphate, tricalcium phosphate, hydrolyzed wheat protein, sodium hexametaphosphate, succinic acid, niacinamide, monosodium phosphate, lactic acid, brewer's yeast, torula yeast, calcium phosphate, soy lecithin, iron (ferrous sulfate), magnesium carbonate, vitamin B1 (thiamin mononitrate), vitamin B6 (pyridoxine hydrochloride), vitamin B2 (riboflavin), vitamin B12	
		MorningStar Farms® Original Sausage Patties	Water, wheat gluten, soy protein concentrate, vegetable oil (corn, canola and/or sunflower), soy flour, soy protein isolate. Contains 2% or less of maltodextrin, methylcellulose, yeast extract, salt, fruit juice for color, natural flavors, spices, sugar, potato starch, konjac flour, onion powder, soy sauce (water, soybeans, salt, wheat)	

(Continued)

TABLE 17.1 (Continued)

ORIGIN	BRAND	PRODUCT	INGREDIENTS	REFERENCE
		Morningstar Farms® Veggie Breakfast Sausage Links	Water, wheat gluten, corn oil, egg whites, soy protein concentrate, contains 2% or less of potato starch, salt, sodium caseinate, soy protein isolate, methylcellulose, sugar, canola oil, spices, hydrolyzed vegetable protein (soy, wheat, and corn), yeast extract, caramel color, guar gum, natural and artificial flavors, autolyzed yeast extract, onion powder, soy sauce powder (soy sauce [soybeans, salt, wheat], maltodextrin), disodium inosinate, disodium guanylate, soybean oil, sunflower oil, xanthan gum, sesame oil. Vitamins and minerals: niacinamide, iron (ferrous sulfate), vitamin B1 (thiamin mononitrate), vitmain B6 (pyridoxine hydrochloride), vitamin B2 (riboflavin), vitamin B12	
		MorningStar Farms® Veggie Dogs	Water, wheat gluten, corn syrup solids. Contains 2% or less of methylcellulose, dextrose, salt, egg whites, natural flavors, brown sugar (sugar, molasses), hydrolyzed vegetable protein (corn protein, soy protein), hydrolyzed corn protein, soy protein isolate, carrageenan, mustard flour, onion powder, maltodextrin, spices, xanthan gum, hydrolyzed soy protein, autolyzed yeast, paprika, garlic powder, soybeans, disodium guanylate, disodium inosinate, hydrolyzed torula and brewers yeast, wheat, gum arabic, hydrolyzed vegetable protein (corn gluten, soy protein, wheat gluten), soybean oil, thiamin hydrochloride, paprika extract for color, autolyzed yeast extract, lactic acid, nonfat milk, red 40, sunflower oil, citric acid, blue 1	https://www.morningstarfarms.com/
	No Evil Foods	El Capitán—Chorizo	Filtered water, non-GMO vital wheat gluten, organic tomato paste, apple cider vinegar, organic chili powder, sea salt, organic smoked paprika, organic onion powder, nutritional yeast, organic ground cumin, organic garlic powder, mexican oregano, organic cayenne, organic shoyu (water, organic soybeans, organic wheat, salt)	https://www.noevilfoods.com/
		The Stallion—Italian Sausage	Vital wheat gluten, filtered water, organic red kidney beans, organic shoyu, organic chickpea flour, nutritional yeast, sea salt, organic smoked paprika, organic garlic powder, organic chili flakes, organic fennel seed, organic thyme, organic rosemary, organic cayenne, organic black pepper	
	The Meatless Farm Co	Plant-based Breakfast Sausages	Water, pea protein, canola oil, shea oil, methylcellulose. Less than 2% of: seasoning (spices (white pepper, nutmeg, coriander), onion powder, sage, ascorbic acid (antioxidant), sun flower oil), coconut oil, pea fiber, potato fiber, rice protein, natural flavor, salt, vegetable and fruit extracts for color (beetroot, radish, tomato), yeast extract, ascorbic acid (acidity regulator), caramelized carrot concentrate, sage, coriander, nutmeg, carrot concentrate. sausages filled into sodium alginate casings	https://meatlessfarm.com/
		Plant-based Sausages Patties	Water, pea protein, canola oil, shea oil, methylcellulose. Less than 2% of: seasoning (spices (white pepper, nutmeg, coriander), onion powder, sage, ascorbic acid (antioxidant), sun flower oil), coconut oil, pea fiber, potato fiber, natural flavor, salt, rice protein, vegetable and fruit extracts for color (beetroot, radish, tomato), yeast extract, ascorbic acid (antioxidant), caramelized carrot concentrate, sage, coriander, nutmeg, carrot concentrate	
	Tofurky	Beer Brats	Water, vital wheat gluten, expeller pressed canola oil, tofu (water, soybeans, magnesium chloride, calcium chloride), onions, soy flour and/or concentrate, amber ale (water, malted barley, hops, yeast), contains less than 2% of sea salt, cane sugar, spices, dehydrated onion, granulated garlic, garlic puree, carrageenan, dextrose, konjac, potassium chloride	https://tofurky.com/
		Spinach Pesto	Water, vital wheat gluten, tofu (water, soybeans, magnesium chloride, calcium chloride), expeller pressed canola oil, soy flour and/or concentrate, soy sauce (water, soybean, wheat, salt), wheat berries, contains less than 2% of spinach, basil, garlic puree, granulated garlic, dried onion, spices, sea salt, oat fiber, natural flavors, potassium chloride, lactic acid	

REFERENCES

Abid, R., Waseem, H., Ali, J., Ghazanfar, S., Ali, G.M., Elasbali, A.M. & Alharethi, S.H. 2022. "Probiotic Yeast Saccharomyces: Back to Nature to Improve Human Health." *Journal of Fungi* 8 (5): 444. https://doi.org/10.3390/jof8050444.

Açik, M., Çakiroğlu, F.P., Altan, M. & Baybo, T. 2020. "Alternative Source of Probiotics for Lactose Intolerance and Vegan Individuals: Sugary Kefir." *Food Science and Technology* 40 (3): 523–31. https://doi.org/10.1590/fst.27919.

Adise, S., Gavdanovich, I. & Zellner, D.A. 2015. "Looks like Chicken: Exploring the Law of Similarity in Evaluation of Foods of Animal Origin and Their Vegan Substitutes." *Food Quality and Preference* 41: 52–59. https://doi.org/10.1016/j.foodqual.2014.10.007.

Agraib, L.M., Yamani, M.I., Rayyan, Y.M., Abu-Sneineh, A.T., Tamimi, T.A. & Tayyem, R.F. 2022. "The Probiotic Supplementation Role in Improving the Immune System among People with Ulcerative Colitis: A Narrative Review." *Drug Metabolism and Personalized Therapy* 37 (1): 7–19. https://doi.org/10.1515/dmpt-2021-0150.

Ahmadian, F., Razmpoosh, E., Ejtahed, H.-S., Javadi, M., Mirmiran, P. & Azizi, F. 2022. "Effects of Probiotic Supplementation on Major Cardiovascular-Related Parameters in Patients with Type-2 Diabetes Mellitus: A Secondary-Data Analysis of a Randomized Double-Blind Controlled Trial." *Diabetology & Metabolic Syndrome* 14 (1): 52. https://doi.org/10.1186/s13098-022-00822-z.

Altmann, B., Neumann, C., Velten, S., Liebert, F. & Mörlein, D. 2018. "Meat Quality Derived from High Inclusion of a Micro-Alga or Insect Meal as an Alternative Protein Source in Poultry Diets: A Pilot Study." *Foods* 7 (3): 34. https://doi.org/10.3390/foods7030034.

Alves, J.L.B., de Sousa, V.P., Neto, M.P.C., Magnani, M., Braga, V.A., Costa-Silva, J.H., Leandro, C.G., Vidal, H. & Pirola, L. 2016. "New Insights on the Use of Dietary Polyphenols or Probiotics for the Management of Arterial Hypertension." *Frontiers in Physiology* 7 (OCT). https://doi.org/10.3389/fphys.2016.00448.

Anzani, C., Boukid, F., Drummond, L., Mullen, A.M. & Álvarez, C. 2020. "Optimising the Use of Proteins from Rich Meat Co-Products and Non-Meat Alternatives: Nutritional, Technological and Allergenicity Challenges." *Food Research International* 137 (May): 109575. https://doi.org/10.1016/j.foodres.2020.109575.

Aquilanti, L., Garofalo, C., Osimani, A. & Clementi, F. 2016. "Ecology of Lactic Acid Bacteria and Coagulase Negative Cocci in Fermented Dry Sausages Manufactured in Italy and Other Mediterranean Countries: An Overview." *International Food Research Journal* 23 (2): 429–45.

Arbulu, S., Gómez-Sala, B., Garcia-Gutierrez, E. & Cotter, P.D. 2022. "Bioprotective Cultures and Bacteriocins for Food." In *Good Microbes in Medicine, Food Production, Biotechnology, Bioremediation, and Agriculture*, 89–112. Wiley. https://doi.org/10.1002/9781119762621.ch9.

Azevedo, I., Barbosa, J., Albano, H. & Teixeira, P. 2020. "Non Meat-Based Alheiras– a Safer Novel Trend?" *Food Control* 113 (July): 107177. https://doi.org/10.1016/J.FOODCONT.2020.107177.

Barbosa, J., Borges, S. & Teixeira, P. 2014. "Selection of Potential Probiotic Enterococcus Faecium Isolated from Portuguese Fermented Food." *International Journal of Food Microbiology* 191: 144–48. https://doi.org/10.1016/j.ijfoodmicro.2014.09.009.

Bayram, M. & Bozkurt, H. 2006. "The Use of Bulgur as a Meat Replacement: Bulgur-Sucuk (a Vegetarian Dry-Fermented Sausage)." *Journal of the Science of Food and Agriculture* 87: 411–19. https://doi.org/10.1002/jsfa.2712.

Bedin, E., Torricelli, C., Gigliano, S., De Leo, R. & Pulvirenti, A. 2018. "Vegan Foods: Mimic Meat Products in the Italian Market." *International Journal of Gastronomy and Food Science* 13 (November 2017): 1–9. https://doi.org/10.1016/j.ijgfs.2018.04.003.

Behera, S.S. & Panda, S.K. 2020. "Ethnic and Industrial Probiotic Foods and Beverages: Efficacy and Acceptance." *Current Opinion in Food Science* 32: 29–36. https://doi.org/10.1016/j.cofs.2020.01.006.

Belleggia, L., Ferrocino, I., Corvaglia, M.R., Cesaro, C., Milanović, V., Cardinali, F., Garofalo, C., Cocolin, L., Aquilanti, L. & Osimani, A. 2022c. "Profiling of Autochthonous Microbiota and Characterization of the Dominant Lactic Acid Bacteria Occurring in Fermented Fish Sausages." *Food Research International* 154 (April): 110990. https://doi.org/10.1016/j.foodres.2022.110990.

Belleggia, L., Ferrocino, I., Reale, A., Corvaglia, M.R., Milanović, V., Cesaro, C., Boscaino, F., Di Renzo, T., Garofalo, C., Cardinali, F., Aquilanti, L., Cocolin, L. & Osimani, A. 2022a. "Unfolding Microbiota and Volatile Organic Compounds of Portuguese Painho de Porco Preto Fermented Sausages." *Food Research International* 155 (May): 111063. https://doi.org/10.1016/j.foodres.2022.111063.

Belleggia, L., Ferrocino, I., Reale, A., Haouet, M.N., Corvaglia, M.R., Milanović, V., Boscaino, F., Di Renzo, T., Di Bella, S., Borghi, M., Farnetti, S., Cesaro, C., Garofalo, C., Cardinali, F., Aquilanti, L., Musari, E., Cocolin, L. & Osimani, A. 2022b. "Unravelling Microbial Populations and Volatile Organic Compounds of Artisan Fermented Liver Sausages Manufactured in Central Italy." *Food Research International* 154 (April): 111019. https://doi.org/10.1016/j.foodres.2022.111019.

Berghout, J.A.M., Boom, R.M. & van der Goot, A.J. 2015. "Understanding the Differences in Gelling Properties between Lupin Protein Isolate and Soy Protein Isolate." *Food Hydrocolloids* 43: 465–72. https://doi.org/10.1016/j.foodhyd.2014.07.003.

Berman, R. 2019. "'Plant-Based Meat' Is All Hat and No Cattle." *The Wall Street Journal*. https://www.wsj.com/articles/plant-based-meat-is-all-hat-and-no-cattle-11572997943/.

Berrios, J.J., Losso J.N. & Albertos, I. 2022. "Extrusion Processing of Dry Beans and Pulses." In *Dry Beans and Pulses*, 225–46. Wiley. https://doi.org/10.1002/9781119776802.ch9.

Birke Rune, C.J., Song, Q., Clausen, M.P. & Giacalone, D. 2022. "Consumer Perception of Plant-Based Burger Recipes Studied by Projective Mapping." *Future Foods* 6 (December): 100168. https://doi.org/10.1016/j.fufo.2022.100168.

Bis-Souza, C.V., Barba, F.J., Lorenzo, J.M., Penna, A.L.B. & Barretto, A.C.S. 2019. "New Strategies for the Development of Innovative Fermented Meat Products: A Review Regarding the Incorporation of Probiotics and Dietary Fibers." *Food Reviews International*. Taylor and Francis Inc. https://doi.org/10.1080/87559129.2019.1584816.

Bloukas, J.G., Paneras, E.D. & Fournitzis, G.C. 1997. "Effect of Replacing Pork Backfat with Olive Oil on Processing and Quality Characteristics of Fermented Sausages." *Meat Science* 45 (2): 133–44. https://doi.org/10.1016/S0309-1740(96)00113-1.

Bohrer, B.M. 2019. "An Investigation of the Formulation and Nutritional Composition of Modern Meat Analogue Products." *Food Science and Human Wellness* 8 (4): 320–29. https://doi.org/10.1016/j.fshw.2019.11.006.

Bond, S. 2010. "After 20 Years of Research, Scientists Still Investigating the Benefits and Risks of Soy and Isoflavones in Human Nutrition." *Journal of Midwifery and Women's Health* 55 (6): 598. https://doi.org/10.1016/j.jmwh.2010.08.006.

Borch, E., Kant-Muermans, M.L. & Blixt, Y. 1996. "Bacterial Spoilage of Meat and Cured Meat Products." *International Journal of Food Microbiology* 33 (1): 103–20. https://doi.org/10.1016/0168-1605(96)01135-X.

Bou, R., Cofrades, S. & Jiménez-Colmenero, F. 2017. *Fermented Meat Sausages. Fermented Foods in Health and Disease Prevention.* Elsevier Inc. https://doi.org/10.1016/B978-0-12-802309-9.00010-8.

Boukid, F. 2021. "Plant-Based Meat Analogues: From Niche to Mainstream." *European Food Research and Technology* 247 (2): 297–308. https://doi.org/10.1007/s00217-020-03630-9.

Câmara, A.K.F.I., Paglarini, C.S., Vidal, V.A.S., dos Santos, M. & Pollonio, M.A.R. 2020. "Meat Products as Prebiotic Food Carrier." *Advances in Food and Nutrition Research* 94: 223–65. https://doi.org/10.1016/bs.afnr.2020.06.009.

Campos, S.D., Alves, R.C., Mendes, E., Costa, A.S.G., Casal, S. & Oliveira, M.B.P.P. 2013. "Nutritional Value and Influence of the Thermal Processing on a Traditional Portuguese Fermented Sausage (Alheira)." *Meat Science* 93 (4): 914–18. https://doi.org/10.1016/j.meatsci.2012.09.016.

Chandler, S.L. & McSweeney, M.B. 2022. "Characterizing the Properties of Hybrid Meat Burgers Made with Pulses and Chicken." *International Journal of Gastronomy and Food Science* 27 (March): 100492. https://doi.org/10.1016/j.ijgfs.2022.100492.

Chen, Y.P., Feng, X., Blank, I. & Liu, Y. 2022. "Strategies to Improve Meat-Like Properties of Meat Analogs Meeting Consumers' Expectations." *Biomaterials* 287 (August): 121648. https://doi.org/10.1016/j.biomaterials.2022.121648.

Cocolin, L., Urso, R., Rantsiou, K., Cantoni, C. & Comi, G. 2006. "Dynamics and Characterization of Yeasts During Natural Fermentation of Italian Sausages." *FEMS Yeast Research* 6 (5): 692–701. https://doi.org/10.1111/j.1567-1364.2006.00050.x

Cosme, F., Inês, A. & Vilela, A. 2022. "Consumer's Acceptability and Health Consciousness of Probiotic and Prebiotic of Non-Dairy Products." *Food Research International* 151 (January): 110842. https://doi.org/10.1016/j.foodres.2021.110842.

Dasari, S., Kathera, C., Janardhan, A., Kumar, A.P. & Viswanath, B. 2017. "Surfacing Role of Probiotics in Cancer Prophylaxis and Therapy: A Systematic Review." *Clinical Nutrition* 36 (6): 1465–72. https://doi.org/10.1016/j.clnu.2016.11.017.

Date, K., Fagan, R., Crossland, S., Maceachern, D., Pyper, B., Bokanyi, R., Houze, Y., Andress, E. & Tauxe, R. 2011. "Three Outbreaks of Foodborne Botulism Caused by Unsafe Home Canning of Vegetables-Ohio and Washington, 2008 and 20093." *Journal of Food Protection* 74 (12): 2090–96. https://doi.org/10.4315/0362-028X.JFP-11-128.

Day, L. 2011. *Wheat Gluten: Production, Properties and Application. Handbook of Food Proteins.* Woodhead Publishing Limited. https://doi.org/10.1533/9780857093639.267.

Day, L., Augustin, M.A., Batey, I.L. & Wrigley, C.W. 2006. "Wheat-Gluten Uses and Industry Needs." *Trends in Food Science and Technology* 17 (2): 82–90. https://doi.org/10.1016/j.tifs.2005.10.003.

Dekkers, B.L., Boom, R.M. & Goot, A.J. 2018. "Structuring Processes for Meat Analogues." *Trends in Food Science and Technology* 81 (May): 25–36. https://doi.org/10.1016/j.tifs.2018.08.011.

Dušková, M., Kameník, J., Lačanin, I., Šedo, O. & Zdráhal, Z. 2016. "Lactic Acid Bacteria in Cooked Hams—Sources of Contamination and Chances of Survival in the Product." *Food Control* 61: 1–5. https://doi.org/10.1016/j.foodcont.2015.09.019.

Dušková, M., Kameník, J., Šedo, O., Zdráhal, Z., Saláková, A., Karpíšková, R. & Lačanin, I. 2015. "Survival and Growth of Lactic Acid Bacteria in Hot Smoked Dry Sausages (Non-Fermented Salami) with and without Sensory Deviations." *Food Control* 50: 804–8. https://doi.org/10.1016/j.foodcont.2014.10.030.

Elli, L., Branchi, F., Tomba, C., Villalta, D., Norsa, L., Ferretti, F., Roncoroni, L. & Bardella, M.T. 2015. "Diagnosis of Gluten Related Disorders: Celiac Disease, Wheat Allergy and Non-Celiac Gluten Sensitivity." *World Journal of Gastroenterology* 21 (23): 7110–19. https://doi.org/10.3748/wjg.v21.i23.7110.

FAO & WHO. 2002. "WHO Working Group Guidelines for the Evaluation of Probiotics in Food." In *Guidelines for the Evaluation of Probiotics in Food*, London Ontario, Canada, April 30 and May 1, 2002. 1–11. https://isappscience.org/wp-content/uploads/2019/04/probiotic_guidelines.pdf.

FAO & WHO. 2006. "Probiotics in Food." In *Health and Nutrition Properties of Probiotics in Food Including Powder Milk with Live Lactic Acid Bacteria.* https://doi.org/10.1201/9781420009613.ch16.

Favaro, L. & Todorov, S.D. 2017. "Bacteriocinogenic LAB Strains for Fermented Meat Preservation: Perspectives, Challenges, and Limitations." *Probiotics and Antimicrobial Proteins* 9 (4): 444–58. https://doi.org/10.1007/s12602-017-9330-6.

Felício, M.T.S., Ramalheira, R., Ferreira, V., Brandão, T., Silva, J., Hogg, T. & Teixeira, P. 2011. "Thermal Inactivation of Listeria Monocytogenes from Alheiras, Traditional Portuguese Sausage During Cooking." *Food Control* 22 (12): 1960–64. https://doi.org/10.1016/j.foodcont.2011.05.011.

Fernando, S. 2022. "Pulse Protein Ingredient Modification." *Journal of the Science of Food and Agriculture* 102 (3): 892–97. https://doi.org/10.1002/jsfa.11548.

Ferrocino, I., Franciosa, I., Rantsiou, K. & Cocolin, L.S. 2022. "The Microbiome of Fermented Sausages." In *Good Microbes in Medicine, Food Production, Biotechnology, Bioremediation, and Agriculture*, 160–68. Wiley. https://doi.org/10.1002/9781119762621.ch13.

Flores, M. & Piornos, J.A. 2021. "Fermented Meat Sausages and the Challenge of Their Plant-Based Alternatives: A Comparative Review on Aroma-Related Aspects." *Meat Science* 182 (December). https://doi.org/10.1016/j.meatsci.2021.108636.

Font-i-Furnols, M. & Guerrero, L. 2022a. "Spanish Perspective on Meat Consumption and Consumer Attitudes." *Meat Science* 191 (September). https://doi.org/10.1016/j.meatsci.2022.108874.

Font-i-Furnols, M. & Guerrero, L. 2022b. "Understanding the Future Meat Consumers." *Meat Science* 193 (November): 108941. https://doi.org/10.1016/j.meatsci.2022.108941.

Franciosa, I., Alessandria, V., Dolci, P., Rantsiou, K. & Cocolin, L. 2018. "Sausage Fermentation and Starter Cultures in the Era of Molecular Biology Methods." *International Journal of Food Microbiology* 279 (February): 26–32. https://doi.org/10.1016/j.ijfoodmicro.2018.04.038.

Franz, C.M.A.P., Huch, M., Abriouel, H., Holzapfel, W. & Gálvez, A. 2011. "Enterococci as Probiotics and Their Implications in Food Safety." *International Journal of Food Microbiology* 151 (2): 125–40. https://doi.org/10.1016/j.ijfoodmicro.2011.08.014.

Fu, J., Sun, C., Chang, Y., Li, S., Zhang, Y. & Fang, Y. 2022. "Structure Analysis and Quality Evaluation of Plant-Based Meat Analogs." *Journal of Texture Studies*, June. https://doi.org/10.1111/jtxs.12705.

Gad EL Rab, Z.A.S., Youssef, M.K.E., Khalifa, A.H., Limam, S.A. & Mostafa, B.M.D. 2019. "Quality Attributes of Beef Sausage

Supplemented by Flaxseeds and Chickpea." *Journal of Food and Dairy Sciences* 10 (6): 201–7. https://doi.org/10.21608/jfds.2019.48286.

Gardini, F., Suzzi, G., Lombardi, A., Galgano, F., Crudele, M.A., Andrighetto, C., Schirone, M. & Tofalo, R. 2001. "A Survey of Yeasts in Traditional Sausages of Southern Italy." *FEMS Yeast Research* 1 (2): 161–67. https://doi.org/10.1016/S1567-1356(01)00024-1.

Gazikalović, I., Mijalković, J., Šekuljica, N., Luković, N., Tanasković, S.J., Culetu, A. & Knežević-Jugović, Z. 2021. "Hydrolysis of Soft Wheat Flour: Enhanced Functional Properties and the Effect of Starch on Allergenicity Reduction." *Journal of Food Processing and Preservation*, September. https://doi.org/10.1111/jfpp.15925.

Geeraerts, W., De Vuyst, L. & Leroy, F. 2020. "Ready-to-Eat Meat Alternatives, a Study of Their Associated Bacterial Communities." *Food Bioscience* 37: 100681. https://doi.org/10.1016/j.fbio.2020.100681.

Geeraerts, W., Pothakos, V., De Vuyst, L. & Leroy, F. 2017. "Diversity of the Dominant Bacterial Species on Sliced Cooked Pork Products at Expiration Date in the Belgian Retail." *Food Microbiology* 65: 236–43. https://doi.org/10.1016/j.fm.2017.03.007.

Geeraerts, W., Pothakos, V., De Vuyst, L. & Leroy, F. 2018. "Variability within the Dominant Microbiota of Sliced Cooked Poultry Products at Expiration Date in the Belgian Retail." *Food Microbiology* 73: 209–15. https://doi.org/10.1016/j.fm.2018.01.019.

Godfray, H.C.J., Aveyard, P., Garnett, T., Hall, J.W., Key, T.J., Lorimer, J., Pierrehumbert, R.T., Scarborough, P., Springmann, M. & Jebb, S.A. 2018. "Meat Consumption, Health, and the Environment." *Science (New York, N.Y.)*. https://doi.org/10.1126/science.aam5324.

Godschalk-Broers, L., Sala, G. & Scholten, E. 2022. "Meat Analogues: Relating Structure to Texture and Sensory Perception." *Foods* 11 (15): 2227. https://doi.org/10.3390/foods11152227.

Gradinarska, D.N., Valkova-Yorgova, K.I. & Danov, K.R. 2022. "Effect of Red Wine in Dry Fermented Sausages Produced with a Starter Culture for Improving Their Quality and Safety." *Food Science and Technology* 42. https://doi.org/10.1590/fst.47722.

Graham, A.F., Mason, D.R., Maxwell, F.J. & Peck, M.W. 1997. "Effect of PH and NaCl on Growth from Spores of Non-Proteolytic Clostridium Botulinum at Chill Temperature." *Letters in Applied Microbiology* 24 (2): 95–100. https://doi.org/10.1046/j.1472-765X.1997.00348.x.

Grasso, N., Lynch, N.L., Arendt, E.K. & O'Mahony, J.A. 2022. "Chickpea Protein Ingredients: A Review of Composition, Functionality, and Applications." *Comprehensive Reviews in Food Science and Food Safety* 21 (1): 435–52. https://doi.org/10.1111/1541-4337.12878.

Grom, L.C., Coutinho, N.M., Guimarães, J.T., Balthazar, C.F., Silva, R., Rocha, R.S., Freitas, M.Q., Duarte, M.C.K.H., Pimentael, T.C., Esmerino, E.A., Silva, M.C. & Cruz, A.G. 2019. "Probiotic Dairy Foods and Postprandial Glycemia: A Mini-Review." *Trends in Food Science and Technology* 101 (December): 165–71. https://doi.org/10.1016/j.tifs.2020.05.012.

Guo, M. 2009. "Soy Food Products and Their Health Benefits." In *Functional Foods*, 237–77. https://doi.org/10.1533/9781845696078.237.

Haque, M.A., Timilsena, Y.P. & Adhikari, B. 2016. *Food Proteins, Structure, and Function. Reference Module in Food Science*. Elsevier. https://doi.org/10.1016/b978-0-08-100596-5.03057-2.

Hashempour-Baltork, F., Khosravi-Darani, K., Hosseini, H., Farshi, P. & Reihani, S.F.S. 2020. "Mycoproteins as Safe Meat Substitutes." *Journal of Cleaner Production* 253: 119958. https://doi.org/10.1016/j.jclepro.2020.119958.

He, J., Evans, N.M., Liu, H. & Shao, S. 2020. "A Review of Research on Plant-Based Meat Alternatives: Driving Forces, History, Manufacturing, and Consumer Attitudes." *Comprehensive Reviews in Food Science and Food Safety* 19 (5): 2639–56. https://doi.org/10.1111/1541-4337.12610.

Holck, A., Axelsson, L., McLeod, A., Rode, T.M. & Heir, E. 2017. "Health and Safety Considerations of Fermented Sausages." *Journal of Food Quality* 2017 (January). https://doi.org/10.1155/2017/9753894.

Hoogstraaten, M., Frenken, K., Vaskelainen, T. & Boon, W. 2022. "Replacing Meat, an Easy Feat? The Role of Strategic Categorizing in the Rise of Meat Substitutes." *SSRN Electronic Journal*. https://doi.org/10.2139/ssrn.4198727.

Jalava, K., Selby, K., Pihlajasaari, A., Kolho, E., Dahlsten, E., Forss, N., Bäcklund, T., Korkeala, H., Honkanen-Buzalsk, T., Hulkko, T., Derman, Y., Järvinen, A., Kotilainen, H., Kultanen, L., Ruutu, P., Lyytikaïnen, O. & Lindström, M. 2011. "Two Cases of Food-Borne Botulism in Finland Caused by Conserved Olives, October 2011." *Eurosurveillance* 16 (49): 5–7. https://doi.org/10.2807/ese.16.49.20034-en.

Jnawali, P., Kumar, V. & Tanwar, B. 2016. "Celiac Disease: Overview and Considerations for Development of Gluten-Free Foods." *Food Science and Human Wellness* 5 (4): 169–76. https://doi.org/10.1016/j.fshw.2016.09.003.

Jooyandeh, H. 2011. "Soy Products as Healthy and Functional Foods." *Middle-East Journal of Scientific Research* 7 (1): 71–80.

Kamani, M.H., Meera, M.S., Bhaskar, N. & Modi, V.K. 2019. "Partial and Total Replacement of Meat by Plant-Based Proteins in Chicken Sausage: Evaluation of Mechanical, Physico-Chemical and Sensory Characteristics." *Journal of Food Science and Technology* 56 (5): 2660–69. https://doi.org/10.1007/s13197-019-03754-1.

Kapoor, R., Pathak, S., Najmi, A.K., Aeri, V. & Panda, B.P. 2014. "Processing of Soy Functional Food Using High Pressure Homogenization for Improved Nutritional and Therapeutic Benefits." *Innovative Food Science and Emerging Technologies* 26: 490–97. https://doi.org/10.1016/j.ifset.2014.05.015.

Kaškonienė, V., Stankevičius, M., Bimbiraitė-Survilienė, K., Naujokaitytė, G., Šernienė, L., Mulkytė, K., Malakauskas, M. & Maruška, A. 2017. "Current State of Purification, Isolation and Analysis of Bacteriocins Produced by Lactic Acid Bacteria." *Applied Microbiology and Biotechnology* 101 (4): 1323–35. https://doi.org/10.1007/s00253-017-8088-9.

Kayaardi, S. & Gök, V. 2004. "Effect of Replacing Beef Fat with Olive Oil on Quality Characteristics of Turkish Soudjouk (Sucuk)." *Meat Science* 66 (1): 249–57. https://doi.org/10.1016/S0309-1740(03)00098-6.

Keerthana Priya, R., Rawson, A., Vidhyalakshmi, R. & Mohan, R.J. 2022. "Development of Vegan Sausage Using Banana Floret (*Musa Paradisiaca*) and Jackfruit (*Artocarpus Heterophyllus* Lam.) as a Meat Substitute: Evaluation of Textural, Physico-chemical and Sensory Characteristics." *Journal of Food Processing and Preservation* 46 (1). https://doi.org/10.1111/jfpp.16118.

Kılıç, B. & Özer, C.O. 2017. "Effects of Replacement of Beef Fat with Interesterified Palm Kernel Oil on the Quality Characteristics of Turkish Dry-Fermented Sausage." *Meat Science* 131 (April): 18–24. https://doi.org/10.1016/j.meatsci.2017.04.020.

Kim, T., Miller, R., Laird, H. & Riaz, M.N. 2021. "Beef Flavor Vegetable Hamburger Patties with High Moisture Meat Analogs (HMMA) with Pulse Proteins-Peas, Lentils, and Faba Beans." *Food Science & Nutrition* 9 (8): 4048–56. https://doi.org/10.1002/fsn3.2172.

King, L.A., Niskanen, T., Junnikkala, M., Moilanen, E., Lindstrom, M., Korkeala, H., Korhonen, T., Popoff, M., Mazuet, C., Callon, H., Pihier, N., Peloux, F., Ichai, C., Quintard, H., Dellamonica, P., Cua, E., Lasfargue, M., Pierre, F. & Valk, H. 2009. "Botulism and Hot-Smoked Whitefish: A Family Cluster of Type E Botulism in France, September 2009." *Euro Surveillance: Bulletin Européen Sur Les Maladies Transmissibles = European Communicable Disease Bulletin* 14 (45): 1–3. https://doi.org/10.2807/ese.14.45.19394-en.

Klupšaitė, D. & Juodeikienė, G. 2015. "Legume: Composition, Protein Extraction and Functional Properties. A Review." *Chemical Technology* 66 (1): 5–12. https://doi.org/10.5755/j01.ct.66.1.12355.

Kristensen, N.B., Bryrup, T., Allin, K.H., Nielsen, T., Hansen, T.H. & Pedersen, O. 2016. "Alterations in Fecal Microbiota Composition by Probiotic Supplementation in Healthy Adults: A Systematic Review of Randomized Controlled Trials." *Genome Medicine* 8 (1). https://doi.org/10.1186/s13073-016-0300-5.

Kumar, M., Tomar, M., Potkule, J., Reetu, Punia, S., Dhakane-Lad, J., Singh, S., Dhumal, S., Pradhan, P.C., Bhushan, B., Anitha, T., Alajil, O., Alhariri, A., Amarowicz, R. & Kennedy, J.F. 2022. "Functional Characterization of Plant-Based Protein to Determine Its Quality for Food Applications." *Food Hydrocolloids* 123 (February): 106986. https://doi.org/10.1016/j.foodhyd.2021.106986.

Kumar, P., Chatli, M.K., Mehta, N., Singh, P., Malav, O.P. & Verma, A.K. 2017. "Meat Analogues: Health Promising Sustainable Meat Substitutes." *Critical Reviews in Food Science and Nutrition* 57 (5): 923–32. https://doi.org/10.1080/10408398.2014.939739.

Kyriakopoulou, K., Dekkers, B. & van der Goot, A.J. 2018. *Plant-Based Meat Analogues. Sustainable Meat Production and Processing*. Elsevier Inc. https://doi.org/10.1016/B978-0-12-814874-7.00006-7.

Laranjo, M., Potes, M.E. & Elias, M. 2019. "Role of Starter Cultures on the Safety of Fermented Meat Products." *Frontiers in Microbiology* 10 (Apr): 1–11. https://doi.org/10.3389/fmicb.2019.00853.

Leonard, W., Hutchings, S.C., Warner, R.D. & Fang, Z. 2019. "Effects of Incorporating Roasted Lupin (Lupinus Angustifolius) Flour on the Physicochemical and Sensory Attributes of Beef Sausage." *International Journal of Food Science and Technology* 54 (5): 1849–57. https://doi.org/10.1111/ijfs.14088.

Leroy, F., Abraini, F., Beal, T., Dominguez-Salas, P., Gregorini, P., Manzano, P., Rowntree, J. & van Vliet, S. 2022. "Animal Board Invited Review: Animal Source Foods in Healthy, Sustainable, and Ethical Diets—An Argument against Drastic Limitation of Livestock in the Food System." In *Animal*. Elsevier B.V. https://doi.org/10.1016/j.animal.2022.100457.

Li, H., Zhu, Q., Chen, X., Zhou, J. & Wu, J. 2022. "Isolation and Characterization of Coagulase Negative Staphylococci with High Proteolytic Activity from Dry Fermented Sausages as a Potential Starter Culture." *Food Research International* 162 (December): 111957. https://doi.org/10.1016/j.foodres.2022.111957.

Li, Y., Cao, Z., Yu, Z., Zhu, Y. & Zhao, K. 2023. "Effect of Inoculating Mixed Starter Cultures of Lactobacillus and Staphylococcus on Bacterial Communities and Volatile Flavor in Fermented Sausages." *Food Science and Human Wellness* 12 (1): 200–11. https://doi.org/10.1016/j.fshw.2022.07.010.

Lima, M., Costa, R., Rodrigues, I., Lameiras, J. & Botelho, G. 2022. "A Narrative Review of Alternative Protein Sources: Highlights on Meat, Fish, Egg and Dairy Analogues." *Foods* 11 (14): 2053. https://doi.org/10.3390/foods11142053.

Lindström, M., Vuorela, M., Hinderink, K. & Korkeala, H. 2006. "Botulism Associated With Vacuum—Packed Smoked Whitefish in Finland, June-July 2006." *Eurosurveillance* 11 (7): E060720.3. http://doi.org/10.2807/esw.11.29.03004-en.

Madempudi, R.S., Neelamraju, J., Mokashe, N.U. & Ahire, J.J. 2022. "Probiotics in the Management of Diabetes." In *Probiotics*, 407–24. Elsevier. https://doi.org/10.1016/B978-0-323-85170-1.00005-1.

McClements, D.J. & Grossmann, L. 2021. "The Science of Plant-Based Foods: Constructing Next-Generation Meat, Fish, Milk, and Egg Analogs." In *Comprehensive Reviews in Food Science and Food Safety*. Blackwell Publishing Inc. https://doi.org/10.1111/1541-4337.12771.

Min, M., Bunt, C.R., Mason, S.L. & Hussain, M.A. 2019. "Non-Dairy Probiotic Food Products: An Emerging Group of Functional Foods." *Critical Reviews in Food Science and Nutrition* 59 (16): 2626–41. https://doi.org/10.1080/10408398.2018.1462760.

Mousavi, L., Razali, N.N.B. & Ishak, W.R.W. 2019. "Nutritional Composition and Physicochemical Properties of Sausages Developed with Non-Meat Ingredients (Tofu)." *Journal of Chemical Health Risks* 9 (4): 275–82. https://doi.org/10.22034/jchr.2019.669292.

Muguerza, E., Fista, G., Ansorena, D., Astiasaran, I. & Bloukas, J.G. 2002. "Effect of Fat Level and Partial Replacement of Pork Backfat with Olive Oil on Processing and Quality Characteristics of Fermented Sausages." *Meat Science* 61 (4): 397–404. https://doi.org/10.1016/S0309-1740(01)00210-8.

Nadeeshani, H., Senevirathne, N., Somaratne, G. & Bandara, N. 2022. "Recent Trends in the Utilization of Pulse Protein in Food and Industrial Applications." *ACS Food Science & Technology* 2 (5): 722–37. https://doi.org/10.1021/acsfoodscitech.1c00448.

Neville, M., Tarrega, A., Hewson, L. & Foster, T. 2017. "Consumer-Orientated Development of Hybrid Beef Burger and Sausage Analogues." *Food Science and Nutrition* 5 (4): 852–64. https://doi.org/10.1002/fsn3.466.

Nguyen, B.T., Bujna, E., Fekete, N., Tran, A.T.M., Rezessy-Szabo, J.M., Prasad, R. & Nguyen, Q.D. 2019. "Probiotic Beverage from Pineapple Juice Fermented with Lactobacillus and Bifidobacterium Strains." *Frontiers in Nutrition* 6 (May): 1–7. https://doi.org/10.3389/fnut.2019.00054.

Onwezen, M.C., Bouwman, E.P., Reinders, M.J. & Dagevos, H. 2021. "A Systematic Review on Consumer Acceptance of Alternative Proteins: Pulses, Algae, Insects, Plant-Based Meat Alternatives, and Cultured Meat." *Appetite* 159 (April): 105058. https://doi.org/10.1016/j.appet.2020.105058.

Osen, R., Toelstede, S., Wild, F., Eisner, P. & Schweiggert-Weisz, U. 2014. "High Moisture Extrusion Cooking of Pea Protein Isolates: Raw Material Characteristics, Extruder Responses, and Texture Properties." *Journal of Food Engineering* 127: 67–74. https://doi.org/10.1016/j.jfoodeng.2013.11.023.

Öztürk-Kerimoğlu, B., Kavuşan, H.S., Gürel, D.B., Çağındı, O. & Serdaroğlu, M. 2021. "Cold-Set or Hot-Set Emulsion Gels Consisted of a Healthy Oil Blend to Replace Beef Fat in Heat-Treated Fermented Sausages." *Meat Science* 176 (December 2020): 108461. https://doi.org/10.1016/j.meatsci.2021.108461.

Pernu, N., Keto-Timonen, R., Lindström, M. & Korkeala, H. 2020. "High Prevalence of Clostridium Botulinum in Vegetarian Sausages." *Food Microbiology* 91 (December 2019): 103512. https://doi.org/10.1016/j.fm.2020.103512.

Pi, X., Yang, Y., Sun, Y., Cui, Q., Wan, Y., Fu, G., Chen, H. & Cheng, J. 2022. "Recent Advances in Alleviating Food Allergenicity through Fermentation." *Critical Reviews in Food Science and Nutrition* 62 (26): 7255–68. https://doi.org/10.1080/10408398.2021.1913093.

Pimentel, T.C., da Costa, W.K.A., Barão, C.E., Rosset, M. & Magnani, M. 2021. "Vegan Probiotic Products: A Modern Tendency or the Newest Challenge in Functional Foods." *Food Research*

International 140 (December 2020). https://doi.org/10.1016/j.foodres.2020.110033.

Pop, O.L., Suharoschi, R. & Gabbianelli, R. 2022. "Biodetoxification and Protective Properties of Probiotics." *Microorganisms* 10 (7): 1278. https://doi.org/10.3390/microorganisms10071278.

Preidis, G.A. & Versalovic, J. 2014. "Targeting the Human Microbiome With Antibiotics, Probiotics, and Prebiotics: Gastroenterology Enters the Metagenomics Era the Metagenomics Era and Gastroenterology." https://doi.org/10.1053/j.gastro.2009.01.072.

Preidis, G.A., Hill, C., Guerrant, R.L., Ramakrishna, B.S., Tannock, G.W. & Versalovic, J. 2011. "Probiotics, Enteric and Diarrheal Diseases, and Global Health." *Gastroenterology* 140 (1): 8–14. e9. https://doi.org/10.1053/j.gastro.2010.11.010.

Qin, P., Wang, T. & Luo, Y. 2022. "A Review on Plant-Based Proteins from Soybean: Health Benefits and Soy Product Development." *Journal of Agriculture and Food Research* 7 (March): 100265. https://doi.org/10.1016/j.jafr.2021.100265.

Ravani, A. & Sharma, H.P. 2022. "Meat Based Functional Foods." In *Functional Foods*, 235–87. Wiley. https://doi.org/10.1002/9781119776345.ch7.

Razmpoosh, E., Javadi, A., Ejtahed, H.S., Mirmiran, P., Javadi, M. & Yousefinejad, A. 2019. "The Effect of Probiotic Supplementation on Glycemic Control and Lipid Profile in Patients with Type 2 Diabetes: A Randomized Placebo Controlled Trial." *Diabetes and Metabolic Syndrome: Clinical Research and Reviews* 13 (1): 175–82. https://doi.org/10.1016/j.dsx.2018.08.008.

Read, S. & Grundy, E. 2012. "Allostatic Load—A Challenge to Measure Multisystem Physiological Dysregulation." *NCRM Working Paper*, no. 4: 1–10.

Ren, H., Deng, Y. & Wang, X. 2022. "Effect of a Compound Starter Cultures Inoculation on Bacterial Profile and Biogenic Amine Accumulation in Chinese Sichuan Sausages." *Food Science and Human Wellness* 11 (2): 341–48. https://doi.org/10.1016/j.fshw.2021.11.009.

Rincon, L., Botelho, R.B.A. & de Alencar, E.R. 2020. "Development of Novel Plant-Based Milk Based on Chickpea and Coconut." *Lwt* 128 (January): 109479. https://doi.org/10.1016/j.lwt.2020.109479.

Sha, L. & Xiong, Y.L. 2020. "Plant Protein-Based Alternatives of Reconstructed Meat: Science, Technology, and Challenges." *Trends in Food Science and Technology* 102 (May): 51–61. https://doi.org/10.1016/j.tifs.2020.05.022.

Silva, J., Barbosa, J., Albano, H., Sequeira, M., Pinto, A., Bonito, C.C., Saraiva, M. and Teixeira, P. 2019. "Microbiological Characterization of Different Formulations of Alheiras (Fermented Sausages)." *AIMS Agriculture and Food* 4 (2): 399–413. https://doi.org/10.3934/AGRFOOD.2019.2.399.

Singh, A. & Sit, N. 2022. "Meat Analogues: Types, Methods of Production and Their Effect on Attributes of Developed Meat Analogues." In *Food and Bioprocess Technology*. Springer. https://doi.org/10.1007/s11947-022-02859-4.

Smart, M.A. & Pontes, N. 2022. "The Role of Consumer Restraint Versus Indulgence on Purchase Intentions of Hybrid Meat Analogues." *Food Quality and Preference* (October), 104738. https://doi.org/10.1016/j.foodqual.2022.104738.

Sobel, J., Tucker, N., Sulka, A., McLaughlin, J. & Maslanka, S. 2004. "Foodborne Botulism in the United States, 1990–2000." *Emerging Infectious Diseases* 10 (9): 1606–11. https://doi.org/10.3201/eid1009.030745.

Starowicz, M., Poznar, K.K. & Zieliński, H. 2022. "What Are the Main Sensory Attributes That Determine the Acceptance of Meat Alternatives?" *Current Opinion in Food Science* 48 (December): 100924. https://doi.org/10.1016/j.cofs.2022.100924.

Sun, C., Ge, J., He, J., Gan, R. & Fang, Y. 2021. "Processing, Quality, Safety, and Acceptance of Meat Analogue Products." In *Engineering*. Elsevier Ltd. https://doi.org/10.1016/j.eng.2020.10.011.

Szpicer, A., Onopiuk, A., Barczak, M. & Kurek, M. 2022. "The Optimization of a Gluten-Free and Soy-Free Plant-Based Meat Analogue Recipe Enriched with Anthocyanins Microcapsules." *LWT* 168 (October): 113849. https://doi.org/10.1016/j.lwt.2022.113849.

Tahmasebi, M., Labbafi, M., Emam-Djomeh, Z. & Yarmand, M.S. 2016. "Manufacturing the Novel Sausages with Reduced Quantity of Meat and Fat: The Product Development, Formulation Optimization, Emulsion Stability and Textural Characterization." *LWT—Food Science and Technology* 68: 76–84. https://doi.org/10.1016/j.lwt.2015.12.011.

Tangyu, M., Muller, J., Bolten, C.J. & Wittmann, C. 2019. "Fermentation of Plant-Based Milk Alternatives for Improved Flavour and Nutritional Value." *Applied Microbiology and Biotechnology* 103 (23–24): 9263–75. https://doi.org/10.1007/s00253-019-10175-9.

Tremlova, B., Havlova, L., Benes, P., Zemancova, J., Buchtova, H., Tesikova, K., Dordevic, S. & Dordevic, D. 2022. "Vegetarian 'Sausages' With the Addition of Grape Flour." *Applied Sciences* 12 (4): 2189. https://doi.org/10.3390/app12042189.

Trotta, T., Porro, C., Cianciulli, A. & Panaro, M.A. 2022. "Beneficial Effects of Spirulina Consumption on Brain Health." *Nutrients* 14 (676). https://doi.org/10.3390/nu14030676.

Tziva, M., Negro, S.O., Kalfagianni, A. & Hekkert, M.P. 2020. "Understanding the Protein Transition: The Rise of Plant-Based Meat Substitutes." *Environmental Innovation and Societal Transitions* 35 (June): 217–31. https://doi.org/10.1016/j.eist.2019.09.004.

Vatansever, S., Tulbek, M.C. & Riaz, M.N. 2020. "Low- and High-Moisture Extrusion of Pulse Proteins as Plant-Based Meat Ingredients: A Review." *Cereal Foods World* 65 (4). https://doi.org/10.1094/CFW-65-4-0038.

Vivar-Vera, M.A., Pérez-Silva, A., Ruiz-López, I.I., Hernández-Cázares, A.S., Solano-Barrera, S., Ruiz-Espinosa, H., Bernardino-Nicanor, A. & González-Cruz, L. 2018. "Chemical, Physical and Sensory Properties of Vienna Sausages Formulated with a Starfruit Dietary Fiber Concentrate." *Journal of Food Science and Technology* 55 (8): 3303–13. https://doi.org/10.1007/s13197-018-3265-0.

Wang, L., Guo, H., Liu, X.C., Jiang, G., Li, C., Li, X. & Li, Y. 2019. "Roles of Lentinula Edodes as the Pork Lean Meat Replacer in Production of the Sausage." *Meat Science* 156 (November 2018): 44–51. https://doi.org/10.1016/j.meatsci.2019.05.016.

WGO. 2017. "Probiotics and Prebiotics." https://doi.org/10.1016/B978-0-12-814330-8.00006-8.

Yadav, M.K., Kumari, I., Singh, B., Sharma, K.K. & Tiwari, S.K. 2022. "Probiotics, Prebiotics and Synbiotics: Safe Options for next-Generation Therapeutics." *Applied Microbiology and Biotechnology* 106 (2): 505–21. https://doi.org/10.1007/s00253-021-11646-8.

Yang, P., Zhong, G., Yang, J., Zhao, L., Sun, D., Tian, Y., Li, R. & Rong, L. 2022. "Metagenomic and Metabolomic Profiling Reveals the Correlation between the Microbiota and Flavor Compounds and Nutrients in Fermented Sausages." *Food Chemistry* 375 (May): 131645. https://doi.org/10.1016/j.foodchem.2021.131645.

Yesuraj, D., Deepika, C., Ravishankar, G.A. & Rao, A.R. 2022. "Seaweed-Based Recipes for Food, Health-Food Applications, and Innovative Products Including Meat and Meat Analogs." In *Sustainable Global Resources of Seaweeds Volume 2*, 267–92. Cham: Springer International Publishing. https://doi.org/10.1007/978-3-030-92174-3_14.

Zahari, I., Östbring, K., Purhagen, J.K. & Rayner, M. 2022. "Plant-Based Meat Analogues from Alternative Protein: A Systematic Literature Review." *Foods* 11 (18): 2870. https://doi.org/10.3390/foods11182870.

Traditional Fermented Meat Products from Portugal

18

Luís Soares, Joana Gomes, Teresa Bento de Carvalho, Mónica Oliveira, Lourenço Pinto de Rezende, Irene Martín, Joana Bastos Barbosa, and Paula Teixeira

18.1 INTRODUCTION

Fermented meat products are important elements in the traditional gastronomy of several Portuguese regions, not only because of their unique flavours and textures, but also due to their impact in the economic stability and development of these territories. With dozens of different products, from cured hams and bacon, to smoked sausages and pork loins, the Portuguese landscape is dotted with small and medium-sized businesses and small non-industrial traditional cuisines, each with its own recipes and processing methods to produce such foods. The aim of this chapter is to raise recognition of these Portuguese products, starting with a glimpse into the origin of these ancient traditions and flavours of Portuguese traditional cuisine with a focus on meat fermented products.

18.1.1 Historical Perspective

The origin and understanding of meat fermentation in and around the Mediterranean Sea is lost in the mists of time and myth (Toldrá 2004). In these early writings, references to these processing techniques appear early in Mediterranean literature and are assumed to be conventional recipes and not innovations. For example, in Homer's *Odyssey* (circa 800 BC), there are mentions of sausages being consumed during the siege of Troy, an event which modern historians point to have happened around the 12th century BC, void of any surprise or amazement (Cornwall and Brooklyn 1937; Homer 2018; Strauss 2006). Since Odysseus (or Ulysses), the hero of this story, is said to have founded the first settlements in what is now Lisbon (de Armas 2013; de Góis 1988), one can speculate on the idea of curing, smoking, or drying of meats being innovations brought to the extreme west of the Iberian Peninsula by the mythical King of Ithaca, that is, if these processes were not already present. This relationship between civilisation and the salting and maturation of meats in the Iberian Peninsula endured throughout the centuries, with salt being an extremely important ingredient of not only Carthaginian gastronomy but also of this maritime empire's economy (Sáez Romero 2014; Pilkington 2013). While the Phoenicians never permanently settled the Atlantic coast of the peninsula, the establishment of trading posts in the area to support the flow of long-distance commerce of cured products, such as salted fish and sausages, was so impactful as to induce innovation in the production of ceramic vases used to transport such foods (Pilkington 2013). If not through Greek or Punic influence, the fermentation of meats undoubtedly became entrenched in Iberian cuisine after the conquest and colonisation of *Hispania* (the Latin name for Iberia) by the Roman Empire (Jarvis 2004).

Fermentation of meats, achieved through salting, smoking, and maturation, remained one of the few methods of meat preservation until the invention and popularisation of refrigerators in the 20th century (Thorarinsdottir et al. 2004). These techniques, which became more complex over the centuries, requiring specific woods, spices, and salts, allowed those among the lower socio-economic status classes to consume meat all year round rather than only during the short period after slaughter and before decay (Jarvis 2004; Lawrie 1995; Seong et al. 2016). The increasing complexity of these processes and the requirement for non-European ingredients, such as ground pepper and cardamom (Verluyten et al. 2004), to achieve improved fermented products was, in part, responsible for the Portuguese quest for the New World and the search for new trade routes from East Asia to Europe (Boileau 2010; O'rourke and Williamson 2009). It was this demand for alternative trade routes to those controlled by the Arab Sultanates of the Levant that led this feudal kingdom on the fringes of Europe in the direction of becoming the first global empire of the Modern Period (Boileau 2010; Mirkovich 1943).

DOI: 10.1201/9781003605300-18

18.1.2 Cultural Impact

Having been a staple of virtually all European cultures' gastronomy for the last 2,000 to 4,000 years, fermented meat products represent national and regional identities as much as any anthems or flags. Portuguese tradition of *charcuterie* is no different, and Protected Denomination of Origin (PDO), Protected Geographical Indication (PGI), and Traditional Speciality Guaranteed (TSG) schemes are put in place all over the country to ensure that these products remain true to tradition (Soeiro 2001a; Soeiro 2001b). Establishing these schemes not only protects traditional producers but also guarantees the preservation of the unique sensorial characteristics of the various Portuguese fermented meat products.

With two protected breeds of pig, the Bísaro Transmontano and the Porco Alentejano, and dozens of PDO and PGI products from these animals, Portuguese fermented meats are heavily represented by the pork sausages and hams, such as Chouriça de Vinhais, Salpicão, Paio Alentejano, and Presunto de Barrancos (DGADR 2023; Boileau 2010). While this culture of pork curing is especially significant in the rural interior, these products are also common in households of the industrialised coastal regions. This results in an average consumption of 41.7 kg of pork meat annually per capita, the 15th highest in the world (*FAOSTAT* 2020; Helgi Library 2023).

Adding to this, innovation and health-focused approaches and trends have forced research for new but familiar meat-like fermented foods. For example, new recipes for alheira have been tried and tested (Azevedo et al. 2020). Traditionally, alheira is a sausage made of chopped pork and poultry meat, lard, wheat bread, olive oil, and pork fat mixed with salt, garlic, and spices (Abrams et al. 2011; Albano et al. 2009a; Macieira et al. 2019; Silva et al. 2019), but these novel recipes attempted to replace pork with fish or to altogether remove any animal product from the formulation, aiming for non-meat and vegan sausages (Azevedo et al. 2020). While results suggest that these products are safe to eat if properly cooked and stored, their economic impact is not yet fully understood since consumers' expectation of quality for true-to-tradition products and distrust felt towards innovative ones might result in economical failure (Leroy et al. 2013). Also, attempts to improve the microbiological safety of chouriço, a traditional pork meat sausage, by adding natural essential oils and macroalgae or lactic acid bacteria as starter cultures to the formulation have been the focus of several studies (Alfaia et al. 2018; Coelho-Fernandes et al. 2021a; Marçal et al. 2021).

18.1.2.1 Other Rich and Impactive Portuguese Cured Products

While the American Meat Science Association includes fish and poultry in the definition of meat, the gastronomical concept excludes the flesh and organs of fish. However, since the use of salt, drying, and maturation of fish and molluscs, such as cod and octopus, is as important to Portuguese cuisine as the fermentation of meat, in the stricter sense, the authors feel the inclusion of a small mention of these products would contribute to the enrichment of the scientific knowledge on this subject and stimulate research in this field of food science.

Portugal, besides its rich tradition in the cure, maturation, and overall fermentation of pork products, has the fourth-highest per capita consumption of fish in the world and the first in countries with a population over 1 million, with 60% of the nearly 60 kg of live weight equivalent per citizen being salted cod (FAOSTAT 2020). Not only this, but salted cod is also the main ingredient of hundreds of different Portuguese dishes and the traditional food of Portuguese Christmas, displaying the relevance this product has in Portuguese gastronomy. While the more conventional salting of codfish (Bacalhau de Cura Tradicional Portuguesa ETG—Portuguese Traditional Cured Codfish TSG) allows for little to no bacterial growth, except for that of extremely halophilic bacteria, yellow-cured codfish, a premium product, possess intense sensorial characteristics, a consequence of bacterial metabolism (Dias et al. 2021). Besides codfish, another Portuguese fermented fish appetiser, muxama, commonly known as the ham of the sea, is a salted artisanal dried tuna ham of significant relevance in the sunny region of the Algarve (Soeiro 2001c). Therefore, although these fermented fish products will not be the focus of this chapter, it should be noted that fish preservation and fermentation are not a rarity and in no way less impactful to Portuguese gastronomy than other meat products.

18.1.3 Traditional Fermentation Process

Traditional Portuguese gastronomy is essentially based on local ingredients obtained from farming and livestock (Talon et al. 2007). By selecting prime pieces of fresh pork cuts that are free of preservatives, desirable flavours and textures for the final product are achieved. After selection, through grinding and cutting, the product is refined to the desired density, and the resulting paste is mixed with various spices and herbs (Talon et al. 2007). These seasonings vary with the cultural customs of each region, giving way to the wide variety of sausages produced in Portugal, such as chouriça, chouriço, salpicão, linguiça, cacholeira, bucho, butelo, or alheira. The resulting paste is then used to fill natural casings, such as sheep and hog intestines, although synthetic collagen and/or cellulose casings have also been used in industrially produced sausages. To guarantee a stable and correctly enclosed product, intestines used as casings in these traditional sausages require visual inspection and storage in heavily salted water (FAO 2007). Before being used for its final purpose, reinspection is performed, and the salt is removed by washing and immersion in cold water for 20 minutes to 1 hour. Because intestines are usually contaminated with pathogenic and spoilage-inducing microorganisms, decontamination with aqueous solutions of acetic acid or, more traditionally, vinegar is a crucial step in ensuring food safety. Similarly, the ground product should be manipulated under controlled conditions, such as temperature and humidity, to inhibit bacterial proliferation and product deterioration (ADIV 2006).

After filling, the sausage is stored at low temperatures, around 5 to 15°C, for 3 to 8 days in high-humidity rooms (82–98%). During this period, the proliferation of natural and desirable microbiota occurs. Alternatively, adding selected starter cultures or sugars can achieve a more controlled and predictable fermentation, though this is not part of the traditional process. During fermentation, parameters such as pH, water activity and/or humidity, and temperature should be controlled. Though it is not mandatory, this fermentation can be accompanied by smoking of the product (20–50°C). Through exposure to the burning of specific woods, ideally not resinous woods, chemicals released from this reaction can change organoleptic characteristics and preserve the product, extending its shelf life (ADIV 2006).

Finally, drying and maturation of the sausage can last for several months. This period of storage at low temperatures helps to stabilise the product, lowering water activity and inhibiting bacterial proliferation and/or metabolism, stopping the fermentation process. This final product, if found satisfactory in evaluation parameters such as microbiological safety, water activity, acidity, and overall sensorial characteristics, is ready for packaging and commercialisation (ADIV 2006).

By contrast, Portuguese ham, known as presunto, is salted for extended periods of time and subsequently fermented and stabilised. This product is sold and consumed with little physical processing, with the bone of the pig leg visible and the layers of fat and meat still discernible. In this case, to achieve the desired organoleptic characteristics, two phases of salting of the cut are performed, with a stabilisation phase in between. This process is considerably longer than the one executed for traditional sausages, taking at least a year before the product is suitable for consumption (Teixeira et al. 2018).

All these processes are regulated by national and international directives from the Portuguese Republic and the European Union, with *Autoridade de Segurança Alimentar e Económica* (ASAE) and the European Food Safety Agency (EFSA) serving as the responsible authorities in defining safety practices. Besides this, the Portuguese *Direção-Geral de Agricultura e Desenvolvimento Rural* (DGADR) certifies a total of four PDO and 39 PGI pork products, and these are presented in Tables 18.1 and 18.2.

TABLE 18.1 Portuguese Pork Products with Protected Denominations of Origin Recognition, by Type of Product and Region of Production (Soeiro 2001a; Door List of European Union [DLEU] 2023a, 2023c)

PDO PRODUCT	*TYPE OF PRODUCT*	*DISTRICT OF PRODUCTION (REGION)*
Carne de Bísaro Transmontano/Carne de Porco Transmontano	Pig meat	Bragança and Vila Real (*Trás-os-Montes e Alto Douro*)
Carne de Porco Alentejano	Pig meat	Évora, Beja, Portalegre and Setúbal (*Alentejo*)
Presunto/Paleta de Barrancos	Pig leg (ham)	Beja (*Alentejo*)
Presunto/Paleta do Alentejo	Pig leg (ham)	Beja, Castelo Branco, Évora, Santarém, and Portalegre (*Alentejo*)

TABLE 18.2 Portuguese Pork Products with Protected Geographical Indication Recognition, by Type of Product and Region of Production (Soerio 2001b; DLEU 2023b).

PGI PRODUCT	*TYPE OF PRODUCT*	*DISTRICT OF PRODUCTION (REGION)*
Alheira de Barroso—Montalegre	Meat/bread sausage	Vila Real (*Trás-os-Montes e Alto Douro*)
Alheira de Mirandela	Meat/bread sausage	Bragança (*Trás-os-Montes e Alto Douro*)
Alheira de Vinhais	Meat/bread sausage	Bragança (*Trás-os-Montes e Alto Douro*)
Butelo de Vinhais/Bucho de Vinhais/Chouriço de Ossos de Vinhais	Meat/cartilage sausage	Bragança (*Trás-os-Montes e Alto Douro*)
Cacholeira Branca de Portalegre	Offal sausage	Portalegre (*Alentejo*)
Chouriça de Carne de Barroso—Montalegre	Meat sausage	Vila Real (*Trás-os-Montes e Alto Douro*)
Chouriça de Carne de Melgaço	Meat sausage	Viana do Castelo (*Alto Minho*)
Chouriça de Sangue de Melgaço	Meat/blood sausage	Viana do Castelo (*Alto Minho*)
Chouriça de Carne de Vinhais/Linguiça de Vinhais	Meat sausage	Bragança (*Trás-os-Montes e Alto Douro*)
Chouriça Doce de Vinhais	Meat/honey sausage	Bragança (*Trás-os-Montes e Alto Douro*)
Chouriço Azedo de Vinhais/Azedo de Vinhais/ Chouriço de Pão de Vinhais	Meat/bread sausage	Bragança (*Trás-os-Montes e Alto Douro*)

(Continued)

TABLE 18.2 (Continued)

PGI PRODUCT	*TYPE OF PRODUCT*	*DISTRICT OF PRODUCTION (REGION)*
Chouriço de Abóbora de Barroso—Montalegre	Meat/pumpkin sausage	Vila Real (*Trás-os-Montes e Alto Douro*)
Chouriço de Carne de Estremoz e Borba	Meat sausage	Évora (*Alentejo*)
Chouriço de Portalegre	Meat sausage	Portalegre (*Alentejo*)
Chouriço Grosso de Estremoz e Borba	Meat sausage	Évora (*Alentejo*)
Chouriço Mouro de Portalegre	Meat/blood sausage	Portalegre (*Alentejo*)
Farinheira de Estremoz e Borba	Fat/bread sausage	Évora (*Alentejo*)
Farinheira de Portalegre	Fat/bread sausage	Portalegre (*Alentejo*)
Linguiça de Portalegre	Meat sausage	Portalegre (*Alentejo*)
Linguiça do Baixo Alentejo/Chouriço de Carne do Baixo Alentejo	Meat sausage	Beja, Évora, Portalegre (*Alentejo*)
Lombo Branco de Portalegre	Pork loin	Portalegre (*Alentejo*)
Lombo Enguitado de Portalegre	Pork loin	Portalegre (*Alentejo*)
Morcela de Assar de Portalegre	Blood/fat sausage	Portalegre (*Alentejo*)
Morcela de Cozer de Portalegre	Blood/fat sausage	Portalegre (*Alentejo*)
Morcela de Estremoz e Borba	Blood/fat sausage	Évora (*Alentejo*)
Paia de Estremoz e Borba	Pork loin	Évora (*Alentejo*)
Paia de Lombo de Estremoz e Borba	Pork loin	Évora (*Alentejo*)
Paia de Toucinho de Estremoz e Borba	Pork loin	Évora (*Alentejo*)
Painho de Portalegre	Pork loin	Portalegre (*Alentejo*)
Paio de Beja	Pork loin	Beja (*Alentejo*)
Presunto de Barroso	Pig leg (ham)	Vila Real (*Trás-os-Montes e Alto Douro*)
Presunto/Paleta de Campo Maior e Elvas	Pig leg (ham)	Portalegre (*Alentejo*)
Presunto de Melgaço	Pig leg (ham)	Viana do Castelo (*Alto Minho*)
Presunto/Paleta de Santana da Serra	Pig leg (ham)	Beja (*Alentejo*)
Presunto de Vinhais/Presunto Bísaro de Vinhais	Pig leg (ham)	Bragança (*Trás-os-Montes e Alto Douro*)
Salpicão de Barroso-Montalegre	Pork loin	Vila Real (*Trás-os-Montes e Alto Douro*)
Salpicão de Melgaço	Pork loin	Viana do Castelo (*Alto Minho*)
Salpicão de Vinhais	Pork loin	Bragança (*Trás-os-Montes e Alto Douro*)
Sangueira de Barroso-Montalegre	Meat/blood sausage	Vila Real (*Trás-os-Montes e Alto Douro*)

18.2 VARIETY OF PORTUGUESE FERMENTED MEAT PRODUCTS BY REGIONS

A wide variety of fermented meat products are produced in some rural regions of Portugal. In this section, a brief description of the ingredients, preparation, and smoking of most traditional Portuguese fermented meat products, especially those with Protected Geographical Indication, will be provided by region. Additionally, the geographical distribution of the different traditional meat products is illustrated in Figure 18.1.

18.2.1 Trás-os-Montes e Alto Douro

The Trás-os-Montes e Alto Douro region, which includes the districts of Bragança and Vila Real, is well known for its rich history of traditional fermented/cured meats. This secular tradition has risen from the need to reduce food waste by allowing the use of the entirety of the pigs' edible parts, from snout to tail, as well as preserving the meat after slaughter for longer periods. The *Bísaro* pig is the most typical breed of this region, being the basic raw material for several PDO and PGI statuses due to its meat characteristics and qualities, mainly its distinctive flavour and marbled meat that confers its texture and tenderness (Amass 2017).

Vinhais, Mirandela and Montalegre are well known for their smoked meat, especially alheira, butelo, chouriço/

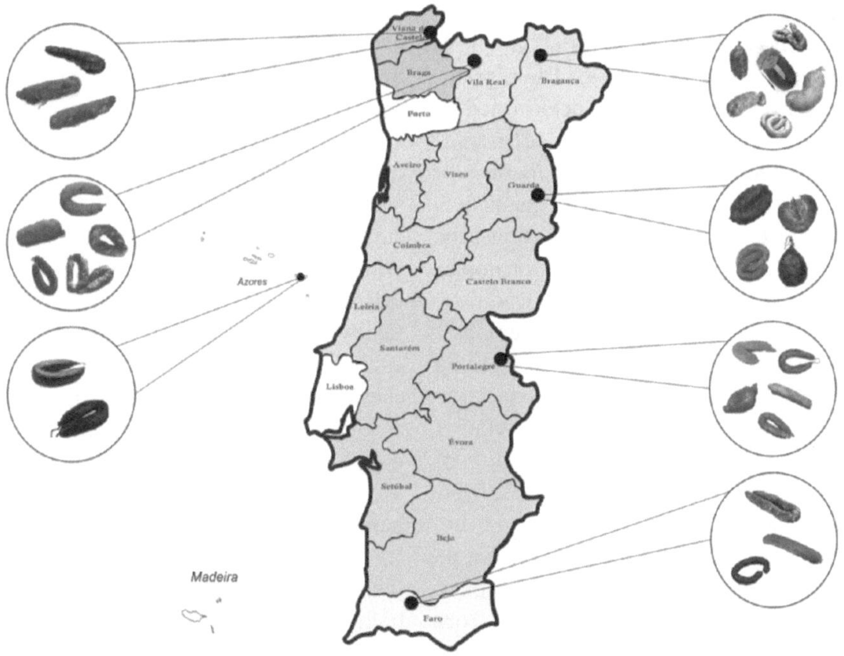

FIGURE 18.1 Illustration of the geographical distribution across the regions of Portugal where the different traditional meat products are produced. (Light green—Algarve; light purple—Alentejo; pink—Centro; light brown—Trás-os-Montes e Alto Douro; blue—Alto Minho).

chouriça and salpicão. Depending on the region, ingredient composition and texture may vary. Alheira is a traditional smoked, naturally fermented horseshoe-shaped sausage with a brownish-yellow colour, an integral element of traditional Portuguese cuisine (Ferreira et al. 2006). Its name derives from the word *alho*, the Portuguese word for garlic, which is part of the recipe (DGADR 2016h). A heterogeneous paste composed of boiled meat, bread, olive oil, garlic, and other seasonings is stuffed in casings made from animal products and is smoked for a maximum of 8 days. Its distinct characteristics are a result of the cold and humid climate of the region, which has always forced fireplaces to be constantly lit, burning autochthonous oak-based firewood, filling the rooms with low-intensity but gradually increasing smoke, as well as the heterogeneous texture of the paste, in which the pieces of meat are clearly visible, the bread used, which is kneaded and baked especially for alheira by bakers from Trás-os-Montes, and the seasonings used (Regulation of European Commission [Reg EC] 2015c).

Alheira de Mirandela PGI (Reg EC 2016) is one of the most produced and consumed products in Portugal. Its origins date back to 15th-century Jewish communities, who arrived in Trás-os-Montes having been expelled from the Spanish kingdoms in 1492 (Gerber 1994). Jews and New Christians did not eat pork and, to hide their identity and avoid persecution, they made the first alheiras without pork, using only bread and poultry, a product identical to a typical pork meat sausage. Eventually, the sausage was adopted by the population of Trás-os-Montes, with pork meat being added to the recipe (Reg EC 2006b). Alheira has since become an important delicacy in various popular and religious festivals of the region, being a commonly offered product in such events. Its ingredients are a combination of pork meat (*Bísaro* pigs) and/or poultry, pork lard, regional wheat bread, and olive oil mixed with salt, garlic, and spices. Weighing between 150 and 200 g, an alheira is approximately 3 cm in diameter and 25 cm in length (Teixeira et al. 2022). Its production begins with cooking seasoned poultry and pork meat in water. Bread is then cut into thin slices and softened in the hot meat broth. The remaining condiments and shredded meats are added to the paste, filling the natural casings, whose ends are connected by cotton yarn. Finally, the sausages are smoked over low heat with firewood from the region (mainly oak and olive) (Reg EC 2015c).

In contrast to Alheira de Mirandela PGI, in Alheira de Vinhais PGI (Reg EC 2023a), the meat mixture is more homogeneous, with only some shredded pieces of meat visible. The history of this traditional fermented product is closely linked to pig breeding in the region. Also, the Alheira de Vinhais PGI recipe differs in the fact that poultry is only used in the broth used for the boiling of the meats and not as stuffing for the sausage. Paprika is also added in this variation from Vinhais. On the other hand, the processing takes place only in the district of Bragança. The sausages are smoked over low heat with chestnut or oak firewood from the region, and a cure or stabilisation ensues for more than 8 days. Alheira de Vinhais PGI is packaged locally and sold in whole pieces, as its texture does not allow it to be sliced or cut. The experience of the local population in pig breeding of the region's climate and the traditional smoking method makes this sausage unique (Reg EC 2007h).

The Alheira de Barroso-Montalegre PGI (Reg EC 2007a) recipe is made with the addition of rabbit and stuffed into thinner casings than other alheiras (Reg EC 2006a). The ingredients of Alheira de Barroso-Montalegre PGI are determined by the traditional *Barrosã* diet, imposed by climatic, geographical, and socio-economic conditions, consisting mainly of bread, potatoes, and pork. To produce this sausage, the meat is cut into pieces of medium size and stuffed in the casings; the sausage is lightly smoked, using peat, oak, willow, or poplar wood, for 4 days in specialised smoking rooms or, traditionally, in a *fumeiro*. Finally, the sausage is placed in a temperature-controlled room to cure until it reaches the desired aroma and flavour (Reg EC 2007a). The municipality of *Montalegre* is the area where the processing and the packaging of Alheira de Barroso-Montalegre PGI take place, with the production of the raw material taking place in the municipalities of Chaves, Boticas, and Montalegre (Reg EC 2006b).

Butelo de Vinhais PGI (Reg EC 2023b), also known as bucho de vinhais or chouriço de ossos de vinhais is a smoked sausage produced with the meat, fat, bone, and cartilage from the ribs and spine of pure- or cross-bred *Bísaro* pigs. The meat is seasoned with garlic, paprika, bay leaves, wine, water, and salt. It weighs between 1 and 2 kg and is 10 to 15 cm in diameter. It is well known for its distinctive smoky flavour and shape with visible bones. Meat, along with bones and cartilages, is cut into small pieces, seasoned, and left for 3 days. Pig casings are then filled and tied in one of the extremities, ensuring that all components are well compacted and smoked for 15 days with oak wood. The curing process is carried out in a cooler environment, with time depending on the specific environmental conditions of the processing facilities (Reg EC 2008b).

Chouriça Doce de Vinhais PGI (Reg EC 2023e) is a cured sausage made from meat from pure or crossbreed *Bísaro,* with local bread, Trás-os-Montes olive oil, pig's blood, honey, and walnuts or almonds. It has a horseshoe shape and is approximately 20 to 25 cm long. Externally, pieces of meat, walnuts, and almonds are visible. The meat is boiled, and bread and the remaining ingredients are then added. The final mixture is stuffed in casings and smoked in an oak or chestnut wood fire with the curing process taking place in cool environments for at least 15 days (Reg EC 2007c).

Chouriça de carne de Vinhais PGI (Reg EC 1997d), also known as linguiça de vinhais, is a long horseshoe-shaped smoked sausage made from the meat and fat of *Bísaro* pigs, seasoned with salt, wine, water, garlic, paprika and/or chilli powder, and bay leaf. It is between 30 to 35 cm long and characteristically of a brown and red hue, with a bright red to white interior.

To achieve its distinctive flavours and textures, the meat, usually from younger pigs between 9 and 18 months, and fat are cut into small pieces, seasoned, and stuffed into pig intestine casings and subsequently smoked for over 35 days in oak and chestnut wood (Reg EC 1997d).

Chouriço Azedo de Vinhais PGI (Reg EC 2008d) is a smoked sausage made from the meat and fat of the *Bísaro* pig, mixed with the local wheat bread, Trás-os-Montes olive oil, paprika, and garlic. It has a cylindrical shape of about 20 to 25 cm long with a typical brownish-yellow tint, weighing on average 300 g. A heterogeneous mixture of boiled meat, bread, and seasonings is stuffed into intestine casings. The smoking process usually lasts for at least 4 weeks under dense oak or chestnut wood firewood, giving this product its typical texture and flavour (Reg EC 2023f).

Salpicão de Vinhais PGI (Reg EC 1998) is a traditional smoked sausage, made from pork meat and wine, traditionally with a dark brown exterior with a marbled interior. It has a cylindrical shape around 15 to 20 cm long. It has a characteristically smoky flavour with wine notes. Pork loin is usually used as the main meat part for this product. Small meat cubes are seasoned and marinated for 4 days. The final mixture is then stuffed into intestine casings and slowly smoked for at least 40 days under oak and chestnut wood fires (Reg EC 1997c).

Presunto de Vinhais PGI (Reg EC 2007i) is a smoked ham produced from the leg of the *Bísaro* pig. Fresh meat, no less than 8 kg, is marinated for 24 hours. The piece of meat is then covered in salt for 30 hours at low temperatures and in a high relative humidity environment. Excess salt is then washed away, and the leg is hung to dry. Before smoking, leg pieces are rubbed with paprika, olive oil, and lard. Smoking is carried out at low temperatures and low relative humidity. The entire process lasts for at least 12 months. Its typical aroma and distinctive flavour are mainly due to the type of meat used and the slow smoking process (Reg EC 2007f).

Salpicão de Barroso-Montalegre PGI (Reg EC 2007b) is a traditional smoked sausage, with a cylindrical shape and about 10 to 20 cm long. It has a brownish-red colour, and its main ingredients are *Bísaro* pig meat, salt, garlic, and paprika. Pork loin is cut into small pieces, seasoned, and marinated for 8 days at temperatures lower than 10°C in low-humidity environments. This mixture is then stuffed into casings and smoked over low fire for 30 days, using mainly oak wood. The low fire gives this product its characteristically smoky flavour and texture (Reg EC 2006c).

Chouriça de Carne de Barroso-Montalegre PGI (Reg EC 2007b) is a 25- to 35-cm-long traditional horseshoe-shaped sausage. It is made from pig meat and seasoned with salt, garlic, wine, and paprika. Fat is visible on the exterior of this brownish-red sausage. Pieces from various parts of the pig are used to manufacture this product. After seasoning, the mixture is marinated for 5 days at temperatures lower than 10°C. It is then stuffed into casings and smoked over a low fire of mainly oak wood for 10 to 15 days. The region's cold environment and low fire smoke give this product its typical flavour and texture (Reg EC 2005a).

Chouriço de Abóbora de Barroso-Montalegre PGI (Reg EC 2007b) is a traditional smoked sausage made with pig meat, pumpkin, salt, garlic, and paprika. It has the typical horseshoe shape and is about 6 cm in diameter. It has a dented exterior with an orange and brown hue. Meat pieces and pumpkin are seasoned and left to marinate for 5 days at temperatures lower than 10°C. After this, the casings are stuffed, and the sausage is smoked over low fire for 10 to 15 days. The pumpkin is the main differential ingredient in this product, conferring its unique flavour (Reg EC 2005b).

Sangueira de Barroso-Montalegre PGI (Reg EC 2007b) is a sausage made of meat, fat, and blood from the *Bísaro* pig. It is seasoned with garlic, salt, onion, parsley, olive oil, wine, paprika, and bread. Meat, salt, and garlic are all mixed and left to marinate for 3 days. After that period, the blood, bread, and remaining seasonings are added, and this new mixture is stuffed into casings and made into the final horseshoe shape. Smoking is carried out with a low fire, mainly of oak wood, for 3 to 4 days. It is typically of a dark colour with a soft interior (Reg EC 2005c).

Presunto de Barroso PGI (Reg EC 2023g) is a smoked ham from the legs of pigs slaughtered between 16 and 18 months of age. The leg is salted for 2 to four 4 and then smoked between t2wo weeks and 3 months. The light and gradual smoking give this product its smoky flavour and tender texture. The final product has a brownish-red hue with some off-white fat streaks (Cooperativa Agrícola dos Produtores de Batata de Semente de Montalegre 1996).

18.2.2 Alto Minho

Fermented and smoked sausages from Alto Minho, the region composed of the districts of Braga and Viana do Castelo, are high-quality products that have come to occupy a prominent place in Portuguese society and gastronomic habits (Município Melgaço, n.d.). The production of these smoked and fermented meat products is mostly performed in the localities of Melgaço, Ponte de Lima, and Arcos de Valdevez (Diagnóstico e Plano de Acção da Agenda 21 Melgaço 2008). In this chapter, the authors will only highlight the Melgaço region with its four PGI products, chouriça de carne de melgaço, chouriça de sangue de melgaço, presunto de melgaço, and salpicão de melgaço.

Chouriça de carne de Melgaço PGI (Reg EC 2023c) is a cylindrical horseshoe-shaped sausage cured by smoke and cold and obtained from pig meat from the shoulder, belly, trimmings, and pork fat (Reg EC 2014a). This traditional product, in which the production area is limited to the municipality of Melgaço, has a light brown colour, with a diameter that varies between 2 to 3 cm, and a length between 30 and 35 cm. It has a smooth appearance and a moderately moist surface. It begins with the cutting of the meat of *Bísaro* pigs (pure or crossed), slaughtered with a minimum live weight of 110 kg and at least 32 weeks old, and fat into small pieces between 2 and 3 cm, mixed and seasoned with onion, garlic, hot or sweet paprika, salt, bay leaves, white wine, and water. A mixture called *sorça* is then formed, which rests in a dry, cool place (8 to 11°C) and is stirred twice a day. Casing filling occurs after up to a week of incubation, and the Chouriça de Carne de Melgaço PGI is placed in the smokehouse, hung by the tied string. Smoking must be slow due to the combustion of dry and non-resinous wood, and the most-used wood comes from oak, broom, and birch trees, which are autochthonous species. After 15 days of smoking, the sausages are placed in a *sequeiro*, a dry, cool place, preserved by processes that avoid changing their characteristics, such as using local cold conditions (Melgaço Rural, n.d.-a). Melgaço has a specific climate (microclimate) characterised by long, cold, and dry winters and hot and short summers that allow ideal conditions for curing meat sausages, favourably conditioning their characteristics (Melgaço Rural, n.d.). The Chouriça de Carne de Melgaço PGI has particular characteristics in its interior appearance, with irregular, clearly visible marbling, and its texture, which is very succulent and tender. This fermented chouriça combines moderately acidic and spicy flavours, salty and with an intense smoky tang (Reg EC 2014a).

Chouriça de sangue de Melgaço (Reg EC 2023d) is a sweet, acidic, and spicy smoked-cured sausage made with bloody meat trimmings, fat, and pig blood (Taste Atlas, n.d.-a). Production is done in the geographical region of the municipality of Melgaço, covering all its parishes. This region offers special natural environments for its production, such as altitude, a particular microclimate, and the firewood used in smoking (Reg EC 2023d). Adding to the meat, blood and fat, onion, garlic, salt, bay leaf and pepper, cumin, and sweet and/or spicy paprika are used for seasoning. This mixture, also known as *sorça*, is preserved in a cool and slightly humid place for up to 3 days. After this period, the gut casing is filled, and occasionally, rye flour can be added at this point. The sausage is then placed in the smokehouse. As mentioned before, the firewood used in smoking is an important traditional custom and comes from species typical of the local flora. Dry and non-resinous wood is used for slow smoking for 10 to 15 days. Afterwards, the sausages are preserved in a *sequeiro* (Reg EC 2023d).

Presunto de Melgaço PGI (Reg EC 2015b) is a smoked ham produced from the hind legs of *Bísaro* pigs weighing more than 10 kg. The first step is trimming the fat and massaging the flesh, followed by salting of the leg piece. Following this step is the drying of the ham and smoking period. The final product presents a mild smoky flavour, tender texture, and intense smoked aroma (Melgaço Rural, n.d.-c; Reg EC 2014c).

Salpicão de Melgaço PGI (Reg EC 2015a) is a sausage made with succulent meat and is distinguished from other sausages by its intense smoky aroma, sweetness, acidity, spiciness, and its moderately salty and smoky aftertaste (Taste Atlas, n.d.-b). Its marbled appearance when cut is unique. When sliced, it is firm and moderately juicy. This is due to the fat content in the intramuscular pieces. Notably, it differs from the other mentioned products by using exclusively specific and noble pork cuts (loin, tenderloin, leg, and shoulder) (Reg EC 2015a). Garlic; onion; salt; bay leaf; and green, red, or white wine, with water and, optionally, cumin are condiments added to the meat (Reg EC 2015d). The mixture is then stuffed in natural casings, with the structure of the sausage supported by hidden knots inside the casing that are tied to the ends of the sausage, which, in turn, also represent a unique characteristic of this sausage (Melgaço Rural, n.d.-b).

18.2.3 Centro

Fermented sausages have always been an integral part of the diet of the central region of Portugal. In this region, located between the Trás-os-Montes e Alto Douro (north) and the Alentejo (south) and represented by the districts of Aveiro, Guarda, Viseu, Castelo Branco, and Leiria, pig farming has consistently played a very important role in the economy of local populations, being an important source of income for most families. Thus, the day of the slaughter of the pigs is usually a day of celebration in this region, with whole communities, families, and friends coming together to enjoy this gastronomic event.

Ancient techniques, from the cutting of selected meats to the slow drying of pork sausages, have been passed down throughout generations, preserving the sausage-making traditions in this region of Portugal (Câmara Municipal Castelo Branco 2023). Castelo Branco, Guarda, Sertã, Seia, Serra da Estrela, and Sabugal are localities well known for their meat smoking tradition and production of fermented sausages, especially bucho da guarda, bucho da sertã, chouriça de bofe da guarda, chouriça de carne da guarda, chouriço raiano, farinheira da guarda, farinheira raiana, morcela da guarda, and morcela doce de jarmelo. While these products do not have Protected Geographical Indication recognition, they are important and traditional fermented products, typical of the region (DGADR 2023). Therefore, the authors consider it appropriate to include some of them in this chapter.

Chouriça de bofe da guarda (DGADR 2016b) is a traditional sausage made from lungs, heart, and pig blood, water, or wine (or both) and the seasonings salt, garlic, sweet paprika, and ground chilli, which gives it a slightly spicy flavour. It is 2.5 to 3.5 cm in diameter and 20 to 25 cm in length (when stretched) and weighs between 150 to 250 g. This is a dark-coloured product, with a rough exterior texture in the form of a horseshoe (https://ocentrofazbem.pt/portfolios/enchidos/). Chouriça de bofe da guarda has a spongy consistency, and the interior is not very cohesive, breaking down when sliced. Nowadays, this product is made with pork meat with traces of blood instead of using the belly and heart, and it is used in regional cuisine dishes (DGADR 2016b).

Meanwhile, chouriço raiano is a sausage made from prime pork meat (loin and leg meat) and some fat, with the pig's small intestine used as a casing. It is blood coloured, has a smooth outer texture, and is firm and compact when sliced. This product is about 15 to 20 cm in length and 4 to 5 cm in width, and the weight is 250 to 350 g (DGADR 2016c). To make this product, the meat is cut into relatively small pieces which are then placed in a container with a sauce made up of local wine, garlic, salt, and sweet and spicy paprika for 3 to 4 days, and following this, the casing is filled. Finally, the ends are tied with cotton thread, and it goes to the smokehouse with oak and chestnut wood for 2 to 3 weeks. Each producer has their own method of seasoning, which can contribute to variations in the product. Like most sausages, it has a slightly spicy flavour. Chouriço raiano can be eaten on its own or in traditional dishes from the region (DGADR 2016c).

Bucho is an emblematic product of the region (Anonymous 2018). Bucho da guarda is a rough sausage with a spherical-elongated shape and a red-orange colour, consisting of non-cohesive pieces of pork meat, which disintegrate when cut, with or without bones. It weighs between 150 and 250 g and is 25 to 35 cm in diameter and 20 to 25 cm in length when stretched out. The flavour of this sausage is slightly spicy. To produce bucho da guarda, the pork meat, namely the head, tail, ribs, cartilage, and snout, is cut into medium-sized pieces and seasoned with salt, garlic, paprika, ground chilli, local red wine, and water (Soeiro 2001b). After the meat has been in this vinha d'alho stage for about 3 days, it is drained and stuffed inside washed pig stomachs and bladders. The casing is tied with a cotton thread, and the sausage is placed in an oak and chestnut wood smoker for a week. After smoking, the product is kept in dry and cool places to prevent mould growth. Bucho da guarda has been around for many years, and it is a typical product of Domingo Gordo (the Sunday before Carnival) eaten after cooking in a linen cloth (DGADR 2016a).

Farinheira da guarda is a regional sausage based on wheat bread, water from cooking bones and meat, rendered pork fat, condiments, and cow gut as a casing. Like most sausages, it has a horseshoe shape and is dark orange in colour, with a slightly rough outer texture, relatively soft, and the inner paste is semi-soft and moderately homogeneous. This product is about 25 to 35 cm long, about 2.5 to 3.5 cm wide, and 150 to 250 g in weight. Production of farinheira da guarda begins with wheat bread being cut into thin slices and softened with the hot water from cooking meat and bones. Then, rendered pork fat is added, and the paste obtained is crushed with the hands and seasoned with salt, paprika, and chilli (Soeiro 2001b). The casing is filled and placed in a smokehouse with oak and chestnut wood. The flavour of farinheira is greasy and slightly spicy (DGADR 2016d). It is usually consumed in typical regional dishes and is marketed throughout the country.

Morcela is also a traditional product of the central region of Portugal. Morcela da guarda is a horseshoe-shaped sausage with a dark colour made with pig's blood and fat and wheat bread (Lusa 2016). It has a medium consistency, the interior dough is homogeneous and cohesive, and it has a smooth outer texture. When cut, the slices have a compact, undeformed appearance. It weighs between 300 to 400 g and has a length of 35 to 45 cm when stretched and a diameter of 5.5 to 6.5 cm (Soeiro 2001b). Morcela da guarda has been made since ancient times and is a derivative of the pig slaughter, traditionally carried out in the winter period, which was the basic structure of food sources for many families throughout the year (Lusa 2016). Its production begins by collecting the blood, and, to prevent it from curdling, a spoonful of coarse salt is added and stirred in; this salt is considered the seasoning of the morcela. Then, the bread, preferably hard and dry, is cut into thin slices and sprinkled with blood. Onion (raw or fried in fat), chopped parsley, and pork fat are added. Then, the mixture is again seasoned with salt, and cumin and ground chilli are added. This paste is stuffed in pig casings prepared on the day of the slaughter or the day before. At the end, the sausage is blanched for 5 minutes and cleaned with a dry cloth before going to the smokehouse, where it remains for 2 to 3 days. This product is usually eaten fried, sliced, accompanied by boiled potatoes and boiled turnip greens, or in the cozido à portuguesa dish.

As a starter, morcela can also be eaten boiled or roasted (when fresh) (DGADR 2016e).

18.2.4 Alentejo

The Alentejo region is known for its rich gastronomy, and fermented meat products are part of its history. Around the 15th century, the technique of fermenting and curing meat emerged in the region as a natural method of food preservation (Quiroga and Olivença, n.d.). The use of Alentejana breed pork meat as the raw material gives the typical sausages of the region a unique and very particular flavour. Over time, the production of fermented meat products became an important economic activity in the region. Producers raised pigs on their own properties and used all the animal meat to make these products that were sold on the local market and consumed by the locals. Currently, Alentejo fermented sausages are not only consumed by the local population but are also enjoyed by people from all over the world who want to experience the traditional flavours of the region (Quiroga and Olivença, n.d.).

Chouriço de carne is typically prepared with meat from Alentejana breed pigs and seasoned with a mash of garlic, paprika, and salt for 1 to 2 days. At the end of this time, white wine and the blood of the pig (collected at the time of the slaughter) are added (Soeiro 2001b). The mixture is put into a thick pork casing, the meats are tightened, and the ends of each sausage are tied together with string to form a bow. Sausages are then smoked between 8 days and 5 to 6 weeks (DGADR 2016g). Depending on the place of production, fermented sausages have specific characteristics, mainly in terms of seasoning and quantity of blood used as well as smoking time.

Chouriço de barrancos is made with pork meat from Alentejana breed pigs, with the addition of some slightly bloody meat pieces and fat from the meat itself. Salt, garlic, paprika, and pepper are the seasonings used to spice the meat (Soeiro 2001b). Water is then added, and the mixture is left to ferment in vats for 24 hours. The fresh pork casing, previously washed, turned, and seasoned with orange slices, is filled with the meat preparation (DGADR 2016f). They are then hung and exposed in an appropriate room and undergo a natural curing process where there is no smoke. The typical microclimate of this region allows for this type of curing.

Chouriço de Carne de Estremoz e Borba PGI (Reg EC 2004), which is produced in the district of Évora, is a sausage cured by slow smoking at a temperature of between 30 and 40°C for a minimum of 3 days (Reg EC 2002a). Acorns are the main ingredient used in the diet of the pigs that make up Chouriço de Carne de Estremoz e Borba PGI and, as a result, give the sausage a very distinctive taste.

Chouriço Grosso de Estremoz e Borba PGI (Reg EC 2004) is a smoke-cured sausage which is between 15 and 25 cm in length and 5 to 8 cm in diameter. The meat preparation is made from Alentejana breed pigs and is seasoned with salt, water, paprika, and garlic. Later, the product undergoes other processes, such as maturation, stuffing, tying, and curing with wood specific to the Alentejo region, such as Holm oak (Reg EC 2002b).

Chouriço de Portalegre PGI (Reg EC 1997b) and Chouriço Mouro de Portalegre PGI (Reg EC 1997b) are part of a large set of traditional sausages from the region of Portalegre. Chouriço de Portalegre PGI is a smoked sausage prepared with pork meat from Alentejana breed pigs cut into pieces of about 2 cm. The meat is seasoned in an appropriate container with salt, dried garlic, sweet bell pepper, paprika, and white wine from the region of Portalegre. The resulting mixture is kept for 1 to 5 days in an environment where the temperature does not exceed 10°C and the humidity is around 80 to 90%. Next, a dry natural casing of bovine and porcine animals is filled, and the product is smoked for a minimum of 3 and a maximum of 10 days (Natural Carnes, n.d.-a). Chouriço Mouro de Portalegre PGI is made with bloody meat trimmings, fats, and entrails from Alentejana breed pigs. Due to the Portalegre region's culture, these smoked sausages are very similar in the process of maturation, filling, curing, and smoking with Holm oak and/or cork wood specific to the region (Reg EC 1997b).

Linguiça de Portalegre PGI (Reg EC 1997b) and Linguiça/Chouriço de Carne do Baixo Alentejo PGI (Reg EC 2007b) are sausages composed mainly of pork from Alentejana breed pigs and hard fats obtained from the cutting of the pig. Like the chouriços of the same region, the meat is seasoned and goes through a process of several stages until the sausage is smoked, during which it acquires a characteristic colour, flavour, and aroma (Natural Carnes, n.d.-b; Reg EC 2006d).

Presunto/Paleta de Barrancos PDO (Reg EC 2014b) is a fermented meat product produced and processed in the municipality of Barrancos which consists of legs/shanks; salted, dried, cured, and unsmoked forelegs; and fore-ends from adult pigs of the Alentejana breed slaughtered between 12 and 20 months old. The weight is over 5 kg (ham) and 3 kg (shoulder). After the slaughtering, the legs and shoulders undergo a transformation process of several stages: salting, curing, drying/maturing, and ageing. The duration of this process is variable and ranges between 12 and 24 months for the presunto and 6 and 18 months for the paletas. It is during the transformation process that the product progressively acquires the appearance, colour, texture, and smooth and delicate flavour that characterise it. Due to the climatic conditions required for the transformation and maturation of the presuntos/paletas de Barrancos, the know-how of the population, and local methods, production is limited to the municipality of Barrancos. It is important to stress that the microclimate typical of this locality allows the drying, maturing, and ageing of the presuntos/paletas to occur naturally, without resorting to smoking, thus maintaining the tradition of the manufacturing technique since ancient times. Presunto/Paleta de Barrancos PDO can be sold whole, in pieces or sliced, with or without bone (Reg EC 2014b).

Presunto/Paleta de Santana da Serra PGI (Reg EC 2008c) is produced in various parishes of the district of Beja. The meat used for production comes from pigs of the Alentejana breed, which are slaughtered between 12 and 24 months, and only carcasses weighing a minimum of 90 kg are used. The legs and shoulders must weigh a minimum of 7.5 kg and 5 kg for presuntos and paletas, respectively. Processing includes several stages, including cutting, salting from 1 day to 1.5 days per kg at a temperature of 0 to 5°C; the ham is washed and hung for 1 to 3 days; then drying takes place for 30 to 60 days at a temperature of 3 to 5°C; and maturation takes 4 to 5 months for presunto and 2 to 3 months for paleta. The maturation and curing process occurs under natural climatic conditions, without resorting to smoking or other types of preservation apart from salting (Reg EC 2007f).

Presunto/Paleta do Alentejo PDO (Reg EC 2007g) is produced in the same way as the other cured hams of the region, with the pork used in its production coming from Alentejana breed pigs. The unique characteristics of this product are a result of the specific feeding of the pigs, which dictates the texture, flavour, and overall sensorial properties. The production of Presunto/Paleta do Alentejo PDO occurs in the districts of Beja, Évora, and Portalegre, extending to some locations in Castelo Branco, Faro, Guarda, Santarém, and Setúbal (Reg EC 2007g).

Presunto/Paleta de Campo Maior e Elvas PGI (Reg EC 2008c) are hams made from pigs of the Alentejana breed. The production stages are very similar to those described previously. However, some differences in the salting, drying, and curing times of these hams are typical of the Campo Maior and Elvas area. Salting is carried out for 15 days at a temperature below 5°C. Then, drying takes longer, approximately 30 days at a temperature between 3 and 5°C. Finally, the curing takes place in two stages, the first lasting 45 days at a temperature between 8 and 12°C and the second lasting around 20 days at a slightly higher temperature between 12 and 18°C. The climate typical of the region is suitable for drying and curing the hams without resorting to smoking, which confers on the product qualities that are very particular and traditional to this region (Reg EC 2007i).

Morcela de Assar de Portalegre PGI (Reg EC 1997b) is a smoked sausage consisting of fat (head, neck, belly, and bacon) and blood of Alentejana breed pigs that must be collected in a suitable container and kept at a temperature that prevents coagulation. Salt, crushed dried garlic, cumin, paprika, and sometimes wine from the Portalegre region are used as seasonings. To facilitate the binding of the seasonings to the meat and fats, the preparation is left to incubate for 12 to 24 hours in environments whose temperature does not exceed 10°C and with humidity between 80 and 90%. Then it is stuffed into a thin or thick pork casing, and the sausage is smoked for a maximum of 10 days (DGADR 2016i).

Morcela de Cozer de Portalegre PGI (Reg EC 1997b) is a sausage made of meat from the Alentejana breed pigs and spiced the same way as morcela de assar. The particularity of this sausage is that, after the filling of the small intestine, large intestine, stomach, or bladder, the sausage is immersed in a container of lightly seasoned water at a high temperature without boiling for 45 minutes to 2 hours (DGADR 2016j).

Morcela de Estremoz e Borba PGI (Reg EC 2004) is a smoked sausage made from the chaps, red meat, and fat obtained from the cutting of Alentejana breed pig carcasses and fat. After cutting and cleaning the pieces, the remaining sausage ingredients are added (salt, water, paprika, garlic, blood, cloves, and sometimes cumin, wine, and pepper). Afterwards, the product undergoes the traditional processes of maturation, filling and curing with autochthonous Holm oak wood. The final product is a horseshoe-shaped sausage, 20 to 35 cm long and 2 to 4 cm in diameter. The sausage is black and somewhat rough and dull. The aroma and taste are pleasant, mild with a garlic and smoky flavour and a balance between sweet and salty taste (Reg EC 2002c).

Cacholeira Branca de Portalegre PGI (Reg EC 1997b) is a horseshoe-shaped sausage, cylindrical, with a length of up to 50 cm and a diameter of 30 to 50 mm. It is prepared from the offal (liver, spleen, heart, and pancreas) and fat of Alentejana breed pigs to which salt, garlic, and cumin are then added. The meat mixture is left to marinate for 12 to 24 hours in a place with a temperature no higher than 10°C and humidity close to 80 to 90%. The sausages are then stuffed and cooked in hot water but without boiling, for half an hour to an hour. At the end, the sausages are immersed in cold water for a period of 2 to 4 minutes (DGADR 2016k). Cacholeira Branca de Portalegre PGI has a greyish-brown colour on the outside and should have a mild, delicate, and not very salty taste.

Farinheira de Portalegre PGI (Reg EC 1997b) is a smoked sausage made from fresh pork fat from the Alentejana pig breed and wheat flour. The resulting paste is seasoned with salt, garlic, pepper, and paprika and is left to stand for 2 to 5 days at a temperature not exceeding 10°C. Afterwards, boiling water is added with wheat flour, and manual kneading is done to homogenise the mixture. Finally, the cow casing is filled, and the product is smoked for 3 to 15 days (DGADR 2016l). Farinheira de Estremoz e Borba PGI is produced in a similar way to the traditional farinheira de portalegre, but with the addition of less fat and more wheat flour, which gives Farinheira de Estremoz e Borba PGI its distinctive flavour.

Paia de Estremoz e Borba PGI (Reg EC 2004), Paia de Lombo de Estremoz e Borba PGI (Reg EC 2004), and Paia de Toucinho de Estremoz e Borba PGI (Reg EC 2004) are smoked sausages produced in the same region of Estremoz e Borba, two small municipalities of the district of Évora, near the border with the Spanish Extremadura, with the production techniques being very similar. The main differences are in the cut of the pork and in the fat used to make the sausage. Paia de Estremoz e Borba PGI is made from pork from the bacon, topside, and leg (Reg EC 2002d). Alternatively, Paia de Lombo de Estremoz e Borba and Paia de Toucinho de Estremoz e Borba PGI are made from the whole loin and the belly of the pig, respectively (Reg EC 2002d, 2002e).

Painho de Portalegre PGI (Reg EC 1997b) and Paio de Beja PGI (Reg EC 2007b) are smoked sausages made from meat from the loin, leg and shoulder, and fat of Alentejana breed pigs. The spices used to season the meat can differ slightly and contribute to the different taste of each sausage. Painho de Portalegre PGI is seasoned with salt, crushed dried garlic, and sweet and garden paprika, while Paio de Beja PGI with salt, red bell pepper paste, crushed dried garlic, cumin, paprika, and pepper. The production of both is similar, and after seasoning, the paste is kept at a temperature not exceeding 10°C for approximately 2 days. The pork casing is then stuffed, and the product is submitted for smoking (IMAIAA 1997; Reg EC 2006e).

Lombo Branco de Portalegre PGI (Reg EC 1997b) is an air-cured sausage made from fresh pork loin from pigs of the Alentejana breed. After being trimmed, the pork loin is sliced to facilitate the penetration of the seasonings. First, the loin is rubbed with a mixture of water and salt and then with crushed garlic and salt (wine may be added). It is placed for 2 to 7 days at a maximum temperature of 10°C. The preparation is stuffed into the natural intestine skin and tied with pure cotton thread. Finally, it is dried in a natural environment. The duration of this process depends on the season in which it is produced (Reg EC 1997a).

Lombo Enguitado de Portalegre PGI (Reg EC 1997b) is produced in a very similar way to lombo branco de portalegre. However, the final stage involves smoking done twice, each lasting for about 6 days, using Holm oak and/or cork oak firewood (IMAIAA 1996).

18.2.5 Algarve—Monchique

The determination to ensure the preservation of the knowledge, history, and costumes of the Monchique region has led to the survival and development of the unique Alentejana breed bBlack pork sausage production units in the Monchique mountain range (DGADR 2016n). Local companies and those devoted to tradition are dedicated to black pig breeding and the production and commercialisation of traditional sausages. The know-how and flavours were kept throughout generations, and the quality standards remained high. While Monchique region fermented sausages do not retain PDO or PGI recognition, the authors would like to highlight the three fermented sausages that are produced and sold in the area—chouriço de carne, morcela de carne and farinheira de milho, due to their social, cultural, and economic significance to this region.

Chouriço de carne da serra de monchique, also known as linguiça da serra de monchique, is a reddish horseshoe-shaped sausage, 4 to 5 cm long, cylindrical and knotted at the ends (DGADR 2016m). This fermented sausage is produced using finely chopped lean pork meat, to which some fatty streaky meat is added. Salt, garlic, peppers and red pepper broth, white wine, and/or vinegar are added, and the resulting meat mixture is left to sit for 3 to 7 days to mature (Soeiro 2001b). It is important to stir or knead it every day with a wooden spoon while adjusting the seasoning. Pork guts, previously washed and seasoned with salt and orange or lemon slices, are filled with this emulsion and left to dry for 15 to 20 days in a traditional smokehouse or kitchen chimney. Preservation in wheat, which provided a fresh and airy environment, was once ideal for stable preservation, but this old custom has been replaced by the immersion of these products in olive oil, a more modern and equally effective preservation method. Typically, chouriço de carne is consumed raw, cut into slices, grilled, or used in other dishes as a flavour enhancement (Soeiro 2001b).

Morcela de carne da serra de monchique is a meat and blood sausage. The leftover meat trimmings from the preparation of other meat sausages are used to produce the morcela (DGADR 2016n). Adding to this, bloody muscle mass from cleaning the carcasses and fried or non-fried fatty meat (*torresmos*) are included and chopped to a fine crumb (Soeiro 2001b). Interestingly, *torresmos* are a layer of skin, meat, and fat removed after frying the bacon and are considered a tasty snack.

Similarly, the appearance of chouriço de carne da serra do monchique and morcela de carne is identical, except for the blood colour of the second fermented sausage. Finely chopped meats are marinated in pig's blood diluted with salt and vinegar, garlic, and various spices (cumin, pepper, cinnamon, and ground nutmeg). After 24 hours, the homogeneous dough is stuffed in fine fresh pork intestines, previously washed and seasoned with lemon or orange. Cooking is carried out in the so-called caldo mouro (Moorish broth), which contains water, whole onion, salt, parsley, and bay leaf. Morcela is left on a slow fire for 30 minutes to 1 hour so the casing does not burst. Afterwards, the fermented sausages are left to dry for 8 to 15 days. Preservation, formerly done by immersion in salted lard or olive oil, is being replaced by refrigeration. Morcela is consumed roasted or cooked as a main meal or snack (DGADR 2016n; Soeiro 2001b).

The production of farinheira de milho, also known as morcela de milho, is interestingly rare and unique, being seasoned with cinnamon and stuffed in raw cloth. The caldo mouro in which the morcela was cooked is reused, and lean, minced pork meat and minced pork mesenteric fat are added (Soeiro 2001b). Likewise, corn flour, salt, garlic, onion, herbs, parsley, pepper, and, with a higher proportion than the other spices, cinnamon are added, contributing to the sensorial characteristics of this product. After cooking, the dough obtained is placed in raw cloth, known as talegos (or sometimes the large intestine of the pig), of variable lengths (20 to 30 cm) tied with normal sewing thread but resistant so as not to open. After cooking for 2 to 4 hours, the wrapped fermented sausages are dried in an airy place and preserved by freezing (Soeiro 2001b). Farinheira should be consumed in a short period of time and is traditionally eaten boiled or fried in lard for breakfast the day after it is prepared.

18.2.6 Região Autónoma dos Açores

The Açores is one of the two autonomous regions of Portugal. Located in the North Atlantic Ocean, Açores is an archipelago of nine volcanic islands with breathtaking mountain views, tropical greenery, large blue lakes, and waterfalls (França et al. 2003). Its traditional heritage in the making of several characteristic products is very present among the locals, and the knowledge has been passed down throughout the centuries. In the Açores, there are two products locally produced: morcela (produced on the island São Miguel) and chouriço de mouro (https://azoresgetaways.com/en-ca/destination/azores/why-visit/food). Although these products do not have PGI recognition due to their relevance as typical fermented products of the region, they will be presented in this section.

Morcela de são miguel is a sausage associated with the winter festivities of São Miguel, patron of the largest island of the archipelago, celebrated on the 29th of September. Being a part of the celebrations, it is also commonly referred to in repositories of popular songs sung around this time. This fermented product is prepared with blood, parsley, onion, cooked rice or rolled wheat, ground cinnamon, ground pepper, pieces of bacon, shreds, toasted corn or wheat flour, fat from the intestines, garlic, vinegar, salt, herbs, and cloves. This mixture is placed in well-washed intestine casings with sour orange. The guts are tied together and placed in a wicker. Morcela is cooked in salted water, hanging over the mouth of the pot, airing it. It is then moved to the smokehouse, or the *chambaril* is placed in the chimney. Later, these products are placed in glass or earthenware jars covered with pig lard. In São Miguel Island, it is customary to serve morcela with pineapple or orange to facilitate its digestion. It can also be eaten fried in lard, oil, or baked in the oven and accompanied by white rice, sweet potato, or boiled yam (Soeiro 2001b).

Chouriço de mouro dos açores, made from the less tender parts of the pig, fat, and offal, is considered a delicious spicy sausage to be eaten as a snack or as a main meal. The meats are seasoned with salt, ground pepper, and garlic and left to marinate for 3 days in wine with garlic (*vinha d'alho* in Portuguese). The mixture is then stuffed in cow guts, smoked for 15 days, and preserved in lard. In some cases, the meat is seasoned with lemon juice (Soeiro 2001b).

18.3 CHARACTERISATION OF DIFFERENT FERMENTED MEAT PRODUCTS

Despite the large number of fermented meat products that are traditionally produced in Portugal, only a few captured the attention of researchers who have dedicated their studies to their characterisation. Such products include alheira (with the largest number of studies) (Albano et al. 2008; Coelho-Fernandes et al. 2021b; Esteves et al. 2008; Ferreira et al. 2006, 2007a; Patarata et al. 2008; Ramalhosa et al. 2012) and other products from the northern region, such as chouriço (Matos et al. 2013; Santos et al. 2011; Silva et al. 2018), chouriça and salpicão (Campelos et al. 2016; Ferreira et al. 2007b, 2009; Pereira et al. 2015), cacholeira, and morcela (Belleggia et al. 2020; Santos et al. 2011). The main results of these studies, especially those related to quality and safety issues, will be summarised and discussed in the upcoming sections.

Regarding microbiological characterisation, the microbiota of fermented meat products fluctuate greatly between producers and regions due to the different raw materials used, different processing techniques, and the heterogeneous nature of the products. In addition, because variables such as temperature and time were not always fully controlled in most artisanal production, these fluctuations were to be expected from batch to batch, even within the same processing unit (Marcos et al. 2016).

According to the Guidelines of the Food Safety Authority of Ireland (Food Safety Authority of Ireland 2001), a fermented meat product can be classified as satisfactory, acceptable, unsatisfactory, or potentially hazardous depending on the level and type of microbiological contamination. Chemical hazards also had a significant impact on the safety of meat products, with a particular concern about polycyclic aromatic hydrocarbons (PAHs) due to the smoking stage during the processing of these products (Santos et al. 2011).

18.3.1 Alheira

The high microbial load of alheira is mainly composed of lactic acid bacteria (LAB) from the *Lactobacillaceae*, *Streptococcaceae*, and *Enterococcaceae* families (Esteves et al. 2008; Fernandes et al. 2022). These bacteria ferment carbohydrates into organic acids which contribute to the taste of these products. However, the growth of some microorganisms with amino genic capacity (e.g., some *Enterococcus* strains) during fermentation and proteolytic activity can contribute to the production and accumulation of biogenic amines (Alves et al. 2017).

Alheira samples from the north of Portugal and tested by Ferreira et al. (2006) showed LAB as the predominant microbiota (with 7 to 8 log colony forming units (CFU)/g). The authors also reported that most samples would be considered unsatisfactory due to high counts of *Enterobacteriaceae* and *Escherichia coli*. Due to the detection of *Listeria monocytogenes* in some samples, these would immediately be reported and considered unacceptable/potentially hazardous. A chemical analysis was also performed, with lead, cadmium, arsenic, and mercury levels being all within the threshold imposed by European legislation. Biogenic amines in alheira samples from the north of Portugal were shown not to represent an obvious or generalised hazard (Ferreira et al. 2006). In 2007, the same authors characterised 37 lots of alheira from 17 producers from the north of Portugal regarding their microbiological safety. Although not finding *Campylobacter* spp. and *Escherichia coli* O157, the authors detected *Salmonella* spp. in two lots and *L. monocytogenes* in levels higher than 100 CFU/g in more than 60% of the lots analysed (Ferreira et al. 2007a).

Albano et al. (2008) studied the microbial ecology of traditional alheira de mirandela samples using PCR-DGGE (denaturing gradient gel electrophoresis) and traditional microbiological analyses. The authors found LAB to be the predominant microbiota, but also 80% of alheira de mirandela samples were considered unsatisfactory due to the high counts of *Enterobacteriaceae*. Results also showed the presence of potential spoilage strains such as *Pseudomonas* and the possibility of product contamination with the pathogen *L. monocytogenes* due to the detection of *Listeria* spp. in some sample points (Albano et al. 2008). In the same year and for alheiras de mirandela samples, Esteves et al. (2008) reported the prevalence of *Staphylococcus aureus,* mostly due to contaminated surfaces and raw materials; *Clostridium perfringens* and *Salmonella* spp. *L. monocytogenes*, *Bacillus cereus*, and *Yersinia enterocolitica* were also found, but in small numbers, with *Campylobacter* spp. not being detected in any samples (Esteves et al. 2008).

Patarata et al. (2008) studied and compared the physicochemical and sensory characteristics of alheira samples from different-sized producers and reported that this product is prone to high variability in its microbiological, chemical, and sensory qualities, with a distinction between alheira samples produced at industrial scale and by traditional producers. The authors also stated that products of industrial origin accomplished higher levels of acceptability due to their higher moisture levels and the presence of a higher percentage of meat (Patarata et al. 2008).

Coelho-Fernandes et al. (2021b) set out to evaluate the association between some physicochemical and microbiological attributes of alheira samples produced by different regional producers of Trás-os-Montes. This study differentiated three clusters of factories: one that was considered safe focused on alheira samples with LAB prevalence and lower pH; a second one with sausages of higher protein content, pH, and meat content; and a third one with alheira samples of questionable quality, with lower meat percentage and higher incidence of pathogens. These findings highlighted the need to implement better hygiene protocols during the processing of these types of products (Coelho-Fernandes et al. 2021a). The same conclusions were reached by Ramalhosa et al. (2012) when studying regional Portuguese kitchens used to produce alheira de vinhais. No *Salmonella* spp., *S. aureus*, or *Listeria* spp. were found. However, occasionally, deficient food processing hygienic conditions were observed in terms of microbiological environment air quality, as well as poor hygiene practices due to the presence of yeasts, moulds, coliforms, and *E. coli* on surfaces and equipment. Once again, proper handling and safe practices were identified as crucial when it comes to producing microbiologically safe products (Ramalhosa et al. 2012).

It should be highlighted that although alheira is not a ready-to-eat sausage, Felício et al. (2011) have shown that the temperatures reached in the product during cooking by the consumer may not be sufficient to destroy pathogenic bacteria if they are present.

18.3.2 Chouriço

As in the case of alheira, the fermentation process also favours the growth of LAB in the chouriço products. Matos et al. (2013) studied two Portuguese chouriço—chouriço alentejano and chouriço ribatejano, and related the predominance of anaerobic microbiota and enterococci to the presence of LAB. However, pathogenic groups such as coagulase-positive staphylococci, sulphite-reducing clostridia, and *Clostridium perfringens* were also represented in the microbiota of both types of chouriço (Matos et al. 2013).

The microbiological safety of chouriço depends on a combination of several hurdles, and the standardisation of the production process is imperative for the quality and safety of the product. In another study, Silva et al. (2018) investigated the risk factors and main microbiological differences in five stages of the production of Portuguese chouriço regarding the levels of *Enterobacteriaceae*, *S. aureus*, and *L. monocytogenes* in the product. The authors found that contamination occurred at the mixing and maceration stages due to the significantly increased numbers of all the microorganisms at the end of these processing stages. Furthermore, the presence of nitrite and polyphosphate exceeded allowable concentrations, slowing fermentation, which is suspected to have contributed to the increased numbers of microorganisms detected during maceration (Silva et al. 2018).

Regarding chemical characterisation, Santos et al. (2011) investigated the contamination levels of PAHs in several traditional dry fermented meat sausages from Alentejo, including chouriço de carne, paínho and paio tradicional and blood sausages chouriço mouro, cacholeira and morcela. The eight PAHs selected as indicators of carcinogenic in foods (PAH8) constituted less than 0.5% of the total contamination: benzo[a] anthracene (0.21 µg/kg) and chrysene (8.9 µg/kg) were the most concentrated compounds regardless of the product. The authors also reported that blood sausages were potentially more hazardous, as the total PAH8 levels were often higher (Santos et al. 2011).

18.3.3 Cacholeira and Morcela

In one of the few studies found in the literature, Santos et al. (2011) reported that cacholeira and morcela had the highest total PAHs content contamination (1701.27 µg/kg and 996.58 µg/kg, respectively) compared to the meat counterpart (chouriço de carne—877.37 µg/kg).

Belleggia et al. (2020) characterised chacoleira blood sausage in terms of physico-chemical, microbiological, and volatile traits. The metataxonomic analysis revealed a simple bacterial composition, dominated by *Latilactobacillus sakei* in the two batches analysed and *Carnobacterium*, *Enterococcus*, *Kluyvera*, *Lactococcus*, and *Serratia* as minor genera. The authors also identified 72 volatile compounds; however, due to differences in raw materials, artisanal production, and seasonality, the presence of these compounds varied significantly between lots (Belleggia et al. 2020).

18.3.4 Salpicão and Chouriça

The chemical and microbiological characterisation of salpicão de vinhais and chouriça de vinhais was evaluated by Ferreira et al. (2007b). The microbiological classification of salpicão and chouriça samples as unsatisfactory was due to high contamination with *Enterobacteriaceae, E. coli* or *S. aureus.* One sample of chouriça was classified as potentially hazardous due to *L. monocytogenes* counts exceeding 100 CFU/g.

Regarding the chemical characterisation, the maximum values for nitrates, nitrites, cadmium, and lead were never surpassed (Reg EC 2006a, 2008a). The chouriça samples presented higher levels of the biogenic amines tyramine and putrescine, while the salpicão samples had higher tyramine levels (Ferreira et al. 2007b). The evolution of the microbiota and microbiological safety during the production and storage of the same type of fermented meat products was investigated by the same authors (Ferreira et al. 2009). During the smoking process, LAB increased over time, becoming the most abundant organisms and consequently lowering the pH. The authors highlighted the importance of working under hygienic conditions and using raw materials of good microbiological quality to ensure the production of safe traditional salpicão de vinhais and chouriça de vinhais (Ferreira et al. 2009). A few years later, Campelos et al. (2016) investigated the microbiological and physico-chemical characteristics of 77 samples of salpicão de vinhais and chouriça de vinhais. The authors found that, despite the absence of pathogens such as *S. aureus*, spores of sulphite-reducing clostridia, *E. coli* 0157:H7, *Yersinia* spp. and *Salmonella* spp., and/or *L. monocytogenes* were found in 14.3% of the samples. This result emphasises the need to implement good manufacturing practices to control *L. monocytogenes* contamination in meat processing plants (Campelos et al. 2016).

Pereira et al. (2015) studied salpicão from Trás-os-Montes at different stages of production, from raw meat to final product. Total viable mesophilic bacteria increased significantly during production. While *Enterobacteriaceae* decreased, *S. aureus* increased throughout processing and was present in the final product of both producers (Pereira et al. 2015). As aalpicão is a ready-to-eat product, it is essential to better monitor its microbiological quality, as well as to improve good hygiene and manufacturing practices in the production of these traditional sausages (Pereira et al. 2015).

18.4 STRATEGIES TO IMPROVE QUALITY AND ENSURE FOOD SAFETY OF FERMENTED MEAT PRODUCTS

The production of fermented meat products is of paramount importance to the meat industry in many countries. Ensuring the quality and safety of food during preparation and cooking is crucial to maintain public health. When good practices and hygiene conditions are not sufficient or not even employed, other methods are necessary. Thus, biopreservation methods, such as adding bacteriocin-producing LAB to raw meat preparations, have gained attention in recent years. These can not only extend the shelf-life of products but also maintain their microbiological safety by inhibiting the growth of certain pathogens and spoilage bacteria (Yost 2014; Ursachi et al. 2020). Proper cooking methods and practices also play a major role in preventing foodborne diseases. Accurate cooking instructions are necessary to ensure that non-ready-to-eat products are properly cooked, achieving high internal temperatures to kill pathogens that may be present due to poor raw product processing or handling (Byrd-Bredbenner et al. 2013).

18.4.1 Autochthonous LAB as Starter and Protective Cultures

As previously mentioned, LAB are the main microbial group found in Portuguese fermented meat products and have been shown to play an important role in the production of these products (Azevedo et al. 2020; Correia Santos et al. 2017). The dominant LAB are lactobacilli and enterococci and can be found in relatively high numbers (7–9 log CFU/g) (Albano et al. 2009b; Barbosa et al. 2009; Dias et al. 2020, 2021). In alheira and chouriça, the numbers of enterococci range from 4 to 8 log CFU/g and 4 to 7 log CFU/g, respectively (Barbosa et al. 2014).

The use of LAB as natural preservatives is a subject of great interest due to the potential production of metabolites with antimicrobial activity, such as organic acids (lactic and acetic), hydrogen peroxide, antimicrobial enzymes and bacteriocins (Abrams et al. 2011; Martín et al. 2022). Biopreservation using LAB through the incorporation of starter cultures is a promising technology due to their generally recognised as safe (GRAS) recognition and environmentally friendly approach (Linares-Morales et al. 2018). These methods are usually used to increase the shelf life and quality of the products by delaying spoilage and pathogen growth (Macieira et al. 2018). Starter cultures are used to improve the quality and safety of fermented products mainly through their bioprotective (inhibition or reduction of the growth of foodborne pathogens), probiotic (production of substances essential for the health of consumers), and fermentative (production of secondary metabolites) capacities (Van Ba et al. 2016). Although the use of starter cultures is not currently a widespread practice in the traditional Portuguese sausage-making industry, several authors have investigated the use of autochthonous bacteria as starter cultures in these products. Lebert et al. (2006) developed a starter consisting of a mixture of bacteria isolated from traditional French sausages (*Lt. sakei*, *Staphylococcus equorum*, and *Staphylococcus succinus*). Its addition to the batter increased the microbiological and chemical characteristics of Saucisson Fermier due to the reduction of *L. monocytogenes*, enterococci, biogenic amines, lipid oxidation, and total cholesterol oxides without altering the flavour of the sausages (Lebert et al. 2006). Vaz-Velho et al. (2013) demonstrated that *Lt. sakei* ST153 decreased *L. monocytogenes* counts (up to 2 log CFU/g) when inoculated into a sterilised paste of alheira. This study demonstrated a promising methodology to improve the safety of Portuguese fermented sausages and showed the possibility of using *Lt. sakei* ST153 as a protective culture (Vaz-Velho et al. 2013). In a study by Alfaia et al. (2018), *Lactiplantibacillus plantarum* and *Lt. sakei* isolated from fermented meat products demonstrated an inhibitory role against *L. monocytogenes* and *S. aureus* and showed a suitable biogenic amine profile to be used as a protective culture (Alfaia et al. 2018). The inoculation of *Lt. sakei* CECT7056 (previously isolated from paio), individually or in combination with *Staphylococcus xylosus,* led to the elimination of *L. monocytogenes* in paio (Dias et al. 2017). In addition, these cultures were associated with a reduction in the levels of putrescine, cadaverine, and tyramine. For all these reasons, *Lt. sakei* CECT7056 should be considered as a protective culture in the production of paio. Dias et al. (2020) also studied the effect of autochthonous starter cultures on the safety and quality of painho da beira baixa and observed that *Lt. sakei* CV3C2 together with *S. equorum* S2M7 contributed to the safety of the smoked sausages by reducing *Enterobacteriaceae* and *L. monocytogenes* and the content of biogenic amines without affecting their sensory quality (Dias et al. 2020).

Bacterial strains that are well adapted to specific matrices, preferably LAB isolated from similar foods, are preferred because they allow less variability in production, thus diversifying the starter/protective culture market. At the same time, they are able to maintain the organoleptic and microbiological characteristics of traditional fermented sausages, thus increasing product's acceptability to consumers (Macieira et al. 2018; Talon et al. 2007). However, selecting the best protective food cultures must be considered on a case-by-case basis. For example, the addition of a bacteriocinogenic culture of *Lb. plantarum* ST153Ch, which showed anti-listerial activity in alheira (Macieira et al. 2019), to chouriço de vinha d'alhos did not show any significant activity in the elimination of *L. monocytogenes* when compared to samples without added culture (Macieira et al. 2018). Therefore, the choice of the most suitable cultures and application methods must be made according to the manufacturing process to ensure the best effect of bioprotective cultures for the industry (Macieira et al. 2018).

The use of nitrite in the curing phase of fermented and cured meats is a commonly used approach for technological needs (improved flavour, taste, aroma, and colour and prevented oxidation) and microbial safety, especially to inhibit the extremely toxigenic bacteria *Clostridium botulinum*. However, this additive has recently been scrutinised due to its association with several blood diseases and cancers (Kalaycıoğlu and Erim 2019; Sebranek and Bacus 2007; Shakil et al. 2022; Ward et al. 2018). Therefore, the reduction, replacement, and/or complete elimination from the curing and overall fermentative process have been the focus of some studies (Fraqueza et al. 2021; Hospital et al. 2015; Patarata et al. 2022). Very recently, Patarata et al. (2022) evaluated the effect of nitrite elimination from cured loins on the colour and inhibition of foodborne pathogens. For cured loins without added nitrite, more than 50% of consumers considered the colour weaker but still acceptable. The authors also reported that the survival of *Clostridium sporogenes*, *L. monocytogenes*, and *Salmonella* spp. was mainly affected by water activity reduction instead of the presence of nitrite. The authors also evaluated the addition of a protective culture of *Staphylococcus xylosus* to the meat. Apart from contributing to the development of appropriate colour, the results did not support a conclusion on the potential use of this culture in the preparation of cured loins (Patarata et al. 2022). Nonetheless, protective food cultures may be a potential solution to control pathogens present in nitrite-free meat products, but more studies are needed to evaluate this potential properly.

18.4.2 Accurate Indications of Safe Food Preparation and Cooking

As mentioned in the previous sections, traditional sausages and hams have a high number of bacteria, but most bacteria in these products are non-pathogenic, with the presence of harmful microorganisms occurring occasionally (Campelos et al. 2016; Ferreira et al. 2006; Matos et al. 2013).

Although most heavily cured and smoked meats can be consumed safely without needing further heat processing (if hygiene practices were followed during production), some, such as alheira, require cooking before use. As observed by Ferreira et al. (2007a) and Matos et al. (2013), *Salmonella* spp., *L. monocytogenes*, coagulase-positive staphylococci, sulphite-reducing clostridia, and *Clostridium perfringens* can be present in alheira and chouriço, so thorough cooking is essential to eliminate such bacteria. To our knowledge, there is only one study in the literature evaluating the thermal inactivation of pathogens during the cooking of traditional sausages (Felício et al. 2011). The authors estimated the survival of *L. monocytogenes* by calculating the D and z values and combining them with the internal temperature profiles of alheira during its cooking by frying, electric grilling, and oven roasting (both gas and wood-fired). Except for roasting, the other cooking methods evaluated may not be sufficient to inactivate this pathogen in alheira, as some of the internal temperature profiles were markedly low, suggesting the persistence of cold spots in these products during consumer processing (Felício et al. 2011). As these traditional products are often exposed and sold at room temperature, this increases the ability of this pathogen (or others) to multiply and reach high numbers in the products. As the most common way consumers cook these sausages is by frying, which usually does not allow the interior of the product to reach sufficient inactivation temperatures, high levels of *L. monocytogenes* or other pathogens can lead to infections and foodborne illness. The authors of this pioneering study advise of the need to warn consumers to fry or grill these products for longer to reduce the likelihood of survival of this pathogen, if present (Felício et al. 2011). To reduce such risks, labelling with concise instructions for adequate cooking conditions, if followed by consumers, will allow safe consumption of these sausages. As labelling is an irreplaceable form of communication of the nutritive and quality characteristics of foods (Hall and Osses 2013; Shangguan et al. 2019; Viola et al. 2016), the authors believe that further attention should be given to the inclusion of highlighted recommendations regarding cooking temperatures, techniques, and times.

18.4.2.1 Concluding Remarks

Traditional fermented meats are important representatives of the cultural heritage and economic fingerprints of different regions in Portugal. Chouriço, alheira, salpicão, and presunto are the main products of many feasts across the country and are products that encourage the reunion of families, friends, and entire communities to come together to share delicious meals. While some of these products are well known by most of the Portuguese population and in some cases are even recognised internationally, some others are relegated to the obscurity of general ignorance. Similarly, scientific research on these products focuses on a limited characteristic of such products, for example, the microbiological, physicochemical, nutritional, and sensory characteristics of alheira and other products being thoroughly studied, while there are practically no such studies for a large number of other products.

Due to the intrinsic microbial activity required to produce fermented products, the microbial balance, responsible for the success or failure of the fermentation process, is fragile. Also, the presence of pathogenic bacteria in traditional and industrially produced fermented products calls for implementing preventive strategies to guarantee food safety. From farm to fork, full compliance with hygienic practices throughout processing raw meat, intestines, fat, and bread to the final fermented and cured meat is essential to achieve a safe product. Furthermore, innovation in biopreservation techniques and the addition of beneficial bacteria as food cultures could help to reduce the risks of pathogens and spoilage microorganisms, promoting food safety while preserving the traditional and essential sensory characteristics of these sausages. Finally, the authors believe that including accurate and understandable labelling on the packages, indicating techniques for the preparation, consumption, and storage of these products in a safety-oriented manner, would positively influence consumer safety practices and thus lead to the prevention of foodborne illnesses.

18.4.2.2 Acknowledgements

The authors would like to thank the scientific collaboration under the scope of the project "cLabel+: Innovative 'clean label' natural, nutritious, and consumer-oriented foods" (POCI-01–0247-FEDER-046080), co-financed by the Fund European Regional Development (ERDF) through the Competitiveness and Internationalisation Operational Program (POCI), and Fundação para a Ciência e a Tecnologia (FCT) through the project GenoPhenoTraits4Persitence – Genomic and phenotypic traits contributing to persistence of *Listeria monocytogenes* in food processing environment (PTDC/BAA-AGR/4194/2021) and project UIDP/00329/2020. Financial support for authors Joana Gomes, Mónica Oliveira, Lourenço Pinto de Rezende, and Teresa Bento de Carvalho, was provided by the doctoral fellowships 2021.06027.BD, 2021.08345.BD, 2022.13871.BD, and 2023.03709.BD, respectively.

REFERENCES

Abrams, D., Barbosa, J., Albano, H., Silva, J., Gibbs, P. A., and P. Teixeira. 2011. Characterization of bacPPK34 a bacteriocin produced by *Pediococcus pentosaceus* strain K34 isolated from "Alheira." *Food Control 22*(6): 940–946. https://doi.org/10.1016/j.foodcont.2010.11.029

ADIV. 2006. Recomendações práticas de higiene para enchidos tradicionais fermentados e secos. Retrieved February 6, 2023, from the World Wide Web: http://maisagro.pt/wp-content/uploads/2017/06/oshcnrp06_recomendacoes_praticas_de_higiene_para_enchidos_tradicionais_fermentados_e_secos.pdf

Albano, H., Henriques, I., Correia, A., Hogg, T., and P. Teixeira. 2008. Characterization of microbial population of alheira (a traditional Portuguese fermented sausage) by PCR-DGGE and traditional cultural microbiological methods. *Journal of Applied Microbiology 105*(6): 2187–2194. https://doi.org/10.1111/j.1365-2672.2008.03947.x

Albano, H., Pinho, C., Leite, D., Barbosa, J., Silva, J., Carneiro, L., Magalhães, R., Hogg, T., and P. Teixeira. 2009a. Evaluation of a bacteriocin-producing strain of *Pediococcus acidilactici* as a biopreservative for alheira, a fermented meat sausage. *Food Control 20*(8): 764–770. https://doi.org/10.1016/j.foodcont.2008.09.021

Albano, H., van Reenen, C. A., Todorov, S. D., Cruz, D., Fraga, L., Hogg, T., Dicks, L., and P. Teixeira. 2009b. Phenotypic and genetic heterogeneity of lactic acid bacteria isolated from alheira, a traditional fermented sausage produced in Portugal. *Meat Science 82*: 389–398. https://doi.org/10.1016/j.meatsci.2009.02.009

Alfaia, C. M., Gouveia, I. M., Fernandes, M. H., Fernandes, M. J., Semedo-Lemsaddek, T., Barreto, A. S., and M. J. Fraqueza. 2018. Assessment of coagulase-negative Staphylococci and lactic acid bacteria isolated from Portuguese dry fermented sausages as potential starters based on their biogenic amine profile. *Journal of Food Science 83*(10): 2544–2549. https://doi.org/10.1111/1750-3841.14298

Alves, S. P., Alfaia, C. M., Škrbić, B. D., Živančev, J. R., Fernandes, M. J., Bessa, R. J. B., and M. J. Fraqueza. 2017. Screening chemical hazards of dry fermented sausages from distinct origins: Biogenic amines, polycyclic aromatic hydrocarbons and heavy elements. *Journal of Food Composition and Analysis 59*: 124–131. https://doi.org/10.1016/j.jfca.2017.02.020

Amass. 2017. Guia dos Enchidos Portugueses. Retrieved February 6, 2023, from the World Wide Web: https://www.amasscook.com/guia-dos-enchidos-portugueses/?lang=pt-pt

Anonymous. 2018. Bucho e enchidos com muita procura nos talhos da Guarda. Jornal A Guarda. Retrieved February 16, 2023, from the World Wide Web: https://www.jornalaguarda.com/index.php/economia/bucho-e-enchidos-com-muita-procura-nos-talhos-da-guarda

Azevedo, I., Barbosa, J., Albano, H., and P. Teixeira. 2020. Non meat-based alheiras– a safer novel trend? *Food Control 113*. https://doi.org/10.1016/j.foodcont.2020.107177

Barbosa, J., Borges, S., and P. Teixeira. 2014. Selection of potential probiotic *Enterococcus faecium* isolated from Portuguese fermented food. *International Journal of Food Microbiology 191*: 144–148. http://dx.doi.org/10.1016/j.ijfoodmicro.2014.09.009

Barbosa, J., Ferreira, V., and P. Teixeira. 2009. Antibiotic susceptibility of enterococci isolated from traditional fermented meat products. *Food Microbiology 26*(5): 527–532. https://doi.org/10.1016/j.fm.2009.03.005

Belleggia, L., Ferrocino, I., Reale, A., Boscaino, F., Di Renzo, T., Corvaglia, M. R., Cocolin, L., Milanović, V., Cardinali, F., Garofalo, C., Clementi, F., Aquilanti, L., and A. Osimani. 2020. Portuguese cacholeira blood sausage: A first taste of its microbiota and volatile organic compounds. *Food Research International 136*: 109567. https://doi.org/10.1016/j.foodres.2020.109567

Boileau, J. P. 2010. A culinary history of the Portuguese Eurasians: The origins of Luso-Asian cuisine in the sixteenth and seventeenth centuries. Unpublished doctoral dissertation. University of Adelaide. https://digital.library.adelaide.edu.au/dspace/handle/2440/77948

Byrd-Bredbenner, C., Berning, J., Martin-Biggers, J., and V. Quick. 2013. Food safety in home kitchens: A synthesis of the literature. *International Journal of Environmental Research and Public Health 10*(9): 4060–4085. https://doi.org/10.3390/ijerph10094060

Câmara Municipal Castelo Branco. 2023. Os Enchidos e os Presuntos de Castelo Branco. Retrieved February 16, 2023, from the World Wide Web: https://www.cm-castelo-branco.pt/visitante/sabores-de-perdi%C3%A7%C3%A3o/os-nossos-enchidos-e-presuntos/

Campelos, M. I., Silva, J., Gibbs, P. A., and P. Teixeira. 2016. Contributing data for risk assessment of traditional fermented sausages: "Salpicão de Vinhais" and "Chouriça de Vinhais". *Cogent Food & Agriculture 2*: 1131886. http://doi.org/10.1080/23311932.2015.1131886

Coelho-Fernandes, S., Rodrigues, G., Faria, A. S., Caleja, C., Pereira, E., Pinela, J., Carocho, M., Barros, L., Cadavez, V., and U. Gonzales-Barron. 2021a. Effect of sage (*Salvia officinalis* L.) extract on the survival of *Staphylococcus aureus* in Portuguese *Alheira* sausage during maturation. *Biology and Life Sciences Forum 6*(1): 125. https://doi.org/10.3390/FOODS2021-11047

Coelho-Fernandes, S., Zefanias, O., Rodrigues, G., Faria, A. S., Fernandes, Â., Barros, L., Cadavez, V., and U. Gonzales-Barron. 2021b. Microbiological and physicochemical assessment of artisanally produced alheira fermented sausages in Northern Portugal. *Proceedings 70*: 16. https://doi.org/10.3390/foods_2020-07627

Cooperativa Agrícola dos Produtores de Batata de Semente de Montalegre. 1996. Caderno de especificações da Indicação Geográfica "Presunto de Barroso". In: Official Journal of the European Union. Directorate-General for Agriculture—Agricultural product quality policy—B-1049 Brussels, Belgium. Retrieved February 23, 2023, from the World Wide Web: https://ec.europa.eu/geographical-indications-register/eambrosia-public-api/api/v1/attachments/59758

Cornwall, E. E., and N. Y. W. Brooklyn. 1937. What the Ancient Greeks Ate. *Annals of Medical History 9*(1): 30. https://www.ncbi.nlm.nih.gov/pmc/articles/PMC7942920/

Correia Santos, S., Fraqueza, M. J., Elias, M., Salvador Barreto, A., and T. Semedo-Lemsaddek. 2017. Traditional dry smoked fermented meat sausages: Characterization of autochthonous enterococci. *Lebensmittel-Wissenschaft & Technologie 79*: 410–415. https://doi.org/10.1016/j.lwt.2017.01.042

de Armas, F. 2013. The geography of imperial deceit: Misplacing Goa and Lisboa in El burlador de Sevilla. *Bulletin of Spanish Studies: Hispanic Studies and Researches on Spain, Portugal and Latin America* 90(4–5): 495–507. https://doi.org/10.1080/14753820.2013.802563

de Góis, D. 1988. *Descrição da cidade de Lisboa* (J. da F. Alves, Ed.). Livros Horizonte.

Diagnóstico e Plano de Acção da Agenda 21 Melgaço. 2008. Retrieved February 16, 2023, from the World Wide Web: https://www.cm-melgaco.pt/wp-content/uploads/docs_cmm/Agenda_21_local/Fases_de_implementacao/Plano_de_accao/Plano_Accao.pdf

Dias, I., Laranjo, M., Fialho, R., Potes, M. E., Véstia, J., Agulheiro-Santos, A. C., Fraqueza, M. J., and M. Elias. 2017. Effect of autochthonous starter cultures in the production of Paio, a traditional Portuguese dry-cured sausage. *Archivos de Zootecnia*: 161–165.

Dias, I., Laranjo, M., Potes, M. E., Agulheiro-Santos, A. C., Ricardo-Rodrigues, S., Fialho, R., Véstia, J., Fraqueza, M. J., Oliveira, M., and M. Elias. 2020. Autochthonous starter cultures are able to reduce biogenic amines in a traditional Portuguese smoked fermented sausages. *Microorganisms 8*: 686. http://doi.org/10.3390/microorganisms8050686

Dias, S., Chambel, L., Tenreiro, R., Nunes, L., and V. Loureiro. 2021. Microbial characterization of yellow curing process of cod-FISH. *International Journal of Food Science* 2021. https://doi.org/10.1155/2021/6072731

Direção-Geral de Agricultura e Desenvolvimento Rural (DGADR). 2016a. Produtos Tradicionais Portugueses—Bucho da Guarda. Retrieved February 23, 2023, from the World Wide Web: https://tradicional.dgadr.gov.pt/pt/cat/salsicharia-fumados-presuntos-e-paletas/248-bucho-da-guarda

Direção-Geral de Agricultura e Desenvolvimento Rural (DGADR). 2016b. Produtos Tradicionais Portugueses—Chouriça de Bofe da Guarda. Retrieved February 22, 2023, from the World Wide Web: https://tradicional.dgadr.gov.pt/pt/cat/salsicharia-fumados-presuntos-e-paletas/776-chourica-de-bofe-da-guarda

Direção-Geral de Agricultura e Desenvolvimento Rural (DGADR). 2016c. Produtos Tradicionais Portugueses—Chouriça Raiano. Retrieved February 22, 2023, from the World Wide Web: https://tradicional.dgadr.gov.pt/pt/cat/salsicharia-fumados-presuntos-e-paletas/251-chourico-raiano

Direção-Geral de Agricultura e Desenvolvimento Rural (DGADR). 2016d. Produtos Tradicionais Portugueses—Farinheira da Guarda. Retrieved February 23, 2023, from the World Wide Web: https://tradicional.dgadr.gov.pt/pt/cat/salsicharia-fumados-presuntos-e-paletas/253-farinheira-da-guarda

Direção-Geral de Agricultura e Desenvolvimento Rural (DGADR). 2016e. Produtos Tradicionais Portugueses—Morcela da Guarda. Retrieved February 22, 2023, from the World Wide Web: https://tradicional.dgadr.gov.pt/pt/cat/salsicharia-fumados-presuntos-e-paletas/409-morcela-da-guarda

Direção-Geral de Agricultura e Desenvolvimento Rural (DGADR). 2016f. Produtos Tradicionais Portugueses—Chouriço de Barrancos. Retrieved February 22, 2023, from the World Wide Web: https://tradicional.dgadr.gov.pt/pt/cat/salsicharia-fumados-presuntos-e-paletas/419-chourico-de-barrancos

Direção-Geral de Agricultura e Desenvolvimento Rural (DGADR). 2016g. Produtos Tradicionais Portugueses—Chouriço de Carne. Retrieved February 22, 2023, from the World Wide Web: https://tradicional.dgadr.gov.pt/pt/cat/salsicharia-fumados-presuntos-e-paletas/778-chourico-de-carne

Direção-Geral de Agricultura e Desenvolvimento Rural (DGADR). 2016h. Produtos Tradicionais Portugueses—Alheira de Barroso—Montalegre IGP. Retrieved February 21, 2023, from the World Wide Web: https://tradicional.dgadr.gov.pt/pt/cat/salsicharia-fumados-presuntos-e-paletas/457-alheira-de-barroso-montalegre-igp

Direção-Geral de Agricultura e Desenvolvimento Rural (DGADR). 2016i. Produtos Tradicionais Portugueses—Morcela de Assar de Portalegre IGP. Retrieved February 22, 2023, from the World Wide Web: https://tradicional.dgadr.gov.pt/pt/cat/salsicharia-fumados-presuntos-e-paletas/622-morcela-de-assar-de-portalegre-igp

Direção-Geral de Agricultura e Desenvolvimento Rural (DGADR). 2016j. Produtos Tradicionais Portugueses—Morcela de Cozer de Portalegre IGP. Retrieved February 22, 2023, from the World Wide Web: https://tradicional.dgadr.gov.pt/pt/cat/salsicharia-fumados-presuntos-e-paletas/623-morcela-de-cozer-de-portalegre-pgi-en

Direção-Geral de Agricultura e Desenvolvimento Rural (DGADR). 2016k. Produtos Tradicionais Portugueses—Cacholeira Branca de Portalegre IGP. Retrieved February 22, 2023, from the World Wide Web: https://tradicional.dgadr.gov.pt/pt/cat/salsicharia-fumados-presuntos-e-paletas/555-cacholeira-branca-de-portalegre-igp

Direção-Geral de Agricultura e Desenvolvimento Rural (DGADR). 2016l. Produtos Tradicionais Portugueses—Farinheira de Portalegre IGP. Retrieved February 22, 2023, from the World Wide Web: https://tradicional.dgadr.gov.pt/pt/cat/salsicharia-fumados-presuntos-e-paletas/606-farinheira-de-portalegre-igp

Direção-Geral de Agricultura e Desenvolvimento Rural (DGADR). 2016m. Produtos Tradicionais Portugueses—Morcela de Carne de Monchique. Retrieved February 24, 2023, from the World Wide Web: https://tradicional.dgadr.gov.pt/pt/cat/salsicharia-fumados-presuntos-e-paletas/399-morcela-de-carne-de-monchique

Direção-Geral de Agricultura e Desenvolvimento Rural (DGADR). 2016n. Produtos Tradicionais Portugueses—Chouriço de Carne da Serra de Monchique. Retrieved February 24, 2023, from the World Wide Web: https://tradicional.dgadr.gov.pt/pt/cat/salsicharia-fumados-presuntos-e-paletas/403-chourico-de-carne-da-serra-de-monchique

Direção-Geral de Agricultura e Desenvolvimento Rural (DGADR). 2023. Produtos Tradicionais Portugueses—Zona Geográfica: Centro. Retrieved February 23, 2023, from the World Wide Web: https://tradicional.dgadr.gov.pt/pt/zona-geografica/centro?start=20

Door List of European Union (DLEU). 2023a. eAmbrosia: The EU geographical indications register—Protected Geographical Indication (PGI) of meat products in Portugal. Retrieved February 21, 2023, from the World Wide Web: https://ec.europa.eu/info/food-farming-fisheries/food-safety-and-quality/certification/quality-labels/geographical-indications-register/

Door List of European Union (DLEU). 2023b. eAmbrosia: The EU geographical indications register—Protected Designation of Origin (PDO) of meat products in Portugal. Retrieved February 21, 2023, from the World Wide Web: https://ec.europa.eu/info/food-farming-fisheries/food-safety-and-quality/certification/quality-labels/geographical-indications-register/

Door List of European Union (DLEU). 2023c. eAmbrosia: The EU geographical indications register—Protected Designation of Origin (PDO) of fresh meat (and offal) in Portugal. Retrieved February 21, 2023, from the World Wide Web: https://ec.europa.eu/info/food-farming-fisheries/food-safety-and-quality/certification/quality-labels/geographical-indications-register/

Esteves, A., Patarata, L., Saraiva, C., and C. Martins. 2008. Assessment of the microbiological characteristics of industrially produced *alheira*, with particular reference to foodborne pathogens. *Journal of Food Safety 28*: 88–102. https://doi.org/10.1111/j.1745-4565.2007.00097.x

FAOSTAT. 2020. Food Balances—Food Balances (2010-). Retrieved February 20, 2023, from the World Wide Web: https://www.fao.org/faostat/en/#compare

Felício, M. T. S., Ramalheira, R., Ferreira, V., Brandão, T., Silva, J., Hogg, T., and P. Teixeira. 2011. Thermal inactivation of *Listeria monocytogenes* from *alheiras*, traditional Portuguese sausage during cooking. *Food Control 22*(12): 1960–1964. https://doi.org/10.1016/j.foodcont.2011.05.011

Fernandes, N., Faria, A. S., Carvalho, L., Choupina, A., Rodrigues, C., Cadavez, V., and U. Gonzales-Barron. 2022. Molecular identification of lactic acid producing bacteria isolated from *Alheira*, a traditional Portuguese fermented sausage. *Biology and Life Sciences Forum 18*: 73. https://doi.org/10.3390/Foods2022-13035

Ferreira, V., Barbosa, J., Silva, J., Felício, M. T., Mena, C., Hogg, T., Gibbs, P., and P. Teixeira. 2007a. Characterisation of *alheiras*, traditional sausages produced in the North of Portugal, with respect to their microbiological safety. *Food Control 18*(5): 436–440. https://doi.org/10.1016/j.foodcont.2005.11.011

Ferreira, V., Barbosa, J., Silva, J., Gibbs, P., Hogg, T., and P. Teixeira. 2009. Microbiological profile of *Salpicão de Vinhais* and *Chouriça de Vinhais* from raw materials to final products: Traditional dry sausages produced in the North of Portugal. *Innovative Food Science and Emerging Technologies 10*(2): 279–283. https://doi.org/10.1016/j.ifset.2008.11.001

Ferreira, V., Barbosa, J., Silva, J., Vendeiro, S., Mota, A., Silva, F., João Monteiro, M., Hogg, T., Gibbs, P., and P. Teixeira. 2007b. Chemical and microbiological characterisation of "Salpicão de Vinhais" and "Chouriça de Vinhais": Traditional dry sausages produced in the North of Portugal. *Food Microbiology 24*(6): 618–623. https://doi.org/10.1016/j.fm.2006.12.007

Ferreira, V., Barbosa, J., Vendeiro, S., Mota, A., Silva, F., Monteiro, M. J., Hogg, T., Gibbs, P., and P. Teixeira. 2006. Chemical and microbiological characterization of *alheira*: A typical Portuguese fermented sausage with particular reference to factors relating to food safety. *Meat Science 73*(4): 570–575. https://doi.org/10.1016/j.meatsci.2006.02.011

Food and Agriculture Organization (FAO). 2007. *Small-Scale Sausage Production*. Food and Agriculture Organization of the United Nation, Rome. ISBN 92-5-102187-2.

Food Safety Authority of Ireland. 2001. Guidelines for the interpretation of results of microbiological analysis of some ready to eat foods sampled at the point of sale. Guidance Note No. 3. Retrieved February 20, 2023, from the World Wide Web: http://www.fsai.ie/publications/index.asp

França, Z., Cruz, J. V., Nunes, J. C., and V. H. Forjaz. 2003. Geologia dos açores: uma perspectiva actual. *Açoreana 10*(1): 11–140.

Fraqueza, M. J., Laranjo, M., Elias, M., and L. Patarata. 2021. Microbiological hazards associated with salt and nitrite reduction in cured meat products: Control strategies based on antimicrobial effect of natural ingredients and protective microbiota. *Current Opinion in Food Science 38*: 32–39. https://doi.org/10.1016/J.COFS.2020.10.027

Gerber, J. S. 1994. *The Jews of Spain: A History of the Sephardic Experience*. Free Press.

Hall, C., and Osses, F. 2013. A review to inform understanding of the use of food safety messages on food labels. *International Journal of Consumer Studies 37*(4): 422–432. https://doi.org/10.1111/IJCS.12010

Helgi Library. 2023. Pork Meat Consumption Per Capita | Helgi Library. Retrieved February 20, 2023, from the World Wide Web: https://www.helgilibrary.com/indicators/pork-meat-consumption-per-capita/

Homer. 2018. *The Odyssey* (F. Lourenço, Ed.; 1st ed.). Quetzal.

Hospital, X. F., Carballo, J., Fernández, M., Arnau, J., Gratacós, M., and E. Hierro. 2015. Technological implications of reducing nitrate and nitrite levels in dry-fermented sausages: Typical microbiota, residual nitrate and nitrite and volatile profile. *Food Control 57*: 275–281. https://doi.org/10.1016/J.FOODCONT.2015.04.024

Instituto dos Mercados Agrícolas e Indústria Agro—Alimentar (IMAIAA). 1996. Regulamento (CEE) N° 2081/92 Do Conselho, 'Lombo Enguitado de Portalegre—Indicação

Geográfica', EC No: G/PT/00018/96.05.02. Ministério da Agricultura. Retrieved February 22, 2023, from the World Wide Web: https://ec.europa.eu/geographical-indications-register/eambrosia-public-api/api/v1/attachments/61204

Instituto dos Mercados Agrícolas e Indústria Agro—Alimentar (IMAIAA). 1997. Regulamento (CEE) N° 2081/92 Do Conselho, 'Painho de Portalegre', EC No: G/PT/00017/96.05.02. Ministério da Agricultura. Retrieved February 22, 2023, from the World Wide Web: https://ec.europa.eu/geographical-indications-register/eambrosia-public-api/api/v1/attachments/61817

Jarvis, N. 2004. Curing and canning of fishery products: A history. *Marine Fisheries Review 50*.

Kalaycıoğlu, Z., and F. B. Erim. 2019. Nitrate and nitrites in foods: Worldwide regional distribution in view of their risks and benefits. *Journal of Agricultural and Food Chemistry 67*(26): 7205–7222. https://doi.org/10.1021/ACS.JAFC.9B01194

Lawrie, R. 1995. The structure, composition and preservation of meat. In G. Campbell-Platt & P. E. Cook (Eds.), *Fermented Meats*: 1–38. https://doi.org/10.1007/978-1-4615-2163-1

Lebert, I., Leroy, S., Lebecque, A., Chacornac, J. P., and R. Talon. 2006. Saucissons secs fermiers du Massif central—Développement d'un ferment indigène pour ameliorer la securite et maintenir la typicité des saucissons. *Viandes et Produits Carnes 25*(5): 177–180.

Leroy, F., Geyzen, A., Janssens, M., de Vuyst, L., and P. Scholliers. 2013. Meat fermentation at the crossroads of innovation and tradition: A historical outlook. *Trends in Food Science & Technology 31*(2): 130–137. https://doi.org/10.1016/J.TIFS.2013.03.008

Linares-Morales, J. R., Gutiérrez-Méndez, N., Rivera-Chavira, B. E., Pérez-Vega, S. B., and G. V. Nevárez-Moorillón. 2018. Biocontrol processes in fruits and fresh produce, the use of lactic acid bacteria as a sustainable option. *Frontiers in Sustainable Food Systems 2*: 50. https://doi.org/10.3389/fsufs.2018.00050

Lusa. 2016. Criada confraria para valorizar e divulgar a morcela típica da Guarda. *Público*. Retrieved February 23, 2023, from the World Wide Web: http://lifestyle.publico.pt/restaurantesebares/357223_criada-confraria-para-valorizar-e-divulgar-a-morcela-tipica-da-guarda

Macieira, A., Albano, H., Pinto, M., Linheiro, R., Barbosa, J., and P. Teixeira. 2019. Evaluation of a bacteriocinogenic *Lactobacillus plantarum* strain on the microbiological characteristics of "Alheira de Vitela." *AIMS Agriculture and Food 4*(2): 223–236. https://doi.org/10.3934/agrfood.2019.2.223

Macieira, A., Barros, D., Vaz-Velho, M., Pinheiro, R., Fonseca, S., Albano, H., and P. Teixeira. 2018. Effects of *Lactobacillus plantarum* bacteriocinogenic culture on physicochemical, microbiological, and sensorial characteristics of "Chouriço Vinha d'Alhos", a traditional Portuguese sausage. *Journal of Food Quality and Hazards Control 5*(4): 118–127. http://jfqhc.ssu.ac.ir/article-1-481-en.html

Marçal, C., Pinto, C. A., Silva, A. M. S., Monteiro, C., Saraiva, J. A., and S. M. Cardoso. 2021. Macroalgae-fortified sausages: Nutritional and quality aspects influenced by non-thermal high-pressure processing. *Foods 10*(2): 209. https://doi.org/10.3390/FOODS10020209

Marcos, C., Viegas, C., Almeida, A. M., and M. M. Guerra. 2016. Portuguese traditional sausages: Different types, nutritional composition and novel trends. *Journal of Ethnic Foods 3*(1): 51–60. https://doi.org/10.1016/j.jef.2016.01.004

Martín, I., Rodríguez, A., Delgado, J., and J. J. Córdoba. 2022. Strategies for biocontrol of *Listeria monocytogenes* using lactic acid bacteria and their metabolites in ready-to-eat meat and dairy-ripening products. *Foods 11*(4): 542. https://doi.org/10.3390/foods11040542

Matos, T. J. S., Bruno-Soares, A., and A. A. Azevedo. 2013. Microbial spoilage of Portuguese chouriço along shelf life period. *Brazilian Journal of Microbiology 44*(1): 105–8. https://doi.org/10.1590/S1517-83822013000100014

Melgaço Rural. n.d.-a. Caderno de especificações da 'Chouriça de carne de Melgaço' IGP. *Melgaço Rural—Associação de Produtos Locais*. Retrieved February 24, 2023, from the World Wide Web: https://tradicional.dgadr.gov.pt/images/prod_imagens/salsicharia/docs/ce_chourica_carne_melgaco.pdf

Melgaço Rural. n.d.-b. Caderno de especificações do 'Salpicão de Melgaço' IGP. *Melgaço Rural—Associação de Produtos Locais*. Retrieved February 25, 2023, from the World Wide Web: https://tradicional.dgadr.gov.pt/images/prod_imagens/salsicharia/docs/CE_salpicao_Melgaco_IGP.pdf

Melgaço Rural. n.d.-c. Caderno de especificações do 'Presunto de Melgaço' IGP. *Melgaço Rural—Associação de Produtos Locais*. Retrieved February 24, 2023, from the World Wide Web: https://tradicional.dgadr.gov.pt/images/prod_imagens/salsicharia/docs/CE_presunto_Melgaco_IGP.pdf

Mirkovich, N. 1943. Ragusa and the Portuguese Spice Trade. Slavonic and East European Review. *American Series 2*(1): 174. https://doi.org/10.2307/3020140

Município Melgaço. (no data). Fumeiro. https://www.cm-melgaco.pt/visitar/gastronomia/fumeiro/#toggle-id-2

Natural Carnes. n.d.-a. Caderno de Especificações: Chouriço de Portalegre—Indicação Geográfica. Agrupamento de Produtores Pecuários do Norte Alentejo, S.A. Retrieved February 24, 2023, from the World Wide Web: https://ec.europa.eu/geographical-indications-register/eambrosia-public-api/api/v1/attachments/59531

Natural Carnes. n.d.-b. Caderno de Especificações: Linguiça de Portalegre—Indicação Geográfica. Agrupamento de Produtores Pecuários do Norte Alentejo, S.A. Retrieved February 22, 2023, from the World Wide Web: https://ec.europa.eu/geographical-indications-register/eambrosia-public-api/api/v1/attachments/59604

O'Rourke, K. H., and J. G. Williamson. 2009. Did Vasco da Gama matter for European markets? *The Economic History Review 62*(3): 655–684. https://doi.org/10.1111/J.1468-0289.2009.00468.X

Patarata, L., Carvalho, F., and M. J. Fraqueza. 2022. Nitrite-free implications on consumer acceptance and the behavior of pathogens in cured pork loins. *Foods 11*(6): 796. https://doi.org/10.3390/FOODS11060796

Patarata, L., Judas, I., Silva, J. A., Esteves, A., and C. Martins. 2008. A comparison of the physicochemical and sensory characteristics of *alheira* samples from different-sized producers. *Meat Science 79*(1): 131–138. https://doi.org/10.1016/j.meatsci.2007.08.009

Pereira, A. P., Gomes, A., Cadavez, V., Estevinho, L. M., Gonzales-Barron, U., and T. Dias. 2015. Microbiological and physicochemical aspects of "salpicão", a traditional dry sausage produced in the northeast of Portugal. *Agriculture & Food 3*: 112–120.

Pilkington, N. L. 2013. An archaeological history of carthaginian imperialism [Doctoral thesis, Columbia University]. https://doi.org/10.7916/D80G3SCF

Quiroga, Z., and S. Olivença. n.d. Alentejo and Algarve: Culinary traditions of the South of Portugal. Retrieved February 28, 2023, from the World Wide Web: https://www.ohmycodtours.com/blog/algarve-alentejo-traditiona-food-best-restaurants/

Ramalhosa, E., Magalhães, A. L., and E. L. Pereira. 2012. Characterization of regional Portuguese kitchens for *alheiras de Vinhais* (PGI) production with respect to the processing conditions, final product quality and legal framing. *Food Control 26*(1): 133–138. https://doi.org/10.1016/j.foodcont.2012.01.007

Reg (EC). 1997a. 2081/92. Protecção das indicações geográficas e denominações de origem—Lombo Branco de Portalegre.

Reg (EC). 1997b. 2081/92. Protected designations of origin and protected geographical indications' provided for in Council Regulation (EEC) on the protection of geographical indications and designations of origin for agricultural products and foodstuff—Lombo Branco de Portalegre (IGP), Lombo Enguitado de Portalegre (IGP), Painho de Portalegre (IGP), Cacholeira Branca de Portalegre (IGP), Chouriço Mouro de Portalegre (IGP), Linguiça de Portalegre (IGP), Morcela de Assar de Portalegre (IGP), Morcela de Cozer de Portalegre (IGP), Farinheira de Portalegre (IGP), Chouriço de Portalegre (IGP).

Reg (EC). 1997c. 2081/92. Protection of geographical indications and designations of origin—Salpicão de Vinhais.

Reg (EC). 1997d. 2081/92. Protection of geographical indications and designations of origin—Chouriça de Carne de Vinhais ou Linguiça de Vinhais, Chouriça de Carne de Vinhais ou Linguiça de Vinhais.

Reg (EC). 1998. 1265/98. Protection of geographical indications and designations of origin for agricultural products and foodstuffs—Salpicão de Vinhais (PGI), Chouriça de Carne de Vinhais or Linguiça de Vinhais (PGI).

Reg (EC). 2002a. 2081/92. Protection of geographical indications and designations of origin—Chouriço de Carne de Estremoz e Borba.

Reg (EC). 2002b. 2081/92. Protection of geographical indications and designations of origin—Chouriço Grosso de Estremoz e Borba.

Reg (EC). 2002c. 2081/92. Protection of geographical indications and designations of origin—Morcela de Estremoz e Borba.

Reg (EC). 2002d. 2081/92. Protection of geographical indications and designations of origin—Paia de Lombo de Estremoz e Borba.

Reg (EC). 2002e. 2081/92. Protection of geographical indications and designations of origin—Paia de Toucinho de Estremoz e Borba.

Reg (EC). 2004. 2400/69. Register of protected designations of origin and protected geographical indications—Paia de Toucinho de Estremoz e Borba, Chouriço de Carne de Estremoz e Borba, Paia de Lombo de Estremoz e Borba, Morcela de Estremoz e Borba, Chouriço grosso de Estremoz e Borba, Paia de Estremoz e Borba.

Reg (EC). 2005a. 2081/92. Protection of geographical indications and designations of origin—Chouriça de Carne de Barroso—Montalegre.

Reg (EC). 2005b. 2081/92. Protection of geographical indications and designations of origin—Chouriço de Abóbora de Barroso—Montalegre.

Reg (EC). 2005c. 2081/92. Protection of geographical indications and designations of origin—Sangueira de Barroso—Montalegre.

Reg (EC). 2006a. 1881/2006. Maximum levels for certain contaminants in foodstuffs.

Reg (EC). 2006b. 2081/92. Protection of geographical indications and designations of origin—Alheira de Barroso—Montalegre.

Reg (EC). 2006c. 2081/92. Protection of geographical indications and designations of origin—Salpicão de Barroso—Montalegre.

Reg (EC). 2006d. 510/2006. Protection of geographical indications and designations of origin—Linguíça do Baixo Alentejo or Chouriço de Carne do Baixo Alentejo.

Reg (EC). 2006e. 510/2006. Protection of geographical indications and designations of origin—Paio de Beja.

Reg (EC). 2007a. 148/2007. Register of protected designation of origin and protected geographical indications.

Reg (EC). 2007b. 148/2007. Register of protected designations of origin and protected geographical indications—Geraardsbergse mattentaart (PGI), Pataca de Galicia or Patata de Galicia (PGI), Poniente de Granada (PDO), Gata-Hurdes (PDO), Patatas de Prades or Patates de Prades (PGI), Mantequilla de Soria (PDO), Huile d'olive de Nîmes (PDO), Huile d'olive de Corse or Huile d'olive de Corse-Oliu di Corsica (PDO), Clémentine de Corse (PGI), Agneau de Sisteron (PGI), Connemara Hill Lamb or Uain Sléibhe Chonamara (PGI), Sardegna (PDO), Carota dell'Altopiano del Fucino (PGI), Stelvio or Stilfser (PDO), Limone Femminello del Gargano (PGI), Azeitonas de Conserva de Elvas e Campo Maior (PDO), Chouriça de Carne de Barroso-Montalegre (PGI), Chouriço de Abóbora de Barroso-Montalegre (PGI), Sangueira de Barroso-Montalegre (PGI), Batata de Trás-os-Montes (PGI), Salpicão de Barroso-Montalegre (PGI), Alheira de Barroso-Montalegre (PGI), Cordeiro de Barroso, Anho de Barroso or Borrego de leite de Barroso (PGI), Azeite do Alentejo Interior (PDO), Paio de Beja (PGI), Linguíça do Baixo Alentejo or Chouriço de carne do Baixo Alentejo (PGI), Ekstra deviško oljčno olje Slovenske Istre (PDO).

Reg (EC). 2007c. 510/2006. Protection of geographical indications and designations of origin—Chouriça Doce de Vinhais.

Reg (EC). 2007f. 510/2006. Protection of geographical indications and designations of origin for agricultural products and foodstuffs—Presunto de Vinhais or Presunto Bísaro de Vinhais.

Reg (EC). 2007g. 510/2006. Protection of geographical indications and designations of origin for agricultural products and foodstuffs—Presunto Do Alentejo or Paleta Do Alentejo.

Reg (EC). 2007h. 510/2006. Relativo à protecção das indicações geográficas e denominações de origem dos produtos agrícolas e dos géneros alimentícios—Alheira de Vinhais.

Reg (EC). 2007i. 676/2008. Register of protected designations of origin and protected geographical indications—Ail de la Drôme (PGI), Všestarská cibule (PDO), Slovenská bryndza (PGI), Ajo Morado de Las Pedroñeras (PGI), Gamoneu or Gamonedo (PDO), Alheira de Vinhais (PGI), Presunto de Vinhais or Presunto Bísaro de Vinhais (PGI).

Reg (EC). 2008a. 1333/2008. Food additives.

Reg (EC). 2008b. 723/2008. Register of protected designations of origin and protected geographical indications—Afuega'l Pitu (PDO), Mazapán de Toledo (PGI), Agneau de Lozère (PGI), Oignon doux des Cévennes (PDO), Butelo de Vinhais or Bucho de Vinhais or Chouriço de Ossos de Vinhais (PGI), Chouriça Doce de Vinhais (PGI).

Reg (EC). 2008c. 943/2008. Register of protected designations of origin and protected geographical indications—Presunto de Campo Maior e Elvas or Paleta de Campo Maior e Elvas (PGI), Presunto de Santana da Serra or Paleta de Santana da Serra (PGI), Slovenský oštiepok (PGI).

Reg (EC). 2008d. 994/2008. Register of protected designations of origin and protected geographical indications—Salame S. Angelo (PGI), Chouriço Azedo de Vinhais or Azedo de Vinhais or Chouriço de Pão de Vinhais (PGI), Presunto do Alentejo or Paleta do Alentejo (PDO).

Reg (EC). 2014a. 1151/2012. Regulation of the European Parliament and of the Council on quality schemes for agricultural products and foodstuff on the protection of geographical indications and designations of origin for agricultural products and foodstuffs—Chouriça de carne de Melgaço.

Reg (EC). 2014b. 1151/2012. Regulation of the European Parliament and of the Council on quality schemes for agricultural products and foodstuff on the protection of geographical indications and designations of origin for agricultural products and foodstuffs—Presunto de Barrancos/Paleta de Barrancos.

Reg (EC). 2014c. 1151/2012. Regulation of the European Parliament and of the Council on quality schemes for agricultural products and foodstuff on the protection of geographical indications and designations of origin for agricultural products and foodstuffs—Presunto de Melgaço.

Reg (EC). 2015a. 1151/587. Protected designations of origin and protected geographical indications—Salpicão de Melgaço (PGI).

Reg (EC). 2015b. 2015/591. Protected designations of origin and protected geographical indications—Presunto de Melgaço (PGI).

Reg (EC). 2015c. 510/2006 do Conselho relativo à proteção das indicações geográficas e denominações de origem dos produtos agrícolas e dos géneros alimentícios—Alheira de Mirandela.

Reg (EC). 2015d. 510/2006. Regulation of the European Parliament and of the Council on quality schemes for agricultural products and foodstuff on the protection of geographical indications and designations of origin for agricultural products and foodstuffs—Salpicão de Melgaço.

Reg (EC). 2016. 1151/2012. Extract from the register of protected designation of origin and protected geographical indications established by article 11 of regulation (eu)—Alheira de Mirandela.

Reg (EC). 2023a. 1151/2012. Extract from the register of protected designation of origin and protected geographical indications established by article 11 of regulation (eu)—Alheira de Vinhais.

Reg (EC). 2023b. 1151/2012. Extract from the register of protected designation of origin and protected geographical indications established by article 11 of regulation (eu)—Butelo de Vinhais/Bucho de Vinhais/Chouriço de Ossos de Vinhais.

Reg (EC). 2023c. 1151/2012. Extract from the register of protected designation of origin and protected geographical indications established by article 11 of regulation (eu)—Chouriça de carne de Melgaço.

Reg (EC). 2023d. 1151/2012. Extract from the register of protected designation of origin and protected geographical indications established by article 11 of regulation (eu)—Chouriça de sangue de Melgaço.

Reg (EC). 2023e. 1151/2012. Extract from the register of protected designation of origin and protected geographical indications established by article 11 of regulation (eu)—Chouriça Doce de Vinhais.

Reg (EC). 2023f. 1151/2012. Extract from the register of protected designation of origin and protected geographical indications established by article 11 of regulation (eu)—Chouriço Azedo de Vinhais/Azedo de Vinhais/Chouriço de Pão de Vinhais.

Reg (EC). 2023g. 1151/2012. Extract from the register of protected designation of origin and protected geographical indications established by article 11 of regulation (eu)—Presunto de Barroso.

Sáez Romero, A. M. 2014. Fish processing and salted-fish trade in the Punic West: New archaeological data and historical evolution. *Fish & Ships:* 159–174. https://doi.org/10.4000/BOOKS.PCCJ.1711pil

Santos, C., Gomes, A., and L. C. Roseiro. 2011. Polycyclic aromatic hydrocarbons incidence in Portuguese traditional smoked meat products. *Food and Chemical Toxicology 49*(9): 2343–2347. https://doi.org/10.1016/j.fct.2011.06.036

Sebranek, J. G., and J. N. Bacus. 2007. Cured meat products without direct addition of nitrate or nitrite: What are the issues? *Meat Science 77*(1): 136–147. https://doi.org/10.1016/J.MEATSCI.2007.03.025

Seong, P. N., Seo, H. W., Kang, S. M., Kim, Y. S., Cho, S. H., Kim, J. H., and V. B. Hoa. 2016. Beneficial effects of traditional seasonings on quality characteristics of fermented sausages. *Asian-Australasian Journal of Animal Sciences 29*(8): 1173. https://doi.org/10.5713/AJAS.15.0738

Shakil, M. H., Trisha, A. T., Rahman, M., Talukdar, S., Kobun, R., Huda, N., and W. Zzaman. 2022. Nitrites in cured meats, health risk issues, alternatives to nitrites: A review. *Foods 11*(21): 3355. https://doi.org/10.3390/FOODS11213355

Shangguan, S., Afshin, A., Shulkin, M., Ma, W., Marsden, D., Smith, J., Saheb-Kashaf, M., Shi, P., Micha, R., Imamura, F., and D. Mozaffarian. 2019. A meta-analysis of food labeling effects on consumer diet behaviors and industry practices. *American Journal of Preventive Medicine 56*(2): 300. https://doi.org/10.1016/J.AMEPRE.2018.09.024

Silva, B. N., Cadavez, V., Pires, P., Dias, T., and U. Gonzales-Barron. 2018. Microbiological safety of Portuguese dry-fermented *Chouriço* sausages as affected by processing and physicochemical factors. *Madridge Journal of Food Science and Technology 3*(2): 137–148. https://doi.org/10.18689/mjft-1000121

Silva, J., Barbosa, J., Albano, H., Sequeira, M., Pinto, A., Bonito, C. C., Saraiva, M., and P. Teixeira. 2019. Microbiological characterization of different formulations of "alheiras" (fermented sausages). *AIMS Agriculture and Food 4*(2): 399–413. https://doi.org/10.3934/AGRFOOD.2019.2.399

Soeiro, A. 2001a. In Produtos Tradicionais Portugueses, Volume I. DGARural-Direção-Geral de Desenvolvimento Rural, Lisboa, Portugal, p. 288.

Soeiro, A. 2001b. In Produtos Tradicionais Portugueses, Volume II. DGARural-Direção-Geral de Desenvolvimento Rural, Lisboa, Portugal, p. 376.

Soeiro, A. 2001c. In Produtos Tradicionais Portugueses, Volume III. DGARural-Direção-Geral de Desenvolvimento Rural, Lisboa, Portugal, p. 352.

Strauss, B. 2006. *The Trojan War: A New History*. Simon and Schuster.

Talon, R., Leroy, S., and I. Lebert. 2007. Microbial ecosystems of traditional fermented meat products: The importance of indigenous starters. *Meat Science 77*(1): 55–62. https://doi.org/10.1016/j.meatsci.2007.04.023

Taste Atlas. n.d.-a. Retrieved February 25, 2023, from the World Wide Web: https://www.tasteatlas.com/chourica-de-sangue-de-melgaco

Taste Atlas. n.d.-b. Retrieved February 25, 2023, from the World Wide Web: https://www.tasteatlas.com/salpicao-de-melgaco

Teixeira, A., Domínguez, R., Reyes, J. F., Lorenzo, J. M., Trindade, M. A., and S. Rodrigues. 2022. Alheira. In: J. M. Lorenzo, R. Domínguez, M. Pateiro, and P. E. Munekata (Eds.), *Production of Traditional Mediterranean Meat Products. Methods and Protocols in Food Science*. Humana, New York, NY. https://doi.org/10.1007/978-1-0716-2103-5_18

Teixeira, A., Pereira, E., Barbosa, N., and S. Rodrigues. 2018. Efeito do tempo de cura nas caracteristicas fisicoquimicas de presunto Bísaro. In A. S. Santos & M. Almeida (Eds.), *XX ZOOTEC—20o Congresso Nacional de Zootecnia | APEZ*. Revista Portuguesa de Zootecnia.

Thorarinsdottir, K. A., Arason, S., Bogason, S. G., and K. Kristbergsson. 2004. The effects of various salt concentrations during brine curing of cod (*Gadus morhua*). *International Journal of Food Science & Technology 39*(1): 79–89. https://doi.org/10.1046/J.0950-5423.2003.00757.X

Toldrá, F. 2004. Meat: Fermented meats. In J. Scott Smith & Y. H. Hui (Eds.), *Food Processing. Principles and Applications* (1st ed.). Blackwell Publishing.

Ursachi, C. Ş., Perţa-Crişan, S., and F. D. Munteanu. 2020. Strategies to improve meat products' quality. *Foods 9*: 1883. https://doi.org/10.3390/foods9121883

Van Ba, H., Seo, H., Kim, J., Cho, S., Kim, Y., Ham, J., and P. Seong. 2016. The effects of starter culture types on the technological quality, lipid oxidation and biogenic amines in fermented sausages. *LWT—Food Science and Technology 74*: 19198.

Vaz-Velho, M., Jácome, S., Noronha, L., Todorov, S., Fonseca, S., Pinheiro, R., Morais, A., Silva, J., and P. Teixeira. 2013. Comparison of antilisterial effects of two strains of lactic acid bacteria during processing and storage of a Portuguese salami-like product "alheira." *Chemical Engineering Transactions 32*: 1807–1812. http://doi.org/10.3303/CET1332302

Verluyten, J., Leroy, F., and L. de Vuyst. 2004. Effects of different spices used in production of fermented sausages on growth of and curvacin A production by *Lactobacillus curvatus* LTH 1174. *Applied and Environmental Microbiology 70*(8): 4807. https://doi.org/10.1128/AEM.70.8.4807-4813.2004

Viola, G. C. V., Bianchi, F., Croce, E., and E. Ceretti. 2016. Are food labels effective as a means of health prevention? *Journal of Public Health Research 5*(3): 139–142. https://doi.org/10.4081/JPHR.2016.768

Ward, M. H., Jones, R. R., Brender, J. D., de Kok, T. M., Weyer, P. J., Nolan, B. T., Villanueva, C. M., and S. G. van Breda. 2018. Drinking water nitrate and human health: An updated review. *International Journal of Environmental Research and Public Health 15*(7): 1557. https://doi.org/10.3390/IJERPH15071557

Yost, C. K. 2014. Biopreservation. *Encyclopaedia of Meat Sciences*, 76–82.

Innovative Potential of Traditional Greek Lactic Acid Fermented Vegetables for Functional Foods

19

Spiros Paramithiotis

19.1 INTRODUCTION

Lactic acid fermentation has been applied for centuries. The currently prevailing belief is that it was principally employed as a method to preserve temporally excess foodstuff, aiming to increase their availability in times of scarcity. With time, such commodities became part of local tradition and food culture. Nowadays, some of them are considered characteristic of specific regions, such as sauerkraut for Northern Europe and kimchi for the Korean peninsula.

The East Mediterranean region is characterized by a constant production of fresh fruits and vegetables throughout the year. Therefore, the food culture that has emerged directed the consumption of fresh commodities, according to their harvest period. An extended variety of fruits and vegetables were consumed at that time, as mentioned in ancient Greek texts such as the *Odyssey* of Homer, the *Symposium* of Xenophon and the *Deipnosophists* of Athenaeus; some types or varieties of which, named after their place of origin, were characterized by a variable degree of acceptance and appreciation. In terms of preservation, the preferred methods seemed to be smoking and dehydration. Fermentation was also extensively applied but only in breadmaking, the production of dairy products, alcoholic beverages and vinegar. Lactic acid fermentation of fruits and vegetables seemed to be rather uncommon.

Nowadays, this situation has been modified; pickling has become an essential part of the local food culture. However, in most cases, this is performed through the addition of vinegar, which seems quite reasonable given its long tradition. Spontaneous fermentation was only marginally exercised, until recently. The trend of the last decade that directs the consumption of healthier food enriched with functional properties led to the rediscovery of old recipes. In addition, some of these products are currently available in the market as a response to the consumer demand for minimally processed preservative-free food. This attempt, although in its infancy, is very promising both in terms of the development of clean-label products and in terms of sustainability through the minimal use of resources and local economy growth.

This chapter aims to collect all available studies on the characterization of spontaneously lactic acid fermented vegetable-based Greek products and discuss and compare them with similar ones throughout the world. In addition, their innovative potential, from a contemporary perspective, is presented and thoroughly discussed.

19.2 LACTIC ACID FERMENTED VEGETABLES IN GREECE

The key characteristics of lactic acid fermentation of vegetables in Greece are the re-invention of old recipes, the diversity of the available raw materials and fermentation temperatures. The common characteristic is the dominance of brine-salting over dry-salting methods. The amount of salt used depends on the perception of the maker, since salt is added to the brine until it is on the borderline between pleasant and unpleasant. This criterion leads to brine of 6–10% NaCl initial content. Incubation is usually performed in cellars, the temperature of which, most of the time, ranges between 10 and 20°C. The products described in the next paragraphs are ones that have attracted attention and therefore been studied to some extent. Many more can be found in rural Greece and will be studied in time to come.

DOI: 10.1201/9781003605300-19

19.2.1 Cauliflower Fermentation

Cauliflower belongs to the *Brassica oleracea* species, which also includes other common types of vegetables such as broccoli, kale and Brussels sprouts that are widely consumed throughout the globe. Cauliflower is considered a good source of fiber and vitamins, particularly C, K, B3 and B9 (https://fdc.nal.usda.gov/). In addition, cauliflower is rich in sulfur-containing phytochemicals, namely glucosinolates and S-methylcysteine sulfoxide, as well as polyphenols and to a lesser extent carotenoids (Picchi et al. 2012).

Cauliflower leaves are used together with the leaves of *B. rapa* subsp. *campestris* var. *cuneifolia* and *B. juncea* to produce gundruk, which is a fermented ethnic product of the Himalayas (Tamang 2010). The availability of lactic acid fermented cauliflower in the European market was mentioned by Buckenhueskes (1993). However, it was not clear whether these products were made from the florets or the leaves. Paramithiotis et al. (2010) was the first to report on the dynamics of the micro community during cauliflower floret fermentation. LAB prevailed in the microecosystem already from the third day of fermentation, resulting in the decrease of the pH value to 3.78 and the increase of total titratable acidity to 0.65 % lactic acid at the end of fermentation (day 26). Enterococci formed the secondary microbiota. SDS-PAGE of whole cell proteins combined with partial 16S-rRNA gene sequencing revealed that *Leuconostoc mesenteroides* group dominated the LAB micro community during the first 5 days, replaced by the *Lactiplantibacillus plantarum*-group, which dominated until the end of fermentation. Early dominance of LAB during cauliflower fermentation was also verified by Wouters et al. (2013a); in that case, the final pH value of the brine was recorded at 3.5. However, a rather diverse LAB microcommunity was revealed by $(GTG)_5$ fingerprinting combined with partial sequencing of the 16S-rRNA gene, which was in general verified by PCR-DGGE analysis. Indeed, *Weissella kimchii* dominated during the first 3 days, with *Levilactobacillus brevis*, *W. viridescens* and *Latilactobacillus sakei/curvatus* also being present. Then, and until the 10th day, *W. kimchi* and *W. viridescens* were gradually replaced by a consortium consisting of *Enterococcus* spp., *Lt. sakei/curvatus*, *Lv. brevis*, *Ln. mesenteroides* and *Pediococcus pentosaceus*. Then, from the 14th until the 35th day of fermentation, *Lv. brevis* dominated, with *Pd. pentosaceus* and *Ln. mesenteroides* occasionally present.

Such differences are quite reasonable to occur in spontaneous fermentations and may be assigned to differences in the recipe and fermentation conditions. To standardize the fermentation, Wouters et al. (2013b) used *Lp. plantarum* strain IMDO 788 as a starter culture. The strain effectively dominated native populations and carried out the fermentation, resulting in a final product with pH value of 3.35; lactic acid and ethanol were reported as the major metabolites. From a safety perspective, spontaneous fermentation was not able to eliminate *Listeria monocytogenes* and *Salmonella* Typhimurium. Indeed, Paramithiotis et al. (2012) reported that although a decline in the population of the pathogens, which were inoculated at 2, 4 and 6 log CFU/mL at the beginning of fermentation, was evident, they remained at enumeration level at the end of fermentation.

19.2.2 Unripe (Green) Tomato Fermentation

Tomatoes (*Solanum lycopersicum* L) are considered a good source of minerals, especially copper, zinc, potassium, phosphorus, magnesium and vitamins (A, C, K, B3) and contain phenolic compounds and carotenoids (https://fdc.nal.usda.gov/). Their amount is highly dependent on the variety and the maturity stage. Indeed, this was verified by Duma et al. (2018), whose study focused on the vitamin C, lycopene, total flavonoid and total phenolic composition of ten varieties of tomatoes at two maturity stages, namely unripe (green) and ripe (red). More accurately, total flavonoids ranged between 2.17 and 6.67 and between 2.87 and 10.03 mg catechin/100 g, in unripe and ripe samples, respectively. Total phenolic content ranged between 4.79 and 8.09 and 6.77 and 14.34 mg GAE/100 g in unripe and ripe samples, respectively; vitamin C ranged between 1.93 and 11.11 and between ca. 9.50 and 22.82 mg/100 g and lycopene between 0.58 and ca. 2.50 and between 3.08 and 27.11 mg/100 g in unripe and ripe samples, respectively.

Fermentation of green tomatoes was first reported by Fleming et al. (1983). In that study, quartered green tomatoes were submerged in brine containing 5% NaCl and 0.1% acetic acid (pH 4.5, adjusted with NaOH) and inoculated with *Lp. plantarum*. Fermentation took place at 24°C and was completed after 16 days. Lactic acid was the major metabolic end product. The pH value was kept constant at 3.8 through continuous monitoring and addition of NaOH. The effect of the ripening stage on the development of the microecosystem during spontaneous fermentation of green tomatoes was assessed by Paramithiotis et al. (2014a). More accurately, tomatoes belonging to the 'immature green' and mature green' stages of gradual pH values (group 1, 3.8–4.2; group 2, 4.3–4.8; group 3, 4.9–5.4), were submerged to 8% NaCl brine and incubated at 20°C. Fermentation was completed after 35 days; the final pH values were 3.78, 4.07 and 3.46, and the final TTA values (% lactic acid) were 0.39, 0.46 and 0.41, for groups 1, 2 and 3, respectively. LAB dominated all fermentations, with yeasts forming a secondary microbiota. Fermentation of tomatoes belonging to groups 1 and 2 were dominated by *Ln. mesenteroides*. On the contrary, fermentation of group 3 tomatoes was only initially dominated by *Ln. mesenteroides*; from the 14th day until the end, the fermentation was driven by a consortium consisting of *Ln. mesenteroides*, *Ln. citreum* and *Lacticaseibacillus casei*. Yeasts and LAB were reported by Bah et al. (2019) as the dominant microbial populations in spontaneous tomato fermentation. Regarding the former, *Meyerozyma caribbica* and *Wickerhamomyces onychis* were detected throughout fermentation and *Candida caprophila* only during the first 2 weeks. *Leuconostoc mesenteroides* dominated the LAB microbiota during the initial stages and *Lp. plantarum* during the final ones. In addition, the tomatoes were also characterized by the increase of flavonoid and phenolic concentration during the first 2 and 3 weeks of fermentation, respectively, and then decreased until the end of fermentation. The increase of lycopene concentration and antioxidant activity throughout fermentation was reported as well.

19.2.3 Asparagus Fermentation

Asparagus (*Asparagus officinalis* L.) is a good source of fiber, vitamins (A, E, K, B9) and minerals (calcium, magnesium, phosphorus, potassium, zinc, copper) (https://fdc.nal.usda.gov/). In addition, it contains several bioactive phytochemicals, such as carotenoids, saponins and phenolic and sulfur-containing compounds (Deli et al. 2000; Jang et al. 2004; Huang and Kong 2006; Fuentes-Alventosa et al. 2008). The phytochemical content is affected by the variety, the section of the spear and the harvesting season (Chitrakar et al. 2019).

Spontaneous asparagus fermentation was studied by Paramithiotis et al. (2014b). It was reported that lactic acid bacteria dominated the fermentation and enterococci formed a secondary microbiota. The fermentation was completed after 26 days; the final pH value of the brine was 4.1, while the TTA was 0.4 (% lactic acid). The LAB micro community was dominated by *Weissella* spp.; *W. viridescens* seemed to be more abundant than *W. cibaria*. Sporadic occurrence of *Enterococcus faecium* and *Lt. sakei* was also recorded.

The effect of fermentation on the nutritional value and the biologically active compounds of asparagus was studied by Tabaszewska et al. (2018) and Zhang et al. (2018). In the first study, the fermentation was spontaneous, driven by LAB, but no further detail other than the final pH value of 3.7 and the final TTA value of 1.02 (% lactic acid) was offered. The fermented asparagus was characterized by total polyphenolic content, total flavonoid content and antioxidative activity that exhibited no statistically significant changes compared to fresh ones. The polyphenolic compounds were further assessed, and the decrease of ferulic acid, p-coumaric acid, caffeic acid and gallic acid in the fermented asparagus was reported. On the other hand, the amount of quercetin, kaempferol and apigenin was increased. Finally, the L-ascorbic acid showed no statistically significant changes, but a decrease in the amounts of dehydroascorbic acid and vitamins B1, B2 and B6 was reported. Zhang et al. (2018) inoculated asparagus pulp with *Lp. plantarum* NCU116 and allowed it to ferment at 37°C for 36 h. Increase of the concentration of functional polysaccharides, which were further reported to exhibit hepatoprotective activity (Zhang et al. 2020), as well as immunoregulatory and enhanced free radical scavenging abilities, were reported.

19.2.4 Fermentation of Globe Artichoke Immature Inflorescence

Globe artichoke (*Cynara cardunculus* var. *scolymus* L. Fiori) is considered a good source of fiber, minerals (calcium, magnesium, phosphorus, potassium, zinc, copper, manganese), vitamins (B3, B9, C, K), lutein and zeaxanthin (https://fdc.nal.usda.gov/). Moreover, it is particularly rich in polyphenols, the amount of which is affected by cultivar, agricultural practices, harvest time and developmental stage (Zayed et al. 2020).

A very interesting and rather unusual fermentation, namely fermentation of globe artichoke immature inflorescence, was presented by Paramithiotis et al. (2016). The preparation involved acid treatment and blanching to prevent browning of the inflorescence and then brine fermentation. As revealed by both culture-dependent and -independent approaches, the fermentation was driven by yeasts and more accurately by *Hanseniaspora* spp. The pH value increased from the initial value of 3.40 to 4.75 and then decreased to 4.50. The low initial pH value was assigned to the citric acid contained in the lemon juice that was used for the acid treatment. The increase of the pH value could be due to the release of proton scavengers by the injured cells of the inflorescence and yeast metabolism. Finally, the decrease that took place after the initial pH equilibration was solely assigned to yeast metabolism.

19.2.5 Radish Root Fermentation

Radish (*Raphanus sativus* L.) belongs to the Brassicaceae family. It is a very popular crop species whose roots and leaves are of high nutritional value. More accurately, the leaves are a good source of proteins and together with the roots are good sources of minerals, particularly potassium, calcium and magnesium, as well as polyphenolic compounds (Goyeneche et al. 2015; Manivannan et al. 2019). Specifically, their antioxidant capacity has been attributed to the high phenolic, flavonoid and anthocyanin content, which, in turn, is affected by the cultivar (Park et al. 2016).

Radish root fermentation is very common in Asia. Indeed, they are used to produce sinki in India (Tamang 2010) and are very often included for the production of kimchi in Korea (Patra et al. 2016). Regarding Europe, the first report on radish root fermentation was made by Pardali et al. (2017). It that study, *Lp. plantarum* was reported to drive spontaneous radish root fermentation at both 20 and 30°C. At both temperatures, *Pd. pentosaceus* was also present during the first days and *Lv. brevis* during the final stage. Glucose and fructose were detected as the available carbon sources, which were depleted during fermentation at 30°C, while lactic acid, acetic acid and ethanol were the main metabolic end products. The final pH and TTA (% lactic acid) values after 17 d of fermentation at 20°C were 3.62 and 0.4, respectively, and after 11 days of fermentation at 30°C were 3.60 and 0.35, respectively. Moreover, Pappa et al. (2018) reported on the effect of garlic, lemon thyme or mint addition on spontaneous fermentation of radish roots along with the fate of two major foodborne pathogens, *L. monocytogenes* and *Escherichia coli* O157:H7. Fermentation took place at 20°C and was completed after 11–17 d, suggesting that the addition of the aromatic plants had no effect on the course of fermentation. In addition, it was reported that *Lp. plantarum* dominated the micro ecosystem, highlighting the aforementioned conclusion. The increase of total phenolic content in the brine during fermentation was also reported, most probably due to their controlled release from the plant material or the hydrolysis of their glycosides due to the decrease of the pH value. The addition of lemon thyme resulted in further increase of total phenolic content. Regarding the fate of the pathogens inoculated into the brine at the onset of fermentation at 4 and 6 log CFU/mL, their absence was verified as the fermentation time was prolonged. However, the initial population of the pathogens seemed to have a more decisive effect on the prolongation of the fermentation time required for complete elimination of the pathogens than the addition of the herbs.

19.3 TOWARDS INNOVATIVE PRODUCTS

Innovation is a term used to describe any improvement in a product through incorporation of temporal trends in a way that is accepted by the consumer. Currently, innovation in fermented products in general is mainly achieved through reformulation of the recipes and the use of selected starter cultures. In the case of the former, in fermented vegetables, research has focused on the reduction of the NaCl used, mostly through partial replacement with different chloride salts. Thus, the effect of a variety of salts, such as $CaCl_2$, $Ca(OH)_2$, KCl and $ZnCl_2$, on the development of the microecosystem in spontaneous fermentations, the growth rate of starter cultures and the physicochemical and organoleptic characteristics of fermented vegetable products has been extensively studied and documented (Bautista-Gallego et al. 2013).

Selection of starter cultures that would retain the traditional character of a product in terms of texture, physicochemical characteristics and organoleptic qualities and at the same time provide additional health benefits has been a target since the first studies on the microecosystem of traditional fermented products took place. These additional health benefits are subjected to the trends of each time; microbial strains have been exhaustively screened for a wide range of properties that extend from a basic probiotic capacity to a sophisticated anti-disease activity and from the production of the L-enantiomer of lactate, instead of the D-enantiomer, to the enrichment of the commodity with vitamins and GABA (γ-aminobutyric acid), through specific metabolic activities.

19.3.1 Development of Probiotic Products

In vitro examination has been extensively applied to perform a preliminary evaluation of the probiotic potential of microbial isolates. The *in vitro* tests employed may be clustered into three broad categories: safety evaluation, stress tolerance and functional potential.

Assessment of hemolytic activity, antibiotic resistance and production of enzymes or metabolites that may directly or indirectly exert adverse health effects, such as gelatinase and amino acid decarboxylases, are usually employed for safety evaluation. Such an evaluation was usually performed through phenotype assessment; nowadays it tends to be replaced by the detection of the respective genes, even though such detection does not necessarily predict the phenotype. The ability of the microorganisms under examination to cope with the stresses encountered during passage through the human gastro-intestinal tract (GIT) is predicted by calculating the surviving population upon exposure to simulated gastric or intestinal conditions. The former necessarily includes exposure to low pH values in the presence of pepsin, while the latter includes exposure to alkaline pH values in the presence of pancreatin and bile salts. The ability to colonize the GIT is also important and may be inferred by studying the hydrophobicity of the cell surface, auto aggregation potential, adhesion to cell lines and binding to mucus. Finally, the functional potential is subjected to constant updating, according to research trends and innovations. In general, the functional properties include activity against a series of diseases, modulation of immune responses and protection of the mucosal barrier.

The probiotic potential of isolates from vegetable fermentation, among other sources, has been extensively assessed. This wealth of knowledge has been combined and organized into a database to assist further research (Zhao et al. 2019). Despite this, application of potentially probiotic microorganisms as starter or adjunct cultures in vegetable fermentation is generally lacking. This may be attributed primarily to the current ongoing trend towards the consumption of juices, which better fits the current on-the-go generation, and secondarily to the diversity of the raw materials and production procedures employed, as well as the presence of autochthonous populations, which demands the use of competitive starter cultures adapted to specific conditions. Although several successful attempts to apply starter cultures in vegetable fermentation are available in the literature (Tolonen et al. 2004; Johanningsmeier et al. 2005; Wiander and Ryhanen 2005; Kim et al. 2019; Lee et al. 2020), in only one, at least as far as the author is concerned, were probiotic strains employed (Beganovic et al. 2011). More accurately, Beganovic et al. (2011) applied a mixture of probiotic strains, namely *Lp. plantarum* L4 and *Ln. mesentroides* LMG 7954, for the fermentation of cabbage heads. The controlled fermentation was completed earlier than the spontaneous one, reaching final pH and titratable acidity values of ca. 3.22 and 35 °SH, respectively. *Lactiplantibacillus plantarum* strain L4 seemed to prevail in the LAB micro community, according to PFGE analysis of isolates.

Another insight to this issue was provided by Campos et al. (2019), where an appetizer was prepared consisting of an homogenized mixture of fruits and vegetables, inoculated with freeze-dried cultures of *Lp. plantarum* LP299v and *La. rhamnosus* GG. The product was incubated at 37°C for 24 h to allow microorganisms to grow and then stored at 8°C for up to 90 d. Storage stability as well as survival of the strains to simulated gastrointestinal conditions were assessed. Both microorganisms retained high populations throughout storage; based on the surviving populations, consumption of 100 g portion would provide with the necessary bacterial population to exert health benefits.

19.3.2 Development of Functional Products

The metabolites produced by LAB during fermentation may affect the organoleptic, technological and functional properties of the final product. The production of these metabolites is generally affected by the fermentation conditions and the ability of the microbial strain to effectively complete the respective metabolic pathway. Among them, research has recently focused on the production of vitamins of the B-complex, γ-aminobutyric acid and sulforaphane.

19.3.2.1 γ-Aminobutyric Acid

GABA production has been reported in vertebrates, plants and microorganisms. In humans it holds a very important physiological role; it is the major inhibitory neurotransmitter in the central nervous system. This fact is fueling the interest in enriching food with GABA through fermentation. Although microbial production of GABA has been intensively studied, especially over the last decade, it is still uncertain whether oral intake indeed affects the physiological processes with which it is implicated (Hepsomali et al. 2020).

A wide variety of yeasts, enterobacteria and lactic acid bacteria have been reported to produce GABA (Han and Lee 2017; Perpetuini et al. 2020). Especially regarding the latter, strains belonging to *La. paracasei*, *Lactococcus lactis* subsp. *lactis*, *Lp. plantarum*, *Lv. brevis*, *Lactobacillus delbrueckii* subsp. *bulgaricus*, *Lentilactobacillus buchneri*, *La. rhamnosus*, *Ln. mesenteroides*, *Ln. pseudomesenteroides*, *W. cibaria*, *Streptococcus salivarius* subsp. *thermophilus*, *Ec. durans* and many more have been reported as GABA-producing (Siragusa et al. 2007; Lim et al. 2017; Han et al. 2020).

Among the factors that affect GABA production, pH value, temperature and nutrient availability are the most studied ones. The pH value that facilitates GABA production has been reported to be at least species-specific. Indeed, the pH value of optimum production has been reported at 4.2 for *Lv. brevis* (Soo et al. 2006), 4.5 for *Str. salivarius* subsp. *thermophilus* (Yang et al. 2008), 5.0 for *La. paracasei* (Komatsuzaki et al. 2005) and 6.0 for *Lp. plantarum* (Di Cagno et al. 2010). Accumulation of GABA at such low pH value is also facilitated by the much higher optimum pH value of GABA transaminase that catalyzes the conversion of GABA to succinic semi-aldehyde using the pyruvate or the α-ketoglutarate as the amino acceptor. Indeed, it has been reported to be ca. 8.5, at least in pseudomonads and *Saccharomyces cerevisiae* (Dhakal et al. 2012). Temperature affects GABA production by affecting population. Thus, more efficient GABA production by LAB has been reported in most of the studies within the range 30–37°C (Komatsuzaki et al. 2005; Cho et al. 2007; Kim et al. 2009; Di Cagno et al. 2010; Li et al. 2010). Glutamate and pyridoxal-5'-phosphate are necessary for GABA production. Thus, exogenous addition may result in the increase of GABA production (Huang et al. 2007; Komatsuzaki et al. 2005; Li et al. 2010). However, this is not always the case, and several studies report that their addition had no such outcome (Di Cagno et al. 2010; Li and Cao 2010; Yang et al. 2008). Finally, it seems that the type and concentration of carbon and nitrogen sources may also affect GABA accumulation but in a strain-dependent manner (Dhakal et al. 2012).

GABA-producing LAB have been used as starter cultures to produce kimchi and fermented cucumber. In the first case, Cho et al. (2011), after studying *in vitro* GABA production by a *Le. buchnerii* strain, applied it in kimchi fermentation. The final product was characterized by eight-times increased GABA concentration without any difference in the sensorial profile compared to the spontaneously fermented one. Improvement in both GABA concentration and sensorial scores was achieved by monosodium glutamate along with a GABA-producing strain (Seok et al. 2008; Lee et al. 2018). Moore et al. (2021) studied the GABA content in fresh, acidified and fermented cucumbers. GABA was detected in all products; however, in fermented cucumbers, it was in more than double the concentration. Moreover, pasteurization and storage at 28°C for 6 months did not affect its concentration.

19.3.2.2 B-Complex Vitamins

Vitamins are essential micronutrients, most of which cannot be synthesized by humans; thus, adequate dietary supply is necessary. Although they are present in many foods, deficiencies are prevalent in both developing and developed countries, mostly because of an imbalanced diet. Therefore, fortification with chemically synthesized vitamins or biofortification through the utilization of microorganisms capable of producing vitamins have been proposed as suitable measures to address this issue.

Lactic acid bacteria have been recognized as producers of vitamins of the B group; thus, biofortification using suitable strains has been the focus of intensive study. Among them, research has focused on the production of vitamins B2 (riboflavin), B9 (folate) and B12 (cobalamin).

Riboflavin is the precursor of flavin coenzymes, namely flavin adenine dinucleotide (FAD) and flavin mononucleotide (FMN), which possess very important physiological roles (Szczuko et al. 2019). Strains belonging to the lactic acid bacteria species *Ec. faecium*, *Lc. lactis* subsp. *lactis*, *Lb. acidophilus*, *Lp. plantarum*, *Limosilactobacillus fermentum*, *La. rhamnosus*, *Lm. mucosae* and *Ln. mesenteroides* have been reported as riboflavin producers (Burgess et al. 2006; Ewe et al. 2010; Capozzi et al. 2011; Juarez del Valle et al. 2014; Thakur et al. 2016; Hati et al. 2019; Yepez et al. 2019; Sabo et al. 2020). Some of them have been effectively used to increase riboflavin content of milk-, soy- and cereal-based products (Capozzi et al. 2011; Juarez del Valle et al. 2014; Waters et al. 2015; Thakur et al. 2016; Yepez et al. 2019).

Folate is a very important cofactor, necessary for many cellular functions, including synthesis of DNA, RNA and proteins (Stover 2004). Strains belonging to several LAB have been reported to produce folate extracellularly, including strains belonging to the species *Str. thermophilus*, *Lb. amylovorus*, *Lp. plantarum*, *Lt. sakei* and *Lc. lactis*. These and many more have been effectively used to increase folate content in dairy and soy products, fruit and vegetable juices and sourdoughs (Gangadharan and Nampoothiri 2011; Laino et al. 2012, 2013, 2014; Kariluoto et al. 2014; Da Silva et al. 2016; Albuquerque et al. 2017; Carrizo et al. 2017, 2020; Meucci et al. 2018; Mosso et al. 2018; Tamene et al. 2019).

Cobalamin is exclusively produced by microorganisms and serves as a cofactor of enzymes that are implicated in the metabolism of amino acids and fatty acids as well as in the synthesis of DNA and hemoglobin (Watanabe and Bito 2018). Strains belonging to the lactic acid bacteria species *Ec. faecium*, *Ec. faecalis*, *La. casei*, *Furfurilactobacillus rossiae*, *Lm. reuteri*, *Lp. plantarum*, *Loigolactobacillus coryniformis*, *Lm. fermentum* and *La. rhamnosus* have been reported as cobalamin producers (Taranto et al. 2003; Masuda et al. 2012; de Angelis et al. 2014; Torres et al. 2016; Li et al. 2017a, 2017b; Hati et al. 2019). Among them, *Lm. reuteri* strains have been reported to increase the cobalamin content of soy products (Gu et al. 2015; Bao et al. 2019). Moreover, *Propionibacterium* spp. are also known as efficient cobalamin producers and have been used for that purpose in dairy fermentations (Van Wyk et al. 2011).

The main sources of folates for humans are fruits and vegetables, especially green vegetables (Delchier et al. 2016). Certain fruits and vegetables, especially dark green vegetables, are a very good source of riboflavin (Powers 2003). However, the capacity of selected starter cultures to increase their content during vegetable fermentation has only been marginally studied. More accurately, Jagerstad et al. (2004) analyzed fresh white cabbage and commercially available canned sauerkraut as well as domestic samples of fermented white cabbage. Moreover, vegetable mixes were subjected to lactic acid fermentation by several mixtures of starter cultures. One canned sample and both domestic ones were reported to contain almost the double amount of folate than the raw white cabbage. The starter culture consisting of *Lp. plantarum*, *Leuconostoc* sp. and *Lc. lactis/cremoris* was defined as superior to the other. In addition, *Propionibacterium freudenreichii* was able to produce vitamin B12. Thompson et al. (2020) used four *Lp. plantarum* strains to ferment cauliflower, white beans and their mixture and reported a statistically significant increase in riboflavin and folate content in the fermented vegetables compared to the unfermented ones. Moreover, *Lp. plantarum* strain 299 resulted in a statistically significant increase of vitamin B12 as well.

19.3.2.3 Sulforaphane Production

Sulforaphane is a highly bioactive molecule that results from the action of myrosinase or myrosinase-like enzymes on glucoraphanin, a glucosinolate present in vegetables of the family Cruciferae. Myrosinase is also present in the same vegetables, but spatially separated; substrate and enzyme are mixed upon tissue disruption. Glucoraphanin hydrolysis yields thiohydroximate-O-sulfinate, which is very unstable and thus is rearranged, mostly to sulforaphane, at pH values between 5 and 8 and in the absence of ferrous ions and the epithiospecifier protein. On the contrary, at more acidic pH values and in the presence of the aforementioned two compounds, thiohydroximate-O-sulfinate is mostly rearranged to sulforaphane nitrile (Wu et al. 2021). The occurrence of myrosinase or enzymes with myrosinase-like activity has been reported in microorganisms of the human gastrointestinal tract and *Bacillus* spp. members of the Enterobacteriaceae family, such as strains of *E. coli*, *Enterobacter clocae*, *Eb. asburiae*, *Eb. lubwigii*, *Eb. xiangfangensis* and *Citrobacter* spp., as well as LAB, such as strains of *Lp. plantarum*, *Lv. brevis*, *Lt. sakei*, *Ligilactobacillus agilis*, *Lc. hircilactis*, *Lc. lactis* subsp. *lactis*, *Ec. faecalis* and *Ec. casseliflavus* (Mullaney et al. 2013; Angelino and Jeffery 2014; Albaser et al. 2016; Luang-In et al. 2016, 2018; Narbad and Rossiter 2018). Sulforaphane has been associated with specific health benefits, namely cancer chemoprotective activity (Juge et al. 2007), indirect antioxidant capacity (Guerrero-Beltrán et al. 2012) and anti-inflammatory activity (Wei et al. 2020).

Several attempts to increase sulforaphane content of fermented vegetables using myrosinase-positive starter cultures have been reported. Luang-In et al. (2018) used myrosinase-positive LAB as starter cultures in cabbage fermentation, and in many cases a statistically significant improvement of sulforaphane content was reported. Cai et al. (2019) studied sulforaphane production after inoculation of broccoli puree with a mixture of five *Lp. plantarum* and two *Ln. mesenteroides* strains and incubation at 30°C for 4 d. The stability of sulforaphane during storage at 4 and 25°C over a period of 90 d was also assessed. Fermentation resulted in nearly doubling the amount of sulforaphane present in raw broccoli puree. However, not all glucoraphanin was converted to sulforaphane. This was attributed to (1) pH value decrease; sulforaphane could be produced as long as the pH value was above 5, and (2) the limited amount of glucoraphanin that was present during the early stages of fermentation, that is, when the pH value was above 5. Ye et al. (2019) reported that LAB fermentation facilitates glucoraphanin release, most likely in a time-dependent manner, which partially coincides with the time the pH value was above 5. Significant improvement in sulforaphane production was obtained by preheating the broccoli florets at 65°C for 3 min before maceration and fermentation (Cai et al. 2020). Zubaidah et al. (2020) reported that the sulforaphane content of sauerkraut made with the addition of *Ln. mesenteroides* FNCC 0023 was higher compared to the spontaneously fermented one. Abellan et al. (2021) provided a new insight and created a broccoli-supplemented American amber ale-type beer. Maximum sulforaphane concentration was observed after 3 days of maceration in beer (18.18 μmol/L). However, this concentration was significantly decreased by 78.1% after 14 days from removal of the broccoli heads from the beer. Finally, after bottling, less than 5 μmol/L remained in the product; this decreased to 0.80 μmol/L at the end of the storage period (150 d).

19.4 CONCLUSIONS AND FUTURE PERSPECTIVES

A variety of interesting spontaneously fermented vegetable-based products can be found currently in Greece. Although most such products are still homemade, some of them are available in the market by small entrepreneurs, and more will be commercialized soon. These can fulfill consumer demand for minimal processing without the addition of preservatives. Moreover, the wide variety of available vegetable raw materials, combined with the wide range of herbs and spices, can lead to an extended diversity of products with distinctive organoleptic properties. In fact, this depicts the innovation that takes place at the level of small entrepreneurs: mixing ingredients towards the creation of new sensorial combinations. From the more science-intensive contemporary perspective, the innovative potential is huge and depends upon the metabolic capacity of the microbial strains employed for fermentation. A single strain or a combination of strains may successfully drive fermentation and provide the desired additional functional properties. As our understanding of microbial physiology and metabolic capacity is increased, our ability to design products with desired traits will be improved. Therefore, research should focus on exposing and understanding the biosynthetic potential of food-grade microorganisms and correlate it with biotic and abiotic growth conditions towards the ultimate aim of improvement of functional ingredients for healthy dietary intake.

REFERENCES

Abellan, A., Domínguez-Perles, R., Gimenez, M.J., Zapata, P.J., Valero, D., and C. García-Viguera. 2021. The development of a broccoli supplemented beer allows obtaining a valuable dietary source of sulforaphane. *Food Bioscience* 39:100814.

Albaser, A., Kazana, E., Bennett, M., Cebeci, F., Luang-In, V., Spanu, P.D., and J.T. Rossiter. 2016. Discovery of a bacterial glycoside hydrolase family 3 (GH3) glucosidase with myrosinase activity from a *Citrobacter* strain isolated from soil. *Journal of Agricultural and Food Chemistry* 64:1520–1527.

Albuquerque, M.A.C., Bedani, R., LeBlanc, J.G., and S.M.I. Saad. 2017. Passion fruit by-product and fructooligosaccharides stimulate the growth and folate production by starter and probiotic cultures in fermented soymilk. *International Journal of Food Microbiology* 261:35–41.

Angelino, D., and E. Jeffery. 2014. Glucosinolate hydrolysis and bioavailability of resulting isothiocyanates: focus on glucoraphanin. *Journal of Functional Foods* 7:67–76.

Bah, A., Ferjani, R., Fhoula, I., Gharbi, Y., Najjari, A., Boudabous, A., and H.I. Ouzari. 2019. Microbial community dynamic in tomato fruit during spontaneous fermentation and biotechnological characterization of indigenous lactic acid bacteria. *Annals of Microbiology* 69:41–49.

Bao, X., Xiang, S.S., Chen, J., Shi, Y., Chen, Y.G., Wang, H.H., and X. Zhu. 2019. Effect of *Lactobacillus reuteri* on vitamin B12 content and microbiota composition of furu fermentation. *LWT-Food Science and Technology* 100:138–143.

Bautista-Gallego, J., Rantsiou, K., Garrido-Fernandez, A., Cocolin, L., and F.N. Arroyo-Lopez. 2013. Salt reduction in vegetable fermentation: reality or desire? *Journal of Food Science* 78:R1095–R1100.

Beganovic, J., Pavunc, A.L., Gjuracic, K., Spoljarec, M., Suskovic, J., and B. Kos. 2011. Improved sauerkraut production with probiotic strain *Lactobacillus plantarum* L4 and *Leuconostoc mesenteroides* LMG 7954. *Journal of Food Science* 76:M124–M129.

Buckenhueskes, H.J. 1993. Selection criteria for lactic acid bacteria to be used as starter cultures for various food commodities. *FEMS Microbiology Reviews* 12:253–272.

Burgess, C.M., Smid, E.J., Rutten, G., and D. van Sinderen. 2006. A general method for selection of riboflavin-overproducing food grade micro-organisms. *Microbial Cell Factories* 5:24.

Cai, Y.X., Augustin, M.A., Jegasothy, H., Wang, J.H., and N.S. Terefe. 2020. Mild heat combined with lactic acid fermentation: a novel approach for enhancing sulforaphane yield in broccoli puree. *Food & Function* 11:779–786.

Cai, Y.X., Wang, J.H., McAuley, C., Augustin, M.A., and N.S. Terefe. 2019. Fermentation for enhancing the bioconversion of glucoraphanin into sulforaphane and improve the functional attributes of broccoli puree. *Journal of Functional Foods* 61:103461.

Campos, P.A., Martins, E.M.F., Martins, M.L., Martins, A.D.O., Júnior, B.R.C.L., Silva, R.R., and L.M. Trevizano. 2019. In vitro resistance of *Lactobacillus plantarum* LP299v or *Lactobacillus rhamnosus* GG carried by vegetable appetizer. *LWT-Food Science and Technology* 116:108512.

Capozzi, V., Menga, V., Digesu, A.M., De Vita, P., Van Sinderen, D., Cattivelli, L., Fares, C., and G. Spano. 2011. Biotechnological production of vitamin B2-enriched bread and pasta. *Journal of Agricultural and Food Chemistry* 59:8013–8020.

Carrizo, S.L., de Moreno de LeBlanc, A., LeBlanc, J.G., and G.C. Rollan. 2020. Quinoa pasta fermented with lactic acid bacteria prevents nutritional deficiencies in mice. *Food Research International* 127:108735.

Carrizo, S.L., Montes de Oca, C.E., Hebert, M.E., Saavedra, L., Vignolo, G., LeBlanc, J.G., and G.C. Rollan. 2017. Lactic acid bacteria from Andean grain amaranth: a source of vitamins and functional value enzymes. *Journal of Molecular Microbiology and Biotechnology* 27:289–298.

Chitrakar, B., Zhang, M., and B. Adhikari. 2019. Asparagus (*Asparagus officinalis*): processing effect on nutritional and phytochemical composition of spear and hard-stem byproducts. *Trends in Food Science and Technology* 93:1–11.

Cho, S.Y., Park, M.J., Kim, K.M., Ryu, J.-H., and H.J. Park. 2011. Production of high γ-aminobutyric acid (GABA) sour kimchi using lactic acid bacteria isolated from mukeunjee kimchi. *Food Science and Biotechnology* 20:403–408.

Cho, Y.R., Chang, J.Y., and H.C. Chang. 2007. Production of gamma aminobutyric acid (GABA) by *Lactobacillus buchneri* isolated from kimchi and its neuroprotective effect on neuronal cells. *Journal of Microbiology and Biotechnology* 17:104–109.

Da Silva, F.F.P., Biscola, V., LeBlanc, J.G., and B.D.G. de Melo Franco. 2016. Effect of indigenous lactic acid bacteria isolated from goat milk and cheeses on folate and riboflavin content of fermented goat milk. *LWT-Food Science and Technology* 71:155–161.

de Angelis, M., Bottacini, F., Fosso, B., Kelleher, P., Calasso, M., Di Cagno, R., Ventura, M., Picardi, E., van Sinderen, D., and M. Gobbetti. 2014. *Lactobacillus rossiae*, a vitamin B12 producer, represents a metabolically versatile species within the genus *Lactobacillus*. *PLoS One* 9:e107232.

Delchier, N., Herbig, A.-L., Rychlik, M., and C.M.G.C. Renard. 2016. Folates in fruits and vegetables: contents, processing, and stability. *Comprehensive Reviews in Food Science and Food Safety* 15:506–528.

Deli, J., Matus, Z., and G. Tóth. 2000. Carotenoid composition in the fruits of *Asparagus officinalis*. *Journal of Agricultural and Food Chemistry* 48:2793–2796.

Dhakal, R., Bajpai, V.K., and K.-H. Baek. 2012. Production of GABA (γ-aminobutyric acid) by microorganisms: a review. *Brazilian Journal of Microbiology* 43:1230–1241

Di Cagno, R., Mazzacane, F., Rizzello, C.G., Angelis, M.D.E., Giuliani, G., Meloni, M., Servi, B.D.E., and G. Marco. 2010. Synthesis of γ-aminobutyric acid (GABA) by *Lactobacillus plantarum* DSM19463: functional grape must beverage and dermatological applications. *Applied Microbiology and Biotechnology* 86:731–741.

Duma, M., Alsina, I., Dubova, L., and I. Erdberga. 2018. Bioactive compounds in tomatoes at different stages of maturity. *Proceedings of the Latvian Academy of Sciences* 72:85–90.

Ewe, J.-A., Wan-Abdullah, W.-N., and M.-T. Liong. 2010. Viability and growth characteristics of *Lactobacillus* in soymilk supplemented with B-vitamins. *International Journal of Food Science and Nutrition* 61:87–107.

Fleming, H.P., McFeeters, R.F., Thompson, R.L., and D.C. Sanders. 1983. Storage stability of vegetables fermented with pH control. *Journal of Food Science* 48:975–981.

Fuentes-Alventosa, J.M., Jaramillo, S., Rodríguez-Gutiérrez, G., Cermeño, P., Espejo, J.A., Jiménez-Araujo, A., Guillen-Bejarano R., Fernandez-Bolanos, J., and R. Rodriguez-Arcos. 2008. Flavonoid profile of green asparagus genotypes. *Journal of Agricultural and Food Chemistry* 56:6977–6984.

Gangadharan, D., and K. Nampoothiri. 2011. Folate production using *Lactococcus lactis* ssp *cremoris* with implications for fortification of skim milk and fruit juices. *LWT-Food Science and Technology* 44:1859–1864.

Goyeneche, R., Roura, S., Ponce, A., Vega-Gálvez, A., Quispe-Fuentes, I., Uribe, E., and K. Di Scala. 2015. Chemical characterization and antioxidant capacity of red radish (*Raphanus sativus* L.) leaves and roots. *Journal of Functional Foods* 16:256–264.

Gu, Q., Zhang, C., Song, D.F., Li, P., and X. Zhu. 2015. Enhancing vitamin B12 content in soy-yogurt by *Lactobacillus reuteri*. *International Journal of Food Microbiology* 206:56–59.

Guerrero-Beltrán, C.E., Calderón-Oliver, M., Pedraza-Chaverri, J., and Y.I. Chirino. 2012. Protective effect of sulforaphane against oxidative stress: recent advances. *Experimental and Toxicologic Pathology* 64:503–508.

Han, M., Liao, W.-Y., Wu, S.-M., Gong, X., and C. Bai. 2020. Use of *Streptococcus thermophilus* for the in situ production of γ-aminobutyric acid-enriched fermented milk. *Journal of Dairy Science* 103:98–105.

Han, S.-M., and J.-S. Lee. 2017. Production and its anti-hyperglycemic effects of γ-aminobutyric acid from the wild yeast strain *Pichia silvicola* UL6–1 and *Sporobolomyces carnicolor* 402-JB–1. *Mycobiology* 45:199–203.

Hati, S., Patel, M., Mishra, B.K., and S. Das. 2019. Short-chain fatty acid and vitamin production potentials of *Lactobacillus* isolated from fermented foods of Khasi Tribes, Meghalaya, India. *Annals of Microbiology* 69:1191–1199.

Hepsomali, P., Groeger, J.A., Nishihira, J., and A. Scholey. 2020. Effects of oral gamma-aminobutyric acid (GABA) administration on stress and sleep in humans: a systematic review. *Frontiers in Neuroscience* 14:923.

Huang, J., Mei, L., Wu, H., and D. Lin. 2007. Biosynthesis of caminobutyric acid (GABA) using immobilized whole cells of *Lactobacillus brevis*. *World Journal of Microbiology and Biotechnology* 23:865–871.

Huang, X., and L. Kong. 2006. Steroidal saponins from roots of *Asparagus officinalis*. *Steroids* 71:171–176.

Jagerstad, M., Jastrebova, J., and U. Svensson. 2004. Folates in fermented vegetables—a pilot study. *Lebensmittel-Wissenschaft und Technologie* 37:603–611.

Jang, D.S., Cuendet, M., Fong, H.H.S., Pezzuto, J.M., and A.D. Kinghorn. 2004. Constituents of *Asparagus officinalis* evaluated for inhibitory activity against cyclooxygenase– 2. *Journal of Agricultural and Food Chemistry* 52:2218–2222.

Johanningsmeier, S.D., Fleming, H.P., Thompson, R.L., and R.F. McFeeters. 2005. Chemical and sensory properties of sauerkraut produced with *Leuconostoc mesenteroides* starter cultures of differing malolactic phenotypes. *Journal of Food Science* 70:343–349.

Juarez del Valle, M., Laino, J.E., Savoy de Giori, G., and J.G. LeBlanc. 2014. Riboflavin producing lactic acid bacteria as a biotechnological strategy to obtain bioenriched soymilk. *Food Research International* 62:1015–1019.

Juge N., Mithen R.F., and M. Traka. 2007. Molecular basis of chemoprevention by sulforaphane: a compressive review. *Cellular and Molecular Life Sciences* 64:1105–1127.

Kariluoto, S., Edelmann, M., Nystrom, L., Sontag-Strohm, T., Salovaara, H., Kivela, R., Herranen, M., Korhola, M., and V. Piironen. 2014. In situ enrichment of folate by microorganisms in beta-glucan rich oat and barley matrices. *International Journal of Food Microbiology* 176:38–48.

Kim, J.Y., Lee, M.Y., Ji, G.E., Lee, Y.S., and K.T. Hwang. 2009. Production of γ-aminobutyric acid in black raspberry juice during fermentation by *Lactobacillus brevis* GABA100. *International Journal of Food Microbiology* 130:12–16.

Kim, M.-J., Lee, H.-W., Lee, M.-E., Roh S.W., and T.-W. Kim. 2019. Mixed starter of *Lactococcus lactis* and *Leuconostoc citreum* for extending kimchi shelf-life. *Journal of Microbiology* 57:479–484.

Komatsuzaki, N., Shima, J., Kawamoto, S., Momosed, H., and T. Kimura. 2005. Production of γ-aminobutyric acid (GABA) by *Lactobacillus paracasei* isolated from traditional fermented foods. *Food Microbiology* 22:497–504.

Laino, J.E., Juarez del Valle, M., Savoy de Giori, G., and J.G. LeBlanc. 2013. Development of a high folate concentration yogurt naturally bio-enriched using selected lactic acid bacteria. *LWT-Food Science and Technology* 54:1–5.

Laino, J.E., Juarez del Valle, M., Savoy de Giori, G., and J.G. LeBlanc. 2014. Applicability of a *Lactobacillus amylovorus* strain as coculture for natural folate bioenrichment of fermented milk. *International Journal of Food Microbiology* 191:10–16.

Laino, J.E., LeBlanc, J.G., and G. Savoy de Giori. 2012. Production of natural folates by lactic acid bacteria starter cultures isolated from artisanal Argentinean yogurts. *Canadian Journal of Microbiology* 58:581–588.

Lee, J.-J., Choi, Y.-J., Lee, M.J., Park, S.J., Oh, S.J., Yun, Y.-R., Min, S.G., Seo, H.-Y., Park, S.-H., and M.-A. Lee. 2020. Effects of combining two lactic acid bacteria as a starter culture on model kimchi fermentation. *Food Research International* 136:109591.

Lee, K.W., Shim, J.M., Yao, Z., Kim, J.A., and J.H. Kim. 2018. Properties of kimchi fermented with GABA-producing lactic acid bacteria as a starter. *Journal of Microbiology and Biotechnology* 28:534–541.

Li, H., and Y. Cao. 2010. Lactic acid bacterial cell factories for gamma-aminobutyric acid. *Amino Acids* 39:1107–1116.

Li, H., Qiu, T., Huang, G., and Y. Cao. 2010. Production of gamma aminobutyric acid by *Lactobacillus brevis* NCL912 using fed-batch fermentation. *Microbial Cell Factories* 9:85.

Li, P., Gu, Q., Wang, Y., Yu, Y., Yang, L., and J.V. Chen. 2017b. Novel vitamin B12-producing *Enterococcus* spp. and preliminary in vitro evaluation of probiotic potentials. *Applied Microbiology and Biotechnology* 101:6155–6164.

Li, P., Gu, Q., Yang, L., Yu, Y., and Y. Wang. 2017a. Characterization of extracellular vitamin B12 producing *Lactobacillus plantarum* strains and assessment of the probiotic potentials. *Food Chemistry* 234:494–501.

Lim, H.S., Cha, I.-T., Roh, S.W., Shin, H.-H., and M.-J. Seo. 2017. Enhanced production of gamma-aminobutyric acid by optimizing culture conditions of *Lactobacillus brevis* HYE1 isolated from kimchi, a Korean fermented food. *Journal of Microbiology and Biotechnology* 27:450–459.

Luang-In, V., Albaser, A.A., Nueno-Palop, C., Narbad, A., Bennett, M., and J.R. Rossiter. 2016. Metabolism of glucosinolates and desulfo-glucosinolates by selected human gut bacteria. *Current Microbiology* 73:442–451.

Luang-In, V., Deeseenthum, S., Udomwong, P., Saengha, W., and M. Gregori. 2018. Formation of sulforaphane and iberin products from Thai cabbage fermented by myrosinase-positive bacteria. *Molecules* 23: 955.

Manivannan, A., Kim, J.-H., Kim, D.-S., Lee, E.-S., and H.-E. Lee. 2019. Deciphering the nutraceutical potential of *Raphanus sativus*—a comprehensive overview. *Nutrients* 11:402.

Masuda, M., Ide, M., Utsumi, H., Niiro, T., Shimamura, Y., and M. Murata. 2012. Production potency of folate, vitamin B12 and thiamine by lactic acid bacteria isolated from Japanese pickles. *Bioscience, Biotechnology, and Biochemistry* 76:2061–2067.

Meucci, A., Rossetti, L., Zago, M., Monti, L., Giraffa, G., Carminati, D., and F. Tidona. 2018. Folates biosynthesis by *Streptococcus thermophilus* during growth in milk. *Food Microbiology* 69:116–122.

Moore, J.F., DuVivier, R., and S.D. Johanningsmeier. 2021. Formation of γ-aminobutyric acid (GABA) during the natural lactic acid fermentation of cucumber. *Journal of Food Composition and Analysis* 96:103711.

Mosso, A.L., Jimenez, M.E., Vignolo, G., LeBlanc, J.G., and N.C. Samman. 2018. Increasing the folate content of tuber-based foods using potentially probiotic lactic acid bacteria. *Food Research International* 109:168–174.

Mullaney, J.A., Kelly, W.J., McGhie, T.K., Ansell, J., and J.A. Heyes. 2013. Lactic acid bacteria convert glucosinolates to nitriles efficiently yet differently from Enterobacteriaceae. *Journal of Agricultural and Food Chemistry* 61:3039–3046.

Narbad, A., and J.T. Rossiter. 2018. Gut glucosinolate metabolism and isothiocyanate production. *Molecular Nutrition & Food Research* 62:1700991.

Pappa, S., Papadelli, M., Paramithiotis, S., Daferera, D., Polissiou, M.G., and E.H. Drosinos. 2018. Effect of herb addition on spontaneous fermentation of radish (*Raphanus sativus* L.) roots in brine and the fate of *L. monocytogenes* and *E. coli* O157:H7. *Journal of Medicinal Plants Studies* 6:32–39.

Paramithiotis, S., Doulgeraki, A.I., Karahasani, A., and E.H. Drosinos. 2014b. Microbial population dynamics during spontaneous fermentation of *Asparagus officinalis* L. young sprouts. *European Food Research and Technology* 239:297–304.

Paramithiotis, S., Doulgeraki, A.I., Tsilikidis, I., Nychas, G.J.E., and E.H. Drosinos. 2012. Fate of *Listeria monocytogenes* and *Salmonella* sp. during spontaneous cauliflower fermentation. *Food Control* 27:178–183.

Paramithiotis, S., Doulgeraki, A.I., Vrelli, A., Nychas, G.J.E., and E.H. Drosinos. 2016. Evolution of the microbial community during traditional fermentation of globe artichoke immature inflorescence. *International Journal of Clinical & Medical Microbiology* 1:117.

Paramithiotis, S., Hondrodimou, O.L., and E.H. Drosinos. 2010. Development of the microbial community during spontaneous cauliflower fermentation. *Food Research International* 43:1098–1103.

Paramithiotis, S., Kouretas, K., and E.H. Drosinos. 2014a. Effect of ripening stage on the development of the microbial community during spontaneous fermentation of green tomatoes. *Journal of the Science of Food and Agriculture* 94:1600–1606.

Pardali, E., Paramithiotis, S., Papadelli, M., Mataragas, M., and E.H. Drosinos. 2017. Lactic acid bacteria population dynamics during spontaneous fermentation of radish (*Raphanus sativus* L.) roots in brine. *World Journal of Microbiology and Biotechnology* 33:110.

Park, C.H., Baskar, T.B., Park, S.-Y., Kim, S.-J., Arasu, M.V., Al-Dhabi, N.A., Kim, J.K., and S.U. Park. 2016. Metabolic profiling and antioxidant assay of metabolites from three radish cultivars (*Raphanus sativus*). *Molecules* 21:157.

Patra, J.K., Das, G., Paramithiotis, S., and H.S. Shin. 2016. Kimchi and other widely consumed traditional fermented foods of Korea: a review. *Frontiers in Microbiology* 7:1493.

Perpetuini, G., Tittarelli, F., Battistelli, N., Suzzi, G., and R. Tofalo. 2020. γ-aminobutyric acid production by *Kluyveromyces marxianus* strains. *Journal of Applied Microbiology* 129:1609–1619.

Picchi, V., Migliori, C., Lo Scalzo, R., Campanelli, G., Ferrari, V., and L.F. Di Cesare. 2012. Phytochemical content in organic and conventionally grown Italian cauliflower. *Food Chemistry* 130:501–509.

Powers, H.J. 2003. Riboflavin (vitamin B-2) and health. *American Journal of Clinical Nutrition* 77:1352–1360.

Sabo, S.S., Mendes, M.A., Araújo, E.S., Muradian, L.B.A., Makiyama, E.N., LeBlanc, J.G., Borelli, P., Fock, R.A., Knobl, T., and R.P.S. Oliveira. 2020. Bioprospecting of probiotics with antimicrobial activities against *Salmonella* Heidelberg and that produce B-complex vitamins as potential supplements in poultry nutrition. *Scientific Reports* 10:7235.

Seok, J.-H., Park, K.-B., Kim, Y.-H., Bae, M.-O., Lee, M.-K., and S.-H. Oh. 2008. Production and characterization of kimchi with enhanced levels of γ-aminobutyric acid. *Food Science and Biotechnology* 17:940–946.

Siragusa, S., De Angelis, M., Di Cagno, R., Rizzello, C.G., Coda, R., and M. Gobbetti. 2007. Synthesis of γ-aminobutyric acid by lactic acid bacteria isolated from a variety of Italian cheeses. *Applied and Environmental Microbiology* 73:7283–7290.

Soo, I.M.C., Lee, J.W., Park, S.M., Lee, M.Y., Ji, G.E., Park, M.S., and T.R. Heo. 2006. Improvement of γ-aminobutyric acid (GABA) production using cell entrapment of *Lactobacillus brevis* GABA 057. *Journal of Microbiology and Biotechnology* 16:562–568.

Stover, P.J. 2004. Physiology of folate and vitamin B12 in health and disease. *Nutrition Reviews* 62:S3–S12.

Szczuko, M., Ziętek, M., Kulpa, D., and T. Seidler. 2019. Riboflavin—properties, occurrence and its use in medicine. *Pteridines* 30:33–47.

Tabaszewska, M., Gabor, A., Jaworska, G., and I. Drozdz. 2018. Effect of fermentation and storage on the nutritional value and contents of biologically-active compounds in lacto-fermented white asparagus (*Asparagus officinalis* L.). *LWT—Food Science and Technology* 92:67–72.

Tamang, J.P. 2010. *Himalayan fermented foods. Microbiology, nutrition and ethnic values*. Boca Raton: CRC Press, pp. 25–31.

Tamene, A., Baye, K., Kariluoto, S., Edelmann, M., Bationo, F., Leconte, N., and C. Humblot. 2019. *Lactobacillus plantarum* P2R3FA isolated from traditional cereal-based fermented food increase folate status in deficient rats. *Nutrients* 11:2819.

Taranto, M., Vera, J., Hugenholtz, J., De Valdez, G., and F. Sesma. 2003. *Lactobacillus reuteri* CRL1098 produces cobalamin. *Journal of Bacteriology* 185:5643–5647.

Thakur, K., Lule, V.K., Rajni, C.S., Kumar, N., Mandal, S., Anand, S., Kumari, V., and S.K. Tomar. 2016. Riboflavin producing probiotic lactobacilli as a biotechnological strategy to obtain riboflavin-enriched fermented foods. *Journal of Pure and Applied Microbiology* 10:161–166.

Thompson, H.O., Onning, G., Holmgren, K., Strandler, H.S., and M. Hultberg. 2020. Fermentation of cauliflower and white beans with *Lactobacillus plantarum*—impact on levels of riboflavin, folate, vitamin B12 and amino acid composition. *Plant Foods for Human Nutrition* 75:236–242.

Tolonen, M., Rajaniemi, S., Pihlava, J.M., Johansson, T., Saris, P.E.J., and E.L. Ryhanen. 2004. Formation of nisin, plant-derived biomolecules and antimicrobial activity in starter culture fermentations of sauerkraut. *Food Microbiology* 21:167–179.

Torres, A.C., Vannini, V., Bonacina, J., Font, G., Saavedra, L., and M.P. Taranto. 2016. Cobalamin production by *Lactobacillus coryniformis*: biochemical identification of the synthetized corrinoid and genomic analysis of the biosynthetic cluster. *BMC Microbiology* 16:240.

Van Wyk, J., Witthuhn, R.C., and T.J. Britz. 2011. Optimisation of vitamin B12 and folate production by *Propionibacterium freudenreichii* strains in kefir. *International Dairy Journal* 21:69–74.

Watanabe, F., and T. Bito. 2018. Vitamin B12 sources and microbial interaction. *Experimental Biology and Medicine* 243:148–158.

Waters, D.M., Mauch, A., Coffey, A., Arendt, E.K., and E. Zannini. 2015. Lactic acid bacteria as cell factory for the delivery of functional biomolecules and ingredients in cereal-based beverages: a review. *Critical Reviews in Food Science and Nutrition* 55:503–520.

Wei, L., Wang, J., Yan, L., Shui, S., Wang, L., Zheng, W., Liu, S., Liu, C., and L. Zheng. 2020. Sulforaphane attenuates 5-fluorouracil induced intestinal injury in mice. *Journal of Functional Foods* 69:103965.

Wiander, B., and E.L. Ryhanen. 2005. Laboratory and large-scale fermentation of white cabbage into sauerkraut and sauerkraut juice by using starters in combination with mineral salt with a low NaCl content. *European Food Research and Technology* 220:191–195.

Wouters, D., Grosu-Tudor, S., Zamfir, M., and L. De Vuyst. 2013a. Bacterial community dynamics, lactic acid bacteria species diversity and metabolite kinetics of traditional Romanian vegetable fermentations. *Journal of the Science of Food and Agriculture* 93:749–760.

Wouters, D., Grosu-Tudor, S., Zamfir, M., and L. De Vuyst. 2013b. Applicability of *Lactobacillus plantarum* IMDO 788 as a starter culture to control vegetable fermentations. *Journal of the Science of Food and Agriculture* 93:3352–3361.

Wu, Y., Lv, C., Zou, L., Sun, J., Song, X., Zhang, Y., and J. Mao. 2021. Approaches for enhancing the stability and formation of sulforaphane. *Food Chemistry* 345:128771.

Yang, S.Y., Lu, F.X., Lu, Z.X., Bie, X.M., Jiao, Y., Sun, L.J., and B. Yu. 2008. Production of gamma-aminobutyric acid by *Streptococcus salivarius* subsp *thermophilus* Y2 under submerged fermentation. *Amino Acids* 34:473–478.

Ye, J.H., Huang, L.Y., Terefe, N.S., and M.A. Augustin. 2019. Fermentation-based biotransformation of glucosinolates, phenolics and sugars in retorted broccoli puree by lactic acid bacteria. *Food Chemistry* 286:616–623.

Yepez, A., Russo, P., Spano, G., Khomenko, I., Biasioli, F., Capozzi, V., and R. Aznar. 2019. In situ riboflavin fortification of different kefir-like cereal-based beverages using selected Andean LAB strains. *Food Microbiology* 77:61–68.

Zayed, A., Serag, A., and M.A. Farag. 2020. *Cynara cardunculus* L.: outgoing and potential trends of phytochemical, industrial, nutritive and medicinal merits. *Journal of Functional Foods* 69:103937.

Zhang, Z., Fan, S., Huang, D., Xiong, T., Nie, S., and M. Xie. 2020. Polysaccharides from fermented *Asparagus officinalis* with *Lactobacillus plantarum* NCU116 alleviated liver injury via modulation of glutathione homeostasis, bile acid metabolism, and SCFA production. *Food & Function* 11:7681–7695.

Zhang, Z.-H., Fan, S., Huang, D.-F., Yu, Q., Liu, X.-Z., Li, C., Wang, S., Xiong T., Nie S.-P., and M.-Y. Xie. 2018. Effect of *Lactobacillus plantarum* NCU116 fermentation on *Asparagus officinalis* polysaccharide: characterization, antioxidative, and immunoregulatory activities. *Journal of Agricultural and Food Chemistry* 66:10703–10711.

Zhao, W., Liu, Y., Latta, M., Ma, W., Wu, Z., and P. Chen. 2019. Probiotics database: a potential source of fermented foods. *International Journal of Food Properties* 22:198–217.

Zubaidah, E., Susanti, I., Yuwono, S.S., Rahayu, A.P., Srianta, I., and P.J. Blanc. 2020. Effect of *Lactobacillus plantarum* and *Leuconostoc mesenteroides* starter cultures in lower salt concentration fermentation on the sauerkraut quality. *Food Research* 4:1038–1044.

Traditional Turkish Fermented Beverages and Their Health Benefits

20

Hasan Tanguler and Hande Baltacioglu

20.1 INTRODUCTION

Fermentation is characterized as the conversion of organic compounds by biochemical modification by microorganisms and their enzymes into organic acids (especially lactic acid) or alcohols (ethyl alcohol) (Aloys and Angeline 2009; Baschali et al. 2017; Garcia et al. 2020). It is the simplest, cheapest, most economical and most reliable way to increase the nutritional value of food, functional properties, shelf life, sensory properties and other qualities associated with the edible nature of these products (Blandino et al. 2003; Achi 2005; Baschali et al. 2017; Erten and Tanguler 2019; Veiga et al. 2019). Fermented food/beverages have occupied a significant place in the nutrition of human beings across the world by different civilizations since prehistoric times (Aloys and Angeline 2009; Garcia et al. 2020).

Furthermore, it is reported that not only fermented products but also some microorganisms associated with these products are effective during fermentation and have important effects on health (Sanlier et al. 2019). Traditionally, in the manufacture of fermented products, symbiotic bacterial mixes (typically lactic acid bacteria [LAB] and acetic acid bacteria), LAB-yeast mixtures, only yeasts or LAB could be effective (Veiga et al. 2019). Some of them are probiotic microorganisms. The uses of probiotics in dairy-based products have been extensively studied while less researched in other products. LAB, which is present in raw materials or added as a starter culture in milk, fruit or vegetable juices increases safety by inhibiting the growth of spoilage as well as pathogenic bacteria and fungi and improves the sensorial and nutritional quality. On the other hand, LAB acts on various intestinal pathogens and positively affects health by improving the human intestinal flora (Swain and Ananadharaj 2018; Garcia et al. 2020).

Numerous biochemical changes occur in the fermentation, leading to flavour components with changes in the texture of the food matrix (Veiga et al. 2019). The aroma of fermented products is caused by the metabolism or enzymatic activity of microorganisms that are effective in fermentation. Hundreds of volatile organic compounds are produced by microorganisms. Produced from food ingredients through a range of biochemical pathways, these aroma agents are significant in improving consumption and advancing the preference for fermented products. The characteristics and aroma of many traditional foods such as wine, beer, olives, pickles, yoghurt and others, where fermentation plays a major role in their production, are formed by the activity of microbial cultures specific to that product. Factors affecting aroma production during fermentation are raw material type, food ingredients, biological factors (presence of desired microorganisms and use of starter culture), pH, temperature, nutrients (nitrogen and carbon sources), water activity, aeration and contamination sources. The raw material and microorganisms utilized in the manufacture of a fermented product are primarily responsible for the aroma of that product (Cabaroglu and Koyuncu 2019).

Higher alcohols are formed by the catabolic (Ehrlich) or anabolic pathways, especially by yeasts in ethyl alcohol fermentation. In the catabolic pathway, amino acids in the fermentation medium are taken into the cell by the yeast. Keto acids are formed from amino acids inside the cell. Then, keto acids are decarboxylated into aldehydes and reduced to alcohols by NAD-linked dehydrogenase from aldehydes. In the anabolic pathway, carbohydrates are taken into the cell instead of amino acids in the production of higher alcohol (Erten and Canbas 2003). From carbohydrate metabolism, keto acids and then higher alcohols are formed by the pyruvate route. The compounds responsible for the aroma formation in fermented products are esters. Esters are compounds that are biochemically produced by yeasts and provide fruity flavours and odours to foods and drinks such as banana, raspberry, pear and pineapple. They arise through interactions between ethyl alcohol and Acetyl-CoA catalysed by alcohol acetyltransferase during fermentation. Lactones are important compounds that give a

DOI: 10.1201/9781003605300-20

variety of fermented products fruity flavours such as apricot, peach, strawberry, and coconut, oily and sweetish. Lactones can be formed by the metabolic activity of yeasts, moulds and bacteria during fermentation (Cabaroglu and Koyuncu 2019). Acetic, lactic, butanoic, hexanoic and octanoic acids formed by microorganisms (yeast and bacteria) are important acids that affect the flavour and aroma of many fermented products. Another flavouring substance in the aroma of fermented food is carbonyl compounds (aldehydes and ketones), and acetaldehyde constitutes the vast majority. Aldehydes synthesized by microorganisms are generally intermediate products formed during the formation of alcohols from keto acids (Erten and Canbas 2003; Cabaroglu and Koyuncu 2019).

Recently, humanity has also realized the nutritional and sensorial features of fermented products (Prajapati and Nair 2008; Baschali et al. 2017). Each fermented product produced has a unique microflora. This unique flora can increase the levels of important substances such as proteins, amino acids and their precursors (Jeyaram et al. 2009; Cabaroglu and Koyuncu 2019). Besides, fermentation has a positive effect on the digestibility of proteins and the bioavailability of minerals and other micronutrients (Sanlier et al. 2019; Demiray 2019; Ignat et al. 2020). Especially during fermentation, microorganisms synthesize vitamins and minerals, produce biologically active peptides with enzymes such as proteinase and peptidase and remove some non-nutrients (Sanlier et al. 2019).

Fermentation also has positive properties such as detoxification, reduction or elimination of unwanted harmful (such as compounds that reduce nutrient absorption) substances in raw foods and even lactose, which are significant for people's health (Gadaga et al. 1999; Aloys and Angeline 2009; Tamer et al. 2019a). Therefore, there are good indications that fermented food/beverages have many important effects on health (Sanlier et al. 2019; Anli 2019). However, although some claims about the health effects of fermented products are based on personal experiences and statements of individuals, experimental evidence is still not sufficient, and therefore more evidence that fermented products have physiological properties is needed (Prajapati and Nair 2008; Baschali et al. 2017; Demiray 2019). There are studies reported in the literature that some fermented products reduce blood pressure due to their various components, help with coping with stress and have a positive effect on liver and colorectal cancer (Gille et al. 2018). For example, the amount of melatonin increases during the fermentation of some products. Melatonin is suggested to have a positive effect on the physical ageing process and cellular atherosclerosis, arthritis, carcinogenesis, diabetes and neurodegenerative disorders (Sanlier et al. 2019; Anli 2019). Moreover, it may also cause a decrease in the formation of free radicals that have an effect on cellular damage that causes many pathologies (Anli 2019). From these points of view, the production of fermented products and functional beverages will become progressively significant for the human diet (Jeyaram et al. 2009). Recently, due to the modern lifestyle, consumers' desire for healthy and diverse quality products has increased the popularity of functional fermented products (Veiga et al. 2019). This popularity has attracted the attention of not only consumers but also scientists, and the number of studies on the fermentation process and the end product for health benefits has increased (Sanlier et al. 2019).

A wide variety of studies have been carried out on important industrial fermented products, and although there is sufficient information, there is still not enough information, especially on those belonging to certain geographical regions (Prajapati and Nair 2008; Kabak and Dobson 2011; Baschali et al. 2017). In traditional Turkish homes, they are more eager to prepare meals at home than Americans and Europeans, and therefore fermented products are usually prepared at home or in small-scale businesses, such as specific traditional foods (Tamer et al. 2019b). For the whole society to benefit from the health benefits of these products, technologies must be advanced and commercialized (Prajapati and Nair 2008). Also, with the increase and spread of information about these products, especially regarding their nutritional, therapeutical and functional features, the opportunities for these products to spread worldwide increase (Sanlier et al. 2019; Nowak and Kuligowski 2017). Therefore, this chapter provides general information, the metabolic process of the fermentation steps and the associated health benefits of these traditional Turkish fermented beverages (*şalgam* beverage, *hardaliye*, *gilaburu*, *boza* and *ayran*).

20.2 FERMENTED *GILABURU*

There is now considerable attention to the positive effects of nutrition on human health (Rakin et al. 2012). In particular, nutritionists recommend raising the consumption of vegetables and fruits for a healthy life and for preventing chronic diseases. This is because fruits and vegetables are inherently high in carbohydrates, fibre, vitamins, minerals, phenolics and phytochemicals (Garcia et al. 2020). Consumption of fruits and vegetables is far below the daily recommended portion in many countries (Rakin et al. 2012). Furthermore, fresh fruits and vegetables are naturally rich in sugar and slightly acidic, causing them to spoil easily without processing (Prajapati and Nair 2008). They are difficult to manage in storage because of these perishable properties. Therefore, the most effective way to protect both fruit and vegetables and increase their intake is through fermentation (Sanlier et al. 2019; Garcia et al. 2020). Various factors, especially during the processing of vegetables, cause undesirable negative qualities. For example, although heat treatment is a significant process of protecting food and preserving functional properties, it can cause changes in properties such as antioxidant activity, taste and odour (Rakin et al. 2012). When it comes to perishable vegetables, people first have extended their shelf life by adding salt or seawater or by drying, and then fermentation has been utilized. Fermented fruit and vegetable products, according to tradition, have played a significant part in human nutrition since ancient times, and these products have been pursued since people started collecting and storing food (Prajapati and Nair 2008; Swain and Ananadharaj

2018). Many traditional Turkish fermented beverages such as *hardaliye*, *boza*, *şalgam* beverage and *ayran* are enjoyed with pleasure in Turkey (Altay et al. 2013; Tanguler 2010). One of these is the fermented *gilaburu* made from the cranberry bush (Sagdic et al. 2014; Tanguler 2014).

Cranberry bush, scientifically called *Viburnum* (*V.*) *opulus* L., is the fruit of a deciduous shrub plant belonging to the Adoxaceae (formerly "Caprifoliaceae") family (Sagdic et al. 2014; Karacelik et al. 2015; Velioglu et al. 2006; Akman et al. 2021) and resembles grapefruit (Cam et al. 2007; Ceylan et al. 2018). Usually, it grows naturally in Anatolia and the Black Sea regions of Turkey as wild fruit (Yilmaztekin and Sislioglu 2015; Ceylan et al. 2018). Cranberry bush is mostly grown on a large scale in Kayseri city in Turkey (Yilmaztekin and Sislioglu 2015; Eren and Hayta 2015). It also grows naturally in metropolises such as Istanbul, Izmir, Ankara and cities such as Konya, Bursa, Sakarya, Trabzon, Çorum, Kahramanmaras, Kirsehir, Erzurum and Samsun (Yapar 2008).

Gilaburu was known as "rose marbling" during the Seljuk and Ottoman Empires. The most important reason for this is the beautiful flowers it has. The name rose marbling changed to *gilaburu* over time (Eren and Hayta 2015). Today, despite being called *gilaburu* in Turkey (Kalyoncu et al. 2013; Baschali et al. 2017), this wild fruit is also known as gildar, dagdagan (Karacelik et al. 2015), gilaboru, gileboru, gileburu, girabolu, gulebru, girebolu and gili gili in different regions. In Turkey, there are distinct species of this wild fruit. These are *V. opulus* L., *V. lantana* L., *V. orientale* Palas and *V. tinus* L. (Kalyoncu et al. 2013; Ceylan et al. 2018). Across the world, this plant is known by alternative names such as "European cranberry bush, American cranberry bush, cranberry tree, guelder rose, wild guelder rose, gueldres-rose, cherry-wood, rose elder, snowball bush, crampbark tree and whitten tree" (Sagdic et al. 2014; Ceylan 2015; Seker et al. 2016).

Cranberry bushes are 1.3–4 m high, and it easily reaches these heights due to its ability to grow rapidly. Cranberry bushes that grow spontaneously in wetlands and water edges are generally used for decorative purposes almost everywhere in the world. Also, the cranberry bush has a remarkable creamy-white flower that appears in the spring season and fruits that are red all fall season (Sagdic et al. 2014; Erten et al. 2017).

Cranberry bush with a dark red colour and stone fruits can be consumed in distinct forms such as fruit drinks, dried fruit, pickles and jam. In Turkey, cranberry bush is harvested in October–November. However, the fruit can not be consumed directly due to its acidic taste (Aktas 2012; Karacelik et al. 2015; Eren and Hayta 2015; Ceylan et al. 2018; Tamer et al. 2019b). This acidic taste of the fruit is due to the valeric acid it contains (Kalyoncu et al. 2013; Yilmaztekin and Sislioglu 2015; Tamer et al. 2019b).

Recently, interest in cranberry bush fruit has increased after it has been revealed that it is a source of various biochemical components, which include triterpenoids, diterpenoids, sesquiterpenes, iridoids, β-carotene, anthocyanins and proanthocyanidins (Zayachkivska et al. 2006; Kalyoncu et al. 2013). The fruit (and fruit juice) also contains carotenoids; phenolic compounds like flavonoids, including catechin, epicatechin, quercetin glycosides, chlorogenic and 3-methyl-butanoic acids; ascorbic acid; and malic and tartaric acid (Velioglu et al. 2006; Erten et al. 2017; Tamer et al. 2019b; Ceylan et al. 2018). Catechin (between 285 to 352 mg/kg) is the most plentiful compound among all the phenolics. Tartaric and malic acid amounts of cranberry bush berries are in the range of 1.24 to 1.41 and 1.21 to 1.37 g/kg, respectively (Ozrenk et al. 2011). The amounts of ascorbic acid and total acidity are 595.2 mg/kg and 17.9 g/kg, respectively (Akbulut et al. 2008). Cranberry bush fruit is a natural mineral source (10,764 ppm potassium, 1289 ppm Mg and 1304 ppm P). Therefore, this fruit can be considered an alternative for the beverage industry (Kalyoncu et al. 2013). Moreover, the energy grade and protein, oil, soluble solid matter, ash, total phenolics and total anthocyanin contents of cranberry bush fruits are 257 kJ/g and 65 mg/kg, 6.7 g/kg, 104 g/kg, 13 g/kg, 3254 mg/kg and 654 mg/kg, respectively (Akbulut et al. 2008). Also, it is stated that the antioxidant properties of the fruit have potential benefits in avoiding and managing illnesses related to reactive oxygen species, especially the potential of reducing the risk of carcinogenesis. Also, consuming *gilaburu* as fruit juice has a potential preventive effect on cancer (Ceylan et al. 2018).

The bark and fruits of this shrub plant have found wide use in pharmacology, and although not all of its positive effects on health have been scientifically proven, these positive effects are mentioned separately in different geographies. These effects are given as follows. It prevents or reduces (1) the formation of kidney stones, (2) uterine infections, (3) hypertension, (4) asthma, (5) digestive problems, (6) cough, (7) abdominal pain and (8) menstrual cramps (Aktas 2012; Sagdic et al. 2014; Ceylan 2015; Yilmaztekin and Sislioglu 2015; Tamer et al. 2019b). Besides, (9) it has antispasmodic properties (Sagdic et al. 2014), and (10) it is good for infections such as the common cold. However, it is a fact that more clinical study is still required to define whether the juice of the fruit can be used as medicine (Ceylan et al. 2018).

Fruit and fruit juice from cranberry bushes have recently been sold in the markets in Turkey (Cam et al. 2007) because of their nutritious and medicinal properties (Ersoy et al. 2017). Therefore, there is a widespread belief that this wild fruit is good for the treatment of high blood pressure; heart ailments; tuberculosis; stomach pain; digestive problems; duodenal ulcers and bleeding; liver, bile, uterine, kidney and bladder ailments; and cold and cough (Kalyoncu et al. 2013; Ersoy et al. 2017). In research performed to investigate the *in vivo* and *in vitro* impacts of *gilaburu* fruit drink on experimentally created Ehrlich ascites tumour (EAT) in mice, it is reported that *gilaburu* extracts in different fractions may have an antitumoral effect on EAT cells (Al 2019). Similarly, Ceylan et al. (2018) reported that *gilaburu* extract exhibited antitumor activity against EAT by modulating lipid peroxidation and enhancing endogenous antioxidant defence systems.

In a research carried out by Zayachkivska et al. (2006), *V. opulus* proanthocyanidins were reported to assist in increased endogenous nitric oxide formation, suppression of lipid peroxidation and mobilization of antioxidant activity and potent gastroduodenal protective activity through changes in the glycoconjugate content of the rat gastroduodenal mucosa. In a study by Zakłos-Szyda and Pawlik (2019), they reported that components of *V. opulus* polyphenolic compounds have high potential. They reported that this could be utilized in the manufacture of dietary supplements or pharmacological preparations for humans with therapeutic potential for those with a raised risk or tendency to develop breast or cervical cancer.

An important product obtained from this cranberry bush fruit, which is a functional and nutritious product, is fermented fruit juice (fermented *gilaburu*) (Sagdic et al. 2014). *Gilaburu*, a non-alcoholic fermented product obtained from naturally fermented cranberry bush fruits (Velioglu et al. 2006; Baschali et al. 2017), is a fermented drink that has been enjoyed for many years in Turkey, especially Central Anatolia. In the production of fermented *gilaburu*, first, harvested cranberry bush fruits are selected and washed. The fruits are then transferred to plastic vessels containing water and left to ferment in the darkness at ambient temperature for spontaneous fermentation. After fermentation, which lasts between 3 and 5 months (Velioglu et al. 2006; Sagdic et al. 2014; Yilmaztekin and Sislioglu 2015; Baschali et al. 2017), the fruits ripen. Fruits that ripen by fermentation lose their bitter taste, albeit partially, but still cannot be consumed at this stage (Eren and Hayta 2015). After the fruits ripen by fermentation, the fermented fruits are crushed and diluted 1:4 with drinkable water. Fermented juice cannot be considered sensorially acceptable due to its astringent flavour that leaves a tactual feel at this stage (Velioglu et al. 2006; Sagdic et al. 2014; Yilmaztekin and Sislioglu 2015). This unwanted taste can be reduced with prolonged fermentation (Sagdic et al. 2014). In addition to the effect of long-lasting fermentation and time, the addition of sugar, if desired, reduces astringency. However, there is still no standard process used for fermented *gilaburu* production, especially in terms of fermentation time and temperature (Yilmaztekin and Sislioglu 2015; Baschali et al. 2017; Erten et al. 2017).

The LAB identified in fermented *gilaburu* are species belonging to the genera Lactobacillus (*L.*) and Leuconostoc (*Ln.*). In a study, 332 isolates were obtained from fermented *gilaburu*. Among these, the more defined types in terms of quantity are as follows: *Lactiplantibacillus (Lpb.) plantarum* (formerly known as *L. plantarum*), *Levilactobacillus brevis* (Basonym: *L. brevis*), *Lacticaseibacillus (Lcb.) casei* (Basonym *L. casei*), *Liquorilactobacillus hordei* (Basonym *L. hordei*) and *Lcb. pantheris* (Basonym *L. pantheris*). The species with only one identified are *Ln. mesenteroides, Lpb. paraplantarum* (Basonym *L. paraplantarum*), *L. cornyformis, Lentilactobacillus (Llb.) buchneri* (Basonym *L. buchneri*), *Llb. parabuchneri, Ln. pseudomesenteroides* and *Schleiferilactobacillus harbinensis* (Basonym *L. harbinensis*) (Sagdic et al. 2014; Garcia et al. 2020; Zheng et al. 2020). In addition to the bacteria mentioned here, in a study by Yapar (2008), *Lcb. paracasei* subsp. *paracasei* (Basonym *L. paracasei* subsp. *paracasei*), Pediococcus (*P.*) spp. and *L. delbrueckii* ssp. *delbrueckii* were isolated from fermented *gilaburu* beverages (Sagdic et al. 2014; Zheng et al. 2020). Moreover, LAB, total mesophilic aerobic bacteria (TMAB) and mould/yeast numbers in fermented *gilaburu* beverage vary from 3.92–8.44 log cfu/mL, 5.73–8.97 and 5.00–7.58 log cfu/g, respectively. On the other hand, pathogenic bacteria such as *Alicyclobacillus* sp., *Staphylococcus aureus*, total coliform and *Escherichia (E.) coli* are not present in fermented *gilaburu* beverage (Sagdic et al. 2014).

The general composition of fermented *gilaburu* is given in Table 20.1. Since sugars in fruits are used in fermentation and decrease, the amount of reducing sugar in *gilaburu* obtained varies between 13.52–55.14 g/kg. On the other hand, there are 58 volatile compounds in fermented *gilaburu*. These are alcohols (27), volatile acids (10), ketones (12), aldehydes (3), esters (3) and lactones (3). 3-methylbutanoic acid is the main volatile compound found in fermented *gilaburu*. The fermentation time and the composition of volatile substances are significantly affected (Yilmaztekin and Sislioglu 2015).

In recent years, the consumption of fermented *gilaburu* juice is thought to have positive effects on health; therefore, its use is increasing in Turkey (Eren and Hayta 2015). In a study

TABLE 20.1 General Composition of Fermented *Gilaburu*

	AMOUNT	*REFERENCES*
Ash (%)	0.39–0.62	Yapar (2008)
Protein (g/Kg)	0.39–0.58	Yapar (2008)
Total acidity (%)*	1.59–2.23	Yapar (2008)
pH	2.40–4.44	Sagdic et al. (2014), Yapar (2008)
Tannic acid (mg/L)	181.6–902.9	Soylak et al. (2002)
Brix	7.90–10.70	Yapar (2008)
Alcohol (g/L)	0.66–6.12	Yapar (2008)
Crude Fiber (%)	0.46–0.68	Yapar (2008)
Fat (g/100mL)	0.3–0.4	Yapar (2008)
Ascorbic acid (mg/100 mL)	28.94–66.1	Yapar (2008), Soylak et al. (2002)
Total phenolic substance (mg/100 mL)	601–731	Yapar (2008)

Note: *: as lactic acid.

in which fruit pomace leftover from fermented *gilaburu* production was evaluated in cake production, it was reported that total phenolic content and radical scavenging activity values increased in cake samples due to increased *gilaburu* pomace addition (Seker et al. 2016).

Fermented *gilaburu* is rich in vitamin precursors such as B group vitamins, carotene and ascorbic acid, mainly due to the *gilaburu* fruit used in its production and the fermentation process (Eren and Hayta 2015). Moreover, fermented *gilaburu* is considered a functional product due to the dense antioxidants in high amounts like phenolic compounds, flavonoids and anthocyanins it contains. In addition to the fact that fermented *gilaburu* is rich in antioxidants, it is a beverage with antimicrobial properties. It may also be chemopreventive in the early stages of colon cancer (Afrin et al. 2016). Also, fermented *gilaburu* is considered a probiotic drink because it contains some LAB strains with probiotic potential (Sagdic et al. 2014; Baschali et al. 2017). Lately, Akman et al. (2021) reported that LAB strains isolated from traditional fermented *gilaburu* have high probiotic potential and can be used in the production or design of various functional foods/beverages.

20.3 *ŞALGAM (SHALGAM)* BEVERAGE

There are many types of traditional Turkish fermented beverages in Turkey (Altay et al. 2013; Tanguler 2010). One of these products originating in Turkey is *şalgam* beverage and has long been an important product that is consumed with pleasure. *Şalgam* beverage, a turbid and sour drink with red and shades of colour, is traditionally made at home or in household-scale businesses. Also, in the last 20–30 years, it has been produced on a small and large scale commercially and sold in the market (Erten and Tanguler 2012; Altay et al. 2013).

A traditional fermented product in the region of the north of India and Pakistan, Kanji has significant similarities to *şalgam* beverage (Sethi 1990; Montet et al. 2015). Similar to *şalgam* beverage, it is manufactured by spontaneous fermentation using purple/black carrot and rock salt. However, what is different from *şalgam* beverage is that the process also uses crushed mustard and red pepper powder as raw materials in production (Berry et al. 1989; Sethi 1990). Another product similar to *şalgam* beverage is lacto-fermented carrot juice. The use of orange-rooted carrots and starter culture in production is a different aspect when compared with *şalgam* beverage. Homofermentative LAB such as *Lactobacillus* strains are used as starter culture in production (Demir et al. 2004, 2006)

It is a product unique to Adana province and its surroundings, famous for its local flavours such as kebap (Turkish meatball), Turkish spicy paste (traditional appetizer) and bici bici (consisting of grated ice, cooked starch, powdered sugar and sherbet). It is stated that 20–25% of tourists coming to the region come to consume specific other products, especially *şalgam* beverage. Being identified with the name of the city of Adana, *şalgam* beverage is one of the leading products of the city in terms of gastronomic tourism. It is also a classic product that local and even some foreign tourists (especially Turks living abroad) coming to the city receive as a gift (Coskun 2019). For *şalgam* beverage produced in Tarsus in 2005, a Geographical Sign Registration Certificate (Registration number: 84) was obtained from the Turkish Patent Institute under the name of "Tarsus *Şalgam*" (Anonymous 2020).

Şalgam beverage is consumed cold in summer and preferably with or without hot chilli pepper in winter. It is recommended to consume *şalgam* beverage without hot pepper powder to perceive its real taste (Erten et al. 2008, Tanguler and Erten 2012; Coskun 2019). To make seasoned *şalgam* beverage, hot pepper (*Capsicum annuum*) powder is used (Tanguler and Erten 2012; Anonymous 2020). For this purpose, pickled hot pepper can also be added to the product. However, if powder or pickled hot pepper is kept too long after mixing, it should be consumed immediately, as it may spoil (Coskun 2019).

Şalgam beverage is produced in large quantities in provinces such as Adana, Mersin, Hatay/Antakya and Maraş and the districts of these provinces. The most important reason for production advantages is that the ambience and circumstances of the area/region are appropriate for the production of *şalgam* beverage and the consumption is more preferred with a preference for regional beverages and foods (Erten and Tanguler 2012; Tanguler et al. 2020). *Şalgam* beverage is consumed with foods such as kebab, pizza, lahmacun and meatballs and also beverages such as raki, which is the conventional Turkish anise-flavoured distillate alcoholic beverage (Tamer et al. 2019b; Tanguler et al. 2020). The popularity of *şalgam* beverage has increased due to the increasing interest in functional foods by conscious consumers in recent times (Incedayi et al. 2008; Coskun 2019). In recent years with the relevance given to functional foods and beverages, *şalgam* beverage has also been sold in some metropolises (Istanbul, Ankara, Konya and Bursa) and is even found in markets in some European capitals (Erten and Tanguler 2019). Adana-London *Şalgam* days were held in 2018 to promote *şalgam* beverage in England. This event was jointly organized by the Adana Metropolitan Municipality, Adana Chamber of Commerce and *şalgam* beverage Producers Association. The reason it is called *Şalgam* days is that it is called *Şalgam* in the city of Adana (Coskun 2019).

The main raw material utilized in the manufacture of *şalgam* beverage is the black (purple) carrot (*Daucus carota* spp. *sativus* var. *atrorubens* Alef.), and sourdough, bulgur flour (setik), salt, turnip, radish and water are also added (Erten et al. 2008; Tanguler et al. 2020). Black carrot provides the red colour to *şalgam* beverage with the anthocyanin substances that are present in black carrots. Cyanidin glycosides are the most abundant anthocyanins in black carrots, and there are also malvidin and peonidin glycosides. The amount of fermentable sugar in black carrots varies between 5.1–6.5%, containing sugars such as sucrose (1.2–3.3%), glucose (1.1–5.6%) and fructose (1.0–4.4) (Erten and Tanguler 2019).

Sourdough or baker's yeast (*Saccharomyces, S. cerevisiae*) is utilized in the manufacture of *şalgam* beverage, and the sourdough used is obtained by keeping the baker's yeast at room temperature overnight (Erten et al. 2008; Erten and Tanguler 2019). Fermented bread dough taken from bakeries is also used in making *şalgam* beverages in homes (Erginkaya and Aksan 2004). Bulgur flour (setik) is the part of the boiled and dried wheat that is formed during the crushing after the outer shells are separated and remain under the sieve. Bulgur flour, which constitutes 2–3% of the wheat grain, contains 4.4–5.8% starch and 2.2–3.3% total sugar. It is a minor raw material used at a rate of 3% in the production of *şalgam* beverage. It provides nutrients to microorganisms during fermentation (Erten et al. 2008; Erten and Tanguler 2012, 2019).

Unrefined rock salt is used at concentrations of 1%–2% in production (Canbas and Deryaoglu 1993; Erten et al. 2008). The salt used in production supports the growth of LAB by inhibiting pathogenic and spoilage organisms during fermentation depending on its concentration (Tanguler et al. 2014). Depending on availability, turnip (*Brassica rapa*) is also used by some companies in production. In turnip, a plant belonging to the cruciferous family (Brassicaceae) with fermentable sugars containing 1.41% glucose, 1.10% fructose and 0.21% sucrose is available (Erten and Tanguler 2019). Turnip is used in *şalgam* beverage production by some producers in specific production in periods when it can be supplied by some producers and in different proportions, or it is not used at all by some producers (Erten et al. 2008).

There is currently no standard flowchart commonly used by the producers in the manufacturing of *şalgam* beverage. Therefore, production can vary from one factory to another or according to the producer. However, commercial production follows two methods: traditional production (dough fermentation and carrot/main fermentation) and direct (fast) production without dough fermentation (Erten et al. 2008; Tanguler 2010; Tanguler et al. 2014, 2020; Erten and Tanguler 2019).

The traditional method is carried out in two stages, where, in the first stage, 3% bulgur flour, 0.2% salt and 0.2% sourdough (or baker's yeast) are mixed by adding water to them (Erten and Tanguler 2019). Lukewarm water is generally used for kneading on cold days (Canbas and Deryaoglu 1993). The mixture is kneaded and pulped and left to ferment in plastic tanks at room conditions or 25°C for 3–5 days or more (Erten and Tanguler 2019). At this stage, LAB and yeasts are enriched, and as a result of their action, the acidity increases significantly and the pH drops. At the end of fermentation, the dough with cracks on its surface is mixed with water 3–4 times and sometimes even 5 times for 5–10 minutes and then extracted (Erten et al. 2008; Incedayi et al. 2008; Tanguler 2010; Coskun 2019). A liquid extract from the dough (sourdough) fermentation is transferred to the tank to perform the second (main) fermentation (Erten et al. 2008). This extract is good for the fast and healthy start and maintenance of the second fermentation. However, extraction is not performed in some productions. After dough fermentation, the fermented product is transferred into a cloth pouch or hung in the fermentation container to start the carrot fermentation (Tanguler and Erten 2012). If extraction is carried out, the extracts combined with 10–20% (according to the manufacturer) cleaned and chopped black carrots are added to the tank (Erten et al. 2008; Erten and Tanguler 2019). While black carrots, which are generally the basic raw materials, are washed, some producers do not require washing in production. However, these producers do not use all carrots that are damaged or defective (Erten and Tanguler 2012). It has been reported that at least 15% black carrot should be used for the production of *şalgam* beverage with an acidity value following Turkish *Şalgam* Standards, TS 11149 (Tanguler et al. 2014). On the other hand, about 1% salt and, if desired, 1% chopped turnip are added to the tank. Water is transferred to the container until it reaches the fill level and left for the second fermentation in room temperature conditions (Erten et al. 2008; Erten and Tanguler 2019). Traditionally, while special barrels made of mulberry wood were used in fermentation (Coskun 2019), today mostly fiberglass, plastic and stainless steel containers are utilized. The second fermentation generally varies between 3 and 25 days at 10–35°C depending on the producer, microflora, ambient temperature and other factors. During fermentation, lactic acid is formed, and natural colouring compounds such as anthocyanins pass into the *şalgam* beverage and give the product its distinctive red colour and pleasant sour taste (Erten et al. 2008; Tanguler et al. 2014; Erten and Tanguler 2019).

The difference between direct *şalgam* beverage production from the traditional production method is that the first fermentation is not done. Only the second fermentation is involved in this production. A content of 10–20% chopped black carrot, 3% bulgur flour, 1% chopped turnip, 1.2% salt, 0.2% sourdough (or baker's yeast) and water are added to the tank and mixed. Fermentation usually takes place at room temperature. Once the fermentation is completed, the procedures specified in the traditional method are applied (Erten et al. 2008; Erten and Tanguler 2019).

As far as is known, any starter culture isolated from the autochthonous or allochthonous environment is not used in *şalgam* beverage production (Tanguler et al. 2020), but some commercial production processes use up to 15% of *şalgam* beverage obtained from the previous production in next batch of fermentation (Erten and Tanguler 2019). However, it has been reported that *Lpb. plantarum, Limosilactobacillus fermentum (Basonym L. fermentum)* and *Lcb. paracasei* subsp. *paracasei* can be used as autochthonous starter cultures in *şalgam* beverage production (Tanguler 2010; Tanguler and Erten 2013; Zheng et al. 2020). Generally, the obtained *şalgam* beverage is not filtered and clarified, but the sediment removal process is applied. The product whose sediment is removed is packaged in airtight glass or plastic bottles or plastic containers and put on the market (Canbas and Deryaoglu 1993; Erten and Tanguler 2019). *Şalgam* beverage is a perishable beverage, and it should be consumed in a short time since generally no preservatives are added to it. Therefore, it should be stored in cool indoor environments until it is consumed, but it is generally kept at room temperature conditions. The general shelf life of a beverage in a closed container with cool storage at 4°C is 3–4 months (Erten et al. 2008; Tanguler 2010; Erten and Tanguler 2012). Tanguler (2010) separated *şalgam* beverages and applied sterile (membrane) filtration in two groups and stored them at 4°C and 20°C for 150–180 days. The study suggested that *şalgam* beverages could be stored to the end of storage without any deterioration in their chemical, physical, microbiological and sensory properties. According to the Turkish Standards Institute (TS 11149 Turkish Standards), the shelf life of *şalgam* beverage can be increased by heat treatment (TSE 2003). However, the pasteurization process causes undesirable sensory properties such as cooked carrot taste and odour in *şalgam* beverages (Tanguler 2010). *Şalgam* beverage produced in this way is less preferred by consumers (Tanguler et al. 2020). Recently, studies have been performed on ultraviolet pasteurization (Dogan 2017), heat + sonication (Avci 2008), ultrasonication, high-pressure process pasteurization (Ates 2019) and pulsed UV light (Karaoglan et al. 2017). Also, the shelf life of the product can be extended by using chemical additives such as salts of benzoic acid (Na-benzoate, K-benzoate) (Erten et al. 2008; TSE 2016; Tanguler et al. 2020) and, although forbidden, natamycin (Avci 2008).

The microbiology of *şalgam* beverages is complicated. Lactic acid fermentation by LAB is the most important reaction. However, yeasts are less effective in *şalgam* beverage fermentation (Erten et al. 2008). These microorganisms originate from the surface of carrot/turnip, the surface of tanks where it is manufactured and also from the sourdough (or baker's yeast) extract (Erginkaya and Hammes 1992; Erten and Tanguler 2012, 2016). LAB ferment hexoses by homofermentative and heterofermentative pathways. *Pediococcus* spp., *Lactococcus* (*Lc.*) spp., *Streptococcus* spp. and some *Lactobacillus* spp. are examples of homofermentative LAB. On the other hand, bacteria belonging to the genera *Leuconostoc*, *Oenococcus* and *Weissella* (*W.*) are examples of heterofermentative LAB. Homofermentative LAB usually generates lactic acid (>90%) as a major or sole end product through glycolysis (EMP, Embden Meyerhof Parnas pathway). Heterofermentative LAB uses the pentose phosphate or phosphoketolase pathway depending on the presence or absence of oxygen in the environment. They produce lactate, carbon dioxide and ethyl alcohol or acetic acid depending on these different routes (Caplice and Fitzgerald 1999; Tanguler 2010).

Tanguler (2010) isolated and identified 18 LABs from *şalgam* beverage samples produced in the laboratory under controlled conditions and by small- and large-scale enterprises in the sector. These belong to the genera *Lactobacillus, Lactococcus, Pediococcus* and *Leuconostoc. Lpb. plantarum* and to a lesser extent *Lcb. paracasei* subsp. *paracasei* dominate *şalgam* beverage fermentations. The species belonging to the genus *Lactobacillus* play an important role in the fermentation of various vegetables due to their acid tolerance. As in cucumber and olive fermentations generated by lactic acid fermentation (Harris 1998; Hutkins 2006), *Lpb. plantarum* is dominant in *şalgam* beverage fermentation, and the fermentation is usually terminated by this bacterium. *Levilactobacillus brevis, Limosilactobacillus fermentum, L. delbrueckii, Lpb. pentosus* (Basonym *L. pentosus*), *Llb. buchneri, P. pentosaceus, Ln. mesenteroides* and *Lc. lactis* were also isolated during *şalgam* beverage fermentation depending on different factories (Tanguler 2010; Zheng et al. 2020). However, these bacteria are generally found in much lower numbers than *Lpb. plantarum*. On the other hand, *Leuconostoc* and *Pediococcus*, which are seen at the beginning of fermentation, cannot be isolated from the environment at the end of fermentation (Tanguler 2010).

In *şalgam* beverage fermentation, there are also total mesophilic aerobic bacteria and coliform bacteria in the environment. With the onset of fermentation, dominant microflora and LAB increase, while coliforms begin to decrease directly as a natural consequence of lowering pH value, and then they cannot be isolated from the environment. CO_2 generated by heterofermentative LAB and yeasts during fermentation replaces the ambient air and in combination with the added rock-salt inhibits unwanted Gram (-) bacteria. Besides, the number of total mesophilic aerobic bacteria, total yeast and non-*Saccharomyces* yeast first increases and then generally decreases (Tanguler 2010; Erten and Tanguler 2012).

While *S. cerevisiae* is particularly effective in fermentation, non-*Saccharomyces* yeasts oxidize sugars to CO_2 and water during fermentation and storage. At the same time, they form small amounts of alcohol and secondary metabolites (Karaoglan et al. 2017). Wild yeasts such as *Candida (C.) krusei, C. pelliculosa, C. lipolytica* (Ozer 2009) and *C. inconspicua*, which are particularly responsible for spoilage in storage, can produce unpleasant and unwanted tastes and odours. Especially if the temperature is slightly higher and the contact area with air is large, this undesirable situation occurs rapidly and is more pronounced (Avci 2008; Karaoglan et al. 2017). They can cause a change in the quality of the product and deterioration over time by causing a decrease in acidity (by breaking down lactic acid) and an increase in the pH of *şalgam* beverage. pH value up to 7.30 and swelling due to gas is a very common problem, especially in products produced in the summer (Avci 2008; Erten and Tanguler 2012; Tanguler et al. 2020).

Flavour compounds found in *şalgam* beverages mainly originate from the raw material (black carrot, setik), yeast and LAB metabolism during fermentation. As a result of the analysis performed by GC-MS in *şalgam* beverages produced by five different methods, 60 aroma compounds, including 20 terpene compounds, 9 ester compounds, 9 higher alcohols, 5 volatile acids, 6 volatile phenols, 5 lactone compounds, 3 naphthalenes, 2 carbonyls and 1 C_{13}-norisoprenoid compound, were identified. On the other hand, the most abundant compounds in the aroma composition are higher alcohols and then terpenes and naphthalenes (Tanguler 2010). Terpenes such as pinene, β-caryophyllene, limonene and terpinolene are compounds derived from black carrots (Kreutzmann et al. 2007) and contribute significantly to the characteristic aroma of *şalgam* beverage (Tanguler 2010).

20.3.1 Benefits of *Şalgam* Beverage on Health

Lactic acid fermentation improves the nutritional and sensory properties, aroma, health aspects, acceptability and shelf life of foods and beverages (Garcia et al. 2020). *Şalgam* beverage is extremely nutritious due to the vitamins, minerals and other

TABLE 20.2 Chemical and Physical Composition of *Şalgam* Beverage

	AMOUNT	*REFERENCES*	*TSE 2003, 2016*
Dry matter (%)	1.69–4.0	Ozler (1995), Cakir (2011)	>2.8
Ash (%)	0.86–2.02	Ozler (1995), Gok (2017)	<2.0
Total acidity (g/L)*	3.42–12.6	Cakir (2011), Gok (2017)	>6.0
pH	3.11–4.31	Ozler (1995), Gok (2017)	<3.8
Lactic acid (g/L)	3.05–8.38	Ozler (1995), Gok (2017)	>4.5
Volatile acid (g/L)**	0.3–3.41	Ozler (1995), Gok (2017)	<1.0
Salt (%)	0.82–2.6	Ozler (1995), Cakir (2011);	<1.7
Carbohydrate (g/L)	0.29	Baser (2013)	–
Sucrose (g/L)	0.03–0.09	Tanguler et al. (2015, 2020)	–
Glucose (g/L)	0.05–0.21	Tanguler et al. (2015, 2020)	–
Fructose (g/L)	0.12–0.17	Tanriseven et al. (2020)	–
Total sugar (g/L)	0.32–0.46	Tanriseven et al. (2020)	–
Protein (g/L)	2.8	Baser (2013)	–
Alcohol (% v/v)	0.13–0.83	Gok (2017)	–
Total phenolic content (mg GA/L)	387–3388	Baser (2013), Cakir (2011)	–
Antioxidant capacity • DPPH (μmol TE/mL) • ABTS (μmol TE/mL) • FRAP (μmol TE/mL)	2.47– 5.96 2.43–5.08 2.01–3.61	Baser (2013), Tanriseven et al. (2020)	–
Iron (mg/L)	0.18–4.66	Gok (2017)	<15
Sn (mg/kg)	nd–0.37	Cakir (2011), Yilmaz-Ersan and Turan (2012)	<200
Zinc (mg/L)	0.079–2.2	Cakir (2011), Gok (2017)	<5.0
Pb (mg/kg)	nd–0.11	Yilmaz-Ersan and Turan (2012)	<0.05
Cu (mg/L)	nd–0.62	Cakir (2011)	<5.0
Cd (mg/L)	nd–0.04	Yilmaz-Ersan and Turan (2012)	–
Arsenic (mg/L)	0.5–1.0	Gok (2017)	<0.2
Chemical preservative (g/L) • Benzoic acid • Sorbic acid	nd–1.1 nd–0.48	Ozturk (2009), Cakir (2011)Cakir (2011)	<0.2 –
Artificial dye-stuff	nd	Cakir (2011), Gok (2017)	***
Ascorbic acid (mg/100g)	0.89–3.37	Ozler (1995)	–
Tryptamine (mg/L)	3.1–66.03	Ozdestan and Uren (2010), Yaldirak (2011)	–
Phenylethylamine (mg/L)	1.82–23.25	Yaldirak (2011)	–
Putrescine (mg/L)	5.0–59.19	Ozdestan and Uren (2010), Yaldirak (2011)	–
Cadaverine (mg/L)	1.5–12.3	Ozdestan and Uren (2010)	–

Notes: *: as lactic acid, **: as acetic acid, ***: should not be found, GA: Gallic acid, nd: not determined.

ingredients it contains (Table 20.1). Besides, although it varies from manufacturer to manufacturer, there is also ethyl alcohol in *şalgam* beverage (Baser et al. 2012; Altay et al. 2013; Sanlier et al. 2019). However, consumption of 300 mL does not significantly affect the blood alcohol level (Tumer et al. 2018). Moreover, it is considered a microbiologically safe beverage. It does not have a suitable environment for pathogens due to its low pH value and high total acidity (Yardimci et al. 2015). Similarly, Cakir (2011) reported that the number of *Salmonella*, *E. coli* and coliform bacteria in *şalgam* beverages is below 10 cfu/mL, and they are generally not detected.

Table 20.2 shows the chemical and physical composition of *şalgam* beverage. As can be seen from Table 20.2, the predominant acid in *şalgam* beverage is lactic acid formed during fermentation by LAB. In addition to lowering the pH value of *şalgam* beverages and giving them a sour taste, lactic acid helps regulate the pH value of the digestive system, facilitates digestion, is refreshing and helps the body to make better use of some minerals (Yardimci et al. 2015). Additionally, lactic acid adds appetizing properties to the product (Coskun 2019). Lactic acid also alters the redox state by reducing the reactive oxygen species load in intestinal enterocytes (Anli 2019).

Black carrot contains a significant amount of antioxidants, vitamins, phenolics and β-carotene, and β-carotene has a positive effect on cardiovascular diseases, cataracts and the immune system (Yardimci et al. 2015; Cirisoglu and Olum 2019). Phenolic compounds found in significant amounts in *şalgam* beverage can protect the human body from free radicals and retard the progress of many chronic diseases due to their antioxidant activities. Baser (2013) reported that the anti-proliferative effect of *şalgam* beverage on cells is higher than black carrots, and the most important reason for this may be the chemical composition (salt, organic acids, sodium benzoate) and microbial flora (LAB) of *şalgam* beverage. Therefore, it has been reported to have potential against colon cancer cells.

Since *şalgam* beverage also contains a significant amount of B group vitamins (Cirisoglu and Olum 2019), it can be said that it calms nerves and the stomach and positively affects liver function (Yardimci et al. 2015). It also contains minerals (sodium, calcium, potassium, magnesium and iron), and they show positive effects on the immune system and bone and dental health. Consumers also use it as a medicine because they assume it contains antiseptic substances and cleans the digestive organs. It is also consumed because it relaxes the intestines (Yilmaz-Ersan and Turan 2012; Yardimci et al. 2015).

Turnip, which is preferred by some producers in the production of *şalgam* beverage, increases the positive effects of *şalgam* beverage on health due to the vitamins (A, B, C) and minerals (calcium, magnesium, iron, sulfur and iodine) it contains. These vitamins and minerals, which pass from raw materials such as carrots and turnips to the product, are suggested to help digestion; support the intestines; prevent constipation; help weight loss; reduce stress; prevent colds; and have a positive effect on heart, kidney and stomach functions (Coskun 2019).

Şalgam beverage has similar antioxidant activity (ABTS or FRAP values) values when compared with other fruit juices (apple juice, orange juice, strawberry juice and cranberry juice) (Tanriseven et al. 2020). Therefore, as it has significant antioxidant capacity, its consumption potentially positively contributes to the plasma lipid profile, antihemolytic defences and platelet function in healthy individuals. Additionally, *şalgam* beverage contains an important quantity of anthocyanins. For this reason, its consumption, because of anti-inflammatory and antioxidant activities, could potentially contribute to reducing the risks of diabetes, arthritis, cardiovascular disease and cancer (Toktas et al. 2018). The phenolic compounds and anthocyanins found in *şalgam* beverage and the antioxidant potential of *şalgam* beverage show that this beverage could potentially be a functional beverage (Tanriseven et al. 2020). It is reported that *şalgam* beverage also has anti-inflammatory and diuretic properties and can be targeted especially for cholesterol, weight control and protection of arthritis (Kabak and Dobson 2011; Tamer et al. 2019b). On the other hand, phenolic compounds and antioxidant properties used in *şalgam* beverage production and found in *şalgam* beverage are also found in *şalgam* beverage residues (fermented black carrot, bulgur flour). Therefore, these residues are utilized in the production of high value-added products (Tanguler and Tatlisoy 2022).

Due to all these potential positive health-relevant properties, it can be considered an important functional beverage and also a probiotic beverage, since it contains many probiotic microorganisms. On the other hand, it is a low-calorie beverage. Therefore, all these features increase the importance of this traditional drink with functional properties (Cirisoglu and Olum 2019).

As mentioned earlier, rock salt used in *şalgam* beverage production supports the growth of LAB during fermentation, inhibits pathogen and spoilage organisms and also has a positive effect on taste (Erten and Tanguler 2012). The high salt content used in its production causes the salt content of *şalgam* beverage to be high, so it is not a recommended product for hypertensive patients (Dikmen and Filazi 2016). Furthermore, the World Health Organization (WHO) reports that high sodium consumption is associated with increased blood pressure and cardiovascular disease (WHO 2003). It also causes water to be drawn from the cells, causing thirst. Thirst can cause an abnormal concentration of sodium in the plasma (hypernatremia), and hypernatremia can cause bloating, high blood pressure, rapid heartbeat, difficulty breathing, coma and death. For these reasons, health concerns about excessive salt consumption have started to emerge in consumers, and as a result, consumer demand for low-sodium food and beverages has increased recently. In this context, it has become important to reduce the amount of salt in *şalgam* beverage. In a study conducted for this purpose, it was stated that NaCl could be replaced by 50% with KCl (Agirman and Erten 2018).

Biogenic amines are compounds that can be found naturally in foods or occur due to microbial activity and can cause toxic effects when consumed in high quantities. Aromatic biogenic amines (vasoactive or psychoactive amines) can cause histamine-related food poisoning, foodborne migraines, severe hypotension or hypertension, dizziness, nausea, heart palpitations, skin rash and vomiting. They can even cause intracerebral haemorrhage, anaphylactic shock syndrome and death in very severe medical cases. The microbial and other circumstances of fermentation have not been standardized for safety to such health challenges. Therefore, it is difficult to prevent the formation of biogenic amines in fermented products (Cosansu 2009; Spano et al. 2010; Tamer et al. 2019b). Biogenic amines such as putrescine, tyramine and histamine can be produced by LAB during *şalgam* beverage fermentation. Putrescine is the most common biogenic amine in *şalgam* beverage, and its amount varies between 5.0 and 42.3 mg/L. The total amount of biogenic amines is between 26.7 and 134.3 mg/L; these values are not at high enough levels to cause adverse health effects on consumers (Ozdestan and Uren 2010). On the other hand, Yaldirak (2011) stated that the total quantity of biogenic amine that was formed in *şalgam* beverage produced at 35°C was lower than in *şalgam* beverage produced at 25°C.

20.4 HARDALIYE

As stated before, fermented herbal products can be used by many different groups of people because they do not contain any allergens (Rakin et al. 2012). One of these fermented products is *hardaliye*. It is a refreshing beverage. *Hardaliye* is an unsalted, low-fat, bitter-tasting and refreshing beverage (Aktop et al. 2023), and it is a non-dairy and low-fat product with an acidic taste that is highly palatable (Amoutzopoulos et al. 2013; Baschali et al. 2017). It is a grape-based traditional fermented non-alcoholic or low-alcoholic beverage specific to the Thrace region of Turkey, although the tradition is disappearing over time (Arici and Coskun 2001; Sanlier et al. 2019; Aksoy et al. 2022).

In the production of *hardaliye*, the main raw material is crushed red grape (*Vitis vinifera* L.) or pomace, which is produced by lactic acid fermentation (Kelebek 2009; Erten et al. 2017). In addition to dark red grapes, different grape varieties are also used in the production of *hardaliye* (Coskun 2017). Grape variety plays an important role in colour, taste and chemical characteristics in the products where the grape is used as raw material. The grape varieties used in production also affect the colour and aroma of the *hardaliye* beverage (Askin and Atik 2016; Coskun 2017). Red grape and their products are rich in antioxidants with high bioavailability, including resveratrol, phenolics (flavonols, tannins and anthocyanins among others), pigments and ascorbic acid (Kelebek 2009; Amoutzopoulos et al. 2013; Aktop et al. 2023). Therefore, there is great interest in these products with potential health benefits (Askin and Atik 2016). Grape juice obtained from dark red grapes contains very high antioxidants, and the amount of antioxidants is higher than orange and tomato juices. The phenolic compounds and sugar content in red grapes can range from 1.61 to 10.85 g/kg and 180 to 260 g/kg, respectively, depending on the cultivar and degree of ripeness. Grapes contain generally quercetin (12–15 mg/kg), kaempferol (<2 mg/kg), myricetin (4.5 mg/kg), catechin (19 mg/kg) and coumaric acid (1–3 mg/kg). The main sugars found in grapes are glucose and fructose and to a lesser extent sucrose (Kelebek 2009; Coskun 2017; Demiray 2019). Resveratrol and flavonoids found in grape skin, seeds and hence grape juice are reported to reduce the risk of arteriosclerosis by inhibiting LDL cholesterol oxidation and have a regulating effect on blood pressure and blood sugar (Coskun 2017).

Besides red grapes, ground and/or heated mustard seeds are other raw materials used (Erten et al. 2017; Coskun 2017; Swain and Ananadharaj 2018). Also for these raw materials, fewer preservatives (such as benzoic acid) and occasionally cherry tree leaves (*Prunus cerasus* L.) are utilized in the production of *hardaliye* (Arici and Coskun 2001; Baschali et al. 2017; Erten et al. 2017). Although white mustard seeds (*Brassica alba* L. Boiss) can be used in production, the use of black mustard seeds (*Brassica nigra* L. Koch) is more suitable due to their higher inhibitory effect. Mustard seeds and, if added during production, cherry leaves give flavour and taste to *hardaliye*. Black mustard seeds contain a significant amount of oil (35–40%) and protein (20%). Mustard oil contains isothiocyanates. Isothiocyanates occur as a result of the hydrolysis of thioglucosides in the presence of water and enzymes. Mustard seeds also contain thioglucosides such as sinigrin and enzymes (myrosinase). Sinigrin breaks down into allyl isothiocyanate, resulting in the characteristic sharpness of *hardaliye* (Coskun and Arici 2006; Erten et al. 2017).

The use of raw black mustard seeds (inhibitory effect of allyl isothiocyanate) and preservatives (sulfur, salts of sorbates and benzoates) inhibits or reduces alcohol formation by affecting the yeasts that come from the grapes during fermentation (Aydogdu et al. 2014; Gurbuz 2018; Aksoy et al. 2022). However, undesirable ethyl alcohol formation may occur during fermentation and storage. The usage of mustard seeds up to 3% (w/v) is not adequate to inhibit the activity of yeasts during fermentation and subsequent storage of the product. However, it has been determined that using more than 2% mustard seeds negatively affects the sensorial characteristics of *hardaliye*. Therefore, in addition to mustard seeds, preservatives (at the industrial scale) prevent the formation of alcohol in the final product by acting on microorganisms (especially yeasts) (Prado et al. 2008; Coskun and Arici 2011; Erten et al. 2017; Gurbuz 2018). It is stated that the number of preservatives (sorbate salts) to be used in beverages except dairy products should be less than 300 mg/L according to the Turkish Food Codex Flavored Beverages Food Additives Regulation. However, some manufacturers can use up to 1 g/L of preservatives in the production of *hardaliye* (Aydogdu et al. 2014). For these reasons, it can be said that the beneficial effects of *hardaliye* are not only related to grapes but also to mustard, which gives the mustard-enriched drink an antibacterial and antifungal effect (Coskun et al. 2018).

Hardaliye is commonly made in homes by the conventional method (Baschali et al. 2017) in September, October and November when the cooler season begins (Coskun and Arici 2011; Demiray 2019). Industrial production has been carried out recently, albeit on a small scale (Coskun 2017). In conventional *hardaliye* production, harvested red grapes are washed, crushed/pressed and placed along with the pulp into stainless steel boilers or oak (or plastic or glass) barrels (Arici and Coskun 2001; Demiray 2019). Approximately 0.2% ground raw black mustard seeds, 0.1% benzoic acid and/or cherry leaves are also added to the barrel (Arici and Coskun 2001; Tamer et al. 2019b). One layer of crushed grapes and one layer of crushed black mustard seeds and preservatives are filled in layers (Coskun and Arici 2011). On the other hand, various additives such as cloves and ginger are also used to add aroma to *hardaliye* in some production processes (Aydogdu et al. 2014). Later on, the mixture is left to allow spontaneous fermentation (Arici and Coskun 2001; Tamer et al. 2019b). Recently, the production of *hardaliye* has been carried out by adding starter culture (*Fructilactobacillus sanfranciscensis, Basonym: L. sanfrancisco, L. acetolerans, Limosilactobacillus pontis* (Basonym *L. pontis*), *L. paracasei* ssp. *pasteurized*) to grape juice (Coskun and Arici 2006; Zheng et al. 2020). During lactic acid fermentation, some space should be left on the tank due to the swelling (CO_2) caused by yeasts and heterofermentative LAB that may be present spontaneously in the environment from grape skins. Also, the tank cover should be kept half-closed during fermentation, and the mixture should be mixed and homogenized every 2 days by taking the liquid from the tap under the barrel and pouring it over the barrel again (Coskun and Arici 2011; Coskun 2017; Demiray 2019). Fermentation is carried out at room temperature (5–10 days) (Tamer et al. 2019b). After fermentation from the obtaining liquid, mustard seeds, vine leaves and grape residues are removed, filtered and then cooled to 4°C (Amoutzopoulos et al. 2013; Tamer et al. 2019b). *Hardaliye*, following fermentation, is cooled and packaged, for which transparent bottles are usually used (Askin 2019; Askin and Atik 2016). At that stage, *hardaliye* can be consumed fresh as well as aged (Prado et al. 2008; Tamer et al. 2019b; Demiray 2019). However, if the bitterness caused by the mustard seeds used in production is at a level that prevents the consumption of the product, it is more appropriate to consume *hardaliye* by adding water (Coskun 2017).

The pH values of *hardaliye* vary between 3.12 and 4.33, and under these acidic conditions, some chemical and physical compositions of *hardaliye* are given in Table 20.3. It is a product that is stable (Askin 2019), and its shelf life is usually 1–2 months. Although the shelf life can be extended further by preserving it in the refrigerator at low temperatures, various studies have been undertaken to extend the shelf life (Coskun 2017; Tamer et al. 2019b). In a study conducted by Demiray (2019), it is reported that *hardaliye* can be pasteurized with technologies such as high hydrostatic pressure and ultrasonication and the use of preservatives in industrial production can be prevented. On the other hand, pasteurization can be considered for longer shelf life, as well as research on the use of starter cultures. If the storage period is prolonged without applying the necessary preservation methods, alcohol may be formed and/or its content may increase in *hardaliye* (Coskun 2017). In a study conducted by Gurbuz (2018), the alcohol content of *hardaliye* samples obtained using mustard seeds and/or starters was found to be between 11% and 13% (v/v) after 2 months of storage in room temperature conditions. However, it has been stated to be between 5% and 13% (v/h) at the end of storage at refrigerator temperature. However, the amount of alcohol was found within legal limits in samples produced using chemical preservatives and stored at refrigerator temperature for the same period. On the other hand, with the storage of *hardaliye*, the brown colour values increase, while the anthocyanin and total phenolic content decrease (Askin 2019; Askin and Atik 2016).

Grapes can naturally be contaminated with different yeast, mould and bacterial strains before, during and after harvest. However, grape microflora may differ depending on ripeness and ambient circumstances. Therefore, this different natural microflora found in grapes causes different microorganisms to be dominant in the production of *hardaliye* and therefore it does not allow the production of a standardized product. However, these diverse strains play a significant role in the manufacture of this product (Coskun 2017). The microorganisms that are effective in the production of *hardaliye* are mainly LAB and unknown fungi. Some LAB species (*Lcb. paracasei* subsp. *paracasei, Lcb. casei* subsp. *pseudoplantarum, Levilactobacillus brevis, Limosilactobacillus pontis, Paucilactobacillus vaccinostercus*, Basonym: *L. vaccinostercus, L. acetotolerans* and *L. sanfransiscensis*) in *hardaliye* obtained by spontaneous fermentation have been identified (Arici and Coskun 2001; Baschali et al. 2017; Zheng et al. 2020; Aktop et al. 2023). Arici et al. (2017) have also isolated species such as *Lpb. plantarum, Lpb. pentosus* and *Secundilactobacillus collinoides* (Basonym *L. collinoides*) (Zheng et al. 2020) in mustard besides these species. The LAB, the total mesophilic aerobic bacteria and yeasts and mould counts ranged from 1.0×10^2 to 8×10^5 cfu/mL, 3.0×10^1 to 3.2×10^6 cfu/mL and 1.0×10^1 to 8.1×10^4 cfu/mL, respectively. *Hardaliye* does not contain bacteria that threaten human health such as coliforms and *E. coli* (Arici and Coskun 2001; Coskun et al. 2018). Low pH is important for allyl isothiocyanates to modulate its antibacterial effect (Coskun 2017). Allyl isothiocyanates can completely inhibit the growth of certain types of bacteria (for example the foodborne pathogen, *E. coli* O157:H7) and moulds (*Aspergillus (Asp.) flavus, Penicillium (Pe.) commune, Pe. corylophilum, Pe. discolor, Pe. polonicum* ve *Pe. roqueforti* among others) (Nielsen and Rios 2000; Luciano and Holley 2009; Coskun 2017). Similarly, in research carried out by Aydogdu et al. (2014), it was stated that coliforms, *E. coli*, yeast and moulds decreased during the fermentation in *hardaliye* production and that they could not be isolated from this environment at the end of fermentation.

20.4.1 Benefits of *Hardaliye* on Health

The characteristic aroma of mustard comes from the skins of grape and mustard seeds, as well as some aromas also arise during the fermentation of grape must and other mixtures (Coskun 2017). *Hardaliye*, besides being a nonfat, very tasty product, contains important nutrients (vitamins and minerals) necessary for the functioning of the body (Bakircioglu et al. 2017; Demiray 2019). It is not easy to show that *hardaliye* has health-promoting effects. However, it has important health effects, and its consumption needs to be expanded (Baschali et al. 2017). *Hardaliye* is also considered a non-dairy probiotic beverage due to the probiotic LAB it contains (Altay et al. 2013; Kilic et al. 2016; Aladeboyeje and Sanli 2021). These make *hardaliye* a more valuable functional product (Coskun 2017). *Hardaliye* can be added to daily nutrition programs as a healthy choice to meet the energy and carbohydrate needs of individuals with high energy needs. However, the fact that no sugar is included in the composition of the beverage and the carbohydrate compounds it contains naturally originate from the raw material such as grapes increases the health value of the product (Faikoglu 2014).

Malvidin-3-O-glucoside is the predominant anthocyanin found in the product. Moreover, there are also high quantities of resveratrol in its composition (Askin and Atik 2016). *Hardaliye*, which provides a high antioxidant effect because it is rich in phenolic compounds, quercetin, gallic acid and

TABLE 20.3 Chemical and Physical Composition of *Hardaliye*

	AMOUNT	*REFERENCES*
Energy (kcal/100 mL)	75.54–316	Amoutzopoulos et al. (2013), Coskun (2017)
Dry matter (%)	5.2–11.49	Faikoglu (2014), Coskun (2017)
Ash (%)	0.23	Coskun (2017)
Fat (g/100 mL)	0.2	Coskun (2017)
Total dietary fiber (g/100 mL)	0.98	Coskun (2017)
Total acidity (g/L)*	3.8– 11.93	Aydogdu et al. (2014), Coskun et al. (2018)
pH	3.12–4.33	Faikoglu (2014), Vardar et al. (2015), Coskun et al. (2018)
Volatile acid (g/L)**	0.24–0.72	Coskun et al. (2018)
Brix	20.5–21.4	Vardar et al. (2015), Askin and Atik (2016)
Carbohydrates (g/100 mL)	17.53	Coskun (2017)
Fructose (g/100 mL)	9.51	Coskun (2017)
Glucose (g/100 mL)	9.43	Coskun (2017)
Total sugar (g/100 mL)	7.2–21.4	Coskun (2017), Coskun et al. (2018)
Alcohol (%)	nd–6.0	Coskun et al. (2018)
Calcium (mg/100 mL)	3.3–13.45	Bakircioglu et al. (2017), Coskun (2017)
Cu (mg/L)	0.10–0.27	Bakircioglu et al. (2017)
Mn (mg/L)	1.0–3.2	Bakircioglu et al. (2017)
Potassium (mg/100 mL)	94.14	Coskun (2017)
Magnesium (mg/100 mL)	0.39–13.59	Bakircioglu et al. (2017), Coskun (2017)
Iron (mg/100 mL)	0.91	Coskun (2017)
Phosphorus (mg/100 mL)	24.5	Coskun (2017)
Zinc (mg/L)	0.39–2.4	Bakircioglu et al. (2017), Coskun (2017)
Sodium (mg/100 mL)	3.2–5.8	Bakircioglu et al. (2017)
Vitamin B1 (mg/100 mL)	0.003	Coskun (2017)
Vitamin B2 (mg/100 mL)	0.01	Coskun (2017)
Vitamin B6 (mg/100 mL)	0.05	Coskun (2017)
Niacin (mg/100 mL)	0.31	Coskun (2017)
Vitamin C (mg/L)	2.35	Amoutzopoulos et al. (2013), Coskun (2017)
Folate (µg)	1.01	Coskun (2017)
Total phenolic substance (mg GA/L)	368.8-1743–3529	Amoutzopoulos et al. (2013), Askin and Atik (2016), Coskun (2017), Coskun et al. (2018)
Total antioxidant ABTS (µmol Trolox/mL) DPPH (µmol Trolox/mL) CUPRAC (µmol Trolox/mL)	53.5–70.4 19.9–32.8 9.1–40.1	Faikoglu (2014)
ORAC (mmol TE/L)	64	Coskun (2017)
Total anthocyanins (mg M3GE/L)	4.5–210.8	Coskun et al. (2018)
trans-Resveratrol (mg/L)	2.72	Coskun (2017)
Quercetin (mg/L)	65.5	Coskun (2017)

Notes: *: as lactic acid, **: as acetic acid, GA: Gallic acid, M3GE; malvidin-3-glucoside equivalent, nd: not determined.

trans-resveratrol (Bakircioglu et al. 2017; Demiray 2019), potentially has relevance to preventing the formation and development of cancer cells by preventing oxidative stress. Also, it has a decreasing effect on plasma lipid peroxidation parameters and serum homocysteine concentrations. The high nutritional, functional features and health benefits of *hardaliye* are due to the raw materials used in its production and also from the fermentation process. The essential oils found in mustard seeds especially potentially have positive roles in human health. Etheric oils, an important ingredient in mustard seeds used in production, are another reason *hardaliye* has a potentially positive effect on health. Sinigrin from thioglucosides, another component of mustard seeds, is a suppressing agent on carcinogenesis. Moreover, allyl isothiocyanate found in mustard seeds has many anti-cancer properties (Coskun 2017; Aladeboyeje and Sanli 2021). On the other hand, *hardaliye* has also a positive effect on coronary heart diseases (Coskun 2017; Aladeboyeje and Sanli 2021).

Some oxidative stress indicators such as diene conjugate and malondialdehyde also may decrease with *hardaliye* consumption. It may also reduce the serum homocysteine level, which poses a significant health risk when its level in the blood rises above normal (Faikoglu 2014). Similarly, Amoutzopoulos et al. (2013) reported that *hardaliye* has antioxidant properties and lowers plasma malondialdehyde, diene conjugate and homocysteine levels. In a study conducted, it was observed that the total antioxidant capacity of the groups consuming *hardaliye* was higher than the control group. Therefore, this result shows that *hardaliye* has an important potential to increase total antioxidant activity (Faikoglu 2014). For all these reasons, *hardaliye* is an important functional beverage potentially appropriate for consumption by vegetarians, children, hypertensive people, those with milk intolerance and people with high cholesterol levels (Aladeboyeje and Sanli 2021; Aktop et al. 2023).

20.5 *BOZA*

Cereal grains are a very important source of human nutrition, as they contain proteins, carbohydrates, vitamins, minerals and fibers. However, the nutritional and sensorial properties of its products are not of very high quality. One of the processes used to enhance the nutritional, sensorial properties and functional properties of grains is fermentation. Therefore, fermented grain products have gained great importance recently (Yegin and Fernández-Lahore 2012). Around one-third of people worldwide use cereal-based fermented foods and beverages in their diets. While beer is the most popular among fermented cereal-based beverages, another important cereal-based fermented beverage consumed in Turkey is *boza* (Tanguler 2014). It is a type of millet beer (Baschali et al. 2017).

Boza, a traditional beverage unique to the Turks dating back to ancient times, is a low-alcohol drink with a thick consistency and slightly sweet or sour flavour (depending on its acidity) (Akpinar-Bayizit et al. 2010; Cabaroglu and Koyuncu 2019; Tamer et al. 2019b; Erten and Tanguler 2019). Although typical *boza* is known for its sour taste, *boza*'s flavour may vary depending on the raw materials utilized in the production, the amounts, preparation method, storage conditions and native properties of yeasts and LAB that are effective in the production process (Akpinar-Bayizit et al. 2010; Yegin and Fernández-Lahore 2012).

It is thought that *boza* came from indigenous people before the Ottoman Empire in Turkey (von Mollendorff et al. 2016). Xenophon, one of the famous Greek historians and philosophers, stated that *boza* was produced in Eastern Anatolia around 400 BC. In time, it spread over a very wide geography with the conquests and migrations of the Ottomans (Yegin and Fernández-Lahore 2012; von Mollendorff et al. 2016; Tamer et al. 2019b). *Boza*, whose colour can range from light beige to dark beige, is a beverage in a colloid suspension (Blandino et al. 2003; Baschali et al. 2017). It is called by different names in different countries such as Busaa (in Kenya) (Yegin and Fernández-Lahore 2012), bousa (in Nigeria) (Akpinar-Bayizit et al. 2010), boza (in Bulgaria, Serbia, Croatia, Hungary and Albania), bozan (in Romania), bozas (in Greece), buza (in Russia and Poland), bouza (in France and Egypt) and busa (in Germany). However, the origin of the word *boza* is "buze", which means millet in Persian (Baschali et al. 2017; Tezel and Sanlibaba 2019). On the other hand, although there are similarities between *boza* in these countries, the raw materials and quantities used as well as the alcohol content in the final product are among the main differences (Yegin and Fernández-Lahore 2012). The alcohol concentration in *boza* produced in Eastern European countries and the Balkans can be up to 7% by volume (Akpinar-Bayizit et al. 2010; Petrova and Petrov 2017). Various cereal flours such as millet, barley, wheat, rye, oat, bulgur, corn and rice are utilized as raw materials in the manufacture of *boza* (Blandino et al. 2003; Tanguler 2014; Sanlier et al. 2019). Also, because whole grains have significant bioactive components and nutritional properties, if they are constantly taken into the body, they have a significant effect on human health (Sanlier et al. 2019; Ignat et al. 2020).

Different raw materials are produced in different amounts with different recipes and methods in the production of *boza* in different regions (Erten and Tanguler 2019; Tezel and Sanlibaba 2019). For example, there is a standard (TS 9778 Turkish Standard) for *boza* making in Turkey, and this standard is defined as the product obtained through the following steps.

> (I) Adding drinking water to one or more of the crushed or flours of millet, wheat, corn and other grains,
> (II) cooking,
> (III) mixing by adding white sugar,
> (IV) Realization of alcohol fermentation and lactic acid fermentation of the mixture obtained according to the technique.
>
> (TSE 1992)

However, generally, *boza* production is carried out as follows. Millet is the basic grain in the manufacture of *boza* (Tanguler 2014). In addition to millet, wheat, bulgur, corn or rice is purified from foreign materials such as stone and soil and is ground (300–800 μm) in crushing mills. Then it is sieved and separated from the crust, semolina and flour. The ground flour and/or semolina is mixed with drinkable water at a rate of 4–6 times its weight and boiled in large copper pots for 2–8 hours depending on the boiling heat and the raw materials. The mixture is constantly mixed during the boiling process to avoid aggregation and to get a homogeneous mash (Yegin and Fernández-Lahore 2012; Erten and Tanguler 2019; Tezel and Sanlibaba 2019). Naturally, as the mixture absorbs water in the boiling process, hot water is added several times until a homogeneous paste is obtained (Tanguler 2014). During this process, a series of physicochemical changes take place in the raw material. Meanwhile in the mixture, starch hydrolyzes to dextrin, glucose concentration increases, protein denatures, and lipids break down to glycerin and butyric acid (Tezel and Sanlibaba 2019). The cooked mash is then cooled to around 20°C within 12 to 24 hours and filtered through a rice sieve to remove the bran, hull and other foreign materials (Yegin and Fernández-Lahore 2012; Erten and Tanguler 2019). The filtered mash is called "sugarless raw *boza*" (Arici and Daglioglu 2002).

The sugarless raw *boza* is allowed to ferment by adding 15–25% sugar (sucrose) to aid fermentation (Yegin and Fernández-Lahore 2012; Erten and Tanguler 2019). The malt used in *boza* production outside of Turkey, the Turkish type *boza* is not used. Healthy *boza* obtained from the previous fermentation containing LAB and yeasts (or sourdough or yoghurt) is added to the raw *boza* taken into barrels as a starter culture for fermentation (Arici and Daglioglu 2002; Arslan et al. 2015; Erten and Tanguler 2019). The inoculation ratio varies depending on the production season and ambient temperature. When sourdough is used as a starter culture, a less viscous and more acidic product is obtained when compared to using the previous *boza*. On the other hand, when yoghurt is used in production, *boza* with high viscosity but more acidic and characteristic yoghurt flavour is obtained. Fermentation is carried out at 15–30°C for 24 hours (Arici and Daglioglu 2002; Erten and Tanguler 2019; Tezel and Sanlibaba 2019; Tamer et al. 2019b). Two different fermentation takes place in production at the same time: ethyl alcohol and lactic acid fermentation. While ethyl alcohol fermentation affects the emergence of CO_2 bubbles and increases in volume, odour and taste properties, lactic acid fermentation is mainly effective in the formation of lactic acid and some other metabolites (Tanguler 2014; Altay et al. 2013).

Fermentation is not completely terminated in *boza* production, usually, it continues for 1 day or longer (Erten and Tanguler 2019). This fermentation time is sufficient to achieve the characteristic sensory qualities of *boza*, but in practice, the fermentation time may be extended (Ignat et al. 2020). At the end of 24 hours, the partially fermented *boza* is cooled below 15°C (generally to +4°C), packaged in 1-kg plastic or glass containers and stored under refrigerated conditions to keep the fermentation under control (Yegin and Fernández-Lahore 2012; Erten and Tanguler 2019). *Boza*, whose shelf life is not long, should be consumed within 3–5 days (Arici and Daglioglu 2002; Tanguler 2014) for the best sensory properties (Yegin and Fernández-Lahore 2012). In a study conducted by Kentel (2001), to extend the shelf life of *boza*, pasteurization (10 minutes at 70°C and 80°C), chemical preservative (Na-benzoate and/or K-sorbate) addition and cooling process (4°C and 20°C) was applied. It has been reported that pasteurized samples can be stored for more than a month without losing their characteristic taste and smell. On the other hand, it has been stated that the shelf life of partially fermented *boza* stored at 4°C and added chemical preservatives can be extended up to 12 days.

Boza, which is a highly nutritious beverage, contains carbohydrates, proteins, vitamins, fiber, thiamine, riboflavin, pyridoxine, niacin and lactic acid (Kabak and Dobson 2011; Altay et al. 2013; Baschali et al. 2017). It can be consumed daily by people of all ages, especially during cold days and nights of winter and autumn, plain or by being aromatized with cinnamon and roasted chickpeas (leblebi) (Blandino et al. 2003; Akpinar-Bayizit et al. 2010; Baschali et al. 2017; Sanlier et al. 2019; Erten and Tanguler 2019). However, it is mostly served plain or with the addition of cocoa and is consumed every season in Bulgaria (Petrova and Petrov 2017; Tezel and Sanlibaba 2019). *Boza* was traditionally made in the family and at home according to the recipes of the family elders. Recently, however, it has been produced and sold in some patisseries and even small-scale factories (Tamer et al. 2019b).

The microbial flora that is effective in the fermentations occurring in *boza* production is quite heterogeneous. While homo- and heterofermentative LAB are effective in lactic acid fermentation, various yeast species play an important role in ethyl alcohol fermentation. The difference in strain and the number of microorganisms is due to the type and/or amount of raw materials utilized in manufacture and the fermentation conditions (Gotcheva et al. 2001; Akpinar-Bayizit et al. 2010; Petrova and Petrov 2017; Erten and Tanguler 2019).

Since the interactions between LAB and yeasts are not checked during spontaneous fermentation in traditional *boza* production, changes in the property and stability of *boza* may arise. However, LAB is always dominant in production (Yegin and Fernández-Lahore 2012). In the study comparing *boza* samples collected from the local market and *boza* prepared under laboratory conditions, it was reported that the heterofermentative LAB was dominant over homofermentative bacteria (Kozat 2000). During fermentation, many species belonging to Lactobacillus, Lactococcus, Leuconostoc, Pediococcus, Enterococcus (*En.*), Oenococcus and Weissella genera were isolated. These species were *Lpb. plantarum, Fructilactobacillus sanfranciscensis,* Basonym: *L. sanfrancisco, Lcb. paracasei* subsp. *paracasei, Limosilactobacillus fermentum, Levilactobacillus brevis, L. confusus, Lcb. rhamnosus* (Basonym: *L. rhamnosus*), *L. acidophilus, Lpb. pentosus, L. coprophilus, Loigolactobacillus coryniformis* (Basonym *L. coryniformis*), *L. sakei, L. maltaromicus, Llb. parabuchneri* (Basonym *L. parabuchneri*), *L. amelibiosum, L. pseudomesenteroides, Ln. mesenteroides* subsp. *mesenteroides, Ln. mesenteroides* subsp. *dextranicum, Ln. raffinolactis, Ln. lactis, Ln. paramesenteroides, Oenococcus oeni* (syn. *Ln. oenos*), *W. confusa, W. paramesenteroides, Lc. lactis* subsp. *lactis, En. faecium* and *P. pentosaceus* (Hancioglu and Karapinar 1997; Zorba et al. 2003; Tamer and Copur 2004; Botes et al. 2007; Moncheva et al. 2003; Zheng et al. 2020). In a study carried out by Kozat (2000), it was stated that the most common LAB strain among 56 isolates was *Levilactobacillus brevis* subsp. *lindneri*.

On the other hand, yeast species such as *C. pelliculosa, C. colliculosa, C. lambica, C. krusei* (Kozat 2000), *C. tropicalis, C. pararugosa, C. diversa, C. boidinii, C. lactiscondes, C. norvegica, C. inconspicua, S. cerevisiae, S. carlsbergensis, S. uvarum, Pichia (Pi.) fermentans, Pi. norvegensis, Pi. membranifaciens, Dekkera bruxellensis, Issatchenkia orientalis, Rhodotorula (R.) mucilaginosa, R. araucariae, Torulaspora delbrueckii* (Hancioglu and Karapinar 1997; Zorba et al. 2003; Tamer and Copur 2004; Botes et al. 2007; Moncheva et al.

2003) and *Cryptococcus laurentii* have also been isolated from *boza* during in fermentation (Kozat 2000). On the other hand, the mould species isolated from *boza* are *Asp. fumigatus, Pe. chrysogenum, Fusarium oxysporum, Geotrichum candidum* and *Geotrichum capitatum*. Moreover, mycotoxin was also found in some *boza* samples (Uysal et al. 2009).

The physicochemical features and nutritive value of the *boza* may vary depending on the raw materials utilized in its manufacture, their amount, size and production process (eg spontaneous fermentation temperature and time) (Hammes et al. 2005; Baschali et al. 2017). For example, when wheat is used instead of millet, the amount of lactic acid can increase approximately twice (Akpinar-Bayizit et al. 2010). Along with the *boza* standard (TS 9778) reported by the Turkish Standards Institute (TSE 1992), some chemical and physical compositions of *boza*, in general, are shown in Table 20.4.

TABLE 20.4 Chemical and Physical Composition of *Boza*

	AMOUNT	*REFERENCES*	*TSE 1992*
Dry matter (%)	5.57–29.82	Altay et al. (2013)	>20
Ash (%)	0.02–0.17	Tamer et al. (2019b)	<0.2
Protein (%)	0.27–2.75	Altay et al. (2013)	–
Total acidity (%)*	0.13–0.61	Tamer et al. (2019b), Akpinar-Bayizit et al. (2010)	0.2–0.5[a] 0.5–1.0[b]
pH	3.16–4.63	Tamer et al. (2019b)	–
Lactic acid (mg/kg)	300	Akpinar-Bayizit et al. (2010)	–
Oxalic acid (mg/kg)	142	Akpinar-Bayizit et al. (2010)	–
Pyruvic acid (mg/kg)	72	Akpinar-Bayizit et al. (2010)	–
Acetic acid (mg/kg)	38	Akpinar-Bayizit et al. (2010)	–
Volatile acid (%)**	0.006–0.13	Arici and Daglioglu (2002), Yegin and Fernández-Lahore (2012)	0.1[a]–0.2[b]
Carbohydrate (%)	12.3–57.5	Yegin and Uren (2008)	>10
Total sugar (%)	10.64–2.59	Tamer et al. (2019b)	>10
Sucrose (%)	0.05	Celik et al. (2016)	–
Alcohol (% v/v)	0%–7.0%	Baschali et al. (2017), Hancioglu and Karapinar (1997), Petrova and Petrov (2017)	< 2
Total phenolic content (mg/100 g)[γ]	124–203	Berktas (2011)	–
Total flavonoid (%)[α]	71.8–106.6	Berktas (2011)	–
Antioxidant activity (μmol TE/100 g)	900–1097.3	Berktas (2011)	–
Fat (g/100mL)	0.5	Guven et al. (2004), Tezel and Sanlibaba (2019)	–
Cholesterol (mg/100 mL)	1	Guven et al. (2004)	–
Calcium (mg/100 mL)	26.8–29	Guven et al. (2004), Celik et al. (2016)	–
Potassium (mg/100 mL)	52.5	Celik et al. (2016)	–
Iron (mg/100 mL)	1.3	Guven et al. (2004)	–
Phosphorus (mg/100 mL)	9.92–97	Guven et al. (2004), Celik et al. (2016)	–
Zinc (mg/100 mL)	0.31–1	Guven et al. (2004), Celik et al. (2016)	–
Sodium (mg/100 mL)	1	Guven et al. (2004)	–
Vitamin A (I.U/100 mL)	6.9	Guven et al. (2004)	–
Thiamine (mg/100 mL)	0.09–0.25	Guven et al. (2004), Yegin and Fernández-Lahore (2012)	–
Riboflavin (mg/100 mL)	0.05–0.21	Guven et al. (2004), Yegin and Fernández-Lahore (2012)	–
Niacin (mg/100 mL)	1.16	Guven et al. (2004)	–
Nicotinamide (mg/100 g)	0.51–0.60	Yegin and Fernández-Lahore (2012)	–
Pyridoxine (mg/100 g)	0.32–0.36	Yegin and Fernández-Lahore (2012)	–

Note: [a]: Sweet Boza, [b]: Sour boza, *: as lactic acid, **: as acetic acid, [γ]: as ferulic acid equivalent, [α]:as catechin equivalent.

On the other hand, volatile esters (such as ethyl acetate and isoamyl acetate), higher alcohols (such as isoamyl alcohol, 1-propanol and isobutyl alcohol) and acetaldehyde are important dominant flavouring substances found in *boza* (Cabaroglu and Koyuncu 2019). As can be seen from the table, the general composition properties of *boza* are within wide limits. The most important reason for this is the production method that can vary from region to region, as stated previously. For example, while alcohol content produced in Turkey ranged from 0.03% to 0.39%, it may rise to 1.5% (w/v) in other parts of the world (Yegin and Uren 2008; Tamer et al. 2019b).

20.5.1 Benefits of *Boza* on Health

Boza is potentially a complete herbal drink. It is an important functional drink, as it is produced from natural raw materials; contains vitamins, minerals, carbohydrates, proteins, fiber, bioactive compounds, prebiotics and probiotic microorganisms; and has protective and therapeutic effects (Arslan et al. 2015; Sanlier et al. 2019; Tezel and Sanlibaba 2019). It has positive effects on the digestive and intestinal flora, especially due to the lactic acid it contains (Arici and Daglioglu 2002). Fermented cereal products such as *boza* have low fat and essential amino acid content compared to fermented dairy products. Furthermore, these cereal products are rich in plant components such as fibers, flavonoids and phenolics, which have a positive effect on oxidative stress, inflammation and hyperglycemia (Tanguler 2014; Tezel and Sanlibaba 2019). The phenolic compounds that are contained in *boza*'s raw materials have antioxidant potential and potentially scavenge harmful free radicals in the body. *Boza* is also rich in mineral (Ca, Fe, P, Zn, Na) and vitamin (vitamin A, vitamins B_1, B_2, B_6) content (Ignat et al. 2020), which change in amount during fermentation. It is a good alternative, especially for vegetarians and vegan diet-preferring individuals (Petrova and Petrov 2017; Tezel and Sanlibaba 2019).

The raw materials used in the production of *boza* contain substances that have positive effects on health. One of them is omega-3, which is found in millets and whose fatty acid is known to reduce the risks of major cardiovascular diseases such as myocardial infarction; sudden death due to cardiac attacks and coronary heart disease; and potentially chemoprevention of various cancers such as breast, prostate, colon and kidney. Millet also contains omega-6 fatty acids. Also, the positive effects of *boza* on health have been proven in some studies, and these effects can be stated as follows. It is suggested to contribute to balancing blood pressure, improving colon health and gastrointestinal health, reducing plasma cholesterol, increasing milk production and quantity in lactating women and facilitating digestion (Petrova and Petrov 2017; Tezel and Sanlibaba 2019; Ignat et al. 2020).

Grains, the major raw material of *boza*, contain both water-soluble and insoluble fibers (Petrova and Petrov 2017; Ignat et al. 2020). Fermentable fibers are non-digestible food constituents that can be consumed by microorganisms found in limited counts in the intestine and are called prebiotics, which have a health-improving effect through improving probiotic support functions (Aladeboyeje and Sanli 2021).

These fibers contained in *boza* are compounds that are not hydrolyzed by human digestive enzymes but have positive effects on host health by selectively stimulating the growth and/or activity of some bacteria in the colon (Petrova and Petrov 2017; Tezel and Sanlibaba 2019). Therefore, another positive effect of *boza* on human health is its symbiosis because it contains prebiotics and probiotic microorganisms together (Arslan et al. 2015; Baschali et al. 2017; Tezel and Sanlibaba 2019). Probiotic microorganisms found in *boza* are *Lpb. plantarum, Lcb. paracasei, Lcb. rhamnosus* and *Lpb. pentosus*. Some of these probiotic bacteria exhibit auto-aggregation properties, antibacterial and antiviral activity (Todorov et al. 2008; Baschali et al. 2017). These microorganisms with probiotic properties are not microorganisms used as a starter in *boza* production. Since spontaneous or uncontrolled fermentation occurs in *boza* production, it is not certain that the same microorganisms and especially probiotics are always present (Tezel and Sanlibaba 2019). Arslan et al. (2015) stated that in their study with certified probiotic starter cultures (*L. acidophilus* LA-5 and *Bifidobacterium bifidum* BB-12 and *S. boulardii*), *boza* samples had more than 10^6 cfu/ml viable microorganisms and therefore could be considered a probiotic product.

Lactic acid fermentation takes place beside ethyl alcohol fermentation in *boza* production, which causes a decrease in the pH value of the product. Microorganisms that are effective in *boza* production form distinct antimicrobials such as organic acids (especially lactic acid), H_2O_2, fatty acid and bacteriocins during fermentation (Arici and Daglioglu 2002; Petrova and Petrov 2017; Charyyev et al. 2019; Tamer et al. 2019b). In a study carried out by Charyyev et al. (2019), the enterocin-producing *En. faecium* YT52 strain isolated from *boza* poses a low risk to consumer health, and this strain was reported to be effective against foodborne pathogens such as *Listeria monocytogenes* and *Bacillus (B.) cereus*. Therefore, it has been stated that this strain can be used as a starter or co-starter culture in fermented products to increase food safety. Guven and Benlikaya (2005) reported that *B. cereus* could not reach potentially toxic levels in fermented *boza* and that *boza* was safe.

Low pH and antimicrobial agents prevent the growth of pathogens and/or spoilage microorganisms, making the product safe for health (Petrova and Petrov 2017; Swain and Ananadharaj 2018). It has been reported that many strains such as *Lpb. plantarum, Lcb. paracasei, Lcb. rhamnosus, Lpb. pentosus, Limosilactobacillus fermentum, Lc. lactis, Ln. mesenteroides, Ln. pseudomesenteroides, En. faecium* and *P. pentosaceus* are capable of producing bacteriocins in studies performed in *boza* produced using different methods and from different regions (Tezel and Sanlibaba 2019).

On the other hand, since there are biogenic amine-producing bacteria (*Fructilactobacillus sanfranciscensis, Basonym: L. sanfrancisco, Oenococcus oeni, Ln. mesenteroides*) in the microflora of *boza*, biogenic amines such as putrescine, tyramine, β-phenylethylamine, histamine and spermidine may be found in *boza* (Yegin ve Uren 2008; Cosansu 2009; Yegin and Fernández-Lahore 2012; Zheng et al. 2020). The sum of the biogenic amines amounts ranges from 25–69 mg/kg (Yegin and Uren 2008), and some researchers state that it can exceed 100 mg/kg. The most abundant biogenic amine is tyramine. Foods and beverages containing high biogenic amine may cause health risks such as hypotension or hypertension, nausea, headache, rash, dizziness, cardiac palpitation and emesis. Especially in patients using medication, serious health problems may occur (Yegin ve Uren 2008; Cosansu 2009; Tamer et al. 2019b). However, less than 0.5 kg per meal seems to be safe for people who are not taking medication (Yegin and Fernández-Lahore 2012).

Compounds that play an important role in regulating blood pressure by reducing the activity of the angiotensin-converting enzyme (ACE), which plays an important role in regulating blood pressure and fluid-salt balance in humans, are known as ACE inhibitors. Kancabas and Karakaya (2013) stated that *boza*, protein hydrolysate, fractionated hydrolysates and dialysates obtained after *in vitro* digestion contain bioactive compounds with different ACE inhibitory activities. Therefore, they stated that this beverage may be taken into account as an important source of ACE-inhibiting peptides and can be consumed instead of carbonated beverages.

20.6 *AYRAN*

Ayran is a traditional fermented milk product that is prepared by adding water (30–50%) and salt (max 1%) to yoghurt or by adding *Streptococcus* (*St.*) *thermophilus* (formerly known as *St. salivarius* subsp. *thermophilus*) and *L. delbrueckii* subsp. *bulgaricus* (formerly known as *L. bulgaricus*) to standardized milk for fermentation (TFC 2009; Zheng et al. 2020; Atik et al. 2023). *Ayran* has been identified by different names in different regions such as *doogh* in Iran, *lassi* in Southern Asia, *than* in Armenia, *shenina* in Jordan, *laban ayran* in Syria and Lebanon, *laban arbil* in Iraq, *moru* in South India, *ayryan* in Bulgaria and *ariani* in Greece (Kabak and Dobson 2011; Baschali et al. 2017; Otles and Nakilcioglu-Tas 2019). The manufacturing processes of *ayran* are shown in Figure 20.1 (Kabak and Dobson 2011).

Ayran has been widely consumed in many countries in the Balkans, Central Asia and the Caucasus, especially in Turkey. The general characteristics of *ayran* obtained from

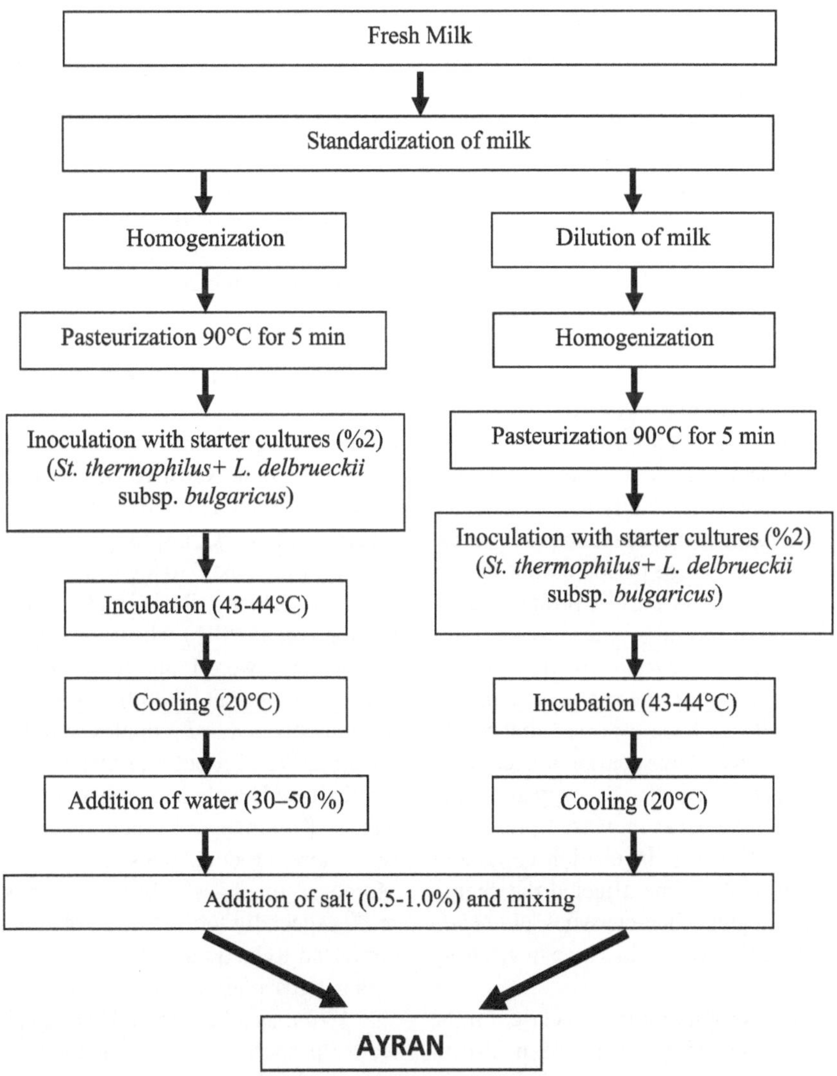

FIGURE 20.1 Manufacturing processes of *ayran*.

different locations in Turkey have been determined in many studies. The chemical compositions of *ayran* marketed in Kars and Ankara in Turkey have been determined in the range between 0.67±0.12% and 0.88±0.10% for acidity (lactic acid), 5.45±1.06% and 8.28±1.98% for total dry solids, 0.88±0.17 and 1.42±0.10 % for salt content and 1.08±0.56% and 2.09±0.34% for fat content (Gulmez et al. 2003). Chemical contents of *ayran* from local dairy plants in Konya, Karaman, Aksaray and Nigde have been reported as in the ranges of 2.4–2.6%, 2.2–2.7% and 8.5–9.0% for protein, fat and total dry matter, respectively (Ayar et al. 2009). It has been reported that production methods have no significant differences in terms of chemical composition (Kocak et al. 2006). However, different thermal treatments (at 75, 85 and 95°C for 5 minutes) of milk significantly affected acetaldehyde, whey separation and viscosity of *ayran* samples during the 1st, 7th and 14th days of storage at 4–5°C, while they had no significant effect on the physical, chemical and microbiological properties of *ayran* such as pH, titratable acidity, lactic acid content, the number of yoghurt bacteria and sensory properties during storage. pH values, titratable acidity and the lactic acid content of *ayran* samples were determined in the ranges of 4.09–4.32, 36.31–41.57 (°SH) and 0.50–0.65 (%), respectively (Tamucay-Ozunlu and Kocak 2010a). The benzoic acid levels in *ayran* samples were found to be in the range of 1.54 to 16.57 mg/L (Yildiz et al. 2012). The strongest aroma-active components of *ayran* were reported as ethyl lactate (whey/creamy), 2-methylbutanal (fruity), acetoin (buttery creamy) and butanoic acid (cheesy-sweet) (Sarhir et al. 2019).

The microbial flora of *ayran* is similar to that of yoghurt. However, the bacterial population in homemade *ayran* is lower than in industrially manufactured products (Otles and Nakilcioglu-Tas 2019). The bacterial diversity of *ayran* has been determined, and dominant species have been reported to belong to *Leuconostoc* and *Enterococcus* genera. Moreover, 8 bacterial strains of LAB showed probiotic characteristics based on bile salt tolerance, acid tolerance, and antibiotic resistance (Zhadyra et al. 2021). The increase in acidity has affected the microbial populations of *ayran*. Ending the incubation process at 4.6, 4.3 and 4.0 pH during the storage period on the 1st, 7th and 14th days had a significant effect on pH, titratable acidity, lactic acid, whey separation, acetaldehyde, viscosity and numbers of *L. delbrueckii* subsp. *bulgaricus*, while it had no significant effect on the numbers of *St. thermophilus* and sensory properties of samples. It has been reported that as the incubation pH decreased, pH and whey separation decreased, titratable acidity, lactic acid, acetaldehyde, viscosity and numbers of *L. delbrueckii* subsp. *bulgaricus* increased (Tamucay-Ozunlu and Kocak 2010b). Moreover, incubation conditions such as incubation temperature and time affected the chemical, rheological and microbiological characteristics and *in situ* exopolysaccharide yield, which caused an increase in viscosity of *ayran* (Yilmaz et al. 2015).

The effect of microbial transglutaminase (TGase) on the chemical and physical properties of *ayran* has been investigated. The addition of enzymes improved the physical properties of *ayran*, while it did not significantly change the chemical properties of *ayran* during the storage period. The viscosity of *ayran* increased, serum separation was prevented and gel permeability decreased with the pretreatment of milk with TGase (Sanli et al. 2013). On the other hand, thermosonication has a positive effect on the physical properties of *ayran*. The viscosity increased while the serum separation of *ayran* decreased with the thermosonication treatment (Erkaya et al. 2015). Furthermore, high-intensity ultrasound was applied with heat treatment (60°C) for the microbial inactivation and a similar reduction for microbial inactivation was observed compared to the heat treatment at 70°C (Kilic-Akyilmaz et al. 2023).

Different levels of water and salt have affected the rheological properties and serum separation of *ayran*. As the levels of water and salt increased, the consistency coefficient decreased, and the flow behaviour index and serum separation increased (Koksoy and Kilic 2003). Different hydrocolloids can be used to give consistency and improve the rheological properties of *ayran*. Usage of various hydrocolloids such as xanthan gum, carboxymethylcellulose, -carrageenan, locust bean gum, agar-agar and modified starch at different rates on the production of *ayran* has the possibility to give consistency to *ayran* which has low dry matter (Engez et al. 2006). On the other hand, fermentation conditions such as fermentation temperature and types of culture can also affect the rheological properties and serum separation of *ayran*. High fermentation temperature (43°C) and a final pH of 4.3 caused higher viscosity, higher consistency coefficient and lower serum separation in *ayran* (Ozdemir and Kilic 2004). Although some of the hydrocolloids such as high methoxyl pectin and gelatin can cause an unpleasant taste and odour in *ayran*, locust bean gum can be used without affecting the taste and the odour in *ayran* (Koksoy and Kilic 2004).

20.6.1 Benefits of *Ayran* on Health

Ayran is a highly nutritional drink due to its high vitamin and mineral content. Nutritive values (approximate per serving) of *ayran* have been reported containing 7.0 g protein, 2.7 g fat, 9.6 g carbohydrate, 0.2 mg iron, 194 mg phosphorus, 18 mg cholesterol, 1 mg zinc, 710 mg sodium, 190 IU vitamin A, 0.06 mg thiamine, 0.30 mg riboflavin, 0.16 mg niacin and 2 mg vitamin C (Gogus et al. 2016). The trace metal contents (Al, Cd, Se and As) of *ayran* were found to be below the maximum contents allowed by the TFC (2009) and Ayar et al. (2009). The phthalate amounts in *ayran* samples were found to be in the range of 38–59 μg/kg di-n-butyl phthalate (DBP) and 26–81 μg/kg di (2-ethylhexyl) phthalate (DEHP), less than the limit-imposed values (Korkmaz ve Kuplulu 2019). It can be said that *ayran*, as a dairy product, does not pose severe health problems. Moreover, the digestibility of *ayran* is very high compared to milk, especially for lactose-intolerance people since lactose is converted to lactic acid by the LAB in the fermentation process (Gogus et al. 2016).

Ayran, as a functional beverage, has antioxidant and protective properties on intestinal cells due to its high amounts of natural LAB, acetic and lactic acid. It has been reported that the dose-dependent effect of *ayran* significantly reduced DNA damage (Grishina et al. 2010). The antiradical power value of *ayran* was determined as 0.092 ±0.063 mM TE/kg (Najgebauer-Lejko and Sady 2015). The antioxidant capacity of *ayran* can be strengthened with the addition of natural extracts and dark-coloured fruits and vegetables. *Ayran* products that have high antioxidant activity can be developed with the addition of black carrot powder (Say et al. 2018). The addition of freeze-dried black carrot fiber to *ayran* can improve its mineral contents (Say et al. 2022). Moreover, the *ayran–şalgam* mixture as a new product has the potential for healthy dietary options (Uzay et al. 2021). The use of the powder form of chitosan-coated liposomes containing cocoa hull waste phenolic extract in powder form can be recommended for *ayran* to increase total phenolics, total flavonoids and total antioxidant capacity (Altin et al. 2018). The addition of *Mentha longifolia* L. essential oil due to its flavour and antioxidant properties to *ayran* increased the viability of *Lcb. casei* without affecting the physicochemical and organoleptic characteristics of *ayran* (Mahmoudi 2014).

To improve functional properties of *ayran*, fat replacers (protein-based Carbelac Dairy-Lo and carbohydrate-based inulin) and probiotic bacteria (*L. acidophilus* ve *Bifidobacterium bifidum*) along with yoghurt bacteria (*L. delbrueckii* subsp. *bulgaricus* and *St. thermophilus*) were used in *ayran* (Tas and Guzel-Seydim 2010). Similarly, monounsaturated/polyunsaturated fatty acid, oleic, linoleic, α-linolenic and folic acid contents and antioxidant activity of *ayran* prepared with commercial probiotic cultures can be increased (Shalabi et al. 2023). In the same way, the addition of whey protein hydrolysate to *ayran* enhanced the growth and activity of probiotic bacteria and the nutritional value of *ayran* (Celekli et al. 2020). On the other hand, viability of probiotic bacteria (*L. acidophilus* LA-5) and physicochemical properties of *ayran* (pH, acidity, viscosity, sedimentation and colour) have also been improved with the addition of aloe vera gel (Rezazadeh-Bari et al. 2019). Quinoa flour can also be used to develop the physical, chemical, microbiological and sensory properties of *ayran* (Akkoyun and Arslan 2020).

Although *ayran* is usually made from cow's milk, goat's milk is also used for the production of *ayran*. Goat's milk has been reported as a valuable raw material for the production of *ayran* due to its fatty acid profile. There have been 12 types of fatty acids in *ayran* samples from goat's milk, and saturated fatty acids (SFAs) such as C10:0, C14:0, C16:0, and C18:0 have been reported to be the most prevalent in *ayran* samples from goat's milk. Moreover, *ayran* samples from goat's milk have a high amount of monounsaturated fatty acid (C18:1n9c) (Shunekeyeva 2021).

20.7 CONCLUSIONS AND FUTURE DIRECTIONS

Today, with the growing world population, coupled with the rapidly aging population in several parts of the world, people have started to become more conscious about healthy eating and therefore wish to adopt healthy dietary habits and consume more functional foods. Also, as high calorie-based diet-linked chronic diseases such as type 2 diabetes, heart disease and cancers increase with multiple impacts of the environment, there are many causes that could be better countered with a healthier diet. Therefore, functional foods and beverages (fermented products) obtained through fermentation, which are thought to have beneficial effects on health, are important, and these concepts in people's diets have been in place since old times. In the future, due to the rapidly increasing population, aging populations, consumer demands and protein and vitamin deficiencies, as well as their positive effects on health, fermented products with established functional properties will become even more important in people's daily nutrition and advancing health protection.

Among the health effects of fermented products, which are thought to be beneficial, include the potential for reducing the risk of hypertension and lowering glycemic index related to diabetes, obesity, high cholesterol and some cancers. In summary, many compounds such as vitamins, melatonin, exopolysaccharides, antimicrobial peptides and bioactive compounds are synthesized and/or enhanced in fermented products due to fermentation. However, the claims expressed regarding the health effects of fermented beverages are mostly based on individuals' personal experiences and statements through empirical evidence, which is not sufficient. Therefore, more extensive studies should be conducted for the industrial production of standard-quality fermented beverages with functional properties. These studies should focus on searching for and selecting the most suitable starter culture and/or cultures (single or combined) to determine and increase their positive effects on health, determining the bacteriocin production coupled with probiotic activities of the cultures and determining the most suitable growth conditions. This can be integrated into advances in the determination of the nutritional properties and important effects of fermented beverages on health to further improve these properties and prevent the development of pathogenic microorganisms.

REFERENCES

Achi, O.K. 2005. The potential for upgrading traditional fermented foods through biotechnology. *Afr. J. Biotechnol.* 4(5):375–380. https://doi.org/10.5897/AJB2005.000–3070

Afrin, S., Giampieri, F., Gasparrini, M., et al. 2016. Chemopreventive and therapeutic effects of edible berries: A focus on colon cancer prevention and treatment. *Molecules* 21(2):169. https://doi.org/10.3390/molecules21020169

Agirman, B., and Erten, H. 2018. The influence of various chloride salts to reduce sodium content on the quality parameters of salgam (şalgam): A traditional Turkish beverage based on black carrot. *J. Food Qual.* 3292185:1–11. https://doi.org/10.1155/2018/3292185

Akbulut, M., Causir, S., Marakoglu, T., et al. 2008. Chemical and technological properties of European cranberrybush. *Asian. J. Chem.* 20:1875–1885.

Akkoyun, Y., and Arslan, S. 2020. The impact of quinoa flour on some properties of ayran. *Food Sci Nutr.* 8:5410–5418. https://doi.org/10.1002/fsn3.1832

Akman, P.K., Ozulku, G., Tornuk, F., et al. 2021. Potential probiotic lactic acid bacteria isolated from fermented gilaburu and shalgam beverages. *LWT Food Sci. Technol.* 149:111705. https://doi.org/10.1016/j.lwt.2021.111705

Akpinar-Bayizit, A., Yilmaz-Ersan, L., and Ozcan, T. 2010. Determination of Boza's organic acid composition as it is affected by raw material and fermentation. *Int. J. Food Prop.* 13:648–656. https://doi.org/10.1080/10942911003604194

Aksoy, A.S., Arici, M., and Yaman, M. 2022. The effect of hardaliye on reducing the formation of malondialdehyde during in vitro gastrointestinal digestion of meat products. *Food Biosci.* 47:101747. https://doi.org/10.1016/j.fbio.2022.101747

Aktas, A. 2012. *Bioactive phenolic composition of gilaburu, red grape, scented grape and black nut fruits determined by LC-ESI-MS and on-line HPLC-ABTS methods.* Karadeniz Technical University, Science Institute, Trabzon, Turkey, MSc Thesis, 139 p. (in Turkish).

Aktop, S., Sanlibaba, P., and Gucer, Y. 2023. Grape-based traditional foods produced in Turkey. *Italian J. Food Sci.* 35(3):55–74.

Al, O. 2019. *The effects of different fractions of gilaburu (Viburnum opulus) juice on the experimentally induced cancer in mice.* Erciyes University, Institute of Health Sciences, Faculty of Medicine, Department of Anatomy, Kayseri, Turkey, PhD Thesis, 95 p.

Aladeboyeje, O., and Sanli, N.O. 2021. Fermented traditional probiotic beverages of Turkish origin: A concise review. *Int. J. Life Sci. Biotechnol.* 4(3):546–564. https://doi.org/10.38001/ijlsb.936982

Aloys, N., and Angeline, N. 2009. Traditional fermented foods and beverages in Burundi. *Food Res. Int.* 42:588–594. https://doi.org/10.1016/j.foodres.2009.02.021

Altay, F., Kabancioglu-Guler F., Daskaya-Dikmen C., et al. 2013. A review on traditional Turkish fermented non-alcoholic beverages: Microbiota, fermentation process and quality characteristics. *Int. J. Food Microbiol.* 167:44–56. https://doi.org/10.1016/j.ijfoodmicro.2013.06.016

Altin, G., Gultekin-Ozguven, M., and Ozcelik, B. 2018. Liposomal dispersion and powder systems for delivery of cocoa hull waste phenolics via ayran (drinking yoghurt): Comparative studies on in-vitro bioaccessibility and antioxidant capacity. *Food Hydrocolloids* 81:364–370. https://doi.org/10.1016/j.foodhyd.2018.02.051

Amoutzopoulos, B., Loker, G.B., Samur, G., et al. 2013. Effects of a traditional fermented grape-based drink 'hardaliye' on antioxidant status of healthy adults: A randomized controlled clinical trial. *J. Sci. Food Agric.* 93(14):3604–3610. https://doi.org/10.1002/jsfa.6158

Anli, S.C. 2019. Fermented food and health. In: *Fermented Food: Microbiology, Technology and Health*, eds. E. Anli and P. Sanlibaba, Vol. 23(691–708). Ankara, Turkey: Nobel Publisher (in Turkish).

Anonymous. 2020. https://www.turkpatent.gov.tr/TURKPATENT/resources/temp/F6EA3F94-C8F2-482B-99EB-36D26AB90970.pdf;jsessionid=F03FEDF4BE16061751BC3C66071044A1. Accessed date: September 20, 2020.

Arici, M., and Coskun, F. 2001. Hardaliye: Fermented grape juice as a traditional Turkish beverage. *Food Microbiol.* 18:417–421. https://doi.org/10.1006/fmic.2001.0413

Arici, M., Coskun, F., Celikyurt, G., et al. 2017. Some technological and functional properties of lactic acid bacteria isolated from Hardaliye. *J. Agric. Sci.* 23:428–437. https://doi.org/10.5072/zenodo.57840

Arici, M., and Daglioglu, O. 2002. Boza: A lactic acid fermented cereal beverage as traditional Turkish food. *Food Rev. Int.* 18(1):39–48. https://doi.org/10.1081/FRI-120003416

Arslan, S., Durak, A.N., Erbas, M., et al. 2015. Determination of microbiological and chemical properties of probiotic boza and its consumer acceptability. *J. Am. Coll. Nutr.* 34(1):56–64. https://doi.org/10.1080/07315724.2014.880661

Askin, B. 2019. Effect of different temperatures on hardaliye quality during storage. *Harran J. Agric. Food Sci.* 23(1):13–21. https://doi.org/10.29050/harranziraat.4094 (in Turkish).

Askin, B., and Atik, A. 2016. Color, phenolic composition, and antioxidant properties of hardaliye (fermented grape beverage) under different storage conditions. *Turk. J. Agric. For.* 40:803–812. https://doi.org/10.3906/tar-1602–111

Ates, C. 2019. *Effectiveness of ultrasonication and high-pressure processing pasteurization on quality characteristics and shelf life of fermented salgham drink.* Bolu Abant İzzet Baysal University, Bolu, Turkey, MSc Thesis, 477 p. (in Turkish).

Atik, I., Atik, A., Akarca, G., et al. 2023. Production of high-mineral content of ayran and kefir: Effect of the fishbone powder obtained from garfish *(Belone belone). Int. J. Gastron. Food Sci.* 33:100786. https://doi.org/10.1016/j.ijgfs.2023.100786

Avci, S. 2008. *Inactivation of Candida inconspicua in shalgam juice by temperature and sonication.* Master's Thesis, Institute of Science, Department of Food Engineering, Van, Turkey (in Turkish).

Ayar, A., Sert, D., and Akın, N. 2009. The trace metal levels in milk and dairy products consumed in middle Anatolia-Turkey. *Environ. Monit. Assess.* 152:1–12. https://doi.org/10.1007/s10661-008-0291-9

Aydogdu, H., Yildirim, S., Halkman, A.K., and Durgun, T. 2014. A study on production and quality criteria of hardaliye; a traditional drink from the Thrace region of Turkey. *The Food (GIDA)* 39(3):139–145. https://doi.org/10.5505/gida.36844

Bakircioglu, D., Topraksever, N., and Kurtulus, Y.B. 2017. Determination of several element levels in hardaliye beverages using flame atomic absorption spectrometry after ultrasound-probe extraction. *J. AOAC Int.* 100(5):1531–1538. https://doi.org/10.5740/jaoacint.17–0117

Baschali, A., Tsakalidou, E., Kyriacou, A., et al. 2017. Traditional low-alcoholic and non-alcoholic fermented beverages consumed in European countries: A neglected food group. *Nutr. Res. Rev.* 30:1–24. https://doi.org/10.1017/S0954422416000202

Baser, G.M. 2013. *Biological and chemical characterization of şalgam juice.* Yeditepe University, Istanbul, Turkey, MSc Thesis, 110 p.

Baser, M., Ekinci, F., and Guclu-Ustundag, O. 2012. The antioxidant capacity of traditional fermented beverage, Shalgam juice. *New Biotech.* 29:S119. https://doi.org/10.1016/j.nbt.2012.08.332

Berktas, I. 2011. *The effect of producing boza by using different raw materials on phenolic content and quality.* Istanbul Technical University, Food Engineering Department, Istanbul, Turkey, MSc Thesis, 51 p.

Berry, S.K., Manan, J.K., Joshi, G.J., et al. 1989. Preparation and evaluation of ready-to-serve black carrot beverage (Kanji). *J. Food Sci. Technol.* 26(6):327–328.

Blandino, A., Al-Aseeri, M.E., Pandiella, S.S., et al. 2003. Cereal-based fermented foods and beverages. *Food Res. Int.* 36:527–543. https://doi.org/10.1016/S0963–9969(03)00009–7

Botes, A., Todorov, S.D., von Mollendorff, J.W., et al. 2007. Identificaation of lactic acid bacteria and yeast from boza.

Proc. Biochem. 42:267–270. https://doi.org/10.1016/j.procbio.2006.07.015

Cabaroglu, T., and Koyuncu, G. 2019. Aroma composition of fermented food. In: *Fermented Food: Microbiology, Technology and Health*, eds. E. Anli and P. Sanlibaba, Vol. 4(83–100). Ankara, Turkey: Nobel Publisher (in Turkish).

Cakir, P. 2011. *A research on the composition of shalgam juices produced in our country and their compliance with food legislation.* Namık Kemal University, Institute of Science, Manisa, MSc Thesis, 53 p.

Cam, M, Hisil, Y., and Kuscu, A. 2007. Organic acid, phenolic content, and antioxidant capacity of fruit flesh and seed of *Viburnum opulus*. *Chem. Nat. Compd.* 43(4):460–461. https://doi.org/10.1007/s10600-007-0161-7

Canbas, A., and Deryaoglu, A. 1993. Research on the production technique and composition of Shalgam juice. *Doga-Turkish J. Agric. For.* 17:119–129 (in Turkish).

Caplice, E., and Fitzgerald, G.F. 1999. Food fermentations: Role of microorganisms in food production and preservation. *Int. J. Food Microbiol.* 50:131–149.

Celekli, A., Alslibi, Z., and Bozkurt, H. 2020. An investigation of effects of whey protein hydrolysate on yoghurt starter cultures and probiotic bacteria in ayran. *Kahramanmaras Sutcu Imam Univ. Tarim ve Doga Derg.* 23(5):1225–1239. https://doi.org/10.18016/ksutarimdoga.vi.625601

Celik, I., Isik, F., and Yilmaz, Y. 2016. Effect of roasted yellow chickpea (leblebi) flour addition on chemical, rheological and sensory properties of Boza. *J. Food Proc. Preserv.* 40:1400–1406. https://doi.org/10.1111/jfpp.12725

Ceylan, D. 2015. *The effects of gilaburu (Viburnum opulus) juice on experimentally induced cancer in mice.* Erciyes University, Institute of Natural and Applied Sciences, Kayseri, Turkey, PhD Thesis, 90 p.

Ceylan, D., Aksoy, A., Ertekin, T., et al. 2018. The effects of gilaburu (*Viburnum opulus*) juice on experimentally induced Ehrlich ascites tumour in mice. *J. Can. Res. Ther.* 14:314–320. https://doi.org/10.4103/0973-1482.181173

Charyyev, M.G., Tuncer, B.O., Kankaya, D.A., et al. 2019. Bacteriocinogenic properties and safety evaluation of *Enterococcus faecium* YT52 isolated from Boza, a traditional cereal based fermented beverage. *J. Consum. Prot. Food S.* 14:41–53. https://doi.org/10.1007/s00003-019-01213-9

Cirisoglu, E., and Olum, E. 2019. The importance of functional foods in Turkish cuisine for gastronomic tourism. *J. Turk. Tour. Res.* 3(4):1659–1680. https://doi.org/10.26677/TR1010.2019.264 (in Turkish)

Cosansu, S. 2009. Determination of biogenic amines in a fermented beverage, Boza. *J. Food Agric. Environ.* 7(2):54–58. https://hdl.handle.net/20.500.12619/67534

Coskun, F. 2017. A traditional Turkish fermented non-alcoholic grape-based beverage, "Hardaliye". *Beverages* 3(2):1–11. https://doi.org/10.3390/beverages3010002

Coskun, F. 2019. Shalgam beverage. In: *Fermented Foods: Microbiology, Technology and Health*, eds. E. Anli and P. Sanlibaba, Vol. 1(569–581). Ankara, Turkey: Atlas Academic Press (in Turkish).

Coskun, F., and Arici, M. 2006. The effects of using different mustard seeds and starter cultures on some properties of hardaliye. *Ann. Microbiol.* 56(4):335–337. https://doi.org/10.1007/BF03175027

Coskun, F., and Arici, M. 2011. Effect of the use of different mustard seeds and grape varieties on some properties of hardaliye. *Acad. Food J.* 9(3):6–11.

Coskun, F., Arici, M., Gulcu, M., et al. 2018. Physicochemical, functional and microbiological properties of hardaliye beverages produced from different grapes and collected from different households. *J. Agric. Sci.* 24:278–285. https://doi.org/10.15832/ankutbd.446458

Demir, N., Acar, J., and Bahceci, K.S. 2004. Effects of storage on quality of carrot juices produced with lactofermentation and acidification. *Eur. Food Res. Technol.* 218:465–468. https://doi.org/10.1007/s00217-004-0883-8

Demir, N., Bahceci, K.S., and Acar, J. 2006. The effects of different initial *Lactobacillus plantarum* concentrations on some properties of fermented carrot juice. *J. Food Proc. Preserv.* 30:352–363. https://doi.org/10.1111/j.1745-4549.2006.00070.x

Demiray, M. 2019. *Pasteurization of traditional hardaliye drink by innovative food processing technologies.* Bolu Abant Izzet Baysal University Graduate School of Natural and Applied Sciences, Bolu, Turkey, MSc Thesis, 154 p.

Dikmen, B.Y., and Filazi, A. 2016. Nutraceuticals: Turkish perspective. In: *Nutraceuticals; Efficacy, Safety and Toxicity*, ed. R.C. Gupta, Vol. 69(971–981). Amsterdam: Elsevier. https://doi.org/10.1016/B978-0-12-802147-7.00069-3

Dogan, K. 2017. *Optimization of ultraviolet pasteurization conditions of salgam (şalgam) juice and determination of its shelf life.* Yildiz Technical University Institute of Science, Istanbul, Turkey, MSc Thesis, 99 p.

Engez, S., Ugur, H., and Karagulle, B. 2006. Investigation of usage possibilities of various thickeners in ayran production. *Celal Bayar Univ. J. Sci.* 2(2):129–134 (in Turkish).

Eren, E.M., and Hayta, M. 2015. Traditional healthy fruit juice of Kayseri, 'Gilaburu'. Paper presented at the 3rd International Symposium on Traditional Foods from Adriatic to the Caucasus, Sarajevo/Bosnia and Herzegovina.

Erginkaya, Z., and Aksan, E. 2004. Adana province traditional drink: Şalgam. Paper presented at Traditional Foods Symposium, Van, Turkiye (in Turkish).

Erginkaya, Z., and Hammes, W.P. 1992. A study on the development of microorganisms during the Shalgam fermentation and identification of isolated lactic acid bacteria. *Gida (Food)* 17(5):311–314 (in Turkish).

Erkaya, T., Baslar, M., Sengul, M., et al. 2015. Effect of thermosonication on physicochemical, microbiological and sensorial characteristics of ayran during storage. *Ultrason. Sonochem.* 23:406–412. https://doi.org/10.1016/j.ultsonch.2014.08.009

Ersoy, N., Ercisli, S., and Gundogdu, M. 2017. Evaluation of European cranberrybush (*Viburnum opulus* L.) genotypes for agro-morphological, biochemical and bioactive characteristics in Turkey. *Folia Hort.* 29(2):181–188. https://doi.org/10.1515/fhort-2017-0017

Erten, H., Agirman, B., Boyaci Gunduz, C.P., et al. 2017. Regional fermented vegetables and fruits in Europe. In: *Lactic Acid Fermentation of Fruits and Vegetables*, ed. S. Paramithiotis, 205–235. Boca Raton: CRC Press.

Erten, H., and Canbas, A. 2003. Flavors formed during alcohol fermentation. *Food* 28(6):615–619 (in Turkish).

Erten, H., and Tanguler, H. 2012. Şalgam (Shalgam). In: *Handbook of Plant-Based Fermented Food and Beverage Technology*, ed. Y.H. Hui, Vol. 40(657–664). Boca Raton: CRS Press.

Erten, H., and Tanguler, H. 2016. Shalgam (şalgam): A traditional Turkish lactic acid fermented beverage obtained mainly from black carrots. In: *Handbook of Vegetable Preservation and Processing, Second Edition*, eds. Y.H. Hui, E.O. Evranuz, H. Erten, G. Bingol, and M.E.J. Flores, Vol. 2(841–849). Boca Raton: CRC Press.

Erten, H., and Tanguler, H. 2019. Fermented herbal products. In: *Food Biotechnology*, ed. N. Aran, Vol. 9(243–278). Ankara: Nobel Publishing House.

Erten, H., Tanguler, H., and Canbas, A. 2008. A traditional Turkish lactic acid fermented beverage: Shalgam (Salgam). *Food Rev. Int.* 24:352–359. https://doi.org/10.1080/87559120802089324

Faikoglu, F. 2014. *Investigation of quality and sensory properties of hardaliye produced with Adakarasi, Papazkarasi and Karecikkarasi grape varieties*. Uludag University Graduate School of Natural and Applied Sciences, Uludag, Turkey, MSc Thesis, 74 p.

Gadaga, T.H., Mutukumira, A.N., Narvhus, J.A., et al. 1999. A review of traditional fermented foods and beverages of Zimbabwe. *Int. J. Food Microbiol.* 53:1–11. https://doi.org/10.1016/s0168–1605(99)00154–3

Garcia, C., Guerin, M., Souidi, K., et al. 2020. Lactic fermented fruit or vegetable juices: Past, present and future. *Beverages* 6(8):1–32. https://doi.org/10.3390/beverages6010008

Gille, D., Schmid, A., Walther, B., et al. 2018. Fermented food and non-communicable chronic diseases: A review. *Nutrients* 10(4):448. https://doi.org/10.3390/nu10040448

Gogus F., Otles S., Erdogdu F., et al. 2016. Functional and nutritional properties of some Turkish traditional foods. In: *Functional Properties of Traditional Foods: Integrating Food Science and Engineering Knowledge in to the Food Chain, Vol 12*, eds. K. Kristbergsson and S. Otles, 87–104. Boston, MA: Springer. https://doi.org/10.1007/978-1-4899-7662-8_6

Gok, S. 2017. *Determining the qualification properties and suitability to the Turkish food codex of şalgam that sold packaged or unpacked in Adana province*. Kahramanmaraş Sutcu Imam University, Institute of Science and Technology, Department of Food Engineering, Kahramanmaras, Turkey, MSc Thesis, 60 p.

Gotcheva, V., Pandiella, S.S., Angelov, A., et al. 2001. Monitoring the fermentation of the traditional Bulgarian beverage Boza. *Int. J. Food Sci. Technol.* 36:129–134. https://doi.org/10.1046/j.1365–2621.2001.00429.x

Grishina, A., Kulikova, I., and Alieva, L. 2010. Antigenotoxic effect of kefir and ayran supernatants on faecal water-induced DNA damage in human colon cells. *Nutr. Cancer* 63(1):73–79. https://doi.org/10.1080/01635581.2010.516873

Gulmez, M., Guven, A., Sezer, C., et al. 2003. Evaluation of microbiological and chemical quality of ayran samples marketed in Kars and Ankara cities in Turkey. *J. Faculty Vet. Med. Univ. Kafkas* 9(1):49–52.

Gurbuz, O. 2018. *Effects of antimicrobial agents used in hardaliye production on the fermentation*. Trakya University, Institute of Science, Tekirdag, Turkey, MSc Thesis, 107 p.

Guven, K., and Benlikaya, N. 2005. Acid pH produced by lactic acid bacteria prevent the growth of *Bacillus cereus* in boza, a traditional fermented Turkish beverage. *J. Food Saf.* 25:98–108. https://doi.org/10.1111/j.1745–4565.2005.00568.x

Guven, S., Vatan, E., and Ogutcu, M. 2004. Some modifications in Boza production. Paper presented at the Traditional Foods Symposium, Gaziantep, Turkey (in Turkish).

Hammes, W.P., Brandt, M.J., Francis, K.L., et al. 2005. Microbial ecology of cereal fermentations. *Trends Food Sci. Technol.* 16(1–3):4–11. https://doi.org/10.1016/j.tifs.2004.02.010

Hancioglu, O., and Karapinar, M. 1997. Microflora of Boza, a traditional fermented Turkish beverage. *Int. J. Food Microbiol.* 35:271–274. https://doi.org/10.1016/S0168–1605(96)01230–5

Harris, L.J. 1998. The microbiology of vegetable fermentations. In: *Microbiology of Fermented Foods*, ed. B.J.B. Wood, 45–72. London: Blackie Academic & Professional.

Hutkins, R.W. 2006. *Microbiology and Technology of Fermented Foods*. Oxford, UK: IFT Press, Blackwell Publishing.

Ignat, M.V., Salantă, L.C., Pop, O.L., et al. 2020. Current functionality and potential improvements of non-alcoholic fermented cereal beverages. *Foods* 9:1–28. https://doi.org/10.3390/foods9081031

Incedayi, B., Uylaser, V., and Copur, U.A. 2008. A traditional Turkish beverage shalgam: Manufacturing technique and nutritional value. *J. Food Agric. Environ.* 6(3):31–34.

Jeyaram, K., Singh, T.H., Romi, W., et al. 2009. Traditional fermented foods of Manipur. *Indian J. Tradit. Knowl.* 8(1):115–121.

Kabak, B., and Dobson, A.D. 2011. An introduction to the traditional fermented foods and beverages of Turkey. *Crit. Rev. Food Sci. Nutr.* 51(3):248–260. https://doi.org/10.1080/10408390903569640

Kalyoncu, I.H., Ersoy, N., Elidemir, A.Y., et al. 2013. Some physicochemical characteristics and mineral contents of Gilaburu (*Viburnum opulus* L.) fruits in Turkey. *Int. J. Biol. Food Vet. Agric. Eng.* 7(6):169–171.

Kancabas, A., and Karakaya, S. 2013. Angiotensin-converting enzyme (ACE)-inhibitory activity of boza, a traditional fermented beverage. *J. Sci. Food Agric.* 93:641–645. https://doi.org/10.1002/jsfa.5883

Karacelik, A.A., Kucuk, M., Iskefiyeli, Z., et al. 2015. Antioxidant components of *Viburnum opulus* L. determined by on-line HPLC–UV–ABTS radical scavenging and LC–UV–ESI-MS methods. *Food Chem.* 175:106–114. https://doi.org/10.1016/j.foodchem.2014.11.085

Karaoglan, H.A., Keklik, N.M., and Isikli, N.D. 2017. Modelling inactivation of *Candida inconspicua* isolated from turnip juice using pulsed UV light. *J. Food Proc. Eng.* 40:1–9. https://doi.org/10.1111/jfpe.12418

Kelebek, H. 2009. *Researches on the phenolic compounds profile of okuzgozu, bogazkere and kalecik karasi cultivars grown in different regions and their wines*. Cukurova University, Institute of Science and Technology, Adana, Turkey, PhD Thesis, 259 p.

Kentel, Z.B. 2001. *A study on prolonging the shelf life of Boza*. Ankara University Graduate School of Natural and Applied Sciences, Ankara, Turkey, MSc Thesis, 56 p (in Turkish).

Kilic, G.B., Agdas, K., Karahan, A., et al. 2016. Effect of *Lactobacillus plantarum* AK4–11 and different grape varieties on the properties of hardaliye. *J. Agric. Sci.* 22(4):512–521. https://doi.org/10.1501/Tarimbil_0000001409

Kilic-Akyilmaz, M., Kurt, C., Uzunoglu, T.P., et al. 2023. Comparison of high intensity ultrasound and heat treatment for extending shelf life of a fermented milk beverage. *Int. Dairy J.* 141:105617. https://doi.org/10.1016/j.idairyj.2023.105617

Kocak, C., Avsar, Y.K., and Tamucay, B. 2006. A comparative study on the production methods of ayran. *Gida (Food)* 31(4):225–231.

Koksoy, A., and Kilic, M. 2003. Effects of water and salt level on rheological properties of ayran, a Turkish yoghurt drink. *Int. Dairy J.* 13(10):835–839.

Koksoy, A., and Kilic, M. 2004. Use of hydrocolloids in textural stabilization of a yoghurt drink, ayran. *Food Hydrocoll.* 18:593–600. https://doi.org/10.1016/j.foodhyd.2003.10.002

Korkmaz, S., and Kuplulu, O. 2019. Determination of phthalates in some milk products by liquid chromatography/tandem mass spectrometry. *Ankara Universitesi Veteriner Fakultesi Dergisi* 66(3):230–236. https://doi.org/10.33988/auvfd.436873

Kozat, P. 2000. *Microbiological and biochemical characterization of Boza a Turkish traditional fermented beverage*. Department of Food Engineering, ODTU, Ankara, Turkey, MSc Thesis, 102 p.

Kreutzmann, S., Thybo, A.K., and Bredie, W.L.P. 2007. Training of a sensory panel and profiling of winter hardy and coloured carrot genotypes. *Food Qual. Prefer.* 18:482–489. https://doi.org/10.1016/j.foodqual.2006.05.009

Luciano, F.B., and Holley, R.A. 2009. Enzymatic inhibition by allyl isothiocyanate and factors affecting its antimicrobial action against *Escherichia coli* O157:H7. *Int. J. Food Microbiol.* 131:240–245. https://doi.org/10.1016/j.ijfoodmicro.2009.03.005

Mahmoudi, R. 2014. Effect of *Mentha longifolia* L. essential oil on physicochemical properties of the bio-ayran. *J. Essent. Oil. Bear. Plants.* 17(1):56–66. https://doi.org/10.1080/0972060X.2013.813205

Moncheva, P., Chipeva, V., Kujumdzieva, A., et al. 2003. The composition of the microflora of boza, an original Bulgarian beverage. *Biotech. Biotechnol. Eq.* 17:164–168. https://doi.org/10.1080/13102818.2003.10819214

Montet, D., Ray, R.C., and Zakhia-Rozis, N. 2015. Lactic acid fermentation of vegetables and fruits. In: *Microorganisms and Fermentation of Traditional Foods Vol. 1*, eds. R.C. Ray and D. Montet, 108–140. Boca Raton: CRC Press.

Najgebauer-Lejko, D., and Sady, M. 2015. Estimation of the antioxidant activity of the commercially available fermented milks. *Acta Sci. Pol. Technol. Aliment.* 14(4):387–396. https://doi.org/10.17306/J.AFS.2015.4.38

Nielsen, P.V., and Rios, R. 2000. Inhibition of fungal growth on bread by volatile components from spices and herbs, and the possible application in active packaging, with special emphasis on mustard essential oil. *Int. J. Food Microbiol.* 60:219–229. https://doi.org/10.1016/S0168–1605(00)00343–3

Nowak, J., and Kuligowski, M. 2017. Functional properties of traditional food products made by mold fermentation. In: *Fermented Foods, Part II, Technological Interventions*, eds. R.C. Ray and D. Montet, 46–73. Boca Raton: CRC Press.

Otles, S., and Nakilcioglu-Tas, E. 2019. Nutritional components of some fermented nonalcoholic beverages. In: *Fermented Beverages*, eds. A.M. Grumezescu and A.M. Holban, 287–319. Woodhead Publishing.

Ozdemir, H., and Kilic, M. 2004. Influence of fermentation conditions on rheological properties and serum separation of ayran. *J. Text. Stud.* 35:415–428. https://doi.org/10.1111/j.1745–4603.2004.tb00604.x

Ozdestan, O., and Uren, A. 2010. Biogenic amine content of shalgam (salgam): A traditional lactic acid fermented Turkish beverage. *J. Agric. Food Chem.* 58:2602–2608. https://doi.org/10.1021/jf903775z

Ozer, N.O. 2009. *Inactivation of spoilage factor wild yeasts by heat treatment in salgam beverage to prolong the shelf-life*. Yuzuncu Yil University, Van, Turkey, MSc Thesis, 158 p (in Turkish).

Ozler, N. 1995. *Research on shalgam juice production.* Uludağ University, Institute of Science, Bursa, Master's Thesis, 55 p (in Turkish).

Ozrenk, K., Gundogdu, M., Keskin, N., et al. 2011. Some physical and chemical characteristics of Gilaburu (*Viburnum opulus* L.) fruits in Erzincan region. *Igdir Univ. J. Inst. Sci. Technol.* 1(4):9–14.

Ozturk, O. 2009. *A research on the composition of turnip juices in Adana market.* Çukurova University, Institute of Science, Department of Food Engineering, Master's Thesis, 43 p (in Turkish).

Petrova, P., and Petrov, K. 2017. Traditional cereal beverage boza fermentation technology, microbial content and healthy effects. In: *Fermented Food Part II: Technological Interventions, First Edition*, eds. C.R. Ramesh and M. Didier, 284–305. USA: CRC Press. ISBN 978-1-1386-3784-9.

Prado, F.C., Parada, J.L., Pandey, A., et al. 2008. Trends in non-dairy probiotic beverages. *Food Res. Int.* 41:111–123. https://doi.org/10.1016/j.foodres.2007.10.010

Prajapati, J.B., and Nair, B.M. 2008. The history of fermented foods. In: *Handbook of Fermented Functional Foods, Second Edition*, ed. E.R. Farnworth, 1–24. Boca Raton: CRC Press.

Rakin, M., Vukašinović Sekulić, M., and Mojović, L. 2012. Health benefits of fermented vegetable juices. In: *Handbook of Plant-Based Fermented Food and Beverage Technology*, ed. Y.H. Hui, 385–406. Florida: CRC Press.

Rezazadeh-Bari, M., Najafi-Darmian, Y., Alizadeh, M., et al. 2019. Numerical optimization of probiotic ayran production based on whey containing transglutaminase and Aloe vera gel. *J. Food Sci. Technol.* 56:3502–3512. https://doi.org/10.1007/s13197-019-03841-3

Sagdic, O., Ozturk, I., Yapar, N., et al. 2014. Diversity and probiotic potentials of lactic acid bacteria isolated from gilaburu, a traditional Turkish fermented European cranberrybush (*Viburnum opulus* L.) fruit drink. *Food Res. Int.* 64:537–545. https://doi.org/10.1016/j.foodres.2014.07.045

Sanli, T., Sezgin, E., Senel, E., et al. 2013. The effect of transglutaminase on some physicochemical and sensory properties of the Turkish drinking yoghurt ayran. *Int. J. Dairy Technol.* 66(3):410–416. https://doi.org/10.1111/1471–0307.12045

Sanlier, N., Gokcen, B.B., and Sezgin, A.C. 2019. Health benefits of fermented foods. *Crit. Rev. Food Sci. Nutr.* 59(3):506–527. https://doi.org/10.1080/10408398.2017.1383355

Sarhir, S.T., Amanpour, A., and Selli, S. 2019. Characterization of ayran aroma active compounds by solvent-assisted flavor evaporation (SAFE) with gas chromatography–mass spectrometry–olfactometry (GC–MS–O) and aroma extract dilution analysis (AEDA). *Anal. Lett.* 52(13):2077–2091. https://doi.org/10.1080/00032719.2019.1594244

Say, D., Saydam, İ.B., and Guzeler, N. 2018. Some properties of ayran fortified with black carrot powder. *J. Adv. VetBio Sci. Technol.* 3(3):54–60. https://doi.org/10.31797/vetbio.499749

Say, D., Saydam, İ.B., and Guzeler, N. 2022. Effects of freeze-dried black carrot fiber addition on the physicochemical, color, sensory attributes, and mineral contents of ayran. *J. Food Process. Preserv.* 46:e17225. https://doi.org/10.1111/jfpp.17225

Seker, I.T., Ertop, M.H., and Hayta, M. 2016. Physicochemical and bioactive properties of cakes incorporated with gilaburu fruit (*Viburnum opulus*) pomace. *Qual. Assur. Saf. Crop. Foods* 8(2):261–266. https://doi.org/10.3920/QAS2014.0542

Sethi, V. 1990. Lactic fermentation of black carrot juice for spiced beverage. *Indian Food Pack.* 44(3):7–12.

Shalabi, O.M.A.K., Hassan, A.M., Ismail, M.M., et al. 2023. Characterization of the ayran made with commercial probiotic cultures for fatty acids, cholesterol, folic acid levels, and anti-oxidative potential. *Probiot. Antimicrob. Prot.* https://doi.org/10.1007/s12602-023-10100-7

Shunekeyeva, A.A. 2021. Influence of starter cultures' type on the microbiological, rheological and sensory properties of ayran samples from goat's milk. *OnLine J. Biol. Sci.* 21(1):154–160. https://doi.org/10.3844/ojbsci.2021.154.160

Soylak, M., Elci, L., Saracoglu, S., et al. 2002. Chemical analysis of fruit juice of European cranberrybush (Viburnum opulus) from Kayseri-Turkey. *Asian J. Chem.* 14(1):135–138.

Spano, G., Russo, P., Lonvaud-Funel, A., et al. 2010. Biogenic amines in fermented foods. *Eur. J. Clin. Nutr.* 64:S95–S100. https://doi.org/10.1038/ejcn.2010.218

Swain, M.R., and Ananadharaj, M. 2018. Regional fermented vegetables and fruits in Asia-pacific. In: *Lactic Acid Fermentation of Fruits and Vegetables*, ed. S. Paramithiotis, 181–203. Boca Raton: CRC Press.

Tamer, C.E., and Copur, O.U. 2004. Our traditional drink: Boza. Paper presented at the Traditional Foods Symposium, 85–89 (in Turkish).

Tamer, C.E., Omeroglu, P.Y., and Copur, O.U. 2019b. Functional and traditional nonalcoholic beverages in Turkey. In: *Non-Alcoholic Beverages, the Science of Beverages*, eds. A.M. Grumezescu and A.M. Holban, 483–521. Duxford, United Kingdom: Elsevier.

Tamer, C.E., Suna, S., and Ozcan-Sinir, G. 2019a. Toxicological aspects of ingredients used in nonalcoholic beverages. In: *Non-Alcoholic Beverages, the Science of Beverages*, eds. A.M. Grumezescu and A.M. Holban, 441–481. Duxford, United Kingdom: Elsevier.

Tamucay-Ozunlu, B., and Kocak, C. 2010a. The effect of different heat treatments of milk on quality of ayran. *Gida (Food)* 35(5):355–362.

Tamucay-Ozunlu, B., and Kocak, C. 2010b. Effect of ending the incubation at different pHs on quality of ayran. *Gida (Food)* 35(2):113–119.

Tanguler, H. 2010. *Determination of lactic acid bacteria effective in shalgam beverage production and development of shalgam beverage production technique*. Cukurova University, Adana, Turkey, PhD Thesis, 367 p (in Turkish).

Tanguler, H. 2014. Traditional Turkish fermented cereal based products: Tarhana, Boza and Chickpea bread. *Turk. J. Agric. Food Sci. Technol.* 2(3):144–149. https://doi.org/10.24925/turjaf.v2i3.144–149.111

Tanguler, H., Dinc, S.O., and Beylikci, S.C. 2020. Problems, suggestions and last trends related to salgam beverage which is traditional product of Turkey. *Turk. J. Agric. Food Sci. Technol.* 8(6):1266–1271. https://doi.org/10.24925/turjaf.v8i6.1266–1271.3266

Tanguler, H., and Erten, H. 2012. Occurrence and growth of lactic acid bacteria species during the fermentation of shalgam (şalgam), a traditional Turkish fermented beverage. *LWT Food Sci. Technol.* 46:36–41. https://doi.org/10.1016/j.lwt.2011.10.026

Tanguler, H., and Erten, H. 2013. Selection of potential autochtonous starter cultures through lactic acid bacteria isolated and identificated from shalgam: A traditional Turkish fermented beverage. *Turk. J. Agric. For.* 37:212–220. https://doi.org/10.3906/sag-1210–82

Tanguler, H., Gunes, G., and Erten, H. 2014. Influence of addition of different amounts of black carrot (*Daucus carota*) on shalgam quality. *J. Food Agric. Environ.* 12(2):60–65.

Tanguler, H., Saris Per, E.J., and Erten, H. 2015. Microbial, chemical and sensory properties of shalgams made using different production methods. *J. Sci. Food Agric.* 95:1008–1015. https://doi.org/10.1002/jsfa.6781

Tanguler, H., and Tatlisoy, A. 2022. Evaluation of shalgam residuals as a partial wheat flour substitution in tarhana production as a green food product. *J. Food Process. Preserv.* 2022:e17267. https://doi.org/10.1111/jfpp.17267

Tanriseven, D., Kadiroglu, P., Selli, S., et al. 2020. LC-DAD-ESI-MS/MS-assisted elucidation of the phenolic compounds in shalgams: Comparison of traditional and direct methods. *Food Chem.* 305:125505. https://doi.org/10.1016/j.foodchem.2019.125505

Tas, T., and Guzel-Seydim, Z. 2010. Determination of the effects of various fat replacers and probiotic use on buttermilk quality criteria. *J. Food.* 35(2):105–112 (in Turkish).

Tezel, B.U., and Sanlibaba, P. 2019. Boza. In: *Fermented Food: Microbiology, Technology and Health*, eds. E. Anli and P. Sanlibaba, 593–608. Ankara, Turkey: Nobel Publisher (in Turkish).

TFC. 2009. *Turkish Food Codex, fermented milk products.* Turkish Food Codex Regulations, Regulation No: 2009/25.Republic of Turkey Ministry of Food, Agriculture and Livestock, Ankara, Turkey.

Todorov, S.D., Botes, M., Guigas, C., et al. 2008. Boza, a natural source of probiotic lactic acid bacteria. *J. Appl. Microbiol.* 104:465–477. https://doi.org/10.1111/j.1365–2672.2007.03558.x

Toktas, B., Bildik, F., and Ozcelik, B. 2018. Effect of fermentation on anthocyanin stability and in vitro bioaccessibility during shalgam (şalgam) beverage production. *J. Sci. Food Agric.* 98:3066–3075. https://doi.org/10.1002/jsfa.8806

TSE. 1992. *TS 9778 the Standard of Boza.* Turkish Standards Institute, Necatibey Caddesi 112 Bakanliklar, Ankara (in Turkish).

TSE. 2003. *TS 11149 Standard of Shalgam Beverage.* Turkish Standards Institute, Necatibey Caddesi 112 Bakanliklar, Ankara (in Turkish).

TSE. 2016. *TS 11149: 2003 Revised of Shalgam Beverage.* Turkish Standards Institute, Necatibey Caddesi 112 Bakanliklar, Ankara (in Turkish).

Tumer, A.R., Lale, A., Gurler, M., et al. 2018. The effects of traditional fermented beverages on ethanol, ethyl glucuronide and ethyl sulphate levels. *Egypt. J. Forensic Sci.* 8:1–5. https://doi.org/10.1186/s41935-018-0064-8

Uysal, U.D., Oncu, E.M., Berikten, D., et al. 2009. Time and temperature-dependent microbiological and mycotoxin (ochratoxin-A) levels in Boza. *Int. J. Food Microbiol.* 130:43–48. https://doi.org/10.1016/j.ijfoodmicro.2008.12.032

Uzay, M., Ozturk, H.İ., Buzrul, S., et al. 2021. A study on rheological properties, sensory evaluation and shelf life of ayran–shalgam mixtures. *J. Food Sci. Technol.* 58:2479–2486. https://doi.org/10.1007/s13197-020-04754-2

Vardar, U.S., Tekin, Z.H., and Ozcelik, G. 2015. Physicochemical and bioactive properties of hardaliye. Paper presented at the 3rd International Symposium on Traditional Foods from Adriatic to Caucasus, Sarajevo, Bosnia and Herzegovina.

Veiga, M., Costa, E.M., Voss, G., et al. 2019. Engineering and health benefits of fruits and vegetable beverages. In: *Non-Alcoholic Beverages, the Science of Beverages*, eds. A.M. Grumezescu and A.M. Holban, 363–440. Duxford, United Kingdom: Elsevier.

Velioglu, Y.S., Ekici, L., and Poyrazoglu, E.S. 2006. Phenolic composition of European cranberry bush (*Viburnum opulus* L.) berries and astringency removal of its commercial juice. *Int. J. Food Sci. Technol.* 41:1011–1015. https://doi.org/10.1111/j.1365–2621.2006.01142.x

Von Mollendorff, J.W., Vaz-Velho, M., and Todorov, S.D. 2016. Boza, a traditional cereal-based fermented beverage: A rich source of probiotics and bacteriocin-producing lactic acid bacteria. In: *Functional Properties of Traditional Foods*, eds. K. Kristbergsson and S. Otles, 157–188. New York: Springer.

WHO. 2003. WHO/FAO, Diet, Nutrition and the Prevention of Chronic Diseases, 2003. Report of the joint WHO/FAO expert consultation, WHO Technical Report Series, No. 916 (TRS 916), WHO Library Cataloguing-in-Publication Data, Geneva, Switzerland. 160 p. ISBN 92 4 120916 X.

Yaldirak, G. 2011. *The effect of different fermentation temperature on biogenic amine formation in shalgam juice produced by natural fermentation.* Sakarya University Institute of Science, Sakarya, Turkey, MSc Thesis, 58 p.

Yapar, N. 2008. *The possibility of using lactic acid bacteria isolated from traditionally fermented European Cranberrybush (Viburnum opulus L.) juice in the manufacture of industrial fruit juice.* Erciyes University, Kayseri, Turkey, MSc Thesis, 96 p (in Turkish).

Yardimci, H., Ozcelik, A.O., and Ozcelik, F. 2015. A special drink: Turnip juice and its effects on health. Paper presented at the 3rd International Symposium on Traditional Foods from Adriatic to Caucasus, Sarajevo, Bosnia and Herzegovina.

Yegin, S., and Fernández-Lahore, M. 2012. Boza: A traditional cereal-based, fermented Turkish beverage. In: *Handbook of Plant-Based Fermented Food and Beverage Technology*, ed. Y.H. Hui, 533–542. Florida: CRC Press.

Yegin, S., and Uren, A. 2008. Biogenic amine content of Boza: A traditional cereal-based, fermented Turkish beverage. *Food Chem.* 111:983–987. https://doi.org/10.1016/j.foodchem.2008.05.020

Yildiz, A., Erdogan, S., Saydut, A., et al. 2012. High-performance liquid chromatography analysis and assessment of benzoic acid in yogurt, ayran, and cheese in Turkey. *Food Anal. Methods* 5:591–595. https://doi.org/10.1007/s12161-011-9288-z

Yilmaz, M.T., Dertli, E., Toker, O.S., et al. 2015. Effect of in situ exopolysaccharide production on physicochemical, rheological, sensory, and microstructural properties of the yogurt drink ayran: An optimization study based on fermentation kinetics. *J. Dairy Sci.* 98(3):1604–1624. https://doi.org/10.3168/jds.2014–8936

Yilmaz-Ersan, L., and Turan, M.A. 2012. Major and minor element concentrations in fermented salgam beverage. *Int. J. Food Prop.* 15:903–911. https://doi.org/10.1080/10942912.2010.506621

Yilmaztekin, M., and Sislioglu, K. 2015. Changes in volatile compounds and some physicochemical properties of European cranberrybush (*Viburnum opulus* L.) during ripening through traditional fermentation. *J. Food Sci.* 80(4):687–694. https://doi.org/10.1111/1750–3841.12836

Zakłos-Szyda, M., and Pawlik, N. 2019. The influence of *Viburnum opulus* polyphenolic compounds on metabolic activity and migration of hela and MCF cells. *Acta Innov.* 31:33–42. https://doi.org/10.32933/ActaInnovations.31.4

Zayachkivska, O.S., Gzhegotsky, M.R., Terletska, O.I., et al. 2006. Influence of *Viburnum opulus* proanthocyanidins on stress-induced gastrointestinal mucosal damage. *J. Physiol. Pharm.* 57(5):155–167.

Zhadyra, S., Han, X., Anapiyayev, B.B., et al. 2021. Bacterial diversity analysis in Kazakh fermented milk Shubat and Ayran by combining culture-dependent and culture-independent methods. *LWT* 141:110877. https://doi.org/10.1016/j.lwt.2021.110877

Zheng, J., Wittouck, S., Salvetti, E., et al. 2020. A taxonomic note on the genus Lactobacillus: Description of 23 novel genera, emended description of the genus Lactobacillus Beijerinck 1901, and union of Lactobacillaceae and Leuconostocaceae. *Int. J. Syst. Evol. Microbiol.* 70(4):2782–2858. https://doi.org/10.1099/ijsem.0.004107

Zorba, M., Hancioglu, O., Genc, M., et al. 2003. The use of starter cultures in the fermentation of Boza, a traditional Turkish beverage. *Process. Biochem.* 38:1405–1411. https://doi.org/10.1016/S0032–9592(03)00033

Traditional Turkish Fermented Foods and Their Health Benefits

21

Hande Baltacioglu and Hasan Tanguler

21.1 INTRODUCTION

Traditional food products that are manufactured by fermentation have been consumed abundantly all over the world during the last few decades. Today, it has been reported that more than 5000 different fermented products are consumed in the world (Ray et al. 2016). These products have been produced in Middle Eastern countries, Asia, and Africa, at home-scale and/or small/large-scale industrial levels (Kabak and Dobson 2011; Tanguler 2014). Fermentation, which is known as one of the oldest food preparation techniques, has been used in food protection for ages to enhance the quality, nutritional value, sensory properties, and shelf life of food by the aid of microorganisms such as yeast, fungi, and bacteria (Sankaranarayanan and Amaresan 2020).

In recent years, fermented foods have gained popularity among consumers due to their beneficial effects on health. With the help of fermentation, the quantity of some carbohydrates, such as lactose in milk, which is not easily digested, is decreased, while vitamins, essential amino acids, and minerals are increased (Blandino et al. 2003). Different organic acids (lactic and acetic acid), carbon dioxide, ethyl alcohol, antimicrobial peptides, and bacteriocins that are formed during fermentation act as preservatives (Aloys and Angeline 2009; Sankaranarayanan and Amaresan 2020). This preservation method has been used by ancient people to store foods derived from plant or animal materials, which are perishable and seasonal. Fermented food products were very popular among people living in the Middle East during 7000–8000 BC. The first report on the health effects of fermented milk on gastrointestinal infections was by the Roman historian Pliny in 76 AD. Then, in light of advances in science, the health impact of fermented foods attracted more attention from consumers (Ray et al. 2016).

Certain components found in milk, cereals/grains, and vegetables can be substrates for the production of fermented foods as the fermenting nutrient-dense matrix. Fermented products, which use these substrate food matrices, are important sources of probiotics and nutritional compounds (Kabak and Dobson 2011). Yoghurt is a well-known fermented milk product both historically and commercially worldwide. Yoghurt, obtained by fermentation of milk with lactic acid bacteria (LAB) such as *Streptococcus thermophilus* (formerly known *St. salivarius* ssp. *thermophilus*) and *Lactobacillus* (*L.*) *delbrueckii* ssp. *bulgaricus* (formerly known *L. bulgaricus*) is a highly nutritious protein-rich product (Prado et al. 2008; Béal and Helinck 2014; Sanlier et al. 2019; Zheng et al. 2020). Besides proteins, it has many compounds that have nutritional value such as minerals and vitamins in high amounts (Konuray and Erginkaya 2020).

The amino acid and vitamin values of cereals and legumes can be increased through fermentation in cereal-based fermented foods (Ozdemir et al. 2007). One of the most well-known cereal and legume-based fermented products specific to Turkey is chickpea bread. It is a conventional product and is specially produced in Central Anatolia and the Mediterranean regions of Turkey. The mixture of wheat flour, salt, and chickpea yeast, obtained by fermentation of chickpea, salt, and water at 37–40°C for 16–18 h, is used to produce chickpea bread (Konuray and Erginkaya 2020). *Tarhana*, prepared with wheat flour, yoghurt, vegetables (such as tomato and onion), and herbs (for example, mint), is another traditional fermented food known in Turkey and mostly consumed as soup in winter. After the fermentation, the *tarhana* mixture is dehydrated in the sunlight. It has an acidic taste due to the pH decrease during fermentation (Karacil and Acar Tek 2013).

These traditional foods that reflect Turkish culture are widely consumed throughout the world due to their good taste. Beyond this, they have several health effects such as antimicrobial, probiotic, and antioxidant activity. There have been many studies about these fermented products due to the increasing

DOI: 10.1201/9781003605300-21

demand for them. In this chapter, the general information, metabolic process of the fermentation steps, and health benefits of these traditional Turkish fermented foods (yoghurt, chickpea bread, and *tarhana*) are highlighted.

21.2 YOGHURT

Yoghurt, traditionally manufactured from milk by fermentation, has been produced since ancient times (Kabak and Dobson 2011; Uzunsoy and Ozer 2019). Although it is not known exactly, it is thought that yoghurt was discovered by nomadic Turkish tribes in Central Asia in 2000 BC by souring the milk that was stored in animal stomachs to transport (Weerathilake et al. 2014; Uzunsoy and Ozer 2019). Roman historian Gaius Plinius Secundus, who lived between AD 23 and 79, used the word "yoghurt" for the first time and wrote down the idea that nomadic communities produced yoghurt by souring milk (Weerathilake et al. 2014). It is stated that yoghurt was consumed in the Far East in the 10th–11th centuries in all of the Uyghur texts in which the word "yoghurt" is included. Similarly, the word yoghurt is used in its present meaning in the works titled "Divan-u Lugati Turk" (written in 1072) written by Kasgarli Mahmut in the 10th century and "Kutadgu Bilig" written by Balasagunlu Yusuf Hacib. According to Kasgarli Mahmud, Turks stated that they were the first community to use yoghurt for medicinal purposes, and he wrote that Turkish communities used yoghurt for the first time for diarrhoea and skin burns (Uzunsoy and Ozer 2019).

Yoghurt has also been described as a cleansing agent and beauty lotion. It is reported that in 1208, Genghis Khan, the founder of the Mongol Empire, added yoghurt to his daily diet to give his soldiers courage and strength. According to the prevailing thought today, yoghurt is a Turkish invention, and it first spread to the Balkans and the Middle East, and then to Europe, through migration routes from the regions living under Turkish rule and culture. The clearest information on the spread of yoghurt in Europe is found in French sources. According to this, yoghurt reached Europe for the first time in 1542 via the Turkish doctor sent to the French King Francis I by Suleiman the Magnificent to perform his treatment and spread to Europe. The transition of yoghurt production to the Americas after Europe was a few centuries later. In 1905, Stamen Grigorov, a 27-year-old student, discovered *L. delbrueckii subsp. bulgaricus* (formerly known as *L. bulgaricus*) in Bulgaria. In 1909, Elie Metchnikoff revealed that ageing is caused by bacteria that are toxic in the intestinal flora, and lactic acid bacteria have a life-prolonging effect. In "Prolongation of Life: Optimistic Studies" published by Metchnikoff, *L. delbrueckii* subsp. *bulgaricus* was reported to have potentially life-prolonging effects. Isaac Carasso (1874–1939), from Thessaloniki, was recorded in 1919 as the first person to market fermented milk products, which are not well known in Europe, as a health product in pharmacies. After Isaac Carasso immigrated to Barcelona, he established one of the first yoghurt factories in Europe. This factory continues its production under the name of Groupe Danone today. The first commercial yoghurt culture was established in the USA in 1932 and produced by Dr. J. M. Rossell (Tamime and Robinson 2007; Fisberg and Machado 2015; Uzunsoy and Ozer 2019).

There have been many yoghurt or yoghurt-like products around the world that are similar to each other but are called by different names such as *kisselmleka/naja/yaourt* in the Balkans, *urgotnic* in Balkan mountains, *yiaourti* in Greece, *cieddu* in Italy, *mezzoradu* in Sicily, *gioddu* in Sardinia, *tarho/taho* in Hungary, *viili* in Finland, *filmjolk/fillbunke/filbunk/surmelk/taettem* in Scandinavia, *skyr* in Iceland, *gruzoviz* in Yugoslavia, *iogurte* in Portugal, *donskaya/varenetes/kurugna/ryzhenka/guslyanka* in Russia, *busa* in Turkestan, *katyk* in Transcaucasia, *mazun/matzoon, matsun, matsoni, madzoon* in Armenia, *leban/laban* in Lebanon and some Arab countries, *zabady/zabade* in Egypt and Sudan, *mast/dough/doogh* in Iran and Afghanistan, *roba/rob* in Iraq, *dahi/dadhi/dahee* in India, *shosim/sho/thara* in Nepal, and *yogurt/yoghurt/yaort/yourt/yaourti/yahourth/yogur/yaghourt* in other countries (Uzunsoy and Ozer 2019; Tamang et al. 2020). These different products have different organoleptic properties. While Turkish-type yoghurt is acidic, savoury, and thick-gel viscous, other yoghurt-like products may be mild-acidic, thick, and sticky and have a sweet taste (Tamang et al. 2020).

Various modifications have been made based on practical observation for centuries to increase the strength of yoghurt. The best known among these modifications is removing the water from the yoghurt and transforming it into a more concentrated and acidic product. This product is known as torba yoghurt (filtered yoghurt, bag yoghurt) and has been manufactured in Anatolia for centuries. In the traditional torba yoghurt production method, yoghurt is filtered through a cheesecloth bag for a long time (10–14 h) to remove whey. At the end of the process, salt can be added to enhance the shelf life of the product, and it is stored at 4°C (Kabak and Dobson 2011). Another modification is the drying of yoghurt. This dried yoghurt is known as kurut and is made by dehydration of yoghurt or ayran. It is especially manufactured in the Eastern part of Turkey and consumed with traditional foods. It is mixed with water to gain a viscous yoghurt texture, so it is easily used instead of yoghurt (Konuray and Erginkaya 2020).

Concerning the Turkish Food Codex (TFC), yoghurt is a coagulated dairy product manufactured by lactic acid fermentation (LAF) caused by the bacteria *L. delbrueckii* subsp. *bulgaricus* and *St. thermophilus*. A more detailed definition is given in the Turkish Standards Institute Yoghurt Standard (TS 1330). According to this, yoghurt is the product obtained after adding yoghurt culture involving *L. delbrueckii* subsp. *bulgaricus* and *St. thermophilus* to pasteurized cow's milk (TS 1018), buffalo milk (TS 11045), sheep's milk (TS 11044), goat's milk (TS 11046), or their combination, with or without homogenization with the addition of powdered milk (TS 1329) if it is required, and then processed in accordance with the Yoghurt Making Rules Standard (TS 10935) (TFC 2009).

In Turkey, yoghurt can be manufactured traditionally at home scale and has been since ancient times. In this production method, the first step is boiling of milk until some of the water has evaporated (it is determined by just looking at it; the milk is boiled until about one-third of its starting volume has evaporated), then it is cooled down to 40–45°C (people check this temperature by dipping their fingers in milk to determine the temperature of milk; if it has reached a level that will not burn their finger, they consider it has reached the appropriate temperature), and then a small piece of yoghurt is used to ferment the milk, and it is kept at stable temperature (a thick cloth is wrapped around the container to maintain its temperature) until the yoghurt structure is formed and then cooled down (Konuray and Erginkaya 2020).

The manufacturing of yoghurt has been made at an industrial scale during the last few decades. This production method includes the standardization of milk (skim milk or whey protein concentrate is used to regulate total solid content, and cream/butterfat is used to adjust fat content), homogenization (55–65°C and 10–20 MPa at first stage pressures and 5 MPa at second stage pressure), milk heat treatment (80–85°C for 30 min or 90–95°C for 5 min), cooling to incubation temperature (~45°C), inoculation of starter culture (2% v/v), packing into individual containers, incubation/fermentation (at 42–43°C for 3–6 h until pH reached to 4.4), cooling (< 4°C), and storage for set-type yoghurt (Lee and Lucey 2010; Kabak and Dobson 2011; Weerathilake et al. 2014). In the stirred-type yoghurt production process, after inoculation, yoghurt is left to ferment without being placed in single containers. After that, it is cooled, stirred, cooled, pumped, and packaged. The final product is stored in a cold (<4°C) place. Many ingredients are added to yoghurt. One of them is stabilizers that increase the texture, prevent syneresis, and distribute ingredients properly. The addition of stabilizers such as gelatin, starch, and pectin increased the pH value and decreased the resistance to the separation of whey and curd of yoghurt (Saadi et al. 2022). Sweeteners, flavours, and sometimes fruits may also be added to increase the flavour of yoghurt (Weerathilake et al. 2014).

The basic chemical quality characteristics of yoghurt are defined in the TFC (TFC 2009). Accordingly, yoghurt should contain at least 3% protein by weight and a maximum of 15% fat by weight. Also, the lowest and highest values of titration acidity must be 0.6% and 1.5% as lactic acid, respectively. Although yoghurt is similar to milk in terms of its chemical composition, it differs depending on the composition of milk, resulting from the processes applied to the milk during yoghurt making, the substances added during production, or the changes that occur during bacterial fermentation. In general, yoghurt contains 80–86% water, 14–20% dry matter, 2–8% fat, 4–8% protein, 2–5% milk sugar, 0.8–1.2% mineral matter, 4.6 g saturated fatty acid, 2.3 g monounsaturated fatty acid, 0.3 g polyunsaturated fatty acid, 28.0 mg cholesterol, 0.1 mg iron, 100 mg sodium, 260 mg calcium, 320 mg potassium, 200 mg phosphorus, and 0.9 mg zinc per 200 g (Kizilaslan and Solak 2016). Moreover, the vitamin content of yoghurt is reported as 0.06 mg thiamin, 0.2 mg niacin, 0.27 mg riboflavin, 0.1 mg vitamin B_6, 0.2 mg vitamin B_{12}, 18 µg folate, and 21 µg carotene per 100 g (Weerathilake et al. 2014).

The general features of yoghurt obtained from different locations in Turkey can be distinctive. The chemical compositions of the yoghurt marketed in Ankara, Turkey, have been determined in the range between 35–70 SH for the acidity, 9.05–14.25% for total dry solids, for 7.25–10.85% nonfat dry-matter, and 1.4–4.4% for fat content (Kochisarli and Ergul 1987). Chemical contents of yoghurt samples produced in Kirklareli, Turkey, have been reported as in the range between 3.90–5.25, 0.10–1.88%, and 1.0–3.8% for pH, acidity, and fat content, respectively (Cetin et al. 2014). The pH, acidity, fat, nonfat dry-matter, protein, and ash contents of yoghurt samples sold in Sanliurfa, Turkey were 3.31–4.16, 0.98–1.56%, 1.40–3.85%, 6.41–9.83%, 2.24–5.44%, and 0.59–0.85%, respectively (Turkoglu et al. 2003). In another type of yoghurt sold in the UK, the fat, protein, and carbohydrate (total sugar) content vary between 1.5–5.2 g/100g, 2.7–5.4 g/100g, and 5.5–19.6 g/100g, respectively (Moore et al. 2018). Yoghurt is known as a good source of calcium owing to both its high content and also the high absorbability of calcium from yoghurt (Pei et al. 2017).

Some nutrients such as vitamins and minerals can be added to yoghurt to reduce nutritional deficiencies due to its high consumption rate (Gahruie et al. 2015). Yoghurt is a suitable product for the fortification of vitamin D_3. It is very stable in set-type yoghurt and does not degrade during processing and long-term storage of 4 weeks (Kazmi et al. 2007). Moreover, vitamin A (10,000 IU per 227 g) and vitamin C (300 mg per 227 g), which supply at least 100% of the Recommended Dietary Allowances (RDA) for both vitamins, are found to be stable in yoghurt during 6 weeks of storage at 3°C (Ilic and Ashoor 1988). Microencapsulated iron and vitamin C can be also used for the fortification of drink yoghurt, which showed high efficiency of 73% and 76%, respectively (Kim et al. 2003a).

Yoghurt is a typical viscoelastic gel, so elastic and viscous characteristics are evaluated together in the measurement of its physical quality. It can act as either a viscoelastic fluid if it is stirred or a viscoelastic solid if it is set-type yoghurt. Dry matter fortification, thermal treatment, incubation temperature, and enzymatic cross-linking of caseins can affect the physical properties of yoghurts (Lee and Lucey 2010). Natural food ingredients such as flaxseed to a certain extent can be added to increase the textural properties of yoghurt (Mousavi et al. 2019).

Yoghurt is described as a coagulated product that is formed by LAF with *L. delbrueckii* subsp. *bulgaricus* and *St. thermophilus* (Sanlier et al. 2019). Yoghurt, as a good carrier of probiotic strains, has been a product preferred by consumers for a long time. In recent years, "bio-yoghurts" that contain AB cultures (*L. acidophilus* and *Bifidobacterium*) and traditional yoghurt organisms (*St. thermophilus* and *L. delbrueckii* subsp. *bulgaricus*) have gained popularity (Stanton et al. 2003). Consumption of this probiotic yoghurt, which is obtained with the addition of AB cultures, improves the normal flora of people (McFarland 2015). Traditional yoghurt processing procedures can be carried out for the manufacturing of probiotic yoghurts, and probiotic bacteria can be added simultaneously with traditional yoghurt cultures before fermentation or after fermentation to the chilled product before packaging (Stanton et al. 2003).

Due to its acidic nature, yoghurt has qualities that can inhibit the growth of many pathogenic bacteria. In particular, *Listeria monocytogenes*, *Salmonella* spp., coliform bacteria, and *Staphylococcus aureus* rarely survive in yoghurt. These bacteria quickly become inactive within the first 24 hours after yoghurt production. The survival time of pathogenic bacteria in yoghurt is largely related to the extent of contamination and the pH of the yoghurt. It is known that these bacteria partially preserve their survival in very severe bacterial contamination or acidity above pH 4.6. While yoghurt starter bacteria inhibit the growth of enterotoxigenic *Staphylococcus aureus* to some extent, staphylococcal enterotoxin-A synthesized by this pathogenic bacteria can persist in yoghurt during storage. Similarly, although yoghurt starter bacteria show strong antagonistic effects against *Listeria monocytogenes*, some strains of this bacterium develop acid adaptation in acidic environments, so they can grow in yoghurt during storage. The presence of coliform bacteria in yoghurt is accepted as an indicator of insufficient sanitation in yoghurt production. Coliform bacteria generally cannot survive in yoghurts with low acidity values. It is also known that these bacteria are not resistant to some antibiotics synthesized by yoghurt starter cultures. On the other hand, the contamination of verotoxigenic *Escherichia* (*E.*) *coli* O157:H7, which has been effective in food poisoning in recent years, in yoghurt after thermal treatment brings a serious health risk. Although *E. coli* O157:H7 is adversely affected by low acidity, it is not completely destroyed during storage (Uzunsoy and Ozer 2019).

In the yoghurt manufacturing process, during the fermentation of milk, LAB convert lactose to lactic acid and produce compounds that give yoghurt its characteristic texture, flavour, and aroma. LAB produce exopolysaccharides (EPS), which improve the texture of yoghurt by increasing the viscosity. Lactic acid provides the acidic taste of yoghurt, while the characteristic flavour of the end product is mainly attributed to acetaldehyde above and beyond that provided by the pungency of the lactic acid (Water 2003; Weerathilake et al. 2014; Uzunsoy and Ozer 2019).

21.2.1 Benefits of Yoghurt on Health

Yoghurt is a very nutritious food product due to its protein, vitamin, and mineral content. It provides the daily requirements of some nutrients indicated by RDA. For instance, a serving of 50 g of yoghurt is suggested to ensure 41% of the RDA value of calcium for a 5-year-old due to its calcium content being high with good bioavailability (Weerathilake et al. 2014).

Yoghurt has a substantial role in the diet owing to its health benefits that are mentioned even in the Bible and the ancient books of Hinduism (Weerathilake et al. 2014). Yoghurt contains proteins with high biological value, and its digestibility is twice as high as milk. A significant part of the lactose in milk is hydrolysed and turned into lactic acid; the amount of lactose decreases to approximately 4%. Therefore, the digestibility of carbohydrates in yoghurt increases compared to milk (Kizilaslan and Solak 2016). Many studies, whose subjects were people whose ages range from 7 months to 53 years, have shown that yoghurt consumption reduces lactose intolerance symptoms and yoghurt cultures significantly increase lactose digestion (Savaiano and Hutkins 2021). In some studies, it has been reported that lactose digestion and tolerance are increased when people who have difficulty digesting food add yoghurt to their diet (Kolars et al. 1984; Hertzler and Clancy 2003). Lactose digestion is significantly improved with fresh yoghurt alone even between different yoghurts (Savaiano et al. 1984; Martini et al. 1987). Moreover, while lactose digestion was only marginally improved with milk fermented with *Bifidobacterium* (*Bi.*) *bifidus*, significant improvements were achieved with milk fermented with LAB (Martini et al. 1991).

Studies have shown that there is a direct relationship between yoghurt consumption and intestinal health (Kok and Hutkins 2018). LAB and probiotic bacteria (*L. acidophilus* and *Bi. bifidum*) that can be highly viable in yoghurt are reported to provide the intestinal balance that is also associated with good health (Adolfsson et al. 2004; Morelli 2014). These strains colonize both the small and large intestines and protect against pathogenic bacteria by providing a physical barrier, producing antimicrobial peptides, bacteriocins, and short-chain fatty acids that lower the pH to be unsuitable for the growth of other microbes (Kok and Hutkins 2018). It is declared that consumption of yoghurt with probiotics is the cure for some gastrointestinal illnesses such as ulcerative colitis and Crohn's disease, preventing urogenital infections, alleviating constipation, counteracting acute gastroenteritis-induced barrier dysfunction, protecting against diarrhoea, preventing infant diarrhoea, and protecting against colon/bladder cancer (Adolfsson et al. 2004; Mckinley 2005; Weerathilake et al. 2014; Pei et al. 2017). Several studies showed that various *Lactobacillus* strains inhibited the growth of *Helicobacter* (*H.*) *pylori*, which is associated with peptic ulcer disease, gastric adenocarcinoma, chronic gastritis, and mucosa-associated lymphoid tissue lymphoma (Adolfsson et al. 2004). Different *Lactobacillus* strains such as *Lacticaseibacillus casei* (Basonym *L. casei*) (Zheng et al. 2020), *L. acidophilus*, and *L. gasseri* produce different amounts of lactic acid that are correlated with inhibitory activity on *H. pylori* (Adolfsson et al. 2004). Moreover, bacteriocins generated by LAB have also anti-*Helicobacter* activity. Lacticins produced by *Lactococcus* (*Lc.*) *lactis* along with many different bacteriocins showed the greatest *Helicobacter* inhibitory activity when used against different strains of *H. pylori* (Kim et al. 2003b). Probiotic strains can be protected from stomach acid by encapsulation so their viability can be improved, and their metabolic effects in the intestine may become apparent (Pei et al. 2017). Yoghurt consumption provides improvement of the plasma concentrations of B groups vitamins such as B_1, B_2, and B_6 in healthy adults and young women due to the beneficial action of yoghurt cultures (Elmadfa et al. 2001; Fabian et al. 2008).

To support the growth of probiotics in the colon, prebiotics is used in yoghurt production, so the gastrointestinal motility and immune system can be improved (Das et al. 2019). Prebiotics such as fructooligosaccharides, xylooligosaccharides, galactooligosaccharides, polydextrose, and some polysaccharides are obtained from different fruit and vegetable sources such as soybeans, Jerusalem artichoke (inulin), oats, wheat, barley, onion, banana, and asparagus by commercial extraction (Fazilah et al. 2018). The addition of guar gum, inulin, β-glucan, dietary fibers from apple and banana, passion fruit, and cooked rice as prebiotics may help increase the number and existence of probiotic bacteria (Mudgil et al. 2016; Shaghaghi et al. 2013; Fernandez-Garcia and McGregor 1997; Espirito Santo et al. 2012; Espirito-Santo et al. 2013; Kumari et al. 2015). Moreover, due to the symbiotic effect of prebiotics and probiotics, some antibacterial factors such as bacteriocin can be released, the absorption of minerals can be improved, diarrhoea can be prevented, and the assimilation of nutrients can be optimized (Fazilah et al. 2018; Das et al. 2019). It is suggested that the antioxidant capacity of human plasma and antioxidant enzymes activities can be positively influenced by symbiotics containing *Lacticaseibacillus casei* (Basonym *L. casei*) and inulin (Kleniewska et al. 2016; Zheng et al. 2020). Adding to children's diet every day with yoghurt including probiotic bacteria (*Bi. animalis* subsp. *lactis*) and inulin significantly enhanced the frequency of bowel movements and reduced fever days (Ringel-Kulka et al. 2015). It was presented that the antioxidant activity of yoghurts was enhanced by the presence of probiotics (Tavakoli et al. 2019). Additionally, the antioxidant activity of yoghurts can be improved with the addition of some phytochemicals such as phenolic compounds, carotenoids and phytosterols in the form of plant extracts, essential oil, or fruit waste (Fazilah et al. 2018, Abdullah et al. 2023). Phytochemicals in spices provide an inhibitory effect against pathogenic bacteria such as *Listeria monocytogenes* and *E. coli* so the growth of probiotics can be enhanced (Azizkhani and Tooryan 2016). Phytochemicals in the form of fruit waste such as pineapple pomace for use in yoghurt production showed antioxidant properties as well as prebiotic effects (Sah et al. 2016).

Many studies have shown that yoghurt consumption improved health and reduced the risk of many diseases such as metabolic syndrome, cancer, cardiovascular disease, obesity, and diabetes (Fernandez et al. 2017; Kok and Hutkins 2018). In an epidemiological study, yoghurt consumption was reported to reduce the risk of metabolic syndrome in older Mediterranean adults (Babio et al. 2015). It was recommended that high consumption of fermented milk (yoghurt) lessened the incidence of cardiovascular disease at a rate of 15% (Sonestedt et al. 2011). Strong evidence was reported for the opposite relationship between yoghurt intake and the risk of type 2 diabetes concerning most observational studies (Salas-Salvado et al. 2017). In two different meta-analysis studies, an opposite relationship was reported between yoghurt intake and the risk of type 2 diabetes. It was reported that consumption of yoghurt between 80–125 g/day and 80–140 g/day declined the risk of diabetes at a rate of 14% and 11–17%, respectively (Aune et al. 2013; Gijsbers et al. 2016). Yoghurt supplementation at a rate of 5% is reported to prevent glucose intolerance; standardize enzyme activities that are liver specific; reduce the amounts of oxidative stress markers in the liver and plasma; prevent collagen deposition, inflammatory cell infiltration, and fibrosis in the liver; and standardize the intestinal lining and brush border in rats with high-fat diet-induced obesity. Therefore, it was suggested that metabolic syndrome was prevented alternatively by supplementation of yoghurt in high-fat diet-fed rats (Lasker et al. 2019). Published studies between 1987 and 2017 showed that the consumption of yoghurt has a favourable effect on gastrointestinal health and prevents diarrhoea and allergies among children aged 4 to 24 months (Donovan and Rao 2019). Similar results were also observed for preschool children aged 3–5 who supplemented with a serving of 125 g of yoghurt 5 days a week. For children, calcium, zinc, and vitamin B_2 intake were improved; the incidence and duration of upper-respiratory infection and diarrhoea were decreased; and the health, growth, and development of the children were promoted (He et al. 2005). Yoghurt provides important macro- and micronutrients such as protein, fatty acids, calcium, zinc, potassium, magnesium, phosphorus, vitamin A, vitamin B_5, riboflavin, and vitamin B_{12} in children's diet that could provide beneficial effects on gut microbiota, obesity, and cardiometabolic risk (Marette and Picard-Deland 2014). Based on the clinical trial it was reported that consumption of yoghurt with probiotics can improve the symptoms of constipation during pregnancy (Mirghafourvand et al. 2016). Supplementation of PUFA in yoghurt provides cardiovascular health benefits without altering body weight or low-density lipoprotein plasma levels (Cortés et al. 2023).

In a meta-analysis of the studies up to November 2019, daily consumption of probiotic yoghurt at a rate below 200 g for more than 8 weeks showed a significant reduction in serum CRP (C-reactive protein) levels (<3 mg/dl) in overweight people (Mousavi et al. 2020). Studies conducted over 4–8 weeks showed that yoghurt consumption decreased total cholesterol at a rate of 4%, LDL cholesterol at a rate of 5%, and significantly reduced blood pressure owing to its high content of protein, calcium, and probiotics. The consumption of yoghurt is also associated with losing weight, which reduced cardiovascular disease in observational studies (Astrup 2014). Similarly, serum total cholesterol levels were lowered in 14 healthy volunteers by consuming probiotic yoghurt containing *L. acidophilus* and *Bi. lactis* compared to ordinary yoghurt (Ataie-Jafari et al. 2009). The consumption of yoghurt containing *L. delbrueckii* subsp. *bulgaricus* and *St. thermophiles* significantly reduced metabolic risk markers such as total cholesterol concentrations and plasma glucose levels compared to a yoghurt-free diet in two studies (Dumas et al. 2017).

21.3 CHICKPEA BREAD

Chickpea (*Cicer arietinum* L.) is one of the most important legume crops intended for human consumption in the world. It belongs to the Fabaceae family and is a basic food crop with high protein (21.70%), minerals (Ca, Mg, Fe, P, K), vitamins (niacin, thiamine, ascorbic acid), essential amino acids, and unsaturated fatty acids (oleic, linoleic), particularly in tropical and subtropical areas. Chickpeas can be cooked both as a meal, and fermented products of chickpeas are produced in Middle Asia, the Middle East, and African countries (Adebo et al. 2017; Aydin and Cebi 2019).

One of the fermented products of chickpeas is chickpea bread, produced in small bakeries or homes in Turkey. Chickpea bread, a legume-cereal-based fermented product, is specific to Turkey. It is produced by chickpea yeast, wheat flour, and salt. To obtain chickpea yeast as a dough-leavening agent, the crushed chickpea seeds, salt, and water at 37–40°C are fermented for 16–18 h, and then the mixture is filtered. Chickpea yeast makes the dough softer, affects the color of the inner and outer crust, and gives it its characteristic taste and aroma (Tanguler 2014; Konuray and Erginkaya 2020). Chickpea yeast is also used to make simit, galeta, Melba toast (peksimet), and kumru (Ozer et al. 2010). The chemical constituents of chickpea bread have been found as carbohydrate 47.45%, moisture 35%, ash 1.75%, and protein 13.60% (Tuncel et al. 2010).

LAB have been isolated and identified in the liquid phase of fermented chickpea. *Lc. lactis* (90%), *Lactiplantibacillus (Lpb.) plantarum* (formerly known as *L. plantarum)* (6.7%), and *Levilactobacillus brevis* (Basonym: *L. brevis*) (3.3%) were phenotypically identified in chickpea fermentation liquid in a total of 60 isolates of LAB (Aydin and Cebi 2019; Zheng et al. 2020). LAB, aerobic spore bacteria, yeast and anaerobe spore bacteria were found as dominant microflora in the liquid phase of fermented chickpea and dough produced from this liquid. *Saccharomyces cerevisiae*, *Candida parapsilosis*, *Meyerozyma guilliermondii*, *Pichia fermentans*, and *Cryptococcus albidosimilis* were identified at the rate of 40.7%, 33.9%, 20.3%, 3.4%, and 1.7%, respectively, from 59 yeast strains in chickpea liquid starter and dough samples (Boyaci Gunduz and Erten 2022). In the manufacturing of bread with the liquid phase of fermented chickpea, not only did LAB and yeast have an important role but also the spore-forming bacteria of *Bacillus* (*B.*) and *Clostridium* species were found to be useful (Erginkaya et al. 2016).

There have been few studies on the microflora of chickpea bread. In a work conducted in Turkey, *Enterococcus* (*En.*) *mundtii*, *En. casseliflavus*, *Fructilactobacillus sanfranciscensis (Basonym: L. sanfrancisco)*, *Saccharomyces* (*S.*) *cerevisiae*, *Lpb. plantarum*, *L. viridescens*, *Loigolactobacillus bifermentans* (Basonym: *L. bifermentans*), *Pediococcus* (*P.*) *urinae-Equi*, *St. thermophilus*, and *Lc. lactis* subsp. *cremoris* were identified in chickpea bread (Hancioglu-Sıkılı 2003; Zheng et al. 2020). In a study in which fermented chickpea liquor and dough samples were picked up from bakeries in Turkey and examined for the microbiological character, *Weissella* (*W.*) *cibaria* and *W. confusa* constituted 12.4% and 47.8%, respectively, of total LAB isolated. *En. faecium*, *En. lactis*, *Levilactobacillus brevis*, *Lpb. plantarum*, *Leuconostoc* (*Ln.*) *mesenteroides*, *Ln. mesenteroides* subsp. *dextranium*, *Pediococcus* (*P.*) *acidilactici*, *Pediococcus* (*P.*) *pentosaceus*, and *Streptococcus* (*St.) lutetiensis* were identified as the other species. Chickpea liquor and dough samples inoculated with *W. confusa* showed a higher formation of butanoic acid and ethyl acetate at a rate of 81.52% and 8.15%, respectively, than the control sample (Boyaci Gunduz et al. 2020).

In another study in which fermented chickpea liquor was formulated into whole-grain wheat bread at a rate of 30% and 45% levels, loaf volume increased by approximately 10–25%, crumb hardness decreased by about 10–15%, and antioxidant activity was enhanced by about 10%. However, the usage of fermented chickpea liquor had a limited effect on flavour and colour (Sayaslan and Sahin 2018). Moreover, the effects of this liquor on gluten-free bread were investigated. Chickpea liquor was put into gluten-free bread formulations as yeast source at a rate of 30% and 45% levels, and the loaf volume increased by about 3–12%, staling was retarded, the flavour was improved, and nutritional quality was increased by about 10% with the addition of yeast (Sahin et al. 2018). To increase the quality and extend the shelf life, it has been suggested to use sour dough and fermented chickpea extract together as a starter in gluten-free bread (Mygdalia et al. 2022).

Studies have investigated the changes in composition and cellulase, amylase, α-galactosidase, invertase, and proteinase activities in chickpea liquid during fermentation. The population of bacilli and clostridia was increased in the liquor throughout fermentation. A gradual decrease was observed in pH, and acidity was increased. Similarly, free fatty acid concentrations were increased. The amount of reducing sugars increased up to 12 h, and then a gradual decrease was observed for reducing sugar. A decline was also detected for many of the soluble carbohydrates with fermentation, and solubilization-degradation of starch was observed after 8 hours. The increasing rate of the free amino acid content was determined during fermentation, and protein degradation was found to be prominent after 8 hours of incubation (Hatzikamari et al. 2007a). In the fermenting liquid of chickpea, *Bacillus* spp. (especially *B. cereus*, *B. licheniformis*, and *B. thuringiensis*) and *Clostridium* spp. (especially *Cl. perfringens* and *Cl. beijerinckii*) were identified and increased during fermentation when the pH of the fermenting liquid had lowered to 5.4. In addition, it was reported that it can be safe for consumption (Hatzikamari et al. 2007b).

Turkey has a wide variety of bread cultures. Nevsehir bagel is a traditional type of chickpea bread consumed in Central Anatolia and produced using natural chickpea yeast. Chickpea yeast is prepared by breaking dry chickpeas and fermenting them in hot water at a constant temperature for about 16 hours (Figure 21.1a). Chickpeas incubated in water are fermented and foamed (Figure 21.1b). The fermented chickpea yeast is filtered, and the dough is kneaded with this filtered water. Nevsehir bagel has an oblong shape and has a little bulging area on the centre of it (Figure 21.1c). Nevsehir bagel has a distinctive smell from the chickpea yeast inside it, and it is salt-free and unadulterated (Sozbilen 2019; Anonymous 2021).

Karahoyuk bread is also a traditional type of chickpea bread known in Denizli province and its surroundings. Chickpea yeast and chickpea flour are used for the production of this bread, which has a unique taste, appearance, and

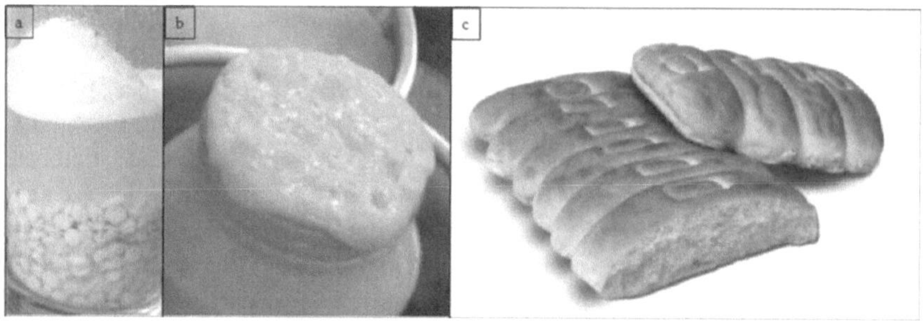

FIGURE 21.1 (a) Soaking chickpeas in water to obtain yeast. (b) Chickpea starter used for a dough. (c) Traditional chickpea bread (Turkish bagel) consumed in Nevsehir, Turkey.

aroma. The hardness increased during 5 days of storage. It is thought that chickpea flour, which is one of the raw materials used in bread production, is effective in the color formation of the bread crumb and the dominance of the yellow color due to the increase in the b value. When color parameters of fresh bread crumb and crust were compared, color parameters L* and b* of bread crumb samples were found to be higher than those of crust samples (Demiray et al. 2015).

21.3.1 Benefits of Chickpea Bread on Health

Chickpea yeast is commonly used as a leavening agent in bread formulations to obtain prominent flavour and taste. It may also develop the functional, rheological, nutritional, and sensorial quality of products (Saad et al. 2015). The nutritional and physicochemical properties and shelf life of wheat bread prepared with chickpea yeast were enhanced compared to the bread prepared with instant yeast. The antioxidant activity and mineral content of the wheat bread prepared with chickpea yeast were found to be significantly higher than the control sample. The mineral content of the samples prepared with chickpea yeast was reported as 57.53% for Na, 3.25% for Mg, 15.56% for K, 11.03% for P, 6.78% for S, 300 ppm for Ca, 151 ppm for Al, 243 ppm for Fe, 107 ppm for Zn, 70 ppm for Mn, and 22 ppm for Cu (Hendek Ertop and Coskun 2018). The chemical contents of kumru, which is a traditional bread made with chickpea yeast and manufactured in Izmir in Turkey, were presented as 6.9–19.4 g/kg for total ash, 633.6–697.1 g/kg for total dry matter, 7.3–16.4 g/kg for salt, 1.01–2.51 g/kg for total free amino acid (as leucine), 0.01–0.02 mg/kg for Cd, 0.00–0.018 mg/kg for Pb, 1.71–10.7 mg/kg for Cu, 5.73–13.18 mg/kg for Zn, 236–396 mg/kg for Ca, and 1027–1410 mg/kg for P (Ozdestan et al. 2012).

Chickpea liquor has a high amount of amino acids. Lysine and serine represented 77.13% and 6.49% of free amino acids, respectively. Arginine and tyrosine were reported as 24% and 20% of total amino acids, respectively. The total carbohydrate level in the freeze-dried chickpea liquor was indicated as 1.47%. Vitamin levels for chickpea liquor were demonstrated as 143 mg/g for nicotinic acid, 31.4 mg/g for pyridoxine, 242 mg/g for thiamin, 11.1 mg/g for riboflavin, 5.9 mg/g for folate, and 245 mg/g for vitamin B_{12}. Chickpea liquor showed very close antioxidant activity compared to the tertiary butyl hydroquinone. *B. megaterium* and *B. subtilis* have been isolated and *B. subtilis* reported as a substantial starter culture for alkaline fermentation in fermented legumes (Saad et al. 2016). Furthermore, many of the LAB in the fermentation liquid of chickpeas and the dough of chickpea bread have been isolated and identified (Aydin and Cebi 2019; Boyaci Gunduz et al. 2020). As is well known, most LAB act as probiotics and can strengthen the immune system of the host even if effects are generally dependent on strain and dose level. In addition to this, LAB have shown antiviral effects that affect viral infections in the upper respiratory and gastrointestinal tract by generating bioactive compounds such as peptides. Besides the living culture, paraprobiotics, non-living cells of probiotics that are formed during the processing and storage of fermented products, also have health benefits. In some studies, it has been suggested that the living and non-living cells of probiotics have similar immuno-stimulatory effects (Ogel and Ozturk 2020). It has been reported that oral intake of heat-killed *Lpb. plantarum* as paraprobiotics has a protective effect against influenza virus infection in mice (Maeda et al. 2009). Therefore, in light of the evidence, chickpea bread can be thought of as not only probiotics but also paraprobiotics and may also have antiviral effects owing to its heat-killed LAB.

21.4 *TARHANA*

Tarhana is a cereal-based fermented product that is manufactured traditionally from a yoghurt and cereal mixture. After fermentation, it is dried and ground to store for a long time and when consumed reconstituted with water and cooked as a soup in winter. Moreover, the dried slices of *tarhana* can also be consumed as nuggets, chips, or snacks in Anatolia, especially in Kahramanmaras. It is a significant part of the diet in Turkey. It is known that *tarhana* was first made by Turkish settlers who lived in Middle Asia, and it was taken to Anatolia, the Middle East, the Balkans, and some European countries by them. It is known by different names in different regions such as "kishk" in Syria, Palestine, Jordan, Lebanon and Egypt; "trahanas" in

Greece, "kushuk" in Iraq and Iran; "talkuna" in Finland, and "thanu" in Hungary (Daglioglu 2000; Tanguler 2014; Tanguler and Tatlısoy 2022).

Tarhana is mainly manufactured at the home scale in Turkey, and several types of *tarhana* have been specified by the Turkish Standardization Institute (TS-2282) such as: (1) flour *tarhana*, (2) cracked wheat (goce) *tarhana*, (3) semolina *tarhana*, and (4) mixed *tarhana* (Anonymous 1981). The major ingredients of *tarhana* are yoghurt (stirred- or set-type yoghurt) and cereals and legumes such as wheat flour, legume flour, chopped wheat, and/or semolina, which can be used separately or in combinations in different quantities in the different types of *tarhana* production (Daglioglu 2000; Maskan and Ibanoglu 2002; Ozdemir et al. 2007). Vegetables that are cooked or raw (tomatoes, green pepper, and onions) and herb and spices such as mint and paprika are put into yoghurt and cereals and fermented for up to 7 days. Sometimes milk, soybean, chickpea, lentil, cornflour, and egg can be also added to formulations in some regions. The proportions of the ingredients may also change from region to region. In most recipes, yoghurt is blended with wheat flour at a rate of ½ (w/w), or in some recipes, an equal amount of yoghurt is added to wheat flour (Daglioglu 2000). This results in the characteristic yeast flavour, the odor of *tarhana*, and an acidic and sour taste due to the organic acid production developed during dough fermentation. It can be stored for more than a year owing to its low moisture content (<10%) and relatively low pH that inhibit growth of pathogenic and spoilage microorganisms (Daglioglu 2000; Tanguler 2014).

Two different commercial-scale productions are defined; the first is the direct or straight method, and the other is the sourdough method (Tanguler 2014). In the straight method (Figure 21.2a), chopped and blended onion is mixed with wheat flour, durum wheat semolina, lentil flour, yoghurt, tomato paste, vegetable oil, red pepper paste, salt, citric acid, and baker's yeast. If desired, tomato juice, chopped red pepper, or various spices may be added in different ratios. Then, water and/or yoghurt base are added and kneaded for dough formation. The dough is laid on a tray in the form of a thin layer subject to fermentation at an average temperature of 35°C for 5 days. Finally, at the end of fermentation, it is sun-dried for home-scale production or oven-dried for commercial production (until a moisture content of 6–10% is reached) at 55°C and ground to a particle size of below 800 μm (Daglioglu 2000; Ozdemir et al. 2007; Tanguler 2014). In the second method (Figure 21.2b), three recipes where each one has distinctive amounts and types of ingredients are used for the production of *tarhana*. Generally, similar ingredients (wheat flour, durum wheat semolina, yoghurt, tomato paste, onion, pepper paste, lentil soup, salt, vegetable oil, and citric acid) used in the direct method are used in different ratios in the first recipe. However, baker's yeast is not used in this method. After mixing and kneading all ingredients in the recipe, the obtained dough is laid on a tray and then fermented at 40–42°C for up to 5 days. For recipe 2, the obtained dough in the first recipe is mixed with flour, tomato paste, semolina, pepper paste, salt, starch, and citric acid and kneaded, then dried to a moisture content

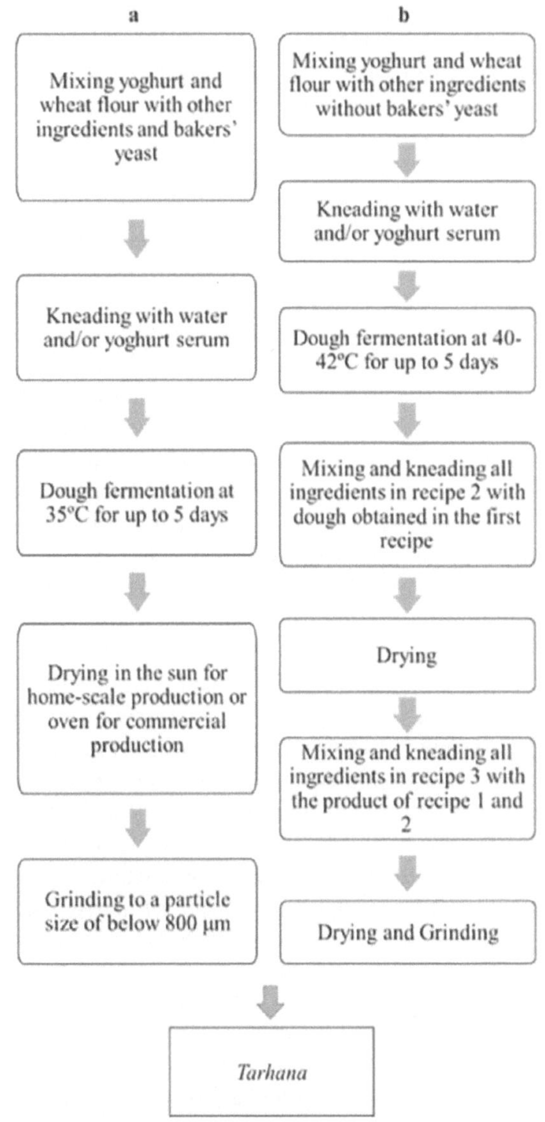

FIGURE 21.2 *Tarhana* production (a) by straight method, (b) by sourdough method.

Source: (Adapted from Daglioglu 2000; Ozdemir et al. 2007; Tanguler 2014).

below 8% at 80°C. For recipe 3, which is used for the production of the final product, the dough obtained from the first and second recipes, tomato paste, and pepper paste are mixed and kneaded. The resulting dough is laid on a tray in the form of a thin layer, dehydrated to a moisture content below 10% at 80°C, and ground to a particle size of 800 μm (Daglioglu 2000; Tamer et al. 2007).

The aromatic taste and flavour of *tarhana* can be formed during fermentation by both LAB and as well as yeast by producing lactic acid, ethanol, CO_2, and organic fermentation products. Many different LAB were isolated and identified in *tarhana*, such as *St. thermophilus*, *Lc. lactis*, *Lc. diacetylactis*, *Lpb. plantarum*, *L. delbrueckii* subsp. *bulgaricus*, *L. acidophilus*, *Ln. cremoris*, *Lacticaseibacillus casei* (Basonym *L. casei*), *L. helveticus*, *Levilactobacillus brevis*, *Ln. mesenteroides* subsp. *mesenteroides, P. pentosaceus*, and *P. acidilactici*. Yeasts identified in *tarhana* are *S. cerevisiae* and *Rhodotorula glutinis* (Daglioglu 2000; Sengun et al. 2009; Settanni et al. 2011; Tanguler 2014). The most produced organic acids in the fermentation of *tarhana* were lactic, succinic, and acetic acids. The amounts of alcohols and esters increased during fermentation, while that of ketones decreased. Ethyl acetate, ethyl lactate, ethanol, 1-butanol-3-methyl, limonene, ethyl caprylate, benzaldehyde, acetic acid, 1,8-cineole, ethyl capronate, neodihydrocarveol, benzene ethanol, and d-carvone are responsible for the formation of the typical aroma of *tarhana* during fermentation (Ozdemir et al. 2019). Drying methods have a great effect on the aroma active volatiles of *tarhana*. In vacuum drying, a greater number of aromatic and active components are retained than the sun-dried method (Gocmen et al. 2004).

At the end of fermentation, the pH (3.8–4.2) becomes lower due to the formation of organic acids, and the moisture content can be lowered below 10% by drying. Those combined processes cause a bacteriostatic effect that inhibits pathogen and spoilage microorganisms so shelf life is extended (Daglioglu 2000; Sengun and Karapinar 2012; Tanguler 2014). The survival and growth of inoculated pathogenic bacteria such as *E. coli* 0157:H7 and *Yersinia enterocolitica* during the fermentation and hot air or microwave oven drying of *tarhana* was investigated. While counts of viable *Yersinia enterocolitica* decreased dramatically after the first day of fermentation, *E. coli* 0157:H7 survived until the end of fermentation, but counts of *E. coli* 0157:H7 decreased significantly after drying for both drying methods (Aytac 1996). Similarly, the viable numbers of pathogenic bacteria of *Staphylococcus aureus* and *Salmonella typhimurium* were significantly lessened during the first and second days of the fermentation for wet *tarhana*, respectively. However, the number of viable microorganisms became less to an undetectable level on the 4th day of the fermentation period of wet *tarhana* for moulds, the 4th day of the fermentation period of wet *tarhana* for *E. coli* O157:H7, and the 16th day of the fermentation period of wet *tarhana* for *B. cereus*. Therefore, it is advised not to eat wet *tarhana* for the first 7 days of fermentation (Turantas and Kemahlioglu 2012).

21.4.1 Benefits of *Tarhana* on Health

Tarhana is a very nutritious product, and its nutritive value and composition vary according to the ingredients and their quantities used in the recipe (Tanguler and Tatlısoy 2022). The compositions of 134 *tarhana* samples collected from dissimilar regions of Turkey reported in the range between 6.4–13.9% for moisture content, 12.0–29.9% for protein, 41.8–77.5% for carbohydrate, 1.6–18.2% for fat, 0.01–3.1% for fiber, 0.56–10.4% for salt, and 1.4–14.2% for ash (Daglioglu 2000). After fermentation, the mean values of water content, protein, fiber, ash and salt were reported as 61.05%, 16.79%, 2.83%, 8.94%, and 6.48% for wet *tarhana*, respectively (Erbas et al. 2005). Moreover, *Lpb. plantarum* strains, isolated from *tarhana*, produced between 120 and 400 mg/L exopolysaccharides that have antioxidant, cholesterol-lowering, antidiabetic, anticancer, and immunomodulatory properties (Zehir Senturk et al. 2019). *Tarhana* has an antimutagenic activity that decreased the sodium azide mutagenicity in *Salmonella typhimurium*, ranging between 22 and 33% (Karakaya and Kavas 1999a).

Organic acids, ascorbic acid, thiamin, riboflavin, niacin, folate, pantothenic acid, calcium, iron, sodium, potassium, zinc, magnesium, copper, and different amino acids can be produced during fermentation (Daglioglu 2000; Maskan and Ibanoglu 2002; Turantas and Kemahlioglu 2012; Tanguler 2014). Lactic acid, pyruvic acid, and acetic acid were detected at levels of 10.6, 1.9, and 2.4 g/kg, respectively for dried *tarhana*, and as the fermentation time increased, lactic acid formation as well as total organic acid content of *tarhana* increased (Bozkurt and Gurbuz 2008). Average values of minerals and vitamins were reported as 1090 mg/kg for calcium, 0.36 g/kg for iron, 6.34 g/kg for sodium, 1.14 g/kg for potassium, 0.78 g/kg for magnesium, 0.018 g/kg for zinc, 4.5 g/kg for copper, 6.12 g/kg for manganese, 0.1 mg/kg for vitamin B_1, and 0.8 mg/kg for vitamin B_2 (Daglioglu 2000). The thiamin, riboflavin, niacin, vitamin B_6, and folate contents of *tarhana* dough increased during fermentation. While the contents of thiamin (vitamin B_1), riboflavin (vitamin B_2), niacin (vitamin B_3) vitamin B_6, and folate (vitamin B_9) were detected as 4.80, 1.85, 7.68, 15.30, and 2.25 mg/kg, respectively, at the initial part of fermentation and increased to 10.15, 5.15, 24.04, 19.35, and 6.10 mg/kg, respectively, at the final part of fermentation (Certel et al. 2007). In another study, riboflavin, niacin, pantothenic acid, ascorbic acid, and folate contents of the *tarhana* samples increased with fermentation, but no differences were observed for thiamine and pyridoxine levels. The highest losses of water-soluble vitamins were reported at 70°C for the 35-h drying period (Ekinci 2005). The effects of extrusion variables (barrel temperature, feed rate, and screw speed) did not have any significant effect on the *in vitro* protein digestibility, thiamin, and riboflavin content of *tarhana*. However, oven-drying (55°C for 48 h) after extrusion decreased thiamin at a rate of ~30% in the samples (Ibanoglu et al. 1997).

Tarhana is a plentiful source of free amino acids (Ozdemir et al. 2007). Average values of the amino acid composition of *tarhana* were indicated as 5.81 g/kg for lysine, 6.1 g/kg for histidine, 5.55 g/kg for arginine, 14.4 g/kg for aspartic acid, 8.56 g/kg for threonine, 11.3 g/kg for serine, 53.05 g/kg for glutamic acid, 60.94 g/kg for proline, 4.57 g/kg for glycine, 5.7 g/kg for alanine, 1.64 g/kg for cysteine, 8.51 g/kg for valine, 3.24 g/kg for methionine, 6.54 g/kg for isoleucine, 11.52 g/kg for leucine, 3.92 g/kg for tyrosine, and 7.33 g/kg for phenylalanine (Daglioglu 2000). The quercetin content of *tarhana* was determined as 59.2 mg/kg (Karakaya and El 1999b).

As a novel product, kefir was used in *tarhana* production instead of yoghurt. The addition of kefir increased the acidity values, protein, ash, and phenolic contents of the product (Demirci et al. 2019). To manufacture *tarhana* with high β-glucan content, barley flours can be chosen and compared to *tarhana* produced from wheat flours (Erkan et al. 2006). The addition of soybean flour raised the protein values of *tarhana*, but the addition of maize flour reduced the protein content of samples (Kose and Cagindi 2002). *Tarhana* made with prebiotics (inulin and lactose each 3%) and different probiotics (*St. thermophilus*, *L. acidophilus*, and *Bi. bifidum*) showed more reduction in low-density lipoproteins, total plasma cholesterol, and triglycerides, while high-density lipoprotein increased (Gabrial et al. 2010). Supplementation of immature wheat grain in *tarhana* formulation increased the moisture; protein; ash; antioxidant activity; total phenolic contents; and Ca, Mg, K, P, and Zn contents of the samples (Aktas et al. 2015). Supplementation of the wheat germ in *tarhana* formulation increased the nutritional content of the product. By the addition of wheat germ, the Mg, Zn, K, Mn, fatty acids, oil, acid, alcohol, aldehyde, ketone, and terpene contents of *tarhana* samples can be enhanced (Tekgul et al. 2021). Cornelian cherry *tarhana*, which has a geographical mark in Turkey, is a traditional, cereal-based fermented product. Usage of cornelian cherry puree instead of tomato and/or paprika puree decreased the pH value at around 3.6 and increased the acidity of the product compared to the traditional *tarhana* (Isik et al. 2014). It was found that the total anthocyanin value was in the range from 12.03 to 41.81 mg/kg, total phenolics was from 357.40 to 762.81 mg/kg, and total antioxidant activity was from 2.79 to 12.47 mM (Koca 2008). The effect of fermentation on quality parameters of cornelian cherry *tarhana* produced with different flours such as wheat, durum wheat clear, buckwheat, or whole-grain hull-less barley flour was investigated. Fermented cornelian cherry *tarhana* samples had a significantly high amount of total phenolic compounds and higher antioxidant activity. However, the ascorbic acid and anthocyanin contents of the *tarhana* samples were found to be lower. Cornelian cherry *tarhana* manufactured with buckwheat flour is suggested as a gluten-free soup for gluten-sensitive and celiac patients (Karademir and Yalcin 2019).

In addition to the mentioned health benefits, *tarhana* shows a hypolipidemic effect and proteolytic activity. It has a beneficial effect on protein digestibility and the bioavailability of minerals. It can also be used as a biological preservative by showing antimicrobial properties (Gok 2023). As a result, *tarhana*, which is usually consumed as a soup in the winter months, is an important and beneficial food for sick individuals, people at risk, the elderly, and children due to its high protein, mineral, organic acid, free amino acid, and vitamin content (Tanguler and Tatlısoy 2022).

21.5 CONCLUSIONS AND FUTURE DIRECTIONS

In this review, the production methods and health benefits of yoghurt, chickpea bread, and *tarhana* were discussed. These three fermented foods, which can be produced both at the home scale and commercial scale, are a very important part of the diet in Turkey due to their delicious taste and health-promoting functions. Today, many domestic fermented foods are preferred owing to their functional, sensorial, and health-enhancing properties, and they are sold all over the world. These increasing demands enhance the popularity of these foods, and these products have become more available on the international markets.

The mentioned fermented products are a good source of vitamins, proteins, antimicrobial peptides, exopolysaccharides, bioactive compounds, and minerals and also have probiotic and paraprobiotic properties. At the same time, consumption of these fermented products is thought to be related to the potential reduction of the risk of many diseases such as metabolic syndrome, cancer, cardiovascular disease, obesity, diabetes, hypertension, and lowering the glycaemic index. Although claims on the health effects of fermented products are based on most observational studies, more prominent evidence must be collected. Therefore, in the future, more *in vivo* studies should be conducted to determine the health effects of fermented foods. Thus, results from *in vivo* studies will draw the attention of people and provide health-relevant rationale about the consumption of fermented foods, and the demand, especially for domestic fermented products, will increase.

REFERENCES

Abdullah, R., Arshad, H., Kaleem, A., Iqtedar, M., Aftab, M., F. Saleem. 2023. Assessment of angiotensin converting enzyme inhibitory activity and quality attributes of yoghurt enriched with *Cinnamomum verum*, *Elettaria cardamomum*, *Beta vulgaris* and *Brassica oleracea*. *Saudi J. Biol. Sci.* 30(2):103556. https://doi.org/10.1016/j.sjbs.2023.103556

Adebo, O. A., Njobeh, P. B., Adebiyi, J. A., Gbashi, S., Phoku, J. Z., E. Kayitesi. 2017. Fermented pulse-based food products in developing nations as functional foods and ingredients. In: *Functional Food-Improve Health through Adequate Food*, ed. M. C. Hueda. InTech. https://doi.org/10.5772/intechopen.69170

Adolfsson, O., Meydani, S. N., R. M. Russell. 2004. Yoghurt and gut function. *Am. J. Clin. Nutr.* 80(2):245–256. https://doi.org/10.1093/ajcn/80.2.245

Aktas, K., Demirci, T., N. Akin. 2015. Chemical composition and microbiological properties of tarhana enriched with immature wheat grain. *J. Food Process. Pres.* 39(6):3014–3021. https://doi.org/10.1111/jfpp.12554

Aloys, N., N. Angeline. 2009. Traditional fermented foods and beverages in Burundi. *Food Res. Int.* 42:588–594. https://doi.org/10.1016/j.foodres.2009.02.021

Anonymous. 1981. *TS 2282 the Standard of Tarhana*. Ankara: Turkish Standards Institute (in Turkish).

Anonymous. 2021. *TR71 Region Regional Products*. https://www.ahika.gov.tr/assets/upload/dosyalar/tr71_bolgesi_yoresel_urunler_katalogu.pdf. Last accessed 6 December 2021.

Astrup, A. 2014. Yoghurt and dairy product consumption to prevent cardiometabolic diseases: Epidemiologic and experimental studies. *Am. J. Clin. Nutr.* 99(5 Suppl):1235S–1242S. https://doi.org/10.3945/ajcn.113.073015

Ataie-Jafari, A., Larijani, B., Alavi Majd, H., F. Tahbaz. 2009. Cholesterol-lowering effect of probiotic yoghurt in comparison with ordinary yoghurt in mildly to moderately hypercholesterolemic subjects. *Ann. Nutr. Metab.* 54(1):22–27. https://doi.org/10.1159/000203284

Aune, D., Norat, T., Romundstad, P., L. J. Vatten. 2013. Dairy products and the risk of type 2 diabetes: A systematic review and dose-response meta-analysis of cohort studies. *Am. J. Clin. Nutr.* 98:1066–1083. https://doi.org/10.3945/ajcn.113.059030

Aydin, F., K. Cebi. 2019. Isolation and identification of lactic acid bacteria from fermented chickpeas. *BSJ-Agri.* 2(2):79–85.

Aytac, S. A. 1996. Fate of *Escherichia coli* O157:H7 and *Yersinia enterocolitica* in the preparation of tarhana. *Adv. Food Sci.* 18:28–31.

Azizkhani, M., F. Tooryan. 2016. Antimicrobial activities of probiotic yoghurts flavoured with peppermint, basil, and zataria against *Escherichia coli* and *Listeria monocytogenes*. *J. Food Qual. Hazards Control* 3:79–86.

Babio, N., Becerra-Tomas, N., Martinez-Gonzalez, M. A., et al. 2015. Consumption of yoghurt, low-fat milk, and other low-fat dairy products is associated with a lower risk of metabolic syndrome incidence in an elderly Mediterranean population. *J. Nutr.* 145:2308–2316. https://doi.org/10.3945/jn.115.214593

Béal C., S. Helinck. 2014. Yoghurt and other fermented milks. In: *Microorganisms and Fermentation of Traditional Foods*, eds. R. C. Ray and D. Montet, 141–187. Boca Raton: CRC Press.

Blandino, A., Al-Aseeri, M. E., Pandiella, S. S., et al. 2003. Cereal-based fermented foods and beverages. *Food Res. Int.* 36:527–543. https://doi.org/10.1016/S0963–9969(03)00009–7

Boyaci Gunduz, C. P., Gaglio, R., Franciosi, E., et al. 2020. Molecular analysis of the dominant lactic acid bacteria of chickpea liquid starters and doughs and propagation of chickpea sourdoughs with selected *Weissella confusa*. *Food Microbiol.* 91:103490. https://doi.org/10.1016/j.fm.2020.103490

Boyaci Gunduz, C. P., H. Erten. 2022. Yeast biodiversity in chickpea sourdoughs and comparison of the microbiological and chemical characteristics of the spontaneous chickpea fermentations. *J. Food Process. Preserv.* 46:e16479. https://doi.org/10.1111/jfpp.16479

Bozkurt, O., O. Gurbuz. 2008. Comparison of lactic acid contents between dried and frozen tarhana. *Food Chem.* 108:198–204. https://doi.org/10.1016/j.foodchem.2007.10.063

Certel, M., Erbas, M., Uslu, M. K., M. O. Erbas. 2007. Effects of fermentation time and storage on the water-soluble vitamin contents of tarhana. *J. Sci. Food Agric.* 87:1215–1218. https://doi.org/10.1002/jsfa.2810

Cetin, B., Atik, A., S. Karasu. 2014. Physicochemical and microbiological quality of yoghurt and Ayran samples produced in Kırklareli, Turkey. *Acad. Food J.* 12(2):57–60 (in Turkish).

Cortés, L. N., Villamil, R. A., L. Y. Cortés. 2023. The impact of PUFA-enriched yogurt consumption on cardiovascular risk markers: A review. *PharmaNutrition* 23:100330. https://doi.org/10.1016/j.phanu.2023.100330

Daglioglu, O. 2000. Tarhana as a traditional Turkish fermented cereal food. Its recipe, production and composition. *Nahrung* 44(2):85–88. https://doi.org/10.1002/(SICI)1521–3803(20000301)44:2<85::AID-FOOD85>3.0.CO;2-H

Das, K., Choudhary, R., K. A. Thompson-Witrick. 2019. Effects of new technology on the current manufacturing process of yoghurt-to increase the overall marketability of yoghurt. *LWT* 108:69–80. https://doi.org/10.1016/j.lwt.2019.03.058

Demiray, E., Celik, İ., Nogay, O., Y. Tulek. 2015. Color and textural properties of Denizli Karahöyük (Turkey) Bread Zıgır. *Acad. Food J.* 13(3):223–228 (in Turkish).

Demirci, A. S., Palabiyik, I., Ozalp, S., G. Tirpanci Sivri. 2019. Effect of using kefir in the formulation of traditional Tarhana. *Food Sci. Technol.* 39(2):358–364. https://doi.org/10.1590/fst.29817

Donovan, S. M., G. Rao. 2019. Health benefits of yoghurt among infants and toddlers aged 4 to 24 months: A systematic review. *Nutr. Rev.* 77(7):478–486. https://doi.org/10.1093/nutrit/nuz009

Dumas, A. A., Lapointe, A., Dugrenier, M., et al. 2017. A systematic review of the effect of yoghurt consumption on chronic diseases risk markers in adults. *Eur. J. Nutr.* 56:1375–1392. https://doi.org/10.1007/s00394-016-1341-7

Ekinci, R. 2005. The effect of fermentation and drying on the water-soluble vitamin content of tarhana, a traditional Turkish cereal food. *Food Chem.* 90:127–132. https://doi.org/10.1016/j.foodchem.2004.03.036

Elmadfa, I., Heinzle, C., Majchrzak, D., H. Foissy. 2001. Influence of probiotic yoghurt on the status of vitamins B(1), B(2) and B(6) in the healthy adult human. *Ann. Nutr. Metab.* 45:13–18. https://doi.org/10.1159/000046700

Erbas, M., Certel, M., M. K. Uslu. 2005. Microbiological and chemical properties of tarhana during fermentation and storage as wet-sensorial properties of tarhana soup. *LWT* 38:409–416. https://doi.org/10.1016/j.lwt.2004.06.009

Erginkaya, Z., Unal Turhan, E., E. A. Ozer. 2016. The production of bread with chickpea ferment and dominant microflora. *J. Agric. Faculty Bursa Uludag Univ.* 30(1):89–99 (in Turkish).

Erkan, H., Celik, S., Bilgi, B., H. Koksel. 2006. A new approach for the utilization of barley in food products: Barley Tarhana. *Food Chem.* 97(1):12–18. https://doi.org/10.1016/j.foodchem.2005.03.018

Espirito Santo, A. P., Cartolano, N. S., Silva, T. F., et al. 2012. Fibers from fruit by-products enhance probiotic viability and fatty acid profile and increase CLA content in yoghurts. *Int. J. Food Microbiol.* 154:135–144. https://doi.org/10.1016/j.ijfoodmicro.2011.12.025

Espirito-Santo, A. P., Lagazzo, A., Sousa, A. L. O. P., et al. 2013. Rheology, spontaneous whey separation, microstructure and sensorial characteristics of probiotic yoghurts enriched with passion fruit fiber. *Food Res. Int.* 50:224–231. https://doi.org/10.1016/j.foodres.2012.09.012

Fabian, E., Majchrzak, D., Dieminger, B., et al. 2008. Influence of probiotic and conventional yoghurt on the status of vitamins B1, B2 and B6 in young healthy women. *Ann. Nutr. Metab.* 52:29–36. https://doi.org/10.1159/000114408

Fazilah, N. F., Ariff, A. B., Khayat, M. E., et al. 2018. Influence of probiotics, prebiotics, synbiotics and bioactive phytochemicals on the formulation of functional yoghurt. *J. Funct. Foods* 48:387–399. https://doi.org/10.1016/j.jff.2018.07.039

Fernandez-Garcia, E., J. U. McGregor. 1997. Fortification of sweetened plain yogurt with insoluble dietary fiber. *Z. Unters. Lebensm.* 204(6):433–437. https://doi.org/10.1007/s002170050108

Fernandez, M. A., Panahi, S., Daniel, N., et al. 2017. Yoghurt and cardiometabolic diseases: A critical review of potential mechanisms. *Adv. Nutr.* 8(6):812–829. https://doi.org/10.3945/an.116.013946

Fisberg, M., R. Machado. 2015. History of yogurt and current patterns of consumption. *Nutr. Rev.* 73(1):4–7. https://doi.org/10.1093/nutrit/nuv020

Gabrial, S. G., Zaghloul, A. H., Khalaf-Allah, A. E., et al. 2010. Synbiotic Tarhana as a functional food. *J. Am. Sci.* 6(12):847–857.

Gahruie, H. H., Eskandari, M. H., Mesbahi, G., et al. 2015. Scientific and technical aspects of yoghurt fortification: A review. *Food Sci. Hum. Wellness* 4(1):1–8. https://doi.org/10.1016/j.fshw.2015.03.002

Gijsbers, L., Ding, E. L., Malik, V. S., et al. 2016. Consumption of dairy foods and diabetes incidence: A doseresponse meta-analysis of observational studies. *Am. J. Clin. Nutr.* 103:1111–1124. https://doi.org/10.3945/ajcn.115.123216

Gocmen, D., Gurbuz, O., Rouseff, R., et al. 2004. Gas chromatographic-olfactometric characterization of aroma active compounds in sun-dried and vacuum-dried tarhana. *Eur.*

Food Res. Technol. 218:573–578. https://doi.org/10.1007/s00217-004-0913-6

Gok, I. 2023. Functional potential of several Turkish fermented traditional foods: Biotic properties, bioactive compounds, and health benefits. *Food Rev. Int.* 39(5):2568–2593. https://doi.org/10.1080/87559129.2021.1962340

Hancioglu-Sıkılı, O. 2003. *Investigation of microbiological and flavour characteristics of chickpea sweet dough*, PhD Thesis, Ege University, Izmir, Turkey, p. 203.

Hatzikamari, M., Kyriakidis, D., Tzanetakis, N., et al. 2007a. Biochemical changes during a submerged chickpea fermentation used as a leavening agent for bread production. *Eur. Food Res. Technol.* 224:715–723. https://doi.org/10.1007/s00217-006-0363-4

Hatzikamari, M., Yiangou, M., Tzanetakis, N., et al. 2007b. Changes in numbers and kinds of bacteria during a chickpea submerged fermentation used as an agent for bread production. *Int. J. Food Microbiol.* 116:37–43. https://doi.org/10.1016/j.ijfoodmicro.2006.12.030

He, M., Yang, Y. X., Han, H., et al. 2005. Effects of yoghurt supplementation on the growth of preschool children in Beijing suburbs. *Biomed. Environ. Sci.* 18(3):192–197.

Hendek Ertop, M., Y. Coskun. 2018. Shelf-life, physicochemical, and nutritional properties of wheat bread with optimized amount of dried chickpea sourdough and yeast by response surface methodology. *J. Food Process. Preserv.* 42(7):e13650. https://doi.org/10.1111/jfpp.13650

Hertzler, S. R., S. M. Clancy. 2003. Kefir improves lactose digestion and tolerance in adults with lactose maldigestion. *J. Am. Diet Assoc.* 103:582–587. https://doi.org/10.1053/jada.2003.50111

Ibanoglu, S., Ainsworth, P., G. D. Hayes. 1997. In vitro protein digestibility and content of thiamin and riboflavin in extruded tarhana, a traditional Turkish cereal food. *Food Chem.* 58:141–144. https://doi.org/10.1016/S0308–8146(96)00233–6

Ilic, D., S. Ashoor. 1988. Stability of vitamins A and C in fortified yoghurt. *J. Dairy Sci.* 71:1492–1498. https://doi.org/10.3168/jds.S0022–0302(88)79712-X

Isik, F., Celik, I., Y. Yilmaz. 2014. Effect of cornelian cherry use on physical and chemical properties of Tarhana. *Acad. Food J.* 12(2):34–40.

Kabak, B., A. D. Dobson. 2011. An introduction to the traditional fermented foods and beverages of Turkey. *Crit. Rev. Food Sci. Nutr.* 51(3):248–260. https://doi.org/10.1080/10408390903569640

Karacil, M. S., N. Acar Tek. 2013. Fermented products produced in the world: Historical process and relationships with health. *J. Agric. Faculty Bursa Uludag Univ.* 27(2):163–173 (in Turkish).

Karademir, E., E. Yalcin. 2019. Effect of fermentation on some quality properties of cornelian cherry tarhana produced from different cereal/pseudocereal flours. *Qual. Assur. Saf. Crop. Foods* 11(2):127–135. https://doi.org/10.3920/QAS2018.1389

Karakaya, S., A. Kavas. 1999a. Antimutagenic activities of some foods. *J. Sci. Food Agric.* 79:237–242. https://doi.org/10.1002/(SICI)1097–0010(199902)79:2<237::AID-JSFA178>3.0.CO;2-K

Karakaya, S., S. N. El. 1999b. Quercetin, luteolin, apigenin and kaempferol contents of some foods. *Food Chem.* 66:289–292. https://doi.org/10.1016/S0308–8146(99)00049–7

Kazmi, S. A., Vieth, R., D. Rousseau. 2007. Vitamin D3 fortification and quantification in processed dairy products. *Int. Dairy J.* 17:753–759. https://doi.org/10.1016/j.idairyj.2006.09.009

Kim, T. S., Ahn, J., Seok, J., H. Kwak. 2003a. Microencapsulated iron for drink yoghurt fortification. Asian-Australas. *J. Anim. Sci.* 16:581–587. https://doi.org/10.5713/ajas.2003.581

Kim, T. S., Hur, J. W., Yu, M. A., et al. 2003b. Antagonism of *Helicobacter pylori* by bacteriocins of lactic acid bacteria. *J. Food Prot.* 66:3–12. https://doi.org/10.4315/0362–028X–66.1.3

Kizilaslan, N., I. Solak. 2016. Yoghurt and effects on human health. *Gaziosmanpasa J. Sci. Res.* 12:52–59 (in Turkish).

Kleniewska, P., Hoffmann, A., Pniewska, E., R. Pawliczak. 2016. The influence of probiotic *Lactobacillus casei* in combination with prebiotic inulin on the antioxidant capacity of human plasma. *Oxid. Med. Cell. Longev.* 1340903:1–10. https://doi.org/10.1155/2016/1340903

Koca, A. F. 2008. Physical, chemical and antioxidant properties of tarhana with cornelian cherry. *Asian J. Chem.* 20(7):5667–5672.

Kochisarli, I., Ergul, E. 1987. Researches on some quality properties of yoghurt samples sold in Ankara market. *J. Food* 12(3):175–177 (in Turkish).

Kok, C. R., R. Hutkins. 2018. Yoghurt and other fermented foods as sources of health-promoting bacteria. *Nutr. Rev.* 76:4–15. https://doi.org/10.1093/nutrit/nuy056

Kolars, J. C., Levitt, M. D., Aouji, M., D. A. Savaiano. 1984. Yoghurt: An autodigesting source of lactose. *N. Engl. J. Med.* 310:1–3.

Konuray, G., Z. Erginkaya. 2020. Traditional fermented food products of Turkey. In: *Fermented Food Products*, eds. A. Sankaranarayanan, N. Amaresan, and D. Dhanasekaran, 55–65. Boca Raton: CRC Press.

Kose, E., O. S. Cagindi. 2002. An investigation into the use of different flours in tarhana. *Int. J. Food Sci. Technol.* 37:219–222. https://doi.org/10.1046/j.1365–2621.2002.00559.x

Kumari, A. G. I. P., Ranadheera, C. S., Prasanna, P. H. P., et al. 2015. Development of a rice incorporated synbiotic yogurt with low retrogradation properties. *Int. Food Res. J.* 22(5):2032–2040.

Lasker, S., Rahman, M. M., Parvez, F., et al. 2019. High-fat diet-induced metabolic syndrome and oxidative stress in obese rats are ameliorated by yoghurt supplementation. *Sci. Rep.* 9(1):20026. https://doi.org/10.1038/s41598-019-56538-0

Lee, W. J., J. A. Lucey. 2010. Formation and physical properties of yogurt. *Asian-Australas. J. Anim. Sci.* 23(9):1127–1136. https://doi.org/10.5713/ajas.2010.r.05

Maeda, N., Nakamura, R., Hirose, Y., et al. 2009. Oral administration of heat-killed *Lactobacillus plantarum* L-137 enhances protection against influenza virus infection by stimulation of type I interferon production in mice. *Int. Immunopharmacol.* 9(9):1122–1125. https://doi.org/10.1016/j.intimp.2009.04.015

Maskan, M., S. Ibanoglu. 2002. Hot air drying of cooked and uncooked Tarhana dough, a wheat flour-yoghurt mixture. *Eur. Food Res. Technol.* 215:413–418. https://doi.org/10.1007/s00217-002-0572-4

Marette, A., E. Picard-Deland. 2014. Yoghurt consumption and impact on health: Focus on children and cardiometabolic risk. *Am. J. Clin. Nutr.* 99(5):1243S–1247S. https://doi.org/10.3945/ajcn.113.073379

Martini, M. C., Smith, D. E., D. A. Savaiano. 1987. Lactose digestion from flavored and frozen yoghurts, ice milk, and ice cream by lactase-deficient persons. *Am. J. Clin. Nutr.* 46:636–640. https://doi.org/10.1093/ajcn/46.4.636

Martini, M. C., Lerebours, E. C., Lin, W. J., et al. 1991. Strains and species of lactic acid bacteria in fermented milks (yoghurts): Effect on in vivo lactose digestion. *Am. J. Clin. Nutr.* 54:1041–1046. https://doi.org/10.1093/ajcn/54.6.1041

McFarland, L. V. 2015. From yaks to yoghurt: The history, development, and current use of probiotics. *Clin. Infect. Dis.* 60(2):S85–S90. https://doi.org/10.1093/cid/civ054

Mckinley, M. C. 2005. The nutrition and health benefits of yoghurt. *Int. J. Dairy Technol.* 58(1):1–12. https://doi.org/10.1111/j.1471–0307.2005.00180.x

Mirghafourvand, M., Homayouni Rad, A., Mohammad Alizadeh Charandabi, S., et al. 2016. The effect of probiotic yogurt on constipation in pregnant women: A randomized controlled clinical trial. *Iran. Red Crescent Med. J.* 18(11):e39870. https://doi.org/10.5812/ircmj.39870

Moore, J. B., Horti, A., B. A. Fielding. 2018. Evaluation of the nutrient content of yoghurts: A comprehensive survey of yoghurt products in the major UK supermarkets. *BMJ Open* 8(8):e021387. https://doi.org/10.1136/bmjopen-2017–021387

Morelli L. 2014. Yoghurt, living cultures, and gut health. *Am. J. Clin. Nutr.* 99:1248S–1250S. https://doi.org/10.3945/ajcn.113.073072

Mousavi, M., Heshmati, A., Garmakhany, A. D., et al. 2019. Texture and sensory characterization of functional yoghurt supplemented with flaxseed during cold storage. *Food Sci. Nutr.* 7:907–917. https://doi.org/10.1002/fsn3.805

Mousavi, S. N., Saboori, S., O. Asbaghi. 2020. Effect of daily probiotic yoghurt consumption on inflammation: A systematic review and meta-analysis of randomized controlled clinical trials. *Obes. Med.* 18:100221. https://doi.org/10.1016/j.obmed.2020.100221

Mudgil, D., Barak, S., B. S. Khatkar. 2016. Development of functional yoghurt via soluble fiber fortification utilizing enzymatically hydrolyzed guar gum. *Food Biosci.* 14:28–33. https://doi.org/10.1016/j.fbio.2016.02.003

Mygdalia, A. S., Nouska, C., Hatzikamari, M., et al. 2022. A sourdough process based on fermented chickpea extract as leavening and anti-staling agent for improving the quality of gluten-free breads. *Food Res. Int.* 159:111593. https://doi.org/10.1016/j.foodres.2022.111593

Ogel, Z. B., H. I. Ozturk. 2020. Antiviral mechanisms related to lactic acid bacteria and fermented food products. *Biotech. Stud.* 29(1):18–28. https://doi.org/10.38042/biost.2020.29.01.03

Ozdemir, S., Gocmen, D., A. Yildirim Kumral. 2007. A traditional Turkish fermented cereal food: Tarhana. *Food Rev. Int.* 23(2):107–121. https://doi.org/10.1080/87559120701224923

Ozdemir, N., Simsek, O., Temiz, H., A. H. Con. 2019. The effect of fermentation time on the volatile aromatic profile of Tarhana dough. *Food Sci. Technol. Int.* 25(3):212–222. https://doi.org/10.1177/1082013218815325

Ozdestan, O., Guven, G., Alpozen, E., et al. 2012. Determination of some chemical properties of Kumru: A traditional fermented cereal food. *J. Food* 37(6):309–315.

Ozer, E. A., Erginkaya, Z., Ozer, S., E. Unal. 2010. One of our traditional breads: Chickpea fermented bread. Paper presented at the 1st International Symposium on Traditional Foods from Adriatic to Caucasus, Tekirdag, Turkey, pp. 455–457.

Pei, R., Martin, D. A., DiMarco, D. M., B. W. Bolling. 2017. Evidence for the effects of yoghurt on gut health and obesity. *Crit. Rev. Food Sci. Nutr.* 57(8):1569–1583. https://doi.org/10.1080/10408398.2014.883356

Prado, F. C., Parada, J. L., Pandey, A., C. R. Soccol. 2008. Trends in non-dairy probiotic beverages. *Food Res. Int.* 41:111–123. https://doi.org/10.1016/j.foodres.2007.10.010

Ray, M., Ghosh, K., Singh, S., K. C. Mondal. 2016. Folk to functional: An explorative overview of rice based fermented foods and beverages in India. *J. Ethn. Foods* 3(1):5–18. https://doi.org/10.1016/j.jef.2016.02.002

Ringel-Kulka, T., Kotch, J. B., Jensen, E. T., et al. 2015. Randomized, double-blind, placebo-controlled study of synbiotic yoghurt effect on the health of children. *J. Pediatr.* 166(6):1475–1481.e3. https://doi.org/10.1016/j.jpeds.2015.02.038

Saad, A. H., Elmassry, R. A., Wahdan, K. M., et al. 2015. Chickpea (*Cicer arietinum*) steep liquor as a leavening agent: Effect on dough rheology and sensory properties of bread. *Acta Period. Technol.* 46:91–102. https://doi.org/10.2298/APT1546091S

Saad, A. M., Elmassry, R. A., Hamed, A. S., et al. 2016. Characterization of composition, antioxidant potential and microbial organisms upon submerged *Cicer arietinum* fermentation. *Food Measure.* 10:319–326. https://doi.org/10.1007/s11694-016-9309-0

Saadi, A. M., Jafar, N. B., M. A. Jassim. 2022. Effect of some types of stabilizers on the quality of yogurt during storage. *J. Hyg. Eng. Des.* 38:125–130.

Sah, B. N. P., Vasilijevic, T., McKenie, S., O. N. Donkor. 2016. Effect of pineapple waste powder on probiotic antioxidant and anti-mutagenic activities of yoghurt. *J. Food Sci. Technol.* 53:1698–1708. https://doi.org/10.1007/s13197-015-2100-0

Sahin, N., Koyuncu, M., A. Sayaslan. 2018. Effects of chickpea yeast utilization on gluten-free bread. *Harran J. Agric. Food Sci.* 22(4):513–524 (in Turkish). https://doi.org/10.29050/harranziraat.382537

Salas-Salvado, J., Guasch-Ferre, M., Diaz-Lopez, A., N. Babio. 2017. Yoghurt and diabetes: Overview of recent observational studies. *J. Nutr.* 147(7):1452S–1461S. https://doi.org/10.3945/jn.117.248229

Sankaranarayanan, A., N. Amaresan. 2020. Diversity of global fermented food products: An overview. In *Fermented Food Products*, eds. A. Sankaranarayanan, N. Amaresan, and D. Dhanasekaran, 3–24. Boca Raton: CRC Press.

Sanlier, N., Gokcen, B. B., A. C. Sezgin. 2019. Health benefits of fermented foods. *Crit. Rev. Food Sci. Nut.* 59(3):506–527. https://doi.org/10.1080/10408398.2017.1383355

Savaiano, D. A., AbouElAnouar, A., Smith, D. E., D. E. Levitt. 1984. Lactose malabsorption from yoghurt, pasteurized yoghurt, sweet acidophilus milk, and cultured milk in lactase deficient individuals. *Am. J. Clin. Nutr.* 40:1219–1223. https://doi.org/10.1093/ajcn/40.6.1219

Savaiano, D. A., R. W. Hutkins. 2021. Yoghurt, cultured fermented milk, and health: A systematic review. *Nutr. Rev.* 79(5):599–614. https://doi.org/10.1093/nutrit/nuaa013

Sayaslan, A., N. Sahin. 2018. Effects of fermented-chickpea liquor (chickpea yeast) on whole-grain wheat flour bread properties. *Qual. Assur. Saf. Crops Foods* 10(2):183–192. https://doi.org/10.3920/QAS2017.1225

Sengun, I. Y., Nielsen, D. S., Karapinar, M., M. Jakobsen. 2009. Identification of lactic acid bacteria isolated from Tarhana, a traditional Turkish fermented food. *Int. J. Food Microbiol.* 135(2):105–111. https://doi.org/10.1016/j.ijfoodmicro.2009.07.033

Sengun, I. Y., M. Karapinar. 2012. Microbiological quality of Tarhana, Turkish cereal based fermented food. *Qual. Assur. Saf. Crops Foods* 4:17–25. https://doi.org/10.1111/j.1757–837X.2011.00118.x

Settanni, L., Tanguler, H., Moschetti, G., et al. 2011. Evolution of fermenting microbiota in tarhana produced under controlled technological conditions. *Food Microbiol.* 28:1367–1373. https://doi.org/10.1016/j.fm.2011.06.008

Shaghaghi, M., Pourahmad, R., H. R. Mahdavi Adeli. 2013. Synbiotic yoghurt production by using prebiotic compounds and probiotic lactobacilli. *J. Int. Res. J. Appl. Basic Sci.* 5:839–846.

Sonestedt, E., Wirfält, E., Wallström, P., et al. 2011. Dairy products and its association with incidence of cardiovascular disease: The Malmö diet and cancer cohort. *Eur. J. Epidemiol.* 26(8):609–618. https://doi.org/10.1007/s10654-011-9589-y

Sozbilen, G. 2019. An alternative bread recommendation for food and beverage businesses: "Nevsehir Simidi". Paper presented at the 4th International Gastronomy Tourism Research Congress, p. 530 (in Turkish).

Stanton, C., Desmond, C., Coakley, M., et al. 2003. Challenges facing the development of probiotic-containing functional foods. In: *Handbook of Fermented Functional Foods*, ed. E. R. Farnworth, 27–58. Boca Raton: CRC Press.

Tamang, J. P., Cotter, P. D., Endo, A., et al. 2020. Fermented foods in a global age: East meets West. *Compr. Rev. Food Sci. Food Saf.* 19:184–217. https://doi.org/10.1111/1541–4337.12520

Tamer, C. E., Kumral, A., Asan, M., I. Sahin. 2007. Chemical composition of traditional tarhana having different formulations. *J. Food Process. Preserv.* 31:116–126. https://doi.org/10.1111/j.1745–4549.2007.00113.x

Tamime, A. Y., Y. K. Robinson. 2007. yoghurt *Science and Technology* (Third Edition). Abington Hall, Abington, Cambridge, England: Woodhead Publishing Limited.

Tanguler, H. 2014. Traditional Turkish fermented cereal based products: Tarhana, Boza and Chickpea bread. *Turk. J. Agric. Food Sci. Technol.* 2(3):144–149. https://doi.org/10.24925/turjaf.v2i3.144–149.111

Tanguler, H., A. Tatlısoy. 2022. Evaluation of shalgam residuals as a partial wheat flour substitution in tarhana production as a green food product. *J. Food Process. Preserv.* e17267:1–12. https://doi.org/10.1111/jfpp.17267

Tavakoli, M., Najafi, M. B. H., M. Mohebbi. 2019. Effect of the milk fat content and starter culture selection on proteolysis and antioxidant activity of probiotic yoghurt. *Heliyon* 5(2):e01204. https://doi.org/10.1016/j.heliyon.2019.e01204

Tekgul, Y., Caliskan Koc, G., Erten, E. S., A. Akdogan. 2021. Determination of the effect of wheat germ on the mineral and fatty acid composition and aroma compounds of Tarhana: A traditional fermented cereal food. *J. Food Process. Preserv.* 45(2):e15144. https://doi.org/10.1111/jfpp.15144

TFC. 2009. Turkish Food Codex, fermented milk products. Turkish Food Codex Regulations, Regulation No: 2009/25. Republic of Turkey Ministry of Food, Agriculture and Livestock, Ankara, Turkey.

Tuncel, N. B., Yilmaz, N., S. Ozcelik. 2010. Production and investigation of some properties of chickpea bread. Paper presented at the 1st International Symposium on Traditional Foods from Adriatic to Caucasus, Tekirdag, Turkey, pp. 447–449.

Turantas, F., K. Kemahlioglu. 2012. Fate of some pathogenic bacteria and molds in Turkish Tarhana during fermentation and storage period. *J. Food Sci. Technol.* 49(5):601–607. https://doi.org/10.1007/s13197-010-0200-4

Turkoglu, H., Atasoy, A. F, H. B. Ozer. 2003. Some micribiological properties of raw milk, yogurt and fresh Urfa cheese produced and sold in Sanliurfa province. *J. Faculty Agric.* 7(3–4):69–76 (in Turkish).

Uzunsoy, I., B. Ozer. 2019. Yoğurt. In: *Fermente gıdalar*, eds. E. Anli and P. Sanlibaba, 187–231. İstanbul: Nobel Yayıncılık (in Turkish).

Water, J. V. 2003. Yoghurt and immunity: The health benefits of fermented milk products that contain lactic acid bacteria. In: *Handbook of Fermented Functional Foods*, ed. E. R. Farnworth, 114–137. London, Boca Raton: CRC Press.

Weerathilake, W. A. D. V., Rasika, D. M. D., Ruwanmali, J. K. U., et al. 2014. The evolution, processing, varieties and health benefits of yoghurt. *Int. J. Sci. Res. Publ.* 4(4):353–362.

Zehir Senturk, D., Dertli, E., Erten, H., O. Simsek. 2019. Structural and technological characterization of ropy exopolysaccharides produced by *Lactobacillus plantarum* strains isolated from Tarhana. *Food Sci. Biotechnol.* 29(1):121–129. https://doi.org/10.1007/s10068-019-00641-5

Zheng, J., Wittouck, S., Salvetti, E., et al. 2020. A taxonomic note on the genus Lactobacillus: Description of 23 novel genera, emended description of the genus *Lactobacillus* Beijerinck 1901, and union of *Lactobacillaceae* and *Leuconostocaceae*. *Int. J. Syst. Evol. Microbiol.* 70(4):2782–2858. https://doi.org/10.1099/ijsem.0.004107

Fermented Foods from Chile and Health Benefits

22

M. Carolina Henríquez

22.1 INTRODUCTION

Chile is in the far SW of South America. It occupies an area of 756,100 km^2, making it the world's longest country and possibly also the narrowest. The agricultural sector represents 8.6% of the country's labor force, and this is marked by a great diversity of climates and soils that permit the cultivation of a broad range of species such as fruits, horticultural crops, cereals, and legumes (Muñoz et al. 2020).

Chile is the principal producer and exporter of fresh fruits in the southern hemisphere and is a leader in the export of more than 50 different fruit species such as apples, table grapes, kiwis, plums, pears, blueberries, mandarins, lemons, and oranges and dried fruits like walnut and almond, and in processed fruits, it is a leader in raisins, dried apples, plums and peppers, frozen raspberries, grape and apple juice, tomato puree, canned peaches, and wine (Muñoz et al. 2020; ODEPA 2022). In Chile, the fermented products that are highlighted are alcoholic beverages from different fresh fruits such as grape and apple or cereals like corn.

In effect, the development of alcoholic beverages from fermentation was a custom in South America before the arrival of the Spanish people. The most widespread approach was *chicha* using corn (*Zea mays*). Chicha is a low-alcohol wine consumed before fermentation is complete (Crowley 2000). Later, with the incorporation of European plants, the custom of making alcoholic beverages from fruits, particularly grape (*Vitis vinifera*), spread to all communities (Lacoste et al. 2015a). Other species utilized were wheat (*Triticum aestivum*) and barley (*Hordeum vulgare*); fruits like apples (*Malus domestica*) and pears (*Pyrus communis*); and native fruits such as Chilean strawberries (*Fragaria chiloensis*), piñones (fruits from *Araucaria araucana*), maqui (*Aristotelia chilensis*), murtilla (*Ugni molinae*), and algarrobo (*Prosopis* spp.), among others. These kinds of alcoholic fermented beverages have accompanied the social and cultural life of the peoples of South America (Lacoste et al. 2015a; Schmeda-Hirschmann et al. 2019).

On other hand, it is important to indicate that in Chile, most of the fermented products are developed and processed using artisan methods, except wine. The processing and associated technology used to produce these fermented beverages are transmitted from generation to generation.

22.2 FERMENTED ALCOHOLIC BEVERAGES DEVELOPED IN CHILE

22.2.1 Wine

Wine is a product of the alcoholic fermentation of the must, or the grape juice obtained from the pulp of the species *Vitis vinifera*; vinification is the process that transforms the grape juice into wine, a product of fermentation with yeast, which converts the sugars in the juice to alcohol (Simunovic 1999).

Chile has a 450-year history of wine production. Wine was introduced to America on the second voyage of Christopher Colombus in 1493. Subsequently the Spanish established it in México (1522), Perú (1542), and Chile (1554). At the same time, the Portuguese colonizers introduced alcoholic fermentation to the east coast from South America (1532) (Crowley 2000; Pszczólkowski and Lacoste 2016). Probably the origin of the wine originally arrived in America to the Canary Islands; one of the autochthonous varieties is Listan Prieto, and from their seed descended many ecotypes founded in America, such a Mission (California), Negra Corriente (Perú), Misionera (Bolivia), Uva del País o País (Chile), and Criolla Chica (Argentina) (Pszczólkowski and Lacoste 2016).

Chile has around 137,000 ha of wine cultivation area distributed between the Coquimbo (30.0° SL) and Araucanía (38.7° SL) regions, although there are some additional expansions to the wine-growing areas in Southern Araucanía and Northern Coquimbo. The Chilean wine industry is rather concentrated and strongly export oriented. Wine exports represent

DOI: 10.1201/9781003605300-22

more than 1,800 million dollars and more than 80% of wine produced, reaching more than 130 countries (Mora 2019). The most cultivated grapevine varieties in Chile are Cabernet-Sauvignon (28.84%), Sauvignon blanc (11.21%), Merlot (8.26%), País (8.05%), Carmenere (7.92%), and Chardonnay (7.91%) (SAG 2022).

Of these grape varieties, Carménère is an emblematic red grape wine variety from Chile originating from the region of Médoc, Bordeaux, France. This was believed to be extinct after the catastrophic effects of *Daktulosphaira vitifoliae* that ravaged the French vineyards in the mid-19th century (Domínguez and Agosin 2010). The Carménère variety was rediscovered and redeveloped by the Chilean wine industry (Gutiérrez-Gamboa et al. 2020). It has high tannin content and is generally soft and round, even though it is sometimes possible to obtain a small degree of bitterness (Pszczólkowski 2004). On other hand, Carménère grape seeds contain a lower concentration of monomeric flavan-3-ols and (+)-catechin and a higher mean degree of polymerization, average molecular weight of the flavanol fraction, (–)-epicatechin, epicatechin-3-O-gallate, gallic acid, and dimeric procyanidins esterified with gallic acid content (Obreque-Slier et al. 2010). Additionally, Carménère seed has a higher concentration of proanthocyanidin B1, B4G, and B2 (Obreque-Slier et al. 2012). Carménère grape skins have higher contents of total anthocyanins, monomeric flavan-3-ols, and total flavonoids; a higher mean degree of polymerization; a higher percentage of galloylation; a higher average molecular weight of the flavanol fraction; and a higher color intensity (Obreque-Slier et al. 2010).

The Listán Prieto or País grape variety was the first variety established in Chile, incorporated by the Spanish in the 16th century (Díaz and Morales 2019). This variety has traditionally been cultivated by small wine-growers (Gutiérrez-Gamboa et al. 2020), and it have been redeveloped in the last decade; in previous decades, this variety not was considered as a priority for wine production, due to it usually resulted in low-quality wines, mainly due to its roughness and taste imbalance (Fernández et al. 2015). However, these characteristics indicate the presence of a wide variety of phenolic compounds, such as the abundance of proanthocyanidins, higher mean degree of polymerization (mDP), and presence of epicatechin-gallate (ECG) (Villarroel 2009). In addition, powder of País skin extract has a higher inhibition against angiotensin I-converting enzyme, which is responsible for increasing blood pressure, and this was associated with a higher mean degree of polymerization (mDP) and the presence of epicatechin-gallate (ECG) (Godoy et al. 2012). On other hand, grape pomace contains a good source of anthocyanins such as malvidin-3-glucoside, while leaf extracts have abundant amounts of flavonols like quercetin-3-hexoside and phenolic acids such as caftaric acid (Aguilar et al. 2016). In relation to the total phenolic content, this varied between 0.1 mg gallic acid equivalent (GAE)/g (juice) to 0.8 (skin) (Lutz et al. 2012).

Carignan is another grape variety that has a unique heritage in Chilean winemaking and, with the País grape variety, has given a new identity to the country in the world wine scenario due to their history and heritage. These two grapevine varieties have been cultivated mostly by small winegrowers located in Maule, Itata, and Bío Bío Valleys; this constitutes the largest wine-producing area of Chile (Gutiérrez-Gamboa et al. 2018). The first vine cuttings were introduced from the south of France around 1939 (Martínez-Gil et al. 2018). Gutiérrez-Gamboa et al. (2018) determined the phenolic compounds in Carignan grapes, and the results obtained indicate that total anthocyanin content varied from 1393.05 to 1856.03 mg kg^{-1}; total flavonol content ranged between 124.17 and 218.93 mg kg^{-1}; total flavanol concentration varied from 86.85 to 158.13 mg kg^{-1}; and total content of hydroxycinnamic acids ranged between 23.91 and 34.84 mg kg^{-1}. Martínez-Gil et al. (2018) determined the phenolic compounds of Carignan noir grapes from six sites of the Maule Valley (Chile). The compounds identified by these authors were flavonols (myricetin, quercetin, laricitrin, kaempferol, isorhamnetin, and syringetin), anthocyanins (delphinidin, cyanidin, petunidin, peonidin, and malvidin), flavanols (catechin, epicatechin, epicatechin-3-gallate, epigalocatechin, and procyanidins B1 and B2), and three individual hydroxycinnamic acids (trans-caftaric acid, trans-+cis-cutaric acid, and caffeic acid).

22.2.2 Pisco

Pisco is the most important distilled beverage produced and consumed in Chile. It corresponds to a non-aged or slightly aged kind of brandy, produced in the semiarid regions of southern Chile and Perú; it is produced by the distillation of fermented grape juice (Bordeu et al. 2012; Pszczólkowski and Lacoste 2016). Chilean pisco was born at the beginning of the 18th century, associated with the varieties introduced by the Spanish colonizers (País grape and Moscatel de Alejandria varieties) and the criollas varieties derived from naturally crossing them (Pszczólkowski and Lacoste 2016). Moscatel de Alejandria is a variety of North African origin, and it is widely distributed in South America, particularly in Perú, Bolivia, Chile, and Argentina. According to Lacoste et al. (2015a), farming of this variety started in La Serena (Region de Coquimbo) in the early 18th century. In this varieties the terpenes such as linalool are the most important compounds (Pszczólkowski and Lacoste 2016).

Pisco has been recognized, delimited, and protected by the State as Denomination of Origin since 1931 and has managed to sustain itself as an emblematic national beverage (Mitchell and Terry 2011). To use the name Pisco, there are specific grape varieties that can be used and the vinification, distillation, packaging, and labeling are carried out within the Pisco production area that corresponds to the Atacama and Coquimbo regions (Simunovic 1999; Bordeu et al. 2012).

It is important to indicate that Chilean pisco represents one of the few typical products of Latin America that maintained its identity for three centuries (Pszczólkowski and Lacoste 2016).

However, even when it is a distillate with unique production in Chile, its per capita consumption is low, 2.1 liters, compared to the consumption of other alcoholic beverages such as wine (16.2 liters) or beer (50 liters) (Araya-Pizarro 2022).

This young distillate is produced predominantly from Muscat wines in the north of Chile (Atacama and Coquimbo regions). Other varieties used for Pisco production in Chile correspond to Moscatel de Alejandría and criollas group of grape varieties such as Moscatel de Austria, Moscatel Rosada, Moscatel Negra, and Pedro Giménez, with low percentages of European varieties (Pszczólkowski and Lacoste 2016).

Muscat grape cultivars are known for the strong aromatic properties that are conferred to wines, mainly due to their high presence of free and bound terpenes, which present low odor thresholds (Belancic et al. 1997).

Pisco processing includes the steps of fermentation, distillation, and maturation. The distillation is based on the simultaneous evaporation and condensation of the fermented grape juice to purify it; this process is done in copper alembics (Bordeu et al. 2012; Tsakiris et al. 2014).

22.2.3 Grape Chicha

Chicha is an alcoholic beverage; this correspond to grape juice or must, with a not completed fermentation process. The taste is very sweet, and the alcohol content is low (Simunovic 1999).

Grape chicha is a typical product of the popular viticulture of the central valley of Chile. The production of grape chicha as a product for mass consumption is documented from the mid-18th century, and the production and marketing were popularized between the 18th and 19th centuries (Lacoste et al. 2015b). This fermented grape beverage is consumed mainly in September due to this month being the celebration of Chilean Independence, and it is less consumed during the rest of the year. In Chile, the importance of grape chicha is recognized as a fundamental part of the identity and culture of the country, and this presence is maintained over time and occupies a privileged place of consumption and appreciation (Castro 2016), due to the cultural importance of being consumed in different celebrations such as during the harvest, carnival, national holidays, and even political events (Castro et al. 2016a).

Despite its deep historical importance and identity, there are no historical studies or figures of production and consumption of this drink. This beverage has been maintained due to the production being mainly artisan and by small producers (Castro 2016). Due the identity of this product, it has been recognized by the Chilean legislation (Decree N° 78, 31-07-1986) (Castro 2016).

In relation to the production, raw grape chicha has very simple preparation. In this product, the must is fermented, but only partially. Only a part of the grape sugar turns into alcohol; therefore, one obtains a product very sweet with low alcohol content. This product is only consumed during the autumn (March–April in the south hemisphere). The processing of cooked grape chicha is more complex. It is made from a must around a density of 1,090 to 1,110 g/L (reached in March–April depending on the area). Then it is boiled and mixed in a copper container with ashes of shoots that provide carbonates that neutralize the high acidity produced by concentrating the volume during the boiling process. The foam is then removed with a horsehair (or plastic) strainer. The process stops at 1,130–1,135 g/L. Then it is packed in 225–500-L barrels or clay pots well covered with mud. The barrels or jars are stored buried in the ground or very cool places. In the first days of September, the cooked grape chicha is packaged in bottles that are closed with corks. This product is ready for distribution and consumption on the national holiday of September 18 (Independence Day) (Lacoste et al. 2015b).

22.2.4 Apple Chicha

Apple chicha is an alcoholic beverage obtained by the incomplete fermentation of apple juice. It is produced mainly in the south of Chile (Región de La Araucanía, Región de Los Lagos, and Región de Los Ríos). Commercial advertisements (from 1870 and 1885) seem to indicate that one of the most important characteristics of apple chicha is its digestive capacity (Castro 2016). At the present, almost 90% of chicha producers do not distinguish apples varieties at the moment of processing of the alcohol, although they have marked preferences for the "Limona" variety; it has yellow skin and small size, is compact, and the taste is bittersweet (Salas 1999).

22.2.5 Pipeño Wine

Pipeño wine is a typical wine of south-central Chile, originating in the second half of the 18th century. It is a crude, unfiltered wine that is not bottled and is drunk soon after harvest (Crowley 2000).

Since then, it has remained a typical wine from the south of the Central Valley, especially between the Maule and Bio-Bio rivers. This wine is made from País grapes in the case of rosés and reds and Moscatel de Alejandría for white wines. It owes its name to the "pipes", Chilean oak barrels, which were first used in the 18th century (Lacoste et al. 2015c). These wines are made from traditional methods; they are foot stepped by man, fermented in open vats and kept in Chilean oak pipes (Lacoste et al. 2015c). For a long time, it was negatively valued by wine experts and specialists and was considered a marginal wine, part of poor farmers' consumption and processing using criollas (native) grapes and was a minor wine of little interest for the elites. However, in recent years their identity, history and social has taken importance. In recent years, there has been a revaluation of Pipeño given a renewed interest in products with identity and heritage values from the national food industry (Lacoste et al. 2015c). This revival is part of a larger process, marked for the appreciation of products with identity. In effect, Pipeño development is integrated with the social, economic, and cultural history of Chile from the colonial period to the present, particularly in the population, that lived in rural areas (Lacoste et al. 2015c).

22.2.6 Pajarete Wine

Pajarete wine is a typical and selected wine produced in the northern area of Chile, particularly in Huasco and Elqui Valley; the records of this wine began in 1790 up to now. This wine is processed by small artisanal winegrowers, and in 1953, it was recognized as a Denomination of Origin wine (Castro et al. 2016b).

Due this early recognition as a Denomination of Origin, this wine has an important symbolic role in the process of strengthening the identity of Chilean and Latin American food products in general (Castro et al. 2016b).

In the 1980s, País grapes were used for the processing of red or colored Pajaretes wines and Moscatel de Alejandría, Moscatel de Austria, Torontel, and Moscatel Rosa for white Pajaretes wines (Sotomayor 1987).

Pajarete wine is characterized as having an alcohol content that ranges between 15% and 20%. This is implemented using various methods, but the most natural way is to sun-dry the grapes after the harvest in such a way to obtain a higher concentration of sugar through natural dehydration. Within the local classification, Pajarete wine belongs to the sweet wines (Castro et al. 2016b).

22.2.7 Mudai

Mudai is a typical alcoholic beverage used in indigenous ceremonies. This beverage is processed with piñon, the seed of *araucaria* (*Araucaria araucana.* (Mol) K. Koch) (common name *araucaria* or *pehuen* in the Mapuche language). It is one of the most important conifer trees, endemic to Chile and Argentina (Hoffmann 1991). It is a species of high conservation value that was declared a national monument by the Convention on International Trade in Endangered Species World Flora and Fauna (CITES). The international trade of its wood and subproducts is forbidden, but the commercialization of piñon and products derived by them is not prohibited (Herrmann 2006). The natural distribution is relatively limited ranging, from latitude 37°20 to 40°20 LS (Herrmann 2005). There are two centers of distribution: the Andes Cordillera, which constitutes the main growing area (37°30 to 40°02 LS), and the coastal Cordillera of Nahuelbuta (between 37°20 and 38°40 LS) (Herrmann 2005; Hoffmann 1991). *Araucaria araucana* can reach a height of 45 m, develop a diameter of up to 2 m, and attain a maximum age of at least 1300 years. Pehuen is considered sacred and spiritual for the Mapuche Pehuenche, the native people that live in the south of Chile, who believe that the tree makes possible their life, and it is a physical expression of culture and life. During the *ngillatun*, one of the most important community ceremonies of the Pehuenches; they offer muday to the tree called pehuen and ask for permission, before starting to collect the piñones to ensure a good harvest. In addition, piñon provides a staple diet for the people, constituting an important source of carbohydrates. This seed has been eaten raw, boiled, or toasted (Aagesen 2004; Herrmann 2005).

In relation to the chemical composition, piñon has a low content of lipids and proteins, and starch is the most important compound (64% of the dry weight). Total dietary fiber represented about 25.4% of the dry weight of peeled piñon seeds, and the total phenolics in the peeled piñon seeds were 0.71 ± 0.02 mg GAE/g DW and in the raw coat were 66.34 ± 1.89 mg GAE/g DW (Henríquez et al. 2008).

Mudai was formerly obtained by chewing the piñon seed, which hydrolyzed part of the starch and facilitated the subsequent alcoholic fermentation of the paste mixed with water (Herrmann 2005).

22.2.8 Aloja

Aloja is an ancient autochthonous alcoholic beverage that can be fermented by the natural microflora of *Prosopis* pods flour; this highlights the sweet taste. It was the social beverage for ethnic groups such as the Wichí, Pilagá, and Diaguitas (Encina 1952).

Trees belonging to the genus *Prosopis* are popularly known as *algarrobo* in South America. This tree has a height between 5 and 15 m, and it grows in the north central region of Argentina and Chile, south region of Peru, and central west region of Bolivia. Algarrobo was sacred for ancient populations, being one of the main food sources (Rodríguez et al. 2020). *Prosopis alba* fruit is used to dissolve gallstones and as an anti-bronchitis treatment and laxative. Its flowers have diuretic properties, and the bark is used as an astringent and to heal eye infections (Pasiecznik et al. 2004). Its fresh pods were consumed directly, and the dry pods were used for many food product preparations such as flour, patay (a kind of bread made from *Prosopis* flour), arrope (syrup), or the beverages aloja and añapa; this is an unfermented beverage processed with *Prosopis* pod flour.

For the traditional aloja preparation, the ripe pods are milled in a mortar or cut into small pieces by hand, put in a vessel, and mixed with water to stimulate the fermentation process. After fermentation, from 3 to 10 d in a fresh, dry, and dark place, the remains of the pods are manually discarded, and a fresh amount of ground algarroba must be incorporated to stimulate the fermentation. The optimum point of the beverage is detected by tasting, and aloja must be consumed in a few days because it has a short shelf life. It is utilized by regional people as a diuretic beverage (Figueroa and Dantas 2006; Sciammaro et al. 2016).

For the preparation of añapa, the ripe pods are ground in a mortar or fragmented by hand, mixed with an excess of cold water, and placed in a vessel pot; dispersion is agitated and filtered, and the extract is then deposited in vessels (Figueroa and Dantas 2006; Sciammaro et al. 2016). Both beverages are drunk for their refreshing quality and perceived "health" benefits.

Rodríguez et al. (2020) studied the chemical composition of añapa and aloja, and they reported that sugar was the main macronutrient found in these two beverages. Total phenolic compounds were significantly higher in aloja (150 to 180 mg GAE/L) than in añapa (21 mg/L), and according to these authors, these effects might be explained by a higher solubility of phenolic compounds in the ethanol produced during fermentation; this process could be effective to extract phenolic compounds bound to cell structures because of microorganism activity during fermentation. Considering the phenolic group, these authors determined that aloja showed more flavonoid, flavone, and flavon-3-ol content than añapa. The main phenolic compounds found in these two beverages were two apigenin C glycosides: vicenin II and vitexin, quercetin, and cinnamic acid. In addition, both beverages were highlighted for their higher antioxidant capacity.

22.2.9 Tecu

Tecu is an alcoholic fermented beverage made by the Araucanian people using maqui fruits (Poblete 1998). The word maqui comes from the Mapudungun language, and its meaning refers to the black berry of this species (Mosbach 1992). These fruits are collected from wild populations between December and February due to the lack of commercial plantations of this species (Hoffmann 1991).

Araucaria chilensis, commonly known as maqui, is an evergreen shrub (perennial), found in Chile from the region of Coquimbo (30°S latitude) to the region of Aysen del General Carlos Ibañez del Campo (45°S latitude), up to 2,500 m of altitude, and it is also found in the Archipelago of Juan Fernández and in Argentina (Hoffmann 1991). Maqui reaches a height of 4 to 5 m (Hoffmann et al. 1992). For the Mapuche culture, the plant is a sacred symbol of "good intentions" (Hoffmann et al. 1992). Ethnopharmacological records recommend using the leaves as an infusion for the treatment of throat diseases, pain, infections such as mouth and stomach ulcers, kidney pain, intestinal tumors, fever, and hemorrhoids, which suggests anti-inflammatory, analgesic, antimicrobial, and antioxidant properties. Powder processed with dried leaves is used for wound healing and scarring, and it has a relaxing action on muscle fiber. Its phytochemicals include alkaloids, flavonoids, and tannins (Montes and Wilkomirsky 1987; Muñoz et al. 2001; Delporte 2006; Suwalsky et al. 2008; Vogel et al. 2008).

The fruit of this species is a small spherical berry, black-purplish and bright in color, and it is recommended to be used as an antidiarrheal treatment and for amygdalitis, pharyngitis, and oral ulcers (Rodríguez et al. 1983; Montes and Wilkomirsky 1987; Muñoz et al. 2001; Vogel et al. 2008).

In addition, the fruit is slightly acidic, refreshing, and astringent; is edible; and is used in the preparation of sweets and jams and in the manufacture of juices and beverages such as tecu (INFOR-CHILE 2003). Maqui has been listed as the plant with the highest content of phenolics when compared with other berries (Misle et al. 2011). The fruit is characterized by the high level of phenolic compounds, mainly anthocyanins (cyaniding glucosides, delfidine, malvidine, and petunidine), flavan-3-ols (catequin and epicatequins), proanthocyanidin B, phenolic acid (ferulic acid, gallic acid, p-coumaric acid, sinapic acid, 4-hydroxybenzoic acid, vanillic acid, and derivates of ellagic acid), and flavonols (quercetin and myricetin) (Escribano-Bailón et al. 2006; Céspedes et al. 2010). These compounds have antioxidant, anti-inflammatory, cardioprotective, anti-diarrheal, anti-aterogenic, hypoglycemic, and astringent properties (Hoffmann et al. 1992; Céspedes et al. 2008; Gironés-Vilaplana et al. 2012; Rojo et al. 2012; Rodríguez et al. 2022). In effect, different studies have shown the capacity of maqui extracts to inhibit low density lipoprotein (LDL) oxidation and protect human endothelial cells against oxidative stress (Miranda-Rottmann et al. 2002), to inhibit α-glucosidase and α-amylase enzymes (Rubilar et al. 2011), to inhibit the adipogenesis and accumulation of lipids (Schreckinger et al. 2010), to stimulate the expression of inflammation mediators (Reyes-Farías et al. 2016; Céspedes et al. 2017; Céspedes-Acuña et al. 2018), and to act as an antidiabetic agent (Rojo et al. 2012).

The highest content in anthocyanin could be attributed to its high surface/volume ratio, especially considering the low fruit weight compared to the other polyphenol-rich fruits (blueberries, raspberries, strawberries, and blackberries) (Fredes et al. 2014).

In addition, maqui fruit can be mixed with corn to prepare an alcoholic beverage, and this has been described in the Mapuche culture (Godoy-Aguirre 2018).

22.2.10 Lahueñe

Lahueñe is a product obtained by the fermentation of endemic and wild Chilean strawberry (*Fragaria chiloensis*) juice (Finn et al. 2013). Labarca (1994) reported that the Mapuche's favorite fermented drink (lahueñe mushca) was made from the small, red-fruited native strawberry called llahue or lahueñe.

In Chile, the origin of *Fragaria chiloenesis* has not been completely determined; it may have been introduced from North America by birds. After making landfall in Chile, the species spread from the coast into the mountains, eventually developing four biotypes. The genotypes grown in Chile have a range of skin color from almost white to slightly red (Finn et al. 2013).

Fragaria chiloensis has been a highly valued fruit for its aromatic characteristics, sweet taste, and pleasant aroma and is considered one the six important foods of Chile. The Chilean strawberry can be found in two botanical forms, the white strawberry (*Fragaria chiloensis* ssp. *chiloensis* f. *chiloensis*) or the red strawberry (*Fragaria chiloensis* ssp. chiloensis f. *patagonica*). The production area for the white strawberry is from Region del Maule (35°30'S) to Region de Araucanía, mainly in the Cordillera of Nahuelbuta, and the red strawberry is usually located on the Andean slopes of Chile and the southern Patagonia in Argentina (Schmeda-Hirschmann et al. 2019). At least two native peoples, the Mapuche, between Rio Bío-Bío and south–central Chile, and the Picunche, between

Rio Itata and Rio Bío-Bío, began the domestication process (Finn et al. 2013).

The chemical composition of this species indicates the presence of proanthocyanidins, hydrolysable tannins, ellagitanins, anthocyanins (cyanindin and pelargonidin), flavonol glycosides (quercetin, kaempferol), phenolic acid such as ellagic acid, E-cinnamic acid glycosides, p-coumaric acid derivative, chlorogenic acid, and caffeic=ferulic acid derivative (Simirgiotis et al. 2009; Morales-Quintana and Ramos 2019). The antioxidant activity of the Chilean strawberry has been related to the anthocyanin and ellagitannin content of the fruits (Cheel et al. 2007). The antioxidant activity of Chilean *Fragaria chiloensis* samples has been related to the anthocyanin and ellagitannin content of the fruits (Cheel et al. 2007). In addition, Chilean white strawberry extracts were effective inhibitors of the enzyme α-glucosidase, a therapeutic strategy for the control of hyperglycemia, and acetylcholinesterase activities (Noriega et al. 2021).

Other berry fruits that have been used to produce the fermented alcoholic beverage chicha are the aromatic fruits of *Amomyrtus luma*, murta (*Ugni molinae*), and *Luma apiculata*; these fruits were macerated into alcohol to obtain liquors. These fruits belong to the Myrtaceae family, and they are highlighted by their high phenolic content (mainly anthocyanins) and antioxidant capacity (Schmeda-Hirschmann et al. 2019)

Murta or murtilla is a forest-based wild native shrub of Chile and Argentina that produces small berries as fruits, which develop a variety of fruit colors like soft green, yellow, fuchsia, and light and dark red. In the Mapuche language, it is called *uñi*. Murta berries are widely used in folk medicine as analgesics for different types of pain and as an anti-inflammatory and are known for their astringent power (Fredes et al. 2020). Different studies have shown that murtilla berries showed a higher content of total beta-carotene, phenolic compounds, and flavonoids (Rodríguez et al. 2014; Junqueira-Gonçalves et al. 2019). The principal phenolic compounds found in the fruits are quercetin, kaempferol isoquercetin, rutin, luteolin, gallic acid, p-coumaric acid, caffeic acid, feruloyl quinic acid, chlorogenic acid, and anthocyanins (petunidin, pelargonidin, delphinidin, malvidin, and peonidin) (Castro et al. 2021). The total phenolic content of murta fruits is significantly higher in comparison to other cultivated berries such as blueberries or strawberries (Arancibia-Avila et al. 2011). In addition, anti-inflammatory and antimicrobial properties have been identified in ethanolic and methanolic extracts processed from this fruit (Junqueira-Gonçalves et al. 2019; Castro et al. 2021)

22.3 CONCLUSION

Chile is an important producer and exporter of fresh fruits in Latin America. However, the production of fermented products is only focused on fermented alcoholic beverages, and of them, only Chilean wine processed and fermented with white and red grapes is recognized around the world.

The fermentation-based processing of other fermented products such as juice or dairy products developed with fruits such as maqui, murtilla, grapes, or apples that possess a rich and diversified chemical composition of bioactive compounds and health-promoting properties has been studied; in addition, this kind of product could be consumed or processed to obtain a freeze-dried powder that can be added in the formulation of other types of healthy food. These new types of products could positively impact the human health of the population, and on the other hand, they can impact the economic resources that the country of origin can generate.

Furthermore, it is necessary to lead a transition from wild collection to the cultivation of native plants such as murtilla, maqui, or Chilean strawberry that have a high concentration of phytochemicals, and this implies that the production must be optimized and enhanced, considering various aspects like production management, technology, and techniques, to obtain consistent homogeneous products that can be incorporated in national production and exports.

REFERENCES

Aagesen, D. L. 2004. Burning monkey-puzzle: Native fire ecology and forest management in northern Patagonia. *Agriculture and Human Values* 21(2–3):233–242. https://doi.org/10.1023/B:AHUM.0000029402.85972.6c

Aguilar, T., C. Loyola, J. de Bruijn, et al. 2016. Effect of thermomaceration and enzymatic maceration on phenolic compounds of grape must enriched by grape pomace, vine leaves and canes. *European Food Research and Technology* 242:1149–1158. https://doi.org/10.1007/s00217-015-2619-3

Arancibia-Avila, P., F. Toledo, E. Werner, et al. 2011. Partial characterization of a new kind of Chilean murtilla-like berries. *Food Research International* 44(7):2054–2062. https://doi.org/10.1016/j.foodres.2011.01.016

Araya-Pizarro, S. (2022). Valor de marca del pisco chileno: Aportes desde la región pisquera de Chile. *RETOS Revista de Ciencias de la Administración y Economía* 12(23):139–159. https://doi.org/10.17163/ret.n23.2022.09

Belancic, A., E. Agosin, A. Ibacache, et al. 1997. Influence of sun exposure on the aromatic composition of Chilean Muscat grape cultivars Moscatel de Alejandria and Moscatel rosada. *American Journal of Enology and Viticulture* 48:181–186.

Bordeu, E., E. Agosín, and G. Casaubon. 2012. Pisco: Production, flavor chemistry, sensory analysis, and product development. In: *Alcoholic Beverages, Sensory Evaluation and Consumer Research, Woodhead Publishing Series in Food Science, Technology and Nutrition*, ed. J. Piggott, 331–347. https://doi.org/10.1533/9780857095176.3.331

Castro, A. 2016. Chicha y sidra de manzana en Chile (1870–1930): Manzanas con identificación de origen. *RIVAR* 3(9):4–26. https://www.redalyc.org/journal/4695/469546924002/html/

Castro, A., A. León, F. Cussen, et al. 2016a. ¡Viva la chicha nueva! La chicha en la vida popular y campesina chilena. *IDESIA* 34(1): 77–83. https://doi.org/10.4067/S0718–34292016000100009

Castro, A., P. Pszczólkowski, F. Mujica, et al. 2016b. El pajarete de Huasco y Elqui (Chile). Historia de un vino escogido. *IDESIA* 34(4): 23–35. http://doi.org/10.4067/S0718–34292016005000020

Castro, R., P. Ramos, C. Parra-Palma, et al. 2021. *Ugni molinae* fruit as a source of bioactive compounds with good quality traits. *BioMed Research International* 6683877. https://doi.org/10.1155/2021/6683877

Céspedes, C. L., M. El-Hafidi, N. Pavon, et al. 2008. Antioxidant and cardioprotective activities of phenolic extracts from fruits of Chilean blackberry *Aristotelia chilensis* (Elaeocarpaceae), Maqui. *Food Chemistry* 107:820–829. https://doi.org/10.1016/j.foodchem.2007.08.092

Céspedes, C. L., M. Valdez-Morales, J. G. Ávila, et al. 2010. Phytochemical profile and the antioxidant activity of Chilean wild black-berry fruits, *Aristotelia chilensis* (Mol) Stuntz (Elaeocarpaceae). *Food Chemistry* 119:886–895. https://doi.org/10.1016/j.foodchem.2009.07.045

Céspedes, C. L., N. Pavón, M. Domínguez, et al. 2017. The Chilean superfruit black-berry *Aristotelia chilensis* (Elaeocarpaceae), Maqui as mediator in inflammation-associated disorders. *Food Chemistry and Toxicology* 108:438–450. https://doi.org/10.1016/j.fct.2016.12.036

Céspedes-Acuña, C. L., J. Xiao, Z. J. Wei, et al. 2018. Antioxidant and anti-inflammatory effects of extracts from Maqui berry *Aristotelia chilensis* in human colon cancer cells. *Journal of Berry Research* 8:275–296. https://doi.org/10.3233/JBR-180356

Cheel, J., C. Theoduloz, J. A Rodríguez, et al. 2007. Free radical scavenging activity and phenolic content in achenes and thalamus from *Fragaria chiloensis* ssp. *chiloensis*, *F. vesca* and *F. x ananassa* cv. Chandler. *Food Chemistry* 102:36–44. https://doi.org/10.1016/j.foodchem.2006.04.036

Crowley, W. K. 2000. Chile's wine industry: Historical character and changing geography yearbook. *Conference of Latin Americanist Geographers* 26:87–101. http://www.jstor.org/stable/25765889

Delporte, C. 2006. *Determinación de las actividades anti-inflamatorias, analgésicas, antioxidantes y antimicrobianas de las hojas de Aristotelia chilensis (Maqui).* Identificación de los compuestos activos. Resúmenes del Primer Simposio Internacional de Plantas Medicinales y Aromáticas: Materia primas, puntos críticos y rol de las Universidades, Boletín Latinoamericano Caribe Plantas Medicinales Aromáticas, pp. 135–136.

Díaz, I., and D. Morales. 2019. *Existencia y diversificación de la cepa País en Chile.* https://www.inia.cl

Domínguez, A. M., and E. Agosin. 2010. Gas chromatography coupled with mass spectrometry detection for the volatile profiling of *Vitis vinifera* cv. Carménère wines. *Journal of the Chilean Chemical Society* 55:385–391. http://doi.org/10.4067/S0717-97072010000300025

Encina, F. A. 1952. *Historia de Chile desde la Prehistoria hasta 1891*, 3rd edition, Santiago, Chile.

Escribano-Bailón, M. T., C. Alcalde-Eon, O. Muñoz, et al. 2006. Anthocyanins in berries of maqui [*Aristotelia chilensis* (Mol.) Stuntz]. *Phytochemical Analysis* 17:8–14. http://doi.org/10.1002/pca.872

Fernández, K., M. Vega, and E. Aspé. 2015. An enzymatic extraction of proanthocyanidins from País grape seeds and skins. *Food Chemistry* 168:7–13. http://doi.org/10.1016/j.foodchem.2014.07.021

Figueroa, G., and M. Dantas. 2006. Recolección, procesamiento y consumo de frutos silvestres en el noroeste semiárido argentino. Casos actuales con implicancias arqueológicas. *La Zaranda de Ideas* 2:35–50. https://plarci.org/index.php/lazarandadeideas/article/view/611

Finn, Ch., J. Retamales, and G. Lobos, et al. 2013. The Chilean strawberry (*Fragaria chiloensis*): over 1000 years of domestication. *Hortscience* 48(4):418–421. https://doi.org/10.21273/HORTSCI.48.4.418

Fredes, C., G. Montenegro, J. Zoffoli, et al. 2014. Comparison of total phenolic, total anthocyanin and antioxidant activity of polyphenol-rich fruits grown in Chile. *Ciencia e Investigación Agraria* 41(1):49–60. http://doi.org/10.4067/S0718-16202014000100005

Fredes, C., A. Parada, J. Salinas, et al. (2020). Phytochemicals and traditional use of two southernmost Chilean berry fruits: Murta (*Ugni molinae* Turcz) and Calafate (*Berberis buxifolia* Lam.). *Foods* 9(1):54. https://doi.org/10.3390/foods9010054

Gironés-Vilaplana, A., P. Valentao, D. A. Moreno, et al. 2012. New beverages of lemon juice enriched with the exotic berries, maqui, acai, and blackthorn: Boactive components and in vitro biological properties. *Journal of Agriculture and Food Chemistry* 60:6571–6580. http://doi.org/10.1021/jf300873k

Godoy, S., M. Roeckel, and K. Fernández. 2012. Influence of the structure and composition of the País grape proanthocyanidins on the inhibition of angiotensin I-converting enzyme (ACE). *Food Chemistry* 134:346–350. https://doi.org/10.1016/j.foodchem.2012.02.171

Godoy-Aguirre, C. 2018. Evaluating vegetable processing and the elaboration of fermented beverages in an El Vergel context in Mocha Island (ad 1000–1300). *Chungara* 50(1):107–120. http://doi.org/10.4067/S0717-73562018005000401

Gutiérrez-Gamboa, G., M. Carrasco-Quiroz, N. Verdugo-Vásquez, et al. 2018. Characterization of grape phenolic compounds of 'Carignan' grapevines grafted onto 'País' rootstock from Maule Valley (Chile): Implications of climate and soil conditions. *Chilean Journal of Agriculture Research* 78(2):310–315. http://doi.org/10.4067/S0718-58392018000200310

Gutiérrez-Gamboa, G., S. Y. Liu, and P. Pszczólkowskic. 2020. Resurgence of minority and autochthonous grapevine varieties in South America: A review of their oenological potential. *Journal of the Sciences of Food and Agriculture* 100:465–482. https://doi.org/10.1002/jsfa.10003

Henríquez, C., B. Escobar, F. Figuerola, et al. 2008. Characterization of piñon seed (*Araucaria araucana* (Mol) K. Koch) and the isolated starch from the seed. *Food Chemistry* 107:592–601. https://doi.org/10.1016/j.foodchem.2007.08.040

Herrmann, M. 2005. Knowledge, values, uses and management of the *Araucaria araucana* forest by the indigenous Mapuche Pewenche people: A basis for collaborative natural resource management in southern Chile. *Natural Resources Forum* 29: 120–134. https://doi.org/10.1111/j.1477-8947.2005.00121.x

Herrmann, T. M. 2006. Indigenous knowledge and management of *Araucaria araucana* forest in the Chilean Andes: Implications for native forest conservation. *Biodiversity and Conservation* 15:647–662. https://doi.org/10.1007/s10531-005-2092-6

Hoffmann, A. 1991. *Native Flora of Chile, Araucana Area*, 4th editions. Santiago, Chile: Claudio Gay Foundation.

Hoffmann, A., C. Farga, J. Lastra, et al. 1992. *Medicinal Plants in Common Use in Chile.* Santiago, Chile: Claudio Gay Foundation.

INFOR-Chile. 2003. Technological and commercial innovation on non-timber forestry products in Chile. Information Bulletin 1 *Aristotelia chilensis* (Mol.) Stuntz, Maqui, Maquei, Clon, Queldrón.

Junqueira-Gonçalves, M. P., L. Yáñez, C. Morales, et al. 2019. Isolation and characterization of phenolic compounds and anthocyanins from murta (*Ugni molinae* Turcz.) fruits. assessment of antioxidant and antibacterial activity. *Molecules* 20:5698–5713. http://doi.org/10.3390/molecules20045698

Labarca, E. 1994. *Butamalón.* Editorial. Santiago, Chile: Universitaria y Fondo de Cultura Económica.

Lacoste, P., A. Castro, F. Briones, et al. 2015c. El pipeño: historia de un vino típico del sur del Valle Central de Chile. *Idesia (Arica)* 33(3):87–96. http://doi.org/10.4067/S0718-34292015000300013

Lacoste, P., P. Pszczolkowski, F. Briones, et al. 2015b. Historia de la chicha de uva: Un producto típico en Chile. *Idesia* 33(2):87–96. http://doi.org/10.4067/S0718–34292015000200011

Lacoste, P., P. Pszczólkowski, and N. Soto. 2015a. Moscatel de Alejandría en Chile y Argentina: Origen y relación con el pisco. *Idesia* 33(3):79–86. http://doi.org/10.4067/S0718–34292015000300012

Lutz, M., Y. Cajas, and C. Henríquez. 2012. Phenolics content and antioxidant capacity of Chilean grapes cv. País and Cabernet Sauvignon. *CYTA: Journal of Food* 10(4):251–257. https://doi.org/10.1080/19476337.2011.633244

Martínez-Gil, A. M., G. Gutiérrez-Gamboa, T. Garde-Cerdán, et al. 2018. Characterization of phenolic composition in Carignan noir grapes (*Vitis vinifera* L.) from six wine-growing sites in Maule Valley, Chile. *Journal of the Sciences of Food and Agriculture* 98:274–282. https://doi.org/10.1002/jsfa.8468

Miranda-Rottmann, S., A. A Aspillaga, D. D. Pérez, et al. 2002. Juice and phenolic fractions of the berry *Aristotelia chilensis* inhibit LDL oxidation in vitro and protect human endothelial cells against oxidative stress. *Journal of Agriculture and Food Chemistry* 50:7542–7547. https://doi.org/10.1021/jf025797n

Misle, E., E. Garrido, H. Contardo, et al. 2011. Maqui [*Aristotelia chilensis* (Mol.) Stuntz]-the amazing Chilean tree: A review. *Journal of Agriculture, Sciences and Technology B* 1:473–482.

Mitchell, J. T., and W. C. Terry. 2011. Contesting pisco: Chile, Perú, and the politics of trade. *Geographical Review* 101:518–535. http://www.jstor.org/stable/23208637

Montes, M., and T. Wilkomirsky. 1987. *Medicina tradicional chilena*. Concepción, Chile: Editorial de la Universidad de Concepción.

Mora, G. M. 2019. The Chilean wine industry. In: *The Palgrave Handbook of Wine Industry Economics*, eds. A. A. Ugaglia, J. M. Cardebat, and A. Corsi, 177–200. Berlin/Heidelberg, Germany: Springer International Publishing.

Morales-Quintana, L., and P. Ramos. 2019. Chilean strawberry (Fragaria chiloensis): An integrative and comprehensive review. *Food Research International* 119:769–776. https://doi.org/10.1016/j.foodres.2018.10.059

Mosbach, E. W. 1992. *Indigenous Botany of Chile*. Santiago, Chile: Editorial Andrés Bello.

Muñoz, C., C. Mattar, R. Neira, et al. 2020. Sustainable agriculture and healthy food in Chile. In: *Challenges and Opportunities for Food and Nutrition Security in the Americas: The View of the Academies of Sciences*, ed. IANAS, 190–210. https://ianas.org/wp-content/uploads/2020/07/Chile.pdf

Muñoz, O., M. Montes, and T. Wilkomirsky. 2001. *Plantas medicinales de uso en Chile*. Química y Farmacología. Santiago, Chile: Editorial Universitaria.

Noriega, F., C. Mardones, S. Fischer, et al. 2021. Seasonal changes in white strawberry: Effect on aroma, phenolic compounds and its biological activity. *Journal of Berry Research* 11:103–118. https://doi.org/10.3233/JBR-200585

Obreque-Slier, E., A. Peña-Neira, R. López-Solís, et al. 2010. Comparative study of the phenolic composition of seeds and skins from Carménère and Cabernet Sauvignon grape varieties (*Vitis vinifera* L.) during ripening. *Journal of Agriculture and Food Chemistry* 58:3591–3599. https://doi.org/10.1021/jf904314u

Obreque-Slier, E., R. López-Solís, L. Castro-Ulloa, et al. 2012. Phenolic composition and physicochemical parameters of Carménère, Cabernet Sauvignon, Merlot and Cabernet Franc grape seeds (*Vitis vinifera* L.) during ripening. *LWT: Food Sciences and Technology* 48:134–141. https://doi.org/10.1016/j.lwt.2012.02.007

ODEPA (Oficina de Estudios y Políticas Agrarias). 2022. *Frutas frescas y procesadas*. https://www.odepa.gob.cl/rubros/frutas-frescas-y-procesadas

Pasiecznik, N. M., P. J. Harris, and S. J. Smith. 2004. *Identifying Tropical Prosopis Species a Field Guide*. Coventry, UK: HDRA Publishing.

Poblete, G. 1998. *Vegetative Propagation of maqui [Aristotelia chilensis (Mol.) Stuntz]*. Thesis, Faculty of Agronomy. Chillán, Chile: University of Concepción, p. 22.

Pszczólkowski, P. 2004. La invención del cv. Carménère (Vitis vinifera L.) en Chile, desde la mirada de uno de sus actores. *Universum* 2:150–165. http://doi.org/10.4067/S0718–23762004000200010

Pszczólkowski, P., and P. Lacoste. 2016. Variedades criollas, una oportunidad para el pisco de Chile. *Revista de la Facultad de Ciencias Agrarias* 48:239–251. https://www.redalyc.org/pdf/3828/382846012001.pdf

Reyes-Farías, M., K. Vasquéz, F. Fuentes, et al. 2016. Extracts of Chilean native fruits inhibit oxidative stress, inflammation and insulin-resistancelinked to the pathogenic interaction between adipocytes and macrophages. *Journal of Functional Foods* 27:69–83. https://doi.org/10.1016/j.jff.2016.08.052

Rodríguez, F., F. Cattaneo, X. V. Zech, et al. 2020. Aloja and añapa, two traditional beverages obtained from *Prosopis alba* pods: Nutritional and functional characterization. *Food Biosciences* 35:100546. https://doi.org/10.1016/j.fbio.2020.100546

Rodríguez, K., K. Ah-Hen, A. Vega-Gálvez, et al. 2014. Changes in bioactive compounds and antioxidant activity during convective drying of murta (*Ugni molinae* T.) berries. *International Journal of Food Sciences and Technology* 49(4):990–1000. https://doi.org/10.1111/ijfs.12392

Rodríguez, L., A. Trostchansky, H. Vogel, et al. 2022. A comprehensive literature review on cardioprotective effects of bioactive compounds present in fruits of *Aristotelia chilensis* Stuntz (Maqui). *Molecules* 27:6147. https://doi.org/10.3390/molecules27196147

Rodríguez, R., O. Mattei, and M. Quezada. 1983. *Arboreal Flora of Chile*. Concepción, Chile: Editorial University of Concepción.

Rojo, L. E., D. Ribnicky, S. Logendra, et al. 2012. In vitro and in vivo anti-diabetic effects of anthocyanins from Maqui berry (*Aristotelia chilensis*). *Food Chemistry* 131(2):387–396. https://doi.org/10.1016/j.foodchem.2011.08.066

Rubilar, M., C. Jara, Y. Poo, et al. 2011. Extracts of Maqui (Aristotelia chilensis) and Murta (Ugni molinae Turcz.): Sources of antioxidant compounds and α-glucosidase/α-amylase inhibitors. *Journal of Agriculture and Food Chemistry* 59:1630–1637. https://doi.org/10.1021/jf103461k

SAG. 2022. *Catastro Vitícola Nacional*. https://www.sag.gob.cl/noticias/sag-presenta-catastro-viticola-nacional-2022

Salas, M. 1999. Procesamiento y calidad de sidra en Valdivia-Chile. *Agro Sur* 27(1):1–9. https://doi.org/10.4206/agrosur.1999.v27n1–01

Schmeda-Hirschmann, G., F. Jiménez-Aspee, C. Theoduloz, et al. 2019. Patagonian berries as native food and medicine. *Journal of Ethnopharmacology* 241:111979. https://doi.org/10.1016/j.jep.2019.111979

Schreckinger, M. E., J. Wang, G. Yousef, et al. 2010. Antioxidant capacity and in vitro inhibition of adipogenesis and inflammation by phenolic extracts of *Vaccinium floribundum* and *Aristotelia chilensis*. *Journal of Agriculture and Food Chemistry* 58:8966–8976. https://doi.org/10.1021/jf100975m

Sciammaro, L. P., D. P. Ribotta, and M. C. Puppo. 2016. Chapter 14: Traditional food products from *Prosopis* sp. flour. In: *Traditional Foods, Integrating Food Science and Engineering Knowledge into the Food Chain 10*, eds. K. Kristbergsson and J. Oliveira, 209–16. New York: Springer Science+Business Media. https://doi.org/10.1007/978–1–4899–7648–2_14

Simirgiotis, M. J., C. Theoduloz, P. Caligari, et al. 2009. Comparison of phenolic composition and antioxidant properties of two native Chilean and one domestic strawberry genotypes. *Food Chemistry* 113:377–385. https://doi.org/10.1016/j.foodchem.2008.07.043

Simunovic, Y. 1999. *Manual de bebidas alcohólicas y vinagres*. Subdepartamento Divulgación Técnica Servicio Agrícola y Ganadero. https://www.sag.gob.cl/sites/default/files/MANUAL_BEBIDAS_ALCOHOLICAS.pdf

Sotomayor, J. P. 1987. *Denominación de Origen: Vino Asoleado y Pajarete*. I Simposio Internacional sobre Denominación de Origen de productos vitivinícolas en América Latina. La Serena, del 22 al 27 de noviembre de 1987. Santiago de Chile, SAG, tomo I, pp. 298–313.

Suwalsky, M., P. Vargas, M. Avello, et al. 2008. Human erythrocytes are affected in vitro by flavonoids of *Aristotelia chilensis* (Maqui) leaves. *International Journal of Pharmaceutics* 363:85–90. https://doi.org/10.1016/j.ijpharm.2008.07.005

Tsakiris, A., S. Kallithraka, and Y. Kourkoutas. 2014. Grape brandy production, composition and sensory evaluation. *Journal of the Sciences of Food and Agriculture* 94:404–414. https://doi.org/10.1002/jsfa.6377

Villarroel, M. 2009. *Caracterización del contenido de proantocianidinas en las uvas País Negra (Vitis vinífera L) del Valle del Itata*. Concepción, Chile: Universidad de Concepción.

Vogel, H., I. Rasmilic, J. San Martin, et al. 2008. *Medicinal plants of Chile*. Talca, Chile: Editorial Universidad de Talca.

Biotechnological Advances in Diverse Brazilian Fruit Fermentation

23

Carolina Saori Ishii Mauro, Maria Thereza Carlos Fernandes, Paulo Terumitsu Saito, Fernanda Silva Farinazzo, Deepak Kumar Verma, and Sandra Garcia

23.1 INTRODUCTION

Brazil boasts the most plant diversity in the world, with studies pointing to 32,797 different angiosperm species (plants with flowers and fruits) (Barros et al. 2017; Teixeira et al. 2019; Barbosa et al. 2020). There is a wealth of natural landscapes where a variety of species exist from the Amazon to the Pampa, passing through the Caatinga, the Cerrado, the Pantanal, and the Atlantic Forest (Infante et al. 2016; Anunciação et al. 2019; Teixeira et al. 2019; da Silva et al. 2019b; Brack et al. 2020; Schulz et al. 2020; de Assis et al. 2021). Although the Brazilian flora are the most diversified in the world, the uses, features, and potentials of most of the plants are unknown, which is reflected in the country's and the world's limited food supply. Only 15 plant species are believed to provide much of the food consumed by people (Teixeira et al. 2019). Food monotony is thus caused by a lack of government incentives and policies that encourage the transition from large monocultures to basic ecological production systems, integrating and valuing resources through farming families and traditional communities that seek to exploit biodiversity resources in a sustainable manner to ensure environmental conservation, income generation, and the promotion of food sovereignty (Köhler and Brack 2016; da Silva et al. 2017; Dantas et al. 2019).

Brazil has continental dimensions and is home to a variety of biomes, some of which are more endangered than others. These lands have been protected by native populations for cultural and survival reasons. There is a wide range of natural wildlife and flora in these biomes, with many at risk of extinction or vulnerability (Neri-Numa et al. 2018; Teixeira et al. 2019; Schulz et al. 2020; de Assis et al. 2021). Native peoples have known and used this biodiversity for many years. The local and indigenous fruits are used by the residents of these areas, but the majority are not widely commercialized or industrialized, making their usage and investigation by other people challenging. Although there has been little scientific research on these fruits, it is known that they contain bioactive compounds (BACs) with potential health advantages (de Araujo et al. 2019; Albuquerque et al. 2020; Barbosa et al. 2020; Dacoreggio et al. 2021; Andrade et al. 2022; Morais et al. 2022). Some authors reported that exotic, wild, and poorly investigated fruits can contribute to innovations and the development of new products in the food, pharmaceutical, and medicinal industries if further studies can be undertaken (Teixeira et al. 2019; Schulz et al. 2020; Dacoreggio et al. 2021). However, with sustained European invasion and pressure on indigenous tribes, the habits and customs of the Old World have been overlaid with the diversity of local traditions and cultures, which have gradually declined. Ingás, cajas, araticuns, araçás, cambucás, pequis, jatobás, and many other species, which were part of the food and culture of indigenous or aboriginal inhabitants, were marginalized and even viewed with disdain. Native fruits are considered unconventional food plants because, for the most part, they go unreported or are unknown to a major portion of the population, particularly urban dwellers, necessitating their presentation and discussion of their applications, qualities, and potentials (Köhler and Brack 2016).

DOI: 10.1201/9781003605300-23

On special occasions such as celebrations, festivities, and religious rites, aborigines drink fermented beverages made from native fruits, grains, honey, starchy tubers, sugar cane, and palm saps. They are also a part of the diet and a source of nutrients, but information about these products is primarily distributed through oral reporting, with few publications. Fermentation takes place because of the action of wild resident microorganisms and substrate contaminants. The alcoholic strength of 10 to 11% v/v obtained in cauim, caxiri, and pajauaru beverages, as well as the inclusion of fruit juices, suggests that these beverages are from fermented fruit (Puerari 2014; Ferreira 2015; Resende et al. 2018; Santos et al. 2021a). Osvaldo Gonsalves de Lima, a microbiologist, published an important book on Brazilian Indigenous drinks in 1975, entitled *Pulque, Balchê, and Pajuaru, on the Ethnobiology of Fermented Beverages and Foods* (Lima 1975). Most fermented indigenous foods have an unknown and complex microbiota, which is characterized by the association of lactic acid bacteria, bacilli, yeasts, and fungi (van Hijum et al. 2013).

Researchers have recently shown an increased interest in native fruits from Brazilian biomes because of their immense diversity, the content of BACs favorable to health, and the possibility of exploration for a more diverse and nutritious diet. Native fruit consumption can also aid local economic development by giving opportunities to people in rural areas and indigenous people. Fermented beverages derived from traditional Brazilian fruits are attractive because they improve food quality by increasing bioactive ingredients after fermentation and improving sensory qualities. Consumers may be more accepting of these changes because of these transformations (Fioravante et al. 2017). Fermentation can also help to extend the shelf life of perishable and seasonal fruits, decreasing post-harvest losses (Da Silva 2014; Verma et al. 2017). Barros et al. (2020) investigated the effect of probiotic fermentation on the bioaccessibility of phenolic components and antioxidant potential of native and exotic fruit by-products, as well as *in vitro* gastrointestinal (GI) digestion. The bioaccessibility of cupuaçu (*Theobroma grandiflorum*) residue increased significantly after digestion, as evaluated by ORAC 339.83±0.06%, and around 15.01±1.54 to 237.77±4.10% of phenolic compounds were bioaccessible after probiotic fermentation. Catechin and epicatechin were found in all the fruit extracts tested. Simple phenolic acids and hydroxybenzoic acid were detected after GI ingestion and fermentation by probiotics, as demonstrated by previous investigators. Among the residues investigated, soursop (*Annona muricata*) showed better bioaccessibility, antioxidant activity, and inhibition of α-amylase, implying a potential anti-diabetic action.

Given the diversity and complexity of fruit species found in various biomes throughout Brazil, the goal of this chapter is to present the most common fruits found in various Brazilian biomes that have been fermented or used to develop fermented products and whose dissemination has been documented in the scientific literature. Furthermore, biochemical and nutritional components, as well as their biological activity for human health, have been reviewed, as these fruits are important and have an impact on food and human health.

23.2 BIOCHEMICAL AND NUTRITIONAL COMPONENTS OF BRAZILIAN FRUITS

Several researchers and scientists have studied the biochemical and nutritional components of Brazilian fruits in recent years due to their high concentrations of sugars, fatty acids, vitamins, minerals, flavonoids, pigments, and sterols (Fujita et al. 2015; Uekane et al. 2017; Anunciação et al. 2019; Bastos et al. 2019; da Silva et al. 2019a; Lazarini et al. 2020; Schulz et al. 2020; Gibbert et al. 2021). Major biochemical and nutritional components of Brazilian fruits, as well as their chemical structure, are shown in Figure 23.1. Researchers and scientists reported a wide variation among these biochemical and nutritional components and stated that intrinsic and extrinsic factors that include fruit species, fruit cultivars, soil and soil conditions, rainfall, environmental conditions, ripening stage, and geographic location influenced the biochemical and nutritional components of Brazilian fruits (Neri-Numa et al. 2018; de Araujo et al. 2019; Teixeira et al. 2019; Schulz et al. 2020).

There are several reasons for looking into the biochemical and nutritional properties of Brazilian fruits, including anti-cancer (Guzzo da Silva et al. 2019; Gibbert et al. 2021), anti-inflammatory (Lazarini et al. 2020; Xiong et al. 2020; Amorim et al. 2021), antimicrobial (Fujita et al. 2015), antioxidants (Barros et al. 2017; Lazarini et al. 2020; de Assis et al. 2021; Fraga et al. 2021; Gibbert et al. 2021; Rudke et al. 2021; Andrade et al. 2022), antiproliferative (Neri-Numa et al.

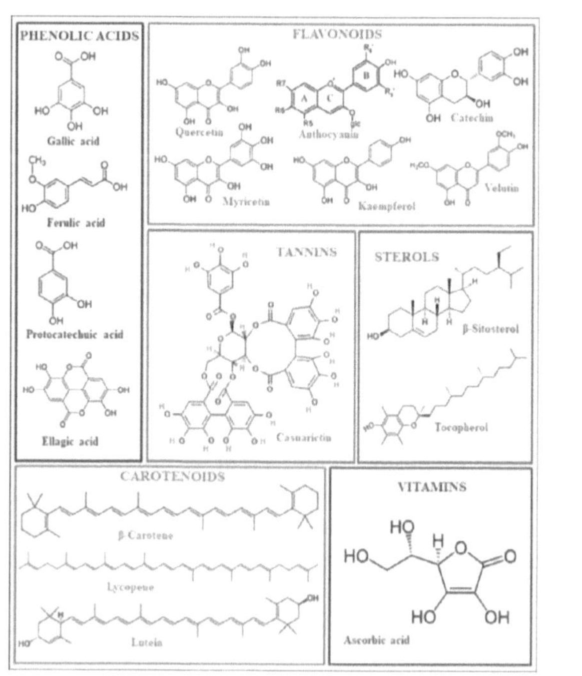

FIGURE 23.1 Chemical structure of major biochemical and nutritional components of Brazilian fruits (Neri-Numa et al. 2018).

2013; Guzzo da Silva et al. 2019), cardiovascular protection (da Silva et al. 2019a), glycemic control, and weight control (Fujita et al. 2015, 2017), as well as neuroprotective properties (Antognoni et al. 2020; Schulz et al. 2019). As a result of malnutrition and an imbalanced diet, a rising number of people are suffering from nutritional deficiencies and chronic illnesses because of eating diets that are poor in nutritional content and diversity (da Silva et al. 2019a; Antognoni et al. 2020; Schulz et al. 2020; Dacoreggio et al. 2021). As a result of this situation, new research must advance the understanding of the composition and health benefits of native Brazilian fruit species that are now underappreciated. Due to reduced food and nutritional security issues, as well as the need to promote biodiversity conservation and management, the usage of these species might boost the number of native foods now consumed in Brazil and other countries. Brazilian native species may be of interest to pharmaceutical enterprises in addition to food, which may help enhance the local economy by encouraging the cultivation and investigation of these fruits (Infante et al. 2016; Neri-Numa et al. 2018; de Araujo et al. 2019; Teixeira et al. 2019; Schulz et al. 2020).

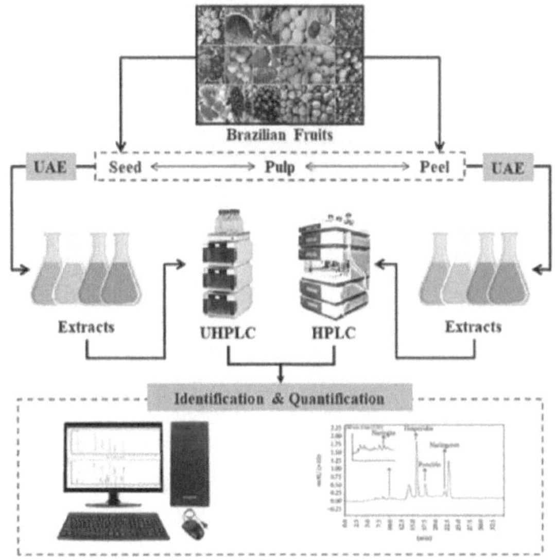

FIGURE 23.2 Schematic illustration of extraction, identification, and quantification of Brazilian fruit bioactive compounds.

23.2.1 Extraction, Identification, and Quantification of Biochemical Components

In terms of genetic diversity, Brazil is one of the most varied countries in the world (Araujo et al. 2020). In terms of fruit production, Brazil ranks third after China and India due to its soil, climate, and geographic location (Araujo et al. 2020). It produces 42 million tonnes of tropical, subtropical, and temperate fruits (Uekane et al. 2017). As a result of the taste and use of naturally occurring BACs in indigenous Brazilian fruits, there has been an increase in consumer interest in these fruits (Neri-Numa et al. 2018; Donado-Pestana et al. 2018). The agro-industry might benefit from many underutilized native and exotic fruit species in Brazil, many of which are wild or only available at local trade fairs (Anunciação et al. 2019; Barbosa et al. 2020). Several fruits rich in BACs, such as phenolic compounds, carotenoids, and ascorbic acid, could be a potential source of nutrients and BACs in Brazil (Araujo et al. 2020; Barbosa et al. 2020). Some of these plants and their bioactivity potential are unknown, and as a result, they are rarely incorporated into everyday meals (Infante et al. 2016; Araujo et al. 2020). Therefore, research on extracting, profiling, and quantifying phytochemical components, primarily phenolic compounds found in Brazilian fruits, is required. A schematic representation of the extraction, identification, and quantification of Brazilian fruit phytochemicals is shown in Figure 23.2. These studies can be useful in future research examining the association between diet and disease. Furthermore, they can aid in the development and subsequent utilization of new products. The next paragraphs discuss various extraction, identification, and quantification methods for Brazilian fruits.

Secondary metabolism-derived molecules that benefit human health are known as phytochemicals or bioactive compounds (Das et al. 2017; Meregalli et al. 2020). In order to acquire BACs from fruits, extraction is a crucial process in which a solvent act on the vegetative cellular structure and removes the ingredient of interest (Gualberto et al. 2021). A variety of extraction procedures and solvents should be examined since plant matrixes contain a wide range of antioxidants (Bezerra et al. 2020; Chávez-González et al. 2020; Verma and Srivastav, 2020b; Gualberto et al. 2021; Tripathy et al. 2021). Maceration and percolation are few standard conventional extraction methods (Oroian et al. 2020b), and on the other hand, the non-conventional methods include ultrasound-assisted (Klein et al. 2018; Asfaram et al. 2019; Fernandez-Barbero et al. 2019; Oroian et al. 2020b; Rudke et al. 2021), supercritical fluid (del Pilar Garcia-Mendoza et al. 2017; Klein et al. 2018; Verma et al. 2018; Santos et al. 2019), ohmic heating assisted extraction (Mercali et al. 2015), microwave-assisted extraction (Oroian et al. 2020b), and liquid-pressurized extraction (del Pilar Garcia-Mendoza et al. 2017), which employs different organic solvents, such as acetone and methanol acidified with hydrochloric acid (Table 23.1).

There are few complications with traditional extraction techniques, and the process is straightforward. However, more time and money are needed to achieve their goals (Meregalli et al. 2020; Chávez-González et al. 2020). Non-conventional extraction methods are more successful in overcoming these difficulties by minimizing these factors (Meregalli et al. 2020). Increased attention has been paid to employing clean methods that simultaneously produce excellent extraction yields and reduce solvent-related toxicities (Kim et al. 2019; Chávez-González et al. 2020). Ultrasound-assisted extraction (UAE) is a cost-effective, simple, and efficient method of extraction. Reduced energy usage and the avoidance of heat degradation of some active chemicals can be achieved by using this method

TABLE 23.1 Recent Studies on the Extraction, Identification, and Quantification of Bioactive Compounds of Brazilian Fruits

			IDENTIFIED BIOACTIVE COMPOUND AND ITS CONCENTRATION		
FRUIT	*PART U=USED*	*EXTRACTION METHOD*	*BIOACTIVE COMPOUNDS*	*CONCENTRATION*	*REFERENCES*
Psidium guajava	Seed, peel	UAE	Naringenin	13.29±7.65 µg/g DW	Gualberto et al. (2021)
			Vanillin	16.08±0.0 µg/g DW	
			ρ-coumaric acid	10.80±0.23 µg/g DW	
			Ferulic acid	11.69±0.24 µg/g DW	
			Vanillic acid	13.27±0.24 µg/g DW	
			Chlorogenic acid	6.97±0.43 µg/g DW	
	Whole fruit	–	Apigenin	0.01–0.032 mg/100 g	Shukla et al. (2021)
			Ascorbic acid	7.45–75.07 mg/100 g	
			Gallic acid	9.46–63.08 mg/100 g	
			Lycopene	0.34–0.92 mg/100 g	
			Myrcetin	0.09–0.034 mg/100 g	
			Quercetin	0.11–2.51 mg/100 g	
Spondias tuberosa	Seed, peel	UAE	Chlorogenic acid	20.28±0.15 µg/g DW	Gualberto et al. (2021)
			Gallic acid	37.82±0.32 µg/g DW	
			Rutin	423.58±0.72 µg/g DW	
			Vanillic acid	39.41±0.24 µg/g DW	
			Vanillin	31.59±0.03 µg/g DW	
			ρ-coumaric acid	30.47±0.23 µg/g DW	
Eugenia calycina	Pulp, seed	UAE	Ellagic acid	NQ	Araujo et al. (2020)
			Epicatechin gallate		
			Myricitrin		
Eugenia involucrata	Fruit, seed	Maceration	Hydroxybenzoate derivates (HBD) and tannins	1.65±0.04 mg/mL of extract	Girardelo et al. (2020)
			Total flavonol derivatives and ellagic acid	0.68±0.01 mg/mL of extract	
Psidium cattleianum	Peel	UAE	–	**NQ**	Meregalli et al. (2020)
Myrciaria dubia	Whole fruit	UAE	Alnusiin, cyanidin-3-glucoside	NQ	Rodrigues et al. (2020)
			Ellagic acid		
			Gallic acid		
			p-coumaric,		
			Quercetin		
Baccharis dracunculifolia	Whole fruit	UAE	Artepillin C	807.6 ± 57.7 mg/g sample	Cavalaro et al. (2019)
			p-coumaric	45.6 ± 0.2 mg/g sample	
Myrciaria cauliflor	Pulp, peel	UAE	Cyanidin chloride	0.573 mg/L	Fernandez-Barbero et al. (2019)
			Cyanidin-3-*O*-glucosid	0.780 mg/L	
			Delphinidin-3-*O*-glucoside	0.826 mg/L	
			Gallic acid	8.07 mg/L	
Acca sellowiana	Peel	SCFE	Ellagic acid Ferulic acid Gallic acid	NQ	Santos et al. (2019)

(Continued)

TABLE 23.1 (Continued)

			IDENTIFIED BIOACTIVE COMPOUND AND ITS CONCENTRATION		
FRUIT	*PART U=USED*	*EXTRACTION METHOD*	*BIOACTIVE COMPOUNDS*	*CONCENTRATION*	*REFERENCES*
Lecythis pisonis	Nuts and shells	UAE	Catechin	NQ	Demoliner et al. (2018)
			Ellagic acid		
			Epicatechin		
			Ferulic acid		
			Myricetin		
Lycium barbarum	Whole fruit	Sequential extraction	4-hydroxybenzoic	**NQ**	Pedro et al. (2018)
			Caffeic		
			Catechin		
			Chlorogenic		
			Ferulic acid		
			Gallic acid		
			Kaempferol		
			Naringin		
			p-coumaric		
			Quercetin		
			Rutin		
			Syringic		
			trans-cinnamic		

for the extraction of various compounds (Cavalaro et al. 2019; Oroian et al. 2020a, 2020b; Chávez-González et al. 2020). As a green technique, ultrasound has less solvent required, has a shorter extraction time, and is easier to use (Klein et al. 2018; Rodrigues et al. 2020; Chávez-González et al. 2020). Acoustic cavitation is formed because of ultrasound-assisted extraction, which is described as the formation, growth, and implosion of bubbles (Cavalaro et al. 2019). BACs are released when bubbles are generated owing to alternating compression and expansion collapse on the surface of the substance (Asfaram et al. 2019; Oroian et al. 2020a; Oroian et al. 2020b). The collapse of cavitation bubbles increases mass transfer, which in turn leads to an increase in reactivity (Chemat et al. 2017; Cavalaro et al. 2019).

The yield and concentration of chemicals in the extract are greatly affected by processing factors such as the solid-to-solvent ratio, extraction duration, and temperature (Cavalaro et al. 2019; Verma and Srivastav 2020b). Significant parameters influencing ultrasound efficiency include ultrasonic amplitude and cycle, solvent type and composition, temperature, and pH (Fernandez-Barbero et al. 2019; Verma and Srivastav 2020b). Process variables can be fixed in a variety of ways; mathematical models, in this sense, minimize the number of trials necessary to analyze parameter effects and interactions (Das et al. 2017; Meregalli et al. 2020; Verma and Srivastav 2020b). The use of response surface methodology (RSM), a practical and vital technique to find the interaction of many processing factors, can minimize the number of tests, enabling computation and assessing the experimental error (Cavalaro et al. 2019). Such an extract may be obtained with optimized features using RSM (Pedro et al. 2018). As a result, the RSM mathematical model may be efficiently used to obtain an extract with high antioxidant content (Pedro et al. 2016). Using RSM, it is feasible to determine the best extraction point, where the responses, such as antioxidant capacity, are the most favorable.

To identify and quantify BACs, various chromatographic methods, including high-performance liquid chromatography (HPLC), ultra-high-performance liquid chromatography (UHPLC), and thin-layer chromatography (TLC) have been used ((Lade et al. 2014; De Freitas et al. 2017; Pedro et al. 2018; Fernandez-Barbero et al. 2019; Girardelo et al. 2020; Rodrigues et al. 2020; Gualberto et al. 2021; Shukla et al. 2021). HPLC is the technique of choice for many BACs because of its ease of use and high sensitivity (Verma and Srivastav 2020a). Reverse-phase C18 columns, a binary solvent gradient, and various detection techniques like diode array detection or mass spectrometry are the most common HPLC conditions used. This approach has been used to identify and quantify BACs in Brazilian fruit (Cavalaro et al. 2019; Verma and Srivastav 2020a; Shukla et al. 2021). Complex BACs in natural products or biological samples may now be identified using chromatographic methods and mass spectrometry (Fernandez-Barbero et al. 2019; Rodrigues et al. 2020; Verma and Srivastav 2020a; Gualberto et al. 2021).

In addition, a recently developed UHPLC connected to a photodiode array detector or quadrupole mass spectrometer is utilized to identify and quantify BACs in Brazilian fruit (Table 23.1) (Fernandez-Barbero et al. 2019; Rodrigues et al. 2020; Gualberto et al. 2021). UHPLC-PDA analysis of phenolic composition revealed a significantly greater phenolic content in *Eugenia involucrata* fruit (Girardelo et al. 2020). Hydroxybenzoate derivatives were the most common kind of identified phenolic, followed by flavonol derivatives, ellagic acid derivatives, and hydroxycinnamate derivatives, in that order (Girardelo et al. 2020). Furthermore, 17 compounds were identified and validated manually in *Myrciaria dubia* using UHPLC-Q–TOF-MS/MS. The bulk of the compounds were from the groups of anthocyanins, ellagitannins, ellagic acid derivatives, flavonols, phenolic acids, and proanthocyanidins (Rodrigues et al. 2020). Recent research effectively employed UPLC-MS to analyze the BACs of *Psidium guajava* cultivars (Shukla et al. 2021).

23.2.2 Biological Activity and Potential Health Benefits of Brazilian Fruit

Tropical fruits have been more popular in recent years since they are recognized as a source of BACs in addition to their nutritional benefits (Dutra et al. 2017). The antioxidant activity of BACs contained in fruits relates to protection against oxidative stress in the body, which is a significant cause of chronic degenerative illness (Antognoni et al. 2020; Dacoreggio et al. 2021). These substances are more prevalent in indigenous and alien fruit species, which are mostly grown in tropical nations, increasing the food industry's interest in these fruits (Sayago-Ayerdi et al. 2021). Brazil's geographical and climatic qualities are favorable for fruit cultivation (da Silva et al. 2019a). The north-eastern area of Brazil provides favorable climatic conditions for growing a diverse range of exotic fruits with high nutritional value and agricultural and industrial potential (Barros et al. 2017; Dutra et al. 2017; Dantas et al. 2019). The detailed biological activity and its importance to human health are shown in Table 23.2.

The pulp of fruits is the major source of antioxidants, such as phenols and flavones, which can protect cells from oxidative stress and prevent illnesses caused by oxidative stress (Stafussa et al. 2018). Various experimental investigations clearly explain the anti-diabetic, anticancer, and anti-neurodegenerative effects of phenolic substances (Figure 23.3) (Lin et al. 2016; Saleem et al. 2017). In this context, the usage of the Brazilian fruit uvaia (*Eugenia pyriformis*) has lately increased due to the fruit's ability to reduce chronic illnesses such as cardiovascular disease, cancer, and diabetes (da Silva et al. 2019a). Furthermore, a balanced diet rich in Brazilian fruits decreases inflammation since they are high in carotenoids, terpenes, phenolic compounds, and soluble and insoluble fibers (Neri-Numa et al. 2018; Donado-Pestana et al. 2018).

TABLE 23.2 Various Biological Activities Performed by Bioactive Compounds of Brazilian Fruits

FRUIT	*MAJOR BIOACTIVE COMPOUNDS*	*EMPLOYED ASSAY*	*IMPORTANT FINDINGS*	*REFERENCES*
Hancornia speciosa	Catechin, chlorogenic acid, epicatechin, epicatechin gallate, hesperidin, kaempferol 3-glucoside, myricetin, narigenin, procyanidin (B1 and B2), resveratrol, rutin, and syringic acid	ORAC assay measures antioxidant capacity	Fermentation of fruit pulp improved the antioxidant activity.	de Assis et al. (2021)
Talisia esculenta	Phenolic compounds, carotenoids, ascorbic acid, and tocopherols	DPPH assay measures antioxidant capacity	Fruit peel (30.98–81.41%) exhibited higher antioxidant activity than pulp (4.26–29.33%).	Fraga et al. (2021)
		ABTS assay measures antioxidant capacity	The antioxidant activity of the different extracts is significantly different depending on the solvent used for extraction (ethanolic extract has higher activity than hexane extract).	
		FRAP assay measures antioxidant capacity	Ethanolic extract showed a higher capacity to reduce iron.	
Mauritia flexuosa	Caffeic acid, ferulic acid, chlorogenic acid, gallic acid, ellagic acid, epicatechin, hispidulin, isoorientin, kaempferol, myricetin, phytosterol, quercetin, and tocopherols	DPPH assay measures antioxidant capacity	Ultrasound and water-ethanol combination showed lower antioxidant activity (1187.17±2.00 μg/mL)	Rudke et al. (2021)
		ABTS assay measures antioxidant capacity	Ultrasound and ethanol combination showed higher antioxidant activity (0.63±0.03 mmol TE g^{-1} of extract).	

(Continued)

TABLE 23.2 (Continued)

FRUIT	*MAJOR BIOACTIVE COMPOUNDS*	*EMPLOYED ASSAY*	*IMPORTANT FINDINGS*	*REFERENCES*
Syzygium cumini	β-carotene, lycopene, 3.4-dihydroxybenzoic acid, acetic acid, succinic acid, quinic acid, and tartaric acid	α-Amylase inhibition	Ethanolic and water extract showed greater inhibition of α-amylase (95.3% and 84.1% respectively).	Andrade et al. (2022)
Caryocar brasiliense	Polyphenols, proanthocyanidins, and carotenoid	*In vitro* α-glucosidase inhibition assay	Ethyl acetate fraction inhibited α-glucosidase with IC_{50} = 4.8±0.1	Caldeira et al. (2021)
		In vitro α-amylase inhibition assay	Fruit extract inhibited α-amylase depending on the sample concentration (0.003–1 mg/mL).	
Campomanesia xanthocarpa	Gallic acid, catechin, hesperidin, epicatechin, quercetin 3-*O*-glucoside, sinapic acid, naringenin, quercetin, kaempferol, and apigenin	α-glucosidase inhibitory inhibition assay	Sample inhibited α-glucosidase enzyme activity with IC_{50} values of 117 and 169 µg/mL of unripe and ripe samples.	Barros et al. (2020)
		α-amylase inhibition assay	Sample inhibited α-amylase enzyme activity with IC_{50} values of 11.3 and 25.6 µg/mL of unripe and ripe samples.	
Syzygium cumini	Catechin, chlorogenic acid, ellagic acid, ferulic acid, syringic acid, gallic acid, vanillic acid, ellagitannins, epicatechin, epigallocatechin, gallacatechin, gallotannins, myricetin, *p*-coumaric acid, quercetin, and rutin	Anticancer activity by cell proliferation assay (lung carcinoma line CP-H460)	*S. cumini* of 2 mg/mL was able to inhibit approximately 80% of the viability of the lung carcinoma cell.	Gibbert et al. (2021)
Psidium guajava	–	Anticancer activity by cell cytotoxicity assay (human myeloid leukemia cell line U937)	At a concentration of 1.25% and 5% v/v, substantial cytotoxic effects were seen. Lower levels of cytotoxicity were found in pulp and pulp-containing peel extracts compared to those seen in the pure peel.	Suwanwong and Boonpangrak (2021)
Schinus terebinthifolius	δ-3-carene, α-phellandrene, limonene, α-pinene, and germacrene D	Anticancer activity by *in vitro* antiproliferative activity assay (Human tumor cell lines)		Guzzo da Silva et al. (2019)
Mauritia flexuosa	Flavonoids, carotenoids, mono- and polyunsaturated fatty acids, tocopherols, phenolic compounds, and ascorbic acid	Anti-inflammatory activity by ELISA	Inflammatory tissue measures such as peritoneal leukocyte count and TNF-α levels were lowered.	Amorim et al. (2021)
Eugenia selloi	Hydroxybenzoic acid, ellagitannins, and flavonoids	Anti-inflammatory activity by ELISA	Cytokine release, neutrophil migration, and NF-κB activation were decreased by fruit extract.	Lazarini et al. (2020)
Euterpe oleracea	–	Anti-inflammatory *in vitro* assay	Extraction of phenolic compounds from fruit pulp inhibited the early inflammation response that was associated with proanthocyanidins and suppressed the expression of pro-inflammatory genes such as interleukin-1β, interleukin-6, nitric oxide synthase, and cyclooxygenase–2.	Xiong et al. (2020)

Abbreviations: ORAC: Oxygen radical absorbance capacity; DPPH: 2,2-diphenyl-1-picrylhydrazyl; ABTS: 2,2-azinobis-(3-ethylbenzothiazoline-6-sulfonate); FRAP: Fluorescence recovery after photobleaching; ELISA: Enzyme-linked immunosorbent assay

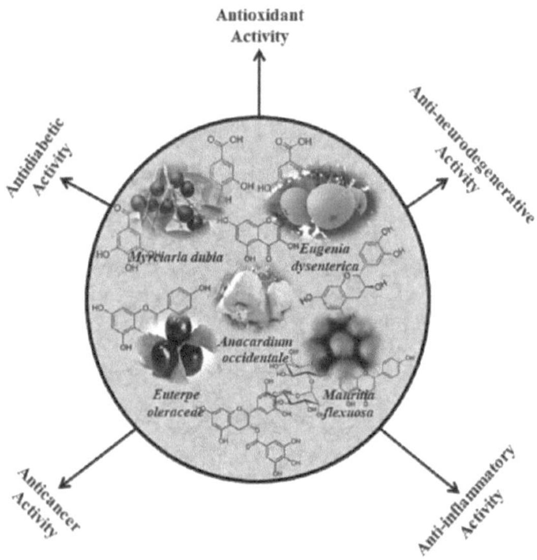

FIGURE 23.3 Various biological activities of Brazilian fruits.

According to published reports, the main BACs of *Myrciaria jaboticaba* include anthocyanins, ellagic acid, ellagitannins, gallic acid, limonene, quercetin, and terpenes (Plaza et al. 2016; da Silva et al. 2017). These chemicals of *M. jaboticaba* are primarily responsible for anti-inflammatory, antiproliferative, and antioxidant activities *in vivo* and *in vitro* (Batista et al. 2014; Plaza et al. 2016). Traditional uses of the Eugenia genus of Brazilian fruits include the treatment of digestive ailments, bronchitis, influenza, cough, and high blood pressure (Araujo et al. 2020). Furthermore, the *Eugenia* genus contains a high concentration of phenols, flavonoids, ellagic acid, and gallic acid derivatives (Teixeira et al. 2018); as a result, they have anti-diabetic, anti-inflammatory, antiglycant, antimutagenic, and antiproliferative activity (Neri-Numa et al. 2013; Bastos et al. 2019; Araujo et al. 2020). Therefore, for the food, pharmaceutical, and cosmetic sectors, BACs of the fruits of the Eugenia genus serve as a natural source of phenolics that can be employed as dietary supplements, ingredients in functional foods, food additives, and a replacement for synthetic antioxidants in cosmetics.

23.3 FRUIT FERMENTATION OF DIVERSE BRAZILIAN BIOMES

Brazil's large region is home to a plethora of edible native fruit plant species that have yet to be discovered (Teixeira et al. 2019; Schulz et al. 2020). As a result, gathering information on native valuable plant species is essential if we are to make better use of them. The preservation of such data from documents and historical bibliography has been the focus of various research groups over the last decade (Teixeira et al. 2019; Schulz et al. 2020). Traditional uses of fruit plant species documented by European scientists/naturalists who visited or resided in Brazil in the late 1800s and early 1900s received special attention (Teixeira et al. 2019). Table 23.3 includes some fruit plants from various Brazilian biomes that are currently underutilized, even though scientists and researchers are working to improve their utilization, as detailed are discussed in the following sections.

23.3.1 Caatinga Biome

The Caatinga biome is home to one of the world's biggest seasonal savannas, stretching across much of northeastern Brazil and a tiny piece of southeast Brazil (Da Silva et al. 2019b). This biome's vegetation is mostly made up of shrubs and succulent trees, the majority of which are equipped with thorns and bromeliads. These traits are the result of numerous adaptations to the year-round low water availability regime (Da Silva et al. 2019b). These natural vegetation species, which can be found throughout the Caatinga, produce a wide range of fruits that have yet to be fully investigated, particularly in terms of food production and integration in diets. However, several researchers have concluded that these fruits have possible functional characteristics, owing to their high concentrations of minerals and BACs such as phenolics and betalains (Stintzing et al. 2005; Dantas et al. 2019; Macêdo et al. 2023). As a result, the development of functional food products, such as beverages, is a viable option for the use of these exotic fruits, and various studies have demonstrated that this is feasible.

The Caatinga wild passion fruit species (*Passiflora cincinnata* Mast.), as depicted in Figure 23.4, is recognized for its tolerance to drought and certain pests, as well as being high in potassium, iron, phosphate, calcium, complex B, and vitamins A and C (Araujo et al. 2020). This species has been studied for its potential use as an ingredient in fermentation processes for the manufacture of alcohol (Santos et al. 2021a), probiotic fermentation (Santos et al. 2017), and other functional drinks such as kefir (Mendes et al. 2021a). Santos et al. (2021a) investigated the ability of *P. cincinnata* to produce an alcoholic beverage fermented by *Saccharomyces cerevisiae var. bayanus Maurivin.*, as well as confirming the presence of 21 phenolic compounds, including phenolic acids, flavonols, flavanols, and stilbenes, which were identified and quantified using HPLC-DAD-FD (Santos et al. 2021a). After the fermentation process of the beverage produced from Caatinga's passion fruit, a rise in the concentration of phenolic compounds was observed, which happened owing to the likely impact on the phenolic fraction caused by the activity of commercial yeast, pectinolytic enzyme. In addition, the duration of the clarifying, fermentation, and stabilization processes, the pH value, temperature, mineral concentrations, quantity of dissolved oxygen, and degree of homogenization were all factors that influenced the whole process (Santos et al. 2021a). Furthermore, Rotta et al. (2019) reported that the phenolic acid concentrations in fermented passion fruit beverages from Caatinga were greater than those observed in the fruits of various *Passiflora* species. Previous research on the biological activity of the species

TABLE 23.3 Various Fruits from diverse Brazilian Biomes

SCIENTIFIC NAME	*FAMILY*	*COMMON/POPULAR NAME*
(1) Caatinga biome		
Passiflora cincinnata	Passifloraceae	Maracujá tubarão
Spondias tuberosa	Anacardiaceae	Umbuzeiro
Spondias bahiensis	Anacardiaceae	Umbu-caja
Anacardium occidentale	Anacardiaceae	Cajueiro, acajaíbaa
Ziziphus joazeiro	Rhamnaceae	Juazeiroa
Cereus jamacaru	Cactaceae	Cardo da praia
(2) Cerrado and Pantanal biomes		
Acrocomia aculeata	Arecaceae	Bocaiuva or macaúba
Dipteryx alata	Fabaceae	Baru or cumbaru
Anacardium humile	Anacardiaceae	Cerrado cashew, Cajuía
Eugenia dysenterica	Myrtaceae	Cagaiteira
Hancornia speciosa	Apocynaceae	Mangabeira, Mangaba fruit
Platonia insignis	Clusiaceae	Bacurizeiro
Caryocar brasiliense	Caryocaraceae	Pequi
(3) Atlantic forest biome		
Psidium guajava	Myrtaceae	Guava, guajava
Psidium cattleianum	Myrtaceae	Araça de coroa
Eugenia pyriformis	Myrtaceae	Uvalha do campo
Eugenia jambolana	Myrtaceae	Jamun, jambolão
Campomanesia xanthocarpa	Myrtaceae	Guabiroba do mato
Euterpe edulis	Arecaceae	Juçara palm
Myrciaria jaboticaba	Myrtaceae	Jabuticaba
Eugenia uniflora	Myrtaceae	Pitangueira vermelha
Eugenia brasiliensis	Myrtaceae	Grumixameira
(4) Amazon biome		
Euterpe oleracea	Arecaceae	Acaizeiro
Mauritia flexuosa	Arecaceae	Buriti do brejo
Eugenia stipitata	Myrtaceae	Araçá-boi
Oenocarpus bacaba	Arecaceae	Bacaba açú
Myrciaria dubia	Myrtaceae	Camu-camu
(5) Pampa biome		
Eugenia involucrata	Myrtaceae	Rio-grande cherry, Cerejeira, Pitanga preta
Eugenia multicostata	Myrtaceae	Araçá-piranga, Pau alazao
Eugenia myrcianthes	Myrtaceae	Pitangatuba
Eugenia rostrifolia	Myrtaceae	Batinga
Butia eriospatha	Arecaceae	Butiazeiro

Sources: Teixeira et al. (2019)

P. edulis and *P. cincinnata* revealed significant antioxidant capacity in *in vitro*, which was linked to phenolic substances found in the fruit pulp, such as flavonoids (Santos et al. 2021b). These components might be directly related to fruits' functional potential and their associated products, such as beverages. According to Santos et al. (2017), Caatinga passion fruit pulp and pectin derived from its bark were excellent carriers of probiotics in fermented drinks. This was observed because the survival of the *Lacticaseibacillus rhamnosus* ATCC 7469 strain after simulated GI simulation for the passion fruit probiotic drink increased in the final storage time, with higher values found when extracted pectin was added; then indicated a

FIGURE 23.4 Various fruits from Caatinga biome.

possible prebiotic effect. This passion fruit has also been used in the production of kefir (Mendes et al. 2021a). Mendes et al. (2021b) recently published a research report in which they described the reductions in survival following the simulated circumstances of the GI system in which bacteria and yeasts maintained the maximum stability with survival always more than 70%. It is worth noting that, in addition to the functional claim, the use of fruits as a substrate for the fermentation of lactic acid bacterial cultures has the added benefit of adding various tastes, resulting in products with varied sensory and physicochemical properties and extended shelf life (Sagdic et al. 2014; Sabidi et al. 2020).

Umbu (*Spondias tuberosa* L.) and umbu-caja (*Spondias bahiensis*) are two Caatinga biome-typical species depicted in Figure 23.4. The fruit of the umbu tree is a drupe that is usually 2 to 4 cm in diameter and weighs 10 to 20 g (Teixeira et al. 2019). The unique flavor of this fruit is produced by the pulp, which is due to the combination of sugars, acids, and phenolic compounds (De Lima et al. 2018). Fruit consumption is mostly prompted due to its sweet-sour flavor. However, vitamin C, carotenoids, minerals, and other nutritional components, as well as its potential as a functional fermented beverage ingredient, have spurred its greater inclusion in the diet (Santos et al. 2006; de Moura Neto et al. 2016; De Lima et al. 2018). Umbu pulp can be processed manually or industrially, with *Saccharomyces cerevisiae* and *Spirulina* spp., producing wines and liqueurs, respectively (Carmo et al. 2012; dos Santos Neto et al. 2021). The investigation of fermentation methods and the physicochemical characterization of the products revealed that the drinks produced met the quality criteria set by Brazilian legislation (Bastos et al. 2017; BMH 2019). According to the Food and Drug Administration (FDA), the fermented beverage made from *Spirulina* spp. has functional qualities since the microalgae are good for the consumer's health (dos Santos Neto et al. 2021). da Silva Júnior et al. (2021) investigated Kombucha analogues using umbu-cajá. They demonstrated that adding the fruit helped to diversify and improve the chemical and biological properties of the fermented product, resulting in a sweeter drink with fruity fragrances. This was possible because acetic, butyric, citric, succinic, and malic acids, in addition to terpenes, which were the principal volatile compounds described, were found in beverages (da Silva Júnior et al. 2021). Kombucha had high antioxidant activity, and the most bioaccessible phenolics identified were caftaric acid, catechin, and hesperidin (da Silva Júnior et al. 2021). Umbu has been reported to have antioxidant properties due to the presence of phenolic compounds (Gualberto et al. 2021). The phenolic acids identified in the fruit were protocatechuic, *p*-hydroxybenzoic, ferulic, gallic, and ellagic, as well as the flavonoids rutin, quercetin, and myricetin (Dutra et al. 2017). According to Dutra et al. (2017), this was due to the high bioaccessibility of the umbu fruit components (73.92%). These fruits can help protect the body from free radical damage because of the synergistic action of the substances in them (Melo and Andrade 2010). As a result, they might be more successfully included in industrialized products as constituents in food formulations, particularly functional beverages.

The cashew (*Anacardium occidentale* L.), illustrated in Figure 23.4, is a member of the Anacardiaceae family that is native to the northeast coast of Brazil and is now grown in many tropical countries where both the nuts and succulent pseudofruit are extracted, processed, and sold (De Alencar et al. 2018). The fruits (chestnut) and pseudofruits (peduncle) of this species are both utilized in the production of food products. Walnut oil has a high nutritional value due to its high level of oleic acid and other BACs such as tocopherols and sitosterol (Lisboa et al. 2020). Fresh cashew pseudofruit can be eaten, although it frequently has an astringent flavor because of its high tannin content. As a result, it is often processed into juice or fermented drinks. Drinks prepared from cashew juice fermented by probiotic bacteria might be an ideal option, transforming it into a functional beverage supplemented with beneficial microbes for human health (Sahu and Samal 2020). According to Pereira et al. (2011), cashew juice fermented by *Lacticaseibacillus casei* NRRL B-442 is a nutritious functional beverage containing probiotics. Cashew juice was shown to be just as effective as dairy products for the growth of *L. casei*, with the microbe remaining alive at 8 log CFU/mL after 42 days at 4°C storage. Kaprasob et al. (2017) also demonstrated the production of a fermented probiotic beverage made from cashew juice. After 48 hours of fermentation, the findings revealed that the lactic acid bacteria *L. plantarum*, *L. casei*, and *L. acidophilus* enhanced phenolic compound content, such as tannins, volatile compounds, and vitamin C in cashew juice. The fermentation process enhanced the content of condensed tannins in cashew juice while decreasing the concentration of hydrolyzable tannins. Changes in the types of flavor volatiles were also examined during juice fermentation in this study. The results showed that lactic acid bacteria changed the flavor profiles of fermented cashew juice, with the fruity odor decreasing while the whiskey and acid odor rose. Shori et al. (2022) investigated the impact of probiotic microorganisms *Lacticaseibacillus rhamnosus*, *Lacticaseibacillus casei*, or *Lactiplantibacillus plantarum* on cashew milk-based yogurt post-acidification after 21 days of storage. Yogurts prepared from cashew milk fermented by lactic acid bacteria demonstrated an increase in cell viability, flavonoids, and antioxidant capacity that was substantially dependent on the strain of *Lactobacillus* spp. utilized (Shori et al. 2022). Some lactic acid bacteria strains are capable of bioconverting phenolic substances into secondary bioactive antioxidant metabolites (Baú et al. 2015).

Other Caatinga fruits represented in Figure 23.4, such as juá (*Ziziphus joazeiro*) and mandacaru (*Cereus jamacaru* P. DC.), have the potential to be utilized as ingredients in functional beverage formulations (De Almeida et al. 2011; Da Silva 2014; Teixeira et al. 2019), but much more study is required. Juá fruits are high in dietary fiber and phenolic compounds and are non-acidic fruits with high moisture content. Carbohydrate (>20%), water (70 to 80%), potassium (165.29 mg/100 g), calcium (18.92 mg/100 g), and magnesium (16.91 mg/100 g), as well as vitamin C (102.58 mg/100kg) are all abundant in them (Diniz 2016; Silva et al. 2017). Because of the presence of biochemical and nutritional components, juá fruits may be widely employed in beverage preparation. They have strong antioxidant activity due to a high quantity of more bioaccessible phenolic compounds and, as a result, a bioaccessible fraction, as described by Oliveira et al. (2020). This strong antioxidant capacity adds to the fruit's functional impact, which has expressive biological characteristics due to various types of flavanol, flavonol, procyanidin, and phenolic acids. The flavonols, which are the primary phenolics in juá fruits, consist mostly of epigallocatechin gallate, catechin, and procyanidin B1 (Oliveira et al. 2020). The utilization of the mandacaru fruit in biotechnological processes is an alternative method of obtaining products with higher added value and, as a result, profit. The production of a fermented drink from the fruit, as well as others found in the Caatinga region, can help to promote the growth of this culture. De Almeida et al. (2006) and Oliveira et al. (2011a) investigated the kinetics of a mandacaru fermented beverage by *Saccharomyces cerevisiae*, both demonstrating that the fruits performed well as a substrate for the fermentation process in the development of beverages. Furthermore, the *C. jamacaru* fruit has excellent nutritional value due to its high total fiber content and the presence of phenolic compounds in sufficient quantities to provide a good supply of phenolic compounds for the diet of the local population. The total antioxidant capacity is caused by the total phenolic component concentration, particularly flavonoids (Vencioneck Dutra et al. 2018; Soares et al. 2021). According to Vencioneck Dutra et al. (2018), this antioxidant activity has the potential to induce a considerable decrease in the viability of sarcoma 180 in 24 and 48 h of *in vitro* treatment, showing good biological activity and suggesting that mandacaru can be beneficial in the production of food products. Although there is research on some typical fruits of the Brazilian Caatinga biome in the literature, it is still insufficient. The bioactive potential of fruits and their nutritional composition are still poorly understood, particularly in terms of accessible carbohydrates and dietary fiber content in the pulp, as well as the bioaccessibility of phenolic compounds. When developing novel product formulations, such as functional fermented drinks, this information is critical for assessing diet quality and functionality.

23.3.2 Cerrado and Pantanal Biomes

The Cerrado is the second biggest Brazilian biome and one of the world's biomes with the greatest biodiversity. This biome is characterized by savanna formation and covers an area of around 2.0 million km^2, accounting for approximately 23% of Brazilian land (Soares et al. 2017). The cerrado has unique edaphoclimatic characteristics, with a strong seasonal tropical climate. In the dry winter, the relative humidity of the air is between 38 and 40%, whereas, in the rainy season, the humidity reaches 97%, and the temperature ranges between 22 and 27°C (Mendes et al. 2012). Plants from the Brazilian Cerrado are adapted to different environmental conditions that require them to develop defense mechanisms to protect themselves against physical, chemical and biological agents during their evolutionary process due to these climatic features. As a result, the existence of BAC in them can be linked. Therefore, this biome has a wide range of fruits with excellent nutritional content, distinct flavor, and scent, making them appealing for exploration, study, and marketing (Reis and Schmiele 2019).

The Pantanal is one of the world's biggest continuous wetlands, with plains covering about 90% of its surface area and accounting for 48.3% of the wetland area in this region of Brazil. Its approximate area in 2018 was 102,953 km^2 (IBGE 2020) The Pantanal is home to a diverse range of vegetation, with over 2,000 plant species recognized (Jorge and Borsato 2009; Pott et al. 2011). The use of Pantanal wild plants in the production of products like drinks, cakes, breads, and jams, among others, can enhance the economic status of local people while also enhancing food security and valorizing biodiversity and local culture (Bortolotto et al. 2017).

Cerrado and Pantanal fruits have a high nutritional value and antioxidants that can be beneficial for human health. A study that evaluated four fruit flours from these biomes (i.e., jatobá pulp, cumbaru almond, bocaiuva pulp, and bocaiuva almond) showed positive effects of the flours on the gut microbiota of individuals after *in vitro* controlled conditions. The promising microbiota-modulating abilities of the fruit flours provide evidence that Cerrado and Pantanal fruits can be incorporated into the diet to exert beneficial health effects on consumers (Mauro et al. 2022).

Fruits from the Cerrado and Pantanal can be substrates for fermentation by microorganisms because they contain a high concentration of carbohydrates and other elements that may be fermented by yeasts and probiotic bacteria (Damasceno-Junior et al. 2010; Mauro et al. 2022). In this context, studies published in scientific publications attempt to generate beneficial fermented drinks and beverages using the edible components of these fruits, which are described throughout this chapter.

Acrocomia aculeata (Jacq.) Lodd. Ex Mart. is a Cerrado and Pantanal palm tree also known as bocaiuva or macaúba. The bocaiuva fruit, represented in Figure 23.5, is green to orange in color and is highly valued and used by the local inhabitants (Damasceno-Junior et al. 2010). The fresh pulp has a sweet flavor and a mucilaginous texture; it may be consumed raw or dehydrated, and it can be used in a variety of

FIGURE 23.5 Various fruits from Cerrado and Pantanal biomes.

products like jams, ice creams, cakes, and beverages (Costa et al. 2018). The inclusion of *A. aculeata* pulp, soy extract, or soy milk has been employed as a matrix for the formation of fermented drinks. Vieira (2017) developed three fermented beverages using soybean extract and bocaiuva pulp. The fermentation was carried out with the help of a traditional yogurt milk culture containing *Streptococcus thermophilus* and *Lactobacillus bulgaricus*, as well as the probiotic strain *Lactobacillus acidophilus* LA–5. The macaúba pulp formulation had greater amounts of total dietary fibers, both insoluble and soluble. After 28 days at 7°C, the drinks fermented by the typical yogurt culture and macaúba pulp had a viability of 7 log CFU/mL. However, following refrigerated storage, the ultimate viability of *L. acidophilus* fermented drinks with macaúba pulp was about 4 log CFU/mL. The author suggested that the use of macaúba pulp could help improve the nutritional value of beverages. On the other hand, Silva et al. (2021a) developed kefir-fermented soymilk enriched with inulin and 3.5 or 7.0% *A. aculeata* powder-pulp. The author fermented kefir grains (4 g/100 mL) for 12 hours at 25°C. This substrate demonstrated survivability over 10^7 CFU/mL of *Lactobacillus, Lactococcus,* and yeasts present in the fermented beverage with inulin and macaúba supplementation after refrigerated storage at 7°C for 16 days. The supplementation enhanced the beverage's technological properties, such as decreased syneresis and phase separation, without affecting the product's viability. Bortolotto et al. (2015) investigated wild plants and food applications known in four Pantanal rural populations. Older individuals from various tribes recounted the food applications of *A. aculeata*, such as cooking oil obtained from the seeds and the production of "chicha," a beverage consumed fermented or unfermented by the indigenous Guató population. However, only a few senior individuals are aware of these applications, which are unknown to younger generations, posing a threat to future understanding of conventional technology.

The baru or cumbaru (*Dipteryx alata* Vogel) is a natural fruit of the Brazilian Cerrado, and the almonds (Figure 23.5) from the baru or cumbaru have a high protein and fatty acid content (Fioravante et al. 2017; Silva et al. 2021b). Although baru is not commonly consumed by the general public, the edible parts of this fruit, such as almonds and pulp, can provide nutrients. The aqueous extract of almonds can be utilized to encourage the production of novel food products through fermentation by beneficial microorganisms. Fioravante et al. (2017) developed a fermented flavored drink with probiotic potential based on a water-soluble extract of baru almond. The extract was fermented by a culture of *Streptococcus thermophilus, Lactobacillus acidophilus,* and *Bifidobacterium.* After fermentation, thickeners, sugar, and plum pulp were added, producing seven formulations. Following sensorial analysis, the average score of total acceptability was 7.03, which falls between the terms "like somewhat" and "like moderately." The beverage had strong nutritional properties since it was a food with high manganese content as well as a source of magnesium and phosphorus. The vitality of *L. acidophilus* and *Bifidobacterium* rose from 10^3 CFU/mL on day 1 to 10^8 CFU/mL on day 28 after 28 days of storage. The scientists speculated that the high soluble fiber content of plum pulp may have contributed to the enhanced growth of beneficial bacteria during storage. Whereas Coutinho et al. (2021) used a different approach to study the influence of green banana starch on the physicochemical and sensory qualities of a fermented product based on baru almond water extract by probiotic bacteria. The addition of 4.5% green banana starch was highly accepted and favored over the control sample, as it produced enhanced firmness, decreased product syneresis, and preserved the probiotic qualities. Baru pulp is a natural source of sugar and provides an advantageous matrix for the development of novel fermented products. To produce alcoholic beverages, researchers studied the fermentation process of baru pulp, correlating microbial growth with substrate consumption and alcohol production. Toledo et al. (2018) studied the population growth of bread yeast (*Saccharomyces cerevisiae*) in response to variations in CO_2 concentration in baru pulp extract. It was determined in this study that a higher concentration of yeast promotes population development, hence shortening the time required for fermentation to generate the alcoholic beverage (Toledo et al. 2018). Recently, a group of researchers characterized a beverage fermented by *Saccharomyces cerevisiae* as having a reduction in sugars, guaranteeing a successful fermentation without interference, and substantial levels of acetic acid in the development of another fermented alcoholic beverage employing baru pulp. The authors presented results that were in accordance with Brazilian fermented fruit regulations (Silva et al. 2021b).

The "Cerrado" cashew (*Anacardium humile*) is a shrub plant with two parts: a nut (the genuine fruit) and the fruit's pedicel, which grows into the cashew. This pseudofruit is golden and smaller in size than common cashew (Damasceno-Junior et al. 2010). Santana et al. (2020) developed a clarified "Cerrado" cashew juice mixed with *Saccharomyces boulardii* and other sweeteners. The probiotic *S. boulardii* (10^8 CFU/mL) was added to the juices, and the cellular viability was higher than 7 log CFU/mL for nearly all treatments after 28 days of storage. It was shown that the juices had significant levels of antioxidant activity (DPPH method: 86–88% discoloration), total phenolic compounds (75–105 mg/100 mL), and total flavonoids (average of 71 mg/100 mL). Due to the inclusion of *S. boulardii*, the beverage was pink in color and effervescent or bubbling to a degree.

Eugenia dysenterica DC (often referred to as cagaiteira or cagaita) is a fruit plant of the Myrtaceae family, shown in Figure 23.5, commonly found in the Cerrado. The fruits of *E. dysenterica* have a peripheral shape and can weigh anywhere from 2.0 g to over 30.0 g, with a diameter ranging from 2.0 to 3.0 cm, with a yellow to reddish skin when mature, with around one to five seeds (Naves et al. 2002). It has high nutritional value, and studies have shown that the leaf extract contains BACs such as triterpenes (gammacer-16-en-3β-ol, neolupenol, physostigmine, taraxerol, α-amyrin, and β-amyrin), phenolic compounds (catechin and quercetin), and carotenoids (Gasca et al. 2017). Traditionally, *E. dysenterica* has been regarded as a valuable source of vitamin C and protein for the treatment of chronic constipation and irritable bowel syndrome (IBS). There has been research linking the cagaita phytoconstituents with antioxidant and antibacterial properties, as well as enzyme inhibitory properties (such as α-amylase and α-glucosidase) and an anti-obesity and gastro-protective impact (Donado-Pestana et al. 2015; Daza et al. 2017).

Alcoholic and acetic fermentation are methods of using and conserving raw resources. Cagaita vinegar is a novel product that adds value to this indigenous fruit. Vinegars are utilized all over the world for their anti-diabetic, anti-tumor, anti-microbial, antifungal, and digestion-stimulating properties (Lima et al. 2010; de Morais Cardoso et al. 2011). Santana (2019) demonstrated the viability of acetic fermentation development by employing cagaita. For fermentation, two types of enzymes, amylase, and pectinase, were utilized, as well as two subspecies of the yeast *Saccharomyces cerevisiae* UFLA CA11 and thermo-resistant LNF Angel, and required simple sugar source of sucrose (16° Brix). First, alcoholic fermentation was carried out, followed by slow acetic fermentation at 25 °C employing acetic bacteria with an acidity of 9% (v/v) and alcoholic fermentation with an alcohol content of 9% (v/v). Oliveira et al. (2011b) developed a wine from cagaita fruits, and their findings demonstrated that the procedures typically employed in fermented beverage production may be modified for cagaita wine production. In an investigation by Oliveira et al. (2011b), the cagaita was fermented by the two yeasts utilized, *Saccharomyces cerevisiae* UFLA CA11 and *Saccharomyces cerevisiae* UFLA CAT–1. Cagaita wines made with immobilized cells and Ca-alginate produced better yields. This wine was highly received (at least 70% for overall features), even by tasters who were unfamiliar with it. The sensory evaluation used criteria including taste, fragrance, and overall characteristics to categorize the beverages. The approach utilized was simple, and it may be implemented in a limited area without incurring significant costs. Considering how well-liked the beverage is, this approach might provide an alternative method of using a tropical fruit's bioactive pulp.

The mangaba fruit (*Hancornia speciosa* Gomes) (Figure 23.5) is a commercial fruit plant that is mostly consumed as pulp, juice, and ice cream in industrial processes, ice cream parlors, supermarket chains, and intermediate products through suppliers (Bailão et al. 2015). Furthermore, the fruits of *H. speciosa* contain important minerals (Ca, P, Zn, Fe) as well as a significant level of vitamin C (33–247 mg/100 g pulp) and iron (2.4–4.1 mg/100 g pulp) (Rodrigues 2019). The fruits of *H. speciosa* are also an important source of antioxidants (Bailão et al. 2015). Recently, mangaba vinegar's total phenolic component concentration and total carotenoid content were found to be 19.2±8.20 mg/100 g and 2.6±0.6 mg/100 g, respectively (Siqueira et al. 2021). The ethanol to acetic acid conversion yield was 90%. According to the results of the principal component analysis, pH and volatile acidity had a significant impact on the product, while color and scent had a strong positive link. The finished product complied with all applicable laws and regulations, proving that mangaba vinegar with antioxidant potential may be produced for the benefit of customers. Furthermore, tasters liked the product in the sensory evaluation, confirming the potential economic value of this Cerrado fruit product (Siqueira et al. 2021).

Bacuri (*Platonia insignis* Mart., Clusiaceae) is a member of the Clusiaceae family that is native to the southern Amazon jungle and the Brazilian Cerrado and was most likely domesticated by the Amazo Amerindians (Rogez et al. 2004). The bacuri fruit (Figure 23.5) contains polyisoprenylated benzophenones, xanthones (responsible for a variety of biological actions), and polyphenolic chemicals with antioxidant potential. Diterpenes, xanthones, and fatty acids were found in the seeds of the bacuri, whereas monoterpenes were found in the pulp (Acuna et al. 2009; Kumar et al. 2013). Mendes et al. (2021b) investigated the viability and growth of *Lacticaseibacillus rhamnosus* ATCC 9595 and *Limosilactobacillus fermentum* ATCC 23271 in bacuri juice, and the findings laid the groundwork for the development of a non-dairy probiotic product derived from bacuri juice. Both strains grew in bacuri juice and retained viability after 28 days of storage. However, the growth of *L. rhamnosus* ATCC 9595 (7.40±0.04 log CFU/mL) was substantially greater than that of *L. fermentum* (6.29±0.12 log CFU/mL). Both strains, *L. fermentum* ATCC 23271 and *L. rhamnosus* ATCC 9595, were able to grow in bacuri juice without any additional nutrients. Both *L. fermentum* ATCC 23271 and *L. rhamnosus* ATCC 9595 grew substantially faster in bacuri juice (without the addition of nutritional additives) than in MRS medium ($p < 0.05$). The mean difference between growth in bacuri juice and MRS for *L. fermentum* ATCC 23271 was 2.84 log CFU/mL ($p < 0.05$), whereas the difference for *L. rhamnosus* ATCC 9595 was 7.39 log ($p < 0.001$). Tenebrio molitor larvae infected with enteroaggregative *Escherichia coli* 042 were used in this study to compare the anti-infective effects of fermented and non-fermented juices. However, pre-treatment with fermented and non-fermented supernatants of both juices considerably enhanced the survival of *E. coli*-infected larvae, according to the data obtained after inoculating them 2 hours after infection (Mendes et al. 2021b).

As a food source, the pequi fruit (*Caryocar brasiliense* Camb), depicted in Figure 23.5, has enormous potential. Anti-clastogenicity has been shown to be improved by using an aqueous extract of pequi fruit (Khouri et al. 2007). Some research has been done on the processing of pequi fruit because of its health advantages as well (Machado et al. 2015; Machado et al. 2016; Caldeira et al. 2021). Ferreira et al. (2021)

aimed to establish, characterize, and assess the stability of an isotonic beverage made from whey permeate and carotenoid extract powder from pequi at two different temperatures (4 and 25°C). The powdered carotenoid from pequi, with a w/v concentration of 3%, was chosen because of its osmolality (314.89 mOsmol/L) and light yellow color, which were both typical of hydro electrolytic drinks. Sodium (662 mg/L) and potassium (1363.73 mg/L) were both found in significant amounts in the beverage. ABTS radicals had an antioxidant capacity of 10.79 mol equivalence of Trolox/100 mL, while DPPH had an antioxidant capacity of 73.38 mol equivalence of Trolox/100 mL. The athletes liked the taste of the hydro electrolytic beverage, and it met microbiological standards during a stability test. Thus, it can be concluded that pequi carotenoids and other natural ingredients in the beverage have bioactive characteristics, making it potentially beneficial for athletes to consume.

23.3.3 Atlantic Forest Biome

The Atlantic Forest is one of the biomes with the most biodiversity on the globe, and it is found in 17 Brazilian states. According to the Atlas of Atlantic Forest Remnants 2014–2015, the present extent of the Atlantic Forest corresponds to less than 8% of its original area, making it one of the most vulnerable biomes (Bastos et al. 2017). The Atlantic Forest biome encompasses 390,869.13 km² in the South Region, accounting for around 68% of the region's land, accounting for nearly 98% of the State of Paraná, the whole State of Santa Catarina, and approximately 42% of Rio Grande do Sul. Figure 23.6 depicts many notable fruit species, including guava (*Psidium guajava*) and jabuticaba (*Plinia trunciflora* O. Berg), as well as juçara (*Euterpe edulis* Martius), which are native to the Atlantic Forest. Figure 23.6 also depicts poorly studied species such as grumixama (*Eugenia brasiliensis*), araçá (*Psidium cattleianum*), and uvaia (*E. pyriformis*) (Silva et al. 2014).

The Myrtaceae plant family contains the genus Eugenia, which is one of the largest genera, with approximately 3,800 edible species distributed across the world's tropical and subtropical climates (Donado-Pestana et al. 2018; Bastos et al. 2019; de Araujo et al. 2019). Many Eugenia species are used medicinally because of the presence of antioxidant substances, including phenols and anthocyanins, which have favorable physiological features like antidepressant, antirheumatic, diuretic, and anti-inflammatory characteristics (Neri-Numa et al. 2013; Aguiar et al. 2016; Chhikara et al. 2018; Donado-Pestana et al. 2018; Bastos et al. 2019; de Araujo et al. 2019).

FIGURE 23.6 Various fruits from Atlantic forest biome.

About half of all species in Brazil are indigenous to the Atlantic Rain Forest Biome, which is home to the Myrtaceae family of plants, which includes native fruit species found throughout the country from the north to the south (Donado-Pestana et al. 2018). The genera Eugenia and Myrciaria contain most of the species found in Brazil (Fujita et al. 2015; Plaza et al. 2016; de Araujo et al. 2019; Albuquerque et al. 2020). These plants are distinguished by the presence of opposite, simple, entire, translucent leaves with glands containing oils and produce fleshy, berry-like fruits with a wide variety of beneficial properties (Plaza et al. 2016; de Araujo et al. 2019). Regional folk medicine uses many parts of these plants extensively as an astringent, anti-inflammatory, and antihypertensive and for the treatment of digestive problems. Polyphenols and carotenoids, monoterpenes, sesquiterpenes, and their derivatives from Eugenia species are multitarget phytochemicals that have antioxidant, anti-inflammatory, and anti-diabetic activities (de Araujo et al. 2019).

Eugenia jambolana Larmark of the Myrtaceae family is a big tree found in Algeria, Bangladesh, Brazil, California, Florida, India, Israel, Philippines, and Thailand that bears fruits as indicated in Figure 23.6, known as black plum, jambolan, jambolão (Brazil), jambul (Hindi), and jamun (Bezerra et al. 2015; Martins et al. 2020). The unripe fruits are green, but as they grow, they turn a faint magenta tint before turning a deep purple or black. When fully mature, the fruit has a sweet, moderately sour, and astringent flavor profile, and it becomes purple on the tongue as a result (Baliga et al. 2011). In animal tests, the leaves and seeds of *E. jambolana* showed hypoglycemic, hypolipidemic, and anti-inflammatory properties, while the pulp had anti-glycemic and anti-diabetic effects (Sharma et al. 2008; Bastos et al. 2019; Dacoreggio et al. 2021). Significant ($p < 0.05$) decreases in LDL (about 27% MD, 29% SD) and triglyceride levels (about 35% MD, 37% SD), as well as an increase in HDL (about 21% MD, 34% SD), were seen the hypolipidemic action after treatment with this extract compared to untreated diabetic mice (Sharma et al. 2008). In 2015, Bezerra et al. (2015) tested four different kinds of caprine frozen yogurt to find out if fresh-dried and spray-dried pulp of jambolan fruit (*E. jambolana*) and a probiotic strain (*B. animalis* subsp. *lactis* BI-07) work together or not. Aside from the added probiotic bacteria surviving in the frozen storage and being accepted by the customers in this context, Bezerra et al. (2015) also examined for physical and chemical properties, overrun, meltdown tests, chosen bioactive-related qualities, and sensory acceptability. Bezerra et al. (2015) found that adding probiotics to frozen yogurt lowered the pH ($p < 0.05$), and all samples had low overrun values (14.2 to 22.6%). Samples that included the probiotic strain had higher levels of phenolic content ($p < 0.05$), indicating that biotransformation might be taking place, resulting in higher levels of phytochemicals.

Additionally, the initial cell counts of the frozen yogurt samples containing probiotics were roughly 9 log CFU/g, and a remarkable probiotic survival rate of 97% was recorded for the course of the 90-day frozen storage period. Samples of frozen yogurt containing probiotic bacteria had lower flavor rankings ($p < 0.05$). When it came to appearance, fragrance, and consistency, frozen yogurt samples were similar. This research showed overall that caprine frozen yogurt containing jambolan fruit was an excellent carrier for the probiotic strain and had favorable biological as well as technological properties (Bezerra et al. 2015).

Guabiroba (*Campomanesia xanthocarpa* O. Berg) is a fruit species native to Brazil's Atlantic Forest that belongs to the Myrtaceae family. It can be found from Minas Gerais to the extreme south of Rio Grande do Sul (Teleginski et al. 2018). Figure 23.6 shows that it is a berry-like fruit that is yellow, globose, and approximately 2.5 cm long and 2–3 cm broad, with a smooth and thin epicarp. The endocarp is succulent, sweet, and fragrant, with 1 to 32 yellowish seeds with little pinkish patches containing essential oil glands (Barbieri et al. 2018; Teleginski et al. 2018; Barros et al. 2019, 2020). It is classified as a functional fruit since it contains high amounts of dietary fiber, phenolic compounds, antioxidants, and vitamin C (Silva et al. 2014; Barbieri et al. 2018; Barbosa et al. 2020; Donado-Pestana et al. 2018; Teixeira et al. 2019; Schulz et al. 2020; Prestes et al. 2021). However, the pulps of *C. xanthocarpa* are not currently commercially available (Barbieri et al. 2018). The ability of fermented milk mixed with guabiroba pulp to preserve *Bifidobacterium* BB-12 cells during GI simulation has been determined (Prestes et al. 2021). Recently, Prestes et al. (2021) investigated the effect of adding guabiroba pulp to fermented milk on the survival of the probiotic *Bifidobacterium animalis* subsp. *lactis* (BB-12), starter cultures *L. bulgaricus* and *S. thermophilus*, and the content of total phenolic compounds and antioxidant activity under simulated *in vitro* GI conditions. Prestes et al. (2021) observed that milk fermented with 10% guabiroba pulp offered the maximum survival rate and viable cell count at all stages of the GI simulation. For a 10% pulp sample, the initial counts before the TGI simulation were 6.47 log CFU/g and 10.37 log CFU/g at the latter (descending colon). Even for sensitive strains of *L. bulgaricus* and *S. thermophilus*, the addition of 10% guabiroba pulp resulted in a considerable survival rate in the digestive stage with severe stomach acidity, with rates of 60.4% and 21.7% higher than the control sample (Prestes et al. 2021). This suggested that guabiroba BACs may have a possible prebiotic influence on bifidobacteria and/or protective effects on the three microorganisms studied. Prestes et al. (2021) demonstrated an increase in polyphenols with the development of *in vitro* GI simulation for the total phenolic content and showed significant increases in these compounds, from 161.45 to 532.24 mg GAE/100 g. The concentration of phenolic compounds in the descending colon was five times higher than in the first stage before digestion. The antioxidant activity increased in tandem with the rise in total phenolic compound concentration. According to the FRAP and DPPH tests, the sample with 10% guabiroba pulp had 275% and 614% higher antioxidant activity than the control sample in the early stage of the simulated TGI (Prestes et al. 2021).

Euterpe edulis Martius is a palm popularly known as juçara palm, native to Brazil's Atlantic Forest, where it ranges from the state of Bahia to the north of Rio Grande do Sul, and is known for the palm heart, which has high commercial value, good sensory qualities, and high yield, but the palm has the disadvantage of not forming new stems from the same tree, which causes the plant to die after cutting the heart of palm (Santos et al. 2008). This has had consequences for the loss of this species to the point that it is now included on the list of endangered species of Brazilian flora, as the natural occurrence of palm heart has significantly diminished as a result of excessive harvesting, and it is included in the "vulnerable" category (BSSE 2016). The Juçara fruits illustrated in Figure 23.6 are raw materials for producing pulp or beverages with a high lipid content, rich in unsaturated fatty acids and phenolic compounds, and high anthocyanin content (Borges et al. 2013; Silva et al. 2014; Brito Junior et al. 2023). According to Guergoletto et al. (2017a), the phenolic compounds in juçara can promote the survival of *Lactobacillus* strains, making it a suitable carrier for probiotics. In 2017, Guergoletto et al. (2017a) investigated the pH and temperature parameters for *Limosilactobacillus reuteri (L. reuteri* LR92), *Lactiplantibacillus plantarum* (*L. plantarum* BG112) fermentation in juçara pulp. The probiotics were grown at 32°C with a pH of 5.7 for *L. reuteri* and 25°C with a pH of 5.0 for *L. plantarum*, with both bacteria fermenting for 30 hours. There was an increase in total phenolic component content and antioxidant activity (DPPH, ABTS) following juçara pulp fermentation. *L. reuteri* ingested all the glucose and fructose in the juçara, but *L. plantarum* only consumed glucose. The authors confirmed cell viability stability in juçara extract with counts of 8.73 and 8.71 log CFU/g for *L. plantarum* BG112 and *L. reuteri* LR92, respectively, after 28 days of storage at 4 °C (Guergoletto et al. 2017a). Simultaneously, Guergoletto et al. (2017b) also studied the influence of gum arabic, maltodextrin, and gelatin on the survival of *L. reuteri* LR92 in fermented juçara pulp after spray-drying. To prepare the powder, 10% carrier agents were added. The authors observed that following dehydration, *L. reuteri* vitality was more than 6.0 log CFU/g for all carriers examined, while gelatin-based powders had higher bacterial survival after drying (8.63±0.11 log CFU/g). Gum arabic and maltodextrin produced higher yields and solubilities. According to the study, these findings show that gelatin is an effective medium for drying probiotic microorganisms (Guergoletto et al. 2017b). In the study by Braga et al. (2018), the capacity to transform juçara anthocyanins has also been investigated. According to them, the average optimal conditions occurred with a mixture of 20% juçara pulp and 10% glucose in MRS medium, in which the microrganism was capable of achieving the highest enzyme production (β-galactosidase, α-galactosidase and β-glucosidase) and also produce the most extensive conversion of major anthocyanins. *Lactobacillus delbruekii* was able to change the anthocyanins of jussara pulp more extensively than other strains of probiotics examined. Study of anthocyanins and phenolic acids was done before and after fermentation to verify changes in phenolic components based on the selection of *L delbruekii* with increased potential for bioconversion of anthocyanins for further analysis. The authors reported the discovery of caffeic acid, which can be converted into 3,4-dihydroxyphenylpropionic acid and then into protocatechuic acid. After 2 hours of fermentation, these three phenolic acids were identified, and they were most likely phenolic acids formed from cyanidin-3-glucoside, cyanidin-3–5-hexose-pentose, and cyanidin-3-rutinoside. Perlagonidin-3-rutinoside and perlagonidin-3-glucoside were bioconverted into *m*-hydroxyphenylacetic acid and *p*-coumaric acid after 2 hours of fermentation. After 8 hours of fermentation, peonidin-3-rutinoside was bioconverted into ferulic or isoferulic acid. Fermented juçara had significantly higher antioxidant activity (ABTS) than fresh juçara pulp, increasing from 142.1 to 212.0 µmol TE/100 g of material (an increase of 49.2%). The ORAC experiment validated this impact, boosting antioxidant activity from 46.3 to 88.3 µmol TE/100 g of material (an increase of 90.7%) (Braga et al. 2018). Miranda et al. (2020) developed probiotic juçara and passion fruit jelly candies with *Bacillus coagulans* GBI-30 6086, assessing the strain's antioxidant activity, probiotic viability, and *in vitro* resistance to the GI system. *B. coagulans* had initial counts over 6.3 log CFU/g and kept the probiotic viable for 90 days with counts above 6.4 log CFU/g. After the simulated TGI, the jelly candy demonstrated a slight rise in *B. coagulans* count and stability, with counts in the gastric phase (initial phase) of 6.57 log CFU/g, increasing to 6.83 log CFU/g and 6.82 log CFU/g in enteric phases I and II, respectively. The probiotic microrganism was present in around 7.82 log CFU/g of candy (30g). During the 90-day storage period, total anthocyanins, and antioxidant capacity (as measured by the ABTS assay) in juçara and passion fruit jelly candies did not change significantly ($p > 0.05$).

Jabuticaba (*Myrciaria jaboticaba* (Vell.) Berg) belongs to the Myrtaceae family and is native to the Brazilian Atlantic Forest, where it grows naturally and produces more than 5,000 tonnes of fruits per year (Albuquerque et al. 2020). The fruit is dark violet in color, as can be seen in Figure 23.6, and is highly valued due to the sweet flavor of the pulp (Leite-Legatti et al. 2012; Inada et al. 2019; Albuquerque et al. 2020). In general, the plant is medium to big in size, growing 6 to 9 meters tall, and is distinguished by blooms that grow directly from the main stems and branches, lending it an aesthetic quality (Albuquerque et al. 2020). The Paulista variety is distinguished by its big size and excellent production capacity, as well as its large and late-maturing fruits. On the other hand, the Sabará variety is the most popular and widely grown, with medium growth and tiny, delicious fruits (Jesus et al. 2004; Lima et al. 2008). The fibers and phenolics in jabuticaba are mostly concentrated in the bark and seeds, which are the primary fractions that form in the jabuticaba by-product (Lima et al. 2008; Inada et al. 2019; Albuquerque et al. 2020). Massa et al. (2020) examined the prebiotic effects of a digested by-product of jabuticaba (*M. jaboticaba*) in a medium containing MRS, confirming the growth and metabolism of probiotics *Lactobacillus acidophilus* LA-05, *Lacticaseibacillus casei* (*Lactobacillus casei* LAFTI L-26), and *Bifidobacterium animalis subsp. lactis*, as well as the effects of fermentation on the alteration of phenolic compounds. A drop in pH, a rise in lactic acid and short-chain fatty acid synthesis (acetic, butyric, and propionic acid), and the consumption of glucose and fructose over time were all signs that probiotics in a medium with digested jabuticaba substrate were engaged in vigorous metabolic processes. According to the results, the counts of probiotics in the medium containing digested jabuticaba by-product were increased by roughly 2 log CFU/mL after 48 hours of fermentation. A reduction in caffeine ($p < 0.05$) was seen after 48 hours of fermentation in the presence of *L. acidophilus* LA-05 or *B. lactis* BB-12, whereas a rise in caffeine was observed in the presence of *L. casei* L–26. Probiotic inoculation had no effect on the levels of chlorogenic acid, catechins, or epicatechins in the culture medium (Massa et al. 2020). Cultivation increased the *trans*-resveratrol content of the three strains, as well as a decrease in the level of gallic acid and the other phenolic compounds such as naringenin, procyanidin B2, myricetin, epicatechin gallate, delphinidin 3-glucoside, and cyanidin 3-glucoside. Growth in mediums containing *L. casei* L-26 or *B. lactis BB*-12 lowered epigallocatechin, but growth in *L. acidophilus* LA-05 media did not.

Pitanga (*Eugenia uniflora*) is a widespread species in Brazil, native to the Atlantic Forest, Pampa, and Cerrado biomes, and found in the Northeast (Bahia), Midwest (Mato Grosso do Sul), Southeast (Minas Gerais, Rio de Janeiro, and São Paulo), and South (Paraná, Rio Grande do Sul, and Santa Catarina) regions (Vieira et al. 2011; Mazine et al. 2020). The fruit of *E. uniflora* is illustrated in Figure 23.6. It is a globose berry with flattened poles, roughly seven to ten longitudinally defined grooves, 1.5 to 5.0 cm in diameter, and capped with persistent sepals. When the epicarp begins to mature, it changes color from green to yellow, orange, red, and dark crimson, eventually reaching virtually black. Sweet and sour are the flavors, with a powerful and distinctive scent. The endocarp is 3.0 to 5.0 mm thick and pink to crimson in color (Vieira et al. 2011). The pitanga fruit contains insoluble fibers that help in the proper functioning of the digestive system. This fruit contains vitamins A, B complex, and C, as well as minerals such as calcium, iron, and phosphorus. The dark red color of the fruit is due to the presence of lycopene, a strong antioxidant that has been shown to be particularly beneficial in avoiding degenerative diseases such as cancer (Vieira et al. 2011). Ferreira (2015) fermented milk with pitanga pulp (3%) and examined the viability for 28 days at 4°C using a probiotic mixed culture consisting of *Lactobacillus acidophilus, Bifidobacterium bifidum* and *Streptococcus thermophilus* (Bio Rich Cristian Hansen). The authors observed that the probiotic and symbiotic milk drinks (with FOS) were viable for 28 days, with final counts of 9.18 and 9.19 log CFU/mL, respectively. According to the author, the inclusion of FOS had no effect on probiotic counts in the symbiotic drink, and the low acidity of the product following fermentation did not affect the decrease in probiotic viability. The sensory study revealed that both symbiotic and probiotic samples were well tolerated, with an acceptability value of greater than 70% (Ferreira 2015). In a study conducted by da Silva Júnior et al. (2021), kombucha was mixed with pitanga (*E. uniflora*) and showed increased acidity and numerous organic acids, except for malic acid, but predominantly acetic acid. Terpenes had a role in the growth or appearance of novel terpenoids such as curzerene and β-caryophyllene. Kombuchas flavored with fruit had a significant level of antioxidant activity, with epigallocatechin gallate as the major phenolic compound (above 63%). Following the GI simulation, the phenolic concentration decreased, which most likely contributed to the decrease in antioxidant activity.

One of the most common trees in the Atlantic Forest, the grumixama (*Eugenia brasiliensis* Lam.) is a tiny tree with enormous, scented leaves and flowers that may be utilized for landscaping (Martins et al. 2020). This species is particularly resistant to climatic change, which occurs from the north of the state of Bahia to the south of the state of Santa Catarina in Brazil. Figure 23.6 depicts the fruit of *E. brasiliensis*. As a result, it has been utilized to make a variety of artisanal commodities, including commercially frozen pulp and fruit, and other products (Nascimento et al. 2017; Teixeira et al. 2018). Cambessèdes (1832–1833) identified three types of grumixama varieties based on their fruit colors as follow: (1) α-variety (purple fruits), (2) β-variety (red fruits), and (3) γ-variety (white fruits described as yellow) (Moreno et al. 2007). As a result, it is also known as the Brazilian cherry because of its light-colored, sweet, and juicy pulp, which melts on the tongue and has a flavor reminiscent of that of a cherry (Nascimento et al. 2017). While purple grumixama has an abundance of anthocyanins (especially cyanidin 3-glucoside), the yellow fruit is dominated by the flavonoid quercetin. It is also rich in ellagitannins, particularly monomeric forms of castalagin/vescalagin and strictinin in the purple or yellow grumixama, respectively (Teixeira et al. 2018). According to Silva et al. (2019), the addition of grumixama pulp to yogurt formulations can serve as a technical strategy and a possible substrate for bacteria during fermentation. Silva et al. (2019) determined the content of biogenic amines (putrescine, cadaverine, spermidine, spermine, and tyramine), proteolysis activity, and bacterial count during the fermentation period using thermophilic bacteria (1% vol/vol; *Lactobacillus delbrueckii ssp. bulgaricus* and *Streptococcus thermophilus*) in sheep milk yogurts containing 10% (wt/vol). They prepared and evaluated four treatments: natural, prebiotic (5% wt/vol; fructooligosaccharide), probiotic (*B. animalis* BB-12), and synbiotic (*B. animal* BB-12 and 5% wt/vol; fructooligosaccharide). On day 0, the yogurt and probiotic bacteria remained alive (≥7 log CFU/mL). Proteolytic activity was consistent during fermentation in all treatments. All treatments demonstrated the same behavior for all biogenic amines, with concentration changes. Putrescine, cadaverine, and spermidine concentrations fell, but spermine concentrations remained stable, and tyramine concentrations increased. The authors concluded that tyramine can be produced during the fermentation of functional sheep milk yogurts.

23.3.4 Amazon Biome

The Amazon biome is the biggest in Brazil, covering around 4.1 million km². More than half of the biome (6.7 million km²) is located inside Brazilian territory, including portions of the states of Acre, Amazonas, Rondônia, Pará, Mato Grosso, Amapá, Tocantins, and Maranhão. The Amazon Biome covers around 49% of the country's land area and is home to a diverse range of ecosystems, including the world's biggest tropical forest, open fields, and huge regions of mangroves (Magnusson et al. 2016; Uekane et al. 2017; Anunciação et al. 2019; Barbosa et al. 2020; Xiong et al. 2020).

The Amazon rainforest is well known for its tremendous biological potential, which is mostly based on several fruit species with high BAC content (De Freitas et al. 2017; Uekane et al. 2017; Barbosa et al. 2020; Xiong et al. 2020). Amazonian fruits include several beneficial components, including phenolic compounds, carotenoids, fatty acids, vitamin C, and dietary fiber (De Freitas et al. 2017; Anunciação et al. 2019; Barbosa et al. 2020; Xiong et al. 2020). By incorporating Amazonian fruits into the development of new food products, we can increase their functionality while also imparting an exotic flavor and aroma that is well received by consumers, thereby adding value to these little-known fruits and expanding their availability to other regions of Brazil (Damiani et al. 2020). Probiotic microbes offer several health advantages and are the best technological approach for enhancing BAC profiles in fruit, which aligns with consumer health and nutrition knowledge (Fujita et al. 2017).

Açai (*Euterpe oleracea* Mart) is a berry native to the Brazilian Amazon biome and is represented in Figure 23.7. For its distinctive flavor and bioactive characteristics, such as antioxidant and anti-inflammatory effects, açai berries have gained international attention (Rufino et al. 2011; Xiong et al. 2020). Açaí is high in anthocyanins, which are water-soluble pigments that give the fruit its distinctive color (Rufino et al. 2011). When ripe, its fruits are globose or somewhat fast drupes with adherent flowers and violet or green color (Oliveira et al. 2002). Recognition as a fruit tree of economic interest has already transcended the Amazon's boundaries, with the fruit tree products being marketed in major Brazilian capitals in a variety of ways (such as ice cream, popsicles, energy food, accompanied by other fruits and cereals, energy drinks, and

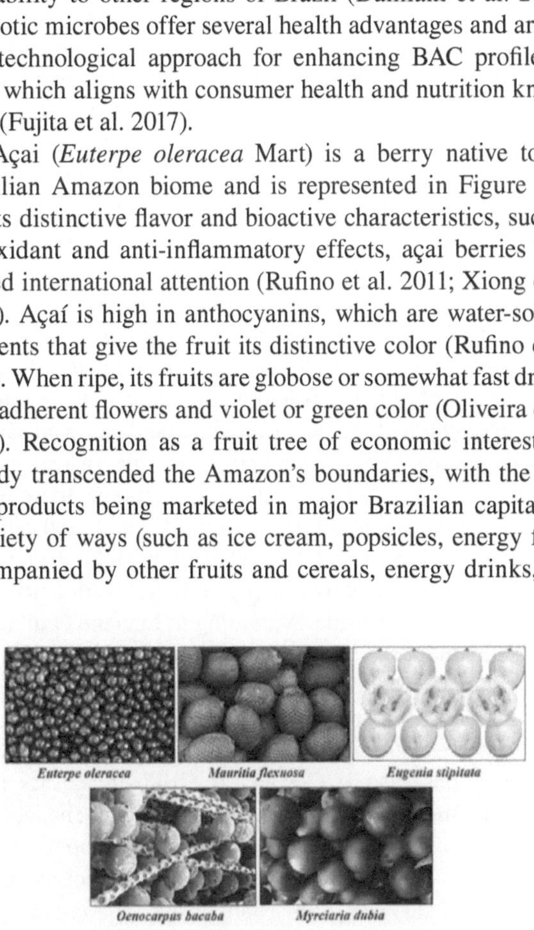

FIGURE 23.7 Various fruits from Amazon biome.

jellies). Because of the commercialization of juice products, numerous Brazilian producers have expressed interest in commercializing their crops (Oliveira et al. 2002). Açaí has a lot of promise in the creation of probiotic products since these foods may combine the health advantages of these bacteria with the nutritional benefits of this fruit. In recent years, several fermented products have been made with açaí, such as ice cream, smoothies, açaí drinks, and blended drinks (Costa et al. 2017; Luciano et al. 2018; Freitas et al. 2021; Baygut et al. 2023). Baygut et al. (2023) investigated the fermentation kinetics of a fermented soy beverage with the addition of açaí powder. The authors found that adding açaí powder influenced the fermentation kinetics based on changes in pH, accelerating more significant fermentation in the initial period. They concluded that the determination of the fermentation kinetics allowed obtaining the development of technology to obtain functional foods. De Almeida et al. (2008) developed a probiotic açaí yogurt with açaí and *B. bifidum* that maintained average populations of 7 log CFU/mL after 21 days of refrigerated storage. do Espírito Santo et al. (2010) observed that adding açai pulp to yogurt increased the counts of *L. acidophilus* L10, *B. longum* Bl05, and *B. animalis ssp. lactis* Bl04 after 28 days of cold storage. They also found that adding açai pulp to probiotic yogurt had boosted the concentration of both fatty acid content (monounsaturated and polyunsaturated) in the yogurt, as well as the formation of conjugated linoleic and α-linolenic acids during the fermentation of skim milk with *B. animalis ssp. lactis* Bl04 and B94 strains, which might have health advantages. In 2018, *L. acidophilus* LA3 was added to açai-based smoothies for 28 days to see how it affected physicochemical parameters, sugar and organic acid content, probiotic survival in a simulated human GI environment, phenolic profile, and sensory attributes (Luciano et al. 2018). After 1 day of storage, *L. acidophilus* LA3 counts were around 8.5 log CFU/mL, and there were no changes ($p < 0.05$) in subsequent storage periods. On days 14 and 28 of storage, viable counts of *L. acidophilus* LA3 were around 7.5 log CFU/mL following *in vitro* digestion. Lower levels of glucose, fructose, and maltose were found in the *L. acidophilus* LA3 açai smoothie, but lactic and succinic acids were found to be greater in the smoothie. In açai smoothies with *L. acidophilus* LA3 on day 1 of storage, there was an increase in the contents of procyanidin B1 and B and catechin, as well as a reduction in the contents of cyanidin 3-glucoside, epicatechin gallate, malvidin 3-glucoside, *p*-coumaric acid, and quercetin 3-glucoside. According to Luciano et al. (2018), these fruit smoothies can be used as a matrix for lactobacilli-enriched non-dairy drinks. *Lacticaseibacillus rhamnosus* GG (*L. rhamnosus* GG) strain viability and tolerance to simulated GI stress were examined and morphological alterations during stress were observed by Costa et al. (2017) in açai ice cream (with 12% pulp açai). There were no significant changes in *L. rhamnosus* GG viability in any formulation for up to 112 days of storage at −18°C temperature. To test the effect of the food matrix on the survival of microorganisms, *L. rhamnosus* GG was inoculated into açai ice cream and found to have a higher survival rate than fresh culture. All formulations presented a decrease of at least 5 log cycles in all storage periods following the GI simulation. When the probiotic was introduced to ice cream, fewer morphological alterations were noticed than when the fresh culture was used (Costa et al. 2017).

Buriti (*Mauritia flexuosa* L.) is a Brazilian palm that is native to the Amazon and Cerrado biomes. The fruits of *M. flexuosa* (Figure 23.7) are elliptical to oval in form, with a pericarp of reddish-brown triangular scales around it. The mesocarp is edible and comprises a coating of fleshy and greasy yellowish or orange substances. It features a spongy, thin, white to yellowish endocarp with a very rigid, ovoid, whitish-colored seed (Sampaio and Carraza 2012; Amorim et al. 2021; Rudke et al. 2021). The fruits range in diameter from 37.24 to 52.83 mm (Rossi et al. 2014). The buriti palm is little known outside of Brazil's northeast region and is notable for its high content of carotenoids, primarily β-carotene, with provitamin A activity (Rosso and Mercadante 2007), as well as being an excellent source of fatty acids, carbohydrates, proteins, mineral salts, B complex vitamins, and phenolic compounds (Carneiro and Carneiro 2011) and can be used in fermented beverages (Morais et al. 2022). Borgonovi (2018) found that buriti pulp at a concentration of 10g/L in MRS medium allowed the growth of *Lacticaseibacillus casei* (*L. casei* SJRP38) and *Limosilactobacillus fermentum* (*L. fermentum* SJRP30) with counts of 8.4 and 8.6 log UFC-1 after 24 hours of incubation. With these results, the research was advanced with the preparation of fermented buriti milk (addition of 1% w/v powdered buriti pulp) with *L. casei* SJRP38 and results indicated that the total phenolic compound content was reduced from 98.48 to 56.14 mg GAE/100 g after 28 days of storage at 4°C but showed stability in the antioxidant activity measured by the DPPH method (with 0.54 μmol Trolox/g). The fermented buriti milk was likewise stable after 28 days of storage (Borgonovi 2018).

Araçá-boi (*Eugenia stipitata* Mcvaugh) is a plant from the western Amazon, a three-metre-high shrub with year-round production. Figure 23.7 depicts its fruits, which are globose yellowish berries with a great volume of pulp ranging from 30 to 800 g (average 200 g). When mature, the pulp is succulent, with a good flavor but is quite acidic, and has a high quantity of vitamin C, proteins, carbohydrates, fiber, and mineral salts (Gentil and Clement 1997; Falcão et al. 2000; Neri-Numa et al. 2013; Barros et al. 2019). The fruit's pericarp is canary yellow, velvety, and slender (Falcão et al. 2000). Souza et al. (2022) investigated the optimization of the production process of fermented araçá-boi wine to increase the tyrosol (metabolite) content, which is related to the antioxidant capacity (DPPH), through a 2^3 factorial design, evaluating the effect of the type of yeast, must filtration (Filtered or Unfiltered) and fermentation temperature (18°C and 28°C). After studying the factorial design, the authors reported that all the factors interfered in a significant way in the concentration of tyrosol and antioxidant capacity. However, the fermentation temperature was the most relevant factor to the antioxidant response, that is, it showed a higher concentration of tyrosol (835.23–920.00 μmol L^{-1}) and higher antioxidant activity (IC50 = 370.7–464.9 μg mL^{-1}) in araçá-boi wine samples at a temperature of 28°C.

Bacaba (*Oenocarpus bacaba* Mart.) is a member of an Arecaceae family native palm found in two Brazilian biomes, the Amazon and Cerrado (Diniz et al. 2020). The fruits (Figure 23.7) occur in bunches and are spherical, subglobose drupes with a highly oleaginous mucilaginous pulp, and the mesocarp is about 1.5 mm thick with a thin and fibrous endocarp (Ferreira and Medeiros 2005; Guimarães 2013). Its fruits are edible and highly valued by the Amazon's indigenous and cabocla people (Queiroz and Bianco 2009). Chicha is a fermented drink made with bacaba palm fruit (*O. bacaba*) that can be eaten fermented or unfermented. Other names for the drink include "bacaba wine" and "bacaba milk." Lima (1975) references a paper on the derivation of Náhuatl from the phrase chicha (or chichia), which means "to render sour or bitter." Chicha also means "spit" in Náhuatl, which corresponds to the essential process in the preparation of the drink. Spontaneous fermentation can occur in the postharvest period, between fruit collection and the production of beverages; and acetic acid (0.90 g/L), succinic acid (2.69 g/L), citric acid (0.49 g/L), glycerol (0.40 g/L), and ethanol (0.52 g/L) were the identified metabolites. They were reported as rich sources of phenolic compounds as well as carbohydrates and fat. It has also been reported that using the bacaba berry (*O. bacaba*), the indigenous Umutina people make a purple-hued beverage called bacaba chicha (Puerari 2014; Puerari et al. 2015; Resende et al. 2018). In a study by Puerari (2014), after 6 hours of fermentation, there was a 5.11 log CFU/mL increase in bacterial population, with LAB as the largest group; increasing concentrations of citric and lactic (1.4 g/L) acids; the average glycerol content in the beverage was 0.42 g/L; and ethanol had been eliminated, making this chicha an acidic, nonalcoholic beverage. At 36 hours of fermentation, the pH levels had dropped from 5.2 to 3.9. *Enterococcus hormaechei*, *Leuconostoc lactis,* and *Pantoea dispersa* were among the LAB strains identified using culture-dependent analysis. *Acetobacter* spp., *Bifidobacterium*, *Pichia caribbica, Propionibacterium*, and *Propionibacterium avidum* were found to be present in the polymerase chain reaction (PCR)–denaturing gradient gel electrophoresis (DGGE) results, as well as other uncultured bacteria and fungi. For the synthesis of BACs, Barros et al. (2019) isolated strains of endophytic bacteria from the achachairu, araçá-boi, and bacaba fruits and examined their liquid fermentation capability. *Candida, Ochrobactrum*, and *Rhodotorula* were among the genera of major microbial strains detected. It was found that *Rhodotorula glutinis*, which was isolated from the araçá-boi, was the most effective because of its high production of BACs. After testing numerous residues for their potential in the culture medium for phenolic compounds, the bacaba fruit residue was found to be the most effective. Using a Plackett-Burman design to optimize the process, Barros et al. (2019) found that the pH of the medium and the yeast extract were the two primary parameters that controlled the fermentation process, resulting in a maximum output of 887.55±3.61 mg GAE/100 g after 48 hours. According to a comparison of fermented and unfermented bacaba residues using UHPLC-QqQ-MS/MS analytical method, chlorogenic acid (460.00%), catechin (94.75%), and rutin (27.53%) were found to be higher in the fermented bacaba. As a result, ABTS (39.25%), FRAP (90.32%), and ORAC (118.62%) revealed an increase in antioxidant capacity of the fermented products (Barros et al. 2019).

An interesting study (González et al. 2023) developed a pigment from black sapote (Diospyros digyna) fermented by lactic acid bacteria and yeasts, from this fruit native and endemic to Mexico. In the Cerrado, Pantanal, Atlantic Forest, Amazon and Caatinga biomes, similar fruits also occur, such as *Diospyros inconstans* lacq. called persimmon of the forest and *Diospyros lasiocalyx* (Mucuíba, a name of Tupi-Guarani origin and meaning "fruit of the field"), both from the Ebenaceae family (Passos, M.A.B., 2023), which could be researched and better explored, like the Mexican species. Additionaly Diospyros (Ebenaceae) pantropical genus, among which 25 species distributed in tropical areas showed significant inhibitory activity on dengue virus replications (Peyrat et al. 2016). The Mexican fruit includes in its composition many vitamins, minerals, and bioactive compounds, some of which are pigments that might possess nutraceutical properties (González et al 2023; Oscar et al., 2023). The pulp of black sapote contains carotenoids (β carotene and lycopene), xanthophylls (lutein and β kryptoxanthin), chlorophylls, anthocyanins, and other phenolic compounds. The authors extracted these pigments from the fermentation of the fruit pulp by yeasts and LAB and added the powder obtained by spray drying to gummy bears. The pigment powders were added at 1, 3, and 5% to the gummy bear formulations, which were evaluated in color and some sensory characteristics. Results showed that the fermentation process increased the antioxidant capacity and total phenolic compounds, particularly when fermentation was performed with *L. rhamnosus.* The addition of these pigments into food matrices might produce highly colored foods with health-benefitting properties. Thus, native Brazilian fruits of the same genus could also be used in this way in processed products and not just in their natural form or in liqueurs.

Camu-camu (*Myrciaria dubia* McVaugh) of the Myrtaceae family is represented in Figure 23.7, a native fruit tree found on the banks of rivers and flooded lakes in the Amazon. The species is found across the Brazilian Amazon, including the states of Pará, Amapá, Amazonas, Rondônia, Roraima, and Mato Grosso (Maués and Couturier 2002). The fruit (Figure 23.7) is a spherical berry with a smooth and shiny surface, purplish-red hue, 2 to 4 cm in diameter, and one to four seeds, valued for the confection of ice cream, juices, and liqueurs, and with high economic potential (Villachica 1996). However, because this fruit spoils quickly, exporting and storing it can be prohibitively expensive (Fujita et al. 2017). Probiotic beverages incorporating LAB fermentation, such as those made from the high-phenolic antioxidant content and anti-diabetic properties of camu-camu fruit, could benefit from the fruit's enhanced shelf life, enhanced sensory qualities, and increased functionality of previously established human health-relevant BACs (Fujita et al. 2015, 2017). Fujita et al. (2017) investigated powdered camu-camu preparation with the goal of improving the antioxidant, anti-hyperglycemic, and anti-hypertensive effects of camu-camu coupled with soybean milk by fermentation with *L. plantarum* and *L. helveticus.* Camu-camu powders were created in two ways: freeze-dried and spray-dried in soy milk with gum arabic as a carrier agent. After 72 hours of fermentation, Fujita et al. (2017) reported that *L. helveticus* outperformed *L. plantarum* in terms of growth and stability ($p < 0.05$). The growth of *L. plantarum* ($p < 0.05$) was not affected by the addition of 0.5% or 1% of camu-camu powder. On the other hand, camu-camu slowed the development of *L. helveticus*, particularly when compared to the control. The fermentation of the camu-camu and soymilk combination with both probiotic strains resulted in greater α-glucosidase and α-amylase inhibitory activities, whereas total phenolic content and antioxidant activity did not change substantially. The camu-camu (0.5 and 1% additions) and soymilk combination fermented with *L. plantarum* also improved the action of ACE enzyme inhibitors. As a food-based beverage for type 2 diabetes and hypertension prevention, the health advantages of camu-camu were rationally enhanced by employing lactic acid fermentation in a soymilk substrate design (Fujita et al. 2017).

23.3.5 Pampa Biome

The Pampa, also known as Pampas, Campanha Gaúcha, Campos Sulinos, or Campos do Sul, is the unique Brazilian biome found in a single federal unit. Pampa was only designated as a biome in 2004. This vital territory extends beyond Brazil's boundaries and has an enormous range of fauna and flora species, bringing together strengths and weaknesses. Areas in the process of sanding, the human transmigration from the countryside; the purchase of vast areas for the exploitation of forestry; massive crops of transgenic grains with their pesticides and intensive agriculture (Morais et al. 2022); increased mining; and many other conflicts point to problems in the use of their natural resources. According to the Ministry of Environment, the Pampa is Brazil's second most destroyed ecosystem, with just 36% left (Winckler and Filho 2020). The Pampa, located in the Rio Grande do Sul, accounts for around 2% of the Brazilian area. This name comes from the Quechua language and means "flat area." This Brazilian biome's environment is largely made up of native fields. The Pampas has a chilly subtropical climate, with an average annual temperature of 19°C (Winckler and Filho 2020; Lopes et al. 2020). The Pampa biome is confined to the Brazilian Pampa, which covers an area of 178,243 km^2 in the state of Rio Grande do Sul, which corresponds 63% of the state's area and 2% of Brazilian land. In South America, the fields that include the Pampas cover parts of Uruguay and Argentina, covering an area of 750,000 km^2 (Winckler and Filho 2020; Lopes et al. 2020).

A survey was carried out in 2020 by Brack et al. (2020), who identified 213 species of natural fruit trees in the Rio Grande do Sul, encompassing 48 families and 102 genera. The plant families with the most species included Annonaceae (9 spp.), Arecaceae (14 spp.), Cactaceae (14 spp.), Myrtaceae (43 spp.), Passifloraceae (10 spp.), and Solanaceae (15 spp.). The genus *Eugenia* (Myrtaceae), with the greatest diversity, included species such as araçá-piranga (*E. multicostata*), batinga (*E. rostrifolia*), pitanga (*E. uniflora*), rio-grande cherry (*E. involucrata*), uvaia (*E. pyriformis*), wild peach (*E. myrcianthes*), and others. Figure 23.8 shows some of the fruit plants

FIGURE 23.8 Various fruits from the Pampa biome.

found in the Pampa biome. Schulz et al. (2020) investigated the potential health benefits of seven underutilized Brazilian dark fruits (*Bactris setosa, Euterpe edulis, Eugenia involucrata, E. brasiliensis, Myrciaria jaboticaba, M. cauliflora,* and *Myrcianthes pungens*) and found that they had high nutritional and bioactive value *in vitro* and *in vivo* studies. These species have economic potential; however, the majority are unknown and only consumed locally. Anthocyanin levels in *Euterpe edulis* and *Eugenia brasiliensis* were found to be high (Schulz et al. 2020).

Butia eriospatha (Martius) Beccari is a member of the Arecaceae family and is native to South America. It is also known as butiá or butiá-da-serra. *Butia* is a small genus of roughly eight species in South America, seven of which exist endemically and natively in Brazil in the states of Paraná, Santa Catarina, and Rio Grande do Sul (Lorenzi 2004; Köhler and Brack 2016). This plant produces globose, succulent fruits with an average diameter of 1.7 to 1.9 cm, which mature into a yellowish epicarp (Henderson et al. 1995). Throughout the year, the concentrated juice may be frozen, stored, and processed until the next harvest (de Souza Cardoso et al. 2014). Butiá is one of the most popular natural fruits in the Rio Grande do Sul, and it may be found in local slang and idioms. Botanically, it is the common name for the fruit of various species of Butia palm trees. The fruits, which can range in color from yellow to orange to pink, are frequently consumed and liked by the general populace and animals. Butiá, butiazeiro, and butiazais (huge regions with a preponderance of butia trees) have traditionally been associated with human communities. Historical and archaeological records show that the fruits and seeds were part of the indigenous diet for over 8000 years, while the leaves were utilized to make baskets. Several organizations are now formed in different districts of the Rio Grande state for the usage and processing of butiá fruit. A local cooperative on the North Coast is benefitting from the fruits of the butiá-da-praia (*Butia catarinensis* Noblick & Lorenzi) and producing frozen pulps and ice creams. *Butia odorata* (Barb. Rodr.) fruit processing is offered informally as frozen pulps in the Central and South Coast areas. Many households in the Serra e Norte region already engage in fruit harvesting and processing (Köhler and Brack 2016). dos Santos Cruxen et al. (2017) utilized butia to make probiotic ice cream with great acceptance and purchase intent. The probiotic counts were 8.51 log CFU/g of *Bifidobacterium lactis* after 90 days of storage. The native fruit containing BACs was mixed into the ice cream, showing product qualities such as 71.2 g/100 g overrun, a density of 575 g/cm^3, and a pH of 4.7.

Various raw materials have been examined as alternatives to ethanol production, and one of these possibilities is palm tree whose fruits contain carbohydrates or sugars that allow fermentation. Palm trees are a huge plant family that may be found in virtually any ecosystem. In Brazil, there are around 119 species belonging to 39 distinct genera (Lorenzi 2004; de Souza Cardoso et al. 2014). Because of the abundance of carbohydrates, Butiá pulp aqueous extract proved an appropriate and promising substrate for ethanol synthesis, making it an alternative to sugar cane. As carbohydrate sources, the original extract contained 29.97 g/L (32.86%) glucose, 39.84 g/L (43.69%) fructose, 3.43 g/L (3.76%) sucrose, and 17.96 g/L (19.69%) fructooligosaccharides (FOS). The simplex method was used to analyze fermented products, and it was found that FOS present in the extract did not undergo fermentation because butiá had not yet reached the required maturity level, and this type of carbohydrate is not fermented by *Saccharomyces cerevisiae*, but 0.75 g/L of fructose, traces of glucose, and the absence of sucrose were also found (de Souza Cardoso et al. 2014).

23.4 CONCLUDING REMARKS

Brazil has a wide range of climates and areas with distinct edaphoclimatic characteristics, making it one of the most diverse countries in the world. The flora are so diverse that many species are undiscovered or understudied. A greater understanding of macro-and micro-nutrient composition, particularly of those with positive and bioactive effects, is required for their sustainable and conscientious use. According to this chapter, new products fermented with non-conventional native or exotic fruits, representative of the six Brazilian biomes and with potential for further studies in the future, can enable the commercialization and consumption of these rich flora that are not yet fully known and enable regional economic exploration by the communities located in these regions.

23.4.1 Acknowledgments

The authors would like to thank the National Council of Technological and Scientific Development (CNPq) and the Coordination for the Improvement of Higher Education Personnel (CAPES), as well as the State University of Londrina, for giving a scholarship and permitting this study.

REFERENCES

Acuna, U. M., Jancovski, N., and Kennelly, E. J. 2009. "Polyisoprenylated benzophenones from Clusiaceae: Potential drugs and lead compounds." *Current Topics in Medicinal Chemistry*, 9, no. 16: 1560–1580. https://doi.org/10.2174/156802609789909830

Aguiar, T. M., Sabaa-Srur, A. U. O., and Smith, R. E. 2016. "Study of Grumixama (*Eugenia brasiliensis*, Lam) fruit pulp and development of a jelly: Rheological, sensorial and colorimetric evaluation." *The Natural Products Journal*, 6, no. 2: 142–151.

Albuquerque, B. R., Pereira, C., Calhelha, R. C., Alves, M. J., Abreu, R. M., Barros, L., and Ferreira, I. C. 2020. "Jabuticaba residues (*Myrciaria jaboticaba* (Vell.) Berg) are rich sources of valuable compounds with bioactive properties." *Food Chemistry*, 309: 125735. https://doi.org/10.1016/j.foodchem.2019.125735

Amorim, V. R., Rodrigues, D. C. D. N., Silva, J. D. N., Ramos, C. L. S., Almeida, L. M. N., Almeida, A. A. C., and Ferreira, P. M. P. 2021. "Anti-inflammatory mechanisms of fruits and by-products from *Mauritia flexuosa*, an exotic plant with functional benefits." *Journal of Toxicology and Environmental Health, Part A*, 84, no. 11: 441–457. https://doi.org/10.1080/15287394.2021.1881672

Andrade, J. K. S., Barros, R. G. C., Pereira, U. C., Nogueira, J. P., Gualberto, N. C., de Oliveira, C. S., and Narain, N. 2022. "Bioaccessibility of bioactive compounds after *in vitro* gastrointestinal digestion and probiotics fermentation of Brazilian fruits residues with antioxidant and antidiabetic potential." *LWT*, 153: 112469. https://doi.org/10.1016/j.lwt.2021.112469

Antognoni, F., Potente, G., Mandrioli, R., Angeloni, C., Freschi, M., Malaguti, M., and Tartarini, S. 2020. "Fruit quality characterization of new sweet cherry cultivars as a good source of bioactive phenolic compounds with antioxidant and neuroprotective potential." *Antioxidants*, 9, no 8: 677.

Anunciação, P. C., Giuffrida, D., Murador, D. C., de Paula Filho, G. X., Dugo, G., and Pinheiro-Sant'Ana, H. M. 2019. "Identification and quantification of the native carotenoid composition in fruits from the Brazilian Amazon by HPLC–DAD–APCI/MS." *Journal of Food Composition and Analysis*, 83: 103296.

Araujo, N. M. P., Arruda, H. S., Dos Santos, F. N., de Morais, D. R., Pereira, G. A., and Pastore, G. M. 2020. "LC-MS/MS screening and identification of bioactive compounds in leaves, pulp and seed from *Eugenia calycina* Cambess." *Food Research International*, 137: 109556.

Asfaram, A., Sadeghi, H., Goudarzi, A., Kokhdan, E. P., and Salehpour, Z. 2019. "Ultrasound combined with manganese-oxide nanoparticles loaded on activated carbon for extraction and pre-concentration of thymol and carvacrol in methanolic extracts of *Thymus daenensis, Salvia officinalis, Stachys pilifera, Satureja khuzistanica*, and mentha, and water samples." *Analyst*, 144, no. 6: 1923–1934.

Bailão, E. F. L. C., Devilla, I. A., Da Conceição, E. C., and Borges, L. L. 2015. "Bioactive compounds found in Brazilian Cerrado fruits." *International Journal of Molecular Sciences*, 16, no. 10: 23760–23783. https://doi.org/10.3390/ijms161023760

Baliga, M. S., Bhat, H. P., Baliga, B. R. V., Wilson, R., and Palatty, P. L. 2011. "Phytochemistry, traditional uses and pharmacology of *Eugenia jambolana* Lam. (black plum): A review." *Food Research International*, 44, no. 7: 1776–1789. https://doi.org/10.1016/j.foodres.2011.02.007

Barbieri, S. F., de Oliveira Petkowicz, C. L., de Godoy, R. C. B., de Azeredo, H. C. M., Franco, C. R. C., and Silveira, J. L. M. 2018. "Pulp and Jam of Gabiroba (*Campomanesia xanthocarpa* Berg): Characterization and rheological properties." *Food Chemistry*, 263: 292–299.

Barbosa, A. P. P., Moraes, A. F., and Chisté, R. C. 2020. "Physicochemical characterization and quantification of bioactive compounds of Antrocaryon amazonicum fruits cultivated in Brazilian Amazonia." *CyTA: Journal of Food*, 18, no. 1: 616–623.

Barros, R. G. C., Andrade, J. K. S., Denadai, M., Nunes, M. L., and Narain, N. 2017. "Evaluation of bioactive compounds potential and antioxidant activity in some Brazilian exotic fruit residues." *Food Research International*, 102: 84–92.

Barros, R. G. C., de Oliveira, C. S., Oliveira, L. T. S., Pereira, U. C., Silva, T. O. M., Denadai, M., and Narain, N. 2019. "Enhancement of phenolic antioxidants production in submerged cultures of endophytic microorganisms isolated from achachairu (*Garcinia humilis*), araçá-boi (*Eugenia stipitata*) and bacaba (*Oenocarpus bacaba*) fruits." *LWT*, 111: 370–377.

Barros, R. G. C., Pereira, U. C., Andrade, J. K. S., de Oliveira, C. S., Vasconcelos, S. V., and Narain, N. 2020. "*In vitro* gastrointestinal digestion and probiotics fermentation impact on bioaccessbility of phenolics compounds and antioxidant capacity of some native and exotic fruit residues with potential antidiabetic effects." *Food Research International*, 136: 109614. https://doi.org/10.1016/j.foodres.2020.109614

Bastos, A., Hirota, M., and Balazina, A. 2017. SOS Mata Atlantica annual activities report–2017. Accessed on October 7, 2021. https://cms.sosma.org.br/wp-content/uploads/2018/10/AF_RA_SOSMA_2017_web.pdf

Bastos, R. G., Salles, B. C. C., Bini, I. F., Castaldini, L. P., Silva, L. C. D., Vilela, A. A., and da Silva, M. A. 2019. "Phytochemical composition, antioxidant and *in vivo* antidiabetic activities of the hydroethanolic extract of *Eugenia florida* DC. (Myrtaceae) leaves." *South African Journal of Botany*, 123: 317–332.

Batista, Â. G., Lenquiste, S. A., Cazarin, C. B. B., da Silva, J. K., Luiz-Ferreira, A., Bogusz, S., Jr., and Maróstica, M. R., Jr. 2014. "Intake of jaboticaba peel attenuates oxidative stress in tissues and reduces circulating saturated lipids of rats with high-fat diet-induced obesity." *Journal of Functional Foods*, 6: 450–461.

Baú, T. R., Garcia, S., and Ida, E. I. 2015. "Changes in soymilk during fermentation with kefir culture: Oligosaccharides hydrolysis and isoflavone aglycone production." *International Journal of Food Sciences and Nutrition*, 66, no. 8: 845–850. https://doi.org/10.3109/09637486.2015.1095861

Baygut, H., Cais-Sokolińska, D., Bielska, P., and Teichert, J. 2023. "Fermentation kinetics, microbiological and physical properties of fermented soy beverage with acai powder." *Fermentation*, 9, no. 4: 324. https://doi.org/10.3390/fermentation9040324

Bezerra, F. D. S., da Costa, D. F., and Koblitz, M. G. B. 2020. "Integral use of oil-based raw materials with "green solvents": review and opportunities." *Research, Society and Development*, 9, no. 8: e372985388.

Bezerra, M., Araujo, A., Santos, K., and Correia, R. 2015. "Caprine frozen yoghurt produced with fresh and spray dried jambolan fruit pulp (*Eugenia jambolana* Lam) and *Bifidobacterium animalis subsp. lactis* BI–07." *LWT*, 62, no. 2: 1099–1104.

BMH (BRAZIL, Ministry of Health). 2019. Health and functional property claims. Updated March 2019. Accessed on October 7, 2021. https://www.gov.br/agricultura/pt-br/assuntos/inspecao/produtos-vegetal/legislacao-1/biblioteca-de-normas-vinhos-e-bebidas/alegacoes-de-propriedade-funcional-aprovadas_anvisa.pdf

Borges, G. D. S. C., Gonzaga, L. V., Jardini, F. A., Mancini Filho, J., Heller, M., Micke, G., and Fett, R. 2013. "Protective effect of *Euterpe edulis* M. on Vero cell culture and antioxidant evaluation based on phenolic composition using HPLC-ESI-MS/MS." *Food Research International*, 51, no. 1: 363–369.

Borgonovi, T. F. 2018. Biocomposites from passion fruit and buriti pulp: characterization and application in milk. Master Dissertation. São Paulo State University, São Paulo, Brazil.

Bortolotto, I. M., Amorozo, M. C. d. M., Neto, G. G., Oldeland, J., and Damasceno-Junior, G. A. 2015. "Knowledge and use of wild edible plants in rural communities along Paraguay River, Pantanal, Brazil." *Journal of Ethnobiology and Ethnomedicine*, 11: 46. https://doi.org/10.1186/s13002-015-0026-2

Bortolotto, I. M., Hiane, P. A., Ishii, I. H., Souza, P. R., Campos, R. P., Gomes, R. J. B., Farias, C. S., Leme, F. M., Arruda, R. C. O., Costa, L. B. L. C., and Damasceno-Junior, G. A. 2017. "A knowledge network to promote the use and valorization of wild food plants in the Pantanal and Cerrado, Brazil." *Regional Environmental Change*, 17: 1329–1341. https://doi.org/10.1007/s10113-016-1088-y

Brack, P., Köhler, M., Corrêa, C. A., Ardissone, R. E., Sobral, M. E. G., and Kinupp, V. F. 2020. "Fruits native to Rio Grande do Sul, Brazil: Food richness and potential." *Rodriguésia*, 71: e03102018. https://doi.org/10.1590/2175-7860202071091

Braga, A. R. C., Mesquita, L. M. de. S., Martins, P. L. G., Habu, S., and de Rosso, V. V. 2018. "*Lactobacillus* fermentation of jussara pulp leads to the enzymatic conversion of anthocyanins increasing antioxidant activity." *Journal of Food Composition and Analysis*, 69: 162–170.

Brazil State Secretariat for the Environment (BSSE). 2016. SMA Resolution No. 57 of June 5, 2016. Publishes the second revision of the official list of endangered flora species in the State of São Paulo. *Official Gazette of the State of São Paulo*, São Paulo, June 7, 2016.

Brito Júnior, M. R. D., Ugalde, F. Z., Gonzaga, L. V., Schulz, M., Fett, R., Costa, A. C. O., and Tribuzi, G. 2023. "Physicochemical characteristics and antioxidant potential of a fruit beer produced with Juçara (*Euterpe edulis Martius*) fruit pulp." *Brazilian Archives of Biology and Technology*, 66: 222–234.

Caldeira, A. S., Mbiakop, U. C., Pádua, R. M., van de Venter, M., Matsabisa, M. G., Campana, P. R., and Braga, F. C. 2021. "Bioguided chemical characterization of pequi (*Caryocar brasiliense*) fruit peels towards an anti-diabetic activity." *Food Chemistry*, 345: 128734.

Carmo, S. K. S., Sá, S. K. C. V. L., Almeida, M. M., and Swarnakar, R. 2012. "Production and characterization of umbu fermented from its commercial pulp." *Revista Brasileira de Produtos Agroindustriais*, 14, no. 1: 15–20.

Carneiro, T. B., and de Mello Carneiro, J. G. 2011. "Buriti fruits and dehydrated pulp (*Mauritia flexuosa* L.): Physical, chemical and technological aspects." *Revista Verde de Agroecologia e Desenvolvimento Sustentável*, 6, no. 2: 22.

Cavalaro, R. I., da Cruz, R. G., Dupont, S., de Moura, J. M. L. N., and de Souza Vieira, T. M. F. 2019. "*In vitro* and *in vivo* antioxidant properties of bioactive compounds from green propolis obtained by ultrasound-assisted extraction." *Food Chemistry*, X, no. 4: 100054.

Chávez-González, M. L., Sepúlveda, L., Verma D. K., Luna-García, H. A., Rodríguez-Durán, L. V., Ilina, A., and Aguilar, C. N. 2020. "Conventional and emerging extraction processes of flavonoids." *Processes*, 8: 434. https://doi.org/10.3390/pr8040434

Chemat, F., Rombaut, N., Sicaire, A. G., Meullemiestre, A., Fabiano-Tixier, A. S., and Abert-Vian, M. 2017. "Ultrasound assisted extraction of food and natural products. Mechanisms, techniques, combinations, protocols and applications. A review." *Ultrasonics Sonochemistry*, 34: 540–560.

Chhikara, N., Kaur, R., Jaglan, S., Sharma, P., Gata, Y., and Panghal, A. 2018. "Bioactive compounds and pharmacological and food applications of Syzygium cumini: A review." *Food & Function*, 9, no. 12: 6096–6115.

Costa, J. M. C., Oliveira, D. M., and Costa, L. E. C. 2018. "Macauba Palm: *Acrocomia aculeata*." In: Rodrigues, S., Oliveira Silva, E., and Brito, E. S. (Eds.), *Exotic Fruits*. Academic Press, pp. 297–304. https://doi.org/10.1016/B978-0-12-803138-4.00039-3

Costa, M. G. M., Ooki, G. N., Vieira, A. D. S., Bedani, R., and Saad, S. M. I. 2017. "Synbiotic Amazonian palm berry (açai, *Euterpe oleracea* Mart.) ice cream improved *Lactobacillus rhamnosus*

GG survival to simulated gastrointestinal stress." *Food & Function*, 8, no. 2: 731–740.

Coutinho, G. S. M., Ribeiro, A. E. C., Prado, P. M. C., Oliveira, É. R., Careli-Gondim, Í., Oliveira, A. R., and Vilas Boas, E. V. D. B. 2021. "Green banana starch enhances physicochemical and sensory quality of baru almond-based fermented product with probiotic bacteria." *International Journal of Food Science & Technology*, 56, no. 10: 5097–5106. https://doi.org/10.1111/ijfs.15260

da Silva, I. L., da Silva, L. A. D. O., and Coelho, L. C. B. B. 2019b. "The Brazilian Caatinga biome and its biotechnological potential." In: Teodor, R. (Ed.), *Advances in Applied Science and Technology*, Vol. 5. B P International Publisher, India, pp. 1–20. https://doi.org/10.9734/bpi/aast/v5

da Silva, A. P. G., Spricigo, P. C., Purgatto, E., de Alencar, S. M., Sartori, S. F., and Jacomino, A. P. 2019a. "Chemical composition, nutritional value and bioactive compounds in six uvaia accessions." *Food Chemistry*, 294: 547–556.

Da Silva, J. B. 2014. Physical-chemical and chemical characterization of the fruit of Juazeiro (*Ziziphus juazeiro* Mart.) and evaluation of its conservation by lactic fermentation. PhD Thesis. Federal University of Rio Grande do Norte, Brazil.

da Silva, J. K., Batista, Â. G., Cazarin, C. B. B., Dionísio, A. P., de Brito, E. S., Marques, A. T. B., and Junior, M. R. M. 2017. "Functional tea from a Brazilian berry: Overview of the bioactives compounds." *LWT*, 76: 292–298.

da Silva Júnior, J. C., Magnani, M., da Costa, W. K. A., Madruga, M. S., Olegário, L. S., Borges, G. D. S. C., Dantas, A. M., dos Santos Lima, M., de Lima, L. C., de Lima Brito, I., and de Magalhães Cordeiro, A. M. T. 2021. "Traditional and flavored kombuchas with pitanga and umbu-cajá pulps: Chemical properties, antioxidants, and bioactive compounds." *Food Bioscience*, 44 (part A): 101380. https://doi.org/10.1016/j.fbio.2021.101380

Dacoreggio, M. V., Santetti, G. S., Inácio, H. P., Kempka, A. P., and de Mello Castanho Amboni, R. D. 2021. "A comprehensive review of *Eugenia pyriformis* Cambess: Reported bioactivities and health effects." *Food Reviews International*, 1–15.

Damasceno-Junior, G. A., Souza, P. R., Bortolotto, I. M., Ramos, M. I. L., Hiane, P. A., Braga Neto, J. A., Ishii, I. H., Costa, D. C., Ramos Filho, M. M., Gomes, R. J. B., Barbosa, M. M., and Rodrigues, R. B. 2010. *Flavors From the Cerrado and Pantanal: Getting to Know Them to Value Native Fruits, Recipes and Good Use Practices* (1st edition). Campo Grande, Brazil: Publisher of the Federal University of Mato Grosso do Sul.

Damiani, C., Silva, J. S., Amorim, K. A., Miranda P. H. S., and dos Santos, A. C. 2020. "Alimentos funcionais desenvolvidos a partir dos frutos da amazônia." In: de Almeida, A. F., and Santos, C. C. A. do A. (Eds.), *Frutos Amazônicos: Biotecnologia e Sustentabilidade*, Vol. 1, no. 24. Palmas, Tocantins, Brazil, pp. 26–37.

Dantas, A. M., Mafaldo, I. M., de Lima Oliveira, P. M., dos Santos Lima, M., Magnani, M., and Borges, G. D. S. C. 2019. "Bioaccessibility of phenolic compounds in native and exotic frozen pulps explored in Brazil using a digestion model coupled with a simulated intestinal barrier." *Food Chemistry*, 274: 202–214.

Das, A. B., Goud, V. V., and Das, C. 2017. "Extraction of phenolic compounds and anthocyanin from black and purple rice bran (*Oryza sativa* L.) using ultrasound: A comparative analysis and phytochemical profiling." *Industrial Crops and Products*, 95: 332–341.

Daza, L. D., Fujita, A., Granato, D., Fávaro-Trindade, C. S., and Genovese, M. I. 2017. "Functional properties of encapsulated Cagaita (*Eugenia dysenterica* DC.) fruit extract." *Food Bioscience*, 18: 15–21. http://doi.org/10.1016/j.fbio.2017.03.003

De Alencar, P. G., De Espindola, G. M., and Carneiro, E. L. N. C. 2018. "Dwarf cashew crop expansion in the Brazilian semiarid region: Assessing policy alternatives in Pio IX, Piauí." *Land Use Policy*, 79: 1–9.

De Almeida, M. H. B., Zoellner, S. S., Da Cruz, A. G., Moura, M. R. L., De Carvalho, L. M. J., Freitas, M. C. J., and De S Sant'ana, A. 2008. "Potentially probiotic açaí yogurt." *International Journal of Dairy Technology*, 61, no. 2: 178–182.

De Almeida, M. M., da Silva, F. L. H., Conrado, L. S., Mota, J. C., and Freire, R. M. M. 2011. "Kinetics and characterization of cereus *Cereus jamacaru* P. DC. Wine." *Revista Verde de Agroecologia e Desenvolvimento Sustentável*, 6, no. 2: 176–183.

De Almeida, M. M., Tavares, D. P. S. A., Rocha, A. S. R., Oliveira, L. S. C., da Silva, F. L. H., and Mota, J. C. 2006. "Cinética da produção do fermentado do fruto do mandacaru." *Revista Brasileira de Produtos Agroindustriais*, 8, no. 1: 35–42.

de Araujo, F. F., Neri-Numa, I. A., de Paulo Farias, D., da Cunha, G. R. M. C., and Pastore, G. M. 2019. "Wild Brazilian species of *Eugenia* genera (Myrtaceae) as an innovation hotspot for food and pharmacological purposes." *Food Research International*, 121: 57–72. https://doi.org/10.1016/j.foodres.2019.03.018

de Assis, B. B. T., Pimentel, T. C., Dantas, A. M., dos Santos Lima, M., Borges, G. D. S. C., and Magnani, M. 2021. "Biotransformation of the Brazilian Caatinga fruit-derived phenolics by *Lactobacillus acidophilus* La-5 and *Lacticaseibacillus casei* 01 impacts bioaccessibility and antioxidant activity." *Food Research International*, 110435.

De Freitas, F. A., Araújo, R. C., Soares, E. R., Nunomura, R. C. S., da Silva, F. M. A., da Silva, S. R. S., de Souza, A. Q. L., de Souza, A. D. L., Franco-Montalbán, F., Acho, L. D. R., Lima, E. S., Bataglion, G. A., and Koolen, H. H. F. 2017. "Biological evaluation and quantitative analysis of antioxidant compounds in pulps of the Amazonian fruits bacuri (*Platonia insignis* Mart.), ingá (*Inga edulis* Mart.), and uchi (*Sacoglottis uchi* Huber) by UHPLC-ESI-MS/MS." *Journal of Food Biochemistry*, 42 no. 1: e12455. https://doi.org/10.1111/jfbc.12455

De Lima, M. A. C., Silva, S. M., and De Oliveira, V. R. 2018. "Umbu: *Spondias tuberosa*." *Exotic Fruits*: 427–433.

de Morais Cardoso, L., Martino, H. S. D., Moreira, A. V. B., Ribeiro, S. M. R., and Pinheiro-Sant'Ana, H. M. 2011. "Cagaita (*Eugenia dysenterica* DC.) of the Cerrado of Minas Gerais, Brazil: Physical and chemical characterization, carotenoids and vitamins." *Food Research International*, 44, no. 7: 2151–2154. https://doi.org/10.1016/j.foodres.2011.03.005

de Moura Neto, L. G., de Sousa Lira, J., da Silva Torres, M. M. F., Barbosa, I. C., do Amarante Melo, G. F., and Soares, D. J. 2016. "Development of a mixed drink made from hydrosoluble soybean extract, coconut water and umbu pulp (*Spondias tuberosa*)." *Acta Scientiarum Technology*, 38, no. 3: 371–376. https://doi.org/10.4025/actascitechnol.v38i3.27064

de Souza Cardoso, C., Spacino, K. R., da Silva, E. T., Gomes, K., Angilelli, O. F. G., Moreira, I., and Borsato, D. 2014. "Application of the super-modified simplex optimization method in the fermentation of Butiá Palm (*Butia eriospatha* (Martius) Beccari) aqueous extract." *International Journal of Environmental Bioenergy*, 9, no. 2: 130–142.

del Pilar Garcia-Mendoza, M., Espinosa-Pardo, F. A., Baseggio, A. M., Barbero, G. F., Junior, M. R. M., Rostagno, M. A., and Martínez, J. 2017. "Extraction of phenolic compounds and anthocyanins from juçara (*Euterpe edulis* Mart.) residues using pressurized liquids and supercritical fluids." *The Journal of Supercritical Fluids*, 119: 9–16.

Demoliner, F., de Britto Policarpi, P., Vasconcelos, L. F. L., Vitali, L., Micke, G. A., and Block, J. M. 2018. "Sapucaia nut (*Lecythis pisonis* Cambess) and its by-products: A promising and underutilized source of bioactive compounds. Part II: Phenolic compounds profile." *Food Research International*, 112: 434–442.

Diniz, F. V., de Lima, Y. M. M., da Paz, F. S., da Silva, A. L. D., Gomes, L. C., Santos, G. S., and Carvalho, C. M. 2020. "Enzymatic activity of bacaba (*Oenocarpus bacaba* Mart.) endophytic fungi." *Biota Amazônia*, 10, no. 3: 7–11.

Diniz, V. M. 2016. Juazeiro fruit processing to obtain flour and oil. Doctorate Thesis in Process Engineering. Federal University of Campina Grande, Paraíba, Brazil.

do Espírito Santo, A. P., Silva, R. C., Soares, F. A., Anjos, D., Gioielli, L. A., and Oliveira, M. N. 2010. "Açai pulp addition improves fatty acid profile and probiotic viability in yoghurt." *International Dairy Journal*, 20, no. 6: 415–422.

do Oliveira, M. S. P., de Carvalho, J. E. U., de Nascimento, W. M. O., and Muller, C. H. 2002. Cultivation of açaí palm for fruit production. Belém: Embrapa Eastern Amazon. 18p. (Embrapa Amazônia Oriental. Circular Técnica, 26). Accessed on October 12, 2021. https://www.agencia.cnptia.embrapa.br/Repositorio/Producaodefrutos+Circ_tec_26_000gbz56rpu02wx5ok01dx91cobm2bes.pdf

do Rufino, M. S. M., Pérez-Jiménez, J., Arranz, S., Alves, R. E., de Brito, E. S., Oliveira, M. S. P., and Saura-Calixto, F. 2011. "Açaí (*Euterpe oleraceae*) "BRS Pará": A tropical fruit source of antioxidant dietary fiber and high antioxidant capacity oil." *Food Research International*, 44, no. 7: 2100–2106.

dos Santos, A. F., Júnior, C. C., and Neves, E. J. M. 2008. *Palm Trees for Palm Heart Production: Juçara, Pupunheira and Palmeira Real*. Columbus: Embrapa Forests, p. 188.

Donado-Pestana, C. M., Belchior, T., and Genovese, M. I. 2015. "Phenolic compounds from cagaita (*Eugenia dysenterica* DC.) fruit prevent body weight and fat mass gain induced by a high-fat, high-sucrose diet." *Food Research International*, 77: 177–185. https://doi.org/10.1016/J.FOODRES.2015.06.044

Donado-Pestana, C. M., Moura, M. H. C., de Araujo, R. L., de Lima Santiago, G., de Moraes Barros, H. R., and Genovese, M. I. 2018. "Polyphenols from Brazilian native Myrtaceae fruits and their potential health benefits against obesity and its associated complications." *Current Opinion in Food Science*, 19: 42–49.

dos Santos Cruxen, C. E., Hoffmann, J. F., Zandoná, G. P., Fiorentini, Â. M., Rombaldi, C. V., and Chaves, F. C. 2017. "Probiotic butiá (*Butia odorata*) ice cream: Development, characterization, stability of bioactive compounds, and viability of *Bifidobacterium lactis* during storage." *LWT*, 75: 379–385. https://doi.org/10.1016/j.lwt.2016.09.011

dos Santos Neto, J. P., dos Santos, Y. M. P., Cunha, P. S. J., do Moraes Gatti, V. C., da Silva Barata, H., Nunes, G. G. T., and de Santana, M. C. C. B. 2021. "Functional liquor based on Umbu (Spondias tuberosa Arruda) and Microalgae (Spirulina spp.)." *Research, Society and Development*, 10, no. 3: e42010313557. http://doi.org/10.33448/rsd-v10i3.13557

Dutra, R. L. T., Dantas, A. M., de Araújo Marques, D., Batista, J. D. F., de Albuquerque Meireles, B. R. L., de Magalhães Cordeiro, Â. M. T., and Borges, G. D. S. C. 2017. "Bioaccessibility and antioxidant activity of phenolic compounds in frozen pulps of brazilian exotic fruits exposed to simulated gastrointestinal conditions." *Food Research International*, 100: 650–657.

Falcão, M. D. A., Galvão, R. D. M. S., Clement, C. R., Ferreira, S. A. D. N., and Sampaio, S. D. G. 2000. "Phenology and productivity of araçá-boi (*Eugenia stipitata*, Myrtaceae) in Central Amazonia." *Acta Amazonica*, 30, no. 1: 9. https://doi.org/10.1590/1809-43922000301021

Fernandez-Barbero, G., Pinedo, C., Espada-Bellido, E., Ferreiro-González, M., Carrera, C., Palma, M., and Garcia-Barroso, C. 2019. "Optimization of ultrasound-assisted extraction of bioactive compounds from jabuticaba (*Myrciaria cauliflora*) fruit through a Box-Behnken experimental design." *Food Science and Technology*, 39: 1018–1029.

Ferreira, L. R., Fontes, E. A. F., Marinho, L. M. G., de Barros, F. A. R., Stringheta, P. C., and Ramos, A. M. 2021. "Elaboration, characterization and color stability of an isotonic beverage based on whey permeate with carotenoid powder from pequi." *Research, Society and Development*, 10, no. 8: e41810817233. https://doi.org/10.33448/rsd-v10i8.17233

Ferreira, M. G. R., and Medeiros, M. S. 2005. Bacaca (*Oenocarpus bacaba* Mart.). embrapa, Porto Velho (Technical information). Accessed on October 7, 2021. https://ainfo.cnptia.embrapa.br/digital/bitstream/item/24783/1/folder-bacaca.pdf

Ferreira, T. D. A. 2015. Development of a fermented milk beverage flavored pitanga (*Eugenia uniflora* L.) with probiotic and symbiotic characteristics. Dissertation (Postgraduate Program in Food Science and Technology). Universidade Federal Rural de Pernambuco, Brazil.

Fioravante, M. B., Hiane, P. A., and Braga Neto, J. A. 2017. "Elaboration, sensorial acceptance and characterization of fermented flavored drink based on water-soluble extract of baru almond." *Ciência Rural*, 47, no. 9: 5–10. https://doi.org/10.1590/0103-8478cr20151646

Fraga, L. N., de Souza Oliveira, A. K., Aragão, B. P., de Souza, D. A., Dos Santos, E. W. P., Melo, J. A., and de Carvalho, I. M. M. 2021. "Mass spectrometry characterization, antioxidant activity, and cytotoxicity of the peel and pulp extracts of Pitomba." *Food Chemistry*, 340: 127929.

Freitas, H. V., Dos Santos Filho, A. L., Rodrigues, S., Abreu, V. K. G., Narain, N., Lemos, T. D. O., Gomes, W. F., and Pereira, A. L. F. 2021. "Synbiotic açaí juice (Euterpe oleracea) containing sucralose as noncaloric sweetener: Processing optimization, bioactive compounds, and acceptance during storage." *Journal of Food Science*, 86: 730–739. https://doi.org.ez78.periodicos.capes.gov.br/10.1111/1750-3841.15617

Fujita, A., Sarkar, D., Genovese, M. I., and Shetty, K. 2017. "Improving anti-hyperglycemic and anti-hypertensive properties of camu-camu (*Myriciaria dubia* Mc. Vaugh) using lactic acid bacterial fermentation." *Process Biochemistry*, 59: 133–140. https://doi.org/10.1016/j.procbio.2017.05.017

Fujita, A., Sarkar, D., Wu, S., Kennelly, E., Shetty, K., and Genovese, M. I. 2015. "Evaluation of phenolic-linked bioactives of camu-camu (*Myrciaria dubia* Mc. Vaugh) for antihyperglycemia, antihypertension, antimicrobial properties and cellular rejuvenation." *Food Research International*, 77: 194–203. https://doi.org/10.1016/j.foodres.2015.07.009

Gasca, C. A., Castillo, W. O., Takahashi, C. S., Fagg, C. W., Magalhães, P. O., Fonseca-Bazzo, Y. M., and Silveira, D. 2017. "Assessment of anti-cholinesterase activity and cytotoxicity of cagaita (*Eugenia dysenterica*) leaves." *Food and Chemical Toxicology*, 109: 996–1002.

Gentil, D. F. O., and C. R. Clement. 1997. "The Araza (Eugenia stipitata): Results and directions." *Acta Horticulturae*, 452: 9–17.

Gibbert, L., Sereno, A. B., de Andrade, M. T. P., da Silva, M. A. B., Miguel, M. D., Montrucchio, D. P., and Dias, J. D. F. G. 2021. "Nutritional composition, antioxidant activity and anticancer potential of *Syzygium cumini* (L.) and *Syzygium malaccense* (L.) fruits." *Research, Society and Development*, 10, no. 4: e5210413743.

Girardelo, J. R., Munari, E. L., Dallorsoleta, J. C., Cechinel, G., Goetten, A. L., Sales, L. R., and Conterato, G. M. 2020. "Bioactive compounds, antioxidant capacity and antitumoral activity of ethanolic extracts from fruits and seeds of *Eugenia involucrata* DC." *Food Research International*, 137: 109615.

González, O. J., González-Pérez, J. E., Mejía-Garibay, B., López-Malo, A. and Guerrero-Beltran, J. A. 2023. Caramel colour pigments from black sapote (*Diospyros digyna*): Obtention and food application. *Sustainable Food Technology*, https://doi.org/10.1039/D2FB00037G

Gualberto, N. C., de Oliveira, C. S., Nogueira, J. P., de Jesus, M. S., Araujo, H. C. S., Rajan, M., and Narain, N. 2021. "Bioactive compounds and antioxidant activities in the agro-industrial residues of acerola (*Malpighia emarginata* L.), guava (*Psidium guajava* L.), genipap (*Genipa americana* L.) and umbu (*Spondias tuberosa* L.) fruits assisted by ultrasonic or shaker extraction." *Food Research International*, 147: 110538.

Guergoletto, K. B., Mauro, C. S. I., and Garcia, S. 2017a. "Juçara (*Euterpe edulis*) pulp as a substrate for probiotic bacteria fermentation: Optimisation process and antioxidant activity." *Emirates Journal of Food and Agriculture*, 29, no. 12: 949–955.

Guergoletto, K. B., Busanello, M., and Garcia, S. 2017b. "Influence of carrier agents on the survival of *Lactobacillus reuteri* LR92 and the physicochemical properties of fermented juçara pulp produced by spray drying." *LWT*, 80: 321–327. https://doi.org/10.1016/j.lwt.2017.02.038

Guimarães, A. C. G. 2013. Functional and nutritional potential of jerivá (*Syagrus romanzoffiana*) and bacaba (*Oenocarpus bacaba*) flour. Dissertation (Food Science). Universidade Federal de Lavras, Lavras, Brazil.

Guzzo da Silva, B., Foglio, M. A., Vieira e Rosa, P. D. T., Taranto, O. P., and Frattini Fileti, A. M. 2019. "Optimization of hydrodistillation and in vitro anticancer activity of essential oil from *Schinus terebinthifolius* Raddi fruits." *Chemical Engineering Communications*, 206, no. 5: 619–629. https://doi.org/10.1080/00986445.2018.1515074

Henderson, A., Galeano, G., and Bernal, R. 1995. *Field Guide to the Palms of the Americas*. New Jersey: Princeton University Press, pp. 1–351.

IBGE (Brazilian Institute of Geography and Statistics). 2020. *Census 2019: Biomes and Coastal-Marine System*. Rio de Janeiro: IBGE.

Inada, K. O. P., Nunes, S., Martínez-Blázquez, J. A., Tomás-Barberán, F. A., Perrone, D., and Monteiro, M. 2019. "Effect of high hydrostatic pressure and drying methods on phenolic compounds profile of jabuticaba (*Myrciaria jaboticaba*) peel and seed." *Food Chemistry*, 309: 125794.

Infante, J., Rosalen, P. L., Lazarini, J. G., Franchin, M., and Alencar, S. M. D. 2016. "Antioxidant and anti-inflammatory activities of unexplored Brazilian native fruits." *PLoS One*, 11, no. 4: e0152974.

Jesus, N. D., Martins, A. B. G., Almeida, E. J. D., Leite, J. B. V., Ganga, R. M. D., Scaloppi Junior, E. J., and Moreira, R. F. C. 2004. "Characterization of four groups of jabuticabeira, under the conditions of Jaboticabal-SP." *Revista Brasileira de Fruticultura*, 26: 482–485.

Jorge, M. H. A., and Borsato, A. V. 2009. *Embrapa Pantanal Actions in Medicinal Plants in the Region of Corumbá-MS*. Corumbá: Brazilian Agricultural Research Corporation—Embrapa Pantanal, p. 3 (ADM, 137).

Kaprasob, R., Kerdchoechuen, O., Laohakunjit, N., Sarkar, D., and Shetty, K. 2017. "Fermentation-based biotransformation of bioactive phenolics and volatile compounds from cashew apple juice by select lactic acid bacteria." *Process Biochemistry*, 59: 141–149. https://doi.org/10.1016/j.procbio.2017.05.019

Khouri, J., Resck, I. S., Poças-Fonseca, M., Sousa, T. M. M., Pereira, L. O., Oliveira, A. B. B., and Grisolia, C. K. 2007. "Anticlastogenic potential and antioxidant effects of an aqueous extract of pulp from the Pequi tree (Caryocar brasiliense Camb)." *Genetics and Molecular Biology*, 30, no. 2: 442–448. https://doi.org/10.1590/s1415-47572007000300024

Kim, S. W., Ko, M. J., and Chung, M. S. 2019. "Extraction of the flavonol quercetin from onion waste by combined treatment with intense pulsed light and subcritical water extraction." *Journal of Cleaner Production*, 231: 1192–1199.

Klein, E. J., Santos, K. A., Palu, F., Vieira, M. G. A., and da Silva, E. A. 2018. "Use of supercritical CO_2 and ultrasound-assisted extractions to obtain α/β-amyrin-rich extracts from uvaia leaves (*Eugenia pyriformis* Cambess.)." *The Journal of Supercritical Fluids*, 137: 1–8.

Köhler, M., and Brack, P. 2016. "Frutas nativas no Rio Grande do Sul: cultivando e valorizando a diversidade." *Revista Agriculturas*, 13, no. 2.

Kumar, S., Sharma, S., and Chattopadhyay, S. K. 2013. "The potential health benefit of poly-isoprenylated benzophenones from Garcinia and related genera: Ethnobotanical and therapeutic importance." *Fitoterapia*, 89: 86–125. https://doi.org/10.1016/j.fitote.2013.05.010

Lade, B. D., Patil, A. S., and Paikrao, H. M. 2014. "Systematic optimization of thin-layer chromatography for wound-induced differential secondary metabolites in *Passiflora foetida*." *Journal of Planar Chromatography-Modern TLC*, 27, no. 5: 385–392.

Lazarini, J. G., Franchin, M., Soares, J. C., Nani, B. D., Massarioli, A. P., de Alencar, S. M., and Rosalen, P. L. 2020. "Anti-inflammatory and antioxidant potential, in vivo toxicity, and polyphenolic composition of *Eugenia selloi* BD Jacks. (pitangatuba), a Brazilian native fruit." *PLoS One*, 15, no. 6: e0234157.

Leite-Legatti, A. V., Batista, Â. G., Dragano, N. R. V., Marques, A. C., Malta, L. G., Riccio, M. F., and Maróstica, M. R. 2012. "Jaboticaba peel: Antioxidant compounds, antiproliferative and antimutagenic activities." *Food Research International*, 49, no. 1: 596–603.

Lima, A. D. J. B., Corrêa, A. D., Alves, A. P. C., Abreu, C. M. P., and Dantas-Barros, A. M. 2008. "Chemical characterization of jabuticaba fruit (*Myrciaria cauliflora* Berg) and its fractions." *Archivos Latinoamericanos de Nutricion*, 58, no. 4: 416–421.

Lima, O. G. (1975). Pulque, Balchê and Pajauaru in the ethnobiology of beverages and fermented foods. Universidade Federal de Pernambuco, São Paulo. Official State Press.

Lima, T. B., Silva, O. N., Oliveira, J. T. A., Vasconcelos, I. M., Scalabrin, F. B., Rocha, T. L., and Franco, O. L. 2010. "Identification of *E. dysenterica* laxative peptide: A novel strategy in the treatment of chronic constipation and irritable bowel syndrome." *Peptides*, 31, no. 8: 1426–1433. https://doi.org/10.1016/J.PEPTIDES.2010.05.003

Lin, S. T., Tu, S. H., Yang, P. S., Hsu, S. P., Lee, W. H., Ho, C. T., and Chen, L. C. 2016. "Apple polyphenol phloretin inhibits colorectal cancer cell growth via inhibition of the type 2 glucose transporter and activation of p53-mediated signaling." *Journal of Agricultural and Food Chemistry*, 64, no. 36: 6826–6837.

Lisboa, M. C., Wiltshire, F. M. S., Fricks, A. T., Dariva, C., Carrière, F., Lima, Á. S., and Soares, C. M. F. 2020. "Oleochemistry potential from Brazil northeastern exotic plants." *Biochimie*, 178: 96–104. https://doi.org/10.1016/j.biochi.2020.09.002

Lopes, R. P., Pereira, J. C., Kerber, L., and Dillenburg, S. R. 2020. "The extinction of the Pleistocene megafauna in the Pampa of southern Brazil." *Quaternary Science Reviews*, 242: 106428. https://doi.org/10.1016/j.quascirev.2020.106428

Lorenzi H. 2004. *Brazilian and Exotic Cultivated Palms*. Nova Odessa, Brazil: Plantarum, pp. 1–416.

Luciano, W. A., Matte, T. C., Portela, I. A., de Medeiros, L. L., dos Santos Lima, M., Maciel, J. F., and Magnani, M. 2018. "Effects of *Lactobacillus acidophilus* LA-3 on physicochemical and sensory parameters of açaí and mango based smoothies and its survival following simulated gastrointestinal conditions." *Food Research International*, 114: 159–168.

Macêdo, E. D. L. C., Pimentel, T. C., de Sousa Melo, D., de Souza, A. C., de Morais, J. S., dos Santos Lima, M., and Magnani, M. 2023. "Yeasts from fermented Brazilian fruits as biotechnological tools for increasing phenolics bioaccessibility and improving the volatile profile in derived pulps." *Food Chemistry*, 401: 134200.

Machado, M. T. C., Mello, B. C. B. S., and Hubinger, M. D. 2015. "Evaluation of pequi (*Caryocar brasiliense* Camb.) aqueous extract quality processed by membranes." *Food and Bioproducts Processing*, 95: 304–312. https://doi.org/10.1016/j.fbp.2014.10.013

Machado, M. T. C., Trevisan, S., Pimentel-Souza, J. D. R., Pastore, G. M., and Hubinger, M. D. 2016. "Clarification and concentration of oligosaccharides from artichoke extract by a sequential process with microfiltration and nanofiltration membranes." *Journal of Food Engineering*, 180: 120–128. https://doi.org/10.1016/j.jfoodeng.2016.02.018

Magnusson, W. E., de Holanda, A. S. S., de Freitas, M. A., Ramalho, E. E., Akama, A., Ferreira, L., Menin, M., Nunez, C. V., de Rodrigues, D. J., Manzatto, A. G., de Paggoto, R. C., and Ishikawa, N. K. 2016. Amazônia biodiversidade incontável. In: Peixoto, A. L., Luz, J. R. P., and de Brito, M. A. (Org.), *Conhecendo a Biodiversidade. Brasília: Ministério da Ciência, Tecnologia, Inovações e Comunicações*. CNPq, PPBio, pp. 112–123.

Martins, D. R. S., Sanches, K. F. M., Sanjines-Argandonã, E. J., and Tobal, T. M. 2020. "Grumixama (*Eugenia brasiliensis*) and jambolan (*Eugenia jambolana*): Source of bioactive compounds and viability in the preparation of jams and yogurts." *Research, Society and Development*, 9, no. 7: 1–16.

Massa, N. M. L., Menezes, F. N. D. D., de Albuquerque, T. M. R., de Oliveira, S. P. A., dos Lima, M. S., Magnani, M., and de Souza, E. L. 2020. "Effects of digested jabuticaba (*Myrciaria jaboticaba* (Vell.) Berg) by-product on growth and metabolism of *Lactobacillus* and *Bifidobacterium* indicate prebiotic properties." *LWT*, 131: 109766.

Maués, M. M., and Couturier, G. 2002. "Biologia floral e fenologia reprodutiva do camu-camu (*Myrciaria dubia* (H.B.K.) McVaugh, Myrtaceae) no Estado Pará, Brasil." *Revista Brasileira de Botânica*, 25, no. 4: 441–448.

Mauro, C. S. I., Hassani, M. K., Barone, M., Esposito, M. T., Calle, Y., Behrends, V., Garcia, S., Brigidi, P., Turroni, S., and Costabile, A. 2022. "Cerrado and Pantanal fruit flours affect gut microbiota composition in healthy and post-COVID-19 individuals: An *in vitro* pilot fermentation study." *International Journal of Food Science and Technology*, 58, no. 8: 4495–4510. https://doi.org/10.1111/ijfs.16274

Mazine, F. F., Valdemarin, K. S., Bünger, M., Faria, J. E. Q., Fernandes, T., Giaretta, A., Santana, K. C., Sobral, M., and Souza, M. A. D. 2020. *Eugenia* in Flora do Brasil 2020. Jardim Botânico do Rio de Janeiro. Accessed on October 7, 2021. http://reflora.jbrj.gov.br/reflora/floradobrasil/FB10499

Melo, E. D. A., and Andrade, R. D. S. 2010. "Bioactive compounds and antioxidant potential from the "umbuzeiro" fruits." *Alimentos e Nutrição*, 21, no. 3: 453–457.

Mendes, M. R. A., Munhoz, C. B. R., Silva Júnior, M. C., and Castro, A. A. J. F. 2012. "Vegetation and soil relationship in moist grassland in the National Park of Sete Cidades, Piauí, Brazil." *Rodriguésia*, 63, no. 4: 971–984. http://doi.org/10.1590/S2175-78602012000400014

Mendes, R. M. L., de Andrade, R. H. C., Marques, M. D. F. F., and de Andrade, E. R. 2021a. "Potential use of the passion fruit from caatinga in kefir." *Food Bioscience*, 39: 100809. https://doi.org/10.1016/j.fbio.2020.100809

Mendes, Y. C., Mesquita, G. P., Costa, G. D. E., Barbosa da Silva, A. C., Gouveia, E., Silva, M. R. C., and Zagmignan, A. 2021b. "Evaluation of growth, viability, lactic acid production and anti-infective effects of *Lacticaseibacillus rhamnosus* ATCC 9595 in Bacuri Juice (*Platonia insignis*)." *Foods*, 10, no. 3: 603. https://doi.org/10.3390/foods10030603

Mercali, G. D., Gurak, P. D., Schmitz, F., and Marczak, L. D. F. 2015. "Evaluation of non-thermal effects of electricity on anthocyanin degradation during ohmic heating of jaboticaba (*Myrciaria cauliflora*) juice." *Food Chemistry*, 171: 200–205. https://doi.org/10.1016/j.foodchem.2014.09.006

Meregalli, M. M., Puton, B. M. S., Camera, F. D. M., Amaral, A. U., Zeni, J., Cansian, R. L., and Backes, G. T. 2020. "Conventional and ultrasound-assisted methods for extraction of bioactive compounds from red araçá peel (*Psidium cattleianum* Sabine)." *Arabian Journal of Chemistry*, 13, no. 6: 5800–5809.

Miranda, J. S., Costa, B. V., de Oliveira, I. V., de Lima, D. C. N., Martins, E. M. F., de Júnior, B. R. C., do Benevenuto, W. C. A. N., de Queiroz, I. C., da Silva, R. R., and Martins, M. L. 2020. "Probiotic jelly candies enriched with native Atlantic Forest fruits and *Bacillus coagulans* GBI-30 6086." *LWT*, 126: 109275.

Morais, R. A., Teixeira, G. L., Ferreira, S. R. S., Cifuentes, A., and Block, J. M. 2022. "Nutritional composition and bioactive compounds of native Brazilian fruits of the arecaceae family and its potential applications for health promotion." *Nutrients*, 14: 4009. https://doi.org/10.3390/nu14194009

Moreno, P. R. H., Lima, M. E. L., Sobral, M., Young, M. C. M., Cordeiro, I., Apel, M. A., and Henriques, A. T. 2007. "Essential oil composition of fruit colour varieties of *Eugenia brasiliensis* Lam." *Scientia Agricola*, 64: 428–432. https://doi.org/10.1590/S0103-90162007000400014

Nascimento, L. S. M., Santiago, M. C. P. A., Oliveira, E. M. M., Borguini, R. G., Braga, E. C. O., Martins, V. C., Pacheco, S., Souza, M. C., and Gogoy, R. L. O. 2017. "Characterization of bioactive compounds in *Eugenia brasiliensis*, Lam. (Grumixama)." *Nutrition and Food Technology: Open Access*, 3, no. 3: 1–7. http://doi.org/10.16966/2470-6086.146

Naves, R. V., Borges, J. D., and Chaves, L. J. 2002. "A Cagaiteira." *Revista Brasileira de Fruticultura*, 24, no. 2: 289–596. https://doi.org/10.1590/S0100-29452002000200001

Neri-Numa, I. A., Carvalho-Silva, L. B., Morales, J. P., Malta, L. G., Muramoto, M. T., Ferreira, J. E. M., and Pastore, G. M. 2013. "Evaluation of the antioxidant, antiproliferative and antimutagenic potential of araçá-boi fruit (*Eugenia stipitata* Mc Vaugh—Myrtaceae) of the Brazilian Amazon Forest." *Food Research International*, 50, no. 1: 70–76.

Neri-Numa, I. A., Sancho, R. A. S., Pereira, A. P. A., and Pastore, G. M. 2018"Small Brazilian wild fruits: Nutrients, bioactive compounds, health-promotion properties and commercial interest." *Food Research International*, 103: 345–360. https://doi.org/10.1016/j.foodres.2017.10.053

Oliveira, A. S., Santos, D. C., de Oliveira, E. N. A., da Silva, F. L. H., and Florentino, E. R. 2011a. "Production of alcoholic fermented fruit of thornless mandacaru (*Cereus jamacaru*)." *Revista Brasileira de Produtos Agroindustriais, Campina Grande*, 13, no 3: 271–277.

Oliveira, M. E. S., Pantoja, L., Duarte, W. F., Collela, C. F., Valarelli, L. T., Schwan, R. F., and Dias, D. R. 2011b. "Fruit wine produced from cagaita (*Eugenia dysenterica* DC) by both free and immobilised yeast cell fermentation." *Food Research International*, 44, no. 7: 2391–2400. https://doi.org/10.1016/j.foodres.2011.02.028

Oliveira, P. M. L., Dantas, A. M., Morais, A. R. S., Gibbert, L., Krüger, C. C. H., Lima, M. S., Magnani, M., and Borges, G. S. C. 2020. "Juá fruit (*Ziziphus joazeiro*) from Caatinga: A source of dietary fiber and bioaccessible flavanols." *Food Research International*, 129: 108745.

Oroian, M., Dranca, F., and Ursachi, F. 2020b. "Comparative evaluation of maceration, microwave and ultrasonic-assisted extraction of phenolic compounds from propolis." *Journal of Food Science and Technology*, 57, no. 1: 70–78.

Oroian, M., Ursachi, F., and Dranca, F. 2020a. "Influence of ultrasonic amplitude, temperature, time and solvent concentration on bioactive compounds extraction from propolis." *Ultrasonics Sonochemistry*, 64: 105021.

Oscar, J. G., Julio, G. P., Beatriz, M. G., Aurelio, L. M., and Ángel, G. B. J. 2023. "Caramel colour pigments from black sapote (*Diospyros digyna*): Obtention and food application." *Sustainable Food Technology*, 1, no. 4: 555–566. https://doi.org/10.1039/D2FB00037G

Passos, M. A. B. 2023. "Plantas Alimentícias Não Convencionais (PANC) no Estado do Maranhão, Brasil." *Revista Foco*, 16, no. 3: e1380. https://doi.org/10.54751/revistafoco.v16n3-096

Pedro, A. C., Granato, D., and Rosso, N. D. 2016. "Extraction of anthocyanins and polyphenols from black rice (*Oryza sativa* L.) by modeling and assessing their reversibility and stability." *Food Chemistry*, 191: 12–20.

Pedro, A. C., Maurer, J. B. B., Zawadzki-Baggio, S. F., Ávila, S., Maciel, G. M., and Haminiuk, C. W. I. 2018. "Bioactive compounds of organic goji berry (*Lycium barbarum* L.) prevents oxidative deterioration of soybean oil." *Industrial Crops and Products*, 112: 90–97.

Pereira, A. L. F., Maciel, T. C., and Rodrigues, S. 2011. "Probiotic beverage from cashew apple juice fermented with *Lactobacillus casei*." *Food Research International*, 44, no. 5: 1276–1283. https://doi.org/10.1016/j.foodres.2010.11.035

Peyrat, L. A., Eparvier, V., Eydoux, C., Guillemot, J. C., Stien, D., and Litaudon, M. 2016. "Chemical diversity and antiviral potential in the Pantropical Diospyros genus." *Fitoterapia*, 112: 9–15.

Plaza, M., Batista, Â. G., Cazarin, C. B. B., Sandahl, M., Turner, C., Östman, E., and Júnior, M. R. M. 2016. "Characterization of antioxidant polyphenols from *Myrciaria jaboticaba* peel and their effects on glucose metabolism and antioxidant status: A pilot clinical study." *Food Chemistry*, 211: 185–197.

Pott, A., Oliveira, A. K., Damasceno-Junior, G. A., and Silva, J. S. 2011. "Plant diversity of the Pantanal wetland." *Brazilian Journal of Biology*, 71: 265–273.

Prestes, A. A., Verruck, S., Vargas M. O., Canella, M. H. M., Silva, C. C., da Barros, E. L. S., Dantas, A., de Oliveira, L. V. A., Maran, B. M., Matos, M., Helm, C. V., and Prudencio, E. S. 2021. "Influence of guabiroba pulp (*Campomanesia xanthocarpa* o. berg) added to fermented milk on probiotic survival under in vitro simulated gastrointestinal conditions." *Food Research International*, 141: 110135. https://doi.org/10.1016/j.foodres.2021.110135

Puerari, C. 2014. Microbial diversity and physicochemical characterization of beverages produced by indigenous Umutina—Brazil: rice and bacaba chicha. Doctoral Thesis, Universidade Federal de Lavras, Lavras, Brazil.

Puerari, C., Magalhães-Guedes, K. T., and Schwan, R. F. 2015. "Bacaba beverage produced by Umutina Brazilian Amerindians: Microbiological and chemical characterization." *Brazilian Journal of Microbiology*, 46: 1207–1216. https://doi.org/10.1590/S1517–838246420140964

Queiroz, M. S. D. M., and Bianco, R. 2009. "Morphology and germination development of *Oenocarpus bacaba* Mart. (Arecaceae) from the western Amazon." *Revista Árvore*, 33: 1037–1042. https://doi.org/10.1590/S0100–67622009000600006

Reis, A. F., and Schmiele, M. 2019. "Características e potencialidades dos frutos do Cerrado na indústria de alimentos." *Brazilian Journal of Food Technology*, 22: e2017150. https://doi.org/10.1590/1981–6723.15017

Resende, L. V., Pinheiro, L. K., Miguel, M. G. D. C. P., Ramos, C. L., Vilela, D. M., and Schwan, R. F. 2018. "Microbial community and physicochemical dynamics during the production of 'Chicha', a traditional beverage of Indigenous people of Brazil." *World Journal of Microbiology and Biotechnology*, 34, no. 3: 1–11. https://doi.org/10.1007/s11274-018-2429-4

Rodrigues, L. F. 2019. Development of sweet breads with added nutritional and functional value from mangaba pulp and flour (*Hancornia speciosa* Gomes) and pequi peel flour (*Caryocar brasiliense* Camb.). Dissertation, Universidade Federal de Lavras, Lavras, Brazil.

Rodrigues, L. M., Romanini, E. B., Silva, E., Pilau, E. J., da Costa, S. C., and Madrona, G. S. 2020. "Camu-camu bioactive compounds extraction by ecofriendly sequential processes (ultrasound assisted extraction and reverse osmosis)." *Ultrasonics Sonochemistry*, 64: 105017.

Rogez, H., Buxant, R., Mignolet, E., Souza, J. N., Silva, E. M., and Larondelle, Y. 2004. Chemical composition of the pulp of three typical Amazonian fruits: araça-boi (*Eugenia stipitata*), bacuri (*Platonia insignis*) and cupuaçu (*Theobroma grandiflorum*). *European Food Research and Technology*, 218, no. 4: 380–384. https://doi.org/10.1007/s00217-003-0853-6

Rossi, A. A., Gomes, A., Silveira, G., Ramalho, A., and Barbosa, R. 2014. "Morphological characterization of fruits and seeds of *Mauritia flexuosa* L. F. (Arecaceae) naturally occurring in Mato Grosso Amazonia." *Enciclopédia Biosfera*, 10, no. 18: 852–862.

Rosso, V. V., and Mercadante, A. Z. 2007. "Identification and quantification of carotenoids, by HPLC-PDA-MS/MS from Amazonian fruits." *Journal of Agricultural and Food Chemistry*, 55, no. 13: 5062–5072.

Rotta, E. M., Rodrigues, C. A., Jardim, I. C. S. F., Maldaner, L., and Visentainer, J. V. 2019. "Determination of phenolic compounds and antioxidant activity in passion fruit pulp (*Passiflora* spp.) using a modified QuEChERS method and UHPLCMS/MS." *Food Science Technology*, 100: 397–403.

Rudke, A. R., Andrade, K. S., Mazzutti, S., Zielinski, A. A. F., Alves, V. R., Vitali, L., and Ferreira, S. R. S. 2021. "A comparative study of phenolic compounds profile and in vitro antioxidant activity from buriti (*Mauritia flexuosa*) by-products extracts." *LWT*: 111941.

Sabidi, S., Koh, S. P., Abd Shukor, S., Adzni Sharifudin, S., and Sew, Y. S. 2020. "Safety assessment of fermented jackfruit (*Artocarpus heterophyllus*) pulp and leaves in Sprague-Dawley rats." *Food Science & Nutrition*, 8: 4370–4378.

Sagdic, O., Ozturk, I., Yapar, N., and Yetim, H. 2014. "Diversity and probiotic potentials of lactic acid bacteria isolated from gilaburu, a traditional Turkish fermented European cranberry bush (*Viburnum opulus* L.) fruit drink." *Food Research International*, 64: 537–545.

Sahu, A., and Samal, K. C. 2020. "Potential application of Cashew Apple (*Anacardium occidentale* L.) as a probiotic beverage fermented with Lactobacillus." *Vigyan Varta*, 1, no. 6: 48–54.

Saleem, F., Sarkar, D., Ankolekar, C., and Shetty, K. 2017. "Phenolic bioactives and associated antioxidant and anti-hyperglycemic functions of select species of Apiaceae family targeting for type 2 diabetes relevant nutraceuticals." *Industrial Crops and Products*, 107: 518–525.

Sampaio, M. B., and Carraza, L. R. 2012. Technological manual for the integral use of the fruit of the buriti leaf (*Mauritia flexuosa*). Society, Population and Nature Institute (ISPN), Brasília-DF. Technological manual No. 4. Accessed on October 12, 2021. https://ispn.org.br/site/wp-content/uploads/2018/10/ManualTecnologicoBuriti.pdf

Santana, J. F. D. S. 2019. Evaluation of Cagaita (*Eugenia dysenterica*) fruits for acetic fermented production. Masters Dissertation (In Agrochemistry), Federal Institute of Education, Science and Technology, Rio Verde, Brazil.

Santana, R. V., Santos, D. C., Santana, A. C. A., Oliveira Filho, J. G., Almeida, A. B., Lima, T. M., Silva, F. G., and Egea, M. B. 2020. "Quality parameters and sensorial profile of clarified "Cerrado" cashew juice supplemented with *Saccharomyces boulardii* and different sweeteners." *LWT*, 128: 109319. https://doi.org/10.1016/j.lwt.2020.109319

Santos, C. T., Marques, G. M. R., Fontan, G. C. R., Fontan, R. D. C. I., Bonomo, R. C. F., and Bonomo, P. 2006. "Elaboration and characterization of a milky drink fermented with Umbu pulp (*Spondias tuberosa* sp.)." *Revista Brasileira de Produtos Agroindustriais*, 8, no. 2: 111–116.

Santos, E., Andrade, R., and Gouveia, E. 2017. "Utilization of the pectin and pulp of the passion fruit from Caatinga as probiotic food carriers." *Food Bioscience*, 20: 56–61. https://doi.org/10.1016/j.fbio.2017.08.005

Santos, P. H., Ribeiro, D. H. B., Micke, G. A., Vitali, L., and Hense, H. 2019. "Extraction of bioactive compounds from feijoa (*Acca sellowiana* (O. Berg) Burret) peel by low and high-pressure techniques." *The Journal of Supercritical Fluids*, 145: 219–227.

Santos, R. T. S., Biasoto, A. C. T., Rybka, A. C. P., Castro, C. D. P. C., Aidar, S. T., Borges, G. S. C., and Silva, F. L. H. 2021a. "Physicochemical characterization, bioactive compounds, *in vitro* antioxidant activity, sensory profile and consumer acceptability of fermented alcoholic beverage obtained from Caatinga passion fruit (*Passiflora cincinnata* Mast.)." *LWT*, 148: 111714.

Santos, T. B., de Araújo, F. P., Neto, A. F., de Freitas, S. T., de Araújo, J. S., de Vilar, S. B. O., Araújo, A. J. B., and Lima, M. S. 2021b. "Phytochemical compounds and antioxidant activity of the pulp of two Brazilian passion fruit species: *Passiflora cincinnata* Mast. and *Passiflora edulis* Sims." *International Journal of Fruit Science*, 22: 1–15.

Sayago-Ayerdi, S., García-Martínez, D. L., Ramírez-Castillo, A. C., Ramírez-Concepción, H. R., and Viuda-Martos, M. 2021. "Tropical fruits and their co-products as bioactive compounds and their health effects: A review." *Foods*, 10, no. 8: 1952.

Schulz, M., Gonzaga, L. V., de Souza, V., Farina, M., Vitali, L., Micke, G. A., and Fett, R. 2019. "Neuroprotective effect of juçara (*Euterpe edulis* Martius) fruits extracts against glutamate-induced oxytosis in HT22 hippocampal cells." *Food Research International*, 120: 114–123. https://doi.org/10.1016/j.foodres.2019.02.030

Schulz, M., Seraglio, S. K. T., Brugnerotto, P., Gonzaga, L. V., Costa, A. C. O., and Fett, R. 2020. "Composition and potential health effects of dark-colored underutilized Brazilian fruits: A review." *Food Research International*, 137: 109744. https://doi.org/10.1016/j.foodres.2020.109744

Sharma, B., Balomajumder, C., and Roy, P. 2008. "Hypoglycemic and hypolipidemic effects of flavonoid rich extract from *Eugenia jambolana* seeds on streptozotocin induced diabetic rats." *Food and Chemical Toxicology*, 46, no. 7: 2376–2383.

Shori, A. B., Aljohani, G. S., Al-zahrani, A. J., Al-sulbi, O. S., and Baba, A. S. 2022. "Viability of probiotics and antioxidant activity of cashew milk-based yogurt fermented with selected strains of probiotic *Lactobacillus* spp." *LWT*, 153: 112482. https://doi.org/10.1016/j.lwt.2021.112482

Shukla, S., Kushwaha, R., Singh, M., Saroj, R., Puranik, V., Agarwal, R., and Kaur, D. 2021. "Quantification of bioactive compounds in guava at different ripening stages." *Food Research*, 5, no. 3: 183–189.

Silva, J., de Fátima Moura, M., Silva, T., Araújo, N., and Barbosa, I. 2017. "Physicochemical characterization and proximate and mineral composition of the juazeiro fruit (*Ziziphus joazeiro* Mart.)." *Enciclopédia Biosfera*, 14, no. 25: 291.

Silva, J. C. M., Santana, R. V., Almeida, A. B., Takeuchi, K. P., and Egea, M. B. 2021a. "Changes in the chemical, technological, and microbiological properties of Kefir-fermented soymilk after supplementation with inulin and *Acrocomia aculeata* pulp." *Applied Sciences*, 11, no. 12: 5575. https://doi.org/10.3390/app11125575

Silva, J. S., Ferreira, N. B. S., Asquieri, E. R., Damiani, C., and Asquieri, E. M. A. R. 2021b. "Chemical monitoring of Baru (Dipteryx alata Vog.) pulp fermented beverage." *Food Science and Technology (Campinas)*, 41, no. 1: 155–162. https://doi.org/10.1590/fst.14420

Silva, N., Rodrigues, E., Mercadante, A., and De Rosso, V. 2014. "Phenolic compounds and carotenoids from four fruits native from the Brazilian Atlantic Forest." *Journal of Agricultural and Food Chemistry*, 62, no. 22: 5072–5084.

Silva, V. L. M., Costa, M. P., Vieira, C. P., and Conte-Junior, C. A. 2019. "Short communication: Biogenic amine formation during fermentation in functional sheep milk yogurts." *Journal of Dairy Science*, 102, no. 10: 8704–8709.

Siqueira, A. E. B., Silva, E. P. D., Santiago, E. D. S., Teixeira, É. T. L. B., Mesquita, J. D. A., Cristo, P. F. R. F. D., and Sousa, D. D. A. 2021. "Elaboration of mangaba vinegar by semi-solid fermentation combined with enzymatic activity: Chemical characterization and sensory evaluation." *Ciência Rural*, 51, no. 10. http://doi.org/10.1590/0103-8478cr20200638

Soares, L. M. N., Silva, G. M., Buriti, F. C. A., and Alves, H. S. 2021. "*Cereus jamacaru* D.C. (Mandacaru): A promising native Brazilian fruit as a source of nutrients and bioactives derived from its pulp and skin." *Plant Foods and Human Nutrition*, 76: 170–178.

Soares, L. V., Melo, R., Oliveira, W. S., Souza, P. M., and Schmiele, M. 2017. Chapter 5: Brazilian Cerrado fruits and their potential use in bakery products. In: Lewis, H. (Ed.), *Bread: Consumption, Cultural Significance and Health Effects*. New York: Nova Publisher, pp. 125–160.

Souza, P. L. L., Ramos, A. S., Dos Santos, A. D. C., Boeira, L. S., de Bezerra, J. A., and Machado, M. B. 2022. "Evaluation of sensory and antioxidant properties of araçá-boi wines as an effect of yeast type, must filtration and fermentation temperature." *Chemical Papers*, 76: 3531–3540. https://doi.org/10.1007/s11696-022-02119-x

Stafussa, A. P., Maciel, G. M., Rampazzo, V., Bona, E., Makara, C. N., Junior, B. D., and Haminiuk, C. W. I. 2018. "Bioactive compounds of 44 traditional and exotic Brazilian fruit pulps: Phenolic compounds and antioxidant activity." *International Journal of Food Properties*, 21, no. 1: 106–118.

Stintzing, F. C., Herbach, K. M., Mosshammer, M. R., Carle, R., Yi, W., Sellappan, S., Akoh, C. C., Bunch, R., and Felker, P. 2005. "Color, betalain pattern, and antioxidant properties of Cactus Pear (*Opuntia* spp.) clones." *Journal of Agricultural and Food Chemistry*, 53, no. 2: 442–451.

Suwanwong, Y., and Boonpangrak, S. 2021. "Phytochemical contents, antioxidant activity, and anticancer activity of three common guava cultivars in Thailand." *European Journal of Integrative Medicine*, 42: 101290.

Teixeira, L. D. L., Hassimotto, N. M., and Lajolo, F. M. 2018. Grumixama—*Eugenia brasiliensis* Lam. In: *Exotic Fruits*. Academic Press, pp. 219–224.

Teixeira, N., Melo, J. C., Batista, L. F., Paula-Souza, J., Fronza, P., and Brandao, M. G. 2019. "Edible fruits from Brazilian biodiversity: A review on their sensorial characteristics versus bioactivity as tool to select research." *Food Research International*, 119: 325–348. https://doi.org/10.1016/j.foodres.2019.01.058

Teleginski, F., Zuffellato-Ribas, K. C., Koehler, H. S., Degenhardt-Goldbach, J., and Teleginski, E. 2018. "Resgate vegetativo de Mart. ex O. Berg por Alporquia." *Ciência Florestal*, 28, no. 2: 820–826.

Toledo, M. A., dos Santos, D. A., Cunha, A. H. N., Brasil, E. P. F., Lemes, G. F., and Vieira, J. A. 2018. "Utilization of a continuous flow system to evaluate the parameters of the baru pulp fermentation process for the production of alcoholic beverage." *International Journal of Development Research*, 8, no. 1: 18534–18539.

Tripathy, S., Verma, D. K., Thakur, M., Patel, A. R., Srivastav, P. P., Singh, S., Gupta, A. K., Chávez-González, M. L., Aguilar, C. N., Chakravorty, N., Verma, H. K., and Utama, G. L. 2021. "Curcumin extraction, isolation, quantification and its application in functional foods: A review with a focus on immune enhancement activities and COVID-19." *Frontiers in Nutrition*, 675. https://doi.org/10.3389/fnut.2021.747956

Uekane, T. M., Nicolotti, L., Griglione, A., Bizzo, H. R., Rubiolo, P., Bicchi, C., and Rezende, C. M. 2017. "Studies on the volatile fraction composition of three native Amazonian-Brazilian fruits: Murici (*Byrsonima crassifolia* L., Malpighiaceae), bacuri (*Platonia insignis* M., Clusiaceae), and sapodilla (*Manilkara sapota* L., Sapotaceae)." *Food Chemistry*, 219: 13–22.

van Hijum, S. A., Vaughan, E. E., and Vogel, R. F. 2013. "Application of state-of-art sequencing technologies to indigenous food fermentations." *Current Opinion in Biotechnology*, 24, no. 2: 178–186. https://doi.org/10.1016/j.copbio.2012.08.004

Vencioneck Dutra, J. C., Moisés Ferreira, J., Costalonga Pereira, P. R., Ben-Hur de Oliveira, J., Vitorino Gervásio, S., Bernardes Xavier, M., and Pimentel Batitucci, M. D. C. 2018. "*Cereus jamacaru* DC hydroalcoholic extract promotes anti-cytotoxic and antitumor activity." *Pharmaceuticals*, 11, no. 4: 130. https://doi.org/10.3390/ph11040130

Verma, D. K., Dhakane, J. P., Mahato, D. K., Billoria, S., Bhattacharjee, P., and Srivastav, P. P. 2018. "Supercritical fluid extraction (SCFE) for rice aroma chemicals: Recent and advance extraction method." In: Verma, D. K., and Srivastav, P. P. (Eds.), *Science and Technology of Aroma, Flavour and Fragrance in Rice*. Apple Academic Press, pp. 179–198.

Verma, D. K., Mahato, D. K., Billoria, S., Kapri, M., Prabhakar, P. K., Ajesh Kumar, V., and Srivastav, P. P. 2017. "Microbial approaches in fermentations for production and preservation of different food." In: Verma, D. K., and Srivastav, P. P. (Eds.), *Microorganisms in Sustainable Agriculture, Food and the Environment*, Vol. 1, as part of book series on Innovations in Agricultural Microbiology, Apple Academic Press, pp. 105–142.

Verma, D. K., and Srivastav, P. P. 2020a. "Bioactive compounds of rice (*Oryza sativa* L.): Review on paradigm and its potential benefit in human health." *Trends in Food Science & Technology*, 97: 355–365. https://doi.org/10.1016/j.tifs.2020.01.007

Verma, D. K., and Srivastav, P. P. 2020b. "A paradigm of volatile aroma compounds in rice and their product with extraction and identification methods: A comprehensive review." *Food Research International*, 130: 1–33. https://doi.org/10.1016/j.foodres.2019.108924

Vieira, M. C. 2017. Production of a fermented beverage based on hydrosoluble soy extract added with macauba pulp (*Acrocomia aculeata* (Jacq.) Lodd. ex Mart.) using traditional yogurt dairy culture and probiotic strain. Masters Dissertation (In Food Science), Faculty of Pharmacy, Federal University of Minas Gerais, Belo Horizonte, p. 127. http://hdl.handle.net/1843/33794

Vieira, N. K., Bourscheid, K., Dickel, M. L., Ritter, M. R., and Barros I. B. I. 2011. "*Eugenia uniflora*." In: Coradin, L., Siminski, A., & Reis, A. (Eds.), *Native Species of the Brazilian Flora of Current or Potential Economic Value: Plants for the Future—Southern Region*. Brasília: Ministry of the Environment, cap. 5, pp. 619–628.

Villachica, H. L. 1996. "The cultivation of camu-camu (*Myrciaria dubia* (H.B.K.) McVaugh) in the Peruvian Amazon." *Mirigraf, Lima*, 1996.

Winckler, L. T., and Teixeira Filho, A. 2020. *Proceedings of the 1st Congress on the Pampa Biome*. Pelotas, RS: Editora UFPel, p. 227. ISBN 978-65-86440-04-1.

Xiong, J., Matta, F. V., Grace, M., Lila, M. A., Ward, N. I., Felipe-Sotelo, M., and Esposito, D. 2020. "Phenolic content, anti-inflammatory properties, and dermal wound repair properties of industrially processed and non-processed acai from the Brazilian Amazon." *Food & Function*, 11, no. 6: 4903–4949.

Food Fermentation in Brazil

24

Historical Aspects, Production Processes, and Health-Relevant Importance

Francieli Begnini, Leidiane Andreia Acordi Menezes, and Cristina Maria Zanette

24.1 INTRODUCTION

If it is true that "The identity of a people is mainly due to their language and food culture" (Sonati et al. 2009), it cannot be denied that in Brazil, fermented foods certainly contribute to shaping what characterizes and differentiates as the Brazilian people from others. Therefore, fermented foods are of great importance to Brazilians. Dozens of beverages, cheese, and other fermented products are spread all over the country. Their consumption and production are part of Brazilian culture and history, and they have significant economic and nutritional value.

All over the American continent, food culture is related to the local populations and people who moved there during colonial times. Besides its indigenous heritage, Brazil, with its continental size, inherited the influence of many immigrants who settled there—Portuguese, Italian, German, Polish, Spanish, Japanese, Arabs, and others (Sonati et al. 2009). In particular, Brazilian fermented products result from the culinary practices of indigenous peoples and the techniques brought by European immigrants. Thus, they represent the mixing between the aboriginal peoples' traditions and the colonization the country went through.

The Brazilian indigenous population, estimated by the Brazilian Institute of Geography and Statistics (*IBGE—Instituto Brasileiro de Geografia e Estatística*, in Brazilian Portuguese), was approximately 2,400,000 people in the 16th century. However, in 2010, it was just over 800,000, with the vast majority living in the north and northeast regions (IBGE 2012), reflecting a current greater cultural and culinary influence. On the other hand, Europeans who immigrated to Brazil from the 19th century onwards created settlements in some states of the northeast, in Minas Gerais (southeast), and in the southern regions, spreading their customs and traditions (Sonati et al. 2009; Pineda et al. 2021). Thus, each region has developed a unique food culture, rooted in the traditions of all cultures that formed the country.

Although different peoples formed the pillars of the Brazilian people's eating habits, one technique unites all the Brazilian regions: fermentation. Fermented products, which have been produced over the centuries, still have substantial involvement in the Brazilian economy and diet, regardless of the country's region. Despite such importance, most of them remain artisanal or small scale, reinforcing their definition as Brazilian cultural heritage.

This chapter will address the cultural and historical aspects of the leading Brazilian fermented products, as well as their production processes, the peculiarities of each type of fermentation, the microorganisms involved, and their nutritional value.

24.2 BRAZILIAN FERMENTED PRODUCTS

Cachaça is one of the leading Brazilian fermented products. This sugar cane spirit is the third most consumed distilled alcoholic beverage in the world and number one in Brazil. Its production started in Brazil when the Portuguese brought in sugar cane and distillation techniques. Annual production is

DOI: 10.1201/9781003605300-24

estimated at 800 million liters. The main producing states are São Paulo, Minas Gerais, Pernambuco, Ceará, and Paraíba (IBRAC 2023a).

Another substrate widely used in the fermentation process is cassava, used for polvilho azedo (sour starch), puba flour (carimã), and beiju production. Among them, sour starch is the best known and most consumed one, used to prepare Brazilian cheese bread and chipa bread throughout the country. An estimated level of 536,600 tons of native and sour cassava starch were produced in 2018 alone. Over the past 30 years, there has been a decline in production by small starch manufacturers in the traditional production states of Minas Gerais (MG) and Santa Catarina (SC) and industrial production growth in the state of Paraná. The leading producing states are Paraná (PR), Mato Grosso do Sul (MS), and São Paulo (SP), with 60%, 27.3%, and 9.54%, respectively (Groxko 2020).

Cassava is also the main substrate in indigenous food and alcoholic and non-alcoholic beverage production. Other substrates such as fruits, rice, sweet potatoes, peanuts, and corn, are also used by indigenous tribes to produce food and beverages for nutritional, stimulating, medicinal, festive, and religious purposes. Production is usually done using a family recipe passed down through generations and carried out empirically and on a small scale (Almeida 2009; Schwan et al. 2017). Most studies on these products are about the microbiota and physico-chemical characterization of Caxiri, Yakupa, Cauim (Kawi), Chicha, Taruba, and Calugi beverages. The microbiota involved in co-culture fermentation are lactic acid bacteria (LAB), yeasts, and filamentous fungi (Ramos and Schwan 2019). These beverages are produced by different indigenous tribes, mainly in Brazil's north and midwest regions, and marketed only in a few places. This characteristic restricts the access to these beverages for the population living in other areas of the country.

On the other hand, Brazil has a large variety of cheeses, which are more widespread in the country compared to the indigenous fermented beverages. Cheese production originated with European settlers and immigrants, but production techniques have been modified over time in different micro-regions and incorporated into the culture of those born in Brazil. Currently, regional artisanal cheeses are produced with unique sensory, microbiological, and physico-chemical characteristics. Some examples of unique cheese are marajó cheese obtained from raw milk from buffaloes and produced in the state of Pará (Marajó island); coalho (Curd cheese) and manteiga (butter cheese), both made in the north and northeast regions; and caipira and cabacinha cheese produced in the midwestern region. In the southeastern region, the state of Minas Gerais stands out as a major artisanal cheese producer, with minas artisanal cheese widely marketed and studied. The southern region also stands out in cheese production with its main products: colonial, kochkäse, and serrano cheese.

Law No. 13860 was passed in Brazil in 2018 to regulate the production and marketing of artisanal foods (Brazil 2018). Artisanal products are identified by a seal with the word ARTE (artisanal) and may be sold nationwide. After technical analysis, the seal is granted by the agriculture and livestock agencies of various states. Certification contributes to the appreciation of artisanal products, preserving tradition, culture, and regionalism. The diversity of fermented foods and beverages produced in different regions of Brazil is shown in Figure 24.1.

FIGURE 24.1 Mapping of the main Brazilian fermented foods and their respective regions of origin.

24.2.1 Sour Cassava Starch

Cassava (*Manihot esculenta* Crantz), also known as *manioc, mandioca*, or *yuca,* is a root widely produced in continents such as Africa, Asia, and Latin America, where it is an important food for more than 500 million people (Nair and Prajapati 2003). The world production of cassava roots in 2021 was 314.8 million tons. Brazil is the largest producer in Latin America, with 18.09 million tons (Food and Agriculture Organization of the United Nations 2021).

Cassava root is a source of carbohydrates and vitamins used as food, animal feed, and processed products (mainly starch and flour) (Balagopalan 2002). Carbohydrates are represented by starch (~27% fresh weight) and smaller amounts of free sugars (Araújo et al. 2019). Besides amylaceous food, raw cassava should not be consumed, as it contains antinutritional substances such as phytate, tannin and oxalate, nitrate, and saponins that reduce nutrient bioavailability (Montagnac et al. 2009; Cereda and dos Santo Brito 2017). Raw cassava, especially the bitter type, contains cyanogenic glucosides (CNGs), such as linamarin. Furthermore, due to the cassava root grinding process, β-glycosidase enzymes hydrolyze CNGs, releasing hydrogen cyanide (HCN), which is toxic (EFSA Panel on Contaminants in the Food Chain 2019).

The inactivation of HCN and anti-nutritional substances occurs both by cooking and the fermentation process. Also, during fermentation, the microbiota consumes non-digestible oligosaccharides such as raffinose, stachyose, and verbascose, increasing the food's digestibility and reducing problems such as abdominal distention and flatulence (Ramos and Schwan 2017).

Wet starch fermentation is a traditional technology commonly used in Latin America (Penido et al. 2018). Sour cassava starch is a modified starch known as *polvilho azedo* in Brazil and *almidon agrio* in Argentina, Colombia, Bolivia, and Paraguay (Cereda and Brito 2017). This flour has characteristic expansion properties and flavor due to the lactic acid produced during fermentation. Therefore, it is considered irreplaceable in traditional bakery products with cheese such as *pão de queijo* (Brazilian cheese bread) in Brazil, *chipa* in Paraguay, *cuñape* in Bolivia, and *pan de yuca* in Colombia and Ecuador (Parada et al. 1996; Cereda and dos Santo Brito 2017; Figueroa et al. 2018). Furthermore, sour cassava starch is a gluten-free ingredient considered a potential wheat flour replacement for use in bakery products (bread, pizza, cookies) and in expanded snacks (Alvarado et al. 2013; Jensen et al. 2015; Díaz et al. 2019).

In Brazil, *polvilho azedo* is mainly produced by small and medium-sized agro-industries. The state of Paraná is the largest cassava starch producer, corresponding to 60% of the total domestic production. The agro-industries have an average daily processing capacity of 300 tons of cassava starch (Alves and Vedovoto 2003). Among the cassava processing stages for sour starch production, fermentation and drying present the greatest variation. Cassava starch production stages are shown in Figure 24.2.

Sour cassava starch production's first stage involves washing and peeling the roots. This is followed by the grating and sieving process to obtain a mass, which is washed and filtered to extract the starch. It is then purified by decantation (18 to 24 h) or centrifugation (Chaves-López et al. 2014). The purified starch is then placed in fermentation tanks (under a 10- to 20-cm layer of water) to carry out a spontaneous semi-solid fermentation for 20 to 90 days, depending on climatic conditions (Chiquiza-Montaño et al. 2016). Fermentation tanks may be open or closed. The construction materials used may be either wood or concrete and covered with ceramic tiles (Aquino et al. 2016).

Sour cassava fermentation is performed by microorganisms naturally found in the raw material, water, utensils, and fermentation tanks (Cereda and dos Santo Brito 2017). Spontaneous fermentation occurs by a succession of microbial populations, with the dominant microorganisms showing greater resistance to acidic conditions, such as lactic acid bacteria and yeast (Ampe et al. 2001).

During fermentation, pH reduction of the supernatant liquid from a 6.0 to 6.5 initial pH to 3.0 to 3.8 final pH occurs (Parada et al. 1996; Marcon et al. 2006) due to organic acid production by homo and heterofermentative LAB. The production of organic acids (mainly lactic, acetic, tartaric, and succinic acids) by LABs is important for product preservation and sensory and pasting properties (Freire et al. 2015; Reboucas et al. 2016). Acids and amylase enzymes attack the starch granule amorphous regions (Marcon et al. 2006)

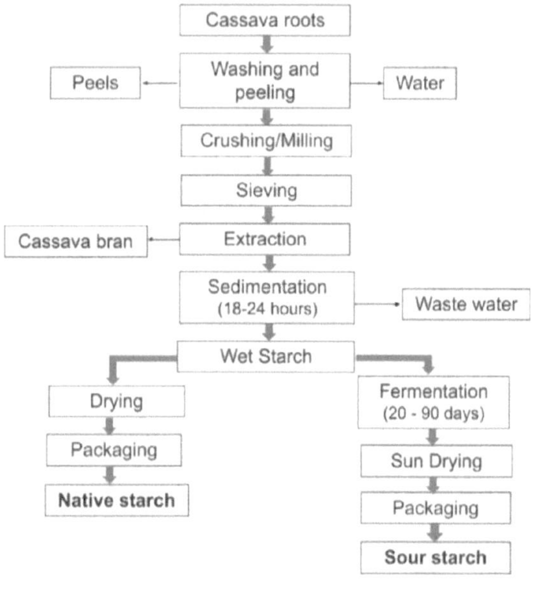

FIGURE 24.2 Schematic diagram of cassava starch production.

Source: (Reprinted and modified with permission from: Chaves-López, C., Serio, A., Grande-Tovar, C. D., Cuervo-Mulet, R., Delgado-Ospina, J., and Paparella, A. 2014. Traditional fermented foods and beverages from a microbiological and nutritional perspective: The Colombian heritage: Colombian fermented foods and beverages. Compr Rev Food Sci Food Saf 13:1031–1048. https://doi.org/10.1111/1541–4337.12098).

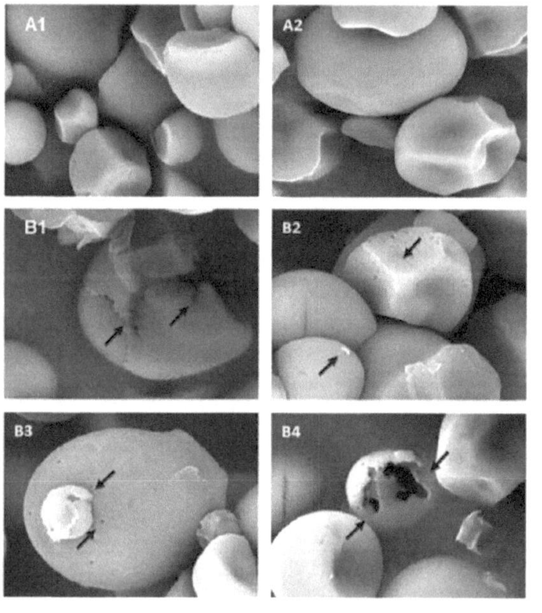

FIGURE 24.3 Scanning electron microscopy micrograph of native and sour cassava starch granules. A1 and A2: industrial native starch; B1, B2, B3 and B4: sour starch.

Source: (Reprinted with permission from: Marcon, M. J. A. 2009. Proposta para o mecanismo de expansão do polvilho azedo com base nas características físico químicas (Portuguese). PhD Thesis. Federal University of Santa Catarina).

(Figure 24.3), changing the starch microstructure. As a result, paste and peak viscosity reduce and solubility increases when compared to native starch (Putri et al. 2012). Furthermore, organic acids inhibit pathogenic bacteria growth, and LABs produce antimicrobial compounds such as bacteriocins and hydrogen peroxide, improving sour cassava starch microbiological safety (Schwan et al. 2007; Ayodeji et al. 2017).

In addition to LAB, yeasts are also involved and isolated during starch fermentation. Low pH during fermentation favors yeast growth. The yeast population is smaller than that of LAB; however, yeasts are believed to produce aromatic compounds that contribute to sour starch sensory characteristics (Lacerda et al. 2005; Penido et al. 2018).

The end of the fermentation stage is crucial for the quality of the final product; however, it is defined empirically in most industries (Penido et al. 2018). Some parameters such as smell, bubbles on the surface, acidic taste, and a bread baking test are used to define the fermentation process endpoint (Aquino et al. 2016).

After the fermentation stage, the sour starch is dried to ensure its expansion property. Sun-drying is traditionally carried out with quite variable exposure time (Figure 24.4b). Solar ultraviolet (UV) radiation is essential to obtain high expansion on baking (Alvarado et al. 2013; Aquino et al. 2016). Sour cassava starch exposure to UV wavelengths causes oxidative depolymerization and the appearance of carboxyl and carbonyl groups in the amylose and amylopectin chains (Dias et al. 2011; Alvarado et al. 2013). Sun-drying requires large areas and presents problems related to the standardization and product physical contamination (insect fragments, soil particles) of the product (Oliveira et al. 2020). Physical contamination could be minimized using greenhouses, as shown in Figure 24.4a. Overall, the product's final moisture is approximately 14% (Cereda and Brito 2017).

Sour cassava starch is considered an artisanal product even when made in medium-sized industries due to spontaneous fermentation and lack of process standardization. Thus, sour cassava starch presents variations in its physico-chemical, rheological, and microbiological characteristics (de Sena Aquino et al. 2013; Aquino et al. 2016). Lack of uniformity negatively impacts the desirable expansion properties in bakery products and becomes a problem when selling them to demanding markets. The use of starter cultures may contribute to greater product uniformity.

24.2.2 Fermented Beverages

Fermented beverages are produced worldwide using different substrates and processes, aiming to preserve and increase nutritional value and flavor. In Brazil, the best-known traditional fermented beverages are cachaça and indigenous beverages, which will be further discussed in the following.

24.2.2.1 Cachaça

Cachaça is a strong spirit obtained by fermented sugar cane juice distillation with an alcoholic content ranging between 38 and 48% at 20°C and may have 6% sugar added.

Cachaça production starts with inoculum production or *pé de cuba,* in Brazilian Portuguese. A small amount of sugar cane juice or wort is mixed with commercial yeasts (*Saccharomyces cerevisiae*) or wild yeasts. Different substrate sources, such as corn or cassava flour, rice, soy, and fruit, are added in the artisanal process, obtaining a wort and cachaça with unique microbiota and sensory characteristics. This inoculum is fermented for 5–7 days with strong aeration and added daily with diluted sugar cane juice. Then, the active inoculum is used for large-scale alcoholic fermentation. Fermentation is carried out in fermentation tanks for 14–24 h under 28–32°C controlled temperature and pH 4.0–4.5. At the end of the fermentation process, a wine with 7–12% alcoholic content is obtained, which is then distilled in copper stills (Morais et al. 1997; Portugal et al. 2017; Alcarde and Ricardo 2017).

The distillation process occurs in three different stages. First, 1–2% of the useful boiler volume, known as the head, is separated, and it is composed of the volatile compounds that are more soluble in ethanol than in water, such as methanol and acetaldehyde. Second, the purest fraction, known as the "heart", is obtained representing 40–45% v/v, which is used

FIGURE 24.4 Sun-drying systems used in Brazil. (a) Greenhouse; (b) outdoor.

Source: (Reprinted with permission from: Kerplas estufas (2021), http://www.kerplasestufas.com.br/ and Catuzo (2021), www.catuzo.com.br).

for commercial purposes, generating 80–85% ethanol and higher alcohols. In the final stage, the fraction known as the "tail" is extracted, and it is composed of water-soluble volatile compounds, such as acetic acid and 5-hydroxymethylfurfural (HMF). Total separation of the head and the tail fractions is essential to produce cachaça with good sensory qualities (Reche et al. 2007; Cardeal et al. 2008; Alcarde and Ricardo 2017).

The aging process using wooden barrels may be done to obtain a spirit with better sensory characteristics. The aging period ranges from 1 year to produce the standard or premium aged cachaça to at least 3 years for the extra premium cachaça. Standard aged cachaça is a non-aged and aged cachaça blend. The pure aged one is designated premium cachaça (Brasil 2005). Cellulose, hemicellulose, lignin, and phenolic compound extraction occurs during the aging process due to the direct contact of the distilled spirit with the barrel, changing its color, odor, and taste. The sensory characteristics and phenolic compounds of cachaça are influenced by the type of wood used and the aging process period (Cardello and Faria 2000; Gabriel et al. 2012; Rodrigues et al. 2016; Carvalho et al. 2020).

According to IBRAC (Brazilian Cachaça Institute), Brazil could currently produce an average 1.2 billion liters of cachaça annually; however, its annual output is lower than 800 million liters. A total of 951 producers, mainly from the states of São Paulo (SP), Pernambuco (PE), Ceará (CE), Minas Gerais (MG), and Paraíba (PB), were registered in 2021. Consumption stood out in the states of São Paulo, Pernambuco, Rio de Janeiro, Ceará, Bahia, and Minas Gerais. In the same year (2022), cachaça was exported to 76 countries, generating a revenue of US$ 20.8 million, equivalent to 9.31 million liters. The leading importing countries were Paraguay, Germany, the USA, Portugal, and France (IBRAC 2023b).

24.2.2.2 Indigenous Beverages

Indigenous fermented beverages, alcoholic or non-alcoholic, may be produced by spontaneous fermentation with different substrates, such as corn, cassava, rice, palm, sweet potatoes, and others (Ramos and Schwan 2017). Some examples of these products with their respective purposes are presented in Table 24.1.

These beverages are produced by artisanal processes taking from 1 to 13 days until they are ready (Figure 24.5). Tarubá is the only one produced by solid fermentation and with the longest process (Figures 24.5 and 24.6). Cauim and calugi (Figures 24.5 and 24.7) are fermented by inoculum obtained

TABLE 24.1 Examples of Brazilian Indigenous Beverages with Their Respective Substrate, Tribe, and Purpose of Utilization

BEVERAGE	*SUBSTRATE*	*TRIBE*	*PURPOSE*	*REGION OF TRIBE*	*REFERENCES*
Chicha	Bacaba, rice, pumpkin, corn, sweet corn, cassava	Umutina	Rituals	Mato Grosso	(Puerari et al. 2015a, 2015b)
Cauim	Cassava, rice, corn, peanut, cotton seed, banana, pumpkin	Tapirapé, Tupinambás	Nutritional and Cauinage Festival	Mato Grosso, North region, Eastern regions	(Almeida et al. 2007; Ramos et al. 2010)
Calugi	Rice, cassava, corn	Javaé	Nutritional	Tocantis	(Miguel et al. 2012)
Tarubá	Cassava	Irurama, Arapyun	Nutritional and religious rituals	Amazonas, Pará	(Lima 2006; Ramos et al. 2015)
Caxiri	Cassava	Juruna, Macuxi	Sacred rituals	Mato Grosso, Roraima	(Lima 2006)
Yakupa	Cassava	Juruna	Nutritional	Mato Grosso	(Lima 2006)

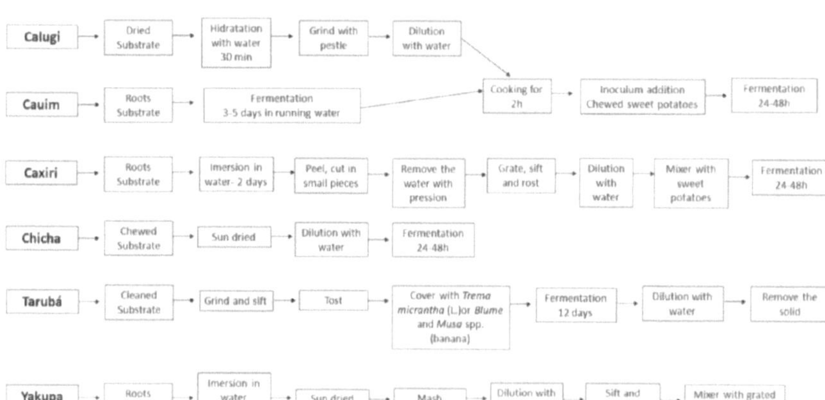

FIGURE 24.5 Flowchart of production of indigenous fermented beverages.

FIGURE 24.6 Taruba preparation. (a, b) cassava mass is placed in wood receptacle (gareira), covered with leaves, and fermented for 12 days. (c) Fermented cassava mass. (d) Fermented cassava mass diluted with water. (e) Filtering the mixture. (f) Final beverage, tarubá.

Source: (Reprinted with permission from: Ramos, C. L., de Sousa, E. S. O., Ribeiro, J., Almeida, T. M. M., do Santos, C. C. A. A., Abegg, M. A., and Schwan, R. F. 2015. Microbiological and chemical characteristics of tarubá, an indigenous beverage produced from solid cassava fermentation. Food Microbiol 49:182–188. https://doi.org/10.1016/j.fm.2015.02.005).

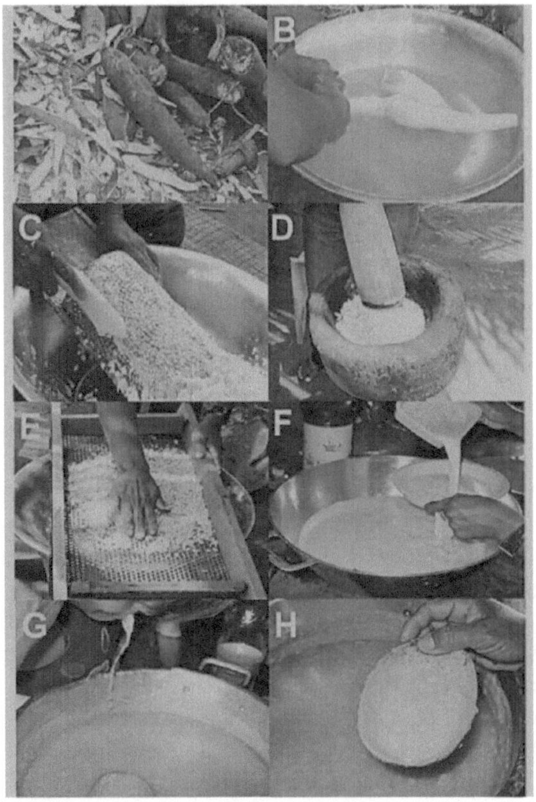

FIGURE 24.7 Steps in the preparation of calugi: peeling and removing the unwanted parts (a), washing the roots (b), grating cassava to obtain a wet mass (c), maceration of corn in a rustic wooden pestle (d), sieving (e), mixing with water and sieving to remove the peel (f), inoculation with mastication juice of the sweet potato (g) and fermented calugi (h).

Source: (Reprinted with permission from: Schwan, R. F., do Santos, C. A., Santos, R. R. M. M., and Duarte, W. F. 2014. Bacterial dynamics and chemical changes during the spontaneous production of the fermented porridge (Calugi) from cassava and corn. Afr J Microbiol Res 8:839–849. https://doi.org/10.5897/AJMR2013.6240).

from a fluid resulting from the chewing of sweet potato by female indigenous Brazilians (Almeida et al. 2007). For chicha production, the substrate is chewed. The chewing process involves the amylolytic enzymes found in saliva that hydrolyze the starch to soluble carbohydrates, which are used as substrate, mainly by lactic acid bacteria and yeast, during the fermentation process (Almeida et al. 2007; Ramos et al. 2010, 2011).

The production process is affected by the environment (humidity and temperature), autochthonous microbiota, and substrate modifying the sensorial attributes and nutritional quality of the products obtained by each indigenous tribe (Chaves-López et al. 2014). The autochthonous microbiota is obtained from the leaves, water, and the environment, which will be discussed in more detail in Section 24.3.1.

Cassava is the most-used substrate (Table 24.1), and it is an essential nutrient source for indigenous people, mainly for its carbohydrates and vitamins. It has also the ability to grow in marginally fertile soils with different harvesting seasons (Uchechukwu-Agua et al. 2015). This root is not considered safe for raw consumption, as discussed in Topic 24.2.1. However, grinding, chewing, and fermentation processes contribute to over 80% reduction of cyanogenic glucoside and antinutritional compounds found in the fermented beverage (Montagnac et al. 2009).

Although indigenous foods are an important source of B-complex vitamins, amino acids, proteins, and soluble carbohydrates, Brazilian indigenous beverages are not commercially or academically explored. Tarubá and cauim are sold in local commerce and at food fairs in northern Brazil. However, the rudimentary production process hinders large-scale commercialization, needing research and investment for improvement.

24.2.3 Artisanal Cheese

In Brazil, cheese has socioeconomic, historical, and nutritional importance due to its significant marketing and high consumption. Cheese production and consumption are widespread in all regions of the country and undoubtedly part of Brazilian culture.

The country has a large cheese production both at industrial and artisanal levels. According to the Institute of Food Technology (ITAL, Instituto de Tecnologia de Alimentos, in Brazilian Portuguese) (ITAL 2020), considering solely cheese using Brazilian production technology, the most industrially produced varieties in certified establishments are the minas, coalho, and colonial cheeses. In addition to industrial production, most Brazilian states have a strong artisanal cheese-producing and consuming tradition. Artisanal cheeses are gaining increasing importance, mainly due to the demands of modern consumers, who seek differentiated products that are less processed and rich in identity and culture (Pineda et al. 2021).

Although the European settlers introduced the original recipes for cheese production in Brazil, each Brazilian region adapted their production to make it unique in its particularities. This has resulted in different artisanal cheeses with inter-regional differences in terms of product types and manufacturing techniques, affected by settlement and migration cultural heritage, the available resources, and each region's climate conditions. These cheeses are characterized using raw milk in most cases, natural starter cultures, and traditional and straightforward manufacturing technology (Kamimura et al. 2019), which is passed down through generations, keeping their unique features. Their production technology involves methods developed by cheesemakers over the centuries, influenced by both social and cultural aspects. These cheeses have typical and distinctive organoleptic characteristics, much appreciated by the local population. They result from the origin and composition of starter cultures, raw materials employed, and ripening conditions (Campagnollo et al. 2018; Kamimura et al. 2019; Margalho et al. 2020).

According to Slow Food Brasil (2020), an association dedicated, among other issues, to the appreciation of regional foods, more than 25 varieties of artisanal Brazilian cheese have been catalogued throughout the country. The typology varies in each Brazilian region and is intrinsically linked to the local culture. Some of them are specific to one solely region, such as marajó and serrano cheese. However, others receive the same denomination but are produced in different regions, such as minas cheese (made in seven different regions of Minas Gerais state, each one with their peculiar characteristics) and coalho and manteiga cheese (produced in some of the northern and northeastern states of Brazil). The most important varieties in terms of production and consumption will be discussed in the next section.

24.2.3.1 Minas Cheese

Minas cheese is one of the most popular cheeses in Brazil, produced both artisanally and industrially due to its great relevance. Artisanal minas cheese is the oldest and most traditional cheese in Brazil and produced in different regions of Minas Gerais state. Today, seven traditional cheese-producing micro-regions are recognized by the *Instituto Mineiro de Agropecuária* IMA (Minas Gerais Institute of Agriculture and Livestock) as Minas cheese producers: Araxá, Serra da Canastra, Cerrado, Serra do Salitre, Triângulo Mineiro, Campo das Vertentes, and Serro.

Its main characteristic is related to the product's sensory attributes that are mainly the result of an endogenous starter culture known as *pingo* (Perin et al. 2017). *Pingo* is a fermented whey portion from the drainage process of cheese produced the previous day, collected for use as a starter for the next day's production. *Pingo* contains a high sodium chloride concentration and many endogenous lactic acid bacteria, responsible for the cheese's typical sensory characteristics (Bachmann et al. 2011; Campagnollo et al. 2018; Kamimura et al. 2019; Margalho et al. 2020). The endogenous microbiota found in *pingo* and raw milk is characteristic of each region, resulting in the typical terroir and artisanal cheese with local production characteristics (Carneiro et al. 2020).

This starter (*pingo*) contains several microbial groups that drive cheese fermentation and maturation (Dores et al. 2013), which varies depending on the region where it is produced. All traditional micro-regions follow the same process, which involves adding *pingo*, using raw milk, and maturing at room temperature, seldom exceeding 8 days (Monteiro 2018). Despite that, the cheese's final characteristics may vary according to the micro-region where it is being produced. Minas cheese has a firm consistency, homogeneous mass, high moisture, and characteristic flavor. It is typically a fresh, soft, slightly salted, and acid white cheese, with a 14–21-day shelf life. The addition of dyes or preservatives is not allowed. The cheeses are cylindrical and weigh approximately 1 kg (Lollo et al. 2015; Kamimura et al. 2019).

Minas artisanal cheese is responsible for generating income for many small-scale family producers. Its century-old production is part of Minas Gerais state's culture and constitutes a heritage to be preserved. Given such importance, minas artisanal cheese was recognized as an intangible cultural heritage by the Historic and Artistic Heritage of Minas Gerais State Institute (IEPHA MG) and by the National Historic and Artistic Heritage Institute (IPHAN) and received the Geographical Indication seal award from the National Institute of Intellectual Property (INPI) in 2002. This title aims to safeguard the traditions and identity of minas cheese, strengthening the product's identity and its artisanal production process as well as stimulating the maintenance of cultural diversity (Monteiro 2018; Pineda et al. 2021).

24.2.3.2 Marajó Cheese

Marajó cheese is so named as it is typical of Marajó Island located in the northern Brazilian state of Pará. It is obtained by the spontaneous coagulation of skimmed, unpasteurized buffalo milk and it is classified as a cooked, unripened, and unpressed cheese. There are two varieties of this cheese, the cream type and the butter type. In the first one, the mass is cooked by adding cream obtained from skimmed milk, with about 50% moisture and 22% lipid content. In the second one, butter is added during the mass cooking process, with 35%

moisture and 42% lipid content (Lourenço et al. 2002; Seixas et al. 2015).

Marajó cheese production is of significant importance to the region for its economic, historical, cultural, and social aspects. It has been produced for over 200 years, originally made from bovine milk. However, over the years, it began to be produced exclusively with buffalo milk, and its consumption became quite traditional among the local population (Figueiredo 2006; Figueiredo et al. 2011). The history of its production is associated with the local people's history and the growing importance of buffalo farming in the region. About 50% of the Brazilian buffalo population is found in the northern region, especially in Marajó island in Pará state, which holds the largest Brazilian buffalo population (Seixas et al. 2015).

Marajó cheese production has a family and artisanal character, contained and advanced in small scale-industries using the know-how accumulated over two centuries, which results in a strong territorial identity. The product's unique features come from the animal breeding system, animals fed on natural pastures, and the use of raw milk with a low-tech manual production system. Artisanal production is not included in official statistics, and the existence of artisanal production units and producing farms in these regions is empirically known (Seixas et al. 2015). Despite this, cheese production moves the local economy, generating income, development, and jobs, allowing families to remain in the region. It should be noted that the two marajó cheese types are highly relevant and unique. In this sense, cheese producers face a great challenge: to maintain tradition and modernize without losing the quality and characteristics that make this cheese so unique (Nascimento et al. 2019).

24.2.3.3 Coalho Cheese (Curd Cheese)

Coalho cheese is the most typical artisanal cheese from the northeast of Brazil. One of the most traditional Brazilian cheeses, it is widely consumed both by the local population and throughout the country. It has been produced for over 150 years, mainly in the states of Pernambuco, Ceará, Rio Grande do Norte, and Paraíba. It is considered a cultural element of the northeastern region, with significant cultural, economic, and historical relevance due to its marketing and consumption. Coalho cheese is produced by both small-scale and large-scale producers. However, most cheeses distributed in the market are produced in a homemade and artisanal manner. It is produced from raw milk, coagulated by rennet addition. Its sensory attributes are quite peculiar. Its main characteristics are the slightly salted and acidic flavor, elastic consistency, compact and soft texture, white color, and melting-by-heat resistance. It may be found in varied sizes and shapes (rectangular or on long wooden sticks, ranging from 100 g to chunks of up to 5 kg), with regular and irregular small eyes or without them. It is mainly consumed in a cooked form such as "roast cheese" and often eaten fresh (Wilkinson et al. 2017; Silva et al. 2012; Kamimura et al. 2019; Brasil 2001; Vidal 2011).

24.2.3.4 Manteiga Cheese (Butter Cheese)

Its processing consists of coagulating whole or skimmed milk, draining the curd obtained by acidification, and adding "butter-from-a-bottle" to the melted curd. "Butter-from-a-bottle" is a type of regional butter produced in northeastern Brazil, which has a semi-liquid consistency at 25°C. It is made by skimming the milk, whipping the cream, and melting at temperatures ranging from 100°C to 130°C, followed by filtration to separate the supernatant (Ambrósio et al. 2001). The main features of manteiga cheese are its salty and rancid aroma and flavor, buttery flavor, oiliness, and cylindrical or rectangular shape ranging from 1 to 15 kg. Its color ranges from straw-colored to caramel (Kamimura et al. 2019).

24.2.3.5 Colonial Cheese

The leading artisanal cheese in the south of Brazil is colonial cheese. This cheese is not identified with a specific geographical location but with a social and cultural category—families of German and Italian immigrants living on small farms and organized in settlements. Initially located in the state of Rio Grande do Sul, these settlers migrated to the two other southern states, Paraná and Santa Catarina, taking with them the colonial cheese production tradition (Wilkinson et al. 2017). Colonial cheese is made from raw milk with industrial rennet added. It has a hard or semi-hard crust; the mass is compact, with some eyes, soft and creamy, with yellow coloration. It is sold in pieces of 0.5 to 1.5 kg, easily sliced, and melted under heat. As there is no legislation regulating the identity and quality standard of colonial cheese, its physical-chemical and sensory characteristics may vary considerably due to variations in the production process. Ripening duration and type may vary between 30 and 75 days, either at room temperature, under refrigeration, or a combination of both (Kamimura et al. 2019; Carvalho et al. 2019).

24.2.3.6 Serrano Cheese

Produced in the Campos de Cima da Serra micro-region located along the borders of the Santa Catarina and Rio Grande do Sul states—northeast of Rio Grande do Sul and Serra do Rio do Rastro in Santa Catarina—serrano cheese, made with raw milk, is a typical mountain cheese. Serrano cheese production is an over 200-year tradition in this region (Wilkinson et al. 2017; Kamimura et al. 2019; Carvalho et al. 2019). According to the producing region's legislation, serrano cheese must be manufactured in a small cheese shop located on a milk-producing property following the region's historical and cultural tradition, defined as a producing one. In addition, it must only be produced with fresh raw whole milk, industrial rennet, and salt. The maturation process must ensure product safety and the biochemical and physical transformations responsible for the development of the typical sensory characteristics of this cheese. It is classified as medium moisture content (semi-hard), semi-fat (fat content in dry matter ranging from 25% to 44%), and with a yellowish crust (2017).

24.2.3.7 Kochkase

The term *Kochkäse* has a German origin, meaning cooking cheese. Kochkäse is an artisanal soft cheese traditionally produced in some regions of the Vale do Itajaí (Santa Catarina state, southern Brazil). The tradition of producing this type of cheese has cultural roots in German colonization and its strong cultural influence in this region. It is produced with raw skimmed milk, naturally coagulated, and desorbed. After that, the fresh cheese is pressed and naturally fermented again for 2 to 7 days. Last, the fermented cheese mass is cooked. Cooking not only provides critical sensory attributes for kochkäse characterization but also improves the product's microbiological quality. The final product is a cheese with over 55% moisture content and lower than 10% fat content in the dry extract (Meisen et al. 2017, 2019; Schlei et al. 2020).

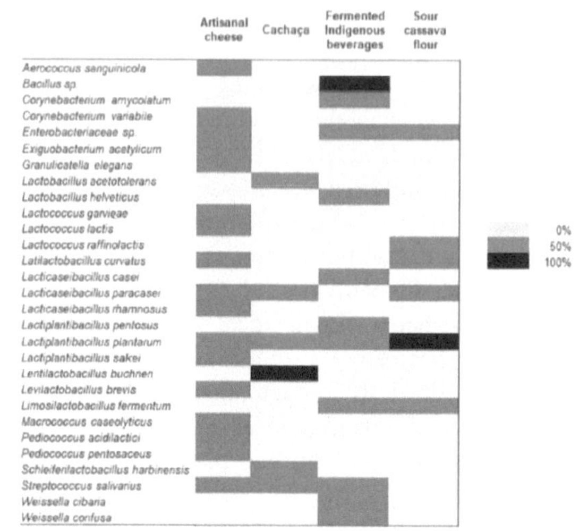

FIGURE 24.8 Heatmap depicting the relative abundance of the most frequently isolated bacteria in Brazilian fermented foods.

24.3 FUNCTIONAL PROPERTIES OF FERMENTED FOODS

In the last decades, the academic community has sought to isolate and identify the microbiota and the compounds produced during the fermentation process, as well as to understand the benefits of consuming these foods. LAB and yeast are the main microorganisms present in fermented foods. Some of them have probiotic properties, contributing to the prevention and treatment of different diseases, such as diabetes, obesity, cancer, hypertension, allergies, and others. In addition, due to LAB and yeast metabolism in fermented food, many compounds with functional properties such as bioactive peptides are produced.

24.3.1 Microbiota

Fermented foods may be produced either by spontaneous fermentation (traditional process) or by adding starter culture (industrial process). In spontaneous fermentation, the microbiota present in the substrate, air, tools, and hands are responsible for the fermentation process (Lindner 2017). Microbiota and the product's features are influenced by factors such as region, weather, substrate, and inoculum used during the process.

In fermented indigenous beverages, *Corynebacterium amycolatum* and *Streptococcus salivarius* were identified as contaminating bacteria from the chewed substrate (Almeida et al. 2007; Ramos et al. 2010; Miguel et al. 2012). *C. amycolatum* produces amylase and protease enzymes. They assist the substrate hydrolysis and release soluble carbohydrates and amino acids that are used by other bacteria and yeasts during fermentation (Almeida et al. 2007). In contrast, *S. salivarius* has a low amylase and proteinase activity (Rul et al. 1994; Whitehead and Cotta 1995). *S. salivarius* is a lactic acid bacteria used as starter culture in fermented foods such as yogurt and cheese due to their ability to use glucose and galactose, producing polysaccharide molecules and increasing the viscosity of the products (Stanley 2003).

Brazilian fermented food has diverse bacterial microbiota with more than 70% represented by the LAB group (Figure 24.8). LABs are classified as generally recognized as safe (GRAS) and are largely responsible for the fermentation of most foods such as cheese, yogurt, fermented milk, beverages, sourdough, fish products, pickled vegetables, and others.

LAB belong to the order *Lactobacillales,* which includes over 50 genera placed in five families (*Aerococcaceae, Carnobacteriaceae, Enterococcaceae, Lactobacillaceae*, and *Streptococcaceae*). The most important bacteria in the fermentation of foods are the *Lactobacillaceae* (Ray and Joshi 2014). Since 2020, with reviewed taxonomic data, 25 different genera belong to this family (Zheng et al. 2020).

For the Brazilian fermented foods evaluated, *Lactiplantibacillus plantarum* was the only bacteria present in all categories (Figure 24.9). *L. plantarum* shows good ecological flexibility and is commonly found in the soil, tools, flours, grain, fruits, and different fermented foods (Guidone et al. 2014; Siepmann et al. 2019). It is an extensively studied bacteria and used as a starter culture in food fermentation due to its exopolysaccharides and bacteriocin production abilities (Bernbom et al. 2006; Wang et al. 2015; Engelhardt et al. 2018). Furthermore, this bacterium has probiotic effects, reducing gastrointestinal tract infection burden symptoms, reducing intestinal heavy metal absorption, and relieving hepatic oxidative stress (Shi et al. 2017; Tong et al. 2017). Additionally, it has been studied for use in the treatments of Alzheimer's, Parkinson's, diabetes, obesity, cancer, hypertension, urinogenital complications, and liver disorders (Arasu et al. 2016).

Two LABs deserve to be highlighted here, *Lacticaseibacillus paracasei* and *Lactococcus lactis*, found in fermented indigenous beverages and cachaça, respectively.

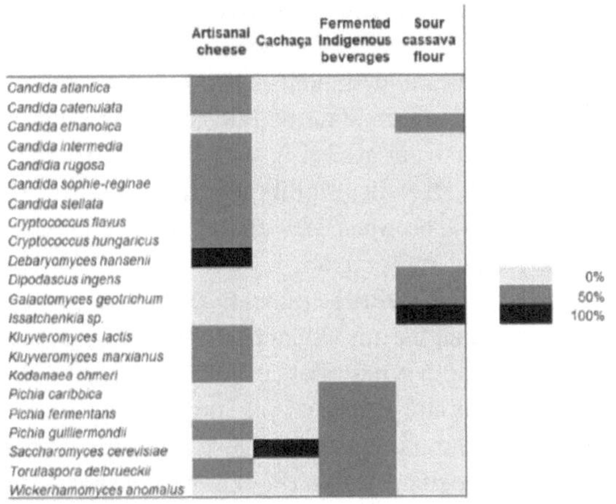

FIGURE 24.9 Heatmap depicting the relative abundance of the most frequently isolated yeasts in Brazilian fermented foods.

Both bacteria have probiotic effects. The first one is commonly isolated in feces of healthy children and has shown promising results in inhibiting breast cancer growth (Biffi et al. 1997), reducing triacylglycerol levels (Bjerg et al. 2015), preventing allergic rhinitis and colitis (Gobbetti and Minervini 2014), and countering infants and young children's diarrheal causes (Kumar and Salminen 2016). *Lc. lactis* showed reduction of DNA damage (Lee et al. 2007) and had beneficial effects against lung, colon, gastric, and breast cancer models (Han et al. 2015; Lee et al. 2015). Also, *Lc. lactis* is known for nisin production, a bacteriocin used in food industries since 1953 when it was deemed safe and approved for food use in England. The US Food and Drug Agency (FDA) approved nisin in 1988 for use in pasteurized products and processed cheese to inhibit the growth of *Clostridium botulinum* (Jozala et al. 2015).

Another key finding is the presence of *Lentilactobacillus buchneri* in 100% of cachaça samples, as seen in Figure 24.8. This LAB has previously been isolated in a bioethanol production plant (Liu et al. 2008), cheese (Fröhlich-Wyder et al. 2013), pickled juice (Zeng et al. 2010), the human intestine, and the oral cavity (Kõll et al. 2010). This bacterium can convert lactic acid into acetic acid under both aerobic and anaerobic conditions, increasing silage conservation time (Oude Elferink et al. 2001).

LAB plays a vital role in the fermentation process, producing organic acids that decrease pH, modifying sensory features of products, and limiting Gram-negative and mesophilic bacteria growth.

Bacillus and *Enterobacteriaceae* sp. were found in fermented indigenous beverages (Figure 24.8). High populations of *Bacillus* genus bacteria are typically found in soils, possibly causing harmful effects. However, they may be beneficial in the fermentation process due to their ability to release a wide range of degradative enzymes, including amylase, protease, and cellulase (Amoa-Awua and Jakobsen 1995; Ramos et al. 2015). Cellulase is an important enzyme in cassava fermentation. It contributes to substrate tissue hydrolysis, facilitating other enzymes action, such as linamarase (Amoa-Awua and Jakobsen 1995; Rusul and Yaacob 1995; Ramos et al. 2015). Linamarase hydrolyzes linamarin (the predominant cyanogenic glycoside found in cassava) to acetone cyanohydrin and glucose, reducing toxicity (Cooke et al. 1978). *Enterobacteriaceae* sp. are related to environmental and hygienic conditions and may be isolated from different foods, such as infants' formula, cereal products, milk powder (Shaker et al. 2007), and Chinese liquor (Zheng et al. 2012) and even in low-pH media, such as yogurt (Hervert et al. 2017).

A fermented product's low pH is essential for aroma development and for selecting the most adaptable microbiota. The reduced pH favors yeast growth that produces some volatile compounds, such as alcohols, esters, aldehydes, and ketones, contributing to the fermented food's sensory features (Freire et al. 2015). In addition, some yeasts may produce amylases, releasing fermentable sugars to bacteria and providing more amino acids and vitamins essential to LAB growth due to their growth and autolysis. Such a characteristic is typical of *Saccharomyces cerevisiae*, present in cachaça and fermented indigenous beverages (Figure 24.9) (Velasco et al. 2004; Alexandre and Guilloux-Benatier 2006; Ramos and Schwan 2017; Oyewole 2001; Ponomarova et al. 2017).

Yeast diversity is lower than bacterial diversity in Brazilian fermented foods (Figure 24.9). During the fermentation process, especially a longer stage, yeast growth inhibition by excessive acids or anti-fungal production may occur. Besides that, we also need to consider the low number of published studies investigating the presence of yeasts in the final products. For cassava flour, 23 samples were considered for bacterial identification, whereas only 2 were used for yeast identification. These results demonstrate the need for further investigation in this field.

The cheese category had 31 samples evaluated (the largest number), where greater yeast diversity was found, mostly by *Candida catenulata*, *Debaryomyces hansenii,* and *Kluyveromyces lactis*. These three yeasts are commonly found in milk, cheese, yogurt, and dairy plant environments, with *C. catenulata* and *D. hansenii* as the dominant yeast species found on cheese crusts (Reps 1993; Mounier et al. 2009; Büchl and Seiler 2011). *C. catenulata* yeast has strong extracellular lipolytic and proteolytic properties (Welthagen and Viljoen 1998; Gkatzionis et al. 2014). *D. hansenii* growth on the cheese surface is favored due to its ability to tolerate high NaCl concentrations and low pH, having lactate as its main carbon source (Petersen et al. 2002). *D. hansenii* plays an essential role in proteolytic activity, and it is responsible for generating the acidic, fruity, and cheesy aroma during cheese ripening (Belloch et al. 2011; Büchl and Seiler 2011). On the other hand, *K. lactis* has weak proteolytic and lipolytic activities; however, together with *Kluyveromyces marxianus*, they are the only two lactose-fermenting species usually present in milk and dairy products (Belloch et al. 2011).

24.3.2 Bioactive Compounds

In addition to being a potential carrier of probiotics, the fermentation process results in many compounds with functional properties due to LAB and yeast metabolism in the fermented food. Bioactive peptides are the most commonly compounds found in cheese. There is a huge potential for Brazilian fermented products from their beneficial health effects; however, scientific studies are scarce and need more research support.

During cheese manufacturing and ripening, proteinases from diverse origins degrade caseins, releasing peptides of different sizes, some of which have interesting functional properties. These peptides, once released, exhibit different bioactivities on the digestive, cardiovascular, immune, and nervous systems (Grappin et al. 1985; Rank et al. 1985).

Regarding the presence of bioactive peptides in Brazilian cheese, antimicrobial peptides are certainly the most-investigated group. Table 24.2 summarizes the studies on these bioactive antimicrobial peptides. Some of these peptides are commonly referred to as bacteriocins and defined as low-molecular-weight proteins or protein complexes with antimicrobial properties. Their inhibitory activity is related to formation of pores in the bacterial membrane, DNA degradation, and inhibition of peptidoglycan synthesis (Favaro et al. 2015).

Very few studies have been found for Brazilian cheese regarding antioxidant peptides. Silva et al. (2012) found about 70 different peptide fractions with antioxidant activity in coalho cheese after digestive process simulation measuring for ABTS radical (2,2'-azinobis-3-ethylbenzothiazoline-6-sulfonic acid) inhibition, reaching nearly 84% oxidation inhibition in *in vitro* tests. According to the authors, the results are comparable to some wines and standard antioxidants such as BHT. More recently, the antioxidant activity of coalho cheese was also reported (Santos et al. 2019), with the antioxidant potential related to ABTS radical inhibition after digestive process simulation, ranging between 38% and 43% for bovine and buffalo coalho cheese, respectively.

Regarding the DPPH radical (2,2-diphenyl-1-picrylhydrazil) inhibition, the antioxidant activity found was approximately 87% for both types of cheese. Santos et al. (2019) also reported the inhibitory activity of angiotensin-I-converting enzyme. After simulating the digestive process, peptides extracted from coalho cheese (bovine milk) showed a 39% anti-hypertensive activity and 14% when extracted from minas cheese. Fontenele et al. (2017) detected by chromatography four peptides known for their antihypertensive properties in coalho cheese. Some peptide sequences with immunomodulatory activity were also found in this study. Antihypertensive and antioxidant activities were also reported in minas cheese samples (Rocha et al. 2020).

24.4 FUTURE PERSPECTIVES

The production of fermented foods in Brazil occurs mainly by spontaneous fermentation. Process and environmental conditions variations provide a wide variety of fermented

TABLE 24.2 Peptide Fractions with Antimicrobial Activity Found in Brazilian Cheeses

TYPE OF CHEESE	SOURCE/TYPE OF PEPTIDE	INHIBITORY ACTION*	REF
Coalho cheese	Water-soluble extract	*Bacillus subtilis* MIC 6,22 µg/mL to 9,37 µg/mL; *Enterococcus faecalis* MIC 199,37 µg/mL.	(Dias et al. 2019)
Coalho cheese	Water-soluble extract	*E. faecalis,* *B. subtilis,* *Escherichia coli, Pseudomonas aeruginosa* Conc. 50 mg/mL	(Silva et al. 2012)
Coalho cheese	Water-soluble extract	*E. faecalis,* *B. subtilis,* *Escherichia coli, Pseudomonas aeruginosa* Conc. 225 mg/mL	(Lima et al. 2017)
Coalho cheese	A segment (193–209) from b-casein A segment (1–23) from aS1-casein	Peptides known for their antimicrobial properties detected by chromatography	(Fontenele et al. 2017)
Minhas cheese	Water-soluble extract	*E. coli* MIC 11–17 µg/mL	(Fialho et al. 2018)
Minas, Coalho and Manteiga cheeses	Proteic cell-free supernatant from 27 LAB strains isolated from cheeses	*Listeria monocytogenes* *Staphylococcus aureus*	(Margalho et al. 2020)
Minas cheese	Proteic cell-free supernatant from *Lactococcus lactis* isolated from Minas cheese	*Listeria monocytogenes* *Staphylococcus aureus*	(Costa et al. 2013)

Note: *MIC: Minimum inhibitory concentration. Conc.: Concentration of the extract used.

foods and beverages, which are little known nationally and internationally.

Although the amount of research on Brazilian fermented products has increased in recent years, most studies aim at isolation and identification of microorganisms. Few studies have verified bioactive compounds, probiotic strains, and nutritional changes during fermentation (increased nutrient digestibility and antinutritional factor degradation). For advancing research on functional bioactive benefits, studies on Brazilian fermented products are of the utmost importance for using rich biodiversity resources and ensuring cultural, social, and economic heritage conservation. Furthermore, these studies may contribute to new applications in the food industry worldwide and the spread of Brazilian products internationally.

REFERENCES

Alcarde, Ricardo A. 2017. *Cachaça: ciência, tecnologia e arte (portuguese)*. Blucher.

Alexandre, H., and M. Guilloux-Benatier. 2006. Yeast autolysis in sparkling wine: A review. *Austr J Grape Wine Res* 12:119–127. https://doi.org/10.1111/j.1755–0238.2006.tb00051.x

Almeida, E. G., C. C. T. C. Rachid, and R. F. Schwan. 2007. Microbial population present in fermented beverage "cauim" produced by Brazilian Amerindians. *Int J Food Microbiol* 120:146–151. https://doi.org/10.1016/j.ijfoodmicro.2007.06.020

Alvarado, P. M., L. Grosmaire, A. G. Dufour, et al. 2013. Combined effect of fermentation, sun-drying and genotype on breadmaking ability of sour cassava starch. *Carbohydr Polym* 98:1137–1146. https://doi.org/10.1016/j.carbpol.2013.07.012

Ambrósio, C. L. B., N. B. Guerra, and J. M. Filho. 2001. Características de identidade, qualidade e estabilidade da manteiga de garrafa: parte I—características de identidade e qualidade (portuguese). *Food Sci Technol* 21:3. https://doi.org/10.1590/S0101–20612001000300011

Amoa-Awua, W. K. A., and M. Jakobsen. 1995. The role of *Bacillus* species in the fermentation of cassava. *J Appl Bacteriol* 79:250–256.

Ampe, F., A. Sirvent, and N. Zakhia. 2001. Dynamics of the microbial community responsible for traditional sour cassava starch fermentation studied by denaturing gradient gel electrophoresis and quantitative rRNA hybridization. *Int J Food Microbiol* 65:45–54. https://doi.org/10.1016/S0168–1605(00)00502-X 10.1590/1981–6723.5515

Aquino, A. C. M. S., V. M. Gervin, and E. R. Amante. 2016. Avaliação do processo produtivo de polvilho azedo em indústrias de Santa Catarina (portuguese). *Braz J Food Technol* 19: 1–8. http://dx.doi.org/10.1590/1981-6723.5515

Arasu, M. V., N. A. Al-Dhabi, S. Ilavenil, K. C. Choi, and S. Srigopalram. 2016. In vitro importance of probiotic *Lactobacillus plantarum* related to medical field. *Saudi J Biol Sci* 23:S6–S10. https://doi.org/10.1016/j.sjbs.2015.09.022

Ayodeji, B. D., C. Piccirillo, V. Ferraro, et al. 2017. Screening and molecular identification of lactic acid bacteria from gari and fufu and gari effluents. *Ann Microbiol* 67:123–133. https://doi.org/10.1007/s13213-016-1243-1

Bachmann, H.-P., M.-T. Fröhlich-Wyder, E. Jakob, et al. 2011. *Cheese|Raw Milk Cheeses*. Encyclopedia of Dairy Sciences, pp. 652–660.

Balagopalan, C. 2002. Cassava utilization in food, feed and industry. In: *Cassava: Biology, Production, and Utilization*, eds. R. J. Hillocks and J. M. Thresh, 301–318. CABI Publishing.

Belloch, C., A. Querol, and E. Barrio. 2011. Yeasts and molds|*Kluyveromyces* spp. In: *Encyclopedia of Dairy Sciences* (Second Edition), ed. J. W. Fuquay, 754–764. Elsevier.

Bernbom, N., T. R. Licht, P. Saadbye, F. K. Vogensen, and B. Nørrung. 2006. *Lactobacillus plantarum* inhibits growth of *Listeria monocytogenes* in an in vitro continuous flow gut model, but promotes invasion of *L. monocytogenes* in the gut of gnotobiotic rats. *Int J Food Microbiol* 108:10–14. https://doi.org/10.1016/j.ijfoodmicro.2005.10.021

Biffi, A., D. Coradini, R. Larsen, L. Riva, and G. Di Fronzo. 1997. Antiproliferative effect of fermented milk on the growth of a human breast cancer cell line. *Nutr Cancer* 28:93–99. https://doi.org/10.1080/01635589709514558

Bjerg, A. T., M. Kristensen, C. Ritz, et al. 2015. Four weeks supplementation with *Lactobacillus paracasei subsp. paracasei L. casei W8®* shows modest effect on triacylglycerol in young healthy adults. *Ben Microbes* 6:29–39. https://doi.org./10.3920/BM2014.0058

Brasil. 2001. *Instrução Normativa nº 30, de 26 de junho de 2000. Regulamento Técnico de Identidade e Qualidade de Manteiga da Terra ou Manteiga de Garrafa, Queijo de Coalho e Queijo de Manteiga*. Diário Oficial da República Federativa do Brasil: seção 1, Brasília, DF.

Brasil. 2005. *Instrucao Normativa MAPA n 13 de 29/06/2005—Federal—LegisWeb*. https://www.legisweb.com.br/legislacao/?id=76202. Accessed January 14, 2024. Accessed January 20, 2021.

Brazil. 2018. http://www.planalto.gov.br/ccivil_03/_ato2015–2018/2018/Lei/L13680.htm. Accessed December 15, 2020.

Büchl, N. R., and H. Seiler. 2011. Yeasts and molds|Yeasts in milk and dairy products. In: *Encyclopedia of Dairy Sciences*, ed. J. W. Fuquay, 744–753. Academic Press.

Campagnollo, F. B., L. P. Margalho, B. A. Kamimura, et al. 2018. Selection of indigenous lactic acid bacteria presenting anti-listerial activity, and their role in reducing the maturation period and assuring the safety of traditional Brazilian cheeses. *Food Microbiol* 73:288–297. https://doi.org/10.1016/j.fm.2018.02.006

Cardeal, Z. L., de Souza, P. P., da Silva, M. D. R. G., and Marriott, P. J. 2008. Comprehensive two-dimensional gas chromatography for fingerprint pattern recognition in cachaça production. *Talanta* 74:793–799. https://doi.org/10.1016/j.talanta.2007.07.021

Cardello, H., and J. B. Faria. 2000. Perfil sensorial e características físico-químicas de aguardentes comerciais brasileiras envelhecidas e sem envelhecer (portuguese). *Braz J Food Technol* 3:41–50.

Carvalho, D. G., L. Ranzan, L. F. Trierweiler, and J. O. Trierweiler. 2020. Determination of the concentration of total phenolic compounds in aged cachaça using two-dimensional fluorescence and mid-infrared spectroscopy. *Food Chem* 329:127142. https://doi.org/10.1016/j.foodchem.2020.127142

Carvalho, M. M., L. O. de Fariña, D. Strongin, C. L. L. F. Ferreira, and J. D. D. Lindner. 2019. Traditional colonial-type cheese from the south of Brazil: A case to support the new Brazilian laws for artisanal cheese production from raw milk. *J Dairy Sci* 102:9711–9720. https://doi.org/10.3168/jds.2019–16373

Catuzo. 2021. http://www.catuzo.com.br. Accessed February 20, 2021.

Cereda, M. P., and V. H. dos Santo Brito. 2017. Fermented foods and beverages from Cassava (Manihot esculenta Crantz) in South America. In: *Fermented Foods of Latin America: From Traditional Knowledge to Innovative Applications*, eds. A. L. B. Penna, L. A. Nero, and S. D. Todorov, 192–223. CRC Press.

Chaves-López, C., A. Serio, C. D. Grande-Tovar, R. Cuervo-Mulet, J. Delgado-Ospina, and A. Paparella. 2014. Traditional fermented foods and beverages from a microbiological and nutritional perspective: The Colombian heritage: Colombian fermented foods and beverages . . . *Compr Rev Food Sci Food Saf* 13:1031–1048. https://doi.org/10.1111/1541-4337.12098

Chiquiza-Montaño, L. N., O. I. Montoya, C. Restrepo, and F. Orozco-Sánchez. 2016. Estudio de la Microbiota del Proceso de Producción de Almidón Agrio de Yuca. *Inf Tec* 27:3–14. https://doi.org/10.4067/S0718-07642016000500002

Costa, H. H. S., M. R. Souza, L. B. Acúrcio, A. E. Cunha, M. F. S. Resende, and Á. C. Nunes. 2013. Potencial probiótico in vitro de bactérias ácido-láticas isoladas de queijo-de-minas artesanal da Serra da Canastra, MG (portuguese). *Arq. Bras. Med. Vet. Zootec* 65:1858–1866. https://doi.org/10.1590/S0102-09352013000600038

Cooke, R. D., G. G. Blake, and J. M. Battershill. 1978. Purification of cassava linamarase. *Phytochemistry* 17:381–383. https://doi.org/10.1016/S0031-9422(00)89320-6

de Almeida, E. G. 2009. Caracterização fisico-química e microbiológica de bebidas fermentadas produzidas pelos índios Tapirapé (portuguese). PhD, Federal University of Lavras, MG (Brazil).

de Alves, E. R. A., and G. L. Vedovoto. 2003. A indústria do amido de mandioca (portuguese). https://www.infoteca.cnptia.embrapa.br/infoteca/handle/doc/109712. February 10, 2021.

das Araújo, F. C. B., E. F. de Moura, R. L. Cunha, J. T. Farias Neto, and R. S. de Silva. 2019. Chemical root traits differentiate "bitter" and "sweet" cassava accessions from the Amazon. *Crop Breed Appl Biotechnol* 19:77–85. https://doi.org/10.1590/1984-70332019v19n1a11

das Dores, M. T., J. E. da Nobrega, and C. L. de Luces Fortes Ferreira. 2013. Room temperature aging to guarantee microbiological safety of Brazilian artisan Canastra cheese. *Food Sci Technol* 33:180–185.

de Carneiro, J. O., A. C. S. D. Chaves, M. P. Stephan, C. A. Boari, and M. G. B. Koblitz. 2020. Artisan minas cheese of Serro: Proteolysis during ripening. *Heliyon* 6:e04446. https://doi.org/10.1016/j.heliyon.2020.e04446

de Lourenço, L. F. H., C. L. Sousa, M. Simão Neto, and J. B. de Lourenço Júnior. 2002. Análise microbiológica do requeijão marajoara no norte do Brasil (portuguese). *Higiene alimentar* 16, no. 96:55–59.

da Miguel, M. G. C. P., M. G. M. da Cruz Pedrozo, M. R. R. Santos, W. F. Duarte, E. G. de Almeida, and R. F. Schwan. 2012. Physicochemical and microbiological characterization of corn and rice "calugi" produced by Brazilian Amerindian people. *Food Res Int* 49:524–532. https://doi.org/10.1016/j.foodres.2012.08.012

de Oliveira, D. I., L. Demogalski, A. H. Dias, L. A. A. Pereira, A. Alberti, P. R. Los, and I. M. Demiate. 2020. Traditional sour cassava starch obtained with alterations in the solar drying stage. *Food Sci Technol* 41. https://doi.org/10.1590/fst.16120

de Sena Aquino, A. C. M., J. M. Pereira, L. B. Watanabe, and E. R. Amante. 2013. Standardisation of the sour cassava starch reduces the processing time by fermentation water monitoring. *Int J Food Sci Technol* 48:1892–1898. https://doi.org/10.1111/ijfs.12167

Dias, A. R. G., E. R. da Zavareze, M. C. Elias, E. Helbig, D. O. da Silva, and C. F. Ciacco. 2011. Pasting, expansion and textural properties of fermented cassava starch oxidised with sodium hypochlorite. *Carbohydr Polym* 84:268–275. https://doi.org/10.1016/j.carbpol.2010.11.033

Dias, G. M. P., A.B. da Silva, N.M.C. Granja, T.N. da Silva, G.V.M. Lima, M.T. Cavalcanti, and A.L.F. Porto. 2019. Can Coalho cheese lactic microbiota be used in dairy fermentation to reduce foodborne pathogens? *Scientia Plena* 15:1–9. https://doi.org/10.14808/sci.plena.2019.021501

Díaz, A., C. Dini, S. Z. Viña, and M. A. García. 2019. Fermentation and drying effects on bread-making potential of sour cassava and ahipa starches. *Food Res Int* 116:620–627. https://doi.org/10.1016/j.foodres.2018.08.081

dos Santos, W. L., E. G. da Silva, M. E. B. da Silva, E. F. T. da Silva, and K. A. Moreira. 2019. Potencial biológico e estudo in vitro da digestão de peptídeos solúveis obtidos de diferentes variedades de queijo (portuguese). *PUBVET* 13:148.

EFSA Panel on Contaminants in the Food Chain, D. Schrenk, M. Bignami, et al. 2019. Evaluation of the health risks related to the presence of cyanogenic glycosides in foods other than raw apricot kernels. *EFSA J* 17:e05662. https://doi.org/10.2903/j.efsa.2019.5662

Engelhardt, T., K. Szakmár, G. Kiskó, C. Mohácsi-Farkas, and O. Reichart. 2018. Combined effect of NaCl and low temperature on antilisterial bacteriocin production of *Lactobacillus plantarum* ST202Ch. *LWT Food Sci Technol* 89:104–109. https://doi.org/10.1016/j.lwt.2017.10.027

Favaro, L., A. L. B. Penna, and S. D. Todorov. 2015. Bacteriocinogenic LAB from cheeses: Application in biopreservation? *Trends Food Sci Technol* 41:37–48. https://doi.org/10.1016/j.tifs.2014.09.001

Fialho, T. L., L. C. Carrijo, M. J. Magalhães Júnior, M. C. Baracat-Pereira, R. H. Piccoli, and L. R. De Abreu. 2018. Extraction and identification of antimicrobial peptides from the Canastra artisanal minas cheese. *Food Res. Int.* 10:406–413. https://doi.org/10.1016/j.foodres.2018.02.009

Figueiredo, E. L. 2006. Elaboração e caracterização do "Queijo Marajó", tipo creme, de leite de búfala, visando sua padronização (portuguese). Master Dissertation, Federal University of Pará.

Figueiredo, E. L., J. de Brito Lourenço Júnior, M. U. Toro, and S. C. G. de Lima. 2011. Queijo do Marajó Tipo Creme: Parâmetros Físico-Químicos e sensoriais (portuguese). *Rev Inst Laticínios Cândido Tostes* 66:26–33.

Figueroa, J. A., E. M. Cadena, C. C. Rincon, and J. G. Salcedo. 2018. Modification of the fermentation process for Sour Cassava starch with expansion properties. *Adv J Sci Technol* 16:55–61. https://doi.org/10.19026/ajfst.16.5937

Fontenele, M. A., M. S. R. do Bastos, K. M. O. Dos Santos, M. P. Bemquerer, and A. S. do Egito. 2017. Peptide profile of Coalho cheese: A contribution for protected designation of origin (PDO). *Food Chem* 219:382–390. https://doi.org/10.1016/j.foodchem.2016.09.171

Food and Agriculture Organization of the United Nations. 2021. *FAOSTAT Statistical Database.*

Freire, A. L., C. L. Ramos, and R. F. Schwan. 2015. Microbiological and chemical parameters during cassava based-substrate fermentation using potential starter cultures of lactic acid bacteria and yeast. *Food Res Int* 76:787–795. https://doi.org/10.1016/j.foodres.2015.07.041

Fröhlich-Wyder, M.-T., D. Guggisberg, R. Badertscher, D. Wechsler, A. Wittwer, and S. Irmler. 2013. The effect of *Lactobacillus buchneri* and *Lactobacillus parabuchneri* on the eye formation of semi-hard cheese. *Int Dairy J* 33:120–128. https://doi.org/10.1016/j.idairyj.2013.03.004

Gabriel, A. V. M. D., M. R. Verruma-Bernardi, L. A. C. Margarido, et al. 2012. Effect of the spontaneous fermentation and the ageing on the chemo-sensory quality of Brazilian organic cachaça. *Ciencia Rural* 42:918–925. https://doi.org/10.1590/S0103-84782012000500026

Gkatzionis, K., D. Yunita, R. S. T. Linforth, M. Dickinson, and C. E. R. Dodd. 2014. Diversity and activities of yeasts from different parts of a Stilton cheese. *Int J Food Microbiol* 177:109–116. https://doi.org/10.1016/j.ijfoodmicro.2014.02.016

Gobbetti, M., and F. Minervini. 2014. Lactobacillus|*Lactobacillus casei*. In: *Encyclopedia of Food Microbiology*, eds. C. A. Batt and M. L. Tortorello, 432–438. Academic Press.

Grappin, R., T. C. Rank, and N. F. Olson. 1985. Primary proteolysis of cheese proteins during ripening: A review. *J Dairy Sci* 68:531–540. https://doi.org/10.3168/jds.S0022–0302(85)80855–9

Groxko, M. 2020. *Mandioca: Análise da Conjuntura (portuguese)*. https://www.agricultura.pr.gov.br/sites/default/arquivos_restritos/files/documento/2019–12/Mandioca%202020.pdf. Accessed January 15, 2021.

Guidone, A., T. Zotta, R. P. Ross, C. Stanton, M. C. Rea, E. Parente, and A. Ricciardi. 2014. Functional properties of *Lactobacillus plantarum* strains: A multivariate screening study. *LWT Food Sci Technol* 56:69–76. https://doi.org/10.1016/j.lwt.2013.10.036

Han, K. J., N.-K. Lee, H. Park, and H.-D. Paik. 2015. Anticancer and anti-inflammatory activity of probiotic *Lactococcus lactis* NK34. *J Microbiol Biotechnol* 25:1697–1701. https://doi.org/10.4014/jmb.1503.03033

Hervert, C. J., N. H. Martin, K. J. Boor, and M. Wiedmann. 2017. Survival and detection of coliforms, *Enterobacteriaceae*, and gram-negative bacteria in Greek yogurt. *J Dairy Sci* 100:950–960. https://doi.org/10.3168/jds.2016–11553

IBGE—Brazilian Institute of Geography and Statistics. 2012. *Os indígenas no Censo Demográfico 2010 primeiras considerações com base no quesito cor ou raça (portuguese)*. https://www.ibge.gov.br/indigenas/indigena_censo2010.pdf. Accessed March 10, 2021.

IBRAC—Brazilian Cachaça Institute. 2023a. https://ibrac.net/servicos/mercado-interno. Accessed January 26, 2023.

IBRAC—Brazilian Cachaça Institute. 2023b. *Mercado Externo*. https://ibrac.net/servicos/mercado-externo. Accessed January 14, 2023.

ITAL—Institute of Food Technology. 2020. *Brazil Dairy Trends 2020*. Secretaria de Agricultura e Abastecimento do Estado de São Paulo. https://ital.agricultura.sp.gov.br/brasilfoodtrends/16/. Accessed June 20, 2024.

Jensen, S., L. H. Skibsted, U. Kidmose, and A. K. Thybo. 2015. Addition of cassava flours in bread-making: Sensory and textural evaluation. *LWT Food Sci Technol* 60:292–299. https://doi.org/10.1016/j.lwt.2014.08.037

Jozala, A. F., L. C. de Lencastre Novaes, and A. P. Junior. 2015. Nisin. In: *Concepts, Compounds and the Alternatives of Antibacterials*, eds. V. Bobbarala. IntechOpen. https://doi.org/10.5772/60932

Kamimura, B. A., M. Magnani, W. A. Luciano, et al. 2019. Brazilian artisanal cheeses: An overview of their characteristics, main types and regulatory aspects. *Compr Rev Food Sci Food Saf* 18:1636–1657. https://doi.org/10.1111/1541–4337.12486

Kerplas estufas. 2021. http://www.kerplasestufas.com.br. Accessed February 20, 2021.

Kõll, P., R. Mändar, I. Smidt, et al. 2010. Screening and evaluation of human intestinal Lactobacilli for the development of novel gastrointestinal probiotics. *Curr Microbiol* 61:560–566. https://doi.org/10.1007/s00284-010-9653-y

Kumar, H., and S. Salminen.2016. Probiotics. In: *Encyclopedia of Food and Health*, eds. B. Caballero, P. M. Finglas, and F. Toldrá, 510–515. Academic Press.

Lacerda, I. C. A., R. L. Miranda, B. M. Borelli, et al. 2005. Lactic acid bacteria and yeasts associated with spontaneous fermentations during the production of sour cassava starch in Brazil. *Int J Food Microbiol* 105:213–219. https://doi.org/10.1016/j.ijfoodmicro.2005.04.010

Lee, N.-K., K. J. Han, S.-H. Son, S. J. Eom, S.-K. Lee, and H.-D. Paik. 2015. Multifunctional effect of probiotic *Lactococcus lactis* KC24 isolated from kimchi. *LWT Food Sci Technol* 64:1036–1041. https://doi.org/10.1016/j.lwt.2015.07.019

Lee, N. K., J. E. Noh, G. H. Choi, E. Park, and H. I. Chang. 2007. Potential probiotic properties of *Lactococcus lactis NK34* isolated from jeotgal. *Food Sci Biotechnol* 16, no. 5:843–847.

Lima, M. dos S. F. de., R. A. da Silva, J. L. de Lima Filho, A. L. F. Porto, and M. T. H. Cavalcanti. 2017. Queijo de coalho artesanal: fonte alternativa de peptídeos antimicrobianos (portuguese). *Braz J Food Technol* 20:1–7. https://doi.org/10.1590/1981-6723.19316

Lima, T. S. 2006. *Um peixe olhou pra mim: o povo Yudjá e a perspectiva* (portuguese). São Paulo: Editora UNESP.

Lindner, J. D. D. 2017. Characteristics and production of microbial cultures. In: *Fermented Foods of Latin America: From Traditional Knowledge to Innovative Applications*, eds. A. L. B. Penna, L. A. Nero, and S. D. Todorove, 279–306. CRC Press.

Liu, S., K. A. Skinner-Nemec, and T. D. Leathers. 2008. *Lactobacillus buchneri* strain NRRL B-30929 converts a concentrated mixture of xylose and glucose into ethanol and other products. *J Ind Microbiol Biotechnol* 35:75–81. https://doi.org/10.1007/s10295-007-0267-8

Lollo, P. C. B., P. N. Morato, C. S. Moura, et al. 2015. Hypertension parameters are attenuated by the continuous consumption of probiotic Minas cheese. *Food Res Int J* 76:611–617. https://doi.org/10.1016/j.foodres.2015.07.015

Marcon, M. J. A. 2009. Proposta para o mecanismo de expansão do polvilho azedo com base nas características físico químicas (portuguese). PhD Thesis. Federal University of Santa Catarina.

Marcon, M. J. A., M. A. Vieira, K. Santos, K. N. de Simas, R. A. Dias de Mello Castanho, and E. R. Amante. 2006. The effect of fermentation on cassava starch microstructure. *J Food Process Eng* 29:362–372. https://doi.org/10.1111/j.1745–4530.2006.00073.x

Margalho, L. P., M. D. Feliciano, C. E. Silva, J. S. Abreu, M. V. F. Piran, and A. S. Sant'Ana. 2020. Brazilian artisanal cheeses are rich and diverse sources of nonstarter lactic acid bacteria regarding technological, biopreservative, and safety properties: Insights through multivariate analysis. *J Dairy Sci* 103:7908–7926. https://doi.org/10.3168/jds.2020–18194

Meisen, M. N., D. S. T. Suarez, R. de Cássia Siqueira Curto Valle, S. K. F. Yamaguchi, C. K. De Souza, and L. F. de Carvalho. 2017. Avaliação Sensorial de queijo Kochkase produzido com leite cru e pasteurizado consumido na região do médio Vale do Itajaí (portuguese)—Santa Catarina. *Engevista* 19:906. https://doi.org/10.22409/engevista.v19i4.912

Meisen, M. N., B. Weizstein, M. C. dos Santos, N. F. Moratelli, B. H. Kipper, A. P. Ferreira, and L. F. de Carvalho. 2019. Análise de características de qualidade durante processamento de queijo *Kochkäse* produzido a partir de leite cru e pasteurizado (portuguese). *Brazil. J. Develop.* 5:13383–13396. https://doi:10.34117/bjdv5n8-140.

Montagnac, J. A., C. R. Davis, and S. A. Tanumihardjo. 2009. Processing techniques to reduce toxicity and antinutrients of cassava for use as a staple food. *Compr Rev Food Sci Food Saf* 8:17–27. https://doi.org/10.1111/j.1541–4337.2008.00064.x

Monteiro, R. P. 2018. O queijo Minas artesanal e seu potencial para a agroindústria familiar (portuguese). In: *Queijo Minas artesanal: valorizando a agroindústria familiar*, eds. R. P. Monteiro and V. M. da Matta. Embrapa.

Morais, P. B., C. A. Rosa, V. R. Linardi, C. Pataro, and A. B. R. A. Maia. 1997. Short communication: Characterization and succession of yeast populations associated with spontaneous fermentations during the production of Brazilian sugar-cane aguardente. *World J Microbiol Biotechnol* 13:241–243. https://doi.org/10.1023/A:1018558302062

Mounier, J., C. Monnet, N. Jacques, A. Antoinette, and F. Irlinger. 2009. Assessment of the microbial diversity at the surface of Livarot cheese using culture-dependent and independent approaches. *Int J Food Microbiol* 133:31–37. https://doi.org/10.1016/j.ijfoodmicro.2009.04.020

Nascimento, E. C., B. E. V. da Cruz, and M. F. Calvi. 2019. Queijos diferentes, origem geográfica comum: história e tradição da produção dos queijos do Marajó (portuguese). *Ateliê Geográfico* 13:190–208. https://doi.org/10.5216/ag.v13i3.57694

Nair, B., and J. Prajapati. 2003. The history of fermented foods. In: *Handbook of Fermented Functional Foods*, ed. E. R. Farnworth, 1–25. CRC Press. https://doi.org/10.5216/ag.v13i3.57694

Oude Elferink, S. J., J. Krooneman, J. C. Gottschal, S. F. Spoelstra, F. Faber, and F. Driehuis. 2001. Anaerobic conversion of lactic acid to acetic acid and 1, 2-propanediol by *Lactobacillus buchneri*. *Appl Environ Microbiol* 67:125–132. https://doi.org/10.1128/AEM.67.1.125–132.2001

Oyewole, O. B. 2001. Characteristics and significance of yeasts' involvement in cassava fermentation for "fufu" production. *Int J Food Microbiol* 65:213–218. https://doi.org/10.1016/S0168–1605(01)00431–7

Parada, J. L., E. Zapata, S. V. de Fabrizio, and A. Martínez. 1996. Microbiological and technological aspects of cassava-starch fermentation. *World J Microbiol Biotechnol* 12:53–56. https://doi.org/10.1007/BF00327801

Penido, F. C. L., F. B. Piló, S. H. de Cicco, et al. 2018. Selection of starter cultures for the production of sour cassava starch in a pilot-scale fermentation process. *Braz J Microbiol* 49:823–831. https://doi.org/10.1016/j.bjm.2018.02.001

Perin, L. M., M. L. S. Sardaro, L. A. Nero, E. Neviani, and M. Gatti. 2017. Bacterial ecology of artisanal Minas cheeses assessed by culture-dependent and -independent methods. *Food Microbiol* 65:160–169. https://doi.org/10.1016/j.fm.2017.02.005

Petersen, K. M., S. Westall, and L. Jespersen. 2002. Microbial succession of *Debaryomyces hansenii* strains during the production of Danish surfaced-ripened cheeses. *J Dairy Sci* 85:478–486. https://doi.org/10.3168/jds.S0022–0302(02)74098–8

Pineda, A. P. A., G. Z. Campos, N. J. Pimentel-Filho, B. D. G. M. de Franco, and U. M. Pinto. 2021. Brazilian artisanal cheeses: Diversity, microbiological safety, and challenges for the sector. *Front Microbiol* 12:666922. https://doi.org/10.3389/fmicb.2021.666922

Ponomarova, O., N. Gabrielli, D. C. Sévin, et al. 2017. Yeast creates a niche for symbiotic lactic acid bacteria through nitrogen overflow. *Cell Syst* 5:345–357.e6. https://doi.org/10.1016/j.cels.2017.09.002

Portugal, C. B., A. P. de Silva, A. M. Bortoletto, and A. R. Alcarde. 2017. How native yeasts may influence the chemical profile of the Brazilian spirit, cachaça? *Food Res Int* 91:18–25. https://doi.org/10.1016/j.foodres.2016.11.022

Puerari, C., K. T. Magalhães-Guedes, and R. F. Schwan. 2015a. Physicochemical and microbiological characterization of chicha, a rice-based fermented beverage produced by Umutina Brazilian Amerindians. *Food Microbiol* 46:210–217. https://doi.org/10.1016/j.fm.2014.08.009

Puerari, C., K. T. Magalhães-Guedes, and R. F. Schwan. 2015b. Bacaba beverage produced by Umutina Brazilian Amerindians: Microbiological and chemical characterization. *Braz J Microbiol* 46:1207–1216. https://doi: 10.1590/S1517-838246420140964

Putri, W. D. R., M. D. W. Haryadi, and M. N. Cahyanto. 2012. Role of lactic acid bacteria on structural and physicochemical properties of sour cassava starch. *APCBEE Proc* 2:104–109. https://doi.org/10.1016/j.apcbee.2012.06.019

Ramos, C. L., E. G. de Almeida, A. L. Freire, and R. Freitas Schwan. 2011. Diversity of bacteria and yeast in the naturally fermented cotton seed and rice beverage produced by Brazilian Amerindians. *Food Microbiol* 28:1380–1386. https://doi.org/10.1016/j.fm.2011.06.012

Ramos, C. L., E. G. de Almeida, G. V. M. de Pereira, P. G. Cardoso, E. S. Dias, and R. F. Schwan. 2010. Determination of dynamic characteristics of microbiota in a fermented beverage produced by Brazilian Amerindians using culture-dependent and culture-independent methods. *Int J Food Microbiol* 140:225–231. https://doi.org/10.1016/j.ijfoodmicro.2010.03.029

Ramos, C. L., E. S. O. de Sousa, J. Ribeiro, et al. 2015. Microbiological and chemical characteristics of tarubá, an indigenous beverage produced from solid cassava fermentation. *Food Microbiol* 49:182–188. https://doi.org/10.1016/j.fm.2015.02.005

Ramos, C. L., and R. F. Schwan. 2017. Technological and nutritional aspects of indigenous Latin America fermented foods. *Curr Opin Food Sci* 13:97–102. https://doi.org/10.1016/j.cofs.2017.07.001

Ramos, C. L., and R. F. Schwan. 2019. Ethnic fermented foods of America. In: *Fermented Food Products*, eds. A. Sankaranarayanan, N. Amaresan, and D. Dhanasekaran, 41–54. CRC Press.

Rank, T. C., R. Grappin, and N. F. Olson. 1985. Secondary proteolysis of cheese during ripening: A review. *J Dairy Sci* 68:801–805.

Ray, R., and V. Joshi. 2014. Fermented foods: Past, present and future. https://doi.org/10.13140/2.1.1849.8241

Reboucas, K. H., L. Paes Gomes, A. M. O. Leite, and T. M. Uekane. 2016. Evaluating physicochemical and rheological characteristics and microbial community dynamics during the natural fermentation of cassava starch. *Food Process Technol J* 7. https://doi.org/10.4172/2157–7110.1000578

Reche, R. V., A. F. L. Neto, A. A. D. Silva, C. A. Galinaro, R. Z. D. Osti, and D. W. Franco. 2007. Influence of type of distillation apparatus on chemical profiles of Brazilian cachaças. *J Agri Food Chem* 55:6603–6608. https://doi.org/10.1021/jf0704110

Reps, A. 1993. Bacterial surface-ripened cheeses. In: *Cheese: Chemistry, Physics and Microbiology: Volume 2 Major Cheese Groups*, eds. P. F. Fox, P. L. H. McSweeney, T. M. Cogan, and T. P. Guinee, 137–172. Boston, MA: Springer US.

Rocha, R. S., R. Silva, J. T. Guimarães, et al. 2020. Possibilities for using ohmic heating in Minas Frescal cheese production. *Food Res Int* 131:109027. https://doi.org/10.1016/j.foodres.2020.109027

Rodrigues, L. M. A., M. das Graças Cardoso, W. D. Santiago, et al. 2016. Phenolic extracts of cachaça aged in different woods and quantifying antioxidant activity and antifungal properties: Phenolic extracts of cachaça aged in different woods. *J Inst Brew* 122:644–652. https://doi.org/10.1002/jib.375

Rul, F., V. Monnet, and J.-C. Gripon. 1994. Purification and characterization of a general aminopeptidase (St-PepN) from *Streptococcus salivarius ssp. thermophilus* CNRZ 302. *J Dairy Sci* 77:2880–2889. https://doi.org/10.3168/jds.S0022–0302(94)77228–3

Rusul, G., and N. H. Yaacob. 1995. Prevalence of *Bacillus cereus* in selected foods and detection of enterotoxin using TECRA-VIA and BCET-RPLA. *Int J Food Microbiol* 25:131–139. https://doi.org/10.1016/0168–1605(94)00086-l

Schlei, K. P., B. L. Angioletti, L. Fernandes de Carvalho, S. L. Bertoli, M. G. Ratto Reiter, and C. Krebs de Souza. 2020. Influence of temperature on microbial growth during processing of kochkäse cheese made from raw and pasteurised milk. *Int Dairy J* 109:104786. https://doi.org/10.1016/j.idairyj.2020.104786

Schwan, R. F., E. G. Almeida, M. A. G. Souza-Dias, and L. Jespersen. 2007. Yeast diversity in rice-cassava fermentations produced by the indigenous Tapirapé people of Brazil. *FEMS Yeast Res* 7:966–972. https://doi.org/10.1111/j.1567–1364.2007.00241.x

Schwan, R. F., C. A. do Santos, R. R. M. M. Santos, and W. F. Duarte. 2014. Bacterial dynamics and chemical changes during the spontaneous production of the fermented porridge (Calugi) from cassava and corn. *Afr J Microbiol* 8:839–849. https://doi.org/10.5897/AJMR2013.6240

Schwan, R. F., C. L. Ramos, E. G. de Almeida, V. F. Alves, and E. C. P. De Martinis. 2017. Brazilian indigenous fermented food. In: *Fermented Foods of Latin America: From Traditional Knowledge to Innovative Applications*, eds. A. L. B. Penna, L. A. Nero, and S. D. Todorove, 224–236. CRC Press Fermented Foods of Latin America.

Seixas, V. N. C., M. R. Félix, G. M. Silva, and I. T. Perrone. 2015. Characterization of Marajó's butter-type cheese made in two seasons of the year (portuguese). *Ciência Rural* 45:730–736. https://doi.org/10.1590/0103–8478cr20140463

Shaker, R., T. Osaili, W. Al-Omary, Z. Jaradat, and M. Al-Zuby. 2007. Isolation of *Enterobacter sakazakii* and other Enterobacter sp. from food and food production environments. *Food Control* 18:1241–1245. https://doi.org/10.1016/j.foodcont.2006.07.020

Shi, Y., Q. Zhai, D. Li, et al. 2017. Restoration of cefixime-induced gut microbiota changes by *Lactobacillus cocktails* and fructooligosaccharides in a mouse model. *Microbiol Res* 200:14–24. https://doi.org/10.1016/j.micres.2017.04.001

Siepmann, F. B., B. S. de Almeida, V. Ripari, et al. 2019. Brazilian sourdough: microbiological, structural, and technological evolution. *Eur Food Res Technol* 245:1583–1594. https://doi.org/10.1007/s00217-019-03254-8

Silva, R. A., M. S. F. Lima, J. B. M. Viana, et al. 2012. Can artisanal "Coalho" cheese from Northeastern Brazil be used as a functional food? *Food Chem* 135:1533–1538. https://doi.org/10.1016/j.foodchem.2012.06.058

Slow Food Brasil. 2020. *Conhecendo e degustando os Queijos Artesanais do Brasil (portuguese)*. https://slowfoodbrasil.org/. Accessed November 30, 2020.

Sonati, J. G., R. Vilarta, C. de Silva. 2009. Influências Culinárias e Diversidade Cultural da Identidade Brasileira: Imigração, Regionalização e suas Comidas (portuguese). In: *Qualidade de vida e cultura*, eds. R. T. Mendes, R. Vilarta, and G. L. Gutierrez, 137–147. IPES.

Stanley, G. 2003. CHEESES|Starter cultures employed in cheesemaking. In: *Encyclopedia of Food Sciences and Nutrition*, ed. B. Caballero, 1051–1056. Academic Press.

Tong, Y., Q. Zhai, W. Lu, et al. 2017. New insights in integrated response mechanism of *Lactobacillus plantarum* under excessive manganese stress. *Food Res Int* 102:323–332. https://doi.org/10.1016/j.foodres.2017.10.014

Uchechukwu-Agua, A. D., O. J. Caleb, and U. L. Opara. 2015. Postharvest handling and storage of fresh Cassava root and products: A review. *Food Bioprocess Technol* 8:729–748. https://doi.org/10.1007/s11947-015-1478-z

Velasco, I., S. Tenreiro, I. L. Calderon, and B. André. 2004. *Saccharomyces cerevisiae* Aqr1 is an internal-membrane transporter involved in excretion of amino acids. *Eukaryot Cell* 3:1492–1503. https://doi.org/10.1128/EC.3.6.1492–1503.2004

Vidal, R. H. L. 2011. Diagnóstico regional do processo de queijo coalho comercializado em Natal/RN (portuguese). Master's Thesis. Federal University of Rio Grande do Norte.

Wang, J., X. Zhao, Z. Tian, Y. Yang, and Z. Yang. 2015. Characterization of an exopolysaccharide produced by *Lactobacillus plantarum* YW11 isolated from Tibet Kefir. *Carbohydr Polym* 125:16–25. https://doi.org/10.1016/j.carbpol.2015.03.003

Welthagen, J. J., and B. C. Viljoen. 1998. Yeast profile in Gouda cheese during processing and ripening. *Int J Food Microbiol* 41:185–194. https://doi.org/10.1016/s0168–1605(98)00042–7

Whitehead, T. R., and M. A. Cotta. 1995. Identification of intracellular amylase activity in *Streptococcus bovis* and *Streptococcus salivarius*. *Curr Microbiol* 30:143–148. https://doi.org/10.1007/BF00296199

Wilkinson, J., C. Cerdan, and C. Dorigon. 2017. Geographical indications and "origin" products in Brazil: The interplay of institutions and networks. *World Dev* 98:82–92. https://doi.org/10.1016/j.worlddev.2015.05.003

Zeng, X. Q., D. D. Pan, and Y. X. Guo. 2010. The probiotic properties of *Lactobacillus buchneri* P2. *J Appl Microbiol* 108:2059–2066. https://doi.org/10.1111/j.1365–2672.2009.04608.x

Zheng, J., S. Wittouck, E. Salvetti, et al. 2020. A taxonomic note on the genusLactobacillus: Description of 23 novel genera, emended description of the genus Lactobacillus Beijerinck 1901, and union of Lactobacillaceae and Leuconostocaceae. *Int J Syst Evol Microbiol* 1:2782–2858. https://https://doi.org/10.1099/ijsem.0.004107

Zheng, X.-W., Z. Yan, B.-Z. Han, et al. 2012. Complex microbiota of a Chinese "Fen" liquor fermentation starter (Fen-Daqu), revealed by culture-dependent and culture-independent methods. *Food Microbiol* 31:293–300. https://doi.org/10.1016/j.fm.2012.03.008

Traditional Andean Potato-Based Fermented Foods and Their Potential for Health

25

Lena Gálvez Ranilla

25.1 INTRODUCTION AND HISTORICAL BACKGROUND

The potato is considered one of the major primary food crops worldwide after sugar cane, maize, rice, and wheat (FAO 2022). Countries such as China, India, and Ukraine have been registered as the main world producers, according to recent statistical data (FAO 2022). This current expansion of potato cultivation across the world is the result of a domestication process that began thousands of years ago in a geographical area different from that of current major production areas. Western South America has been reported as the primary center of origin and diversity of the potato (De Haan and Rodriguez 2016). The origin and domestication of the cultivated potato (*Solanum tuberosum*) has been proposed to occur more than 8000 years ago in the region encompassing the northern and southern Andes in Peru (Garzón 2007; De Jong 2016). Archaeological excavations in the central coast of Peru revealed remains of this tuber dating from about 8000–6000 years BC (Lumbreras 2006), indicating the expansion of potato intake from Andes to lowland areas. However, its intensive cultivation seems to have taken place in the region close to Lake Titicaca (located on the border of current countries of Peru and Bolivia) (Garzón 2007). This was the base of the rise of important Pre-Columbian societies such as the *Tiwanaku* civilization (500–1000 AD) that settled down near Lake Titicaca and spread from southern Peru to northern regions of Bolivia, Argentina, and Chile (Garzón 2007; De Jong 2016). After that, the Inca civilization settled in Cuzco and expanded its territory around 4000 km from Ecuador in the north to central Chile in the south, creating the *Tawantinsuyo* empire (1430 AD) (Garzón 2007). Incan society improved the intensive agriculture of potato, and it therefore became a staple food among the Andean population (Garzón 2007; FAO 2009).

Currently, cultivated species such as *Solanum tuberosum*, *Solanum curtilobum*, *Solanum juzepczukii*, and *Solanum ajanhuiri* have been recognized, along with two *S. tuberosum* groups: *S. tuberosum* Andigenum and Chilotatum (De Haan and Rodriguez 2016). The *S. tuberosum* Andigenum group is distributed from northern Chile and Argentina to Colombia; includes diploid, triploid, and tetraploid subgroups; and has the highest intraspecific diversity, generally embracing non-bitter (with low glycoalkaloid contents), floury varieties and landraces (Ovchinnikova et al. 2011; De Haan and Rodriguez 2016; De Haan et al. 2019). *S. curtilobum* and *S. juzepczukii* species include bitter landraces and are distributed from central Peru to southern Bolivia, whereas *S. ajanhuiri* (also a bitter species) is exclusively cultivated in the Peruvian–Bolivian Altiplano close to Lake Titicaca (De Haan and Rodriguez 2016; De Haan et al. 2019).

The process of potato domestication in the Andean region involved not only the adaptation and improvement of agronomic traits but also the selection of non-bitter characteristics in potato to make this food edible (Lumbreras 2006). Bitter potato species are cold resistant and naturally have high glycoalkaloid contents, which give them protection against pathogens and pests in highlands (Ginzberg et al. 2009; Dahlin et al. 2017). However, glycoalkaloids confer a bitter taste and at certain doses are toxic to humans (Langkilde et al. 2008; De Jong 2016).

Ancient Andean people found innovative ways to overcome such drawbacks through the application of different processes adapted to the Andean environmental conditions. The use of a natural freeze-drying method to process bitter potatoes to obtain a product called *chuño* (term used in the Aymara and Quechua native languages) has been reported to occur between 2000–3000 years ago (Horkheimer 2004; Peñarrieta et al. 2011a). The processing of *chuño* would have been applied even before the Tiwanaku period (400 BC) according to archaeological evidence, showing a maximum expansion in the Inca

DOI: 10.1201/9781003605300-25

period (Goldstein 2003; Horkheimer 2004). During *chuño* processing and use, most of the glycoalkaloid compounds are eliminated, and the final freeze-dried product can be stored long term. In fact, *chuño* played an important role in the development of the Tiwanaku and Inca civilizations since it was the base of food security in the Andean environments (Fonseca and Ordinola 2011).

Another technique used by Andean communities for processing potatoes while increasing their shelf life was fermentation. *Tocosh* is a naturally potato-based fermented product prepared since pre-Columbian times and used as food and a medicine (Velasco-Chong et al. 2020). This product is also referred to as the "antibiotic of the Incas" because it is believed to contain a penicillin-type effect, according to traditional Andean medicine (De Moreno et al. 2015; De Haan and Rodriguez 2016).

Both *chuño* and *tocosh* are still produced and consumed among the Andean population; however, most of the information about their nutritional and health-relevant functional properties is based on traditional knowledge inherited from ancient times.

Some publications have shown that *chuño* and *tocosh* may be important sources of beneficial probiotic microorganisms; however, scientific information about these Andean fermented foods is still very limited (Jiménez et al. 2018; Ramos et al. 2018). The objective of the present chapter is to analyze the current knowledge about the traditional processing of *chuño* and *tocosh*, their composition, health benefits, microbiome diversity, and associated potential for health and biotechnological applications. The implications of the valorization of these fermented foods for the development of Andean food systems are also emphasized through suggestions about future research.

25.2 TRADITIONAL PROCESSING OF ANDEAN POTATO-BASED FERMENTED FOODS

25.2.1 Andean Freeze-Dried Potato, *Chuño*

Freeze-dried Andean potatoes are locally named *chuño* and are classified in two types: white and black *chuño*. White *chuño*, also known as *tunta* in the Aymara native language, is the common name given to the freeze-dried Andean potato which has a snow-white color and a rough surface with slight grooves depending on the origin and quality of the potato variety (Peruvian Ministry of Production 2007; Fonseca et al. 2008). This product is generally made from medium to large potatoes (from 3.9 to >5.1 cm diameter) (Fonseca et al. 2008). Black *chuño* is also a freeze-dried potato but with dark brown color and commonly processed from small potatoes (around 3 cm diameter or less) (Peruvian Ministry of Production 2007).

White and black *chuño* production is usually spread throughout the Andean highland regions of Peru and Bolivia countries at altitudes between 3800 and 4600 meters above sea level (masl) (Yoshikawa and Apaza 2020). During the dry season (from May to the end of July), the temperatures are commonly below the freezing point at night (from −1.4 to −5°C), whereas high daily solar radiation and low relative humidity (average 27% in the southern Peruvian Andes) have been registered (Fonseca et al. 2008; De Haan et al. 2010; Peñarrieta et al. 2011a). Andean inhabitants have taken advantage of these temperature alternations to naturally freeze-dry potatoes and produce *chuño* to ensure food security in periods of scarcity in this region. It has been reported that this food can last up to 10 years, but currently the product is stored at most 1 year for commercial purposes (Fonseca et al. 2008; De Haan et al. 2010). In addition to such natural adaptation of potato processing, *chuño* processing does not require the use of fuel, which makes this technology a sustainable food production model in Andean systems (Yamamoto 1988).

Another important objective of the production of *chuño* is to decrease the glycoalkaloid content from bitter native potato species and make them available as food. Glycoalkaloids are steroidal alkaloids and toxic compounds commonly found in the *Solanum* genus and are responsible for the bitterness in wild-type potato varieties where glycoalkaloids are found at higher concentrations (Friedman and Levin 2016). Acute toxicity with glycoalkaloids has been associated with adverse effects on the central nervous system due to the inhibition of acetylcholinesterase (a neurotransmitter), whereas gastrointestinal disorders and effects on other organs by disruption of the cell membranes have been observed for low toxicity (Keukens et al. 1995; Friedman et al. 1997; Langkilde et al. 2008).

Traditionally, the raw material (fresh potatoes) used to produce *chuño* was exclusively wild-type or bitter species. However, sweet or non-bitter and improved varieties such as those from the *S. tuberosum* ssp. *andigena* group are also currently used at the commercial level (Table 25.1) (Fonseca et al.

TABLE 25.1 Potato Species Used for Processing *chuño* in Peru and Bolivia

SPECIE	PLOIDY	CHARACTERISTIC
Solanum tuberosum ssp. *andigena*	$2n = 4x = 48$	Non-bitter Includes floury and improved varieties
Solanum stenotomum	$2n = 2x = 24$	Non-bitter Includes floury varieties
Solanum juzepczukii	$2n = 3x = 36$	Bitter
Solanum curtilobum	$2n = 5x = 60$	Bitter
Solanum ajanhuiri	$2n = 2x = 24$	Bitter

Source: Adapted from Terrazas et al. (2008), Fonseca et al. (2008), De Haan (2009), and De Haan et al. (2010).

2008; Terrazas et al. 2008; De Haan et al. 2010; Peñarrieta et al. 2011a). Among the bitter species, *S. juzepzukii* is extensively used to produce *chuño* in Peru and Bolivia.

The processing of white *chuño* has several variations across the Andean region, but basically two general procedures have been reported. Figure 25.1 shows the traditional technology applied to produce white *chuño* in the highlands of Peru (Figure 25.1A) and Bolivia (Figure 25.1B). Potatoes are first selected according to their size and variety to be processed in different groups to ensure a uniform freeze-drying process. Then potatoes are evenly spread on the ground in a single layer and then frozen for 3–4 days, taking advantage of cold temperatures at night. Tubers are piled in the early morning and protected during the day with straw, thick blankets, or nets to avoid the potatoes browning due to sunlight exposure (Fonseca et al. 2008). This procedure is repeated for 3–4 days until the tubers reach a stone-like texture (Fonseca et al. 2008; De Haan et al. 2010; Peñarrieta et al. 2011a). This process is generally performed in flat areas or *pampas* close to a river or running water. Yoshikawa and Apaza (2020) determined that at the beginning of the 3–4-day period, potatoes are exposed to freeze–thaw cycles where water inside the potatoes reaches supercooling temperatures (1–3°C colder than the freezing point), followed by a sudden increase of temperature to the freezing point (from −0.17 to −1°C, depending on the potato variety). After several cycles of potato freeze–thaw, both the supercooling and freezing temperature points decrease, creating a temperature gradient that causes water evaporation from the thawed potatoes during high temperatures in the daytime (Yoshikawa and Apaza 2020).

Differences between the Peruvian and Bolivian methods are observed from this point (Figure 25.1). In case of the Peruvian method, potatoes are fermented by soaking the frozen tubers in special net-built cages in a running water river for around 20–30 days, with regular rotation every 5–7 days (Fonseca et al. 2008; De Haan et al. 2010). In contrast, a pressing step is applied before fermentation in the Bolivian method. Frozen potatoes are pounded barefoot, allowing the mechanical removal of the peel and water (Peñarrieta et al. 2011a). The fermentation stage in the Bolivian case consists of storing the semidried tubers in 5-m-deep holes, or *tajanas* (in the Aymara language), located close to a river or a pond for 21–30 days. *Tajanas* are filled with running water to maintain a cold and humid environment, and potatoes are covered with straw and flat stones (Paz-Hidalgo 1995; Peñarrieta et al. 2011a).

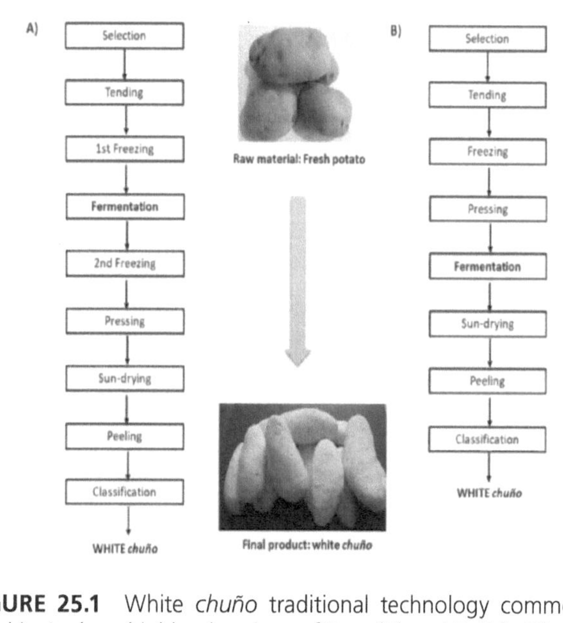

FIGURE 25.1 White *chuño* traditional technology commonly used in Andean highland regions of Peru (A) and Bolivia (B).

Source: Adapted from Paz-Hidalgo (1995), Fonseca et al. (2008), De Haan et al. (2010), and Peñarrieta et al. (2011a).

Following the fermentation, tubers are spread in a field on straw or nets and exposed to a second freezing for one night in the Peruvian method (Figure 25.1A). Next, potatoes are placed within net bags and the peel is removed by stepping on tubers (pressing) in the water during the day (Fonseca et al. 2008; De Haan et al. 2010). Tubers are then spread on clean nets and sun-dried for 7–10 days. The remaining skin is removed by abrasive manual separation until a white product is obtained (Fonseca et al. 2008; De Haan et al. 2010). In case of the Bolivian process, potatoes are sun-dried after fermentation, and a final peeling is applied to remove remaining skin (Peñarrieta et al. 2011a). Finally, white *chuño* is classified based on quality parameters such as color, size, and physical appearance (Fonseca et al. 2008). The average total time of processing white *chuño* varies between 38 and 42 days, and the tubers are always protected from the sunlight exposure until the foot-pounding step in both cases (Fonseca et al. 2008; De Haan et al. 2010; Peñarrieta et al. 2011a).

Depending on the Andean region, variations in the production of white *chuño* have been observed and are generally related to the sequence and duration of each processing step. In some localities of the central Peruvian Andes, white *chuño* is obtained in a similar way as in Bolivia but with different fermentation times (from 5–13 to 7–21 days) (Werge 1979; De Haan et al. 2010). There is another product called *moraya* which is generally reported as white *chuño*. However, it corresponds to a slightly different product, according to Peruvian regulations, since the fermentation process is carried out in semi-stagnant water instead of the running water usually used in the white *chuño* preparation (Peruvian Ministry of Production 2007).

Previous publications have commonly referred to the fermentation phase in white chuño processing as a "washing or soaking step", highlighting that its main objective is the reduction of glycoalkaloid contents in bitter potato varieties (Yamamoto 1988; Fonseca et al. 2008; De Haan et al. 2010; Peñarrieta et al. 2011a). Investigations by Ramos et al. (2018) and Santos-Mendoza et al. (2018) have pointed out that a submerged fermentation process also occurs in potatoes at this stage; however, more research is needed to determine the variability of the microbiome composition under the different technological parameters applied to obtain *chuño*.

The traditional processing of black *chuño* is shown in Figure 25.2. Potatoes are also spread on the ground for about 1–4 days, then frozen, stomped barefoot, and sun-dried,

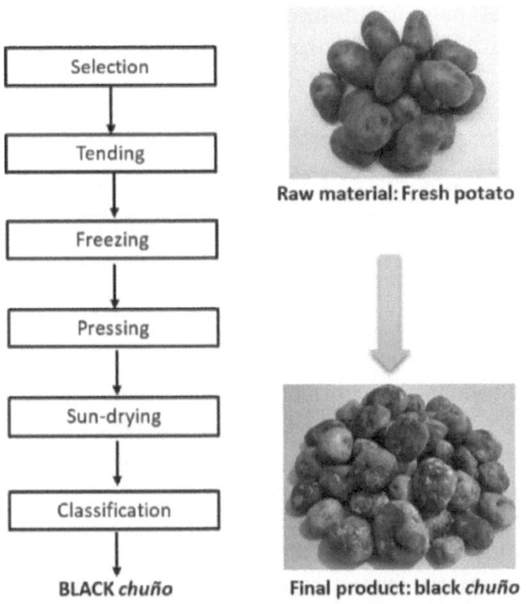

FIGURE 25.2 Black *chuño* traditional technology used in the Andean highland regions from Peru and Bolivia.

Source: Adapted from De Haan et al. (2010) and Peñarrieta et al. (2011a).

similarly to the white *chuño* process. However, in the black *chuño* procedure, tubers are exposed to sunlight during the freezing and treading steps and are not soaked into water. The production time lasts around 14 days, and the technology is almost the same across the Andean Peruvian and Bolivian regions (De Haan et al. 2010; Peñarrieta et al. 2011a). A fermentation process has not been clearly reported to take place in black *chuño*. Zvietcovich et al. (1985) stated that in some central and southern Peruvian Andean communities, a maturation period may follow the treading or pressing underfoot during the processing of black *chuño*. In this phase, moist tubers are piled up and fermented for 2 or 3 days, giving *chuño* its characteristic final flavor (Zvietcovich et al. 1985). Different lactic acid bacterial species have been found in unprocessed Andean potatoes and were variable depending on the tuber variety (Ramos et al. 2018). This microbiota may play a role in the maturation time; however, more research should be undertaken to understand the biochemical and microbiological changes during black *chuño* preparation.

Currently, white and black *chuño* are still processed under artisanal conditions following traditional techniques. In the case of white *chuño*, some local initiatives to improve the final quality of the product such as the use of disinfected tools in the process and the pressing of frozen potatoes using rubber boots instead of barefoot have increased its commercial value (Fonseca et al. 2008). Although it is common to find black *chuño* in local Andean markets, its production is mostly targeted to domestic consumption among Andean communities. *Chuño* should be rehydrated for 10 to 30 minutes with water and boiled before consumption (Fonseca et al. 2008). It is primarily prepared in soups or stews, combined with grains and meat, while in other preparations, *chuño* is mixed with cheese or fruits (Werge 1979).

25.2.2 Fermented Andean Potato, *Tocosh*

Tocosh (also known as *togosh* or *toqosh*) is an ancestral potato-based fermented product currently prepared by small communities located in the northern and central Andean regions from Peru such as Ancash, La Libertad, Huanuco, and Junin (De Haan and Rodriguez 2016; Velasco-Chong et al. 2020). The wider processing of *chuño* in these regions is not common since the weather is more humid and minimum temperatures are higher than those reached in the dry southern Andean and Altiplano areas (Schindler and De Haan 2021). Thus, the production and processing of *tocosh* is an adaptation of potato processing according to the different Andean environmental conditions to guarantee food security in places where wider *chuño* processing was not possible (Schindler and De Haan 2021).

The general procedure to obtain *tocosh* is shown in Figure 25.3. The process begins with the building of pools protected with straw or mesh (0.70 × 1.50 m deep) near a river or a stream (Jiménez et al. 2018; Velasco-Chong et al. 2020). The raw material commonly consists of discarded whole bitter potato varieties, which are placed into the pools, piled between straw layers, and finally covered on top with straw and stones to secure them (Yamamoto 1988; De Moreno et al. 2015). Then the potatoes are left to ferment for variable times and conditions. De Moreno et al. (2015) reported that fermentation time

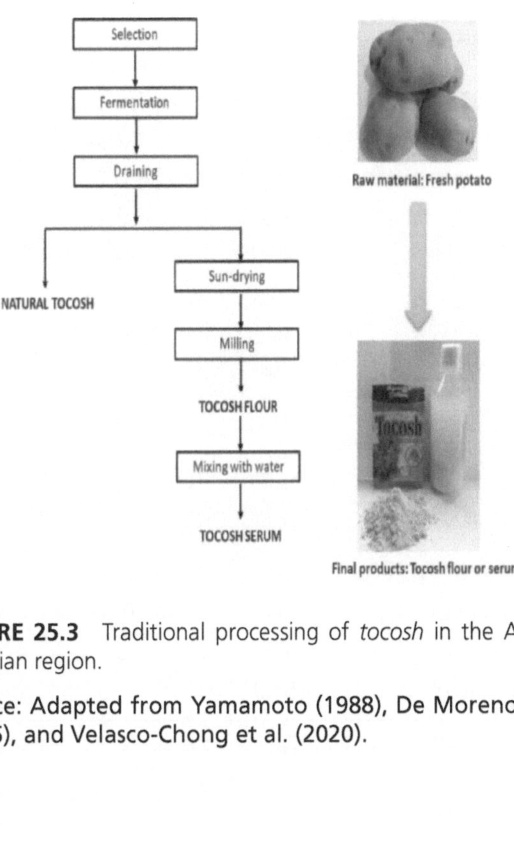

FIGURE 25.3 Traditional processing of *tocosh* in the Andean Peruvian region.

Source: Adapted from Yamamoto (1988), De Moreno et al. (2015), and Velasco-Chong et al. (2020).

may vary from 1 to 12 months in slight running water conditions, whereas fermentation times ranging from 2 weeks to 12 months in stagnated water have been mentioned by Yamamoto (1988).

During the fermentation stage, potatoes darken due to enzymatic browning, and several biochemical and microbiological processes occur, as will be further analyzed in later sections (Ali et al. 2016). The fermentation is primarily carried out by lactic acid bacteria, which may play a role in glycoalkaloid degradation in bitter potato varieties (Veselá et al. 2002; Jiménez et al. 2018).

After the fermentation time, the potatoes are taken out of the pools and placed in a dry shaded area for water draining (De Moreno et al. 2015). No references have been found for the duration of this step, but it may allow obtaining a partially dried, fermented, unpeeled potato with reduced size and a strong and unpleasant odor (Velasco-Chong et al. 2020). This final product is also known as "natural *tocosh*" and is typically placed within sacks and covered with straw for self-consumption or commercial purposes. Other further processing includes sun-drying and milling to get a pulverized product, or "*tocosh* flour" (Figure 25.3) (De Moreno et al. 2015). This form of *tocosh* has longer shelf life (around 1 year) that allows its distribution to other Andean regions. *Tocosh* is usually consumed after a boiling process. The natural *tocosh* or the flour may be used in the preparation of soups, stews, or *api* (term in the Quechua native language for *mazamorra*, which is a sort of sweet porridge). Another common presentation of *tocosh* is like a "serum" which is prepared by diluting 1 teaspoon of *tocosh* flour (around 1–3 g) in 100 mL of water (Velasco-Chong et al. 2020). This preparation is generally targeted for medicinal uses, and it is recommended to be consumed before food intake (Velasco-Chong et al. 2020).

25.3 NUTRITIONAL AND BIOACTIVE COMPOSITION

25.3.1 *Chuño*

The nutritional composition of raw *chuño* is shown in Table 25.2. According to the Peruvian Food Composition Table, the macronutrient, mineral, and vitamin contents of both types of raw *chuño* have an overall decrease (except in the total carbohydrates, calcium, and iron) in comparison to the unprocessed material (raw potato) (MINSA-INS-CENAN 2017). However, these comparisons are only referential since the information about the potato variety used for the processing of *chuño* has not been registered by MINSA-INS-CENAN (2017).

Table 25.2 also includes the nutritional composition of boiled potato and *chuño*, which reflects their common forms of preparation and consumption. Burgos et al. (2009) compared the protein concentrations of nine cooked potato varieties from the central Peruvian Andes with the corresponding boiled white *chuño* products. A significant decrease of the protein contents (around 48–83% in fresh weight) was observed following *chuño* processing, and the contents of this macronutrient varied depending on the potato variety (Burgos et al. 2009).

TABLE 25.2 Nutritional Composition of Raw and Boiled *Chuño* per 100 g in Dry Weight (dw)

	RAW			BOILED		
COMPOUND	*POTATO*[a]	*WHITE CHUÑO*[a]	*BLACK CHUÑO*[a]	*POTATO*	*WHITE CHUÑO*	*BLACK CHUÑO*
Protein (g)	8.2	2.3	4.7	9.1–11.7[b]	1.8–4.2[b]	NR
Lipids (g)	0.4	0.6	0.2	NR	NR	NR
Ash (g)	3.9	2.2	2.7	NR	NR	NR
Total carbohydrates (g)	87.5	94.9	92.4	NR	NR	NR
Calcium (mg)	35.3	112.3	51.2	27–52.3[c]	46.3–123.6[c]	19.1–29.3[c]
Phosphorus (mg)	184.3	65.9	236.3	198.6–303.3[c]	81–102.3[c]	99.7–186.0[c]
Potassium (mg)	NR	NR	NR	1400.0–1576.6[c]	9.8–53.0[c]	440.0–876.7[c]
Magnesium (mg)	NR	NR	NR	70.7–91.7[c]	18.3–34.0[c]	23.7–60.0[c]
Sodium (mg)	NR	NR	NR	1.6–12.4[c]	11.3–19.0[c]	0.5–3.2[c]
Zinc (mg)	1.1	NR	NR	1.0–1.3[c]	0.3–0.6[c]	0.4–0.8[c]
Iron (mg)	2.0	4.0	1.0	1.7–2.2[c]	1.4–2.5[c]	2.3–3.1[c]
Thiamin (mg)	0.4	0.04	0.2	NR	NR	NR
Riboflavin (mg)	0.4	0.05	0.2	NR	NR	NR
Niacin (mg)	6.5	0.5	4.0	NR	NR	NR
Vitamin C (mg)	54.9	1.3	2.0	NR	NR	NR

Notes: [a]MINSA-INS-CENAN (2017). [b]Burgos et al. (2009). [c]De Haan et al. (2010). NR: not reported.

Variables such as the genotype, the production environment, and the type of process significantly influenced the mineral composition of boiled white *chuño* (De Haan et al. 2012). The concentrations of phosphorus, potassium, magnesium, and zinc generally decreased in both boiled *chuño* types in comparison with potato; however, white *chuño* shows the highest losses in comparison with black *chuño* (Table 25.2). Conversely, the ranges of calcium and iron concentrations increased following the *chuño* process. Minerals may decrease by leaching during the treading and the soaking/fermentation stages, whereas calcium and iron are likely absorbed from water in the case of white *chuño* (De Haan et al. 2010; De Haan et al. 2012). The effect of traditional potato freeze-drying on the contents of the other macronutrients such as lipids, carbohydrates, and dietary fiber along with vitamin composition need further investigation considering variables such as the potato variety, the type of processing, and other agroclimatic variables.

Regarding the bioactive composition of *chuño*, a study reported changes of phenolic antioxidants of several *Solanum* potato species from Bolivia during black *chuño* processing (Peñarrieta et al. 2011b). *Chuño* processed with different *S. tuberosum* varieties had 5.0–7.7 µmol gallic acid equivalents per gram in dry weight (µmol GAE/g dw) of total phenolic contents, whereas the total flavonoids varied from 0.2 to 0.6 µmol catechin equivalents per gram in dw (µmol CE/g dw) (Peñarrieta et al. 2011b). The total phenolic compounds in *chuño* decreased around 30–70% when compared with its corresponding unprocessed potato variety; however, specific phenolic compounds such as chlorogenic and protocatechuic acids remained almost unchanged after *chuño* processing (0.1–0.2 and 0.01–0.3 µmol/g dw of protocatechuic acid in *chuño* and potato, respectively; 0.6–1.0 and 0.7–1.0 µmol/g dw of chlorogenic acid in *chuño* and potato, respectively) (Peñarrieta et al. 2011b). Epicatechin (0.2–7.0 µmol/g dw) and gallic acid (0.06–1.6 µmol/g dw) were the major phenolic compounds in *chuño*, while syringaldehyde was detected at lower concentrations (<0.01–0.1 µmol/g dw) (Peñarrieta et al. 2011b). Despite the losses of some antioxidants during the *chuño* process such as epicatechin, gallic acid, and syringaldehyde, the total antioxidant capacity of 100 g of *chuño* (0.15 mmol trolox equivalents) was comparable to the value shown by 200 g of cauliflower (Nilsson et al. 2006; Peñarrieta et al. 2011b).

25.3.2 *Tocosh*

Table 25.3 shows the macronutrient composition of *tocosh* made with the Peruvian white potato variety *Hualash* in comparison with the corresponding unprocessed potato. A decrease in the protein, lipid, crude fiber, and ash contents is observed in potato following *tocosh* processing (Mercado 2020). A reduction of protein, fat, and mineral (sodium, potassium, calcium, magnesium, and phosphorus) contents in naturally fermented raw potato in water for 48 h with respect to unfermented potato was also found by Aregbe et al. (2019). These losses were likely attributed to a mixed effect of the metabolic activity of microorganisms and the leaching of derived soluble compounds into the fermenting medium (Aregbe et al. 2019). On the other hand, the total carbohydrate concentration in *tocosh* increased around 12% with respect to the value in raw potato (Table 25.3). Similar carbohydrate (80%) and protein (3.91%) concentrations with low contents of lipids were reported by De Moreno et al. (2015) in *tocosh*.

TABLE 25.3 Nutritional Composition of *Tocosh* Made from the Peruvian Potato Variety *Hualash* (g/100 g dw)

COMPOUND	*POTATO*	*NATURAL TOCOSH*
Protein	9.3	3.6
Lipids	1.6	1.0
Crude fiber	18.1	1.7
Ash	5.9	1.9
Total carbohydrates	83.3	93.4

Source: Adapted from Mercado (2020).

The bioactive composition of *tocosh* has been evaluated mostly at the qualitative level. Sugars, phenols, alkaloids, saponins, and steroids have been detected in a solution of *tocosh* flour from the region of Huanuco in Peru (Velasco-Chong et al. 2020). In another study, different metabolites such as phenolic compounds, alkaloids, triterpenoids, steroids, reducing sugars, and free amino acids were reported in a decoction of *tocosh* flour (*Solanum tuberosum*) processed in Junin (Peru) (Sandoval et al. 2015). The specific identification and contents of the previous compounds have not been investigated yet. Yábar et al. (2019) identified different organic acids by gas chromatography with potential antimicrobial properties such as butanoic or butyric acid, octadecanoic acid, and methyl formate in *tocosh* with 30 days of fermentation.

25.4 BIOCHEMICAL AND MICROBIOLOGICAL CHANGES IN ANDEAN POTATO-BASED FERMENTED FOODS

Scientific information about the biochemical and microbiological changes in Andean potato-based fermented foods is scarce in comparison with other traditional fermented foods worldwide (Negrete-Romero et al. 2021). Based on the technological process of both *chuño* and *tocosh* described previously, carbohydrates (mainly as a starch), which are found in high concentrations in raw potatoes (Tables 25.2 and 25.3), may be anaerobically fermented by the native microorganisms in succession with initially hydrolysis and subsequent fermentation as energy source during the soaking stage, likely stimulating the development of a diversity of microbiota and leading to diverse biochemical changes (Aregbe et al. 2019).

In the case of *chuño*, Ramos et al. (2018) evaluated the changes in the microorganism concentrations at different processing stages of white *chuño* processed in the Puno region (southern Peruvian Andes) from non-bitter potato varieties (*Solanum tuberosum* ssp. *andigena*, varieties Imilla and Chaska). The lactic acid bacteria (LAB) population in fresh potatoes (raw material) was higher in the Chaska variety (~4 log CFU/g) than the Imilla variety (~3 log CFU/g), revealing the influence of the potato variety and associated agronomical and growth environmental conditions on the autochthonous microbiota (Ramos et al. 2018). Independent of the potato variety, LAB counts increased to around 8 log CFU/g after 8 and 21 days of water immersion (fermentation stage) and decreased to approximately 4 log CFU/g in the final *chuño* product (after peeling and sun-drying), while the yeast (with lower counts than LAB) overall decreased at the end of the process (from 3 to 1 log CFU/g) (Ramos et al. 2018). Major LAB species identified using molecular methods were *Latilactobacillus curvatus*, *Leuconostoc mesenteroides*, *Latilactobacillus sakei*, *Levilactobacillus brevis*, and *Enterococcus mundtii* (Ramos et al. 2018). No information about the identification of yeast species in *chuño* has been found. However, Rodriguez and Chambi (2019) determined that white *chuño* may be a suitable substrate for the growth of *Saccharomyces boulardii*, showing high potential as a probiotic carrier.

Microbiological biotransformations in potato might be different in *tocosh* as compared with *chuño* due to differences in their technological process. The physicochemical parameters have been shown to vary along the fermentation time in *tocosh* according to some studies. Yábar and Reyes (2017) reported initial pH values of 5.5–5.8 (no fermentation) in Peruvian fresh potatoes (Yungay and Peruanita varieties from the Huancayo region) that decreased to 3.6–3.8 after 60 days of fermentation, while a simultaneous increase of the titratable acidity (from 0.09% to 0.48% lactic acid) was observed. In another study, the pH values varied from 5.5 (in fresh potatoes) to 3.8 and 4.2 after 1 and 8 months of fermentation, respectively (Jiménez et al. 2018). These chemical changes reflect the growth of organic acid-producing microorganisms such as lactic acid bacteria.

LAB are gram positive, rod- and coccus-shaped bacteria able to ferment carbohydrates, producing mostly lactic acid (over 85%) in the case of homofermentative microorganisms (e.g., *Lactobacillus* sp., *Pediococcus* sp.) and other final metabolites such as carbon dioxide, ethanol, organic acids, and volatile compounds in the case of heterofermentative bacteria (e.g., *Leuconostoc* sp.) (John et al. 2007; Reis et al. 2012; Barus et al. 2021). An increase in the LAB counts from 4 to 8 log CFU/g has been observed in *tocosh* after 60 days of fermentation (Yábar and Reyes 2017). Likewise, the LAB concentration increased (from ~4 to ~8 log CFU/g) in *tocosh* after 8 months of fermentation, whereas the total yeast and mold counts remained almost invariable until the first month (4 log CFU/g), decreasing to 2 log CFU/g at the end of the fermentation period (Jiménez et al. 2018).

In relation to the LAB identification, preliminary studies by Quillama et al. (2012) in *tocosh* from the northern and central Peruvian Andes revealed that over the 250 isolated bacterial strains, 80% and 20% corresponded to the *Lactobacillus* and *Pediococcus* genera, respectively. LAB species such as *Lactiplantibacillus plantarum*, *Companilactobacillus alimentarius*, *Lacticaseibacillus casei*, *Levilactobacillus brevis*, *Limosilactobacillus reuteri*, and *Limosilactobacillus fermentum* were identified in the same study using biochemical and culturing methods (Quillama et al. 2012).

High-throughput techniques, including DNA isolation, 16S rRNA gene amplification, and pyrosequencing, allowed the identification of a wide microbiological diversity in *tocosh*, which was variable depending on the fermentation time (Jiménez et al. 2018). The relative abundance (%) of the *Lactobacillus* genus was high (~95%) in fresh potatoes, with major predominance of *Latilactobacillus sakei*, *Leuconostoc gasicomitatum*, *Leuconostoc gelidum*, and *Leuconostoc mesenteroides* (Jiménez et al. 2018). Non-LAB genera such as *Clostridium* (non-pathogenic), *Propionispira*, and *Prevotella* predominated over LAB after 1 and 8 months of fermentation; however, *Lactobacillus* species diversity increased along the fermentation time (Jiménez et al. 2018). *Clostridium acetobutylicum* and *Clostridium tyrobutyricum* were the major species in 1-month fermented *tocosh*, followed by different LAB species (*Leuconostoc gasocomitatum*, *Latilactobacillus sakei*, *Lactiplantibacillus plantarum*, *Paucilactobacillus oligofermentans*, and *Levilactobacillus senmaizukei*), whereas *Lacticaseibacillus casei*, *Loigolactobacillus backii*, and *Levilactobacillus senmaizukei* were more abundant in the 8-month *tocosh* (Jiménez et al. 2018). In contrast to these results, the genus *Weissella*, followed by *Gluconobacter*, *Bacillus*, and *Klebsiella*, occurred in high abundance in Chinese potatoes previously steamed before a submerged 48-h fermentation, whereas the *Lactobacillus* genus was the less predominant (~3%) (Aregbe et al. 2019). These results reveal significant effects of the processing method on the microbial composition in potato-based fermented foods.

Yeasts are another group of microorganisms also found in *tocosh* but in lower numbers than LAB (Jiménez et al. 2018). However, no studies have been reported about the identification of yeast species as in the case of *chuño*. Aregbe et al. (2020) isolated strains of *Pichia* sp. strain AQGWD7 and *Saccharomyces cerevisiae* strain YJM1592 from potato fermented for 48 h and highlighted the ability of *S. cerevisiae* to produce gas and volatile compounds, which are important characteristics for potential technological applications. It is most likely that different strains of *S. cerevisiae* may be also present in *tocosh*, but this needs to be confirmed with further research.

A diversity of LAB isolated from *tocosh* and white *chuño* has shown different biotechnological properties, exhibiting strain-specific behavior (Table 25.4). The antimicrobial activity and the capacity of producing vitamin B compounds have been detected in several LAB isolated from Peruvian samples of *tocosh* and *chuño* (Yépez et al. 2017; Mosso et al. 2018; Ramos et al. 2018). It is widely known that LAB inhibit the growth of pathogenic and spoilage microorganisms through the production of several antimicrobial active metabolites such as organic acids, hydrogen peroxide, diacetyl, ethanol, fatty acids, antifungal compounds, and bacteriocins (Reis et al. 2012). In the case of some isolated strains from *tocosh*,

TABLE 25.4 Biotechnological Potential of Lactic Acid Bacteria from *Tocosh* and *Chuño*

ORIGIN	*BIOTECHNOLOGICAL PROPERTY*	*ISOLATED MICROORGANISMS*	*ANTIMICROBIAL ASSOCIATED COMPOUND*	*SOURCE*
Tocosh (fermented in 1–8 months), elaborated in the Huanuco region. Central Peruvian Andes.	Folate (vitamin B9) production	*Leuconostoc mesenteroides, Latilactobacillus sakei*	–	Jiménez et al. (2018)
	Riboflavin (vitamin B2) production	*Leuconostoc mesenteroides, Latilactobacillus sakei*	–	
	Exopolysaccharides production from sucrose	*Leuconostoc mesenteroides, Latilactobacillus sakei*	–	
	Phytate-degrading activity	*Leuconostoc mesenteroides, Latilactobacillus sakei, Lacticaseibacillus casei*	–	
	Amylolytic activity	*Leuconostoc mesenteroides, Latilactobacillus sakei*	–	
	Biogenic amines (tyramine) production	*Leuconostoc mesenteroides, Latilactobacillus sakei, Lacticaseibacillus casei*	–	
	Antibacterial activity	*Leuconostoc mesenteroides* (against *Listeria monocytogenes* FBUNT and *Bacillus subtilis* 168), *Latilactobacillus sakei* (against *Listeria monocytogenes* FBUNT and *Bacillus subtilis* 168), *Lacticaseibacillus casei* (against *Bacillus subtilis* 168)	Proteinaceous compound	
Tocosh (fermented in 1–8 months), elaborated in the Huanuco region. Central Peruvian Andes.	Folate production and probiotic activity (tolerance to gastrointestinal conditions)	*Latilactobacillus sakei* 2T3MS8, 2T2MM10 *Limosilactobacillus fermentum* T3M3 *Lacticaseibacillus casei* 3T3MS2, 3T3M3	–	Mosso et al. (2018)
Tocosh No reference for the origin	Antifungal activity (spoiler fungi)	*Latilactobacillus sakei* T3M7, *Leuconostoc mesenteroides* T2MM3, 2T3Y9 (All strains against *Aspergillus oryzae* CECT 2094[NT], *Penicillium roquerfoti* CECT 2905[NT], *Meyerozyma guilliermondii* CECT 1021)	–	Yépez et al. (2017)
	Antifungal activity (mycotoxigenic fungi)	*Limosilactobacillus fermentum* T3M3, *Leuconostoc mesenteroides* T1M3 (All strains against *Aspergillus parasiticus* CECT 2681, *Penicillium expansum* CECT 2278, *Fusarium verticilloides* CECT 2987)	• 3,5 Di-O-caffeoylquinic acid • Cyclo (L-Tyr-L-Pro) • Caffeoylhexose-deoxyhexoside • Protocatechuic acid hexoside • Phenylpyruvic acid • All obtained from *L. fermentum* T3M3	

(*Continued*)

TABLE 25.4 (Continued)

ORIGIN	*BIOTECHNOLOGICAL PROPERTY*	*ISOLATED MICROORGANISMS*	*ANTIMICROBIAL ASSOCIATED COMPOUND*	*SOURCE*
	Antibacterial activity	*Companilactobacillus farciminis* T3Y6, *Limosilactobacillus fermentum* T3M3, *Lacticaseibacillus casei* 3T3MS2, *Latilactobacillus sakei* T3Y3, *Leuconostoc mesenteroides* T1M3, T1M5, T2MM3, T2MM5, T2MM6, T2MS1. (All strains against *Listeria innocua* CECT 910^T, *Escherichia coli* O157:H7 CECT 5947; *Salmonella enterica* ssp. *enterica* serovar Typhi CECT 4138)	–	
Chuño (*Solanum tuberosum* ssp. *andigena*, Imilla and Chaska varieties) with 21 days of fermentation elaborated in El Collao province, Puno region. Southern Peruvian Andes	Antibacterial activity	*Latilactobacillus curvatus* JCM 1096^T *Leuconostoc mesenteroides* *Enterococcus mundtii* DSM 4838^T (All strains against *Listeria innocua* 7)	Bacteriocin-like protein compound (in *Enterococcus mundtii* cell-free supernatant)	Ramos et al. (2018)
Chuño elaborated in El Collao province, Puno region. Southern Peruvian Andes.	Riboflavin (vitamin B2) production	*Leuconostoc mesenteroides* 2TP10-BAL18, 2TP08-BAL04, 2TP08-BAL09, 2TP02-BAL01	–	Santos-Mendoza et al. (2018)

the antifungal activity shown by *Limosilactobacillus fermentum* against the mycotoxigenic fungi *Aspergillus parasiticus*, *Penicillium expansum*, and *Fusarium verticillioides* has been associated with the production of several phenolic acid derivatives, whereas bacteriocin-like protein compounds produced by *Enterococcus mundtii* isolated from white *chuño* have been linked to the inhibition of *Listeria innocua* (Yépez et al. 2017; Ramos et al. 2018).

Additionally, exopolysaccharide (EPS) production and phytate-degrading capacity, relevant for improving the mineral availability (e.g., iron and zinc), have been also observed in strains of *Leuconostoc mesenteroides* and *Latilactobacillus sakei* isolated from *tocosh* (Jiménez et al. 2018). LAB strains of *Leuconostoc*, *Lactobacillus*, *Pediococcus*, *Weissella*, and *Streptococcus* have shown the capacity to synthetize EPS, which may play a role in different ecological functions, technological applications, and health benefits, including prebiotic potential (Korcz et al. 2018). Some *Lactobacillus* species isolated from Peruvian *tocosh* have also exhibited *in vitro* probiotic activity due to their ability to maintain the minimum recommended bacteria concentrations for a probiotic candidate (8 log CFU/mL) after their exposure to simulated gastrointestinal conditions (Mosso et al. 2018).

Based on this scientific evidence, *chuño* and *tocosh* might be considered natural sources of beneficial microorganisms, mainly LAB, with potential biotechnological and health applications. However, the influence on their microbiota diversity and associated functionalities due to several factors such as the potato variety, the fermentation time, and the environmental conditions of processing, which are generally variable along the Andean region, remain unknown and require further investigation.

25.5 HEALTH-RELEVANT FUNCTIONAL PROPERTIES ASSOCIATED WITH ANDEAN POTATO-BASED FERMENTED FOODS

The health-relevant functional properties attributed to *tocosh* and *chuño* according to some *in vitro* and *in vivo* studies are summarized in Table 25.5.

Tocosh has been used not only as a food but also as a medicine among Andean communities based on their traditional knowledge inherited since pre-Columbian times (Velasco-Chong et al. 2020). Popular health applications of *tocosh* include pneumonia, colds, postpartum treatment, gastric ulcers, wound healing, stimulation of the immune system, protection of gastric mucosa against inflammation, and as an antimicrobial agent, among others (De Moreno et al. 2015). It has been established that the no-observed-adverse-effect levels (NOAEL) of *tocosh* flour is 1000 mg/kg body weight (w) (in a rat model), which is within the regular dose traditionally consumed by Andean people for medicinal purposes (500–1000 mg/kg daily) (Velasco-Chong et al. 2020).

Despite its importance in traditional Andean medicine, few scientific reports that support such health claims have been published to date. The health benefits of *tocosh* have been investigated in its natural form or prepared as a porridge. Hydroalcoholic extracts from natural *tocosh* have shown *in vitro* antibacterial activity against Gram-positive bacteria such as *Staphylococcus aureus*, *Streptococcus mutans*, and *Streptococcus mitis*, showing potential for the protection against oral infections (Yábar et al. 2019; Mayta-Tavalino et al. 2019; Enciso et al. 2020). Based on previous results, Mayta-Tovalino et al. (2019) developed an experimental *tocosh* toothpaste which showed antibacterial effect against *S. aureus* and *S. mutans*. The authors suggested that an antimicrobial protein homologous of acid phosphatase previously isolated from potato could be involved (Feng et al. 2003). The antimicrobial compounds biosynthesized by LAB during *tocosh* fermentation may also play an important role and would partly explain its common use as "antibiotic" for the treatment of infections in traditional Peruvian medicine (Yépez et al. 2017).

The protection and regenerative capacity of the gastric mucosa has also been investigated in *tocosh* through *in vivo* studies. *Tocosh* porridge (cooked product) administered at doses from 900 to 2700 mg/kg weight (w) in rats before an ethanol-induced gastric ulcer treatment showed cytoprotective properties on rat gastric mucosa. The dose of 1800 mg/kg weight (w) had the best results (97.1% protection), even higher than sucralfate drug treatment (92.7%) (Sandoval et al. 2015). When the ethanol was first administered before the *tocosh* porridge treatment at a dose of 800 mg/kg, the gastric ulcers induced by ethanol were reduced from 29.5% to 2.96% at 48 h and to 0.89% after 72 h, which reveals the regenerative capacity of *tocosh* (Loli et al. 2016). An orogastric intubation for *tocosh* administration was used in both studies, therefore results under regular oral intake, and extrapolated to humans may be different and should be investigated.

The mechanisms involved in the reported *in vitro* and *in vivo* beneficial effects of *tocosh* have not been completely studied. The cytoprotective effect of *tocosh* seems to be linked to its antioxidant capacity, since the gastric mucosa tissue subjected to the *tocosh* treatment (at a dose of 2700 mg/kg w) showed lower lipoperoxidation than the ethanol-treated gastric mucosa (Sandoval et al. 2015). The *in vivo* attenuation of HCl/ethanol-induced gastric damage by probiotic *Limosilactobacillus fermentum* has been reported and was associated with different antioxidant mechanisms (Suo et al. 2016). Antioxidant peptides derived from a potato protein hydrolysate have also shown protection against gastric mucosal injury due to ethanol (Kudoh et al. 2003). Furthermore, Chintha et al. (2021) observed that the *in vitro* free radical scavenging antioxidant capacity of unfermented sweet potatoes was retained or improved following a 72-h fermentation with *Lactiplantibacillus plantarum* and varied depending on the tuber variety. Consequently, biologically active peptides, among other antioxidant metabolites likely derived from the LAB biotransformation reactions,

TABLE 25.5 Reported Health-Relevant Functional Properties in *Tocosh* and *Chuño*

SAMPLE	*ORIGIN*	*STUDIED HEALTH-RELEVANT FUNCTIONAL PROPERTY*	*MODEL USED AND EXPERIMENTAL DESIGN*	*MAIN OUTCOMES*	*SOURCE*
Powdered *tocosh* from *Solanum* spp. No reference about the *tocosh* fermentation time.	Commercial product purchased in Lima (Peru)	Antioxidant capacity	*In vitro* A: Ethanol: water (80:20) extract B: Methanol: water: acetic acid (50:42:8) extract Methods: Total phenolic contents (TPC) ABTS[a], DPPH[b], and FRAP[c] antioxidant capacity methods	• TPC: 2.5 and 2.8 mg gallic acid equivalents/kg dw for A and B extracts, respectively. • 2.1 and 1.0 mmol trolox equivalents/kg dw for A and B extracts, respectively (ABTS[a]). • No response with the DPPH[b] and FRAP[c] methods.	Brizzolari et al. (2019)
Natural *tocosh* from *Solanum tuberosum* prepared under artisanal and laboratory conditions with 30 days of fermentation. Concentrated *tocosh* serum at 10, 20, and 30% was also evaluated.	No specified	Antioxidant and antibacterial capacity	*In vitro* • The antibacterial capacity with the Kirby-Bauer inhibition halo method (% of inhibition in relation to an antibiotic) T1: *Escherichia coli* ATCC 25922 with the artisanal, laboratory *tocosh* and serum at 10, 20, and 30% T2: *Staphylococcus aureus* ATCC 25923 with the artisanal, laboratory *tocosh* and serum at 10, 20, and 30% • The minimum inhibitory concentration (MIC) of different dilutions of *tocosh* against *E. coli* and *S. aureus* • The antioxidant capacity by the DPPH[b] antioxidant method (artisanal versus laboratory *tocosh*)	• *Tocosh* samples better inhibited both bacteria than serum and showed the highest inhibition values at 30 days of fermentation. • *S. aureus* was more sensitive to *tocosh* than *E. coli* (144% and 51–57% of inhibition, respectively, and MIC: 0.78% versus 25%, for *S. aureus* and *E. coli*, respectively) • The laboratory *tocosh* had higher antioxidant capacity than the artisanal sample showing the highest value at 30 days of fermentation (55%).	Yábar et al. (2019)
Natural freeze-dried *tocosh* from *Solanum tuberosum* with 6–9 months of fermentation. *Tocosh* ethanolic extract (TEE) (100 g/500 mL ethanol and macerated for 7 days).	Palcamayo district, Tarma province, Junin region. Central Peruvian Andes	Antibacterial and cytotoxic capacity	*In vitro* • The antibacterial capacity with the Kirby-Bauer inhibition halo method The TEE alone and a *tocosh* dentifrice formula with TEE were evaluated. A treatment with chlorhexidine (positive control) and distilled water (negative control) were included T1: *Streptococcus mutans* ATCC 25175 with TEE T2: *Streptococcus mitis* ATCC 49456 with TEE T3: *Staphylococcus aureus* ATCC 25923 with TEE T4: *Candida albicans* ATCC 10231 with TEE • Cytotoxic capacity with the MTT[d] method, using ET3 and, DU145 cell lines	• TEE showed antibacterial capacity on *S. aureus*, *S. mitis*, and *S. mutans* (mean inhibition halo of 19.6, 16.5, 13.7 mm, respectively). TEE did not inhibit *C. albicans*. • The inhibition of TEE against *S. aureus* and *S. mitis* was similar to the positive control. • The *tocosh* dentifrice also showed antibacterial action on *S. aureus, S. mitis*, and *S. mutans* (14.7 mm, 14.4 mm, and 16.2 mm, respectively, versus 14.4 mm shown by a control dentifrice). • No cytotoxic effect of TEE on ET3 and CU145 cell lines was found.	Mayta-Tovalino et al. (2019)

(Continued)

TABLE 25.5 (Continued)

SAMPLE	*ORIGIN*	*STUDIED HEALTH-RELEVANT FUNCTIONAL PROPERTY*	*MODEL USED AND EXPERIMENTAL DESIGN*	*MAIN OUTCOMES*	*SOURCE*
Natural *tocosh* from *Solanum tuberosum*. *Tocosh* hydroalcoholic extract (THE) (1 kg/10 L 60% ethanol and macerated for 10 days) and concentrated in a water bath at 40°C. No reference about the *tocosh* fermentation time.	Tarma province, Junin region. Central Peruvian Andes	Antibacterial capacity	*In vitro* The Kirby-Bauer inhibition halo method was used T1: *Streptococcus mutans* ATCC 25175 with 25% THE T2: *Streptococcus mutans* ATCC 25175 with 50% THE T3: *Streptococcus mutans* ATCC 25175 with 75% THE T4: *Streptococcus mutans* ATCC 25175 with 100% THE T5: *Streptococcus mutans* ATCC 25175 with 0.12% chlorhexidine (positive control) T6: *Streptococcus mutans* ATCC 25175 with distilled water (negative control)	• All THE concentrations showed antimicrobial inhibition on *S. mutans*. • 75 and 100% THE treatments showed the highest mean inhibition halo (29.7 and 33.1 mm, respectively). • Both treatments were better than the positive control (mean value: 26 mm).	Enciso et al. (2020)
Tocosh as a porridge. No reference about the *tocosh* fermentation time.	Not specified	Probiotic potential	*In vivo* (rats with oral daily intake for 7 days) T1: no *tocosh* treatment (negative control) T2: whole *tocosh* T3: *tocosh* only with peel T4: *Lactobacillus acidophilus* (positive control)	• T2 correlated with higher tail growth than T1. • T3 allowed lower stool weight and coliforms than T1.	Prentice (2005)
Tocosh flour from *Solanum tuberosum* diluted in water and cooked to obtain a gel (porridge). No reference about the *tocosh* fermentation time.	Raez Patiño local market Huancayo district and province, Junin region. Central Peruvian Andes	Cytoprotective capacity of the gastric mucosa	*In vivo* (male Sprague Dawley rats) Administration through orogastric intubation T1: negative control, 0.9% NaCl (10 mL/kg weight, w) T2: positive control, ethanol T3: *tocosh* 900 mg/kg w + ethanol T4: *tocosh* 1800 mg/kg w + ethanol T5: *tocosh* 2700 mg/kg w + ethanol T6: sucralfate 30 mg/kg w + ethanol The *tocosh* and sucralfate treatments were first applied and then the ethanol after 1 h (70%, 10 mL/kg weight, w) for inducing gastric ulcers The evaluation of the gastric mucosa was performed after 1 h	• *Tocosh* at all doses had cytoprotective properties on the rat gastric mucosa. The dose of 1800 mg/kg showed the best protection (97.1%) higher than sucralfate (92.7%). • Antioxidant effect of *tocosh* at all doses. The gastric mucosa with the lowest lipoperoxidation at the *tocosh* dose of 2700 mg/kg.	Sandoval et al. (2015)

(Continued)

TABLE 25.5 (Continued)

SAMPLE	*ORIGIN*	*STUDIED HEALTH-RELEVANT FUNCTIONAL PROPERTY*	*MODEL USED AND EXPERIMENTAL DESIGN*	*MAIN OUTCOMES*	*SOURCE*
Tocosh gel (porridge). No reference about the *tocosh* fermentation time.	Obtained from a local market. Central Peruvian Andes	Regenerative capacity of the gastric mucosa	*In vivo* (male albino rats) Administration through orogastric intubation T1: negative control (no ethanol) T2-T4: positive control (with ethanol for 24, 48, 72 h) T5: *tocosh* 800 mg/kg w, 24 h + ethanol T6: *tocosh* 800 mg/kg w, 48 h + ethanol T7: *tocosh* 800 mg/kg w, 72 h + ethanol T8: omeprazole 40 mg/kg w, 72 h T9: ranitidine 50 mg/kg w, 72 h The ethanol (50%, 10 mL/kg w) was first applied to induce gastric ulcers before the *tocosh* treatments The mucosa gastric was evaluated at 24, 48 and 72 h after each alcohol and *tocosh* treatments	• The treatments (T6-T7) with *tocosh* 800 mg/kg w (after 48 h and 72 h) reduced the erosive ulcers caused by the ethanol from 29.5% to 2.96% (48 h) and 0.89% (72 h). • *Tocosh* protection was comparable to omeprazole and ranitidine treatments and showed a regenerative capacity on the injured mucosa gastric.	Loli et al. (2016)
White *chuño* from *Solanum juzepczukii* (Piñasa variety) with 15–30 days of fermentation plus a second fermentation period (21 days) after the pressing stage. An extract of 0.8 g/mL of rehydrated *chuño* in sterile water was used.	Yunguyo province, Puno region. Southern Peruvian Andes	Antibacterial capacity	*In vitro* The antibacterial capacity with the Kirby-Bauer inhibition halo method T1: *Salmonella enterica*[a] with *chuño* extract (24 mg) T2: *Salmonella enterica*[a] with chloramphenicol (30 μg) (positive control) T3: *Salmonella enterica*[a] with nitrofurantoin (300 μg) (positive control) T4: *Salmonella enterica*[a] with thermally denatured nitrofurantoin (negative control)	• White *chuño* extract showed an antibacterial capacity against *S. enterica*[a] at a level of limit sensitivity according to the Duraffourd scale (mean inhibition halo of 10.4 mm). • *S. enterica*[1] was highly sensitive to nitrofurantoin (mean inhibition halo of 22 mm), whereas no inhibition was observed with chloramphenicol (antibiotic resistance).	Huayhua and García (2021)

Notes: [a]ABTS: 2,2′-azino-bis(3-ethylbenzothiazoline-6-sulphonic acid); [b]DPPH: 2,2-diphenyl-1-picrylhydrazyl; [c]FRAP: ferric reducing antioxidant power; [d]MTT, (3–(4,5-dimethylthiazol-2-yl)-2,5-diphenyl-tetrazolium bromide). T: treatment.

Note: [a]*S. enterica* ssp. *enterica* serovar Typhimurium ATCC 13311. T: treatment.

would explain the properties of *tocosh* for gastric mucosa protection, but more *in vivo* and clinical studies are necessary to fully understand the implicated mechanisms.

Unlike *tocosh*, *chuño* has received little attention with respect to its potential health-relevant functional qualities likely associated with its bioactive and beneficial microbiome composition (Table 25.5). Only one report was found about the *in vitro* antibacterial capacity of an aqueous extract from *chuño* which showed a level of limit sensitivity on *Samonella enterica* ssp. *enterica* serovar Typhimurium (Huayhua and García 2021).

25.6 FINAL PERSPECTIVES AND FUTURE RESEARCH

The Andes encompass a geographical area with a great diversity of ecosystems and climates, especially the tropical Andes (Colombia, Ecuador, Peru, and Bolivia), which comprise unique biodiversity hotspots in the world (Tognelli et al. 2016; Espinoza et al. 2020). Food diversity in the Andes is therefore rich and has long been the base of food security for Andean populations (Horkheimer 2004). Paradoxically, people from this region face a double burden of malnutrition and obesity-linked chronic diseases, mainly due to several socio-economic factors and a lack of awareness about the nutritional benefits of traditional foods over modern ultra-processed products (Palacios et al. 2021; Sanchez-Samaniego et al. 2021). For instance, high overweight prevalence rates were reported in 2017–2018 among Peruvian adults, especially from rural areas (40.2%), whereas the iron deficiency prevalence in children aged 6–59 months has been qualified as "moderate almost limiting with severe deficiency" in Peru and Bolivia (mean values in 2019 of 29.6% and 36.9%, respectively) (Peruvian Ministry of Health 2018; World Health Organization 2023). De Haan et al. (2019) determined that potato varietal diversity, including freeze-dried derived products, significantly sustains food security in Andean food systems, but current diets are limited in animal food sources, vegetables, and fruits, leading to a final imbalance of critical micronutrients such as iron and zinc. Thus, the improvement of food diversity intake considering traditional preparations may contribute to global strategies to overcome the current health situation.

Traditional fermented foods and beverages have been present in the diet of different cultures worldwide and have lately received high scientific attention due to their nutritional and health benefits (Şanlier et al. 2019). Potato-based fermented foods such as *chuño* and *tocosh* have contributed to Andean food security since ancient times and constitute good examples of highland food system adaptation to severe Andean climate conditions. In this chapter, the traditional processing of *chuño* (white and black) and *tocosh* was presented, along with scientific knowledge about their nutritional and bioactive composition, microbiome diversity, and health-relevant properties based on *in vitro* and *in vivo* studies. The potential of both *chuño* and *tocosh* as natural sources of beneficial microorganisms, mainly LAB, with a variety of biotechnological functional qualities was also discussed. Based on this knowledge, *chuño* and *tocosh* may be considered potential functional foods by themselves. However, the current scientific information about both products is very limited in comparison with dairy or other vegetable-derived fermented foods (Negrete-Romero et al. 2021; You et al. 2022).

Some topics for future research are summarized as follows:

- Understanding the variability of *chuño* and *tocosh* quality by the effects of their different regional processing variants, the impact of Andean ecosystems (abiotic and biotic factors), and the influence of potato variety diversity. This information would help to find the best conditions for optimizing the nutritional, bioactive composition, and potential health benefits properly adapted to specific Andean localities. There is evidence that some commercial potato varieties have buffering capacity, which limits the expected pH decrease in lactic acid fermentations in contrast to native potatoes (Mosso et al. 2016). Hence, the selection of potato variety seems to be relevant for an adequate fermentation process in *chuño* and *tocosh*.
- The application of integrated omics-based technologies for the identification of microbiome diversity in *chuño* and *tocosh* and the elucidation of microbiome-derived metabolite associations relevant for nutrition and health. Based on this knowledge, proper curing and characterization of native microbiota with different biotechnological applications could be targeted.
- Among isolated microorganisms, vitamin and phytate-degrading LAB species should be targeted for their potential application in the development of potato- or vegetable-based fermented foods with better micronutrient availability. A similar strategy has been proposed by Gabaza et al. (2017) to improve the iron and zinc availability of African sorghum and milled fermented foods.
- The characterized beneficial LAB and yeast from *tocosh* and *chuño* can be applied for the design of a wide array of functional foods with different health-relevant characteristics either from Andean potatoes or other vegetable substrates. Mosso et al. (2016) reported the feasibility of developing probiotic foods based on Andean tubers through LAB fermentation, whereas Panda et al. (2009) and Kim et al. (2012) developed lactic acid–fermented functional beverages from potato juices (probiotic product) and sweet potato (enriched with anthocyanin antioxidants), respectively. Andean native potatoes could be used as suitable matrices for the design of gluten-free (Roman et al. 2019) and non-dairy fermented products with probiotic properties, but

further *in vivo* and clinical trials are then necessary to support potential health claims.

- *Tocosh* and *chuño* are commonly produced under artisanal conditions, which can be enhanced with sustainable modern technology adapted to Andean systems to improve yields and guarantee food safety without compromising the beneficial microflora diversity.
- Climate conditions have changed over the last 20 years, and minimum temperatures ranges have increased in several Andean areas, which is leading to the processing of *chuño* at higher altitudes (Yoshikawa and Apaza 2020). This is negatively impacting some Andean communities which depend on *chuño* production and consumption; research should also consider such current challenges.

Chuño and *tocosh* constitute important cultural and biological heritage that should be preserved and characterized. More comprehensive and multidisciplinary research is necessary for future *in situ* initiatives aimed at improving Andean food systems based on the production of fermented potato food products.

REFERENCES

Ali, H. M., A. M. El-Gizawy, R. E. El-Bassiouny, M. A. Saleh. 2016. The role of various amino acids in enzymatic browning process in potato tubers, and identifying the browning products. *Food Chemistry* 192:879–885. https://doi.org/10.1016/j.foodchem.2015.07.100

Aregbe, A. Y., T. Mu, H. Sun. 2019. Effect of different pretreatment on the microbial diversity of fermented potato revealed by high-throughput sequencing. *Food Chemistry* 290:125–134. https://doi.org/10.1016/j.foodchem.2019.03.100

Aregbe, A. Y., T. Mu, H. Sun. 2020. Isolation and selection of technologically important lactic acid bacteria and yeast from fermented potato. *International Journal of Food Science & Technology* 55:1735–1743. https://doi.org/10.1111/ijfs.14432

Barus, W. L., Z. Bachruddin, C. Hanim, L. M. Yusiati. 2021. Effect of yellow sweet potato extract (*Ipomoea batatas* L.) as a prebiotic source for the kinetics of fermentation and the production of lactic acid by *Lactobacillus paracasei*. *IOP Conference Series: Earth and Environmental Science* 686:012047. https://doi.org/10.1088/1755-1315/686/1/012047

Brizzolari, A., A. Brandolini, P. Glorio-Paulet, A. Hidalgo. 2019. Antioxidant capacity and heat damage of powder products from South American plants with functional properties. *Italian Journal of Food Science* 31:731–748. https://dx.doi.org/110.14674/IJFS-1521

Burgos, G., S. De Haan, E. Salas, M. Bonierbale. 2009. Protein, iron, zinc and calcium concentrations of potatoes following traditional processing as "chuño." *Journal of Food Composition and Analysis* 22:617–619. https://doi.org/10.1016/j.jfca.2008.09.001

Chintha, P., D. Sarkar, K. Pecota, M. Dogramaci, K. Shetty. 2021. Improving phenolic bioactive-linked functional qualities of sweet potatoes using beneficial lactic acid bacteria-based biotransformation strategy. *Horticulturae* 7:367. https://doi.org/10.3390/horticulturae7100367

Dahlin, P., M. C. Müller, S. Ekengren, L. S. McKee, V. Bulone. 2017. The impact of steroidal glycoalkaloids on the physiology of *Phytophthora infestans*, the causative agent of potato late blight. *Molecular Plant-Microbe Interactions* 30:531–542. https://doi.org/10.1094/MPMI-09-16-0186-R

De Haan, S. 2009. Potato diversity at height: multiple dimensions of farmer-driven in-situ conservation in the Andes. PhD Thesis. Wageningen University, The Netherlands. https://research.wur.nl/en/publications/potato-diversity-at-height-multiple-dimensions-of-farmer-driven-i (accessed January 15, 2023).

De Haan, S., F. Rodriguez. 2016. Potato origin and production. In: *Advances in Potato Chemistry and Technology*, eds. J. Singh and L. Kaur, 1–32. Cambridge: Academic Press. https://doi.org/10.1016/B978-0-12-800002-1.00001-7

De Haan, S., G. Burgos, J. Arcos, Raul Ccanto, Maria Scurrah, Elisa Salas, Merideth Bonierbale. 2010. Traditional processing of black and white *chuño* in the Peruvian Andes: Regional variants and effect on the mineral content of native potato cultivars. *Economic Botany* 64:217–234. https://doi.org/10.1007/s12231-010-9128-x

De Haan, S., G. Burgos, R. Ccanto, Jesus Arcos, Maria Scurrah, Elisa Salas, Merideth Bonierbale. 2012. Effect of production environment, genotype and process on the mineral content of native bitter potato cultivars converted into white *chuño*. *Journal of the Science of Food and Agriculture* 92:2098–2105. https://doi.org/10.1002/jsfa.5589

De Haan, S., G. Burgos, R. Liria, F. Rodriguez, H. M. Creed-Kanashiro, M. Bonierbale. 2019. The nutritional contribution of potato varietal diversity in Andean food systems: A case study. *American Journal of Potato Research* 96:151–163. https://doi.org/10.1007/s12230-018-09707-2

De Jong, H. 2016. Impact of the potato on society. *American Journal of Potato Research* 93:415–429. https://doi.org/10.1007/s12230-016-9529-1

De Moreno, M. A., S. D. Todorov, G. M. Vignolo, G. Savoy, J. G. Leblanc. 2015. Indigenous fermented foods of Latin America. In: *Microorganisms and Fermentation of Traditional Foods*, eds. R. C. Ray, D. Montet, 312–340. Boca Raton: CRC Press. https://doi.org/10.1201/b17307

Enciso, S., J. Medina, F. Mauricio, Cesar Mauricio-Vilchez, Daniel Alvitez-Temoche, Luzmila Vilchez, Frank Mayta-Tovalino. 2020. Antibacterial effectiveness of four concentrations of the hydroalcoholic extract of *Solanum tuberosum* (tocosh) against *Streptococcus mutans* ATCC 25175TM: A comparative *in vitro* study. *International Journal of Dentistry* 8856382:1–5. https://doi.org/10.1155/2020/8856382

Espinoza, J. C., R. Garreaud, G. Poveda, Chenggang Liang, Jin Xu, Changling Zhang, Liyuan He. 2020. Hydroclimate of the Andes part I: Main climatic features. *Frontiers in Earth Science* 8:64. https://doi.org/10.3389/feart.2020.00064

FAO (Food and Agriculture Organization of the United Nations). 2009. *Section 5 Potato Fact Sheets*. https://www.fao.org/4/i1127e/i1127e.pdf (accessed March 15, 2023).

FAO (Food and Agriculture Organization of the United Nations). 2022. *World Food and Agriculture: Statistical Yearbook 2022*. Rome, Italy. https://www.fao.org/3/cc2211en/cc2211en.pdf (accessed March 15, 2023).

Feng, J., F. Yuan, Y. Gao, Chenggang Liang, Jin Xu, Changling Zhang, Liyuan He. 2003. A novel antimicrobial protein isolated from potato (*Solanum tuberosum*) shares homology with an acid phosphatase. *Biochemical Journal* 376:481–487. https://doi.org/10.1042/bj20030806

Fonseca, C., M. Ordinola. 2011. *Mejorando la competitividad de la agroindustria rural: El caso de la tunta en el altiplano peruano*. Centro Internacional de la Papa, Lima, Peru. https://cgspace.cgiar.org/server/api/core/bitstreams/098ffefa-b1c7-4113-84b2-73190dfd6fcb/content (accessed March 31, 2023).

Fonseca, G., E. Huarachi, W. Chura, G. Cotrado. 2008. *Guía de las buenas prácticas de procesamiento para la producción artesanal de la tunta*. http://cipotato.org/wp-content/uploads/2014/09/004295.pdf (accessed March 31, 2024).

Friedman, M., C. E. Levin. 2016. Glycoalkaloids and calystegine alkaloids in potatoes. In: *Advances in Potato Chemistry and Technology*, eds. J. Singh, L. Kaur, 167–194. Cambridge: Academic Press. https://doi.org/10.1016/B978-0-12-800002-1.00007-8

Friedman, M., C. F. Burns, C. A. Butchko, J. T. Blankemeyer. 1997. Folic acid protects against potato glycoalkaloid R-chaconine-induced disruption of frog embryo cell membranes and developmental toxicity. *Journal of Agricultural and Food Chemistry* 45:3991–3994. https://doi.org/10.1021/jf9702205

Gabaza, M., M. Muchuweti, P. Vandamme, K. Raes. 2017. Can fermentation be used as a sustainable strategy to reduce iron and zinc binders in traditional African fermented cereal porridges or gruels? *Food Reviews International* 33:561–586. https://doi.org/10.1080/87559129.2016.1196491

Garzón, F. J. M. 2007. Sociedades precolombinas asociadas a la domesticación y cultivo de la papa (*Solanum tuberosum*) en Sudamérica. *Revista Latinoamericana de la Papa* 14:1–9. https://doi.org/10.37066/ralap.v14i1.139

Ginzberg, I., J. G. Tokuhisa, R. E. Veilleux. 2009. Potato steroidal glycoalkaloids: biosynthesis and genetic manipulation. *Potato Research* 52:1–15. https://doi.org/10.1007/s11540-008-9103-4

Goldstein, P. S. 2003. From stew-eaters to maize-drinkers. In: *The Archaeology and Politics of Food and Feasting in Early States and Empires*, eds. T. L. Bray, 143–172. Boston: Kluwer Academic/Plenum Publishers. https://doi.org/10.1007/978-0-306-48246-5_6

Horkheimer, H. 2004. *Alimentación y obtención de alimentos en el Perú Prehispánico*. Peru: Instituto Nacional de Cultura del Perú.

Huayhua, H. J., R. A. García. 2021. Efecto antimicrobiano de la tunta (*Solanum juzepczukii*) sobre la *Salmonella enterica* subespecie *enterica* serovar Typhimurium. *Revista de Investigaciones Altoandinas* 23:37–46. http://dx.doi.org/10.18271/ria.2021.224

Jiménez, E., A. Yépez, A. Pérez-Cataluña, Elena Ramos Vásquez, Doris Zúñiga Dávila, Graciela Vignolo, Rosa Aznar. 2018. Exploring diversity and biotechnological potential of lactic acid bacteria from tocosh-traditional Peruvian fermented potatoes-by high throughput sequencing (HTS) and culturing. *LWT: Food Science and Technology* 87:567–574. https://doi.org/10.1016/j.lwt.2017.09.033

John, R. P., K. M. Nampoothiri, A. Pandey. 2007. Fermentative production of lactic acid from biomass: An overview on process developments and future perspectives. *Applied Microbiology and Biotechnology* 74:524–534. https://doi.org/10.1007/s00253-006-0779-6

Keukens, E. A., T. de Vrije, C. Van den Boom, Pieter de Waard, Henk H. Plasman, Felix Thiel, Vladimir Chupin, Wim M. F. Jongen, Ben de Kruijff. 1995. Molecular basis of glycoalkaloid induced membrane disruption. *Biochimica et Biophysica Acta (BBA)-Biomembranes* 1240:216–228. https://doi.org/10.1016/0005–2736(95)00186–7

Kim, N. J., H. L. Jang, K. Y. Yoon. 2012. Potato juice fermented with *Lactobacillus casei* as a probiotic functional beverage. *Food Science and Biotechnology* 21:1301–1307. https://doi.org/10.1007/s10068-012-0171-5

Korcz, E., Z. Kerényi, L. Varga. 2018. Dietary fibers, prebiotics, and exopolysaccharides produced by lactic acid bacteria: Potential health benefits with special regard to cholesterol-lowering effects. *Food & Function* 9:3057–3068. https://doi.org/10.1039/C8FO00118A

Kudoh, K., M. Matsumoto, S. Onodera, Y. Takeda, K. Ando, N. Shiomi. 2003. Antioxidative activity and protective effect against ethanol-induced gastric mucosal damage of a potato protein hydrolysate. *Journal of Nutritional Science and Vitaminology* 49:451–455. https://doi.org/10.3177/jnsv.49.451

Langkilde, S., M. Schrøder, D. Stewart, Otto Meyer, Sean Conner, Howard Davies, Morten Poulsen. 2008. Acute toxicity of high doses of the glycoalkaloids, α-solanine and α-chaconine, in the Syrian golden hamster. *Journal of Agricultural and Food Chemistry* 56:8753–8760. https://doi.org/10.1021/jf8012794

Loli, R. A., M. H. Sandoval, R. Callohuari, L. A. Mundaca. 2016. Tratamiento regenerativo de la mucosa gástrica con la mazamorra de tocosh de papa, en animales de experimentación. *Theorēma* 3:91–97. https://revistasinvestigacion.unmsm.edu.pe/index.php/Theo/article/view/12607

Lumbreras, L. G. 2006. Una labor ancestral. In: *La papa, tesoro de los Andes*, ed. C. Graves, 54–55. Lima, Peru: Centro Internacional de la Papa. https://cipotato.org/wp-content/uploads/2014/05/la_papa_tesoro_de_los_andess.pdf

Mayta-Tovalino, F., G. Sedano-Balbin, P. Romero-Tapia, Daniel Alvítez-Temoche, María Álvarez-Paucar, Luis Gálvez-Calla, Sonia Sacsaquispe-Contreras. 2019. Development of new experimental dentifrice of Peruvian *Solanum tuberosum* (tocosh) fermented by water stress: Antibacterial and cytotoxic activity. *The Journal of Contemporary Dental Practice* 20:1206–1211. https://thejcdp.com/doi/JCDP/pdf/10.5005/jp-journals-10024–2681

Mercado, L. G. 2020. Evaluación de la fermentación y del secado en la elaboración de tocosh fresco y harina de tocosh desodorizado. Undergraduate thesis. National University of the Center of Peru. https://repositorio.uncp.edu.pe/handle/20.500.12894/7041 (accessed October 20, 2022).

MINSA-INS-CENAN (Peruvian Ministry of Health-National Institute of Health-National Center of Food and Health). 2017. *Tablas peruanas de composición de alimentos*. https://repositorio.ins.gob.pe/bitstream/handle/20.500.14196/1034/tablas-peruanas-QR.pdf?sequence=3&isAllowed=y (accessed October 12, 2022).

Mosso, A. L., M. E. Jimenez, G. Vignolo, J. G. LeBlanc, N. C. Samman. 2018. Increasing the folate content of tuber based foods using potentially probiotic lactic acid bacteria. *Food Research International* 109:168–174. https://doi.org/10.1016/j.foodres.2018.03.073

Mosso, A. L., M. O. Lobo, N. C. Sammán. 2016. Development of a potentially probiotic food through fermentation of Andean tubers. *LWT-Food Science and Technology* 71:184–189. https://doi.org/10.1016/j.lwt.2016.03.008

Negrete-Romero, B., C. Valencia-Olivares, G. A. Baños-Dossetti, B. Pérez-Armendáriz, G. A. Cardoso-Ugarte. 2021. Nutritional contributions and health associations of traditional fermented foods. *Fermentation* 7:289. https://doi.org/10.3390/fermentation7040289

Nilsson, J., K. Olsson, G. Engqvist, Jimmy Ekvall, Marie Olsson, Margareta Nyman, Björn Åkesson. 2006. Variation in the content of glucosinolates, hydroxycinnamic acids, carotenoids, total antioxidant capacity and low-molecular-weight carbohydrates in Brassica vegetables. *Journal of the Science of Food and Agriculture* 86:528–538. https://doi.org/10.1002/jsfa.2355

Ovchinnikova, A., E. Krylova, T. Gavrilenko, Tamara Smekalova, Mikhail Zhuk, Sandra Knapp, David M. Spooner. 2011. Taxonomy of cultivated potatoes (*Solanum* section *Petota*: Solanaceae). *Botanical Journal of the Linnean Society* 165:107–155. https://doi.org/10.1111/j.1095-8339.2010.01107.x

Palacios, C., M. Magnus, A. Arrieta, H. Gallardo, R. Tapia, C. Espinal. 2021. Obesity in Latin America, a scoping review of public health prevention strategies and an overview of their impact on obesity prevention. *Public Health Nutrition* 24:5142–5155. https://doi.org/10.1017/S1368980021001403

Panda, S. H., S. K. Naskar, P. S. Sivakumar, R. C. Ray. 2009. Lactic acid fermentation of anthocyanin-rich sweet potato (*Ipomoea batatas* L.) into lacto-juice. *International Journal of Food Science & Technology* 44:288–296. https://doi.org/10.1111/j.1365-2621.2007.01692.x

Paz-Hidalgo, L. 1995. *Elaboración de la tunta*. https://es.scribd.com/document/23623143/Elaboracion-de-la-tunta (accessed March 31, 2023).

Peñarrieta, J. M., J. A. Alvarado, J. A. Bravo, B. Bergenståhl. 2011a. Chuño and tunta; the traditional Andean sun-dried potatoes. In: *Potatoes: Production, Consumption and Health Benefits*, ed. C. Caprara, 1–12. New York: Nova Science Publishers.

Peñarrieta, J. M., T. Salluca, L. Tejeda, J. A. Alvarado, B. Bergenståhl. 2011b. Changes in phenolic antioxidants during *chuño* production (traditional Andean freeze and sun-dried potato). *Journal of Food Composition and Analysis* 24:580–587. https://doi.org/10.1016/j.jfca.2010.10.006

Peruvian Ministry of Production. 2007. Norma técnica peruana NTP 011.400:2007. Tubérculos procesados. Papa deshidratada. Tunta. Requisitos y definiciones. https://cipotato.org/wp-content/uploads/publication%20files/books/004294.pdf

Peruvian Ministry of Health. 2018. *Estado Nutricional en adultos de 18 a 59 años, Perú: 2017–2018*. https://web.ins.gob.pe/sites/default/files/Archivos/cenan/van/sala_nutricional/sala_3/2021/Informe%20Tecnico-%20Estado%20nutricional%20en%20adultos%20de%2018%20a%2059%20a%C3%B1os%2CVIANEV%202017–2018.pdf (accessed January 10, 2023).

Prentice, M. M. 2005. Estudio del efecto de tocosh de papa como probiótico en el control del peso corporal y mayor crecimiento en ratas jóvenes frente a cultivo de *Lactobacillus acidophillus*. V Congreso mundial de medicina tradicional, Lima-Perú. https://nanopdf.com/download/medicina-holistica-universidad-norbert-wiener_pdf (accessed March 31, 2023).

Quillama, E., S. Dávila, A. Medina, S. Avalos, D. Paredes. 2012. Evaluación de la biodiversidad láctica de "tocosh", alimento fermentado tradicional de Perú. XXI Reunión científica Instituto de ciencias biológicas Antonio Raimondi (ICBAR): encuentro de las ciencias biológicas, Lima-Peru. https://sisbib.unmsm.edu.pe/bibvirtualdata/congres/basic/icbar_xxi/cap05.pdf (accessed March 31, 2023).

Ramos, E. R., R. A. Santos, E. Velázquez, C. E. Velezmoro, D. E. Zúñiga. 2018. Genetic diversity and antimicrobial activity of lactic acid bacteria in the preparation of traditional fermented potato product 'tunta'. *World Journal of Microbiology and Biotechnology* 34:1–10. https://doi.org/10.1007/s11274-018-2525-5

Reis, J. A., A. T. Paula, S. N. Casarotti, A. L. B. Penna. 2012. Lactic acid bacteria antimicrobial compounds: characteristics and applications. *Food Engineering Reviews* 4:124–140. https://doi.org/10.1007/s12393-012-9051-2

Rodriguez, M., A. D. Chambi. 2019. Determinación de la curva de crecimiento microbiano *Saccharomyces boulardii* en tunta variedades chaska y negra. *Fides et Ratio-Revista de Difusión cultural y científica de la Universidad La Salle en Bolivia* 18:201–214. http://www.scielo.org.bo/pdf/rfer/v18n18/v18n18_a11.pdf

Roman, L., M. Belorio, M. Gomez. 2019. Gluten-free breads: The gap between research and commercial reality. *Comprehensive Reviews in Food Science and Food Safety* 18:690–702. https://doi.org/10.1111/1541-4337.12437

Sanchez-Samaniego, G., S. M. Hartinger, P. S. Tallman, D. Mäusezahl. 2021. Cardiovascular disease in the Peruvian Andes: Local perceptions, barriers, and paths to preventing chronic diseases in the Cajamarca region. *International Journal of Public Health* 66:1–8. https://doi.org/10.3389/ijph.2021.1604117

Sandoval, M. H., J. Tenorio, A. Tinco, R. A. Loli, S. Calderón. 2015. Efecto antioxidante y citoprotector del tocosh de *Solanum tuberosum* papa en la mucosa gástrica de animales de experimentación. *Anales de la Facultad de Medicina* 76:15–20. http://dx.doi.org/10.15381/anales.v76i1.11070

Şanlier, N., B. B. Gökcen, A. C. Sezgin. 2019. Health benefits of fermented foods. *Critical Reviews in Food Science and Nutrition* 59:506–527. https://doi.org/10.1080/10408398.2017.1383355

Santos-Mendoza, R., E. Ramos-Vásquez, A. Iris Zavaleta, D. Zúñiga-Dávila, C. Velezmoro-Sánchez. 2018. Bacterias ácido lácticas productoras de riboflavina aisladas del proceso de elaboración de la "tunta". *Ecología Aplicada* 17:185–189. http://dx.doi.org/10.21704/rea.v17i2.1238

Schindler, B., S. De Haan. 2021. Fermentation: tocosh and chicha. In: *Los andes y los alimentos del futuro. 50 Andean future foods*, ed. International Potato Center (CIP), 302–305. Lima, Peru: International Potato Center (CIP). https://hdl.handle.net/10568/116283

Suo, H., X. Zhao, Y. Qian, Peng Sun, Kai Zhu, Jian Li, Baozhong Sun. 2016. *Lactobacillus fermentum* Suo attenuates HCl/ethanol induced gastric injury in mice through its antioxidant effects. *Nutrients* 8:155. https://doi.org/10.3390/nu8030155

Terrazas, F., X. Cadima, R. García, J. Zeballos. 2008. *Catálogo etnobotánico de papas nativas*. https://www.bivica.org/files/papas-nativas.pdf (accessed March 31, 2023).

Tognelli, M. F., C. A. Lasso, C. A. Bota-Sierra. 2016. *Estado de conservación y distribución de la biodiversidad de agua dulce en los andes tropicales. Gland, Suiza, Cambridge, UK, and Arlington, USA: UICN (Unión Internacional para la Conservación de la Naturaleza y los Recursos Naturales)*. https://www.iucn.org/es/resources/publicacion/estado-de-conservacion-y-distribucion-de-la-biodiversidad-de-agua-dulce-en

Velasco-Chong, J. R., O. Herrera-Calderón, J. P. Rojas-Armas, Renán Dilton Hañari-Quispe, Linder Figueroa-Salvador, Gilmar Peña-Rojas, Vidalina Andía-Ayme, Ricardo Ángel Yuli-Posadas, Andres F. Yepes-Perez, Cristian Aguilar. 2020. Tocosh flour (*Solanum tuberosum* L.): A toxicological assessment of traditional Peruvian fermented potatoes. *Foods* 9:719. https://doi.org/10.3390/foods9060719

Veselá, M., M. Drdák, P. Šimon, M. Veselý. 2002. Utilization of *Lactobacillus* sp. for steroid glycoalkaloids degradation by lactic acid fermentation. *Food/Nahrung* 46:251–255. https://doi.org/10.1002/1521-3803(20020701)46:4<251::AID-FOOD251>3.0.CO;2-U

Werge, R. W. 1979. Potato processing in the central highlands of Peru. *Ecology of Food and Nutrition* 7:229–234. https://doi.org/10.1080/03670244.1979.9990534

World Health Organization (WHO). 2023. *Global health observatory (GHO)*. https://www.who.int/data/gho (accessed January 12, 2023).

Yábar, E., V. Reyes, J. Casas. 2019. Evaluación de la actividad antioxidante y antibacteriana del tocosh de papa (*Solanum tuberosum*). *Journal of Agri-Food Science* 1:47–52. https://revistas.uncp.edu.pe/index.php/jafs/article/view/545

Yábar, E. F., V. J. Reyes. 2017. Caracterización fisicoquímica y microbiológica del tocosh de papa (*Solanum tuberosum*) durante su elaboración. *Prospectiva Universitaria* 14:31–36. http://hdl.handle.net/20.500.12894/4688

Yamamoto, N. 1988. Potato processing learning from a traditional Andean system. In: *The Social Sciences at CIP*, ed. International Potato Center (CIP), 160–181. Lima, Peru: International Potato Center (CIP). https://pdf.usaid.gov/pdf_docs/PNABB136.pdf

Yépez, A., C. Luz, G. Meca, G. Vignolo, J. Mañes, R. Aznar. 2017. Biopreservation potential of lactic acid bacteria from Andean fermented food of vegetal origin. *Food Control* 78:393–400. https://doi.org/10.1016/j.foodcont.2017.03.009

Yoshikawa, K., F. Apaza. 2020. Unfrozen state by the supercooling of *chuño* for traditional agriculture in altiplano Andes. *Environmental and Sustainability Indicators* 8(100063):1–4. https://doi.org/10.1016/j.indic.2020.100063

You, L., C. Yang, H. Jin, L. Y. Kwok, Z. Sun, H. Zhang. 2022. Metagenomic features of traditional fermented milk products. *LWT-Food Science and Technology* 155:112945. https://doi.org/10.1016/j.lwt.2021.112945

Zvietcovich, G., W. Salas, M. Vega. 1985. *Inventario tecnológico de los sistemas poscosecha en la sierra del Perú.* Lima, Peru: Instituto Interamericano de Cooperación para la Agricultura (IICA). https://repositorio.iica.int/handle/11324/18467

Cacao Fermentation Process in Mexico and Targeting Functional Food Products

26

María J. Paris, Leonor Lastra-Vargas, Enrique Palou, Emma Mani-López, and Aurelio López-Malo

26.1 INTRODUCTION

Food fermentation is a process used since ancient times to preserve and enhance the shelf-life of food (Ojeda-Linares et al. 2020). Although in the beginning it might have been discovered accidentally, it indeed preserves food and brings it new flavors and sensory properties through the generation of a large variety of volatile compounds (De Roos and De Vuyst 2018; Rezac et al. 2018; Tamang et al. 2020). Furthermore, it is considered an enriching process since it brings nutritional enrichment to the food through reactions that occur during the process, making nutrients more available. Additionally, some microorganisms found in fermented food have probiotic effects, contributing to additional health benefits (García-Caballero et al. 2018).

Fermented foods are, in general, prepared using an artisanal approach, and certain ethnic groups usually consume them as part of their gastronomy (Romero-Luna et al. 2017). This pattern is observed in Mexico, where the gastronomy domain has a wide variety of fermented foods and beverages, normally prepared in localized households, with empirical knowledge passed from generation through generation. Only a small portion is produced on a larger or even industrial scale (Ramírez-Guzmán et al. 2019).

Mexican fermented foods are prepared from a variety of vegetable types, including cereals such as corn, agave, dasylirion, palm plant sap; seeds such as cacao; and fruits such as pineapple and prickly pears. From animal origin, both milk and meat are used as raw materials (González-Tenorio et al. 2013; Cooper-Bribiesca et al. 2018; Córdova Lázaro et al. 2018; Ramírez-Guzmán et al. 2019).

Several fermented foods and beverages of Mexican origin are worth further study due to their proven and potential nutritious, healthy, and functional benefits, given the nature of their ingredients or components and participating microorganisms. However, one of these products that might not come to mind quickly is cacao and the several ways Mexicans consume it.

Regardless of the current production areas and uses of cacao, Mexicans have great pride in the Mexican origins of chocolate. Mexico had a solid connection to cacao throughout its early history. Cacao has played an important role in shaping Mexican culture and traditions. In pre-Hispanic times, cacao was used for curing, rituals, ceremonies, and trade. Since then, cacao has been used in various recipies of foods and beverages (McNeil 2006; Arias-González 2014). In 2019, the Mexican Senate approved the second day of September as the "National Day of Cacao and Chocolate", with the justification that it is one of the representative products of the Mexican culture (SADER Yucatán 2020). Additionally, according to the nutritional value, cacao has different components with bioactive properties such as polyphenols and methylxanthines worth studying and exploring (Hu et al. 2016).

Despite the very important position that cacao bears in Mexican history, culture, and gastronomy, Mexico cacao consumption surpasses the domestic production of the seed. In fact, cacao production in Mexico has been suffering a noticeable decrease due to agroeconomic issues. Consequently, reductions in yields, productivity, and profits have been observed (Hipólito-Romero et al. 2017).

Throughout this chapter, we will look at chocolate from a different perspective from that of the high-sugar, high-fat candy we are most familiar with by diving into its history and the conserved traditional practices of cacao production and consumption. In addition, the health-promoting properties of its components and the efforts to conserve and enhance its bioactive components will be presented. These insights might motivate research and development on Mexican cacao and its

DOI: 10.1201/9781003605300-26

products, aiming to preserve the traditions, improve the quality, and better the lives of the small producers.

26.2 SELECTED FERMENTED MEXICAN FOODS

Mexican eating habits are embedded with tradition. Thus, the consumption of fermented foods and beverages fits many purposes among the Mexican population. Some are important sources of nutrients for indigenous people who consume them for breakfast, nutritious snacks, or even medicine. Others, such as spirits and other alcoholic beverages, are suitable for gathering and socialization (Soleri and Cleveland 2007; Sotelo et al. 2012; Giles-Gómez et al. 2016; Ramos and Schwan 2017; Cooper-Bribiesca et al. 2018).

Mexican fermented food products are primarily beverages, but some semi-solid or solid foods can also be found. The fermentation is typically dominated by a consortium of microorganisms rather than by the activity of a single microorganism. Commonly, lactic acid bacteria (LAB), yeasts, or even acetic acid bacteria (AAB) are involved. As such, lactic, alcoholic, and acetic fermentation are common, and depending on the food, one, two, or all of these types of fermentation can occur to obtain the product (Ramírez-Guzmán et al. 2019). Some products may be left with the microorganism used in the fermentation process, while others might be distilled, boiled, heated, or baked, eliminating the organisms responsible for the fermentation. When the microorganisms are left in the food, they usually work as probiotics or as food preservatives, preventing further contamination with other microorganisms (Romero-Luna et al. 2017). On the other hand, the combination of the different microorganisms participating in the fermentation contributes to the particular sensory properties developed in fermented foods, from flavor to texture (Alcántara-Hernández et al. 2010; Crafack et al. 2013; Santiago-Urbina and Ruíz-Terán 2014; Kirchmayr et al. 2017; Ramos and Schwan 2017; Pérez-Armendáriz and Cardoso-Ugarte 2020).

Mexican traditional fermented beverages can be divided into alcoholic and non-alcoholic. As for non-alcoholic drinks, *atole agrio* and pozol *agrio* are some examples. The alcoholic beverages are either non-distilled such as tesgüino, pulque, colonche, tepache, and tuba or distilled like tequila, mezcal, bacarona, and sotol (Ramírez-Guzmán et al. 2019).

26.3 CACAO

The raw material from which chocolate is made are the seeds of the *Theobroma cacao* L. tree. This tree belongs to the Malvaceae family and is one of the 20 to 22 species of the genus *Theobroma*. Broadly, two subspecies are recognized, *T. cacao* spp. *cacao* and *T. cacao* spp. *sphaerocarpum,* which correspond to the cultivars known as *Criollo* and *Forastero,* respectively. Other cultivars resulting from the mixing of these two are known as *Trinitario* (Ozturk and Young 2017; Gibson and Newsham 2018; Figueroa-Hernández et al. 2019). The shape, color, size, and number of beans of the fruit pods, as well as the color of the cotyledons, are characteristics that, in theory, could be used to differentiate between varieties: *Criollo* with pointed, furrowed, warty pods and white cotyledons, and *Forastero* with smooth round pods and purple cotyledons. However, in practice, these characteristics are highly variable (Ogata et al. 2006; Kongor et al. 2016). Based on genetic information, Motamayor et al. (2008) have proposed that *T. cacao* be classified into ten groups (*Marañon, Curaray, Criollo, Iquitos, Nanay, Contamana, Amelonado* (*Forastero*), *Purús, Nacional,* and *Guiana*) to reflect the genetic diversity of the species more accurately.

Criollo is native to Central and South America, the Caribbean islands, and Sri Lanka, and it is considered of fine quality. The seed is white, ivory, or pale purple in color and has a thin shell and low pH. It is rich in pyrazines, the most important volatile components in cacao (Marseglia et al. 2020), and poor in tannins, presenting slight bitterness and astringency. Its flavor is delicate, complex, soft, spicy, fruity, and nutty, with chocolate and caramel notes (Kongor et al. 2016; Ozturk and Young 2017; Utrilla-Vázquez et al. 2020). However, this variety is very fragile and susceptible to diseases, providing low production yields and costing five times more than *Forastero*, a robust cacao variety (Arana-Sánchez et al. 2015; Gibson and Newsham 2018; Utrilla-Vázquez et al. 2020). The cacao *Criollo* variety was introduced in Mesoamerica from the Amazon, and now *Forastero* and *Trinitario* varieties can also be found (Soleri and Cleveland 2007). *Forastero*, on the other hand, is a very resistant plant, not prone to diseases, and as such, it is the most used (95% of world production), providing high yields. It is native to the Amazon, and its beans are purple and flat, with a thick shell; it has a strong acidic and chocolate flavor and is high in phenolic compounds, including tannins (Kongor et al. 2016; Ozturk and Young 2017; Figueroa-Hernández et al. 2019; Utrilla-Vázquez et al. 2020). It is responsible for the full chocolate body, although it is not very aromatic. It is somewhat bitter and is typically used as a cacao base and blended with other cacaos to prepare different chocolate profiles (Gibson and Newsham 2018). *Trinitario* is a hybrid between *Criollo* and *Forastero*, and due to this, it has the high quality and delicate flavor of *Criollo* and the hardiness of *Forastero*. Its quality can be between average and high, and it is finer than *Forastero*. Its susceptibility to diseases is between that of *Forastero* and *Criollo*, and thus it is produced in lower amounts that *Forastero* (Arana-Sánchez et al. 2015; Gibson and Newsham 2018; Utrilla-Vázquez et al. 2020). *Trinitario* beans are variable in color and have strong chocolate features and exclusive wine notes. Together with *Criollo*, it is used to produce high-quality chocolate, providing fruity, flowery, herbal, wooden, nutty, and candy notes to chocolate, and they account for 5% of world's cacao production (Kongor et al. 2016; Castro-Alayo et al. 2019; Utrilla-Vázquez et al. 2020). Although cacao originates from Mexico, most of the currently commercialized cacao is produced in African countries such as the Ivory Coast and Ghana, which are the leading providers of *Forastero* cacao. On the other hand, Mexico is among the few *Criollo* variety producers (CacaoMexico 2020).

In general, cacao pods are oval; they measure between 12 to 30 cm in length and contain 30 to 40 seeds (Figueroa-Hernández et al. 2019; Ordoñez-Araque et al. 2020). The seeds, composed of embryo, cotyledon, and testa, are surrounded by a white, sweet, acid, aromatic, edible, thick, and viscous mucilage. The mucilage is attached to the testa and is rich in fermentable sugars such as glucose, fructose, and sucrose that are important in cacao fermentation, as will be mentioned later in this chapter (Afoakwa et al. 2013; Chin et al. 2013; Ho et al. 2014; Kongor et al. 2016; Castro-Alayo et al. 2019; Ordoñez-Araque et al. 2020; Utrilla-Vázquez et al. 2020). Cacao seeds are around 2 to 2.5 cm in length and approximately 1 to 1.5 cm in width (Ha et al. 2016). They are formed by the embryo and two cotyledons (Schwan and Wheals 2004). The seeds contain a shell or testa, and the bean is called the nib after the shell is removed (Gibson and Newsham 2018). When cacao beans are freshly removed from the pods, they are composed of approximately 35.5% water, 31% fat, 12.5% proteins, 12% polysaccharides, 3% low molecular weight carbohydrates, 5.5% polyphenols, 1% acids, 1.5% theobromine, and 1% caffeine (Afoakwa et al. 2008; Ho et al. 2014). Cacao fat is mostly triacylglycerols (95%), while albumins, globulins (including vicilins), prolamins, and glutelins are the most abundant proteins (Kongor et al. 2016; Barišić et al. 2019; Rawel et al. 2019). Starch accounts for 5% of cacao composition, and pectin and cellulose correspond to 2 to 3%. Sucrose, making up 2.5% of cacao composition, is the most abundant sugar in cacao beans (90%), followed by fructose and glucose. Galactose, sorbose, arabinose, xylose, mannitol, and inositol can also be found (Afoakwa et al. 2008; Ho et al. 2014; Barišić et al. 2019). Concerning cacao seed polyphenol content, catechins or flavan-3-ols, anthocyanins, and proanthocyanidins are found, with catechins the most abundant (37%) (Kongor et al. 2016). Last, the primary organic acids found in cacao beans are citric, oxalic, malic, lactic, and succinic. Although citric acid is the most predominant, its concentration decreases throughout fermentation and drying, being converted to lactic acid, acetic acid, and carbon dioxide (Rodriguez-Campos et al. 2011; Ozturk and Young 2017). Cacao pulp has an average of 84.5% water and 12.5% sugar and other macronutrients and micronutrients that favor microbial growth. The sugars present (glucose, fructose, and sucrose), as well as the citric acid content and a pH varying from 3.0 to 3.5, are ideal for yeast and lactic acid bacteria growth during cacao fermentation. The concentration of glucose and fructose might also increase in the pulp inside the pod with the post-harvesting period (Schwan and Wheals 2004; Rodriguez-Campos et al. 2011; Ordoñez-Araque et al. 2020).

T. cacao plantations are situated mainly 10 degrees north and south of the Equator at altitudes between 4 and 800 mamsl. However, they can also be found 22 degrees north or 21 degrees south of the Equator and at altitudes between 1,000 and 1,300 mamsl. The optimum conditions for growth and development of the tree are an annual temperature between 23 and 28°C; rainfall between 1,200 and 2,500 mm distributed throughout the year; terrain inclination of less than 30%; and nutrient-rich, loamy, clay-loam, and sandy-loam soils with a pH between 6.0 and 7.0 (Arrazate et al. 2011). *T. cacao* tree grows in the shade, protected from wind (Figueroa-Hernández et al. 2019; CacaoMexico 2020).

In 2016, a protected designation of origin was given to cacao grown in the Grijalva region of Tabasco, denominated Cacao Grijalva. Cacao has been produced in the Grijalva region since pre-Hispanic times. Its characteristic quality is given by the region's unique geographic and climatic conditions and the interaction between human and natural factors. Cacao Grijalva is cooperatively produced in domesticated jungles with balance between man and nature, where the flower's pollination is achieved mainly by tiny flies of the genus *Forcipomyia* (Instituto Mexicano de la Propiedad Industrial 2016).

In 2017, Mexico was the 13th highest cacao-producing country in the world, with 27,430 tons in an area of around 60,000 ha, with the states of Tabasco, Chiapas, and Guerrero the leading producers, contributing 17,430, 9612, and 245 tons in the year 2017, respectively (SIAP 2018). Oaxaca is also a cacao-producing state with considerably less input; in 2007 and 2008, the state generated 25 and 14 tons, respectively. Cacao has also been found in the states of Yucatán, Quintana Roo, Veracruz, and San Luis Potosí (Arrazate et al. 2011). Cacao production in Mexico results from around 40,000 small producers (SADER Yucatán 2020). Given the soil and climate conditions necessary for its growth, areas in the states of Veracruz and Nayarit have great potential for this crop's farming (CEDRSSA 2020; SIAP 2018).

The cacao-shaded plantations under a diversified agroforestry system in Tabasco and Chiapas closely resemble a natural jungle. The small producers integrate into their plantation around 66 different species of trees chosen to provide shade for the cacao plants and a source of nitrogen fixation and providers of organic material for the conservation of the soil. They can also be used as wood for construction or firewood for energy and their fruits for food as an added source of income for the farmers. Some examples are the leguminous plants cocohite or cocuite (*Gliricidia sepium)* and gasparito (*Erythrina americana*), which also serve as supporting system for the growth of pitahaya (*Hylocereus undatus*) and vanilla (*Vanilla planifolia* Jacks ex Andrews), fruits such as plums (*Spondias purpurea* L.), oranges and lemons (*Citrus* spp.), sapodilla (*Manilkara zapota* L.), mamey (*Pouteria sapota*), chestnut (*Artocarpus altilis*), pepper (*Pimenta dioica*), cedar (*Cedrela odorata*), mahogany (*Swietenia macrophylla*), and oak (*Tabebuia rosea*), among others. These plantations can also work as natural corridors for birds, mammals, reptiles, amphibians, and insects, connecting natural vegetation spots (Ogata 2017).

26.4 CACAO HISTORY

Today Mexico is known as the source of chocolate, where the seeds of *T. cacao* once played an important role for the Olmec, Maya and Mexica or Aztecs, and other pre-Columbian cultures of the Americas. They were not only used in the culinary traditions and rituals of the elite but also used for trade and viewed as a symbol of outstanding social status and wealth (McNeil 2006).

It was Linnaeus who created the term *Theobroma cacao* in 1753, under his taxonomic binomial system for genus and species (Dillinger et al. 2000), bringing together *Theobroma,* from Greek origins, meaning food of the gods, and *cacao,* the New World name for cacao (Coe and Coe 2013). This is a strange coincidence when one looks at the Maya and Mexica cultures and the divine origins of cacao in their religions (Dillinger et al. 2000).

Although there is still debate on a Mesoamerican origin, the plant *T. cacao* most likely originated in South America, in the northwest Amazon basin, from where it was dispersed either naturally by rodents, bats, and monkeys or by humans (McNeil 2006). Similarly, there is still no consensus on the origin of the domestication of the *T. cacao* tree. It has been proposed that *T. cacao* could have been brought to Mesoamerica as a cultigen and, once there, further domesticated. However, independent of the domestication origin, unlike the South Americans, who used the pulp of the seeds to create a fruit beverage or an alcoholic drink after fermentation, Mesoamericans would use the seeds to produce chocolate (McNeil2006; Powis et al. 2011). Bletter and Daly (2006) theorize the reason behind this dual use of *T. cacao* in pre-Columbian cultures was due to the high availability of other stimulant beverage plant species in South America such as yerba mate (*Ilex paraguariensis* A. St.-Hil.), guayusa (*Ilex guayusa* Loes.), guaraná (*Paullinia cupana* Kunth), and yoco (*Paullinia yoco* R. E. Schult. And Killip), as well as yaupon (*Ilex vomitoria* Aiton), coca (*Erythroxylum coca* Lam. and *E. novogranatense* O. E. Schulz), ili-ga-wa-yik (*Virola sebifera* Aubl.), and tobacco (*Nicotiana rustica* L. and *N. tabacum* L.). These contained higher levels of caffeinated sources and/or other stimulating compounds and required significantly less processing.

There is strong evidence for the Mesoamerican origins of the word "cacao". It is known from linguist studies that the Olmecs (from the lowlands of the Mexican Gulf Coast, 1500–400 BC.) spoke an ancestral form of language from the Mixe-Zoquean or Mije-Sokean family, similar to what the Popoluca people of southeastern Veracruz speak today. It is thought that the word "cacao", which was originally pronounced *kakawa,* was part of the vocabulary of the Olmec civilization as far back as 1000 BC (Kaufman and Justeson 2008). This leads to thinking that the discovery of the processing of *T. cacao* into chocolate occurred in Mesoamerica by the Olmecs (Coe and Coe 2013). These findings are supported by high-performance liquid chromatography coupled to mass spectrometry (HPLC-MS); with this technique, the presence of theobromine in archeological pottery from the Olmec capital San Lorenzo, and thus the use of chocolate between 1800 and 1000 BC, could be proven (Powis et al. 2011; Coe and Coe 2013). Theories on the dissemination of cacao use as chocolate through Olmec to other Mesoamerican cultures such as the Maya could be somewhat sustained, although not confirmed. HPLC-MS studies have important relevance thanks to the unique composition of *T. cacao,* since it is the only Mesoamerican plant that contains theobromine as the primary methylxanthine, which can then be used as a marker for the presence of cacao (Hurst et al. 2002; Hurst 2006). These investigations have revealed that chocolate was processed and consumed by the Mokaya, one of Mesoamerica's earliest sedentary inhabitants from the Pacific Coast of Chiapas, Mexico, as far back as 1900 BC, and by pre-Olmec people in Mexico's Gulf Coast area as far back as 1750 BC (Powis et al. 2007). The presence of theobromine, caffeine, and theophylline in fragments of ceramic drinking vessels also suggests the consumption of cacao in the US southwest and northwest Mexico (A.D. 750–1400). Since the plants are not native to the areas, these findings strengthen the evidence of the importance of cacao among other traded species in social and economic relationships between distant populations through the acquisition of resources and the sharing of cultural practices (Crown et al. 2015).

During the research for this chapter, many references were found that claim that chocolate is Mexico's gift to the world. This may be because the first time the Spanish conquerors Hernán Cortéz and his men witnessed the preparation and consumption of the beverage made from cacao was at Moctezuma's court in Mexica capital Tenochtitlan, now Mexico City, even though they were already familiar with the seeds from Cuba and Haiti (Dillinger et al. 2000). However, Coe and Coe (2013) argue that there is no historical evidence for the argument of the introduction of chocolate to Spain (and thus Europe and the world) via Cortés and his encounters with Moctezuma in Mexico. Instead, the authors point to the Kekchi Maya of Guatemala, who in 1544 was taken to Spain by Dominican friars in the recorded list of presents offered to Prince Philip; there were receptacles of beaten chocolate. Furthermore, documents for the first official shipment of cacao beans are dated from 1585 from Veracruz, Mexico, to Seville, Spain (Coe and Coe 2013).

Whatever the origin of cacao or the place of its domestication and from which it was introduced to the rest of the world, there is no doubt about the importance it has had in Mexican history, culture, and economy. In this sense, the next section will focus on the many uses of cacao in Mexico, many of which are rooted in tradition.

26.5 TRADITIONAL USES OF CACAO IN MEXICO

Cacao fermentation and its use in Mexican recipes are part of the country's culture and traditions, and local knowledge on the subject has been passed on through generations. Cacao fermentation has been kept on an artisanal scale, and the guidelines for its use as a traditional food ingredient are in the hands of Mexican people who prepare the recipes for household consumption or sale in street markets (Córdova Lázaro et al. 2018). As previously mentioned in this chapter, cacao has been used as food since pre-Hispanic times, and some ancient recipes prepared with cacao have contributed to the current Mexican gastronomy. Some recipes did not change much since ancient times, while others suffered considerable modification. Some additions were, indeed, made to recipes containing cacao with the arrival of the Spaniards; in Oaxaca, for instance, cacao is used in the formulation of a foamy beverage containing sugar, cinnamon, and almonds, ingredients that did not exist in pre-Hispanic Mexico and that were brought by the Spaniards (Soleri and Cleveland 2007).

Several beverages, foams, or pastes prepared with cacao are consumed throughout Mexico with Oaxaca and Tabasco being examples of states that stand out for their traditional uses. In Oaxaca state, a beverage called chocolate is consumed among the population. Despite its name, it is not necessarily related to what is commonly known in other countries as chocolate. This beverage, which can be either water or milk based, contains, apart from cacao, cinnamon and sugar. Toasted cacao beans are ground together with cinnamon and sugar. A paste is formed due to cacao butter melting during the grinding. A small amount of it is mixed with water to prepare a cup of chocolate (Mission chocolate 2015). Chocolate can be prepared using Mexican chocolate bars (*tablillas*), which should first be well dissolved in hot water before the addition of more warm water to prepare *chocolate de agua* (water-based chocolate) or milk to prepare *chocolate de leche* (milk-based chocolate). A *tablilla* is a chocolate bar made with cacao, cinnamon, and sugar. Mixing the blend with *molinillo* (a kitchen tool) is very important to prepare chocolate with some foam on it (Barros and Buenrostro 2011). *Mole* is another important Mexican recipe prepared in Oaxaca and Puebla. It is a relatively thick dark brown paste or sauce made with cacao; chili; nuts; seeds such as pumpkin seeds and almonds; herbs such as coriander; and spices such as cinnamon, clove, anise, and peppercorn; raisins; tortilla; and bread. The sauce is traditionally served with turkey or chicken (Laudan and Pilcher 1999; Soleri and Cleveland 2007).

On the other hand, *popo* is a refreshing foam prepared with cacao, rice, yucca peels, cocolmeca (*Milleria quinqueflora*), chupipe roots (*Gonolobus niger*), and sugar. After milling all the ingredients together, a paste is obtained. This is mixed with water and sheared into a foam (*popo*). *Popo* is consumed in Tlacotalpan (Veracruz) (Barros and Buenrostro 2011).

An interesting observation regarding Mexican recipes with cacao is that cacao is sometimes combined with maize, one of the most important crops in the Mexican diet. This practice finds its root in pre-Hispanic times when civilizations such as the Mayas and the Aztecs included cacao and maize in the same dish (Chocolate Class 2016). Today, beverages such as *tejate, tecalate, atole de chocolate*, and *champurrado* containing cacao and maize are consumed throughout the country. *Tejate*, for instance, is a beverage prepared and consumed in the valley of Oaxaca and with cultural and social meaning for the community that consumes it. It is a source of energy, fat, methylxanthines, and minerals such as K and Fe. It is used as a refresher beverage among people working in maize cultivation (Soleri and Cleveland 2007; Sotelo et al. 2012). It is prepared using Mexican maize (*Criollo* variety) dough. This is prepared from grains boiled in water with ash (nixtamalization), which are then recovered, washed, and milled until a smooth paste is obtained. The interesting fact about this nixtamalization process is that it is done with ash. In contrast, the majority of nixtamalization processes are done with lime (calcium hydroxide). On the other hand, fermented cacao seeds, mamey (*Pouteria sapota*) seeds, and *rosita de cacao* (*Quararibea funebris*) flowers are toasted and ground with a *metate* to form a paste. The two doughs are mixed and ground again to guarantee a smooth texture. Sometimes brown sugar is added (Soleri and Cleveland 2007; Sotelo et al. 2012). The ingredients are mixed by hand in a large clay bowl, to which cold water is gradually added with constant stirring, which promotes the separation of the cacao butter. The rapid movement of the arms and the force of the added water help incorporate air into the liquid and form a foam characteristic of the beverage (Mission Chocolate 2015). Although the cacao seeds used in this process are obtained from *T. cacao*, seeds from *Theobroma bicolor* (commonly known as *cacao blanco,* meaning white cacao in English, or *pataxte*) are sometimes used together with *T. cacao* seeds in this recipe. *T. bicolor* seeds are not fermented or roasted and are known to provide cacao aromas and flavors that are not as high quality as *T. cacao* seeds. However, they yield good-quality cacao butter (Soleri and Cleveland 2007). Chocolate *atole* and *champurrado* are also prepared with maize and cacao; *atole* is a Mexican thick drink prepared with milled nixtamalized maize. This is cooked in boiling water until the beverage gets thick. Sugar is normally added to sweeten the beverage. To prepare chocolate *atole*, a paste is made using fermented *T. bicolor* seeds, rice, wheat grains, and cinnamon. The paste is mixed with water to form a foam. This is added on top of common *atole* to obtain chocolate *atole* (Mission Chocolate 2015). On the other hand, *champurrado* is a mixture of nixtamalized maize dough, cacao, cinnamon, cloves, brown sugar (or piloncillo, a sugarcane sweetener), and milk or water to make a thick, homogeneous warm beverage. The preparation process involves heating cacao bars (*tablillas*), cinnamon, and brown sugar (or piloncillo), followed by the addition of corn dough, previously dissolved in water or milk. The whole mixture is left to cook until it is thick (Mission Chocolate 2015). In Sonora state, *champurrado* is prepared with wheat flour instead of nixtamalized maize dough and sweetened with piloncillo (Barros and Buenrostro 2011). Another interesting *atole* is *bapu*. This is maize *atole* with cacao foam and flower petals on top, and it is consumed in Oaxaca (Barros and Buenrostro 2011). *Tascalate*, on the other hand, is a thin drink, colored red or orange, made with toasted and ground tortilla, roasted cacao beans, cinnamon powder, anatto (responsible for the drink color), and sugar or panela (a sugarcane sweetener). These ingredients are ground to obtain a powder mixed in small portions with cold water to yield *tascalate* (Mission Chocolate 2015). *Tanchuca* is a beverage consumed in Tabasco and other states in the south of México, prepared with cacao, maize, and anise (Barros and Buenrostro 2011). *Chilate* is a Mexican beverage prepared in the state of Guerrero with cacao, rice, cinnamon, and piloncillo or brown sugar. Cacao is roasted in the *comal*, while rice and cinnamon are soaked in water. All these ingredients are milled to prepare a paste, which is then mixed with water to yield the beverage. Finally, the sweetener, previously dissolved in water, as well as ice is added to the drink. The beverage is served very foamy, characteristic that is achieved by generous stirring of the beverage with an appropriate household mixer tool called *molinillo* or serving it from a height of approximately 30 to 50 cm to promote aeration and consequently foam formation (Instituto Mexicano de la Propiedad Industrial (IMPI) 2015; Gobierno de México 2018a, 2018b). If preferred, milk can be used in *chilate* preparation. *Chileatole* is a maize-based beverage or soup for which chili, among other ingredients, is used for its preparation (IMPI 2015). However, the designation *chileatole* is also linked to a beverage prepared with maize dough, cacao, and chili (Mother Earth Gardener 2020). It is said that in pre-Hispanic times, *chileatole* was a beverage prepared with cacao, vanilla, honey, and chili. As such, a Spaniard described it as fortifying, though bitter and hot (Ramírez González 2006; Escenario Tlaxcala 2019). In another beverage, *cacahuatole* (or called *agua de barranca*), cacao, and maize are also combined; broad beans, anise, and cinnamon are also used. It is originally from Tlaxcala and presents a strong cacao flavor (Inforural 2018; Gobierno de México 2019; Slow Food 2020). The preparation process implies roasting and milling all the ingredients to prepare a paste that will later be mixed with water, sugar, and ice. The beverage is served foamy, and the foam is generated with the *molinillo*.

Finally, *chorote*, as previously mentioned in this chapter, is a water-based drink prepared with maize and roasted cacao (Córdova Lázaro et al. 2018). Its roots come from pre-Hispanic times, and it is typically consumed in southeast Mexico in states such as Tabasco, Chiapas, Campeche, Yucatan, and Quintana Roo, as well as in Oaxaca by indigenous people (Ben Omar and Ampe 2000; Mendez-Albores et al. 2004; Castillo-Morales et al. 2005; Cooper-Bribiesca et al. 2018; Pérez-Armendáriz and Cardoso-Ugarte 2020). For its preparation, milled nixtamalized maize is mixed with fermented, roasted, and milled cacao. The obtained dough is molded into small balls and fermented for 9 days at 28°C, increasing the LAB concentration (the most predominant bacteria), yeasts, nitrogen-fixing bacteria, and molds (Castillo-Morales et al. 2005). Some examples of LAB present in *chorote* belong to the *Streptococcus* (*Streptococcus infantarius* ssp. *infantarius*), *Enterococcus*, and *Weissella* genera (Ben Omar and Ampe 2000; Romero-Luna et al. 2017; Cooper-Bribiesca et al. 2018; López-Hernández et al. 2018; Ramírez-Guzmán et al. 2019; Pérez-Armendáriz and Cardoso-Ugarte 2020).

26.6 CACAO PROCESSING

26.6.1 Preconditioning

To start cacao processing, pods must be first harvested. The pods are cut from the trees using a sickle and cut open with a machete. The seeds and mucilage are removed with a spoon-like tool, piled up in banana leaves and covered or put in lidded wooden boxes (CacaoMexico 2020). In certain parts of the globe, pre-conditioning of the pulp aims to change pulp properties to improve cacao quality. Pod storage for some days before their opening (Ozturk and Young 2017) has been suggested as a preconditioning procedure, aiming to speed up fermentation and sucrose conversion and reduce bean acidity. According to Crafack et al. (2013), rapid pulp degradation limits the amount of substrate available for organic acid production. It improves pulp aeration, which overall improves the chemical composition and flavor profile of the beans. Excessive pulp removal to reduce sugar content and avoid over-acidification has also been suggested (Afoakwa et al. 2013; Crafack et al. 2013; Kongor et al. 2016). Bean spreading is another preconditioning technique that also reduces the water and sugar content of the pulp, as well as acetic acid formation. Spreading also removes the initial anaerobic phase due to rapid pulp draining, which increases bean pH and reduces cacao acidification (Biehl et al. 1990).

26.6.2 Fermentation

Fermentation is a very important step in cacao processing (Ordoñez-Araque et al. 2020) due to its effect on flavor and color development of the beans (Figueroa-Hernández et al. 2019). It occurs in a rather empirical or uncontrolled way, and as such, the process is subjected to fluctuations or variation. For fermentation to occur, cacao beans and pulp should be kept in piles in a warm environment on banana leaves or large wooden boxes. The piles should be small and often turned to facilitate the aeration of the mucilage and help the occurrence of important concomitant reactions. Cacao fermentation is carried out by a succession of yeast, LAB, and AAB in which peak timing and concentration depend on changes in nutrition, oxygen availability, temperature, and pH (Crafack et al. 2013; Ramos et al. 2014; Kongor et al. 2016; De Roos and De Vuyst 2018). Yeasts are the first type of microorganism to participate in cacao fermentation due to the sugar content of the pulp. This raises the possibility of pulp fermentation and sugar conversion into ethanol and carbon dioxide (Ordoñez-Araque et al. 2020). Apart from the sugar content, at the beginning of the process, the cacao fermentation environment is acidic, anaerobic, and has a low pH of approximately 3.6, which are suitable conditions for the development of yeasts (Rodriguez-Campos et al. 2011; Crafack et al. 2013; Ozturk and Young 2017; Ordoñez-Araque et al. 2020). Due to the reaction caused by yeasts, the temperature in the pulp rises. Crafack et al. (2013) reported a temperature increase from 25°C to 32°C during glucose and fructose conversion to ethanol by yeasts, which later increased to 40°C together with the appearance of a strong acetic acid smell by the time glucose and fructose were consumed entirely. Other authors have also suggested cacao fermentation temperature between 22 to 30°C (Papalexandratou et al. 2011; Crafack et al. 2013; Ordoñez-Araque et al. 2020). Yeast population increases in the first 12 hours; then it remains constant for the next 12 hours and later decreases drastically with pulp sugar depletion and increasing temperature, with a total duration of 36 hours (Crafack et al. 2013; Castro-Alayo et al. 2019; Ordoñez-Araque et al. 2020). At this stage, the oxygen content of the pulp is low, and the pH is below 4 (Castro-Alayo et al. 2019).

As yeasts are inhibited, bacteria take over. Both lactic acid bacteria and acetic acid bacteria begin to grow. LAB, which are microaerophilic and tolerate acidity (Ordoñez-Araque et al. 2020), are present from the beginning of the fermentation but take over when the medium turns from anaerobic to microaerobic (Castro-Alayo et al. 2019; Figueroa-Hernández et al. 2019). They are the most active between 16 and 48 hours of fermentation, with their peak at around 36 hours. While LAB reach their peak, the yeast population decreases (Schwan and Wheals 2004; Crafack et al. 2013; John et al. 2020; Ordoñez-Araque et al. 2020).

AAB, which are strictly aerobic, only appear after the first day of fermentation. Their presence is promoted by pulp degradation by pectinolytic enzymes, making the pulp less viscous for the entrance of oxygen. The turn-over of the heaps to facilitate aeration, and keeping temperatures between 28 and 30°C and an optimum pH of 4.5 is also favorable to AAB growth (Ordoñez-Araque et al. 2020). They become the most predominant microorganisms around 48 to 112 hours of fermentation and reach their peak at 88 hours. After reaching their peak, AAB counts start to decrease, probably due to the high temperatures (50°C) generated towards the end of fermentation by their own ethanol oxidizing activity, until they are non-viable (Papalexandratou et al. 2011; Crafack et al. 2013; Ordoñez-Araque et al. 2020).

Different fermentation methods and durations are used in other countries, which generate different flavor profiles (Marseglia et al. 2020). A duration of 2 to 8 days is typically used for cocoa bean fermentation (Ozturk and Young 2017; Castro-Alayo et al. 2019; Pacheco-Montealegre et al. 2020), in general longer when done on plantations and shorter on smallholder farms (Schwan and Wheals 2004). The duration also depends on the variety; *Forastero* and *Trinitario* take 5 to 8 days to be fermented, while 2 to 4 days are enough for *Criollo* fermentation (Ozturk and Young 2017; Castro-Alayo et al. 2019). Either way, knowing when to stop the fermentation is important to avoid poor fermentation or overfermentation. While the former may allow mold infestation, poor chocolate flavor development upon roasting, or even unpleasant flavors such as a potato flavor on the beans, the latter can increase the amount of microorganisms associated with off-flavor such as aerobic spore-forming *Bacillus* and filamentous fungi, and generate hammy or rotten flavors (Schwan and Wheals 2004; Afoakwa et al. 2008; Kongor et al. 2016; Castro-Alayo et al. 2019). Romero-Cortes et al. (2013) suggested that since fermentation is usually ended empirically by workers' perceptions, it is important to define parameters that analytically allow one to determine the end of the fermentation and standardize the process. The authors referred to the fermentation index as a measurement of anthocyanin content and an indirect measurement of the degree of fermentation. The beans change color from purple to brown due to anthocyanin decomposition, indicating progression of the fermentation process (Sunoj et al. 2016; Barišić et al. 2019). A fermentation index ≥ 1 indicated sufficiently fermented cacao, and principal component analysis results suggested that this could be reached after 48 hours under Mexican environmental conditions (Romero-Cortes et al. 2013). The fermentation index did not seem to be enough to predict the conclusion of fermentation because fermentation finished after 144 hours. However, the authors (Romero-Cortes et al. 2013) suggested that the fermentation index, together with pH, could constitute a way to detect the end of fermentation. Indeed, Afoakwa et al. (2008) mentioned that pH values around 5.5 and 5.8 were associated with unfermented beans, while pH values of 4.75 to 5.19 were characteristic of fermented ones. However, these pH intervals were both associated with poorer chocolate flavor and higher notes of off-flavor by Rodriguez-Campos et al. (2011), who found that intermediate pH (5.20–5.49) generated higher notes of chocolate flavor compared to the previously mentioned pH intervals. Crafack et al. (2013) also suggested that a pH of 4.0–4.5 in the cotyledon leads to lower flavor potential than a pH of 5.0–5.5. In general, beans with low pH are associated with low quality and a poorer flavor profile. The ratio of 3-methyl-1-butyl acetate:methyl-1-butanol can also indicate the degree of fermentation; a ratio higher than 1.5 is related to overfermentation (Rodriguez-Campos et al. 2011). The number of peptides present in dried cacao beans has also been used to assess the degree of fermentation. An increasing number of peptides was observed when cacao changed from underfermented and fairly fermented to well fermented. The proportion of hydrophilic peptides and free hydrophobic amino acids in cacao fermentation also varied depending on the degree of fermentation (Kumari et al. 2018).

26.6.2.1 Microorganisms Involved

The microorganisms participating in cacao fermentation have a significant influence on the speed and success of the process and the quality of the aromas and textures obtained (Ramos et al. 2014; Mota-Gutierrez et al. 2019). Microorganisms shape fermentation based on their proportion, the timing of their distribution during fermentation, and the specific combination of the different species that might be present during the process. On the other hand, the study of the microorganisms involved in the fermentation of cacao could be important for process standardization, which, for instance, allows evaluating new possibilities for small- and large-scale cacao fermentation, manipulating flavor and texture profiles, developing functional nutrients, or simply defining starter cultures that allow cleaner process conditions (Mota-Gutierrez et al. 2019). Pods have been considered the main source of important microorganisms participating in fermentation (Lefeber et al. 2011; Ordoñez-Araque et al. 2020), although they can also be picked up from tools and other utensils, banana leaves used to hold the pulp and the seeds, or from workers' hands (Schwan and Wheals 2004; Lefeber et al. 2011; Crafack et al. 2013; Ozturk and Young 2017; John et al. 2020; Ordoñez-Araque et al. 2020).

In general, the yeast population is more diverse than the bacteria group and plays an indispensable role in cacao fermentation (Figueroa-Hernández et al. 2019). Yeasts belonging to the genera *Saccharomyces, Pichia, Hanseniaspora, Kluyveromyces,* and *Candida* are the most commonly found in cacao fermentation (Figueroa-Hernández et al. 2019; Ordoñez-Araque et al. 2020). *Kloeckera, Trichosporon, Schizosaccharomyces, Kloeckera, Torulospora, Debaromyces, Geotrichum, Rhodotorula, Saccharomycopsis,* and *Wickerhamomyces* are other yeast genera that have also been found in cacao fermentation in different countries. *S. cerevisiae* is among the most dominant yeast species in cacao fermentation (Ho et al. 2014; Mota-Gutierrez et al. 2019; Ordoñez-Araque et al. 2020), and together with *Hanseniaspora opuntiae*, they are among the most desirable in fermentation (Lefeber et al. 2012). *H. guilliermondii, H. opuntiae,* and *Kloeckera apiculata* are examples of yeasts found in fermentation's early stages, declining relatively rapidly, while *S. cerevisiae, K. marxianus, P. membranifaciens, P. kudriavzevii,* and *Candida* spp. dominate the later stage of fermentation. Some *Candida* found during cacao fermentation are *Candida pelliculosa, Candida tropicalis, Candida zeylanoides Candida rajasthanensis, Candida diversa, Candida jaroonii, Candida quercitrusa, Candida humilis,* and *Candida stellimalicola.* Other yeasts identified in cacao fermentation are *Hanseniaspora thailandica, H. guilliermondii, H. uvarum, Pichia manshurica, Pichia galeiformis, P. fermentans, P. kluyveri, P. guilliermondii, Saccharomycopsis crataegensis, Kluveromyces thermotolerans, Torulopsis candida, Torulopsis castelli, Torulopsis holmii, Torulopsis delbrueckii, Torulopsis pretoriensis, Schizosaccharomyces pombe, Issatchenkia hanoiensis, Issatchenkia terricola, Rhodosporidium fluviale, Dekkera anomala, Zygosaccharomyces bailii, Lodderomyces ellongisporus,* and *Kloeckera apis* (Schwan and Wheals 2004; Crafack et al. 2013; Ho et al. 2014; Arana-Sánchez et al. 2015; Figueroa-Hernández et al. 2019; Ordoñez-Araque et al. 2020). In Mexican cacao fermentation, the species *S. cerevisiae, H. guilliermondii, P. kudriavzevii,* and *S. crataegensis* were the most predominant (Arana-Sánchez et al. 2015; Hernández-Hernández et al. 2016).

Although yeasts are mainly responsible for ethanol production during fermentation, Ho et al. (2014) suggested that ethanol presence is not crucial to achieving the amount of acetic acid needed to kill the bean. It might, however, speed up the rate of acetic acid formation. On the other hand, pulp sugars were more rapidly used in the presence of yeasts. Nonetheless, the rate of consumption of cacao bean sugar did not seem to be affected by their presence, which suggests that yeasts do not contribute much to cacao bean sugar conversion. Cacao fermentation benefits from yeasts in many ways, being especially important for developing cacao sensory properties. When yeasts were present in cacao fermentation, the authors obtained fully brown-colored and less acidic cacao, with characteristic chocolate flavor. The brown color was an indicator of successful fermentation. In the absence of yeasts, bean fermentation was unsuccessful; they showed purplish color and made chocolate lighter in color. Yeasts might also play a role in glycerol production and contribute to a drier, less sticky, and better-degraded pulp. In their absence, beans kept a higher amount of their shell material and part of the pulp attached to the seed coat (Ho et al. 2014). Yeasts also help to buffer the pH through the production of organic acids as acetic, malic, oxalic, and phosphoric (Ordoñez-Araque et al. 2020). Yeasts, rather than LAB or AAB, are responsible for producing flavor compounds such as alcohols (candy and flowery notes) and esters (fruity flavors) during fermentation (Ho et al. 2014).

Additionally, although pyrazines are mainly formed during roasting, some are formed during fermentation. These were significantly less when yeasts were not present in fermentation, suggesting the role of these microorganisms in pyrazine production (Lefeber et al. 2011; Crafack et al. 2013; Ho et al. 2014; Ramos et al. 2014; Ozturk and Young 2017; Ordoñez-Araque et al. 2020). Aldehydes, ketones, and organic acids are also formed due to yeast activity (Mota-Gutierrez et al. 2019). Some yeast such as *Kloeckera apiculata, S. cerevisiae, Candida* spp., and *K. marxianus* were found in cacao fermentation and produced volatiles belonging to fuselols, fatty acids, and fatty acid esters (Schwan and Wheals 2004; Ho et al. 2014). *Kloeckera apiculata* and *S. cerevisiae* var. *chevalieri* have also been associated with producing desirable compounds found in high-quality cacao, such as isopropyl acetate, ethyl acetate, 1-propanol, isoamyl alcohol, 2,3-butanediol, diethyl succinate, and phenylethanol (Rodriguez-Campos et al. 2012). Finally, Figueroa-Hernández et al. (2019) obtained fruity, acidic, yogurt-like, and balsamic flavored cacao in the presence of *P. kluyveri.*

LAB appear as the number of yeasts diminish and the pulp drains away, and they mainly belong to *Lactobacillus* and *Leuconostoc* genera (Mota-Gutierrez et al. 2019). The participation of LAB in cacao fermentation might serve different functions. It has been suggested that the production of lactic acid and the consumption of citric acid by LAB helps balance the process's pH. By decreasing pH from 7 to 5.3, they also allow proteolytic enzymes to break cacao bean proteins to produce flavor precursors (Ho et al. 2014). *L. fermentum* (new name *Limosilactobacillus fermentum*) and *L. plantarum* (new name *Lactiplantibacillus plantarum*) are among the most important LAB species participating in cacao fermentation. *L. fermentum* produces aroma compounds such as diacetyl, acetoin, and 2,3-butanediol, apart from effectively consuming citric acid and increasing the pH of the pulp (Crafack et al. 2013; Figueroa-Hernández et al. 2019). Some LAB are found more at the beginning of the fermentation, and others appear gradually over time before the yeast population declines (Ordoñez-Araque et al. 2020). *Leuconostoc mesenteroides* and

Lactococcus lactis were present from the first day of fermentation (Papalexandratou et al. 2011; Lefeber et al. 2012; Ho et al. 2014; Oliveira et al. 2018; Ordoñez-Araque et al. 2020). *Lactobacillus* sp. (*L. delbrueckii, Lactobacillus cellobiosus, Lactobacillus amylovorus, Lactobacillus pentosus* (new name *Lactiplantibacillus pentosus*), *Lactobacillus curieae* (new name *Lentilactobacillus curieae*) and *Lactobacillus casei* (new name *Lacticaseibacillus casei*), *Leuconostoc pseudomesenteroides, Weissella* sp. (*W. paramesenteroides* and *W. cibaria*), *Pediococcus acidilactici, Fructobacillus pseudoficulneus, Streptococcus salivarius*, and *Enterococcus* sp. (*Enterococcus casseliflavus* and *Enterococcus faecium*) are also examples of LAB found during cacao fermentation (Schwan and Wheals 2004; Lefeber et al. 2011; Papalexandratou et al. 2011; Ho et al. 2014; Ordoñez-Araque et al. 2020).

When oxygen is more available, acetic acid bacteria proliferate with the decline of yeasts and lactic acid bacteria. AAB activity and the acetic acid produced therefrom are associated with the breakdown of internal bean structure and the triggering of the biochemical reactions involved in flavor precursors and color formation (Crafack et al. 2013; Arana-Sánchez et al. 2015). Both ethanol (produced by yeasts) and lactic acid (produced by LAB) function as substrates for acetic acid production by AAB, demonstrating the close interaction occurring among the three groups of microorganisms participating in cacao fermentation. Although acetic acid is desirable in fermentation, its early production by either heterofermentative LAB or AAB can be associated with acidic beans and off-flavors. Early lactic acid production by LAB is also associated with off-flavors and bean sourness (Papalexandratou et al. 2011; Rodriguez-Campos et al. 2011; Crafack et al. 2013; De Roos and De Vuyst 2018). The concentration of lactic acid at the beginning of the fermentation is, in effect, low and starts to increase throughout the fermentation, acidifying the structure of the beans, together with acetic acid (Papalexandratou et al. 2011; Rodriguez-Campos et al. 2011; Ho et al. 2014; Ramos et al. 2014; Arana-Sánchez et al. 2015; Figueroa-Hernández et al. 2019; Ordoñez-Araque et al. 2020). Glucose and mannitol can also be oxidized by AAB, mediated by the presence of ethanol (Moens et al. 2014). *Acetobacter* and *Gluconobacter* are the predominant AAB genera in cacao fermentation. *Acetobacter pasteurianus* stands out among the AAB species while *L. plantarum* (new name *Lactiplantibacillus plantarum*) and *L. fermentum* (new name *Limosilactobacillus fermentum*) are prevailing microbiota during cacao beans processing (Papalexandratou et al. 2011; Lefeber et al. 2011; Lefeber et al. 2012; Ho et al. 2014; Ramos et al. 2014; De Roos and De Vuyst 2018; Figueroa-Hernández et al. 2019; Mota-Gutierrez et al. 2019; Ordoñez-Araque et al. 2020). Other *Acetobacter* found during cacao fermentation are *A. aceti, A. fabarum, A. ghanensis, A. senegalensis, A. indonesiensis, A. malorum, A. cerevisiae, A. lovaniensis, A. orientalis, A. peroxydans, A. pomorum, A. ascendens, A. rancens, A. xylinum, A. lovaniensis, A. xylinum, A. peroxydans, A. tropicalis*, and *A. senegalensis*. Some *Acetobacter* species can over-oxidize acetic and lactic acids, decreasing the overall consequences of over-acidification on the flavor profile. *Gluconobacter* such as *G. liquefaciens, G. saccharivorans, G frateurii*, and *G. oxydans* are also among the AAB participating in cacao fermentation (Lefeber et al. 2011; Papalexandratou et al. 2011; Crafack et al. 2013; Ho et al. 2014; Ramos et al. 2014; Ordoñez-Araque et al. 2020).

Filamentous fungi, *Bacillus, Enterobacteriaceae*, and other Gram-positive and Gram-negative bacteria have been found towards the end of fermentation when the temperature is around its maximum (approximately 50°C). Their appearance is favored by prolonged fermentation time, contributing off-flavor to cacao (Schwan and Wheals 2004; Ozturk and Young 2017; Figueroa-Hernández et al. 2019; Ordoñez-Araque et al. 2020). Additionally, filamentous fungi, which can also appear during drying, are a source of mycotoxins, which might increase the amount of free fatty acids and be involved in pulp hydrolysis (Schwan and Wheals 2004; Romanens et al. 2019). In general, *Aspergillus, Eremothecium, Lichtheimia, Geotrichum, Penicillium, Rhizopus, Fusarium, Mucor, Neurospora,* and *Syncephalastrum* are common at the end of cacao fermentation. Accordingly, it is not surprising that aflatoxin B1, ochratoxin A, fumonisin, and citrinin have been found in fermented cacao (Ozturk and Young 2017; Mota-Gutierrez et al. 2019; Romanens et al. 2019; Ordoñez-Araque et al. 2020). *Aspergillus fumigatus* and *Mucor racemosus* has been found as predominant species, although other species such as *Aspergillus niger, Fusarium moniliforme, Fusarium oxysporum, Lasiodiplodia theobromae, Paecilomyces varioti, Penicillium citrinum, Penicillium implicatus, Penicillium spinosum, Thielaviopsis ethaceticus*, and *Trichoderma viridae* have also been identified (Schwan and Wheals 2004). Bacteria of the genus *Bacillus* appear with increased aeration, pH, and temperature at the final stage of fermentation (Castro-Alayo et al. 2019). *Bacillus cereus, Bacillus flexus,* and *B. megaterium* generate off-flavors through acetic acid, lactic acids, 2,3-butanediol, tetramethylpyrazine, and other pyrazine production. They produce proteolytic, lipolytic, amylolytic, and pectinolytic enzymes, which, by degrading biomolecules, produce undesirable flavor compounds. *Bacillus pumilus, Bacillus fusiformis*, and *B. subtilis* were also suggested to produce pectin lyase and, as such, to have a role in bean pectin degradation. Other *Bacillus* species that have been identified in cacao fermentation are *Bacillus coagulans, Bacillus circulans*, and *B. licheniformis* (Papalexandratou et al. 2011; Castro-Alayo et al. 2019; Figueroa-Hernández et al. 2019; Ordoñez-Araque et al. 2020). Other non-LAB Gram-positive genera of bacteria found in cacao fermentation are *Cellulomonas, Micrococcus*, and *Staphylococcus* (Ozturk and Young 2017). Non-AAB Gram-negative bacteria include the enterobacterial family (probably related to pectin degradation, citric acid and glucose assimilation, as well as gluconic acid production), *Zymomonas, Acinobacter, Erwinia, Phyllophora, Pantoea,* and *Tatumella* (Lefeber et al. 2011; Papalexandratou et al. 2011; Ozturk and Young 2017).

26.6.2.2 Starter Cultures

Starter culture application in cacao fermentation aims towards a higher-quality product with an improved flavor profile and a more standardized and controlled fermentation process. The chosen microorganism can also provide desirable characteristics such as heat, acid, ethanol and phenolic tolerance, and citric or lactic acid conversion capability, as well as mannitol or acetic acid production capacity (Lefeber et al. 2012; Crafack et al. 2013; Figueroa-Hernández et al. 2019). Starter cultures can combine different yeasts, LAB or AAB, where two or all three groups of microorganisms can be used (Romanens et al. 2019). Highly aromatic and pectinolytic yeasts have been proposed as starters, namely *Pichia kluyveri* and *Kluyveromyces marxianus* strains, respectively (Crafack et al. 2013). Significant differences in volatile cacao composition were observed when either yeast was used in cacao fermentation. In both cases, the obtained chocolate presented acidic, bittery, fruity, berry, yoghurt-like, and balsamic aromas. *P. kluyveri* contributed higher concentrations of phenylacetaldehyde and 2-methoxyphenol and less pentanal, 1-pentanol, ethylphenyl acetate, and 2,3-octanedione in comparison to microorganisms present in spontaneous fermentation. On the other hand, cacao fermentation inoculated with *K. marxianus* showed higher concentrations of benzyl alcohol, phenethyl alcohol, benzyl acetate, phenethyl acetate, and 2-methoxyphenol compared to a spontaneously fermented cacao (Crafack et al. 2014). Certain *S. cerevisiae* strains have also been proposed as cacao fermentation starter cultures. These include *S. cerevisiae* UFLA CA11, due to the rapid rate of carbohydrate consumption and their conversion to ethanol, avoiding the growth of undesirable microorganisms (Ramos et al. 2014). Romanens et al. (2019) suggested that using the yeasts *S. cerevisiae* and *Haseniaspora opuntiae* and the LAB *L. fermentum* (new name *Limosilactobacillus fermentum*) in cacao fermentation could reduce the incidence of filamentous fungi such as *Aspergillus*, *Penicillium,* and *Gibberella* genera (Romanens et al. 2019). *L. fermentum* was also advantageous for its fructose and citrate conversion capacity, mannitol production, acid, heat, and ethanol tolerance (Figueroa-Hernández et al. 2019). On the other hand, specific characteristics of *A. pasteurianus* such as heat tolerance and ethanol, mannitol, and lactic acid oxidizing capacity were found convenient in a cacao fermentation as starter culture (Figueroa-Hernández et al. 2019).

L. fermentum (new name *Limosilactobacillus fermentum*) 222, which had, apart from citric acid conversion and mannitol production capacity, the ability to produce 2,3-butanediol, was combined with *A. pasteurianus* 386B, a strain capable of producing acetic acid and acetoin through ethanol and lactic acid oxidation, respectively. Moens et al. (2014) also found that *A. pasteurianus* 386B oxidized mannitol and glucose into fructose and gluconic acid, respectively, and was tolerant to high ethanol and acetic acid concentrations. *L. fermentum* 222 and *A. pasteurianus* 386B were used together as a starter culture. Spontaneous cacao bean fermentation enabled a more rapid fermentation process with enhanced citric acid consumption and lactic acid production from carbohydrates. Yeasts and other AAB growth were favored as well. On the other hand, different metabolites were obtained in the cacao pulp when *L. fermentum* 222 and *A. pasteurianus* 386B were used alone or combined with *S. cerevisiae* H5S5K23, an ethanol-tolerant and non-citric acid-converting yeast. Additionally, in the presence of *S. cerevisiae* H5S5K23, more consistent ethanol production by yeasts and acetic acid production by AAB were obtained (Lefeber et al. 2012; Moens et al. 2014). In a study carried out by Crafack et al. (2013), fermentation using a starter culture formed by *P. kluyveri* CH, *L. fermentum* (new name *Limosilactobacillus fermentum*) *L18* and *A. pasteurianus* A149 acquired a higher score for sweetness, fruitiness, cocoa aroma, and general liking compared to spontaneous fermentation. Species such as *Torulaspora delbrueckii, L. lactis, L. plantarum* (new name *Lactiplantibacillus plantarum*), *A. aceti*, and *G. oxydans* are also examples of microorganisms that have been used in starter culture mixtures (Lefeber et al. 2012; Figueroa-Hernández et al. 2019).

26.6.2.3 Reactions

The importance of fermentation in cacao processing is mainly due to the flavor precursors formed during this step (Ordoñez-Araque et al. 2020). Unfermented cacao beans do not contain enough aroma precursors for flavor development during roasting. Such precursors, namely amino acids, peptides, and sugars, are generally formed during fermentation through a series of reactions affected by cacao composition, fermentation duration, types of microorganisms, and, consequently, by ethanol, lactic acid, and acetic acid concentrations in the medium (Utrilla-Vázquez et al. 2020). Fermentation-associated reactions start in the pulp. Yeasts start growing due to the sugar content in the pulp, in part generated by their own invertase activity over sucrose, generating glucose and fructose (Castro-Alayo et al. 2019). These microorganisms (*K. marxianus*, *S. cerevisiae* var. *chevalieri*, *Candida rugopelliculosa*, and *K. thermotolerans* as some examples), secrete pectinolytic enzymes that degrade the pulp, making it thinner and allowing it to drain away (Lefeber et al. 2011; Crafack et al. 2013; Ho et al. 2014; Ramos et al. 2014; Castro-Alayo et al. 2019; Figueroa-Hernández et al. 2019; Ordoñez-Araque et al. 2020). Ethanol, as well as some organic acids such as acetic, oxalic, succinic, and malic, are formed as a result of pulp sugar breakdown (Kongor et al. 2016). Conversely, some yeasts are also responsible for metabolizing citric acid, increasing pH, which, together with the increased ethanol levels, inhibits them (Schwan and Wheals 2004; Papalexandratou et al. 2011; Ramos et al. 2014).

LAB consume glucose, producing lactic acid and, to a lesser extent, acetic acid and alcohol. They also metabolize citric acid present in the pulp (Papalexandratou et al. 2011; Oliveira et al. 2018) producing lactic acid, acetic acid, acetaldehyde, diacetyl, and mannitol, while they can also produce mannitol through the reduction of fructose. Although responsible for metabolizing citric acid, resulting in the reduction of acidity and an increase of pH (which favor AAB appearance), they can also produce citric acid, increasing acidity and lowering pH. Contrary to yeast, LAB have no proteolytic activity (Ordoñez-Araque et al. 2020).

AAB oxidize ethanol to acetaldehyde and later to acetic acid. Eventually, throughout the course of reactions, both acetic and lactic acids can be converted to carbon dioxide and water, which helps avoid over-acidification of the beans (Crafack et al. 2013). AAB can also convert lactic acid into acetic acid and acetoin (Figueroa-Hernández et al. 2019). Moens et al. (2014) observed that both ethanol and lactic acid could be enzymatically converted to acetic acid, either through fast or slow oxidation by *Acetobacter*. Other products of AAB metabolism are ethyl acetate, butanol, and isopropanol and organic acids, as well as other important volatiles for cacao aroma (Schwan and Wheals 2004; Afoakwa et al. 2008; Rodriguez-Campos et al. 2011; Crafack et al. 2013; Ramos et al. 2014; Kongor et al. 2016; Ozturk and Young 2017; Figueroa-Hernández et al. 2019; Ordoñez-Araque et al. 2020). The formation of ethanol and lactic acid through the middle of the fermentation and the increasing temperature and acetic acid concentration towards the end triggers biochemical and enzymatic reactions first on the bean's surface and later inside the cotyledon. While the first phase of cacao reactions occurs in the pulp at 25 to 45°C, subsequent enzymatic reactions in the cotyledon occur at 42–52°C (Castro-Alayo et al. 2019). As acid penetrates the bean, its pH decreases from 6.5 to 4.5, which leads to reactions that cause the breakdown of storage cells, bean swelling, germination hindering, and the death of the embryos. The breakdown of polyphenols within cells decreases astringency and bitterness (Barišić et al. 2019). The lower pH also favors proteolysis reactions which, together with other hydrolytic reactions, are important to produce flavor compound precursors (Crafack et al. 2013; Ho et al. 2014; Ramos et al. 2014; Figueroa-Hernández et al. 2019; Ordoñez-Araque et al. 2020). Enzymatic reactions that break storage proteins such as vicilin (7S)-class globulin into hydrophilic short-chain peptides and hydrophobic free amino acids entail the participation of aspartic endoprotease (optimal temperature around 47°C and pH of 3.8) and serine carboxy-(exo) peptidase (optimal temperature between 46°C and 48°C and pH pf 5.8). Peptides are also the product of albumin degradation (Afoakwa et al. 2008; Kongor et al. 2016; D'Souza et al. 2018; Kumari et al. 2018; Barišić et al. 2019; Figueroa-Hernández et al. 2019).

Peptides are almost nonexistent in unfermented cacao. They are produced during fermentation, and their size gets smaller throughout the process until a variety of peptides with a length between 2 and 8 amino acids are obtained at the end of the process (D'Souza et al. 2018). The obtained peptides and amino acids (alanine, leucine, phenylalanine, and tyrosine) are flavor precursors that generate flavor compounds during roasting (Afoakwa et al. 2008; De Roos and De Vuyst 2018; D'Souza et al. 2018; Figueroa-Hernández et al. 2019). Peptides and amino acids obtained during fermentation are known to contribute nutty and cocoa flavor notes, among others (John et al. 2020). De Britol et al. (2004) used an exogenous proteolytic enzyme for cacao protein breakdown and observed an increase in chocolate flavor and bitter notes in cacao products. Sucrose concentration in the beans, which is higher than monosaccharides at the beginning of fermentation, decreases due to its conversion to glucose and fructose through acid hydrolysis and enzymatic activity invertase (Crafack et al. 2013; Kongor et al. 2016). These obtained reducing sugars are also flavor precursors that will be involved in the Maillard reaction together with amino acids during roasting. The period of enzymatic activity in cacao beans is short, and the enzymes are rapidly inactivated by heat and high concentrations of acetic acid, ethanol, and polyphenols (Hansen et al. 1998). Other enzymes such as polyphenol oxidases are also important for flavor precursor development. Polyphenols such as catechins, epicatechin, and procyanidins are, in fact, cacao flavor precursors. They go through oxidation by the action of polyphenol oxidases, being converted to quinones. Some anthocyanidins are hydrolyzed by glycosidases and later oxidized to quinones by polyphenol oxidase. These can form complexes with amino acids or proteins to form tannins. The beans become more flavorful, less bitter, and less astringent, and their color changes from purple to brown (Gibson and Newsham 2018; Figueroa-Hernández et al. 2019; Ordoñez-Araque et al. 2020). Polyphenols are found in different proportions among cacao varieties, contributing to the other flavor profiles observed. The *Criollo* variety, for instance, seems to have fewer amounts and types than *Forastero* (Afoakwa et al. 2008).

26.6.2.4 Flavor Components

The cacao flavor profile is affected by different factors, from the variety of beans and the conditions of the plant's environment to farming practices. These certainly influence bean and pulp composition, as does pod preconditioning before fermentation. Post-harvesting procedures, including fermentation (and the microorganisms involved), drying, roasting, alkalization, and conching, lead to transformations in cacao that shape its flavor (Crafack et al. 2013; Kongor et al. 2016; Barišić et al. 2019; Figueroa-Hernández et al. 2019; Valverde García et al. 2020). It has been reported that the volatile cacao profile is affected more by processing than by the cacao type (Utrilla-Vázquez et al. 2020). In this regard, fermentation plays an important role in flavor development. Unfermented cocoa beans develop little flavor during roasting, and their quality might depend mostly on aspects such as plant genotype and botanical origin, the age of the tree, and climate and soil conditions (Kongor et al. 2016). The cacao flavor profile is formed by over 500 volatile compounds, including esters, aldehydes, alcohols, polyphenols, pyrazines, and alkaloids (Kongor et al. 2016; Mota-Gutierrez et al. 2019; Marseglia et al. 2020). Non-volatile components responsible for cacao flavor include amino acids, sugars, organic acids, proanthocyanidins, and diketopiperazines (D'Souza et al. 2018). Cacao polyphenols (such as anthocyanins) and alkaloids (theobromine and caffeine) contribute to the widely known cacao bitterness and astringency (Ho et al. 2014; Kongor et al. 2016). These could be considered the primordial cacao flavor components considering that they are always present in raw cacao and undergo several changes during cacao production (Barišić et al. 2019). However, due to reactions that start during fermentation, cacao bitterness and astringency significantly decrease and give way to the development of new flavors or flavor precursors (Chin et al. 2013; Kongor et al. 2016; Gibson and Newsham 2018; Ordoñez-Araque et al. 2020). Other than flavor precursors obtained from bean storage proteins and carbohydrate degradation, the pulp might also contribute to flavor compounds that migrate to the bean (Castro-Alayo et al. 2019). On the other hand, microorganisms participating in cacao fermentation produce volatile aroma compounds such as alcohols, organic acids, esters, and aldehydes that contribute to the cacao flavor profile (Pacheco-Montealegre et al. 2020). Yeasts have been reported to produce phenylethylacetate (fruit aroma), 2-phenylethanol (floral), and acetoin (butter flavor). The yeasts *Kloeckera apiculata* and *S. cerevisiae* var. *chevalieri* produce high-quality compounds, namely esters such as ethyl acetate, isopropyl acetate, and diethyl succinate and alcohols including 1-propanol, phenylethanol, and isoamyl alcohol, as well as 2,3-butanediol (Rodriguez-Campos et al. 2011). LAB are associated with aldehyde and ketone production (Castro-Alayo et al. 2019).

Among volatile aroma compounds found in cacao, pyrazines are among the most important and are associated with earthy, roasted, sweet chocolate, and nutty flavors. Pyrazines are not found in unfermented beans, and some are formed during fermentation. At the same time, they continue to be produced during drying and roasting. Some examples of pyrazines that have been found before roasting of cacao are trimethylpyrazine, tetramethylpyrazine, 2,6-dimethylpyrazine, 2,3-dimethyl-5-ethyl pyrazine, and 2,3,5-trimethyl-6-ethyl pyrazine. Studies suggest that they can result from the genus *Bacillus*'s metabolism. An example is tetramethylpyrazine, which has been associated with *Bacillus subtilis* found during fermentation. High amounts of 2,3,5-trimethylpyrazine, which is responsible for toasted notes, have been found in Mexican cacao (Rodriguez-Campos et al. 2011; Ho et al. 2014; Barišić et al. 2019; Marseglia et al. 2020; Valverde García et al. 2020).

Other important components are aldehydes, ketones, alcohols, and esters. Aldehydes and ketones are desirable molecules in cacao aroma. They are produced during fermentation through reactions occurring within the beans, contributing sweet and fruity flavors, respectively (Valverde García et al. 2020). Aldehydes can be found at the beginning and end of fermentation (Utrilla-Vázquez et al. 2020). Some aldehydes present in cacao are benzaldehyde and phenylacetaldehyde. 2-and 3-methylbutanal can also be found in cacao; they provide malt and chocolate aromas and are obtained by the action of lactic acid bacteria due to the degradation of isoleucine and leucine. On the other hand, ketones such as diacetyl are obtained from yeast and bacterial metabolism and are responsible for buttery flavor. Acetoin (buttery flavor) is the main ketone in roasted, fermented beans, although others such as 2-heptanone, 2-pentanone, 2-nonanone, and acetophenone are also produced during fermentation (Ho et al. 2014; Barišić et al. 2019; Marseglia et al. 2020; Utrilla-Vázquez et al. 2020). Ketones are typically produced from the beginning of fermentation and tend to increase over time (Utrilla-Vázquez et al. 2020).

Alcohol compounds, which are usually obtained by yeast metabolism, are found from the beginning of the fermentation and are associated with long fermentation times. They give cacao flowery, candy, and sweet chocolate aromas and decrease with drying due to evaporation. 2-phenylethanol, 2,3-butanediol, 1, 3-butanediol, benzyl alcohol, 3-methyl-1-butanol, 3-methyl-2-butanol, 2-methyl-1-propanol, and 2-heptanol, among other alcohols, are found during cacao fermentation and are desirable in the cacao flavor profile. Alcohols can be oxidized to acetic acid but also to esters (Rodriguez-Campos et al. 2011; Barišić et al. 2019; Marseglia et al. 2020; Utrilla-Vázquez et al. 2020; Valverde García et al. 2020).

Although not as important as pyrazines, esters are obtained from the reaction of alcohols with organic acids during fermentation; give honey, fruity, and flowery notes to cacao; and are present from the beginning of the fermentation (Kongor et al. 2016; Barišić et al. 2019; Marseglia et al. 2020; Utrilla-Vázquez et al. 2020). Some esters formed during fermentation are ethyl acetate (formed by esterification of ethanol and acetic acid); methyl isovalerate; methyl isopentanoate; and isobutyl-, isoamyl-, and phenylethyl acetates (Barišić et al. 2019). Esters of acetic acid, such as 3-methyl-1-butanol acetate, 2-phenylethyl acetate, methyl acetate, 1-butanol-3-methylbenzoate, and butyrolactone, are probably the result of yeast metabolism (Marseglia et al. 2020).

Acids, such as lactic and acetic, can produce desirable notes such as floral (2-methylpropionic acid), although they can also generate undesirable rancid, sour, and fatty aromas (Lefeber et al. 2011; Crafack et al. 2013; Ramos et al. 2014; Ozturk and Young 2017; Barišić et al. 2019; Ordoñez-Araque et al. 2020; Valverde García et al. 2020).

Finally, phenols and methyl phenols are also found in cacao and are responsible for smoky notes, while the sweetness of the pulp contributes to caramel and walnut flavors (Castro-Alayo et al. 2019; Marseglia et al. 2020).

Cacao from different origins has different flavor profiles due to the different genotypes and fermentation and roasting processes. Flavor profiles vary between cacao varieties as well. Mexican *Criollo* variety showed an interesting flavor profile, with high levels of amino acids and peptides and a high amount of volatile compounds (Marseglia et al. 2020). In general, *Criollo* cacao has a fruity and sugary flavor profile given by compounds such as 3-methylbutanal, 2-methylbutanal, 2-methylpropanal, and 2-phenylethyl acetate. The basic cacao or chocolate aromas in *Criollo* are given by 2,3,5,6-tetramethylpyrazine and 2,3,5-trimethylpyrazine and are generally low due to shorter fermentation compared to the less fine cacao varieties. Fine cacao has also been associated with the presence of compounds such as valencene, aromadendrene, germacrene, 1,3 cyclohexadiene, and octacosane (Castro-Alayo et al. 2019).

26.6.3 Drying

Drying follows fermentation. The fermented cacao beans are left to sun dry, spread on a tray, with frequent raking of the beans. They can also be dried by air-blowing, shade drying, and oven drying at a temperature not higher than 60°C for at least 2 days, although sun drying shows greater benefits regarding off-flavor avoidance (Schwan and Wheals 2004; Afoakwa et al. 2008; Gibson and Newsham 2018; CacaoMexico 2020). Drying is performed until the water content is approximately 7% of cacao weight to avoid subsequent mold contamination, and beans are often turned for better and more even drying (Kongor et al. 2016; Barišić et al. 2019). Drying should not be too rapid or acetic acid evaporation will be inefficient due to bean shell hardening, nor too slow to prevent molding, low acidity, and defects in color (Kongor et al. 2016). While the concentration of volatile acids is reduced due to oxidation and evaporation, and the concentration of compounds such as acetoin and phenylacetaldehyde decreases, there is an increase of other compounds. Aldehydes and pyrazines such as tetramethylpyrazine are examples of important classes of components formed during drying. Additionally, the content of esters such as 3-methyl-1-butanol acetate, and isobuthyl acetate and secondary alcohols, including 2,3-butanediol and 1,3-butanediol, increase during drying (Afoakwa et al. 2008; Rodriguez-Campos et al. 2011; Kongor et al. 2016; Utrilla-Vázquez et al. 2020). Flavor precursors continue to be formed during drying through the oxidation of polyphenols, although not to the same extent as during fermentation. These reactions lead to the formation of new flavor components and reduction of bitterness and astringency (Rodriguez-Campos et al. 2011; Barišić et al. 2019; Utrilla-Vázquez et al. 2020).

26.6.4 Roasting

Roasting is one of the most important processes affecting the cacao flavor profile. It is vital to transform flavor precursors formed during fermentation and drying into flavor components. Cacao beans are roasted at a temperature range of 120 to 150°C for 15 to 45 min, depending on the cacao variety. The *Criollo* variety, for instance, needs a lower temperature compared to other varieties (Kongor et al. 2016; Barišić et al. 2019). When roasting occurs at high temperatures such as 135–150°C, melanoidinds are formed through Maillard reactions, intensifying the cacao's brown color (Barišić et al. 2019). Most of the flavors generated during roasting are due to the Maillard reaction between peptides or amino acids obtained during fermentation and reducing sugars such as glucose and fructose. These sugars are present in harvested cacao beans but are also formed during fermentation (D'Souza et al. 2018). Some examples of amino acids participating in the Maillard reaction are leucine, alanine, phenylalanine, and tyrosine (Afoakwa et al. 2008; Barišić et al. 2019). Generally, the Maillard reaction is the condensation between an α-amino group of amino acids, peptides, or proteins and the carbonyl group of reducing sugars. It is favored by heat, pH above 3, and low moisture content and brings specific chocolate aromas to cacao (Schwan and Wheals 2004; Afoakwa et al. 2008; Kongor et al. 2016; Barišić et al. 2019). It has been proposed that glucose reaction with leucine, glutamine, and valine can form sweet chocolate, chocolate, and penetrating chocolate notes, respectively (Afoakwa et al. 2008). Flavor compounds formed during roasting include pyrazines, Strecker aldehydes, alcohols, ethers, furans, pyranes, pyrroles, acids, esters, and amines. In effect, pyrazine, aldehyde, and alcohol concentrations increase sharply during roasting (Kongor et al. 2016). Indeed, although some pyrazines are formed before, the majority are formed during roasting (Marseglia et al. 2020). They are mainly formed from the reaction between α-dicarbonyl compounds and amino acids during Strecker degradation (Afoakwa et al. 2008; Rizzi 2008). Strecker degradation refers to the oxidation of amino acids by α-dicarbonyl, forming an imine. Due to its origin, the imine is hydrolyzed to ammonia or a primary amine and an aldehyde called a Strecker aldehyde (Yaylayan 2003). 2-ethylpyrazine (musty and nutty aromas), 2,3-dimethylpyrazine (caramel, baked, cacao, sweet notes), and 3-ethyl-2,5-dimethylpyrazine (roasted and smoky tones) are examples of pyrazines formed during roasting (Marseglia et al. 2020). On the other hand, Strecker aldehydes (2-ethylpropanal, 2-methylbutanal, and 3-methylbutanal) produced during roasting are responsible for strong chocolate flavor (Barišić et al. 2019; Marseglia et al. 2020). The reaction of carbonyl derived from the Maillard reaction with free amino acids degrades specific amino acids into specific Strecker aldehydes, generating characteristic aromas (Kongor et al. 2016, Barišić et al. 2019). Strecker aldehydes can be formed directly from free amino acids or from Amadori products. (Yaylayan 2003; Rizzi 2008). Amadori compounds are those formed by the condensation of amino acids with reducing sugars in the Millard reaction and go through further reactions until turning into flavor components (D'Souza et al. 2018; Barišić et al. 2019). At the same time, the concentration of aldehydes such as benzaldeyde and benzenacetaldehyde increases during roasting, certain aldehydes such as 2-methylpropanal and 5-methyl-2-phenyl-2-hexenal are produced for the first time during roasting. The latter contributes to intense bitterness and strong cacao notes (Marseglia et al. 2020). Some alcohols, such as 2-heptanol formed during roasting, are a product of the thermal degradation of amino acids and provide green flavor and sweet aromas to cacao (Marseglia et al. 2020). Hydroxymethylfurfurals and other furfurals are formed during roasting due to sugar dehydration in an acidic environment (Afoakwa et al. 2008; Barišić et al. 2019). Some examples are 3-furfuryl alcohol (bready flavor) and 4-hydroxy-2,5-dimethyl-3(2H)-furanone (caramel notes). Pyrroles, such as (1-ethyl-2-pyrrolidinyl)-methanol, obtained from Strecker degradation, and pyranes, such as 2,3-dihydro-3,5-dihydroxy-6-methyl-4H-pyran-4-one, also result from cacao roasting (Rizzi 2008; Marseglia et al. 2020). Due to high temperatures, roasting also leads to sugar caramelization, bringing additional flavors to cacao (Barišić et al. 2019). The cacao flavor profile becomes more refined with roasting, less acidic due to the evaporation of certain volatile acids, less bitter, and less astringent, and some undesirable flavors produced during fermentation are eliminated (Kongor et al. 2016; Gibson and Newsham 2018; Febrianto and Zhu 2022). Some volatile organic acids such as lactic, citric, oxalic, citric, tartaric, and succinic acids do not seem to be affected by roasting (Kongor et al. 2016).

26.6.5 Alkalization

Cacao alkalization, or also called Dutching, before or after roasting brings several benefits to cacao. It reduces the astringency, bitterness, and acidity of the product and modifies its flavor, making it milder and improved. On the other hand, alkalization also turns the cacao darker in color and improves its solubility (Afoakwa et al. 2008; Miller et al. 2008; Valverde García et al. 2020). The alkalis commonly used are NaOH, Na_2CO_3, $NaHCO_3$, KOH, K_2CO_3, $KHCO_3$, $(NaH_4)_2CO_3$, $Ca(OH)_2$, and $CaCO_3$, alone or combined, at concentrations between 1 and 6%, depending on the desired final product. The higher the temperature, pressure, and alkali concentration, the longer the reaction and the darker the chocolate. Likewise, the higher the pH, the darker, less astringent, less bitter, less acidic, and more soluble the product. The darkness of the cacao is also favored by the presence of oxygen from injected air, and it results from a combination of caramelization, Millard reactions, and polyphenol oxidation (Li et al. 2014; Valverde García et al. 2020). The changes in color and the reduction in bitterness and astringency are related to the decrease of polyphenol compounds observed during Dutching (Miller et al. 2008; Li et al. 2014). Other aroma compounds are also affected during the process; some such as ketones and furans might be formed, and others such as pyrazines might be lost during the process. In addition, nutrients such as amino acids, proteins, sucrose, fat, and flavonoids might decrease as well (Valverde García et al. 2020).

26.6.6 Conching

Conching consists of shearing cocoa liquor (a thick, dark fluid obtained from nib crushing), sugar, and milk solids against a surface, which generates heat and helps break down small particles. It makes the chocolate smoother, with better color, and the flavor becomes softer and improved. Some flavors such as caramel, roasted, and malted increase with heating. During this process, excess moisture evaporates, as well as some undesirable compounds such as volatile acids. As a result, the liquid achieves suitable viscosity and fine crystallization. To get fine crystals, it is important to run cycles of heating and cooling for hours or days. Cocoa butter and lecithin are added at the end of conching to lubricate and emulsify, respectively (Schwan and Wheals 2004; Gibson and Newsham 2018; Barišić et al. 2019).

26.6.7 Tempering

Chocolate gets its glossy appearance by tempering. This process aims at achieving a consistent amount and size of β2 cacao butter crystals through cycles of heating and cooling and correct stirring of the product. As a result, small crystals and a tight crystalline cocoa butter matrix are obtained, giving the chocolate a smooth and glossy surface. The process also improves cacao color and melting properties and provides resistance against liquid oil migration (Gibson and Newsham 2018; Barišić et al. 2019; Pirouzian et al. 2020).

26.7 CACAO PROCESSING IN MEXICO

In Mexico, cacao is produced mainly in Tabasco and Chiapas but also in Guerrero and Oaxaca states, in quantities that approach the 27,000 tons per year (Soleri and Cleveland 2007; Gobierno de México 2018b; Utrilla-Vázquez et al. 2020). All three of the most important cacao varieties (*Forastero, Criollo*, and *Trinitario*) are produced in Mexico, and the plant is one of the most important agricultural products of the country. Cacao fermentation is an artisanal and uncontrolled process that is traditionally carried out in the household (Córdova Lázaro et al. 2018; Utrilla-Vázquez et al. 2020). The communities of the producing states appreciate chocolate, and it could be important to promote the product outside the boundaries of the communities to give it added value (Córdova Lázaro et al. 2018). A study done among people from Chontalpa (Tabasco state) who use cacao as part of their diet revealed that 92% of the interviewers are willing to pay more than standard price for the presence of traditional elements in semi-dark and dark chocolate (Jaramillo et al. 2018). This reveals the importance that communities give to artisanal products. Cacao beans consumed in Oaxaca are fermented in the pod, together with its mucilage. After fermentation, beans are dried and roasted before being used for *tejate*, chocolate (beverage), or mole preparation. However, most of the cacao consumed in Oaxaca is not fermented, dried, or roasted in the state; beans that were previously sent through processing are brought to Oaxaca from other states (Soleri and Cleveland 2007). In Chontalpa (Tabasco), cacao bean fermentation is carried out in an artisanal fashion and for household consumption. Cacao pods are harvested from October to June and piled and opened the next day. This procedure is believed to concentrate the flavor of the beans. The pods are opened, and beans and mucilage are removed and put in a plastic bucket and then in baskets made with *Smilax domingensis*. After a couple of hours, some leakage of the mucilage is observed and beans are transferred to boxes made with wood of *Tabebuia rosea*, of different sizes, depending on the amount of beans. Fermentation lasts 5 days with occasional turn-over, after which beans are sun-dried spread on *Agave fourcroydes* or plastic sacks for 3 to 5 days, from sunrise to sunset. Afterwards, beans of approximately 1 gram each are selected to be roasted, while beans that are deformed, broken, humid, spoiled, or have thin or inexistent cotyledon are removed (Córdova Lázaro et al. 2018). The cacao drying, roasting, and grinding methods have been used since pre-Hispanic times but are now adapted to modern machinery (Chin et al. 2013). Heat for roasting is generated using wood charcoal and a wooden or stone structure on top of which a typical clay tray called a *comal* is laid. Cacao beans are roasted in the hot earthenware *comal* at low flame for approximately 30–40 minutes. The beans are constantly stirred with a wooden tool that resembles a flat-fronted spoon. Roasting is complete when beans have swelled up; the shell becomes brittle and starts to break; they gain dark tones; are easily peeled; and present a bitter, sour taste (Chin et al. 2013; Córdova Lázaro et al. 2018). Beans are peeled through shaking inside a container with an opening through which the peels are discharged or by breaking them with a stone, a plastic or enamel cup, or the wrist. Peeled cacao (nib) can be left exposed in the sun from morning to afternoon, with some sugar to facilitate milling (Córdova Lázaro et al. 2018). This is done using a traditional manual stone grinder called a *metate* or a metal machine called a star grinder, which can be manual or electric (Chin et al. 2013; Córdova Lázaro et al. 2018). A sandy paste (5 kg cacao) is obtained from the first milling. For the second milling, the paste is added to the grinder with cinnamon (100 g). In the third, the sun-dried sugar (1 kg) is added. In the fourth milling, milk powder (250 g) is added. Shape will be given to the chocolate by adding it to a mold, forming small chocolate tablets. These are left to dry and harden at room temperature and are ready to be used (Córdova Lázaro et al. 2018). The more modern and less traditional method involves roasting with a drum roaster and mechanical grinding using a ball mill (Chin et al. 2013). Finally, although cacao in Mexico can be fermented before consumption, in some cases, cacao is only washed with water to remove most of its pulp. This process is called *lavado* (meaning washing in Spanish), and it is followed by sun-drying to yield terracotta-colored cacao (Chin et al. 2013).

26.8 CACAO AND HEALTH

26.8.1 Cacao's Introduction to Europe: Before and After

The importance of cacao for pre-Colombian civilizations can be observed from the evidence of its uses. Henderson et al. (2007) explain that even from its earliest uses (fermenting the pulp to generate an alcoholic beverage) in non-stratified communities like Early Formative Puerto Escondido (lower Río Ulúa Valley, 1100 B.C., present-day Honduras), the beverage was used in special ceremonies like marriages and births, which created social distinctions and obligations, thus arguing that cacao beverages had a part to play in the emergence of the elites. Later, cacao was consumed in the form of drinks made from roasted and ground beans, which were consumed especially by the elite in ceremonies like Mayan naming rituals and marriage ceremonies. It was also given to sacrificial victims in rituals and was used as a form of currency and tax payment, even in civilizations like the Aztec and the Chacoan that were established in climatic regions where cacao trees could not be grown and therefore had to procure cacao through trade (Ozturk and Young 2017). It is unknown if ordinary Mayan people could afford to consume cacao and, if so, how they consumed it, since the evidence for the use of cacao in Classic Maya comes only from the painted or carved vessels found in the tombs of the elite. For the Mexica, cacao was an intoxicating food reserved only for priests, the highest government officials, military officers, and distinguished warriors, all adult males (Coe and Coe 2013).

The use of cacao with medicinal applications by ancient civilizations is particularly interesting, especially in light of the increased interest in the consumption and study of dark chocolate for its potential functional properties.

In "The Ritual of the Bacabs", a codex written in the Mayan language and discovered in Yucatan in 1914, a series of illnesses and what were thought to be their causal origins are described, along with the chants to be spoken over the patients who suffered from different ailments. These included skin eruptions, fever, and seizures. At the conclusion of the chants, the indication was for the patients to drink *chacah* (medical chocolate), which also contained peppers, honey, and tobacco juice (Dillinger et al. 2000). More complete information on the use of the cacao plant and the "chocolate" product by the pre-Colombian peoples of the Americas, from a medical perspective, can be obtained from the documents written following the Spanish conquest, in which extensive information on the native cultures was recorded.

Of these manuscripts, the earliest treatise on Mexican medicinal plants and native remedies is the Badianus Manuscript (a 16th-century Aztec Herbal manuscript), showing no trace of European influence. The manuscript was written in the Aztec language (Nahuatl) by a native physician, Martinus de la Cruz, and translated to Latin by another Aztec Indian, Juannes Badianus (World Research Foundation 2020). Among the plants identified in the text as having healing properties is a color painting of a cacao tree that describes the use of cacao flowers in baths to reduce fatigue suffered by government administrators (Grivetti 2009). Lippi (2013) identified references in the manuscript to using cacao derivatives as nutrients or as remedies against angina, constipation, tartar-related dental problems, dysentery, dyspepsia, indigestion, fatigue, gout, and hemorrhoids. Another of these important manuscripts is The Florentine Codex, written in 1590 by a Spanish priest, Bernardino de Sahagún, who reflected in his writings his deep interest in Mexica/Aztec medical knowledge and the time he took to learn their culture (Grivetti 2009). In the Codex, Sahagún warns about the effects of excessive use of green cacao, which can lead to drunkenness and confusion, and gives a detailed description of how various cacao brews were to be prepared. Sahagún also identifies the illnesses that were treated using cacao and recipes of the corresponding cacao-based remedies, for example, a chocolate beverage consumed to treat complaints of the bowel and stomach and said to cure infections when combined with liquid from silk cotton tree bark (*Castilla elastica*). Five cacao beans ground and blended with the root of *tlayapoloni xiuitl* (an unknown plant) were consumed to treat childhood diarrhea, while a mixture of 8–10 ground cacao beans, dried maize kernels, and *tlacoxochitl* (*Calliandra anomala*) was consumed to relieve fever and fainting spells. Finally, after the ingestion of an infusion of opossum tail, people consumed a medicinal chocolate beverage that also contained *mecaxochitl* (*Piper sanctum*), *uey nacaztli* (*Chiranthodendron pentadactylon*), and *tlilixochitl* (*Vanilla planifolia*) to treat coughs with phlegm. Cacao was also recorded as an enhancer of flavor for other medicinal remedies (Dillinger et al. 2000).

Upon reaching Europe in the 16th century, chocolate continued to show its medicinal properties; however, there was no consensus on what type of maladies it was suitable for. Through the late 16th and early 17th centuries, the effort to classify chocolate for its ability to heat, cool, moisten, or dry the body, under the theory of four humors, complicated things further since it seemed to be both cold and dry, due to its astringent flavor, and hot and moist, coming from its fat component and its ability to nourish and fatten the body. This dual property was unthought-of at the time and led to confusion, speculation, experimentation, discussion, and much theorizing, especially fueled by mid-17th-century physicians who sold chocolate and therefore had the incentive to promote its therapeutic virtues. At the same time, others warned about excessive consumption, generating yet another debate in which the increasing list of ingredients added to chocolate was not forgotten. Eventually, chocolate was recommended to treat almost any illness, becoming a "universal medicine" that could be taken under any pretext (Albala 2007). The extent of this universality is well exemplified by Dillinger et al. (2000), who presented a summary of the positive claims and use of cacao and chocolate from 1590 to 1902 and then from 1956 to 1985, comprising around 127 claims for the use of the cacao seed prepared as chocolate, 2 claims for the cacao bark, 19 claims for cacao butter, 5 claims for the cacao flower, 1 claim for cacao fruit pulp, and 2 claims for cacao leaf. The authors identify the following consistent cacao/chocolate medicine-related uses: prescribed to gain weight, to stimulate the nervous system (in feeble or exhausted patients with lack of energy or apathy symptoms), to patients suffering from hyperactivity (since physicians thought it produced a tranquilizing effect), to improve digestion, to stimulate the kidneys, and as a vehicle to administer other drugs.

Skipping forward to the 20th century, excessively processed chocolate with added sugar and milk and other ingredients gained popularity as a high-fat and high-calorie food and was associated with obesity, dental problems, and an overall unhealthy lifestyle, most likely because the added ingredients counteracted cacao's health benefits (Lippi 2013).

Some traditional cacao beverages and products, as well as the production techniques from ancient cultures, remained in communities from the Americas, as did some of its medicinal uses. From fieldwork done in different regions from the state of Oaxaca in Mexico (Central Valley, Sierra Norte Juárez, and La Mixteca Alta), it is known that chocolate is one of the principal foods consumed daily owing to beliefs about its nutritional properties and sacred nature. Cacao continues to be used as a sacred plant in ceremonies performed in sacred places such as caves and mountain entrances. The team found a traditional healer or *curandero* from the Sierra Juárez who prepares a cure for bronchitis using cacao beans. It was also learned from these expeditions that chocolate is eaten to diminish the effectiveness of scorpion, bee, or wasp venom and that *curanderos* from La Mixteca Alta cure *espanto* (illness from fright) by "feeding" (planting/burying) cacao beans and other products (tobacco, fermented beverages, and herbs) in the earth at the place where the fright was experienced. This is thought to be both payment to the earth in exchange for healing the patient and a distractor for mother earth to release the patient's spirit. A Mixtec community from Madera residing in the San Joaquin Valley in California, United States (which retains Oaxacan origins), reported combating fatigue with a blend of chocolate and freshly beaten eggs added to hot water. When added to a *manzanilla* (chamomile) infusion, chocolate was used to relieve pain, and when mixed with cinnamon and *ruda* (rue), it helped to alleviate stomach ache. Chocolate beverages were also used to treat the common cold. The same community reported the use of chocolate to lower high blood pressure and, interestingly, also to raise low blood pressure (Dillinger et al. 2000; Grivetti 2009).

26.8.2 Cacao Compounds with Proven or Potential Health Benefits

The following section will focus on the reconciliation of cacao, cocoa, and dark chocolate as contributors to a healthy diet. However, it is important to first describe some terminology since cacao, cocoa, and chocolate are sometimes used interchangeably. This distinction is important because the composition of the three products is very different, which can result in different conclusions from the studies and great difficulty when comparing results.

Chocolate and cocoa are products derived from processing the seeds of the cacao (*T. cacao*) tree (Mellor 2016; Petyaev and Bashmakov 2017). After the seeds have been harvested, fermented, dried, and roasted, they are milled to create what is known as chocolate liquor (also known as cocoa liquor, cacao liquor, cocoa mass (Borchers et al. 2000), or cacao paste in Mexico). When that liquor is pressed, the fat, known as cocoa butter, separates, leaving behind the cocoa powder (Crozier et al. 2011; Petyaev and Bashmakov 2017; Mustiga et al. 2019), or cocoa solids (Petyaev and Bashmakov 2017). Thus, chocolate is the mixture of cocoa butter and cocoa powder (Mellor 2016) that has been mixed with refined sugar, conched, tempered, and molded (Borchers et al. 2000). Three basic types of chocolate can be distinguished; dark chocolate, which contains mostly cocoa solids (80% of total weight or less than 35% by weight of cocoa liquor (Seem et al. 2019)); milk chocolate, which contains less than 10% total weight of cocoa solids as well as high-fat milk and sugar; and white chocolate, which is a mixture of cocoa butter, milk, and sugar and contains no cocoa solids (Petyaev and Bashmakov 2017). On the other hand, in Mexico, the beverage made from cocoa mass, water or milk, and other ingredients is also known as chocolate.

From the compilation of information done by Borchers et al. (2000), Katz et al. (2011), Ellam and Williamson (2013), Petyaev and Bashmakov (2017), and Campos-Bossacoma et al. (2019), we learn that the main components found in cacao, cocoa, and chocolate with bioactive properties are polyphenols, methylxanthines, and selected minerals. Miller et al. (2009) and Belščak-Cvitanović et al. (2012) found a strong correlation between the concentration of flavanols and the percentage of nonfat cocoa solids present in different cacao products. As Miller et al. (2009) pointed out, it is important to know a more detailed composition of cacao products since some of their compounds have been related to the health benefits of cacao, cocoa, and chocolate. As shown in the examples from Table 26.1, the differences in composition between cacao

TABLE 26.1 Reported Concentration of Functional Compounds in Different Cacao Products

PRODUCTS	*COMPOUNDS MEASURED*			*REFERENCE*
	Total polyphenol content (mg gallic acid equivalents/g)		Total flavanol content (mg/g)	Crozier et al. (2011)
Dark chocolate	24.8		13.4	
Cocoa powder	48.2		30.1	
	Total flavonoid content (mg gallic acid equivalents/g)	Flavan-3-ols (mg catechin equivalents/g)	Procyanidins (mg cyanidin chloride equivalents/g)	Belščak-Cvitanović et al. (2012)
Milk chocolate (30 % cocoa)	0.83	1.05	1.16	
Semisweet chocolate (38% cocoa)	1.30	1.23	1.38	
Dark chocolate (72% cocoa)	1.26	1.50	1.46	
Cocoa liquor (100% cocoa)	2.31	1.83	1.95	
	Total polyphenol content average range (mg epicatechin equivalent/g)			Niemenak et al. (2006)
Freshly harvested beans from different clones	from 67 to 149.2			
2-day-old fermented bean from different clones	from 101.3 to 143.6			
	Total flavanols (mg/g)	Epicatechin (mg/g)	Catechin (mg/g)	Miller et al. (2009)
Cocoa powder	22.73	1.85	0.58	
Baking chocolate	15.10	1.14	0.49	
Dark chocolate	3.65	0.90	0.16	
Semisweet chocolate baking chips	4.89	0.48	0.19	
Milk chocolate	0.68	0.10	0.04	
Chocolate syrup	0.66	0.07	0.04	
	(-)-Epicatechin (mg/100g)		(+)-Catechin (mg/100g)	Seem et al. (2019)
Cacao beans	99.18		88.45	
Cocoa, dry powder, unsweetened	196.43		64.82	
Cocoa, dry powder, unsweetened, processed with alkali	56.60		36.71	
Baking chocolate, unsweetened	141.83		64.33	
Dark chocolate	84.40		24.20	
Milk chocolate	10.88		4.16	
	Total phenolic content (mg catechin equivalent/100g)	Total flavonoid content (mg catechin equivalent/100g)	Theobromine (mg/100g)	Meng et al. (2009)
Dark chocolate	578.64	28.30	883.11	
Milk chocolate	160.46	13.48	125.15	
White chocolate	126.39	7.70	Below detection level	
	Theobromine (mg/100g)		Caffeine (mg/100g)	Srdjenovic et al. (2008)
Cocoa	462.1		48.9	
Baking chocolate	1004.1		158.0	
Milka chocolate	100.4		5.6	
	Magnesium			Borchers et al. (2000)
Natural cocoa powder blends	593.64 mg/100g			
Dark chocolate (Dove)	33.06 mg/serving*			
	20.07 mg/serving*			
Milk chocolate (Dove)	*1.4 oz = 39.7 g			
	Caffeine (mg/100g)	Theobromine (mg/100g)	Magnesium (mg/100g)	USDA (2020)
Cocoa powder, not reconstituted (FDC ID: 781462)	230	2057	499	
Chocolate, sweet or dark (FDC ID: 789231)	56	470	230	
Milk chocolate candy, plain (FDC ID: 789222)	20	205	63	

products translate into differences in the concentration of bioactive compounds. In contrast, Alañón et al. (2016) studied and compared the composition of commercial chocolate samples (24 dark chocolate and 17 milk chocolate) and found no correlation between the percentage of cocoa solids and the epicatechin content of the chocolates. This, as the authors mention, brings out the importance of labeling products with information that could be more relevant to the consumer, such as flavanol content.

26.8.2.1 Polyphenols

After an extensive revision of the term, Quideau et al. (2011) propose the following definition of true plant polyphenols:

> Plant secondary metabolites derived exclusively from the shikimate-derived phenylpropanoid and/or the polyketide pathway(s), featuring more than one phenolic ring and being devoid of any nitrogen-based functional group in their most basic structural expression.

Although this group consists of compounds with functions related to plant structure, growth and survival, plant polyphenols are most commonly known for their role in plant defense against oxidative stress. This is caused by excess reactive oxygen species triggered by light (UV), low temperatures, pathogen or herbivore attacks, and/or nutrient deficiency (Tsao 2010; Lattanzio 2013). Furthermore, they are well recognized for their antioxidant capacity and other less-understood biological actions related to their role in disease prevention (Manach et al. 2004). This has led to the appreciation of certain foods and beverages rich in plant polyphenols like tea, coffee, red wine, fruits, berries, and chocolate (Quideau et al. 2011). For example, in contrast to other products, chocolate has been shown to preserve its polyphenol content 2 years after manufacture while maintaining low levels of microbiological growth (Mellor 2016).

According to a structural classification, the polyphenols present in cacao belong mainly to the flavanol (flavan-3-ols or catechins) and proanthocyanidin groups (dimers, trimers, and other polymers of flavan-3-ols) and second to the anthocyanidins group (commonly known as anthocyanins), both belonging to the larger flavonoid group (Tsao 2010). In unfermented cacao beans, 12–18% of the total dry weight is attributed to polyphenols, of which approximately 58% are proanthocyanidins, 37% are flavan-3-ols, and 4% are anthocyanins. The polyphenols in cacao beans are essential for chocolate's flavor and color formation, which occur through different chemical reactions during various processing stages (fermentation, drying, roasting, and conching). These, in turn, change the polyphenol content in cacao beans to a different content in cocoa and chocolate (e.g., more than 90% of the initial catechin content is lost during fermentation) (Barišić et al. 2019). It is the flavan-3-ol monomers (-)-epicatechin and (+)-catechin as well as several procyanidin oligomers which have been associated with the health benefits of cacao and its derived products (Niemenak et al. 2006). This has been attributed mainly to their antioxidant capacity, which is determined by their structure, where the delocalization of electrons over the aromatic system is responsible for the neutralization of free radicals and chelation of redox-active metals (Fe^{2+} and Cu^{+}) (Jalil and Ismail 2008; Katz et al. 2011). The lowered susceptibility of low-density lipoprotein (LDL) to oxidation after cocoa ingestion has been attributed to the scavenging of LDL surface oxidants and the prevention of free radical formation by chelation of metal catalysts (Allgrove and Davison 2018).

Ellam and Williamson (2013) reviewed human intervention studies on the effect of cocoa published between the years 2000 and 2012 and described that the major reported impacts were attributed to the flavanol content and were improved endothelial function (measured in biomarkers related to nitric oxide metabolism, flow-mediated dilation, angiotensin-converting enzyme activity and nitric oxide (NO)/nitrite levels), decreased LDL oxidation susceptibility, platelet aggregation inhibition and activation, lower levels of F2-isoprostanes, and changes in blood pressure and cholesterol levels.

Borchers et al. (2000) point out that epidemiological studies cannot report causal relationships between the compound of interest and the observed diminished disease risk since there is the possibility that the associations are given by another element or elements contained in the food used as the delivery mechanism. Instead, the authors turn to the review of the studies of individual compounds as evidence that these can affect human health. The authors conclude that cocoa flavonoids, particularly (-)-epicatechin and quercetin, could help decrease the risk of coronary artery disease mainly through their antioxidant capacity, thus preventing oxidative modifications of LDL and through inhibition of macrophage LDL uptake and reduction of platelet activity. This antioxidant capacity of flavonoids may also be responsible for their beneficial effects in inflammatory processes by inhibiting important enzymes for eicosanoid production (cyclooxygenases and lipoxygenases), which are in turn mediators of inflammation. These effects may be better manifested in the population at risk with high levels of inflammatory and oxidative stress (Allgrove and Davison 2018).

Due to its epicatechin content, cocoa has been shown to upregulate nitric oxide production both acutely and chronically by inhibiting NADPH oxidase and therefore preventing NADPH oxidase from producing reactive oxygen species which can cause endothelial dysfunction, in part by scavenging nitric oxide. Decreased nitric oxide production, jointly with abnormal signaling and increased oxidative stress, has been shown to develop endothelial dysfunction in the cardiac risk population. Endothelial dysfunction has been related to the initiation of the atherosclerosis process. Katz et al. (2011) and Ellam and Williamson (2013) describe this as one of the main reasons cocoa epicatechin has been associated with a reduced risk of afflictions related to endothelial dysfunction such as coronary artery disease incidence and mortality. From the randomized, placebo-controlled, and crossover or double-blinded studies reviewed by the same authors (Katz et al. 2011), as well as from their own research, it has been shown that cocoa consumption (especially flavanol-rich sugar-free cocoa) can improve flow-mediated dilation of the brachial artery, increase microvascular circulation, induce vasodilation, and reduce blood pressure. Additionally, flavonoid-rich dark chocolate consumption improved flow-mediated dilation of the brachial artery, systolic and diastolic blood pressure, insulin sensitivity, serum LDL, and β-cell function and decreased total cholesterol. In 2012 the European Food Safety Authority (EFSA Panel) concluded that there was enough evidence to support the health claim that "cocoa flavanols help maintain endothelium-dependent vasodilation, which contributes to normal blood flow" and that the general population can obtain the effects by consuming, within a balanced diet, a daily dose of high-flavanol cocoa powder (2.5 g) or dark chocolate (10 g), which are the equivalent to 200 mg of cocoa flavanols/day (EFSA Panel 2012).

Petrone et al. (2013) reviewed the available evidence that tested the effects of cocoa consumption on flow-mediated dilation. The analysis included studies with a controlled parallel-arm or crossover design, which investigated the effects of flavonoid-rich cocoa products in subjects ≥18 years of age and which had available information on baseline and postintervention mean and standard deviation values. The authors found that consumption of dark chocolate or cocoa products significantly increased flow-mediated dilation, with a more substantial effect on participants who had one or more cardiovascular disease risk factors. It is important to mention that the cocoa quantities utilized were usually larger than what is generally consumed (flavonoid content from experimental groups: 88 to 963 mg) and that the most comprehensive study lasted 12 weeks. Additionally, the formulation of the cocoa products used had wide variations. These led researchers to conclude that more studies need to be done to properly determine optimal dosage (flavonoid content and type), usage frequency, and other consumption patterns. Another meta-analysis by Jafarnejad et al. (2020) on the effect of cocoa consumption on blood pressure of middle-aged and elderly subjects (between 45 and 70 years of age) searched for randomized control trial human studies published between 2007 and March 2019, which included sufficient detail, had the appropriate design or placebo/control treatment, and used cocoa as the primary or secondary intervention. Most of the studies compared a flavanol-rich treatment with a low-flavanol control, but the observed effects are not attributed to any one component in cocoa. It was concluded that longer-term consumption of cocoa was associated significantly with a reduction in systolic and diastolic blood pressure where the mean changes found were comparable to clinically relevant values (improvement of 4.00 mm Hg for systolic and 2.04 mm Hg for diastolic blood pressure). However, the beneficial effects were observed only in the elderly population with an existing hypertensive condition.

The effects of the intake of cocoa flavanols on exercise performance and recovery have also been addressed by several research groups. Decroix et al. (2018) and Massaro et al. (2019) conducted independent systematic reviews of the studies of this type published before 2017 and 2019, respectively, and reached similar conclusions. Intake of cocoa flavanols, either acute, sub-chronic, or chronic, can reduce oxidative stress induced by exercise, improve vascular function, and lower blood pressure. It may also positively affect fat and carbohydrate metabolism during exercise but does not seem to affect the enhancement of physical performance. As stated by many others after this type of meta-analysis, the authors report a lack of consistency between studies, complicating comparisons between results and undermining the possible conclusions.

It has also been observed that the antioxidant activity of cocoa polyphenols influences insulin resistance and reduces the risk of diabetes (Katz et al. 2011). Up to 2015, the evidence of the beneficial effects of cocoa on fatty liver and pancreatic function, whose pathophysiology is associated with type 2 diabetes mellitus, came from *in vitro* and animal model studies, along with support for the theory that the mechanism may work beyond the antioxidant capacity of polyphenols and that it involves their action as signaling molecules, altering gene expression and in turn modulating endogenous antioxidant capacity. However, confirmation studies from human clinical trials are still needed (Mellor and Naumovski 2016). The lack of evidence from human intervention studies may come from the controversial approach of giving cocoa or chocolate to diabetics since, although flavonoid-rich, they also have a high sugar and high calorie content (Allgrove and Davison 2018). A comprehensive review on the effects of cacao flavanols on the improvement of risk factors for metabolic syndrome, a group of metabolic disorders that increase the risk of diabetes type II and cardiovascular disease, can be found in Flores (2019).

Some researchers have observed a reduction in weight gain after cocoa consumption. For example, Massot-Cladera et al. (2015) relate their findings to cocoa flavonoids and their reported capacity to inhibit lipid digestion and absorption, reduce fat deposition, down-regulate lipid metabolism-related genes, and their effects on the composition of gut microbiota. However, many authors have questioned the feasibility of these and other effects of cocoa on health by considering the high amounts of sugar and fat that can be found in chocolate (Katz et al. 2011; Ackar et al. 2013; Zugravu and Otelea 2019; MacDonald et al. 2020).

From the antioxidant activity of cocoa polyphenols, it has been proposed that they might affect the immune response and other cell functions by affecting the expression of associated genes through changes in redox-sensitive signaling pathways. In the same way, they could prevent DNA oxidative damage affecting carcinogenesis (Katz et al. 2011). The idea that cocoa flavanols may aid in the prevention of certain types of cancer has been concluded through the positive results of *in vitro* studies using lung carcinoma and colon adenocarcinoma cell lines, as well as *in vivo* with mouse models and chemically induced carcinogenesis in rats. However, extrapolation to human effects needs to be done with caution since human intervention studies are still lacking (Allgrove and Davison 2018).

Preliminary results showed that catechin and epicatechin derivatives' antioxidant capacity could have protective effects in neuronal injury and neuroinflammation, while epicatechin can be involved in long-term memory formation (Katz et al. 2011). Cocoa's neuroprotective effects, some of which in a few studies were reflected as a modest but significant improvement in cognitive function biomarkers (cerebral blood flow, decreased fatigue, and increased task processing), can also be the result of an increase in cerebral and nervous system blood flow as well as vasodilation (Katz et al. 2011; Ellam and Williamson 2013; Allgrove and Davison 2018). Other possible mechanisms through which cocoa polyphenols affect the brain may be the scavenging of reactive oxygen species and changes in cellular signals and gene expression mediated by catechins. These could lead to long-term memory improvement and affect other cellular mechanisms such as plasticity, apoptosis, differentiation, multiplication, and survival (Zugravu and Otelea 2019). The current data from clinical trials show promise regarding the effect of cacao polyphenols on cognitive performance such as memory and executive function. However, the evidence is still weak since there are many sources of variation between studies (Barrera-Reyes et al. 2020).

Despite the widely referenced antioxidant activity of cacao flavanols, Barrera-Reyes et al. (2019) suggest that a direct antioxidant mechanism is probably not the reason for the observed biological effects of (-)-epicatechin. Instead, they may modulate the antioxidant response through the modulation of intracellular signaling pathways. The team observed that, after consuming a single dose of procyanidin and (-)-epicatechin rich cocoa powder, there was a moderate change in the expression of genes related to the decrease in reactive oxygen species production, calcium, and inflammatory response modulation.

Cocoa consumption has also been associated with a positive effect against pain. In contrast to the proposed mechanism for improvement of the endothelial function, the pain-relieving action of cocoa has been associated with the effect of its polyphenols in the reduction of nitric oxide, as well as cytokines, chemokines, and reactive oxygen species. These are molecules involved in the inflammatory response, the modulation of opioidergic mechanisms, and interference with GABAergic systems. However, these conclusions come from preclinical studies. Further research is needed since the mechanisms could come from other compounds present in cocoa, such as methylxanthines or other alkaloids, N-acylethanolamines, tryptophan, or biogenic amines (De Feo et al. 2020).

26.8.2.2 Methylxanthines

The methylxanthines theobromine, caffeine, and theophylline are purine alkaloids derived from purine nucleotides, another type of secondary metabolite found in cacao seeds. Theobromine is the most abundant of the methylxanthines (2.2 to 2.7%), while caffeine is present in lower concentrations (0.6%). Their accumulation in selected tissues or organs (stamen or male part of the flower and seeds of the cacao tree) supports the hypothesis that, as for other alkaloids, purine alkaloids help keep predators away from nutrient-rich tissues (Anaya et al. 2006). However, the bitter taste that methylxanthines confer to the products in which they are found is probably not the main reason for consuming these products; this is more likely due to their psychoactive properties (Oñatibia-Astibia et al. 2017). Caffeine, a well-known component of coffee, tea, and cola, is a recognized central nervous system stimulant that increases alertness, attention, and psychomotor performance at doses around 2 mg/kg of body weight. It is also known to improve an athlete's endurance as a modulator of cardiovascular function (Gammone et al. 2018). An 8-ounce cup of brewed coffee, black tea, and green tea contain a typical amount of 95 mg, 47 mg, and 25 mg of caffeine, respectively, with a range between 75 and 165 mg for coffee, 14 and 70 mg for black tea, and 24 and 45 mg for green tea. On the other hand, 8 ounces of a cocoa beverage can contain between 3 and 32 mg of caffeine, with the typical amount being 6 mg (IFIC 2018).

It has been observed from animal studies using methylxanthine concentrations at levels present in methylxanthine-containing food that these compounds act as antagonists of non-selective adenosine receptors. The level of affinity of methylxanthines to the adenosine receptor subtypes present in the brain (A1 and A2A) is highest for paraxanthine (a product of the metabolism of caffeine) and theophylline, followed by caffeine, and finally theobromine with the lowest affinity (three times lower than caffeine (Gammone et al. 2018)). Studies with knock-out mice have shown the importance of these receptors in sleep, motor activity, heart rate, body temperature, and oxygen consumption regulation. When pharmacological or genetic interventions block A2A receptors in the hippocampus, a positive effect on memory in normal and aged animals and animal models of Parkinson's and Alzheimer's diseases has been observed. Since caffeine, theophylline, and theobromine can penetrate the central nervous system by crossing the blood-brain barrier (by facilitated diffusion in the case of caffeine), these effects, added to the actions of flavanols, could help explain some of the reported effects of cacao (and other methylxanthine-containing products like coffee and tea) on the improvement of brain health (Oñatibia-Astibia et al. 2017; Camandola et al. 2019). Other studies, with caffeine in particular, suggest that an increase in the expression of granulocyte colony-stimulating factor, a reduction in markers of inflammation and oxidative stress, and changes in the regulation of mitochondrial output may also be responsible for the neuroprotective effects of methylxanthines, which are very likely to be multifactorial (Oñatibia-Astibia et al. 2017). Other results reported for caffeine have not been included in this review since the concentrations needed were several times higher than what can be reached from dietary intake and thus unlikely to be a factor for the reported effects of caffeine from coffee, tea, or cocoa consumption (Oñatibia-Astibia et al. 2017).

Travassos et al. (2015) found a significant association between theobromine and $A\beta_{42}$ levels in cerebrospinal fluid. $A\beta_{42}$ is an important biomarker for Alzheimer's disease pathology since it reflects amyloid-β deposition in the brain. The researchers relate this finding to a possible protective role of theobromine in Alzheimer's disease. Interestingly, theobromine concentration in cerebrospinal fluid did not correlate with intake from dietary sources or other methylxanthines in the plasma (Janitschke et al. 2021). This led to the speculation that theobromine could be formed from caffeine metabolism in the brain, which, together with its potential role in neurodegenerative disease, needs further investigation. Another mechanism by which theobromine could potentially aid in the prevention of neurological disorders, as well as liver diseases, is the suppression of mTOR (mammalian target of rapamycin, a serine/threonine protein kinase) activity in the brain and liver since hyperactivity of mTOR has been related to a variety of neurodegenerative diseases such as Alzheimer's and tuberous sclerosis. This hypothesis was proposed by Sugimoto et al. (2018), who measured the levels of phosphorylated mTOR (used as an index of mTOR activity) in Winston rats after oral administration of a theobromine supplemented diet (0.05%) for 40 days and observed a significant reduction. mTOR has a role in regulating protein synthesis (which promotes cell growth and proliferation) and degradation, longevity, and cytoskeletal formation. When the protein homeostasis mTOR-mediated balance is disrupted, the production of key proteins related to synaptic plasticity is altered, leading to detrimental effects on cognitive function (Talboom et al. 2015).

There is a substantial difference between the research done on the health-related effects of cocoa flavanols and those of cocoa methylxanthines. Furthermore, most of what is known about the potential health effects of cocoa methylxanthines comes from the research done on caffeine and its relationship to coffee consumption. Given the high concentration of theobromine in cacao, cocoa, and its products and the potential it has shown, there is a lot to be learned from its study as part of the cocoa matrix where, as Jalil and Ismail (2008) reflect, it can aid or interfere with flavanols (and other components) and consequently their beneficial health effects. More importantly, since it has also been reported that theobromine can increase concentrations of high-density lipoprotein cholesterol and reduce low-density lipoprotein cholesterol concentrations in the serum, thus possibly decreasing fatal heart attack risk (Oñatibia-Astibia et al. 2017; Gammone et al. 2018), effects on the cardiovascular system have regularly been associated with cocoa flavanol intake.

26.8.2.3 Minerals

Cocoa powder is considered a rich source of minerals such as Fe (13.9 mg per 100 g), Cu (4.6 mg per 100 g), Mg (593.6 mg per 100 g), K (2,058 mg per 100 g), Zn (7.9 mg per 100 g), Mn (4.7 mg per 100 g), P (795.3 mg per 100 g), Na (9.0 mg per 100 g), and Ca (169.4 mg per 100 g). Despite also containing oxalic acid, phytic acid, procyanidins, and fiber, which can interfere with the absorption of minerals, their high content in cocoa powder may be enough to consider it a significant source. This is the case with magnesium (Borchers et al. 2000). Approximately 10% of the recommended daily allowance of Mg can be obtained from a 40-g portion of dark chocolate with 70% of cocoa (Ellam and Williamson 2013).

Although the effect of decreased platelet reactivity after cocoa consumption has been highly attributed to cocoa flavanols (platelet activation contributes to the inflammation and thrombosis that lead to cardiovascular disease), the minerals potassium, magnesium, and calcium, as well as stearic acid present in cocoa, have been proposed to be partly responsible (Katz et al. 2011).

One must keep in mind that cacao and its derived products can also contain trace elements such as Cd, Ni, and Pb, which are toxic. From the analysis of 139 chocolates from different regions of the world, it was found that six samples (from India, Trinidad, Ecuador, and Peru) contained amounts of Pb that exceeded the limit set for edible fats, which includes cacao butter, by the Codex Alimentarius (>0.10 mg/kg). The study also found that one sample from Central America and 15 from South America exceeded the European Union limits for Cd (0.80 mg/kg in chocolates with ≥50% cacao solids) (Vanderschueren et al. 2019).

26.8.2.4 Other Potential Bioactive Components of Cacao

According to Borchers et al. (2000), the evidence suggests that regular consumption of chocolate does not have a negative effect on cardiovascular health despite the high content of saturated fat (around 60%) in cocoa butter. This is attributed to the type of saturated fat present, primarily stearic acid (18:0) and palmitic acid (16:0) in lesser amounts. Concerning cholesterol levels, stearic acid has proven to be neutral, while palmitic acid can be neutral in normolipidemic subjects. Thus, in healthy subjects, cocoa butter does not seem to be atherogenic. There is some evidence indicating that epicatechin and the fatty acids present in chocolate may be responsible for the observed reduction in low density lipoprotein oxidation, since this effect has also been observed after consumption of milk and white chocolate, which are low in flavonoids. This is important since oxidized LDL can affect the endothelium and lead to coronary artery disease (Katz et al. 2011). Melo et al. (2020) found that oleic acid (C18:1 n9), considered to have LDL-cholesterol lowering activity, was the most common of the unsaturated fatty acids in chocolate samples (70% cocoa solids) from different cacao varieties cultivated in the southern region of Bahia, Brazil.

It has recently been reported that dry fermented cacao beans (León-Perez et al. 2018), cacao husk, and cacao pod husk (Ruesgas-Ramón et al. 2019), as well as dark chocolate (León-Perez et al. 2020), contain enough phytoprostanes and phytofurans to be considered for further study: 221.46–1589.83 ng/g dw and 1.18–13.13 ng/g dw, respectively, in dry fermented cacao beans; 474.3 ng/g dw and 278.0 ng/g dw, respectively, in cacao husk; 179.9 ng/g dw and 393.8 ng/g dw, respectively, in cacao pod husk; and 2.45–1192.43 ng/g and 144.57–840.19 ng/g, respectively, in dark chocolate. They could be an important factor for the explanation of the reported health benefits of cacao and dark chocolate as well as a source of nutritional supplements for the enrichment of other food products. Phytoprostanes and phytofurans are formed in plants by the non-enzymatic oxidation of polyunsaturated fatty acids, such as α-linolenic acid (ALA, C18:3, n-3), as a result of high concentrations of reactive oxygen species. The interest in the search for these components in cacao and its products arose from their reported anti-inflammatory and apoptosis-inducing activities, cytoprotective response in immature brain cells, and role in the regulation of immune function. The interest also arose from their possible activity as mediators of oxidative stress, involvement in the development of the brain and central nervous system, and potential role in preventing metabolic disorders. This new information will open new research opportunities to study their physiological functions *in vivo* (León-Perez et al. 2018, 2020; Ruesgas-Ramón et al. 2019) for a better understanding of the health benefits of cacao and its products.

Martorell et al. (2013) found that non-treated and protease-treated "Barquillo", a protein-rich (20–27%) cocoa by-product from the cocoa butter extraction process, have a protective antioxidant effect on *Caenorhabditis elegans* (wild type N2 strain) treated with H_2O_2 to induce oxidative stress and can reduce paralysis in *C. elegans* (transgenic Alzheimer's disease model, strain CL4176) where the protease-treated solutions had significantly higher effects. These results indicate that cocoa's antioxidant activity can be attributed not only to polyphenols but also to bioactive peptides present in cocoa. To test this hypothesis, the researchers isolated and identified bioactive peptides from hydrolyzed Barquillo fractions with antioxidant and paralysis-delaying activity. The most active peptide, peptide 13L, was further characterized, and it was found that its activity is most likely due to an inhibition of amyloid-β peptide oligomerization, which is directly related to the amyloid-β peptide toxicity in *C. elegans* CL4176, as well as through the modulation of synaptic and proteosomal functions. Although not proposed as an explanation for the reported neuroprotective effects of cocoa, the bioactivity found for the discovered 13L peptide is suggested for further study in murine models and human trials as a therapeutic agent in preventing aging-related diseases.

Another compound of interest recently identified in high concentrations in fermented cacao powder (4486.2±165.6 μg/kg dw) is kynurenic acid. Kynurenic acid in cacao and other fermented foods is formed as a product of tryptophan catabolism in yeast through the kynurenine pathway. The relevance for further studies related to the health properties of cacao as a result of kynurenic acid content and changes in concentration through cacao's processing stages comes from reports of kynurenic acid's neuroprotective and antioxidant activities as well as positive effects on gastrointestinal diseases (Yilmaz and Gökmen 2018).

Interestingly, while the National Institutes of Health of the United States report that there are only a few foods that naturally contain vitamin D (NIH 2020), Kühn et al. (2018) hypothesized that cacao beans might contain vitamin D because of their high content of ergosterol (vitamin D precursor) as a result of fungal contamination which would be then converted to vitamin D_2 after the bean's exposure to sun while drying. The authors measured the vitamin D_2 and D_3 (synthesized by some plants) content in cacao and its products. The cacao product analyzed contained varying amounts of vitamin D_2 and trace or non-detectable amounts of vitamin D_3: raw cacao beans (0.2 μg/100 g), cocoa powder (1.63 μg/100 g), cocoa butter (2.98 μg/100 g), dark chocolate (3.95 μg/100 g), milk chocolate (1.96 μg/100 g), white chocolate (1.35 μg/100 g). This suggested that vitamin D content might result from fungal contamination and explains the differences in vitamin D_2 content from the products since they all had differing raw material origins. Therefore, they concluded that cacao, cocoa, and their products are dietary sources of vitamin D.

Finally, melanoidins are high-molecular-weight compounds formed during the roasting of cacao beans through the Maillard reaction and are responsible for the brown color and aroma formations of cocoa and other thermally processed foods such as roasted coffee, bread, cooked meat, malt, roasted potatoes, soy sauces, and any other thermally processed food containing amino groups (from amino acids, proteins, or vitamins) and carbonyl groups (from reducing sugars or oxidized lipids) (Langer and Rzeski 2014; Pérez-Burillo et al. 2020). Given the variety of substrates where melanoidins can be formed, their detailed structure is still unknown. In coffee and cocoa, they are generally bound to phenolic compounds (Pérez-Burillo et al. 2020). Some authors have associated the presence of phenolic compounds with the reported biological functions of melanoidins (Oracz and Zyzelewicz 2019). Knowledge of the antimicrobial, prebiotic, antihypertensive, anti-adhesive, metal-chelating, and antioxidant properties of melanoidins has come mainly from studies on those formed in coffee (Quiroz-Reyes and Fogliano 2018). Melanoidins from cocoa have been shown to have antioxidant proprieties (Oracz and Zyzelewicz 2019; Zhang et al. 2019) and to increase the production of short-chain fatty acids favoring *Bifidobacterium* growth in simulated digestion and fermentation processes (Pérez-Burillo et al. 2020). To the best of our knowledge, the study of cocoa melanoidins and their potential health-promoting effects is still highly unexplored and worth further investigation.

26.8.2.5 Health Effects Not Related to Other Components

Polyphenols, fiber, and theobromine present in cocoa can reach the colon, where they can enhance or inhibit the growth of a particular bacterial group, thereby significantly changing the composition and functionality of the gut microbiota (Campos-Bossacoma et al. 2019). At the same time, the bacteria present in the gut can metabolize cocoa's bioactive compounds and turn them into new and simpler metabolites, which can be absorbed into blood circulation. From the studies done on this subject with cocoa as a whole and its individual components, Campos-Bossacoma et al. (2019) conclude that the changes in microbiota due to cocoa consumption depend on its fiber content and that polyphenols and theobromine can influence the effect of the fiber by neutralizing, adding to, or synergizing with its effect. A whole-cocoa diet seems to suppress the prebiotic effect of the cocoa fiber in female Wistar rats. In contrast to a cocoa fiber diet, it did not increase *Bifidobacterium* counts, resulted in a lower production of short-chain fatty acids, and decreased the proportion of IgA-coated bacteria (Massot-Cladera et al. 2015). In summary, it was reported that cocoa consumption diminishes total bacteria, *Bacteroides* spp., *Clostridium histolyticum*,

Clostridium perfringens, *Staphylococcus* spp., *Streptococcus* spp., *Bifidobacterium* spp., and the *Firmicutes/Bacteroidetes* ratio. In turn, *Eubacterium rectale*, *Clostridium coccoides*, *Prevotella*, *Bacterioides* spp., and total short-chain fatty acids (acetic and butyric acids) increase. There have also been reports of new species of *Actinobacteria*, *Cyanobacteria*, *Firmicutes,* and *Proteobacteria* phyla. Others have also reported lower numbers of *Clostridium* spp. and an increase in *Lactobacillus* and *Bifidobacterium* species after a month of intervention with a high-flavanol cocoa beverage (Campos-Bossacoma et al. 2019). *Lactobacillus* and *Bifidobacterium* species are recognized by their anti-inflammatory activity in the intestines and have antimicrobial effects on harmful bacteria (Zugravu and Otelea 2019).

26.8.2.6 Some Considerations

Although many of the health claims attributed to cacao and its derived products have been associated with its high polyphenol content and fewer to the presence of other compounds such as theobromine, Mellor (2016) explains that there is a difficulty in identifying if the health benefits come from one active compound or a synergistic effect from various compounds from cocoa (referring to cacao and its derived products), since it contains other compounds that may have an impact on human health. Some examples of these compounds are those found among its unique fatty acid and amino acid profile or micronutrients content (Ca, Cu, Fe, Mg, Mn, P, K, Na, Zn (Borchers et al. 2000)). Massot-Cladera et al. (2017) observed that the metabolic hormones and urinary metabolic profiles of Wistar rats fed with whole-cocoa or a cocoa fiber diet differed and thus concluded that the non-fiber cocoa compounds and/or the interaction between compounds present in cocoa must play a role in these effects.

As we have seen, there are many reports on the effects of cocoa polyphenols and other components on health, from studies on cell lines to animal models, epidemiological studies, and clinical trials. However, there is still no consensus regarding dosage, delivery, duration, or even target population for an effective usage or treatment. The concerns presented next by the following authors may shed some light on the reasons.

After reviewing some of the literature published before 2013 on the health effects of chocolate polyphenols on health, Ackar et al. (2013) conclude that cocoa and dark chocolate could benefit human health, particularly with impact on the cardiovascular system. However, chocolate should be further questioned due to its high sugar and saturated fat content. Nevertheless, the authors do call for more systematic approaches to human studies since they found no consensus on the number of examinees in clinical studies, and information on the type of chocolate used in the study (specific polyphenol content and composition) and individual nutritive preferences was lacking or unclear. Zugravu and Otelea (2019) reach a similar conclusion and exemplify how dark chocolate consumers may also prefer a low-sugar diet. This could be the hidden reason for the observed benefits seen in epidemiological studies. A similar insight is given in Ellam and Williamson (2013). The authors point to the variation in intervention times and the type of cocoa used in the study as factors contributing to a difficult rigorous interpretation of results. The authors also noticed the role of sugar and observed that cocoa's positive changes (concerning flow-mediated dilation) are greatest in the absence of sugar. When considering cocoa intake from various sources using a dietary questionnaire given to 45,653 French women, it was concluded that moderate consumption of plain chocolate (2.6 g cocoa/day) was associated with a slightly reduced risk of hypertension. However, when consumed in deserts and drinks (foods high in fat and sugar), cocoa consumption was associated with the opposite effect (MacDonald et al. 2020). Garcia-Yu et al. (2018) conclude that the studies available so far on the impact of chocolate flavanols on blood pressure reduction are not enough to conclude a clear dosage effect relationship. The authors also show concern that while several studies show that moderate consumption of dark chocolate (50 g/day or every other day) relates to a beneficial outcome for a variety of cardiovascular-related conditions (hypertension, cholesterol, coronary heart disease, heart failure, cerebrovascular accidents, and peripheral vascular disease), these studies are focused to a limited population (middle-aged and older Caucasian men and women), and many derive their data from self-reported serving amounts. These are important difficulties when trying to draw firm conclusions.

An interesting example of conflicting results can be found in Katz et al. (2011), where the authors compare reports from two studies done with coronary artery disease patients. Although there were some differences regarding the characteristics of the studied populations, the major difference between the studies was the flavanol delivery method, which may have affected the results, since the amount of accompanying kcal and fat per day varied greatly. One study found favorable effects using a cocoa beverage providing either 375 or 9 mg of cacao flavanols, 50 kcal, and <1 g of fat per day. The other study did not find significant differences in measures between treatment groups and used a beverage/solid chocolate combination that provided 444 mg or 19.6 mg of cacao flavanols, 275 kcal, and 18 g of fat per day. It might be that other ingredients present in each delivered product affect the availability of absorption and rate of *in vivo* efficacy of the polyphenols (or other functional components) (Allgrove and Davison 2018), as well as exerting effects of their own that might oppose, hide, or overtake those of the studied molecules. Looking at these results and the questions from researchers, it is understandable why the debate on the health-related properties of cocoa and chocolate has lasted for so long.

The information surrounding cacao's components and their health benefits is vast, and many studies and reviews were not included in this section. However, the general ideas on the information obtained have been presented and, more importantly, the gaps in knowledge that need to be addressed.

26.8.3 Factors That Change the Composition of Cacao and Cacao Products

As we learned in the previous section, not all cacao products have the same concentration of functional compounds, which affects their health-promoting potential. However, it is not only the different proportions of ingredients that influence the composition of the products; when Alañón et al. (2016) found no correlation between epicatechin content and percentage of cocoa solids in different dark and milk commercial chocolates, they explained that the observed results are a consequence of the many factors that influence flavanol content (and other components) in the final chocolate product. These factors include bean variety, origin, and processing treatments. An interesting review of the factors that may be responsible for the differences in polyphenolic compound concentration in cacao beans and its products is given by Oracz et al. (2013) and Tee et al. (2021), where they again point to cultivation area, bean characteristics at harvest, and climate conditions during development, as well as the processing steps of cacao beans.

Just as factors such as genotype, fermentation, drying, roasting, and probably soil type, the tree's age, and storage and transportation of the beans are critical for the formation of the many flavor compounds in cocoa (Kongor et al. 2016), the composition of bioactive compounds seems to be modified by the intrinsic and extrinsic factors affecting the cacao bean. Several authors have found significant differences in the epicatechin content from cacao beans of different origins (Othman et al. 2010) and in total phenolic content, flavonoids, anthocyanins, catechin, epicatechin, theobromine, caffeine, and caffeic acid from different cacao clones of the same region (Maciel et al. 2017). Differences in the fatty acid profile of chocolates (70% cocoa solids) made with different varieties of cacao from the same region (Melo et al. 2020) and in the polyphenol composition of chocolates made with cacao from different regions and varieties (Cambrai et al. 2017) have also been observed. The mineral content, fatty acids, total polyphenols, catechin, epicatechin, theobromine, and caffeine from chocolate bars with varying amounts of cocoa solids from different origins and varieties as well as from cacao beans from different origins and varieties have differed as well (Grassia et al. 2019; Ramos-Escudero et al. 2021).

Cambrai et al. (2017), who analyzed the composition of 47 different chocolates by principal component analysis, concluded that the geographical origin and variety of the cacao used to manufacture the chocolates have an important influence on the polyphenol composition of the chocolate samples. As such, the polyphenol profile of the chocolate can be used to confirm the origin and variety of cacao stated in the labeling. This study can be taken as a first approach to understanding the different factors that influence the composition of the cacao seed and its derived products. However, direct relationships between genetic, ecological, and processing factors and cacao seed composition must be fully understood. This information can help find the best strategies to produce cacao products of better quality and/or higher functional value.

26.8.3.1 Ecological, Geographic, and Genetic Aspects

Genetic analysis of the tree *T. cacao* L. has revealed a reduction in genetic diversity over its evolutionary history caused by inbreeding. Before its domestication, this reduction could result from environmental and/or ecological events and later the result of artificial selection through domestication. Genetic evidence suggests that during the domestication of the cacao tree, the bitterness of the cacao beans was intentionally reduced. This can be observed in the resulting selection of genes related to the metabolism of polyphenols and methylxanthines, which are responsible for the astringency and bitterness in cacao beans (Cornejo et al. 2018). This finding supports the reported differences in composition between cacao varieties. Febrianto and Zhu (2019) found significant differences in the composition of bioactive compounds (flavan-3-ols, phenolic acids, flavonols, anthocyanins, methylxanthines) between 26 cacao genotypes from Indonesia grown in the same plantation area and processed under the same conditions. The authors propose that those variants with high phenolic content and low theobromine/caffeine ratio could be used as fine-flavor cocoa beans with high nutritional value. A higher theobromine/caffeine ratio has been related to the *Forastero* genotype followed by *Trinitario* and, finally, *Criollo* related to the lower ratios. However, some authors have found false classifications when using this ratio (Pedan et al. 2018). Given morphological and anatomical characteristics, the *T. cacao* species has three genetic groups, *Criollo, Forastero*, and *Trinitario*. The *Criollo* genotype is known to produce higher-quality chocolates than the *Forastero* genotype.

However, unlike *Forastero, Criollo* presents low yields and high susceptibility to diseases. The increased productivity and resistance to pests and diseases make *Forastero* the most-cultivated cacao tree, making up to 95% of the world's cocoa production. *Trinitario* is a hybrid of *Criollo* and *Forastero* for high quality, high yield, and pest resistance. *Criollo* and *Trinitario* are the cultivars used to produce "fine" cacaos used to make high-quality dark chocolate. *Nacional* cacao trees are also known as a variety of fine cacao trees found only in northern Ecuador, which, from a botanical perspective, are part of the *Forastero* variety (Oracz et al. 2013; Kongor et al. 2016; Wickramasuriya and Dunwell 2018; Borja Fajardo et al. 2022).

On the other hand, Elwers et al. (2009) found that the differences in the amounts of total polyphenols and (-)-epicatechin found in seed samples of different cacao varieties (*Forastero, Nacional, Criollo*, and *Trinitario*) from different origins could not be explained by a genetic influence but were affected by the nutrient supply of the mother plant. Cacao seeds from fertilized sites had significantly lower amounts of total polyphenols, flavan-3-ols, and anthocyanins. Although seeds from the

Criollo variety contain significantly lower or no anthocyanins, the differences in catechin content observed by other authors between *Criollo* and other varieties can be explained by the stronger diminution of catechin content during fermentation and drying of *Criollo* seeds compared to the other types.

The genetic and environmental factors that affect the fatty acid composition of cacao seeds were studied by Mustiga et al. (2019). The team created a genetic map from DNA extracted from the leaves of 420 offspring cacao plants and 2 heterozygous parents searching for quantitative trait loci related to palmitic, stearic, and arachidic levels acid in cacao beans. All individuals studied were planted in the same year under controlled soil and shade conditions. The bean samples analyzed for fatty acid composition were processed in the same way. To study the effects of pollen type on fatty acid profile, three pollen sources were used to pollinate the flowers: open pollination, self-pollination, and pollen from clone SIAL 70. Since there were two harvest seasons, the authors could incorporate data on temperature and rainfall amounts into their analysis. Although the parent plants had a similar fatty acid profile, their genetic makeup is different enough that, through recombination, the resulting progeny received different combinations of alleles that resulted in a high variation of fatty acid compositions (phenotypic trait), which in turn allowed for the identification of the corresponding genomic regions. This made it clear that genetic factors play a role in the fatty acid profile of cacao beans, and the authors propose that a clone with low saturated (palmitic acid) and high unsaturated (stearic acid) fatty acids could be bred. The effect of temperature and amount of rainfall during the development of the pod (4 to 6 months before harvest) was confirmed. High temperatures were weakly correlated with arachidic acid and strongly correlated with palmitic and stearic acid. Low temperatures were strongly correlated with oleic and linoleic acid. The amount of rain was associated with a reduction in stearic acid and increased linoleic acid and total fat content. When compared to an open pollination, the pollen donor SIAL 70 was related to a decrease in palmitic acid, and when compared to the self-pollination treatment, it was related to an increase in stearic and oleic acid. All this is strong evidence of the role of environmental and genetic factors on the composition of cacao seed fatty acids.

Ecological conditions may also have an impact on the composition of cacao beans. Samaniego et al. (2020) measured the theobromine, caffeine, total phenolic, epicatechin, and procyanidin content of cacao trees of the *Nacional* cultivar from two different regions in Ecuador. The authors found a higher content of theobromine and caffeine content in trees from the Amazonian region than from the Pacific Coast region. This suggests an effect of the production area on the methylxanthine content. However, no effect was observed for the theobromine/caffeine ratio, which has been related to cacao genotypes, confirming that all samples came from the same variety.

Regarding total phenolic content, the authors found the same trend between regions but also statistically significant differences among different provinces in each region. Epicatechin and procyanidin content was only found to be statistically different at the province level as well as at the farm level. This supports the previous evidence that the production area significantly affects the phenolic content of the cacao beans. The authors briefly explain that the higher altitude of the Amazonian region could be responsible for the higher methylxanthine content. The same suggestion was made by Peláez et al. (2016) after analyzing the theobromine content in fresh cacao beans from three Peruvian regions (Tingo María, San Alejandro, and Curimana), where the highest theobromine content was found in the Tingo María region, with the highest altitude. However, the areas are not fully characterized in either study in terms of climate, soil composition, or other factors that could also explain the observed differences between production areas. This is an interesting research approach that could help intentionally direct the composition of the cacao seeds towards a desirable concentration of bioactive components.

Based on the composition of single-origin chocolates, Vanderschueren et al. (2019) were able to trace the geographical origin of the chocolates by generating a type of elemental fingerprint using the concentration of Ba, Cd, Mo, Sr, and Zn. The elements with the highest concentration in chocolates (Ca, K, Mg, P, and S) were similar for all chocolates and could not be used in the fingerprint. Ca presented a negative correlation with cacao content that could be explained by the milk powder content of chocolates. Similarly, Pb could not be correlated to either origin or cacao content, which suggests it originates in the production process. For the rest of the elements, their concentration was positively correlated with the chocolate's cacao content, indicating that the elements are present in the raw material. In the case of Cd, the authors explain that hotspot areas have been reported where the cacao plant can have an elevated Cd uptake, which then accumulates in the beans. This can be true for the rest of the elements, which reveals the importance of the soil in the concentration of elements in the beans.

There is no single factor that can explain the differences in composition between fresh and processed cacao seeds, and it is very likely that the variability comes from the interaction of several factors. This could at least explain the observed quantitative differences in polyphenols between cacao seeds from different pods of the same tree where genetic, physiological, environmental, and agronomic factors could be interacting and modifying the polyphenols in each individual pod (Niemenak et al. 2006; Mihai et al. 2022). A simple example of this can be found in Dang and Nguyen (2019), where the effects of pod maturity stage and fermentation time and type (unaided and pectinase-aided) are shown to have an impact on proanthocyanidins, caffeine, and theobromine content of the resulting cocoa powders. With increasing maturity of the pods, from immature to mature but not after (from mature to half and fully ripe), the amounts of proanthocyanidins, caffeine, and theobromine of the corresponding cocoa powder increase. After a 5-day fermentation of the beans from each stage of maturity, the proanthocyanidin concentration was significantly lower in the cocoa powder from fully ripe pods. This is explained as a result of the structural changes of the cacao cells where the cell walls become softer and easier to break by external factors, leading to the leaking of cell components and their consequent biochemical changes. Theobromine was found to increase significantly compared to cocoa powder from unfermented beans, except for the half-ripe stage, the stage at which the pods are usually harvested. When fermentation for beans of half-ripe pods was extended to 7 days, a more pronounced loss of proanthocyanidin content was observed, from 0.8% on day 5 to 24.0% on day 7. The reduction was even stronger when the fermentation was aided with the pectinase enzyme Pectinex Ultra SP-L, where on day 5 and 7, the proanthocyanidin contents were 20.4 and 43.9% less, respectively. A similar but less pronounced effect was observed for caffeine and theobromine content during prolonged and aided fermentation.

26.8.3.2 Processing Steps and Conditions

Cacao seeds must be fermented, dried, and roasted to generate the characteristic flavor of cocoa liquor. Flavor precursors are generated during fermentation and are transformed into a wide variety of flavor compounds during roasting (Kongor et al. 2016; Goya et al. 2022). Furthermore, chocolate manufacturing includes further processing steps and mixing of ingredients. All of these have an important impact on the composition of the final products, influencing their health-promoting potential. Agus et al. (2018) analyzed the proximal composition, total phenolic content, antioxidant activity, and phytosterol content of unprocessed, fermented, and roasted Peninsular Malaysian cacao beans. The observed significant differences led to the recommendation of a higher control of the processing (fermentation and roasting temperature) to preserve the quality of cacao beans.

26.8.3.2.1 Fermentation

Some of the most important changes that occur during fermentation are those that modify the polyphenol composition of the cacao beans, since they are essential for the formation of color and the reduction of astringency and bitterness. During this process, polyphenolic containing cells, where polyphenols are stored, are destroyed, and the released polyphenols undergo aerobic oxidation by the enzyme polyphenol oxidase and are converted to insoluble forms (tannins) by reacting with proteins (Kongor et al. 2016; Barišić et al. 2019). Measuring anthocyanin and total polyphenol content at the start and end of the fermentation (132 h) as well as at the end of the drying process (132 h and 7% moisture content) for a mixture of five cacao clones grown in the Urabá region (Antioquia, Colombia) shows a clear reduction of these components during processing, more markedly for total polyphenol content. A positive relationship between polyphenol content and a sensory descriptor for bitter flavor was also observed. The decrease in anthocyanin concentration is related to the hydrolyzation of the glycosidic bond of cyanidin-3-O-galactoside and cyanidin-3-O-orabinoside, which also cause the beans to change from violet to brown in color. Besides the oxidation of polyphenols to insoluble tannins, the decrease in concentration in polyphenol content is associated to their diffusion out of the bean during fermentation (Barrientos et al. 2019; Cortez et al. 2023). On the other hand, *trans*-resveratrol (3,5,4′-trihydroxy-*trans*-stilbene), a stilbene aglycone that belongs to the stilbene class of phenolic compounds, can be found only after cacao bean fermentation. This is because, in raw beans, resveratrol is present in its glycosylated form, which is hydrolyzed to *trans*-resveratrol during fermentation through the action of the enzyme β-glucosidase. *Trans*-resveratrol is not affected by the following high-temperature processes of chocolate production like roasting or alkalization (Salvador et al. 2019).

Theobromine levels are also affected by the fermentation process. Acetic acid can hydrolyze the bonds between theobromine and tannins present in unfermented beans, releasing part of the theobromine, which then diffuses to the cacao shell (Barišić et al. 2019). Peláez et al. (2016) observed that fermented and dried bean samples from the Tingo María, San Alejandro, and Curimana areas of Peru had a reduction of 21, 23, and 29% of theobromine and 40, 38, and 26% of caffeine, respectively, compared to fresh cacao beans from the same areas.

26.8.3.2.2 Alkalization

The alkalization of cocoa, also known as Dutching, was first introduced in the 19th century to enhance solubility of the cocoa powder; now it is also used to reduce the acetic acid that is formed during fermentation, raising the pH and producing darker-colored cocoa with a milder flavor. During alkalization, the natural cocoa material is mixed with an alkali solution, generally K_2CO_3 and/or NaOH at concentrations between 1 and 6%, and treated with temperature (from 60–130°C) and pressure (between 1 and 12 atm). Depending on the final pH, which relates to the harshness of the process, the resulting product will have different color, flavor, and solubility characteristics, which are helpful for different purposes (Valverde García et al. 2020). Other researchers have shown that this process also reduces the polyphenol content and antioxidant activity in cocoa. Adding to these findings, Alasti et al. (2020) demonstrated that the dosage and type of alkaline treatment significantly affect this reduction. The authors found that total polyphenol content was highest for the non-alkalized cocoa powder and was reduced considerably as the alkali solution concentration increased (0.5, 1.0, 1.5, and 3.0% (w/w)). Alkalization with a K_2CO_3 solution, either on its own or in combination with NH_4HCO_3, resulted in lower polyphenol content than alkalization with a NaOH solution, either on its own or combined with NH_4HCO_3. The lower polyphenol content was also associated with a darker cocoa color, which was thought to result from catechins, anthocyanins, and procyanidins changing into quinones and linking to proteins to form brown insoluble compounds. According to the comprehensive review on the alkalization process of cocoa by Valverde García et al. (2020) the reduction of the total phenol and flavanol contents is the result of the oxidation of polyphenols and their interaction with other molecules (such as proteins, amino acids, peptides, products of Maillard reactions and other flavonoids), as well as the glycosylation of flavanols and increased activity of polyphenol oxidase. (-)-epicathechin is isomerized to (-)-catechin, and (+)-catechin is epimerized to (-)-catechin, lowering their bioaccessibility. Caffeine and theobromine are also reduced later, interacting with bases to form salts.

26.8.3.2.3 Roasting

The composition of cacao beans is significantly affected by roasting. Increasing temperatures from 110 to 150°C causes a reduction in total carbohydrate, free sugars, and protein content, while higher roasting air humidity seems to reduce the entrance of oxygen into the kernel, slowing down lipid oxidation. Even at the lowest temperature (110°C), a reduction in total flavan-3-ol content can be observed, coupled with increased catechin content compared to initial levels. The degradation of polyphenols is slower at lower temperatures and higher air humidity (Oracz and Nebesny 2019). Polyphenols are thermolabile molecules affected by the high temperatures used during roasting (between 120 and 140°C) (Barišić et al. 2019) and drying (Urbańska et al. 2019). This, as well as the heat-related epimerization of (-)-epicatechin to (-)-catechin, explain the observed significant reduction of total polyphenols and procyanidin content, as well as the lower epicatechin/catechin ratio after roasting of the cacao beans (Spizzirri et al. 2019). A strategy to preserve the polyphenols is to apply lower temperatures during shorter times (Urbańska et al. 2019). On the other hand, Stanley et al. (2018) observed that while roasting temperatures of 150°C or higher reduce (-)-epicatechin and dimer-pentamer proanthocyanidin content, they also increase proanthocyanidin hexamers and heptamers, which increased *in vitro* inhibition of pancreatic lipase activity. This effect was previously correlated with fatty liver disease prevention and inflammation mitigation in high fat-fed mice.

Applying higher temperatures (>135°C) and relative air humidity of 0.3% seems to aid the formation of melanoidins, which have also been attributed health-promoting benefits, possibly due to the presence of soluble free phenolics and insoluble bound phenolics present in the melanoidin structure. However, the composition of free and bound phenolic compounds in the high-molecular-weight melanoidin fraction of roasted cacao beans also changes depending on roasting conditions. The increase in roasting temperatures and decrease in relative air humidity leads to a reduction in the free phenolics content, as well as in the carbohydrates and proteins content, while bound phenolic compounds content increases at higher temperatures. It might be possible, then, that by tinkering with roasting conditions, the levels of bioactive compounds in high-molecular-weight melanoidin fractions could be increased, which in turn might change their functional properties (Oracz and Nebesny 2019; Oracz et al. 2019). As can be seen through these examples and suggested by the previous authors, the optimization of roasting parameters can lead to the formation of a product of the desired composition aimed at maximizing its functional potential.

26.8.3.3 The Role of Biotechnology

One way to obtain a cacao product with the desirable flavor characteristics, cost, and compliance with reduced toxic trace element regulations is to mix cacao from different origins (Vanderschueren et al. 2019). Bean selection and controlled fermentation, roasting, and alkalization treatments have also been applied to produce dark chocolate with high flavonoid contents (Jalil and Ismail 2008). High-quality beans from the *Criollo* variety can be used to make raw chocolate, which helps conserve the functional components of the beans that would otherwise be lost during roasting (Spizzirri et al. 2019). Additionally, it would also be possible to increase the content of cocoa liquor from unroasted beans into chocolates to increase the content of bioactive compounds with health beneficial effects (Żyżelewicz et al. 2018).

Some researchers are trying to understand the inner workings of the fermentation process by analyzing the microbial diversity and structure and even the metabolic pathways involved, aiming at identifying the types of microorganisms that contribute to a desired aroma and flavor profile for use in directed fermentations with targeted inoculations (Agyirifo et al. 2019). Therefore, fermentation with a starter culture of selected single genus yeast, isolated from cacao pulp, was used to analyze the antioxidant properties (total polyphenols, total flavonoids, and 2, 2-diphenyl-1-picrylhydrazyl (DPPH) free radical scavenging activity) of the resulting fermented cacao beans. Compared to the natural fermentation, 2 (*Hanseniaspora thailandica* and *Pichia kudriavzevii*) out of 13 starter cultures derived from yeast isolated from cacao pulp led to fermented beans with higher total phenolic and flavonoid content at certain fermentation points, while *P. hudriavzevii* culture resulted in the most increased DPPH free radical scavenging activity (Ooi et al. 2020).

Lee et al. (2019) have developed a system that can simulate on-farm cacao fermentation (presenting similar chemical changes) but under laboratory settings that allow the control of microorganisms, pulp composition, external temperature, and aeration. The system can also be used at any time of the year and in non-tropical regions since the dried unfermented beans used as starting material can be preserved if properly stored. This system has also been demonstrated to work on a larger pilot scale (Racine et al. 2019). Both are thought to develop experiments for the study of flavanol composition during fermentation and potentially optimize the parameters to obtain fermented cacao beans with a higher flavanol content. However, they can also be used to study aroma and flavor compounds and other sensory characteristics.

Another approach for manipulated fermentations to obtain the desired product comes from the study of simulated fermentations. John et al. (2020) used a submerged fermentation system in which temperature, time, and the concentration of the main microbial metabolites generated during cacao fermentation, ethanol, lactic acid, and acetic acid, were controlled under aseptic conditions. A design of experiments was applied to study how these individual factors and their combinations affect bean pH, the diversity and quantity of peptides, and flavan-3-ols dynamics during a simulated fermentation environment. The researchers explain that this is an initial approach to an experimental model of cacao fermentation that is simple, since only a few of the many factors influencing the changes in bean composition during spontaneous fermentation are considered. Moreover, as in the case of polyphenol content, bean composition is also influenced by factors such as genetics, environment, and individual bean matrix, among others. However, by generating fermentation models, researchers might have an idea of what needs to be modified during a spontaneous fermentation to obtain a certain result. In this case, it was shown that the greatest effects came from temperature, acetic acid, and incubation time.

Other approaches for obtaining high-quality cacao products may come from genetic research, where the relatively small genome of the cacao plant facilitates the search for molecular markers that could be associated with specific traits and then applied in breeding programs to obtain cacao plants with the desired genetic background that could ensure certain flavor profile, bean size, color, cacao butter content, and yield, among others. Some markers have already been associated with butter content, disease resistance, and yield. This information on genome sequence coupled to proteomic and transcriptomic studies during cacao embryogenesis could help improve somatic embryogenesis propagation efforts, which would be of great help for the required replanting rate that may be needed to avoid future production shortages (Wickramasuriya and Dunwell 2018).

Two US patents describing methods for the modulation of flavonoid content (U.S. Patent No. 8,969,655 B1 (Preston and Murphy 2015)) and for increasing the production of phenolic compounds (U.S. Patent No. 9,428,759 B2 (Rengifo 2016)) in cacao plants have been published. The strategy is to transform the cacao plant cell with either a nucleic acid vector containing a sequence that encodes a polypeptide, enzyme, or transcription factor, capable of modulating the level of one or more flavonoids (U.S. Patent No. 8,969,655 B1) or with a DNA plasmid that promotes the production of phenolic compounds (U.S. Patent No. 9,428,759 B2).

However useful it might be for improving cacao varieties, the genetic modification of cacao faces the same challenges as other genetically modified crops in terms of consumer acceptance and surrounding controversies, which have maintained resulting studies in laboratories and greenhouses (Wickramasuriya and Dunwell 2018).

26.9 CONCLUSIONS AND FUTURE DIRECTIONS

Mexico is a country recognized for its rich culinary tradition and diversity. Among this broad spectrum of food ingredients and recipes, Mexican cacao shares a strong historical, cultural, and culinary connection with Mexican people. Mexico bears pride in being recognized as the place for the original introduction of cacao and chocolate beverages to the world. There has been a considerable amount of research done surrounding cacao in general, from its genetics, cultivation, and composition, to its processing, flavor profiles, and functional properties. However, cacao that is grown and processed in Mexico has been highly overlooked, leaving many gaps in knowledge. This results in an unexploited area for discovering added value for Mexican cacao and its products.

Arrazate et al. (2011) propose that the promotion of Mexican *Criollo* in international market varieties that have unique aroma characteristics is a good opportunity to consolidate new markets and reorient production goals. In this regard, documentation of the chemical profile of Mexican cacao at different stages of processing is important, and until now, little is reported on the topic.

Currently, consumers seek artisanal products with a marked identity. There is an opportunity for Mexican cacao to meet this demand if it is promoted as a unique product that is still produced in an artisanal fashion, keeping an ancestral pre-Hispanic imprint in its processing. As Arrazate et al. (2011), Ogata (2007, 2017), and CEDRSSA (2020) point out, the development of high-quality artisanal products that maintain the cultural richness of the production areas can generate added value for Mexican cacao. These high-quality products would have to be accompanied by high-quality control standards during fermentation, drying, and chocolate manufacturing to bring important economic value and income to small producers (Ogata 2007). For this, it is important to study the cacao biota of Mexican fermentation since consistency in production is achieved with the aid of scientific knowledge of the process.

It is also important that farmers obtain incentives or sponsorships to boost productivity and yields to make Mexican cacao production an economically sustainable activity. This could come in the form of new trees and training on the maintenance of the plantation. Carmona and Cruz (2017) mention that many cacao plantations in Mexico are supported by old trees that have surpassed their optimum production age.

Ogata (2017) also points to the conservation of the jungle and diversification of sources of income for small farmers since cacao production can be integrated in a diversified system that closely resembles a natural jungle. This could also add value to the product since it is accompanied by efforts to conserve high-diversity ecosystems and will also avoid the use of fertilizers and pesticides, since the plantation would be part of a healthy ecosystem. Furthermore, the authors hope to prevent the monoculture cultivation of cacao under non-shaded systems since these systems fall out of the cultural context of the farmers and lead to biological collapse.

There is also the opportunity of taking advantage of the increasing demand from consumers for healthy and functional products. More research should be done on the composition of Mexican cacao and chocolate and production processes that retain health-promoting compounds. In this way, Mexican chocolate could be promoted as a healthy natural alternative.

With everything we know about cacao, there is a great potential for Mexico to exploit the traditional origins of cacao cultivation and processing in an environmentally friendly manner directed towards a product with the potential for possessing attractive health-enhancing properties. However, there is still much to be done in terms of research if this objective is to be accomplished.

REFERENCES

Ackar, D., Lendić, K. V., Valek, M., Šubarić, D., Miličević, B., Babić, J., and I. Nedić. 2013. Cocoa polyphenols: Can we consider cocoa and chocolate as potential functional food? *Journal of Chemistry* 2013(1):289392. https://doi.org/10.1155/2013/289392

Afoakwa, E. O., Kongor, J. E., Takrama, J. F., and A. S. Budu. 2013. Changes in acidification, sugars and mineral composition of cocoa pulp during fermentation of pulp pre-conditioned cocoa (*Theobroma cacao*) beans. *International Food Research Journal* 20(3):1215–1222.

Afoakwa, E. O., Paterson, A., Fowler, M., and A. Ryan. 2008. Flavor formation and character in cocoa and chocolate: A critical review. *Critical Reviews in Food Science and Nutrition* 48(9):840–857. https://doi.org/10.1080/10408390701719272

Agus, B. A. P., Mohamad, N. N., and N. Hussain. 2018. Composition of unfermented, unroasted, roasted cocoa beans and cocoa shells from Peninsular Malaysia. *Journal of Food Measurement and Characterization* 12:2581–2589. https://doi.org/10.1007/s11694-018-9875-4

Agyirifo, D. S., Wamalwa, M., Otwe, E. P., Galyuon, I., Runo, S., Takrama, J., and J. Ngeranwa. 2019. Metagenomics analysis of cocoa bean fermentation microbiome identifying species diversity and putative functional capabilities. *Heliyon* 5:e02170. https://doi.org/10.1016/j.heliyon.2019.e02170

Alañón, M. E., Castle, S. M., Siswanto, P. J., Cifuentes-Gómez, T., and J. P. E. Spencer. 2016. Assessment of flavanol stereoisomers and caffeine and theobromine content in commercial chocolates. *Food Chemistry* 208:177–184. https://doi.org/10.1016/j.foodchem.2016.03.116

Alasti, F. M., Asefi, N., Maleki, R., and S. S. SeiiedlouHeris. 2020. The influence of three different types and dosage of alkaline on the inherent properties in cocoa powder. *Journal of Food Science and Technology* 57(7):2561–2571. https://doi.org/10.1007/s13197-020-04293-w

Albala, K. 2007. The use and abuse of chocolate in 17th century medical theory. *Food and Foodways* 15:53–74. https://doi.org/10.1080/07409710701193381

Alcántara-Hernández, R. J., Rodríguez-Álvarez, J. A., Valenzuela-Encinas, C., Gutiérrez-Miceli, F. A., Castañón-González, H., Marsch, R., Ayora-Talavera, T., and L. Dendooven. 2010. The bacterial community in 'taberna' a traditional beverage of Southern Mexico. *Letters in Applied Microbiology* 51(5):558–563. https://doi.org/10.1111/j.1472-765X.2010.02934.x

Allgrove, J. E., and G. Davison. 2018. Chocolate/cocoa polyphenols and oxidative stress. In: *Polyphenols: Mechanisms of Action in Human Health and Disease*, eds. R. R. Watson, V. R. Preede, and S. Zibadi, 207–219. London: Academic Press. https://doi.org/10.1016/C2016-0-04277-8

Anaya, A. L., Cruz-Ortega, R., and G. R. Waller. 2006. Metabolism and ecology of purine alkaloids. *Frontiers in Bioscience* 11:2354–2370. https://doi.org/10.2741/1975

Arana-Sánchez, A., Segura-García, L. E., Kirchmayr, M., Orozco-Ávila, I., Lugo-Cervantes, E., and A. Gschaedler-Mathis. 2015. Identification of predominant yeasts associated with artisan Mexican cocoa fermentations using culture-dependent and culture-independent approaches. *World Journal of Microbiology and Biotechnology* 31(2):359–369. https://doi.org/10.1007/s11274-014-1788-8

Arias-González, J. 2014. Un vertiginoso viaje etnohistórico dentro de los "imaginarios alimentarios" en el simbolismo del cacao en México. *Anales de Antropología* 48(1):79–95. https://doi.org/10.1016/s0185-1225(14)70490-4

Arrazate, C. H. A., Fuentes, J. M. V., Rojas, E. C., Méndez, R. A. G., López, A. M., Medina, J. F. A., Esquivez, A. S., and S. E. Zaragoza. 2011. *Diagnóstico del cacao en México*. Estado de México: Universidad Autónoma Chapingo.

Barišić, V., Kopjar, M., Jozinović, A., Flanjak, I., Ačkar, Đ., Miličević, B., Šubarić, D., Jokić, S., and J. Babić. 2019. The chemistry behind chocolate production. *Molecules* 24(17):3163. https://doi.org/10.3390/molecules24173163

Barrera-Reyes, P. K., Cortés-Fernández de Lara, J., González-Soto, M., and M. E. Tejero. 2020. Effects of cocoa-derived polyphenols on cognitive function in humans. Systematic review and analysis of methodological aspects. *Plant Foods for Human Nutrition* 75:1–11. https://doi.org/10.1007/s11130-019-00779-x

Barrera-Reyes, P. K., Hernández-Ramírez, N., Cortés, J., Poquet, L., Redeuil, K., Rangel-Escareño, C., Kussmann, M., Silva-Zolezzi, I., and M. E. Tejero. 2019. Gene expression changes by high-polyphenols cocoa powder intake: A randomized crossover clinical study. *European Journal of Nutrition* 58:1887–1898. https://doi.org/10.1007/s00394-018-1736-8

Barrientos, L. D. P., Oquendo, J. D. T., Garzón, M. A. G., and O. L. M. Álvarez. 2019. Effect of the solar drying process on the sensory and chemical quality of cocoa (*Theobroma cacao* L.) cultivated in Antioquia, Colombia. *Food Research International* 115:259–267. https://doi.org/10.1016/j.foodres.2018.08.084

Barros, C., and M. Buenrostro. 2011. Pozol, popo, champurrado. *Revista Digital Universitaria* 12(4):1–9.

Belščak-Cvitanović, A., Durgo, K., Gačina, T., Horžić, D., Franekić, J., and D. Komes. 2012. Comparative study of cytotoxic and cytoprotective activities of cocoa products affected by their cocoa solids content and bioactive composition. *European Food Research and Technology* 234:173–186. https://doi.org/10.1007/s00217-011-1632-4

Ben Omar, N., and F. Ampe. 2000. Microbial community dynamics during production of the Mexican fermented maize dough pozol. *Applied and Environmental Microbiology* 66(9):3664–3673. https://doi.org/10.1128/aem.66.9.3664-3673.2000

Biehl, B., Meyer, B., Said, M. B., and R. J. Samarakoddy. 1990. Bean spreading: A method for pulp preconditioning to impair strong nib acidification during cocoa fermentation in Malaysia. *Journal of the Science of Food and Agriculture* 51(1):35–45. https://doi.org/10.1002/jsfa.2740510105

Bletter, N., and D. C. Daly. 2006. Cacao and its relatives in South America: An overview of taxonomy, ecology, biogeography, chemistry, and ethnobotany. In: *Chocolate in Mesoamerica: A Cultural History of Cacao*, ed. C. L. McNeil, 31–68. Florida: University Press of Florida.

Borchers, A. T., Keen, C. L., Hannum, S. M., and M. E. Greshwin. 2000. Cocoa and chocolate: Composition, bioavailability, and health implications. *Journal of Medicinal Food* 3(2):77–105. https://doi.org/10.1089/109662000416285

Borja Fajardo, J. G., Horta, T. H. B., Peñaloza, A. G. C., Sandoval, A. A. P., and A. J. J. Mendez. 2022. Antioxidant activity, total polyphenol content and methylxanthine ratio in four materials of *Theobroma cacao* L., from Tolima, Colombia. *Heliyon* 8:e09402. https://doi.org/10.1016/j.heliyon.2022.e09402

CacaoMexico. 2020. *Principales países productores de cacao*. https://cacaomexico.org/?page_id=201

Camandola, S., Plick, N., and M. P. Mattson. 2019. Impact of coffee and cacao purine metabolites on neuroplasticity and neurodegenerative disease. *Neurochemical Research* 44:214–227. https://doi.org/10.1007/s11064-018-2492-0

Cambrai, A., Marchioni, E., Julien-David, D., and C. Marcic. 2017. Discrimination of coca bean origin by chocolate polyphenol chromatographic analysis and chemometrics. *Food Analytical Methods* 10:1991–2000. https://doi.org/10.1007/s12161-016-0744-7

Campos-Bossacoma, M., Massot-Cladera, M., Pérez-Cano, F. J., and M. Castell. 2019. Influence of a cocoa-enriched diet on the intestinal immune system and microbiota. In: *Dietary Interventions in Gastrointestinal Diseases: Foods, Nutrients, and Dietary Supplements*, eds. R. R. Watson and V. R. Preedy, 213–225. London: Academic Press.

Carmona, O., and R. Cruz. 2017. El cacao, un ingrediente de alto potencial. In: *El cacao alimento divino*, ed. G. Castillo, 1762–2021. Ciudad de Mexico: Fundación Herdez.

Castillo-Morales, M., Wacher-Rodarte, M. C., and H. Hernández-Sánchez. 2005. Preliminary studies on chorote—A traditional Mexican fermented product. *World Journal of Microbiology and Biotechnology* 21(3):293–296. https://doi.org/10.1007/s11274-004-3634-x

Castro-Alayo, E. M., Idrogo-Vásquez, G., Siche, R., and F. P. Cardenas-Toro. 2019. Formation of aromatic compounds precursors during fermentation of Criollo and Forastero cocoa. *Heliyon* 5(1):e01157. https://doi.org/10.1016/j.heliyon.2019.e01157

CEDRSSA. 2020. *Industria del cacao en México*. http://www.cedrssa.gob.mx/post_industria_del_-n-cacao_en_mn-xico-n.htm

Chin, E., Miller, K. B., Payne, M. J., Hurst, W. J., and D. A. Stuart. 2013. Comparison of antioxidant activity and flavanol content of cacao beans processed by modern and traditional Mesoamerican methods. *Heritage Science* 1:9. https://doi.org/10.1186/2050-7445-1-9

Chocolate Class. (2016). *Cacao and corn in the Mayan world*. https://chocolateclass.wordpress.com/tag/corn/

Coe, S. D., and M. D. Coe. 2013. *The True History of Chocolate*. London: Thames & Hudson.

Cooper-Bribiesca, B., Navarro-Ocaña, A., Díaz-Ruiz, G., Aguilar-Osorio, G., Rodríguez-Sanoja, R., and C. Wacher. 2018. Lactic acid fermentation of arabinoxylan from nejayote by *Streptococcus infantarius* ssp. *infantarius* 25124 isolated from pozol. *Frontiers in Microbiology* 18(9):3061. https://doi.org/10.3389/fmicb.2018.03061

Córdova Lázaro, C. E., Jaramillo Villanueva, J. L., Córdova Ávalos, V., Carranza Cerda, I., and J. Morales Jiménez. 2018. Chocolate casero tradicional en la región de la Chontalpa Tabasco, México: actores y saberes locales. *Estudios sociales. Revista de Alimentación Contemporánea y Desarrollo Regional* 28(52):1–27. https://doi.org/10.24836/es.v28i52.577

Cornejo, O. E., Yee, M.-C., Dominguez, V., Andrews, M., Sockell, A., Strandberg, E., Livingstone, D., III, Stack, C., Romero, A., Umaharan, P., Royaert, S., Tawari, N. R., Ng, P., Gutierrez, O., Phillips, W., Mockaitis, K., Bustamante, C. D., and J. C. Motamayor. 2018. Population genomic analyses of the chocolate tree, *Theobroma cacao* L., provide insights into its domestication process. *Nature Communications Biology* 1:167. https://doi.org/10.1038/s42003-018-0168-6

Cortez, D., Quispe-Sanchez, L., Mestanza, M., Oliva-Cruz, M., Yoplac, I., Torres, C., and S. G. Chavez. 2023. Changes in bioactive compounds during fermentation of cocoa (*Theobroma cacao*) harvested in Amazonas-Peru. *Current Research in Food Science* 6:100494. https://doi.org/10.1016/j.crfs.2023.100494

Crafack, M., Keul, H., Eskildsen, C. E., Petersen, M. A., Saerens, S., Blennow, A., Skovmand-Larsen, M., Swiegers, J. H., Petersen, G. B., Heimdal, H., and D. S. Nielsen. 2014. Impact of starter cultures and fermentation techniques on the volatile aroma and sensory profile of chocolate. *Food Research International* 63 Part C:306–316. https://doi.org/10.1016/j.foodres.2014.04.032

Crafack, M., Mikkelsen, M. B., Saerens, S., Knudsen, M., Blennow, A., Lowor, S., Takrama, J., Swiegers, J. H., Petersen, G. B., Heimdal, H., and D. S. Nielsen. 2013. Influencing cocoa flavour using *Pichia kluyveri* and *Kluyveromyces marxianus* in a defined mixed starter culture for cocoa fermentation. *International Journal of Food Microbiology* 167(1):103–116. https://doi.org/10.1016/j.ijfoodmicro.2013.06.024

Crown, P. L., Gu, J., Hurst, W. J., Ward, T. J., Bravenec, A. D., Ali, S., Kebert, L., Berch, M., Redman, E., Lyons, P. D., Merewether, J., Phillips, D. A., Reed, L. S., and K. Woodson. 2015. Ritual drinks in the pre-Hispanic US Southwest and Mexican Northwest. *PNAS* 112(37):11436–11442. https://doi.org/10.1073/pnas.1511799112

Crozier, S. J., Preston, A. G., Hurst, J. W., Payne, M. J., Mann, J., Hainly, L., and D. L. Miller. 2011. Cacao seeds are a "Super fruit": A comparative analysis of various fruit powders and products. *Chemistry Central Journal* 5:5. https://doi.org/10.1186/1752–153X-5–5

Dang, Y. K. T., and H. V. H. Nguyen. 2019. Effects of maturity at harvest and fermentation conditions on bioactive compounds of cocoa beans. *Plant Foods for Human Nutrition* 74:54–60. https://doi.org/10.1007/s11130-018-0700-3

De Britol, E. S., García, N. H. P., and A. C. Amancio. 2004. Use of a proteolytic enzyme in cocoa (*Theobroma cacao* L.) processing. *Brazilian Archives of Biology and Technology* 47(4):553–558. https://doi.org/10.1590/S1516–89132004000400008

De Feo, M., Paladini, A., Ferri, C., Carducci, A., Del Pinto, R., Varrassi, G., and D. Grassi. 2020. Anti-inflammatory and antinociceptive effects of cocoa: A review on future perspectives in treatment of pain. *Pain and Therapy* 9(1):231–240. https://doi.org/10.1007/s40122–020–00165–5

De Roos, J., and L. De Vuyst. 2018. Acetic acid bacteria in fermented foods and beverages. *Current Opinion in Biotechnology* 49:115–119. https://doi.org/10.1016/j.copbio.2017.08.007

Decroix, L., Soares, D. D., Meeusen, R., Heyman, E., and C. Tonoli. 2018. Cocoa flavanol supplementation and exercise: A systematic review. *Sports Medicine* 48(4):867–892. https://doi.org/10.1007/s40279-017-0849-1

Dillinger, T. L., Barriga, P., Escárcega, S., Jimenez, M., Lowe, D. S., and L. E. Grivetti. 2000. Food of the gods: Cure for humanity? A cultural history of the medicinal and ritual use of chocolate. *The Journal of Nutrition* 130:2057S–2072S. https://doi.org/10.1093/jn/130.8.2057S

D'Souza, R. N., Grimbs, A., Grimbs, S., Behrends, B., Corno, M., Ullrich, M. S., and N. Kuhnert. 2018. Degradation of cocoa proteins into oligopeptides during spontaneous fermentation of cocoa beans. *Food Research International* 109:506–516. https://doi.org/10.1016/j.foodres.2018.04.068

EFSA Panel. 2012. Scientific opinion on the substantiation of a health claim related to cocoa flavanols and maintenance of normal endothelium-dependent vasodilation pursuant to article 13(5) of regulation (EC) No 1924/2006. *European Food Safety Authority Journal* 10(7):2809. https://doi.org/10.2903/j.efsa.2012.2809

Ellam, S., and G. Williamson. 2013. Cocoa and human health. *Annual Review of Nutrition* 33:105–128. https://doi.org/10.1146/annurev-nutr-071811–150642

Elwers, S., Zambrano, A., Rohsius, C., and R. Lieberei. 2009. Differences between the content of phenolic compounds in Criollo, Forastero and Trinitario cooca seed (*Theobroma cacao*

L.). *European Food Research and Technology* 229:937–948. https://doi.org/10.1007/s00217-009-1132-y

Escenario Tlaxcala. 2019. *6 cosas interesantes del "chileatole", una riqueza prehispánica de Tlaxcala.* https://escenariotlx.com/6-cosas-interesantes-del-chileatole-una-riqueza-prehispanica-de-tlaxcala/

Febrianto, N. A., and F. Zhu. 2019. Diversity in composition of bioactive compounds among 26 cocoa genotypes. *Journal of Agricultural and Food Chemistry* 67(34):9501–9509. https://doi.org/10.1021/acs.jafc.9b03448

Febrianto, N. A., and F. Zhu. 2022. Comparison of bioactive components and flavor volatiles of diverse cocoa genotypes of *Theobroma grandiflorum, Theobroma bicolor, Theobroma subincanum* and *Theobroma cacao. Foor Research International* 161:111764. https://doi.org/10.1016/j.foodres.2022.111764

Figueroa-Hernández, C., Mota-Gutierrez, J., Ferrocino, I., Hernández-Estrada, Z. J., González-Ríos, O., Cocolin, L., and M. L. Suárez-Quiroz. 2019. The challenges and perspectives of the selection of starter cultures for fermented cocoa beans. *International Journal of Food Microbiology* 301:41–50. https://doi.org/10.1016/j.ijfoodmicro.2019.05.002

Flores, M. E. J. 2019. Cocoa flavanols: Natural agents with attenuating effects on metabolic syndrome risk factors. *Nutrients* 11(4):751. https://doi.org/10.3390/nu11040751

Gammone, M. A., Efthymakis, K., Pluchinotta, F. R., Bergante, S., Tettamanti, G., Riccioni, G., and N. D´Orazio. 2018. Impact of chocolate on the cardiovascular health. *Frontiers in Bioscience* 23(5):852–864. https://doi.org/10.2741/4620

García-Caballero, B. E., Muñoz-Ríos, R., Pensaben, M., Villarreal-Carrera, M. L., Reyes-Vázquez, Z. C., González-Herrera, S. M., Rodríguez-Herrera, R., and O. M. Rutiaga-Quiñones. 2018. Jocoque: A traditional fermented dairy food with functional potential. *Agroproductividad* 11(7):107–111.

Garcia-Yu, I. A., Garcia-Ortiz, L., Gómez-Marcos, M. A., Alonso-Dominguez, R., Gonzalez-Snachez, J., Mora-Simon, S., González-Manzano, S., Rodriguez-Sanchez, E., Maderuelo-Fernandez, J. A., and J. I. Recio-Rodriguez. 2018. Vascular and cognitive effects of cocoa-rich chocolate in postmenopausal women: A study protocol for a randomized clinical trial. *BMJ Open* 8(12):e024095. https://doi.org/10.1136/bmjopen-2018-024095

Gibson, M., and P. Newsham. 2018. *Food Science and the Culinary Arts.* London: Academic Press. https://doi.org/10.1016/C2016-0-01642-X

Giles-Gómez, M., García, J. G. S., Matus, V., Quintana, I. C., Bolívar, F., and A. Escalante. 2016. In vitro and in vivo probiotic assessment of *Leuconostoc mesenteroides* P45 isolated from pulque, a Mexican traditional alcoholic beverage. *SpringerPlus* 5(1):708. https://doi.org/10.1186/s40064–016–2370–7

Gobierno de México. 2018a. *Chilate: bebida prehispánica refrescante.* https://www.gob.mx/siap/articulos/chilate-bebida-prehispanica-refrescante?idiom=es

Gobierno de México. 2018b. *El Chilate Bebida Tradicional del Estado de Guerrero.* https://www.gob.mx/agricultura/guerrero/articulos/el-chilate-bebida-tradicional-del-estado-de-guerrero?idiom=es

Gobierno de México. 2019. *Agua de barranca, espuma para el alma, documental sobre el patrimonio cultural de Tlaxcala.* https://www.gob.mx/cultura/prensa/agua-de-barranca-espuma-para-el-alma-documental-sobre-el-patrimonio-cultural-de-tlaxcala

González-Tenorio, R., Fonseca, B., Caro, I., Fernández-Diez, A., Kuri, V., Soto, S., and J. Mateo. 2013. Changes in biogenic amine levels during storage of Mexican-style soft and Spanish-style dry-ripened sausages with different aw values under modified atmosphere. *Meat Science* 94(3):369–375. https://doi.org/10.1016/j.meatsci.2013.03.017

Goya, L., Kongor, J. E., and D. Pascual-Teresa. 2022. From cocoa to chocolate: effect of processing on flavanols and methylxanthines and their mechanisms of action. *International Journal of Molecular Sciences* 23(22):14365. https://doi.org/10.3390/ijms232214365

Grassia, M., Salvatori, G., Roberti, M., Planeta, D., and L. Cinquanta. 2019. Polyphenols, methylxanthines, fatty acids and minerals in cocoa beans and cocoa products. *Journal of Food Measurement and Characterization* 13:1721–1728. https://doi.org/10.1007/s11694-019-00089-5

Grivetti, L. E. 2009. Medicinal chocolate in new Spain, Western Europe, and North America. In: *Chocolate: History, Culture, and Heritage*, eds. L. E. Grivetti and H. Shapiro, 67–88. New Jersey: John Wiley & Sons, Inc.

Ha, L. T. V., Nusantoro, B., Phillip, R. A., Anh, P. H., Toan, H. T., Messens, K., and K. Dewettinck. 2016. Physico-chemical properties of fourteen popular cocoa bean varieties in Dongnai-highland Vietnam. *Can Tho University Journal of Science* 4:81–86. https://doi.org/10.22144/ctu.jen.2016.046

Hansen, C. E., del Olmo, M., and C. Burri. 1998. Enzyme activities in cocoa beans during fermentation. *Journal of the Science of Food and Agriculture* 77(2):273–281.

Henderson, J. S., Joyce, R. A., Hall, G. R., Hurst, W. J., and P. E. McGovern. 2007. Chemical and archeological evidence for the earliest cacao beverages. *PNAS* 104(48):18937–18940. https://doi.org/10.1073/pnas.0708815104

Hernández-Hernández, C., López-Andrade, P. A., Ramírez-Guillermo, M. A., Guerra Ramirez, D., and J. F. Caballero Perez. 2016. Evaluation of different fermentation processes for use by small cocoa growers in Mexico. *Food Science and Nutrition* 4(5):690–695. https://doi.org/10.1002/fsn3.333

Hipólito-Romero, E., Carcaño-Montiel, M. G., Ramos-Prado, J. M., Vázquez-Cabañas, E. A., López-Reyes, L., and J. Ricaño-Rodríguez. 2017. Efecto de inoculantes bacterianos edáficos mixtos en el desarrollo temprano de cultivares mejorados de cacao (*Theobroma cacao* L.) en un sistema agroforestal tradicional del norte de Oaxaca, México. *Revista Argentina de Microbiología* 49(4):356–365. https://doi.org/10.1016/j.ram.2017.04.003

Ho, V. T. T., Zhao, J., and G. Fleet. 2014. Yeasts are essential for cocoa bean fermentation. *International Journal of Food Microbiology* 174:72–87. https://doi.org/10.1016/j.ijfoodmicro.2013.12.014

Hu, S., Kim, B.-Y., and M.-Y. Baik. 2016. Physicochemical properties and antioxidant capacity of raw, roasted and puffed cacao beans. *Food Chemistry* 194:1089–1094. https://doi.org/10.1016/j.foodchem.2015.08.126

Hurst, W. J. 2006. The determination of cacao in samples of archeological interest. In: *Chocolate in Mesoamerica: A Cultural History of Cacao*, ed. C. L. McNeil, 105–116. Florida: University Press of Florida.

Hurst, W. J., Tarka, S. M., Powis, T. G., Valdez, F., and T. R. Hester 2002. Cacao usage by the earliest Maya civilization. *Nature* 418:289–290. https://doi.org/10.1038/418289a

IFIC. 2018. *Fact Sheet: Everything You Need to Know About Caffeine.* International Food Information Council Foundation. https://foodinsight.org/wp-content/uploads/2015/08/IFIC_Caffeine_v5.pdf

Inforural. 2018. *Agua de barranca, bebida prehispánica que preserva Zacatelco, Tlaxcala.* https://www.inforural.com.mx/agua-de-barranca-bebida-prehispanica-que-preserva-zacatelco-tlaxcala/

Instituto Mexicano de la Propiedad Industrial. 2015. *Lista Complementaria a la Clasificación Internacional de Productos y Servicios.* https://www.gob.mx/cms/uploads/attachment/file/94310/Lista_Complementaria_Dic15.pdf

Instituto Mexicano de la Propiedad Industrial. 2016. *"Cacao Grijalva", se convierte en la 15ª denominación de origen mexicana*. https://www.gob.mx/impi/prensa/cacao-grijalva-se-convierte-en-la-15-denominacion-de-origen-mexicana

Jafarnejad, S., Salek, M., and C. C. T. Clark. 2020. Cocoa consumption and blood pressure in middle-aged and elderly subjects: A meta-analysis. *Current Hypertension Reports* 22:1. https://doi.org/10.1007/s11906-019-1005-0

Jalil, A. M. M., and A. Ismail. 2008. Polyphenols in cocoa and cocoa products: Is there a link between antioxidant properties and health? *Molecules* 13:2190–2219. https://doi.org/10.3390/molecules13092190

Janitschke, D., Lauer, A. A., Bachmann, C. M., Grimm, H. S., Hartmann, T., and M. O. W. Grimm. 2021. Methylxanthines and neurodegenerative diseases: An update. *Nutrients* 13(3):803. https://doi.org/10.3390/nu13030803

Jaramillo, V. J. L., Córdova, L. C. E., and Á. V. Córdova. 2018. Disponibilidad a pagar por atributos culturales en chocolates caseros artesanales de la región de la Chontalpa, Tabasco, México. *Economía Agraria y Recursos Naturales* 18(2):53–73.

John, W. A., Böttcher, N. L., Behrends, B., Corno, M., D'souza, R. N., Kuhnert, N., and M. S. Ullrich. 2020. Experimentally modeling cocoa bean fermentation reveals key factors and their influences. *Food Chemistry* 302:125335. https://doi.org/10.1016/j.foodchem.2019.125335

Katz, D. L., Doughty, K., and A. Ali. 2011. Cocoa and chocolate in human health and disease. *Antioxidants & Redox Signaling* 15(10):2779–2811. https://doi.org/10.1089/ars.2010.3697

Kaufman, T., and J. Justeson. 2008. The history of the word for cacao in ancient Mesoamerica. *Ancient Mesoamerica* 18(2):193–237.

Kirchmayr, M. R., Segura-García, L. E., Lappe-Oliveras, P., Moreno-Terrazas, R., de la Rosa, M., and A. G. Mathis. 2017. Impact of environmental conditions and process modifications on microbial diversity, fermentation efficiency and chemical profile during the fermentation of Mezcal in Oaxaca. *LWT: Food Science and Technology* 79:160–169. https://doi.org/10.1016/j.lwt.2016.12.052

Kongor, J. E., Hinneh, M., Van de Walle, D., Afoakwa, E. O., Boeckx, P., and K. Dewettinck. 2016. Factors influencing quality variation in cocoa (*Theoborma cacao*) bean flavour profile: A review. *Food Research International* 82:44–52. https://doi.org/10.1016/j.foodres.2016.01.012

Kühn, J., Schröter, A., Hartmann, B. M., and G. I. Stangl. 2018. Cocoa and chocolate are sources of vitamin D_2. *Food Chemistry* 269:318–320. https://doi.org/10.1016/j.foodchem.2018.06.098

Kumari, N., Grimbs, A., D'Souza, R. N., Verma, S. K., Corno, M., Kuhnert, N., and M. S. Ullrich. 2018. Origin and varietal based proteomic and peptidomic fingerprinting of *Theobroma cacao* in non-fermented and fermented cocoa beans. *Food Research International* 111:137–147. https://doi.org/10.1016/j.foodres.2018.05.010

Langer, E., and W. Rzeski. 2014. Biological properties of melanoidins: A review. *International Journal of Food Properties* 17(2):344–353. https://doi.org/10.1080/10942912.2011.631253

Lattanzio, V. 2013. Phenolic compounds: Introduction. In: *Natural Products: Phytochemistry, Botany, and Metabolism of Alkaloids, Phenolics and Terpenes*, eds. K. G. Ramawat and J. M. Mérillon, 1543–1580. Berlin Heidelberg: Springer-Verlag.

Laudan, R., and J. M. Pilcher. 1999. Chiles, chocolate, and race in New Spain: Glancing backward to Spain or looking forward to Mexico? *Eighteenth-Century Life* 23(2):59–70. https://www.muse.jhu.edu/article/10489

Lee, A. H., Neilson, A. P., O'Keefe, S. F., Ogejo, J. A., Huang, H., Ponder, M., Chu, H. S. S., Jin, Q., Pilot, G., and A. C. Stewart. 2019. A laboratory-scale model cocoa fermentation using dried, unfermented beans and artificial pulp can simulate the microbial and chemical changes of on-farm cocoa fermentation. *European Food Research and Technology* 245:511–519. https://doi.org/10.1007/s00217-018-3171-8

Lefeber, T., Gobert, W., Vrancken, G., Camu, N., and L. De Vuyst. 2011. Dynamics and species diversity of communities of lactic acid bacteria and acetic acid bacteria during spontaneous cocoa bean fermentation in vessels. *Food Microbiology* 28(3):457–464. https://doi.org/10.1016/j.fm.2010.10.010

Lefeber, T., Papalexandratou, Z., Gobert, W., Camu, N., and L. De Vuyst. 2012. On-farm implementation of a starter culture for improved cocoa bean fermentation and its influence on the flavour of chocolates produced thereof. *Food Microbiology* 30(2):379–392. https://doi.org/10.1016/j.fm.2011.12.021

León-Perez, D., Domínguez-Perles, R., Collado-González, J., Cano-Lamadrid, M., Durand, T., Guy, A., Galano, J.-M., Carbonell-Barrachina, A., Londoño-Londoño, J., Ferreres, F., Jiménez-Cartagena, C., Gil-Izquierdo, A., and S. Medina. 2020. Bioactive plant oxylipins-based lipidomics in eighty worldwide comercial dark chocolates: Effect of cocoa and fatty acid composition on their dietary burden. *Microchemical Journal* 157:105083. https://doi.org/10.1016/j.microc.2020.105083

León-Perez, D., Medina, S., Londoño-Londoño, J., Cano-Lamadrid, M., Carbonell-Barrachina, A., Durand, T., Guy, A., Oger, C., Galano, J.-M., Ferreres, F., Jiménez-Cartagena, C., Restrepo-Osorno, J., and A. Gil-Izquierdo. 2018. Comparative study of different cocoa (*Theobroma cacao* L.) clones in terms of their phytoprostanes and phytofurans contents. *Food Chemistry* 280:231–239. https://doi.org/10.1016/j.foodchem.2018.12.072

Li, Y., Zhu, S., Feng, Y., Xu, F., Ma, J., and F. Zhong. 2014. Influence of alkalization treatment on the color quality and the total phenolic and anthocyanin contents in cocoa powder. *Food Science and Biotechnology* 23(1):59–63. https://doi.org/10.1007/s10068-014-0008-5

Lippi, D. 2013. Chocolate in history: Food, medicine, medi-food. *Nutrients* 5(5):1573–1584. https://doi.org/10.3390/nu5051573

López-Hernández, M., Rodríguez-Alegría, M. E., López-Munguía, A., and C. Wacher. 2018. Evaluation of xylan as carbon source for *Weissella* spp., a predominant strain in pozol fermentation. *LWT: Food Science and Technology* 89:192–197. https://doi.org/10.1016/j.lwt.2017.10.030

MacDonald, C.-J., Madika, A.-L., Bonnet, F., Fagherazzi, G., Lajous, M., and M.-C. Boutron-Ruault. 2020. Consumption of cocoa-containing foods and risk of hypertension in French women. *European Journal of Epidemiology* 35(5):465–469. https://doi.org/10.1007/s10654-020-00603-w

Maciel, L. F., Felício, A. L. S. M., and E. Y. Hirooka. 2017. Bioactive compounds by UPLC-PDA in different cocoa colnes (*Theobroma cacao* L.) developed in the southern region of Bahia, Brazil. *British Food Journal* 119(9): 2117–2127. https://doi.org/10.1108/BFJ-09-2016-0423

Manach, C., Scalbert, A., Morand, C., Rémésy, C., and L. Jiménez. 2004. Polyphenols: Food sources and bioavailability. *The American Journal of Clinical Nutrition* 79(5):727–747. https://doi.org/10.1093/ajcn/79.5.727

Marseglia, A., Musci, M., Rinaldi, M., Palla, G., and A. Caligiani. 2020. Volatile fingerprint of unroasted and roasted cocoa beans (*Theobroma cacao* L.) from different geographical origins. *Food Research International* 132:109101. https://doi.org/10.1016/j.foodres.2020.109101

Martorell, P., Bataller, E., Llopis, S., Gonzalez, N., Alvarez, B., Montón, F., Ortiz, P., Ramón, D., and S. Genovés. 2013. A cocoa peptide protects *Caenorhabditis elegans* form oxidative stress and β-amyloid peptide toxicity. *PLoS One* 8(5):e63283. https://doi.org/10.1371/journal.pone.0063283

Massaro, M., Scoditti, E., Carluccio, M. A., Kaltsatou, A., and A. Cicchella. 2019. Effect of cocoa products and its polyphenolic constituents on exercise performance and exercise-induced muscle damage and inflammation: A review of clinical trials. *Nutrients* 11(7):1471. https://doi.org/10.3390/nu11071471

Massot-Cladera, M., Costabile, A., Childs, C. E., Yaqoob, P., Franch, À. Castell, M., and F. J. Pérez-Cano. 2015. Prebiotic effects of cocoa fiber on rats. *Journal of Functional Foods* 19(Part A):341–352. https://doi.org/10.1016/j.jff.2015.09.021

Massot-Cladera, M., Mayneris-Perxachs, J., Costabile, A., Swann, J. R., Franch, A., Pérez-Cano, F.J., and M. Castell. 2017. Association between urinary metabolic profile and the intestinal effects of cocoa in rats. *British Journal of Nutrition* 117(5):623–634. https://doi.org/10.1017/S0007114517000496

McNeil, C. L. 2006. Introduction: The biology, antiquity, and modern uses of the chocolate tree (*Theobroma cacao* L.). In: *Chocolate in Mesoamerica: A Cultural History of Cacao*, ed. C. L. McNeil, 1–28. Florida: University Press of Florida.

Mellor, D. D. 2016. Cocoa: Composition and health effects. *Encyclopedia of Food and Health* 2016:179–184. https://doi.org/10.1016/b978–0–12–384947–2.00178–1

Mellor, D. D., and N. Naumovski. 2016. Effect of cocoa in diabetes: The potential of the pancreas and liver as key target organs, more than an antioxidant effect? *International Journal of Food Science & Technology* 51(4):829–841. https://doi.org/10.1111/ijfs.13075

Melo, C. W. B., Bandeira, M. J., Maciel, L. F., Bispo, E. S., Souza, C. O., and S. E. Soares. 2020. Chemical composition and fatty acids profile of chocolates produced with different cocoa (*Theobroma cacao* L.) cultivars. *Food Science and Technology Campinas* 40(2):326–333. https://doi.org/10.1590/fst.43018

Mendez-Albores, J. A., Arámbula-Villa, G., Preciado-Ortíz, R. E., and E. Moreno-Martínez. 2004. Aflatoxins in pozol, a nixtamalized, maize-based food. *International Journal of Food Microbiology* 94(2):211–215. https://doi.org/10.1016/j.ijfoodmicro.2004.02.009

Meng, C. C., Jalil, A. M. M., and A. Ismail. 2009. Phenolic and theobromine contents of commercial dark, milk and white chocolates on the Malaysian market. *Molecules* 14:200–209. https://doi.org/10.3390/molecules14010200

Mihai, R. A., Landazuri, A. P. A., Tinizaray, R. B. A., Florescu, L. I., Catană, R., and A. Kosakyan. 2022. Abiotic factors from different Ecuadorian Regions and their contribution to antioxidant, metabolomic and organoleptic quality of *Theobroma cacao* L. beans, variety "Arriba Nacional". *Plants* 11(7):976. https://doi.org/10.3390/plants11070976

Miller, K. B., Hurst, W. J., Flanningan, N., Ou, B., Lee, C. Y., Smith, N., and D. A. Stuart. 2009. Survey of commercially available chocolate- and cocoa-containing products in the United States. 2. Comparison of Flavan-3-ol content with nongat cocoa solids, total polyphenols, and percent cacao. *Journal of Agricultural and Food Chemistry* 57(19):9169–9180. https://doi.org/10.1021/jf901821x

Miller, K. B., Hurst, W. J., Payne, M. J., Stuart, D. A., Apgar, J., Sweigart, D. S., and B. Ou. 2008. Impact of alkalization on the antioxidant and flavanol content of commercial cocoa powders. *Journal of Agricultural and Food Chemistry* 56(18):8527–8533. https://doi.org/10.1021/jf801670p

Mission chocolate. 2015. *5 best chocolate drinks of Mexico*. https://missionchocolate.com/blogs/recipes/87911299–5-cacao-drink-recipes-5-recetas-bebidas-de-cacao

Moens, F., Lefeber, T., and L. De Vuyst. 2014. Oxidation of metabolites highlights the microbial interactions and role of *Acetobacter pasteurianus* during cocoa bean fermentation. *Applied and Environmental Microbiology* 80(6):1848–1857. https://doi.org/10.1128/AEM.03344–13

Mota-Gutierrez, J., Barbosa-Pereira, L., Ferrocino, I., and L. Cocolin. 2019. Traceability of functional volatile compounds generated on inoculated cocoa fermentation and its potential health benefits. *Nutrients* 11(4):884. https://doi.org/10.3390/nu11040884

Motamayor, J. C., Lachenaud, P., da Silva e Mota, J. W., Loor, R., Kuhn, D. N., Brown, J. S., and R. J. Schnell. 2008. Geographic and genetic population differentiation of the Amazonian chocolate tree (*Theobroma cacao* L). *PLoS One* 3(10):e3311. https://doi.org/10.1371/journal.pone.0003311

Mother Earth Gardener. 2020. *Chileatole recipe*. https://www.motherearthgardener.com/food/recipes/chileatole-recipe-zmrz17szcwil

Mustiga, G. M., Morrissey, J., Stack, J. C., DuVal, A., Royaert, S., Jansen, J., Bizzotto, C., Villela-Dias, C., Mei, L., Cahoon, E. B., Seguine, E., Marelli, J. P., and J. C. Motamayor. 2019. Identification of climate and genetic factors that control fat content and fatty acid composition of *Theobroma cacao* L. beans. *Frontiers in Plant Science* 10:1159. https://doi.org/10.3389/fpls.2019.01159

Niemenak, N., Rohsius, C., Elwers, S., Ndoumon, D. O., and R. Lieberei. 2006. Comparative study of different cocoa (*Theobroma cacao* L.) clones in terms of their phenolics and anthocyanins contents. *Journal of Food Composition and Analysis* 19:612–619. https://doi.org/10.1016/j.jfca.2005.02.006

NIH. (2020). *Vitamin D. Fact sheet for health professionals*. U.S. Department of Health and Human Services, National Institutes of Health, Office of Dietary Supplements. https://ods.od.nih.gov/factsheets/VitaminD-HealthProfessional/#h3

Ogata, N. 2007. El cacao. CONABIO. *Biodiversitas* 72:1–5.

Ogata, N. 2017. El cacao como sistema de diversificación productiva para el desarrollo del Sureste de México. In: *El cacao alimento divino*, ed. G. Castillo, 954–1443. Ciudad de Mexico: Fundación Herdez.

Ogata, N., Gómez-Pompa, A., and K. A. Taube. 2006. The domestication and distribution of *Theobroma cacao* L. in the neotropics. In: *Chocolate in Mesoamerica: A Cultural History of Cacao*, ed. C. L. McNeil, 69–89. Florida: University Press of Florida.

Ojeda-Linares, C. I., Vallejo, M., Lappe-Oliveras, P., and A. Casas. 2020. Traditional management of microorganisms in fermented beverages from cactus fruits in Mexico: An ethnobiological approach. *Journal of Ethnobiology and Ethnomedicine* 16:1. https://doi.org/10.1186/s13002-019-0351-y

Oliveira, J. S., Costa, K., Acurcio, L. B., Sandes, S. H. C., Cassali, G. D., Uetanabaro, A. P. T., Costa, A. M., Nicoli, J. R., Neumann, E., and A. L. F. Porto. 2018. In vitro and in vivo evaluation of two potential probiotic lactobacilli isolated from cocoa fermentation (*Theobroma cacao* L.). *Journal of Functional Foods* 47:184–191. https://doi.org/10.1016/j.jff.2018.05.055

Oñatibia-Astibia, A., Franco, R., and E. Martínez-Pinilla. 2017. Health benefits of methylxanthines in neurodegenerative diseases. *Molecular Nutrition and Food Research* 61(6):1600670. https://doi.org/10.1002/mnfr.201600670

Ooi, T. S., Ting, A. S. Y., and L. F. Siow. 2020. Influence of selected native yeast starter cultures on the antioxidant activities, fermentation index and total soluble solids of Malaysia cocoa beans: A simulation study. *LWT: Food Science and Technology* 122:108977. https://doi.org/10.1016/j.lwt.2019.108977

Oracz, J., and E. Nebesny. 2019. Effect of roasting parameters on the physicochemical characteristics of high-molecular-weight Maillard reaction products isolated from cocoa beans of different *Theobroma cacao* L. groups. *European Food Research and Technology* 245:111–128. https://doi.org/10.1007/s00217-018-3144-y

Oracz, J., and D. Zyzelewicz. 2019. In vitro antioxidant activity and FTIR characterization of high-molecular weight melanoidin fractions from different types of cocoa beans. *Antioxidants* 8(11):560. https://doi.org/10.3390/antiox8110560

Oracz, J., Nebesny, E., and D. Żyżelewicz. 2019. Identification and quantification of free and bound phenolic compounds contained in the high-molecular weight melanoidin fractions derived from two different types of cocoa beans by UHPLC-DAD-ESI-HR-MS^n. *Food Research International* 115:135–149. https://doi.org/10.1016/j.foodres.2018.08.028

Oracz, J., Żyżelewicz, D., and E. Nebesny. 2013. The content of polyphenolic compounds in cocoa beans (*Theobroma cacao* L.) depending on variety, growing region and processing operations: A review. *Critical Reviews in Food Science and Nutrition* 55(9):1176–1192. https://doi.org/10.1080/10408398.2012.686934

Ordoñez-Araque, R. H., Landines-Vera, E. F., Urresto-Villegas, J. C., and C. F. Caicedo-Jaramillo. 2020. Microorganisms during cocoa fermentation: systematic review. *Foods and Raw Materials* 8(1):155–162. http://doi.org/10.21603/2308-4057-2020-1-155-162

Othman, A., Jalil, A. M. M., Weng, K. K., Ismail, A., Ghani, N. A., and I. Adenan. 2010. Epicatechin content and antioxidant capacity of cocoa beans from four different countries. *African Journal of Biotechnology* 9(7):1052–1059. https://doi.org/10.5897/AJB09.1219

Ozturk, G., and G. M. Young. 2017. Food evolution: The impact of society and science on the fermentation of cocoa beans. *Comprehensive Reviews in Food Science and Food Safety* 16(3):431–455. https://doi.org/10.1111/1541–4337.12264

Pacheco-Montealegre, M. E., Dávila-Mora, L. L., Botero-Rute, L. M., Reyes, A., and A. Caro-Quintero. 2020. Fine resolution analysis of microbial communities provides insights into the variability of cocoa bean fermentation. *Frontiers in Microbiology* 11:650. https://doi.org/10.3389/fmicb.2020.00650

Papalexandratou, Z., Vrancken, G., De Bruyne, K., Vandamme, P., and L. De Vuyst. 2011. Spontaneous organic cocoa bean box fermentations in Brazil are characterized by a restricted species diversity of lactic acid bacteria and acetic acid bacteria. *Food Microbiology* 28(7):1326–1338. https://doi.org/10.1016/j.fm.2011.06.003

Pedan, V., Weber, C., Do, T., Fisher, N., Reich, E., and S. Rohn. 2018. HPTLC fingerprint profile analysis of cocoa proanthocyanidins depending on origin and genotype. *Food Chemistry* 267:277–287. https://doi.org/10.1016/j.foodchem.2017.08.109

Peláez, P. P., Bardón, I., and P. Camasca. 2016. Methylxanthine and catechin content of fresh and fermented cocoa beans, dried cocoa beans, and cocoa liquor. *Scientia Agropecuaria* 7(4):355–365. https://doi.org/10.17268/sci.agropecu.2016.04.01

Pérez-Armendáriz, B., and G. A. Cardoso-Ugarte. 2020. Traditional fermented beverages in Mexico: Biotechnological, nutritional, and functional approaches. *Food Research International* 136:109307. https://doi.org/10.1016/j.foodres.2020.109307

Pérez-Burillo, S., Rajakaruna, S., Pastoriza, S., Paliy, O., and J. A. Rufián-Henares. 2020. Bioactivity of food melanoidins is mediated by gut microbiota. *Food Chemistry* 316:126309. https://doi.org/10.1016/j.foodchem.2020.126309

Petrone, A. B., Gaziano, J. M., and L. Djoussé. 2013. Effects of dark chocolate and cocoa products on endothelial function: a meta-analysis. *Current Nutrition Reports* 2:267–273. https://doi.org/10.1007/s13668-013-0058-y

Petyaev, I. M., and Y. K. Bashmakov. 2017. Dark chocolate: Opportunity for an alliance between medical science and the food industry? *Frontiers in Nutrition* 4:43. https://doi.org/10.3389/fnut.2017.00043

Pirouzian, H. R., Konar, N., Palabiyik, I., Oba, S., and O. S. Toker. 2020. Pre-crystallization process in chocolate: Mechanism, importance and novel aspects. *Food Chemistry* 321:126718. https://doi.org/10.1016/j.foodchem.2020.126718

Powis, T. G., Cyphers, A., Gaikwad, N. W., Grivetti, L., and K. Cheong. 2011. Cacao use and the San Lorenzo Olmec. *PNAS* 108(21):8595–8600. https://doi.org/10.1073/pnas.1100620108

Powis, T. G., Hurst, W. J., Rodríguez, M. C., Ortíz, P., Blake, M., Cheetham, D., Coe, M. D., and J. G. Hodgson. 2007. Oldest chocolate in the New World. *Antiquity Project Gallery* 81(314).

Preston, D., and R. B. Murphy. 2015. *U.S. Patent No. 8,969,655 B1.* Washington, DC: U.S. Patent and Trademark Office.

Quideau, S., Deffieux, D., Douat-Casassus, C., and L. Pouységu. 2011. Plant polyphenols: Chemical properties, biological activities, and synthesis. *Angewandte Chemie International Edition* 50(3):586–621. https://doi.org/10.1002/anie.201000044

Quiroz-Reyes, C. N., and V. Fogliano. 2018. Design cocoa processing towards healthy cocoa products: The role of phenolics and melanoidins. *Journal of Functional Foods* 45:480–490. https://doi.org/10.1016/j.jff.2018.04.031

Racine, K. C., Lee, A. H., Wiersema, B. D., Huang, H., Lambert, J. D., Stewart, A. C., and A. P. Neilson. 2019. Development and characterization of a pilot-scale model cocoa fermentation system suitable for studying the impact of fermentation on putative bioactive compounds and bioactivity of cocoa. *Foods* 8(3):102. https://doi.org/10.3390/foods8030102

Ramírez González, B. 2006. Patrimonio arqueológico, histórico, intangible y natural de la delegación de Iztapalapa. Universidad Autónoma Metropolitana—Iztapalapa. Distrito Federal, Mexico. https://issuu.com/josemartinez34/docs/patrimonio_de_iztapalapa

Ramírez-Guzmán, K. N., Torres-León, C., Martinez-Medina, G. A., de la Rosa, O., Hernández-Almanza, A., Alvarez-Perez, O. B., Araujo, R., Rodríguez, G. L., Londoño, L., Ventura, J., Rodríguez, R., Martinez, J. L., and C. N. Aguilar. 2019. Traditional fermented beverages in Mexico. In: *Fermented Beverages*, eds. A. M. Grumezescu and A. M. Holban, 605–635. Duxford: Woodhead Publishing. https://doi.org/10.1016/B978-0-12-815271-3.00015-4

Ramos, C. L., Dias, D. R., Miguel, M. G. C. P., and R. F. Schwan. 2014. Impact of different cocoa hybrids (*Theobroma cacao* L.) and *S. cerevisiae* UFLA CA11 inoculation on microbial communities and volatile compounds of cocoa fermentation. *Food Research International* 64:908–918. https://doi.org/10.1016/j.foodres.2014.08.033

Ramos, C. L., and R. F. Schwan. 2017. Technological and nutritional aspects of indigenous Latin America fermented foods. *Current Opinion in Food Science* 13:97–102. https://doi.org/10.1016/j.cofs.2017.07.001

Ramos-Escudero, F., Casimiro-Gonzales, S., Fernández-Prior, Á., Chávez, K. C., Gómez-Mendoza, J., Fuente-Carmelino, L., and A. M. Muñoz. 2021. Colour, fatty acids, bioactive compounds and total antioxidant capacity in commercial cocoa beans (*Theobroma cacao* L.). *LWT: Food Science and Technology* 147:111629. https://doi.org/10.1016/j.lwt.2021.111629

Rawel, H. M., Huschek, G., Sagu, S. T., and T. Homann. 2019. Cocoa bean proteins: Characterization, changes and modifications due to ripening and post-harvest processing. *Nutrients* 11(2):428. https://doi.org/10.3390/nu11020428

Rengifo, R. C. 2016. *U.S. Patent No. 9,428,759 B2.* Washington, DC: U.S. Patent and Trademark Office.

Rezac, S., Kok, C. R., Heermann, M., and R. Hutkins. 2018. Fermented foods as a dietary source of live organisms. *Frontiers in Microbiology* 9:1785. https://doi.org/10.3389/fmicb.2018.01785

Rizzi, G. P. 2008. The Strecker degradation of amino acids: Newer avenues for flavor formation. *Food Reviews International* 24(4):416–435. https://doi.org/10.1080/87559120802306058

Rodriguez-Campos, J., Escalona-Buendía, H. B., Contreras-Ramos, S. M., Orozco-Avila, I., Jaramillo-Flores, E., and E. Lugo-Cervantes. 2012. Effect of fermentation time and drying temperature on volatile compounds in cocoa. *Food Chemistry* 132(1):277–288. https://doi.org/10.1016/j.foodchem.2011.10.078

Rodriguez-Campos, J., Escalona-Buendía, H. B., Orozco-Avila, I., Lugo-Cervantes, E., and M. E. Jaramillo-Flores. 2011. Dynamics of volatile and non-volatile compounds in cocoa (*Theobroma cacao* L.) during fermentation and drying processes using principal components analysis. *Food Research International* 44(1):250–258. https://doi.org/10.1016/j.foodres.2010.10.028

Romanens, E., Leischtfeld, S. F., Volland, A., Stevens, M. J. A., Krähenmann, U., Isele, D., Fischer, B., Meile, L., and S. M. Schwenninger. 2019. Screening of lactic acid bacteria and yeast strains to select adapted anti-fungal co-cultures for cocoa bean fermentation. *International Journal of Food Microbiology* 290:262–272. https://doi.org/10.1016/j.ijfoodmicro.2018.10.001

Romero-Cortes, T., Salgado-Cervantes, M. A., García-Alamilla, P., García-Alvarado, M. A., Rodríguez-Jimenes, G. C., Hidalgo-Morales, M., and V. Robles-Olvera. 2013. Relationship between fermentation index and other biochemical changes evaluated during the fermentation of Mexican cocoa (*Theobroma cacao*) beans. *Journal of the Science of Food and Agriculture* 93(10):2596–2604. https://doi.org/10.1002/jsfa.6088

Romero-Luna, H. E., Hernández-Sánchez, H., and G. Dávila-Ortiz. 2017. Traditional fermented beverages from Mexico as a potential probiotic source. *Annals of Microbiology* 67:577–586. https://doi.org/10.1007/s13213-017-1290-2

Ruesgas-Ramón, M., Figueroa-Espinoza, M. C., Durand, E., Suárez-Quiroz, M. L., González-Ríos, O., Rocher, A., Reversat, G., Vercauteren, J., Oger, C., Galano, J.-M., Durand, T., and C. Vigor. 2019. Identification and quantification of phytoprostanes and phytofurans of coffee and cocoa by- and co-products. *Food and Function* 10(10):6882–6891. https://doi.org/10.1039/c9fo01528k

SADER Yucatán. 2020. *Este 2 de septiembre, México celebra pro segunda ocasión el "Día Nacional del Cacao y el Chocolate"*. https://www.gob.mx/agricultura/yucatan/articulos/este-2-de-septiembre-mexico-celebra-por-segunda-ocasion-el-dia-nacional-del-cacao-y-el-chocolate?idiom=es

Salvador, I., Massarioli, A. P., Silva, A. P. S., Malaguetta, H., Melo, P. S., and S. M. Alencar. 2019. Can we conserve *trans*-resveratrol content and antioxidant activity during industrial production of chocolate? *Journal of the Science of Food and Agriculture* 99:83–89. https://doi.org/10.1002/jsfa.9146

Samaniego, I., Espín, S., Quiroz, J., Ortiz, B., Carrillo, W., García-Viguera, C., and P. Mena. 2020. Effect of the growing area on the methylxanthines and flavan-3-ols content in cocoa beans from Ecuador. *Journal of Food Composition and Analysis* 88:103448. https://doi.org/10.1016/j.jfca.2020.103448

Santiago-Urbina, J. A., and F. Ruíz-Terán. 2014. Microbiology and biochemistry of traditional palm wine produced around the world. *International Food Research Journal* 21(4):1261–1269.

Schwan, R. F., and A. E. Wheals. 2004. The microbiology of cocoa fermentation and its role in chocolate quality. *Critical Reviews in Food Science and Nutrition* 44(4):205–221. https://doi.org/10.1080/10408690490464104

Seem, S. A., Yuan, Y. V., and J. C. Tou. 2019. Chocolate and chocolate constituents influence bone health and osteoporosis risk. *Nutrition* 65:74–84. https://doi.org/10.1016/j.nut.2019.02.011

SIAP. 2018. Cacao. In: *Atlas agroalimentario 2012–2018*, ed. SIAP, Servicio de Información Agroalimentaria y Pesquera, 46–47. Ciudad de México: Servicio de Información Agroalimentaria y Pesquera.

Slow Food. 2020. *Zacatelco Agua de barranca*. https://www.fondazioneslowfood.com/en/ark-of-taste-slow-food/zacatelco-agua-de-barranca/

Soleri, D., and D. A. Cleveland. 2007. Tejate: *Theobroma cacao* and *T. bicolor* in a traditional beverage from Oaxaca, Mexico. *Food and Foodways* 15(1–2):107–118. https://doi.org/10.1080/07409710701260131

Sotelo, A., Soleri, D., Wacher, C., Sánchez-Chinchillas, A., and R. M. Argote. 2012. Chemical and nutritional composition of tejate, a traditional maize and cacao beverage from the Central Valleys of Oaxaca, Mexico. *Plant Foods for Human Nutrition* 67(2):148–155. https://doi.org/10.1007/s11130-012-0281-5

Spizzirri, U. G., Ieri, F., Campo, M., Paolino, D., Restuccia, D., and A. Romani. 2019. Biogenic amines, phenolic, and aroma-related compounds of unroasted and roasted cocoa beans with different origin. *Foods* 8(8):306. https://doi.org/10.3390/foods8080306

Srdjenovic, B., Djordjevic-Milic, V., Grujic, N., Inajc, R., and Z. Lepojevic. 2008. Simultaneous HPLC determination of caffeine, theobromine and theophylline in food, drinks and herbal products. *Journal of Chromatographic Science* 46(2):144–149. https://doi.org/10.1093/chromsci/46.2.144

Stanley, T. H., Van Buiten, C. B., Baker, S. A., Elias, R. J., Anantheswaran, R. C., and J. D. Lambert. 2018. Impact of roasting on the flavan-3-ol composition, sensory-related chemistry, and *in vitro* pancreatic lipase inhibitory activity of cocoa beans. *Food Chemistry* 255:414–420. https://doi.org/10.1016/j.foodchem.2018.02.036

Sugimoto, N., Katakura, M., Matsuzaki, K., Sumiyoshi, E., Yachie, A., and O. Shido. 2018. Chronic administration of theobromine inhibits mTOR signal in rats. *Basic & Clinical Pharmacology & Toxicology* 124(5):575–581. https://doi.org/10.1111/bcpt.13175

Sunoj, S., Igathinathane, C., and R. Visvanathan. 2016. Nondestructive determination of cocoa bean quality using FT-NIR spectroscopy. *Computers and Electronics in Agriculture* 124:234–242. https://doi.org/10.1016/j.compag.2016.04.012

Talboom, J. S., Velazquez, R., and S. Oddo. 2015. The mammalian target of rapamycin at the crossroad between cognitive aging and Alzheimer's disease. *Nature Partner Journals, Aging and Mechanisms of Disease* 1:15008. https://doi.org/10.1038/npjamd.2015.8

Tamang, J. P., Cotter, P. D., Endo, A., Han, N. S., Kort, R., Liu, S. Q., Mayo, B., Westerik, N., and R. Hutkins. 2020. Fermented foods in a global age: East meets West. *Comprehensive Reviews in Food Science and Food Safety* 19(1):184–217. https://doi.org/10.1111/1541-4337.12520

Tee, Y.-K., Bariah, K., Zainudin, B. H., Yap, K.-C. S., and N.-G. Ong. 2021. Impacts of cocoa pod maturity at harvest and bean fermentation period on the production of chocolate with potential health benefits. *Journal of the Science of Food and Agriculture* 102(4):1576–1585. https://doi.org/10.1002/jsfa.11494

Travassos, M., Santana, I., Baldeiras, I., Tsolaki, M., Gkatzima, O., Sermin, G., Yener, G. G., Simonsen, A., Hasselbalch, S. G., Kapaki, E., Mara, B., Cunha, R. A., Agostinho, P., Blennow, K., Zetterberg, H., Mendes, V. M., Manadas, B., and A. de Mendon. 2015. Does caffeine consumption modify cerebrospinal fluid Amyloid-β levels in patients with Alzheimer's disease? *Journal of Alzheimer's Disease* 47(4):1069–1078. https://doi.org/10.3233/JAD-150374

Tsao, R. 2010. Chemistry and biochemistry of dietary polyphenols. *Nutrients* 2(12):1231–1246. https://doi.org/10.3390/nu2121231

Urbańska, B., Derewaika, D., Lenart, A., and J. Kowalska. 2019. Changes in the composition and content of polyphenols in chocolate resulting from pre-treatment method of cocoa beans and thechonolgical process. *European Food Research and technology* 245:2101–2112. https://doi.org/10.1007/s00217-019-03333-w

USDA. 2020. *FoodData Central*. https://fdc.nal.usda.gov/fdc-app.html#/food-details/789222/nutrients

Utrilla-Vázquez, M., Rodríguez-Campos, J., Avendaño-Arazate, C. H., Gschaedler, A., and E. Lugo-Cervantes. 2020. Analysis of volatile compounds of five varieties of Maya cocoa during fermentation and drying processes by Venn diagram and PCA. *Food Research International* 129:108834. https://doi.org/10.1016/j.foodres.2019.108834

Valverde García, D., Pérez Esteve, É., and J. M. Barat Baviera. 2020. Changes in cocoa properties induced by the alkalization process: A review. *Comprehensive Reviews in Food Science and Food Safety* 19(4):2200–2221. https://doi.org/10.1111/1541–4337.12581

Vanderschueren, R., Montalvo, D., De Ketelaere, B., Delcour, J. A., and E. Smolders. 2019. The elemental composition of chocolates is related to cacao content and origin: A multi-element fingerprinting analysis of single origin chocolates. *Journal of Food Composition and Analysis* 83:103277. https://doi.org/10.1016/j.jfca.2019.103277

Wickramasuriya, A. M., and J. M. Dunwell. 2018. Cacao biotechnology: Current status and future prospects. *Plant Biotechnology Journal* 16:4–17. https://doi.org/10.1111/pbi.12848

World Research Foundation. 2020. *The Badianus Manuscript (Codex Barberini, Latin 241) translated by Emily Walcott Emmart, Johns Hopkins Press, 1940*. https://www.wrf.org/ancient-medicine/badianus-manuscript-americas-earliest-medical-book.php

Yaylayan, V. A. 2003. Recent advances in the chemistry of Strecker degradation and Amadori rearrangement: Implications to aroma and color formation. *Food Science and Technology Research* 9(1):1–6. https://doi.org/10.3136/fstr.9.1

Yilmaz, C., and V. Gökmen. 2018. Determination of tryptophan derivatives in khynurenine pathway in fermented foods using liquid chromatography tandem mass spectrometry. *Food Chemistry* 243:420–427. https://doi.org/10.1016/j.foodchem.2017.10.004

Zhang, H., Zhang, H., Troise, A. D., and V. Fogliano. 2019. Melanoidins from coffee, cocoa, and bread are able to scavenge α-dicarbonyl compounds under simulated physiological conditions. *Journal of Agricultural and Food Chemistry* 67(39):10921–10929. https://doi.org/10.1021/acs.jafc.9b03744

Zugravu, C., and M. R. Otelea. 2019. Dark chocolate: To eat or not to eat? A review. *Journal of AOAC International* 102(5):1388–1396. https://doi.org/10.5740/jaoacint.19–0132

Żyżelewicz, D., Budryn, G., Oracz, J., Antolak, H., Kręgiel, D., and M. Kaczmarska. 2018. The effect on bioactive components and characteristics of chocolate by functionalization with raw cocoa beans. *Food Research International* 113:234–244. https://doi.org/10.1016/j.foodres.2018.07.017

Traditional Fermented Foods of Mexico and Potential Functional Benefits

27

Emma Mani-López and Aurelio López-Malo

27.1 INTRODUCTION

Fermentation is a natural and low-cost food preservation process used in foods and beverages, destroying unwanted compounds, inhibiting undesirable microorganisms, enhancing the nutritional value, improving digestibility, enhancing sensory properties, and providing bioactive compounds with functional properties. Major benefits of fermentation are the development of lactic acid bacteria (LAB) with potential probiotic properties (hypolipidemic, hepatoprotective, and antibacterial), removal of anti-nutrient compounds (phytate, enzyme inhibitors, and polyphenols), and increasing the bioavailability of components associated with starch (fiber, vitamins, minerals, proteins) (Sangwan et al. 2014). In addition, metabolites produced during fermentation also contribute to health benefits. The primary metabolites with health-promoting properties are bioactive peptides (with immunomodulatory, antithrombic, and antihypertensive properties) (Tamang et al. 2016), polysaccharides (as antioxidants and radical scavenging properties), and short-chain fatty acids (SCFAs). SCFA (particularly butyrate) prevents or reduces gastrointestinal disorders and colon cancer, promotes maturation of the enteric neurons, and modulates the immune system (Cuvas-Limon et al. 2020). Microorganisms of fermented foods with potential probiotic properties must demonstrate their resistance to conditions that simulate the digestive process, such as low pH, bile salts, and enzymatic action. In addition, they should be able to colonize and proliferate in the digestive tract. It is not easy to designate a traditional fermented food as a probiotic because health benefits must be demonstrated to specific species, and consumption should be in large amounts. However, cohort studies show that health benefits are associated with the intake of fermented foods in various regions worldwide.

Perception of fermented products is evolving from simple preserved foods to nutritional and functional foods nowadays. Traditional fermented foods and beverages provide beneficial effects via microbial (bacteria, yeast, and molds) or probiotic action or by microbial metabolites improving the balance of the intestinal host, resulting in a low risk of gastrointestinal diseases (Romero-Luna et al. 2017). Taylor et al. (2020) conducted a cross-sectional study about plant fermented food intake in the American population. They observed that the overall microbiota was significantly different between non-consumers and consumers with regular intake (3 to 5 times/week) of plant fermented foods. In the same report, using a longitudinal study, the authors discovered various species related to fermented foods in the microbiota composition of consumers of fermented foods after 4 weeks, such as *Lactobacillus acidophilus*, *Lactobacillus brevis* (new name *Levilactobacillus brevis*), *Lactobacillus kefiranofaciens*, *Lactobacillus parabuchneri* (new name *Lentilactobacillus parabuchneri*), *Lactobacillus helveticus*, and *Lactobacillus sakei* (new name *Latilactobacillus sakei*). In other reports, fermented foods such as yogurt are related to better metabolic parameters (Chen et al. 2014), Korean fermented foods have been associated with a low prevalence of dermatitis in their consumers (Park and Bae 2016), and miso and natto were related to typical values of blood pressure in the Japanese population (Nozue et al. 2017).

Properties of fermented food depend directly on the microorganisms and substrate involved in the fermentation product since metabolites produced are derived from them. In addition, native Mexican raw materials used by pre-Hispanic people to prepare fermented products have specific properties and consequently health benefits. Therefore, this chapter provides information about the functional properties and bioactive compounds of traditional Mexican fermented foods and beverages, including *pulque*, *xaxtle*, *pozol*, vanilla, *tesgüino*, *tepache*, *colonche*, and *atole agrio*.

DOI: 10.1201/9781003605300-27

27.2 MEXICAN FERMENTED FOODS

The wide diversity of traditional Mexican fermented foods was due to natural resources available in the wild and those domesticated in the region, providing several raw materials. Maize, one of the primary ingredients in the Mexican diet, was also used to formulate various traditional fermented products. Since maize domestication around 3500 BC in the region of Coxcatlán cave in Tehuacán and Ocampo cave in Tamaulipas localized in central and north Mexico (Long et al. 1989), the development of derived foodstuffs began. Meanwhile, opuntia and agave were endemic plants from semi-arid zones in Mexico consumed abundantly by the ancient Mexicans. After native settlers learned how to make edible opuntia and agave, they extracted the agave sap and prepared prickly pear juices. Then, derived fermented products were probably developed accidentally. Archeological evidence of *pulque* production was reported in the Teotihuacan valley from organic residue analysis of pottery vessels. This demonstrated its production from 550–200 B.C. (Correa-Ascencio et al. 2014).

There are ~200 fermented foods and beverages in Mexico (Romero-Luna et al. 2017). The vast diversity is derived from minor variations in the preparation process and the ingredients used in the relevant foodstuff matrix. Most traditional Mexican fermented foods are produced by spontaneous fermentation; hence a consortium of microorganisms is commonly responsible for the fermentation. Microbial consortia primarily consist of bacteria from different species, fungi, and/or yeast incorporated from raw material, utensils, handling, and the environment. Even though several traditional fermented foods are recognized, very few are studied and documented. Cacao, vanilla, *pulque*, and *pozol* are the most studied. Recently, some researchers have been interested in describing the fermentation process of *atole agrio* and *tepache*. Selected traditional Mexican fermented products such as *pozol*, *atole agrio*, vanilla, and *pulque* have been recognized for their medicinal properties since ancient times. The health benefits of traditional fermented products are associated with their microorganisms and metabolites. In addition, bioactive compounds such as polyphenols, betalains, and saponins also contribute to health benefits.

27.3 FUNCTIONAL PROPERTIES OF FERMENTED FOODS

27.3.1 *Atole Agrio*

Atole agrio is a maize-based nonalcoholic, acidic fermented beverage traditionally prepared in some areas of southeastern Mexico (Oaxaca, Chiapas, Tabasco, and South Veracruz). Inhabitants consumed the beverage for their nutritional and medicinal properties, especially to treat diarrhea (Sánchez-Cortés et al. 2019). *Atole agrio* preparation differs in each region; however, the typical process uses tender maize or corn grains without nixtamalization (cooking grains in limewater). Tender maize is preferred in the *atole agrio* preparation. Grains are cut off from the corn cobs using a knife, ground or grated, mixed with water to obtain a dough, and allowed to ferment for 6 to 12 h at room temperature (~30°C). Meanwhile, the grains are carefully washed and soaked for mature maize kernel for 2 days, draining the water each day (Manzo-Fuentes and Gómez-Gutierrez 2019) before the grinding step. The fermentation takes place in a liquid or solid state (Díaz-Ruiz et al. 2013). In the solid state, the dough is molded into balls. In contrast, for the liquid state, equal portions of dough and water are mixed before fermentation. When fermentation is concluded, the mixture is boiled for 10 min. Typically, the end product is flavored with sugar, honey, cinnamon, or cocoa. Fermentation in a solid state allows the development of more LAB species diversity compared to liquid fermentation. The dominant genera were *Weissella* (36.3%) or *Pediococcus* (29.8%) (Väkeväinen et al. 2018; Peréz-Cataluña et al. 2018). Fermentation time ranges from 6 to 12 h at ~34°C, and the end product reaches a pH of 3.9–4.1 (Chaves-López et al. 2020). Indeed, the low pH of *atole agrio* is associated with antibacterial activity against Gram-negative bacteria (Cuvas-Limon et al. 2020).

The *atole agrio* microbiota is constituted by *Weissella* (~29%, *W. confusa* 18%, *W. cibaria* ~7%, *W. paramesenteroides* 4.5%), *Pediococcus* (*pentosaceus*, 24%), *Lactococcus* (*lactis*, ~18%), *Lactobacillus* (~16%, *Lactoplantibacillus plantarum* (formerly *Lactobacillus plantarum*) was the most abundant at 10%), *Leuconostoc* (6.7%), and *Enterococcus* (5.8%) (Väkeväinen et al. 2018). Väkeväinen et al. (2018) isolated 359 strains from *atole agrio,* and 88 were selected for further tests. A total of 12 strains of *Lb. plantarum* and 8 strains of *Lc. lactis* were capable of producing folate extracellularly, whereas 1 strain of *Lb. plantarum* degraded phytates extracellularly. A total of 9 strains (*Lb. plantarum* (2), *Lc. lactis* (5), and *P. pentosaceus* (2)) from *atole agrio* were amylolytic. Three strains of *Lb. plantarum* and one strain of *P. pentosaceus* were capable of producing EPS from glucose, maltose, raffinose, and sucrose. Also, yeasts have been isolated and identified by sequence analysis of the 26S rDNA D1/D2 domain from *atole agrio* (*Candida intermedia*, *Pichia kudriavzevii*, *Clavispora lusitaniae*, *Candida parapsilosis*, and some maize saprophytes) (Manrique-Donrronsoro et al. 2016). *P. kudriavzevii* and *C. lusitaniae* showed probiotic potential since they could grow at 37°C, survive at low pH (pH = 2.5, 24 h) and in the presence of bile salts (1.5%, 24 h), and form biofilms *in vitro*. Yeasts' concentration in the human gastrointestinal tract (GIT) microbiota is low (0.1%) and detected in ~70% of healthy adults (Lara Hidalgo et al. 2017). They are important in molecular interactions owing to their size (10 times bigger than bacteria) (Czerucka et al. 2007) and as nitrogen suppliers that stimulate LAB growth (Ponomarova et al. 2017).

Regardless, *atole agrio* has been little studied; its microbiota has potential as probiotics and mainly as postbiotics since it is consumed after being boiled.

As with *pozol*, various bacteria isolated from *atole agrio* showed excellent properties. However, there are particular concerns about its benefits since few LAB were resistant to ≥2 antibiotics, 14 strains at least to one antibiotic, and only 5 were susceptible (Väkeväinen et al. 2018). Certainly, *atole agrio* is consumed boiled; hence, most bacteria are inactivated.

27.3.2 *Tesgüino*

Tesgüino is a maize-based alcoholic fermented beverage traditionally prepared from germinated maize or maize stalk juice in northern and northwestern Mexico. *Tesgüino* is prepared and consumed by several ethnic groups of this region during family reunions and celebrations. The typical process consists of steeping the corn kernels for several days; the grains are drained and placed in baskets to germinate (in a dark space or hole in the ground until uncolored sweet seedlings are developed: malted corn). Subsequently, the sprouted grains are ground in a manual metal or stone mill (such as *metate*), boiled in water (~8 h), and fermented in a clay pot with regional herbs and barks added (such as *Bromus arizonicus*, *Phaseolus metcalfei*, *Ariocarpus fissuratus*, *Lophophora williamsii*, *Agave schotti*, *Randia echinocarpa*, *R. watsonii*, *R. laevigata*, or *Coutarea pterosperma*, among others) as a catalyst (Herrera and Ulloa 1973; Romo 1996; Lorence-Quiñones et al. 1999). *Tesgüino* requires between 2 and 3 days at room temperature to complete the fermentation and produces ~3.7% of ethanol, ~0.4 of lactic acid, and 0.1 % of acetic acid (Lorence-Quiñones et al. 1999; Chaves-López et al. 2020). *Tesgüino* should be distinguished from *tejuino*, which is developed with dough from nixtamalized maize (a mestizo version of *tesgüino*) (Lappe et al. 1989).

In *tesgüino* production, a lactic-alcoholic fermentation followed by alcoholic-acetic fermentation was identified. Homofermentative and heterofermentative LAB (*Lactobacillus*, *Leuconostoc*, *Pediococcus*, and *Streptococcus*) start the fermentation providing the refreshing, acidic, and acrid flavor of the beverage. *Saccharomyces cerevisiae*, *Saccharomyces kluyveri*, *Candida guilliermondii*, and *Hansenula anomala* are responsible for alcohol production in *tesgüino*. The yeasts are preserved on the surfaces of clay pots and utensils, which are uniquely used to prepare *tesgüino* (Herrera and Ulloa 1973; Lorence-Quiñones et al. 1999).

Nava-Arenas et al. (2008) used blue corn grains to study a germination process similar to those used to prepare *tesgüino*, and phenolics and anthocyanins were determined in the malted grains. After 96 h of germination, phenolics and anthocyanins increased 253% and 155% with values of 3800 and 775 mg/kg, respectively, whereas antioxidant activity increased 56%. These findings revealed the potential contribution of antioxidants to *tesguïno* if blue corn is used to prepare it.

27.3.3 Vanilla

Vanilla is one of the most-used natural flavors in ice creams, chocolates, cakes, confectionery, soft drinks, and liquors. Vanilla is extracted from cured pods of *Vanilla planifolia*, *Vanilla fragans*, or *Vanilla tahitensis*. The extract or cured pods provide a fruity and floral flavor combined with a deep and aromatic body (Sinha et al. 2008). *V. planifolia* is most valued due to its quality and yield. The Aztecs developed the domestication of crops and the curing process of vanilla pods. They used vanilla to flavor their drinks.

The total changes from raw material to finished product, which means the entire vanilla process, is determined by curing. Curing steps involve various natural vegetative processes in the raw material to favor metabolic reactions related to the production of aromatic flavoring. Specifically, there are two main curing steps in the vanilla process: (1) changes through loss of water and (2) changes related to chemical transformation (De la Cruz et al. 2009). Metabolic changes are closely linked to microbiota developed in vanilla beans, hence fermentation. The traditional curing process included four steps: killing, sweating, drying, and conditioning. The killing step aims to destroy the cell tissue structure of vanilla beans by heating (using the sun, oven, or hot water) or freezing. If hot water (blanching) is used during the killing, the process takes place under high humidity and temperature conditions (1.5–5 min in hot water or air below 80°C) (Xu et al. 2020). In the sweating step (36–48 h in a wooden box), high moisture content should be retained to allow the enzyme-catalyzed reaction to develop the primary flavor compounds. Then, beans are sun- or air-dried in the hot sun (from 12 noon to 15 h) from 5–7 to 12–14 days depending on the vanilla bean grade (De la Cruz et al. 2009). Following this, slow drying consists of spreading the beans in racks in a well-ventilated room at ~35°C (RH ~70%) for 2–8 days to 20–35 days, according to the bean vanilla grade. Finally, vanilla beans are stored for 2–3 months to develop the unique aroma of vanilla (De la Cruz et al. 2009; Xu et al. 2020). Vanilla pods or extract (chopped vanilla pods percolating or macerating with ethyl alcohol and water) are commercially available for flavoring. Commercial natural vanilla extract contains ~1.0 g/L vanillin.

The microbial genus and species result from the handling and environmental conditions of the process, similar to occurance in other spontaneous fermentation processes. Recently, Xu et al. (2020) isolated and identified the microbiota developed during the four steps of the vanilla process. They found *Bacillus* strains in all stages, increasing or decreasing in relative abundance from the blanching to conditioning stage. In addition, *Lactococcus*, *Streptococcus*, and *Pseudomonas* contributed to the fermenting process of vanilla, since they were relatively intermediate in abundance. Regarding the fungal community, *Aspergillus* dominated the curing stages (including fermentation) of vanilla beans. *Aspergillus* increased its relative abundance from blanching to conditioning step, reaching ~90%. *Tausonia*, *Nectria*, and *Mrakia* were also present in relatively high abundance in the conditioning stage; hence, they contributed to flavor development. Other bacteria and fungi genera isolated and identified during the vanilla process were *Ralstonia, Methylobacterium, Pelagibacterium, Kluyvera, Acinetobacter, Acetobacter, Pantoea, Caulobacter, Halomonas, Gluconobacter, Jatrophihabitants, Nesterenkonia, Bradyrhizobium, Sphingomonas, Actinomycetospora, Lachnoclostricium, Meyerozyma, Fusarium, Schizophyllum, Wickerhamomyces, Cladosporium, Aeremonium, Ochroconis, Zasmidium, Pseudocercospora, Cyphellophora, Hyalocladosporiella, Mycosphaerella, Candida, Penicillium, Toxicocladosporium*, and *Pseudoteratosphaeria* (Xu et al. 2020).

Various medicinal properties were attributed to vanilla pods in Mesoamerica. Vanilla pods were used to treat fever, spasm, dysmenorrhea, and hysteria in Venezuela and Argentina (Sinha et al. 2008). The Aztecs considered vanilla an aphrodisiac and stimulant. Also, the cholesterol-lowering activity of vanilla had been reported, and it was related to its regulatory effect on genes involved in cholesterol metabolism (low-density lipoprotein receptor and 3-hydroxy-3-methylglutaryl-coenzyme A reductase) (Krishnasree and Andallu 2013). Vanilla contains phytochemicals (vanillin, p-hydroxybenzaldehyde, vanillic acid, and p-hydroxybenzoic acid), which have redox properties and act as free radical scavengers, chelators of metal ions, and single oxygen quenchers; hence, vanilla is considered an antioxidant (Kaur et al. 2020). *In vitro* tests of DPPH (1,1-diphenyl-2-picrylhydrazyl), hydroxyl, superoxide, and nitric oxide radical scavenging activity of methanol extracts from vanilla pods (500 µg/mL) demonstrated its antioxidant activity (Krishnasree and Andallu 2013). This scavenging activity was attributed to its compounds, such as eugenol, tannins, vanillic acid, p-hydroxy benzoic acid, and vanillin, which acted in synergistic form. Despite vanillin exhibiting antioxidant and great medicinal properties, the natural extract has shown higher antioxidant properties than synthetic vanillin (Kaur et al. 2020).

Several studies have investigated the medicinal properties of vanillin and vanillic acid in animal models and humans. However, these properties are briefly mentioned since vanillin is primarily produced by chemical synthesis; hence, it is not the focus of this chapter. Vanillin has been recognized as pain relief, anti-anxiety, antidepressant, protecting against nerve damage and neurodegeneration, and correcting blood glucose and lipid dysregulation in animal models (Xu et al. 2015; Singletary 2020). Vanillin (100 mg/kg) had antidepressant activity in mice (Shoeb et al. 2013). Vanillin showed antifungal activity against *Candida albicans* (clinical isolates) and *Cryptococcus neoformans* (isolated from human or avian excretes) (Boonchird and Flegel 1982). However, high doses were required to inhibit the fungal growth, excluding its use as a therapeutic agent for systemic mycoses due to its short-term toxicity in mice. In a rat model, vanillin reduced the number of colon tumors induced by various agents (Akagi et al. 1995). Vanillin also has shown antioxidant activity in rat liver mitochondria, protecting proteins and lipids from oxidation caused by photosensitization. Thus, vanillin may prevent oxidative damage to membranes in mammalian tissues (Kamat et al. 2000).

In humans, vanillin had a significant soothing effect on the expression of distress after blood draws in premature and full-term infants after olfactory stimulation (Rattaz et al. 2005; Sadathosseini et al. 2013; Jebreili et al. 2015). Vanillin inhibited chromosomal aberrations caused by ultraviolet radiation on the human chromosome by suppressing the mutation of the *CD59* locus (Keshava et al. 1998). o-vanillin is capable of interacting directly with hemoglobin S (in red blood cells, RBC) to inhibit its polymerization or on hemoglobin to reduce cation loss (maintained RBC hydration), leading lower concentration of hemoglobin S. Thus, *o*-vanillin can contribute to ameliorating sickle cell disease (Hannemann et al. 2014). Although several benefits for vanillin are described, its rapid metabolism resulting in vanillic acid in humans and the short half-life of vanillic acid (0.95 h) in the blood (Syngletare 2020) move it away from use as a therapeutic agent.

27.3.4 *Pozol*

Pozol is a maize-based fermented dough used to prepare a nonalcoholic beverage traditionally produced in the southeastern region of Mexico such as Chiapas, Campeche, Tabasco, and Yucatán (Maya region). The beverage is an important component of the diet and traditional medicine of the inhabitants of this region. Mayans used *pozol* to treat fever, intestinal complaints, diarrhea, and skin infections (Moran et al. 2010). Kernels of white maize are cooked in limewater (nixtamalization), washed to remove pericarps, and coarsely ground. Then the heterogeneous dough is kneaded and shaped into balls (5–8 cm diameter) wrapped in banana leaves to allow fermentation. Its fermentation time ranges from 2 to 7 days at room temperature (27–34°C), and the end product pH varies from 4.7–3.6 (Nuraida et al. 1995). The beverage is prepared in water with a ratio of *pozol* dough:water of 1:2 and 1:3 (Lorence-Quiñones et al. 1999). Figure 27.1 shows nixtamalized corn kernels, fresh dough, and the *pozol* beverage.

According to Ampe et al. (1999), lactic acid bacteria dominate the fermentation process of *pozol*; *Lactococcus*, *Leuconostoc*, and *Streptococcus* are the main genera in the early stage of fermentation (12 h), whereas *Lactobacillus* predominates (~93%) at the end of the process (5 days). Subsequently, other studies confirmed that the predominant microbiota are LAB using 16S rDNA sequence analysis (Escalante et al. 2001) or a metaproteomic approach (Cárdenas et al. 2014). Recently, interesting findings were reported about *pozol* microbiota since increments in nitrogen content (8–9%) and protein were observed during the early 9 to 24 h of fermentation. The increments were related to the activity of nitrogen-fixing bacteria. Bacteria responsible for nitrogen fixation were mainly from the genus *Klebsiella* (*Klebsiella variicola*, *Klebsiella pneumoniae*, *Klebsiella quasipneumoniae*). Although *Klebsiella* is enterobacteria and its presence in pozol indicates concerns, *K. pneumoniae* from *pozol* had different capabilities than a clinical isolate (Rizo et al. 2020). The analysis of similar enterobacteria isolates in other fermented foods has demonstrated their safety. For instance, *K. pneumoniae* and *Citrobacter freundii* (vitamin B_{12} producers isolated from tempeh) were analyzed with PCR to assess

FIGURE 27.1 Raw materials to prepare *pozol* and *pulque* beverages and their potential functional properties (health benefits).

their safety. The results showed these bacteria lacked genes that encode selected enterotoxins; hence, they are safe to use in food (Keuth and Bisping 1994).

The main component of maize kernel is starch, which provides up to ~70% of kernel weight, and simple sugars such as glucose, sucrose, and fructose represent 1–3% (FAO 1992). Complex carbohydrates are in the pericarp, the tip cap, endosperm cell walls, and the germ cell walls (FAO 1992). Therefore, the first energy source for microorganisms is starch in corn dough at the beginning of the fermentation. Amylolytic LAB was reported during the fermentation process, maintaining counts ~10^8 during 96 h (Ben Omar and Ampe 2000). Amylolytic LAB content at the beginning of the fermentation is ~40% and decreases to 3% after 72 h (Guyot 2010). Díaz-Ruiz et al. (2003) isolated 257 LAB strains with amylolytic activity from *pozol*, of which 40 strains were conspicuous. *Streptococcus bovis* was the most abundant amylolytic bacteria and produced high amounts of low-molecular-weight maltooligosaccharides from dough in the early fermentation stage. *Streptococcus infantarius* ssp. *infantarius* 25124 isolated from *pozol* also presented amylolytic activity and production of maltooligosaccharides (Bolaños-Núñez et al. 2021). Following this, amylolytic LAB and non-amylolytic microorganisms degraded the starch, and complex carbohydrates resulted in short-chain carbohydrates or sugars available for other bacteria to synthesize exopolysaccharides (EPSs). EPSs are microbial biopolymers secreted into the extracellular environment. Their cellular function is protected against environmental stresses during fermentation, and their production is regulated transcriptionally to mediate the induction or repression of genes (Chaves-López et al. 2020). Ben Omar and Ampe (2000) observed exopolysaccharide-producing bacteria during the *pozol* fermentation process (96 h) growing in the periphery of dough balls that reached counts of 10^9 CFU/g. *Leuconostoc citreum* CW28 (a bacteria isolated from *pozol*) produced an inulin-like polysaccharide with β-21 and β-26 linkages in a sucrose medium. The carbon atoms' properties were similar to those of inulin, and the polysaccharide was digested with inulinase; hence, it has potential prebiotic activity (Olivares-Illana et al. 2002; Olvera et al. 2017). *Weissella confusa* L9 hydrolyzed xylan to form xylooligosaccharides during fermentation and then used them as a carbon source to grow (López-Hernández et al. 2018). Xylooligosaccharides promote colon and metabolic function and reduce food intake, resulting in a tool to manage weight in humans (Carvalho et al. 2013). *Weissella* strains with the capability to produce xylooligosaccharides can confer multiple health benefits to consumers (López-Hernández et al. 2018). Although no quantification of maltooligosaccharides, inulin, and xylooligosaccharides in pozol dough or beverage has been performed, they are probably present and exert benefits as prebiotics.

On the other hand, selected strains of *pozol* microbiota produced various antimicrobial compounds. Bacilysin (an antibiotic of proteinaceous nature), chlorotetaine (a similar compound to bacilysin), and iturin A (complex compound) are antibacterial (Gram-positive and Gram-negative bacteria) and antifungal substances that are produced by *Bacillus* ssp. CS93 isolated from *pozol* (Phister et al. 2004). The proteinaceous compounds were stable at a wide range of pH (3 to 11) and heat-stable at 121°C (115 min) (Ray et al. 2000). In 2010, Moran et al. detected 12 lipopeptides in the cell-free supernatant of *Bacillus* sp. CS93 with antibacterial activity, which probably contributed to the medicinal properties of *pozol*.

The traditional recognition of *pozol* benefits and the production of polysaccharides (with prebiotic potential) and antimicrobial substances have led researchers to study the potential probiotic properties of bacteria isolated from *pozol*. Rodríguez et al. (2011) evaluated the probiotic potential of eight strains of LAB isolated from *pozol* (three *Streptococcus* strains, two strains of *Leuconostoc. pseudomesenteroides*, one *Weissella paramesenteroides*, one *Lactococcus lactis* ssp. *lactis*, and one *Lactobacillus* spp.). The eight strains were resistant to bile salts, survived simulated gastrointestinal tract conditions, and produced H_2O_2 in different amounts. A total of 75% of the strains survived low pH, 50% tolerated phenol (0.4%), and 87.5% showed adherence to Hep2 cells (localized, diffuse, and aggregative were the adherence patterns observed). Another study investigated four LAB isolated from *pozol* and tested for growing at acidic pHs (4.0–5.0, at 37°C for 24 h), in bile salts (0.1–0.3%, at 37°C for 24 h), and in sodium chloride (NaCl; 2, 4, 7, and 10%, at 37°C for 24 h) and evaluated their antagonistic activity against *Salmonella* sp. to demonstrate their probiotic potential (Velázquez-López et al. 2018). The four bacteria survived at least 55% after 24 h at low pH (pH = 4.0), maintained their viability (~85%) after bile salt treatment, and preserved 5.6×10^4 CFU/mL counts at 4% of NaCl. Regarding antimicrobial activity, the four isolates had a positive inhibition effect against *Salmonella*. Although it is difficult to establish probiotic effects from fermented food, previous studies showed selected *pozol* microbiota fulfill various probiotic properties.

In vivo studies on *pozol* benefits are very scarce. However, Muñoz Cano et al. (2013) evaluated the supplementation of *pozol* (as a traditional food, 30%) in rats. They observed *pozol* reduced significantly plasma triglycerides and alanine amino transferase in rats induced with metabolic syndrome in a long-term study (28 weeks, fed with 30% sucrose). Despite the interesting results, the functional component was not identified or related to these effects.

One concern about *pozol* microbiota is their antibiotic resistance. Some researchers have investigated the resistance of LAB isolated from *pozol* to antibiotics and their capability to encode multidrug-resistant efflux pumps genes that can be transferred to pathogens. For example, Wacher-Rodarte et al. (2016) assessed the multi-antibiotic resistance of LAB isolated from *pozol* and discovered several strains with multi-antibiotic resistance. However, more studies focused on the phenotypic and molecular characterization of the resistance mechanism are necessary to identify their risk.

27.3.5 *Pulque* and *Xaxtle*

27.3.5.1 Pulque

Pulque is a traditional white, viscous, fermented alcoholic beverage (alcohol content ranged from 4% to 7%) obtained from the mead called *aguamiel* (*Agave* sap). *Pulque* pH ranges from 3.5 to 4.2. Mead is extracted from young (6–12 years) maguey agaves such as *Agave salmiana*, *A. mapisaga*, *A. feroz*, *A. atrovirens*, and *A. americana* (Gómez-Aldapa et al. 2012). For mead extraction, an immature floral stem is removed to form a cavity where mead excretes from the maguey leaves and is stored until collected (commonly every ~12 h). The period of mead collection can extend by 2–6 months. After mead is removed, the cavity walls are scraped with a sharp tool to remove 0.5 cm of wall thickness of tissue and promote mead secretion. Fermentation of mead begins slowly in the collection cavity where microorganisms are present. The fermentation step continues in containers where mead is placed, inoculated with a portion (20 to 80% of the mixture) of *pulque* known as "seed", and stored at 16–30°C for 24–48 h (Gómez-Aldapa et al. 2012). During fermentation, acetic acid (71.5 mM) and lactic acid (19 mM) also were produced (Escalante et al. 2008). Recently, a slight variation in *pulque* preparation was reported in one community of Michoacan where mead was boiled (1 min) and cooled (~1 h) before *pulque* "seed" was added (Álvarez-Ríos et al. 2020). According to the manufacturers, this step process (boiled) improved the flavor and *pulque* safety. Figure 27.1 shows agave plants and *pulque*.

Pulque is consumed in religious ceremonies and familiar celebrations for its nutritional and health-promoting properties. In addition, the beverage has been recognized for its medicinal properties since ancient times. Currently, it is part of the traditional pharmacopeia (Rojas-Rivas et al. 2020). According to Backstrand et al. (2002) *pulque* is a good source of iron (1.1 mg/100 g) and ascorbic acid (5.7 mg/100 g). In rural areas, this beverage is the primary source of ascorbic acid. At the same time, it is the third source of non-heme iron (only following tortillas and legumes).

Furthermore, *pulque* consumption was related to high plasma ferritin in nonpregnant Mexican women. In addition, *pulque* provides riboflavin, other B vitamins, and steroidal saponins (selected as bioactive compounds) (Backstrand et al. 2002). In addition, the combination of *pulque* and corn tortilla intake increased the bioavailability *in vitro* of iron and zinc due to *pulque* phytase activity (Tovar et al. 2008).

The *pulque* microbial community has been studied for almost 70 years. Although microbiota development in *pulque* is influenced by environmental factors such as season of the year, *Agave* species (Villarreal Morales et al. 2019), place of production, and handling, among others, a typical profile of microbiota dominates in *pulque*. Recent findings showed the microbial core for *pulque* (*Lactobacillus, Leuconostoc, Weissella, Lactococcus, Acetobacter, Gluconobacter, Zymomonas, Obesumbacterium, Kazachstania, Kluyveromyces, Saccharomyces*, and *Hanseniaspora*) is present in the maguey tissue (*metzal*), mead, and *pulque* fermentation stages, and these selected bacteria are recognized as maguey endophytes (Astudillo-Melgar et al. 2023). Table 27.1 shows the main microorganisms reported in *pulque*. Bacteria are the predominant group, with a wide variety of genera and strains (extensively studied). In contrast, fungi are in minor proportion, and they were less investigated. However, the cultivated counts of both groups are similar, 1.90×10^8 CFU/mL for bacteria and 1.47×10^8 CFU/mL for yeasts (Álvarez-Ríos et al. 2020). *Pulque* microbiota is capable of growing and proliferating in foods when it was incorporated as an ingredient. Wheat dough formulated and fermented with *pulque* promoted the growth of *Lactobacillus* strains (*Lb. paracasei* ssp. *paracasei* (new name *Lacticaseibacillus paracasei*) and *Lb. plantarum*; (Quintero et al. 2014) that can provide beneficial metabolites or postbiotics in the resulting bread. However, further studies are necessary to verify and identify these compounds and their benefits.

Selected bacteria from *pulque* have been studied to show their functional properties and probiotic potential since *pulque* is recognized as health promoting and used to treat gastrointestinal disorders. Table 27.2 displays the properties of selected bacteria isolated from *pulque* with probiotic potential. *Ln. mesenteroides* P45 highlights the potential probiotic microorganisms isolated from *pulque* since its genome was sequenced and published (Riveros-Mckay et al. 2014). Therefore, the analysis of specific genes can be related to probiotic properties in its

TABLE 27.1 Microorganisms Isolated and Identified in *Pulque*

ANALYSIS TECHNIQUE	*PLACE OF PULQUE PRODUCTION AND TIME OF FERMENTATION*	*MICROORGANISMS ISOLATED AND IDENTIFIED*	*REFERENCE*
16S rDNA	Tizayuca, Hidalgo State, Mexico. 6 h of fermentation.	*Lactobacillus* spp., *Lb. acidophilus*, *Lb. kefir*, *Lb. acetotolerans*, *Lb. plantarum*, *Leuconostoc mesenteroides* ssp. *mesenteroides*, *Microbacterium arborescens*, *Flavobacterium johnsoniae*, *Zymomonas mobilis*, *Hafnia alvei*	Escalante et al. (2004)
16S rDNA	Huitzilac, Morelos State, Mexico. 6 h of fermentation.	*Lactobacillus* spp., *Ln. mesenteroides*, *Lb. acidophilus*, *Enterobacter agglomerans*, *Lb. hilgardii*, *Lb. sanfranciscensis*, *Acetobacter malorum*, *Lactococcus lactis* spp. *cremoris*, *Leuconostoc citreum*, *Saccharomyces cerevisiae*, *Z. mobilis*, *Acetobacter orientalis*, *Providencia* spp., *Chryseobacterium* spp., *Bacillus* sp., *Klyuvera ascorbata*	Escalante et al. (2008)
Shotgun metagenomic sequencing	Huitzilac, Morelos State, Mexico. ~12 h of fermentation.	*Zymomonas* (21.48%), *Leuconostoc* (14.3%), *Saccharomyces* (13.51%), *Lactococcus* (13.03%), *Lactobacillus* (7.53%), and *Acinetobacter* (2.85%)	Chacón-Vargas et al. (2020)
Metagenomic DNA analysis	Huitzilac, Morelos State, Mexico. 24 h of fermentation.	*Proteobacteria* (~70% in relative abundance, *Zymomonas* (~36%), *Blastomonas*, *Sphingobium*, *Sphingopyxis*, *Sphingomonas*, *Sphingosinicella*, *Bartonella*, *Novosphingobium*, *Telmatospirillu*, *Acinetobacter*): *Z. mobilis*, *S. japonicum*, *S. chlorophenolicumm*, *S. alaskensis*, *S. wittichii*, *B. australis*, *B. quintana*, *B. bacilliformis*, *B. tribocorum*, *B. clarridgeiae*, *N. aromaticivorans*, *A. baumannii*, *A. oleivorans*, *A. pittii*. *Firmicutes* (~28%, genus *Leuconostoc*, *Lactococcus*, *Lactobacillus*, *Fructobacillus*): *Ln. mesenteroides*, *Ln. gelidum*, *Ln. carnosum*, *Ln. citreum*, *Ln. kimchii*, *Lc. lactis*, *Lc. garvieae*, *Lb. acidophilus*, *Lb. brevis*, *Lb. buchneri*, *Lb. casei* group, *Lb. crispatus*, *Lb. delbrueckii*, *Lb. fermentum*, *Lb. gasseri*, *Lb. helveticus*, *Lb. johnsonii*, *Lb. kefiranofaciens*, *Lb. plantarum*, *Lb. reuteri*, *Lb. rhamnosus*, *Lb. ruminis*, *Lb. sakei*, *Lb. salivarius*, *Lb. sanfranciscensis*. Minority groups (<1%) belong to Cyanobacteria, such as *Actinobacteria*, *Fusobacteria*, *Acidobacteria,* and *Nitrospira*	Escobar-Zepeda et al. (2020)
16S rDNA	Hidalgo State, Mexico. 24 h of fermentation.	Bacteria genus: *Sphingomonas*, *Lactobacillus*, *Acetobacter*, *Leuconostoc*, *Lactococcus*, *Acinetobacter*, *Serratia*, *Weissella*, *Synechococcus*, *Streptomyces*, *Sphingopyxis*, *Gluconobacter*, *Euzebya*, *Dysgonomonas*, *Arcobacter*. Fungi: *Kluyveromyces marxianus*, *Kluyveromyces* sp., *Saccharomyces cerevisiae*, *Kazachstania gamospora*, *Dekkera anomala*, *Westerdykella globosa*, *Verrucaria baldensis*, *Pezizomycotina* sp. *Candida ethanolica*, *Arthroascus schoenii*, *Alternaria porri*, *Aureobasidium leucospermi*, *Galactomyces geotrichum*, *Pyrenochaeta inflorescentiae*, *Penicillium carneum*, *Clavispora lusitaniae*, *Candida zemplinina*	Rocha-Arriaga et al. (2020)
16S rRNA	Huitzilac, Morelos, Mexico. 12–14 h fermentation.	Bacteria: *Zymomonas, Acetobacter, Weissella, Leuconostoc, Lactococcus, Obesumbacterium Gluconobacter*, and *Lactobacillus*. Fungi: *S. cerevisiae, K. marxianus, K. gamospora.*, *Hanseniaspora* sp., and O-Saccharomycetales	Astudillo-Melgar et al. (2023)

TABLE 27.2 Specific Capabilities and Functional Properties Identified in Microorganisms Isolated from *Pulque*

MICROORGANISM AND PLACE OF PULQUE PRODUCTION	*SPECIFIC ABILITIES AND FUNCTIONAL PROPERTIES*	*REFERENCE*
Kluyveromyces marxianus Mineral de la Reforma, Hidalgo State, Mexico	Survival to bile salts (0.05 to 0.30%, 24 h at 37°C): 99% Survival to low pH (pH=1.5 to 2.5, 24 h at 37°C): ~47% to 71% Survival to simulated gastric juice (0.2% NaCl, pH 1.5, 0.32% pepsine, 2 h): 92% Antimicrobial activity: inhibition of *Klebsiella pneumoniae* (clinical isolated, 50 μL of 10^8 CFU/mL by spot-on-lawn method) *K. marxianus* was capable to implant in the small and large intestine obtained from mice model CD-1 *K. marxianus* colonized the gastrointestinal tract of mice model CD-1 (10^8 CFU in 200 μL was administer via intraesophageally for 7 days) reached counts of 10^2 to 10^3 CFU/mL in the small and large intestine	Mendoza-Gardeazábal (2013)
Lactobacillus casei (new name *Lacticasiebacillus casei*) subsp. *paracasei* J57 Jocotitlan, State of Mexico, Mexico	Non-hemolytic activity Survival to acidity (water pH 1.5 for 4 h): 60% Survival to simulated gastric juice (0.3% pepsin, 0.5% NaCl, pH 2.0, 4 h): 55% Survival to bile salts (0.3% salt, 4 h): taurocholic acid (70%), glycocholic acid (70%), deoxycholic acid (60%), oxgall (60%) Bile salt hydrolase activity on glycocholic acid (44.9 U/mg of protein), glycodeoxycholic acid (45.2 U/mg of protein), taurocholate (671.7 U/mg of protein), and taurodeoxycholate (61.5 U/mg of protein)	González-Vázquez et al. (2015)
Leuconostoc mesenteroides P45 Huitzilac in Morelos State, Mexico	Survival to lysozyme: ~90% and ~71% after 30 and 120 min at 37°C, respectively Survival to low pH (pH = 2.5, 24 h at 37°C): ~75% Survival to bile salts (0.3 to 1%, 24 h at 37°C): 100% Antimicrobial activity (supernatants, cell-to cell contact, and cell-to cell contact in favorable conditions of exopolysaccharides production) against: enteropathogenic *Escherichia coli* 2348/69, *Listeria monocytogenes*, and *Salmonella* Typhi ATCC 9992 and Typhimurium ATCC 14028 *In vivo* activity against the invasion of *S.* Typhimurium L1334 (str^r) in mouse liver and spleen: Anti-infective effect was observed in counts from dissected liver and spleen after oral administration ($2x10^9$ CFU/mL) 7 days before *Salmonella* infection A pre-bacteriocin and 6 peptidoglycan hydrolase enzymes with antimicrobial activity were identified in the genome of the bacteria	Giles-Gómez et al. (2016)
Lactobacillus plantarum (new name *Lactiplantibacillus plantarum*) and *Lactobacillus brevis* (new name *Levilactobacillus brevis*) El Venado, Mineral de la Reforma, Hidalgo, Mexico	Strains P24–7 (*Lb. plantarum*) and P-24–8 (*Lb. brevis*) Survival to simulated gastric juice (pH 2, 1% pepsine, 2 h):95–96% Survival to bile salts (0.3%) and pancreatin for 24 h (after gastric juice exposition): 52.5% and 65.3% Strains P24–1, P24–2, P24–3, and P24–4 (*Lb. plantarum*) had antibacterial activity *in vitro* against *E. coli* ATCC 25922 and *Staphylococcus aureus* ATCC 29213 Strains P24–2 (*Lb. plantarum*) and P24–8 (*Lb. brevis*) showed *in vitro* total inhibition growth of *Helicobacter pylori* ATCC 43504 and inhibited its urease activity Strain P24–6 (*Lb. plantarum*) inhibited fully *H. pylori* ATCC 43504 and one clinical isolate (1L) *in vitro*	Cervantes-Elizarrarás et al. (2019)
Lacticaseibacillus paracasei (RVG1, UTMB4), *Lactiplantibacillus plantarum* (RVG2, UTMB1), *Liquorilactobacillus ghanensis* (RVG6), *Levilactobacillus brevis* (UTMB2) San Jerónimo Zoyatitlanapan, Juan N. Méndez, Puebla, Mexico Santa Catarina Otzolotepec, Tehuacán, Puebla, Mexico Tamazulapan. Oaxaca, Mexico	The bacteria subjected to a pH 2 for 3 h showed survival rates of 61–87% (log reduction (CFU/mL) from 3.5 to 1.1). The strains RVG6, UTMB1, UTMB2 y UTMB4 were more resistant than the probiotic reference (*Lb. acidophilus* NCFM) LAB exposed to 0.3% or 0.5% bile salts (37°C for 24 h) presented survival rates from 77 to 97% and 67 to 92%, respectively; the reductions oscillated between 3 and 0.7 log. The strain UTMB2 exhibited the highest survival (92%) at 0.5% bile salts All the bacteria presented antimicrobial activity against different pathogens (*S.* Typhi, *L. monocytogenes*, and *E. coli*) LAB RVG3 y UTMB2 showed good antimicrobial peptide production with antimicrobial activity against *Lactobacillus sakei* ATCC 15521 Al the bacteria presented peptidoglycan hydrolase activity against non-viable cells of *Micrococcus lysodeikticus* *Liquorilactobacillus ghanensis* RVG6 presented the best performance for tests	Ruíz-Ramírez et al. (2022)

genome. Recently, the relative abundance of *Ln. mesenteroides* (0.0427%) were corroborated in *pulque* by metagenomic approach, highlighted its contribution as probiotic (Escobar-Zepeda et al. 2020). *Leuconostoc citreum* and *Lb. brevis*, producers of bacteriocins and gut-brain-related metabolites (as potential probiotics), were also detected in significant relative abundance in *pulque* using a metagenomic approach (Escobar-Zepeda et al. 2020).

The production of EPS in *pulque* is primordial, since it aids the development of textural properties, inherent viscosity, and cloudiness of the final product (which is its distinctive characteristic). *Ln. mesenteroides* IBT-PQ and *Lb. plantarum* isolated from *pulque* were able to produce viscosity in the presence of sucrose (Chellapandian et al. 1998). *Lb. plantarum* was a levansucrase producer as well as *Zymomonas mobilis*, whereas *Ln. mesenteroides* IBT-PQ was a dextransucrase producer. The dextran produced by *Ln. mesenteroides* has a linear backbone (α16 D-Glcp) with branching (α13 D-Glcp). The capability of dextransucrase associated with *Ln. mesenteroides* IBT-PQ related to oligosaccharide yield and dextran yield was similar to the industrial strain (*Ln. mesenteroides* NRRL-B512F) (Chellapandian et al. 1998). Torres-Rodríguez et al. (2014) also reported dextran and levan polymers produced by two strains of *Leuconostoc kimchii* isolated from *pulque*. One strain of *Ln. kimchii* EPSA produced dextran (soluble and cell-associated) with a linear backbone (α16 D-Glcp) and branching (α12 and α13), whereas *Ln. kimchii* EPSB synthetized a polymer mixture composed of a linear dextran (soluble) and a polymer from the cell-associated fraction consist of 79% levan and 21% dextran. *Kluyveromyces lactis* var. *lactis* P7 isolated from *pulque* is an excellent inulinase producer able to form fructooligosaccharides from inulin used as prebiotics (Cruz-Guerrero et al. 2006). *Pulque* analysis revealed that 5.3 g/L of fructose constituted the polymeric structure (Escalante et al. 2008). The presence of enzymes involved in polysaccharide synthesis (dextran, levan) or hydrolysis of agave's inulin in *pulque* that produced polysaccharides (prebiotics) was recently confirmed by Escobar-Zepeda et al. (2020) using metagenomic analysis. Dextrans of different molecular weights and degrees of branching had shown increasing *Bifidobacterium* spp., demonstrating their potential prebiotic activity (Sarbini et al. 2014). These polysaccharides have also been related to other beneficial properties such as antitumor agents, immuno-stimulators, and blood cholesterol-lowering benefits (Patel et al. 2012).

LAB resistance in fermented foods is of concern since the resistance can horizontally transfer to pathogens. González-Vázquez et al. (2015) isolated and identified various LAB from *pulque* and tested against common antibiotics. *Lb. casei* subsp. *paracasei* (new name *Lacticaseibacillus casei* subsp. *paracasei*) J57 was resistant only to cefepime and sensitive to 11 antibiotics. In another study, eight LAB isolated from *pulque* were susceptible to various antibiotics (amikacin, amoxicillin, chloramphenicol, gentamicin, and spectinomycin), three of them were moderately resistant to levofloxacin, one was moderately resistant to tetracycline, and all were resistant to vancomycin (Cervantes-Elizarrarás et al. 2019). Most LAB have been demonstrated an intrinsic vancomycin-resistant which is associated with their typical peptidoglycan which poorly binds to vancomycin (Zhang et al. 2018), instead of acquired resistance.

One concern about *pulque* safety is its content of enterobacteria due to handling. However, the studies that assessed the *pulque* microbiota have found few enterobacteria genera or low relative abundance. Escalante et al. (2008) reported Enterobacteriales in *pulque* such as *Enterobacter agglomerans*, *Enterobacter* spp., *Erwinia rhapontici*, *Kluyvera ascorbate*, *Kluyvera cochleae*, *Providencia* spp., and *Serratia grimensii*. Low relative abundance of *Serratia* spp. and *Enterobacter* genera was detected in *pulque* microbiota as analyzed by Rocha-Arriaga et al. (2020). Bacterial pathogens such as *Pseudomonas aeruginosa*, *Salmonella enterica*, *Yersinia* spp., coliform bacteria (*Klebsiella* spp., *Citrobacter* spp., *Serratia* spp., *Enterobacter* spp., and *E. coli*), and potential pathogens (*Streptococcus mutans* and *Streptococcus salivarius*) were found in very low relative abundance in *pulque* using a metagenomic approach, confirming the safety of *pulque* for human consumption (Escobar-Zepeda et al. 2020).

27.3.5.2 Xaxtle

The sediments accumulated during *pulque* fermentation are called *xaxtle*. It is rich in LAB and yeasts; one gram of lyophilized *xaxtle* had 7×10^6 CFU/g and 4×10^5 CFU/g, respectively (Torres-Maravilla et al. 2016a). Thus, *xaxtle*, as well as pulque, can be a source of potential probiotics. In addition, *xaxtle* could be a consumption option to overcome the disadvantages of ethanol intake from *pulque*.

Torres-Maravilla et al. (2016b) isolated 14 *Lactobacillus* and evaluated their survival to lysozymes, low pH, and bile salts and their hydrophobicity, adherence to intestinal epithelial cells, immunomodulatory secretion of IL-8, and anti-inflammatory activity *in vivo*. Table 27.3 exhibits the main findings of Torres-Maravilla et al. (2016b) study. Four LAB showed immunomodulatory secretion IL-8 activity by HT-29 cells; even more UV-killed bacteria maintained their anti-inflammatory activity, blocking up to 30% IL-8 secretion (showing their postbiotic activity). Three strains, *Lactobacillus sanfranciscensis* LBH1068 (new name *Fructilactobacillus sanfranciscensis*), *Lactobacillus composti* LBH1072, and *Lb. plantarum* LBH1064, showed anti-inflammatory properties *in vivo* in a murine DNBS-induced colitis model, limiting mice weight loss and even increasing body weight (8–13%). The strains also significantly reduced intestinal hyper-permeability, demonstrating their barrier function, protecting from enteric bacteria. The strains had a systemic modulation of cytokine production in the tested model; the three isolates exhibited significant decreases in IL-5 secretion. Moreover, the systematic modulation of cytokine production was analyzed in the splenocytes from NDBS-treated mice and the results showed that secretion of IFN-γ was reduced only by *Lb. sanfaranciscensis* LBH1068 and *Lb. composti* LBH1072, IL-17A was inhibited by *Lb. sanfranciscensis* LBH1068, whereas IL-22 (cytokine responsible for the pro-inflammatory gene expression in patients with inflammatory bowel disease) secretion was reduced only by *Lb. composti* LBH1072. In addition, the same strains capable

TABLE 27.3 Specific Capabilities and Functional Properties Identified in Microorganisms Isolated from *Xaxtle*

PLACE OF XAXTLE PRODUCTION AND ANALYSIS TECHNIQUE TO IDENTIFY THE BACTERIA	*MICROORGANISMS ISOLATED AND IDENTIFIED*	*SPECIFIC ABILITY OF LACTIC ACID BACTERIA STRAINS*	*REFERENCE*
Milpa Alta (Cd. de Mexico), Santiago Villalta and Nanacamilpa (Tlaxcala) API 50CHL and 16 rRNA	*Lactobacillus plantarum* (new name *Lactiplantibacillus plantarum*) LBH1062, *Lb. plantarum* LBH1063, *Lb. plantarum* LBH1064, *Lactobacillus paracasei* ssp. *paracasei* (new name *Lacticaseibacillus paracasei*) LBH1065, *Lb. paracasei* ssp. *paracasei* LBH1066, *Lb. paracasei* ssp. *paracasei* LBH1067 *Lactobacillus sanfranciscensis* (new name *Fructilactobacillus sanfranciscensis*) LBH1068, *Lactobacillus brevis* (new name *Levilactobacillus brevis*) LBH1069, *Lb. brevis* LBH1070, *Lb. brevis* LBH11071, *Lactobacillus composti* (new name *Agrilactobacillus composti*) LBH1072, *Lb. brevis* LBH1073, *Lb. composti* LBH1074, *Lb. plantarum* LBH1075	Survival to lysozyme (≥95%): *Lb. plantarum* LBH1062 and *Lb. brevis* LBH1069. Survival to low pH (90 min at pH 2.0, ~85%): *Lb. brevis* LBH1073 and *Lb. sanfranciscensis* LBH1068. Survival to bile salt (0.3%) ≥85%: *Lb. plantarum* LBH1064, *Lb. sanfranciscensis* LBH1068, and *Lb. brevis* LBH1069. Hydrophobicity (bacterial adherence to solvents): *Lb. plantarum* LBH1064, *Lb. paracasei* ssp. *paracasei* LBH1065, *Lb. composti* LBH1072, and *Lb. composti* LBH1074. Adhesion to intestinal epithelial cells (HT-29): *Lb. brevis* LBH1070. Adhesion to intestinal epithelial cells HT-29-MTX: *Lb. plantarum* LBH1062, *Lb. plantarum* LBH1063, *Lb. paracasei* ssp. *paracasei* LBH1065. Immunomodulation of IL-8 secretion by HT-29 cells: *Lb. plantarum* LBH1062, *Lb. plantarum* LBH1064, *Lb. composti* LBH1072, and *Lb. plantarum* LBH1075.	Torres-Maravilla et al. (2016b)

of tolerating 0.4% of phenol showed reductions of 2–3 log cycles CFU/mL (Torres-Maravilla et al. 2016c). Phenol resistance is important since phenols can be derived from the deamination of aromatic amino acids (from dietary sources) in the gut and exert a bacteriostatic effect against LAB. *Lb. plantarum* (LBH1062, LBH1063, LBH1064, LBH1075), *Lb. paracasei* ssp. *paracasei* (LBH1065), *Lb. sanfranciscensis* (LBH1068), and *Lb. composti* (LBH1072) exhibited strong inhibition against *Escherichia coli* O157:H7 and *Salmonella* Typhimurium ATCC 14028. *Lb. brevis* (LBH11071) strongly inhibited the growth of *Candida albicans* ATCC 10231. Five strains (*Lb. paracasei* ssp. *paracasei* LBH1066 and LBH1067, *Lb. composti* LBH1072, *Lb. brevis* LBH1070 and LBH1073) had a significant ($p < 0.05$) anti-proliferative effect on HT-29 cells. Meanwhile, *Lb. composti* LBH1072 was capable of immune-modulating IL-2 (~20 pg/mL) and IL-6 (~60 pg/mL) on macrophage cell line RAW 264.7. Recently, the relative abundance of *Lb. sanfranciscensis* (0.41%) was corroborated in *pulque* by a metagenomic approach (Escobar-Zepeda et al. 2020) and consequently in *xaxtle*.

Xaxtle can partially replace the yeast in bread formulation. Torres-Maravilla et al. (2016a) evaluated the quality of bread produced with *xaxtle* (addition 1–10 g of dried *xaxtle*, fermentation time from 30–90 min). They obtained a good bread using 3 g of freeze-dried *xaxtle* and 90 min of fermentation. Bread formulated with *xaxtle* had lower values of total starch (10%) and a lower glycemic index (predicted values) compared with bread fermented by *Saccharomyces cerevisiae* or commercial bread. In another study, three samples of *xaxtle* (Santiago Villalta, Tlaxcala State; Nanacamilpa, Tlaxcala State; Milpa Alta, Mexico City, Mexico) were analyzed and used to formulate Italian bread straight from dough. Three main species of LAB were identified (using PCR and API 50CH techniques) in the *xaxtle* samples (*Lactobacillus pentosus* (new name *Lactiplantibacillus pentosus*), *Lb. paracasei* ssp. *paracasei*, and *Lb. brevis*) and a yeast (*S. cerevisiae*) (Sánchez-Pardo et al. 2016). Meanwhile, five LAB species were identified in bread doughs formulated with *xaxtle* (*Lb. paracasei* ssp. *paracasei*, *Lb. plantarum*, *Lb. brevis*, *Lb curvatus* ssp. *curvatus* (new name *Lactilactobacillus curvatus* ssp. *curvatus*), and *Ln. mesenteroides*). *Xaxtle* may improve bread with LAB metabolites secreted during fermentation and as postbiotics. Kuligowski et al. (2019) analyzed the bacteria and yeasts of *xaxtle* using 16S rRNA and 26S rRNA methods, respectively. The sequences matched to *Bacillus subtilis*, *Lactobacillus* spp. (*buchneri*, *kefiri*, *otakinesis*), *Ln. pseudoesenteroides*, *Saccharomyces* spp. (*paradoxus*, *cariocanus*, *mikatae*, *cerevisiae*), and *Z. mobilis*.

Bacteria with probiotic properties isolated from *pulque* and *xaxtle* exhibited greater survival and beneficial properties than commercial probiotics, as shown by studies by González-Vázquez et al. (2015), Torres-Maravilla et al. (2016b, 2016c), Giles-Gómez et al. (2016), and Cervantes-Elizarrarás et al. (2019).

27.3.6 *Tepache*

Currently, *tepache* is prepared by fermenting pineapple *(Ananas comosus)* peel submerged in water with piloncillo (unprocessed sugar cane similar to molasses) or sugar and spiced with cinnamon or pepper. Fermentation takes place in wooden barrels at room temperature ranging from 1 to 5 days, obtaining a refreshing sweet beverage. The first microorganisms isolated from *tepache* were the yeasts *Candida queretana*, *Torulopsis inconspicua*, and *S. cerevisiae* (Herrera and Ulloa 1978). *S. cerevisiae* is the predominant yeast (70%), followed by *Candida apicola* (9%), *Cryptococcus skinneri* (9%), and *Hanseniaspora* (6%) (Corona-González et al. 2013). Recently, Gutiérrez-Sarmiento et al. (2022) analyzed the microbial community structure of *tepache* (using 16S rRN and ITS–5.8 rDNA methods) and recorded that *Firmicutes* phylum (*Leuconostoc* (29%), *Lactococcus* (39%), *Lactobacillus* and *Weisella* (29%)) bacteria dominate the microbiota at the end of fermentation (72 h). Fungal community of *tepache* after 72 h of fermentation was dominated for *Ascomycota* phylum which include *Gibberella* (5%), *Candida* (2%), *Saccharomycetaceae*, *Zygosaccharomyces* (5%), *Saccharomyces* (30–50%), *Meyerozyma* (<2%), *Trichocomaceae*, *Talaromyces* (10%), *Epicoccum* and *Kabatiella* (>0.5%)) (Gutiérrez-Sarmiento et al. 2022). Corona-González et al. (2013) evaluated different temperature levels, initial pH, percentage of piloncillo, and fermentation time to optimize the overall acceptance (sensory analysis) of *tepache* using a central composite design. The formulation with 13% piloncillo, initial pH of 4.0, fermented by 54 h at 22°C were the optimal conditions predicted by the model to obtain a *tepache* with a better level of acceptance.

Lb. plantarum (TPI13) isolated from homemade *tepache* exhibited moderate antimicrobial activity against *Listeria monocytogenes* 7644, Scott A, G19; *Vibrio cholerae* 5921, and *Vibrio parahaemolyticus* 10691 (Alvarado et al. 2006). De la Fuente-Salcido et al. (2015) isolated several LAB from *tepache*. They evaluated their resistance to acidic pH (4.0 and 5.0), growth at 4.0% and 6.5% of NaCl, growth at 15°C or 45°C, antibacterial activity, and bacteriocin production to demonstrate their probiotic potential. More than 60% of the isolated strains grew at the tested conditions (pH, NaCl, and temperature). Regarding the antibacterial activity, ≥42 strains inhibited Gram-positive bacteria (*L. monocytogenes*, *Listeria innocua*, and *Bacillus subtilis*), whereas only 6 strains restricted Gram-negative bacteria (*Samonella* spp., *Salmonella* Typhimurium, *E. coli*, and *K. pneumoniae*). Two bacteria were identified to harbor genes to synthesize nisin and enterocin (*Lc. lactis* TeA1 and *Enterococcus faecium* TeA2, respectively). Moreover, the production of bacteriocins was demonstrated as well as their antibacterial activity against *Micrococcus luteus*, *L. monocytogenes*, *L. innocua*, *Staphylococcus aureus*, *Bacillus cereus*, *B. subtilis*, *Enterococcus faecalis*, and *K. pneumoniae*. Escorza et al. (2016) isolated and identified five LAB strains from *tepache* (*Streptococcus* spp., two bacillus producers of polysaccharides, and two bacillus). One strain survived after 2 h of exposure to pH = 2.0 (83%), 0.3% bile salts (95.92%), and artificial gastric juice formulated with pH = 2.5 and 1000 units of pepsin at 37°C (97.73%). Recently, an isolated bacteria designed as *Lactobacillus pentosus* ABHEAU-05 from *tepache* was also tested to grow in bile salts (0.3%), acid pH (pH = 2.0), resistance to *in vitro* digestion (pH = 2.5, 1000 IU of pepsin incubated for 2 h at 37°C), and the ability to grow and survive on milk during 21 days at 5°C (Escobar-Ramírez et al. 2020). The tested bacteria survived 92.3% in bile salts, 82.8% in acidic conditions, and 97.6% in enzymatic activity. Regarding its growth in milk, the counts increased two logarithmic cycles after 28 h of incubation and maintained counts of 8.3 $\log_{10}$ CFU/mL after 21 days at 5°C.

27.3.7 *Colonche*

Colonche is a red, sweet, low-alcohol beverage obtained from the spontaneous fermentation of the red prickly pear, mainly produced by prickly pear Cardona (*Opuntia streptacantha*). If *O. streptacantha* is not available, other species can be used, such as *O. leucotricha*, *O. orbiculata*, and *O. robusta* (Ulloa and Herrera 1978). It is typically produced in the central region of Mexico (Aguascalientes, Guanajuato, San Luis Potosí, and Zacatecas). In the south region, *O. pilifera* and various cacti species (*Pachycereus weberi*, *Escontria chiotilla*, *Stenocereus* spp., and *Polaskia* spp.) are also utilized (Ojeda-Linares et al. 2023). The traditional process begins by collecting the prickly pears, removing the peel, squeezing the juice, straining the juice through an ixtle sieve (to remove the seeds), then boiling the juice 2–3 h and allowing fermentation for a few days in clay pots (used exclusively for *colonche* production). The fermentation occurs at room temperature (25 to 30°C) (Ojeda-Linares et al. 2020). A recent alternative process introduced by young women is through the inoculated fermentation which includes the addition of previous batches to boil juice and ferment for 4 h in plastic containers (Ojeda-Linares et al. 2022). *Colonche,* after a few hours of fermentation, is sweet and gaseous, with a pleasant taste. The fermentation of *colonche* is achieved by LAB, *Candida valida*, *S. cerevisiae*, and *Torulopsis taboadae* (Godoy et al. 2003). In addition, *Pichia fermentans* were isolated from *colonche* and were responsible for producing several aromatic compounds (Gutiérrez-Moreno et al. 2010). A current microbial analysis of spontaneous *colonche*, *colonche* using inoculum from previous batches, and clay pots showed that clay pots are a rich microbial reservoir for the fermentation process. Spontaneous *colonche* has a great microbial diversity compared with inoculated *colonche*. Spontaneous *colonche* microbiota include bacteria *Lactobacillus, Leuconostoc, Weissella, Tanticharoenia, Tatumella, Rheinheimera, Pediococcus, Gluconacetobacter, Faecalibacterium, Dorea, Aeromonas, Massilia, Elizabethkingia, Clostridium, Bacillus,* and *Alkanindiges* and fungi *Dekkera, Kazachstania, Kluyveromyces, Knufia, Aureobasidium, Candida, Cladosporium, Pichia, Saccharomyces, Torulaspora, Rhodotorula*, and *Wickerhamomyces*. For *conlonche* fermented using inoculum (previous batches), the microbial richness was limited to seven bacteria and two yeast genera dominated by

Lactobacillus, Leuconostoc, Alkanindiges, Pediococcus, Saccharomyces, and *Kazachstani* (Ojeda-Linares et al. 2022).

Although no functional properties for *colonche* have been studied, prickly pears are recognized for their traditional medicinal properties to treat ulcers, dyspnea, glaucoma, liver illnesses, wounds, and fatigue (Ojeda-Linares et al. 2020). Red prickly pears contain betalains responsible for the red-yellow color, which are considered antioxidants (heat and pH stable). *O. streptacantha* are rich in betalains (betacyanins, betaxanthins) and total phenols, with values of ~3500 μg and 115 mg of equivalent gallic acid/100 g in fresh fruit, respectively (Yahia and Mondragon-Jacobo 2011). Therefore, *colonche* could provide antioxidant properties such as in red wine due to betalains. Tesoriere et al. (2004) evaluated the bioavailability, absorption, and antioxidant effects on humans after consuming *Opuntia ficus-indica* pulp fruit (500 g, 28 mg indicaxanthin, 16 mg betanin). Betanin and indicaxanthin were detectable in plasma extracts after 60 min of ingestion, whereas the maximum concentration was reached after 3 h, and the compounds disappeared after 12 h. Indicaxanthin bound to low-density lipoproteins (LDLs) after 3 h and until 5 h after the ingestion of fruit. In contrast, betanin only was detected in LDL after 3 h. Moreover, LDL-proteins isolated 3 h after fruit ingestion showed a marked elongation of the time preceding the onset of oxidation (~30%) when exposed to an oxidant environment *ex vivo* (copper-induced) than those LDL-proteins isolated before the ingestion of the fruit (Tesoriere et al. 2004). Thus, indicaxanthin and betanin are probably involved in LDL protection against induced oxidative modifications. Madrigal-Santillán et al. (2013) demonstrated that red-purple prickly pear juice had antioxidant activity, resulting in the anticlastogenic effect in mouse erythrocytes.

27.4 CONCLUDING REMARKS

Pulque, *xaxtle*, vanilla, and *pozol* are widely studied, and their microbiota and fermentation process have been well established. However, studies related to *pulque* and *pozol* health benefits are only for *in vitro* models and scarcely *in vivo* (mainly culture cells and animals). The health benefits of vanilla extracts also have been investigated poorly. The investigation of medicinal properties has been primarily focused on vanillin, the main component of vanilla extract (mostly its synthetic form). The fermentation process and the microbiota of *atole agrio* and *tepache* were recently investigated and reported on. *Tesgüino* and *colonche* are poorly studied, and their fermentation process is not well known. Therefore, the health benefits of *atole agrio*, *tepache*, *tesgüino*, and *colonche* are unknown and only related to empirical theories. In addition, studies to assess the benefits between the intake of traditional Mexican fermented foods and health benefits in consumers are necessary to verify their benefits.

REFERENCES

Akagi, K., Hirose, M., Hoshiya, T., Mizoguchi, Y., Ito, N., and T. Shirai. 1995. Modulating effects of ellagic acid, vanillin and quercetin in a rat medium term multi-organ carcinogenesis model. *Cancer Letters* 94(1):113–121. https://doi.org/10.1016/0304-3835(95)03833-I

Alvarado, C., García, A. B. E., Martin, S. E., and C. Regalado. 2006. Food-associated lactic acid bacteria with antimicrobial potential from traditional Mexican foods. *Revista Latinoamericana de Microbiología* 48(3–4):260–268.

Álvarez-Ríos, G. D., Figueredo-Urbina, C. J., and A. Casas. 2020. Physical, chemical, and microbiological characteristics of pulque: Management of a fermented beverage in Michoacán, Mexico. *Foods* 9:361. https://doi.org/10.3390/foods9030361

Ampe, F., ben Omar, N., and J.-P. Guyot. 1999. Culture-independent quantification of physiologically-active microbial groups in fermented foods using rRNA-targeted oligonucleotide probes: Application to pozol, a Mexican lactic acid fermented maize dough. *Journal of Applied Microbiology* 87:131–140. https://doi.org/10.1046/j.1365-2672.1999.00803.x

Astudillo-Melgar, F., Hernández-Chávez, G., Rodríguez-Alegría, M. A., Bolívar, F., and A. Escalante. 2023. Analysis of the microbial diversity and population dynamics during the pulque fermentation process. *Fermentation* 9:342. https://doi.org/10.3390/fermentation9040342

Backstrand, J. R., Allen, L. H., Black, A. K., de Mata, M., and G. H. Pelto. 2002. Diet and iron status of nonpregnant women in rural Central Mexico. *The American Journal of Clinical Nutrition* 76:156–164. https://doi.org/10.1093/ajcn/76.1.156

Ben Omar, N., and F. Ampe. 2000. Microbial community dynamics during production of the Mexican fermented maize dough pozol. *Applied and Environmental Microbiology* 66(9):3664–3673. https://doi.org/10.1128/AEM.66.9.3664-3673.2000

Bolaños-Núñez, S., Santiago-Urbina, J. A., Guyot, J.-P., Díaz-Ruiz, G., and C. Wacher. 2021. Microbial interactions between amylolytic and non-amylolytic lactic acid bacteria strains isolated during the fermentation of pozol. *Foods* 10(11):2607. https://doi.org/10.3390/foods10112607

Boonchird, C., and T. W. Flegel. 1982. *In vitro* antifungal activity of eugenol and vanillin against *Candida albicans* and *Cryptococcus neoformans*. *Canadian Journal of Microbiology* 28:1235–1241. https://doi.org/10.1139/m82-184

Cárdenas, C., Barkla, B. J., Wacher, C., Delgado-Olivares, L., and R. Rodríguez-Sanoja. 2014. Protein extraction for the proteomic study of a Mexican traditional fermented starchy food. *Journal of Proteomics* 111:139–147. https://doi.org/10.1016/j.jprot.2014.06.028

Carvalho, A. F. A., Neto, P. de O., da Silva, D. F., and G. M. Pastore. 2013. Xylo-oligosaccharides from lignocellulosic materials: Chemical structure, health benefits and production by chemical and enzymatic hydrolysis. *Food Research International* 51(1):75–85. https://doi.org/10.1016/j.foodres.2012.11.021

Cervantes-Elizarrarás, A., Cruz-Cansino, N. S., Ramírez-Moreno, E., Vega-Sánchez, V., Velázquez-Guadarrama, N., Zafra-Rojas, Q. Y., and J. Piloni-Martini. 2019. *In vitro* probiotic potential of lactic acid bacteria isolated from aguamiel and pulque and antibacterial activity against pathogens. *Applied Sciences* 9(3):601. https://doi.org/10.3390/app9030601

Chacón-Vargas, K., Torres, J., Giles-Gómez, M., Escalante, A., and J. G. Gibbons. 2020. Genomic profiling of bacterial and fungal communities and their predictive functionality during pulque fermentation by whole-genome shotgun sequencing. *Scientific Reports* 10:15115. https://doi.org/10.1038/s41598-020-71864-4

Chaves-López, C., Rossi, C., Maggio, F., Paparella, A., and A. Serio. 2020. Changes occurring in spontaneous maize fermentation: An overview. *Fermentation* 6(1):36. https://doi.org/10.3390/fermentation6010036

Chellapandian, M., Larios, C., Sanchez-Gonzalez, M., and A. Lopez-Munguia. 1998. Production and properties of a dextransucrase from *Leuconostoc mesenteroides* IBT-PQ isolated from *pulque*, a traditional Aztec alcoholic beverage. *Journal of Industrial Microbiology & Biotechnology* 21:51–56. https://doi.org/10.1038/sj.jim.2900560

Chen, M., Sun, Q., Giovannucci, E., Mozaffarian, D., Manson, J. E., Willet, W. C., and F. B. Hu. 2014. Dairy consumption and risk of type 2 diabetes: 3 cohorts of US adults and an updated meta-analysis. *BMC Medicine* 12:215. https://doi.org/10.1186/s12916-014-0215-1

Corona-González, R. I., Ramos-Ibarra, J. R., Gutiérrez-González, P., Pelayo-Ortiz, C., Guatemala-Morales, G. M., and E. Arriola-Guevara. 2013. The use of response surface methodology to evaluate the fermentation conditions in the production of tepache. *Revista Mexicana de Ingeniería Química* 12(1):19–28.

Correa-Ascencio, M., Robertson, I. G., Cabrera-Cortés, O., Cabrera-Castro, R., and R. P. Evershed. 2014. Pulque production from fermented agave sap as a dietary supplement in Prehispanic Mesoamerica. *Proceedings of the National Academy of Sciences* 111(39):14223–14228. https://doi.org/10.1073/pnas.1408339111

Cruz-Guerrero, A. E., Olvera, J. L., García-Garibay, M., and L. Gómez-Ruiz. 2006. Inulinase-hyperproducing strains of *Kluyveromyces* sp. isolated from aguamiel (*Agave* sap) and pulque. *World Journal of Microbiology & Biotechnology* 22:115–117. https://doi.org/10.1007/s11274-005-9005-4

Cuvas-Limon, R. B., Nobre, C., Cruz, M., Rodriguez-Jasso, R. M., Ruíz, H. A., Loredo-Treviño, A., Texeira, J. A., and R. Belmares. 2020. Spontaneously fermented traditional beverages as a source of bioactive compounds: An overview. *Critical Reviews in Food Science and Nutrition* 61(18):2984–3006. https://doi.org/10.1080/10408398.2020.1791050

Czerucka, D., Piche, T., and P. Rampal. 2007. Review article: Yeast as probiotics—*Saccharomyces boulardii*. *Alimentary Pharmacology & Therapeutics* 26:767–778. https://doi.org/10.1111/j.1365–2036.2007.03442.x

De la Cruz, M. J., Rodriguez, J. G. C., García, H. S., Rosado, Z. T. L., García, A. M. A., and O. V. J. Robles. 2009. *Vanilla: Post-Harvest Operations*. Rome: Food and Agriculture Organization of the United Nations.

De la Fuente-Salcido, N. M., Castañeda-Ramírez, J. C., García-Almendárez, B. E., Bideshi, D. K., Salcedo-Hernández, R., and J. E. Barboza-Corona. 2015. Isolation and characterization of bacteriocinogenic lactic acid bacteria from M-Tuba and Tepache, two traditional fermented beverages in Mexico. *Food Science & Nutrition* 3(5):434–442. https://doi.org/10.1002/fsn3.236

Díaz-Ruiz, G., Guyot, J. P., Ruiz-Teran, F., Morlon-Guyot, J., and C. Wacher. 2003. Microbial and physiological characterization of weakly amylolytic but fast-growing lactic acid bacteria: A functional role in supporting microbial diversity in pozol, a Mexican fermented maize beverage. *Applied and Environmental Microbiology* 69(8):4367–4374. https://doi.org/10.1128/AEM.69.8.4367–4374.2003

Díaz-Ruiz, G., Valderrama, A., Esquivel, K., Espinosa, J., Centurión, D., Reyes-Duarte, D., and C. Wacher. 2013. Microbial populations in atole agrio: A traditional Mexican fermented maize beverage. Proceedings of the XVII National Congress of Biotechnology and Bioengineering, Cancún, Mexico.

Escalante, A., Giles-Gómez, M., Hernández, G., Córdova-Aguilar, M. S., López-Munguía, A., Gosset, G., and F. Bolívar. 2008. Analysis of bacterial community during the fermentation of pulque, a traditional Mexican alcoholic beverage, using a polyphasic approach. *International Journal of Food Microbiology* 124:126–134. https://doi.org/10.1016/j.ijfoodmicro.2008.03.003

Escalante, A., Rodríguez, M. E., Martínez, A., López-Munguía, A., Bolívar, F., and G. Gosset. 2004. Characterization of bacterial diversity in Pulque, a traditional Mexican alcoholic fermented beverage, as determined by 16S rDNA analysis. *FEMS Microbiology Letters* 235(2):273–279. https://doi.org/10.1111/j.1574–6968.2004.tb09599.x

Escalante, A., Wacher, C., and A. Farrés. 2001. Lactic acid bacterial diversity in the traditional Mexican fermented dough pozol as determined by 16S rDNA sequence analysis. *International Journal of Food Microbiology* 64(1–2):21–31. https://doi.org/10.1016/S0168–1605(00)00428–1

Escobar-Ramírez, M. C., Jaimez-Ordaz, J., Escorza-Iglesias, V. A., Rodríguez-Serrano, G. M., Contreras-López, E., Ramírez-Godínez, J., Castañeda-Ovando, A., Morales-Estrada, A. I., Felix-Reyes, N., and L. G. González-Olivares. 2020. *Lactobacillus pentosus* ABHEAU-05: An in vitro digestion resistant lactic acid bacterium isolated from a traditional fermented Mexican beverage. *Revista Argentina de Microbiología* 52(4):305–315. https://doi.org/10.1016/j.ram.2019.10.005

Escobar-Zepeda, A., Montor, J. J., Olvera C., Sanchez-Flores, A., and A. Lopez-Munguia. 2020. An extended taxonomic profile and metabolic potential analysis of pulque microbial community using metagenomics. *Journal of Food Science & Technology* 5(2):83–97. DOI:*10.25177/JFST.5.2.RA.10637*

Escorza, I. V. A., Jaimez, O. J., Contreras, L. E., Castañeda, O. A., Añorve, M. J., Ramírez, G. J., and O. L. G. González. 2016. Lactic acid bacteria isolation from a pineapple fermented drink (*Tepache*) and determination of their potential probiotic ability. In Proceedings of the 7th International Symposium on Probiotics, Ciudad de México, México.

FAO, Food and Agriculture Organization of the United Nations. 1992. *Maize in Human Nutrition*. Rome: FAO.

Giles-Gómez, M., Sandoval, G. J. G., Matus, V., Campos, Q. I., Bolívar, F., and A. Escalante. 2016. In vitro and in vivo probiotic assessment of *Leuconostoc mesenteroides* P45 isolated from pulque, a Mexican traditional alcoholic beverage. *SpringerPlus* 5:708. https://doi.org/10.1186/s40064-016-2370-7

Godoy, A., Herrera, T., and M. Ulloa. 2003. *Más allá del pulque y el tepache. Las bebidas alcohólicas no destiladas indígenas de México*. México: UNAM. http://hdl.handle.net/11154/149501

Gómez-Aldapa, C. A., Díaz-Cruz, C. A., Villarruel-López, A., Torres-Vitela, M. R., Rangel-Vargas, E., and J. Castro-Rosas. 2012. Acid and alcohol tolerance of *Escherichia coli* O157:H7 in pulque, a typical Mexican beverage. *International Journal of Food Microbiology* 154:79–84. https://doi.org/10.1016/j.ijfoodmicro.2011.12.027

González-Vázquez, R., Azaola-Espinosa, A., Mayorga-Reyes, L., Reyes-Nava, L. A., Shah, N. P., and Y. Rivera-Espinoza. 2015. Isolation, identification and partial characterization of a *Lactobacillus casei* strain with bile salt hydrolase activity from pulque. *Probiotics and Antimicrobial Proteins* 7:242–248. https://doi.org/10.1007/s12602-015-9202-x

Gutiérrez-Moreno, K., Rodríguez-Lerma, G. K., Cárdenas-Marríquez, M., Botello-Álvarez, E., Jiménez-Islas, H., Rico-Martínez, R., and J. L. Navarrete-Bolaños. 2010. A strategy for biotechnological processes design: Prickly pear (*Opuntia ficus-indica*) wine production. *Chemical Engineering Transactions* 20:315–320. https://doi.org/10.3303/CET1020053

Gutiérrez-Sarmiento, W., Peña-Ocaña, B. A., Lam-Gutiérrez, A., Guzmán-Albores, M., Jasso-Chávez, R., and V. M. Ruíz-Valdiviezo. 2022. Microbial community structure, physicochemical characteristics and predictive functionalities of the Mexican tepache fermented beverage. *Microbiological Research* 260:127045. https://doi.org/10.1016/j.micres.2022.127045

Guyot, J.-P. 2010. Fermented cereal products. In *Fermented Foods and Beverages of the World*, eds. J. P. Tamang and K. Kailasapathy, 247–261. Boca Raton, FL: CRC Press.

Hannemann, A., Cytlak, U. M. C., Gbotosho, O. T., Rees, D. C., Tewari, S., and J. S. Gibson. 2014. Effects of *o*-vanillin on K^+ transport of red blood cells from patients with sickle cell disease. *Blood Cells, Molecules and Diseases* 53:21–26. https://doi.org/10.1016/j.bcmd.2014.02.004

Herrera, T., and M. Ulloa. 1973. *Saccharomyces cerevisiae* una levadura fermentadora del tesgüino de los indios Tarahumaras. *Boletín de la Sociedad Mexicana de Microbiología* 7:33–38. https://doi.org/10.33885/sf.1973.2.427

Herrera, T., and M. Ulloa. 1978. Descripción de una especie nueva de levadura, *Candida queretana*, aislada del tepache de Querétaro, México. *Boletín de la Sociedad Mexicana de Microbiología* 12:13–18. https://doi.org/10.33885/sf.1978.2.477

Jebreili, M., Neshat, H., Seyyedrasouli, A., Ghojazade, M., Hosseini, M. B., and H. Hamishehkar. 2015. Comparison of breastmilk odor and vanilla odor on mitigating premature infants' response to pain during and after venipuncture. *Breastfeeding Medicine* 10(7):362–365. https://doi.org/10.1089/bfm.2015.0060

Kamat, J. P., Ghosh, A., and T. P. A. Devasagayam. 2000. Vanillin as an antioxidant in rat liver mitochondria: Inhibition of protein oxidation and lipid peroxidation induced by photosensitization. *Molecular and Cellular Biochemistry* 209:47–53. https://doi.org/10.1023/A:1007048313556

Kaur, K., Sharma, R., and S. Singh. 2020. Bioactive composition and promising health benefits of natural food flavors and colorants: Potential beyond their basic functions. *Pigment & Resin Technology* 49(2):110–118. https://doi.org/10.1108/PRT-02-2019-0009

Keshava, C., Keshava, N., Ong, T. M., and J. Nath. 1998. Protective effect of vanillin on radiation-induced micronuclei and chromosomal aberrations in V79 cells. *Mutation Research* 397(2):149–159. https://doi.org/10.1016/S0027-5107(97)00203-0

Keuth, S., and B. Bisping. 1994. Vitamin B12 production by *Citrobacter freundii* or *Klebsiella pneumoniae* during tempeh fermentation and proof of enterotoxin absence by PCR. *Applied and Environmental Microbiology* 60(5):1495–1499. https://doi.org/10.1128/aem.60.5.1495-1499.1994

Krishnasree, V., and B. Andallu. 2013. Radical scavenging activity of vanilla (*Vanilla fragans*) pods and commercial vanilla essence. *International Journal of Scientific and Research Publications* 3(11):1–5.

Kuligowski, M., López, O. R., Polanowska, K., Montet, D., Jasinska-Kuligowska, I., and J. Nowak. 2019. Influence of fermentation by different microflora consortia on pulque and pulque bread properties. *Journal of the Science of Food and Agriculture* 99(14):6307–6314. https://doi.org/10.1002/jsfa.9907

Lappe, P., Ulloa, M., and J. Gómez. 1989. Estudio microbiano y cromatográfico del tejuino de Jalisco, México. *Revista Mexicana de Micología* 5:181–203. https://doi.org/10.33885/sf.1989.3.748

Lara Hidalgo, C. E., Hernández-Sánchez, H., Hernández Rodríguez, C., and L. Dorantes-Álvarez. 2017. Yeasts in fermented foods and their probiotic potential. *Austin Journal of Nutrition & Metabolism* 4(1):1045.

Long, A., Benz, B. F., Donahue, D. J., Jull, A. J. T., and L. J. Toolin. 1989. First direct AMS dates on early maize from Tehuacán, Mexico. *Radiocarbon* 3(3):1035–1040. https://doi.org/10.1017/S0033822200012686

López-Hernández, M., Rodríguez-Alegría, M. E., López-Munguía, A., and C. Wacher. 2018. Evaluation of xylan as carbon source for *Weissella* spp. A predominant strain in pozol fermentation. *LWT: Food Science and Technology* 89:192–197. https://doi.org/10.1016/j.lwt.2017.10.030

Lorence-Quiñones, A., Wacher-Rodarte, C., and R. Quintero-Ramírez. 1999. Cereal fermentations in Latin American countries. In *Fermented Cereals, a Global Perspective*, ed. FAO Agricultural Services Bulletin No. 138. Rome: Food and Agriculture Organization of the United Nations.

Madrigal-Santillán, E., García-Melo, F., Morales-González, J. A., Vázquez-Alvarado, P., Muñoz-Juárez, S., Zuñiga-Pérez, C., Sumaya-Martínez, M. T., Madrigal-Bujaidar, E., and A. Hernández-Ceruelos. 2013. Antioxidant and anticlastogenic capacity of prickly pear juice. *Nutrients* 5(10):4145–4158. https://doi.org/10.3390/nu5104145

Manrique-Donrronsoro, R., Díaz-Ruíz, G., and M. C. Wacher-Rodarte. 2016. Probiotic potential of yeasts isolated from atole agrio from Villahermosa, Tabasco. In Proceedings of the 7th International Symposium on Probiotics, Ciudad de México, México.

Manzo-Fuentes, M. I., and D. L. Gómez-Gutierrez. 2019. Difusión de técnica artesanal del atole agrio de Cuauhtémoc, Villaflores. In *El maíz: conocimiento de su patrimonio gastronómico y cultura*, eds. L. Álvarez Rateike and G. Palacios Pola, 123–132. Tuxtla Gutiérrez, Chiapas: Universidad de Ciencias y Artes de Chiapas.

Mendoza-Gardeazábal, A. S. 2013. Caracterización de la levadura *Kluyveromyces marxianus* como microorganismo probiótico. Master dissertation, Universidad Autónoma del Estado de Hidalgo, San Agustín Tlaxiaca, Hidalgo. http://dgsa.uaeh.edu.mx:8080/bibliotecadigital/bitstream/handle/231104/1862/TESIS%20SEL.pdf?sequence=1&isAllowed=y (Accessed on October 26, 2020).

Moran, S., Robertson, K., Paradisi, F., Rai, D. K., and C. D. Murphy. 2010. Production of lipopeptides in *Bacillus* sp. CS93 isolated from pozol. *FEMS Microbiology Letters* 304(1):69–73. https://doi.org/10.1111/j.1574-6968.2009.01882.x

Muñoz Cano, J. M., Aguilar, A. C., and J. C. Hernández. 2013. Lipid-lowering effect of maize-based traditional Mexican food on a metabolic syndrome model in rats. *Lipids in Health and Disease* 12:35. https://doi.org/10.1186/1476-511X-12-35

Nava-Arenas, D., Jiménez-Aparicio, A., and H. Hernández-Sánchez. 2008. Optimization of germination conditions of blue corn (*Zea mays* L.) by Taguchi orthogonal array methodology. *Asian Journal of Plant Sciences* 7(7):682–686. https://doi.org/10.3923/ajps.2008.682.686

Nozue, M., Shimazu, T., Sasazuki, S., Charvat, H., Mori, N., Mutoh, M., Sawada, N., Iwasaki, M., Yamaji, T., Inoue, M., Kokubo, Y., Yamagishi, K., Iso, H., and S. Tsugane. 2017. Fermented soy product intake is inversely associated with the development of high blood pressure: The Japan Public Health Center-Based prospective study. *The Journal of Nutrition* 147(9):1749–1756. https://doi.org/10.3945/jn.117.250282

Nuraida, L., Wacher, M. C., and J. D. Owens. 1995. Microbiology of pozol, a Mexican fermented maize dough. *World Journal of Microbiology & Biotechnology* 11(5):567–571. https://doi.org/10.1007/BF00286375

Ojeda-Linares, C. I., Solís-García, I. A., and A. Casas. 2022. Constructing micro-landscapes: Management and selection practices on microbial communities in a traditional traditional fermented beverage. *Frontiers in Ecology and Evolution* 10:821268. https://doi.org/10.3389/fevo.2022.821268

Ojeda-Linares, C. I., Vallejo, M., and A. Casas. 2023. Disappearance and survival of fermented beverages in the biosphere reserve Tehuacán-Cuicatlán, Mexico: The cases of *tolonche* and *lapo*. *Frontiers in Sustainable Food Systems* 6:1067598. https://doi.org/10.3389/fsufs.2022.1067598

Ojeda-Linares, C. I., Vallejo, M., Lappe-Oliveras, P., and A. Casas. 2020. Traditional management of microorganisms in fermented beverages from cactus fruits in Mexico: An ethnobiological approach. *Journal of Ethnobiology and Ethnomedicine* 16:1. https://doi.org/10.1186/s13002-019-0351-y

Olivares-Illana, V., Wacher-Rodarte, C., Le Borgne, S., and A. López-Munguía. 2002. Characterization of a cell-associated inulosucrase from a novel source: A *Leuconostoc citreum* strain isolated from pozol, a fermented corn beverage of Mayan origin. *Journal of Industrial Microbiology & Biotechnology* 28(2):112–117. https://doi.org/10.1038/sj/jim/7000224

Olvera, C., Santamaría, R. I., Bustos, P., Vallejo, C., Montor, J. J., Wacher, C., and A. Lopez Munguia. 2017. Draft genome sequence of *Leuconostoc citreum* CW28 isolated from pozol, a pre-Hispanic fermented corn beverage. *Genome Announcements* 5:e01283–17. https://doi.org/10.1128/genomea.01283–17

Park, S., and J.-H. Bae. 2016. Fermented food intake is associated with a reduced likelihood of atopic dermatitis is an adult population (Korean National Health and Nutrition Examination Survey 2012–2013). *Nutrition Research* 36(2):125–133. https://doi.org/10.1016/j.nutres.2015.11.011

Patel, S., Majumder, A., and A. Goyal. 2012. Potentials of exopolysaccharides from lactic acid bacteria. *Indian Journal of Microbiology* 52(1):3–12. https://doi.org/10.1007/s12088–011–0148–8

Peréz-Cataluña, A., Elizaquível, P., Carrasco, P., Espinosa, J., Reyes, D., Wacher, C., and R. Aznar. 2018. Diversity and dynamics of lactic acid bacteria in Atole agrio, a traditional maize-based fermented beverage from South-Eastern Mexico, analysed by high throughput sequencing and culturing. *Antonie van Leeuwenhoek* 111:385–399. https://doi.org/10.1007/s10482-017-0960-1

Phister, T. G., O'Sullivan, D. J., and L. L. McKay. 2004. Identification of bacilysin, chlorotetaine, and iturin A produced by *Bacillus* sp. strain CS93 isolated from pozol, a Mexican fermented maize dough. *Applied and Environmental Microbiology* 70(1):631–634. https://doi.org/10.1128/AEM.70.1.631–634.2004

Ponomarova, O., Gabrielli, N., Sévin, D. C., Mülleder, M., Zimgibl, K., Bulyha, K., Andrejev, S., Kafkia, E., Typas, A., Sauer, U., Ralser, M., and K. R. Patil. 2017. Yeast creates a niche for symbiotic lactic acid bacteria through nitrogen overflow. *Cell Systems* 5(4):345–357. https://doi.org/10.1016/j.cels.2017.09.002

Quintero, L. A., Alvarado-Resendiz, M. G., Soto, S. S., Piloni, M. J., Reyes-Santamaria, M. I., and N. Guemes-Vera. 2014. Use of *Lactobacillus* from pulque in sourdough. *Advances in Microbiology* 4:969–977. http://dx.doi.org/10.4236/aim.2014.414108

Rattaz, C., Goubet, N., and A. Bullinger. 2005. The calming effect of a familiar odor on full term newborns. *Journal of Developmental & Behavioral Pediatrics* 26(2):86–92. https://doi.org/10.1097/00004703–200504000–00003

Ray, P., Sanchez, C., O'Sullivan, D. J., and L. L. McKay. 2000. Classification of a bacterial isolate, from pozol, exhibiting antimicrobial activity against several Gram-positive and Gram-negative bacteria, yeasts, and molds. *Journal of Food Protection* 63(8):1123–1132. https://doi.org/10.4315/0362–028x–63.8.1123

Riveros-Mckay, F., Campos, I., Giles-Gómez, M., Bolívar, F., and A. Escalante. 2014. Draft genome sequence of *Leuconostoc mesenteroides* P45 isolated from pulque, a traditional Mexican alcoholic fermented beverage. *Genome Announcements* 2(6):e01130–14. https://doi.org/10.1128/genomea.01130–14

Rizo, J., Rogel, M. A., Guillén, D., Wacher, C., Martinez-Romero, E., Encarnación S., Sánchez, S., and R. Rodríguez-Sanoja. 2020. Nitrogen fixation in pozol, a traditional fermented beverage. *Applied and Environmental Microbiology* 86(16):e00588. https://doi.org/10.1128/AEM.00588–20

Rocha-Arriaga, C., Espinal-Centeno, A., Martinez-Sánchez, S., Caballero-Peréz, J., Alcaraz, L. D., and A. Cruz-Ramírez. 2020. Deep microbial community profiling along the fermentation process of pulque, a biocultural resource of Mexico. *Microbiological Research* 241:126593. https://doi.org/10.1016/j.micres.2020.126593

Rodríguez, A., Villalva, B. Flores, M. T., Sainz, T., Eslava, C., Díaz-Ruiz, G., and M. C. Wacher-Rodarte. 2011. Estudio del potencial probiótico de bacterias ácido lácticas aisladas del pozol. Proceedings of the XIV National Congress of Biotechnology and Bioengineering, Queretaro, Mexico.

Rojas-Rivas, E., Viesca-González, F. C., Favila-Cisneros, H. J., and F. Cuffia. 2020. Consumer's perception of a traditional fermented beverage in Central Mexico. *British Food Journal* 122(2):708–721. https://doi.org/10.1108/BFJ-05-2019-0317

Romero-Luna, H. E., Hernández-Sánchez, H., and G. Dávila-Ortiz. 2017. Traditional fermented beverages from Mexico as a potential probiotic source. *Annals of Microbiology* 67:577–586. https://doi.org/10.1007/s13213-017-1290-2

Romo, A. 1996. *Química, universo, tierra y vida*. D. F., México: Fondo de Cultura Económica.

Ruíz-Ramírez, Y., Guadarrama-Mendoza, P. C., Escalante, A., Giles-Gómez, M., and R. Valadez-Blanco. 2022. Probiotic activity traits in vitro and production of antimicrobial peptides by *Lactobacillaceae* isolates from pulque using *Lactobacillus acidophilus* NCFM as control. *Brazilian Journal of Microbiology* 53(2):921–933. https://doi.org/10.1007/s42770-022-00684-7

Sadathosseini, A. S., Negarandeh, R., and Z. Movahedi. 2013. The effect of a familiar scent on the behavioral and physiological pain responses in neonates. *Pain Management Nursing* 14(4):e196–e203. https://doi.org/10.1016/j.pmn.2011.10.003

Sánchez-Cortés, M. S., Verdugo Valdez, A. G., López Zuñiga, E. J., and G. Vela Gutiérrez. 2019. El atole agrio: su presencia y modo de preparación en la cultura alimentaria de Chiapas. In *El maíz: conocimiento de su patrimonio gastronómico y cultura*, eds. L. Álvarez Rateike and G. Palacios Pola, 87–94. Tuxtla Gutiérrez, Chiapas: Universidad de Ciencias y Artes de Chiapas.

Sánchez-Pardo, M. E., Blancas-Nápoles, J. A., Vázquez-Landaverde, P. A., Nari, A., Taglieri, I., Ortiz-Moreno, A., Mayorga-Reyes, L., Sanmartín, C., Bermúdez-Humarán L. G., and E. Torres-Maravilla. 2016. The use of Mexican xaxtle as leavening agent in Italian straight dough bread making to produce pulque bread. *Agrochimica* 60(4):329–342. https://doi.org/10.12871/00021857201647

Sangwan, S., Kumar, S., and S. Goyal. 2014. Maize utilization in food bioprocessing: An overview. In *Maize: Nutrition Dynamics and Novel Uses*, eds. D. P. Chaudhary, S. Kumar, and S. Langyan, 119–134. New Delhi, India: Springer.

Sarbini, S. R., Kolida, S., Deaville, E. R., Gibson, G. R., and R. A. Rastall. 2014. Potential of novel dextran oligosaccharides as prebiotics for obesity management through *in vitro* experimentation. *British Journal of Nutrition* 112(8):1303–1314. https://doi.org/10.1017/S0007114514002177

Shoeb, A., Chowta, M., Pallempati, G., Rai, A., and A. Singh. 2013. Evaluation of antidepressant activity of vanillin in mice. *Indian Journal of Pharmacology* 45(2):141–144. https://doi.org/10.4103/0253–7613.108292

Singletary, K. W. 2020. Vanilla: Potential health benefits. *Nutrition Today* 55(4):186–196. https://doi.org/10.1097/NT.0000000000000412

Sinha, A. K., Sharma, U. K., and N. Sharma. 2008. A comprehensive review on vanilla flavor: Extraction, isolation and quantification of vanillin and others constituents. *International Journal of Food Sciences and Nutrition* 59(4):299–326. https://doi.org/10.1080/09687630701539350

Tamang, J. P., Shin, D.-H., Jung, S.-J., and S.-W. Chae. 2016. Functional properties of microorganisms in fermented foods. *Frontiers in Microbiology* 7:578. https://doi.org/10.3389/fmicb.2016.00578

Taylor, B. C., Lejzerowicz, F., Poirel, M., Shaffer, J. P., Jiang, L., Aksenov, A., Litwin, N., Humphrey, G., Martino, C., Miller-Montgomery, S., Dorrestein, P. C., Veiga, P., Song, S. J., McDonald, D., Derrien, M., and R. Knight. 2020. Consumption of fermented foods is associated with systemic differences in the gut microbiome and metabolome. *mSystems* 5(2):e00901–19. https://doi.org/10.1128/msystems.00901–19

Tesoriere, L., Allegra, M., Butera, D., and M. A. Livrea. 2004. Absorption, excretion, and distribution of dietary antioxidant betalains in LDLs: Potential health effects of betalains in humans. *The American Journal of Clinical Nutrition* 80(4):941–945. https://doi.org/10.1093/ajcn/80.4.941

Torres-Maravilla, E., Blancas-Nápoles, J. A., Vázquez-Landaverde, P. A., Cristiani-Urbina, E., Mayorga-Reyes, L., and M. E. Sánchez-Pardo. 2016a. Evaluation of pulque sediment (xaxtle) as a starter culture in order to obtain a low glycemic index baked product. *Agrociencia* 50(2):183–200.

Torres-Maravilla, E., Lenoir, M., Mayorga-Reyes, L., Allain, T., Sokol, H., Langella, P., Sánchez-Pardo, M. E., and L. G. Bermúdez-Humarán. 2016b. Identification of novel anti-inflammatory probiotic strains isolated from pulque. *Applied Microbiology and Biotechnology* 100:385–396. https://doi.org/10.1007/s00253-015-7049-4

Torres-Maravilla, E., Muñoz-Provencio, D. M., Curiel-Gómez, E., Azaola, A., Sánchez-Pardo, M. E., Mayorga-Reyes, L., and L. G. Bermúdez-Humarán. 2016c. Anti-proliferative effect on HT-29 cells by novel probiotic strains isolated from pulque. In Proceedings of the 7th International Symposium on Probiotics, Ciudad de México, México.

Torres-Rodríguez, I., Rodríguez-Alegría, M. E., Miranda-Molina, A., Giles-Gómez, M., Conca, M. R., López-Munguía, A., Bolívar, F., and A. Escalante. 2014. Screening and characterization of extracellular polysaccharides produced by *Leuconostoc kimchii* isolated from traditional fermented *pulque* beverage. *SpringerPlus* 3:583. https://doi.org/10.1186/2193-1801-3-583

Tovar, L. R., Olivos, M., and M. E. Gutierrez. 2008. Pulque, an alcoholic drink from rural Mexico, contains phytase. Its in vitro effects on corn tortilla. *Plant Foods and Human Nutrition* 63:189–194. https://doi.org/10.1007/s11130-008-0089-5

Ulloa, M., and T. Herrera. 1978. *Torulopsis taboadae* una nueva especie de levadura aislada del colonche de Zacatecas, México. *Boletín de la Sociedad Mexicana de Microbiología* 12:5–12. https://doi.org/10.33885/sf.1978.2.476

Väkeväinen, K., Valderrama, A., Espinosa, J., Centurión, D., Rizo, J., Reyes-Duarte, D., Díaz-Ruiz, G., von Wright, A., Elizaquível, P., Esquivel, K., Simontaival, A.-I., Aznar, R., Wacher, C., and C. Plumed-Ferrer. 2018. Characterization of lactic acid bacteria recovered from atole agrio, a traditional Mexican fermented beverage. *LWT: Food Science and Technology* 88:109–118. https://doi.org/10.1016/j.lwt.2017.10.004

Velázquez-López, A., Covatzin-Jirón, D., Toledo-Meza, M. D., and G. Vela-Gutiérrez. 2018. Bebida fermentada elaborada con bacterias ácido lácticas aisladas del pozol tradicional chiapaneco. *Ciencia UAT* 13(1):165–178. https://doi.org/10.29059/cienciauat.v13i1.871

Villarreal Morales, S. L., Enríquez Salazar, M. I., Michel Michel, M. R., Flores Gallegos, A. C., Montañez-Saens, J., Aguilar, C. N., and R. R. Herrera. 2019. Metagenomic microbial diversity in aguamiel from two *Agave* species during 4-year seasons. *Food Biotechnology* 33:1. https://doi.org/10.1080/08905436.2018.1547200

Wacher-Rodarte, M. C., Trejo-Muñúzuri, T. P., Montiel-Aguirre, J. F., Drago-Serrano, M. E., Gutiérrez-Lucas, R. L., Castañeda-Sánchez, J. I., and T. Sainz-Espuñes. 2016. Antibiotic resistance and multidrug-resistance efflux pumps expression in lactic acid bacteria isolated from pozol, a nonalcoholic Mayan maize fermented beverage. *Food Science & Nutrition* 4(3):423–430. https://doi.org/10.1002/fsn3.304

Xu, F., Chen, Y., Cai, Y., Gu, F., and K. An. 2020. Distinct roles for bacterial and fungal communities during the curing of vanilla. *Frontiers in Microbiology* 11:552388. https://doi.org/10.3389/fmicb.2020.552388

Xu, J., Xu, H., Liu, Y., He, H., and G. Li. 2015. Vanillin-induced amelioration of depression-like behaviors in rats by modulating manoamine neurotransmitters in the brain. *Psychiatry Research* 225(3):509–514. https://doi.org/10.1016/j.psychres.2014.11.056

Yahia, E. M., and C. Mondragon-Jacobo. 2011. Nutritional components and antioxidant capacity of ten cultivars and lines of cactus pear fruit (*Opuntia* spp.). *Food Research International* 44(7):2311–2318. https://doi.org/10.1016/j.foodres.2011.02.042

Zhang, S., Oh, J.-H., Alexander, L. M., Özçam, M., and J.-P. van Pijkeren. 2018. D-Alanyl-D-Alanine ligase as a broad-host-range counterselection marker in vancomycin-resistant lactic acid bacteria. *Journal of Bacteriology* 200(13):e00607–17. https://doi.org/10.1128/jb.00607–17

Index

Note: Page numbers in *italics* indicate a figure and page numbers in **bold** indicate a table on the corresponding page.

A

aakhone, 123, 126
açai berry, 367–368
acetic acid bacteria (AAB), 51, 52, 419, 420–424
Actinomucor elegans, 117
akhuni, 123, 126, 133, 181
alcoholic beverages, 162, 186, 197, 293, 362
caatinga biome, 358
cacao, 428
cachaça, 380–381
chicha, 163
Chilean
aloja, 345–346
apple chicha, 344
grape chicha, 344
lahueñe, 346–347
mudai, 345
pajarete wine, 345
pipeño wine, 344
pisco, 343–344
tecu, 346
wine, 342–343
mahia, 49–50, *50*
Mexican
pozol, 453
pulque, 455
Northeast Indian
atingba, 168
bhaati jaanr, 168
chhang, 168
kodo ko jaanr, 168–169
sujen, 169
şalgam, 307
sufu, 142, 143, 146–147, 150
tuak, 148–149
alcohols, 102, 105–106, 153, 164
Algerian fermented foods, 3
bacteria, 6
biotechnology application, 13–14
cereals, 5–6
cheese
bouhezza, 4
dhan, 5
ighounane, 5
jben, 4
klila, 4–5
rigouta, 5
smen, 5
zebda, 5
dairy products
lben, 4
lfrik, 4
raib, 4
meat products
gueddid, 5
kadid, 5
ossban, dried, 5
mold, 6
risks of, 13
studies, **8–12**
vegetables, 5–6
yeast, 6
alheira, 241–245, 275, 282–283
alkalization, 424, 427, 439–440
aloja, 345–346
Amazon biome, *367*, 367–370
amine, *see* biogenic amines
amine-degrading enzymes (ADEs), 117
amino acid composition, 98–99, 104, 205, 207
amou, 165
amylolytic starters, 197
amou/perokkushi, 165
bakhar, 165–166
hamei, 166
huamo, 166
ipoh, 166
keem, 166
khekhrii, 166
marcha, 167
mod pitha, 166–167
ranu dabai, 167
ranu goti, 167
thiat, 167
vekur pitha, 167–168
xaj-pitha, 168
Andean potato fermentation, 396
biochemical and microbiological changes, 401–402, **403–404**, 405
chuño, 397–399, *398*, *399*
nutritional composition, 400–401, *401*
future research, 409–410
health benefits, 405, **406–408**, 409
historical background, 396–397
tocosh, 399–400, *400*
nutritional composition, 401
angiotensin-converting enzyme (ACE), 61, 116, 128, 319, 370
animal farming, sweet potato, 38
anti-allergenic effects, 171
antibacterial properties, 6, 52, 170, 204, 312, 318
allyl isothiocyanates, 313
atole agrio, 415
antibiotic resistance/resistant, 13, 131, 133, 156, 246, 320
anti-cancer properties, 114, 132, 146, 170, 172, 315
bioactive peptide, 207
soyasaponins, 116
soybean isoflavone, 116
anti-diabetic properties, 123, 129, 131, 352, 364
antifungal properties, 6, 86, 312, 363, 405
fatty acids, 183
Northeast Indian fermented foods, 170
anti-HIV properties, 130, 135
antihypertensives, 183, 207, 364, 390, 450
soybean, 133
anti-inflammatory properties, 30, 45, 129, 358, 436
Amazon biome, 367
Atlantic Forest biome, 364
Brazilian fruits, 352
cacao, 434
kimchi products, 195
lahueñe, 347
şalgam, 311
soyasaponins, 116
soybean isoflavone, 116
sulforaphane, 298
tecu, 346
xaxtle, 458
antioxidants, 130, 132, 172, 182, 195, 242
anti-thrombotic properties, 129, 136
apple chicha, 344
apple vinegar, 51, *see also* cider vinegar
araçá-boi, 368–369
asparagus fermentation, 295
Aspergillus egyptiacus, 111
Aspergillus oryzae, 111
atingba, 168
Atlantic forest biome, *364*, 364–367
atole agrio, 415, 450–452
axone, 123, 126
ayran, *319*, 319–320
benefits of, 320–321

B

bacaba, 368–369
Bacillus cereus, 22, 125, 135, 245, 283
backslopping, 123
bacuri, 363
bakhar, 165–166
barbari bread, 87
baru/cumbaru, 362
bean curd, *see* sufu
bekang, 126, 128, 129, 181, 183
beverage/s, 20, 142, 428, 450, *see also* alcoholic beverages
Brazilian indigenous, 384–385
Iranian
doogh, 78–79
yogurt, 78–79
mahia, 49
Malaysian
fruit vinegar, 152–153
lassi (yoghurt drink), 147–148, *148*
tuak (palm wine), 148–150, *149*
Northeast Indian, *see* alcoholic beverages
teas, 217–218
Turkish
ayran, 319–321
boza, 315–319
gilaburu, 304–307
hardaliye, 312–315
şalgam (shalgam), 307–311
bhaati jaanr, 168, 183
bile salt hydrolase enzyme (BSH), 53, 170
bioactive peptides, 131–132

Brazilian fermented foods, 390
Moroccan fermented foods, 52
Pakistani fermented foods, 193–194
soybean foods, 131–132
terasi, 114–117
yu-lu fish sauce, 207
bioethanol production, 35
biogenic amines, 13, 90, 102–106, 117, 135
yu-lu fish sauce, 206–207
biomes, 351, 352, 358, **359**
Amazon, *367*, 367–370
Atlantic forest, *364*, 364–367
caatinga, 358–361, *360*
cerrado, 361–364, *362*
Pampa, *370*, 370–371
pantanal, 361–364, *362*
biotechnology, 3, 440–441
blood cholesterol, *see* cholesterol
bone health, 25, 59, 61, 147
boubanita/boubanite, 49
bouhezza, 4
boza, 185, 305, 315–318
benefits of, 318–319
chemical and physical composition, **317**
Brazilian fermented foods, 380, *381*
antimicrobial activity, **390**
cheese, 385–386
coalho, 387
colonial, 387
kochkase, 388
manteiga, 387
marajó, 386–387
minas, 386
serrano, 387
bioactive compounds, 390
cachaça, 383–384
future perspectives, 390–391
indigenous beverages, 384–385
microbiota, 388–389, *389*
products, 380–381, *381*
sour cassava starch, 381–383, *382*
Brazilian fruit fermentation, 351–352
biochemical and nutritional components, *352*, 352–353
biological activity and health benefits, 356, **356–357**, 358
extraction, identification, and quantification, *353*, 353–356, **354–356**
biomes, 358, **359**
Amazon, *367*, 367–370
Atlantic forest, *364*, 364–367
caatinga, 358–361, *360*
cerrado, 361–364, *362*
Pampa, *370*, 370–371
pantanal, 361–364, *362*
bread, 6, 25, 239, 308, 435
chickpea, 332–334, *334*
farinheira da guarda, 278
flat bread/naan, 197–198
Iranian, 86–87
morcela, 278
Pakistani, 195
Pantanal wild plants, 361
pulque microbiota, 455
sourdough, 51
Sweet potato, 20
wheat, 275
xaxtle, 459
yeast, 362
buriti, 368
butter, 5, 165, 386, 427, 441
manteiga cheese, 387
smen, 47
zabda baldya, 47

C

caatinga biomes, 358–361, *360*
Caatinga wild passion fruit, 358
cabbage, 69, 71, 195, 296, 298
cacao, *see* Mexican cacao fermentation
cachaça, 380, 383–384, 388, 389
cacholeira, 272, 280, 282, 283
camu-camu, 36–370
cancer, 25, 32, 89, 321, 366
bioactive peptides, 131
bladder, 196, 331
boza, 318
breast, 127, 306, 389
cervical, 306
cocoa flavanols, 432
colon, 52, 127, 135, 331
colorectal, 304
gastric, 170
green tea, 220
liver, 304
prostate, 127
saponins, 129
sufu, 147
sulforaphane, 298
sweet potato, 25
tempeh, 60
Candida albicans, 7
capers, *48*, 50
cardiovascular disease (CVD), 129, 169–170
cardiovascular health, 58–60, 147, 183, 220
carrots, 50, 69, 195, 307–308
cashew, 360
Cerrado, 362
cassava, 381–383
cauliflower, 69, 71, 195, 197, 401
Greek vegetable fermentation, 294
cereal fermentation
Algerian, 5–6
India, 175–176
compilation and dissemination, 188
economic value, 188
enterococci, 185
functionality and nature, 182–185
LAB, 181–182, **184**
microbiological processes, 186–187
microbiology, 180–181
microflora, **177–180**
millet, 185–186
substrate and pre-fermentation treatments, 187–188
substrate for, 176, 180
Iranian, 86–87
Pakistani, 195
cerrado biomes, 361–364, *362*
cheese, 47, 75, 122, 175, 192
Algerian
bouhezza, 4
dhan, 5
ighounane, 5
jben, 4
klila, 4–5
rigouta, 5
smen, 5
zebda, 5
Brazilian, 385–386
coalho, 387
colonial, 387
kochkase, 388
manteiga, 387
marajó, 386–387
minas, 386
serrano, 387
chhurpi/kachcha chhurpi, 165
colonial, 387
Iranian
khiki, 79, 81
koozeh, *84*, 84–85
lighvan, *85*, 85–86, **86**
poosti, 79, 81, *82*, **83**, *83*
siahmazgi, 81–84, *83*, **84**
Moroccan
jben, 47
klila, 47
Pakistani, 197
sufu, 142, 143, 146–147, 150
chhang, 168
chickpea, 145, 187, 228, 240, 246
Turkish fermented foods, 332–334, *334*
benefits of, 334
Chilean fermented foods, 342
aloja, 345–346
apple chicha, 344
grape chicha, 344
lahueñe, 346–347
mudai, 345
pajarete wine, 345
Pipeño wine, 344
pisco, 343–344
tecu, 346
wine, 342–343
China, 17, 57, 105, 110, 122, *see also* douchi; yu-lu fish sauce
Chingshey purple sweet potato (CPSP), 30, 33, 35
cholesterol, 25, 53, 99, 320, 330, 431
atherosclerosis, 132
cardiovascular health, 60
kinema, 130
lactic acid bacteria, 169–170
liver health, 61
non-meat sausages, 239
and obesity, 61
octopamine, 208
polypeptides, 116
saponins, 128–129
soy protein, 240
taurine, 208
vinegar, 152
chongkukjang, 110, 225, 227, 230–231
agricultural background, 231
birth of, 230
healthy benefits, 234–236
historical backgrounds, 230–231
key words, 229
literature, 229–230
literature backgrounds, 231
misconceptions, 231–232
misinformation, 232–234
'si' character, 234
stories of, 227–228
chouriça, 272, 283–284
chouriço, 272, 276, 278, 283, 285
chuño, 397–399, *398*, *399*
nutritional composition, 400–401, *401*
cider vinegar, see apple vinegar
Citrobacter freundii, 58
Clostridium acetobulyticum, 23
Clostridium beijerinckii, 23

Clostridium perfringes, 24
coalho cheese, 387, 390
CODEX, 36, 55
cognitive function, 58–60, 220, 433–434
colonche, 460–461
colorectal cancer (CRC), 13
cooked non-fermented (CNF), 130
critical control point (CCP), **37**, 37–38
curd cheese, *see* coalho cheese
cyclic dipeptides, 52
Cylas formicarius, 19

D

dahi, 147, 148, 164, 165, 172
dairy products, 142–143, 175
  Algerian
    lben, 4
    lfrik, 4
    raib, 4
  cashew juice, 360
  Iranian, 75
    cheeses, 79–86, *82*–85, **83**, **86**
    doogh, 78–79, *79*
    kashk, 76–77, *77*, **78**
    yogurt, 78
  Moroccan, **46**, *46*
    jben, 47
    klila, 47
    lben, 46–47
    raib, 46
    smen, 47
    zabda baldya, 47
  nonmeat fermented sausages, 246
  Northeast Indian, 164
    dahi, 164
    ghee, 165
    hard chhurpi, 165
    soft chhurpi, 165
    somar, 165
  yoghurt, *see* yogurt/yoghurt
dhan, 5
diabetes mellitus (DM), 81, 131
diarrhea, 32
dimethyl trisulfide, 105–106
docosahexaenoic acid (DHA), 99
doogh, 319, 329
  manufacturing, 79
  microbiology, 78–79
  preparation, 78
douchi fermentation, 110, **112**, 129, 135, 227
  development strategies, 118
  diversity, 110–112
  fungal diversity, 113–114
  health benefits, 114, *115*
    biogenic amines, 117
    melanoidins, 117
    polypeptides, 116
    soyasaponins, 116
    soybean isoflavone, 116
    soybean oligosaccharides, 116
    streptokinase, 116
  nutritional characteristics, 113
  production, 110–112, *111*
  research status of, 112–113
durian pulp, *see* tempoyak

E

eicosapentaenoic acid (EPA), 99
el-hammoum, 6, 13
enterococci, 7, 13, 176, 185, 242
Enterococcus avium, 4
Enterococcus faecalis, 4, 47
Enterococcus faecium, 4, 47
Escherichia coli, 24, 242, 282, 295, 363, 459
exopolysaccharides (EPSs), 13, 52, 148, 454

F

fatty acids, 33, 60, 99–100, 106–107, 131, 224, 247, 321, 421
  in chocolate, 434
fermentation history, 1
  Africa, 1
  The Americas, 2
  Asia, 1–2
  Europe, 2
fibrinolytic enzyme, 102, 104, 112, 129
fish, fermented, *see* ikan pekasam
fish sauce, see yu-lu fish sauce
fish sauce, mahyaveh, 90
flat bread/naan, 197–198
Food and Agriculture Organization of the United Nations Statistics (FAOSTAT), 17, 20, 39
food quality, 35–38
food safety, 35–38
fructooligosaccharide, 31, 182, 246, 367, 458
fruits, 1, 305–306, 342, 351, 381, 414, *see also* Brazilian fruit fermentation; Chilean fermented foods
  lemons, 50–51
  Pakistan, 197
  tempoyak, 142, 143, 151–152, **152**
  Pakistan, 197
  Southeast Asia, 153–154
  wine, *see* wine
fruit vinegar, 148, *149*, 152–153
fungi, 1, 21, 81, 219, 303
  amylolytic, 165
  atingba, 168
  bhatti jaanr, 168
  diversity, 113–114
  douchi, 113–114
  xaj-pitha, 168

G

galactooligosaccharide, 31, 246, 332
gamma-aminobutyric acid (GABA), 32, 127, 130, 187, 296
  Greek vegetable fermentation, 297
gastrointestinal health, 59–60, 64, 99, 169, 184, 193, *see also* microbiome; microbiota; prebiotics; probiotics
  Boza and, 318
  infection, 52, 334, 338, 435, 450
  yoghurt and, 331–332
gastro-intestinal tract (GIT), 32, 296, 451
generally recognized as safe (GRAS), 7, 13, 34, 103, 156
ghee, 165
gherkins, 50
Gibberella fujikuroi, 24
gilaburu, 304–307, **306**
globe artichoke, 295
Gluconacetobacter entanii, 153
Gluconobacter oxygens, 153
glycosidic isoflavones, 130
grapes, 73, 150, 197, 312
  chicha, 344
  pisco, 343–344
  wine, 342–343
Greek vegetable fermentation, 293
  asparagus, 295
  B-complex vitamins, 297–298
  cauliflower, 294
  globe artichoke immature inflorescence, 295
  lactic acid, 293
  probiotic products, 296–297
  radish root, 295–296
  sulforaphane production, 298–299
  unripe (green) tomato, 294–295
greenhouse gas (GHG), 61, *62*
green tomato fermentation, 294–295
guabiroba, 365
gundruk, 165, 172, 197, 294
gut health, *see* gastrointestinal health

H

hamei, 163, 166, 168
hamoum, 6
hardaliye, 312–313
  benefits of, 313–315
  chemical and physical composition, **314**
hard chhurpi, 165
hawaijar, 125–126, 130, 131, 133
health benefits, 51
  Andean potato fermentation, 405, **406–408**, 409
  chongkukjang, 234–236
  douchi fermentation, 114, *115*
    biogenic amines, 117
    melanoidins, 117
    polypeptides, 116
    soyasaponins, 116
    soybean isoflavone, 116
    soybean oligosaccharides, 116
    streptokinase, 116
  Mexican cacao fermentation, 428–429
    bioactive components, 434–435
    considerations, 436
    methylxanthines, 433–434
    minerals, 434
    polyphenols, 431–433
  Moroccan fermented foods
    bioactive compounds, 52
    food constituents, 51–52
    probiotic effects, 52–53
  soybean, 127–128
    anti-cancer effects, 132
    anti-diabetic effects, 131
    antihypertensive effects, 132–133
    antioxidant properties, 130
    anti-thrombotic activity, 129
    bioactive peptides, 131–132
    γ-PGA, 128
    menaquinone-7, 133
    NAFLD, 131
    nutritional value, 130–131
    probiotic properties, 132–133
    saponins, 128–129
    vitamin K, 133
  sweet potato (SP) fermentation, 25, 27, 30–31
  tea fermentation, 220–221
  tempeh fermentation, 58–59, *59*
    bone, 61
    cancer, 60
    cardiovascular, 60
    cognitive function, 60
    gut, 59–60

liver, 61
lung, 61
malnutrition, 60
obesity, 61
skeletal muscle recovery, 61
high-density polyethylene (HDPE), 38, 96
high-fat diet (HFD), 131
horreh, 89
microbiology, 89–90
preparation, 89
huamo, 166

I

ighounane, 5
ikan pekasam, 144–145
immune system, 25, 32, 52–53, 311, 405
anti-allergenic effects, 171
antitumor effects, 170
polypeptides, 116
probiotics, 196
immunomodulatory properties, 147, 171, 183, 235, 458
India, 1–2, 17, 57, 122, 175, *see also* cereal fermentation; legume fermentation
Indonesia, 17, 55, *see also* tempeh fermentation; terasi
inulin, 31, 32
ipoh, 166
Iranian fermented foods and beverages, 69, **70–71**
cereals, 86–87
dairy products, 75
cheeses, 79–86, *82*–85, **83**, **86**
doogh, 78–79, *79*
kashk, 76–77, *77*, **78**
yogurt, 78
fish products, 90
horreh, *89*, 89–90
kashk-e zard, *87*, 87–88
tarkhineh, *88*, 88–89
vegetables, 69
pickled garlic, 73–75
verjuice, 73
winter salad, 69–72
isoflavone/s, 58–61, 112
aglycones, 146
chongkukjang, 235
glycosidic, 130, 132, 147
soybean, 116

J

jabuticaba, 366
jang, 225–226, *226*
role of, 227
jben, 4, 47
juá, 361

K

kashk, 76, *77*
chemical properties, 77, **78**
preparation
industrial dried kashks, 76–77
industrial liquid kashks, 77
kashk-e zard, 87
microbiology, 88
preparation, 87
keddid, 48
keem, 166
kefir, 123–124, 170, 358, 366
khekhrii, 163, 166
khiki cheese, 79, 81
microbiology, 81
physicochemical and micro-structural properties, 81
preparation, 81
khlii, 48, *49*
classic, 48
express, 49
light/diet, 48
kimchi, 14, 73, 123–124
baik, 195
chongkukjang, 230
dongchimi, 195
jang, 225
kakdugi, 195
nabak, 195
radish root fermentation, 295
tongbaechu, 195
yeolmoo, 195
kinema, 110, 122, 125, 126–135
Northeast Indian fermented foods, 164
Klebsiella pneumoniae, 58, 102
klila, 4–5, 47
Kluyveromyces marxianus, 47
kochkase, 381, 388
kodo ko jaanr, 168–169
Komagataeibacter europaeus, 153
koozeh cheese, 84
manufacturing, *84*
microbiology, 85
preparation, 84
Korean soy fermentation, 225
chongkukjang, 227
agricultural backgrounds, 231
birth of, 230
healthy benefits, 234–236
historical backgrounds, 230–231
key words, 229
literature, 229–230
literature backgrounds, 231
misconceptions, 231–232
misinformation, 232–234
'si' character, 234
stories of, 227–228
jang, 225–226, *226*
role of, 227
soybean, 228
kourdass, 49

L

lactate, 52, 181, 296
lactic acid bacteria (LAB), 3, 6, 13, 144, 164
cereal fermentation, 181–182
Greek vegetable fermentation, 293
health and nutritional benefits, **184**
legume fermentation, 181–182
Northeast Indian fermented foods, 169–170
Portuguese meat fermentation, 284–285
Lacticaseibacillus casei, 4, 47
Lactiplantibacillus pentosus, 50
Lactiplantibacillus plantarum, 4, 25, 50
Lactobacillus acidophilus, 23
Lactobacillus confusus, 47
Lactobacillus delbrueckii, 79
Lactococcus lactis, 4, 47
lahueñe, 346–347
Laos, 2, 110, 197, 212, *see also* tea fermentation
L-asparaginase-producing bacteria, 102
lassi, 147–148, *148*, 198
lavash bread, 87
lben, 46–47
Algerian fermented foods, 4
legume fermentation, 122, 128, 164
India, 175–176
compilation and dissemination, 188
economic value, 188
enterococci, 185
functionality and nature, 182–185
LAB, 181–182, **184**
microbiological processes, 186–187
microbiology, 180–181
microflora, **177–180**
millet, 185–186
substrate and pre-fermentation treatments, 187–188
substrate for, 176, 180
kinema, 164
tungrymbai/tung rymbái, 164
lemons, 50–51
lemzeiet, 6
Leuconostoc lactis, 4
Leuconostoc mesenteroides, 4, 23, 47
Levilactobacillus brevis, 4
lfrik, 4
lighvan cheese, 85
microbiology, 85
physicochemical changes, 85–86, **86**
preparation, 85
safety, 86
Listeria monocytogenes, 24
liver health, 58, 59, 61
Luar ethnic group, 219–220
Lue-Yuan ethnic group, 219
lung health, 58, 59, 61
lysine, 98–99, 103, 182–183, 186, 205, 334

M

mahia, 49, *50*
mahyaveh, 90, 204
Malaysian fermented foods, 142, 153–154
food regulation, 154, **155–156**, 156
fruit vinegar, 152–153
ikan pekasam (fermented fish), 144–145
lassi (yoghurt drink), 147–148, *148*
sufu (fermented bean curd), 142, 143, 146–147, 150
taucu (fermented soybean paste), *150*, 150–151
techniques, 143, **144**
tempeh (fermented soybean cake), *145*, 145–146
tempoyak (fermented durian pulp), 142, 143, 151–152, **152**
tuak (palm wine), 148–150, *149*
types of, 142–143, **143**
malnutrition, 25, 38, 60
mandacaru, 360
mangaba fruit, 362
manteiga cheese, 386, 387
marajó cheese, 386–387
marcha, 163, 166–169, 180
meat products, 1, 49, 192, 195, 242
dried ossban, 5
gueddid/kadid, 5
Moroccan
boubanita/boubanite, 49
keddid, 48

khlii, 48–49, *49*
kourdass, 49
Pakistani, 195
Portuguese, 284–285
melanoidins, 114, 117, 118, 426, 440
Meliodogyne incognita, 19
methylxanthines, 414, 418, 433–434
Mexican cacao fermentation, 414–416
alkalization, 427
biotechnology, 440–441
conching, 427
drying, 426
eating habits, 415
ecological, geographic, and genetic aspects, 437–439
flavor components, 424–425
functional compounds, **430**
future directions, 441
health benefits, 428–429
bioactive components, 434–435
considerations, 436
methylxanthines, 433–434
minerals, 434
polyphenols, 431–433
history, 417
microorganisms, 420–422
preconditioning, 419
processing, 427–428
processing steps and conditions, 439–440
reactions, 423–424
roasting, 426–427
starter cultures, 422
tempering, 427
traditional uses of, 417–419
Mexican fermented foods, 450
atole agrio, 451–452
colonche, 460–461
pozol, 453–455, *454*
pulque, 455, **456–458**, 458
tepache, 460
tesgüino, 452
vanilla, 452–453
xaxtle, 458–459, **459**
miang, 197, 212, 213, 218, *219*
Laos ethnic groups, 219
Luar ethnic group, 219–220
Lue-Yuan ethnic group, 219
microbiome, 31, 38, 235, 246
antibacterial and antifungal activities, 170
chuño, 397
next-generation sequencing, 171
tocosh, 397
microbiota, 6, 170, 180, 241, 401
alcoholic beverages, 163
alheira, 282
asparagus fermentation, 295
atole agrio, 451–452
autochthonous, 385
Brazilian fruit fermentation, *388,* 388–389, *389*
cacao, 422
cachaça, 383
cassava, 381
Cerrado and Pantanal fruits, 361
chouriço, 283
colonche, 460
enterococci, 7
gluten free (GF), 186
gut health, 59
kefir and kefiran, 171
kinema, 164
minas cheese, 386
pozol, 453
pulque, 455
tempeh, 58
tepache, 460
vanilla, 452
microorganisms, 1, 21, 46, 78, 102
Mexican cacao fermentation, 420–422
protease, 205–206
sourdough, 6
sweet potato, 21–23, *23*
yu-lu fish sauce, 204
milk, *see* dairy products
milk solids non-fat (MSNF), 147–148
millet, 145, 168, 169, 175, 176
minas cheese, 386
minerals, 30, 52, 87, 118, 122
cacao, 434
mod pitha, 166–167, 169
mold, 7, 146, 163, 195, 227, 232, 420, 426, 428
monounsaturated fatty acids (MUFAs), 99
morcela, 278–283
Moroccan fermented foods, 45
apple vinegar, 51
dairy products
jben, 47
klila, 47
lben, 46–47
raib, 46
smen, 47
zabda baldya, 47
fruits and vegetables, 50
capers, 50
carrots, 50
gherkins, 50
lemons, 50–51
olives, 50
onions, 50
pickled peppers, 50
health benefits
bioactive compounds, 52
food constituents, 51–52
probiotic effects, 52–53
mahia, 49, *50*
meat products
boubanita/boubanite, 49
keddid, 48
khlii, 48–49, *49*
kourdass, 49
sourdough, 51
mudai, 345
mycotoxins, 33, 52, 135, 182, 187, 422

N

nattokinase (NK), 129, 132
next-generation sequencing (NGS), 71, 79, 91, 163, 168
non-alcoholic fatty liver disease (NAFLD), 131
Northeast India, 2, *see also* soybean foods
Northeast Indian fermented foods, 162–163, *see also* soybean foods
alcoholic beverages
atingba, 168
bhaati jaanr, 168
chhang, 168
kodo ko jaanr, 168–169
sujen, 169
amylolytic starters
amou/perokkushi, 165
bakhar, 165–166
hamei, 166
huamo, 166
ipoh, 166
keem, 166
khekhrii, 166
marcha, 167
mod pitha, 166–167
ranu dabai, 167
ranu goti, 167
thiat, 167
vekur pitha, 167–168
xaj-pitha, 168
dairy products, 164
dahi, 164
ghee, 165
hard chhurpi, 165
soft chhurpi, 165
somar, 165
health benefits, 169
anti-allergenic effects, 171
antibacterial and antifungal properties, 170
antitumor effects, 170
immunomodulatory effects, 171
lactic acid bacteria, 169–170
pathogens, 170
wound-healing activity, 170
yeast, 171
legume products, 164
vegetables, 165
nutrition
Brazilian fruit fermentation, 352–353
douchi fermentation, 113
soybean, 130–131
sweet potato (SP) fermentation, 25, 27, 30–31
boiled sweet potato, **28–29**
dietary allowance/adequate intake, **30**
orange sweet potato leaves/vines, **31**
tempeh fermentation, 58

O

obesity, 59, 124, 131, 321, 388
high-fat and high-calorie food, 429
octopamine, 208
phytochemicals in, 129
taurine, 208
tempeh, 61
vinegar, 152
octopamine, 206–208
oligosaccharides, 105, 116, 118, 146, 246
olive oil, 171, 239, 244, 281
boubanita/boubanite, 49
diet khlii, 48
khlii/khlea, 48
olive/s, 50, 171, 244, 309
green, 6
onions, 31, 50, 125, 126
orange fleshed sweet potato (OFSP), 20, 27
oudiouan oulli, 4

P

pajarete wine, 345
Pakistani fermented foods, 192, 197–198, **198**
benefits of, 193
bioactive compounds, 193–194
breads, 195
cereals, 195

food products, 194
historical perspective, 192
meats, 195
microorganisms, 192, **193**
probiotics, 196
seafood, 195
soybeans, 195
vegetable products, 195
yogurt, 195, *196*
advantages of, 196
nutrient profiles, 195
proto-cooperation, 195–196
palm wine, *see* tuak
Pampa biome, *370*, 370–371
paneer, 197
pathogens, 182, 187, 196, 242, 245
inhibition, 32
suppression of, 170
pectin, 31, 78, 330, 359, 416
Pediococcus acidilactis, 47
Pediococcus pentosaceus, 5, 50
peptic ulcers, 32
pequi fruit, 363–364
perokkushi, 165
peruyyan, 123, 126, 128, 129
pH, 52, 97, 102, 122, 144
Pichia kudriavzevii, 25
pickled garlic, 73
chemical properties, 75
microbiology, 75
preparation, 75
pickled peppers, 50
pineapple vinegar, 153
Pipeño wine, 344
pisco, 343–344
pitanga, 366
poly-γ-polyglutamic acid (γ-PGA), 128, 132, 235
polyglutamic acid (PGA), 128
polyphenols, 107, 132, 431–439, 451
polyunsaturated fatty acids (PUFAs), 99, 332
poosti cheese, 79, 81
during Aging Period, **82**
chemical properties, **83**
making process, *82*
microbiology, 81
physicochemical and micro-structural properties, 81
preparation, 81
Portuguese meat fermentation, 271
Alentejo region, 279–281
Algarve region, 281
Alto Minho region, 277
cacholeira, 283
Centro region, 277–279
chouriça, 283–284
chouriço, 283
cultural impact, 271
food preparation and cooking, 285
historical perspective, 271
LAB, 284–285
Monchique region, 281
morcela, 283
pork products, **273–274**
Região Autónoma dos Açores region, 281–282
salpicão, 283–284
traditional process, 272–273
Trás-os-Montes e Alto Douro region, 274–277, *275*
pozol, 185, 453–455, *454*, 461
prebiotic/s, 153, 169, 182, 360, 435
boza, 318
nonmeat fermented sausages, 246–247
Sweet potato fermentation, 31–32, **32,** *33*
yoghurt production, 331
probiotics, 31, 124, 148, 164, 169
Algerian fermented foods, 7, 13
cholesterol, 53
hyperlipidemia, 53
immunity system, 52–53
mycotoxin absorption, 52
pH balance, 52
protection against infection, 52
Greek vegetable fermentation, 296–297
Moroccan fermented foods, 52
non-meat sausages, 245–246
Pakistani fermented foods, 196
soybean, 132–133
sweet potato, 31–33, **32, 33**
tarkhineh, 89
Propionibacterium freudenreichii, 58
propionic acid bacteria (PAB), 33
protease, 23, 100–103, 110–113
from fish, 205
from microorganisms, 205–206
protein, 25, 27, 197, 231, 304
animal-sourced, 61
cacao bean, 421
fish, 204–205
hydrolysis, 118
kashk, 76
milk, 246
pickled garlic, 75
plant-based, 122
potato, 405
pozol, 453
pulse, 241
soy, 240
soybean polypeptides, 116
sweet potato (SP) products, 31
tarkhineh, 88
tempeh, 55, *62*
verjuice, 73
wheat, 240–241
pulque, 455, **456–457**, 458, 461
pulse proteins, 228, 239, 241

R

radish, 165, 197, 295–296
raib, 4, 46
ranu dabai, 167
ranu goti, 167
Rhizopus oligosporus, 55
Rhizopus oryzae, 55
Rhizopus stolonifer, 19, 55
rigouta, 5
roasting, 36, 188, 285, 426–427

S

Saccharomyces cerevisiae, 24, 47
Saccharomyces rouxii, 112, 206
şalgam (shalgam), 307–309
benefits of, 309–311
chemical and physical composition, **310**
Salmonella enterica, 57
salpicão, 272, 277, 283–284, 286
salt, 57, 77, 97–99
douchi, 118
yu-lu fish sauce, 204–205
sangak bread, 87
saturated fatty acids (SFAs), 99
sausages, non-meat, 239
functional characteristics, 245
microbiological characteristics, 241–243
physicochemical characteristics, 243–244
prebiotics, 246–247
probiotics, 245–246
pulse proteins, 241
sensorial traits, 244–245
soy protein, 240
trends, 239–240
wheat protein, 240–241
worldwide, 240, **248–265**
seafood, 99, 101, 104, 106, 195
serrano cheese, 381, 386, 387
siahmazgi cheese, 81
making process, *83*
microbiology, 83–84
physicochemical and sensory properties, 84
preparation, 81, 83
sinki, 165, 172, 197, 295
skeletal muscle recovery, 58, 59, 61
smen, 5, 47
soft chhurpi, 164, 165
somar, 164, 165
sour cassava starch, 381–383
electron microscopy micrograph, *382*
schematic diagram of, *382*
sun-drying systems, *383*
sourdough, 123, 298, 388
bread samples, 87
soybean flour in, 25
tarhana, 335
tarkhineh, 88
soyasaponins, 114, 116, 118, 128–129
soybean foods, 20, 25, 55, 58, 106, 110–114
biogenic amines, *see* biogenic amines
cake, *see* tempeh
isoflavone, 116
Korean soy fermentation, 228
melanoidins, 117
milk, 369
in Northeast India, 122–123
akhuni/aakhone/axone, 126
bekang, 126
fermented foods, types of, 123–124, **124**
hawaijar, 125–126
health benefits, 127–133
improvement, 133–134
kinema, 125
peruyyan, 126
products, 126–127, **127**
tungrymbai, 126
oligosaccharides, 116
Pakistani fermented foods, 195
paste, *see* taucu
polypeptides, 116
seeds, *123*
streptokinase, 116
soy protein, 58, 61, 110, 240, 243
spontaneous ferments, 123
Staphylococcus aureus, 24
Staphylococcus saprophyticus, 102
starter culture,

cacao fermentation, 423
sweet potato (SP), 33–35, *34*, *35*
terasi, 102–104
Streptococcus lactis, 23
Streptococcus thermophilus, 5
streptokinase, 114, 116, 118
Streptomyces ipomoeae, 19
sufu, 142, 143, 146–147, 150
sujen, 166, 169
sulforaphane production, 297–299
sweet potato (SP) fermentation, 17–21
animal farming, 38
array of, 24–25
global production, **18–19**, *19*
health benefits, 25, 27, 30–31
leaves, **26–27**
microorganisms, 21–23, *23*
nutrition, 25, 27, 30–31
boiled sweet potato, **28–29**
dietary allowance/adequate intake, **30**
orange sweet potato leaves/vines, **31**
potato tuber, **26–27**
prebiotic, probiotic, and synbiotic, 31–33
product flow process, *33*
production process, *21*, 23–25, *24*, *39*
product safety, 35–38
raw, 21–23, **26–27**
root, **26–27**, 38
starter culture, 33–35, *34*, *35*
yam tuber, **26–27**
vine, **31**, 38

T

taftoon bread, 87
tarhana, 87, 328, 334–336, *335*
benefits of, 336–337
tarkhineh, 88
chemical characteristics, 88–89
functionality and probiotic properties, 89
microbiology, 88
preparation, 88
taucu, 142, 143, 150–151, *151*
taurine, 99, 207, 208
tea fermentation, 212–213
beverage, 217–218
brands, *217*
future prospects, 221–222
health benefits, 220–221
history of, 213–214
miang, 218, *219*
Laos ethnic groups, 219
Luar ethnic group, 219–220
Lue-Yuan ethnic group, 219
processing, 216–217
production and structure, 214–216, **215**, *216*
role and sociocultural relations, 216
sustainable development, 221
tea resources, *214*
varieties, *213*
wild tea processing, *214*
tecu, 346
tempeh fermentation, 55, 145–146
affordability, 62, *63*
in banana leaf, *56*
challenges, 62–65
definition, 55
evolution, 56–58
fried, *57*
health benefits, 58–59, *59*
bone, 61
cancer, 60
cardiovascular, 60
cognitive function, 60
gut, 59–60
liver, 61
lung, 61
malnutrition, 60
obesity, 61
skeletal muscle recovery, 61
history, 56
nutrition, 58
opportunities, 62–65
spread, 56–58
sustainability, 61, *62*
white rice drizzled with sweet soy sauce, *57*
tempoyak, 142, 143, 151–152, **152**
tepache, 415, 450, 451, 460, 461
terasi, 95–96
chemical composition
amino acid composition, 98–99
fatty acid, 99–100
organic acid composition, 99–100
proximate composition, 97–98
salt, 97–98
water activity, 97–98
flavor compounds, 104–106
microbiology
diversity, 100–101
microbial flora, 101–102
starters, 102–104
packaging, 96–97
production, *96*, 96–97
shelf life, 96–97
terasi udang, 95–99, **98**, 105
tesgüino, 415, 452
thiat, 163, 167
tocosh, 399–401, *400*
tofu, 122, 124, 146, 147, 227
tuak, 142, 148–150, *149*
tungrymbai, 110, 123, 126, 128, 129
Northeast Indian fermented foods, 164
tung rymbái, 164
Turkish fermented beverages, 303–304
ayran, *319*, 319–320
benefits of, 320–321
boza, 315–318
benefits of, 318–319
chemical and physical composition, **317**
future directions, 321
gilaburu, 304–307, **306**
hardaliye, 312–313
benefits of, 313–315
chemical and physical composition, **314**
şalgam (shalgam), 307–309
benefits of, 309–311
chemical and physical composition, **310**
Turkish fermented foods, 328–329
chickpea bread, 332–334, *334*
benefits of, 334
future directions, 337
tarhana, 334–336, *335*
benefits of, 336–337
yoghurt, 329–331
benefits of, 331–332
type 2 diabetes, 58, 124, 131, 146, 169, 185

U

umbu, 360
umbu-caja, 360

V

vanilla, 416, 419, 450–453
vegetables
Algerian, 5–6
Iranian, 69
pickled garlic, 73–75
verjuice, 73
winter salad, 69–72
Moroccan, 50
capers, 50
carrots, 50
gherkins, 50
olives, 50
onions, 50
pickled peppers, 50
Pakistani, 195
vekur pitha, 167–168
verjuice, 73, *75*
Vibrio parahaemolyticus, 24
vinegar, 50, 149, 192, 272
apple, 51
cagaita, 363
fruit, 152–153
vitamin B2, 127, 130
vitamin K, 52, 73, 133, 135, 234
vitamins, B-complex, 297–298, 385

W

water activity, 97–98, 101, 135, 243
wheat protein, 240–242
wine, 25, 34, 110, 143, *see also* alcoholic beverages
Chilean, 343–344
pajarete, 345
pipeño, 344
Malaysian, 148–150
rice, 142
sweet potato red, 30
winter salad, 69, 70
16S rRNA gene sequence analysis, *74*
microbial dynamics and acidity, *73*
microbiology, 70
physicochemical properties, 72, *74*
preparation, 70
production, *72*
during ripening period, 71–72
World Food Program (WFP), 20
World Health Organization (WHO), 20, 129, 311

X

xaj-pitha, 168
xaxtle, 450, 458–459, **459**, 461
xylooligosaccharide, 31, 332, 454

Y

yeast, 7, 77–79, 122, 195
Northeast Indian fermented foods, 171
yogurt/yoghurt
açaí, 368
antihypertensive drink, 34

antitumor effects, 170
ayran, 319–321
boza and, 316
cashew, 360, 362
frozen, 364–365
Iranian, 78
doogh, 78–79
kashk, 75–78
kashk-e zard, 87
Malaysian
lassi, 147–148, *148*
Pakistani, 195, *196*
advantages of, 196
nutrient profiles, 195
proto-cooperation, 195–196
prebiotics, 331
tarhana, 334–337
Turkish, 329–332
sheep milk, 367
sweet potato, 33
yu-lu fish sauce, 203
bioactive peptides, 207
biogenic amine, 206–207
flavor, 206
low-salt, 204–205
microorganisms, 204
octopamine, 208
protease, 205–206
taurine, 208

Z

zabda baldya, 47
zebda, 5